Electronic Design Automation for IC Implementation, Circuit Design, and Process Technology

CRC Press
Taylor & Francis Group
Boca Raton London New York

CRC Press is an imprint of the
Taylor & Francis Group, an **informa** business

Electronic Design Automation for IC Implementation, Circuit Design, and Process Technology

Edited by

Luciano Lavagno
Politecnico di Torino
Torino, Italy

Grant Martin
Cadence Design Systems, Inc.
San Jose, California, USA

Igor L. Markov
University of Michigan
Ann Arbor, Michigan, USA

Louis K. Scheffer
Howard Hughes Medical Institute
Ashburn, Virginia, USA

CRC Press
Taylor & Francis Group
Boca Raton London New York

CRC Press is an imprint of the
Taylor & Francis Group, an **informa** business

MATLAB® and Simulink® are trademarks of The MathWorks, Inc. and are used with permission. The MathWorks does not warrant the accuracy of the text or exercises in this book. This book's use or discussion of MATLAB® and Simulink® software or related products does not constitute endorsement or sponsorship by The MathWorks of a particular pedagogical approach or particular use of the MATLAB® and Simulink® software.

CRC Press
Taylor & Francis Group
6000 Broken Sound Parkway NW, Suite 300
Boca Raton, FL 33487-2742

First issued in paperback 2018

© 2016 by Taylor & Francis Group, LLC
CRC Press is an imprint of Taylor & Francis Group, an Informa business

No claim to original U.S. Government works

ISBN-13: 978-1-4822-5460-0 (hbk)
ISBN-13: 978-1-138-58601-7 (pbk)

Library of Congress Cataloging-in-Publication Data

Names: Lavagno, Luciano, 1959- editor. | Markov, Igor L. (Igor Leonidovich),
1973- editor. | Martin, Grant (Grant Edmund) editor. | Scheffer, Lou,
editor.
Title: Electronic design automation for IC implementation, circuit design,
and process technology / edited by Luciano Lavagno, Igor L. Markov, Grant
E. Martin, and Louis K. Scheffer.
Other titles: EDA for IC implementation, circuit design, and process
technology.
Description: Second edition. | Boca Raton : Taylor & Francis, CRC Press,
2016. | Revised edition of: EDA for IC implementation, circuit design,
and process technology. 2006. | Includes bibliographical references and
index.
Identifiers: LCCN 2015034516 | ISBN 9781482254600
Subjects: LCSH: Integrated circuits--Computer-aided design. | Integrated
circuits--Design and construction. | Manufacturing processes.
Classification: LCC TK7874 .E257 2016 | DDC 621.3815--dc23
LC record available at http://lccn.loc.gov/2015034516

Visit the Taylor & Francis Web site at
http://www.taylorandfrancis.com

and the CRC Press Web site at
http://www.crcpress.com

The editors would like to acknowledge the unsung heroes of electronic design automation (EDA), who work to advance the field in addition to their own personal, corporate, or academic agendas. These men and women serve in a variety of ways—they run the smaller conferences, they edit technical journals, and they serve on standards committees, just to name a few. These largely volunteer jobs won't make anyone rich or famous despite the time and effort that goes into them, but they do contribute mightily to the remarkable and sustained advancement of EDA. Our kudos to these folks, who don't get the credit they deserve.

Contents

SECTION I — RTL to GDSII, or Synthesis, Place, and Route

SECTION II — Analog and Mixed-Signal Design

SECTION III — Physical Verification

SECTION IV — Technology Computer-Aided Design

Preface to the Second Edition

When Taylor & Francis Group (CRC Press) first suggested an update to the 2006 first edition of the *Electronic Design Automation for Integrated Circuits Handbook*, we realized that almost a decade had passed in the electronics industry, almost a complete era in the geological sense. We agreed that the changes in electronic design automation (EDA) and design methods warranted a once-in-a-decade update and asked our original group of authors to update their chapters. We also solicited some new authors for new topics that seemed of particular relevance for a second edition. In addition, we added a new coeditor, Igor L. Markov, especially since Louis K. Scheffer has moved out of the EDA industry.

Finding all the original authors was a challenge. Some had retired or moved out of the industry, some had moved to completely new roles, and some were just too busy to contemplate a revision. However, many were still available and happy to revise their chapters. Where appropriate, we recruited new coauthors to revise, update, or replace a chapter, highlighting the major changes that occurred during the last decade.

It seems appropriate to quote from our original 2006 preface: "As we look at the state of electronics and IC design in 2005–2006, we see that we may soon enter a major period of change in the discipline." And "Upon further consideration, it is clear that the current EDA approaches have a lot of life left in them." This has been our finding in doing the revision. Some rather revolutionary changes are still coming; but to a great extent, most of the EDA algorithms, tools, and methods have gone through evolutionary, rather than revolutionary, change over the last decade. Most of the major updates have occurred both in the initial phases of the design flow, where the level of abstraction keeps rising in order to support more functionality with lower NRE costs, and in the final phases, where the complexity due to smaller and smaller geometries is compounded by the slow progress of shorter wavelength lithography.

Major challenges faced by the EDA industry and researchers do not so much require revising previously accumulated knowledge but rather stimulate applying it in new ways and, in some cases, developing new approaches. This is illustrated, for example, by a new chapter on 3D circuit integration— an exciting and promising development that is starting to gain traction in the industry. Two more new chapters cover gate sizing and clock tree synthesis, two areas that had been nascent at the time of the first edition but experienced strong growth and solid industry adoption in recent years.

We hope that the readers enjoy the improved depth and the broader topic coverage offered in the second edition.

Luciano Lavagno
Torino, Italy

Igor L. Markov
Mountain View, California

Grant E. Martin
San Jose, California

Louis K. Scheffer
Washington, DC

Preface to the First Edition

Electronic design automation (EDA) is a spectacular success in the art of engineering. Over the last quarter of a century, improved tools have raised designers' productivity by a factor of more than a thousand. Without EDA, Moore's law would remain a useless curiosity. Not a single billion-transistor chip could be designed or debugged without these sophisticated tools, so without EDA we would have no laptops, cell phones, video games, or any of the other electronic devices we take for granted.

Spurred by the ability to build bigger chips, EDA developers have largely kept pace, and these enormous chips can still be designed, debugged, and tested, and in fact, with decreasing time to market.

The story of EDA is much more complex than the progression of integrated circuit (IC) manufacturing, which is based on simple physical scaling of critical dimensions. EDA, on the other hand, evolves by a series of paradigm shifts. Every chapter in this book, all 49 of them, was just a gleam in some expert's eye just a few decades ago. Then it became a research topic, then an academic tool, and then the focus of a start-up or two. Within a few years, it was supported by large commercial EDA vendors and is now part of the conventional wisdom. Although users always complain that today's tools are not quite adequate for today's designs, the overall improvements in productivity have been remarkable. After all, in what other field do people complain of *only* a 21% compound annual growth in productivity, sustained over three decades, as did the *International Technology Roadmap for Semiconductors* in 1999?

And what is the future of EDA tools? As we look at the state of electronics and integrated circuit design in the 2005–2006 time frame, we see that we may soon enter a major period of change in the discipline. The classical scaling approach to integrated circuits, spanning multiple orders of magnitude in the size of devices over the last 401 years, looks set to last only a few more generations or process nodes (though this has been argued many times in the past and has invariably been proved to be too pessimistic a projection). Conventional transistors and wiring may well be replaced by new nano and biologically based technologies that we are currently only beginning to experiment with. This profound change will surely have a considerable impact on the tools and methodologies used to design integrated circuits. Should we be spending our efforts looking at CAD for these future technologies or continue to improve the tools we currently use?

Upon further consideration, it is clear that the current EDA approaches have a lot of life left in them. With at least a decade remaining in the evolution of current design approaches, and hundreds of thousands or millions of designs left that must either craft new ICs or use programmable versions of them, it is far too soon to forget about today's EDA approaches. And even if the technology changes to radically new forms and structures, many of today's EDA concepts will be reused and evolved for design into technologies well beyond the current scope and thinking.

The field of EDA for ICs has grown well beyond the point where any single individual can master it all, or even be aware of the progress on all fronts. Therefore, there is a pressing need to create a snapshot of this extremely broad and diverse subject. Students need a way of learning about the many disciplines and topics involved in the design tools in widespread use today. As design becomes more and more multidisciplinary, electronics designers and EDA tool developers need to broaden their scope. The methods used in one subtopic may well have applicability to new topics as they arise. All of electronics design can utilize a comprehensive reference work in this field.

With this in mind, we invited many experts from across all the disciplines involved in EDA to contribute chapters summarizing and giving a comprehensive overview of their particular topic or field. As might be appreciated, such chapters, written in 2004–2005, represent a snapshot of the state of the art. However, as surveys and overviews, they retain a lasting educational and reference value that will be useful to students and practitioners for many years to come.

With a large number of topics to cover, we decided to split the handbook into two books. *Electronic Design Automation for IC System Design, Verification, and Testing* is Volume 1 and covers system-level design, micro-architectural design, and verification and test. *Electronic Design Automation for IC Implementation, Circuit Design, and Process Technology* is Volume 2 and covers the classical "RTL to GDSII" design flow, incorporating synthesis, placement, and routing, along with analog and mixed-signal design, physical verification, analysis and extraction, and technology CAD topics. These roughly correspond to the classical "front-end/back-end" split in IC design, where the front end (or logical design) focuses on ensuring that the design does the right thing, assuming it can be implemented, and the back end (or physical design) concentrates on generating the detailed tooling required, while taking the logical function as given. Despite limitations, this split has persisted through the years—a complete and correct logical design, independent of implementation, remains an excellent handoff point between the two major portions of an IC design flow.

Since IC designers and EDA developers often concentrate on one side of this logical/physical split, this seemed to be a good place to divide the book as well.

This volume, *Electronic Design Automation for IC Implementation, Circuit Design, and Process Technology* opens with an overview of the classical RTL to GDSII design flows and then steps immediately into the logic synthesis aspect of "synthesis, place, and route." Power analysis and optimization methods recur at several stages in the flow. Recently, equivalence checking has become more reliable and automated in standard IC flows. We then see chapters on placement and routing and associated topics of static timing analysis and structured digital design. The standard back-end flow relies on standard digital libraries and design databases and must produce IC designs that fit well into packages and onto boards and hybrids. The relatively new emphasis on design closure knits many aspects of the flow together. Indeed, Chapter 13, on design closure, is a good one to read right after Chapter 1 on design flows.

Before diving into the area of analog and mixed-signal design, the handbook looks at the special methods appropriate to FPGA design—a growing area for rapid IC design using underlying fixed but reprogrammable platforms. Then we turn to analog design, where we cover simulation methods, advanced modeling, and layout tools. Physical verification, analysis, and extraction covers design rule checking, transformation of designs for manufacturability, analysis of power supply noise and other noise issues, and layout extraction. Finally, the handbook looks at process simulation and device modeling and advanced parasitic extraction as aspects of technology CAD for ICs.

This handbook with its two books constitutes a valuable learning and reference work for everyone involved and interested in learning about electronic design and its associated tools and methods. We hope that all readers will find it of interest and that it will become a well-thumbed resource.

Louis K. Scheffer

Luciano Lavagno

Grant E. Martin

MATLAB® is a registered trademark of The MathWorks, Inc. For product information, please contact:

The MathWorks, Inc.
3 Apple Hill Drive
Natick, MA 01760-2098 USA
Tel: 508-647-7000
Fax: 508-647-7001
E-mail: info@mathworks.com
Web: www.mathworks.com

Acknowledgments

Louis K. Scheffer acknowledges the love, support, encouragement, and help of his wife, Lynde, his daughter, Lucynda, and his son, Loukos. Without them, this project would not have been possible.

Luciano Lavagno thanks his wife, Paola, and his daughter, Alessandra Chiara, for making his life so wonderful.

Grant E. Martin acknowledges, as always, the love and support of his wife, Margaret Steele, and his two daughters, Jennifer and Fiona.

Igor L. Markov thanks his parents, Leonid and Nataly, for encouragement and support.

Editors

Luciano Lavagno received his PhD in EECS from the University of California at Berkeley, Berkeley, California, in 1992 and from Politecnico di Torino, Torino, Italy, in 1993. He is a coauthor of two books on asynchronous circuit design, a book on hardware/software codesign of embedded systems, and more than 200 scientific papers. He is an inventor of 15 US patents. Between 1993 and 2000, he was the architect of the POLIS project, a cooperation between the University of California at Berkeley, Cadence Design Systems, Magneti Marelli, and Politecnico di Torino, which developed a complete hardware/software codesign environment for control-dominated embedded systems. Between 2003 and 2014, he was one of the creators and architects of the Cadence C-to-Silicon high-level synthesis system.

Since 2011, he is a full professor with Politecnico di Torino, Italy. He has been serving on the technical committees of several international conferences in his field (e.g., DAC, DATE, ICCAD, ICCD, ASYNC, CODES) and of various workshops and symposia. He has been the technical program chair of DAC and the TPC and general chair of CODES. He has been an associate editor of IEEE TCAS and ACM TECS. He is a senior member of the IEEE.

His research interests include the synthesis of asynchronous low-power circuits, the concurrent design of mixed hardware and software embedded systems, the high-level synthesis of digital circuits, the design and optimization of hardware components and protocols for wireless sensor networks, and design tools for WSNs.

Igor L. Markov is currently on leave from the University of Michigan, Ann Arbor, Michigan, where he taught for many years. He joined Google in 2014. He also teaches VLSI design at Stanford University. He researches computers that make computers, including algorithms and optimization techniques for electronic design automation, secure hardware, and emerging technologies. He is an IEEE fellow and an ACM distinguished scientist. He has coauthored five books and has four U.S. patents and more than 200 refereed publications, some of which were honored by best-paper awards. Professor Markov is a recipient of the DAC Fellowship, the ACM SIGDA Outstanding New Faculty award, the NSF CAREER award, the IBM Partnership Award, the Microsoft A. Richard Newton Breakthrough Research Award, and the inaugural IEEE CEDA Early Career Award. During the 2011 redesign of the ACM Computing Classification System, Professor Markov led the effort on the Hardware tree. Twelve doctoral dissertations were defended under his supervision; three of them received outstanding dissertation awards.

Grant E. Martin is a distinguished engineer at Cadence Design Systems Inc. in San Jose, California. Before that, Grant worked for Burroughs in Scotland for 6 years; Nortel/BNR in Canada for 10 years; Cadence Design Systems for 9 years, eventually becoming a Cadence fellow in their Labs; and Tensilica for 9 years. He rejoined Cadence in 2013 when it acquired Tensilica and has been there since, working in the Tensilica part of the Cadence IP group. He received his bachelor's and master's degrees in mathematics (combinatorics and optimization) from the University of Waterloo, Canada, in 1977 and 1978.

Grant is a coauthor of *Surviving the SOC Revolution: A Guide to Platform-Based Design*, 1999, and *System Design with SystemC*, 2002, and a coeditor of the books *Winning the SoC Revolution: Experiences in Real Design* and *UML for Real: Design of Embedded Real-Time Systems*, June 2003, all published by Springer (originally by Kluwer). In 2004, he cowrote, with Vladimir Nemudrov, the first book on SoC design published in Russian by Technosphera, Moscow. In the middle of the last decade, he coedited *Taxonomies for the Development and Verification of Digital Systems* (Springer, 2005) and *UML for SoC Design* (Springer, 2005), and toward the end of the decade cowrote *ESL Design and Verification: A Prescription for Electronic System-Level Methodology* (Elsevier Morgan Kaufmann, 2007) and *ESL Models and Their Application: Electronic System Level Design in Practice* (Springer, 2010).

He has also presented many papers, talks, and tutorials and participated in panels at a number of major conferences. He cochaired the VSI Alliance Embedded Systems study group in the summer of 2001 and was cochair of the DAC Technical Programme Committee for Methods for 2005 and 2006. He is also a coeditor of the Springer Embedded System Series. His particular areas of interest include system-level design, IP-based design of system-on-chip, platform-based design, DSP, baseband and image processing, and embedded software. He is a senior member of the IEEE.

Louis K. Scheffer received his BS and MS from the California Institute of Technology, Pasadena, California, in 1974 and 1975, and a PhD from Stanford University, Palo Alto, California, in 1984. He worked at Hewlett Packard from 1975 to 1981 as a chip designer and CAD tool developer. In 1981, he joined Valid Logic Systems, where he did hardware design, developed a schematic editor, and built an IC layout, routing, and verification system. In 1991, Valid merged with Cadence Design Systems, after which Dr. Scheffer worked on place and route, floorplanning systems, and signal integrity issues until 2008.

In 2008, Dr. Scheffer switched fields to neurobiology, studying the structure and function of the brain by using electron microscope images to reconstruct its circuits. As EDA is no longer his daily staple (though his research uses a number of algorithms derived from EDA), he is extremely grateful to Igor Markov for taking on this portion of these books. Lou is also interested in the search for extraterrestrial intelligence (SETI), serves on the technical advisory board for the Allen Telescope Array at the SETI Institute, and is a coauthor of the book *SETI-2020*, in addition to several technical articles in the field.

Contributors

Jason Anderson
Department of Electrical and Computer Engineering
University of Toronto
Toronto, Ontario, Canada

Mark Bales
Synopsys, Inc.
Mountain View, California

Mark Basel
Fairlead Investments
Portland, Oregon

Faik Baskaya
Department of Electrical and Electronics Engineering
Boğaziçi University
Istanbul, Turkey

Scott T. Becker
Tela Innovations, Inc.
Los Gatos, California

Vaughn Betz
Department of Electrical and Computer Engineering
University of Toronto
Toronto, Ontario, Canada

David Blaauw
Electrical Engineering and Computer Science
University of Michigan
Ann Arbor, Michigan

Robert K. Brayton
Electrical Engineering and Computer Sciences
University of California, Berkeley
Berkeley, California

Doron Bustan
Cadence Design Systems
Haifa, Israel

Neal Carney
Tela Innovations, Inc.
Los Gatos, California

Rajat Chaudhry
Avago Technologies
Austin, Texas

David Chinnery
Mentor Graphics, Inc.
Fremont, California

Minsik Cho
IBM Thomas J. Watson Research Center
Yorktown Heights, New York

Chang-Hoon Choi
Stanford University
Palo Alto, California

Mihir Choudhury
IBM Thomas J. Watson Research Center
Yorktown Heights, New York

John M. Cohn
IBM Systems and Technology Group
Essex Junction, Vermont

Jordi Cortadella
Department of Computer Science
Polytechnic University of Catalonia
Barcelona, Spain

Nicola Dragone
HiQE Capital
Brescia, Italy

Robert W. Dutton
Electrical Engineering
Stanford University
Palo Alto, California

Katherine Fetty
Mentor Graphics, Inc.
Wilsonville, Oregon

Paul D. Franzon
Electrical and Computer Engineering
North Carolina State University
Raleigh, North Carolina

Georges G.E. Gielen
Department of Elektrotechniek ESAT-MICAS
Katholieke Universiteit Leuven
Leuven, Belgium

Jean-Charles Giomi
Cadence Design Systems, Inc.
San Jose, California

Laurence Grodd
Mentor Graphics, Inc.
Wilsonville, Oregon

Carlo Guardiani
HiQE Capital
Verona, Italy

Matthew R. Guthaus
Department of Computer Engineering
University of California, Santa Cruz
Santa Cruz, California

David Hathaway
Independent
Underhill, Vermont

Stephan Held
Research Institute for Discrete Mathematics
University of Bonn
Bonn, Germany

James Hogan
VistaVentures
Santa Cruz, California

Chih-Jen Hsu
Cadence Design Systems
Hsinchu, Taiwan, Republic of Chinax

Jiang Hu
Computer Engineering & Systems Group
Texas A&M University
College Station, Texas

Jin Hu
IBM
New York, New York

Mike Hutton
Altera Corp.
San Jose, California

Ralph Iverson
Synopsys, Inc.
Arlington, Massachusetts

Mark D. Johnson
Synopsys, Inc.
Mountain View, California

Andrew B. Kahng
Computer Science and Engineering
University of California, San Diego
San Diego, California

Mattan Kamon
Coventor, Inc.
Cambridge, Massachusetts

Edwin C. Kan
Department of Electrical and Computer Engineering
Cornell University
Ithaca, New York

William Kao
Consultant
San Jose, California

Vinod Kariat
Cadence Design Systems, Inc.
San Jose, California

Igor Keller
Cadence Design Systems, Inc.
San Jose, California

Kurt Keutzer
Electrical Engineering and Computer Sciences
University of California, Berkeley
Berkeley, California

Sunil P. Khatri
Department of Electrical and Computer Engineering
Texas A&M University
College Station, Texas

Adel Khouja
Cadence Design Systems, Inc.
San Jose, California

Andreas Kuehlmann
Synopsys, Inc.
San Francisco, California

Daniel Liddell
Mentor Graphics, Inc.
Redmond, Washington

Sung-Kyu Lim
School of Electrical and Computer Engineering
Georgia Institute of Technology
Atlanta, Georgia

Mark Po-Hung Lin
Department of Electrical Engineering
National Chung Cheng University
Chiayi, Taiwan, Republic of China

Chi-Yuan Lo
Cadence Design Systems, Inc.
New Providence, New Jersey

Alan Mantooth
Electrical Engineering
University of Arkansas
Fayetteville, Arkansas

Fan Mo
Synopsys, Inc.
Sunnyvale, California

José Monteiro
Embedded Electronic Systems
INESC-ID/Técnico
Lisboa, Portugal

Makoto Nagata
Department of Information Science
Kobe University
Kobe, Japan

Gi-Joon Nam
IBM Thomas J. Watson Research Center
Yorktown Heights, New York

Peter J. Osler
Cadence Design Systems, Inc.
Chelmsford, Massachusetts

Rajendran Panda
Oracle, Inc.
Austin, Texas

Sanjay Pant
Apple Computers
Cupertino, California

Rakesh Patel
Apple Computers
Cupertino, California

Joel R. Phillips
Cadence Design Systems, Inc.
San Jose, California

Ruchir Puri
IBM Thomas J. Watson Research Center
Yorktown Heights, New York

Sherief Reda
School of Engineering
Brown University
Providence, Rhode Island

Haoxing Ren
IBM Thomas J. Watson Research Center
Yorktown Heights, New York

Jaijeet Roychowdhury
Electrical Engineering and Computer Sciences
University of California, Berkeley
Berkeley, California

Rob A. Rutenbar
Department of Computer Science
University of Illinois at Urbana–Champaign
Urbana, Illinois

Sachin S. Sapatnekar
Department of Electrical and Computer Engineering
University of Minnesota
St. Paul, Minnesota

Louis K. Scheffer
Howard Hughes Medical Institute
Washington, DC

Franklin M. Schellenberg
TBD Technologies
Palo Alto, California

Narendra V. Shenoy
Synopsys, Inc.
Mountain View, California

Raminderpal Singh
IBM Research
New York, New York

Fabio Somenzi
Department of Electrical, Computer, and Energy
    Engineering
University of Colorado, Boulder
Boulder, Colorado

Peter Spink
Cadence Design Systems, Inc.
Berkeley, California

Leon Stok
IBM Thomas J. Watson Research Center
New York, New York

Andrzej J. Strojwas
PDF Solutions, Inc.
San Jose, California

Madhavan Swaminathan
School of Electrical and Computer Engineering
Georgia Institute of Technology
Atlanta, Georgia

Gustavo E. Téllez
IBM Thomas J. Watson Research Center
Yorktown Heights, New York

Vivek Tiwari
Intel Corp.
Santa Clara, California

Robert Todd
Mentor Graphics, Inc.
Wilsonville, Oregon

Jimmy Tomblin
Mentor Graphics, Inc.
Wilsonville, Oregon

Nishath Verghese
Synopsys, Inc.
Mountain View, California

Yaoguang Wei
IBM
Austin, Texas

Hua Xiang
IBM Thomas J. Watson Research Center
Yorktown Heights, New York

I

RTL to GDSII, or Synthesis, Place, and Route

Design Flows

1

David Chinnery, Leon Stok, David Hathaway, and Kurt Keutzer

CONTENTS

1.1 EVOLUTION OF DESIGN FLOWS

Scaling has driven digital integrated circuit (IC) implementation from a design flow that uses primarily stand-alone synthesis, placement, and routing algorithms to an integrated construction and analysis flow for design closure. This chapter will outline how the challenges of rising interconnect delay led to a new way of thinking about and integrating design closure tools (see Chapter 13). Scaling challenges such as power, routability, variability, reliability, ever-increasing design size, and increasing analysis complexity will keep on challenging the current state of the art in design closure.

A modern electronic design automation (EDA) flow starts with a high-level description of the design in a register-transfer-level (RTL) language, such as Verilog or VHDL, reducing design work by abstracting circuit implementation issues. Automated tools synthesize the RTL to logic gates from a standard cell library along with custom-designed macro cells, place the logic gates on a floor plan, and route the wires connecting them. The layout of the various diffusion, polysilicon, and metal layers composing the circuitry are specified in GDSII database format for fabrication.

The RTL-to-GDSII flow has undergone significant changes in the past 30 years. The continued scaling of CMOS technologies significantly changed the objectives of the various design steps. The lack of good predictors for delay has led to significant changes in recent design flows. In this chapter, we will describe what drove the design flow from a set of separate design steps to a fully integrated approach and what further changes we see are coming to address the latest challenges.

Similar to the eras of EDA identified by Alberto Sangiovanni-Vincentelli in "The Tides of EDA" [1], we distinguish three main eras in the development of the RTL-to-GDSII computer-aided design flow: the Age of Invention, the Age of Implementation, and the Age of Integration. During the invention era, logic synthesis, placement, routing, and static timing analysis were invented. In the age of implementation, they were drastically improved by designing sophisticated data structures and advanced algorithms. This allowed the software tools in each of these design steps to keep pace with the rapidly increasing design sizes. However, due to the lack of good predictive cost functions, it became impossible to execute a design flow by a set of discrete steps, no matter how efficiently implemented each of the steps was, requiring multiple iterations through the flow to close a design. This led to the age of integration where most of the design steps are performed in an integrated environment, driven by a set of incremental cost analyzers.

Let us look at each of the eras in more detail and describe some of their characteristics and changes to steps within the EDA design flow.

1.2 THE AGE OF INVENTION

In the early days, basic algorithms for routing, placement, timing analysis, and synthesis were *invented*. Most of the early invention in physical design algorithms was driven by package and board designs. Real estate was at a premium, and only a few routing layers were available and pins were limited. Relatively few discrete components needed to be placed. Optimal algorithms of high complexity were not a problem since we were dealing with few components.

In this era, basic partitioning, placement, and routing algorithms were invented. A fundamental step in the physical design flow is partitioning to subdivide a circuit into smaller portions to simplify floor planning and placement. Minimizing the wires crossing between circuit partitions is important to allow focus on faster local optimizations within a partition, with fewer limiting constraints between partitions, and to minimize the use of limited global routing resources. In 1970, Kernighan and Lin [2] developed a minimum cut partitioning heuristic to divide a circuit into two equal sets of gates with the minimum number of nets crossing between the sets. Simulated annealing [3] algorithms were pioneered for placement and allowed for a wide range of optimization criteria to be deployed. Basic algorithms for channel, switch box, and maze routing [4] were invented. By taking advantage of restricted topologies and design sizes, optimal algorithms could be devised to deal with these particular situations.

1.3 THE AGE OF IMPLEMENTATION

With the advent of ICs, more and more focus shifted to design automation algorithms to deal with them rather than boards. Traditional CMOS scaling allowed the sizes of these designs to grow very rapidly. As design sizes grew, design tool implementation became extremely important to keep up with the increasingly larger designs and to keep design time under control. New implementations and data structures were pioneered and algorithms that scaled most efficiently became the standard.

As design sizes started to pick up, new layers of abstraction were invented. The invention of standard cells allowed one to separate the detailed physical implementation of the cells from the footprint image that is seen by the placement and routing algorithms. Large-scale application of routing, placement, and later synthesis algorithms took off with the introduction of the concept of standard cells.

The invention of the standard cell can be compared to the invention of the printing press. While manual book writing was known before, it was a labor-intensive process. Significant automation in the development of printing was enabled by keeping the height of the letters fixed and letting the width of the base of each of the letters vary according to the letter's size. Similarly, in standard cell application-specific IC (ASIC) design, one uses standard cells of common height but varying widths depending on the complexity of the single standard cell. These libraries (Chapter 12) created significant levels of standardization and enabled large degrees of automation. The invention of the first gate arrays took the standardization to an even higher level.

This standardization allowed the creation of the ASIC business model, which created a huge market opportunity for automation tools and spawned a number of innovations. Logic synthesis [5] was invented to bridge the gap from language descriptions to these standard cell implementations.

In the implementation era, a design flow could be pasted together from a sequence of discrete steps (see Figure 1.1). RTL logic synthesis translated a Verilog or VHDL description, performing technology-independent optimizations of the netlist and then mapping it into a netlist with technology gates, followed by further technology-dependent gate-level netlist optimization.

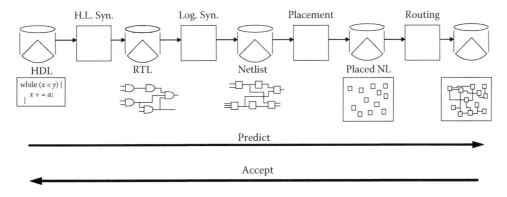

FIGURE 1.1 Sequential design flow.

This was followed by placement of the gates and routing to connect them together. Finally, a timing simulator was used to verify the timing using a limited amount of extracted wiring parasitic capacitance data.

Digital circuit sizes kept on increasing rapidly in this era, doubling in size every 2 years per Moore's law [6]. Logic synthesis has allowed rapid creation of netlists with millions of gates from high-level RTL designs. ICs have grown from 40,000 gates in 1984 to 40,000,000 gates in 2000 to billion gate designs in 2014.

New data structures like Quadtrees [7] and R-trees [8] allowed very efficient searches in the geometric space. Applications of Boolean Decision Diagrams [9] enabled efficient Boolean reasoning on significantly larger logic partitions.

Much progress was made in implementing partitioning algorithms. A more efficient version of Kernighan and Lin's partitioning algorithm was given by Fiduccia and Mattheyses [10]. They used a specific algorithm for selecting vertices to move across the cut that saved runtime and allowed for the handling of unbalanced partitions and nonuniform edge weights. An implementation using spectral methods [11] proved to be very effective for certain problems. Yang [12] demonstrated results that outperformed the two aforementioned methods by applying a network flow algorithm iteratively.

Optimizing quadratic wire length became the holy grail in placement. Quadratic algorithms took full advantage of this by deploying efficient quadratic optimization algorithms, intermixed with various types of partitioning schemes [13].

Original techniques in logic synthesis, such as kernel and cube factoring, were applied to small partitions of the network at a time. More efficient algorithms like global flow [14] and redundancy removal [15] based on test generation could be applied to much larger designs. Complete coverage of all timing paths by timing simulation became too impractical due to its exponential dependence on design size, and static timing analysis [16] based on early work in [17] was invented.

With larger designs came more routing layers, allowing over-the-cell routing and sharing of intracell and intercell routing areas. Gridded routing abstractions matched the standard cell templates well and became the base for much improvement in routing speed. Hierarchical routing abstractions such as global routing, switch box, and area routing were pioneered as effective ways to decouple the routing problem.

Algorithms that are applicable to large-scale designs without partitioning must have order of complexity less than $O(n^2)$ and preferably not more than $O(n \log n)$. These complexities were met using the aforementioned advances in data structures and algorithms, allowing design tools to handle large real problems. However, it became increasingly difficult to find appropriate cost functions for such algorithms. Accurate prediction of the physical effects earlier in the design flow became more difficult.

Let us discuss how the prediction of important design metrics evolved over time during the implementation era. In the beginning of the implementation era, most of the important design metrics such as area and delay were quite easy to predict. (Performance, power, and area are the corresponding key metrics for today's designs.) The optimization algorithms in each of the discrete design steps were guided by objective functions that relied on these predictions. As long as the final values of these metrics could be predicted with good accuracy, the RTL-to-GDSII flow could indeed be executed in a sequence of fairly independent steps. However, the prediction of important design metrics was becoming increasingly difficult. As we will see in the following sections, this led to fundamental changes in the design closure flow. The simple sequencing of design steps was not sufficient anymore.

Let us look at one class of the prediction functions, estimates of circuit delay, in more detail. In the early technologies, the delay along a path was dominated by the delay of the gates. In addition, the delay of most gates was very similar. As a result, as long as one knew how many gates there were on the critical path, one could reasonably predict the delay of the path by counting the number of levels of gates on a path and multiplying that with a typical gate delay. The delay of a circuit was therefore known as soon as logic synthesis had determined the number of logic levels on each path. In fact, in early timing optimization, multiple gate sizes were used to keep delays reasonably constant across different loadings rather than to actually improve the delay of a gate. Right after the mapping to technology-dependent standard cells, the area could be reasonably

TABLE 1.1   Gate Delays

Gate	Logical Effort	Intrinsic Delay	FO4 Delay
INV	1.00	1.00	5.00
NAND2	1.18	1.34	6.06
NAND3	1.39	2.01	7.57
NAND4	1.62	2.36	8.84
NOR2	1.54	1.83	7.99
NOR3	2.08	2.78	11.10
NOR4	2.63	3.53	14.05

well predicted by adding up the cell areas. Neither subsequent placement nor routing steps would change these two quantities significantly. Power and noise were not of very much concern in these times.

In newer libraries, the delays of gates with different logic complexities started to vary significantly. Table 1.1 shows the relative delays of different types of gates. The logical effort characteristic indicates how the delay of the gate increases with load, and the intrinsic delay is the load-independent contribution of the gate delay [18]. The fourth column of the table shows that the fanout-of-4 (FO4) delay of a more complex NOR4 logic gate can be three times the delay of a simple inverter. (The FO4 delay is the delay of an inverter driving a load capacitance that has four times the inverter's input pin capacitance. An FO4 delay is a very useful metric to measure gate delay as it is mostly independent of process technology and operating conditions. Static gates vary only 20% in FO4 delay over a variety of such conditions [19].)

Simple addition of logic levels is therefore insufficient to estimate path delay. Predicting the delay of a design with reasonable accuracy requires knowing what gates the logic is actually mapped to. It became necessary to include a static timing analysis engine (Chapter 6) in the synthesis system to calculate these delays. The combination of timing and synthesis was the first step on the way to the era of integration. This trend started gradually in the 1980s; but by the beginning of the 1990s, integrated static timing analysis tools were essential to predict delays accurately. Once a netlist was mapped to a particular technology and the gate loads could be approximated, a pretty accurate prediction of the delay could be made by the timing analyzer.

At that time, approximating the gate load was relatively easy. The load was dominated by the input capacitances of the gates that were driven. The fact that the capacitance of the wire was estimated by an inaccurate wire load model was hardly an issue. Therefore, as long as the netlist was not modified in the subsequent steps, the delay prediction was quite reliable.

Toward the mid-1990s, these predictions based on gate delays started losing accuracy. Gate delays became increasingly dependent on the load capacitance driven as well as on the rise and fall rates of the input signals (input slew) to the gates. At the same time, the fraction of loads due to wire capacitance started to increase. Knowledge of the physical design became essential to reasonably predict the delay of a path. Initially, it was mostly just the placement of the cells that was needed. The placement affected the delay, but the wire route had much less impact, since any route close to the minimum length would have similar load.

In newer technologies, more and more of the delay started to shift toward the interconnect. Both gate and wire (RC) delay really began to matter. Figure 1.2 shows how the gate delay and interconnect delay compare over a series of technology generations. With a Steiner tree approximation of the global routing, the lengths of the nets could be reasonably predicted. Using these lengths, delays could be calculated for the longer nets. The loads from the short nets were not very significant, and a guesstimate of these was still appropriate. Rapidly, it became clear that it was very important to buffer the long nets really well. In Figure 1.2, we see that around the 130 nm node, the difference between a net with repeaters inserted at the right places and an unbuffered net starts to have a significant impact on the delay. In automated place and route, buffering of long wires became an integral part of physical design. Today's standard approach for repeater (buffer or inverter) insertion on long wires is van Ginneken's dynamic programming buffering algorithm [20] and derivatives thereof. Van Ginneken's buffering algorithm has

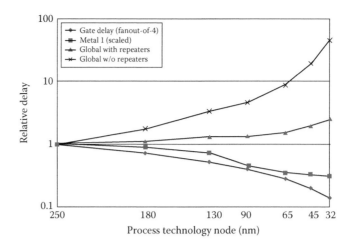

FIGURE 1.2   Gate and interconnect delay.

runtime complexity of $O(n^2)$, where n is the number of possible buffer positions, but this has been improved to near-linear [21].

Recently, a wire's physical environment has become more significant. Elmore RC wire delay models are no longer accurate, and distributed RC network models must be used, accounting for resistive shielding [22] and cross coupling. The cross-coupling capacitance between wires has increased as the ratio between wire spacing and wire height decreases. Furthermore, optical proximity effects when exposing the wafer vary depending on the distance between neighboring wires and the resulting impact when etching the wires also affects their resistance. Therefore, the actual routes for the wires from detailed routing are now very significant in predicting the delays. Moreover, global routing estimates of routing congestion can differ significantly from the actual routes taken, and global routing may mispredict where routing violations occur due to severe detailed routing congestion [23].

Netlist optimizations, traditionally done in logic synthesis, became an important part of place and route. Placement and routing systems that were designed to deal with static (not changing) netlists had to be reinvented.

Until recently, postroute optimizations were limited to minor changes that would cause minimal perturbation to avoid requiring a full reroute. These included footprint compatible cell swaps, particularly for critical path delay fixing by swapping cells to low threshold voltage, and leakage power minimization by swapping to high threshold voltage for paths with timing slack; manual reroutes to overcome severe congestion; manual engineering change orders (ECOs); and the use of metal programmable spare cells for fixes very late in the design cycle, avoiding the need to redo base layers. However, the gap between global routing estimates and detailed routing extracted capacitance and resistance have forced additional conservatism before detailed routing, for example, avoidance of small drive strength cells that will be badly affected by additional wire capacitance, leaving significant optimization opportunities in postroute.

1.4 THE AGE OF INTEGRATION

This decrease in predictability continued and firmly planted us in the age of *integration*. The following are some of the characteristics of this era:

- The impact of later design steps is increasing.
- Prediction is difficult.
- Larger designs allow less manual intervention.
- New cost functions are becoming more important.
- Design decisions interact.
- Aggressive designs allow less guardbanding.

The iterations between sequential design steps such as repeater insertion, gate sizing, and placement steps not only became cumbersome and slow but often did not even converge. EDA researchers and developers have explored several possible solutions to this convergence problem:

- Insert controls for later design steps into the design source.
- Fix problems at the end.
- Improve prediction.
- Concurrently design in different domains.

The insertion of controls proved to be very difficult. As illustrated in Figure 1.3, the path through which source modifications influence the final design result can be very indirect, and it can be very hard to understand the effect of particular controls with respect to a specific design and tools methodology. Controls inserted early in the design flow might have a very different effect than desired or anticipated on the final result. A late fix in the design flow requires an enormous increase in manual design effort. Improving predictions has proven to be very difficult. Gain-based delay models traded off area predictability for significant delay predictability and gave some temporary relief, but in general it has been extremely hard to improve predictions. The main lever seems to be concurrent design by integrating the synthesis, placement, and routing domains and coupling them with the appropriate design analyzers.

After timing/synthesis integration, placement-driven (physical) synthesis was the next major step on the integration path. Placement algorithms were added to the integrated static timing analysis and logic synthesis environments. Well-integrated physical synthesis systems became essential to tackle the design closure problems of the increasingly larger chips.

This integration trend is continuing. Gate-to-gate delay depends on the wire length (unknown during synthesis), the layer of the wire (determined during routing), the configuration of the neighboring wires (e.g., distance—near/far—which is unknown before detailed routing), and the signal arrival times and slope of signals on the neighboring wires. Therefore, in the latest technologies, we see that most of the routing needs to be completed to have a good handle on the timing of a design. Local congestion issues might force a significant number of routing detours. This needs to be accounted for and requires a significantly larger number of nets to be routed earlier for design flow convergence. Coupling between nets affects both noise and delay in larger portions of the design. Therefore, knowledge of the neighbors of a particular net is essential to carry out the analysis to the required level of detail and requires a significant portion of the local routing to be completed. In addition, power has become a very important design metric, and noise issues are starting to significantly affect delays and even make designs function incorrectly. In addition to integrated static timing analysis, power and noise analyses need to be included as well.

For example, white space postroute optimization to fix timing violations after routing is now in common use. Added cells and resized cells are placed at a legal location in available white space to avoid perturbing placement of the other cells and to minimize any changes to detailed routing

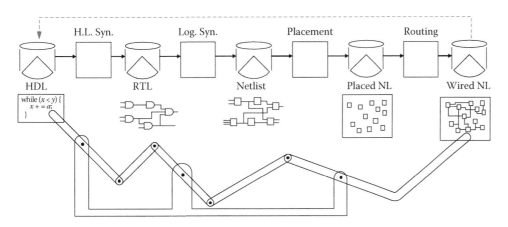

FIGURE 1.3 Controlled design flow.

between the other cells but allowing significant rerouting of nets connected to the modified cells. Larger cells, particularly those of multiple standard cell row heights, may still not be permitted to move in postroute optimization to limit the perturbation.

Consider the analysis of upsizing a cell to reduce short circuit power, which is determined by input slews and load capacitances, in postroute white space dynamic power optimization. This requires incremental evaluation of a new cell location with corresponding changes to detailed routing. The setup and hold timing impact is checked across multiple corners and modes with signal integrity cross talk analysis. Input slews to the cell are determined accounting for the impact on fanin load capacitance and the cell's output slew that affects its fanouts. The changed slews and extracted wire loads impact dynamic power consumption, and the tool must also ensure no degradation in total power, which includes both dynamic and leakage power in active mode. Such automated postroute optimizations can reduce cell area by up to 5%, helping yield and reducing fabrication cost; postroute dynamic power reductions of up to 12% have been achieved with Mentor Graphics' Olympus-SoC™ place-and-route tool [24], reducing total power consumption and extending battery life. Significant setup and hold timing violations can also be fixed in an automated fashion. These are remarkable improvements given how late in the design flow they are achieved, providing significant value to design teams.

These analysis and optimization algorithms need to work in an incremental fashion, because runtime constraints prevent us from recalculating all the analysis results when frequent design changes are made by the other algorithms. In the age of integration, not only are the individual algorithms important, but the way they are integrated to reach closure on the objectives of the design has become the differentiating factor. A fine-grained interaction between these algorithms, guided by fast incremental timing, power, and area calculators, has become essential.

In the era of integration, most progress has been made in the way the algorithms cooperate with each other. Most principles of the original synthesis, placement, and routing algorithms and their efficient implementations are still applicable, but their use in EDA tools has changed significantly. In other cases, the required incrementality has led to interesting new algorithms— for example, trialing different local optimizations with placement and global routing to pick the best viable solution that speeds up a timing critical path [25]. While focusing on the integration, we must retain our ability to focus on advanced problems in individual tool domains in order to address new problems posed by technology and to continue to advance the capabilities of design algorithms.

To achieve this, we have seen a shift to the development of EDA tools guided by four interrelated principles:

1. Tools must be integrated.
2. Tools must be modular.
3. Tools must operate incrementally.
4. Tools must sparsely access only the data that they need to reduce memory usage and runtime.

Let us look at each of these in more detail in the following sections.

1.4.1 TOOL INTEGRATION

Tool integration allows them to directly communicate with each other. A superficial form of integration can be provided by initiating, from a common user interface, the execution of traditional point tools that read input from files and save output to files. A tighter form of integration allows tools to communicate while concurrently executing rather than only through files. This tight integration is generally accomplished by building the tools on a common database (see Chapter 15) or through a standardized message passing protocol.

Tool integration enables the reuse of functions in different domains because the overhead of repeated file reading and writing is eliminated. This helps to reduce development resource requirements and improve consistency between applications. Although tool integration enables incremental tool operation, it does not require it. For example, one could integrate placement and routing programs and still have the placement run to completion before starting routing. Careful design of an application can make it easier to integrate with other applications on a common database model, even if it was originally written as a stand-alone application.

Achieving tool integration requires an agreed-upon set of semantic elements in the design representation in terms of which the tools can communicate. These elements generally consist of cells, pins, and nets and their connectivity, cell placement locations, and wire routes. Cells include standard cells, macros, and hierarchical design blocks. Individual applications will augment this common data model with domain-specific information. For example, a static timing analysis tool will typically include delay and test edge data structures. In order for an integrated tool to accept queries in terms of the common data model elements, it must be able to efficiently find the components of the domain-specific data associated with these elements. Although this can be accomplished by name lookup, when integrated tools operate within a common memory space, it is more efficient to use direct pointers. This, in turn, requires that the common data model provide methods for applications to attach private pointers to the common model elements and callbacks to keep them consistent when there are updates to the elements (see Section 1.4.3.3).

1.4.2 MODULARITY

Modular tools are developed in small, independent pieces. This gives several benefits. It simplifies incremental development because new algorithms can more easily be substituted for old ones if the original implementation was modular. It facilitates reuse. Smaller modular utilities are easier to reuse, since they have fewer side effects. It simplifies code development and maintenance, making problems easier to isolate. Modules are easier to test independently. Tool modularity should be made visible to and usable by application engineers and sophisticated users, allowing them to integrate modular utilities through an extension language.

In the past, some projects have failed largely due to the lack of modularity [26]. To collect all behavior associated with the common data model in one place, they also concentrated control of what should be application-specific data. This made the data model too large and complicated and inhibited the tuning and reorganization of data structures by individual applications.

1.4.3 INCREMENTAL TOOLS

Tools that operate incrementally update design information, or the design itself, without revisiting or reprocessing the entire design. This enables fine-grained interaction between integrated tools. For example, incremental processing capability in static timing analysis helps logic synthesis to change a design and evaluate the effect of that change on timing, without requiring a complete timing analysis to be performed.

Incremental processing can reduce the time required to cycle between analysis and optimization tools. As a result, it can improve the frequency of tool interaction, allowing a better understanding of the consequences of each optimization decision.

The ordering of actions between a set of incremental applications is important. If a tool like synthesis invokes another incremental tool like timing analysis, it needs immediate access to the effects of its actions as reported by the invoked tool. Therefore, the incremental invocation must behave as if every incremental update occurs immediately after the event that precipitates it.

An example of incremental operation is fast what-if local timing analysis for a cell resize that can be quickly rolled back to the original design state and timing state if the change is detrimental. The local region of timing analysis may be limited to the cell, the cell's fanouts because of output slew impacting their delay, the cell's fanins as their load is affected, and the fanins' fanouts, as fanin slews changed with the change in load. Timing changes are not allowed to propagate beyond that region until the resize is committed or until global costing analysis is performed. Fully accurate analysis would require propagating arrival times to the timing endpoints and then propagating back required times, which can be very runtime expensive. Even in a *full* incremental timing analysis update, only those affected paths are traversed and updated, and timing is not propagated further where the change is determined to have no impact, for example, if it remains a subcritical side path.

There are four basic characteristics desirable in an incremental tool:

1. Autonomy to avoid explicit invocation and control of other engines, for example, by registering callbacks so that an engine will be notified of events that impact it
2. Lazy evaluation to defer and minimize additional computation needed
3. Change notification to inform other engines of changes so that they may make associated updates where necessary
4. Reversibility to quickly undo changes, save, and restore

We shall explain these further in the following sections.

1.4.3.1 AUTONOMY

Autonomy means that applications initiating events that precipitate changes in another incremental tool engine do not need to notify explicitly the incremental tool of those events and that applications using results from an incremental tool do not need to initiate explicitly or control the incremental processing in that tool. Avoiding explicit change notification is important because it simplifies making changes to a design and eliminates the need to update the application when new incremental tools are added to the design tool environment. After registering callbacks, incremental applications can be notified of relevant events.

Avoiding explicit control of other incremental tools is important so that the caller does not need to understand the details of the incremental algorithm, facilitating future algorithmic improvements and making mistakes less likely. It also simplifies the use of the incremental tool by other applications.

1.4.3.2 LAZY EVALUATION (FULL AND PARTIAL)

Lazy evaluation means that an incremental tool should try, to the greatest extent possible, to defer processing until the results are needed. This can save considerable processing time when some results are never used. For example, consider a logic synthesis application making a series of individual disconnections and reconnections of pins to accomplish some logic transformation. If an incremental timing analyzer updates the timing results after each of these individual actions, it will end up recomputing time values many times, while only the last values computed are actually used.

Lazy evaluation simplifies the flow of control when recalculating values. If we have interdependent analysis functions that all try to update results when notified of a netlist change, then the callbacks would have to be ordered to ensure that updates occur in the correct order. For example, if timing updates are made before extraction updates, the timing results will be incorrect as stale extraction results are being used. With lazy evaluation, each application performs only invalidation when notified of a design change, then updates are ordered correctly through demand-driven recomputation. After a netlist change, when timing is requested, the timing engine requests extraction results before performing the delay computation. The extraction engine finds that the existing extraction results have been invalidated by the callbacks from the netlist change, so it updates those results, and only then is the timing analysis updated by the timing engine. An example of this is shown in Figure 1.4.

Lazy evaluation may be full, if all pending updates are performed as soon as any information is requested, or partial, if only those values needed to give the requested result are updated. If partial lazy evaluation is used, the application must still retain enough information to be able to determine which information has not yet been updated, since this information may be requested later. Partial lazy evaluation is employed in some timing analyzers [27] by levelizing the design and limiting the propagation of arrival and required times based on this levelization, providing significant benefits in the runtime of the tool.

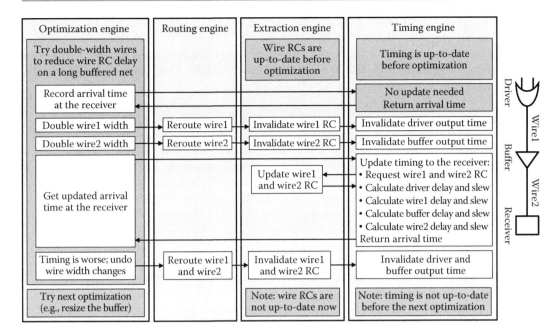

FIGURE 1.4 This example shows change notification callbacks from the routing engine to the extraction and timing analysis engines. There are lazy updates to the wire RC extraction and the timing analysis upon request from the timing engine and optimization engine, respectively.

1.4.3.3 CHANGE NOTIFICATION (CALLBACKS AND UNDIRECTED QUERIES)

With change notification, the incremental tool notifies other applications of changes that concern them. This is more than just providing a means to query specific pieces of information from the incremental tool, since the application needing the information may not be aware of all changes that have occurred. In the simplest situations, a tool initiating a design change can assume that it knows where consequent changes to analysis results (e.g., timing) will occur. In this case, no change notification is required. But in other situations, a tool may need to respond not only to direct changes in the semantic elements of the common data model but also to secondary changes within specific application domains (e.g., changes in timing).

Change notification is important because the applications may not know where to find all the incremental changes that affect them. For example, consider an incremental placement tool used by logic synthesis. Logic synthesis might be able to determine the blocks that will be directly replaced as a consequence of some logic change. But if the replacement of these blocks has a ripple effect, which causes the replacement of other blocks (e.g., to open spaces for the first set of replaced blocks), it would be much more difficult for logic synthesis to determine which blocks are in this second set. Without change notification, the logic synthesis system would need to examine all blocks in the design before and after every transformation to ensure that it has accounted for all consequences of that transformation.

Change notification may occur immediately after the precipitating event or may be deferred until requested. Immediate change notification can be provided through callback routines that applications can register with and that are called whenever certain design changes occur.

Deferred change notification requires the incremental tool to accumulate change information until a requesting application is ready for it. This requesting application will issue an undirected query to ask for all changes of a particular type that have occurred since some checkpoint (the same sort of checkpoint required for undoing design changes). It is particularly important for analysis results, since an evaluation routine may be interested only in the cumulative effect of a series of changes and may neither need nor want to pay the price (in nonlazy evaluation) of immediate change notification.

A typical use of undirected queries is to get information about changes that have occurred in the design, in order to decide whether or not to reverse the actions that caused the changes.

For this purpose, information is needed not only about the resultant state of the design but also about the original state so that the delta may be determined. (Did things get better or worse?) Thus, the application making an undirected query needs to specify the starting point from which changes will be measured. This should use the same checkpoint capability used to provide reversibility.

1.4.3.4 REVERSIBILITY (SAVE/RESTORE AND RECOMPUTE)

Trial evaluation of candidate changes is important because of the many subtle effects any particular design change may cause and because of the order of complexity of most design problems. An application makes a trial design change, examines the effects of that change as determined in part by the incremental tools with which it is integrated, and then decides whether to accept or reject the change. If such a change is rejected, we need to make sure that we can accurately recover the previous design state, which means that all incremental tool results must be reversible.

Each incremental tool should handle the reversal of changes to the data for which it is responsible. In some cases such as timing analysis, the changed data (e.g., arrival and required times) can be deterministically derived from other design data, and it may be more efficient to recompute them rather than to store and recover them. In other cases such as incremental placement, changes involve heuristics that are not reversible, and previous state data must be saved, for example, in a local stack. The incremental tool that handles the reversal of changes should be the one actually responsible for storing affected data. Thus, changes to the netlist initiated by logic synthesis should be reversed by the data model (or an independent layer built on top of it) and not by logic synthesis itself.

All such model changes should be coordinated through a central undo facility. Such a facility allows an application to set checkpoints to which it could return. Applications, which might have information to undo, register callbacks with the facility to be called when such a checkpoint is established or when a request is made to revert to a previous checkpoint.

Ordering of callbacks to various applications to undo changes requires care. One approach is to examine the dependencies between incremental applications and decide on an overall ordering that would eliminate conflicts. A simpler solution is to ensure that each atomic change can be reversed and to have the central undo facility call for the reversal of all changes in the opposite order from that in which they were originally made.

Applications can undo changes in their data in one of two ways, as outlined earlier. Save/restore applications (e.g., placement) may merely store the previous state and restore it without requiring calls to other applications. Recompute applications (e.g., timing analysis) may recompute previous state data that can be uniquely determined from the state of the model. Recompute applications generally do not have to register specific undo callbacks, but to allow them to operate, model change callbacks (and all other change notification callbacks) must be issued for the reversal of model changes just as they are for the original changes.

Such a central undo facility can also be used to help capture model change information. This can be either to save to a file, for example, an ECO file, or to transmit to a concurrently executing parallel process, on the same machine or another one, with which the current applications need to synchronize.

Even when we choose to keep the results of a change, we may need to undo it and then redo it. A typical incremental processing environment might first identify a timing problem area and then try and evaluate several alternative transformations. The original model state must be restored before each alternative is tried, and after all alternatives have been evaluated, the one that gives the best result is then reapplied.

To make sure that we get the same result when we redo a transformation that we got when we did it originally, we also need to be able to reverse an undo. Even for deterministic transformations, the exact result may depend on the order in which certain objects are visited, and such orderings may not be semantic invariants and thus may be altered when undoing a change. Also, a considerable amount of analysis may be required by some transformations to determine the exact changes that are to be made, and we would like to skip this analysis when we redo the transformation. Avoiding nondeterministic behavior is also highly desirable; otherwise, reproducing faulty behavior during debug is fraught with complications.

For these reasons, the central undo/ECO facility should be able to store and manage a tree of checkpoints rather than just a single checkpoint. Trial transformations to a circuit may also be nested, so we need to be able to stack multiple checkpoints.

It is important to remember that there is no magic bullet that will turn a nonincremental tool into an incremental one. In addition to having a common infrastructure, including such things as a common data model and a callback mechanism, appropriate incremental algorithms need to be developed.

Following some of these guidelines also encourages incremental design automation tool development. Rewriting tools is an expensive proposition. Few new ideas change all aspects of a tool. Incremental development allows more stability in the tool interface, which is particularly important for integrated applications. It also allows new ideas to be implemented and evaluated more cheaply, and it can make a required function available more quickly.

1.4.4 SPARSE ACCESS

Sparse access refers to only loading the portion of the design and associated files that are necessary to perform a given task and deferring loading additional data until needed. This can reduce both memory usage and runtime by an order of magnitude in some cases.

For example, a design check to verify correct connectivity to sleep header cells for power gating needs to only traverse the cell pin and net connections and may avoid loading libraries if cell types are identified by a naming convention. The netlist portion traversed might be further limited to just the nets connecting to the sleep header cells and the cells driving the sleep control signal, ensuring that logic is connected to the always-on power supply.

Sparse access to netlist or parasitic data typically requires paging memory to disk with an index to specify where to find data for a given cell or net. Likewise, cells within a library can be indexed to provide individual access rather than loading the entire library, which can reduce the memory overhead for loading composite current source (CCS) libraries by an order of magnitude or more as only a few hundred cells are used out of several thousand. This may require precharacterization of libraries, for example, to provide a cached list of which delay buffers are available to fix hold violations and the subset of those that are Pareto optimal in terms of area or power versus delay trade-off.

1.4.4.1 MONOLITHIC EDA TOOLS AND MEMORY LIMITS

In-house tools at design companies are often fast point tools that load only the necessary portion of the design and associated technology files or libraries to minimize the start-up runtime overhead, as the task being performed may run in a minute to an hour. This can be illustrated by the example of running tens of design checks across all the design blocks in parallel on a server farm on a nightly basis. Such point tools are very useful to analyze and check designs for any significant issues before launching further runs that will take significant time.

In contrast, commercial EDA tools have typically been designed for use in a monolithic manner. One or more major flow steps would be performed within the tool, such as synthesis, placement, and optimization; clock tree synthesis (CTS); post-CTS optimization; detailed routing; and postroute optimization. The entire design, technology files, and libraries are all loaded taking significant memory. The initial loading runtime of several minutes is smaller compared to that of the larger flow step being performed.

Commercial EDA tools are now facing tight memory limits when analyzing and optimizing large designs across multiple processes and operating corners. A combination of 30 corners and modes or more is not uncommon—for example, slow, typical, and fast process corners; low temperature, room temperature, and high temperature; –5% voltage, typical voltage, and +5% voltage; supply voltage modes such as 0.7, 0.9, and 1.1 V; and asleep, standby, awake, and scan operating modes. CCS libraries can be 10 GB each, so just loading the libraries for all these corners can be 100 GB or more of memory. Today's designs are typically anywhere between blocks of several hundred thousand gates to hundreds of millions of gates when analyzed in a flat manner at the top level and can also take 100 GB or more of memory, particularly with standard parasitic exchange format (SPEF)–annotated parasitic capacitances. When running multithreaded across a high-end 16 CPU core server to reduce runtimes that would otherwise take days, some data must also be replicated to run multithreaded to avoid memory read/write collisions and to limit pauses needed to synchronize data. Today's server farms only have a few machines with 256GB of memory, and servers with higher memory than that are very expensive.

EDA developers use a variety of standard programming techniques to reduce memory usage. For example, a vector data structure takes less memory than a linked list (to store the same data), whereas multiple Boolean variables may be packed as bits in a single CPU word and extracted with bitmasks when needed. Duplicated data can be compressed, for example, during layout versus schematic verification [28] where hierarchical representations of layout data represent individual library cells containing many repeated geometric shapes. Memory usage, runtime, and quality of results, such as circuit timing, area, and power, are carefully monitored across regression suites to ensure that code changes in the EDA tool do not degrade results or tool performance.

1.4.4.2 PARALLEL AND DISTRIBUTED COMPUTING FOR EDA

EDA vendors are migrating tools to work in a distributed manner with a controller on one machine distributing task to workers on other machines to perform in parallel. Multithreading and fork–join [28] are also common parallelism techniques used in EDA software, and they can be combined with distributed computation. These methods can reduce runtime for analysis, optimization, and verification by an order of magnitude. Shared resources can also be more effectively used with appropriate queuing.

Many of the algorithms used in the RTL-to-GDSII flow are graph algorithms, branch-and-bound search, or linear algebra matrix computations [29]. While most of these algorithms can be parallelized, Amdahl's law [30] limits the speedup due to the serial portion of the computation and overheads for communication, cache coherency, and bottlenecks where synchronization between processes is needed. Performance improvements with parallelism are also often overstated [31].

Much of the development time in parallelizing software actually focuses on reducing the runtime of the serial portion, for example, avoiding mutually exclusive locks on shared resources to minimize stalls where one process has to wait for another to complete. Fast and memory-efficient implementations of serial algorithms, such as Dijkstra's shortest-path algorithm [32], are still very important in EDA as there is often no better parallel algorithm [31].

As an example of how distributed computing is used in EDA, optimization at several important timing corners may require a server with a large amount of memory, but lower-memory servers can verify that timing violations do not occur at other corners. Analysis at dominant corners causing the majority of setup and hold timing violations reduces the need for analysis at other corners. However, there are still cases where a fix at a dominant corner can cause a violation at another corner, so it cannot be entirely avoided. The controller and the workers cannot afford the memory overhead of loading libraries for all the corners, nor the additional analysis runtime. Distributed optimization allows staying within the server memory limits. One worker performs optimization with analysis at dominant corners only, with other workers verifying that a set of changes do not violate constraints at the other corners.

Partitioning large designs into smaller portions that can be processed separately is a common approach to enabling parallelism [29]. This may be done with min-cut partitioning [12] to minimize cross-border constraints where there will be suboptimality. However, some regions may be overly large due to reconvergent timing paths, latch-based timing, cross coupling between wires, and other factors that limit a simple min-cut partitioning approach. The regions need to be sized so as to balance loads between workers. How best to subpartition a design is a critical problem in EDA [29] and varies by application. For example, timing optimization may consider just critical timing paths, whereas area optimization may try to resize every cell that is not fixed.

Portions of a design can be rapidly optimized in parallel in a distributed manner. Border constraints imposed to ensure consistency between workers' changes to the design do reduce the optimality. Individual workers use sparse access to just read the portion of the design that they are optimizing, though logical connectivity may not be sufficient as cross-coupling analysis considers physically nearby wires.

Parallel computation can be a source of nondeterminism when multiple changes are performed in parallel. To avoid this nondeterminism, incremental analysis on distributed workers may need to use a consistent snapshot of the circuit with periodic synchronizations points at which changes and analysis updates are done or undone in a fixed deterministic order. Some operations may also need to be serial to ensure consistent behavior.

1.5 FUTURE SCALING CHALLENGES

In the previous sections, we mainly focused on how continued scaling changed the way we are able to predict delay throughout the design flow. This has been one of the main drivers of the design flow changes over the past two decades. However, new challenges require rethinking the way we automate the design flow. In the following sections, we will describe some of these in more detail and argue that they are leading to a design closure that requires an integrated, incremental analysis of routability, power, noise, and variability (with required incremental extraction tools) in addition to the well-established incremental timing analysis.

1.5.1 DYNAMIC AND LEAKAGE POWER

Tools have traditionally focused on minimizing both critical path delay and circuit area. As technology dimensions have scaled down, the density of transistors has increased, mitigating the constraints on area, but increasing the power density (power dissipated per unit area). Heat dissipation limits the maximum chip power, which, in turn, limits switching frequency and hence how fast a chip can run. By 90 nm technology, even some high-end microprocessor designers found power to be a major constraint on performance [33]. Chapter 3 provides further detail on power analysis and optimization beyond the discussion in this subsection.

The price of increasing circuit speed with Moore's law has been an even faster increase in dynamic power. The dynamic power due to switching a capacitance C with supply voltage V_{dd} is $fCV^2_{dd}/2$. Increasing circuit speed increases switching frequency f proportionally, and capacitance per unit area also increases.

Transistor capacitance varies inversely to transistor gate oxide thickness t_{ox} ($C_{gate} = \varepsilon_{ox}WL/t_{ox}$, where ε_{ox} is the gate dielectric permittivity and W and L are transistor width and length). Gate oxide thickness has been scaled down linearly with transistor length so as to limit short channel effects and maintain gate drive strength to increase circuit speed [34]. As device dimensions scale down linearly, transistor density increases quadratically and the capacitance per unit area increases linearly. Additionally, wire capacitance has increased relative to gate capacitance, as wires are spaced closer together with taller aspect ratios to limit wire resistance and thus to limit corresponding wire RC delays.

If the supply voltage is kept constant, the dynamic power per unit area increases slower than quadratically due to increasing switching frequency and increasing circuit capacitance. To reduce dynamic power, supply voltage has been scaled down. However, this reduces the drive current, which reduces the circuit speed. The saturation drive current $I_{D,sat} = kW(V_{dd} - 2V_{th})^a/Lt_{ox}$, where V_{th} is the transistor threshold voltage and the exponent a is between 1.2 and 1.3 for recent technologies [35]. To avoid reducing speed, threshold voltage has been scaled down with supply voltage.

Static power has increased with reductions in transistor threshold voltage and gate oxide thickness. The subthreshold leakage current, which flows when the transistor gate-to-source voltage is below V_{th} and the transistor is nominally off, depends exponentially on V_{th} ($I_{subthreshold} = ke^{-qV_{th}/nkT}$, where T is the temperature and n, q, and k are constants). As the gate oxide becomes very thin, electrons have a nonzero probability of quantum tunneling through the thin gate oxide. While gate tunneling current was smaller by orders of magnitudes than subthreshold leakage, it was increasing much faster due to the reduction in gate oxide thickness [36].

Automated logic synthesis and place-and-route tools have focused primarily on dynamic power when logic evaluates within the circuit, as static power was a minimal portion of the total power when a circuit was active. For example, automated clock gating reduces unnecessary switching of logic. Managing standby static power was left for the designers to deal with by using methods such as powering down modules that are not in use or choosing standard cell libraries with lower leakage.

1.5.1.1 LEAKAGE POWER

Standby power may be reduced by using high threshold voltage sleep transistors to connect to the power rails [37]. In standby, these transistors are switched off to stop the subthreshold leakage path from supply to ground rails. This technique is known as power gating. When the circuit is active, these sleep transistors are on, and they must be sized sufficiently wide to cause only a small drop in voltage swing, but not so wide as to consume excessive power. To support this technique, place-and-route software must support connection to the *virtual* power rail provided by the sleep transistors and clustering of cells that enter standby at the same time so that they can share a common sleep transistor to reduce overheads.

In recent 45–28 nm process technologies, leakage can contribute up to about 40% of the total power when the circuit is active. There is a trade-off between dynamic power and leakage power. Reducing threshold voltage and gate oxide thickness allows the same drive current to be achieved with narrower transistors, with correspondingly lower capacitance and reduced dynamic power, but this increases leakage power. Increasing transistor channel length provides an alternate way to reduce leakage at the cost of increased gate delay, with somewhat higher gate capacitance and thus higher dynamic power. Alternate channel lengths are cheaper as alternate threshold voltages require separate implant masks that increase fabrication expense.

Designers now commonly use standard cell libraries with multiple threshold voltages and multiple channel lengths, for example, a combination of six threshold voltage and channel length alternatives: low/nominal/high V_{th}, with regular and +2 nm channel lengths. Low threshold voltage and shorter channel length transistors reduce the delay of critical paths. High threshold voltage and longer channel length transistors reduce the leakage power of gates that have slack. EDA tools must support a choice between cells with smaller and larger leakage as appropriate for delay requirements on the gate within the circuit context.

Significant improvements to leakage power minimization have been achieved in the last couple of years using fast global optimization approaches such as Lagrangian relaxation. This has made it possible to optimize a million-gate netlist in an hour [38], and significant further speedups may be achieved by multithreading.

It is essential that tools treat leakage power on equal terms with dynamic power when trying to minimize the total power. Preferentially optimizing leakage power or dynamic power is detrimental to the total power consumption in active mode. Other operating corners and modes must also be considered during power optimization, as there are also design limits imposed on standby power consumption as not all gates can be power-gated off.

1.5.1.2 DYNAMIC POWER

Leakage power has been significantly reduced with multigated devices, such as triple-gated FinFETs [39]. With 22–14 nm FinFET process technologies, we have seen that leakage power contributes 10%–30% of the total power when the circuit is active, depending on switching activity and threshold voltage choices. So, designers have increased the use of dynamic power minimization techniques.

There has been significant development in industry and academia on register clumping and swapping of multibit flops to reduce the clock load and clock power. Other research has focused on optimizing the clock trees to trade-off latency, clock skew, and clock power. Clock skews in the range of 30–50 ps can be achieved with multisource CTS and in the range of 10–30 ps with automated clock mesh synthesis [40], providing further opportunity for designs to achieve higher performance with trade-offs between clock skew and clock power.

Some industry designs have used fine-grained voltage islands to reduce power consumption [41], though this is yet to be supported by EDA vendor tools due to the complicated design methodology and limited market thus far.

There is further opportunity to automatically optimize wire width and wire spacing in conjunction with gate sizing/V_{th} assignment and buffering to optimize delay versus power consumption. To reduce the dynamic power of high-activity wires, they can be preferentially routed to shorten them, and wider spacing can be used to reduce their coupling capacitance to neighbors. Downsizing a gate reduces the switching power, but in some cases, it may be preferential to upsize the gate or insert a buffer to reduce slew and short circuit power. Vendor EDA tools still perform little optimization of wire spacing or wire width, overlooking trade-offs between (1) increased wire capacitance and (2) reduced wire resistance and reduced RC delay. Nondefault rules (NDRs) for wire width and spacing are typically manually provided by designers, with some limited tool support for choosing between the available NDRs.

Techniques to reduce peak power and glitching, for both peak and average power reductions, are also of interest to designers with little automation support as yet. Glitching can be reduced by insertion of deglitch latches or by balancing path delays, but dynamic timing analysis is runtime expensive, so simplified fast approaches are needed. Logic could also be remapped to better balance path delays or reduce switching activity within the logic, but remapping is quite runtime expensive.

1.5.2 PLACEMENT AND ROUTABILITY WITH COMPLEX DESIGN RULES

Multiple patterning to fabricate smaller feature sizes has added significant design rule complexity that has complicated both placement and routing. The design rules are more complex due to interaction between features on the same layer that must be on different masks to achieve higher resolution and due to potential misalignment of the multipatterning masks. For further details on design rules, please see Chapter 20.

The wavelength of the lithography light source imposes a classical Rayleigh limit on the smallest linewidth of features that can be resolved in IC fabrication, as detailed in Chapter 21, which provides a deeper discussion on multiple patterning. Resolution beyond 65 nm process technologies is limited by the 193 nm wavelength provided by argon fluoride lasers. To scale further, either immersion lithography or double patterning is needed for 45 nm devices, and both immersion lithography and double patterning are required for technologies from 32 to 14 nm [42]. Triple patterning will first be used for 10 nm process technology [43]. Multipatterning has additional fabrication costs for the additional masks to expose layers requiring smaller resolutions. Higher metal layers, which are wider, are more cheaply added without multipatterning or immersion lithography.

Other solutions for fabricating smaller features have not yet been practical for large-scale production. Extreme ultraviolet (EUV) lithography with 13.5 nm wavelength holds the promise of being able to print much smaller transistors at lower cost with simpler design rules [44]. However, the introduction of EUV has been further delayed to at least 7 nm process technology due to a variety of issues, in particular difficulty in achieving a 100 W power source to provide good exposure throughput [45]. Recently, record throughput of 1022 wafers in 24 hours at Taiwan Semiconductor Manufacturing Company and 110 W EUV power capability has been reported by ASML [46]. Electron beams can be used to write small patterns but are far too slow to be practical, except that they have been used extensively in testing fabricated ICs [47]. Double patterning may be needed with EUV at 7 nm and such resolution enhancement techniques will be required with EUV for 5 nm technology [43]. Consequently, foundries are committed to the use of multipatterning to fabricate smaller process technologies, and they are preparing alternative solutions such as self-aligned quadruple patterning in case EUV is not ready in time for 7 nm process technology. The multipatterning design-rule constraints for placement and routing are now major issues.

Since 32nm, horizontal edge-type placement restrictions [48] have been required to improve yield and for design for manufacturability. For example, if the layout shapes of polysilicon on adjacent cell edges will be detrimental to yield, spacing may be required to allow the insertion of dummy poly to minimize the optical interference between the adjacent cells [49]. This results in different horizontal edge-spacing restrictions depending on the layout of the cells.

In 10 nm process technology, we now also see vertical edge abutment restrictions. Such restrictions are introduced to comply with more conservative design rules, for example, to provide sufficient space for nearby vias to connect to cell pins or the power/ground metal stripes. Another objective is to prevent conflict between color choices for wires within cells as adjacent wires may not have the same coloring assignment. (In the context of multipatterning, coloring refers to which pattern a portion of routing is assigned to.) Vertical restrictions are more complicated, specifying distance ranges so that portions of cells may not vertically abut. Early tool support has been provided for these vertical restrictions, but further updates are needed to placers and routers to account for these better during optimization. The Library Exchange Format (LEF) [48] also needs to be updated to add specification of these vertical placement constraints.

In global placement, temporarily increasing the size of (bloating) the cells with an estimate of the additional spacing needed can help account for edge-type placement restrictions and reduce routing congestion [50,51]. Routing congestion is increased by routing blockages, routing NDRs, and preroutes such as the power, ground, and clock grids. Detailed placement must also consider these during placement legalization, as cell pins may be inaccessible to connect to if they are next to or under a preroute, or due to the additional spacing needed for an NDR [51].

With increasing design rule complexity for multipatterning, patterns must be more regular and may be restricted to unidirectional segments at each layer to improve manufacturability [52], disallowing wrong-way routing and requiring more vias, adding to routing congestion. Where wrong-way routing is allowed, there are different design rules with much wider spacing constraints between wires for the nonpreferred direction. Routers have been updated to support these more complex design rules, and both the resistance and the capacitance of vias must also now be considered for accurate parasitic extraction.

There are also rules to insert dummy metal fill between cells to reduce variation in dielectric thickness, but metal fill adds to wire coupling capacitance and thus increases wire delay [53]. Metal fill adds the complication that downsizing a cell can introduce a spacing violation as a gap may be introduced between cells with insufficient space for metal fill. Similarly, a downsized cell may have different edge types, due to different layouts within the cell, which can introduce a placement violation with neighboring cells if the different edge type requires wider spacing.

When performing multiple rounds of patterning (usually two or three), the shapes for each round are identified by *color*. There is different wire coupling capacitance in the same metal layer with the *color choice* due to misalignment error between the masks of different colors for multipatterning [54]. Depending on the foundry and fabrication process, some EDA flows preassign routing tracks a given color, whereas other design flows are colorless with the foundry coloring the provided GDSII layout. When wire color is not determined, as in a colorless design flow, it can be considered in timing analysis by adding two additional minimum and maximum corners for wire resistance and capacitance extraction [55]. The color assignment for wires within cells affects the cell delay characteristics, so EDA tools may select different cell implementations that differ only in the color assignment of wires within the cells, if standard cell libraries provide the different coloring alternatives. Global and detailed routers also need to be aware of wire coloring, either conservatively estimating capacitance and resistance for the worst-case color choices or preferably optimizing the wire color choice based on timing criticality.

Some spacing restrictions do not need to be strict but are there to improve yield, so they may be satisfied where opportunity permits to do so without degrading critical path delay and hence timing yield. Today's EDA tools strictly enforce spacing restrictions. Design groups have built fast incremental tools to provide this capability using internal database implementations with sparse access or scripted slower TCL implementations in vendor tools [56]. There remain further opportunities for native EDA tool support of yield analysis and optimization during placement.

1.5.3 VARIABILITY IN PROCESS AND OPERATING CONDITIONS

Historically, it was assumed that static timing analysis at one slow corner and one fast corner was sufficient to ensure that timing constraints were met. Worst-case circuit delay was estimated from the slow corner for the standard cell library: This was the slow process corner, with a slow operating corner with lower supply voltage (e.g., −10% V_{dd}) and high temperature (e.g., 100°C). Hold-time violations by fast timing paths and worst-case dynamic power were estimated from the fast process corner, with a fast operating corner with higher supply voltage (e.g., +10% V_{dd}) and low temperature (e.g., 0°C). The slow (fast) process corner would be for both p-type and n-type MOSFETs due to higher (lower) threshold voltages and longer (shorter) channel lengths due to process variability, which also reduces (increases) subthreshold leakage current. Worst-case leakage power can be estimated at the fast process and high-temperature corner.

In practice, a mix of *slow* and *fast* process variations and differences in spot temperatures on a chip may lead to unforeseen failures not predicted by a single process corner. Also in 90 nm and smaller process technologies, inverted temperature dependence of delay has added an additional slower operating corner at lower temperature and further complications for static timing analysis [57]. Temperature inversion is due to the competing temperature dependence of drain current with carrier mobility and threshold voltage. The impact of mobility is more significant at higher supply voltage, and the impact of threshold voltage is more significant at lower supply voltage. Mobility decreases with increasing temperature that reduces drain current slowing devices. Threshold voltage decreases with increasing temperature, making the circuit faster. Due to these competing effects, the path delay may no longer monotonically increase with temperature and may instead first decrease and then increase with temperature [57].

Both optimization and timing sign-off with a combination of many corners and modes have become common today. Additionally, the majority of chips fabricated may be substantially faster and have lower average power under normal circuit conditions. The design costs are significant for this overly conservative analysis.

Although some circuit elements are under tighter control in today's processes, the variation of other elements has increased. For example, a small reduction in transistor threshold voltage or channel length can cause a large increase in leakage. The layout of a logic cell's wires and transistors has become more complicated. Optical interference changes the resulting layout. Phase-shift lithography attempts to correct this. Varying etch rates have a greater impact on narrower wires.

Certain cell layouts are more likely to reduce yield, for example, due to increased gate delay or even a complete failure such as a disconnected wire. Ideally, cell layouts with lower yield would not be used, but these cells may be of higher speed. A small decrease in yield may be acceptable to meet delay requirements. To support yield trade-offs, foundries must share some yield information with customers along with corresponding price points.

Yield and variability data can be annotated to standard cell libraries, enabling software to perform statistical analysis of timing, power, and yield. This requires a detailed breakdown of process variability into systematic (e.g., spatially correlated) and random components. With this information, tools can estimate the yield of chips that will satisfy delay and power constraints. This is not straightforward, as timing paths are statistically correlated, and variability is also spatially correlated. However, these correlations can be accounted for in a conservative fashion that is still less conservative than worst-case corner analysis.

Variation in process and operating conditions as they pertain to static timing analysis are examined in Chapter 6. Chapter 22 discusses in detail process variation and design for manufacturability.

1.5.4 CIRCUIT RELIABILITY AND AGING

Transient glitches in a circuit's operation can be caused by cross-coupling noise as well as alpha-particle and neutron bombardment. There are also circuit aging issues that are of increasing concern for reliability, including electromigration, hot-carrier injection, and negative-bias threshold instability. These are discussed in more detail in Chapter 13.

Cross-coupling noise has become more significant with higher circuit frequencies and greater coupling capacitance between wires. Wire cross-coupling capacitance has increased with wires spaced closer together and with higher aspect ratios, which have been used to reduce wire RC delays as dimensions scale down. Wires can be laid out to reduce cross-coupling noise (e.g., by twizzling or shielding with ground wires). There are tools for analyzing cross-coupling noise, and some support for half shielding or full shielding of wires by routing them with one or both sides next to a power or ground net at the cost of additional routing congestion and detours.

As gate capacitance decreases with device dimensions and supply voltages are reduced, smaller amounts of charge are stored and are more easily disrupted by an alpha-particle or neutron strike.

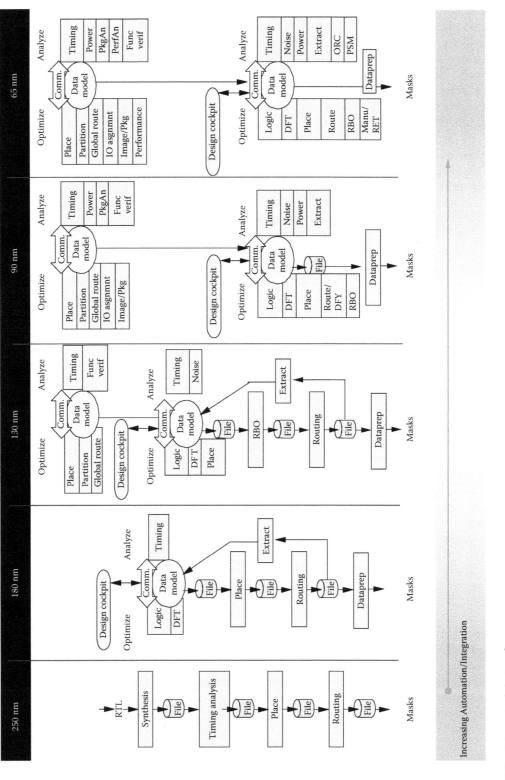

FIGURE 1.5 Integrated design flow.

Soft error rates due to alpha particles increase by a large factor as device dimensions are scaled down [58]. Soft error rates can be reduced by using silicon on insulator technology and other manufacturing methods or by using latches that are more tolerant to transient pulses [58].

For fault tolerance to glitches, circuits can have error detection and correction or additional redundancy. Tool support for logic synthesis and mapping to such circuits will simplify a designer's task.

Electromigration can cause open or short circuits [59,60] preventing correct circuit operation, which is becoming more problematic as current densities increase with narrower wires. Designers now need greater EDA tool support for the analysis of electromigration [61] and automated solutions, such as widening wires with high switching activity and additional spacing around clock cells. In a 10 nm technology, clock cells may also require decoupling capacitors inserted next to them to limit voltage drop.

1.6 CONCLUSION

In this chapter, we looked at how the RTL-to-GDSII design flow has changed over time and will continue to evolve. As depicted in Figure 1.5, we foresee continuing integration of more analysis functions into the integrated environment to cope with design for robustness and design for power. We provided an overview of a typical EDA flow and its steps and outlined the motivation for further EDA tool development.

There is high demand from IC design companies for faster turnaround to iterate through the EDA flow. Designs continue to increase rapidly in number of transistors, integrating more components, with smaller 10 and 7 nm process technologies actively being researched and developed, with 3D ICs on the horizon. More complex design rules for these newer technologies drive major EDA tool development in place and route. To achieve higher performance and lower power, the correlation must be improved between interrelated flow steps from RTL estimation and synthesis through to postroute optimization. This necessitates further tool integration, with analysis across more process corners and more operating modes and choices as to what are appropriate levels of abstraction to achieve better accuracy earlier in the design flow. All of these increase the computational demand, which motivates increasing the parallelism of EDA tools with support for larger sets of data, but still within relatively tight memory constraints. Much research and development continues within the EDA industry and academia to meet these difficult design challenges, as the EDA market continues to grow, albeit maturing with further consolidation within the industry.

The following chapters describe the individual RTL-to-GDSII flow steps in more detail, examining the algorithms and data structures used in each flow step. The additional infrastructure for design databases, cell libraries, and process technology is also discussed in the later chapters. We hope these encourage readers to innovate further with a deeper understanding of EDA.

ACKNOWLEDGMENT

The authors thank Vladimir Yutsis for his helpful feedback on Section 1.5.2.

REFERENCES

1. A. Sangiovanni-Vincentelli, The tides of EDA, *IEEE Design and Test of Computers*, 20, 59–75, 2003.
2. B.W. Kernighan and S. Lin, An efficient heuristic procedure for partitioning graphs, *Bell System Technical Journal*, 49, 291–308, 1970.
3. S. Kirkpatrick, C.D. Gelatt Jr., and M.P. Vevvhi, Optimization by simulated annealing, *Science*, 220, 671–680, 1983.
4. K.A. Chen, M. Feuer, K.H. Khokhani, N. Nan, and S. Schmidt, The chip layout problem: An automatic wiring procedure, *Proceedings of the 14th Design Automation Conference*, New Orleans, LA, 1977, pp. 298–302.
5. J.A. Darringer and W.H. Joyner, A new look at logic synthesis, *Proceedings of the 17th Design Automation Conference*, Minneapolis, MN, 1980, pp. 543–549.
6. D.C. Brock, *Understanding Moore's Law: Four Decades of Innovation*, Chemical Heritage Foundation, Philadelphia, PA, 2006.

7. J.L. Bentley, Multidimensional binary search trees used for associative searching, *Communication of the ACM*, 18, 509–517, 1975.

8. A. Guttman, R-trees: A dynamic index structure for spatial searching, *SIGMOD International Conference on Management of Data*, Boston, MA, 1984, pp. 47–57.

9. R.E. Bryant, Graph-based algorithms for Boolean function manipulation, *IEEE Transactions on Computer*, C-35, 677–691, 1986.

10. C.M. Fiduccia and R.M. Mattheyses, A linear time heuristics for improving network partitions, *Proceedings of the 19th Design Automation Conference*, Las Vegas, NV, 1982, pp. 175–181.

11. L. Hagen and A.B. Kahng, Fast spectral methods for ratio cut partitioning and clustering, *Proceedings of the International Conference on Computer Aided Design*, Santa Clara, CA, 1991, pp. 10–13.

12. H. Yang and D.F. Wong, Efficient network flow based min-cut balanced partitioning, *Proceedings of the International Conference on Computer Aided Design*, San Jose, CA, 1994, pp. 50–55.

13. J.M. Kleinhans, G. Sigl, and F.M. Johannes, Gordian: A new global optimization/rectangle dissection method for cell placement, *Proceedings of the International Conference on Computer Aided Design*, San Jose, CA, 1999, pp. 506–509.

14. C.L. Berman, L. Trevillyan, and W.H. Joyner, Global flow analysis in automatic logic design, *IEEE Transactions on Computer*, C-35, 77–81, 1986.

15. D. Brand, Redundancy and don't cares in logic synthesis, *IEEE Transactions on Computer*, C-32, 947–952, 1983.

16. R.B. Hitchcock Sr., Timing verification and the timing analysis program, *Proceedings of the 19th Design Automation Conference*, Las Vegas, NV, 1982, pp. 594–604.

17. T.I. Kirkpatrick and N.R. Clark, PERT as an aid to logic design, *IBM Journal of Research and Development*, 10, 135–141, 1966.

18. I. Sutherland, and R. Sproull, Logical effort: Designing for speed on the back of an envelope, *Proceedings of the 1991 University of California/Santa Cruz conference on Advanced Research in VLSI*, Santa Cruz, CA, 1991, pp. 1–16.

19. D. Harris et al., The fanout-of-4 inverter delay metric, unpublished manuscript, 1997. http://odin.ac.hmc.edu/~harris/research/FO4.pdf.

20. L.P.P.P. van Ginneken, Buffer placement in distributed RC-tree networks for minimal Elmore delay, *International Symposium on Circuits and Systems*, New Orleans, LA, 1990, pp. 865–868.

21. W. Shi and Z. Li, A fast algorithm for optimal buffer insertion, *IEEE Transactions on Computer-Aided Design*, 24(6), 879–891, June 2005.

22. C. Kashyap et al., An "effective" capacitance based delay metric for RC interconnect, *International Conference on Computer-Aided Design*, 2000, pp. 229–234.

23. I. Bustany et al., ISPD 2014 Benchmarks with sub-45 nm technology rules for detailed-routing-driven placement, *Proceedings of the 2014 International Symposium on Physical Design*, Petaluma, CA, 2014, pp. 161–168.

24. V. Lebars, Data path optimization: The newest answer to dynamic power reduction, *EDN Magazine*, May 2015. http://www.edn.com/electronics-blogs/eda-power-up/4439458/Data-path-optimization--The-newest-answer-to-dynamic-power-reduction.

25. Y. Luo, P.V. Srinivas, and S. Krishnamoorthy, Method and apparatus for optimization of digital integrated circuits using detection of bottlenecks, US Patent 7191417, 2007.

26. J. Lakos, *Large-Scale C++ Software Design*, Addison-Wesley, Reading, MA, 1996.

27. R.P. Abato, A.D. Drumm, D.J. Hathaway, and L.P.P.P. van Ginneken, Incremental timing analysis, US Patent 5 508 937, 1996.

28. S. Hirsch, and U. Finkler, To Thread or not to thread, *IEEE Design and Test*, 30(1), 17–25, February 2013.

29. B. Catanzaro, K. Keutzer, and B. Su, Parallelizing CAD: A timely research agenda for EDA, *Proceedings of the Design Automation Conference*, Anaheim, CA, 2008, pp. 12–17.

30. G. Amdahl, Validity of the single-processor approach to achieving large scale computing capabilities, *Proceedings of the AFIPS Spring Joint Computer Conference*, Atlantic City, NJ, 1967, pp. 483–485.

31. P. Madden, Dispelling the myths of parallel computing, *IEEE Design and Test*, 30(1), 58–64, February 2013.

32. E. Dijkstra, A note on two problems in connexion with graphs, *Numerische Mathematik*, 1, 269–271, 1959.

33. A. Wolfe, Intel clears up post-Tejas confusion, *CRN Magazine*, May 17, 2004. http://www.crn.com/news/channel-programs/18842588/intel-clears-up-post-tejas-confusion.htm.

34. S. Thompson, P. Packan, and M. Bohr, MOS scaling: Transistor challenges for the 21st century, *Intel Technology Journal*, Q3, 1–19, 1998.

35. T. Sakurai and R. Newton, Delay analysis of series-connected MOSFET circuits, *IEEE Journal Solid-State Circuits*, 26, 122–131, 1991.

36. D. Lee, W. Kwong, D. Blaauw, and D. Sylvester, Analysis and minimization techniques for total leakage considering gate oxide leakage, *Proceedings of the 40th Design Automation Conference*, Anaheim, CA, 2003, pp. 175–180.

37. S. Mutoh et al., 1-V power supply high-speed digital circuit technology with multithreshold—voltage CMOS, *IEEE Journal Solid-State Circuits*, 30, 847–854, 1995.

38. G. Flach et al., Effective method for simultaneous gate sizing and Vth assignment using Lagrangian relaxation, *IEEE Transactions on Computer-Aided Design of Integrated Circuits and Systems*, 33(4), 546–557, April 2014.

39. C. Jan et al., A 22 nm SoC platform technology featuring 3-D tri-gate and high-k/metal gate, optimized for ultra low power, high performance and high density SoC applications, *Proceedings of the IEEE International Electron Devices Meeting*, San Francisco, CA, 2012, pp. 3.1.1–3.1.4.

40. D. Chinnery et al., Gater expansion with a fixed number of levels to minimize skew, *Proceedings of SNUG Silicon Valley*, Santa Clara, CA, 2012.

41. H. Xiang et al., Row based dual-VDD island generation and placement, *Proceedings of the 51st Design Automation Conference*, San Francisco, CA, 2014, pp. 1–6.

42. S. Natarajan et al., A 14 nm logic technology featuring 2nd-generation FinFET transistors, air-gapped interconnects, self-aligned double patterning and a 0.0588 μm^2 SRAM cell size, *International Electron Devices Meeting*, San Francisco, CA, 2014.

43. W. Joyner et al., The many challenges of triple patterning, *IEEE Design and Test*, 31(4), 52–58, August 2014.

44. M. van der Brink, Many ways to shrink: The right moves to 10 nanometer and beyond, *SPIE Photomask Technology Keynote*, Monterey, CA, 2014.

45. M. Martini, The long and tortuous path of EUV lithography to full production, *Nanowerk Nanotechnology News*, April 2014. http://www.nanowerk.com/spotlight/spotid=35314.php.

46. A. Pirati et al., Performance overview and outlook of EUV lithography systems, *Proceedings of SPIE, Extreme Ultraviolet (EUV) Lithography VI*, San Jose, CA, Vol. 9422, March 1–18, 2015.

47. L. Wu, Electron beam testing of integrated circuits, US Patent 3969670, 1976.

48. Cadence, *LEF/DEF Language Reference*, Product Version 5.7, December 2010.

49. D. Pan, B. Yu, and J. Gao, Design for manufacturing with emerging lithography, *IEEE Transactions on Computer-Aided Design of Integrated Circuits and Systems*, 32(10), 1453–1472, October 2013.

50. C.-K. Wang et al., Closing the gap between global and detailed placement: Techniques for improving routability, *Proceedings of the International Symposium on Physical Design*, Monterey, CA, 2015.

51. A. Kennings, N. Darav, and L. Behjat, Detailed placement accounting for technology constraints, *Proceedings of the International Conference on Very Large Scale Integration*, Playa del Carmen, Mexico 2014.

52. K. Vaidyanathan, Rethinking ASIC design with next generation lithography and process integration, *Proceedings of SPIE, Design for Manufacturability through Design-Process Integration VII*, San Jose, CA, 2013.

53. J. Subramanian, Performance impact from metal fill insertion, *CDNLive!*, 2007.

54. Y. Ma et al., Self-aligned double patterning (SADP) compliant design flow, *Proceedings of SPIE, Design for Manufacturability through Design-Process Integration VI*, San Jose, CA, 2012.

55. D. Petranovic, J. Falbo, and N. Kurt-Karsilayan, Double patterning: Solutions in parasitic extraction, *Proceedings of SPIE, Design for Manufacturability through Design-Process Integration VII*, San Jose, CA, 2013.

56. D. Chinnery, High performance and low power design techniques for ASIC and custom in nanometer technologies, *Proceedings of the International Symposium on Physical Design*, Stateline, NV, 2013, pp. 25–32.

57. A. Dasdan, and I. Hom, Handling inverted temperature dependence in static timing analysis, *ACM Transactions on Design Automation of Electronic Systems*, 11(2), 306–324, April 2006.

58. C. Constantinescu, Trends and challenges in VLSI circuit reliability, *IEEE Micro*, 23, 14–19, 2003.

59. J. Lienig, Electromigration and its impact on physical design in future technologies, *International Symposium on Physical Design*, Stateline, NV, 2013, pp. 33–40.

60. Y. Zhang, L. Liang, and Y. Liu, Investigation for electromigration-induced hillock in a wafer level interconnect device, *Electronic Components and Technology Conference*, Las Vegas, NV, 2010, pp. 617–624.

61. A. Abbasinasab and M. Marek-Sadowska, Blech effect in interconnects: Applications and design guidelines, *International Symposium on Physical Design*, Monterey, CA, 2015, pp. 111–118.

Logic Synthesis

Sunil P. Khatri, Narendra V. Shenoy,* Jean-Charles Giomi, and Adel Khouja

CONTENTS

---

* Narendra Shenoy did not contribute to the material set forth in Section 2.4.3.2.

2.1 INTRODUCTION

The roots of logic synthesis can be traced to the treatment of logic by George Boole (1815–1865), in what is now termed *Boolean algebra*. Shannon's [1] discovery in 1938 showed that two-valued Boolean algebra can describe the operation of switching circuits. In the early days, logic design involved manipulating the truth table representations as Karnaugh maps [2,3]. The Karnaugh map–based minimization of logic is guided by a set of rules on how entries in the maps can be combined. A human designer can only work with Karnaugh maps containing four to six variables. The first step toward the automation of logic minimization was the introduction of the Quine–McCluskey [4,5] procedure that could be implemented on a computer. This exact minimization technique presented the notion of prime implicants and minimum cost covers that would become the cornerstone of two-level minimization. Another area of early research was in state minimization and encoding of finite-state machines (FSMs) [6–8], a task that was the bane of designers. The applications for logic synthesis lay primarily in digital computer design. Hence, IBM and Bell Laboratories played a pivotal role in the early automation of logic synthesis. The evolution from discrete logic components to programmable logic arrays (PLAs) hastened the need for efficient two-level minimization, since minimizing terms in a two-level representation reduces the area of the corresponding PLA. MINI [9] was an early two-level minimizer based on heuristics. Espresso [10] is an improvement over MINI; it uses the unate recursive paradigm (URP) as a central theme for many optimization steps.

However, two-level logic circuits are of limited importance in very large-scale integrated (VLSI) design; most designs use multiple levels of logic. An early system that was used to design multilevel circuits was logic synthesis system (LSS) [11] from IBM. It used local transformations to simplify logic. Work on LSS and the Yorktown Silicon Compiler [12] spurred rapid research progress in logic synthesis in the 1980s. Several universities contributed by making their research available to the public—most notably, MIS [13] from the University of California, Berkeley, and BOLD [14] from the University of Colorado, Boulder. Within a decade, the technology migrated to commercial logic synthesis products offered by electronic design automation companies.

The last two decades have seen tremendous progress in the field of logic synthesis. It has provided a dramatic productivity boost to digital circuit design, enabling teams to fully utilize the large number of transistors made available by decreasing feature sizes. This chapter provides a brief survey of the rapid advances made in this area. A more comprehensive treatment can be found in books by De Micheli [15], Devadas et al. [16], Hassoun and Sasao [17], and Hachtel and Somenzi [18].

2.2 BEHAVIORAL AND REGISTER-TRANSFER-LEVEL SYNTHESIS

To increase designer productivity, it is crucial to be able to specify and optimize designs at higher levels of abstraction. There has been a significant amount of research on the synthesis of circuits that are behaviorally specified using a hardware description language (HDL). The goal of behavioral synthesis is to transform a behavioral HDL specification into a register-transfer-level (RTL) specification, which can be used as input to a gate-level logic synthesis flow. A general overview of behavioral synthesis can be found in [19–23].

Behavioral optimization decisions are guided by cost functions that are based on the number of hardware resources and states required. These cost functions provide a coarse estimate of the combinational and sequential circuitry required to implement the design.

In a behavioral optimization tool, a front-end parser translates the behavioral HDL description of the design into a control and data flow graph (CDFG) [23]. Sometimes, separate control flow graphs (CFGs) and data flow graphs are created. This CDFG is subjected to high-level transformations, many of which are based on compiler optimizations [24], such as constant propagation, loop unrolling, dead code elimination, common subexpression elimination, code motion [25], and data flow analysis. Some hardware-specific optimizations performed in this step include making hardware-specific transformations (shifting to perform multiplication by a power of 2), minimizing the number of logic levels to achieve speedup, and increasing parallelism. The tasks of *scheduling* and *resource allocation and sharing* generate the FSM and the datapath of the RTL description of the design.

Scheduling [26] assigns operations to points in time, while allocation assigns each operation or variable to a hardware resource. Scheduling identifies places in the CFG where states begin and end, yielding the FSM for the design. Scheduling usually precedes allocation, although they can be intertwined. Among the scheduling algorithms in use are as soon as possible and, its counterpart, as late as possible. Other algorithms include force-directed scheduling [27], list scheduling [28,29], path-based scheduling [30], and symbolic scheduling [31].

Given a schedule, the allocation operation optimizes the amount of hardware required to implement the design. Allocation consists of three parts: functional unit allocation, register allocation, and bus allocation. The goal during allocation is to share hardware units maximally. Allocation for low power has been studied in [32], while [33] reports joint scheduling and allocation for low power. Behavioral synthesis for low power has been studied in [34–36].

Behavioral synthesis typically ignores the size and delay of the required control logic. The CDFG representation is significantly different from the representation used in the RTL network. As a result, a behavioral network graph (BNG) is utilized in [37]. The BNG is an RTL network, with the ability to represent unscheduled behavioral descriptions. Since wiring delays are becoming increasingly important in VLSI design, early estimation of such delays is very helpful. The fast bus delay predictor for high-level synthesis [38] fulfills this need. In [39], the notion of "don't cares" has been exploited in behavioral optimization. Much research has been conducted in the area of high-level synthesis for testability. For details, we refer the interested reader to a survey paper [40] on this topic. Examples of behavioral synthesis systems are found in [41,42]. The Olympus [43] system combines behavioral and logic syntheses.

2.3 TWO-LEVEL MINIMIZATION

Two-level logic minimization is arguably the workhorse of logic synthesis. Its early and most direct application included logic minimization for PLA-based designs. Since then, it has been extensively used in performing node optimization in multilevel technology-independent logic optimization.

Two-level logic minimization is an instance of the classical *unate covering problem* (UCP) [44–48]. For a logic function with n inputs and 1 output, the covering matrix has $O(2^n)$ rows (corresponding to minterms) and $O(3^n/n)$ columns (corresponding to primes of the function). In a typical solution to the UCP, we iterate the steps of row and column dominance and extraction of row singletons (essential primes) until the cover matrix cannot be further reduced (referred to as the "cyclic core"). At this point, techniques such as branch and bound are used to solve the cyclic core. Early solutions include the Quine–McCluskey approach [4,5]. The maximum independent set of primes can be used to bound the solution cost. In the early 1990s, two new exact techniques based on the computation of *signature cubes* were developed. In one of these methods [49], the size of the covering matrix is reduced (both rows and columns) yielding more efficient solutions. The other method [50] is based on the use of reduced ordered binary decision diagrams (ROBDDs) [51]. Characteristic ROBDDs of all primes and minterms are created, and dominance steps are formulated in terms of ROBDD operations. Similarly, in [52], the authors implicitly create the cyclic core, with a significantly lower computational cost than the previous approaches. In [44–46], improved bounding and pruning approaches are introduced and implemented in SCHERZO. The approach of Goldberg et al. [47,48] is based on performing branch and bound with a goal of proving that a given subspace cannot yield a better solution (negative thinking) as opposed to trying to find a better solution via branching (positive thinking). In [53], the authors provide a technique that combines the use of zero-suppressed binary decision diagrams (BDDs) [54] (to represent the data) and Lagrangian relaxation to solve the integer formulation of the UCP, yielding significant improvements.

Unate covering can be an expensive (though exact) approach to solving the two-level minimization problem. Several heuristic approaches have been developed as well. MINI [9] was one of the early heuristic two-level minimizers. ESPRESSO [10] improved on MINI, utilizing the URP at the core of the operations. In this heuristic approach, primes are never enumerated. Rather, operations are performed on a subset of primes. In ESPRESSO, the operations of *Reduce* (which reduces cubes in an ordered manner, such that the new set of cubes is still a cover), *Expand* (which expands cubes into primes, removing cubes that are covered by some other cubes in the cover), and *Irredundant* (which removes redundant cubes in the cover) are iterated until no further improvement is possible. These algorithms are based on cofactoring with respect to the most binate variable in the cover until unate leaves are obtained. The operation is efficiently performed on unate leaves, and then the results are recursively merged upward until the result of the operation on the original cover is obtained. When the reduce, expand, and irredundant iteration encounters a local minimum, a LASTGASP algorithm is called, which attempts to add more primes in a selective manner, in an attempt to escape the local minimum. After running LASTGASP, we again iterate on reduce, expand, and irredundant until no improvement is possible. ESPRESSO requires the computation of the complement of a function, which can become unmanageable for functions such as the Achilles heel function. In [55], a reduced offset computation is proposed to avoid this potential cube explosion.

ESPRESSO yields close to optimum results, with significant speedup compared to exact techniques. To bridge further the gap between ESPRESSO and exact approaches, iterative applications of ESPRESSO [56] can be used. In this technique, after an ESPRESSO iteration, cubes are selectively extracted from the onset and treated as don't-care cubes. Iterating ESPRESSO in this manner has been shown to result in bridging the (albeit small) optimality gap between ESPRESSO and exact techniques, with a modest runtime penalty.

ESPRESSO-MV [57] is the generalization of ESPRESSO to multivalued minimization. Two-level logic minimization has applications in minimizing Internet Protocol (IP) routing tables [58,59]. Hardware implementations of ESPRESSO tailored to this application [60,61] have reported significant speedups.

Boolean relations are a generalization of incompletely specified functions (ISFs), in that they are one-to-many multioutput Boolean mappings. Minimizing such relations involves finding the best two-level logic function that is *compatible* with the relation. The minimization of Boolean relations can be cast as a binate covering problem (BCP). In [62], a Quine–McCluskey-like procedure is provided to find an optimum two-level representation for a Boolean relation. The solution is based on a branch-and-bound covering method, applied to the BCP. Implicit techniques to solve the BCP are provided in [63], along with a comparison with explicit BCP solvers.

In [64], a branch-and-bound BCP algorithm is provided, with the input specified as a conjunction of several ROBDDs. Heuristic minimization of multivalued relations, using the two-level logic minimization paradigm of Brayton et al. [10], is presented in [65]. This multivalued decision diagram [66] (MDD)-based approach is implemented in a tool called GYOCRO [65].

2.4 MULTILEVEL LOGIC MINIMIZATION

2.4.1 TECHNOLOGY-INDEPENDENT OPTIMIZATION

Typical practical implementations of a logic function utilize a multilevel network of logic elements. A standard cell–based logic netlist, for example, utilizes such a network. A multilevel logic network can be abstracted as a directed acyclic graph (DAG), with edges representing wires and nodes representing memory elements or combinational logic primitives. Typically, we consider each node to have a single output, although this can be generalized as well. The combinational logic of a node can be represented in several ways; however, a two-level cover is the most commonly utilized method.

From an RTL description of a design, we can construct a corresponding multilevel Boolean network. This network is optimized using several technology-independent techniques before technology-dependent optimizations are performed. The typical cost function during technology-independent optimizations is total literal* count of the factored representation of the logic function (which correlates quite well with circuit area).

Many technology-independent optimizations can be performed on a multilevel Boolean network. We present several such optimizations, with a brief discussion of the salient techniques for each.

2.4.1.1 DIVISION

A function g is a *divisor* for f if $f = gh + r$, where r is a *remainder* function and h a *quotient* function. Note that h and r may not be unique. If $r = 0$, we refer to the process as "factoring." Division can be of two types: Boolean [13,67,68] and algebraic [13,69,70]. In general, Boolean division explores all possible functions that divide f. It is, therefore, computationally expensive. Algebraic division is less flexible but extremely fast since no Boolean operations are performed. Division can be performed recursively on g, h, and r to obtain an initial multilevel network.

2.4.1.2 KERNELING

One important decision in division is the choice of divisor g. Kernels [69] are used for this purpose. Kernels are cube-free† primary divisors.‡ Kernels can be computed efficiently using algebraic operations. The use of two-cube kernels [71] results in significant speedup in kerneling, with a minimal penalty in quality over the full-fledged kernel extraction.

The division techniques explained earlier can be utilized to optimize a multilevel network in several ways:

- If there exists a node g that can be divided into another node f in the network, we perform *substitution* of g into f.
- By *extracting* common subexpressions among one or more nodes, we may be able to reimplement a network with fewer literals. We find kernels of several nodes, choose a best kernel, and create a new node with the logic function of the kernel. Now, we substitute the new node into all remaining nodes in the design.
- We may *eliminate* a node by *collapsing* it into its fanouts to get out of a local minimum and enable further multilevel optimizations.

---

* A *literal* is a variable or its complement.
† A cube is a conjunction of literals. An expression is cube-free if no single cube can divide it evenly.
‡ Primary divisors are expressions obtained by algebraically dividing f with some cube c.

2.4.2 MULTILEVEL DON'T CARES

The use of multilevel don't cares can be very effective in reducing the literal count of a network. Multilevel don't-care-based optimization is typically invoked at the end of the aforementioned structural optimizations. Multilevel don't cares are first computed as functions of primary input variables as well as internal node variables. Then, by performing an (usually ROBDD-based) image computation, we find the image of the multilevel don't cares of a node in terms of the fanin variables of that node. This allows us to perform two-level minimization on the node using the newly computed don't cares, thereby reducing the literal count.

Early multilevel don't-care computation methods [72] were restricted to networks of NOR gates. The corresponding don't-care computation technique was referred to as the "transduction" method. Two sets of *permissible functions**—the maximum set of permissible functions (MSPFs) and the compatible set of permissible functions (CSPFs)—were defined. The downside of both is that they are defined to be global functions. Further, MSPFs are not compatible in the sense that if a function at some node j is changed, it may require the recomputation of the MSPFs of other nodes in the network. This limitation is removed for CSPFs.

More general multilevel don't-care techniques [73,74] were developed shortly after the transduction method. For these methods, the multilevel don't cares were computed as a disjunction of external don't cares (XDCs), satisfiability don't cares (SDCs), and observability don't cares (ODCs). XDCs for each output are typically specified as global functions, while SDCs for the circuit encode local logical conditions that cannot occur in the network. The network SDC is $\sum_i (y_i \oplus f_i)$, where f_i is the logic function of node y_i of the network. ODCs are computed as $\prod_k (\partial z_k / \partial y_i)$, where $\partial z_k / \partial y_i$ is the Boolean difference of primary output z_k with respect to y_i and encodes the minterms in the global space for which z_k is sensitive to changes in y_i.

The disjunction of the aforementioned don't cares is then imaged back to the local fanins of the node y_i, and the cover f_i is then minimized with respect to these local don't cares. Two-level minimization is used for this purpose.

If a node is minimized with respect to these don't cares (which include the ODC), it results in changes to the ODCs of other nodes in the design. This may be unimportant when the nodes are minimized one after another. However, if the optimization context is one in which the don't cares of all the nodes are utilized simultaneously, this does not guarantee correct results. For such applications, compatible output don't cares (CODCs) [75] can be used. After CODCs are computed for any design, a node can be optimized against its CODCs without the need to recompute CODCs of other nodes.

The don't-care computations outlined earlier rely on an image computation, which requires the computation of global ROBDDs of the circuit nodes. This limits the effectiveness of the don't-care computation. Recently, approximate CODC [76] and ODC [77] computations have been introduced. It was shown that by using a *window* of small topological depth in the forward and reverse directions from the node under consideration, a robust and effective don't-care computation can be devised. In [76], a depth of $k = 4$ resulted in a $30\times$ speedup and a $25\times$ reduction in memory requirements, while 80% of the literal reduction of the traditional CODC technique was obtained.

Another recent development in this area is the introduction of the sets of pairs of functions to be distinguished (SPFDs) [78]. SPFDs were first introduced in the context of Field Programmable Gate Array (FPGA) optimization. Their applicability to general logic network optimization was identified in [79]. The SPFD of a node is a set of ISFs.** As a result, the SPFD of a node encapsulates more information than the don't cares of that node. SPFDs to be distinguished can be represented as bipartite graphs. The SPFD of a node encodes the set of pairs of points in the primary input space, which must be *distinguished* or assigned different functional values. These pairs are redistributed among the fanins of the node, resulting in new SPFDs at the fanins of the node. Coloring these graphs results in new implementations of the fanin functions. In [80], an ROBDD-based implementation of an SPFD-based network optimization package was demonstrated. Results for wire replacement and fanin minimization were provided, with about 10% improvement over CODC-based optimizations. It was also shown that the flexibility offered by SPFDs contains that

---

* An ISF partitions the set of points in B_n into onset, offset, and don't care set points.

of traditional don't-care-based optimizations. Subsequently, SPFD-based optimizations have been demonstrated in other contexts as well, including rewiring [81], power, and delay minimization for FPGA synthesis [82], wire removal for PLA networks using multivalued SPFDs [83], and topologically constrained logic synthesis [84]. Sequential extensions to SPFDs were reported in [85].

2.4.3 TECHNOLOGY-DEPENDENT OPTIMIZATION

The multilevel technology-independent logic optimizations discussed so far in Section 2.4 have utilized a simple cost function, such as literal count. Technology-dependent optimization transforms an optimized technology-independent circuit into a network of gates in a given technology. The simple cost estimates are replaced by more concrete, implementation-driven estimates during and after technology mapping. The circuit structure created by mapping is crucial for the final quality of optimization. Technology mapping and several local optimizations available after technology mapping are described in this section.

2.4.3.1 TECHNOLOGY MAPPING

Mapping is constrained by factors such as the available gates (logic functions) in the technology library, the drive sizes for each gate, and the delay, power, and area characteristics of each gate. Further details on a technology library are given in Section 2.5.1. The initial work on technology mapping relied on rule-based heuristic transforms [86,87]. These were later replaced by rigorous algorithmic approaches. The process of technology mapping is conceptually best described by three generic steps. In the first step, called "subject graph construction," an optimized technology-independent circuit is decomposed into a DAG, consisting of a set of primitive gates (such as nand gates and inverters). There are many logically equivalent decompositions for a given circuit, and an appropriate choice is critical for good quality of mapping. The logic gates in the library are also decomposed into the same set of primitive gates. These are known as "patterns." The second step, called "pattern matching," generates matches of logic gates in the library to each primitive gate in the subject graph. The third step, called "covering," constructs a tiling of the subject graph with a selection of a subset of patterns generated in the previous step. The first constraint of covering is that each primitive gate of the subject graph is contained in at least one selected pattern. The second constraint requires the inputs to a selected pattern to be outputs of another selected pattern. The objective of covering is to find a minimum cost covering. The cost function can be area, delay, power, or a combination of these.

Matching can be performed by a variety of methods: structural matching using graph isomorphism or Boolean matching using BDDs. In the case of structural matching for trees, efficient string matching techniques can be used [88]. Boolean matching [89,90] is more powerful as it can detect matches independent of the local decomposition of the subject graph, under arbitrary input/output phase assignments and input permutations (of the library gate).* To avoid the combinatorial explosion due to permutations and phases, signatures [91,92] have been proposed.

The techniques utilized for the covering step depend on whether we are operating on a DAG or a tree. DAG covering can be formulated as a BCP [93]. It is not viable for very large circuits. Keutzer [94] describes an approach in which the subject graph is split into a set of fanout-free trees. Each tree is mapped optimally (for area). Thus, DAG covering is approximated with a sequence of tree mapping steps. Tree covering is solved using dynamic programming, which leads to optimum coverings [95] under certain conditions on the cost function. Minimum area tree-based technology mapping is solvable in time polynomial in the number of nodes in the subject graph and in the number of patterns in the library. Rudell [93] extends technology mapping to include timing using a binning technique. Touati [96] considers optimal delay mapping using a linear delay model. When the delay models are complex, the covering process requires an area–delay cost curve to be stored at each node in the subject graph. This can result in a high memory consumption. Roy et al. [97] present a compression algorithm that yields a solution bounded from the optimal solution by a user-specified threshold. Minimum area mapping under delay constraints

---

* This is often termed as NPN equivalence.

is more difficult and there is no known optimal algorithm [98]. The main disadvantages with applying tree covering are twofold. First, decomposing DAGs into trees leads to too many small trees, thereby losing any optimality with respect to the original DAG. The unpredictability of load estimation at the multifanout points, which are cut in order to convert DAGs into trees, causes errors in delay estimation. DAG covering is feasible if the delay models are modified to be load independent [99]. Stok et al. [100] and Kukimoto et al. [101] present DAG covering approaches under gain-based and load-independent delay models. Since the quality of technology mapping is very sensitive to the initial decomposition of the subject graph, Lehman et al. [102] propose decomposition within the covering step. A simple way to improve the quality of circuits constructed by tree covering is to use the inverter-pair heuristic [16].

To illustrate tree-based technology mapping for minimum area, consider a library of 10 combinational logic gates decomposed into patterns consisting of two-input nand gates and inverters as shown in Figure 2.1. Each gate may have multiple decompositions (only one is shown for each gate). Consider a tree of logic and its associated decomposition into a subject graph using two-input nand gates and inverters as shown in Figure 2.2. Observe that two inverters are added on a wire to enable the detection of more patterns as suggested by the aforementioned inverter-pair heuristic. A subset of matches of library patterns on the subject graph is indicated in Figure 2.3. Note that multiple matches can occur at the output of a gate. Technology mapping selects the best match of a pattern at a gate from a set of candidate matches using dynamic programming. In this approach, the least area cost of implementing a pattern at a gate

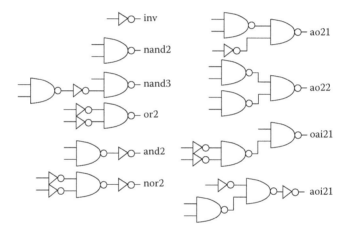

FIGURE 2.1 Decomposition of library gates into patterns.

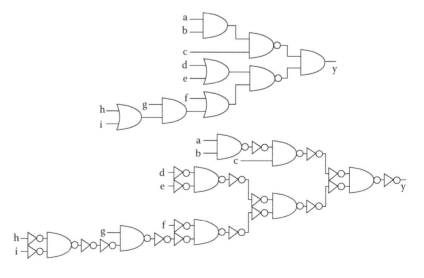

FIGURE 2.2 Logic undergoing mapping and its associated subject graph.

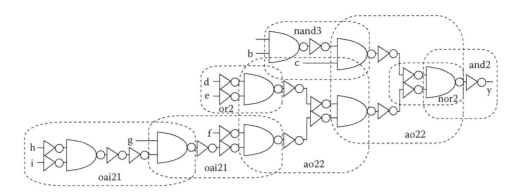

FIGURE 2.3   *Subset of pattern matches on the subject graph.*

is computed from the inputs toward the output of the tree. The selection of a pattern implies that the inputs to the pattern must be selected as well in the final implementation. Hence, the least area cost of selecting a pattern can be computed as the sum of the area of the pattern and the least area costs of implementing the signals at the inputs to the pattern. Since the solution is constructed from inputs toward the output, the latter quantities are readily available. Figure 2.4 shows two options of selecting a gate to implement the logic at *y*. Selecting a nor2 gate (area of 3) requires the inputs to the pattern to be implemented with area costs of 3 and 12, respectively, yielding a total area cost of 18. Selecting an aoi21 gate (area of three) requires the inputs to the pattern to be implemented with area costs of 2 and 12, respectively, yielding a total area cost of 17.

Although area and delay have been traditional objective functions in mapping, power is now a central concern as well. The tree-covering approach for low power under a delay constraint mimics the minimum area covering under a delay constraint. Tiwari et al. [103] and Tsui et al. [104] discuss technology mapping with a low power focus.

2.4.3.2 LOGICAL EFFORT-BASED OPTIMIZATIONS

The delay of a circuit can be expressed using the notions of logical and electrical effort [105]. The delay of a gate can be expressed as

$$d = \tau(p + gh)$$

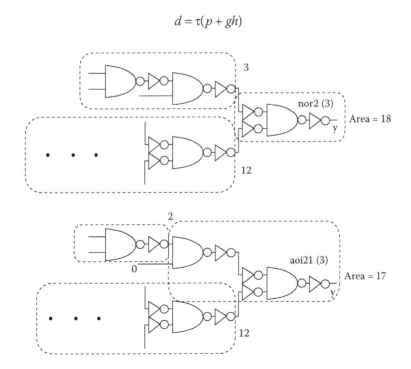

FIGURE 2.4   *Dynamic programming during technology mapping for area.*

Here, τ is a process-dependent parameter, while p is the parasitic delay of the gate (which arises due to the source/drain capacitances of the gate output). The parameter g is the *logical effort* of the gate, and it characterizes the ability of the gate to drive an output current. We assume that an inverter has a logical effort of unity. The logical effort of other gates depends on their topology and describes how much worse these gates are (in comparison with an inverter) in producing an output current, assuming that the input capacitances are identical to that of an inverter. The parameter h is the *electrical effort* (or gain) of the gate and is the ratio of the output capacitance to the input capacitance of a pin of the gate. Note that p and g are independent of the size of the gate, while h is sizing dependent.

In [105], the authors minimize delay along a circuit path by assigning an equal delay budget for each topological level along the circuit path. The resulting solution minimizes delay, with a solution that is not necessarily area minimal. In [106], the authors solve the fanout optimization problem to minimize the input capacitance of the source gate, subject to sink timing constraints, without considering area. In [107], the authors address the problem of minimizing buffer area subject to driver capacitance constraints. In [108], the idea of logical effort is applied to technology mapping, addressing the problem of load distribution at multifanout points. An analogous notion of interconnect effort is introduced in [109,110], combining the tasks of logic sizing and buffer insertion. Logical effort can also be applied to the design of regular structures such as adders [111,112].

2.4.3.3 OTHER TECHNOLOGY-DEPENDENT OPTIMIZATIONS

After technology mapping, a set of technology-dependent optimizations is carried out. Gate sizing, gate replication, and de Morgan transformations make local changes to the circuit. The impact of fanout optimization is more widespread. Critical path restructuring can make significant nonlocal changes. These approaches have the following common strategy. The first phase identifies portions of the circuit that need to be modified for possible benefit. The second phase selects a subset of these portions for change, modifies the circuit, and accepts the change if it results in an improvement. These steps rely on accurate incremental timing analysis to make this decision (see Section 2.5.2.1).

The early work on sizing focused on transistors and used a simple delay model (such as RC trees). This makes the objective function convex and easy to optimize [113–115]. Nonlinear programming techniques have also been used [116–118]. However in the ASIC context, logic optimization can only size gates and not individual transistors. Hence, the next wave of research focused on this aspect [119,120]. Also, the simple delay models cease to be accurate in modern technologies. Coudert et al. [121] provide an excellent survey on gate sizing.

Fanout optimization seeks to distribute optimally a signal to various input pins. This is critical for gates driving a large number of fanouts. Berman et al. [122], Hoover et al. [123], and Singh and Sangiovanni-Vincentelli [124] describe solutions to the fanout problem. Fanout optimization for libraries rich in buffer sizes is addressed by Kung [106]. Using the logical effort model, Rezvani et al. [107] discuss a fanout optimization tool for area and delay.

Bartlett et al. [125] were the first to focus on minimizing the area of a circuit under a delay constraint in the SOCRATES system. Singh et al. [126] describe a strategy for identifying regions for resynthesis using a weighted min-cut heuristic. The weights are determined based on the potential for speedup and an estimate of the area penalty incurred during optimization. Fishburn [127] describes a heuristic for speeding up combinational logic. Further refinements have been made by Yoshikawa et al. [128].

2.5 ENABLING TECHNOLOGIES FOR LOGIC SYNTHESIS

Some technologies have played a key role in the rapid advances in logic synthesis. We discuss some of the enabling research in this section.

2.5.1 LIBRARY MODELING

A key reason for the commercial success of logic synthesis is the availability of technology library models for various semiconductor foundries. Two popular commercial formats for library description are Liberty from Synopsys [129] and Advanced Library Format [130]. The library data can be categorized into two distinct sections:

1. *Technology data*: This includes information such as operating conditions (power supply and temperature) and wire load models. A wire load model is a statistical estimate of the capacitances observed on *typical* nets as a function of the number of pins and the size of the design (in terms of area). It is used as a proxy for net capacitance during synthesis when the physical data required for accurate estimation of net capacitance are unavailable (see discussion in Section 2.7 for more information).
2. *Library cell data*: Each cell in the library is accurately characterized for timing, power, and area. The delay is a function of circuit parameters (such as the input slew, output load, threshold voltage, and critical dimension) and operating conditions (such as local power and ground levels, temperature). For synthesis purposes, the delay is typically modeled as a nonlinear function of input slew and output load. It is specified as a 2D table. The move to smaller geometries is forcing the use of polynomials to capture the dependence on many parameters. The table-based approach does not scale in terms of memory usage as the number of parameters increases. The timing information also includes constraints such as setup and hold constraints. The power data include internal switching power and leakage power.

An issue of debate in the last decade was the nature and number of different logic functions in a standard cell library [131,132]. We currently have libraries with hundreds of logic functions, with multiple sizes for each function. More recent approaches have modeled gate delays using the concept of logical effort [105,133]. This approach assumes that the area of a gate is a continuous function of its load. Thus, a gate can be sized to keep the delay constant. This was first proposed in [99], and its use in technology mapping was explored in [102]. For this methodology to work, each gate in the library needs to have sufficiently many sizes so that the error in discretization in the final step is minimal [134,135].

2.5.2 TIMING ANALYSIS

A good logic optimization system requires access to an efficient timing analysis engine. In this section, we briefly address some of the work in this area.

2.5.2.1 INCREMENTAL STATIC TIMING ANALYSIS

Static timing analysis is covered in detail in Chapter 6. A key requirement for efficient timing optimization in a logic synthesis system is to have a fast incremental static timing analyzer coupled with the netlist representation. Technology-dependent optimizations (Section 2.4.3) operate on mapped circuits. Typically, local changes are made to the circuit. After each change, the design is queried to check if the timing has improved. Invoking static timing analysis on the whole design is a waste of computational resources if the impact of the changes is local. For example, when a gate is sized up, all gates in the transitive fanout cone of the gate and the immediate fanin gates are candidates for a change in arrival times. The actual update in timing is triggered by a query for timing information and is also dictated by the location of the query. This is known as *level limiting* [137]. However, not all arrival times may change as other paths can dominate the arrival. This is known as *dominance limiting*. The objective of incremental static timing analysis is to perform the minimal computation required to update the timing information that is invalidated by logic optimization changes. Timing information includes net capacitances, arrival times, required times, slack, transition times, and pin-to-pin delays. One of the published works on incremental static timing analysis in the context of netlist changes during synthesis is [136].

2.5.2.2 FALSE PATHS

False paths are covered in detail in Chapter 6. Our interest is in logic optimization performed in the context of handling false paths. Keutzer et al. [138] demonstrate that redundancy is not necessary to reduce delay. They provide an algorithm (called the KMS algorithm after the authors) that derives an equivalent irredundant circuit with no increase in delay. Their approach converts the well-known carry-skip adder into a novel irredundant design. Saldanha et al. [139] study the relationship of circuit structure to redundancy and delay. Kukimoto and Brayton [140] extend the notion of false paths to safe replaceability under all input arrival considerations.

2.6 SEQUENTIAL OPTIMIZATION

Historically, the first logic synthesis approaches were combinational in nature. Then, sequential techniques were developed to manipulate and optimize single FSMs. Following this, sequential techniques to manipulate a hierarchy of FSMs were developed. FSMs are similar to finite automata [141], except that FSMs produce output signals, while automata produce no outputs and simply accept or reject input sequences. Moore machines [142] are FSMs whose output functions depend only on the present state, while Mealy machines [142] are FSMs whose outputs depend on the present state as well as the applied input.

A combinational circuit implements a Boolean function, which depends only on the inputs applied to the circuit. Acyclic circuits are necessarily combinational while cyclic circuits may be combinational [144–147]. However, such circuits are not commonly used in practice.

The rest of this section deals with synchronous sequential circuits. Such circuits typically implement state elements using clocked latches or flip-flops. Asynchronous sequential circuits, in contrast, are not clocked. Synchronous sequential behavior can be specified in the form of a netlist consisting of memory elements and combinational logic, as state transition graphs (STGs) or as transition relations. Transition relations are typically represented implicitly using ROBDDs. To avoid memory explosion, the ROBDDs of a transition relation can be represented in a partitioned manner [148]. Often, the FSM behavior is expressed *nondeterministically* by allowing transitions to different states (with different outputs) from a given input state and input combination. Nondeterminism allows for compact representations of the FSM behavior. Also, the transition and output functions may be *incompletely specified.* In such a case, any behavior is considered to be allowed. Of course, any implementation must be deterministic and completely specified; however, the relaxation of these restrictions during optimization allows us to express a multitude of allowable behaviors in a compact fashion.

2.6.1 STATE MINIMIZATION

Given an STG, with symbolic states and transitions, we perform state minimization to combine *equivalent states* (states that produce identical output sequences, given identical input sequences). This is cast as a fixed-point computation. Typically, for completely specified machines, at step i, we construct a set of sets of equivalent states that form a partition of the original state space and refine the set by separating the states that can be distinguished by an additional step. This is continued until convergence, starting with an initial set that consists of all states in the state space. For completely specified machines, the problem has polynomial complexity.

For incompletely specified machines, we compute the set of *prime compatibles*, and for those machines, the problem is Non-deterministic Polinomial (NP)-hard [148]. One of the early approaches for deterministic incompletely specified FSMs was given in [149]. In [150], it was shown that a minimum state cover for an incompletely specified machine could be found using prime compatibles. In [151], an efficient minimization method was introduced, using the notion of compatibles. In [152], exact as well as heuristic minimization results were reported and implemented in a tool called STAMINA. An implicit (using ROBDDs) computation of the compatibles was reported in [153], allowing the technique to handle extremely large numbers of compatibles. In [154], the authors address the minimization of nondeterministic FSMs based on the notion of generalized compatibles, which are computed implicitly.

One of the early works in the minimization of a network of FSMs based on input don't-care sequences was by Kim and Newborn [155], which computed the input don't-care sequences for the driven machine. Later, Rho and Somenzi [156] showed that for an incompletely specified machine, there exists an *analogous* machine, which has to be minimized for the optimization of two interacting completely specified machines. In [157], an implicit procedure to compute all input don't-care sequences in a general FSM network was introduced. The work of Watanabe and Brayton [158] computes the maximum set of permissible behaviors for any FSM in a network of FSMs. The resulting nondeterministic machine is referred to as the "E-machine." State minimization of this machine is reported in [159].

In [160], the problem of minimizing a network of interacting FSMs in the context of language containment-based formal property verification is addressed. The problem of synthesizing low-power FSMs is addressed in [161], while don't-care-based minimization of extended FSMs is reported in [162].

The decomposition of an FSM is addressed in [163], where the objective is power reduction. A Kernighan–Lin-style [164] partitioning procedure is utilized. In [165], the goal of the FSM decomposition is I/O minimization, and it is achieved using a Fiduccia–Mattheyses-based [166] partitioning approach.

2.6.2 STATE ASSIGNMENT

Given a minimized STG, we perform state assignment to assign binary codes to each symbolic state. The choice of binary codes has a great impact on the area and power of the final synthesized design. State assignment can be viewed as an encoding problem, using either input encoding, output encoding, or input–output encoding.

The encoding problem is split into two steps. In the first step, a multivalued representation is minimized, along with a set of constraints on the codes used for the symbolic states. The next step involves finding an encoding that satisfies these constraints. The encoded representation has the same size as that of the minimized multivalued representation. Encoding can be performed for either two- or multilevel implementations of the FSM. In two-level implementations, we attempt to minimize the number of cubes, while in multilevel implementations, we attempt to minimize the number of literals in the design.

The input constraints are *face-embedding* constraints—they specify that any symbol can be assigned to a single face of the Boolean n cube, without that face being shared by other symbols. Output constraints are *dominance** or *disjunctive*† constraints. In [167], it was shown that finding a minimum-length code satisfying input constraints is NP-hard [148].

In [168,169], input-encoding-based state assignment for minimum two-level realizations is reported. A solution producing multilevel implementations is given in [170]. The output-encoding procedure of Devadas and Newton [171] generates disjunctive constraints with nested conjunctive terms. The NOVA algorithm of Villa and Sangiovanni-Vincentelli [172] exploits dominance constraints. In [167,173], both input-encoding and output-encoding constraints are handled simultaneously. Multilevel encoding is reported in [174–176]. State assignment with the goal of minimizing the size of the ROBDD of the FSM transition relation is reported in [177]. Parallel state assignment is explored in [178], while state assignment for low power is studied in [179–183]. State assignment for testability is studied in [183].

2.6.3 RETIMING

Once a sequential netlist has been generated, further optimization can be achieved by *retiming*, which involves moving registers across logic elements, in a manner such that the sequential behavior of the circuit is maintained. The goal of retiming may be to minimize clock period (*min-period retiming*), or the number of registers (*min-area retiming*), or to minimize the number of registers

---

* *Dominance constraints* express the condition in which the code for any symbol covers that of another.
† *Disjunctive constraints* express the condition in which the code of any symbol is the bit-wise OR of other symbols.

subject to a maximum clock period constraint (*constrained min-area retiming*). Retiming may be coupled with resynthesis as well. In all retiming approaches, the circuit is represented as a graph, with edges representing wires and vertices representing gates. The weight of a vertex corresponds to the gate delay, and the weight of an edge corresponds to the number of registers on that edge.

Early retiming efforts [184] were based on mixed integer linear programming approaches. Later, relaxation-based [185] approaches were reported. In general, retimed circuits require a new initial state to be computed [186]. Resettable circuits [187] may be utilized to implement retimed designs so that an initializing sequence, which can be used to bring the machine into a known state after power-up, can be easily computed. In [188], techniques to minimize the effort of finding new equivalent initial states after retiming are integrated in the retiming approach. In [189], a min-area retiming methodology, which guarantees the existence of equivalent initial states, is described.

Retiming for level-sensitive designs was studied in [190–193]. Efficient implementations of min-period and constrained min-area retiming were studied in [194], while Maheshwari and Sapatnekar [195] demonstrated efficient min-area retiming.

In [196], the approach is to combine retiming and resynthesis by moving registers to the circuit boundary, then optimizing the combinational logic and moving registers back into the design. Peripheral retiming for pipelined logic [197] involves retiming such that internal edges have zero weight, although peripheral edges may have negative weights. Once the retiming is legalized (all weights are made greater than or equal to zero), we obtain a functionally equivalent circuit (equivalence is exhibited after an initializing sequence is applied). Another resynthesis approach [198] utilizes the retiming-induced register equivalence to optimize logic.

Retiming has been studied in the FPGA context [199,200] as well as in the context of testability preservation and enhancement [201,202]. Retiming for low power is described in [203]. Recent retiming approaches account for DSM delay constraints [204], interconnect and gate delay [205], and clock distribution [206,207]. Retiming is used to guide state assignment in [208].

2.7 PHYSICAL SYNTHESIS

With the advance of technology in the late 1990s, it became evident that physical effects could not be ignored during logic synthesis. We briefly summarize some of the technology issues that led to this crisis. Decreasing transistor sizes translate to smaller and faster gates. Scaling down gate sizes implies that capacitances of input pins of gates decrease. To incorporate more logic, the routing resources cannot be scaled in a commensurate manner. Consequently, the number of metal interconnect layers increased from three to a much larger number (8–10 is common in current technologies). In order to lower the resistance of interconnect, the cross-sectional area has to be increased. If the width of the wire is increased, then routability suffers as fewer wires can be packed in a given area. Consequently, the height of metal interconnect needs to be increased. This raises the lateral capacitance of an interconnect line to neighboring interconnect lines. Until the mid-1990s, synthesis and physical design (placement and routing) had been very weakly coupled. A simple notion of net weights to indicate timing criticality was sufficient for placement. The concept of net list sign-off served as a viable business model among the foundries, design teams, and tool vendors. The delay estimate of a gate during synthesis relies on an accurate estimate of the capacitance on a net driven by the gate. The net capacitance is the sum of the pin capacitance and the wire capacitance (the loading from the interconnect used to connect electrically all the pins on the net). Synthesis uses wire load models (Section 2.5.1) as a proxy for wire capacitances. As long as the postrouted wire capacitance does not account for a significant fraction of the total net capacitance, the inaccuracies in delay calculation during synthesis do not cause a problem. At 0.25 mm, wire capacitances began to dominate pin capacitances. Consequently, technology-dependent optimizations were less effective without a means of correctly estimating wire capacitances. This, in turn, required access to the physical placement of cells and pins. The placement algorithms had relatively weak timing models and analysis capabilities. Recall that static timing analysis was primarily developed in the context of synthesis. These factors led to nonconvergence of a design through the steps of synthesis, placement, and routing in the design flow. Moreover, effects such as signal integrity issues, which were ignored in previous technology generations, could no longer be ignored due to the strong lateral coupling capacitances.

Today, physical synthesis is a key step that has replaced placement and late timing correction in synthesis. The early work in this area focused on incorporating physical effects during logic optimization [209–212]. Keutzer et al. [213], in an invited paper, discussed the technical and business issues of this discontinuity. The second-generation efforts looked toward a closer integration of synthesis and placement [214–216]. The third-generation research [217,218] enabled commercial products to enter the market. Gosti et al. [219] improved the companion placement model proposed earlier [209] for better area and delay performance. The concept of logical effort [133] enables an alternative view of deferring technology-dependent optimizations until some placement information is available. Physical synthesis uses two sets of distinct optimizations to achieve convergence of the final design. Logic synthesis during placement is able to change the netlist structure undergoing placement. This enables operations such as effective buffering of long wires and sizing up of weak gates. Placement will penalize cell congestion, cell overlaps, and nets with long wire lengths. Working in tandem, the two techniques produce much better results than either one alone.

Carragher et al. [220] discuss layout-driven logic optimization strategies. Logic restructuring [221] and technology mapping [222] with physical information have also been studied. Murgai [225] investigates area recovery using layout-driven buffer optimization. Kudva et al. [224] discuss metrics for measuring routability in the technology-independent optimization phase when the structure of the netlist is defined. Saxena and Halpin [225] discuss incorporating repeaters within the placement formulation.

2.8 MULTIVALUED LOGIC SYNTHESIS

Multivalued logic synthesis has received much attention in recent times. Although research on multivalued circuit [226–228] as well as memory [55] implementations has been reported, such circuits are hard to design and fabricate, as evidenced by the significantly greater attention given to binary-valued circuits.

The role of multivalued techniques is arguably during the early stages of synthesis. After multivalued logic optimizations have been performed, the design can be encoded into binary. At this point, the traditional binary-valued logic optimization flow can be applied.

Early multivalued techniques included the generalization of ESPRESSO to multiple values [57]. The reduced offset computation, which is used in ESPRESSO, also has a multivalued counterpart [230]. ROBDDs have also been generalized to multiple values [66]. Many other binary-valued logic optimization techniques have been generalized to multiple values, such as algebraic division [231], factorization [232], generalized cofactoring [233], don't cares [234], redundancy removal [236], wire removal [237], and satisfiability (SAT) [238]. In [239], nondeterministic multivalued network simplification is addressed. A nondeterministic multivalued relation is used to express the maximum flexibility of a node. In [240], a technique to reduce multivalued algebraic operations into binary is reported, yielding significant speedups in operations on multivalued networks. Multivalued, multilevel synthesis techniques are discussed in [241,242]. MVSIS [242] implements many of the earlier techniques, in addition to encoding.

Static [241] and dynamic [242] MDD variable reordering have been implemented, in addition to dynamic reencoding [243]. MDD-based synthesis [244] and functional decomposition [245] approaches have also been reported. The D-algorithm for ATPG has been generalized to multiple values [246]. In [247], the authors describe fault simulation for sequential multivalued networks.

2.9 SUMMARY

In this chapter, we have explored some of the key state-of-the-art logic synthesis techniques. Our focus has been on mainstream synthesis approaches, based on the CMOS standard cell design paradigm.

We covered high-level, two-level, and multilevel synthesis as well as sequential synthesis in this chapter. Related enabling technologies were covered, along with recent approaches that combine physical design and logic synthesis.

Starting with early logic synthesis techniques, there has been a steady push toward an increased level of abstraction. At the same time, technology considerations have forced logic synthesis to become physically aware. In the future, we will see several challenges. Improved multivalued logic synthesis techniques will enable more effective algorithmic approaches to high-level synthesis. At the same time, the quality of the netlists produced by logic synthesis (from a physical design perspective) will continue to be important. The capacity and speed of optimization will continue to be hurdles for flat multimillion gate synthesis.

A change in the underlying implementation style, such as a diversion from CMOS, can be a driver for new research in logic synthesis approaches. Currently, there is much excitement about logic synthesis for quantum computing.

2.10 UPDATE FOR THE SECOND EDITION

In the past decade, the field of logic synthesis has faced several new challenges. It remains clear that as technology nodes advance, logic synthesis optimizations have a major impact on physical design feasibility. Newer technology nodes (22, 16/14, and 10 nm) have increased the need for logic synthesis to accurately estimate new physical effects during optimization.

Power has grown as an important concern alongside delay and area. Both leakage and dynamic power optimizations have been researched extensively and are performed to varying degrees of accuracy by commercial synthesis tools.

The industry has also evolved to require that all logic synthesis optimizations provided by commercial synthesis tools can be formally verified by independent tools to ensure functional correctness. This new requirement has triggered research interests in both formal verification and logic synthesis, especially for sequential optimizations.

As design sizes continue to increase by more than an order of magnitude in the last decade, logic synthesis capacity improvement has become an active area of research and development leveraging parallelization and multicore CPUs.

2.10.1 PHYSICAL EFFECTS

Physical synthesis continues to be a critical step in the design flow and has been actively enhanced for new technology nodes. Interconnect routing estimation remains the main challenge to properly model interconnect delay and congestion during logic synthesis. First, as interconnect delay continues to become more significant with new technology nodes, delay estimation has been enhanced. More accurate CMOS cell and interconnect delay calculation replaced nonlinear delay models with current source models [248,249]. Second, interconnect routing estimation during logic synthesis remains challenging due to inaccuracies of early placement, blockage constraints affecting final routing, metal layer assignment, and VIA estimation. That being said, Steiner-based routing estimation has been enhanced extensively [250,251] and continues to be the main technique used for routing estimation in logic synthesis.

Physical synthesis has focused on technology-dependent optimizations like remapping, sizing, and buffering linked closely with placement, legalization, and routing [252–254]. At newer technology nodes, the different metal layers in the stack have very different resistive properties per unit length. For a given distance, the interconnect delay varies significantly depending on the metal layer used (higher metal layers providing smaller delay over longer wire lengths). Physical synthesis has been enhanced to determine which delay-critical net connections should be assigned to a higher metal layer—also known as layer promotion [255]. More recent work [256] has explored different congestion models and metrics for each optimization in the physical synthesis flow. As technology nodes have advanced, nondefault rules (NDRs) have recently been added to physical synthesis offerings to address electromigration constraints and reduce parasitic variations by allowing wider wire widths and spacings. Estimating the impacts of NDRs is another growing challenge.

Area and delay optimized logic structures from traditional logic synthesis often exhibit poor physical properties vis-à-vis routability and pin accessibility. Poor pin accessibility at lower geometries can contribute to significant congestion. Various techniques have been proposed to make logic synthesis generate logic structures that simplify placement and routing, especially for datapath and mux/select logic [257–259]. Moreover, logic synthesis faces the added challenge to modeling physical effects even earlier in the flow, at the technology-independent stage, when the design has not yet been mapped.

2.10.2 BOOLEAN OPTIMIZATION AND TECHNOLOGY MAPPING

As designs have increased in complexity and size and are reusing more IP cores, redundant and duplicated logic is becoming more common. To address this, logic optimization has borrowed from ATPG and advanced formal verification techniques including SAT [260], structural hashing, simulation, and recent hybrid techniques known as SAT sweeping that leverages an underlying and-inverter graph as the foundation for the various logic transformations. Collectively, these new optimization techniques enable reductions in the final gate count and circuit depths of the synthesized circuits. Similarly, Mishchenko introduced flexibility computation techniques in Boolean networks [261], and Chatterjee described improvements in technology mapping [262] introducing new Boolean matching algorithms and leveraging advances in sequential verification.

Along with retiming and clock gating, sequential optimizations like sequential merging, sequential duplication, and sequential phase inversion have become commonly available in commercial logic synthesis tools and have initiated advanced research in verification [263,264].

2.10.3 LOW POWER

Clock gating has been the most commonly used design technique to reduce dynamic power. Clock gating elements can be inferred either (1) manually by designers based on design knowledge or (2) automatically by logic synthesis tools. Automatic clock gating has been initially performed by analyzing the next-state function to identify conditions under which the clock can be gated [265,266]. Recently, clock gating has been extended to identify conditions under which the clock does not need to toggle using a combination of simulation and SAT techniques [267]. Sequential clock gating extends clock gating further based on sequential analysis [268–270], creating new challenges for formal sequential verification [271]. In addition, power-aware technology mapping approaches have been researched along with efficient switching-activity computation [272]. Register-transfer-level analysis and optimization like operand isolation or logic gating have been effective to reduce dynamic power in high-toggling logic like datapaths [273]. Standard cell libraries have been enhanced to include multibit flops (e.g., 2 bit or 4 bit) with a shared clock pin, providing good dynamic power savings compared to equivalent single-bit elements. Mapping to these cells with physical awareness reduces power without degrading timing and congestion.

On the other hand, leakage power has been reduced by providing a broader spectrum of library cells with delay/area/leakage trade-offs at various voltage thresholds (Vts) [274–277]. Optimization during logic synthesis can utilize low-Vt cells (with high leakage power and fast delay) on timing critical paths and high-Vt cells in nontiming critical paths (with low leakage power and slow delay).

Finally, commercial logic synthesis tools are now supporting multiple-supply voltage design technique with automatic isolation and level shifter cell insertion. The industry has also converged on a power intent design specification language standard [278].

2.10.4 CAPACITY AND SPEED

Capacity requirements in terms of runtime and memory usage have increased significantly due to an industry trend of designing larger and more complex SoCs leveraging IP reuse. Distributed parallelization based on coarse-grain partitioning has been proposed [279]. Parallelization based on multithreading is best for low-level parallelizable algorithms. Catanzaro discussed the challenges in parallelizing current graph-based algorithms used in logic synthesis and introduced a parallel logic optimization framework based on low-level partitioning [280]. Over the years, distributed synthesis techniques combined with low-level multithreaded algorithms have incrementally delivered tangible speedups.

2.10.5 ECO SYNTHESIS

In the IC industry, it is common practice to make *engineering change orders* (ECOs) to fix functional design errors and timing closure errors or even make small functional changes on an already implemented design. As design size and complexity have increased significantly, it became more difficult to manually implement ECOs. Therefore, ECO synthesis has been recently introduced and delivers automatic techniques to implement ECO later in the design cycle on an existing pre- or postmask layout [281]. ECO synthesis will first determine the minimum functional difference introduced by an ECO change. Extensive research has been done to formally identify a minimum set of logic gates between two implementations using satisfiability-based proof and interpolation techniques [282]. This functional logic difference (patch logic) will be physically optimized and appropriately mapped onto available cells. Specific ECO-aware logic and physical synthesis techniques have been developed to ensure an optimal implementation [283–285].

2.10.6 CONCLUSION OF THE UPDATE

In this update, we have summarized key technology advances in the area of logic synthesis over the last decade. We fully expect that the quality of netlists produced by logic synthesis will remain critical to ensure convergence in the physical design flow. As new technology requirements are identified [286,287], logic synthesis will continue to face new challenges that drive exciting research and create opportunities for innovation in commercial tools.

REFERENCES

1. C. Shannon, A symbolic analysis of relay and switching circuits, *Trans. Am. Inst. Electr. Eng.*, 57, 713–723, 1938.
2. M. Karnaugh, The map method for synthesis of combinational logic circuits, *Trans. Am. Inst. Electr. Eng.*, 72, 593–599, 1953.
3. E. Veitch, A chart method for simplifying truth functions, *Proceedings of the Association for Computing Machinery*, 1952, pp. 127–133.
4. W. Quine, The problem of simplifying truth functions, *Am. Math. Mon.*, 59, 521–531, 1952.
5. E. McCluskey, Minimization of Boolean functions, *Bell Syst. Tech. J.*, 35, 1417–1444, 1956.
6. D. Huffman, The synthesis of sequential switching circuits, *J. Frankl. Inst.*, 257, 161–190, 1954.
7. G. Mealy, A method for synthesizing sequential circuits, *Bell Syst. Tech. J.*, 34, 1045–1079, 1955.
8. E. Moore, Gedanken-experiments on sequential machines, in *Automata Studies*, Princeton University Press, Princeton, NJ, 1956, Vol. 34, pp. 129–153.
9. S. Hong, R. Cain, and D. Ostapko, MINI: A heuristic approach for logic minimization, *IBM J. Res. Dev.*, 18, 443–458, 1974.
10. R. Brayton, A. Sangiovanni-Vincentelli, C. McMullen, and G. Hachtel, *Logic Minimization Algorithms for VLSI Synthesis*, Kluwer Academic Publishers, Dordrecht, the Netherlands, 1984.
11. J. Darringer, D. Brand, J. Gerbi, W. Joyner, and L. Trevillyan, LSS: A system for production logic synthesis, *IBM J. Res. Develop.*, 28, 272–280, 1984.
12. R. Brayton, R. Camposano, G. DeMicheli, R.H.J.M. Otten, and J.T.J. van Eijndhoven, The Yorktown silicon compiler system, in *Silicon Compilation*, D. Gajski, Ed., Addison-Wesley, Reading, MA, 1988.
13. R. Brayton, R. Rudell, A. Sangiovanni-Vincentelli, and A. Wang, MIS: A multiple-level logic optimization system, *IEEE Trans. Comput. Aided Des. Integr. Circ. Syst.*, 6, 1062–1081, 1987.
14. D. Bostick, G.D. Hachtel, R. Jacoby, M.R. Lightner, P. Moceyunas, C.R. Morrison, and D. Ravenscroft, The Boulder optimal logic design system, *Proceedings of IEEE International Conference on Computer-Aided Design*, Santa Clara, CA, 1987, pp. 62–65.
15. G. De Micheli, *Synthesis and Optimization of Digital Circuits*, McGraw-Hill, New York, 1994.
16. S. Devadas, A. Ghosh, and K. Keutzer, *Logic Synthesis*, McGraw-Hill, New York, 1994.

17. S. Hassoun and T. Sasao, Eds., *Logic Synthesis and Verification*, Kluwer Academic Publishers, Dordrecht, the Netherlands, 2002.
18. G. Hachtel and F. Somenzi, *Logic Synthesis and Verification Algorithms*, Kluwer Academic Publishers, Dordrecht, the Netherlands, 2000.
19. R. Camposano, From behavior to structure: High-level synthesis, *IEEE Des. Test Comput.*, 7, 8–19, 1990.
20. M. McFarland, A. Parker, and R. Camposano, The high-level synthesis of digital systems, *Proc. IEEE*, 78, 301–318, 1990.
21. R. Camposano and W. Wolf, *High Level VLSI Synthesis*, Kluwer Academic Publishers, Boston, MA, 1991.
22. D. Gajski and L. Ramachandran, Introduction to high-level synthesis, *IEEE Des. Test Comput.*, 11, 44–54, 1994.
23. R. Camposano and R. Tebet, Design representation for the synthesis of behavioral VHDL models, *Proceedings of the International Symposium on Computer Hardware Description Languages and their Applications*, Washington, DC, 1989, pp. 49–58.
24. A. Aho, R. Sethi, and J. Ullman, *Compiler Principles, Techniques and Tools*, Addison-Wesley, Reading, MA, 1986.
25. S. Gupta, N. Savoiu, N. Dutt, R. Gupta, and A. Nicolau, Using global code motions to improve the quality of results for high-level synthesis, *IEEE Trans. Comput. Aided Des. Integr. Circ. Syst.*, 23, 302–312, 2004.
26. S. Amellal and B. Kaminska, Scheduling of a control data flow graph, *Proceedings of the IEEE International Symposium on Circuits and Systems*, Portland, OR, Vol. 3, 1993, pp. 1666–1669.
27. P. Paulin and J. Knight, Force-directed scheduling for the behavioral synthesis of ASICs, *IEEE Trans. Comput. Aided Des. Integr. Circ. Syst.*, 8, 661–679, 1989.
28. A. Parker, J. Pizarro, and M. Mlinar, MAHA: A program for datapath synthesis, *Proceedings of ACM/IEEE Design Automation Conference*, Las Vegas, NV, 1986, pp. 461–466.
29. S. Davidson, D. Landskov, B. Shriver, and P. Mallet, Some experiments in local microcode compaction for horizontal machines, *IEEE Trans. Comput.*, C-30, 460–477, 1981.
30. R. Camposano, Path-based scheduling for synthesis, *IEEE Trans. Comput. Aided Des. Integr. Circ. Syst.*, 10, 85–93, 1991.
31. I. Radivojevic and F. Brewer, Symbolic techniques for optimal scheduling, *Proceedings of SASIMI Workshop*, Nara, Japan, 1993, pp. 145–154.
32. A. Raghunathan and N. Jha, Behavioral synthesis for low power, *Proceedings of IEEE International Conference on Computer Design*, Cambridge, MA, 1994, pp. 318–322.
33. Y. Fang and A. Albicki, Joint scheduling and allocation for low power, *Proceedings of the IEEE International Symposium on Circuits and Systems*, Atlanta, GA, Vol. 4, 1996, pp. 556–559.
34. C. Gopalakrishnan and S. Katkoori, Behavioral synthesis of datapaths with low leakage power, *Proceedings of IEEE International Symposium on Circuits and Systems*, Scottsdale, AZ, 2002, Vol. 4, pp. 699–702.
35. R. San Martin and J. Knight, Optimizing power in ASIC behavioral synthesis, *IEEE Des. Test Comput.*, 13, 58–70, 1996.
36. K. Khouri and N. Jha, Leakage power analysis and reduction during behavioral synthesis, *IEEE Trans. Very Large Scale Integr. (VLSI) Syst.*, 10, 876–885, 2002.
37. R. Bergamaschi, Bridging the domains of high-level and logic synthesis, *IEEE Trans. Comput. Aided Des. Integr. Circ. Syst.*, Santa Clara, CA, 21, 582–596, 2002.
38. R. Pomerleau, P. Franzon, and G. Bilbro, Improved delay prediction for on-chip buses, *Proceedings of ACM/IEEE Design Automation Conference*, New Orleans, LA, 1999, pp. 497–501.
39. R. Gupta and J. Li, Control optimizations using behavioral don't cares, *Proceedings of IEEE International Symposium on Circuits and Systems*, Atlanta, GA, 1996, Vol. 4, pp. 404–407.
40. K. Wagner and S. Dey, High-level synthesis for testability: A survey and perspective, *Proceedings of ACM/IEEE Design Automation Conference*, Las Vegas, NV, 1996, pp. 131–136.
41. W. Wolf, A. Takach, C.-Y. Huang, R. Manno, and E. Wu, The Princeton University behavioral synthesis system, *Proceedings of ACM/IEEE Design Automation Conference*, Anaheim, CA, 1992, pp. 182–187.
42. G. DeMicheli and D. Ku, HERCULES: A system for high-level synthesis, *Proceedings of ACM/IEEE Design Automation Conference*, Atlantic City, NJ, 1988, pp. 483–488.
43. G. DeMicheli, D. Ku, F. Mailhot, and T. Truong, The Olympus synthesis system, *IEEE Des. Test Comput.*, 7, 37–53, October 1990.
44. O. Coudert, Two-level minimization: An overview, *Integration*, 17, 97–140, 1994.
45. O. Coudert and J.-C. Madre, New ideas for solving covering problems, *Proceedings of ACM/IEEE Design Automation Conference*, San Francisco, CA, 1995, pp. 641–646.
46. O. Coudert, On solving covering problems, *Proceedings of ACM/IEEE Design Automation Conference*, Las Vegas, NV, 1996, pp. 197–202.

47. E. Goldberg, L. Carloni, T. Villa, R. Brayton, and A. Sangiovanni-Vincentelli, Negative thinking by incremental problem solving: Application to unate covering, *Proceedings of IEEE International Conference on Computer-Aided Design*, Santa Clara, CA, 1997, pp. 91–99.

48. E. Goldberg, L. Carloni, T. Villa, R. Brayton, and A. Sangiovanni-Vincentelli, Negative thinking in branch-and-bound: The case of unate covering, *IEEE Trans. Comput. Aided Des. Integr. Circ. Syst.*, 19, 281–294, 2000.

49. P. McGeer, J. Sanghavi, R. Brayton, and A. Sangiovanni-Vicentelli, ESPRESSO-SIGNATURE: A new exact minimizer for logic functions, *IEEE Trans. Very Large Scale Integr. (VLSI) Syst.*, 1, 432–440, 1993.

50. G. Swamy, R. Brayton, and P. McGeer, A fully implicit Quine-McCluskey procedure using BDDs, *Proceedings of the International Workshop on Logic Synthesis*, Tahoe City, CA, 1993.

51. R. Bryant, Graph-based algorithms for Boolean function manipulation, *IEEE Trans. Comput.*, C-35, 677–691, 1986.

52. O. Coudert, J.-C. Madre, and H. Fraisse, A new viewpoint on two-level logic minimization, *Proceedings of ACM/IEEE Design Automation Conference*, Dallas, TX, 1993, pp. 625–630.

53. R. Cordone, F. Ferrandi, D. Sciuto, and R. Calvo, An efficient heuristic approach to solve the unate covering problem, *Proceedings, Design, Automation and Test in Europe Conference*, Paris, France, 2000, pp. 364–371.

54. S. Minato, Implicit manipulation of polynomials using zero-suppressed BDDs, *Proceedings of European Design and Test Conference*, Paris, France, 1995, pp. 449–454.

55. A. Malik, R. Brayton, A.R. Newton, and A. Sangiovanni-Vincentelli, Reduced offsets for two-level multi-valued logic minimization, *Proceedings of ACM/IEEE Design Automation Conference*, Orlando, FL, 1990, pp. 290–296.

56. K. Shenoy, N. Saluja, and S. Khatri, An iterative technique for improved two-level logic minimization, *Proceedings of the International Workshop on Logic Synthesis*, Temecula, CA, 2004, pp. 119–126.

57. R. Rudell and A. Sangiovanni-Vincentelli, Espresso-MV: Algorithms for multiple-valued logic minimization, *Proceedings of IEEE Custom Integrated Circuit Conference*, Portland, OR, 1985, pp. 230–234.

58. H. Liu, Reducing routing table size using ternary CAM, *Proceedings of Hot Interconnects*, Stanford, CA, 2001, pp. 69–73.

59. J. Bian and S. Khatri, IP routing table compression using ESPRESSO-MV, *Proceedings of 11th IEEE International Conference on Networks*, Sydney, Australia, 2003, pp. 167–172.

60. S. Ahmed and R. Mahapatra, m-Trie: An efficient approach to on-chip logic minimization, *Proceedings of IEEE International Conference on Computer-Aided Design*, Santa Clara, CA, 2004, pp. 428–435.

61. R. Lysecky and F. Vahid, On-chip logic minimization, *Proceedings of ACM/IEEE Design Automation Conference*, Anaheim, CA, 2003, pp. 334–337.

62. R. Brayton and F. Somenzi, An exact minimizer for Boolean relations, *Proceedings of IEEE International Conference on Computer-Aided Design*, Santa Clara, CA, 1989, pp. 316–319.

63. T. Villa, T. Kam, R. Brayton, and A. Sangiovanni-Vincentelli, Explicit and implicit algorithms for binate covering problems, *IEEE Trans. Comput. Aided Des. Integr. Circ. Syst.*, 16, 671–691, 1997.

64. S.-W. Jeong and F. Somenzi, A new algorithm for the binate covering problem and its application to the minimization of Boolean relations, *Proceedings of IEEE International Conference on Computer-Aided Design*, Santa Clara, CA, 1992, pp. 417–420.

65. Y. Watanabe and R. Brayton, Heuristic minimization of multiple-valued relations, *IEEE Trans. Comput. Aided Des. Integr. Circ. Syst.*, 12, 1458–1472, 1993.

66. A. Srinivasan, T. Kam, S. Malik, and R. Brayton, Algorithms for discrete function manipulation, *Proceedings of IEEE International Conference on Computer-Aided Design*, Santa Clara, CA, 1990, pp. 92–95.

67. R. Ashenhurst, The decomposition of switching functions, *Proceedings of International Symposium on the Theory of Switching*, Cambridge, MA, 1957, pp. 74–116.

68. E. Lawler, An approach to multilevel Boolean minimization, *J. Assoc. Comput. Mach.*, 11, 283–295, 1964.

69. R. Brayton and C. McMullen, The decomposition and factorization of Boolean expressions, *Proceedings of IEEE International Symposium on Circuits and Systems*, Rome, Italy, 1982, pp. 49–54.

70. R. Brayton and C. McMullen, Synthesis and optimization of multistage logic, *Proceedings of IEEE International Conference on Computer Design*, Port Chester, NY, 1984, pp. 23–38.

71. J. Vasudevamurthy and J. Rajski, A Method for concurrent decomposition and factorization of Boolean expressions, *Proceedings of IEEE International Conference on Computer-Aided Design*, 1990, Santa Clara, CA, pp. 510–513.

72. S. Muroga, Y. Kambayashi, H. Lai, and J. Culliney, The transduction method-design of logic networks based on permissible functions, *IEEE Trans. Comput.*, 38, 1404–1424, 1989.

73. H. Savoj and R. Brayton, The use of observability and external don't cares for the simplification of multilevel networks, *Proceedings of ACM/IEEE Design Automation Conference*, Orlando, FL, 1990, pp. 297–301.

74. H. Savoj, R. Brayton, and H. Touati, Extracting local don't cares for network optimization, *Proceedings of IEEE International Conference on Computer-Aided Design*, Santa Clara, CA, 1991, pp. 514–517.

75. H. Savoj, Don't cares in multilevel network optimization, PhD thesis, Electronics Research Laboratory, College of Engineering, University of California, Berkeley, CA, 1992, p. 94720.

76. N. Saluja and S. Khatri, A robust algorithm for approximate compatible observability don't care (CODC) computation, *Proceedings of ACM/IEEE Design Automation Conference*, San Diego, CA, 2004, pp. 422–427.

77. A. Mishchenko and R. Brayton, SAT-based complete don't care computation for network optimization, *Proceedings of the International Workshop on Logic Synthesis*, Temecula, CA, 2004, pp. 353–360.

78. S. Yamashita, H. Sawada, and A. Nagoya, A new method to express functional permissibilities for LUT based FPGAs and its applications, *Proceedings of IEEE International Conference on Computer-Aided Design*, 1996, pp. 254–261.

79. R. Brayton, Understanding SPFDs: A new method for specifying flexibility, *Proceedings of the International Workshop on Logic Synthesis*, Tahoe City, CA, 1997.

80. S. Sinha and R. Brayton, Implementation and use of SPFDs in optimizing Boolean networks, *Proceedings of IEEE International Conference on Computer-Aided Design*, Santa Clara, CA, 1998, pp. 103–110.

81. J. Cong, J. Lin, and W. Long, A new enhanced SPFD rewiring algorithm, *Proceedings of IEEE International Conference on Computer-Aided Design*, Santa Clara, CA, 2002, pp. 672–678.

82. B. Kumthekar and F. Somenzi, Power and delay reduction via simultaneous logic and placement optimization in FPGAs, *Proceedings, Design, Automation and Test in Europe Conference*, Paris, France, 2000, pp. 202–207.

83. S. Khatri, S. Sinha, R. Brayton, and A. Sangiovanni-Vincentelli, SPFD-based wire removal in standard-cell and network-of-PLA circuits, *IEEE Trans. Comput. Aided Des. Integr. Circ. Syst.*, 23, 1020–1030, July 2004.

84. S. Sinha, A. Mishchenko, and R. Brayton, Topologically constrained logic synthesis, *Proceedings of IEEE International Conference on Computer-Aided Design*, Santa Clara, CA, 2002, pp. 679–686.

85. S. Sinha, A. Kuehlmann, and R. Brayton, Sequential SPFDs, *Proceedings of IEEE International Conference on Computer-Aided Design*, Santa Clara, CA, 2001, pp. 84–90.

86. D. Gregory, K. Bartlett, A. de Geus, and G. Hachtel, SOCRATES: A system for automatically synthesizing and optimizing combinational logic, *Proceedings of ACM/IEEE Design Automation Conference*, Las Vegas, NV, 1986, pp. 79–85.

87. W. Joyner, L. Trevillyan, D. Brand, T. Nix, and S. Gunderson, Technology adaptation in logic synthesis, *Proceedings of ACM/IEEE Design Automation Conference*, Las Vegas, NV, 1986, pp. 94–100.

88. A. Aho and M. Corasick, Efficient string matching: An aid to bibliographic search, *Commun. Assoc. Comput. Mach.*, 333–340, 1975.

89. F. Mailhot and G. De Micheli, Algorithms for technology mapping based on binary decision diagrams and on Boolean operations, *IEEE Trans. Comput. Aided Des. Integr. Circ. Syst.*, 559–620, 1993.

90. J. Burch and D. Long, Efficient Boolean function matching, *Proceedings of IEEE International Conference on Computer-Aided Design*, Santa Clara, CA, 1992, pp. 408–411.

91. J. Mohnke, P. Molitor, and S. Malik, Limits of using signatures for permutation independent Boolean comparison, *Proceedings of the Asia and South Pacific Design Automation Conference*, Makuhari, Japan, 1995.

92. U. Schlichtman and F. Brglez, Efficient Boolean matching in technology mapping with very large cell libraries, *Proceedings of IEEE Custom Integrated Circuit Conference*, San Diego, CA, 1993.

93. R. Rudell, Logic synthesis for VLSI design, PhD thesis, University of California, Berkeley, CA, 1989.

94. K. Keutzer, DAGON: Technology binding and local optimization by DAG matching, *Proceedings of ACM/IEEE Design Automation Conference*, Miami Beach, FL, 1987, pp. 341–347.

95. A. Aho and S. Johnson, Optimal code generation for expression trees, *J. Assoc. Comput. Mach.*, 23(3), 488–501, 1976.

96. H. Touati, Performance oriented technology mapping, PhD thesis, University of California, Berkeley, CA, 1990.

97. S. Roy, K. Belkhale, and P. Banerjee, An a-approximate algorithm for delay-constraint technology mapping, *Proceedings of ACM/IEEE Design Automation Conference*, New Orleans, LA, 1999, pp. 367–372.

98. K. Chaudhary and M. Pedram, A near-optimal algorithm for technology mapping minimizing area under delay constraints, *Proceedings of ACM/IEEE Design Automation Conference*, Anaheim, CA, 1992, pp. 492–498.

99. J. Grodstein, E. Lehman, H. Harkness, B. Grundmann, and Y. Watanabe, A delay model for logic synthesis of continuously sized networks, *Proceedings of IEEE International Conference on Computer-Aided Design*, San Jose, CA, 1995, pp. 458–462.

100. L. Stok, M.A. Iyer, and A. Sullivan, Wavefront technology mapping, *Proceedings, Design, Automation and Test in Europe Conference*, Munich, Germany, 1999.

101. Y. Kukimoto, R. Brayton, and P. Sawkar, Delay-optimal technology mapping by DAG covering, *Proceedings of ACM/IEEE Design Automation Conference*, San Francisco, CA, 1998.

102. E. Lehman, E. Watanabe, J. Grodstein, and H. Harkness, Logic decomposition during technology mapping, *Proceedings of IEEE International Conference on Computer-Aided Design*, San Jose, CA, 1995, pp. 264–271.

103. V. Tiwari, P. Ashar, and S. Malik, Technology mapping for low power in logic synthesis, *Integr. VLSI J.*, 3, 243–268, 1996.

104. C. Tsui, M. Pedram, and A. Despain, Technology decomposition and mapping targeting low power dissipation, *Proceedings of ACM/IEEE Design Automation Conference*, Dallas, TX, 1993, pp. 68–73.

105. I. Sutherland and R. Sproull, Logical effort: Designing for speed on the back of an envelope, *Advanced Research in VLSI*, Santa Cruz, CA, 1991, pp. 1–16.

106. D. Kung, A fast fanout optimization algorithm for near-continuous buffer libraries, *Proceedings of ACM/IEEE Design Automation Conference*, San Francisco, CA, 1998, pp. 352–355.

107. P. Rezvani, A. Ajami, M. Pedram, and H. Savoj, LEOPARD: A logical effort-based fanout optimizer for area and delay, *Proceedings of IEEE International Conference on Computer-Aided Design*, Santa Clara, CA, 1999, pp. 516–519.

108. S. Karandikar and S. Sapatnekar, Logical effort based technology mapping, *Proceedings of IEEE International Conference on Computer-Aided Design*, Santa Clara, CA, 2004, pp. 419–422.

109. S. Srinivasaraghavan and W. Burleson, Interconnect effort—A unification of repeater insertion and logical effort, *Proceedings of IEEE Computer Society Annual Symposium on VLSI*, Tampa, FL, 2003, pp. 55–61.

110. K. Venkat, Generalized delay optimization of resistive interconnections through an extension of logical effort, *IEEE International Symposium on Circuits and Systems (ISCAS)*, Portland, OR, 1993, Vol. 3, pp. 2106–2109.

111. D. Harris and I. Sutherland, Logical effort of carry propagate adders, *Conference Record of the Thirty-Seventh Asilomar Conference on Signals, Systems and Computers*, Pacific Grove, CA, 2003, Vol. 1, pp. 873–878.

112. H. Dao and V. Oklobdzija, Application of logical effort techniques for speed optimization and analysis of representative adders, *Conference Record of the Thirty-Fifth Asilomar Conference on Signals, Systems and Computers*, Pacific Grove, CA, 2001, Vol. 2, pp. 1666–1669.

113. J. Shyu, A. Sangiovanni-Vincentelli, J. Fishburn, and A. Dunlop, Optimization based transistor sizing, *IEEE J. Solid State Circ.*, 23(2), 400–499, 1988.

114. J. Fishburn and A. Dunlop, TILOS: A posynomial programming approach to transistor sizing, *Proceedings of IEEE International Conference on Computer-Aided Design*, Santa Clara, CA, 1985, pp. 326–328.

115. S. Sapatnekar, V. Rao, P. Vaidya, and S. Kang, An exact solution to the transistor sizing problem for CMOS circuits using convex optimization, *IEEE Trans. Comput.-Aided Des. Integr. Circ. Syst.*, 12, 1621–1634, 1993.

116. M. Cirit, Transistor sizing in CMOS circuits, *Proceedings of ACM/IEEE Design Automation Conference*, Miami Beach, FL, 1987, pp. 121–124.

117. K. Hedlund, AESOP: A tool for automated transistor sizing, *Proceedings of ACM/IEEE Design Automation Conference*, Miami Beach, FL, 1987, pp. 114–120.

118. D. Marple, Transistor size optimization in the tailor layout system, *Proceedings of ACM/IEEE Design Automation Conference*, Las Vegas, NV, 1989, pp. 43–48.

119. M. Berkelaar and J. Jess, Gate sizing in MOS digital circuits with linear programming, *Proceedings of European Design Automation Conference*, Glasgow, U.K., 1990.

120. O. Coudert, Gate sizing: A general purpose optimization approach, *Proceedings of European Design and Test Conference*, Paris, France, 1996.

121. O. Coudert, R. Haddad, and S. Manne, New algorithms for gate sizing: A comparative study, *Proceedings of ACM/IEEE Design Automation Conference*, Las Vegas, NV, 1996, pp. 734–739.

122. C. Berman, J. Carter, and K. Day, The fanout problem: From theory to practice, in *Advanced Research in VLSI: Proceedings of the Decennial Caltech Conference*, C.L. Seitz, Ed., MIT Press, Cambridge, MA, 1989, pp. 66–99.

123. H. Hoover, M. Klawe, and N. Pippenger, Bounding fanout in logical networks, *J. Assoc. Comput. Mach.*, 31, 13–18, 1984.

124. K. Singh and A. Sangiovanni-Vincentelli, A heuristic algorithm for the fanout problem, *Proceedings of ACM/IEEE Design Automation Conference*, Orlando, FL, 1990, pp. 357–360.

125. K. Bartlett, W. Cohen, A. De Geus, and G. Hachtel, Synthesis and optimization of multilevel logic under timing constraints, *IEEE Trans. Comput. Aided Des. Integr. Circ. Syst.*, 5, 582–596, 1986.

126. K. Singh, A. Wang, R. Brayton, and A. Sangiovanni-Vincentelli, Timing optimization of combinational logic, *Proceedings of IEEE International Conference on Computer-Aided Design*, Santa Clara, CA, 1988, pp. 282–285.

127. J. Fishburn, A depth-decreasing heuristic for combinational logic, *Proceedings of ACM/IEEE Design Automation Conference*, Orlando, FL, 1990, pp. 361–364.

128. K. Yoshikawa, H. Ichiryu, H. Tanishita, S. Suzuki, N. Nomizu, and A. Kondoh, Timing optimization on mapped circuits, *Proceedings of ACM/IEEE Design Automation Conference*, San Francisco, CA, 1991, pp. 112–117.

129. http://www.synopsys.com/partners/tapin/lib_info.html. Liberty.

130. http://www.eda.org/alf/. Advanced Library Format (alf) home page.

131. K. Keutzer, K. Kolwicz, and M. Lega, Impact of library size on the quality of automated synthesis, *Proceedings of IEEE International Conference on Computer-Aided Design*, Santa Clara, CA, 1987, pp. 120–123.

132. K. Scott and K. Keutzer, Improving cell libraries for synthesis, *Proceedings of IEEE Custom Integrated Circuit Conference*, San Diego, CA, 1994, pp. 128–131.

133 I. Sutherland, R. Sproull, and D. Harris, *Logical Effort*, Morgan-Kaufmann, San Francisco, CA, 1999.

134. R. Haddad, L.P.P.P. van Ginneken, and N. Shenoy, Discrete drive selection for continuous sizing, *Proceedings of IEEE International Conference on Computer Design*, Austin, TX, 1997, pp. 110–115.

135. F. Beeftink, P. Kudva, D. Kung, and L. Stok, Gate-size selection for standard cell libraries, *Proceedings of IEEE International Conference on Computer-Aided Design*, San Jose, CA, 1998, pp. 545–550.

136 R. Abato, A. Drumm, D. Hathaway, and L.P.P.P. van Ginneken, Incremental timing analysis, US Patent 5,508,937, 1993.

137. A.R. Wang, Algorithms for multilevel logic optimization, PhD thesis, University of California, Berkeley, CA, 1989.

138. K. Keutzer, S. Malik, and A. Saldanha, Is redundancy necessary to reduce delay? *IEEE Trans. Comput. Aided Des. Integr. Circ. Syst.*, 10, 427–435, 1991.

139. A. Saldanha, R. Brayton, and A. Sangiovanni-Vincentelli, Circuit structure relations to redundancy and delay, *IEEE Trans. Comput. Aided Des. Integr. Circ. Syst.*, 13, 875–883, 1994.

140. Y. Kukimoto and R. Brayton, Timing-safe false path removal for combinational modules, *Proceedings of IEEE International Conference on Computer-Aided Design*, San Jose, CA, 1999.

141. J. Hopcroft and J. Ullman, *Introduction to Automata Theory, Languages and Computation*, Addison-Wesley, Reading, MA, 1979.

142. Z. Kohavi, *Switching and Finite Automata Theory*, Computer Science Series, McGraw-Hill, New York, 1970.

143. W. Kautz, The necessity of closed circuit loops in minimal combinational circuits, *IEEE Trans. Comput.*, C-19, 162–166, 1971.

144. S. Malik, Analysis of cyclic combinational circuits, *IEEE Trans. Comput. Aided Des. Integr. Circ. Syst.*, 13, 950–956, 1994.

145. T. Shiple, G. Berry, and H. Touati, Constructive analysis of cyclic circuits, *Proceedings of European Design and Test Conference*, Paris, France, 1996, pp. 328–333.

146. M. Riedel and J. Bruck, The synthesis of cyclic combinational circuits, *Proceedings of ACM/IEEE Design Automation Conference*, Anaheim, CA, 2003, pp. 163–168.

147. The VIS Group, VIS: A system for verification and synthesis, in *Proceedings of the 8th International Conference on Computer Aided Verification*, Vol. 1102, R. Alur and T. Henzinger, Eds., New Brunswick, NJ, 1996, pp. 428–432.

148. M.R. Garey and D.S. Johnson, *Computers and Intractability: A Guide to the Theory of NP-Completeness*, W. H. Freeman and Company, New York, 1979.

149. M. Paull and S. Unger, Minimizing the number of states in incompletely specified sequential switching functions, *IRE Trans. Electron. Comput.*, EC-8, 356–367, 1959.

150. A. Grasselli and F. Luccio, A method for minimizing the number of internal states in incompletely specified sequential networks, *IEEE Trans. Electron. Comput.*, EC-14, 350–359, 1965.

151. L. Kannan and D. Sarma, Fast heuristic algorithms for finite state machine minimization, *Proceedings of European Conference on Design Automation*, Amsterdam, the Netherlands, 1991, pp. 192–196.

152. J.-K. Rho, G. Hachtel, F. Somenzi, and R. Jacoby, Exact and heuristic algorithms for the minimization of incompletely specified state machines, *IEEE Trans. Comput. Aided Des. Integr. Circ. Syst.*, 13, 167–177, 1994.

153. T. Kam, T. Villa, R. Brayton, and A. Sangiovanni-Vincentelli, Implicit computation of compatible sets for state minimization of ISFSMs, *IEEE Trans. Comput. Aided Des. Integr. Circ. Syst.*, 16, 657–676, 1997.

154. T. Kam, T. Villa, R. Brayton, and A. Sangiovanni-Vincentelli, Theory and algorithms for state minimization of nondeterministic FSMs, *IEEE Trans. Comput. Aided Des. Integr. Circ. Syst.*, 16, 1311–1322, 1997.

155. J. Kim and M. Newborn, The simplification of sequential machines with input restrictions, *IEEE Trans. Comput.*, 1440–1443, 1972.

156. J.-K. Rho and F. Somenzi, The role of prime compatibles in the minimization of finite state machines, *Proceedings of IEEE International Conference on Computer Design*, Cambridge, MA, 1992, pp. 324–327.

157. H.-Y. Wang and R. Brayton, Input don't care sequences in FSM networks, *Proceedings of IEEE International Conference on Computer-Aided Design*, Santa Clara, CA, 1993, pp. 321–328.

158. Y. Watanabe and R. Brayton, The maximum set of permissible behaviors for FSM networks, *Proceedings of IEEE International Conference on Computer-Aided Design*, Santa Clara, CA, 1993, pp. 136–320.

159. Y. Watanabe and R. Brayton, State minimization of pseudo non-deterministic FSMs, *Proceedings of European Design and Test Conference*, Paris, France, 1994, pp. 184–191.

160. A. Aziz, V. Singhal, R. Brayton, and G. Swamy, Minimizing interacting finite state machines: A compositional approach to language containment, *Proceedings of IEEE International Conference on Computer Design*, Cambridge, MA, 1994, pp. 255–261.

161. A. Dasgupta and S. Ganguly, Divide and conquer: A strategy for synthesis of low power finite state machines, *Proceedings of IEEE International Conference on Computer Design*, Austin, TX, 1997, pp. 740–745.

162. Y. Jiang and R. Brayton, Don't cares in logic minimization of extended finite state machines, *Proceedings of the Asia and South Pacific Design Automation Conference*, Kitakyushu, Japan, 2003, pp. 809–815.

163. J. Monteiro and A. Oliveira, Finite state machine decomposition for low power, *Proceedings of ACM/IEEE Design, Automation Conference*, San Francisco, CA, 1998, pp. 758–763.

164. B.W. Kernighan and S. Lin, An efficient heuristic procedure for partitioning graphs, *Bell Syst. Tech. J.*, 49, 291–307, 1970.

165. M.-T. Kuo, L.-T. Liu, and C.-K. Cheng, Finite state machine decomposition for I/O minimization, *Proceedings of IEEE International Symposium on Circuits and Systems*, Seattle, WA, 1995, Vol. 2, pp. 1061–1064.

166. C.M. Fiduccia and R.M. Mattheyses, A linear-time heuristic for improving network partitions, *Proceedings of ACM/IEEE Design, Automation Conference*, Las Vegas, NV, 1982, pp. 175–181.

167. A. Saldanha, T. Villa, R. Brayton, and A. Sangiovanni-Vincentelli, Satisfaction of input and output encoding constraints, *IEEE Trans. Comput. Aided Des. Integr. Circ. Syst.*, 13, 589–602, 1994.

168. G. DeMicheli, R. Brayton, and A. Sangiovanni-Vincentelli, Optimal state assignment for finite state machines, *IEEE Trans. Comput. Aided Des. Integr. Circ. Syst.*, 4, 269–285, 1985.

169. G. DeMicheli, R. Brayton, and A. Sangiovanni-Vincentelli, Correction to "Optimal state assignment for finite state machines", *IEEE Trans. Comput. Aided Des. Integr. Circ. Syst.*, 5, 239–239, 1986.

170. L. Lavagno, S. Malik, R. Brayton, and A. Sangiovanni-Vincentelli, Symbolic minimization of multilevel logic and the input encoding problem, *IEEE Trans. Comput. Aided Des. Integr. Circ. Syst.*, 11, 825–843, 1992.

171. S. Devadas and A.R. Newton, Exact algorithms for output encoding, state assignment, and four-level Boolean minimization, *IEEE Trans. Comput. Aided Des. Integr. Circ. Syst.*, 10, 13–27, 1991.

172. T. Villa and A. Sangiovanni-Vincentelli, NOVA: State assignment of finite state machines for optimal two-level logic implementation, *IEEE Trans. Comput. Aided Des. Integr. Circ. Syst.*, 9, 905–924, 1990.

173. M. Ciesielski, J.-J. Shen, and M. Davio, A unified approach to input-output encoding for FSM state assignment, *Proceedings of ACM/IEEE Design Automation Conference*, San Francisco, CA, 1991, pp. 176–181.

174. S. Devadas, H.-K. Ma, A.R. Newton, and A. Sangiovanni-Vincentelli, MUSTANG: State assignment of finite state machines targeting multilevel logic implementations, *IEEE Trans. Comput. Aided Des. Integr. Circ. Syst.*, 7, 1290–1300, 1988.

175. B. Lin and A.R. Newton, Synthesis of multiple level logic from symbolic high-level description languages, *Proceedings of the International Conference on VLSI*, Munich, Germany, 1989, pp. 187–196.

176. X. Du, G. Hachtel, and P. Moceyunas, MUSE: A multilevel symbolic encoding algorithm for state assignment, *Proceedings of the 23rd Annual Hawaii International Conference on System Sciences*, Vol. 1, Waikoloa, HI, 1990, pp. 367–376.

177. R. Forth and P. Molitor, An efficient heuristic for state encoding minimizing the BDD representations of the transition relations of finite state machines, *Proceedings of the Asia and South Pacific Design Automation Conference*, Yokohama, Japan, 2000, pp. 61–66.

178. G. Hasteer and P. Banerjee, A parallel algorithm for state assignment of finite state machines, *IEEE Trans. Comput.*, 47, 242–246, 1998.

179. L. Benini and G. DeMicheli, State assignment for low power dissipation, *IEEE J. Solid-State Circ.*, 30, 258–268, 1995.

180. K.-H. Wang, W.-S. Wang, T. Hwang, A. Wu, and Y.-L. Lin, State assignment for power and area minimization, *Proceedings of IEEE International Conference on Computer Design*, Cambridge, MA, 1994, pp. 250–254.

181 C.-Y. Tsui, M. Pedram, and A. Despain, Low-power state assignment targeting two- and multilevel logic implementations, *IEEE Trans. Comput. Aided Des. Integr. Circ. Syst.*, 17, 1281–1291, 1998.

182. X. Wu, M. Pedram, and L. Wang, Multi-code state assignment for low power design, *IEEE Proc. G-Circ. Dev. Syst.*, 147, 271–275, 2000.

183. S. Park, S. Cho, S. Yang, and M. Ciesielski, A new state assignment technique for testing and low power, *Proceedings of ACM/IEEE Design Automation Conference*, San Diego, CA, 2004, pp. 510–513.

184. C. Leiserson and J. Saxe, Optimizing Synchronous Systems, *J. VLSI Comput. Syst.*, 1, 41–67, 1983.

185. J. Saxe, *Decomposable Searching Problems and Circuit Optimization by Retiming: Two Studies in General Transformations of Computational Structures*, PhD thesis, Carnegie Mellon University, Pittsburgh, PA, 1985, CMU-CS-85-162.

186. H. Touati and R. Brayton, Computing the initial states of retimed circuits, *IEEE Trans. Comput. Aided Des. Integr. Circ. Syst.*, 12, 157–162, 1993.

187. C. Pixley and G. Beihl, Calculating resetability and reset sequences, *Proceedings of IEEE International Conference on Computer-Aided Design*, Santa Clara, CA, 1991, pp. 376–379.

188. G. Even, I. Spillinger, and L. Stok, Retiming revisited and reversed, *IEEE Trans. Comput. Aided Des. Integr. Circ. Syst.*, 15, 348–357, 1996.

189. N. Maheshwari and S. Sapatnekar, Minimum area retiming with equivalent initial states, *Proceedings of IEEE International Conference on Computer-Aided Design*, Santa Clara, CA, 1997, pp. 216–219.

190. N. Shenoy, R. Brayton, and A. Sangiovanni-Vincentelli, Retiming of circuits with single phase transparent latches, *Proceedings of IEEE International Conference on Computer Design*, Cambridge, MA, 1991, pp. 86–89.

191. A. Ishii, C. Leiserson, and M. Papaefthymiou, Optimizing two-phase, level-clocked circuitry, *Proceedings of the Brown/MIT Conference on Advanced Research in VLSI and Parallel Systems*, MIT Press, Cambridge, MA, 1992, pp. 245–264.

192. B. Lockyear and C. Ebeling, Optimal retiming of multi-phase, level-clocked circuits, *Proceedings of the Brown/MIT Conference on Advanced Research in VLSI and Parallel Systems*, MIT Press, Cambridge, MA, 1992, pp. 265–280.

193. N. Maheshwari and S. Sapatnekar, Efficient minarea retiming of large level-clocked circuits, *Proceedings, Design, Automation and Test in Europe Conference*, Paris, France, 1998, pp. 840–845.

194. N. Shenoy and R. Rudell, Efficient implementation of retiming, *Proceedings of IEEE International Conference on Computer-Aided Design*, Santa Clara, CA, 1994, pp. 226–233.

195. N. Maheshwari and S. Sapatnekar, Efficient retiming of large circuits, *IEEE Trans. Very Large Scale Integr. (VLSI) Syst.*, 6, 74–83, 1998.

196. S. Malik, E. Sentovich, R. Brayton, and A. Sangiovanni-Vincentelli, Retiming and resynthesis: Optimizing sequential networks with combinational techniques, *IEEE Trans. Comput. Aided Des. Integr. Circ. Syst.*, 10, 74–84, 1991.

197. S. Malik, K. Singh, R. Brayton, and A. Sangiovanni-Vincentelli, Performance optimization of pipelined logic circuits using peripheral retiming and resynthesis, *IEEE Trans. Comput. Aided Des. Integr. Circ. Syst.*, 12, 568–578, 1993.

198. P. Kalla and M. Ciesielski, Performance driven resynthesis by exploiting retiming-induced state register equivalence, *Proceedings, Design, Automation and Test in Europe Conference*, Munich, Germany, 1999, pp. 638–642.

199. J. Cong and C. Wu, Optimal FPGA mapping and retiming with efficient initial state computation, *IEEE Trans. Comput. Aided Des. Integr. Circ. Syst.*, 18, 1595–1607, 1999.

200. J. Cong and C. Wu, An efficient algorithm for performance-optimal FPGA technology mapping with re-timing, *IEEE Trans. Comput. Aided Des. Integr. Circ. Syst.*, 17, 738–748, 1998.

201. A. El-Maleh, T. Marchok, J. Rajski, and W. Maly, Behavior and testability preservation under the retiming transformation, *IEEE Trans. Comput. Aided Des. Integr. Circ. Syst.*, 16, 528–543, 1997.

202. S. Dey and S. Chakradhar, Retiming sequential circuits to enhance testability, *Proceedings of IEEE VLSI Test Symposium*, Cherry Hill, NJ, 1994, pp. 28–33.

203. J. Monteiro, S. Devadas, and A. Ghosh, Retiming sequential circuits for low power, *Proceedings of IEEE International Conference on Computer-Aided Design*, Santa Clara, CA, 1993, pp. 398–402.

204. A. Tabarra, R. Brayton, and A.R. Newton, Retiming for DSM with area-delay trade-offs and delay constraints, *Proceedings of ACM/IEEE Design Automation Conference*, New Orleans, LA, 1999, pp. 725–730.

205. C. Chu, E. Young, D. Tong, and S. Dechu, Retiming with interconnect and gate delay, *Proceedings of IEEE International Conference on Computer-Aided Design*, Santa Clara, CA, 2003, pp. 221–226.

206. T. Soyata, E. Friedman, and J. Mulligan, Incorporating interconnect, register, and clock distribution delays into the retiming process, *IEEE Trans. Comput. Aided Des. Integr. Circ. Syst.*, 16, 105–120, 1997.

207. X. Liu, M. Papaefthymiou, and E. Friedman, Retiming and clock scheduling for digital circuit optimization, *IEEE Trans. Comput. Aided Des. Integr. Circ. Syst.*, 21, 184–203, 2002.

208. B. Iyer and M. Ciesielski, Metamorphosis: State assignment by retiming and re-encoding, *Proceedings of IEEE International Conference on Computer-Aided Design*, Santa Clara, CA, 1996, pp. 614–617.

209. M. Pedram and N. Bhat, Layout driven technology mapping, *Proceedings of ACM/IEEE Design Automation Conference*, San Francisco, CA, 1991, pp. 99–105.

210. M. Pedram and N. Bhat, Layout driven logic restructuring/decomposition, *Proceedings of IEEE International Conference on Computer-Aided Design*, Santa Clara, CA, 1991, pp. 134–137.

211. P. Abouzeid, K. Sakouti, G. Saucier, and F. Poirot, Multilevel synthesis minimizing the routing factor, *Proceedings of ACM/IEEE Design Automation Conference*, Orlando, FL, 1990, pp. 365–368.

212. H. Vaishnav and M. Pedram, Routability driven fanout optimization, *Proceedings of ACM/IEEE Design Automation Conference*, Dallas, TX, 1993, pp. 230–235.

213. K. Keutzer, A.R. Newton, and N. Shenoy, The future of logic synthesis and physical design in deep sub-micron process geometries, *Proceedings of the International Symposium on Physical Design*, Napa Valley, CA, 1997.

214. L. Kannan, P. Suaris, and H. Fang, A methodology and algorithms for post-placement delay optimization, *Proceedings of ACM/IEEE Design Automation Conference*, San Diego, CA, 1994.

215. T. Ishioka, M. Murofushi, and M. Murakata, Layout driven delay optimization with logic resynthesis, *Proceedings of the International Workshop on Logic Synthesis*, Tahoe City, CA, 1997.

216. G. Stenz, B. Riess, B. Rohfleisch, and F. Johannes, Timing driven placement in interaction with netlist transformations, *Proceedings of the International Symposium on Physical Design*, Napa Valley, CA, 1997, pp. 36–41.

217. S. Hojat and P. Villarubia, An integrated placement and synthesis approach for timing closure of PowerPC microprocessor, *Proceedings of IEEE International Conference on Computer Design*, Austin, TX, 1997, pp. 206–210.

218. N. Shenoy, M. Iyer, R. Damiano, K. Harer, H.-K. Ma, and P. Thilking, A robust solution to the timing convergence problem in high performance design, *Proceedings of IEEE International Conference on Computer Design*, Austin, TX, 1999, pp. 250–257.

219. W. Gosti, S. Khatri, and A. Sangiovanni-Vincentelli, Addressing the timing closure problem by integrating logic optimization and placement, *Proceedings of IEEE International Conference on Computer-Aided Design*, San Jose, CA, 2001, pp. 224–231.

220. R. Carragher, R. Murgai, S. Chakraborty, M. Prasad, A. Srivastava, and N. Vemure, Layout driven logic synthesis, *Proceedings of the International Workshop on Logic Synthesis*, Dana Point, CA, 2000.

221. J. Lou, A. Salek, and M. Pedram, Concurrent logic restructuring and placement for timing closure, *Proceedings of IEEE International Conference on Computer-Aided Design*, San Jose, CA, 1999, pp. 31–35.

222. J. Lou, W. Chen, and M. Pedram, An exact solution to simultaneous technology mapping and linear placement problem, *Proceedings of IEEE International Conference on Computer-Aided Design*, San Jose, CA, 1997, pp. 671–675.

223. R. Murgai, Delay constrained area recovery via layout-driven buffer optimization, *Proceedings of the International Workshop on Logic Synthesis*, Tahoe City, CA, 1999, pp. 217–221.

224. P. Kudva, A. Sullivan, and W. Dougherty, Metrics for structural logic synthesis, *Proceedings of IEEE International Conference on Computer-Aided Design*, San Jose, CA, 2002, pp. 551–556.

225. P. Saxena and B. Halpin, Modeling repeaters explicitly within analytical placement, *Proceedings of ACM/IEEE Design Automation Conference*, San Diego, CA, 2004, pp. 699–704.

226. Y.-J. Chang and C. Lee, Synthesis of multi-variable MVL functions using hybrid mode CMOS logic, *Proceedings of International Symposium on Multiple-Valued Logic*, Boston, MA, 1994, pp. 35–41.

227. M. Syuto, J. Shen, K. Tanno, and O. Ishizuka, Multi-input variable-threshold circuits for multi-valued logic functions, *Proceedings of International Symposium on Multiple-Valued Logic*, Portland, OR, 2000, pp. 27–32.

228. T. Temel and A. Morgul, Implementation of multi-valued logic, simultaneous literal operations with full CMOS current-mode threshold circuits, *Electron. Lett.*, 38, 160–161, 2002.

229. E. Lee and P. Gulak, Dynamic current-mode multi-valued MOS memory with error correction, *Proceedings of International Symposium on Multiple-Valued Logic*, Sendai, Japan, 1992, pp. 208–215.

230. H. Wang, C. Lee, and J. Chen, Algebraic division for multilevel logic synthesis of multi-valued logic circuits, *Proceedings of International Symposium on Multiple-Valued Logic*, Boston, MA, 1994, pp. 44–51.

231. H. Wang, C. Lee, and J. Chen, Factorization of multi-valued logic functions, *Proceedings of International Symposium on Multiple-Valued Logic*, Bloomington, IN, 1995, pp. 164–169.

232. Y. Jiang, S. Matic, and R. Brayton, Generalized cofactoring for logic function evaluation, *Proceedings of ACM/IEEE Design Automation Conference*, Anahein, CA, 2003, pp. 155–158.

233. Y. Jiang and R. Brayton, Don't cares and multi-valued logic network minimization, *Proceedings of IEEE International Conference on Computer-Aided Design*, Santa Clara, CA, 2000, pp. 520–525.

234. S. Khatri, R. Brayton, and A. Sangiovanni-Vincentelli, Sequential multi-valued network simplification using redundancy removal, *Proceedings of International Conference on VLSI Design*, Goa, India, 1999, pp. 206–211.

235. S. Sinha, S. Khatri, R. Brayton, and A. Sangiovanni-Vincentelli, Binary and multi-valued SPFD-based wire removal in PLA networks, *Proceedings of IEEE International Conference on Computer Design*, Austin, TX, 2000, pp. 494–503.

236. C. Liu, A. Kuehlmann, and M. Moskewicz, CAMA: A multi-valued satisfiability solver, *Proceedings of IEEE International Conference on Computer-Aided Design*, Santa Clara, CA, 2003, pp. 326–333.

237. A. Mishchenko and R. Brayton, Simplification of non-deterministic multi-valued networks, *Proceedings of IEEE International Conference on Computer-Aided Design*, Santa Clara, CA, 2002, pp. 557–562.

238. J. Jiang, A. Mishchenko, and R. Brayton, Reducing multi-valued algebraic operations to binary, *Proceedings of Design, Automation and Test in Europe Conference*, Munich, Germany, 2003, pp. 752–757.

239. R. Brayton and S. Khatri, Multi-valued logic synthesis, *Proceedings of International Conference on VLSI Design*, Goa, India, 1999, pp. 196–205.

240. M. Gao, J. Jiang, Y. Jiang, Y. Li, A. Mishchenko, S. Sinha, T. Villa, and R. Brayton, Optimization of multivalued multi-level networks, *Proceedings of International Symposium on Multiple-Valued Logic*, Boston, MA, 2002, pp. 168–177.

241. R. Drechsler, Evaluation of static variable ordering heuristics for MDD construction, *Proceedings of International Symposium on Multiple-Valued Logic*, Boston, MA, 2002, pp. 254–260.

242. F. Schmiedle, W. Gunther, and R. Drechsler, Selection of efficient re-ordering heuristics for MDD construction, *Proceedings of International Symposium on Multiple-Valued Logic*, Warsaw, Poland, 2001, pp. 299–304.

243. F. Schmiedle, W. Gunther, and R. Drechsler, Dynamic re-encoding during MDD minimization, *Proceedings of International Symposium on Multiple-Valued Logic*, Portland, OR, 2000, pp. 239–244.

244. R. Drechsler, M. Thornton, and D. Wessels, MDD-based synthesis of multi-valued logic networks, *Proceedings of International Symposium on Multiple-Valued Logic*, Portland, OR, 2000, pp. 41–46.

245. C. Files, R. Drechsler, and M. Perkowski, Functional decomposition of MVL functions using multi-valued decision diagrams, *Proceedings of International Symposium on Multiple-Valued Logic*, Antigonish, Nova Scotia, Canada, 1997, pp. 27–32.

246. V. Shmerko, S. Yanushkevich, and V. Levashenko, Test pattern generation for combinatorial multi-valued networks based on generalized D-algorithm, *Proceedings of International Symposium on Multiple-Valued Logic*, Antigonish, Nova Scotia, Canada, 1997, pp. 139–144.

247. R. Drechsler, M. Keim, and B. Becker, Fault simulation in sequential multi-valued logic networks, *Proceedings of International Symposium on Multiple-Valued Logic*, Antigonish, Nova Scotia, Canada, 1997, pp. 145–150.

248. Effective Current Source Modeling, Cadence Design Systems Inc., San Jose, CA, 2001, http://www.cadence.com.

249. Composite Current Source Modeling, Synopsys Inc., 2004, Mountain View, CA, http://www.synopsys.com.

250. C.J. Alpert, A.B. Kahng, C.N. Sze, and W. Qinke, Timing-driven Steiner trees are (practically) free, *Proceedings of ACM/IEEE Design Automation Conference*, San Francisco, CA, 2006, pp. 389–392.

251. C. Chu and Y. Wong, FLUTE: Fast lookup table based rectilinear Steiner minimal tree algorithm for VLSI design, *IEEE Trans. Comput. Aided Des. Integr. Circ. Syst.*, 27, 70–83, 2008.

252. C.J. Alpert, S.K. Karandikar, Z. Li, G.-J. Nam, S.T. Quay, H. Ren, C.N. Sze, P.G. Villarrubia, and M.C. Yildiz, Techniques for fast physical synthesis, *Proceedings of the IEEE*, 95, 573–599, 2007.

253. C.J. Alpert, C. Chu, and P.G. Villarrubia, The coming of age of physical synthesis, *Proceedings of IEEE International Conference on Computer-Aided Design*, San Jose, CA, 2007, pp. 246–249.

254. K.-H. Chang, I.L. Markov, and V. Bertacco, Safe delay optimization for physical synthesis, *Proceedings of Asian and South-Pacific Design Automation Conference*, Yokohama, Japan, 2007, pp. 628–633.

255. Z. Li, C.J. Alpert, S. Hu, T. Muhmud, S.T. Quay, and P.G. Villarrubia, Fast interconnect synthesis with layer assignment, *Proceedings of International Symposium on Physical Design*, Portland, OR, 2008, pp. 71–77.

256. Z. Li, C.J. Alpert, G.-J. Nam, C.N. Sze, N. Viswanathan, and N.Y. Zhou, Guiding a physical design closure system to produce easier-to-route designs with more predictable timing, *Proceedings of ACM/IEEE Design Automation Conference*, San Francisco, CA, 2012, pp. 465–470.

257. S.I. Ward, Keep it straight: Teaching placement how to better handle designs with datapaths, *Proceedings of International Symposium on Physical Design*, Napa Valley, CA, 2012, pp. 79–86.

258. T. Jindal, C.J. Alpert, H. Jiang, L. Zhuo, G.-J. Nam, and C.B. Winn, Detecting tangled logic structures in VLSI netlists, *Proceedings of ACM/IEEE Design Automation Conference*, Anaheim, CA, 2010, pp. 603–608.

259. H. Xiang, H. Ren, L. Trevillyan, L. Reddy, R. Puri, and M. Cho, Logical and physical restructuring of fan-in trees, *Proceedings of International Symposium on Physical Design*, San Francisco, CA, 2010, pp. 67–74.

260. Q. Zhu, N. Kitchen, A. Kuehlmann, and A. Sangiovanni-Vincentelli, SAT sweeping with local observability don't-cares, *Proceedings of ACM/IEEE Design Automation Conference*, San Francisco, CA, 2006, pp. 229–234.

261. A. Mishchenko, J.S. Zhang, S. Sinha, J.R. Burch, R. Brayton, and M. Chrzanowska-Jeske, Using simulation and satisfiability to compute flexibilities in Boolean networks, *IEEE Trans. Comput. Aided Des. Integr. Circ. Syst.*, 25, 743–755, 2006.

262. S. Chatterjee, A. Mishchenko, R. Brayton, X. Wang, and T. Kam, Reducing structural bias in technology mapping, *IEEE Trans. Comput. Aided Des. Integr. Circ. Syst.*, 25, 2894–2903, 2006.

263. Berkeley Logic Synthesis and Verification Group, ABC: A system for sequential synthesis and verification, http://www.eecs.berkeley.edu/.

264. A. Mishchenko, M. Case, R. Brayton, and S. Jang, Scalable and scalably-verifiable sequential synthesis, *Proceedings of IEEE International Conference on Computer-Aided Design*, San Jose, CA, 2008, pp. 234–241.

265. L. Benini, G. De Micheli, E. Macii, M. Poncino, and R. Scarsi, Symbolic synthesis of clock-gating logic for power optimization of control-oriented synchronous networks, *Proceedings of European Design and Test Conference*, Paris, France, 1997, pp. 514–520.

266. W. Qing, M. Pedram, and W. Xunwei, Clock-gating and its application to low power design of sequential circuits, *IEEE Trans. Circ. Syst. I: Fundam. Theory Appl.*, 47, 415–420, 2000.

267. P. Hurst, Automatic synthesis of clock gating logic with controlled netlist perturbation, *Proceedings of ACM/IEEE Design Automation Conference*, Anaheim, CA, 2008, pp. 654–657.

268. H. Li, S. Bhunia, Y. Chen, T.N. Vijaykumar, and K. Roy, Deterministic clock gating for microprocessor power reduction, *Proceedings of the Ninth International Symposium on High-Performance Computer Architecture*, Washington, DC, 2003, p. 113.

269. S. Ahuja and S. Shukla, MCBCG: Model checking based sequential clock-gating, *Proceedings of IEEE International High Level Design Validation and Test Workshop*, San Francisco, CA, 2009, pp. 20–25.

270. A. Mathur and Q. Wang, Power reduction techniques and flows at RTL and system level, *Proceedings of the 22nd International Conference on VLSI Design*, New Delhi, India, 2009, pp. 28–29.

271. H. Savoj, A. Mishchenko, and R. Brayton, Sequential equivalence checking for clock-gated circuits, *IEEE Trans. Comput. Aided Des. Integr. Circ. Syst.*, 33, 305–317, 2014.

272. S. Jang, K. Chung, A. Mishchenko, and R. Brayton, A power optimization toolbox for logic synthesis and mapping, *Proceedings of International Workshop of Logic and Synthesis*, Berkeley, CA, 2009, pp. 1–8.

273. M. Munch, B. Wurth, R. Mehra, J. Sproch, and N. Wehn, Automating RT-Level operand isolation to minimizing power consumption in datapaths, *Proceedings of IEEE Design Automation and Test in Europe*, Paris, France, 2000, pp. 624–631.

274. H. Chou, Y.-H Wang, and C.P. Chen, Fast and effective gate-sizing with multiple-Vt assignment using generalized Lagrangian relaxation, *Proceedings of Asian and South-Pacific Design Automation Conference*, Shanghai, China, 2005, pp. 381–386.

275. S. Shah, A. Srivastava, D. Sharma, D. Sylvester, and D. Blaauw, Discrete Vt assignment and gate sizing using a self-snapping continuous formulation, *Proceedings of IEEE International Conference on Computer-Aided Design*, San Jose, CA, 2005, pp. 705–712.

276. T. Wu, L. Xie and A. Davoodi, A parallel and randomized algorithm for large-scale dual-Vt assignment and continuous gate sizing, *Proceedings of International Symposium on Low Power Electronics and Design*, Bangalore, India, 2008, pp. 45–50.

277. Y. Liu and J. Hu, A new algorithm for simultaneous gate sizing and threshold voltage assignment, *IEEE Trans. Comput. Aided Des. Integr. Circ. Syst.*, 29, 223–234, 2010.

278. IEEE Standard for Design and Verification of Low-Power Integrated Circuits, IEEE Std 1801™-2013, http://standards.ieee.org/.

279. K. De and P. Banerjee, Parallel logic synthesis using partitioning, *Proceedings of International Conference on Parallel Processing*, Raleigh, NC, 1994, pp. 135–142.

280. B. Catanzaro, K. Keutzer, and S. Bor-Yiing, Parallelizing CAD: A timely research agenda for EDA, *Proceedings of ACM/IEEE Design Automation Conference*, Anaheim, CA, 2008, pp. 12–17.

281. C-C. Lin, K-C. Chen, and M. Marek-Sadowska, Logic synthesis for engineering change, *IEEE Trans. Comput. Aided Des. Integr. Circ. Syst.*, 18, 282–292, 1999.

282. B.-H. Wu, C.-J. Yang, C.-Y. Huang, and J.-H.R. Jiang, A robust functional ECO engine by SAT proof minimization and interpolation techniques, *Proceedings of IEEE International Conference on Computer-Aided Design*, San Jose, CA, 2010, pp. 729–734.

283. S. Krishnaswamy, H. Ren, N. Modi, and R. Puri, DeltaSyn: An efficient logic difference optimizer for ECO synthesis, *Proceedings of IEEE International Conference on Computer-Aided Design*, San Jose, CA, 2009, pp. 789–796.

284. H. Ren, R. Puri, L.N. Reddy, S. Krishnaswamy, C. Washburn, J. Earl, and J. Keinert, Intuitive ECO synthesis for high performance circuits, *Proceedings of IEEE Design Automation and Test in Europe*, Grenoble, France, 2013, pp. 1002–1007.

285. Y. Chen, J. Fang, and Y. Chang, ECO timing optimization using spare cells, *Proceedings of IEEE International Conference on Computer-Aided Design*, San Jose, CA, 2007, pp. 530–535.

286. K.M. Svore, A.V. Aho, A.W. Cross, I.L. Chuang, and I.L. Markov, A layered software architecture for quantum computing design tools, *IEEE Trans. Comput.*, 39, 74–83, 2006.

287. G. De Micheli, Logic synthesis and physical design: Quo vadis?, *Proceedings of IEEE Design Automation and Test in Europe*, Grenoble, France, 2011, p. 1.

Power Analysis and Optimization from Circuit to Register-Transfer Levels

José Monteiro, Rakesh Patel, and Vivek Tiwari

3

CONTENTS

The complexity and speed of today's VLSI designs entail a level of power consumption that, if not addressed, causes an unbearable problem of heat dissipation. The operation of these circuits is only possible due to aggressive techniques for power reduction at different levels of design abstraction. The trends of mobile devices and the Internet of Things, on the other hand, drive the need for energy-efficient circuits and the requirement to maximize battery life. To meet these challenges, sophisticated design methodologies and algorithms for electronic design automation (EDA) have been developed.

One of the key features that led to the success of CMOS technology was its intrinsic low power consumption. It allowed circuit designers and EDA tools to concentrate on maximizing circuit performance and minimizing circuit area. Another interesting feature of CMOS technology is its nice scaling properties, which permitted a steady decrease in the feature size, allowing for numerous and exceptionally complex systems on a single chip, working at high clock frequencies.

Power consumption concerns came into play with the appearance of the first portable electronic systems in the late 1980s. In this market, battery lifetime is a decisive factor for the commercial success of the product. It also became apparent that the increasing integration of active elements per die area would lead to prohibitively large energy consumption of an integrated circuit. High power consumption is undesirable for economic and environmental reasons and also leads to high heat dissipation. In order to keep such a circuit working at acceptable temperature levels, expensive heat removal systems may be required.

In addition to the full-chip power consumption, and perhaps even more importantly, excessive heat is often dissipated at localized areas in the circuit, the so-called hot spots. This problem can be mitigated by selectively turning off unused sections of the circuit when such conditions are detected. The term *dark silicon* has been used to describe this situation where many available computational elements in an integrated circuit cannot be used at the same time [1]. These factors have contributed to the rise of power consumption as a major design parameter on par with performance and die size and a limitation of the continuing scaling of CMOS technology.

To respond to this challenge, intensive research has been invested in the past two decades in developing EDA tools for power optimization. Initial efforts focused on circuit- and logic-level tools, because at these levels EDA tools were more mature and malleable. Today, a large fraction of EDA research targets system- or architectural-level power optimization (Chapters 7 and 13 of *Electronic Design Automation for IC System Design, Verification, and Testing*, respectively), which promise a higher overall impact given the breadth of their application. Together with optimization tools, efficient techniques for power estimation are required, both as an absolute indicator that the circuits' consumption meets some target value and as a relative indicator of the power merits of different alternatives during design space exploration.

This chapter provides an overview of key CAD techniques proposed for low power design and synthesis. We start in Section 3.1 by describing the issues and methods for power estimation at different levels of abstraction, thus defining the targets for the tools presented in the following sections.

In Sections 3.2 and 3.3, we review power optimization techniques at the circuit and logic levels of abstraction, respectively.

3.1 POWER ANALYSIS

Given the importance of power consumption in circuit design, EDA tools are required to provide power estimates for a circuit. When evaluating different designs, these estimates are needed to help identify the most power-efficient alternative. Since power estimates may be required for multiple alternatives, accuracy is sometimes sacrificed for tool response speed when the relative fidelity of estimates can be preserved. Second, an accurate power consumption estimate is required before fabrication to guarantee that the circuit meets the allocated power budget.

Obtaining a power estimate is significantly more complex than circuit area and delay estimates, because power depends not only on the circuit topology but also on the activity of the signals.

Typically, design exploration is performed at each level of abstraction, motivating power estimation tools at different levels. The higher the abstraction level, the less information there is about the actual circuit implementation, implying less assurance about the power estimate accuracy.

In this section, we first discuss the components of power consumption in CMOS circuits. We then discuss how each of these components is estimated at the different design abstraction levels.

3.1.1 POWER COMPONENTS IN CMOS CIRCUITS

The power consumption of digital CMOS circuits is generally divided into three components [2]:

1. Dynamic power (P_{dyn})
2. Short-circuit power (P_{short})
3. Static power (P_{static})

The total power consumption is given by the sum of these components:

$$(3.1) \qquad P = P_{dyn} + P_{short} + P_{static}$$

The dynamic power component, P_{dyn}, is related to the charging and discharging of the load capacitance at the gate output, C_{out}. This is a parasitic capacitance that can be lumped at the output of the gate. Today, this component is still the dominant source of power consumption in a CMOS gate.

As an illustrative example, consider the inverter circuit depicted in Figure 3.1 (to form a generic CMOS gate, the bottom transistor, nMOS, can be replaced by a network of nMOS transistors, and the top transistor, pMOS, by a complementary network of pMOS transistors). When the input goes low, the nMOS transistor is cut off and the pMOS transistor conducts. This creates a direct path between the voltage supply and C_{out}. Current I_p flows from the supply to charge C_{out} up to the voltage level V_{dd}. The amount of charge drawn from the supply is $C_{out}V_{dd}$ and the energy drawn from the supply equals $C_{out}V_{dd}{}^2$. The energy actually stored in the capacitor, E_c, is only half of this, $E_c = \frac{1}{2}C_{out}V_{dd}{}^2$. The other half is dissipated in the resistance represented by the pMOS transistor. During the subsequent low-to-high input transition, the pMOS transistor is cut off and the nMOS transistor conducts. This connects the capacitor C_{out} to the ground, leading to the flow of current I_n. C_{out} discharges and its stored energy, E_c, is dissipated in the resistance represented by the nMOS transistor. Therefore, an amount of energy equal to E_c is dissipated every time the output makes a transition. Given N gate transitions within time T, its dynamic power consumption during that time period is given by

$$(3.2) \qquad P_{dyn} = E_c \times N/T = \frac{1}{2} C_{out}V_{dd}{}^2 \frac{N}{T}$$

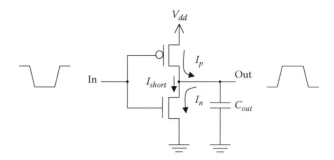

FIGURE 3.1   Illustration of the dynamic and short-circuit power components.

In the case of synchronous circuits, an estimate, α, of the average number of transitions the gate makes per clock cycle, $T_{clk} = 1/f_{clk}$, can be used to compute average dynamic power

$$(3.3) \qquad\qquad P_{dyn} = E_c \times a \times f_{clk} = \tfrac{1}{2}C_{out}V_{dd}^2af_{clk}$$

C_{out} is the sum of the three components C_{int}, C_{wire}, and C_{load}. Of these, C_{int} represents the internal capacitance of the gate. This includes the diffusion capacitance of the drain regions connected to the output. C_{load} represents the sum of gate capacitances of the transistors this logic gate is driving. C_{wire} is the parasitic capacitance of the wiring used to interconnect the gates, including the capacitance between the wire and the substrate, the capacitance between neighboring wires, and the capacitance due to the fringe effect of electric fields. The term αC_{out} is generally called the switched capacitance, which measures the amount of capacitance that is charged or discharged in one clock cycle.

The short-circuit power component, P_{short}, is also related to the switching activity of the gate. During the transition of the input signal from one voltage level to the other, there is a period of time when both the pMOS and the nMOS transistors are on, thus creating a path from V_{dd} to ground. Thus, each time a gate switches, some amount of energy is consumed by the current that flows through both transistors during this period, indicated as I_{short} in Figure 3.1. The short-circuit power is determined by the time the input voltage V_{in} remains between V_{Tn} and $V_{dd}-V_{Tp}$, where V_{Tn} and V_{Tp} are the threshold voltages of the nMOS and the pMOS transistors, respectively. Careful design to minimize low-slope-input ramps, namely, through the appropriate sizing of the transistors, can limit this component to a small fraction of total power; hence, it is generally considered only a second-order effect. Given an estimate of the average amount of charge, Q_{short}, that is carried by the short-circuit current per output transition, the short-circuit power is obtained by

$$(3.4) \qquad\qquad P_{short} = Q_{short}V_{dd}af_{clk}$$

The static power component, P_{static}, is due to leakage currents in the MOS transistors. As the name indicates, this component is not related to the circuit activity and exists as long as the circuit is powered. The source and drain regions of a MOS transistor (MOSFET) can form reverse-biased parasitic diodes with the substrate. There is leakage current associated with these diodes. This current is very small and is usually negligible compared to dynamic power consumption. Another type of leakage current occurs due to the diffusion of carriers between the source and drain even when the MOSFET is in the cutoff region, that is, when the magnitude of the gate-source voltage, V_{GS}, is below the threshold voltage, V_T. In this region, the MOSFET behaves like a bipolar transistor and the subthreshold current is exponentially dependent on $V_{GS}-V_T$. With the reduction of transistor size, leakage current tends to increase for each new technology node, driving up the relative weight of static power consumption. This problem has been mitigated through the introduction of high-κ dielectric materials and new gate geometry architectures [3].

Another situation that can lead to static power dissipation in CMOS is when a degraded voltage level (e.g., the *high* output level of an nMOS pass transistor) is applied to the inputs of a CMOS gate. A degraded voltage level may leave both the nMOS and pMOS transistors in a conducting state, leading to continuous flow of short-circuit current. This again is undesirable and should be avoided in practice.

This condition is true for pure CMOS design styles. In certain specialized circuits, namely, for performance reasons, alternative design styles may be used. Some design styles produce a current when the output is constant at one voltage level, thus contributing to the increase in static power consumption. One example is the domino design style, where a precharged node needs to be recharged on every clock cycle if the output of the gate happens to be the opposite of the precharged value. Another example is the pseudo-nMOS logic family, where the pMOS network of a CMOS gate is replaced by a single pMOS transistor that always conducts. This logic style exhibits a constant current flowing whenever the output is at logic 0, that is, when there is a direct path to ground through the nMOS network.

3.1.2 ANALYSIS AT THE CIRCUIT LEVEL

Power estimates at the circuit level are generally obtained using a circuit-level simulator, such as SPICE [4]. Given a user-specified representative sequence of input values, the simulator solves the circuit equations to compute voltage and current waveforms at all nodes in the electrical circuit. By averaging the current values drawn from the source, I_{avg}, the simulator can output the average power consumed by the circuit, $P = I_{avg}V_{dd}$ (if multiple power sources are used, the total average power will be the sum of the power drawn from all power sources).

At this level, complex models for the circuit devices can be used. These models permit the accurate computation of the three components of power—dynamic, short-circuit, and static power. Since the circuit is described at the transistor level, correct estimates can be computed not only for CMOS but also for any logic design style and even analog modules. After placement and routing of the circuit, simulation can handle back-annotated circuit descriptions, that is, with realistic interconnect capacitive and resistive values. The power estimates thus obtained can be very close to the power consumption of the actual fabricated circuit.

The problem is that such detailed simulation requires the solution of complex systems of equations and is only practical for small circuits. Another limitation is that the input sequences must necessarily be very short since simulation is time consuming; hence, the resulting power estimates may poorly reflect the real statistics of the inputs. For these reasons, full-fledged circuit-level power estimation is typically only performed for the accurate characterization of small-circuit modules. To apply circuit-level simulation to larger designs, one can resort to very simple models for the active devices. Naturally, this simplification implies accuracy loss. On the other hand, massively parallel computers extend the applicability of these methods to even larger designs [5].

Switch-level simulation is a limiting case, where transistor models are simply reduced to switches, which can be either opened or closed, with some associated parasitic resistive and capacitive values. This simplified model allows for the estimation of significantly larger circuit modules under much longer input sequences. Switch-level simulation can still model with fair accuracy the dynamic and short-circuit components of power, but this is no longer true for leakage power. At early technology nodes, designers were willing to ignore this power component since it accounted for a negligible fraction of total power, but now its relative importance is increasing. Leakage power estimation must then be performed independently using specifically tailored tools. Many different approaches have been proposed, some of which are presented in the next section.

Among intermediate-complexity solutions, PrimeTime PX, an add-on to Synopsys' static timing analysis tool [6], offers power estimates with accuracy close to SPICE. This tool employs table lookup of current models for given transistor sizes and uses circuit partitioning to solve the circuit equations independently on each partition. Although some error is introduced by not accounting for interactions between different partitions, this technique greatly simplifies the problem to be solved, allowing for fast circuit-level estimates of large designs.

3.1.3 STATIC POWER ESTIMATION

Static power analysis is typically performed using the subthreshold model to estimate *leakage per unit micron*, which is then extrapolated to estimate leakage over the entire chip. Typically, the stacking factor (leakage reduction from stacking of devices) is a first-order component of this extension and serves to modify the total effective width of devices under analysis [7]. Analysis can be viewed as the modification of this total width by the stacking factor.

Most analytical works on leakage have used the BSIM2 subthreshold current model [8]:

$$(3.5) \qquad I_{sub} = Ae^{\frac{\left(V_{GS} - V_T - \gamma'V_{SB} + \eta V_{DS}\right)}{nV_{TH}}} \left(1 - e^{-\frac{V_{DS}}{V_{TH}}}\right)$$

where

V_{GS}, V_{DS}, and V_{SB} are the gate-source, drain-source, and source-bulk voltages, respectively
V_T is the zero-bias threshold voltage
V_{TH} is the thermal voltage (kT/q)
γ' is the linearized body-effect coefficient
η is the drain-induced barrier lowering (DIBL) coefficient

$$A = \mu_0 C_{ox}(W/L_{eff})V_{TH}^2 e^{1.8}$$

The BSIM2 leakage model incorporates all the leakage behavior that we are presently concerned with. In summary, it accounts for the exponential increase in leakage with reduction in threshold voltage and gate-source voltage. It also accounts for the temperature dependence of leakage.

Calculating leakage current by applying Equation 3.5 to every single transistor in the chip can be very time consuming. To overcome this barrier, empirical models for dealing with leakage at a higher level of abstraction have been studied [9,10]. For example, a simple empirical model is as follows [10]:

$$(3.6) \qquad I_{leak} = I_{off}\frac{W_{tot}}{X_s}X_t$$

where

I_{off} is the leakage current per micron of a single transistor measured from actual silicon at a given temperature
W_{tot} is the total transistor width (sum of all N and P devices)
X_s is an empirical stacking factor based on the observation that transistor stacks leak less than single devices
X_t is the temperature factor and is used to scale I_{off} to the appropriate junction temperature of interest

The I_{off} value is typically specified at room temperature (therefore the need for a temperature factor to translate to the temperature of interest).

The other major component of static power is gate leakage. Gate leakage effectively becomes a first-order effect only when the gate oxide is thin enough such that direct quantum tunneling through the oxide becomes a significant quantity. The amount of gate leakage current is directly proportional to transistor width, and thus, the main effort for gate leakage estimation is to estimate the total transistor width, W_{tot}, similar to what is required for subthreshold current. The exact value of gate leakage depends on the gate-to-source and drain-to-source voltages, V_{GS} and V_{DS}, and these depend on gate input values. State-based weighting of the transistor width can therefore be used for more accurate gate leakage estimation. However, this entails the additional effort of estimating the state probabilities for each gate [11].

At present, V_T is high enough such that subthreshold current is dominated by the dynamic component of the total active current. On the other hand, subthreshold current dominates the total standby current when compared to gate and well leakage components. As oxide thickness continued to scale down, it was feared that gate leakage would become the dominant source of leakage. However, the introduction of new technologies like metal gates and 3D FinFETs has decelerated the trend toward thinner oxides, and therefore, subthreshold leakage will continue to dominate gate leakage for at least a couple of more technology nodes.

3.1.4 LOGIC-LEVEL POWER ESTIMATION

A key observation from Section 3.1.1 that facilitates power estimation at the logic level is that, if the input of the gate rises fast enough, the energy consumed by each output transition does not depend on the resistive characteristics of the transistors and is simply a function of the capacitive load, C_{out}, the gate is driving, $E_c = ½C_{out}V_{dd}^2$. Given parasitic gate and wire capacitance models that allow the computation of C_{out_i} for each gate i in a gate-level description of the circuit, power estimation at the logic level reduces to computing the number of transitions that each gate makes in a given period of time, that is, the switching activity of the gate. This corresponds to either parameter N or α, and we need to only apply Equation 3.2 or 3.3, respectively, to obtain power.

Naturally, this estimate refers only to the dynamic power component. For total power consumption, we must take leakage power into account, meaning that the methods described in the previous section must complement the logic-level estimate. In many cases, power estimates at the logic level serve as indicators for guiding logic-level power optimization techniques, which typically target the dynamic power reduction, and hence, only an estimate for this component is required. There are two classes of techniques for the switching activity computation, namely, simulation-based and probabilistic analyses (also known as dynamic and static techniques, respectively).

3.1.4.1 SIMULATION-BASED TECHNIQUES

In simulation-based switching activity estimation, highly optimized logic simulators are used, allowing for fast simulation of a large number of input vectors. This approach raises two main issues: the number of input vectors to simulate and the delay model to use for the logic gates.

The simplest approach to model the gate delay is to assume zero delay for all the gates and wires, meaning that all transitions in the circuit occur at the same time instant. Hence, each gate makes at most one transition per input vector. In reality, logic gates have nonzero transport delay, which may lead to different arrival times of transitions at the inputs of a logic gate due to different signal propagation paths. As a consequence, the output of the gate may switch multiple times in response to a single input vector. An illustrative example is shown in Figure 3.2.

Consider that initially signal x is set to 1 and signal y is set to 0, implying that both signals w and z are set to 1. If y makes a transition to 1, then z will first respond to this transition by switching to 0. However, at about the same time, w switches to 0, thus causing z to switch back to 1.

This spurious activity can make for a significant fraction of the overall switching activity, which in the case of circuits with a high degree of reconvergent signals, such as multipliers, may be more than 50% [12]. The modeling of gate delays in logic-level power estimation is, thus, of crucial significance. For an accurate switching activity estimate, the simulation must use a general

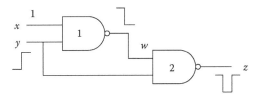

FIGURE 3.2   Example of a logic circuit with glitching and spatial correlation.

delay model where gate delays are retrieved from a precharacterized library of gates. Process variation introduces another level of complexity, motivating a statistical analysis for delay, and consequently of the spurious activity [13].

The second issue is determining the number of input vectors to simulate. If the objective is to obtain a power estimate of a logic circuit under a user-specified, potentially long, sequence of input vectors, then the switching activity can be easily obtained through logic simulation. When only input statistics are given, a sequence of input vectors needs to be generated. One option is to generate a sequence of input vectors that approximates the given input statistics and simulate until the average power converges, that is, until this value stays within a margin ε during the last n input vectors, where ε and n are user-defined parameters.

An alternative is to compute beforehand the number of input vectors required for a given allowed percentage error ε and confidence level θ. Under a basic assumption that the power consumed by a circuit over a period of time T has a normal distribution, the approach described in [14] uses the central limit theorem to determine the number of input vectors that must be simulated:

$$N \geq \left(\frac{z_{\theta/2} s}{\bar{p} \varepsilon} \right)^2$$

where
  N is the number of input vectors
  $\bar{p}$ and s are the measured average and standard deviation of the power
  $z_{\theta/2}$ is obtained from the normal distribution

In practice, for typical combinational circuits and reasonable error and confidence levels, the number of input vectors needed to obtain the overall average switching activity is typically very small (thousands) even for complex logic circuits. However, in many situations, accurate average switching activity for each node in the circuit is required. A high level of accuracy for low-switching nodes may require a prohibitively large number of input vectors. The designer may need to relax the accuracy for these nodes, based on the argument that these are the nodes that have less impact on the dynamic power consumption of the circuit.

Still, today's highly parallel architectures facilitate fast simulation of a large number of input vectors, thus improving the accuracy of this type of Monte Carlo–based estimation methods [15].

3.1.4.2 PROBABILISTIC TECHNIQUES

The idea behind probabilistic techniques is to propagate directly the input statistics to obtain the switching probability of each node in the circuit. This approach is potentially very efficient, as only a single pass through the circuit is needed. However, it requires a new simulation engine with a set of rules for propagating the signal statistics. For example, the probability that the output of an AND gate evaluates to 1 is associated with the intersection of the conditions that set each of its inputs to 1. If the inputs are independent, then this is just the multiplication of the probabilities that each input evaluates to 1. Similar rules can be derived for any logic gate and for different statistics, namely, transition probabilities. Although all of these rules are simple, there is a new set of complex issues to be solved. One of them is the delay model, as mentioned earlier. Under a general delay model, each gate may switch at different time instants in response to a single input change. Thus, we need to compute switching probabilities for each of these time instants. Assuming the transport delays to be Δ_1 and Δ_2 for the gates in the circuit of Figure 3.2 means that signal z will have some probability of making a transition at instant Δ_2 and some other probability of making a transition at instant $\Delta_1 + \Delta_2$. Naturally, the total switching activity of signal z will be the sum of these two probabilities.

Another issue is spatial correlation. When two logic signals are analyzed together, they can only be assumed to be independent if they do not have any common input signal in their support. If there is one or more common input, we say that these signals are spatially correlated. To illustrate this

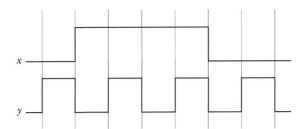

FIGURE 3.3   Example signal to illustrate the concept of temporal correlation.

point, consider again the logic circuit of Figure 3.2 and assume that both input signals, x and y, are independent and have a $p_x = p_y = 0.5$ probability of being at 1. Then, p_w, the probability that w is 1, is simply $p_w = 1 - p_x p_y = 0.75$. However, it is not true that $p_z = 1 - p_w p_y = 0.625$ because signals w and y are not independent: $p_w p_y = (1 - p_x p_y).p_y = p_y - p_x p_y$ (note that $p_y p_y = p_y$), giving $p_z = (1 - p_y + p_x p_y) = 0.75$. This indicates that not accounting for spatial correlation can lead to significant errors in the calculations.

Input signals may also be spatially correlated. Yet, in many practical cases, this correlation is ignored, either because it is simply not known or because of the difficulty in modeling this correlation. For a method that is able to account for all types of correlations among signals see [16], but it cannot be applied to very large designs due to its high complexity.

A third important issue is temporal correlation. In probabilistic methods, the average switching activity is computed from the probability of a signal making a transition 0 to 1 or 1 to 0. Temporal correlation measures the probability that a signal is 0 or 1 in the next instant given that its present value is 0 or 1. This means that computing the static probability of a signal being 1 is not sufficient, and we need to calculate the transition probabilities directly so that temporal correlation is taken into account. Consider signals x and y in Figure 3.3, where the vertical lines indicate clock periods. The number of periods where these two signals are 0 or 1 is the same, and hence, the probability of the signals being at 1 is $p_x^1 = p_y^1 = 0.5$ (and the probability being at 0 is $p_x^0 = p_y^0 = 0.5$). If we only consider this parameter, thus ignoring temporal correlation, the transition probability for both signals is the same and can be computed as $\alpha = p^{01} + p^{10} = p^0 p^1 + p^1 p^0 = 0.5$. However, we can see that, during the depicted time interval, signal x remains low for three clock cycles, remains high for another three cycles, and has a single clock cycle with a rising transition and another with a falling transition. Averaging over the number of clock periods, we have $p_x^{00} = \frac{3}{8} = 0.375$, $p_x^{01} = \frac{1}{8} = 0.125$, $p_x^{10} = \frac{1}{8} = 0.125$, and $p_x^{11} = \frac{3}{8} = 0.375$. Therefore, the actual average switching activity of x is $\alpha_x = p_x^{01} + p_x^{10} = 0.25$. As for signal y, it never remains low or high, making a transition on every clock cycle. Hence, $p_y^{00} = p_y^{11} = 0$ and $p_y^{01} = p_y^{10} = \frac{4}{8} = 0.5$, and the actual average switching activity of y is $\alpha_y = p_y^{01} + p_y^{10} = 1.0$. This example illustrates the importance of modeling temporal correlation and indicates that probabilistic techniques need to work with transition probabilities for accurate switching activity estimates.

It has been shown that exact modeling of these issues makes the computation of the average switching activity an NP-hard problem, meaning that exact methods are only applicable to small circuits and thus are of little practical interest. Many different approximation schemes have been proposed [17].

3.1.4.3 SEQUENTIAL CIRCUITS

Computing the switching activity for sequential circuits is significantly more difficult, because the state space must be visited in a representative manner to ensure the accuracy of the state signal probabilities. For simulation-based methods, this requirement may imply too large an input sequence and, in practice, convergence is hard to guarantee.

Probabilistic methods can be effectively applied to sequential circuits, as the statistics for the state signals can be derived from the circuit. The exact solution would require the computation of the transition probabilities between all pairs of states in the sequential circuit. In many cases, enumerating the states of the circuit is not possible, since these are exponential in the number of sequential elements in the circuit. A common approximation is to compute the transition

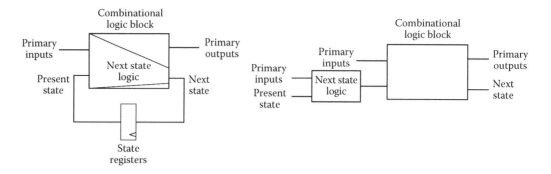

FIGURE 3.4   Creating temporal and spatial correlation among state signals.

probabilities for each state signal [18]. To partially recover the spatial correlation between state signals, a typical approach is to duplicate the subset of logic that generates the next state signals and append it to the present state signals, as is illustrated in Figure 3.4. Then the method for combinational circuits is applied to this modified network, ignoring the switching activity in the duplicated next state logic block.

3.1.5 ANALYSIS AT THE REGISTER-TRANSFER LEVEL

At the register-transfer level (RTL), the circuit is described in terms of interconnected modules of varied complexity from simple logic gates to full-blown multipliers. Power estimation at this level determines the signal statistics at the input of each of these modules and then feeds these values to the module's power model to evaluate its power dissipation. These models are normally available with the library of modules. One way to obtain these power models is to characterize the module using logic- or circuit-level estimators, a process known as macromodeling. We refer to Chapter 13 of *Electronic Design Automation for IC System Design, Verification, and Testing*, where this topic is discussed in more detail.

3.2 CIRCUIT-LEVEL POWER OPTIMIZATION

From the equations that model power consumption, one sees that a reduction of the supply voltage has the largest impact on power reduction, given its quadratic effect. This has been the largest source of power reductions and is widely applied across the semiconductor industry. However, unless accompanied by the appropriate process scaling, reducing V_{dd} comes at the cost of increased propagation delays, necessitating the use of techniques to recover the lost performance.

Lowering the frequency of operation, f_{clk}, also reduces power consumption. This may be an attractive option in situations with low-performance requirements. Yet, the power efficiency of the circuit is not improved, as the amount of energy per operation remains constant.

A more interesting option is to address the switched capacitance term, αC_{out}, by redesigning the circuit such that the overall switching activity is reduced or the overall circuit capacitance is reduced or by a combination of both, where the switching activity in high-capacitance nodes is reduced, possibly by exchanging it with higher switching in nodes with lower capacitance.

Static power due to leakage current, however, presents a different set of challenges. As Equation 3.5 shows, reducing V_{dd} will reduce leakage current as well. However, reducing the number of transitions to reduce switched capacitance has little benefit, since leakage power is consumed whether or not there is a transition in the output of a gate. The most effective way to reduce leakage is to effectively shut off the power to a circuit—this is called power gating. Other techniques are motivated by the relationship of leakage current to the threshold voltage V_T. Increasing the threshold voltage reduces the leakage current. Equation 3.6 motivates other techniques that exploit the relationship of leakage current to circuit topology.

In the following, we briefly discuss the key circuit-level techniques that have been developed to address each of these points. There is a vast amount of published work that covers these and other techniques in great detail, and the interested reader is recommended to start with books [19] and overview papers [20] that cover these topics in greater depth.

3.2.1 TRANSISTOR SIZING

The propagation delay (usually just referred to as delay) of a gate is dependent on the gate output resistance and the total capacitance (interconnect and load) [2]. Transistor sizing (or gate sizing) helps reduce delay by increasing gate strength at the cost of increased area and power consumption. Conversely by reducing gate strength, the switched capacitance, and therefore, the power, can be reduced at the cost of increased delay. This trade-off can be performed manually for custom designs or through the use of automated tools.

Up until the recent past, large parts of high-performance CPUs were typically custom designed. Even now, the most performance critical parts of high-performance designs have a mix of synthesized and custom designed parts. Such designs may involve manual tweaking of transistors to upsize drivers along critical paths. If too many transistors are upsized unnecessarily, certain designs can operate on the steep part of a circuit's power–delay curve. In addition, the choice of logic family used (e.g., static vs. dynamic logic) can also greatly influence the circuit's power consumption. The traditional emphasis on performance often leads to overdesign that is wasteful of power. An emphasis on lower power, however, motivates the identification of such sources of power waste. An example of such waste is circuit paths that are designed faster than they need to be. For synthesized blocks, the synthesis tool can automatically reduce power by downsizing devices in such paths. For manually designed blocks, on the other hand, downsizing may not always be done. Automated downsizing tools can thus have a big impact. The benefit of such tools is power savings as well as productivity gains over manual design methodologies.

The use of multiple-threshold voltages ("multi-V_T") to reduce leakage power in conjunction with traditional transistor sizing is now a widely used design technique. The main idea here is to use lower-V_T transistors in critical paths rather than large high-V_T transistors. However, this technique increases subthreshold leakage due to low V_T. So, it is very important to use low-V_T transistor selectively and optimize their usage to achieve a good balance between capacitive current and leakage current in order to minimize the total current. This consideration is now part of the postsynthesis or postlayout automated tools and flows that recognize both low-V_T and high-V_T substitution. For example, after postlayout timing analysis, a layout tool can operate in incremental mode to do two things: insert low-V_T cells into critical paths to improve speed and insert higher-V_T cells into noncritical paths to bring leakage back down again.

Custom designers may have the flexibility to manually choose the transistor parameters to generate custom cells. Most synthesized designs, however, only have the choice of picking from different gates or cells in a cell library. These libraries typically have a selection of cells ranging from high performance (high power) to low power (low performance). In this case, the transistor-sizing problem reduces to the problem of optimal cell selection either during the initial synthesis flow or of tweaking the initial selection in a postsynthesis flow. This has been an area of active academic research [21] as well as a key optimization option in commercial tools [19].

3.2.2 VOLTAGE SCALING, VOLTAGE ISLANDS, AND VARIABLE V_{DD}

As mentioned earlier, the reduction of V_{dd} is the most effective way of reducing power. The industry has thus steadily moved to lower V_{dd}. Indeed, reducing the supply voltage is the best for low-power operation, even after taking into account the modifications to the system architecture, which are required to maintain the computational throughput. Another issue with voltage scaling is that to maintain performance, threshold voltage also needs to be scaled down since circuit speed is roughly inversely proportional to $(V_{dd}-V_T)$. Typically, V_{dd} should be larger than $4V_T$ if speed is not to suffer excessively. As the threshold voltage decreases, subthreshold leakage current increases exponentially. With every 0.1 V reduction in V_T, subthreshold current increases by 10 times.

In the nanometer technologies, with further V_T reduction, subthreshold current has become a significant portion of the overall chip current. At 0.18 m feature size and less, leakage power starts eating into the benefits of lower V_{dd}. In addition, the design of dynamic circuits, caches, sense amps, PLAs, etc., becomes difficult at higher subthreshold leakage currents. Lower V_{dd} also exacerbates noise and reliability concerns. To combat the subthreshold current increase, various techniques have been developed, as mentioned in the Section 3.2.5.

Voltage islands and variable V_{dd} are variations of voltage scaling that can be used at the circuit level. Voltage scaling is mainly technology dependent and typically applied to the whole chip. Voltage islands are more suitable for system-on-chip design, which integrates different functional modules with various performance requirements onto a single chip. We refer to the chapter on RTL power analysis and optimization techniques for more details on voltage islands. The variable voltage and voltage island techniques are complementary and can be implemented on the same block to be used simultaneously. In the variable voltage technique, the supply voltage is varied based on throughput requirements. For higher-throughput applications, the supply voltage is increased along with operating frequency and vice versa for the lower-throughput application. Sometimes, this technique is also used to control power consumption and surface temperature. On-chip sensors measure temperature or current requirements and lower the supply voltage to reduce power consumption. Leakage power mitigation can be achieved at the device level by applying multithreshold voltage devices, multichannel length devices, and stacking and parking state techniques. The following section gives details on these techniques.

3.2.3 MULTIPLE-THRESHOLD VOLTAGES

Multiple-threshold voltages (most often a high-V_T and a low-V_T option) have been available on many, if not most, CMOS processes for a number of years. For any given circuit block, the designer may choose to use one or the other V_T or a mixture of the two. For example, use high-V_T transistor as the default and then selectively insert low-V_T transistors. Since the standby power is so sensitive to the number of low-V_T transistors, their usage, in the order of 5%–10% of the total number of transistors, is generally limited to only fixing critical timing paths, or else leakage power could increase dramatically. For instance, if the low-V_T value is 110 mV less than the high-V_T value, 20% usage of the former will increase the chip standby power by nearly 500%. Low-V_T insertion does not impact the active power component or design size, and it is often the easiest option in the postlayout stage, leading to the least layout perturbation. Obvious candidate circuits for using high-V_T transistors as the default and only using selectively low-V_T transistors are SRAMs, whose power is dominated by leakage, and a higher V_T generally also improves SRAM stability (as does a longer channel). The main drawbacks of low-V_T transistors are that delay variations due to doping are uncorrelated between the high- and low-threshold transistors, thus requiring larger timing margins, and that extra mask steps are needed, which incur additional process cost.

3.2.4 LONG-CHANNEL TRANSISTORS

The use of transistors that have longer than nominal channel length is another method of reducing leakage power [22]. For example, by drawing a transistor 10 nm longer (long-L) than a minimum sized one, the DIBL is attenuated and the leakage can be reduced by 7×–10× on a 90 nm process. With this one change, nearly 20% of the total SRAM leakage component can be eliminated while maintaining performance. The loss in drive current due to increased channel resistance, on the order of 10%–20%, can be compensated by an increase in width or since the impact is on a single gate stage, it can be ignored for most of the designs [22]. The use of long-L is especially useful for SRAMs, since their overall performance is relatively insensitive to transistor delay. It can also be applied to other circuits, if used judiciously. Compared with multiple-threshold voltages, long-channel insertion has similar or lower process cost—it manifests as size increases rather than mask cost. It allows lower process complexity and the different channel lengths track over process variation. It can be applied opportunistically to an existing design to limit leakage. A potential penalty is the increase in gate capacitance. Overall active power does not increase significantly if the activity factor of the affected gates is low, so this should also be considered when choosing target gates.

The target gate selection is driven by two main criteria. First, transistors must lie on paths with sufficient timing margin. Second, the highest leakage transistors should be chosen first from the selected paths. The first criterion ensures that the performance goals are met. The second criterion helps in maximizing leakage power reduction. In order to use all of the available positive timing slack and avoid errors, long-L insertion is most advisable at the late design stages.

The long-L insertion can be performed by using standard cells designed using long-L transistors or by selecting individual transistors from the transistor-level design. Only the latter is applicable to full custom design. There are advantages and disadvantages to both methods. For the cell-level method, low-performance cells are designed with long-L transistors. For leakage reduction, high-performance cells on noncritical paths are replaced with lower-performance cells with long-L. If the footprint and port locations are identical, then this method simplifies the physical convergence. Unfortunately, this method requires a much larger cell library. It also requires a fine-tuned synthesis methodology to ensure long-L cell selection rather than lower-performance nominal channel length cells. The transistor-level flow has its own benefits. A unified flow can be used for custom blocks and auto placed-and-routed blocks. Only a single nominal cell library is needed, albeit with space for long-L as mentioned.

3.2.5 TOPOLOGICAL TECHNIQUES: STACKING AND PARKING STATES

Another class of techniques exploits the dependence of leakage power on the topology of logic gates. Two examples of such techniques are stacking and parking states. These techniques are based on the fact that a stack of "OFF" transistors leaks less than when only a single device in a stack is OFF. This is primarily due to the self-reverse biasing of the gate-to-source voltage V_{GS} in the OFF transistors in the stack. Figure 3.5 illustrates the voltage allocation of four transistors in series [10]. As one can see, V_{GS} is more negative when a transistor is closer to the top of the stack. The transistor with the most negative V_{GS} is the limiter for the leakage of the stack. In addition, the threshold voltages for the top three transistors are increased because of the reverse-biased body-to-source voltage (body effect).

Both the self-reverse biasing and the body effects reduce leakage exponentially as shown in Equation 3.5. Finally, the overall leakage is also modulated by the DIBL effect for submicron MOSFETs. As V_{DS} increases, the channel energy barrier between the source and the drain is lowered. Therefore, leakage current increases exponentially with V_{DS}.

The combination of these three effects results in a progressively reduced V_{DS} distribution from the top to the bottom of the stack, since all of the transistors in series must have the same leakage current. As a result, significantly reduced V_{DS}, the effective leakage of stacked transistors, is much lower than that of a single transistor.

Table 3.1 quantifies the basic characteristics of the subthreshold leakage current for a fully static four-input NAND gate. The minimum leakage condition occurs for the "0000" input vector (i.e., all inputs a, b, c, and d are at logic zero). In this case, all the PMOS devices are "ON" and

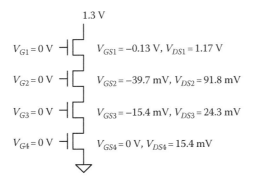

FIGURE 3.5 Voltage distribution of stacked transistors in OFF state.

TABLE 3.1   Stacking Factors of Four-Input NAND

	Minimum	Maximum	Average
Stacking factor Xs	1.75	70.02	9.95
Input vector (a b c d)	(1 1 1 1)	(0 0 0 0)	—

the leakage path exists between the output node and the ground through a stack of four NMOS devices. The maximum leakage current occurs for the "1111" input case, when all the NMOS devices are ON and the leakage path, consisting of four parallel PMOS devices, exists between the supply and the output node. The stacking factor variation between the minimum and maximum leakage conditions reflects the magnitude of leakage dependence on the input vector. In the four-input NAND case, we can conclude that the leakage variation between the minimum and maximum cases is a factor of about 40 (see Table 3.1). The values were measured using an accurate SPICE-like circuit simulator on a 0.18 μm technology library. The average leakage current was computed based on the assumption that all the 16 input vectors were equally probable.

Stacking techniques take advantage of the effects described earlier to increase the stack depth [23]. One of the examples is the sleep transistor technique. This technique inserts an extra series-connected device in the stack and turns it OFF during the cycles when the stack will be OFF as a whole. This comes at the cost of the extra logic to detect the OFF state, as well as the extra delay, area, and dynamic power cost of the extra device. Therefore, this technique is typically applied at a much higher level of granularity, using a sleep transistor that is shared across a larger block of logic. Most practical applications in fact apply this technique at a very high level of granularity, where the sleep state (i.e., inactive state) of large circuit blocks such as memory and ALUs can be easily determined. At that level, this technique can be viewed as analogous to power gating, since it isolates the circuit block from the power rails when the circuit output is not needed, that is, inactive. Power gating is a very effective and increasingly popular technique for leakage reduction, and it is supported by commercial EDA tools [24], but it is mostly applied at the microarchitectural or architectural level and therefore not discussed further in here.

The main idea behind the parking state technique is to force the gates in the circuit to the low-leakage logic state when not in use [25]. As described earlier, leakage current is highly dependent on the topological relationship between ON and OFF transistors in a stack, and thus, leakage depends on the input values. This technique avoids the overhead of extra stacking devices, but additional logic is needed to generate the desirable state, which has an area and switching power cost. This technique is not advisable for random logic, but with careful implementation for structured datapath and memory arrays, it can save significant leakage power in the OFF state.

One needs to be careful about using these techniques, given the area and switching overheads of the introduced devices. Stacking is beneficial in cases where a small number of transistors can add extra stack length to a wide cone of logic or gate the power supply to it. The delays introduced by the sleep transistors or by power gating also imply that these techniques are beneficial only when the targeted circuit blocks remain in the OFF state for long enough to make up for the overhead of driving the transitions in and out of the OFF states. These limitations can be overcome with careful manual intervention or by appropriate design intent hints to automated tools.

Leakage power reduction will remain an active area of research, since leakage power is essentially what limits the reduction of dynamic power through voltage scaling. As transistor technology scales down to smaller feature sizes, making it possible to integrate greater numbers of devices on the same chip, additional advances in materials and transistor designs can be expected to allow for finer-grained control on the power (dynamic and leakage) and performance trade-offs. This will need to be coupled with advances in power analysis to understand nanometer-scale effects that have so far not been significant enough to warrant detailed power models. In conjunction with these models, new circuit techniques to address these effects will need to be developed. As these circuit techniques gain wider acceptability and applicability, algorithmic research to incorporate these techniques in automated synthesis flows will continue.

3.2.6 LOGIC STYLES

Dynamic circuits are generally regarded as dissipating more power than their static counterparts. While the power consumption of a static CMOS gate with constant inputs is limited to leakage power, dynamic gates may be continually precharging and discharging their output capacitance under certain input conditions.

For instance, if the inputs to the NAND gate in Figure 3.6a are stable, the output is stable. On the other hand, the dynamic NAND gate of Figure 3.6b, under constant inputs A = B = 1, will keep raising and lowering the output node, thus leading to high energy consumption.

For several reasons, dynamic logic families are preferred in many high-speed, high-density designs (such as microprocessors). First, dynamic gates require fewer transistors, which means not only that they take up less space but also that they exhibit a lower capacitive load, hence allowing for increased operation speed and for reduced dynamic power dissipation. Second, the evaluation of the output node can be performed solely through N-type MOSFET transistors, which further contributes to the improvement in performance. Third, there is never a direct path from V_{dd} to ground, thus effectively eliminating the short-circuit power component. Finally, dynamic circuits intrinsically do not create any spurious activity, which can make for a significant reduction in power consumption. However, the design of dynamic circuits presents several issues that have been addressed through different design families [26].

Pass-transistor logic is another design style whose merits for low power have been pointed out, mainly due to the lower capacitance load of the input signal path. The problem is that this design style may imply a significantly larger circuit.

Sequential circuit elements are of particular interest with respect to their chosen logic style, given their contribution to the power consumption of a logic chip. These storage elements—flip-flops or latches—are the end points of the clock network and constitute the biggest portion of the switched capacitance of a chip because of both the rate at which their inputs switch (every clock edge) and their total number (especially in high-speed circuits with shallow pipeline depths). For this reason, these storage elements received a lot of attention [27]. For example, dual-edge-triggered flip-flops have been proposed as a lower-power alternative to the traditional single-edge-triggered flip-flops, since they provide an opportunity to reduce the effective clock frequency by half. The trade-offs between ease of design, design portability, scalability, robustness, and noise sensitivity, not to mention the basics trade-offs of area and performance, require these choices to be made only after a careful consideration of the particular design application. These trade-offs also vary with technology node, as the leakage power consumption must be factored into the choice.

In general, one can expect research and innovation in circuit styles to continue as long as the fundamental circuit design techniques evolve to overcome the limitations or exploit the opportunities provided by technology scaling.

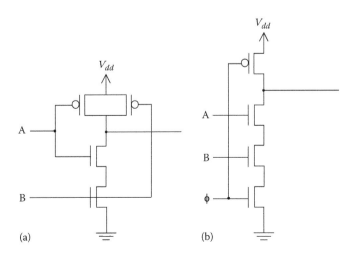

FIGURE 3.6　NAND gate: (a) static CMOS and (b) dynamic domino.

3.3 LOGIC SYNTHESIS FOR LOW POWER

A significant amount of CAD research has been carried out in the area of low power logic synthesis. By adding power consumption as a parameter for the synthesis tools, it is possible to save power with no, or minimal, delay penalty.

3.3.1 LOGIC FACTORIZATION

A primary means of technology-independent optimization is the factoring of logical expressions. For example, the expression $xy \lor xz \lor wy \lor wz$ can be factored into $(x \lor w)(y \lor z)$, reducing transistor count considerably. Common subexpressions can be found across multiple functions and reused. For area optimization, several candidate divisors (e.g., kernels) of the given expressions are generated and those that maximally reduce literal count are selected. Even though minimizing transistor count may, in general, reduce power consumption, in some cases the total effective switched capacitance actually increases. When targeting power dissipation, the cost function must take into account switching activity. The algorithms proposed for low power kernel extraction compute the switching activity associated with the selection of each kernel. Kernel selection is based on the reduction of both area and switching activity [28].

3.3.2 DON'T-CARE OPTIMIZATION

Multilevel circuits are optimized taking into account appropriate don't-care sets. The structure of the logic circuit may imply that some input combinations of a given logic gate never occur. These combinations form the controllability or satisfiability don't-care set of the gate. Similarly, there may be some input combinations for which the output value of the gate is not used in the computation of any of the outputs of the circuit. The set of these combinations is called the observability don't-care set. Although initially don't-care sets were used for area minimization, techniques have been proposed for the use of don't-care sets to reduce the switching activity at the output of a logic gate [29]. The transition probability of a static CMOS gate is given by $\alpha_x = 2p_x^0 p_x^1 = 2p_x^1(1 - p_x^1)$ (ignoring temporal correlation). The maximum for this function occurs for $p_x^1 = 0.5$. Therefore, in order to minimize the switching activity, the strategy is to include minterms from the don't-care set in the onset of the function if $p_x^1 > 0.5$ or in the off-set if $p_x^1 < 0.5$.

3.3.3 PATH BALANCING

Spurious transitions account for a significant fraction of the switching activity power in typical combinational logic circuits [30]. In order to reduce spurious switching activity, the delay of paths that converge at each gate in the circuit should be roughly equal, a problem known as path balancing. In the previous section, we discussed that transistor sizing can be tailored to minimize power primarily at the cost of delaying signals not on the critical path. This approach has the additional feature of contributing to path balancing. Alternatively, path balancing can be achieved through the restructuring of the logic circuit, as illustrated in Figure 3.7.

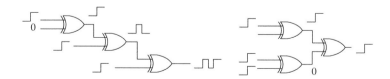

FIGURE 3.7 Path balancing through logic restructuring to reduce spurious transitions.

Path balancing is extremely sensitive to propagation delays, becoming a more difficult problem when process variations are considered. The work in [30] addresses path balancing through a statistical approach for delay and spurious activity estimation.

3.3.4 TECHNOLOGY MAPPING

Technology mapping is the process by which a logic circuit is realized in terms of the logic elements available in a particular technology library. Associated with each logic element is the information about its area, delay, and internal and external capacitances. The optimization problem is to find the implementation that meets the delay constraint while minimizing a cost function that is a function of area and power consumption [31,32]. To minimize power dissipation, nodes with high switching activity are mapped to internal nodes of complex logic elements, as capacitances internal to gates are generally much smaller.

In many cases, the inputs of a logic gate are commutative in the Boolean sense. However, in a particular gate implementation, equivalent pins may present different input capacitance loads. In these cases, gate input assignment should be performed such that signals with high switching activity map to the inputs that have lower input capacitance.

Additionally, most technology libraries include the same logic elements with different sizes (i.e., driving capability). Thus, in technology mapping for low power, the size of each logic element is chosen so that the delay constraints are met with minimum power consumption. This problem is the discrete counterpart of the transistor-sizing problem described in the previous section.

3.3.5 STATE ENCODING

The synthesis of sequential circuits offers new avenues for power optimization. State encoding is the process by which a unique binary code is assigned to each state in a finite-state machine (FSM). Although this assignment does not influence the functionality of the FSM, it determines the complexity of the combinational logic block in the FSM implementation. State encoding for low power uses heuristics that assign minimum Hamming distance codes to states that are connected by edges that have larger probability of being traversed [33]. The probability that a given edge in the state transition graph (STG) is traversed is given by the steady-state probability of the STG being in the start state of the edge, multiplied by the static probability of the input combination associated with that edge. Whenever this edge is exercised, only a small number of state signals (ideally one) will change, leading to reduced overall switching activity in the combinational logic block.

3.3.6 FSM DECOMPOSITION

FSM decomposition has been proposed for low power implementation of an FSM. The basic idea is to decompose the STG of the original FSM into two coupled STGs that together have the same functionality as the original FSM. Except for transitions that involve going from one state in one sub-FSM to a state in the other, only one of the sub-FSMs needs to be clocked. The strategy for state selection is such that only a small number of states is selected for one of the sub-FSMs. This selection consists of searching for a small cluster of states such that summation of the probabilities of transitions between states in the cluster is high, and there is a very low probability of transition to and from states outside of the cluster. The aim is to have a small sub-FSM that is active most of the time, disabling the larger sub-FSM. Having a small number of transitions to/from the other sub-FSM corresponds to the worst case, when both sub-FSMs are active. Each sub-FSM has an extra output that disables the state registers of the other sub-FSM, as shown in Figure 3.8. This extra output is also used to stop transitions at the inputs of the large sub-FSM. An approach to perform this decomposition solely using circuit techniques, thus without any derivation of the STG, was proposed in [34].

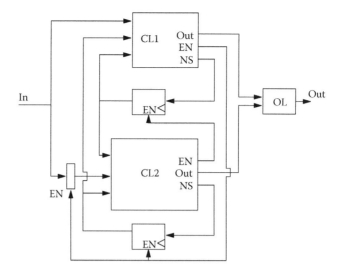

FIGURE 3.8 Implementation diagram of a decomposed FSM for low power.

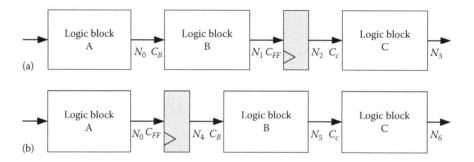

FIGURE 3.9 Two retimed versions, (a) and (b), of a network to illustrate the impact of this operation on the switched capacitance of a circuit.

Other techniques based on blocking input signal propagation and clock gating, such as pre-computation, are covered in some detail in Chapter 13 of *Electronic Design Automation for IC System Design, Verification, and Testing.*

3.3.7 RETIMING

Retiming was first proposed as a technique to improve throughput by moving the registers in a circuit while maintaining input–output functionality. The use of retiming to minimize switching activity is based on the observation that the output of a register has significantly fewer transitions than its input. In particular, no glitching is present. Moving registers across nodes through retiming may change the switching activity at several nodes in the circuit. In the circuit shown in Figure 3.9a, the switched capacitance is given by $N_0C_B + N_1C_{FF} + N_2C_C$, and the switched capacitance in its retimed version, shown in Figure 3.9b, is $N_0C_{FF} + N_4C_B + N_5C_C$. One of these two circuits may have significantly less switched capacitance. Heuristics to place registers such that nodes driving large capacitances have a reduced switching activity, subject to a given throughput constraint, have been proposed [35].

3.4 SUMMARY

This chapter has covered methodologies for the reduction of power dissipation of digital circuits at the lower levels of design abstraction. The reduction of supply voltage has a large impact on power; however, it also reduces performance. Some of the techniques we described apply local voltage reduction and dynamic voltage control to minimize the impact of lost performance.

For most designs, the principal component of power consumption is related to the switching activity of the circuit during normal operation (dynamic power). The main strategy here is to reduce the overall average switched capacitance, that is, the average amount of capacitance that is charged or discharged during circuit operation. The techniques we presented address this issue by selectively reducing the switching activity of high-capacitance nodes, possibly at the expense of increasing the activity of other less capacitive nodes. Design automation tools using these approaches can save 10%–50% in power consumption with little area and delay overhead.

The static power component has been rising in importance with the reduction of feature size due to increased leakage and subthreshold currents. Key methods, mostly at the circuit level, to minimize this power component have been presented.

Also covered in this chapter are power analysis tools. The power estimates provided can be used not only to indicate the absolute level of power consumption of the circuit but also to direct the optimization process by indicating the most power-efficient design alternatives.

REFERENCES

1. M. Taylor, A landscape of the new dark silicon design regime, *IEEE Micro*, 33(5):8–19, 2013.
2. N. Weste and K. Eshraghian. *Principles of CMOS VLSI Design: A Systems Perspective*. Addison-Wesley Publishing Company, Reading, MA, 1985.
3. E. Shauly, CMOS leakage and power reduction in transistors and circuits: Process and layout considerations, *Journal of Low Power Electronics and Applications*, 2(1):1–29, 2012.
4. T. Quarles, *The SPICE3 Implementation Guide*, ERL M89/44, University of California, Berkeley, CA, 1989.
5. L. Han, X. Zhao, and Z. Feng, TinySPICE: A parallel SPICE simulator on GPU for massively repeated small circuit simulations, *Design Automation Conference (DAC)*, Austin, TX, 2013, pp. 89:1–89:8.
6. PrimeTime, Synopsys, http://www.synopsys.com/Tools/Implementation/SignOff/Pages/PrimeTime.aspx (Accessed November 19, 2015.).
7. M. Johnson, D. Somasekhar, and K. Roy, Models and algorithms for bounds on leakage in CMOS circuits, *IEEE Transactions on Computer-Aided Design of Integrated Circuits*, 18(6):714–725, June 1999.
8. B. Sheu et al., BSIM: Berkeley short-channel IGFET model for MOS transistors, *IEEE Journal of Solid-State Circuits*, 22:558–566, August 1987.
9. J. Butts and G. Sohi, A static power model for architects, *Proceedings of MICRO-33*, Monterey, CA, December 2000, pp. 191–201.
10. W. Jiang, V. Tiwari, E. la Iglesia, and A. Sinha, Topological analysis for leakage prediction of digital circuits, *Proceedings of the International Conference on VLSI Design*, Bangalore, India, January 2002, pp. 39–44.
11. R. Rao, J. Burns, A. Devgan, and R. Brown, Efficient techniques for gate leakage estimation, *Proceedings of the International Symposium on Low Power Electronics and Design*, Seoul, South Korea, August 2003, pp. 100–103.
12. L. Zhong and N. Jha, Interconnect-aware low-power high-level synthesis, *IEEE Transactions on Computer-Aided Design of Integrated Circuits and Systems*, 24(3):336–351, March 2005.
13. D. Quang, C. Deming, and M. Wong, Dynamic power estimation for deep submicron circuits with process variation, *Asia and South Pacific Design Automation Conference (ASP-DAC)*, Taipei, Taiwan, 2010, pp. 587–592.
14. R. Burch, F. Najm, P. Yang, and T. Trick. A Monte Carlo approach to power estimation, *IEEE Transactions on VLSI Systems*, 1(1):63–71, March 1993.
15. D. Chatterjee, A. DeOrio, and V. Bertacco, GCS: High-performance gate-level simulation with GPGPUs, *Design, Automation & Test in Europe Conference & Exhibition, DATE '09*, Nice, France, 2009, pp. 1332–1337.
16. A. Freitas and A. L. Oliveira, Implicit resolution of the Chapman-Kolmogorov equations for sequential circuits: An application in power estimation, *Design, Automation and Test in Europe*, Munich, Germany, March 2003, pp. 764–769.
17. M. Pedram, Power minimization in IC design: Principles and applications, *ACM Transactions on Design Automation of Electronic Systems*, 1(1):3–56, 1996.
18. C-Y. Tsui, J. Monteiro, M. Pedram, S. Devadas, A. Despain, and B. Lin, Power estimation methods for sequential logic circuits, *IEEE Transactions on VLSI Systems*, 3(3):404–416, September 1995.
19. C. Piguet, *Low-Power CMOS Circuits: Technology, Logic Design and CAD Tools*, CRC Press, Boca Raton, FL, 2005.
20. M. Alioto, Ultra-low power VLSI circuit design demystified and explained: A tutorial, *IEEE Transactions on Circuits and Systems*, 59(1):3–29, January 2012.

21. P. Kang, Y. Lu, and H. Zhou, An efficient algorithm for library-based cell-type selection in high-performance low-power designs, *Proceedings of the International Conference on Computer Aided Design*, San Jose, CA, November 2012, pp 226–232.

22. L. Clark, R. Patel, and T. Beatty, Managing standby and active mode leakage power in deep sub-micron design, *International Symposium on Low Power Electronics and Design*, August 2005, pp. 274–279.

23. A. Agarwal, S. Mukhopadhyay, A. Raychowdhury, K. Roy, and C. Kim, Leakage power analysis and reduction for nanoscale circuits, *IEEE Mirco*, 26:68–80, March–April 2006.

24. M. Keating, D. Flynn, R. Aitken, A. Gibbons, and K. Shi, *Low Power Methodology Manual: For System-on-Chip Design*, Springer Science Publishing, New York, 2007.

25. D. Lee and D. Blaauw, Static leakage reduction through simultaneous threshold voltage and state assignment, *Proceedings of the Design Automation Conference*, Anaheim, CA, June 2003, pp. 191–194.

26. J. Yuan and C. Svensson, New single-clock CMOS latches and flipflops with improved speed and power savings, *IEEE Journal of Solid-State Circuits*, 32(1):62–69, January 1997.

27. M. Alioto, E. Consoli, and G. Palumbo, Analysis and comparison in the energy-delay-area domain of nanometer CMOS flip-flops: Part I and II, *IEEE Transactions on VLSI Systems*, 19(5):725–750, May 2011.

28. C.-Y. Tsui, M. Pedram, and A. Despain, Power-efficient technology decomposition and mapping under an extended power consumption model, *IEEE Transactions on CAD*, 13(9):1110–1122, September 1994.

29. A. Shen, S. Devadas, A. Ghosh, and K. Keutzer, On average power dissipation and random pattern testability of combinational logic circuits, *Proceedings of the International Conference on Computer-Aided Design*, Santa Clara, CA, November 1992, pp. 402–407.

30. S. Hosun, Z. Naeun, and K. Juho Kim, Stochastic glitch estimation and path balancing for statistical optimization, *IEEE International SOC Conference*, Austin, TX, September 2006, pp. 85–88.

31. V. Tiwari, P. Ashar, and S. Malik, Technology mapping for low power in logic synthesis, *Integration, The VLSI Journal*, 20:243–268, July 1996.

32. C. Tsui, M. Pedram, and A. Despain, Technology decomposition and mapping targeting low power dissipation, *Proceedings of the Design Automation Conference*, Dallas, TX, June 1993, pp. 68–73.

33. L. Benini and G. Micheli, State assignment for low power dissipation, *IEEE Journal of Solid-State Circuits*, 30(3):258–268, March 1995.

34. J. Monteiro and A. Oliveira, Implicit FSM decomposition applied to low power design, *IEEE Transactions on Very Large Scale Integration Systems*, 10(5):560–565, October 2002.

35. J. Monteiro, S. Devadas, and A. Ghosh, Retiming sequential circuits for low power, *Proceedings of the International Conference on Computer-Aided Design*, Santa Clara, CA, November 1993, pp. 398–402.

Equivalence Checking

Andreas Kuehlmann, Fabio Somenzi, Chih-Jen Hsu, and Doron Bustan

CONTENTS

4.1 INTRODUCTION

This chapter covers the challenge of formally checking whether two design specifications are functionally equivalent. In general, there is a wide range of possible definitions of functional equivalence covering comparisons between different levels of abstraction and varying granularity of timing details. For example, one common approach is to consider the problem of machine equivalence, which defines two synchronous design specifications functionally equivalent if, clock by clock, they produce exactly the same sequence of output signals for any valid sequence of input signals. A more general notion of functional equivalence is of importance in the area of microprocessor design. Here, a crucial check is to compare functionally the specification of the instruction set architecture with the register-transfer-level (RTL) implementation, ensuring that any program executed on both models will cause an identical update of the main memory content. A system design flow provides a third example for yet another notion of functional equivalence to be checked between a transaction-level model, for example, written in SystemC, and its corresponding RTL specification or implementation. Such a check has been of increasing interest in a system-on-a-chip design environment, and with the arrival of commercial tools such as Calypto's SLEC, this additional check has become more mainstream. The focus of this chapter will be on the aforementioned machine equivalence, which in practical design flows is currently the most established application of functionally comparing two design specifications. Following common terminology, we will use the term *equivalence checking* synonymously with this narrow view of the problem.

The key to the practical success of formal equivalence checking is the separation of functional and temporal concerns in contemporary design flows. By adopting a synchronous circuit style in combination with a combinational design and verification paradigm, it enables an efficient validation of the overall design correctness by statically and independently checking the timing and function. This approach is crucial for many steps in automated design optimization and analysis, for example, logic synthesis, static timing analysis, test pattern generation, and test synthesis. As outlined in later sections, this paradigm allows equivalence checking to reduce the problem of general machine comparison to the problem of checking pairs of combinational circuits for Boolean equivalence. Such checks can be performed efficiently for many practical designs by exploiting structural similarities between the two circuits.

The need for formal equivalence checking methods was triggered by the necessity to use multiple design models in hardware design flows. Each of these models is tuned for a distinct purpose to address conflicting requirements. For example, in a typical case of using two models, the first would be designed to support fast functional simulation, whereas the second would describe in detail the envisioned circuit implementation, thus ensuring high performance of the final hardware. IBM was one of the first companies to use such a multimodel design flow. Beginning in the late 1970s, the design flow of IBM's mainframe computer chips included RTL modeling to facilitate fast and cycle-accurate functional simulation at the early design stages. In the initial methodology that predated automatic logic synthesis, the actual logic implementation was designed independently and verified against the RTL by formal equivalence checking [1]. The capability of performing equivalence checking on large, industrial-scale designs was key for the success of complex methodologies that use a multitude of models, each specifically tuned for a distinct purpose.

In the early 1980s, IBM's mainframe design flow introduced automatic logic synthesis that largely replaced the manual design step. This did not diminish the importance of equivalence checking that became instrumental in verifying the absence of functional discrepancies that are introduced by tool or methodology bugs or through frequent manual intervention. In the early 1990s, IBM began to complement synthesis-based design flows with custom design circuit blocks. Here again, formal equivalence checking between the RTL specification and manually designed transistor-level implementation provided a critical component to ensure functional correctness [2]. A similar approach was followed in the design flow of later versions of the Alpha microprocessor [3] and Intel's microprocessors [4]. In the mid-1990s, after logic synthesis became broadly adopted in typical ASIC design flows, equivalence checking technology moved into the domain of computer-aided design (CAD) tool vendors. Today, in many ASIC and custom design methodologies, formal equivalence checking is a key component of the overall chip validation procedure, similar to static timing analysis and design rule checking.

Equivalence checking has multiple important applications in hardware design flows. The most common use in ASIC flows is equivalence checking as part of the *sign-off* verification step, which compares the presynthesis RTL specification with the postsynthesis gate-level implementation. This check can catch errors introduced by *engineering changes*, which are regularly applied to the gate-level circuit for implementing late updates of timing or functionality. It can further uncover errors introduced by bugs in the logic synthesis tool, test synthesis step, or the overall composition of the tool flow. As mentioned earlier, another important application can be found in custom design styles typically utilized for high-speed designs such as microprocessors and video image processors. Here, equivalence checking is employed as a design aid, helping the circuit designer to implement a custom circuit at the transistor level according to its RTL specification. Other common uses of equivalence checking include verification of library cells or larger blocks implementing intellectual property, validation of equivalent but more compact simulation models, or the validation of hardware emulation models.

The aforementioned application spectrum of equivalence checking is reflected in differences in the use mode, the tool integration into an overall flow, and the tool interface in terms of debugging feedback and control options. For example, in an ASIC sign-off application, equivalence checking is typically applied in batch mode, verifying an entire chip design in a flat manner. In contrast, in a custom design environment, equivalence checking can provide an interactive design aid for which a tight integration into the schematic editor is highly valuable. Interestingly, such a convenient setup has sometimes extended the use of equivalence checking as a *verification tool* (which just confirms the correctness of an independently designed circuit) to a *design tool* used in a loop to check quick circuit manipulations in a *trial-and-error* manner. Clearly, the importance of debugging support is significantly different for a sign-off verification mode and a design aid mode. The less frequent occurrence of miscompares in the first case requires less sophisticated debugging support—a simple counterexample often suffices. In the latter cases, design miscompares are more common and advanced debugging can offer significant gains in productivity.

4.2 THE EQUIVALENCE CHECKING PROBLEM

Informally, two designs are defined to be functionally equivalent (in the context of this chapter) if they produce identical output sequences for all valid input sequences. For a more formal definition of machine equivalence, we assume that both designs under comparison are modeled as finite-state machines (FSMs). In the following notation, we generally use uppercase and lowercase letters for sets and variables, respectively. Furthermore, we use underlined letters for vectors of variables and denote their individual members by superscripts. We will refer to the original and complemented value of a function as its positive and negative phases, respectively.

Let $M = (X,Y,S,S_0,\delta,\lambda)$ denote a (Mealy-type) FSM in the common manner, where X, Y, S, and S_0 are the sets of inputs, outputs, states, and initial states, respectively. $\delta: X \times S \to S$, S denotes the next-state function and $\lambda: X \times S \to Y$, Y is the output function. Furthermore, let $\underline{s} = (s^1,\ldots,s^u)$, $\underline{x} = (x^1,\ldots,x^v)$, and $\underline{y} = \left(y^1,\ldots,y^w\right)$ be variable vectors holding the values for the encoded states S, inputs X, and outputs Y, respectively, of an FSM M. We will refer to the encoding variables for the states as state bits or registers and to the variables for the inputs and outputs as input and output signals, respectively. Furthermore, let $\underline{\delta}(\underline{x},\underline{s}) = (\underline{\delta}^1(\underline{x},\underline{s}),\ldots,\delta^u(\underline{x},\underline{s}))$ denote the vector of next-state functions for the state bits. A state $\underline{s}_n$ is considered *reachable* if there exists a sequence of transitions from an initial state $\underline{s}_0$ to $\underline{s}_n$, that is, $(\underline{s}_0, \underline{s}_1), (\underline{s}_0, \underline{s}_2), \ldots, (\underline{s}_{n-1}, \underline{s}_n)$, where $\exists \underline{x}. \left(\underline{s}_{i+1} = \delta \left(\underline{x}, \underline{s}_i \right) \right)$ for $0 \leq i < n$.

In the sequel, we assume the following simplifications. First, we restrict comparison to machines that have exactly one initial state, that is, $S_0 = \{\underline{s}_0\}$. Second, we require a one-to-one mapping between the inputs of the two machines and similarly a one-to-one mapping between their outputs. Furthermore, we assume that the timing of both machines is identical, that is, they run at the same clock speed, consume inputs at the same rate, and similarly produce outputs at identical rates. All three assumptions can be lifted in a straightforward manner.

For checking functional equivalence of two machines M_1 and M_2 with pairs of mapped input and outputs, we construct a *product machine* $M = (X, Y, S, S_0, \delta, \lambda) = M_1 \times M_2$ in the following manner:

$$X = X_1 = X_2$$
$$S = S_1 \times S_2$$
$$S_0 = S_{0,1} \times S_{0,2}$$
$$Y = \{0,1\}$$
$$\underline{\delta}(\underline{x}(\underline{s}_1, \underline{s}_2)) = (\underline{\delta}_1(\underline{x}, \underline{s}_1), \underline{\delta}_2(\underline{x}, \underline{s}_2))$$
$$\lambda(\underline{x}, (\underline{s}_1, \underline{s}_2)) = \begin{cases} 1 & \text{if } \underline{\lambda}_1(\underline{x}, \underline{s}_1) = \underline{\lambda}_2(\underline{x}, \underline{s}_2) \\ 0 & \text{otherwise} \end{cases}$$

A schematic view of the construction of the product machine is given in Figure 4.1.

The two machines, M_1 and M_2, are said to be functionally equivalent, iff the output function of M produces "1" for all of its reachable states and input vectors, that is, $\lambda \leftrightarrow 1$.

In general, verifying that the product machine M produces $\lambda \leftrightarrow 1$ for all reachable states is achieved by identifying a characteristic function $\varrho(\underline{s})$ that relates the states of M_1 and M_2 and partitions the state space of M in two parts as depicted in Figure 4.2.

Informally, this characteristic function splits the state space of M into two halves such that the following applies:

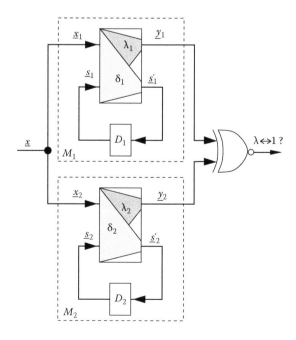

FIGURE 4.1   Product machine for comparing two finite-state machines.

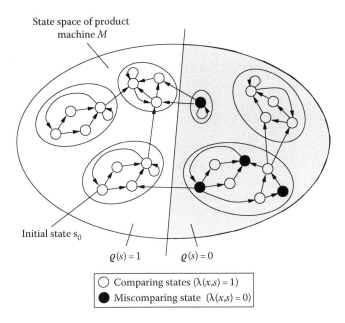

FIGURE 4.2   State-space structure of the product machine.

1. All initial states are in the first part ($\varrho = 1$).
2. All *miscomparing states*, that is, $\lambda \leftrightarrow 0$, are in the second part ($\varrho = 0$).

There is no transition from the first half to the second half.

   If such a characteristic function exists, then one can use an inductive argument to reason that no miscomparing state can be reached from any of the initial states, that is, both machines are functionally equivalent. In fact, one can show that the inverse also holds true, namely, equivalence of M_1 and M_2 implies the existence of such a characteristic function.

   Formally, the product machine M produces $\lambda \leftrightarrow 1$ for all reachable states iff there exists a function $\varrho{:}S \rightarrow \{0,1\}$ such that

$$\varrho(\underline{s}_0) = 1$$
$$\varrho(\underline{s}) = 1 \Rightarrow \forall \underline{x}.\varrho(\delta(\underline{x}, \underline{s})) = 1$$
(4.1)
$$\varrho(\underline{s}) = 1 \Rightarrow \forall \underline{x}.\lambda(\underline{x}, \underline{s}) = 1$$

The task of equivalence checking consists of two parts. First, a candidate for the characteristic function ϱ must be identified, and second, for the given ϱ conditions, Equation 4.1 must be checked for validity. The latter check requires efficient methods for Boolean reasoning, which are described in Section 4.3. Various methods for equivalence checking differ in the way candidates for ϱ are obtained. The most common approach assumes a combinational equivalence checking paradigm and uses a known or *guessed* correspondence of the registers of M_1 and M_2 to derive ϱ implicitly. In this case, the general check for functional equivalence can be reduced to verifying Boolean equivalence of a set of combinational circuits. More details on combinational equivalence checking can be found in Section 4.4, whereas Section 4.5 outlines the methods for general sequential equivalence checking (SEC).

4.3 BOOLEAN REASONING

In this section, we review the techniques for Boolean reasoning that form the foundation for equivalence checking algorithms. We first present binary decision diagrams (BDDs) [5], a popular data structure for Boolean functions, which are especially suited for the manipulation of the characteristic function ϱ. We then review algorithms for the propositional (or Boolean) satisfiability problem, which can be used to decide the validity of Equation 4.1 without explicitly building a representation of ϱ. In particular, we first restrict our attention to the satisfiability of conjunctive normal form (CNF) formulae and then briefly consider the general case.

4.3.1 BINARY DECISION DIAGRAMS

Besides equivalence checking, many algorithms in synthesis and verification manipulate complex logic function. A data structure to support this task is, therefore, of great practical usefulness. BDDs [5] have become highly popular because of their efficiency and versatility.

A BDD is an acyclic graph with two leaves representing the constant functions 0 and 1. Each internal node n is labeled with a variable n and has two children t (*then*) and e (*else*). Its function is inductively defined as follows:

$$(4.2) \qquad f(n) = \left(v \wedge f(t) \right) \vee \left(\neg v \wedge f(e) \right)$$

Three restrictions are customarily imposed on BDDs:

1. There may not be isomorphic subgraphs.
2. For all internal nodes, $t \neq e$.
3. The variables are totally ordered. The variables labeling the nonconstant children of a node must follow in the order of the variable labeling the node itself.

Under these restrictions, BDDs provide a canonical representation of functions, in the sense that for a fixed variable order, there is a bijection between logic functions and BDDs. Unless otherwise stated, BDDs will be assumed to be reduced and ordered. The BDD for $F = (x_1 \wedge x_3) \vee (x_2 \wedge x_3)$ with variable order $x_1 < x_2 < x_3$ is shown in Figure 4.3.

Canonicity is important in two main respects: it makes equivalence tests easy, and it increases the efficiency of *memoization* (the recording of results to avoid their recomputation). On the other hand, canonicity makes BDDs less concise than general circuits. The best-known case is that of multipliers, for which circuits are polynomial and BDDs exponential [6]. Several variants of BDDs have been devised to address this limitation. Some of them have been quite successful for certain classes of problems (for instance, binary moment diagrams [7] for multipliers). Other variants of BDDs have been motivated by the desire to represent functions that map $\{0, 1\}^n$ to some arbitrary set (e.g., the real numbers) [8].

For ordered BDDs, $f(t)$ and $f(e)$ do not depend on v; hence, comparison of Equation 4.2 to Boole's expansion theorem

$$(4.3) \qquad f = (x \wedge f_x) \vee (\neg x \wedge f_{\neg x})$$

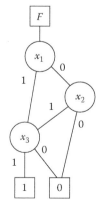

FIGURE 4.3 Binary decision diagram for $F = (x_1 \wedge x_3) \vee (x_1 \wedge x_3)$.

shows that $f(t) = f(n)_v$, the *positive cofactor* (or *restriction*) of $f(n)$ with respect to v, and $f(e) = f(n)_{\neg v}$, the *negative cofactor*. This is the basis of most algorithms that manipulate BDDs, because for a generic Boolean connective $\langle op \rangle$,

$$(4.4) \qquad f\langle op \rangle g = \left(x \wedge \left(f_x \langle op \rangle g_x \right) \right) \vee \left(\neg x \wedge \left(f_{\neg x} \langle op \rangle g_{\neg x} \right) \right)$$

This is applied with x chosen as the first variable in the order that appears in either f or g. This guarantees that the cofactors can be computed easily. If x does not appear in f, then $f_x = f_{\neg x} = f$; otherwise, f_x is the *then* child of f and $f_{\neg x}$ is the *else* child. It is likewise for g. The terminal cases of the recursion depend on the specific operator. For instance, when computing the conjunction of two BDDs, the result is immediately known if either operand is constant or if the two operands are identical or complementary. All these conditions can be checked in constant time if the right provisions are made in the data structures [9].

Two tables are used by most BDD algorithms. The *unique table* allows the algorithm to access all nodes using the triple (v, t, e) as key. The unique table is consulted before creating a new node. If a node with the desired key already exists, it is reused. This approach guarantees that equivalent functions will share the same BDD rather than having isomorphic BDDs; therefore, equivalence checks are performed in constant time.

The *computed table* stores recent operations and their results. Without the computed table, most operations on BDDs would take time exponential in the number of variables. With a lossless computed table (one that records the results of all previous computations), the time for most common operations is polynomial in the size of the operands. The details of the implementation of the unique and computed tables dramatically affect the performance of BDD manipulation algorithms and have been the subject of careful study [10].

The order of variables may have a large impact on the size of the BDD for a given function. For adders, for instance, the optimal orders give BDDs of linear size, while bad orders lead to exponential BDDs. The optimal ordering problem for BDDs is hard [11]. Hence, various methods have been proposed to either derive a variable order from the inspection of the circuit for which BDDs must be built or dynamically compute a good order while the BDDs are built [12].

An exhaustive list of applications of BDDs to problems in CAD is too long to be attempted here. Besides equivalence checking and symbolic model checking, BDD-based algorithms have been proposed for most synthesis tasks, including two-level minimization, local optimization, factorization, and technology mapping. Despite their undeniable success, BDDs are not a panacea; their use is most profitable when the algorithms capitalize on their strengths [13] and avoid their weaknesses by combining BDDs with other representations [14–16], for instance, with satisfiability solvers for CNF expressions.

4.3.2 CONJUNCTIVE NORMAL FORM SATISFIABILITY

A SAT solver returns an assignment to the variables of a propositional formula that satisfies it if such an assignment exists. A *literal* is either a variable or its negation, a *clause* is a disjunction of literals from distinct variables, and a CNF formula is a conjunction of clauses. We will use the following simple CNF formula as a running example:

$$(4.5) \qquad (\neg a \vee b \vee c) \wedge (a \wedge \neg b \vee c) \wedge (\neg c \vee d) \wedge (\neg c \vee \neg d)$$

The variables are the letters a through d; a and $\neg a$ are the two literals (positive and negative) of variable a. The formula is the conjunction of four clauses, each consisting of two or three literals. The Boolean connectives $\vee$, $\wedge$, and $\neg$ stand for disjunction (OR), conjunction (AND), and negation (NOT), respectively.

The CNF satisfiability problem (CNF SAT for short) is the one of finding an assignment (a mapping of each variable to either 0 or 1) such that all clauses contain at least one true literal. In our example Equation 4.5, one such *satisfying assignment* is $a = b = c = d = 0$. A CNF formula with no clauses is trivially satisfiable, whereas a formula containing a clause with no literals (the *empty clause*) is not satisfiable.

The CNF SAT problem is hard, in the sense that no worst-case polynomial-time algorithm is known for it. It is in fact the archetypal nondeterministic polynomial (NP)-complete problem [17]. It can be solved in polynomial time by a nondeterministic machine that can guess a satisfying assignment and then verify it in linear time. However, if a deterministic polynomial-time algorithm were known for CNF SAT, then a polynomial-time algorithm would also exist for *all* problems that a nondeterministic machine can solve in polynomial time.

Practical algorithms for CNF SAT rely on search and on the notion of *resolution*, which says that $(x \vee \neg y) \wedge (y \vee z)$ implies the *resolvent* of the two clauses $(x \vee z)$. A CNF formula is not satisfiable iff the empty clause can be resolved from its clauses. A search algorithm for SAT finds a satisfying assignment to a given formula if it exists and may produce a resolution of the empty clause otherwise. The approach was first described in work by Davis, Putnam, Logemann, and Loveland [18,19]. Hence, it is commonly called the DPLL procedure. However, important details have changed in the last 40 years. The GRASP solver [20], in particular, has popularized the approach to search based on *clause recording* and *nonchronological backtracking* that we outline. Even if we present the algorithm in the context of CNF SAT, it is of wider applicability. In particular, it works on a generic Boolean circuit.

Figure 4.4 shows the pseudocode for the DPLL procedure, which works on three major data structures. The first is the *clause database*, which at an abstract level is just a list of clauses. The second data structure is the *assignment stack* in which all variable assignments currently in effect are kept together with their *levels* and *causes*, that is, with information about when and why the assignment was made. The third data structure is the *assignment queue*, which stores the assignments that have been identified, but not yet affected.

Before they are handed to the DPLL procedure, the clauses are reprocessed to identify *unit clauses* and *pure literals*. A unit clause consists of one literal only, which consequently must be true in all satisfying assignments. The assignments for the unit literals are entered in the assignment queue. Throughout the algorithm, if all the literals of a clause except one are false and the remaining one is unassigned, then that last literal is implied to 1 and the corresponding assignment is *implied*. The implication is added to the assignment queue. A *conflict* occurs when both literals of a variable are implied to true.

A pure literal is such that the opposite literal does not appear in any clause. If there is a satisfying assignment with the pure literal set to 0, then the assignment obtained from it by changing that literal to 1 is also satisfying. Hence, assignments that make pure literals true are also entered in the assignment queue. The handling of unit clauses and pure literals comprises preprocessing.

The procedure "ChooseNextAssignment" checks the assignment queue. If the queue is empty, the procedure makes a *decision*: it chooses one unassigned variable and a value for it and adds the assignment to the queue. Every time a decision is made, the *decision level* is increased. The assignments made during preprocessing are at level 0 and each subsequent decision increases the decision level by 1. The decision level decreases when backtracking occurs. If no unassigned variable can be found, "ChooseNextAssignment" returns false. This causes DPLL to return an affirmative answer, because a complete assignment to the variables has been found that causes no conflict. Since it causes no conflict, it is satisfying.

If the queue was not empty or a decision has been added to it, one assignment is removed from it and added to the assignment stack for the current decision level. The stack also records the cause for the assignment: either a decision or an implication. In the latter case, the clause that caused the implication is recorded.

```
1        DPLL() {
2            while (ChooseNextAssignment()) {
3                while (Deduce() == CONFLICT) {
4                    blevel = AnalyzeConflict();
5                    if (blevel < 0) return UNSATISFIABLE;
6                    else Backtrack(blevel);
7                }
8            }
9            return SATISFIABLE;
10       }
```

FIGURE 4.4 DPLL algorithm.

When a new assignment is added to the stack, its implications are added by "Deduce" to the assignment queue. An implication occurs in a clause when all literals except one are false and the remaining literal is unassigned. Efficient computation of implications for clauses is discussed in [21].

If the implications yield a conflict (both literals of a variable are implied), ANALYZECONFLICT() is launched. Conflict analysis relies on the (implicit) construction of an *implication graph*. Each literal in the conflicting clause has been assigned at some level either by a decision or by an implication. If there are multiple literals from the current decision level, at least one of them is implied. Conflict analysis locates the clause that is the source of that implication and extends the implication graph by adding arcs from the antecedents of the implication to the consequent one. This process is carried out until there is exactly one assignment for the current level among the leaves of the tree. The disjunction of the negation of the leaf assignments gives then the *conflict clause*.

For the example Equation 4.5, preprocessing produces no assignment. Suppose that the first decision is to set a to 0. This is the level-1 decision and is denoted by $\neg a@1$. It is not the choice an SAT solver would likely make but is convenient for illustrative purposes. No assignments are deduced from $\neg a@1$. Suppose the second decision is $b@2$; from it we deduce $c@2$ from the second clause. This, in turn, yields $d@2$ and $\neg d@2$ through the last two clauses. The two opposite assignments to d signal a conflict, which is therefore analyzed.

From Figure 4.5, one sees that $\neg a \wedge b$ is enough to cause the conflict. Hence, every satisfying assignment must satisfy the clause $(a \vee \neg b)$. One also sees that c is sufficient to cause the conflict and that, consequently, $\neg c$ is a (unit) clause that can be added. Both $(a \vee \neg b)$ and c contain exactly one literal assigned at the current level (i.e., at level 2).

The highest level of the assignments in the conflict clause, excluding the current one, is the backtracking level. Backtracking over some decisions that did not contribute to the conflict goes under the name of *nonchronological backtracking*. The single assignment at the current level is known as *unique implication point* (UIP). Conflict clauses based on the first UIP (the one closest to the conflict, like $\neg c$ in the example) have been empirically found to work well [22]. When the conflict clause contains a UIP—not necessarily the first UIP—the effect of backtracking is to make that clause *asserting*. That is, all literals in the clause are false except for the UIP literal. This causes that literal to be asserted to the opposite value that it had when the conflict was detected. The procedure of Figure 4.4 relies on asserting conflict clauses to guarantee termination [23].

Returning to our example, if $(a \vee \neg b)$ is used as a conflict clause, the backtracking level is 1, where the assignment $\neg b@1$ is deduced. If the next decisions are $\neg c@2$ and $d@3$, all variables are assigned, and the search terminates with a satisfying assignment. If, on the other hand, $\neg c$ is added as conflict clause, the backtracking level is 0 (since there is no literal in the clause except the UIP) and $\neg c@0$ is deduced. If the next decision is again $\neg a@1$, we get $\neg b@1$ by implication. At this point, all clauses are satisfied, and no matter what value is chosen for the only unassigned variable, d, we obtain a satisfying assignment at level 2.

In our example, the computation of the first UIP proceeds as follows. Suppose the conflict is detected when examining the fourth clause; that is, $(\neg c \vee \neg d)$ is the *conflicting clause*. Resolution is applied to this clause and to the clause that caused the other implication for d, $(\neg c \vee d)$. The result is $\neg c$. Since this clause contains exactly one literal at level 2, the first UIP has been found and the process stops. If one desires another UIP, one would continue by resolving $\neg c$ with the clause that caused the implication $c@2$, namely, $(a \vee \neg b \vee c)$. The result, $(a \vee \neg b)$, is the other conflict clause that we have examined and contains the remaining UIP for this example.

The clauses that result from conflict analysis are used to steer the search away from a current partial assignment that cannot be extended to a complete assignment. They may also prevent the

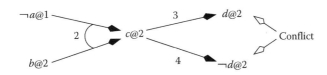

FIGURE 4.5 Implication graph. Each implication is annotated with the ordinal of the clause in Equation 4.5 that caused it.

exploration of equally fruitless regions of the search space. For that reason, conflict clauses are *recorded* in the clause database. Periodically, those that are deemed less useful may be discarded to conserve space. Conflict clauses fulfill three important additional tasks: they are used to decide which decisions to make [21]; they are passed from one problem instance to the next to implement efficient *incremental* SAT solvers [24]; and they are recorded, together with the implication graphs from which they were derived, to produce the resolution proofs of unsatisfiable instances [23]. For the sake of conciseness, the reader can refer to the works in the "Reference" section for the details of these and other tasks. It would suffice to say that a skillful blend of the techniques we have briefly outlined has produced SAT solvers that can deal with many problems with hundreds of thousands of variables and millions of clauses. This kind of capacity is crucial in applications of SAT to equivalence checking.

4.3.3 HYBRID SATISFIABILITY METHODS

Given a generic combinational Boolean circuit, it is possible to translate it into an equisatisfiable CNF formula in linear time. For each logic gate, one generates a set of clauses relating the gate inputs to the gate output. For instance, the clauses that express the behavior of the AND gate $z = a \wedge b$ are $(a \vee \neg z),(b \vee \neg z)$, and $(\neg a \vee \neg b \vee z)$. Thanks to this translation, one can, at least in theory, rely exclusively on a CNF SAT solver when reasoning about Boolean circuits. However, there are advantages to be had from dealing directly with the original circuit as well as from allowing mixed representations of the formulae whose satisfiability is investigated.

When trying to prove that combinational circuits are equivalent, reasoning on the circuits themselves allows one to detect structurally identical subcircuits, which can be merged to simplify the proof of equivalence of the remaining parts. This approach can be extended in two important ways. First, one can use a *semicanonical* circuit representation to increase the chances that equivalent circuits will have the same structure. One example of this semicanonical representation is the *and–inverter graph* (AIG) [14], in which every node of the circuit is an AND gate whose inputs may be inverted. One can quickly identify isomorphic subcircuits in an AIG by hashing the gates. The second extension to structural comparison consists of computing signatures for the nodes of a circuit so that nodes with different signatures are known not to be equivalent. The few nodes with identical signatures can be subjected to further, more expensive, analysis to establish whether they are equivalent. Signature computation is usually based on simulation [25], while the more expensive analysis techniques can be based on either BDDs or CNF SAT.

In summary, direct manipulation of a circuit representation often leads to significant simplification in the task handed to the complete decision procedures. The combination of Boolean circuits, CNF clauses, and BDDs also finds application in general SAT solvers [26–28], which can leverage the strengths of the individual representations.

4.3.4 RELATED APPLICATIONS OF BOOLEAN REASONING IN COMPUTER-AIDED DESIGN

The application of Boolean reasoning in CAD is not limited to equivalence checking. Rather, Boolean reasoning is one of the fundamental tools for analysis and optimization in a wide range of problems in verification, synthesis, and testing. Many CAD problems can be modeled as discrete decision problem with a finite domain and therefore can be expressed as (a sequence of) satisfiability checks. The recent rapid improvements in the size of the SAT problems that can be efficiently handled make such an approach practical for surprisingly large instances.

Satisfiability checking was first introduced to the CAD domain for automatic test pattern generation (ATPG). ATPG computes a suitable set of test vectors that can demonstrate the absence of certain manufacturing faults [29]. Owing to its early use and algorithmic advances, the application of ATPG-style Boolean reasoning was adopted in many areas such as equivalence checking [30,31], logic synthesis [32–34], and verification [35,36]. Later, the introduction and improvement of BDDs, CNF SAT, circuit-based SAT, and hybrid methods substituted or complemented classical ATPG techniques in these domains.

The chapters on assertion-based verification (Chapter 18 of *Electronic Design Automation for IC System Design, Verification, and Testing*), formal property verification (Chapter 20 of

Electronic Design Automation for IC System Design, Verification, and Testing), ATPG (Chapter 22 of *Electronic Design Automation for IC System Design, Verification, and Testing*), and logic synthesis (Chapter 2 of this volume) provide detailed summaries of corresponding CAD areas that make extensive use of Boolean reasoning.

4.4 COMBINATIONAL EQUIVALENCE CHECKING

Most practical equivalence checking approaches assume a *combinational verification paradigm*, that is, they require that the general problem of checking the product machine shown in Figure 4.1 can be reduced to comparing pairs of combinational circuits. In the simplest (and also most common case), they expect a one-to-one correspondence between the registers of both designs under comparison. Such requirements are enforced by an overall design methodology that strikes for a practical compromise, which minimizes the restrictions on the designers, yet ensures an efficient verification flow. For example, by using only combinational logic optimization methods, a one-to-one register correspondence is automatically achieved in a synthesis-based ASIC design flow. Similarly, in a custom design flow, the circuit designer can be required to use the same state encoding for the implementation as given in the RTL specification. This can be further restricted by also requiring the use of identical register names, thereby facilitating an efficient computation of the correspondence.

4.4.1 BASIC APPROACH

In case of a one-to-one register correspondence, the equivalence of the two machines is shown if all corresponding next-state functions and output functions are proved to be functionally equal. For this, all pairs of corresponding register inputs and primary inputs are connected and driven by a unique variable each. This setup corresponds to a specific characteristic function % that is tested for validity. Figure 4.6 shows the corresponding setup, and Figure 4.7 gives the derived *Miter** structure that was first introduced in [31].

Equivalence checking of two designs with corresponding registers is done in two steps:

1. In the first step, the register correspondence is either guessed using simple heuristics or computed exactly. The spectrum of methods ranges from simple name-based methods to systematic methods (e.g., the ones outlined in Section 4.4.2) and often utilizes combinations of both. Name-based methods simply match up instance names of registers or names of signals connected to registers and design ports. This process is typically controlled by user-defined name-matching rules that allow the use of partial name matches, substitution of strings in names, etc.
2. The second step involves the actual functional comparison of the individual combinational circuits. For this, a variety of methods for Boolean reasoning are applied, including the base methods outlined in Section 4.3.

It is important to note that an equivalence check based on register correspondence is *sound* but not *complete*. Soundness states that if all combinational checks pass, the two circuits are indeed functionally equivalent. The lack of completeness means that if any check fails, it cannot be decided whether the two machines are truly inequivalent or if only the given register equivalence is incorrect. In other words, referring to the concept of ϱ introduced in Section 4.2, the failing of check (Equation 4.1) cannot distinguish between an incorrectly chosen candidate for the characteristic function ϱ and true machine inequivalence.

In a practical design methodology, the incompleteness is addressed by rejecting all designs for which the comparison fails and demands a design update with an equivalent input/output behavior and a valid register correspondence. This effectively tightens the correctness criteria for

---

* According to Brand [31], the "Logic Constituting a Miter" resembles two slabs of wood connected with a miter joint.

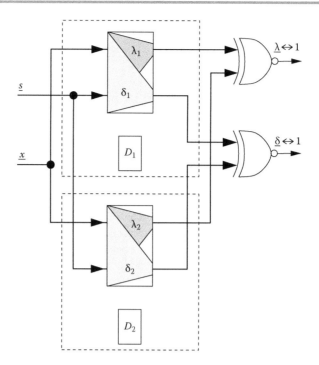

FIGURE 4.6   Combinational equivalence check based on register correspondence.

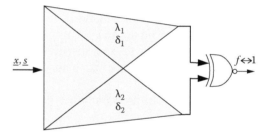

FIGURE 4.7   Miter circuit derived from the configuration of Figure 4.6 for combinational equivalence checking.

design equivalence by requiring identical input/output behavior *and* a valid register correspondence that can prove equivalence by a combinational check.

4.4.2 REGISTER CORRESPONDENCE

In many practical design flows, a candidate register correspondence is derived from naming conventions. In the absence of naming rules, or if registers are only partially matched by names, the register correspondence can be computed automatically as a greatest fixed point using the algorithm outlined in Figure 4.8. The algorithm starts with one equivalence class (bucket) for the register correspondence, containing all registers (line 2). During each iteration, a unique variable is introduced for the outputs of all registers of each bucket (line 5) and all next-state functions are computed (line 8) based on these variables. Next, the buckets are partitioned into pieces that have identical next-state functions (line 11). This process is repeated until a fixed point is reached (line 10). The partitioning of the register buckets is monotonic, that is, the number of buckets increases strictly from iteration to iteration until a fixed point is reached. Thus, the algorithm will terminate after at most u iterations, where u is the total number of registers in the product machines.

```
 1          REGISTER_CORRESPONDENCE() {
 2                  put all registers r into bucket[0]
 3                  do {
 4                          forall buckets i do {
 5                                  initialize output of all registers r ∈ i with variable v[i]
 6                          }
 7                          forall registers r do {
 8                                  compute next state functions δ[r] based on inputs v
 9                          }
10                          if ∀ buckets i: r₁,r₂ ∈ i ⇔ δ[r₁] = δ[r₂] return
11                          split all buckets i into multiple buckets iⱼ s.t. r₁,r₂ ∈ iⱼ ⇔ δ[r₁] = δ[r₂]
12                  }
13          }
```

FIGURE 4.8   Algorithm for computing functional register correspondence.

If a canonical or semicanonical representation is used for the next-state functions, the comparison and refinement in lines 10 and 11 can simply be implemented by array sorting and splitting. For example, if BDDs (see Section 4.3.1) are applied, sorting the array of next-state functions by BDD pointer values moves identical functions to adjacent index positions. A single-array sweep can then perform the partitioning step. Furthermore, if the representation allows efficient function complementation, this approach can also handle complemented registers: the construction of the next-state functions uses the complement of the bucket variable for these registers, and similarly, the resulting functions for them are complemented before comparison. Other examples of function representations that are applicable in this setting are the "and/inverter" graph [14] or Boolean expression diagrams [37].

If a register correspondence between two machines exists, the algorithm of Figure 4.8 is guaranteed to find it. However, in case of discrepancies in any of the next-state functions, the algorithm generally fails. In the extreme case, the effect of a single incorrect next-state function may ripple through each iteration of the algorithm and split all buckets until they contain only one register.

A more relaxed criterion under which registers are considered equivalent can make a correspondence algorithm more robust. For example, an algorithm may consider two registers equivalent if the support sets of their next-state functions are equivalent with respect to the register correspondence. The algorithm given in Figure 4.8 can be modified to accommodate such relaxed criteria by adjusting the criteria in lines 10 and 11. Instead of performing a functional comparison, the support sets need to be simply compared. Clearly, this criterion may result in incorrect matching results. For example, two pairs of registers that represent two distinct state bits may have identical support sets and would get matched by such a criterion. Similarly, a register function may be functionally independent of particular inputs of its support set. If another functionally corresponding register omits any of these inputs, the matching based on input support would fail. Despite the possibility of producing incorrect results, a structural corresponding algorithm based on support sets works well in practice. It is significantly faster than the full functional correspondence algorithm and provides valuable information, even if next-state functions are erroneous. For the application in an equivalence checking tool, the structural correspondence algorithm can be used first, possibly in combination with complete or partial name matching. Only for register pairs where this algorithm produces incorrect matchings, not only the precise but also the more expensive functional correspondence is applied.

More details and improvements of register correspondence algorithms can be found in the literature [38,39]. The automated detection of a register correspondence can be extended to handle don't cares [40].

4.4.3 EQUIVALENCE CHECKING OF RETIMED CIRCUITS

Sequential logic synthesis techniques have been researched for many years, and there are a number of efficient approaches available that are applicable to practical designs. However, thus far their adoption in contemporary tool flows has been slow. The lack of an efficient equivalence checking flow is often cited as one of the reasons for this limited use. In the following, we focus the discussion on verifying circuits that have been optimized by retiming.

Retiming [41,42] is commonly referred to as a structural transformation of a sequential circuit that changes the positions of registers and latches without modifying the design's input–output behavior. The objective of retiming is to decrease the overall design cycle time by balancing the combinational path delays through a realignment of the synchronization grid. Figure 4.9 illustrates the impact of a retiming transformation on the FSM structure of a circuit. The registers are moved either forward or backward, without changing the structure of the remaining logic. As shown in the figure, the register repositioning effectively moves the corresponding logic from one side of the combination logic block to the other.

As mentioned earlier, the most robust and scalable application of equivalence checking is based on a combinational verification paradigm. In the case of retiming, the next-state functions are not comparable, and therefore, a straightforward application of equivalence checking methods mentioned in the previous subsections is not possible. However, by preserving the retime logic from the synthesis flow and applying it to make both designs comparable, the general sequential verification problem can also be reduced to a combinational case. Figure 4.10 illustrates this approach. In essence, both machines are *patched* with pieces of the retime logic to make the interfaces comparable. For this, the forward retime logic is added to the inputs of s_r^f to make the new inputs comparable with s^f. It is also added to the outputs of $s^{f'}$ to make them comparable with the outputs $s_r^{f'}$ of the retimed machine. Similarly, the backward retime logic is added to $s_r^{b'}$ and s^b. As a result, both machines have compatible current- and next-state variables that allow the application of a combinational verification paradigm. Note that this scheme is sound and cannot produce false-positive verification results.

Adding retime logic for equivalence checking can be done in multiple iterations, for example, to patch a sequence of retiming transformations interleaved with combinational optimization. At each step, the number of logic levels increases, which makes the combinational comparison more difficult. Furthermore, efficient equivalence checking algorithms heavily exploit structural similarities of the two designs under comparison [43]. For too many retiming *shells* with little or no structural similarity, the equivalence checking problem may become intractable. However, a targeted and localized application of retiming for alleviating timing problems will typically result in small pieces of retime logic and therefore will not impact the performance of equivalence checking in the given setting.

Further generalizations to verify circuits that have been optimized by retiming can be found in [44–46].

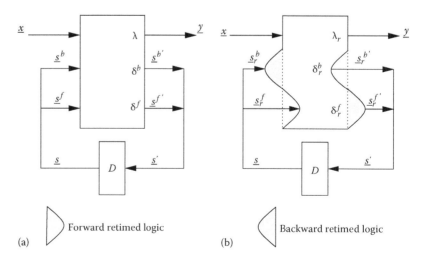

FIGURE 4.9   Effect of retiming on the finite-state machine structure of a design: (a) Forward retimed logic and (b) backward retimed logic.

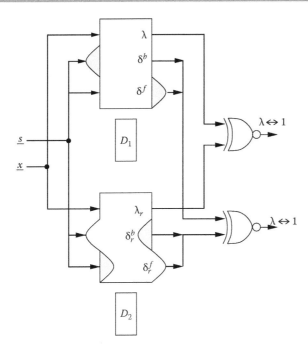

FIGURE 4.10   Application of the retime functions for equivalence checking based on a combinational verification paradigm.

4.5 SEQUENTIAL EQUIVALENCE CHECKING

When the two sequential models to be tested for equivalence are substantially different, in particular when they have not been obtained one from the other through well-understood transformations like retiming or combinational optimizations, one has to resort to a fully general approach to verification that checks that the outputs of the two models are the same for the same inputs in all reachable states. In terms of the characteristic function introduced in Section 4.2, this corresponds to not relying on a guess for ϱ derived from the structure of the two machines.

Instead, one can take ϱ such that it holds for all the states, where $\forall \underline{x} \cdot \lambda(\underline{x}, \underline{s}) = 1$, that is, ϱ is the characteristic functions of all the states of the product machine M, where the outputs of the two machines M_1 and M_2 are identical. Equivalence is established if the initial states are in ϱ and so are all the successors of the states in ϱ. In this case, equivalence is an *inductive invariant* for M. However, if the check on the successor states fails, the result is inconclusive, because the states that caused the failure may not be reachable from the initial states of M. In case of failure, the general approach is to *strengthen* ϱ, that is, to remove some states from it. The strongest ϱ for a given M is the characteristic functions of all its reachable states. The weakest ϱ, on the other hand, that guarantees completeness of the method is the characteristic functions of all the states of M from which no miscomparing state can be reached. Both these functions can be computed by reachability analysis as described in Section 4.5.1. It should be noted that this analysis can be quite expensive for large sequential circuits. Hence, it is sometimes useful to manually strengthen ϱ.

4.5.1 REACHABILITY ANALYSIS

Reachability analysis computes the states of a sequential machine M that either are reachable from a set of designated states or from which the designated states are reachable. In equivalence verification, the states reachable from the initial states form the strongest possible ϱ—the outputs of M_1 and M_2 agree in all reachable states of $M = M_1 \times M_2$ if and only if M_1 is equivalent to M_2. Conversely, the complement of the states from which miscomparing states can be reached forms the weakest ϱ for which the "if and only if" condition holds. The two cases correspond to *forward* and *backward* reachability analyses, respectively. We first discuss forward analysis and then briefly outline the differences.

Forward reachability analysis of a graph can be accomplished by the algorithm of Figure 4.11. The algorithm receives as input the graph and the set of initial states. It uses two main set variables: Q, to hold the states to be processed, and R, to hold the states found to be reachable. At each iteration of the while loop, some unprocessed states are removed from Q, their successors are computed by function *Image*, and the states thus found that were not previously reached are added to both sets. The policy for choosing the states that are extracted from Q determines the type of search. If only the most recently added state is extracted, then one obtains *depth-first search* (DFS). Conversely, if only the least recently added states are extracted, the resulting search is *breadth-first search* (BFS). Other policies are also possible. For hardware verification, the usefulness of DFS is limited by the state-explosion problem: in most practical examples, there are too many states, and each state has too many successors for a search that considers each state and each transition individually.

If all the states in Q are extracted at once, one obtains *symbolic* BFS, in which states are reached in an order consistent with their distance from the initial states. In symbolic reachability analysis, it is customary to represent the sets of states as BDDs. The transitions of the sequential machine are also represented by a BDD. Image computation, that is, the computation of the successors of the states in F, can then be performed by BDD operations that do not enumerate explicitly the states or the transitions. This approach can sometimes deal with huge numbers of states—in excess of 10^{100}. However, this ability depends on the BDDs remaining compact. This, in turn, depends on the state encoding of M, on the order of the state variables in the BDDs, and, of course, on the subsets of states that must be represented. Departure from strict BFS processing [13,47–49] may result in sets of states that have much smaller BDDs; this may considerably speed up the analysis.

Besides the order of traversal, the specifics of the computation of the successors of a set of states have profound influence on the performance of the symbolic reachability analysis algorithm. There are two main approaches: one based on recursive case analysis, which normally employs the transition *functions* [50], and the other based on conjunction and quantification, which employs a properly clustered transition *relation* [51,52]. The conjunction method is often faster, mostly because it is more efficient at caching intermediate computations, but neither is superior; their combination, which increases robustness, is described in [53]. The key problem in the image computation method that relies on conjunction and quantification is the order in which the components of the transition relation are combined to obtain the final result [54,55].

Backward reachability differs from forward reachability in two respects: one is that the starting states are the miscomparing states instead of the initial states of M and the other is that the computation of the successors (image computation) is replaced by the computation of the predecessors of a set of states (preimage computation). There are small differences in the algorithms used for image and preimage computations, due to the fact that the model of sequential circuits expresses the next states directly as a function of the present states and not vice versa.

```
1     FORWARD_REACHABILITY_ANALYSIS() {
2     initialize Q and R to the initial states
3         while (Q ≠ Ø) {
4             let F be nonempty subset of Q
5             Q= Q\F // remove F from Q
6             S = Image(F) // compute successors
7             N = S\R // identity new states
8             Q = Q U N // add new states to Q ...
9             R = R U N //... and to R
10        }
11        return R
12    }
```

FIGURE 4.11   Algorithm for computing forward reachable states.

Forward reachability analysis is more efficient than backward analysis on some models and less efficient on others. As an extreme case, consider checking the equivalence of two modulo-n counters. The algorithm of Figure 4.11 requires n iterations to converge. However, if the two counters are indeed equivalent, backward analysis converges after one iteration. In fact, in this example, equivalence is an inductive invariant, in which case backward analysis always converges in one iteration. Each step of backward reachability corresponds to a strengthening of the invariant, until it becomes inductive or equivalence is refuted.

4.5.2 DEPENDENT VARIABLES IN REACHABILITY ANALYSIS

Suppose a model contains three state variables s_1, s_2, and s_3 such that

$$s_1' = x_1 \qquad s_2' = s_1 \qquad s_3' = x_1 \wedge s_1$$

Suppose that in the initial states, $s_1 = s_2 = s_3$. Then it is easy to see that $s_3 = s_1 \wedge s_2$ in all reachable states. It is, therefore, possible to remove s_3 from the model and replace any usage of this variable with $s_1 \wedge s_2$. The resulting model is typically easier to deal with. We say that s_3 is a dependent variable in the given model. In this section, we study on how we can identify dependent variables and on how we can use this knowledge to improve reachability analysis. The identification and extraction of *dependent variables* are based on the observation that for a Boolean function f,

$$f = (s \leftrightarrow f_s) \wedge (\exists s \cdot f) \qquad \text{if and only if} \qquad \forall s \cdot f = 0$$

Suppose the set $R(s)$ of the reachable states of a model has been computed. If s_i is functionally dependent on R, then the state variable corresponding to s_i can be removed from the model and replaced by a combinational function of other state variables. The resulting model satisfies the same properties as the original model. The advantage of removing dependent variables is that the BDDs are likely to be smaller [56].

For equivalence checking, however, the detection of functional dependencies would be more profitable if performed during fixpoint computation. In particular, one could extract dependencies from the set of states $P(s)$ whose image is being computed. The resulting functions can be used to eliminate the dependent variables from the transition relation used to compute the successors of the states in $P(s)$.

The problem with this approach is that the overhead required to detect the dependencies can be substantial. A more practical approach, albeit a less powerful one, concentrates on special forms of functional dependence, namely, variable equivalence and complementarity. Variables a and b are equivalent in Boolean function f if $f \leq (a \leftrightarrow b)$. They are complementary in f if $f \leq (a \oplus b)$ This means that two variables are equivalent (complementary) if they have the same value (opposite values) in all assignments for which $f = 1$.

When computing the image of $P(s)$, if s_i is equivalent to s_j in P, we can use the same BDD variable for s_j and s_i in both P and the transition relation T. If there is just one initial state, then for the first image computation in reachability analysis only one BDD variable is needed for the current-state variables. In successive image computations, the number of required BDD variables will likely increase, but for some circuits, it will remain substantially lower than the number of state variables. In particular, corresponding register pairs in equivalent circuits will remain represented by the same variable.

The effectiveness of this approach relies on the ability to compute variable equivalence classes efficiently. A partition refinement algorithm for reachability analysis based on forward traversal was implemented in the TiGeR tool in the early 1990s [57] but not published. The main reason why equivalence detection can be made efficient is that it boils down to comparing Boolean functions for equivalence or complementarity—operations that are well supported by BDDs. Later, a method similar to that of TiGeR was discussed by van Eijk [58]; his thesis, in particular, discusses the identification of equivalent variables before reachability analysis, which is described in Figure 4.8.

4.6 NEW DEVELOPMENTS IN 2006–2015

In this section, we discuss new developments in both combinational and SEC during the last decade, since the first edition of the EDA handbook in 2006.

4.6.1 COMBINATIONAL EQUIVALENCE CHECKING

Technology advances facilitate more sophisticated designs and the role of verification becomes more important. In particular, combinational equivalence checking plays a crucial role in the whole implementation process. In order to verify correctness of a design after modification during some implementation stages, the logical equivalence of the current description and the design prior to the implementation stages must be ascertained to make sure all of the verified properties are preserved. If all of the verified properties are preserved, we can therefore save a significant amount of time and cost in verifying those properties again on the implementation. This is why equivalence checking can be adopted at many points throughout design flows. Due to NP-completeness, it is unclear *a priori* that equivalence checking can be performed efficiently in practice. Such efficiency is facilitated by a number of implicit properties, such as the fact that internal equivalent signals are also preserved at the same time during the implementation process, as proof lemmas, which can help prove that equivalence holds. As a result, the success of equivalence checking is based on learning the design intent and effectively finding the design correspondence. Furthermore, state variables after implementation processes are complete and could often be matched such that *combinational* equivalence checking is feasible to be used as a sign-off tool to effectively validate the functionality of implementations. We will review the techniques used to tackle industrial problems from 2006 to 2015.

Over the past decade, the framework of combinational equivalence checking was established by pioneering studies [14,59,60], and the role of Boolean reasoning was almost completely dominated by state-of-the-art SAT solvers [61–65], which effectively deduce the useful correspondences between a design specification and its implementation, as processed by Boolean optimization. However, the relationships between word-level components (datapath) and their implementation are difficult to express and deduce by Boolean arithmetic; thus, the equivalence of implementation with datapath blocks is hard to establish by using conventional Boolean reasoning. Several studies use alternative approaches to prove the equivalence between a word-level specification and its bit-level implementation. Over the past decade, developments in combinational equivalence checking focused not only on demonstrating equivalence but also on dealing with nonequivalence. From generating the evidence of differences to diagnosing the root cause for those differences, and going further to generating the patches to the design that will correct the differences, the direction of nonequivalence analysis has shifted to handle the need for engineering change orders (ECOs). In the following sections, we will first discuss several techniques that strengthen Boolean reasoning. Second, we will revisit the solutions for tackling equivalence checking for datapath blocks. Finally, we will review the techniques used in functional correction, that is, functional ECOs.

As mentioned earlier, the fundamental framework of combinational equivalence checking was well established by [14,59,60], whose basic concept was to establish internal corresponding equivalences, merge and cut the equivalent nodes as new Boolean variables, and then prove internal equivalence, iteratively. However, the property of internal equivalence with respect to internal equivalent cuts may be falsified when the implementation is optimized by using satisfiability don't cares (SDCs) and observability don't cares (ODCs).

Figure 4.12 shows an example in which internal equivalence with respect to the equivalent cuts x, y is *false equivalent* if x and y cannot be 1 simultaneously, that is, $f_1(x_1, y_1) \neq f_2(x_2, y_2)$ where (x_1, x_2) and (y_1, y_2) are the internal equivalent cuts. Goldberg et al. [59] discussed this problem and demonstrated that a SAT solver can robustly solve the problem by learning the block SDC clauses $(\neg x_1 \vee \neg y_1) \wedge (\neg x_2 \vee \neg y_2)$. In this example, with the SAT conflict analysis on the assignment containing $x_1 = 1$ and $y_1 = 1$, it may successfully learn the blocking clause $(\neg x_1 \vee \neg y_1)$. Later, researchers identified effective and efficient algorithms by which a SAT solver can assign variables and learn the blocking clause with respect to SDC optimization. Chaff [21] proposed

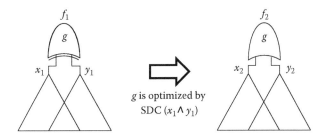

FIGURE 4.12   The example that satisfiability don't-care optimization would introduce nonequivalence in local corresponding cones. If $x_1 = x_2$ and $y_1 = y_2$ and x_1, y_1 cannot be 1 simultaneously, the function $f_1(x_1, y_1) \mathrel{!}= f_2(x_2, y_2)$.

the variable state independent decaying sum heuristic, and MiniSat [61] adjusted the quantity of parameters to assign a variable properly similar to a manual local search on a hotspot region during conflict analysis. In addition, MiniSat [61] provides an efficient data structure to speed up Boolean constraint propagation (BCP). Furthermore, thanks to annual software contests in the SAT community, the state-of-the-art SAT solvers [62–65] have significantly improved, based on enhancements such as CNF optimization [63] and better decision heuristics [62] to tune the practical strategies for tacking large sets of benchmarks.

While combinational equivalence checking relied heavily on state-of-the-art SAT solvers, researchers started to investigate a more useful approach to translate the gate-level implementation into a CNF formula. Velev [66] proposed a systematic way to translate the CNF by reducing intermediate variables and avoiding clause explosion. Mishchenko et al. [67] used a similar idea but a more effective cost-driven approach, borrowing from LUT-based technology mapping, for speeding up SAT solving. Eliminating unimportant variables in a CNF formula not only reduces the time for BCP but also improves variable scoring to speed up Boolean reasoning.

In addition to optimization by SDC, ODC optimization would also destroy internal equivalence, and the most important issue is that the constraints of ODC cannot be encoded by the conventional CNF translation, which is a one-to-one mapping on gates and variables. Figure 4.13 shows an example where f' was optimized by the select signal on output mux s, where the ODC of f is $s = 1$ and f is not equal to f'. Fu et al. [68] used the extra don't-care variable to encode the ODC space and translated gate-level netlists into CNF formulae while considering the ODCs. Although this translation can handle the relation between the implementation optimized by ODCs, the framework [14] cannot simply adopt ODC equivalence in the flow, and in addition, the extra variables would make a SAT solver inefficient or require a customized SAT solver [68]. Equivalence checking on a design after ODC optimization still needs further investigation of both the Boolean reasoning and the corresponding merging framework.

Over the past few years, there have been two approaches, symbolic and signature based, to improve the effectiveness and capability of combinational equivalence checking in handling datapath components with aggressive word-level optimizations. Such optimizations are heavily

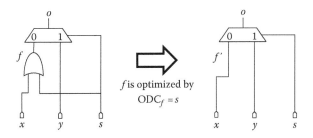

FIGURE 4.13   The example that observability don't-care optimization would introduce nonequivalence for corresponding nets.

used by contemporary synthesizers. Although Boolean reasoning is effective in deducing the correspondence of designs with SDC optimization, it is inefficient in checking the property on datapaths with word-level optimization. A typical example is to check multiplier equivalence. A multiplier can be implemented and optimized with different encodings and allows multiple orderings of addends for implementing the addition circuit. Unfortunately, a BDD cannot construct this function due to memory explosion, and a SAT solver tries to block all the input assignments and thus has problems when deducing equivalence. Therefore, research has tried to resolve this issue either by investigating the theoretical basis for dealing with the problem completely or by adopting a sound approach to handle such large designs. Some of the complete methods use symbolic approaches [7,69–71] in normal form for checking equivalence, but they may encounter an explosion in their representation. Some of the practical approaches [72,73] identify the logical decomposition from the specification of a datapath. Although such practical approaches can handle larger designs, they are incomplete since there are numerous implementations for a given arithmetic operation. Whether using a symbolic or a signature-based approach, researchers are looking forward to some new or improved framework that will combine both advantages.

Symbolic approaches prove equivalence by checking that the calculated canonical form is identical, but they need to avoid BDD explosion while calculating the expression. Therefore, they encode the Boolean function into a more compact representation than a BDD to express the integer arithmetic formula. Bryant and Chen [7,69] proposed the *BMD and *PHDD diagrams, which are *decision diagrams* specifically extended for representing integer arithmetic. Figure 4.14 shows the representation of a 3×3 integer multiplier that has linear complexity compared to the exponential complexity for a BDD representation. However, their work still encountered the memory explosion problem while calculating the expression for internal signals of a gate-level implementation. Pruss et al. [71] encoded Boolean expressions as shown in Table 4.1 and evaluated the formulae by utilizing Gröbner basis and polynomial reduction, which can successfully handle a Montgomery multiplier, not represented compactly in a BDD. Furthermore, Watanabe et al. [70] used the encoding shown in Table 4.2 and combined it with *BMD to verify a hierarchical arithmetic circuit, such as a multiplier and an FIR filter.

Signature-based approaches identify the arithmetic decomposition schemes used by contemporary synthesizers. They rewrite and handle the most common structures in the gate-level implementation and perform the same decomposition to enhance internal equivalence. Figure 4.15 shows the most common components used to decompose an integer multiplier, including the partial product generator and an addition circuit. Stoffel and Kunz [72] presented a reverse-engineering technique that focused on learning the structure of addition circuits. Lai et al. [73] presented

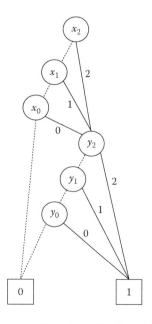

FIGURE 4.14   The PHDD* [69] representation about 3×3 multiplier.

TABLE 4.1   Construct to Encode Boolean Network

Logic Type	Representation Operation
and(z, a, b)	$z = ab$
or(z, a, b)	$z = a + b + ab$
xor(z, a, b)	$z = a + b$
not(z, a)	$z = a + 1$

Source: Pruss, T. et al., Equivalence verification of large Galois field arithmetic circuits using word-level abstraction via Gröbner bases, *Design Automation Conference*, June 2014, pp. 1–6.

TABLE 4.2   Construct to Encode Boolean Network

Logic Type	Representation Operation
and(z, a, b)	$z = ab$
or(z, a, b)	$z = a + b - ab$
xor(z, a, b)	$z = a + b - 2ab$
not(z, a)	$z = 1 - a$

Source: Watanabe, N. et al., Application of symbolic computer algebra to arithmetic circuit verification, *International Conference on Computer Design*, October 2007, pp. 25–32.

FIGURE 4.15 The common components to compose an integer multiplier, including partial product generator and adder tree.

a robust algorithm to identify partial products from a constant multiplier. Both approaches are practical enough to identify a decomposition from the gate-level implementation. However, they are unable to learn such signatures from an implementation if the datapath components are optimized aggressively via SDC and ODC.

The development of combinational equivalence checking not only focuses on deducing an equivalence but, as discussed at the beginning of this section, also concentrates on how to handle nonequivalence in order to support functional ECO processes. In the remainder of this section, we will discuss the theoretical basis of functional ECOs and the SAT-based solutions developed in recent years.

The theoretical basis for ECO was established in pioneering work by Lin et al. [74]. The problem of functional ECO with a single error can be formulated as follows: we assume that the implementation function is $f_1: X \rightarrow Y$ and a new specification is $f_2: X \rightarrow Y$. Thus, we need to determine whether there is some function $h: X \rightarrow B$, also called a *patch*, which can make the revised implementation equivalent to the new specification by replacing some internal logic $g: X \rightarrow B$ in the implementation. This is shown in the following equation, where f_1' is the function that includes g as the extra variable:

$$f_1(X) = f_1'(g(X), X) \quad \text{and} \quad f_2(X) = f_1'(h(X), X) \tag{4.6}$$

Lin et al. further proved the following:

1. The function h exists *iff on* and *off* is a disjunction, where *on*: $X \to B$: $= \neg(f_1'(0, X) = f_2(X))$ and *off*: $X \to B$: $= \neg(f_1'(1, X) = f_2(X))$.
2. The function h is some realization for an incompletely specified function with $h_{on} = on$ and $h_{off} = off$.

Lin et al. used the BDD approach to check the existence of a single fix and create the patch. This approach is not scalable for a large design even with a small patch. Fortunately, Lee et al. [75] presented a SAT-based solution that solved the conjunction of constraints specifying the onset and offset in the incompletely specified function. This provided opportunities to perform SAT-based ECO. Furthermore, Chang et al. [76,77] used diagnosis techniques to efficiently correct bugs inside the design.

In more detail, Lee et al. [75] used SAT solving to check the functional dependency from the constraints specifying the onset $f(X, Y)$ and offset $g(X, Z)$. The function h exists, where $\neg h(X) \wedge f(X, Y) = 0$ and $h(X) \wedge g(X, Z) = 0$, *iff* the SAT solver returns unsatisfiable and can generate a Craig interpolation, the by-product of unsatisfiability, which is one of the realizations of function h. Combining the SAT-based function realization and the existence condition (4.6) of a patch, Wu et al. [78] present a SAT-based ECO solution that tackles the limitations on BDD reasoning. This robust solution can effectively find the single error and fix it. However, the fixable condition (4.6) is only valid for the single error problem. Tang et al. [79] extended the existence condition to a multiple error model and generated the patch by checking the validation of a quantified Boolean formula instead of SAT solving. Although Wu [78] and Tang [79] proposed robust solutions to resolve design changes, the success of ECO is determined by the quality and the size of the patch. However, the McMillian and Pudlák interpolation generated from a resolution graph is hard to control; the interpolant is often unlikely to be a valid patch due to size and timing violations. In recent years, studies such as by Chockler et al. [80] and Petkovska et al. [81] tried to control the size of interpolation, and we are looking forward to more research to generate minimal and predictable patches for ECO.

4.6.2 SEQUENTIAL EQUIVALENCE CHECKING

In the 2006–2015 period, the use of SEC has shifted from comparing RTL versus transistor level to RTL versus RTL. New synthesis flows made RTL versus transistor level more suitable for combinational equivalence checking, where only small parts of the current design RTL are manually customized or modified and thus require SEC.

On the other hand, new design practices such as clock gating and low power design create layered RTL where the first layer defines the design functionality and additional layers define additional control for saving power without changing functionality. Layered RTL is naturally checked with SEC where the basic functionality is checked by comparing two versions of the design: one without the additional layers and the other with the additional layers. In most cases, the additional control logic acts at the block level, which enables using formal methods.

The progress of SEC in the 2006–2015 period has been twofold. On the one hand, SAT solvers made the core reachability analysis much more powerful, and on the other hand, higher level methodologies such as decomposition and alignability have matured and become integral parts of industrial tools. In the following sections, we will review both.

4.6.2.1 SAT-BASED REACHABILITY ANALYSIS

The heart of SEC is *reachability analysis*, and a breakthrough in this field was essential to make sequential equivalence practical on industrial designs. In Section 4.5.1, reachability analysis was discussed using BDDs. Moving from BDDs to SAT-based reachability analysis created the breakthrough.

Recall that reachability analysis is a procedure that takes as inputs a FSM, a set of initial state, and a set of erroneous states and decides whether there is a trace from the set of initial states to the set of erroneous states. The building block of SAT-based reachability analysis is an algorithm that looks for a trace of a specific size with specific properties in the design FSM. Let $\underline{V} = v_0 v_1 \ldots v_n$ be the state variables of the design, which encode the states of S. Let $\delta_0 \delta_1 \ldots \delta_n$ be the next-state functions for the variables of $\underline{V}$. For a given trace length k, the algorithm makes k copies of the design's variables:

$$v_0^0 v_1^0 \ldots v_n^0 v_0^1 v_1^1 \ldots v_n^1 \ldots v_0^{k-1} v_1^{k-1} \ldots v_n^{k-1}$$

and then it creates a SAT formula of the form

$$\wedge_{i=0}^{n} \wedge_{j=0}^{k-2} \left(v_i^{j+1} = \delta_i \left(v_0^j, v_1^j, \ldots, v_n^j \right) \right)$$

Every assignment to

$$v_0^0 v_1^0 \ldots v_n^0 v_0^1 v_1^1 \ldots v_n^1 \ldots v_0^{k-1} v_1^{k-1} \ldots v_n^{k-1}$$

that satisfies the SAT formula $\rho(k)$ represents a trace $s^0 s^1 \ldots s^{k-1}$ such that for every $0 \leq i < k$, the assignment to $v_0^i v_1^i \ldots v_n^i$ encodes the state si.

An immediate usage for the trace formula is bounded model checking [16]. Assume that $I(v_0 v_1 \ldots v_n)$ is a predicate over the variables of the design that define its initial state and $E(v_0 v_1 \ldots v_n)$ is a predicate defining the erroneous states in the design that violate some desired property. Then, the formula $\rho(k) \wedge I \left(v_0^0 v_1^0 \ldots v_n^0 \right) \wedge E \left(v_0^{k-1} v_1^{k-1} \ldots v_n^{k-1} \right)$ is satisfiable if and only if there exists a trace of length k in the design, which starts at an initial state and terminates at an erroneous state. Bounded model checking does a partial proof in the sense that it checks if there is a trace from initial to erroneous state of length smaller or equal to k. However, it does not provide a full proof, in which the trace's length is not bounded. Bounded model checking is useful for bug hunting, that is, showing that an erroneous state is reachable from an initial state. However, it is not very useful for proving that a property holds in a design, because in the general case, such a proof requires performing the bounded satisfiability check for exponentially many bounds. Since the reachability analysis problem is PSPACE (polynomial space) complete [82] in the number of design variables, we do not expect a polynomial reduction to SAT.

Nevertheless, there are a few algorithms that use trace formulae for full proof. Although these algorithms create exponential SAT formulae in the worst case, they often need small polynomial SAT formulae in practice. The most common algorithms for full proof are listed here:

1. *Induction* [83]: In k induction, two trace formulae are tested for satisfiability—the first is the induction base and is the same as bounded model checking, $\rho(k) \wedge I \left(v_0^0 v_1^0 \ldots v_n^0 \right) \wedge E \left(v_0^{k-1} v_1^{k-1} \ldots v_n^{k-1} \right)$. The second formula is the induction step and is of the form $\rho(k) \wedge E \left(v_0^{k-1} v_1^{k-1} \ldots v_n^{k-1} \right) \wedge not \, E \left(v_0^0 v_1^0 \ldots v_n^0 \right) \wedge not \, E \left(v_0^1 v_1^1 \ldots v_n^1 \right) \wedge \ldots \wedge not \, E \left(v_0^{k-2} v_1^{k-2} \ldots v_n^{k-2} \right) \wedge U \left(\underline{v}^0 \underline{v}^1 \ldots \underline{v}^{k-1} \right)$, where the U predicate requires that the encoded states $s^0 s^1 \ldots s^{k-1}$ are pairwise different. An unsatisfiable induction step implies that there is no trace that contains $k - 1$ nonerroneous states and ends with an erroneous state. Proving that both formulae are unsatisfiable for some k implies that there is no trace from an initial state to an erroneous state. To see that, consider a trace from an initial state to an erroneous state. If the trace is shorter than k, then it satisfies the induction base. Otherwise, the first k states are nonerroneous and so there is an interval of the trace that satisfies the induction step. Note that the uniqueness predicate is not needed for the soundness of the algorithm but only for its completeness.
2. *Interpolation* [84]: In interpolation, a proof that there is no trace of length k from the initial state to an erroneous state is used to create a SAT formula over $v_0^0 v_1^0 \ldots v_n^0$ that encodes an overapproximation of the set of states reachable from the initial state in k steps. This approximation replaces $I \left(v_0^0 v_1^0 \ldots v_n^0 \right)$ in the next bounded search, which checks states with distance up to $2k$ steps from the initial state. This procedure of computing

a bounded trace and an overapproximation is repeated until either an erroneous state is found or the overapproximation reaches a fixpoint. This technique is sound; however, since it uses overapproximation, it may find spurious traces to erroneous states. When a spurious trace is found, the overapproximation needs to be refined and this is done by increasing k and rerunning.

3. *IC3* [85]: The algorithm searches for an inductive property P such that the states that satisfy P are not erroneous. A property P is inductive with respect to $M = (X, Y, S, S_0, \delta, \lambda)$ if S_0 satisfies P and $\varpi = P\left(v_0^0 v_1^0 \ldots v_n^0\right) \wedge \left(\wedge_{i=0}^n v_i^1 = \delta_i\left(v_0^0 v_1^0 \ldots v_0^n\right)\right) \wedge not\, P\left(v_0^1 v_1^1 \ldots v_n^1\right)$ is unsatisfiable. That is, P is closed under δ. The algorithm starts its search at $P = \{s:s$ is not erroneous$\}$ and strengthens it iteratively using transitions from a state that satisfies P to a state that does not, that is, an assignment to ϖ. Finding such an inductive P implies that there is no trace from the initial state to an erroneous state.

4.6.2.2 DECOMPOSITION AS A WAY TO REDUCE COMPLEXITY

Applying SEC to big designs with hundreds of thousands of flip-flops requires methods to avoid state explosion. The most common method to achieve this is decomposition. For two designs D_1 and D_2 and their respective Mealy machines $M_1 = (X_1, Y_1, S_1, S_{01}, \delta_1, \lambda_1)$ and $M_2 = (X_2, Y_2, S_2, S_{02}, \delta_2, \lambda_2)$, the decision whether D_1 is sequentially equivalent to D_2 ($D_1 \equiv D_2$) is done in two steps:

1. Decompose

$D_1 = D_1^1 D_1^2 \ldots D_1^n$ and $D_2 = D_2^1 D_2^2 \ldots D_2^n$ into n slices$(M_1 = M_1^1 M_1^2 \ldots M_1^n$ and $M_2 = M_2^1 M_2^2 \ldots M_2^n$.

2. Prove for every pair D_1^i and D_2^i that $D_1^i \equiv D_2^i \left(M_1^i \equiv M_2^i\right)$.

In contrast to combinational equivalency discussed in Section 4.4, the slices here are sequential and include state variables like flip-flops and latches. In most cases, the time $T(D_1 \equiv D_2)$ of checking $(D_1 \equiv D_2)$ is substantially greater than $\sum_{i=0}^n T\left(D_1^i \equiv D_2^i\right)$.

Note that when a circuit is decomposed as $M = M^1{}^\circ M^2$, some of the internal signals become inputs and outputs of the slices. Formally, for

$$M^1 = \left(\underline{x_1}, \underline{y_1}, S_1, S_{01}, \delta_1, \lambda_1\right) \text{ and } M^2 = \left(\underline{x_2}, \underline{y_2}, S_2, S_{02}, \delta_2, \lambda_2\right)$$

the composition $M = M^1{}^\circ M^2$ is

$$M = \left(\underline{x_1 \backslash y_2} \cup \underline{x_2 \backslash y_1}, \underline{y_1} \cup \underline{y_2}, S_1 \times S_2, S_{01} \times S_{02}, \delta, \lambda\right)$$

where

$$\delta\left(x, (s_1, s_2)\right) = \left(\delta_1\left(\left(x_1, \lambda_2\left(x_2, s_2\right)\right), s_1\right), \delta_2\left(\left(x_2, \lambda_1\left(x_1, s_1\right)\right), s_2\right)\right)$$

and $\lambda\left(x, (s_1, s_2)\right) = \left(\lambda_1\left(\left(x_1, \lambda_2\left(x_2, s_2\right)\right), s_1\right), \lambda_2\left(\left(x_2, \lambda_1\left(x_1, s_1\right)\right), s_2\right)\right)$

The composition of λ is valid only when no combinational loops are created in the composition. A composition $M = M^1{}^\circ M^2{}^\circ \ldots{}^\circ M^n$ is defined iteratively as $(M = M^1{}^\circ M^2){}^\circ \ldots){}^\circ M^n)$.

For example, consider the design in Figure 4.16. On the left, the composed design is shown with inputs $\{in\}$, outputs $\{o\}$, and state-holding registers $\{c0, c1, q0, q1\}$. On the right, a decomposition into subcomponents M^1, M^2 is shown, where the inputs, outputs, and registers for M^1 are $\{in\}$, $\{c0, c1\}$, and $\{c0, c1\}$, respectively, and $M^2\{c0, c1\}$, $\{o\}$, $\{q0, q1\}$. Each FSM is shown below its respective design. The output function for M is $\lambda(in, c0, c1, q0, q1) = q0 \oplus q1$, M^1, $\lambda_1(in, c0, c1) = (c0, c1)$; and M^2, $\lambda_2(q0, q1) = q0 \oplus q1$.

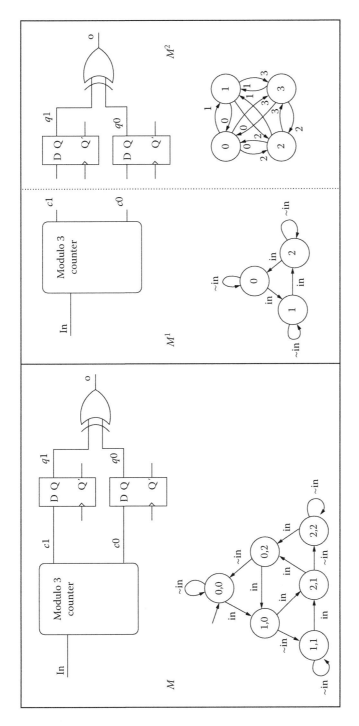

FIGURE 4.16   Decomposition.

How useful a decompositional algorithm is depends on the existence of a good partition, which reduces the equivalence checking to small enough pairs that can be handled by formal techniques. While in theory there is no guarantee that such a partition exists, in practice it often does. Even when a good partition exists, finding it is a challenge. This is usually done by mapping pairs of signals in the designs that are supposed to behave in the same way. Then the mapped signal can be split into two parts. The mapping can be performed automatically, based on signal name or circuit structure, or semiautomatically, where the user gives hints to the tool about possible mapping points. In most cases, compositional algorithms are adaptive. If a slice pair is too hard for the formal reachability engine, the algorithm tries to further partition it. If some slice pair is found to be not equal, then an analysis needs to select one of the following cases:

1. The whole designs are not equal, and a counterexample should be issued.
2. The partitioning is inadequate and needs to be coarser.
3. The partitioning is adequate, but some environment constraints are missing and need to be added.

For example, consider Figure 4.17, where two equal designs are compared. The implementation design is similar to the one in Figure 4.16, but the specification model differs at the output function $\lambda_s(\text{in}, c0, c1, q0, q1) = q0 + q1$. The partition is similar to the one in Figure 4.16. M_s^1 and M_i^1 are the same and proved equal. However, M_s^2 and M_i^2 differ when $(q0, q1) = 3$, $\lambda_i^2(3) = 0$, and $\lambda_s^2(3) = 1$. This is a spurious counterexample because the counter on the left-hand side counts only up to 2, which makes the state $(q0, q1) = 3$ unreachable. Therefore, to complete the proof, an assumption $(c0, c1) \leq 2$ must be added.

4.6.2.3 ALIGNABILITY

Decomposition helps designing and verifying different components of the design locally, sometimes at different design sites at different times in the overall design and verification process. However, while reachability can be done locally, an initial state is global to the whole integration of the design and is usually constructed at integration time. The initial state of the design is represented by an assignment to the design's variables set by the reset sequence. Since the reset sequence is applied to the inputs of the integrated components, it needs to propagate through the design until it reaches all the components. Thus, a component may be verified with one reset sequence but operates with a different reset sequence when integrated into the whole design. *Alignability* is a condition under which SEC is independent of the reset sequence.

A reset sequence is a sequence of input assignments that takes a design from an arbitrary state to a single specific state. For a reset sequence α, $M(\alpha) = (X, Y, S, \alpha, \delta, \lambda)$ is the Mealy machine, where the initial state is the state set by α. This is the initial state from which the reachability analysis is done. In [86], the following observation was made: Let M_s, M_i be designs and let α, β be two different reset sequences. Then, $M_s(\alpha) \equiv M_i(\alpha)$ iff $M_s(\beta) \equiv M_i(\beta)$.

To understand this observation, recall that two designs are equivalent from a pair of initial states (s_{0s}, s_{0i}) if for every sequence of inputs they produce the same sequence of outputs. If we suppose that this is true from the initial states resulting from applying α, then in particular it is true for every input sequence prefixed by β. Since β takes the designs to its initial states, every sequence of inputs produces the equal sequences of outputs from $\beta's$ initial states as well.

The observation made by [86] suggests the following flow for verifying given components M_s, M_i:

1. Find an arbitrary reset sequence α for M_s, M_i.
2. Prove $M_s(\alpha) \equiv M_i(\alpha)$.

This flow ignores the *real* reset sequence that will be used for the design. While this seems a useful flow, it must be integrated with the compositional flow to become practical. However, in [87], it was shown that a simple integration of the two flows is not sound. Thus, there are some

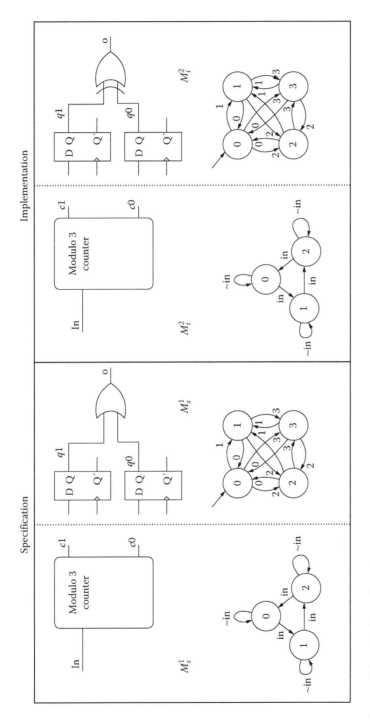

FIGURE 4.17   Sequential equivalence in parts.

limitations on the reset sequence and on the assumptions that allow this integration. The conditions are as follows:

1. The reset sequence must be X initializing.
2. Assumptions used in the proof of subcomponents $M_s^j \overset{?}{\equiv} M_i^j$ must be on the inputs of M_s^j and M_i^j and not on internal signals.

X arithmetic (also known as three-value logic [88]) is an extension of Boolean arithmetic over $\{0, 1, x\}$, where x represents the unknown value. The truth tables of the basic operators AND, OR, and NOT are shown in Figure 4.18. This logic can be extended to any Boolean function using the basic operators, in particular to the next-state function δ and the output function λ. A reset sequence made of $\{0, 1\}$ assignments to a design M is X initializing for M, if it brings M from a state where all state-holding variables have value x to a state where no state variable has value x.

Every X-initializing sequence is a reset sequence, that is, it brings the design to a specific single state, regardless of its previous state. However, the opposite does not hold. In fact, some designs admit reset sequences but no X-initializing sequences. For example, consider Figure 4.19, where the "counter modulo 3" component is enhanced with a reset signal, which, when asserted, increases the counter until it reaches 0. The state machine with reset asserted is on the left-hand side of the figure. Note that the state machine includes the "3" state, which is unreachable from the initial state.

A reset sequence that asserts the reset signal for two cycles brings the counter to state 0. However, there is no X-initializing sequence for the design.

AND	0	1	X
0	0	0	0
1	0	1	X
X	0	X	X

OR	0	1	X
0	0	1	X
1	1	1	1
X	X	1	X

NOT	
0	1
1	0
X	X

FIGURE 4.18   Three-value logic arithmetic.

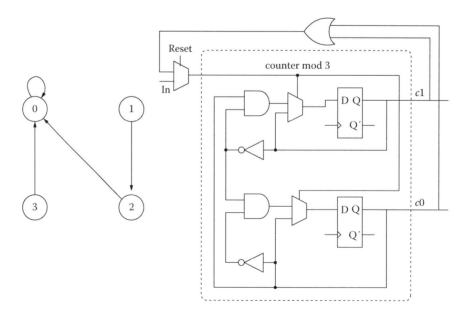

FIGURE 4.19   Extending the counter with reset.

4.7 SUMMARY

In this chapter, we have examined the role of equivalence checking in the design cycle, and we have reviewed the major approaches and techniques applied to this problem. Highly efficient algorithms have been developed for combinational equivalence checking; they rely on structural analysis and on sophisticated decision procedures that can establish the satisfiability of large propositional formulae. Since SEC is a much less tractable problem, every effort is made to reduce it to combinational equivalence by constraining the design methodology and by applying algorithms that deal with unknown register correspondence and retiming. Equivalence checking based on combinational techniques has become an integral part of most design flows and can deal with large-scale designs. In the general case, however, one must analyze the so-called product machine to determine whether miscomparing states are reachable from the initial states. Techniques for this task are based on reachability analysis. Although capacity is a significant concern whenever such an analysis is undertaken, symbolic algorithms can deal with circuits with hundreds of state variables and can therefore be applied to blocks of a design that have undergone deep transformations.

The update sections have covered the major changes in combinational and SEC over the decade since the first edition of the EDA handbook in 2006.

REFERENCES

1. G.L. Smith, R.J. Bahnsen, and H. Halliwell, Boolean comparison of hardware and flowcharts, *IBM J. Res. Dev.*, 26, 106–116, 1982.
2. A. Kuehlmann, A. Srinivasan, and D.P. LaPotin, Verity—A formal verification program for custom CMOS circuits, *IBM J. Res. Dev.*, 39, 149–165, 1995.
3. G.P. Bischoff, K.S. Brace, S. Jain, and R. Razdan, Formal implementation verification of the bus interface unit for the Alpha 21264 microprocessor, *Proceedings of the IEEE International Conference on Computer Design*, 1997, pp. 16–24.
4. J. Moondanos, C.H. Seger, Z. Hanna, and D. Kaiss, CLEVER: Divide and conquer combinational logic equivalence verification with false negative elimination, *Computer Aided Verification (CAV'01)*, Paris, France, 2001, pp. 131–143.
5. R.E. Bryant, Graph-based algorithms for Boolean function manipulation, *IEEE Trans. Comput.*, C-35, 677–691, 1986.
6. R.E. Bryant, On the complexity of VLSI implementations and graph representations of Boolean functions with application to integer multiplication, *IEEE Trans. Comput.*, 40, 205–213, 1991.
7. R. Bryant and Y.-A. Chen, Verification of arithmetic circuits with binary moment diagrams, *Proceedings of the Design Automation Conference*, San Francisco, CA, 1995, pp. 535–541.
8. Y.-T. Lai and S. Sastry, Edge-valued binary decision diagrams for multi-level hierarchical verification, *Proceedings of the Design Automation Conference*, Anaheim, CA, 1992, pp. 608–613.
9. F. Somenzi, Binary decision diagrams, in *Calculational System Design*, M. Broy and R. Steinbrüggen, Eds., IOS Press, Amsterdam, the Netherlands, 1999, pp. 303–366.
10. S. Manne, D.C. Grunwald, and F. Somenzi, Remembrance of things past: Locality and memory in BDDs, *Proceedings of the Design Automation Conference*, Anaheim, CA, 1997, pp. 196–201.
11. D. Sieling, The nonapproximability of OBDD minimization, Technical Report 663, University of Dortmund, Dortmund, Germany, 1998.
12. R. Rudell, Dynamic variable ordering for ordered binary decision diagrams, *Proceedings of the International Conference on Computer-Aided Design*, Santa Clara, CA, 1993, pp. 42–47.
13. K. Ravi and F. Somenzi, Hints to accelerate symbolic traversal, *Correct Hardware Design and Verification Methods (CHARME'99)*, Lecture Notes in Computer Science, Vol. 1703, Springer, Berlin, Germany, 1999, pp. 250–264.
14. A. Kuehlmann and F. Krohm, Equivalence checking using cuts and heaps, *Proceedings of the 34th ACM/IEEE Design Automation Conference*, Anaheim, CA, 1997, pp. 263–268.
15. J.R. Burch and V. Singhal, Tight integration of combinational verification methods, *Proceedings of the International Conference on Computer-Aided Design*, San Jose, CA, 1998, pp. 570–576.
16. A. Biere, A. Cimatti, E. Clarke, and Y. Zhu, Symbolic model checking without BDDs, *Fifth International Conference on Tools and Algorithms for Construction and Analysis of Systems (TACAS'99)*, Lecture Notes in Computer Science, Vol. 1579, Amsterdam, the Netherlands, 1999, pp. 193–207.
17. M.R. Garey and D.S. Johnson, *Computers and Intractability: A Guide to the Theory of NP-Completeness*, 2nd edn., W.H. Freeman and Company, New York, 1979.

18. M. Davis and H. Putnam, A computing procedure for quantification theory, *J. Assoc. Comput. Mach.*, 7, 201–215, 1960.

19. M. Davis, G. Logemann, and D. Loveland, A machine program for theorem proving, *Commn. ACM*, 5, 394–397, 1962.

20. J.P.M. Silva and K.A. Sakallah, Grasp—A new search algorithm for satisfiability, *Proceedings of the International Conference on Computer-Aided Design*, San Jose, CA, 1996, pp. 220–227.

21. M. Moskewicz, C.F. Madigan, Y. Zhao, L. Zhang, and S. Malik, Chaff: Engineering an efficient SAT solver, *Proceedings of the Design Automation Conference*, Las Vegas, NV, 2001, pp. 530–535.

22. L. Zhang, C. Madigan, M. Moskewicz, and S. Malik, Efficient conflict driven learning in Boolean satisfiability solver, *Proceedings of the International Conference on Computer-Aided Design*, San Jose, CA, 2001, pp. 279–285.

23. L. Zhang and S. Malik, Validating SAT solvers using an independent resolution-based checker: Practical implementations and other applications, *Design, Automation and Test in Europe (DATE'03)*, Munich, Germany, 2003, pp. 880–885.

24. H. Jin and F. Somenzi, An incremental algorithm to check satisfiability for bounded model checking, *Electronic Notes in Theoretical Computer Science*, 2004, *Second International Workshop on Bounded Model Checking*, http://www.elsevier.nl/locate/entcs/.

25. C.L. Berman and L.H. Trevillyan, Functional comparison of logic designs for VLSI circuits, *Proceedings of the International Conference on Computer-Aided Design*, Santa Clara, CA, 1989, pp. 456–459.

26. A. Kuehlmann, V. Paruthi, F. Krohm, and M.K. Ganai, Robust Boolean reasoning for equivalence checking and functional property verification, *IEEE Trans. Comput. Aided Des.*, 21, 1377–1394, 2002.

27. M.K. Ganai, P. Ashar, A. Gupta, L. Zhang, and S. Malik, Combining strengths of circuit-based and CNF-based algorithms for a high-performance SAT solver, *Proceedings of the Design Automation Conference*, New Orleans, LA, 2002, pp. 747–750.

28. H. Jin and F. Somenzi, CirCUs: A hybrid satisfiability solver, *International Conference on Theory and Applications of Satisfiability Testing (SAT 2004)*, Vancouver, British Columbia, Canada, 2004.

29. J.P. Roth, Diagnosis of automata failures: A calculus & method, *IBM J. Res. Dev.*, 10, 278–291, 1966.

30. J.P. Roth, Hardware verification, *IEEE Trans. Comput.*, C-26, 1292–1294, 1977.

31. D. Brand, Verification of large synthesized designs, *Digest of Technical Papers of the IEEE/ACM International Conference on Computer-Aided Design*, Santa Clara, CA, 1993, pp. 534–537.

32. M.H. Schulz and E. Auth, Improved deterministic test pattern generation with applications to redundancy identification, *IEEE Trans. Comput. Aided Des.*, 8, 811–816, 1989.

33. D. Brand, Incremental synthesis, *Digest of Technical Papers of the IEEE/ACM International Conference on Computer-Aided Design*, San Jose, CA, 1994, pp. 14–18.

34. L.A. Entrena and K.-T. Cheng, Combinational and sequential logic optimization by redundancy addition and removal, *IEEE Trans. Comput. Aided Des.*, 14, 909–916, 1995.

35. H. Cho, G. Hachtel, S.-W. Jeong, B. Plessier, E. Schwarz, and F. Somenzi, ATPG aspects of FSM verification, *Digest of Technical Papers of the IEEE International Conference on Computer-Aided Design*, 1990, pp. 134–137.

36. M.K. Ganai, A. Aziz, and A. Kuehlmann, Enhancing simulation with BDDs and ATPG, *Proceedings of the 36th ACM/IEEE Design Automation Conference*, New Orleans, LA, 1999, pp. 385–390.

37. H. Hulgaard, P.F. Williams, and H.R. Andersen, Equivalence checking of combinational circuits using Boolean expression diagrams, *IEEE Trans. Comput. Aided Des.*, 18, 903–917, 1999.

38. T. Filkorn, *Symbolische Methoden für die Verifikation endlicher Zustandssysteme*, Dissertation, Technische Universität, München, Germany, 1992.

39. C.A.J. van Eijk and J.A.G. Jess, Detection of equivalent state variables in finite state machine verification, *1995 ACM/IEEE International Workshop on Logic Synthesis*, Tahoe City, CA, 1995, pp. 3-35–3-44.

40. J.R. Burch and V. Singhal, Robust latch mapping for combinational equivalence checking, *Digest of Technical Papers of the IEEE/ACM International Conference on Computer-Aided Design*, San Jose, CA, 1998, pp. 563–569.

41. C. Leiserson and J. Saxe, Optimizing synchronous systems, *J. VLSI Comput. Syst.*, 1, 41–67, 1983.

42. C. Leiserson and J. Saxe, Retiming synchronous circuitry, *Algorithmica*, 6, 5–35, 1991.

43. A. Kuehlmann and C.A. van Eijk, Chapter 13: Combinational and sequential equivalence checking, in *Logic Synthesis and Verification*, S. Hassoun and T. Sasao, Eds., Kluwer Academic Publisher, Boston, MA, 2001.

44. D. Stoffel and W. Kunz, A structural fixpoint iteration for sequential logic equivalence checking based on retiming, *International Workshop on Logic Synthesis*, Tahoe City, CA, 1997.

45. C. van Eijk, Sequential equivalence checking based on structural similarities, *IEEE Trans. Comput. Aided Des.*, 19, 814–819, 2000.

46. S.-Y. Huang, K.-T. Cheng, K.-C. Chen, C.-Y. Huang, and F. Brewer, AQUILA: An equivalence checking system for large sequential designs, *IEEE Trans. Comput.*, 49, 443–464, 2000.

47. K. Ravi and F. Somenzi, High-density reachability analysis, *Proceedings of the International Conference on Computer-Aided Design*, San Jose, CA, 1995, pp. 154–158.

48. G. Cabodi, P. Camurati, L. Lavagno, and S. Quer, Disjunctive partitioning and partial iterative squaring: An effective approach for symbolic traversal of large circuits, *Proceedings of the Design Automation Conference*, Anaheim, CA, 1997, pp. 728–733.

49. R. Fraer, G. Kamhi, B. Ziv, M.Y. Vardi, and L. Fix, Prioritized traversal: Efficient reachability analysis for verification and falsification, *Twelfth Conference on Computer Aided Verification (CAV'00)*, E.A. Emerson and A.P. Sistla, Eds., Lecture Notes in Computer Science, Vol. 1855, Springer, Berlin, Germany, 2000, pp. 389–402.

50. O. Coudert, C. Berthet, and J.C. Madre, Verification of sequential machines using Boolean functional vectors, *Proceedings IFIP International Workshop on Applied Formal Methods for Correct VLSI Design*, L. Claesen, Ed., Leuven, Belgium, 1989, pp. 111–128.

51. H. Touati, H. Savoj, B. Lin, R.K. Brayton, and A. Sangiovanni-Vincentelli, Implicit enumeration of finite state machines using BDD's, *Proceedings of the International Conference on Computer-Aided Design*, 1990, pp. 130–133.

52. J.R. Burch, E.M. Clarke, and D.E. Long, Representing circuits more efficiently in symbolic model checking, *Proceedings of the Design Automation Conference*, San Francisco, CA, 1991, pp. 403–407.

53. I.-H. Moon, J.H. Kukula, K. Ravi, and F. Somenzi, To split or to conjoin: The question in image computation, *Proceedings of the Design Automation Conference*, Los Angeles, CA, 2000, pp. 23–28.

54. D. Geist and I. Beer, Efficient model checking by automated ordering of transition relation partitions, *Sixth Conference on Computer Aided Verification (CAV'94)* D.L. Dill, Ed., Lecture Notes in Computer Science, Vol. 818, Springer, Berlin, Germany, 1994, pp. 299–310.

55. I.-H. Moon, G.D. Hachtel, and F. Somenzi, Border-block triangular form and conjunction schedule in image computation, *Formal Methods in Computer Aided Design*, W.A. Hunt, Jr. and S.D. Johnson, Eds., Lecture Notes in Computer Science, Vol. 1954, Springer, Austin, TX, 2000, pp. 73–90.

56. A.J. Hu and D. Dill, Reducing BDD size by exploiting functional dependencies, *Proceedings of the Design Automation Conference*, Dallas, TX, 1993, pp. 266–271.

57. J.C. Madre, Private communication, 1996.

58. C.A.J. van Eijk, Formal methods for the verification of digital circuits, PhD thesis, Eindhoven University of Technology, Eindhoven, the Netherlands, 1997.

59. E.I. Goldberg, M.K. Prasad, and R.K.Brayton, Using SAT for combinational equivalence checking, *Design, Automation and Test in Europe Conference and Exhibition*, 2001, pp. 114–121.

60. F. Lu, L.-C. Wang, K.-T. Cheng, and C.-Y. Huang, A circuit SAT solver with signal correlation guided learning, *Design, Automation, and Test in Europe*, 2003, pp. 10892–10897.

61. N. Een, and N. Sörensson, MiniSat: A SAT solver with conflict-clause minimization, *The International Conference on Theory and Applications of Satisfiability Testing*, 2005.

62. G. Audemard and L. Simon, Predicting learnt clauses quality in modern SAT solvers, *International Joint Conference on Artificial Intelligence*, 2009, pp. 399–404.

63. A. Biere, Lingeling, plingeling and treengeling entering the SAT competition 2013, *SAT Competition*, 2013, pp. 51–52.

64. M.W. Moskewicz, C.F. Madigan, Y. Zhao, L. Zhang, and S. Malik, Chaff: Engineering an efficient SAT solver, *Design Automation Conference*, pp. 530–535.

65. K. Pipatsrisawat and A. Darwiche, A lightweight component caching scheme for satisfiability solvers, *Theory and Applications of Satisfiability Testing*, 2007, pp. 294–299.

66. M.N. Velev, Efficient translation of Boolean formulas to CNF in formal verification of microprocessors, *Asia and South Pacific Design Automation Conference*, 2004, January 2004, pp. 310–315.

67. A. Mishchenko, S. Chatterjee, R. Brayton, and N. Een, Improvements to combinational equivalence checking, *International Conference on Computer-Aided Design*, November 2006, pp. 836–843.

68. Z. Fu, Y. Yu, and S. Malik, Considering circuit observability don't cares in CNF satisfiability, *Design, Automation and Test in Europe*, Vol. 2, 2005, pp. 1108–1113.

69. Y-A. Chen and R.E. Bryant, An efficient graph representation for arithmetic circuit verification, *IEEE Trans. Comput. Aided Des. Integr. Circuits Syst.*, 20(12), 1443–1454, December 2001.

70. Y. Watanabe, N. Homma, T. Aoki, and T. Higuchi, Application of symbolic computer algebra to arithmetic circuit verification, *International Conference on Computer Design*, October 2007, pp. 25–32.

71. T. Pruss, P. Kalla, and F. Enescu, Equivalence verification of large Galois field arithmetic circuits using word-level abstraction via Gröbner bases, *Design Automation Conference*, June 2014, pp. 1–6.

72. D. Stoffel and W. Kunz, Verification of integer multiplier on arithmetic bit level, *International Conference on Computer-Aided Design*, pp. 183–189, 2001.

73. C.-Y. Lai, C.-Y. Huang, and K.-Y. Khoo, Improving constant-coefficient multiplier verification by partial product identification, *Design, Automation and Test in Europe*, March 2008, pp. 813–818.

74. C.C. Lin, K.C. Chen, S.C. Chang, M. Marek-Sadowska, and K.T. Cheng, Logic synthesis for engineering change, *Design Automation Conference*, 1995, pp. 647–652.

75. C.-C. Lee, J.R. Jiang, Huang C.-Y., and A. Mishchenko, Scalable exploration of functional dependency by interpolation and incremental SAT solving, *International Conference on Computer-Aided Design*, November 2007, pp. 227–233.

76. K.H. Chang, I.L. Markov, and V. Bertacco, Fixing design errors with counterexamples and resynthesis, *Asia and South Pacific Design Automation Conference*, 2007, pp. 944–949.

77. K.H. Chang, I.L. Markov, and V. Bertacco, Automating post-silicon debugging and repair, *International Conference on Computer-Aided Design*, 2007, pp. 91–98.

78. B.-H. Wu, C.-J. Yang, C.-Y. Huang, and J.-H.R. Jiang, A robust functional ECO engine by SAT proof minimization and interpolation techniques, *International Conference on Computer-Aided Design*, November 2010, pp. 729–734.

79. K.-F. Tang, P.-K. Huang, C.-N. Chou, and C.-Y. Huang, Multi-patch generation for multi-error logic rectification by interpolation with cofactor reduction, *Design, Automation and Test in Europe*, 2012.

80. H. Chockler, A. Ivrii, and A. Matsliah, Computing interpolants without proofs, *Haifa Verification Conference*, 2013.

81. A. Petkovska, D. Novo, A. Mishchenko, and P. Ienne, Constrained interpolation for guided logic synthesis, *International Conference on Computer-Aided Design*, November 2014, pp. 462–469.

82. C. Papadimitriou, Chapter 19: Polynomial space, *Computational Complexity*, 1st edn., Addison Wesley, Reading, MA, 1994, pp. 455–490.

83. M. Sheeran, S. Singh, and G. Stålmarck, Checking safety properties using induction and a SAT-solver, *Proceedings of FMCAD*, 2000.

84. K.L. McMillan, Interpolation and SAT-based model checking, *Fifteenth International Conference, CAV* Boulder, CO, 2003.

85. A.R. Bradley, SAT-based model checking without unrolling. *VMCAI*, 2011.

86. C. Pixley, A theory and implementation of sequential hardware equivalence, *IEEE Trans. Comput. Aided Des. Integr. Circuits Syst.*, 11(12), 1469–1478, December 1992.

87. Z. Khasidashvili, D. Kaiss, and D. Bustan, A compositional theory for post-reboot observational equivalence checking of hardware, *FMCAD 2009*, 2009, pp. 136–143.

88. J.S. Jephson, R.P. McQuarrie, and R.E. Vogelsberg, 1969. A three-value computer design verification system. *IBM Syst. J.*, 8(3), 178–188, September 1969.

Digital Layout

Placement

Andrew B. Kahng and Sherief Reda

5

CONTENTS

5.1 INTRODUCTION: PLACEMENT PROBLEM AND CONTEXTS

Placement is an essential step in the physical design flow since it assigns exact locations for various circuit components within the chip's core area. An inferior placement assignment will not only affect the chip's performance but might also make it nonmanufacturable by producing excessive wirelength, which is beyond available routing resources. Consequently, a placer must perform the assignment while optimizing a number of objectives to ensure that a circuit meets its performance demands. Typical placement objectives include total wirelength, timing, congestion, and power. In this chapter, we survey the main algorithmic methods used in state-of-the-art placers.

Figure 5.1 shows the position of placement within the EDA design flow. A placer takes a given synthesized circuit netlist together with a technology library and produces a valid placement layout. The layout is optimized according to the aforementioned objectives and ready for cell resizing and buffering—a step essential for timing and signal integrity satisfaction. Clock-tree synthesis and routing follow completing the physical design process. In many cases, the entire physical design flow or parts of it are iterated a number of times until timing closure is achieved.

A circuit netlist is composed of a number of *components* and a number of *nets* representing the required electrical connectivity between the various components, where a net connects two

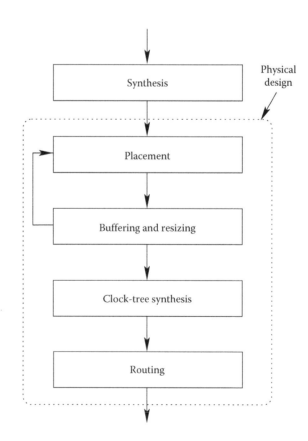

FIGURE 5.1   Placement position in a physical design flow.

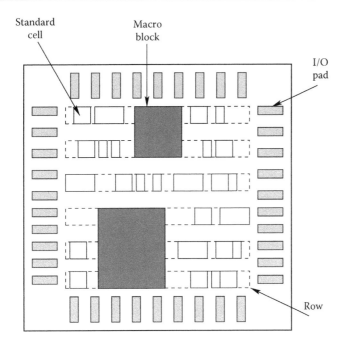

FIGURE 5.2 Placement layout area.

or more components. In the case of application specific integrated circuits (ASIC), the chip's core layout area is comprised of a number of fixed height *rows*, with either some or no space between them. Each row consists of a number of *sites* which can be occupied by the circuit components. A *free site* is a site that is not occupied by any component. Circuit components are either *standard cells*, *macro blocks*, or *I/O pads*.* Standard cells have a fixed height equal to a row's height, but have variable widths. The width of a cell is an integral number of sites. On the other hand, blocks are typically larger than cells and have variable heights that can stretch a multiple number of rows. Figure 5.2 gives a view of a typical placement layout. Blocks can have preassigned locations—say from a previous floorplanning process—which limit the placer's task to assigning locations for just the cells. In this case, the blocks are typically referred to by *fixed blocks*. Alternatively, some or all of the blocks may not have preassigned locations. In this case, they have to be placed with the cells in what is commonly referred to as *mixed-mode* placement.

In addition to ASICs, placement retains its prime importance in gate array structures such as field programmable gate arrays (FPGAs). In FPGAs, placement maps the circuit's subcircuits into programmable FPGA logic blocks in a manner that guarantees the completion of the subsequent stage of routing.

The major placement objectives can be summarized as follows:

- *Wirelength*: Minimizing the total wirelength, or just wirelength, is the primary objective of most existing placers. This is no surprise given that the wirelength must be less than the limited total routing supply of the chip, and that power and delay are proportional to the wirelength and wirelength square, respectively. Consequently, minimizing the wirelength improves the chip's performance. Wirelength is measured by the sum of minimum Steiner tree costs of the various nets, where the Steiner tree cost of a given point set is the minimum cost, or length, of a tree that spans all the given points as well as any subset of additional points (Steiner points) [1]. Routed wirelength is typically slightly larger than the Steiner tree cost since contention on routing resources by different nets might lead to detours, eventually increasing the wirelength. Given that the Steiner tree problem is NP-hard, placers typically minimize and report other metrics that are

---

* We will refer to standard cells by just *cells* and macro blocks by just *blocks*.

faster and easier to compute. These include half-perimeter wirelength (HPWL) and minimum spanning tree (MST). Half-perimeter wirelength is half the perimeter of the smallest bounding box enclosing a given net's components. Half-perimeter wirelength is equivalent to the Steiner minimum tree cost for two- and three-pin nets, and is well-correlated for multi-pin (≥ 4) nets [2]. Minimum spanning tree is also equivalent to the Steiner minimium tree cost for two- and three-pin nets, and within a constant factor for multi-pin (≥ 4) nets [1].

■ *Timing*: The clock cycle of a chip is determined by the delay of its longest path, usually referred to as the *critical path*. Given a performance specification, a placer must ensure that no path exists with delay exceeding the maximum specified delay. Any delay in excess over such specified value is considered *negative slack*, and timing-driven placers must minimize the worst negative slack and the total negative slack to meet performance requirements.

■ *Congestion*: While it is necessary to minimize the total wirelength to meet the total routing resources, it is also necessary to meet the routing resources within various local regions of the chip's core area. A congested region might lead to excessive routing detours. Such detours can ultimately lead to excessive increase in wirelength, adversely impacting both timing and power.

■ *Power*: With increasing clock frequencies and demand for battery-powered mobile devices, minimizing power is becoming an increasingly important objective. Power minimization typically involves distributing the locations of cell components so as to reduce the overall power consumption, alleviate hot spots, and smooth temperature gradients.

Another secondary objective is placement *runtime* minimization. For a given netlist, placement is a one-time effort, and consequently it is usually tolerable to have increased runtimes if this has a positive impact on the placement quality. However, for state-of-the-art designs with multimillion components, placement can take a few days, which is deemed unacceptable for fast prototyping or in timing-closure iterations. For such cases, it is important to seek methods that minimize the total placement runtime with little or no impact to the placement quality.

Given that placement is one of the oldest and first problems in EDA, placement has a rich history of solutions. More than four decades ago, Steinberg [3] considered placement of circuit components on a back board such that the total wirelength is minimized. Steinberg solution starts with an initial placement, obtained say via random placement. Steinberg then selects an *independent set* of components, that is, a set of components that do not share any connections, and optimally reembeds these components within their pool of locations via optimal linear assignment (OLA). The process of selection and optimal reembedding is iterated until no further improvement in solution quality is possible.

Analytical techniques approximate the wirelength objective using quadratic [4–8] or nonlinear formulations [9–12]. The first proposal to use such methods is due to Hall [4], where he suggested minimizing the squared length and devised an eigenvalue approach to solve the problem. Since that point, the central problems in analytical techniques are how to better approximate the wirelength objective, how to numerically solve the nonlinear objective and how to spread out the components that typically heavily overlap in analytical solutions. Approaches for solving the nonlinear objectives include sparse matrix and conjugate-gradient (CG) methods. Cell-spreading techniques include the use of partitioners in top-down frameworks [5,6,13], network flows [7], or additional repelling forces [8,14].

The advent of min-cut partitioners [15] paved the way to the introduction of min-cut placers [16]. The introduction of a linear-time min-cut partitioning heuristic by Fiduccia and Mattheyses [17] and terminal propagation mechanisms [18] further bolstered min-cut placement as an attractive solution. Finally, development of multilevel hypergraph partitioners [19] has initiated a revival in min-cut placement [20–22].

Another thread of placement techniques started with the proposal of simulated annealing as a general combinatorial optimization technique [23]. As a matter of fact, placement along with the Traveling Salesman Problem were the original two problems experimented by Kirkpatrick and Vecchi [23]. Simulated annealing was quickly adopted as a leading technique in placement

[24,25], with methods such as clustering used to improve its execution time [26]. Simulated annealing can also be used with other placement methods such as min-cut to improve their performance [21].

Placement approaches typically differentiate between a *global placement* phase and a *detailed placement* phase. At the beginning of the global phase, all cells belong to one rectangular *bin* that spans the entire chip's core area. As global placement proceeds, cells are spread over the chip's core area into a number of smaller bins. By the end of this phase, each bin will typically contain few cells. In detailed placement, cells are assigned exact locations and all overlaps are removed.

In this chapter, we will a give a brief survey of the various placement approaches. We describe global placement algorithms in Section 5.2, detailed placement algorithms in Section 5.3, and recent (as of 2006) placement trends in Section 5.4. Section 5.5 gives a brief of view of state-of-the-art academic and industrial placers, and Section 5.6 summarizes this chapter content.

5.2 GLOBAL PLACEMENT

The goal of global placement is to find a well-spread, ideally with no overlaps, placement for the given netlist that attains required objectives such as wirelength minimization or timing specifications. Formally, a circuit netlist is represented as a hypergraph $H = (V, E)$, where $V = \{v_1, v_2, ..., v_n\}$ is the set of nodes with each node representing a circuit component, and $E = \{e_1, e_2, ..., e_m\}$ is the set of hyperedges corresponding to the nets, where a hyperedge $e_i \in E$ is a subset of nodes. Traditionally placers are HPWL driven with the main objective to minimize

$$(5.1) \qquad z = \sum_{e_i \in E} \left(\max_{v_j \in e_i} x_j - \min_{v_j \in e_i} x_j + \max_{v_j \in e_i} y_j - \min_{v_j \in e_i} y_j \right)$$

where x_j and y_j are the vertical and horizontal coordinates of a node v_j, and such that all nodes are placed on sites with no overlaps. Fundamentally, z is the HPWL sum of all hyperedges. The placement problem is NP-hard [27], and solutions for the problem rely on heuristics to achieve this objective suboptimally. In this section we survey some of the algorithmic solutions to global placement. Specifically, we describe (1) min-cut placers, (2) simulated-annealing placers, and (3) analytical placers. We also survey some of the main techniques to handle other objectives such as timing, congestion, and placement runtime.

5.2.1 MIN-CUT PLACERS

Min-cut placers operate in a top-down hierarchical fashion by recursively partitioning a given netlist into 2^k, $k \geq 1$, partitions. When $k > 1$, partitioning is commonly referred to as *multiway* partitioning. In the case $k = 1$, partitioning is called *bisection*, and when $k = 2$, partitioning is called *quadrisection*. In addition to netlist partitioning, the placer also recursively divides the layout area into a number of *bins*, and assigns each of the netlist partitions into one of the bins. Min-cut placement is essentially a top-down refinement process, where each bin gets divided into two or four smaller bins with fewer number of cells. Bin dividing is achieved through either a horizontal or a vertical cut in case of bisection [16], or through simultaneous horizontal and vertical cuts in case of quadrisection [28]. Thus the outcome of recursive bin division is a slicing floorplan as illustrated in Figure 5.3. The process of partitioning and dividing all bins exactly once is called *placement level.* Placement levels created by simultaneous netlist partitioning and bin division continue until each bin has a few cells, beyond which detailed placers are used to assign final locations for all nodes within their corresponding bins [20,21,29].

The key concerns of min-cut placement are as follows. How to partition a netlist?, and into how many partitions, for example, bisection or quadrisection? Given a bin, should it be divided horizontally or vertically in case of bisection? Finally, how to capture the global netlist connectivity information when partitioning local netlist instances inside the bins. We tackle each of the aforementioned concerns in the next subsections.

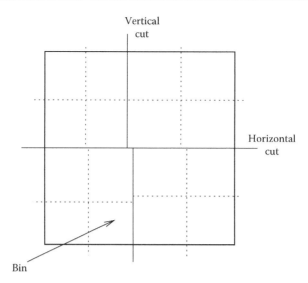

FIGURE 5.3 Slicing outline.

5.2.1.1 MIN-CUT PARTITIONERS

Given a number of netlist partitions, a hyperedge is considered *cut* if its nodes span or reside in more than one partition. The *k-way min-cut partitioning* problem is defined as follows. Given a hypergraph with an associated cost to each hyperedge, partition the nodes of the hypergraph into balanced 2^k subsets while minimizing the total cost of cut hyperedges. Subset balancing is achieved by imposing the constraint that no subset exceeds a given maximum size bound. The min-cut partitioning problem is NP-hard [1].

Kernighan and Lin [15] suggested the first heuristic solution (KL) to this problem within the context of graph partitioning. Assuming $k = 1$, the KL algorithm starts with a random initial balanced partitioning. For each pair of nodes—a pair is comprised of a node from each partition—the algorithm calculates the pair's *score*, or impact on cut size if the pair is swapped. The pair with the largest score or equivalently largest reduction in cut cost is identified, swapped, and *locked* to be prevented from any future swapping. Furthermore, their score is entered in a *record* and the swap score of their neighbors is updated. This process of identifying the best pair to swap, swapping the pair, locking, recording, and updating the neighbors' score is repeated until all pairs of cells are exchanged, essentially yielding the original partitioning. Using the record, the KL algorithm then executes the sequence of swaps that yields the overall minimum cut cost and unlocks all nodes. This constitutes a *KL iteration.* The KL algorithm keeps on iterating until an iteration causes no further reduction in cut size. At this point the algorithm stops.

The KL algorithm suffers from large runtimes due to the need to calculate and update the scores of all pairs of nodes. To overcome these limitations, Fiduccia and Mattheyses proposed a linear time heuristic (FM) based on the KL algorithm [17]. The key differences between the FM and KL algorithms are (1) moves instead of swaps are only considered, (2) hyperedge handling, and (3) a bucket data structure that allows a total update of move scores in linear time.

Despite the improvement in runtime of the FM heuristic in comparison to the KL heuristic, the performance of the FM heuristic, as measured by the cut cost, typically tends to degrade as the problem size increases. To address this problem, modern min-cut partitioners use the FM heuristic in a *multilevel* framework [19,30], where *clustering* or *coarsening* is used to reduce the size of a given flat hypergraph. A given hypergraph H_0 is coarsened k times until a hypergraph H_k of suitable size is attained as shown in Figure 5.4. The FM partitioner is then executed on H_k. Instead of projecting the partitioning results of H_k directly back to H_0, the unclustering is carried out in a controlled *refined fashion*, where the results of H_k are projected *to* H_{k-1} and followed by refinement using the FM partitioner. The process of uncoarsening and refinement is repeated until the original flat hypergraph H_0 is obtained.

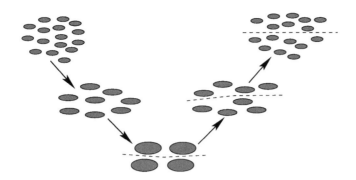

FIGURE 5.4   Coarsening and uncoarsening in a multilevel framework.

In many cases a bin capacity is larger than the total cell area that belongs to it, thus creating an amount of *freespace.* Freespace can be distributed by the partitioner to ensure "smooth" partitioning by allowing tolerances in the specified partition sizes [20,31]. In addition, distributing freespace in a uniform hierarchical manner ensures smooth partitioning till the late placement levels, which results in a reduced amount of final overlaps. The issue of freespace distribution will be further investigated in Section 5.4.2.

5.2.1.2 CUT SEQUENCES

Another important ingredient in min-cut placement is the cut direction. In the case of quadrisection, cut direction selection is trivial; each bin will be simultaneously divided with both a vertical and a horizontal cut [28]. In case of bisection, a cut sequence has to be determined. An *alternating sequence* that strictly alternates between horizontal and vertical cuts is a favorite choice [16]. However, recent (as of 2006) approaches [20,32] suggest selecting a direction based on the bin's aspect ratio. For example, if the height of a bin exceeds its width by a certain ratio then division is carried out horizontally. Such selection is typically justified by wirelength estimates which exhibit an increase if the wrong division direction is chosen.

When comparing multiway partitioning, for example, quadrisection, to bisection, we find out that (1) multiway partitioning fixes the shape of the partitions in contrast to bisection which allows more flexible outlines, and furthermore (2) the added computational complexity of multiway partitioning does not seem to translate to improvements in cut values in comparison to recursive bisection [33].

Another important consideration is whether to allow horizontal cuts that are only aligned with row boundaries or allow "fractional" horizontal cuts that run through the rows [34]. The latter approach might improve placement results but requires an extra effort from the detailed placer to snap the cells to the rows.

5.2.1.3 CAPTURING GLOBAL CONNECTIVITY

While a partitioner might deliver partitions with close to minimum cut costs, this result might not translate to total placement HPWL minimization if each bin is partitioned locally and in isolation of other bin results. Thus it is essential to capture global connectivity while partitioning a particular local bin instance. *Terminal propagation* [18,35] is a mechanism through which connectivity between cells residing in different bins is propagated into local partitioning instances. With terminal propagation, nodes external to a bin being partitioned are propagated as fixed terminals (nodes) to the bin. These terminals bias the partitioner toward placing movable nodes close to their terminals, reducing the overall placement wirelength. Given a bin being partitioned into two child bins and an external node connected to the cells of this bin, it is critical to determine which child bin the node will be propagated to. This is determined according to the distance between the node's position and the centers of the two new child bins as illustrated by the following example.

Example 5.1

Assume a bin B being partitioned into child bins U and L as shown in Figure 5.5. Given a number of external nodes a, b, c, and d that are connected to nodes in block B via a number of nets, it is necessary to determine to which block, U or L, where these nodes should propagate. The distances between each cell location and the centers of U and L are calculated, and each cell is propagated to the closest bin. For example, node a propagates to U; node b is not propagated, or propagated to both bins, since it is equally close to both U and L; node c is propagated to L; and node d is propagated to L.

Another possible terminal propagation case occurs when the nodes of a net are propagated as terminal to both child bins. In this case, the net will be cut anyway, and consequently it is labeled *inessential* and ignored during partitioning [35].

Since the partitioning results of one bin affect the terminal propagation decisions of that bin cells to other bins, a placement level result depends on the order of bin processing. Furthermore, there might exist cyclic dependencies between the different bins [36,37]. For example, one bin B_i results might affect another bin B_j results, with B_j results affecting B_i results in turn. To improve placement results and alleviate these effects, a number of approaches suggest iterating the partitioning process of a placement level until a certain stopping criterion is satisfied [36,37]. It is also possible to use a *sliding window* as shown in Figure 5.6, where at each position of the sliding window, the bins that underlie the window are merged and repartitioned [33]. Since the window positions overlap in their movement, it is possible to move cells out of their assigned bins to improve wirelength.

It is also possible to capture global connectivity in a more direct manner by driving the partitioner to improve the global exact HPWL or MST objective [36,38]. In this case, the score of moving a node across a partitioning line is the exact amount of reduction in placement HPWL due to such a move. These exact approaches, however, incur an increased amount of runtime and are not typically used in practice.

5.2.2 SIMULATED ANNEALING-BASED PLACERS

Simulated annealing is a successful generic combinatorial optimization technique that has its origins in statistical mechanics [23]. Given a combinatorial optimization problem, simulated annealing starts from an initial configuration, and iteratively moves to new configurations until some stopping criterion is satisfied. In each iteration, a random transition from the current configuration to a new configuration is generated. If the new configuration yields an improvement in the cost function or the objective being optimized then the transition is accepted; otherwise the transition is probabilistically rejected, where the rejection probability increases as iterations unfold. The ability to transit to new configurations with worse cost value allows simulated annealing to escape local minima, and is the essential ingredient to its success.

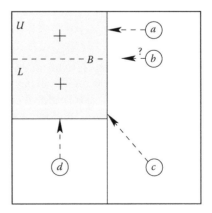

FIGURE 5.5   Terminal propagation. Bin B is being partioned into two bins U and L. Cell a is propagated to U. Cell b is not propagated at all. Cell c is propagated to L. Cell d is propagated to L.

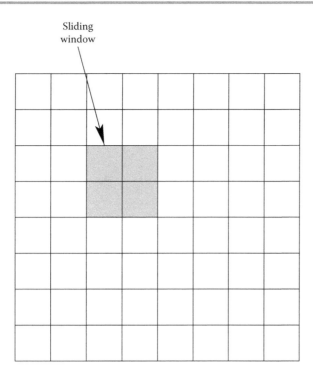

Sliding
window

FIGURE 5.6   A 2 × 2 sliding window over a grid of 8 × 8.

In placement, a new configuration can be generated by either (1) displacing a cell to a new location, (2) swapping two cells, or (3) changing a cell orientation [24,25]. If the change in cost function, ΔC, due to a transition from the current configuration to a new configuration is greater than or equal to zero, that is, deteriorating a minimization objective, then the rejection probability is given by $1 - e^{-(\Delta C/T)}$, where T, commonly referred to as the *temperature*, is a parameter that controls the rejection probability. To control the value of T, a *cooling scheduling* is set up, where T is initially set to a high value and then gradually decreased so that transitions to configurations of worse cost value are less likely. At the end of simulated annealing, $T = 0$, and the algorithm reduces to a greedy improvement iterator which stops when there is no possible further improvement in the solution.

To obtain solutions with good cost values, the cooling scheduling has to be slow. Consequently, simulated annealing typically suffers from excessive runtimes. To improve the runtime, it is possible to cluster or condense the netlist before the start of the simulated annealing process [26]. A three-stage approach to solve the placement problem can be as follows: (1) the netlist is condensed by clustering into a suitable size, (2) simulated annealing is applied on the condensed netlist, and (3) the netlist is unclustered and simulated annealing is applied on the flattened netlist for refinement.

Simulated annealing can be also used in combination with other placement techniques as an improvement operator. For example, the min-cut placer Dragon [21] uses simulated annealing to improve its results.

5.2.3 ANALYTICAL PLACERS

Since their introduction more than three decades ago [4], analytical methods have become one of the most successful techniques in solving the placement problem. This success can be attributed to a number of factors including: (1) the ability to capture mathematically the placement problem in a concise set of equations; (2) the availability of efficient numerical solvers that are able to solve the mathematical formulation in practical runtimes; and (3) the possibility to include analytical solvers within top-down frameworks to produce quality solutions. In this subsection, we survey the major analytical approaches suggested to solve the placement problem.

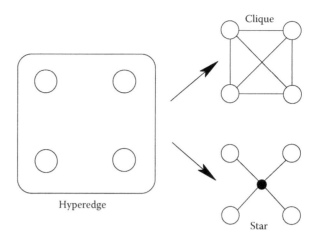

FIGURE 5.7   Net models are used to convert hyperedges into two-pin edges using either cliques or stars.

Most analytical placers typically start by converting a given hypergraph to a graph. Such conversion allows a more convenient mathematical formulation. A *k-pin* hyperedge can be converted into a k clique or a $k + 1$ star using weighted two-pin edges as shown in Figure 5.7. The problem of total HPWL minimization of a graph can be formulated as follows.

Given n objects, let c_{ij} denote the number, or weight, of connections between a pair of nodes v_i and v_j. The placement problem can be solved by minimizing the following equation:

$$(5.2) \qquad z_1 = \frac{1}{2}\sum_{i=1}\sum_{j=1} c_{ij}\left(|x_i - x_j| + |y_i - y_j|\right)$$

under the constraint that no cells overlap. Since the absolute distance $|x_i-x_j| + |y_i-y_j|$ in Equation 5.2 is mathematically inconvenient, it is possible to approximate it by a number of methods. A possible approximation is to replace it by a quadratic term that leads to minimizing the following objective:

$$(5.3) \qquad z_2 = \frac{1}{2}\sum_{i=1}\sum_{j=1} c_{ij}\left(\left(x_i - x_j\right)^2 + \left(y_i - y_j\right)^2\right)$$

which minimizes the total squared edge length measured in Euclidean geometry. z_2 is a continuously differentiable convex function and thus can be minimized by equating its derivative to zero, which reduces to solving a system of linear equations. The approximation of the linear objective by a quadratic one will likely lead to an increase in the total wirelength. As a better approximation, it is possible to use the following objective:

$$(5.4) \qquad z_G = \frac{1}{2}\sum_{i=1}\sum_{j=1} c_{ij}\left(\frac{\left(x_i - x_j\right)^2}{|x_i - x_j|} + \frac{\left(y_i - y_j\right)^2}{|y_i - y_j|}\right) = \frac{1}{2}\sum_{i=1}\sum_{j=1} c_{ij}\left(\frac{\left(x_i - x_j\right)^2}{\hat{x}_{ij}} + \frac{\left(y_i - y_j\right)^2}{\hat{y}_{ij}}\right)$$

where $\hat{x}_{ij} = |x_i - x_j|$ and $\hat{y}_{ij} = |y_i - y_j|$ as proposed in the GORDIANL placer [39]. Since numerical solvers typically operate in an iterative manner, it is possible to incoporate $\hat{x}_{ij}$ and $\hat{y}_{ij}$ by calculating their values from the results of one iteration and using them as constants in the next iteration. A recent (as of 2006) approximation proposed by Naylor [12] and advocated by Kahng and Wang [11] in their APlace placer, skips the hypergraph to graph conversion step and directly approximates Equation 5.1 as follows:

$$(5.5) \qquad z_A = \alpha\sum_{e_i \in E}\left(\ln\left(\sum_{v_j \in e_i} e^{x_j/\alpha}\right) + \ln\left(\sum_{v_j \in e_i} e^{-x_j/\alpha}\right) + \ln\left(\sum_{v_j \in e_i} e^{y_j/\alpha}\right) + \ln\left(\sum_{v_j \in e_i} e^{-y_j/\alpha}\right)\right)$$

FIGURE 5.8 Results of analytical solvers are typically overlapping as shown by this 400 cell example solved using the quadratic objective.

where α is defined as a "smoothing parameter." More accurate nonlinear formulations based on substituting the quadratic terms by higher-degree terms are also possible, but not as computationally efficient as Equation 5.3 or Equation 5.5 [10]. Note that the presence of fixed components, for example, I/O pads, prevents the trivial solution $x_i = y_i = 0$ from being a possible minimum for the previous equations. Equation 5.3 can also be written in a matrix formulation as follows. Let C denote a connection matrix, where c_{ij} gives the number of connections between nodes v_i and v_j. Define a diagonal matrix D as follows: $d_{ij} = 0$ if $i \neq j$ else if $i = j$ then $d_{ii} = \Sigma_{j=1}^{n} c_{ij}$. In addition, define the matrix $B = D - C$, and the row vectors $X^{\mathrm{T}} = (x_1, x_2, \ldots, x_n)$ and $Y^{\mathrm{T}} = (y_1, y_2, \ldots, y_n)$. Using matrix B, Equation 5.3 can be rewritten as

(5.6)
$$z_2 = X^{\mathrm{T}}BX + Y^{\mathrm{T}}BY$$

For illustration of the possible outcome of just numerically solving the analytical objectives, we plot the placement results of a 400 cell netlist in a 20 × 20 grid layout in Figure 5.8 solved using the quadratic objective of Equation 5.3, or equivalently Equation 5.6. It is quite clear that just solving the previous equations yields quite an overlapping placement. Research in analytical techniques mainly focuses on how to numerically minimize either of the previous equations, while spreading out circuit components with minimum component overlap.

5.2.3.1 SPREADING OUT CIRCUIT COMPONENTS

There are two possible strategies in removing component overlap in analytical methods. In the first strategy, the overlap constraint is ignored while solving the previous objectives, leading to a highly overlapped placement. This overlap is then removed by another means. We refer to this strategy as "Remove after Solving." In the second strategy, a measure of overlap is incorporated into the previous objectives while minimizing them. This incorporation can be done via an additional set of terms to minimize or through a number of linear constraints. Thus the outcome solution is minimized with respect to both wirelength and overlaps. We refer to this strategy as "Remove while Solving." We next discuss a number of methods that can be used to implement both strategies.

5.2.3.1.1 Remove after Solving Perhaps one of the simplest methods to remove overlaps—assuming all cells have the same width—is to move the cells from their placement locations to legal placement sites with the minimum total displacement. This can be solved optimally in polynomial time using optimal linear assignment, where the cost of assigning a cell to a placement site is equal to the Manhattan distance between the cell location and the site position. Such an approach requires a runtime of $O(n^3)$ which is deemed expensive, and typically does not yield good solutions if the placement is highly overlapped.

A better approach for overlap removal is through the use of the top-down framework in a manner similar to min-cut placement. Recall that in top-down placement, the layout area is recursively divided by means of bisection or quadrisection, into finer and finer bins, where the min-cut partitioner takes the role of determining the cells that belong to each bin. This approach can be similarly used for analytical placement, where the solution of the analytical program rather than the min-cut partitioner is used to determine the cells that belong to each bin. We next explore a number of methods where analytical placement is used to partition the set of cells among the bin's child bins.

In setting up the analytical program of a subset of cells belonging to a particular bin, the placer is faced with the problem of handling the connections between those cells and the components that lie outside the bin, for example, other cells or I/O pads. To solve this problem, cells outside the bin can be propagated as I/O pads to the periphery of the bin in a manner resembling that of terminal propagation in min-cut placement [6,7]. Such propagation, or *splitting*, of the connection mitigates the mismatch between quadratic and linear wirelength objectives since it leads to shorter connections from the fixed cells to the movable cells.

Once the analytical program is solved, the cells are partitioned among the child bins—produced from either bisection or quadrisection—using the spatial information provided from the analytical solution. In case of bisection, one possibility is to divide the bin into two equal child bins in a direction perpendicular to its longest side. The cell are then sorted along the direction of the longest side and the median cell is identified. All cells before and including the median cell are assigned to the closest child bin, while the other cells are assigned to the other child bin [6]. In case of quadrisection, Vygen [7] has devised an almost optimum linear-time method that reassigns cells to child bins to satisfy their capacities with the minimum total displacement from the analytical solution spatial results.

While the top-down framework offers a good method to remove overlap, it has the drawback of restricting cell movement since cells remain confined within their assigned bin boundaries. This restriction may adversely impact the wirelength. To allow cells to move from their bins, there are two possible solutions. First, it is possible to incorporate the top-down partitioning results "softly" as follows. Instead of solving the analytical placement of each bin separately, the analytical program of the whole netlist can be resolved, while imposing a number of additional linear constraints that coerce the center of gravity of a bin's cells to coincide with the center of the bin they belong to [13]. The additional constraints allow relative flexibility in cell movement while maintaining the global spatial cell distribution. Second, it is possible to use repartitioning via a sliding window moving over all bins in a manner similar to min-cut placement [7,36]. At each position of the sliding window, the bins that lie under the window are merged, analytically resolved, and repartitioned. The overlap in the sliding window positions allows cells to move out of their assigned bins to improve wirelength. While both previous solutions offer a good method to improve the wirelength, they incur an increase in placement runtime.

Another possible improvement to top-down analytical methods is to equip the analytical solver with a min-cut partitioner to improve its results [13]. In this latter scenario, the analytical solution can be thought of as a "seed" for the subsequent min-cut partitioner.

5.2.3.1.2 Remove while Solving
In this approach, the wirelength objective is replaced by an objective that simultaneously minimizes wirelength and overlap. The analytical program is then solved in an iterative manner where every *iteration* yields a better spread placement than the previous iteration.

One possibility to achieve this objective is through the use of a set of *repelling forces* that push cells from high-density regions to low-density ones [8,14]. These forces are calculated and added to the wirelength minimization function in an iterative manner as follows. After solving

Equation 5.3, the additional forces are calculated for each cell and inserted as linear terms in the quadratic program of Equation 5.3. The quadratic program is then resolved resulting in better-spread placement. This constitutes one *transformation iteration* that can be iterated a number of times until a desired even cell distribution is attained.

A recent (as of 2006) approach that has yielded fast runtimes utilizes the concept of spreading forces in a different manner [40]. In each iteration, the utilization of all bins or regions is first calculated, and cells are shifted from bins of high utilization to new locations in bins of low utilization. To encourage the placer to keep the cells close to their newly assigned locations, spreading forces are applied to the shifted cells. These forces are realized via the addition of pseudo-nets connecting every shifted cell to the boundary of the layout region along the direction of its spreading force.

Recently, (as of 2006), Naylor [12] proposed to divide the layout area into a grid and to minimize the uneven cell distribution in the grid to achieve uniform cell spreading. This can be achieved by adding a differentiable penalty function as a weighted term to the wirelength minimization function. The penalty function decreases in value when cells are spread more uniformly. Thus, minimizing the analytical objective function simultaneously minimizes both wirelength and overlaps.

5.2.3.2 NONLINEAR NUMERICAL OPTIMIZATION

Minimization of Equations 5.3 and 5.4, or Equation 5.5 is carried out using numerical techniques [41]. Placement researchers have typically used numerical optimization methods as "black boxes" to provide the required solutions. We next give a glimpse of some of the possible techniques; we refer the interested reader to [41].

One of the first methods suggested to solve Equation 5.6 is through the use of eigenvalues [4]. However, eigenvalue calculation is typically slow. Instead, Cheng and Kuh [5] noticed that matrix B is sparse since a circuit component is typically connected to very few components, and thus sparse-matrix techniques represent a runtime-efficient method to solve the analytical program. Later, Tsay et al. [6] suggested the use of a generalized Gauss–Seidel method, called successive over-relaxation, to solve the set of linear equations.

Given that nonlinear programs described by Equations 5.3 and 5.4, or Equation 5.5 are convex—since matrix B of Equation 5.6 is positive-semidefinite—and typically sparse, gradient methods offer a fast method to find the local minimum. Given a convex function $f(\mathbf{x})$ that has a minimum at $\mathbf{x}^*$, gradient methods find a sequence of solutions $\mathbf{x}_0, \mathbf{x}_1, \mathbf{x}_2, ..., \mathbf{x}_k = \mathbf{x}^*$, where each solution better approximates the minimum. One method is to proceed in the direction of *steepest descent*, which is in the direction of the gradient $\nabla f(\mathbf{x})$. Thus $\mathbf{x}_i = \mathbf{x}_{i-1} + \lambda_{i-1}\nabla f(\mathbf{x}_{i-1})$, where $\lambda_{i-1} < 0$ is the step that minimizes the function $f(x_{i-1} + \lambda\nabla f(\mathbf{x}_{i-1}))$ with respect to λ. The main disadvantage of steepest descent methods is that they might take an unnecessarily long time to converge to the optimal solution.

To improve the convergence rate, the Conjugate Gradient method is typically used [13,41]. Assume that $\mathbf{x}_0$ is the first approximation to $\mathbf{x}^*$, and define $\mathbf{v}_0 = \nabla f(\mathbf{x}_0)$. Conjugate-gradient proceeds by executing the following n times:

1. Calculate $\mathbf{x}_i = \mathbf{x}_{i-1} + \lambda_{i-1}\mathbf{v}_{i-1}$, where again $\lambda_{i-1} < 0$ is the step that minimizes the function $f(\mathbf{x}_{i-1} + \lambda\mathbf{v}_{i-1})$
2. Calculate $\mathbf{v}_i = -\nabla f(\mathbf{x}_i) + \dfrac{\left\|\nabla f(\mathbf{x}_i)\right\|^2}{\left\|\nabla f(\mathbf{x}_{i-1})\right\|^2}\mathbf{v}_{i-1}$
3. Increment i

It can be proven that the vectors $\mathbf{v}_i$ are mutually conjugate, that is, $\mathbf{v}_i^\top B\mathbf{v} = 0$, and that the CG method offers a fast convergence rate to the optimal solution in comparison with the regular steepest descent method.

5.2.4 HANDLING OTHER OBJECTIVES

In addition to total wirelength minimization, placers also have to minimize a number of additional objectives such as the longest path delay, congestion and power. We discuss next how to handle these objectives within the different placement solvers discussed earlier.

5.2.4.1 TIMING-DRIVEN PLACEMENT

Circuit signal paths start at the input pads and flip–flip outputs, and end at the output pads and flip–flop inputs. The delay of a path is the sum of interconnect and gate delays that make up the path. *Critical paths* are all signal paths of delay larger than the specified maximum delay, which is typically the clock period. By controlling the proximity of interconnected components so that no signal path exceeds the specified timing constraint, timing-driven placement can achieve its goal of eliminating all critical paths if feasible.

Timing-driven placement methods depend on the underlying placement solver. Analytical methods can incorporate timing constraints in a number of ways. One way is to formulate the delay of critical paths as a number of linear constraints and include these constraints in the analytical placement program [42]. The linear constraints eliminate placement solutions that do not satisfy the timing constraints, and consequently solving the constrained analytical program automatically leads—if feasible—to solutions with no critical paths. Another way is to multiply the net connectivity coefficients, c_{ij}, in Equations 5.2 and 5.3 by additional weights to reflect their timing criticality [8,43]. Nets included in critical paths receive higher weights than other nets, causing the analytical placer to reduce their length. Linear timing constraints can be indirectly incorporated in a recursive top-down paradigm—whether using a min-cut or analytical solver—as follows [44]. After each placement level, a linear program is constructed where the objective is to minimize the reassignment of cells to bins such that all timing delay constraints are satisfied. Solving the linear program leads to a minimum perturbation from an existing illegal-timing placement to a legal-timing placement.

One possibility to achieve timing-driven placement in min-cut placers is to perform static timing analysis at each placement level and then translate timing slacks to net weights, where nets participating in critical paths receive larger weights [45]. Another solution seeks to control the number of cuts experienced by a critical path, since the more cuts a path experiences, the longer the path tends to be [46]. To control the number of cuts, it is possible to give larger weights to critical nets and to impose an upper bound on the maximum number of times a path can be cut. These weights bias the min-cut partitioner to avoid cutting critical nets eventually reducing critical paths delay. Another recent (as of 2006) method prioritizes cells that belong to critical paths [47]. Given a bin under partition, cells that belong to critical paths are preassigned and fixed to the child bins so as to reduce the negative slack of critical paths. After this cell preassignment, the hypergraph partitioner is invoked on the remaining cells.

5.2.4.2 CONGESTION-DRIVEN PLACEMENT

Minimizing total wirelength ensures meeting the total routing supply. Nevertheless, congestion might occur in localized regions where local routing demand exceeds local routing supply. Congestion impacts routability in two ways. First, congestion might lead to the failure of routing some nets compromising the integrity of the complete physical design stage. Second, congestion forces routers to detour nets away from congested regions, leading to an increase in their wirelength. This increase in wirelength adversely impacts performance if the detoured nets are part of the critical paths.

Given a placement, congestion–reduction methods typically first divide the layout area into rectangular regions. A fast global routing method is then used to estimate the number of wires crossing each region boundary. This estimate is compared against each region's routing supply. If the demand exceeds the supply then the region is either expanded to increase the supply or cells are moved out of the region to reduce the demand. Congestion–reduction methods can be applied during placement or after placement as a postprocessing step. We next discuss briefly fast routing estimates and a number of methods to reduce the congestion.

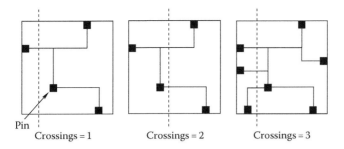

Pin

Crossings = 1   Crossings = 2   Crossings = 3

FIGURE 5.9 Number of wire crossings depends on cut location and number of pins. (From Vorwerk, K. et al., Engineering details of a stable force-directed placer, *Proceedings of IEEE International Conference on Computer Aided Design*, 2004, pp. 573–580.)

Routing-demand estimation methods are numerous and there is no space to discuss them in detail. However, we discuss an attractive routing demand model that was developed from a placement perspective [2]. Cheng [2] observed that the number of wire crossings of a cut line through the bounding box of a net depends on the number of pins of the net as well as the cut line location as shown in Figure 5.9. Given a k-pin net, it is possible to calculate the average number of wire crossings at any location inside the net bounding box as follows: (1) place the k pins of the net randomly; (2) construct a Steiner minimum tree over the k pins and calculate the number of wire crossings at all possible locations; and (3) execute (1) and (2) a number of times and report the average results. The average results for different values of k are calculated once and stored in a lookup table. The table is then used for fast routing estimation for all future placement runs.

As for congestion reduction methods, one simple way to reduce the demand within a congested region is to *inflate* the cells within the region, and utilize repartitioning (see min-cut placers)—originally used to improve wirelength—to move cells out of the congested regions [48]. For example, using a sliding window of size 2 × 2 as shown in Figure 5.6 enables the cells to leave the regions they are currently placed in, thus reducing their congestion.

Alternatively, to increase the supply of a region, Yang et al. [49] consider two possible expansion areas for each region. Since expanded areas of different congested regions can overlap creating new congested regions, an integer linear program is constructed that finds an expansion of all regions that minimizes the maximum congested region. After all congested regions are expanded, cells within each region are placed within the region boundaries using a detailed placer. Another approach [50] expands regions during quadratic placement either vertically or horizontally depending on the demand.

5.2.4.3 REDUCING PLACEMENT RUNTIME

Circuit clustering is an attractive solution to manage the runtime and quality of placement results on a large scale VLSI designs [26,33,51–54]. Clustering takes an input netlist and coarsens or condenses it by merging nodes together to form larger nodes or clusters, and adjusts the nets or hyperedges accordingly as shown in Figure 5.10. Clustering methods have a long history and they are out of the scope of the discussion. We will only focus on the use of clustering in placement.

The interaction between clustering and placement can be classified into two categories. In the first category, clustering is used as a core part of the placement solver [20–22], such as clustering and unclustering within the multilevel partitioner of min-cut placers [19]. In this case, a cluster hierarchy is first generated, followed by a sequence of partitioning and unclustering. Partitioning results of prior clustered netlists are projected to the next level by unclustering, which become the seed for the subsequent partitioning. In the second category of clustering for VLSI placement, the cluster hierarchy is generated at the beginning of the placement as a preprocessing step in order to reduce the netlist size. The coarsened netlist is then passed on to the placer [26,53,54]. Usually, the clustered objects will be dissolved at or near the end of the placement process [54], with a "clean-up" operation applied to the fully uncoarsened netlist to improve the results. In some cases, unclustering and reclustering are executed during intermediate points in the placement flow to allow the placer escape any earlier bad clustering decisions [26,55].

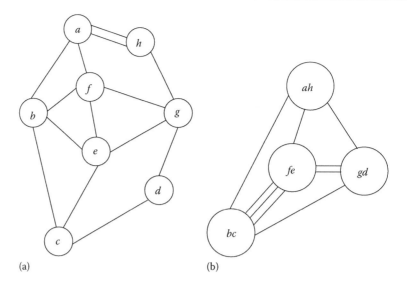

(a)　　　　　　　　(b)

FIGURE 5.10 Clustering of netlist: (a) produces netlist; (b) reducing the number of nodes from 8 to 4 and the number of edges from 14 to 9.

5.3 DETAILED PLACEMENT AND LEGALIZERS

A placement is illegal if cells/blocks overlap and/or occupy illegal sites, for example, between placement rows. Illegal placements might be an outcome of global (twice) placement, or produced from *incremental* changes to legal placements. Such changes include cell resizing and buffer insertion, which are necessary for signal integrity. Figure 5.11 shows a possible outcome from global placement where cells are quite spread out, yet the placement is illegal since there is a small amount of overlap and some cells are not snapped to sites. A placement *legalizer* snaps cells to the sites of rows such that no cells overlap. This has to be done with minimum adverse impact on the quality of the placement. A *detailed* placer takes a legal placement and improves some placement objective such as wirelength or congestion, and produces a new legal placement. In many cases, a detailed placer includes a legalizer that is invoked to legalize any given illegal placement

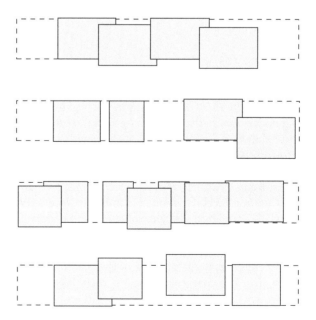

FIGURE 5.11 A possible outcome of global placement. A legalizer legalizes a placement by snapping cells to sites with no overlap.

or to legalize the detailed placer's own placement. Detailed placers are also used in *incremental placement* to legalize the placement after netlist changes. In this case, they are called *engineering Change Order* placers. We can summarize the goals of a detailed placer/legalizer as follows:

1. Remove all overlaps, and snap cells to sites with the minimum impact on wirelength, timing, and congestion.
2. Improve wirelength by reordering groups of cells. Such groups typically include spatially close cells.
3. Improve routability by carefully redistributing free sites.

We classify detailed placement methods as either heuristic or exact. Heuristic methods typically achieve good results in fast runtimes. Exact methods are only applicable for a few cases and usually take longer runtimes.

5.3.1 HEURISTIC TECHNIQUES

A possible legalizing approach starts by dividing the layout area into a number of regions [56]. Regions are then traversed from left to right, where the cells of each region are legalized and placed after preceding regions. During the course of legalization, a *borderline* can be envisioned where cells of regions to its left are legalized, but cells of regions to its right are yet to be legalized. A region is legalized by assigning its cells to new positions by solving a transportation problem, where cells are transported to their new legal positions with the minimum impact to wirelength.

A relatively recent (as of 2006) approach that also divides the layout into regions and balances region utilization, is due to Hur and Lillis [57]. In their approach, an overutilized region, say S, and its nearest underutilized region, say T, are identified, and cells are ripple-moved from S to T along a *monotone path*, that is, a sequence of adjacent bins. Since there are possibly many paths between S and T and many cells to move, the sequence of moves that minimizes the increase in wirelength is identified and executed. In addition, if row-size constraints are not met, then the utilized bin from the largest row, say S, and the most least utilized bin from the smallest row, say T, are identified, and cells are again ripple-moved along a monotone path from S to T.

A recent (as of 2006) detailed placement approach divides the legalization effort into two stages [7,58,59]. In the first stage, the layout area is divided into *zones*. Since some of these zones might be over capacity and consequently produce overlaps, the detailed placer first balances these zones. Such balancing can be achieved while minimizing the total displacement using network flow programs. After all zones are balanced, cells are legalized within rows and free sites are distributed to further minimize the HPWL.

A simple greedy legalization heuristic that empirically proved useful is as follows [34,60]. Initially all cells are sorted by their horizontal coordinate. Cells are then visited in that order and assigned to the location which results in the minimal displacement from their current location, such that the row capacity is not exceeded. The heuristic can also be slightly modified to handle mixed cell and block legalization.

5.3.2 EXACT TECHNIQUES

As for exact techniques in detailed placement, optimal end-case placers can use branch and bound algorithms to place optimally a set of few cells within a row [35]. The algorithm can be suboptimally scaled by sliding a rectangular *window* over the layout area, and placing cells within it optimally [20]. Sliding over the layout area can be iterated a number of times until improvement drops below a certain threshold. Branch and bound techniques are typically plagued with excessive runtime requirements, and window sizes are usually restricted to 7–8 cells to keep runtime manageable. However, if the subset of cells selected are *independent*, that is, they do not share any common nets, and are of the same size, then it is possible to place them optimally in polynomial runtime using OLA [3,61].

Another detailed placement problem that is getting increasingly important due to the ubiquitous presence of free sites in most modern designs is as follows. Given a number of cells placed in a placement row, shift the cells to minimize the wirelength, such that the ordering of cells before and after the shifting is the same [62]. Kahng et al. [62] propose a solution that solves the problem optimally in polynomial time using dynamic programming. The dynamic programming algorithm uses the piecewise-linearity attribute of HPWL to distribute optimally the freespace for wirelength minimization purposes. A correction of the complexity bounds, and further speed up through efficient data structures are realized by Brenner and Vygen [63]. Recently, (as of 2006) Kahng et al. [64] proposed a generic dynamic programming method to remove overlaps within a row given a fixed cell ordering while optimally minimizing either HPWL, total cell displacement, or maximum cell displacement.

5.4 PLACEMENT TRENDS

Technological advances introduce challenging extensions to the placement problem. For example, the recent (as of 2006) proliferation of Intellectual Property cores transformed the placement problem from just cell placement to mixed cell and core, or block, placement. Additionally, once these blocks are placed, they represent a *blockage* where routing cannot be conducted on layers on top of them. Such blockage imposes additional challenges to congestion-driven placement methods.

Increasing system complexity of recent (as of 2006) designs coupled with diminishing feature sizes created further demand for routing resources. This increased demand forced physical design engineers to expand layout areas to be more than the total components area creating *free space* for additional routing resources. Such *free* or *white* space must be distributed by the placer in a manner that improves the design's performance.

Another active research area in placement is that of placer *suboptimality evaluation*. Suboptimality evaluation enables the quantification of existing placers' performance on arbitrary benchmarks. Essentially, suboptimality evaluation reveals how much room for improvement still exists, which guides the continued investment of attention to the placement problem. We next examine a number of recent (as of 2006) solutions for the aforementioned placement challenges, starting with mixed-size placement.

5.4.1 MIXED-SIZE PLACEMENT

Rather than placing blocks before placement either manually or using automatic floorplanning, mixed-size placement simultaneously places cells and blocks, while considering a number of fixed blocks. A number of solutions have been recently (as of 2006) proposed.

One solution clusters the cells in a bottom–up fashion condensing the netlist into blocks and clusters of cells [65]. Quadratic placement in a top–down partitioning framework is then used to globally place the netlist. After global placement, clusters are flattened and their cells are placed without overlap.

Instead of clustering cells, Adya and Markov [66] suggest "shredding" every block into a number of connected cells and then placing the shredded blocks and cells. Afterwards, the initial locations of the blocks are calculated by averaging the locations of the cells created during the shredding process, and the cells are clustered into soft blocks (soft blocks have adjustable aspect ratio) based on their spatial locations in the placement. A fixed outline floorplanner is then used to place both the blocks and the soft blocks without overlaps. Finally, blocks are fixed into place and detailed placers locate positions for the cells of each soft block.

A new (as of 2006) simple approach that has demonstrated strong empirical results treats cells and blocks transparently in min-cut placement [67]. That is, the min-cut placer does not distinguish between cells and blocks. The output of min-cut placement in this case typically contains overlap and thus it is necessary to use a legalizer.

5.4.2 WHITESPACE DISTRIBUTION

Whitespace or *freespace* is the percentage of placement sites not occupied by cells and blocks. Whitespace enlarges the core layout area more than necessary for placement, in order to provide larger routing area. Placement algorithms can allocate whitespace to improve performance in a number of ways including congestion reduction, overlap minimization, and timing improvement.

One of the first approaches sought whitespace allocation in a hierarchical uniform fashion to improve smooth operation of the min-cut partitioner [20,31]. With careful whitespace allocation, bins would have enough whitespace to avoid cell overlaps resulting from unbalanced partitioning. Such overlaps typically occur in min-cut placers due to lack of whitespace, variations in cell dimensions, as well as the choice of cut sequences [64].

Instead of allocating whitespace uniformly, it can be allocated according to congestion to improve the final routability of the placement. In a two-step approach [68], whitespace is first allocated to each placement row proportional to its degree of congestion, with the least congested row receiving a fixed minimum amount of whitespace. In the second step, rows are divided into bins, and bins are assigned whitespace proportional to their congestion, where a bin can be assigned zero whitespace if it is not congested. More recently, (as of 2006) Li et al. [69] improved routability by simultaneously migrating cells connected to nets responsible for the overflow over routing demand, as well as by redistributing whitespace, allocating more whitespace to regions with a deficit in routing supplies.

When large amounts of whitespace is present, it can also be allocated to improve timing [70]. By observing that analytical placers better capture the global placement view than min-cut placers, it is possible to utilize analytical solvers in distributing the whitespace by determining the balance constraints for each partitioning step. Empirical results show that such methods lead to better timing results. In addition, whitespace can be compacted by inserting disconnected *free cells*, only to remove them after placement [71].

5.4.3 PLACEMENT BENCHMARKING

Placement suboptimality quantification is concerned with estimating the performance gap between reported results, for example, wirelength, of placers on a given benchmark and the optimal results of the benchmark. Since the placement problem is NP-hard, it is unlikely that there will exist optimal algorithms for arbitrary instances. Thus researchers attempt to indirectly estimate the performance gap. Existing approaches either use scaling [72] or synthetic constructive benchmarks [73,74] to quantify the suboptimality of existing placers.

Hagen et al. [72] scale a given placement instance by creating a number of identical instances and "patching" them into a single larger scaled instance. The performance of the placer is then compared against a value precalculated from the initial instance.

A recent (as of 2006) paper by Chang et al. [73] uses an overlooked construction method by Hagen et al. [72] to construct optimally a number of benchmarks (PEKO) with known optimal HPWL. Such optimal construction is possible if each k-pin net is constructed with the least possible HPWL. For example, a two-pin net takes a HPWL of 1 unit, a three-pin net takes a HPWL of 2 units, a four-pin net takes a HPWL of 2 units, and so forth. A sample construction is given in Figure 5.12. Empirical results on such artificially constructed benchmarks show that there exists a significant gap between the HPWL from the placer results and the optimal placement HPWL. However, these benchmarks are unrealistic since they only consider local hyperedges. To mitigate this drawback, Cong et al. [74] added global hyperedges to the PEKO benchmarks producing the PEKU benchmarks. The optimal placement of the PEKU benchmarks is unknown; however, it is possible to calculate an upper bound to their optimal placement. By using the optimal placement of the original PEKO benchmarks, an upper bound to the optimal HPWL of the added global hyperedges is readily calculated. Experimental results on the PEKU benchmarks show that the performance of available placers approaches the upper bounds as the percentage of global edges increases. It is not known whether the calculated upper bounds are tight or loose.

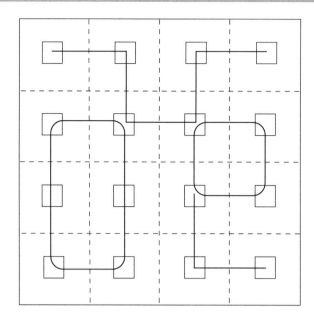

FIGURE 5.12 A benchmark constructed with known optimal HPWL. There are seven two-pin edges, one four-pin hyperedge, and one six-pin hyperedge with a total HPWL of 12.

Another recent (as of 2006) approach quantifies placer suboptimality using arbitrary netlists [75]. Given a netlist and its placement, a number of *zero-change* netlist transformations are applied to the netlist. These transformations do not alter the placement HPWL and can only increase the unknown optimal HPWL value. Executing the placer on the new transformed netlist produces a new suboptimal HPWL result. The difference between the new and old HPWL results is a lower bound on the suboptimality gap.

5.4.4 OTHER TRENDS AND PRACTICAL CONCERNS

In addition to the previous trends, a number of practical issues arise in modern day designs that should be handled by placers. These include:

- Flat placement of 10 million components in mixed-mode designs. Modern designs typically require a number of placement iterations to achieve timing closure. These iterations routinely consume enormous amounts of time—a number of days—till completion. Thus it is necessary to have fast and efficient placement algorithms. Furthermore, components of modern designs include movable cells and movable and fixed blocks. All these need to be smoothly and efficiently handled by the placer.
- Given the ubiquitous presence of mobile devices, temperature- and power-driven placement where the objective is to reduce the total power consumption as well as to smooth temperature differences between spatially proximate regions, is getting increasingly important.
- Placement-driven synthesis, where placement is used to derive logic synthesis optimizations [76–78].

5.5 ACADEMIC AND INDUSTRIAL PLACERS

A number of competitive academic and industrial placers exist to satisfy academic research and commercial needs. We now give a brief overview, as of 2006, of the main academic and industrial tools. The algorithmic techniques used by these placers have been already covered in Section 5.2. Academic placers include:

- *APlace* [11]: An analytical placer that is algorithmically similar to the Physical Compiler from Synopsys. It uses a nonlinear wirelength formulation (Equation 5.5) while penalizing overlaps in a "remove while solving" fashion.
- *BonnPlacer* [7]: An analytical placer that uses a quadratic formulation in a top-down framework. It also uses repartitioning to improve its results.
- *Capo* [20]: A top-down min-cut placer that is based on recursive bisection.
- *Dragon* [21]: A top-down min-cut placer that is based on recursive quadrisection. The placer also uses simulated annealing to improve its results.
- *FengShui* [22]: A top-down min-cut placer that is based on recursive bisection.
- *Kraftwerk* [8]: An analytical placer that uses repelling forces in a "remove while solving" fashion.
- *mPL* [55]: An analytical placer that uses a nonlinear wirelength formulation, and uses clustering in a multilevel framework as a means to reduce instance sizes.
- *Timberwolf* [24]: A simulated annealing-based placer.

While details of academic placers are readily available in academic publication, published data on industrial placers are typically less detailed.

- *IBM CPlace* [70]: An analytical placer that uses the quadratic formulation in a top-down framework. It also distributes whitespace to improve timing and wirelength.
- *Cadence QPlace*: It is believed that QPlace uses the quadratic formulation in a top-down framework.
- *Synopsys Physical Compiler* [12]: Similar to APlace.
- *InternetCAD iTools*: A commercial package based on Timberwolf.

5.6 CONCLUSIONS

Placement is—and will likely remain—one of the essential steps in the physical design part of any integrated circuit's life. In this chapter, we have surveyed the main algorithmic approaches to solve the placement problem. We have also explained how different placement objectives can be handled within those various approaches. Detailed placement as well as placement trends have also been covered. For the future, it seems that the core algorithmic approaches are likely to stay the same; however, they have to be extended to handle smoothly and efficiently multi-million gate mixed-size designs and to achieve timing closure in fewer turn-around cycles.

REFERENCES

1. V.V. Vazirani, *Approximation Algorithms*, 1st ed., Springer, Heidelberg, Germany, 2001.
2. C.E. Cheng, RISA: Accurate and efficient placement routability modeling, *Proceedings of IEEE International Conference on Computer Aided Design*, 1994, San Jose, CA, pp. 690–695.
3. L. Steinberg, The backboard wiring problem: A placement algorithm, *SIAM Rev.*, 3, 37–50, 1961.
4. K.M. Hall, An *r*-dimensional quadratic placement algorithm, *Manage. Sci.*, 17, 219–229, 1970.
5. C.K. Cheng and E.S. Kuh, Module placement based on resistive network optimization, *IEEE Trans. Comput. Aided Des. Integr. Circuits Syst.*, 4, 115–122, 1984.
6. R.S. Tsay, E.S. Kuh, and C.P. Hsu, PROUD: A sea-of-gates placement algorithm, *IEEE Des. Test Comput.*, 5, 44–56, 1988.
7. J. Vygen, Algorithms for large-scale flat placement, *Proceedings of ACM/IEEE Design Automation Conference*, 1997, Anaheim, CA, pp. 746–751.
8. H. Eisenmann and F.M. Johannes, Generic global placement and floorplanning, *Proceedings of ACM/IEEE Design Automation Conference*, 1998, San Francisco, CA, pp. 269–274.
9. B.X. Weis and D.A. Mlynski, A graph theoretical approach to the relative placement problem, *IEEE Trans. Circuits Syst.*, 35, 286–293, 1988.

10. A.A. Kennings and I.L. Markov, Analytical minimization of half-perimeter wirelength, *Proceedings of IEEE Asia and South Pacific Design Automation Conference*, 2000, Yokohama, Japan, pp. 179–184.

11. A. Kahng and Q. Wang, Implementation and extensibility of an analytical placer, *Proceedings of ACM/IEEE International Symposium on Physical Design*, 2004, Phoenix, AZ, pp. 18–25.

12. W. Naylor, Non-linear optimization system and method for wirelength and density within an automatic electronic circuit placer, US Patent 6282693, 2001.

13. J.M. Kleinhans, G. Sigl, F.M. Johannes, and K.J. Antreich, GORDIAN: VLSI placement by quadratic programming and slicing optimization, *IEEE Trans. Comput. Aided Des. Integr. Circuits Syst.*, 10, 356–365, 1991.

14. K. Vorwerk, A. Kennings, and A. Vannelli, Engineering details of a stable force-directed placer, *Proceedings of IEEE International Conference on Computer Aided Design*, 2004, San Jose, CA, pp. 573–580.

15. B. Kernighan and S. Lin, An efficient heuristic procedure for partitioning graphs, *Bell Syst. Tech. J.*, 49, 291–307, February 1970.

16. M.A. Breuer, Min-cut placement, *J. Des. Automat. Fault Tolerant Comput.*, 1, 343–362, 1977.

17. C.M. Fiduccia and R.M. Mattheyses, A linear-time heuristic for improving network partitions, *Proceedings of ACM/IEEE Design Automation Conference*, 1982, Las Vegas, NV, pp. 175–181.

18. A.E. Dunlop and B.W. Kernighan, A procedure for placement of standard-cell VLSI circuits, *IEEE Trans. Comput. Aided Des. Integr. Circuits Syst.*, 4, 92–98, 1985.

19. G. Karypis, R. Aggarwal, V. Kumar, and S. Shekhar, Multilevel hypergraph partitioning: Application in VLSI domain, *Proceedings of ACM/IEEE Design Automation Conference*, 1997, Anaheim, CA, pp. 526–529.

20. A.E. Caldwell, A.B. Kahng, and I.L. Markov, Can recursive bisection alone produce routable placements? *Proceedings of ACM/IEEE Design Automation Conference*, 2000, Los Angeles, CA, pp. 477–482.

21. M. Wang, X. Yang, and M. Sarrafzadeh, DRAGON2000: Standard-cell placement tool for large industry circuits, *Proceedings of IEEE International Conference on Computer Aided Design*, 2001, San Jose, CA, pp. 260–263.

22. M. Yildiz and P. Madden, Global objectives for standard-cell placement, *Proceedings of IEEE Great Lakes Symposium on VLSI*, 2001, West Lafayette, IN, pp. 68–72.

23. S. Kirkpatrick, C.D. Gelatt, Jr., and M.P. Vecchi, Optimization by simulated annealing, *Science*, 220, 671–680, 1983.

24. C. Sechen and A. Sangiovanni-Vincentelli, TimberWolf3.2: A new standard cell placement and global routing package, *Proceedings of ACM/IEEE Design Automation Conference*, 1986, Las Vegas, NV, pp. 432–439.

25. C. Sechen and K.W. Lee, An improved simulated annealing algorithm for row-based placement, *Proceedings of IEEE International Conference on Computer Aided Design*, 1987, Santa Clara, CA, pp. 478–481.

26. W.-J. Sun and C. Sechen, Efficient and effective placement for very large circuits, *IEEE Trans. Comput. Aided Des. Integr. Circuits Syst.*, 14, 349–359, 1995.

27. S. Sahni and T. Gonzalez, P-complete approximation problems, *J. ACM*, 23, 555–565, 1976.

28. P.R. Suaris and G. Kedem, A quadrisection-based combined place and route scheme for standard cells, *IEEE Trans. Comput. Aided Des. Integr. Circuits Syst.*, 8, 234–244, 1989.

29. A. Caldwell, A. Kahng, and I. Markov, End-case placers for standard-cell layout, *Proceedings of ACM/IEEE International Symposium on Physical Design*, 1999, Monterey, CA, pp. 90–96.

30. C.J. Alpert, J.H. Huang, and A.B. Kahng, Multilevel circuit partitioning, *Proceedings of ACM/IEEE Design Automation Conference*, 1997, Anaheim, CA, pp. 530–533.

31. A. Caldwell, I. Markov, and A. Kahng, Hierarchical whitespace allocation in top-down placement, *IEEE Trans. Comput. Aided Des. Integr. Circuits Syst.*, 22, 1550–1556, 2003.

32. M.C. Yildiz and P.H. Madden, Improved cut sequences for partitioning based placement, *Proceedings of ACM/IEEE Design Automation Conference*, 2001, Las Vegas, NV, pp. 776–779.

33. G. Karypis and V. Kumar, Multilevel k-way hypergraph partitioning, *Proceedings of ACM/IEEE Design Automation Conference*, 1999, New Orleans, LA, pp. 343–348.

34. A. Agnihotri, M. Yildiz, A. Khatkhate, A. Mathur, S. Ono, and P. Madden, Fractional cut: Improved recursive bisection placement, *Proceedings of IEEE International Conference on Computer Aided Design*, 2003, San Jose, CA.

35. A.E. Caldwell, A.B. Kahng, and I.L. Markov, Optimal partitioners and end-case placers for standard-cell layout, *IEEE Trans. Comput. Aided Des. Integr. Circuits Syst.*, 19, 1304–1313, 2000.

36. D.J.-H. Huang and A.B. Kahng, Partitioning-based standard-cell global placement with an exact objective, *Proceedings of ACM/IEEE International Symposium on Physical Design*, 1997, Napa, CA, pp. 18–25.

37. A.B. Kahng and S. Reda, Placement feedback: A concept and method for better min-cut placement, *Proceedings of ACM/IEEE Design Automation Conference*, 2004, San Diego, CA, pp. 357–362.

38. K. Zhong and S. Dutt, Effective partition-driven placement with simultaneous level processing and global net views, *Proceedings of IEEE International Conference on Computer Aided Design*, 2000, San Jose, CA, pp. 171–176.

39. G. Sigl, K. Doll, and F.M. Johannes, Analytical placement: A linear or a quadratic objective function? *Proceedings of ACM/IEEE Design Automation Conference*, 1991, San Francisco, CA, pp. 427–431.

40. N. Viswanathan and C. Chu, FastPlace: Efficient analytical placement using cell shifting, iterative local refinement and a hybrid net model, *Proceedings of ACM/IEEE International Symposium on Physical Design*, 2004, Phoenix, AZ, pp. 26–33.

41. D. Wismer and R. Chattergy, *Introduction to Nonlinear Optimization: A Problem Solving Approach*, 1st ed., Elsevier, Amsterdam, the Netherlands, 1978.

42. A. Srinivasan, K. Chaudhary, and E.S. Kuh, RITUAL: A performance driven placement algorithm for small cell ICs, *Proceedings of IEEE International Conference on Computer Aided Design*, 1991, Santa Clara, CA, pp. 48–51.

43. B. Riess and G. Ettelt, SPEED: Fast and efficient timing driven placement, *IEEE International Symposium on Circuits and Systems*, 1995, Seattle, WA, pp. 377–380.

44. B. Halpin, C.Y.R. Chen, and N. Sehgal, Timing driven placement using physical net constraints, *Proceedings of ACM/IEEE Design Automation Conference*, 2001, Las Vegas, NV, pp. 780–783.

45. M. Marek-Sadowska and S. Lin, Timing driven placement, *Proceedings of IEEE International Conference on Computer Aided Design*, 1989, Santa Clara, CA, pp. 94–97.

46. S.L. Ou and M. Pedram, Timing-driven placement based on partitioning with dynamic cut-net control, *Proceedings of ACM/IEEE Design Automation Conference*, 2000, Los Angeles, CA, pp. 472–476.

47. A.B. Kahng, S. Mantik, and I.L. Markov, Min-max placement for large-scale timing optimization, *Proceedings of ACM/IEEE International Symposium on Physical Design*, 2002, Del Mar, CA, pp. 143–148.

48. U. Brenner and A. Rohe, An effective congestion driven placement framework, *Proceedings of ACM/IEEE International Symposium on Physical Design*, 2002, Del Mar, CA, pp. 6–11.

49. X. Yang, R. Kastner, and M. Sarrafzadeh, Congestion reduction during placement based on integer programming, *Proceedings of IEEE International Conference on Computer Aided Design*, 2001, San Jose, CA, pp. 573–576.

50. P.N. Parakh, R.B. Brown, and K.A. Sakallah, Congestion driven quadratic placement, *Proceedings of ACM/IEEE Design Automation Conference*, 1998, San Francisco, CA, pp. 275–278.

51. D.M. Schuler and E.G. Ulrich, Clustering and linear placement, *Proceedings of ACM/IEEE Design Automation Conference*, 1972, Dallas, TX, pp. 50–56.

52. J. Cong and S.K. Lim, Edge separability-based circuit clustering with application to multilevel circuit partitioning, *IEEE Trans. Comput. Aided Des. Integr. Circuits Syst.*, 23, 346–357, 2004.

53. B. Hu and M.M. Sadowska, Fine granularity clustering-based placement, *IEEE Trans. Comput. Aided Des. Integr. Circuits Syst.*, 23, 527–536, 2004.

54. C. Alpert, A. Kahng, G.-J. Nam, S. Reda, and P. Villarrubia, A semi-persistent clustering technique for VLSI circuit placement, *Proceedings of ACM/IEEE International Symposium on Physical Design*, 2005, San Francisco, CA, pp. 200–207.

55. C.-C. Chang, J. Cong, D. Pan, and X. Yuan, Multilevel global placement with congestion control, *IEEE Trans. Comput. Aided Des. Integr. Circuits Syst.*, 22, 395–409, 2003.

56. K. Doll, F. Johannes, and K. Antreich, Iterative placement improvement by network flow methods, *IEEE Trans. Comput. Aided Des. Integr. Circuits Syst.*, 13, 1189–1200, 1994.

57. S.W. Hur and J. Lillis, Mongrel: Hybrid techniques for standard cell placement, *Proceedings of IEEE International Conference on Computer Aided Design*, 2000, San Jose, CA, pp. 165–170.

58. J. Vygen, Algorithms for detailed placement of standard cells, *Design, Automation and Test in Europe*, 1998, Paris, France, pp. 321–324.

59. U. Brenner, A. Pauli, and J. Vygen, Almost optimum placement legalization by minimum cost flow and dynamic programming, *Proceedings of ACM/IEEE International Symposium on Physical Design*, 2004, Phoenix, AZ, pp. 2–9.

60. D. Hill, Method and system for high speed detailed placement of cells within an integrated circuit design, US Patent 6370673, 2001.

61. S. Akers, On the use of the linear assignment algorithm in module placement, *Proceedings of ACM/IEEE Design Automation Conference*, 1981, Nashville, TN, pp. 13–144.

62. A.B. Kahng, P. Tucker, and A. Zelikovsky, Optimization of linear placements for wirelength minimization with free sites, *Proceedings of IEEE Asia and South Pacific Design Automation Conference*, 1999, Hong Kong, pp. 241–244.

63. U. Brenner and J. Vygen, Faster optimal single-row placement with fixed ordering, *Design, Automation and Test in Europe*, 2000, Paris, France, pp. 117–122.

64. A.B. Kahng, I. Markov, and S. Reda, On legalization of row-based placements, *Proceedings of IEEE Great Lakes Symposium on VLSI*, 2004, Boston, MA, pp. 214–219.

65. H. Yu, X. Hing, and Y. Cai, MMP: A novel placement algorithm for combined macro block and standard cell layout design, *Proceedings of IEEE Asia and South Pacific Design Automation Conference*, 2000, Yokohama, Japan, pp. 271–276.

66. S. Adya and I. Markov, Consistent placement of macro-blocks using floorplanning and standard cell placement, *Proceedings of ACM/IEEE International Symposium on Physical Design*, 2002, San Diego, CA, pp. 12–17.

67. A. Khatkhate, C. Li, A.R. Agnihotri, M.C. Yildiz, S. Ono, C.-K. Koh, and P.H. Madden, Recursive bisection based mixed block placement, *Proceedings of ACM/IEEE International Symposium on Physical Design*, 2004, Phoenix, AZ, pp. 84–89.

68. X. Yan, B.K. Choi, and M. Sarrafzadeh, Routability driven whitespace allocation for fixed-die standard-cell placement, *Proceedings of ACM/IEEE International Symposium on Physical Design*, 2002, San Diego, CA, pp. 42–47.

69. C. Li, M. Xie, C.-K. Koh, J. Cong, and P.H. Madden, Routability-driven placement and whitespace allocation, *Proceedings of IEEE International Conference on Computer Aided Design*, 2004, San Jose, CA, pp. 394–401.

70. C. Alpert, G.-J. Nam, and P. Villarrubia, Free space management for cut-based placement, *Proceedings of IEEE International Conference on Computer Aided Design*, 2002, San Jose, CA, pp. 746–751.

71. S. Adya, I. Markov, and P. Villarrubia, On whitespace and stability in mixed-size placement, *Proceedings of IEEE International Conference on Computer Aided Design*, 2003, San Jose, CA, pp. 311–318.

72. L.W. Hagen, D.J.H. Huang, and A.B. Kahng, Quantified suboptimality of VLSI layout heuristics, *Proceedings of ACM/IEEE Design Automation Conference*, 1995, San Francisco, CA, pp. 216–221.

73. C. Chang, J. Cong, and M. Xie, Optimality and scalability study of existing placement algorithms, *Proceedings of IEEE Asia and South Pacific Design Automation Conference*, 2003, Kitakyushu, Japan, pp. 621–627.

74. J. Cong, M. Romesis, and M. Xie, Optimality and scalability study of partitioning and placement algorithms, *Proceedings of ACM/IEEE International Symposium on Physical Design*, 2003, Monterey, CA, pp. 88–94.

75. A.B. Kahng and S. Reda, Evaluation of placer suboptimality via zero-change netlist transformations, *Proceedings of ACM/IEEE International Symposium on Physical Design*, 2005, San Francisco, CA, pp. 208–215.

76. J. Lou, W. Chen, and M. Pedram, Concurrent logic restructuring and placement for timing closure, *Proceedings of IEEE International Conference on Computer Aided Design*, 1999, San Jose, CA, pp. 31–36.

77. W. Gosti, S.P. Khatri, and A.L. Sangiovanni-Vincentelli, Addressing the timing closure problem by intgerating logic optimization and placement, *Proceedings of IEEE International Conference on Computer Aided Design*, 2001, San Jose, CA, pp. 224–231.

78. M. Hrkić, J. Lillis, and G. Beraudo, An approach to placement-coupled logic replication, *Proceedings of ACM/IEEE Design Automation Conference*, 2004, San Diego, CA, pp. 711–716.

Static Timing Analysis

6

Jordi Cortadella and Sachin S. Sapatnekar

CONTENTS

6.1 INTRODUCTION

High-performance circuits have traditionally been characterized by the clock frequency at which they operate. Gauging the ability of a circuit to operate at the specified speed requires timing verification capabilities that measure its delay at numerous steps during the design process. Moreover, delay calculation must be incorporated into the inner loop of timing optimizers at various phases of design, such as synthesis, layout (placement and routing), and in-place optimizations performed late in the design cycle. While both types of timing measurements can theoretically be performed using rigorous circuit simulations, such an approach is liable to be too slow to be practical. Static timing analysis (STA) plays a vital role in facilitating the fast, and reasonably accurate, measurement of circuit timing by finding the minimum or maximum delay over all input logic patterns. The speedup is due to the use of simplified delay models and the limited ability of STA to consider the effects of logical interactions between signals. By construction, delay estimates from STA are built to be pessimistic, thus providing guarantees that an STA sign-off can deliver a working circuit.

STA has become a mainstay of design over the past few decades. This chapter will first overview a set of basic techniques for modeling gate and circuit delays and then define the constraints that must be satisfied to ensure timing correctness. Finally, two classes of STA techniques that comprehend the effects of variation in process and environmental parameters will be presented: corner-based methods and statistical STA (SSTA) techniques.

6.2 REPRESENTATION OF COMBINATIONAL AND SEQUENTIAL CIRCUITS

A combinational logic circuit may be represented as a timing graph $G = (V, E)$, where the elements of V, the vertex set, are the inputs and outputs of the logic gates in the circuit. The vertices are connected by two types of edges: one set of edges connects each input of a gate to its output, which represents the maximum delay paths from the input pin to the output pin, while another set of edges connects the output of each gate to the inputs of its fanout gates and corresponds to the interconnect delays. A simple logic circuit and its corresponding graph are illustrated in Figure 6.1a and b, respectively. We refer to fanin-free nodes as primary inputs and nodes whose values are observed (which may or may not be fanout free) as primary outputs.

A simple transform that converts the graph into one that has a single source s and a single sink t is often useful. If all primary inputs are connected to flip-flops (*FFs*) and transition at the same time, edges are added from the node s to each primary input and from each primary output to the sink t. The case where the primary inputs arrive at different times $a_i \geq 0$ can be handled with a minor transformation to this graph,* adding a dummy node with a delay of a_i along each edge from s to the primary input i.

---

* If some arrival times are negative, the time variable can be shifted to ensure that $a_i \geq 0$ at each primary input.

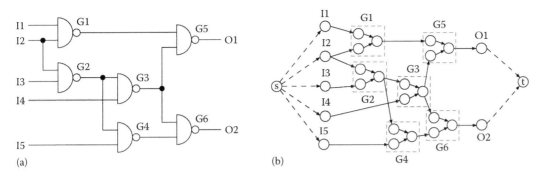

FIGURE 6.1   (a) An example combinational circuit and (b) its timing graph. (From Sapatnekar, S.S., *Timing*, Kluwer Academic Publishers, Boston, MA, 2004.)

This construction generally results in a combinational circuit that is represented as a directed acyclic graph (DAG). This is because mainstream combinational circuits do not have any cycles (though some implementations of combinational circuits are not acyclic). Many timing analyzers handle combinational feedback loops (which translate to cycles in the timing graph) by breaking them either in an *ad hoc* manner or according to a heuristic so that the circuit to be analyzed may be represented as a DAG.

A sequential circuit that consists both of combinational and sequential elements (*FFs* and latches) may be represented as a set of combinational blocks that lie between latches, and a timing graph may be constructed for each of these blocks. For any such block, the sequential elements or circuit inputs that fan out to a gate in the block constitute its primary inputs; similarly, the sequential elements or circuit outputs for which a fanin gate belongs to the block together represent its primary outputs. It is rather easy to find these blocks: to begin with, we construct a graph in which each vertex corresponds to a combinational element, an undirected edge is drawn between a combinational element and the combinational elements that it fans out to, and sequential elements are left unrepresented (this is substantially similar to the directed graph described earlier for a combinational circuit, except that it is undirected). The connected components of this graph correspond to the combinational blocks in the circuit.

The computation of the delay of such a combinational block is an important step in timing analysis. For an edge-triggered circuit, the signals at the primary inputs of such a block make a transition at exactly the same time. The clock period must then satisfy the constraint that it is no smaller than the maximum delay of any path through the logic block, plus the setup time for a *FF* at the primary output at the end of that path. Therefore, finding the maximum delay through a combinational block is a vital subtask. Alternatively, a combinational block that is to be treated as a black box must be characterized in terms of its delays from each input to each output; this problem arises, for example, when the analysis involves hierarchically defined blocks with such timing abstractions. For level-clocked circuits, the timing relations are more complex but are nevertheless based on the machinery developed in this chapter [31].

6.3 STAGE DELAY MODELS

Consider a logic stage, consisting of a gate that drives an interconnect net to one or more sinks. The circuit graph contains an edge, also referred to as a timing arc, from each input pin to the gate output. The delays associated with each timing arc are separately characterized, typically under the single input switching case, where only one cell input is assumed to switch, and all other cell inputs are set so that they enable the maximum delay switching path. For example, during fall delay calculations for a NAND gate, all transistors in the pull-down other than the switching transistor are assumed to be fully on; while analyzing the rise delay, all transistors in the pull-up, except the switching transistor, are assumed to be fully off. For each timing arc, four characterizations are required to represent the output delay and transition time (slew), each computed for the rise and fall cases.

6.3.1 MODELING THE LOAD AS A LUMPED CAPACITANCE

For short interconnects, the wire resistance is negligible as compared to the impedance of the driving transistor(s), and it is reasonable to model the wire and its loads as a lumped capacitive load. The problem of determining the delay of the logic stage then reduces to that of finding the delay of the gate for a specific capacitive load. If the gate is a cell from a library, its delay is typically precharacterized under various loads, C_L, and input transition times (also known as slew), τ_{in}. Two classes of models may be employed:

Equation-based models: In this case, the delay may be characterized in the form of a closed-form equation. Some common closed-form equations are of the following forms:

$$(6.1) \qquad\qquad \alpha\tau_{in} + \beta C_L + \gamma\tau_{in}C_L + \delta$$

and

$$(6.2) \qquad\qquad (a_0 + a_1 C_L + \cdots + a_m C_L^m) \cdot (b_0 + b_1\tau_{in} + \cdots + b_n\tau_{in}^n)$$

where α, β, γ, and δ and the a_is and b_is correspond to fitted coefficients.

Lookup table models: The most commonly used models today employ table lookup, based on a paradigm that is referred to as the nonlinear delay model [4]. A typical use case for this model characterizes the delay of a cell as a 7 × 7 lookup table, with seven points each corresponding to the input slew and output load: this size is a reasonable trade-off between the accuracy of the model and simplicity, in terms of limiting the size and characterization effort. During STA, given a specific input slew and output load, the delay value can be obtained using interpolation.

6.3.2 MODELING THE LOAD AS A DISTRIBUTED RC STRUCTURE

6.3.2.1 IMPACT OF WIRE RESISTANCE

The lumped capacitance model for interconnects is only accurate when the driver resistance overwhelms the wire resistance; when the two are comparable, such a model could have significant errors. A class of approaches models the stage as a linear system and estimates the delay using linear system theory: one example of such a method is the Elmore delay model [31]. However, since the driver is quite nonlinear, such simplistic methods are far too inaccurate for use in STA.

A significant phenomenon that is ignored by linear system–based models is that of *resistive shielding*, which causes the delay at the driver output (referred to as the "driving point") to be equivalent to a situation where it drives a lumped load that is less than the total capacitance of the interconnect. In effect, the interconnect resistance shields a part of the total capacitance from the driving point.

6.3.2.2 CURRENT SOURCE MODELS

The concept of an *effective capacitance* captures the impact of resistive shielding by estimating the capacitance seen at the driving point [28]. This approach begins by setting the effective capacitance to the lumped capacitance and iteratively updates its value.

An enhancement of this approach is the widely used method based on current source models (CSMs) [2,12]. The idea is illustrated in Figure 6.2 for a two-input NAND gate. The CSM is a gate-level black box abstraction of a cell in a library, with the same input and output ports as the original cell. The output port p is replaced by a nonlinear voltage-controlled current source (VCCS), I_p, in parallel with a nonlinear capacitance, C_p. The VCCS model enables the CSM to be load independent and permits it to handle an arbitrary electrical waveform at its inputs. The CSM is characterized in terms of the value of I_p and the charge, Q_p, stored on the capacitor, C_p. The variables, I_p and Q_p, are functions of all input and

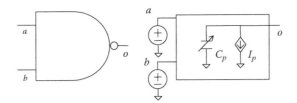

FIGURE 6.2 Example of a current source model: the output port is modeled as a nonlinear voltage-controlled current source dependent on all input port voltages, in parallel with a nonlinear capacitance.

output port voltages, V_i and V_o, respectively. This modeling framework can also be efficiently extended to capture dependencies on the body bias voltages, v_{bp} and v_{bn}, and the temperature T [14]. For the most commonly used version of this model, under the single input switching assumption, the current and charge functions are determined by characterizing the cell at various port voltages as follows:

(6.3)
$$I_p = F(V_i, V_o)$$

(6.4)
$$Q_p = G(V_i, V_o)$$

Here, the functions F and G may be stored as lookup tables. For a cell, I_p characterization involves DC simulations over multiple combinations of DC values of (V_i, V_o), while Q_p is characterized through a set of transient simulations [2].

To use the CSM, the interconnect tree driven by a gate is first reduced to an equivalent π model [27]. Next, a small circuit consisting of the CSM structure driving a π load is simulated to obtain the driving point waveform at the output of the gate. Finally, the full interconnect tree is simulated using the driving point waveform to determine the waveform and delay at the sinks of interest.

6.3.3 INTERCONNECT CROSS TALK EFFECTS ON TIMING

Coupling capacitance between adjacent wires is a significant factor that permits the injection of noise into a victim line from one or more adjacent aggressor lines. If this noise event occurs on the victim during a signal transition, it may increase the delay of the transition, as will be shown in the discussion in Figure 6.3.

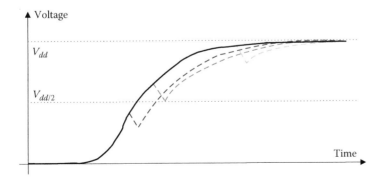

FIGURE 6.3 Three aggressor alignment scenarios to demonstrate that the timing of the noise event affects the delay of the signal.

The analysis of cross talk effects involves several factors:

- *Switching windows* for an aggressor or victim determine the times at which they may switch inject noise into a wire: if the switching window for a victim and a potential aggressor can be shown to be disjoint, the potential aggressor can never inject cross talk noise during switching, and therefore, cross talk is purely a noise margin issue. In other words, the impact of cross talk on delay can be ruled out.
- *Aggressor alignment* determines the worst-case switching time in each aggressor that can induce a maximal delay change in the victim: this involves determining the precise alignment of transition times that can induce the largest noise bump during the victim transition, thus causing the largest delay increase.

Figure 6.3 illustrates the impact of cross talk noise on signal delay. A noiseless waveform rises monotonically from 0 to V_{dd}, but if a noise bump is injected during the transition, the latest point at which the waveform reaches its 50% point (or in general, the voltage threshold used to measure delay) may be pushed to the right. The results of three different aggressor alignment scenarios are shown in the figure: in two of these, the noise effects are seen during the transition and affect the 50% delay point, while in the third, cross talk occurs when the signal is nearly at its final value, and this is treated as a noise issue on an otherwise stable logic value, rather than anything that impacts the transition delay as computed by a timing analyzer.

For a given set of overlaps between timing windows, the concept of Miller capacitances is often used to replace a coupling capacitance by a capacitor to ground. Specifically, if the capacitance between two lines is C_c, then the coupling capacitor is replaced by a capacitance from the victim wire to ground. Depending on whether the potential aggressor is silent, switching in the same direction, or switching in opposite direction during the timing window of the victim, the value of this grounded capacitor is C_c, 0, or $2C_c$, respectively (some variants of this method use the values of C_c, $-C_c$, and $3C_c$, respectively [10]).

For lumped loads, the entire coupling capacitance is replaced by the Miller capacitance, and the gate delay models in Section 6.3.1 are used for delay evaluation. For distributed loads, the coupling capacitances at internal nodes of the RC wire load are replaced by Miller capacitances. Techniques such as those in Section 6.3.2 can then be used to evaluate the delay of the RC load network [19].

Note that the determination of switching windows is nontrivial since the delays of a line and each of its neighbors are interdependent. Breaking this dependency cycle requires an iterative procedure [30]: it can be shown that this solution corresponds to finding the fixpoint of a lattice [39]. For a given allowable timing window, the worst-case aggressor alignment is used to determine the delay of the logic stage.

6.4 TIMING ANALYSIS FOR COMBINATIONAL CIRCUITS

6.4.1 DELAY CALCULATION FOR A COMBINATIONAL LOGIC BLOCK

In this section, we will present techniques that are used for the STA of digital combinational circuits. The word "static" alludes to the fact that this timing analysis is carried out in an input-independent manner and purports to find the worst-case delay of the circuit over all possible input combinations. The computational efficiency (linear in the number of edges in the graph) of this approach has resulted in its widespread use, even though it has some limitations associated with false paths, which will be described in Section 6.4.5.

A method that is commonly referred to as Program Evaluation and Review Technique (PERT) is popularly used in STA. In fact, this is a misnomer, and the so-called PERT method discussed in most of the literature on timing analysis refers to the critical path method (CPM) that is widely used in project management.

Before proceeding, it is worth pointing out that while CPM-based methods are dominantly in use today, other methods for traversing circuit graphs have been used by various timing analyzers. For example, depth-first techniques have been presented in [18].

```
Algorithm CRITICAL_PATH_METHOD
Q = Ø; /* Initialize queue Q */
for all vertices i ∈ V
    n_visited_inputs [i] = 0;
/* Add a vertex to the tail of Q if all inputs are ready */
for all primary inputs i    /* Fanout gates of i */
    for all vertices j such that (i → j) ∈ E
        if (++n_visited_inputs [j] == n_inputs [j]) addQ(j,Q);
while (Q ≠ Ø)
    g = top(Q);
    remove (g,Q);
    compute_delay [g];
    for all vertices k such that (g → k) ∈ E /* Fanout gates of g */
        if (++n_visited_inputs [k] == n_inputs[k]) addQ(k,Q);
```

FIGURE 6.4 Pseudocode for the critical path method.

The algorithm, applied to a timing graph $G = (V, E)$, can be summarized by the pseudocode shown in Figure 6.4. The procedure is best illustrated by means of a simple example. Consider the circuit in Figure 6.5, which shows an interconnection of blocks. Each of these blocks could be as simple as a logic gate or could be a more complex combinational block and is characterized by the delay from each input pin to each output pin. For simplicity, this example will assume that for each block, the delay from any input to the output is identical. Moreover, we will assume that each block is an inverting logic gate such as a NAND or a NOR, as shown by the *bubble* at the output. The two numbers, d_r/d_f, inside each gate represent the delay corresponding to the delay of the output rising transition, d_r, and that of the output fall transition, d_f, respectively. We assume that all primary inputs are available at time zero so that the numbers "0/0" against each primary input represent the worst-case rise and fall arrival times, respectively, at each of these nodes. The CPM proceeds from the primary inputs to the primary outputs in topological order, computing the worst-case rise and fall arrival times at each intermediate node and eventually at the output(s) of the circuit.

A block is said to be *ready for processing* when the signal arrival time information is available for all of its inputs; in other words, when the number of processed inputs of a gate g, n_visited_inputs[g], equals the number of inputs of the gate, n_inputs[g]. Notationally, we will

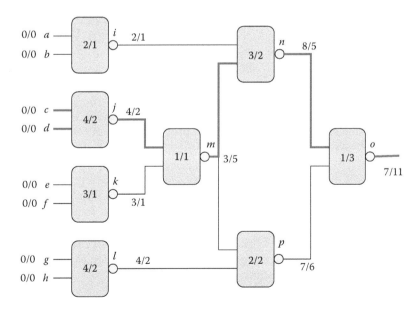

FIGURE 6.5 An example illustrating the application of the critical path method on a circuit with inverting gates. The numbers within the gates correspond to the rise/fall delay of the block, and the italicized bold numbers at each block output represent the rise/fall arrival times at that point. The primary inputs are assumed to have arrival times of zero, as shown. (From Sapatnekar, S.S., *Timing*, Kluwer Academic Publishers, Boston, MA, 2004.)

refer to each block by the symbol of its output node. Initially, since the signal arrival times are known only at the primary inputs, only those blocks that are fed solely by primary inputs are ready for processing. In the example, these correspond to the gates *i*, *j*, *k*, and *l*. These are placed in a queue Q using the function addQ and are processed in the order in which they appear in the queue.

In the iterative process, the block at the head of the queue Q is taken off the queue and scheduled for processing. Each processing step consists of the following:

- Finding the latest arriving input to the block that triggers the output transition (this involves finding the maximum of all worst-case arrival times of inputs to the block) and then adding the delay of the block to the latest arriving input time to obtain the worst-case arrival time at the output. This is represented by compute_delay in the pseudocode.
- Checking all of the blocks that the current block fans out to, to determine whether they are ready for processing. If so, the block is added to the tail of the queue using addQ.

The iterations end when the queue is empty. In the example, the algorithm is executed by processing the gates in the sequence *i*, *j*, *k*, *l*, *m*, *n*, *p*, *o*, and the worst-case delay for the entire block is found to be max(7, 11) = 11 units, as illustrated in the figure.

6.4.2 CRITICAL PATHS, REQUIRED TIMES, AND SLACKS

The *critical path*, defined as the path between an input and an output with the maximum delay as determined by STA, can now easily be found by using a traceback method. We begin with the block whose output is the primary output with the latest arrival time: this is the last block on the critical path. Next, the latest arriving input to this block is identified, and the block that causes this transition is the preceding block on the critical path. The process is repeated recursively until a primary input is reached.

In the example, we begin with Gate *o* at the output, whose falling transition corresponds to the maximum delay. This transition is caused by the rising transition at the output of gate *n*, which must therefore precede *o* on the critical path. Similarly, the transition at *n* is affected by the falling transition at the output of *m* and so on. By continuing this process, the critical path from the input to the output is identified as being caused by a falling transition at either input *c* or input *d* and then progressing as follows: rising *j* → falling *m* → rising *n* → falling *o*.

A useful concept is the notion of the *required time* at a node in the circuit. If a circuit is provided with a set of arrival time constraints at each primary output node, then on the completion of the CPM traversal, one may check whether those constraints are met or not. If they are not met, it is useful to know which parts of the circuit should be sped up and how, and if they are, it may be possible to save design resources by resynthesizing the circuit to slow it down. In either case, the required time at each node in the circuit provides useful information. At any node in the circuit, if the arrival time exceeds the required time, then the path that this node lies on must be sped up.

The computation of the required time proceeds as follows. At each primary output, the rise/fall required times are set according to the specifications provided to the circuit. Next, a backward topological traversal is carried out, processing each gate when the required times at all of its fanouts are known. In essence, this is equivalent to performing a CPM traversal with the directions of the edges in $G(V, E)$ reversed, the block delays negated, the roles of the primary inputs and outputs exchanged, and the required time at the primary outputs of the original graph (now considered primary inputs) set to the timing specification.

The *slack* associated with each node in the circuit graph is then simply computed as the difference between the required time and the arrival time. A positive slack *s* at a node implies that the arrival time at that node may be increased by *s* without affecting the overall delay of the circuit. Such a delay increase will only eat into the slack and may provide the potential to free up excessive design resources used to build the circuit.

For example, in Figure 6.5, if the required time at output *o* is 12 units, then the slack for all signals on the critical path is 1; for an off-critical path signal such as *p*, the slack can be computed as 2 units. If the required time at o is changed to 9 units, then the slack on the critical path is –2 and the slack at p is –1.

6.4.3 EXTENSIONS TO MORE COMPLEX CASES

For ease of exposition, the example in the previous section contained a number of simplifying assumptions. The CPM can work under more general problem formulations and a few of these that are commonly encountered are listed here.

Nonzero arrival times at the primary inputs: If the combinational block is a part of a larger circuit, we may have nonzero rise/fall arrival times at the primary inputs. If so, the CPM traversal can be carried out by simply using these values instead of zero as the arrival times at the primary inputs.

Minimum delay calculations: When the gate delays are fixed, the method described earlier can easily be adapted to find the minimum, instead of the maximum delay from any input to any output. The only changes in the procedure involve the manner in which an individual block is processed—the earliest arrival time over all inputs is now added to the delay of the block to find the earliest arrival time at its output.

Minmax delay calculations: If the gate delay is specified in terms of an interval, $[d_{\min}, d_{\max}]$, then the minimum and maximum arrival time intervals can be propagated in a similar manner; again, these values may be maintained separately for the rising and falling output transitions. The values of $d_{\min}$ and $d_{\max}$ can be computed on the fly while processing a gate.

Generalized block delay models: If the delay from input pin p to output pin q of a blocks is d_{pq} and the values of d_{pq} are not all uniform for a block, then the arrival time at the output q can be computed as $\max_{\text{all inputs } p}(t_p + d_{pq})$, where t_p is the arrival time (of the appropriate polarity) at input p. Note that if $d_{pq} = d$, for every p, then this simply reduces to the expression, $\max_{\text{all inputs } p}(t_p) + d$, which was used in the example in Figure 6.5.

Incorporating input signal transition times: In the example, the delay of each of the individual gates was prespecified as an input to the problem. However, in practice, the delay of a logic gate depends on factors such as the input transition times (i.e., the 10%–90% rise or fall times), which are unknown until the predecessor gate has been processed. This problem is overcome by propagating not only the worst-case rise and fall arrival times but also the input signal transition times corresponding to those arrival times.

Note that this fits in well with the fact that the delay of a gate is only required at the time when it is marked as ready for processing during the CPM traversal. At this time, the input signal arrival times as well as the corresponding signal transition times are available, which implies that the delay of the gate may be computed on the fly, just before it is needed. More complex techniques for handling slope effects are presented in [6].

6.4.4 INCREMENTAL TIMING ANALYSIS

During the process of circuit design or optimization, it is frequently the case that a designer or a CAD tool may alter a small part of a circuit and consider the effects of this change on the timing behavior of the circuit. Since the circuit remains largely unchanged, performing a full STA entails needless computation, since many of the arrival times remain unchanged. The notion of incremental timing analysis provides a way of propagating the changes in circuit timing caused by this alteration only to those parts of the circuit that require it. Generally, incremental analysis is cheaper than carrying out a complete timing analysis.

An *event-driven* propagation scheme may be used: an event is said to occur when the timing information at the input to a gate is changed, and the gate is then processed so that the event is propagated to its outputs. Unlike CPM, in an event-driven approach, it is possible that a gate may be processed more than once. However, when only a small change is made to the circuit, and a small fraction of the gates have their arrival times altered, such a method is highly efficient.

Incremental approaches have been used in timing analysis for decades, and further details on incremental timing analysis are available in [1] and in Chapter 1, which addresses design flows.

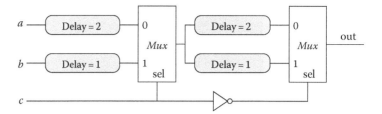

FIGURE 6.6   An illustration of a circuit with false paths.

6.4.5 FALSE PATHS

The true delay of a combinational circuit corresponds to the worst case over all possible logic values that may be applied at the primary inputs. Given that each input can take on one of four values (a steady 0, a steady 1, a $0 \rightarrow 1$ transition, and a $1 \rightarrow 0$ transition), the number of possible combinations for a circuit with m inputs is 4^m, which shows an exponential trend. However, it can be verified that the CPM for finding the delay of a combinational circuit can be completed in $O(|V| + |E|)$ time for a timing graph $G = (V, E)$, and in practice, this computation is considerably cheaper on large circuits than a full enumeration. The difference arises because of a specific assumption made by CPM, namely, that the logic function implemented by a gate is inconsequential and only its delay is relevant.

This may result in estimation errors that are pessimistic. As an example, consider the circuit shown in Figure 6.6, with three inputs, a, b, and c, and one output, out. Assume, for simplicity, that the multiplexer and inverter have zero delays and that the four blocks whose delays are shown are purely combinational. It can be easily verified that the worst-case delay for this circuit computed using the CPM is 4 units. However, by enumerating both possible logic values for the multiplexer, namely, $c = 0$ and $c = 1$, it can be seen that the delay in both cases is 3 units, implying that the circuit delay is 3 units. The reason for this discrepancy is simple—the path with a delay of 4 units can never be sensitized because of the restrictions placed by the Boolean dependencies between the inputs.

While many approaches to false path analysis have been proposed, most are rather too complex to be applied in practice. The identification of false paths includes numerous subtleties. Some paths may not be statically sensitizable under zero delay assumptions but may be dynamically sensitizable. For instance, if the inverter has a delay of 3 units, then the path of delay 4 units is indeed dynamically sensitizable. Various definitions have been proposed in the literature, and an excellent survey, written at a time when research on false paths was at its most active state, is presented in [24].

However, most practical approaches use some designer input and case enumeration, as in the case of Figure 6.6, where the cases of $c = 0$ and $c = 1$ were enumerated. Approaches for pruning user-specified false paths from timing graphs have been presented in [3,5,13,15]. Some approaches have attempted to use satisfiability-based methods for detecting false paths, but their overhead and complexity are generally too large for use within STA and simpler methods are found to be more effective.

6.5 TIMING ANALYSIS FOR SEQUENTIAL CIRCUITS

6.5.1 CLOCKING DISCIPLINES

A general sequential circuit is a network of computational nodes (gates) and memory elements (registers). The computational nodes may be conceptualized as being clustered together in an acyclic network of gates that forms a combinational logic circuit. A cyclic path in the direction of signal propagation is permitted in the sequential circuit only if it contains at least one register. In general, it is possible to represent any sequential circuit in terms of the schematic shown in Figure 6.7, which has I inputs, O outputs, and M registers. The registers' outputs feed into the combinational logic that, in turn, feeds the register inputs. Thus, the combinational logic has $I + M$ inputs and $O + M$ outputs.

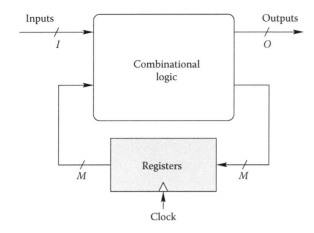

FIGURE 6.7   A general sequential circuit.

The functionality of registers plays a key role in determining the behavior of any sequential circuit, and there are several choices of register type that are available to a designer. The behavior of each register is controlled by the clock signal, and depending on the state of the clock, the data at the register input are either isolated from its output or transmitted to the output. The types of registers in a synchronous system are differentiated by the manner in which they use the clock to transmit data from the input to the output.

Level-clocked latches: These are commonly referred to merely as latches, and these permit data to be transmitted from the input to the output whenever the clock is high. During this time, the latch is said to be *transparent*.

Edge-triggered FFs: These are commonly called "flip-flops" and they use the clock edge to determine when the data are transmitted to the output. In a positive [negative] edge-triggered *FF*, data are transmitted from the input to the output when the clock makes a transition from 0 to 1 [1 to 0]. *FF*s can typically be constructed by connecting two level-clocked latches in a master–slave fashion.

A circuit consisting only of level-clocked latches and gates is referred to as a level-clocked circuit, and a circuit composed of edge-triggered *FF*s and gates is called an edge-triggered circuit. Since a level-clocked latch is transparent during the active period of the clock, the analysis and design of level-clocked circuits are more complex than that of edge-triggered circuits. In level-clocked circuits, combinational blocks are not insulated from each other by the memory elements, and cycle borrowing is possible when the data are latched during the transparent phase of the clock. In such cases, a signal passes through multiple combinational stages before seeing a capturing edge, and timing analysis becomes much more involved since it must perform checks across latch boundaries, accounting for cycle borrowing. Methods for handling multicycle paths are described in [7,21,29,35,36]. A simplifying strategy for level-clocked circuits forces all output signals arrive in time for the active clock edge. This has the advantage of avoiding the complications of cycle borrowing but significantly reduces the benefits achievable from the use of level clocking. In our discussion, we will primarily deal with edge-triggered *D FF*s, which are the most widely used memory elements.

6.5.2 TIMING CONSTRAINTS FOR EDGE-TRIGGERED CIRCUITS

We will now overview the timing requirements for edge-triggered sequential circuits, which consist of combinational blocks that lie between *D FF*s. The basic parameters associated with an *FF* can be summarized as follows:

- The data input of the register, commonly referred to as the *D* input, must receive incoming data at a time that is at least T_s units before the onset of the latching edge of the clock. The data will then be available at the output node, *Q*, after the latching edge. The quantity, T_s, is referred to as the setup time of the *FF*.

■ The input, D, must be kept stable for a time of T_h units, where T_h is called the hold time, so that the data can be stored correctly in the *FF*.

■ For each *FF*, there is a delay between the time the data and clock are both available at the input and the time when the data are latched; this is referred to as the clock-to-Q delay, T_q.

In the edge-triggered scenario, let us consider two *FF*s, FF_i and FF_j, connected only by purely combinational paths (see Figure 6.9). Over all such paths i->j, let the largest delay from FF_i to FF_j be $D(i, j)$ and the smallest delay be $d(i, j)$. Therefore, for any path i->j with delay $\hat{d}$, it must be true that

$$d(i, j) \le \hat{d} \le D(i, j).$$

We will denote the setup time, hold time, and the maximum and minimum clock-to-Q delays of any arbitrary *FF* FF_k as T_{sk}, T_{hk}, and Δ_k and δ_k, respectively. For a negative edge-triggered *FF*, the setup and hold time requirements are illustrated in Figure 6.8. The clock is a periodic waveform that repeats after every P units of time, which is called as the clock period or the cycle time.

Another important component in timing analysis is the clock network that is used to distribute the clock signal throughout the chip. In early stages of the design, it is often considered that the clock network is ideal, that is, there is zero skew between the arrival times of the clock at each memory element as the clock signal arrives at each memory element at exactly the same time. After physical design, the latency of the clock network must also be incorporated in the analysis. For every *FF* FF_k, let CLK_k represent the delay of the clock network between the clock generator and FF_k.

The timing constraints for edge-triggered circuits guarantee that data transfers between pairs of *FF*s are neither too slow (setup constraint) nor too fast (hold constraint). In every constraint, two competing paths are always involved (represented by dotted lines in Figure 6.9):

1. *Launching path*, which is the path starting at the clock generator and covering the logic from launching *FF* (FF_i) to the capturing *FF* (FF_j)
2. *Capturing path*, which is the path starting at the clock generator and delivering the clock to the capturing *FF* (FF_j)

Let us now analyze the two timing constraints that guarantee correct data transfers.

6.5.2.1 SETUP CONSTRAINT

For any pair of *FF*s, FF_i and FF_j, connected by a combinational path, the setup constraint guarantees that data transfers i->j are not too slow. In a generic form, it can be stated as

$$\text{Delay(Launching path)} + T_{s_j} < P + \text{Delay(Capturing path)}$$

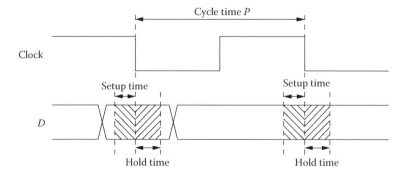

FIGURE 6.8   Illustration of the clocking parameters for a register. (From Sapatnekar, S.S., *Timing*, Kluwer Academic Publishers, Boston, MA, 2004.)

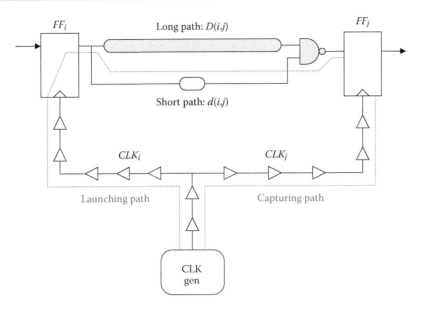

FIGURE 6.9 Timing paths for setup and hold constraints.

given that data launched at FF_i must arrive at FF_j one cycle later, preserving the setup time (T_{s_j}) of the capturing FF. If we consider the components of each path, the constraint is as follows:

(6.5)
$$CLK_i + \Delta_i + D(i,j) + T_{s_j} < P + CLK_j$$

It is interesting to realize that the terms CLK_i and CLK_j cancel each other out when an ideal clock with zero skew is assumed.

Since this constraint places an upper bound on the delay of a combinational path, it is also called the "long-path constraint." Alternatively, this is also referred to as the "zero clocking constraint," because the data will not arrive in time to be latched at the next clock period if the combinational delay does not satisfy this constraint.

6.5.2.2 HOLD CONSTRAINT

This constraint guarantees that data transfers are not too fast. The constraint is necessary to prevent a fast transfer from overriding the previous data before the capturing FF has stored it. The generic form of this constraint is

$$\text{Delay(Capturing path)} + T_{h_j} < \text{Delay(Launching path)}$$

Under this constraint, the data must be stable for an interval, that is, at least as long as the hold time after the clock edge (T_{h_j}), if it is to be correctly captured by the capturing FF (FF_j). Given that the launching path is at the right-hand side of the inequality, the shortest possible delay, $d(i, j)$, must be considered for the combinational paths. More specifically,

(6.6)
$$CLK_j + T_{h_j} < CLK_i + \delta_i + d(i,j)$$

Since this constraint puts a lower bound on the combinational delay on a path, it is referred to as a "short-path constraint." If this constraint is violated, then the data in the current clock cycle are corrupted by the data from the next clock cycle. As a result, data are latched twice instead of once in a clock cycle, and hence, it is also called the "double clocking constraint."

Notice that for an ideal clock network with zero skew ($CLK_i = CLK_j$), if the minimum clock-to-Q delay of FF_i is greater than the hold time of FF_j, that is, $\delta_i \geq T_{h_j}$ (this condition is not always true in practice), then the short-path constraint is always satisfied.

An important observation is that both the long-path and the short-path constraints refer to combinational paths that lie between *FF*s. Therefore, for timing verification of edge-triggered circuits, it is possible to decompose the circuit into combinational blocks and to verify the validity of the constraints on each such block independently. This is not so for level-clocked circuits, which present a greater complexity to the timing verifier.

6.6 TIMING ANALYSIS IN THE PRESENCE OF VARIATION

6.6.1 SOURCES OF VARIATION

Integrated circuits are afflicted with a wide variety of variations that affect their performance. Essentially, under true operating conditions, the parameters chosen by the circuit designer are perturbed from their nominal values due to various types of variations. As a consequence, a single SPICE-level transistor or interconnect model (or an abstraction thereof) is seldom an adequate predictor of the exact behavior of a circuit. These sources of variation can broadly be categorized into two classes, both of which can result in changes in the timing characteristics of a circuit.

Process variations are one-time shifts that result from perturbations in the fabrication process, which cause shifts from the nominal values of parameters such as the effective channel length (L_{eff}), the oxide thickness (t_{ox}), the dopant concentration (N_a), the transistor width (w), the interlayer dielectric (ILD) thickness (t_{ILD}), and the interconnect height and width (h_{int} and w_{int}, respectively).

Environmental variations are run-time changes that arise due to changes in the operating environment of the circuit, such as the temperature or variations in the supply voltage (V_{dd} and ground) levels, or due to changes in the behavior of the circuit as it ages. The chief aging effects that impact circuit timing are bias temperature instability and hot carrier injection, both of which tend to cause transistor threshold voltages to increase with time, thus increasing gate delays.

Variations can also be classified into the following categories:

Interdie variations are the process variations from die to die and affect all the devices on the same chip in the same way. For example, they may cause all of the transistor gate lengths of devices on the same chip to be larger or all of them to be smaller. These may be across-die variations in the same wafer, or across-wafer variations within a lot, or variations across lots of wafers.

Within-die variations correspond to variability within a single chip and may affect different devices differently on the same chip. For example, they may result in some devices having smaller oxide thicknesses than the nominal, while others may have larger oxide thicknesses. In addition to process factors, within-die variations may also be caused by environmental variations such as temperature and V_{dd} factors.

Interdie variations have been a long-standing design issue, and for several decades, designers have striven to make their circuits robust under the unpredictability of such variations. This has typically been achieved by analyzing the design not just at one design point but at multiple *corners*. These corners are chosen to encapsulate the behavior of the circuit under worst-case variations and have served designers well in the past.

Unlike interdie process variations, whose effects can be captured by a small number of STA runs at the process corners, a more sophisticated approach is called for in dealing with within-die process variations. This requires an extension of traditional STA techniques and can be addressed in two ways: either by adding *on-chip variation* (*OCV*) parameters to corner-based approaches (Section 6.6.2) or by using SSTA, which treats delays not as fixed numbers, but as probability density functions (PDFs), taking the statistical distribution of parametric variations into consideration while analyzing the circuit (Section 6.6.3).

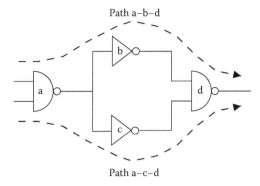

Path a–b–d

Path a–c–d

FIGURE 6.10   An example to illustrate structural correlations in a circuit.

Within-die variations may show some correlations due to the following:

Spatial correlations: To model the within-die spatial correlations of parameters, the die region may be tessellated into n gridded rectangles. Since devices or wires close to each other are more likely to have similar characteristics than those placed far away, it is reasonable to assume perfect correlations among the devices (wires) in the same rectangle, high correlations among those in nearby rectangles, and low or zero correlations in far-away rectangles. Under this model, a parameter variation in a single rectangle at location (x, y) can be modeled using a single random variable $p(x, y)$. For each type of parameter, n random variables are needed, each representing the value of a parameter in one of the n rectangles.

Structural correlations: The structure of the circuit can also lead to correlations that must be incorporated in SSTA. Consider the reconvergent fanout structure shown in Figure 6.10. The circuit has two paths, a–b–d and a–c–d. If, for example, we assume that each gate delay is a Gaussian random variable, then the PDF of the delay of each path is easy to compute, since it is the sum of Gaussians, which admits a closed form. However, the circuit delay is the maximum of the delays of these two paths, and these are correlated since the delays of a and d contribute to both paths. It is important to take such structural correlations, which arise due to reconvergences in the circuit, into account while performing SSTA.

Process variation effects fall into one of the following two categories:

1. *Random variations* depict random behavior that can be characterized in terms of a distribution. This distribution may either be explicit, in terms of a large number of samples provided from fabrication line measurements, or implicit, in terms of a known PDF (such as a Gaussian or a lognormal distribution) that has been fitted to the measurements. Random variations in some process or environmental parameters (such as those in the temperature, supply voltage, or L_{eff}) can often show some degree of local *spatial correlation*, whereby variations in one transistor in a chip are remarkably similar in nature to those in spatially neighboring transistors, but may differ significantly from those that are far away. Other process parameters (such as t_{ox} and N_a) do not show much spatial correlation at all so that for all practical purposes, variations in neighboring transistors are uncorrelated.

2. *Systematic variations* show predictable variational trends across a chip and are caused by known physical phenomena during manufacturing. Strictly speaking, environmental changes are entirely predictable, but practically, due to the fact that these may change under a large number (potentially exponential in the number of inputs and internal states) of operating modes of a circuit, it is easier to capture them in terms of random variations. Examples of systematic variations include those due to spatial within-chip gate length variability, which observes systematic changes in the value of L_{eff} across a reticle due to effects such as changes in the stepper-induced illumination and imaging nonuniformity due to lens aberrations or ILD variations due to the effects of chemical–mechanical polishing on metal density patterns.

6.6.2 CORNER-BASED TIMING ANALYSIS

Given the infeasibility of modeling the behavior of a circuit at all points of the space–time continuum, timing analysis is performed over a discrete set of points of that space.

For an edge-triggered circuit, the first level of discretization is in the time dimension reducing the problem to a one-cycle timing analysis under the assumption that the synchronization of the circuit is done with a constrained clock signal that has a fixed period with small fluctuations (jitter). In this way, timing analysis is only required for paths between sequential elements during one cycle, using an assume-guarantee paradigm for timing verification, that is,

> *if the setup and hold constraints were met for all sequential elements at cycle i, then they must be met at cycle i + 1.*

With this inductive reasoning, timing correctness is guaranteed for any unbounded sequence of cycles.

6.6.2.1 CORNERS

The second level of discretization is on process (P) and environmental variations, that is, voltage (V) and temperature (T). For each parameter, a discrete set of representative values is defined, and a subset of points in the corresponding axis of the discrete grid is selected.

A typical example is shown in Figure 6.11. For process variability, the selected values represent different carrier mobilities for the NMOS and PMOS devices, respectively. The *S* (*slow*) and *F* (*fast*) values usually correspond to variations of the carrier mobilities that produce delays at ±3σ or ±6σ from the typical (*T*) values.*

The voltage dimension has a central point representing the nominal voltage (e.g., 1.0 V) and two extra points representing the maximum and minimum values achievable by the potential fluctuations around the nominal voltage, typically ±10%. Additional values can be included if voltage scaling can be applied to the circuit.

For temperature, manufacturers define different grades according to the application domain in which the circuit is going to be used, for example, industrial, automotive, and military. Each grade has a different range of operating temperatures, for example, [−40°C, 125°C] and [0°C, 85°C]. The extreme values of different grades are selected for the discrete grid, along with some values representing a typical ambient temperature (25°C).

A *corner* is a point in the discrete Process/Voltage/Temperature (PVT) grid of variability, for example, (*FF*, 1.1 V, −40°C). Manufacturers select a set of corners in the PVT grid and provide a characterization of the devices and wires at each corner.

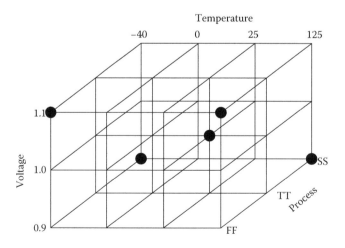

FIGURE 6.11 Discrete representation of corners.

---

* After physical synthesis and parasitic extraction, a new dimension with a discrete set of values for the RC interconnect delays is also incorporated into the grid.

6.6.2.2 GLOBAL AND LOCAL VARIABILITIES

Given the high correlation manifested in the variability of devices and wires located in the same chip, it would be overly conservative and unrealistic to assume that two devices of the same chip may operate at opposite corners of the PVT space. As previously mentioned, interdie variations affect all devices on the same chip in the same way. Additionally, environmental variations also have a high spatial correlation and a similar impact on neighboring devices.

Corner-based timing analysis is an approach to incorporate variability information in a way that is both computationally affordable and not overly pessimistic. It is the most widely used approach in current design flows. The mechanisms for incorporating variability in timing analysis are as follows:

Library corners to model global variations. With this mechanism, all devices and wires are assumed to experience a uniform variability across the die. Timing analysis is performed independently for a selected subset of corners.

Derating factors to model local variations. This is often referred as "on-chip variability." With this mechanism, the delay of each component in a timing path is scaled by a derating factor that models the variability with regard to other competing paths in the timing constraints.

Clock uncertainty to model the variations of the clock period with regard to the nominal period (jitter). Given that the nominal period is a global constant during timing analysis, clock uncertainty is specified as a fixed margin subtracted from the nominal period, for example, 100 ps.

Guardband margins to model any pessimism derived from variations and uncertainties not covered by the previous mechanisms. Usually, these margins are used to cover the inaccuracy of the modeling and analysis tools, unknown variability parameters, imperfect coverage of the test vectors, etc.

In practice, clock uncertainty and guardband margins can be consolidated and specified as only one margin that covers both concepts.

6.6.2.3 MULTICORNER MULTIMODE STA

Analyzing the timing of a circuit under any possible operating scenario is an impossible task since most of the technological and environmental parameters can take on a large number of allowable values. Instead, designers only consider a discrete subset of representative scenarios. The conjecture (or hope) is that a valid timing for these scenarios guarantees a valid timing for any scenario using certain guardband margins. These scenarios correspond to combinations of two conditions: a *mode* (e.g., normal, test, and sleep) in which a circuit operates and a *corner* that captures PVT conditions. A subset of representative PVT conditions is defined for the FEOL layers (transistors). Similarly, a subset of RC conditions are defined for the BEOL layers (interconnect).

Multicorner multimode (MCMM) STA performs multiple timing analyses at a selected subset of combinations of modes, PVT corners, and RC corners, where corners model the global variability of the circuit. OCV is modeled by applying derating factors at the launching/capturing paths. Thus, the setup and hold constraints, (6.5) and (6.6), can be rewritten as (6.7) and (6.8), respectively:

$$(6.7) \qquad \delta_{slow} \cdot (CLK_i + \Delta_i + D(i,j) + T_{s_j}) + T_{guard} < P - T_{jitter} + \delta_{fast} \cdot CLK_j$$

$$(6.8) \qquad \delta_{slow} \cdot (CLK_j + T_{h_j}) + T_{guard} < \delta_{fast} \cdot (CLK_i + \delta_i + d(i,j))$$

The derating factors δ_{slow} and δ_{fast} make the paths slower and faster, respectively. For example, to make the launching path 10% slower and the capturing path 5% faster in the setup constraint, the derating factors should be defined as $\delta_{slow} = 1.1$ and $\delta_{fast} = 0.95$. The derating factors may take different values depending on the constraint, the corner, and the component (gate or wire).

The previous inequalities also include a guardband margin (T_{guard}) often used by designers to account for the inaccuracy of timing analysis. Finally, the setup constraint also includes the clock uncertainty (T_{jitter}) to account for the variations of the clock period.

6.6.2.4 REDUCING PESSIMISM

The timing constraints (6.7) and (6.8) assume that the same derating factor is applied to all components of the launching or capturing paths. This assumption is often pessimistic since it does not consider the correlations of the delays between different timing paths or the mitigation of random variability in long paths.

One interesting observation when analyzing Figure 6.9 is that the paths for the launching and capturing clocks, CLK_i and CLK_j, have a common part. Applying different derating factors to the common part results in a pessimistic analysis that may lead to an overconservative clock period. This phenomenon is called "common path pessimism" [4].

One way to reduce this pessimism is to remove the common part of the clock tree from timing analysis in such a way that the derating factors are only applied to the unshared parts of the launching and capturing clock paths. This approach is called "common path pessimism removal" or "clock reconvergence pessimism removal" [4].

Some techniques propose a more advanced application of derating factors. Using the same derating factor for all paths ignores the spatial correlation between paths located close to each other. From statistics, we also know that the standard deviation of a sum of random variables is significantly smaller than the sum of their standard deviations. For this reason, the local variability of long paths is smaller than that of short paths.

Advanced on-chip variability [34] has been proposed to take into account spatial correlation and path length. With this approach, the derating factor applied to each path depends on the physical distance of the launching and capturing paths and on their length. A simpler approach is *location-based OCV* [20] that only derives derating factors based on the distance between paths. A more complex concept is *parameterized OCV* [26] that analyzes path delays using a more sophisticated statistical timing analysis framework.

6.6.3 STATISTICAL STATIC TIMING ANALYSIS

If the statistics of process variations is known, then the distribution of delays can be determined through an analysis that treats gate delays not as fixed numbers, but as PDFs, and determines the distribution of circuit delay. One way to achieve this is through Monte Carlo analysis, but this approach is computationally expensive. A class of approaches, referred to as SSTA, determines this distribution analytically, with environmental variations being typically captured by corners.

To build a gate delay model that captures the underlying variations in process parameters, we observe that the delay function $d = f(P)$, where P is a set of process parameters, can be approximated linearly using a first-order Taylor expansion:

$$(6.9) \qquad d = d_0 + \sum_{\forall \text{ parameters } p_i} \left[\frac{\partial f}{\partial p_i} \right]_0 \Delta p_i$$

where d_0 is the nominal value of d, calculated at the nominal values of parameters in the set P, $\left[\dfrac{\partial f}{\partial p_i} \right]_0$ is computed at the nominal values of p_i, $\Delta p_i = p_i - \mu_{p_i} \theta$ is a normally distributed random variable, and expression $\Delta p_i \sim N(0, \sigma_{p_i})$. The delay function here can be arbitrarily complex and may include, for example, the effects of the input transition time on the gate delay.

If all parameters in P can be modeled by Gaussian distributions, this approximation implies that d is a linear combination of Gaussians, which is therefore Gaussian. Its mean μ_d and variance σ_d^2 are

$$(6.10) \qquad \mu_d = d_0$$

$$(6.11) \qquad \sigma_d^2 = \sum_{\forall i} \left[\frac{\partial f}{\partial p_i} \right]_0^2 \sigma_{p_i}^2 + 2 \sum_{\forall i \neq j} \left[\frac{\partial f}{\partial p_i} \right]_0 \left[\frac{\partial f}{\partial p_j} \right]_0 cov(p_i, p_j)$$

where $cov(p_i, p_j)$ is the covariance of p_i and p_j.

This approximation is valid when Δp_i has relatively small variations, and hence, the first-order Taylor expansion is adequate and the approximation is acceptable with little loss of accuracy. This is generally true of the impact of within-die variations on delay, where the process parameter variations are relatively small in comparison with the nominal values, and the function changes by a small amount under this perturbation. Hence, the delays, as functions of the process parameters, can be approximated as normal distributions when the parameter variations are assumed to be normal. Second-order expansions have also been explored to cover cases where the variations are larger.

The task of SSTA is to translate these gate delay distributions to circuit delay probabilities while performing an STA-like traversal. In other words, STA must be generalized to handle probabilistic delay functions. The operations in STA are of two types:

1. A gate is being processed in STA when the arrival times of all inputs are known, at which time the candidate delay values at the output are computed using the *sum* operation that adds the delay at each input with the input-to-output pin delay.
2. Once these candidate delays have been found, the *max* operation is applied to determine the maximum arrival time at the output.

For uncorrelated Gaussian parameter variations, the approach in [17] maintains the invariant that the arrival time at each gate output is given by a Gaussian random variable. Since the gate delays are Gaussian, the *sum* operation is merely an addition of Gaussians, which is well known to be a Gaussian. The computation of the max function, however, poses greater problems. The set of candidate delays are all Gaussian so that this function must find the maximum of Gaussians. Such a maximum may be reasonably approximated using a Gaussian [11].

For spatially correlated variations, the analysis can be significantly more involved. The approach in [8] presented a computationally efficient method that uses principal component analysis (PCA) [25] to convert the set of underlying correlated random variables, representing process parameters, into a set of uncorrelated variables in a transformed space; the PCA step can be performed as a preprocessing step for a design. Such as in [10], a PCA-based approach maintains the output arrival time at each gate as a Gaussian variable but represents it as

$$(6.12) \qquad a_i(p_1, \ldots, p_n) = a_i^0 + \sum_{i=1}^{n} k_i p_i' + k_{n+1} p_{n+1}'$$

where the primed variables correspond to the principal components of the unprimed variables and maintain the form of the arrival time after each sum and max operation. The broad outline of the approach is shown in Table 6.1; for details, see [8,9]. The cost of this method corresponds to running a bounded number of deterministic STAs, and it is demonstrated to be accurate, given the statistics of P. A PCA-based approach that has been used extensively in industry is described in [37].

Further generalizations of this approach correspond to the use of nonlinear Taylor series for nonlinear delay functions, and the extension to the case where the underlying parameter variations are non-Gaussian. For the nonlinear Gaussian case, a moment-based approach can be employed [22,23,38]. For the linear case with general parameter distributions, the approach in [32,33] orthogonalizes Gaussian parameters using PCA, and non-Gaussian parameters using independent component analysis [16], and propagates arrival times in the circuit to obtain its delay PDF.

TABLE 6.1   Overall Flow of a PCA-Based SSTA Framework

Input: Process Parameter Variations

Output: Distribution of Circuit Delays

1.	Partition the chip into $n = nrow \times ncol$ grid rectangles, each modeled by spatially correlated variables.
2.	For each type of parameter, determine the n jointly normally distributed random variables and the corresponding covariance matrix.
3.	Perform an orthogonal transformation to represent each random variable with a set of principal components.
4.	For each gate and net connection, model their delays as linear combinations of the principal components generated in step 3.
5.	Using *sum* and *max* functions on Gaussian random variables, perform a CPM-like traversal on the graph to find the distribution of the statistical longest path. This distribution achieved is the circuit delay distribution.

6.7 CONCLUSION

This chapter has presented an overview of techniques used for STA. The exposition has focused mainly on edge-triggered circuits, and similar principles are used, with greater computation, to handle level-clocked circuits where multicycle paths are possible.

STA provides diagnostics that predict the timing correctness of a design. If some timing constraints are violated, the design must be changed to fix those violations. For setup constraints, common strategies include the reduction of T_{long} by transformations such as logic restructuring, gate sizing, placement modification, retiming, and clock skew optimization. If such modifications do not help in meeting the timing constraint, then lengthening the clock period (P) is always a choice that involves a loss of performance. Hold constraints are inherently independent of the clock period, and the usual method to resolve them is to make T_{short} longer by undersizing gates or adding buffers. In doing so, care must be taken not to affect the critical paths that determine the clock period.

Compared to the introductory view of STA outlined in this chapter, real-life design problems involve further complications, such as cycle borrowing across clock cycles when latches or deliberate skew (or both) are used, setup constraints for clock gating, data-to-data constraints, and minimum pulse width checks. Although these issues are not explicitly treated here, the techniques used to address them are all based on methods discussed in this chapter.

REFERENCES

1. R. P. Abato, A. D. Drumm, D. J. Hathaway, and L. P. P. P. van Ginneken. Incremental timing analysis. US Patent 5508937.
2. C. Amin, C. Kashyap, N. Menezes, K. Killpack, and E. Chiprout. A multi-port current source model for multiple-input switching effects in CMOS library cells. In *Proceedings of the ACM/IEEE Design Automation Conference*, San Francisco, CA, 2006, pp. 247–252.
3. K. P. Belkhale. Timing analysis with known false sub-graphs. In *Proceedings of the IEEE/ACM International Conference on Computer-Aided Design*, San Jose, CA, 1995, pp. 736–740.
4. J. Bhasker and R. Chadha. *Static Timing Analysis for Nanometer Designs.* Springer, New York, 2009.
5. D. Blaauw, R. Panda, and A. Das. Removing user specified false paths from timing graphs. In *Proceedings of the ACM/IEEE Design Automation Conference*, Los Angeles, CA, 2000, pp. 270–273.
6. D. Blaauw, V. Zolotov, and S. Sundareswaran. Slope propagation in static timing analysis. *IEEE Transactions on Computer-Aided Design of Integrated Circuits and Systems*, 21(10):1180–1195, October 2002.
7. T. M. Burks, K. A. Sakallah, and T. N. Mudge. Critical paths in circuits with level-sensitive latches. *IEEE Transactions on VLSI Systems*, 3(2):273–291, June 1995.
8. H. Chang and S. S. Sapatnekar. Statistical timing analysis considering spatial correlations using a single PERT-like traversal. In *Proceedings of the IEEE/ACM International Conference on Computer-Aided Design*, San Jose, CA, November 2003, pp. 621–625.

9. H. Chang and S. S. Sapatnekar. Statistical timing analysis under spatial correlations. *IEEE Transactions on Computer-Aided Design of Integrated Circuits and Systems*, 24(9):1467–1482, September 2005.

10. P. Chen, D. A. Kirkpatrick, and K. Keutzer. Miller factor for gate-level coupling delay calculation. In *Proceedings of the IEEE/ACM International Conference on Computer-Aided Design*, San Jose, CA, 2000, pp. 68–75.

11. C. E. Clark. The greatest of a finite set of random variables. *Operations Research*, 9:85–91, 1961.

12. J. F. Croix and D. F. Wong. Blade and Razor: Cell and interconnect delay analysis using current-based models. In *Proceedings of the ACM/IEEE Design Automation Conference*, Anaheim, CA, 2003, pp. 386–389.

13. E. Goldberg and A. Saldanha. Timing analysis with implicitly specified false paths. In *Workshop Notes, International Workshop on Timing Issues in the Specification and Synthesis of Digital Systems*, 1999, pp. 157–164.

14. S. Gupta and S. S. Sapatnekar. Current source modeling in the presence of body bias. In *Proceedings of the Asia/South Pacific Design Automation Conference*, Taipei, Taiwan, 2010, pp. 199–204.

15. D. J. Hathaway, J. P. Alvarez, and K. P. Belkhale. Network timing analysis method which eliminates timing variations between signals traversing a common circuit path. US Patent 5636372.

16. A. Hyvärinen and E. Oja. Independent component analysis: Algorithms and applications. *Neural Networks*, 13:411–430, 2000.

17. E. Jacobs and M. R. C. M. Berkelaar. Gate sizing using a statistical delay model. In *Proceedings of Design and Test in Europe*, Paris, France, 2000, pp. 283–290.

18. N. P. Jouppi. Timing analysis and performance improvement of MOS VLSI design. *IEEE Transactions on Computer-Aided Design of Integrated Circuits and Systems*, CAD-4(4):650–665, July 1987.

19. I. Keller, K. Tseng, and N. Verghese. A robust cell-level crosstalk delay change analysis. In *Proceedings of the IEEE/ACM International Conference on Computer-Aided Design*, San Jose, CA, 2004, pp. 147–154.

20. S. Kobayashi and K. Horiuchi. An LOCV-based static timing analysis considering spatial correlations of power supply variations. In *Proceedings of Design and Test in Europe*, Grenoble, France, March 2011, pp. 1–4.

21. J.-f. Lee, D. T. Tang, and C. K. Wong. A timing analysis algorithm for circuits with level-sensitive latches. In *Proceedings of the IEEE/ACM International Conference on Computer-Aided Design*, San Jose, CA, 1994, pp. 743–748.

22. X. Li, J. Le, P. Gopalakrishnan, and L. T. Pileggi. Asymptotic probability extraction for non-normal distributions of circuit performance. In *Proceedings of the IEEE/ACM International Conference on Computer-Aided Design*, San Jose, CA, 2004, pp. 2–9.

23. X. Li, J. Le, P. Gopalakrishnan, and L. T. Pileggi. Asymptotic probability extraction for nonnormal performance distributions. *IEEE Transactions on Computer-Aided Design of Integrated Circuits and Systems*, 26(1):16–37, January 2007.

24. P. C. McGeer and R. K. Brayton. *Integrating Functional and Temporal Domains in Logic Design.* Kluwer Academic Publishers, Boston, MA, 1991.

25. D. F. Morrison. *Multivariate Statistical Methods.* McGraw-Hill, New York, 1976.

26. A. Mutlu, J. Le, R. Molina, and M. Celik. A parametric approach for handling local variation effects in timing analysis. In *Proceedings of the ACM/IEEE Design Automation Conference*, San Francisco, CA, July 2009, pp. 126–129.

27. P. R. O'Brien and D. T. Savarino. Modeling the driving-point characteristic of resistive interconnect for accurate delay estimation. In *Proceedings of the IEEE/ACM International Conference on Computer-Aided Design*, Santa Clara, CA, 1989, pp. 512–515.

28. J. Qian, S. Pullela, and L. T. Pillage. Modeling the "effective capacitance" for the RC interconnect of CMOS gates. *IEEE Transactions on Computer-Aided Design of Integrated Circuits and Systems*, 13(12):1526–1535, December 1994.

29. K. A. Sakallah, T. N. Mudge, and O. A. Olukotun. Analysis and design of latch-controlled synchronous digital circuits. *IEEE Transactions on Computer-Aided Design of Integrated Circuits and Systems*, 11(3):322–333, March 1992.

30. S. S. Sapatnekar. A timing model incorporating the effect of crosstalk on delay and its application to optimal channel routing. *IEEE Transactions on Computer-Aided Design of Integrated Circuits and Systems*, 19(5):550–559, May 2000.

31. S. S. Sapatnekar. *Timing.* Kluwer Academic Publishers, Boston, MA, 2004.

32. J. Singh and S. S. Sapatnekar. Statistical timing analysis with correlated non-Gaussian parameters using independent component analysis. In *Proceedings of the ACM/IEEE Design Automation Conference*, San Francisco, CA, 2006, pp. 155–160.

33. J. Singh and S. S. Sapatnekar. A scalable statistical static timing analyzer incorporating correlated non-Gaussian and Gaussian parameter variations. *IEEE Transactions on Computer-Aided Design of Integrated Circuits and Systems*, 27(1):160–173, January 2008.

34. Synopsys, Inc. PrimeTime® Advanced OCV Technology. www.synopsys.com/Tools/Implementation/SignOff/CapsuleModule/PrimeTime_AdvancedOCV_WP.pdf, 2009. Accessed on November 24, 2015.

35. T. G. Szymanski. Computing optimal clock schedules. In *Proceedings of the ACM/IEEE Design Automation Conference*, Anaheim, CA, 1992, pp. 399–404.

36. T. G. Szymanski and N. Shenoy. Verifying clock schedules. In *Proceedings of the IEEE/ACM International Conference on Computer-Aided Design*, Santa Clara, CA, 1992, pp. 124–131.

37. C. Visweswariah, K. Ravindran, K. Kalafala, S. G. Walker, and S. Narayan. First-order incremental block-based statistical timing analysis. In *Proceedings of the ACM/IEEE Design Automation Conference*, San Diego, CA, June 2004, pp. 331–336.

38. Y. Zhan, A. J. Strojwas, X. Li, L. T. Pileggi, D. Newmark, and M. Sharma. Correlation-aware statistical timing analysis with non-Gaussian delay distributions. In *Proceedings of the ACM/IEEE Design Automation Conference*, Anaheim, CA, 2005, pp. 77–82.

39. H. Zhou. Timing analysis with crosstalk is a fixpoint on a complete lattice. *IEEE Transactions on Computer-Aided Design of Integrated Circuits and Systems*, 22(9):1261–1269, September 2003.

Structured Digital Design

**Minsik Cho, Mihir Choudhury, Ruchir Puri, Haoxing Ren,
Hua Xiang, Gi-Joon Nam, Fan Mo, and Robert K. Brayton**

CONTENTS

7.1 INTRODUCTION

Structured digital circuits, such as datapaths, programmable logic arrays (PLAs), and memories, are regular in both netlist and layout. Structured digital designs usually involve multibit signals and multibit operations, and hence, regularity occurs naturally. Regularity in designs is desirable because regular circuit structures offer compact layout and good delay estimation.

One motivation for using regularity is the so-called timing closure problem, which arises because design flows are sequential and iterative; early steps need to predict what the later steps will accomplish. Inaccurate prediction leads to wrong decisions that can only be diagnosed later, resulting in many design iterations. In the deep submicron (DSM) domain, wiring delays dominate gate delays [1], thereby increasing the importance of obtaining good estimates of wiring effects early in the flow. Structured circuits, because of their layout regularity, allow more accurate area and timing estimation. Also, regularity provides better guarantees that the structured design layout is faithfully replicated in the fabrication. The reasons for this are subtle but can be summarized as follows. As the mask-making system approaches the physical limits of fabrication, the actual layout patterns on the wafer become different from those during design phases. To compensate these gaps, techniques like optical proximity correction (OPC) are used. However, the number of layout patterns generated by a conventional design flow can lead to OPC requiring an unreasonable amount of time and generating an enormous data set. Regular patterns from structured digital designs can reduce these requirements as well as reduce the discrepancies between the design and fabrication results by using fewer and simpler layout patterns.

Unfortunately, today, structured digital designs are supported by less automation than the unstructured designs. Manual design is common, with computer aids limited to layout editing, design rule checking, and parasitic parameter extraction and simulation. Automatic structured digital design is challenging because a complete structured design flow involves almost every step of a conventional flow, with important differences at every step. This means that an entirely new design flow and a whole set of new algorithms need to be developed to explore important aspects of structural regularity. This subject area is still far from being mature. However, several promising datapath detection and automated layout techniques have been reported, shedding light on possible fully automated structured digital design methodology. In this chapter, we focus on automation techniques for structured design and review some of the algorithms and methodologies being developed. These can be individually employed, supplementing existing conventional flows, or they can be organized in a new flow specific to structured digital design.

The use of regularity is not limited to structured circuit modules. Regular full-chip structures also exist. Gate arrays (GAs), a regular array of transistors and routing channels, have been used for years. New regular chip architectures have emerged recently, including structured application-specific integrated circuits (SASICs) and via-patterned GAs (VPGAs). These technologies will be reviewed as well.

7.2 DATAPATHS

A datapath is a type of circuit structure that performs multibit operations and exhibits logical and physical regularity. It is often used in the design of arithmetic units. A datapath is composed of a set of datapath components, control logic, and their interconnections.

A datapath example is illustrated in Figure 7.1. The vertical data flow and the horizontal control flow are orthogonal. Along the data flow, which consists of multiple slices, lie multibit functional components including multiplexers, adders, multipliers, and registers, called the datapath components. Each datapath component forms a stage. The logic parts within a functional component that are separated by registers are also treated as stages. Due to pipelining, some datapath components, like the top component in the figure, are partitioned into stages. Note that the data flow represents the main direction in which multiple data bits transit (vertical in the figure). Some components, like those of $stage_3$, can have their own internal data direction that is orthogonal to the main data flow. It is also possible that some data bits shift as they enter the next stage, like those at $stage_2$. The regular schematic can be easily translated to a regular layout keeping the arrangement of the stages and slices intact. In some cases, the data flow might be too long to implement in one single set of slices, requiring datapath partitioning and floorplanning techniques, which will be discussed in Section 7.2.2.3.

7.2.1 DATAPATH COMPONENTS

Datapath components are function blocks that realize certain multibit arithmetic or logic operations. There are many references about the details of the structures and functionalities of these components [2]. Here, we briefly describe adders and multipliers, focusing on the trade-offs between area, delay, and regularity. Some complex datapath components are often pipelined to reduce the clock cycle. Pipelining is not described in this chapter.

7.2.1.1 ARITHMETIC OPERATORS

Datapath logic constitutes a significant portion of a general-purpose microprocessor and frequently occurs on the timing-critical paths in high-performance designs. Basic arithmetic

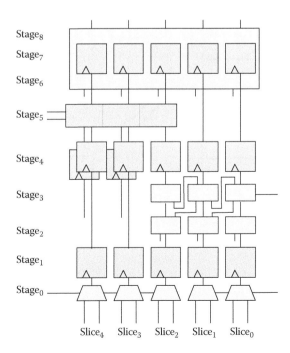

FIGURE 7.1 A datapath example.

operations, such as addition, subtraction, comparison, and multiplication, are most frequently used in datapath logic and often dictate the performance of the entire chip. This section describes how a comparator, a subtracter, and a multiplier can be implemented using an adder, thus facilitating the logic and physical design benefits of parallel prefix adder expansion that will be described in Section 7.2.2.1. For simplicity, integer operands are assumed. The implementation of floating-point operations can be derived from integer operation implementation.

Comparator: Comparators perform one of the four possible operations—$(a > b)$, $(a \geq b)$, $(a < b)$, $(a \leq b)$—on two n-bit integers. The logic implementation of all four comparisons can be derived from the operation $(a > b)$. The operation $(a < b)$ can be implemented using $(a > b)$ by interchanging the operands a and b. The operation $(a \geq b) \equiv !(a < b)$. Similarly, $(a \leq b) \equiv !(a > b)$. An n-bit operation $(a_{n-1},..., a_0 > b_{n-1},..., b_0)$, where $n - 1$ is the most significant bit (MSB) and 0 is the least significant bit (LSB), can be recursively described as follows: the value of $(a_{n-1},..., a_0 > b_{n-1},..., b_0)$ is true when either $(a_{n-1} > b_{n-1})$ or $[(a_{n-1} = b_{n-1})$ and $(a_{n-2},..., a_0 > b_{n-2},..., b_0)]$. Thus,

$$(a_{n-1},...,a_0 > b_{n-1},...,b_0) = (a_{n-1}!b_{n-1}) \vee [(a_{n-1} \oplus b_{n-1}) \cdot (a_{n-2},...,a_0 > b_{n-2},...,b_0)]$$

or

(7.1) $\quad (a_{n-1},...,a_0 > b_{n-1},...,b_0) = (a_{n-1}!b_{n-1}) \vee [(a_{n-1} \vee b_{n-1}) \cdot (a_{n-2},...,a_0 > b_{n-2},...,b_0)]$

Note that this equation is the same as the expression for the carry-out signal of a parallel prefix adder with operand b inverted. Hence, the comparator operation $(a > b)$ can be expressed using an adder, as shown in Figure 7.2.

Subtracter: Subtracter performs the $z = a - b$ operation on two n-bit numbers a and b, producing an n-bit number z.[*] The subtract operation, $(a - b)$, can be expressed as an addition operation as $(a + -b)$. In the widely used two's complement notation for representing negative binary numbers, the value of $-b$ is $(!b + 1)$. The implementation of a subtracter using an adder is shown in Figure 7.2.

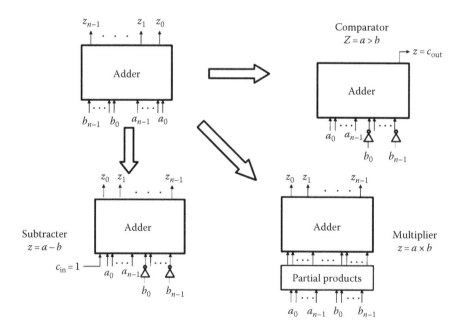

FIGURE 7.2 The implementation of comparator, subtracter, and multiplier using an adder.

---

[*] A carry (borrow)-out is needed, the HDL would pad the operands with an extra bit making it an $(n + 1)$-bit subtract operation.

Multiplier: Multiplier performs the $z = a \times b$ operation on two n-bit numbers a (multiplicand) and b (multiplier), producing a $2n$-bit number z. One of the most commonly used implementations of a multiplier is the array multiplier. The array multiplier consists of a partial product array, where each row is obtained by multiplying a bit from the multiplier to the multiplicand (Figure 7.3). In each row of the multiplier, there is no carry propagation as in a normal multibit adder. Instead, the carry signals are passed on to the next level. Such an adder organization is called a carry-save adder (CSA). The final multibit adder is a conventional adder or a carry-propagate adder. An array multiplier schematic is illustrated in Figure 7.4a. To make the layout a rectangle, the schematic is reshaped, maintaining the array structure and the signal flow, as illustrated in Figure 7.4b. The two input multiplicands enter the module from the left and bottom, and the product leaves from the right and top. The diagonal connections can be implemented with dogleg-shaped wires.

The array multiplier has a speed of $N = 2n$ (bit width of the multibit adder) and is quite regular. Several multiplier architectures have been proposed to optimally implement the addition of the partial products. Improvement in speed can be achieved through rearranging the CSAs such that the longest path is reduced from N to log N. An example is the so-called Wallace tree multiplier illustrated in Figure 7.5 [2]. The notation Ab_i in the figure means a partial product vector. The big disadvantage of a Wallace tree multiplier is the loss of regularity in the layout. Other techniques like Booth encoding can also be used to improve the performance of the multiplier [2].

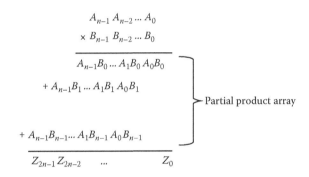

FIGURE 7.3   The implementation of a multiplier using a partial product array and an adder.

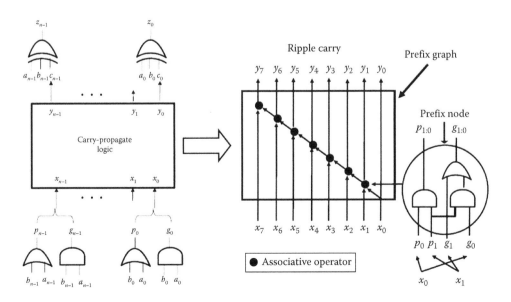

FIGURE 7.4   The construction of a parallel prefix adder.

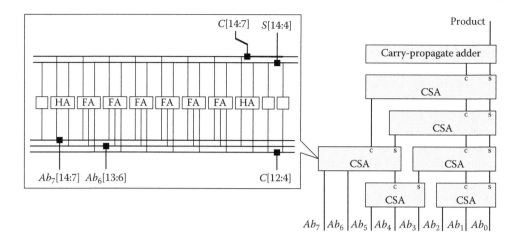

FIGURE 7.5 An 8-bit Wallace tree multiplier.

Complex operations: Other more complex arithmetic operations, such as division, modules, and exponentiation, can be derived from the basic operations described earlier [3].

7.2.1.2 REGISTERS

A datapath often has many pipeline stages, each separated by either register files or simply registers. These registers connect stages of the datapath. Here, we do not refer to registers in the control logic but to those in the main datapath flow. They are also different from the register files in that they are not addressable architecture registers and are not designed as custom circuits. Because large vectors of data must be moved between pipeline stages, these registers can be 64B wide or even wider. The efficient design of these registers is a critical aspect of the overall datapath design.

Two factors are important for datapath register design:

1. Each register bit must be carefully aligned to the data bus to minimize wire length and wiring congestion. In a wide datapath, unaligned register allocation would result in wire crossing between registers and other datapath components. Such unnecessary wire crossings lead to extra congestion and delay on the datapath. Typically, the registers should be placed in the same order as other datapath components. Figure 7.1 shows a datapath with good alignment according to the data flow.
2. The clock driver of the registers should be designed together with the registers and placed to minimize clock load and skew. The conventional clock-tree synthesis techniques create a clock driver to drive registers in different stages or datapath vectors resulting in large skews. For a wide datapath, multiple clock drivers are required to drive the registers. The allocation and placement of those clock drivers should be planned together with the placement and alignment of the registers.

We will discuss techniques to accomplish these two objectives in the later sections of this chapter.

7.2.1.3 OTHER DATAPATH COMPONENTS

Other datapath components, such as barrel shifters and bit-wise logic operators, are simpler than adders, multipliers, and their variations. Unlike the carry chain in the adders, they seldom include long combinational paths through all bits. Thus, most of these components contain a vector of identical elements, which are individually optimized. Area and delay trade-offs for the entire component are often unnecessary, and regularity is well preserved. Decoders and selectors also often occur in vector form, and a regular placement of individual bits can significantly reduce congestion and improve routability of the design.

7.2.2 DATAPATH DESIGN FLOW

A typical design flow for a datapath is illustrated in Figure 7.6. The design input is written in a hardware description language (HDL). Problems like resource scheduling, resource sharing, and pipelining are not described in this chapter but are treated in depth in Chapter 11 of *Electronic Design Automation for Integrated Circuits Handbook*. Datapath components can be identified from the register-transfer-level (RTL) description, and a module generator creates appropriate implementations. The rest of the circuit is compiled to a conventional RTL netlist. At this point, the design is implemented with a list of interconnects linking gates and blocks fully available from the design library. When a module generator is not available, those parts of the circuit that have a certain level of regularity can be extracted from the RTL netlist by a regularity detection algorithm [4]. Regularity extraction can also be applied to a general netlist with or without specific datapath components. The datapath components are floorplanned such that their relative positions in the datapath are determined and circuit timing can be estimated based on the generated floorplan. The RTL netlist is synthesized into a gate-level netlist, at which point datapath components are resynthesized to further improve area and timing. The synthesis process also accounts for timing estimates based on the datapath floorplan. Physical design begins after the gate-level netlist is finalized. The datapath floorplan is used during placement in order to maintain the layout regularity of the datapath and the fidelity of the timing estimation made earlier.

In the following sections, module generation, regularity extraction, floorplanning, resynthesis, and datapath physical design are detailed. Some of these steps can be replaced or enhanced in practice by manual design. Design automation does not conflict with manual intervention; rather, it can serve as a means of quick evaluation and fast feasibility checking while the user is making design changes. The design flow may also require iterations that are triggered when design violations occur. Some commercial synthesis tools advertise "datapath awareness" and integrate module generation, regularity extraction, and resynthesis into conventional design flows.

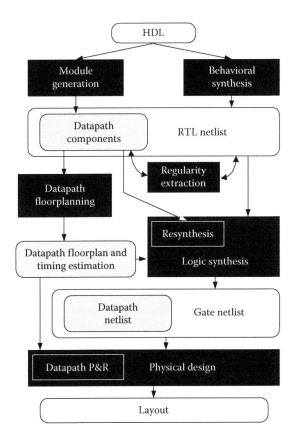

FIGURE 7.6   Datapath design flow.

7.2.2.1 MODULE GENERATION FOR ARITHMETIC LOGIC OPTIMIZATION

Module generation is an automatic method of creating parameterized datapath components. The input is an operator, such as addition, multiplication, or comparison with parameters like data width and speed requirements. The output is a gate-level netlist of the datapath component fully implemented with library cells available in the target technology. Generation of a hard datapath component with its layout requires a library of predesigned modules. This can happen if components laid out in previous designs are kept for reuse, provided that the fabrication technology is unchanged. The generation of an adder, for instance, may involve a set of known structures such as ripple-carry adder, carry-lookahead adder (CLA), carry-bypass adder, carry-select adder (CSA), or ling adder. The type of structure, the area, and delay estimates are chosen for the required bit width and timing, using empirical equations or lookup tables (LUTs). It is preferable to generate the smallest module that meets the timing requirement. Module generation is highly technology dependent, and switching to a different technology may favor entirely different modules under the same functional and timing requirements. The parallel prefix form of CLA is widely used because of the flexibility it offers in the logic structure. Therefore, we further discuss parallel prefix adders in this section.

Parallel prefix adders: Figure 7.4 shows the construction of a parallel prefix adder. The input operands a and b are two n-bit numbers and the output operand is z. At the input side, a propagate p_i and a generate g_i signals are obtained for each bit of the adder. The pair of signals $\{p_i, g_i\}$ for all the bits are then used to obtain the carry outputs y_i of the adder. The carry outputs are then combined with the half sum $(a_i \oplus b_i)$ to obtain the outputs z_i of the adder. The carry-generation logic accounts for most of the area, delay, and power of an adder.

The carry-generation logic is a prefix computation with several choices of structure. A prefix computation on an ordered set of n inputs $x_0, x_1, ..., x_{n-1}$ (where x_{n-1} is the MSB and x_0 is the LSB) and an associative operation o produce an ordered set of n outputs defined as follows:

$$y_i = x_i o \; x_{i-1} o ... o \; x_0 \; \forall i \in [0, \, n-1]$$

Note that the ith output depends on all previous inputs x_j ($j \leq i$). A prefix graph of width n is a graphical representation of a prefix computation. It is a directed acyclic graph (with n inputs and n outputs) whose nodes correspond to the associative operation o in the prefix computation, and there is an edge from node v_i to node v_j if v_i is an input operand for v_j. Figure 7.4b shows the prefix graph for a ripple-carry adder. An introduction to the theory of prefix graphs can be found in [5].

The required bit width and timing determine the choice of structure as well as the area and delay characteristics. The appropriate structures are generated either based on a set of known structures or using algorithmic techniques. Adder design is also highly technology dependent, and switching to a different technology may result in entirely different modules under the same functional and timing requirements. Figure 7.7 shows a few possibilities for 8-bit prefix graphs. Each subfigure is annotated with the level, fanout, and prefix graph size information. Broadly, in the physical implementation, the number of logic levels translates to performance, the size translates to area/power, and the fanout translates to congestion and routability. Notice the trade-off between level, fanout, and size in Figure 7.7.

Regular prefix graph generation: Regular prefix graphs are generated by defining a pattern or rule of connectivity within the prefix graph. The regular adders that are proposed in the literature, such as those of Kogge and Stone [6], Sklansky [7], Brent and Kung [8], Ladner and Fischer [9], Han-Carlson, Knowles, and hybrid power-efficient implementation [10] have different trade-offs in terms of fanouts, logic levels, and sizes (or wire tracks). The trade-offs can be represented using a taxonomy cube [2] illustrated in Figure 7.8.

Algorithmic prefix graph generation: Automated prefix graph generation techniques use algorithms to search the space of prefix graphs to come up with a structure that meets the specified size, fanout, and level constraints. Several algorithms [12–17] have been proposed in the literature. The work on algorithmic prefix graph generation has shown that in the state-of-the-art semiconductor technologies, reducing fanout on the gates in the carry logic can improve the

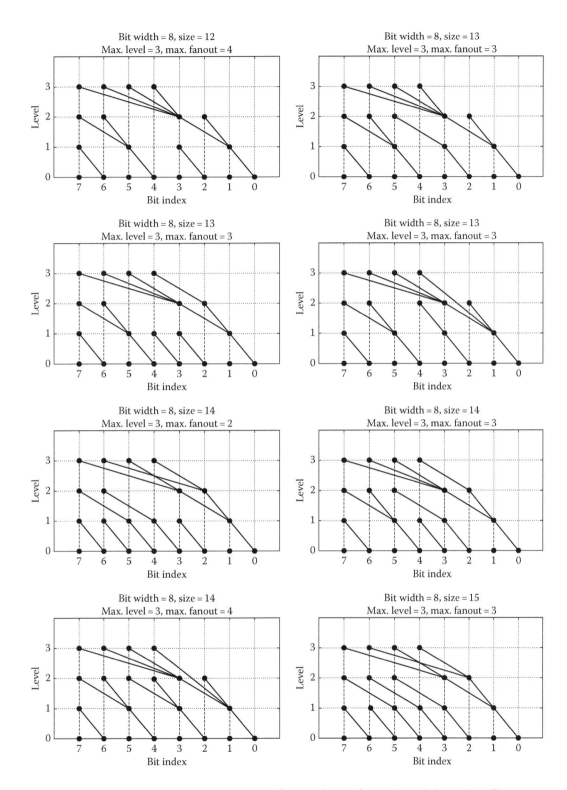

FIGURE 7.7   The implementation alternatives for an 8-bit prefix graph and the trade-off between level, size, and fanout.

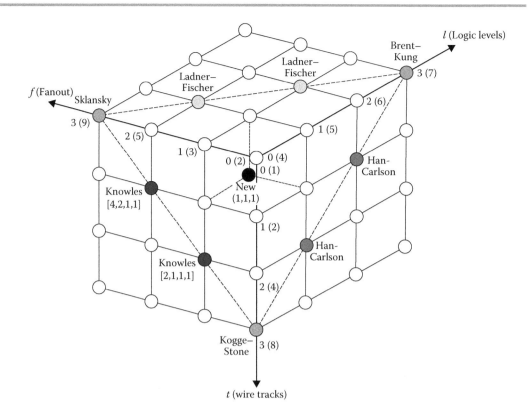

FIGURE 7.8   The taxonomy of regular prefix graph structures. (From Harris, D., A taxonomy of parallel prefix networks, *Proceedings of Asilomar Conference on Signals, Systems and Computers*, 2004.)

performance of adders [16] and algorithmic techniques can generate adders with better performance-area trade-off than regular adders, as illustrated in Figure 7.9b. Each point in the plot represents an adder with a unique prefix graph. Points p1, p2, and p3 correspond to adders generated with techniques presented in [16], DP (dynamic programming based area minimization) corresponds to the adder generated with [17], and BK and KS are Brent–Kung and Kogge–Stone adders, respectively.

Besides the selection of basic logic structures, various optimization techniques can be applied during module generation. These techniques increase the complexity of a module generator but can achieve more economical implementations. The generation process becomes an iterative approach that gradually improves the initial choice of structure. In the following, we list several such optimization methods.

7.2.2.1.1 Arithmetic Transformation   Applying arithmetic transformations to the expression may simplify the operations, resulting in smaller area and delay. For instance, an input expression $Z = A \times B + A \times C$ can be transformed to $Z = A \times (B + C)$, saving one multiplier. Additive operands can swap their positions to change the topology of the addition network without changing the number of adders and the functionality [18]. The transformation allows later inputs to feed adders closer to the end of the network, thus shortening the critical paths.

7.2.2.1.2 Operator Merge   Some complex functions involve several operations, each of which corresponds to a known datapath component like an adder or a multiplier. However, merging the operations may enable resource sharing among different components, reducing the area and the number of levels of logic required. For instance, consider multiplication and accumulation $Z = A \times B + C$, which could be directly implemented as a multiplier followed by an adder. However, it is possible to let operand C merge into the CSA tree of the multiplier to reduce the area and delay. Such operator-merging possibilities are hard to recognize later in the resynthesis once separate multiplier and adder are generated.

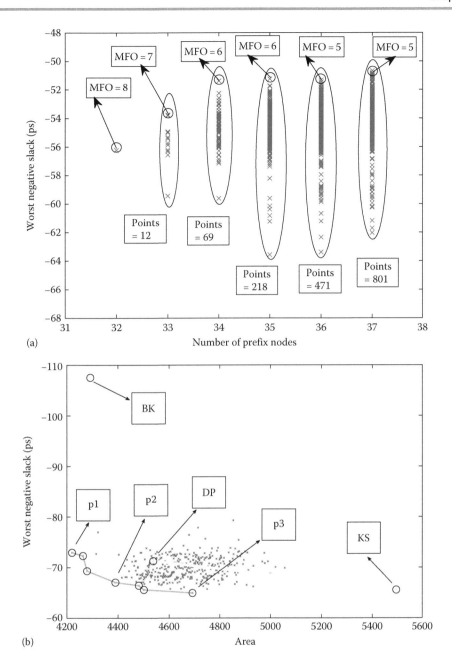

FIGURE 7.9 (a) The adder performance improvement by reducing max. fanout on internal gates. (b) The adder area and performance improvement with algorithmic prefix graph generation over regular prefix graphs such as Brent–Kung and Kogge–Stone adders.

7.2.2.1.3 Constant Propagation and Redundancy Removal   Some operations involve constants. For example, consider $Z = A + 3$. Generating an adder with one input tied to constant "3" can be a starting point. Then the bits of the constant are injected into the adder component, and redundant logic gates are removed along the propagation. Similarly, unused outputs can trigger a backward redundancy removal.

7.2.2.1.4 Special Coefficients   Operations with special coefficients can be replaced by other operations giving the same results but saving area and reducing delay. An example is $Z = 2^M \times X$. Building a multiplier and then using constant propagation can achieve some optimization, but a simpler way is to build an M shifter directly.

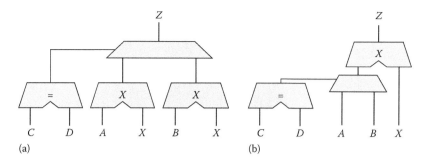

FIGURE 7.10   Restructuring through multiplexing. (a) Faster implementation (b) Smaller implementation.

7.2.2.1.5 Restructuring through Multiplexing   The complex operation $Z = A \times X$ if $(C = D)$ or $B \times X$ if $(C \neq D)$ can be implemented in two ways, as illustrated in Figure 7.10a and b. Obviously implementation (a) is faster but larger than (b). This could be very useful for datapath design, providing alternatives with different area and delay trade-offs.

7.2.2.1.6 Restructuring through Topology Transformations   As shown in Figure 7.11a, consider a complex operation involving three sequential operations. Assume input A takes one among three values A_1, A_2, or A_3. The operation chain can be triplicated, each copy of which has the A input fixed to one of its values. Through constant propagation, OP_1 can be optimized and speeded up. The results of the three copies are selected through a multiplexer controlled by A. The CSA is an application of this kind of generalized select transformation. Another kind is the generalized bypass transformation [19], the application of which is the carry-skip adder. Figure 7.11c shows an example. Inputs C and D are used to produce the *bypass* or *skip* signal, indicating whether Z can be solely represented by the output of OP_1.

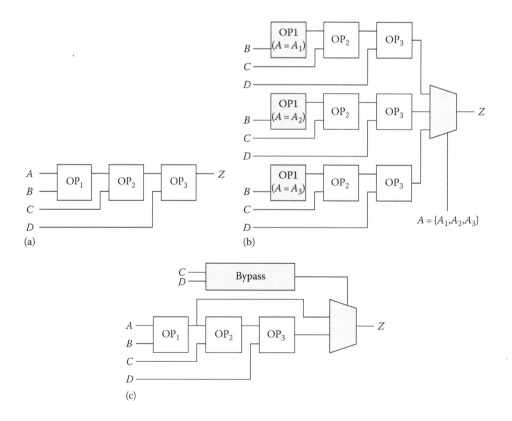

FIGURE 7.11   Restructuring through topology transformations. (a) Original operation (b) Restructuring example with triplication (c) Restructuring example with bypass transformation.

7.2.2.2 REGULARITY EXTRACTION

Regularity extraction can serve as a supplement to module generation. It can collect logic that is not described in terms of datapath components but is regular in nature, for example, it performs multibit operations. Usually, such regular logic appears in datapath components previously designed by hand. There are a few main approaches to regularity extraction. One is to cover the circuit with templates. The other is to expand from *seeds* of multibit structures. Recently, a network flow–based bit-slicing technique has been reported in the literature [4,20].

7.2.2.2.1 Covering Approach   The covering approach is composed of two steps: template generation and covering. A template is a subcircuit that can be instantiated multiple times in the circuit. In contrast to standard cells, templates can be as large as an entire column of the array multiplier. Templates can be given by the user or generated automatically from the circuit. The covering step is similar to the technology mapping step in standard cell designs.

The example shown in Figure 7.12a contains 14 instances where library cells are used as templates. Obviously, this is not what regularity extraction is meant to do. Figure 7.12b and c shows two possible coverings of the circuit with larger templates. Even this simple example discloses two interesting aspects: first, the choice of a template set greatly affects the covering result; second, the size of a template and the number of its instantiations are contradictory. Assuming that only one instantiation of the same template is further optimized and the rest follows the optimization result, a larger number of instantiations are preferred. Smaller templates, which are easier to instantiate more frequently, are more difficult to optimize. In contrast, a large template has more possibilities for optimization. The large template in Figure 7.12c, for example, can be optimized into a smaller and faster one as shown in Figure 7.12d. Therefore, generating a good set of templates is essential to regularity extraction.

Of course, templates can be generated manually, especially when the designer knows what subcircuits need to be used frequently in the circuit. However, an automatic template-generation

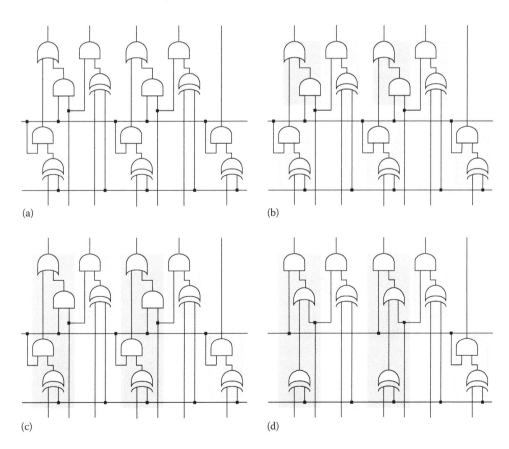

FIGURE 7.12   Regularity extraction. (a) Library-cell based template example. (b) Two library cell template example. (c) Larger template example. (d) Optimization to a larger template.

algorithm was developed [21] in which each pair of gates in the circuit is compared. If identical, then the largest identical (functional and structural) fanin trees are generated for each pair of their input pins. The gate and all its fanin trees form a template. Chowdhary et al. [21] also developed an algorithm that generates templates with multiple fanouts.

Given a set of templates, the next step is to cover the circuit. An iterative approach with greedy covering can be used, based on a combination of the largest-fit-first and most-frequent-fit-first heuristics [21]. In each iteration, the template with the best covering effect—either covering the maximum area or the maximum number of nodes of circuit subgraph—is chosen, the subcircuits that are covered are removed, and the template set is updated. The benefit of this method is twofold. First, it identifies a set of templates and gradually expands it per iteration. Second, it covers the circuit by a subset of these templates and extracts regular circuit components. The iteration is terminated when either the whole circuit is covered or no additional covering is possible.

7.2.2.2.2 Expansion Approach

The expansion approach uses a template set. It starts with a *seed* netlist structure or signal name that indicates regularity and then expands outward until an entire regular structure has been identified. For instance, the net names of a bus in the input netlist usually have the same prefix, so the driver gates of the bus can serve as seeds. Signals like *reset* and *preset* span identical gates in multiple slices, and these gates can serve as seeds as well.

When a set of seeds is collected, the expansion starts. A seed containing one gate per slice is chosen. The expansion in each slice must be exactly identical, with respect to gate functionality, gate pin order, and gate interconnection. The expansion is stopped if mismatches in the slices occur. At this point, the number of mismatched slices is compared against a threshold. If too many slices mismatch, then the expansion is terminated. Otherwise, the slices are partitioned into two groups, with all the matching slices combined in one group and this group is expanded further [22,23].

7.2.2.2.3 Network Flow–Based Datapath Bit Slicing

The network flow approach is to derive the bit slices, based on flow values calculated on a network graph representation of the netlist. A flow network is a directed graph with a source node and a sink node. Besides, each edge is assigned with a capacity and a cost, and the flow on each edge cannot exceed the edge capacity [24,25]. Each gate is represented by a graph node, and the connections between gates are edges. All starting vector gates are connected to the source node and all the ending vector gates to the sink node. The min-cost max-flow algorithm [24] is to find the maximum number of flows from the source to the sink with the minimum cost. An augmenting path indicates the connection path from one starting vector gate to the associated ending vector gate, which in turn, corresponds to a connection path along one-bit slice. The maximum flow in a network corresponds to the number of identified bit slices. This step determines the matching of the starting points and ending points per slice. For example, when the algorithm is applied to a datapath starting from $A[0 \ldots 15]$ to $B[0 \ldots 15]$, the bits are matched as $A[0] \rightarrow B[0], \ldots, A[15] \rightarrow B[15]$. The full bit slice can be extracted with a two-way search approach. The two-way search is to get the fanout/fanin cones of the starting/ending gates of each slice. Only gates identified in both searches are considered as bit-slice gates. This approach tolerates more variations in bit-slice internal structure [4].

7.2.2.3 PHYSICAL DESIGN

Once the datapath netlist is optimized, datapath physical design (including layout) begins. The key ideas of the datapath-driven physical design are (1) how the datapath elements are identified (or defined) and (2) how those elements are placed and routed. The bit slice and stage information is either maintained from the very beginning of the flow or captured during regularity extraction. If the datapath has to be partitioned into several sets of slices during floorplanning, physical design is performed for each set and the connections between the sets are made according to the floorplan.

7.2.2.3.1 Floorplanning

As mentioned, components of a large datapath may be impossible to align in a single set of slices. Thus, smaller datapath slices are constructed whenever the data flow signals change the direction perpendicularly. The datapath floorplanning problem is to determine the relative locations of the datapath components and thus the data flow directions. The general floorplanning problem dealing with the placement of the blocks has been well studied [26]. The objective is to achieve a nonoverlapping placement (locations and orientations) of all the blocks such that the total layout area and interconnection lengths are minimized. The datapath floorplanning problem is unique in that the majority of the nets in a datapath are buses and follow the data flow direction.

One method is to split a long set of slices into several sets and abut them side by side, as shown in Figure 7.13a. With restriction on the total width or height, the goal is to find an optimal partitioning and pack the slice sets such that the total area is minimized. A dynamic programming method has been used to solve this problem [27]. Another method, as shown in Figure 7.13b, is to use a general block-packing algorithm but with additional data flow constraints, that is, adjacent stages should be abutted with the interconnections between them crossing no other stages. Sequence pairs, a representation used in nonoverlapping block placement [26], can be modified to take such constraints into account [28].

It is possible to regenerate some datapath components during floorplanning based on more accurate wiring estimation [29,30]. Initially, the module generator produces implementations of the datapath components that just meet the timing requirements, with the wiring delays ignored. The floorplanning of the components makes available the wiring estimation, in which the datapath components can be replaced by faster implementations if necessary. Swapping implementations of datapath components is not the only way to improve timing. Other methods, especially when wire delays become significant, include component replication and reducing depths (number of stages) of critical paths [31].

7.2.2.3.2 Datapath-Driven Placement and Routing

We will focus on the placement and routing problem of one set of slices in this section. The orthogonal arrangement of the data flow and control flow allows us to cope with two placement problems independently: the ordering of the stages within one slice and the ordering of slices. The slices usually take their natural order, that is, from the LSB to the MSB, so we focus on the ordering of stages.

The ordering of the stages for a set of slices might have been determined by the floorplanning. However, it can be adjusted because, after resynthesis, the internal structure of the stages as well as the interconnections between the stages may have changed. Moreover, the details like the wiring resources, omitted during the floorplanning, could result in inaccurate estimation of routability and timing satisfiability. The stage-ordering problem takes these aspects into account.

As illustrated in Figure 7.14a, a slice contains four stages, labeled from G_A to G_D. The black squares in the figure are the inputs of the stage and the white ones are the outputs. At the bottom of the slice are the external inputs and at the top are the external outputs. The geometric width of a bit of a stage is denoted by $w(stage)$. A stage may have internal routing that occupies $R(stage)$ vertical routing tracks, for example, $R(G_C) = 1$ in the figure. A signal, denoted by l_i, may span several stages. The number of signals crossing a stage is denoted by $r(stage)$, which varies with the ordering of the stages. As mentioned, a signal can come from an adjacent slice in an orthogonal direction, such as the carry signal in an adder. Such special signals are also counted, although this could be an overestimation for the leftmost and rightmost slices. The geometric width of the slice is thus

$$\max\{\max[w(stage)], P \times \max[r(stage) + R(stage)]\}$$

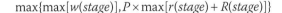

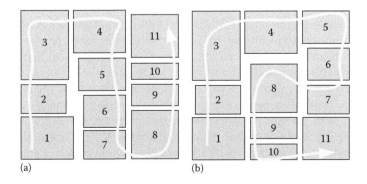

(a)   (b)

FIGURE 7.13 Data flow–aware floorplanning. (a) Datapath floorplanning Example (b) General block packing-based floorplanning for datapath blocks.

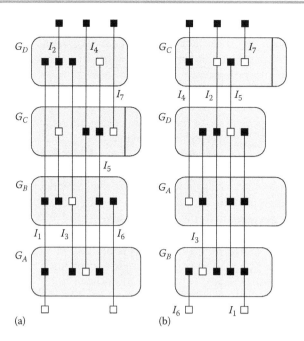

FIGURE 7.14 Ordering of the stages. (a) Original stage ordering (b) After stage ordering optimization.

where P is the vertical wiring pitch. Timing constraints are normally transformed to signal length constraints. The goal of stage ordering is to minimize the slice width while meeting all signal length constraints. Several algorithms can be used to solve this problem. Starting from an initial ordering, usually given by the floorplanning, the ordering of the stages can be altered through simulated annealing [32]; an analytical approach is also feasible [33].

Datapaths can be specifically targeted during placement as the ordering and alignment of gates are the key to successful datapath optimization. Traditional analytical placement techniques are extended to handle datapath structures in [34–36], where the datapath is placed using mixed-block placement techniques. Once the relevant datapath gates are clustered into multiple macros, an existing mixed-block placement engine places datapath gates together, followed by the within-macro datapath detailed placement. Graph automorphism–based datapath extraction and integer linear programming-based bit-stacking selection are proposed to amend the existing half-perimeter wirelength-driven placement [35]. With pseudonets and artificial net weights, global placement is guided to generate a more datapath-friendly output, which is further tuned by bit-stack-based cell swapping.

While Chou et al. [34] and Wang et al. [35] address the datapath placement problem explicitly, Cho et al. [20] take an indirect approach to datapath placement. It first preplaces latches in the datapath with alignment taken into consideration, and then a conventional placer can output a more datapath-friendly placement by taking the preplaced latches as constraints. Also, extracting latches from the datapath is much simpler than finding combinational datapath gates [34,35] with the datapath bit-slicing technique in [4]. Therefore, such latch placement–based datapath optimization can be easily incorporated into the conventional physical design flow and it has been shown to be highly effective [20].

General-purpose routing algorithms can be employed to lay out the interconnections within and between slices. The routing quality depends on the floorplanning and placement. A datapath-specific router [35] uses pin rails that are small predefined horizontal segments crossing several vertical tracks. The routing of a pin pair within a slice seeks to find a path between the two pin rails being used to shift the path from one track to another. The algorithm starts by collecting possible paths for each two-pin connection and splitting the routing probability among them. Then the path probabilities of all the connections are added to the tracks. If routing congestion occurs on a certain segment of a track, some path choices (their probabilities) occupying it are removed. Finally, each path has one and only one choice left, realizing the final routing.

7.2.2.4 RESYNTHESIS

The module generation step generates datapath components based on rough or no wiring delay estimation. Although module regeneration during floorplanning receives more realistic timing information, it still chooses from a fixed set of classical implementations. It is often hard to generate a datapath component that meets timing requirements and has a small area because of the limited number of implementations created from a few classical structures. Therefore, after floorplanning, the datapath components may still have room for further optimization. On the other hand, the regularity of the datapath components should be maintained somehow; complete structural change of the datapath components is not desirable. Another reason for datapath resynthesis is that a datapath component created by a module generator is tagged with special information and forms a rigid boundary, preventing any optimization involving other components or random logic. Breaking the boundary may allow further optimization. However, it loses the boundary information and prevents the identification of the original datapath components. These two factors imply that the datapath resynthesis should focus on the boundary between the existing datapath components and the remaining logic.

Common logic optimization techniques can be adopted [38,39]. However, resynthesis is limited to a subset of the netlist and only small structural changes should take place. Peephole optimization is a useful method for this task [40]. It is a kind of pattern-driven improvement borrowed from compiler techniques, which searches through the netlist under optimization using a small *window*. The subcircuit falling into the window is compared with a set of known patterns. Each identified pattern is replaced by another one that is better in area or delay.

Gate sizing can also be useful in this context [41]. In some cases, upsizing gates along the critical paths may reduce path delays, while downsizing along noncritical paths may reduce area and power. Buffer insertion can help shield heavy loads for driver gates and improve timing.

7.2.2.5 SUMMARY OF THE DATAPATH DESIGN FLOW

The datapath design flow can be summarized as follows. The input HDL description is translated into datapath components using module generators and random logic synthesis. Random logic can be further processed for regularity extraction. A floorplanner determines the relative locations of datapath components. Resynthesis is applied to explore optimization opportunities that lie across component boundaries. Physical design is the last step of the datapath flow where either datapath-oriented constraints are imposed on gates or datapath latches are preplaced. The algorithms involved in the flow take advantage of the regularity of the datapath structure.

7.3 PROGRAMMABLE LOGIC ARRAYS

7.3.1 PROGRAMMABLE LOGIC ARRAY STRUCTURES AND THEIR DESIGN METHODOLOGY

A PLA [52] is a regular structure that is widely used to implement control logic. In the following, the PLA structure is briefly described, and then its design methodology is discussed.

As illustrated in Figure 7.15, the main elements of a PLA are two NOR arrays [42]. At each cross point of a NOR array is either an NMOS transistor or a void (an unconnected or deactivated transistor). The first array is called the AND plane and its output lines called products. The second array is called the OR plane. A PLA represents the logic output functions in a sum-of-products (SOP) form. The *start–done* chain controls the timing of the dynamic PLAs. Owing to its structural regularity, the area and speed of a PLA can be precisely formulated from the SOP expressions that are implemented on it.

The layout generation of a PLA is straightforward. Since the horizontal (product) and vertical (input and output) lines are orthogonal, they can be independently permuted. This permutation acts as placement and routing of the transistors. Hence, the simple PLA structure shown in

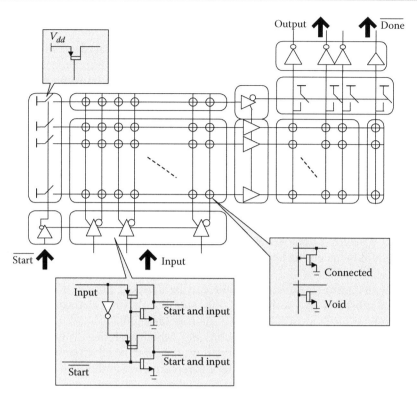

FIGURE 7.15   Programmable logic array structure (dynamic).

Figure 7.15 does not need physical design. A PLA layout–generation program takes the transistor bit maps of the two arrays and outputs a PLA layout. Note that although the PLA itself does not require physical design, a circuit being instantiated by multiple PLAs requires placement and routing for the PLA blocks. The synthesis of PLAs has been well studied and efficient algorithms exist [43,38] for their optimization.

The ratio of the void transistors in the AND and OR planes to the actual transistors can be high in many designs, wasting area and slowing down the PLA because of longer wires. As shown in the simplified PLA diagram in Figure 7.16, the pair of inputs i_1 and i_3 can share the same column (a pair of complementary vertical lines). The same thing applies to outputs o_1 and o_2. Such horizontal compaction is called column folding of the PLA [44]. A restricted version of PLA column folding is discussed in the following. The two complementary lines of an input signal should remain together so that there is only one input pin per input signal. At most, two signals can share one column; hence, pins are always on the top and bottom boundaries of the PLA. The column foldings of the AND and OR planes are separate, so the basic dual-plane structure of the PLA is preserved. With these restrictions, the PLA column-folding problem is reduced to separate interval-packing problems for the AND and OR planes [45]. In the literature, there are also mentions of row-folding methods and simultaneous column-/row-folding methods [46,47], removing the aforementioned restrictions. However, in reality, it is hard for a dynamic PLA structure to support these more aggressive folding methods because of the arrangement of the

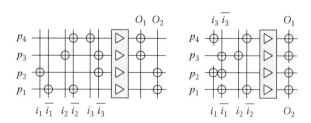

FIGURE 7.16   Programmable logic array folding.

start–done chain and the precharging circuits. In addition, too much folding may cause the input and output pins to spread irregularly on the PLA, which in turn generates placement and routing problems when integrating the PLA onto the chip. These techniques were popular in earlier times when NMOS static PLAs were widely used.

The PLA technology has not seen much change for a while. Recent focus has been on (1) the device density increase with process technology, (2) the importance of low power, and (3) the inclusion of a broad range of intellectual property blocks such as analog-to-digital converters and digital signal processing units.

7.3.2 OTHER PROGRAMMABLE LOGIC ARRAY STRUCTURES

The regular array structure of PLAs is attractive. A single PLA implements several two-level logic functions. However, many logic functions cannot be efficiently implemented using only two levels. In general, multilevel logic synthesis techniques must be adopted to represent various logic functions [39]. A straightforward approach is to have a multilevel Boolean network, with each node in the network being an SOP. After the minimization of the network, the nodes can be mapped to the PLAs. However, this *network of PLA* approach [48] requires block-level placement and routing, causing a loss of global regularity.

Newer PLA-based structures have been proposed. These structures not only maintain the internal regularity of the PLA but also maintain a global regularity when putting multiple PLAs together. The "Whirlpool PLA" is a four-level cyclic structure [49]. The Whirlpool PLA can be synthesized with a four-level minimizer called "Doppio Espresso," which iteratively calls *Espresso* [43], a two-level minimizer, to minimize a pair of adjacent NOR arrays, that is, an SOP. Another PLA-based structure is the "River PLA" [50], which is a stack of PLAs. Interconnections are only present between pairs of adjacent PLAs, and these are implemented with a simple river routing owing to the full permutability of the vertical lines in the PLAs. The synthesis of a River PLA involves multilevel logic minimization, clustering nodes into PLAs in the stack, two-level minimization for each PLA, and river routing with vertical line permutation. A checkerboard is an array of NOR planes [51]. The synthesis algorithms developed for the checkerboard make use of its structural regularity and flexibility. The multilevel logic network is optimized and decomposed into a netlist of NOR gates. Then the gates are placed into the checkerboard. Input pin sharing can occur between gates placed in the same NOR plane. Routing between the NOR planes uses a simple spine topology, which is fast enough to allow the placement and routing to be done at the same time. A detailed description and comparison of all these PLA-based structures can be found in [50,51].

7.4 MEMORY AND REGISTER FILES

7.4.1 MEMORY

Memory components are usually generated by a memory compiler that comes with the technology library [54]. A static RAM (SRAM) compiler is discussed here. Although other memory structures like dynamic RAM and read-only memory have different memory units and surrounding logic, they use very similar compilation techniques.

Large SRAM blocks can be manually partitioned into smaller SRAM banks and floorplanned, or the compiler can be given some predefined partitioning schemes and floorplans [55]. Each bank has a small set of choices of implementations, which differ in area, performance, and power. When input parameters like the address space and the choice of area and speed are given, the memory compiler simply tiles an array of predefined memory unit cells and places peripheral elements like the row decoder and sense amplifiers around the memory array. The output of a memory block compilation includes a layout (which could be an abstracted view of the actual layout), a behavioral model, and a Simulation Program with Integrated Circuit Emphasis (SPICE) model. In the following discussion, we focus on how the memory compiler itself is created; in other words, how the memory unit cells and the peripheral elements are systematically, and, to

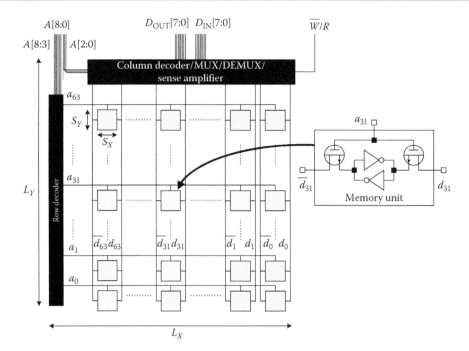

FIGURE 7.17 A 512 × 8-bit static RAM example.

some extent, automatically designed. Although the work can still be done manually, it is much more complicated than the design of a standard cell.

The problem can be stated as follows. Given the address space 2^N, data bit width 2^B, and speed requirement F, determine s_X and s_Y, the X/Y size of the memory unit, the number of rows (2^r), and the number of columns (2^c) such that the speed requirement is satisfied while the total area is minimized. In the statement, the speed requirement F is abstracted, which in reality could represent a set of timing requirements, such as read delay, write delay, and so on.

In Figure 7.17, a simplified 512 × 8-bit SRAM structure is shown. The main body is an array of 2^r-row-by-2^c-column memory units. Note that $2c = 2^{N-r}B$. Figure 7.17 also shows the structure of a typical six-transistor SRAM unit. The high part of the address $A[N-1:N-r]$ is decoded, and one of the 2^r rows is activated. The activated row enables all the memory units in that row, which are connected to the complementary vertical bit-line pairs. If the SRAM is operating in *read* mode, then the lower address $A[N-r-1:0]$ selects B data bits by the multiplexers and the signals are regenerated by the sense amplifiers. If the SRAM is operating in *write* mode, then the data input D_{IN} is demuxed to the corresponding complementary bit-line pairs by $A[N-r-1:0]$.

As most of the SRAM area is occupied by the memory bit units as shown in Figure 7.17, as most of the SRAM area is occupied by the memory bit units, it is reasonable to let s_X and s_Y, rather than the sizes of row decoder, sense amplifiers, etc., to determine the column width and row height. The area of the SRAM block is approximately $2^{r+c}s_X s_Y$, and this is what we want to minimize. For simplicity, we will assume that s_X and s_Y are equal and are written as a single variable s. The sizes of the row decoder and column peripheral circuits are denoted by s_1 and s_2, respectively. The problem, with extensive abstraction, can be formulated as

$$\text{variables}: s, s_1, s_2, r, c$$

$$\text{minimizing} \quad s$$

$$\text{subject to}$$

$$s \geq S_0$$

$$\text{aspect\_ratio}\big(s, r, c\big) \leq ASP_0$$

$$f\big(s, r, c, h_1\big(s, s_1, c\big), h_2\big(s, s_2, r\big)\big) > F$$

The first constraint means that the memory unit cannot be so small that, for example, the sense amplifiers cannot be properly laid out. S_0 is the lower bound for s. The second constraint limits the aspect ratio of the block by ASP_0. The third constraint represents the performance requirement. Performance f, a set of various speeds, is a function of the memory unit size, row and column numbers, row decoder performance h_1, and column periphery performance h_2. Given memory unit size s, the word line capacitance can be extracted. Together with the load capacitance of a memory unit cell, which is also a function of s, the total load of a row on the row decoder can be computed. When the row decoder is sized to s_1, its performance h_1 can be calculated. Similarly, h_2, the performance of the column periphery, can be calculated. The overall performance f depends on r, c, h_1, h_2, and the cell performance. Strictly speaking, memory unit performance is state dependent. For example, when computing the write delay for a memory unit, the previous value of its memory state matters. As a result, f is actually the minimum performance among all possible state transitions (from 0 to 1, 1 to 1, and so on) [56]. The evaluation of function f, which involves the constructions of the unit cell, row decoder and column peripheral circuits, parameter extraction, and SPICE simulation, is time consuming. The constrained optimization problem, although nonlinear or even discrete in nature, could be solved with various techniques. Runtime spent on finding a solution is rarely an issue, because this optimization only needs to be run once for a given technology when configuring the memory compiler. Gradient search is a simple method to use [56]. Exhaustive search is also possible because the number of variables (roughly the number of transistors that need to be sized) is not large, and their ranges are usually small. Even if the sizes of the transistors are manually adjusted, the evaluation of f can still be programmed and automated, which greatly eases the burden of complicated extraction and simulation.

7.4.2 REGISTER FILES

A register file is a special kind of SRAM, which is usually small in memory size but has more than one read/write port. It is widely used in microprocessors and digital signal processors as a fast and versatile storage element. The circuit structure of the register file may differ from that of the SRAM in order to achieve higher performance. However, the basic organization of the memory unit array and peripheral circuits remains the same. Although register files are usually designed by hand to achieve the highest performance, the automatic or semiautomatic optimization similar to SRAM compilation can still be used. The optimization would now focus on the performance rather than the area.

7.4.3 SPECIAL APPLICATIONS OF MEMORY

Memory can be used not only to store data but also to implement logic evaluation. A LUT is a memory block that stores a truth table in the memory cells. The input, feeding the address of the memory, gives the index of the memory cell that stores the output value.

A field-programmable logic array (FPGA) is a LUT-based structure, using SRAM cells to configure logic functions and the interconnections. The design flow for FPGAs is similar to that for standard cells, but the major difference is that the technology mapping targets LUTs as opposed to individual gates. The output of the flow is not a layout but a bit stream that configures the LUTs.

Another application of memory is the cascaded realization of logic function evaluation [57]. As shown in Figure 7.18, the input vector I of logic function $f(I)$ is partitioned into M disjoint subsets labeled as $I_0, I_1, I_2, \dots, I_{M-1}$, and the evaluation of f becomes a sequence of M evaluations, each involving only one subset of the input. The intermediate output of an evaluation is an input to the next evaluation, joined by another subset of I. A sequential version of this approach includes a memory containing M LUTs, a multiplexer selecting a subset of the external inputs, a register storing the intermediate output, an output register storing the output, and a counter sequencing the whole operation. The LUTs, which can be of different sizes, need to be packed into the memory. The partitioning of the input and the realization of the function at each stage can be implemented by binary decision diagram manipulation. The structure is regular and very flexible but runs at low speed.

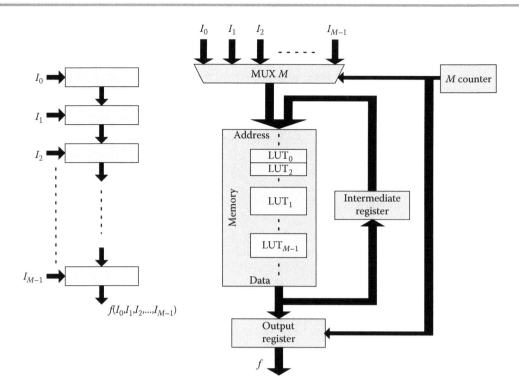

FIGURE 7.18 Cascaded lookup table structure.

A noticeable trend in FPGAs is that vendors start to offer a more comprehensive platform for extensive functionality. It is common to have embedded processor core(s), sensors, and some specialized functional units such as specialized multipliers in modern FPGAs [60,61]. Also, the size of embedded memory (the number of bits) grows significantly with every generation allowing one to avoid unnecessary Input/Output (IO) communications to outside.

7.5 STRUCTURED CHIP DESIGN

In this section, we extend our view from regular circuit modules like datapaths and PLAs to regular full-chip structures [53]. Several structures, such as GAs, SASICs, and VPGAs, as well as their design automation are discussed. The guidelines for using these structures are also given.

7.5.1 GATE ARRAYS

Although these structures are called GAs, the name "transistor array" might be a more appropriate term. An example of a GA structure is illustrated in Figure 7.19a. The core of a GA is composed of an array of basic cells, also called "bases" interleaved by routing channels. The most widely used base structure in CMOS technology is a four-transistor block, two NMOS transistors, and two PMOS transistors. As shown in Figure 7.19b, a base can be configured as a two-input NAND gate by connecting the terminals of the transistors. To configure more complex gates, such as the four-input NAND gate example in Figure 7.19b, multiple adjacent bases can be used. All the metal layers of the GA structure are programmable. If channels exist, the metal layers on top of the bases are used to build the internal connections of the gates; the interconnections between the gates are built within the channels. When more metal layers are available, channels are used to construct larger modules, and the GA becomes a sea of gates with intragate routing taking place on lower metal layers and intergate routing on the upper ones. Masks like diffusion layers are unchanged from design to design if the same GA template is used. The patterns on the fixed layers are regular. Therefore, the GA structure, compared to standard cells, is more cost-effective and manufacturable.

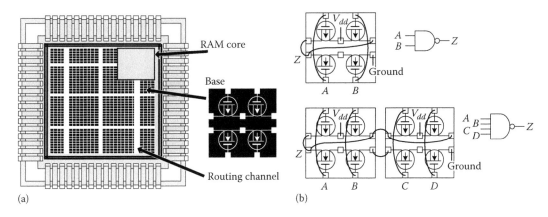

(a)   (b)

FIGURE 7.19   Gate array example. (a) GA structure (b) Four input NAND gate implementation with two input NAND bases.

The synthesis for GAs is almost identical to standard cells, except that the library cells become master cells like the ones shown in Figure 7.19b. The area of a gate is measured by the number of bases it occupies. The physical design for the GA modules is also very similar to standard cells. If the GA structure contains channels, a channel router is needed; if the GA does not contain any channel, and routing takes place over the cell, then an area router is needed.

7.5.2 STRUCTURED APPLICATION-SPECIFIC INTEGRATED CIRCUITS

Cell-based chip structures that dominate today's integrated circuits market provide high area utilization and performance. However, several disadvantages such as the timing closure problem, long turnaround time, and expensive mask costs jeopardize the usefulness of the standard cell structure to meet the fast-growing design complexity and DSM challenges. The ultimate reason for these disadvantages is that standard cells are not regular enough, leading to timing unpredictability and manufacturability problems. SASIC is a new type of chip structure [58,59] introduced to address some of these problems. In contrast to the transistor array structure in the GA, it embeds a cell array on the chip. The SASIC inherits from the GA the benefits of fewer design-dependent masks, enhancing manufacturability and saving mask costs. Power, clock, and scan chain networks are often optimized and fixed for a particular SASIC by the vendor. This greatly simplifies the design flow and reduces the turnaround time. From small- to medium-volume productions, SASIC could be a good alternative to standard cells.

The core area of a typical SASIC, as illustrated in Figure 7.20, consists of arrays of SRAM and gate sites. A site is a physical location on the die where a few types (usually only one) of logic primitives can be implemented. In some SASIC architectures, if an SRAM site is not used, it can be treated as a set of gate sites. The fixed site locations reduce the number of design-dependent masks; hence, the placement problem of logic gates and SRAM modules becomes that of assigning objects to valid and vacant sites. The types of SASIC gate sites are limited compared to the types of gates in a typical standard cell library. The lower metal layers of a SASIC are mostly occupied by the internal connections and configurations of the gates and SRAM, as well as power, ground, and clock networks. A few higher metal layers can be used for signal routing, but some regions on these layers might be occupied by prerouted nets, such as clocks, and some of the internal connections of the SRAMs.

The design methodology for SASICs is very similar to that for standard cells from logic synthesis to physical design. The discussion in this chapter focuses on the unique features of the SASIC design flow. The technology mapping for SASICs deals with a relatively small gate library. Although this implies that some degree of optimization in technology mapping is lost, runtime can be reduced because of fewer gate types or the algorithm can afford to use more CPU-intensive algorithms. Gate sizing and buffering for SASICs have the same limitations and

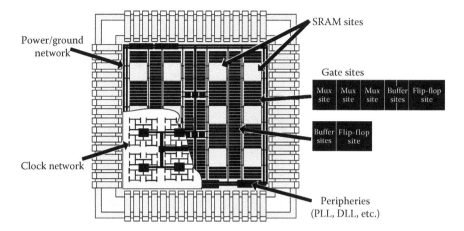

FIGURE 7.20   A structured application-specific integrated circuit example.

benefits due to the restricted gate choices. The placement of the cells and macroblocks must satisfy the site constraints, that is, an object must be placed in an appropriate predefined site on the chip. This is somewhat similar to GA placement, but the difference is that the sites of a SASIC are heterogeneous. A general-purpose cell placer, which is totally site unaware, would have difficulty in translating its placement result to an assignment of gates to legal sites. The placement algorithm needs to control site utilization during placement. An analytical cell placer could be modified so that it alleviates not only the overall row utilization but also the utilization of different site types. A partitioning-based placer could consider similar criteria when swapping cells across the cuts. The power/ground routing for SASICs is normally unnecessary. Clock routing, if not fully predesigned by the SASIC vendor, may still be needed, but it is greatly simplified because of the regularly arranged register sites.

7.5.3 VIA-PATTERNED GATE ARRAY

Both GAs and SASICs require some metal and via layers for configuring the interconnections. Making only via layers configurable leads to a regular chip structure called "via-patterned GA" [60,61]. VPGAs adopt the same logic unit used in FPGAs, the configurable logic block (CLB). A CLB consists of one or two LUTs, one or two flip-flops, and a few other elements like multiplexers. All the metal layers are fixed. The functionalities of the CLBs are configured using lower via layers. On top of each CLB, higher metal layers form a fixed crossbar structure, and signals are routed through vias between these layers. *Jumpers* between the CLBs and the crossbars relay the signal wires from one crossbar to another. A similar routing structure is used in the checkerboard [49], a PLA-based regular structure described in Section 7.3.1. Using a crossbar is not the only choice for VPGAs; switch boxes or other topologies can also be employed [60]. The only design-dependent masks of a VPGA structure are the via layers, reducing mask costs significantly. Most other layers are laid out in a regular and optimal way, so the manufacturability is enhanced. An example of a VPGA is illustrated in Figure 7.21, with the internal structure of a CLB detailed. The three-input LUT is implemented as a symmetric tree rooted at the output terminal Z. The input signals A, B, and C and their complements control the conductive paths from Z, realizing any logic function of the three binary inputs. The via configuration of the LUT in the example implements the logic function $Z = AB\overline{C} + \overline{A}BC$.

Synthesis for VPGAs, including technology-independent optimization and LUT mapping, can be borrowed largely from FPGA methodologies. Routing for VPGAs is more flexible than for FPGAs because of the *sea-of-wires* crossbars. Area routing methods can be tailored to work with the crossbar structure. Modifications come from the fact that two signals cannot share the same segment in a crossbar of the VPGA.

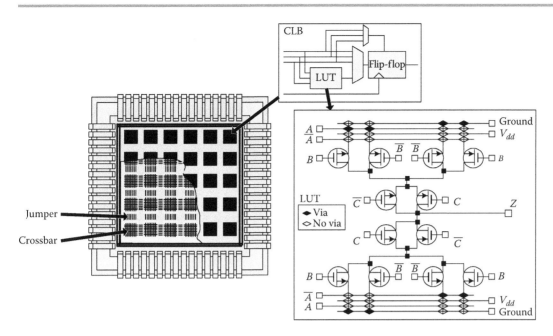

FIGURE 7.21   An example of the via-patterned gate array.

7.5.4 GENERAL GUIDELINES FOR STRUCTURED CHIP DESIGN

Different methodologies can be compared with respect to a number of metrics such as area, performance, power, simpler design methodologies, lower mask costs, faster time to market, and improved manufacturability. The last concern is difficult to quantify and motivates additional research. It is difficult to claim that one metric is more important than the others because it depends on the design targets and constraints. The last four metrics are becoming increasingly important as the technology advances. GA, SASIC, and VPGA are all competitors of standard-cell-based design, but they enjoy advantages of simpler design methodologies, lower mask costs, faster time to market, and improved manufacturability. On the other hand, area and performance disadvantages would prevent them from being used in high-volume and high-performance applications. If any one of the criteria such as area or performance or power is critical (e.g., in microprocessors or wireless applications), standard cells and manual techniques are still the best choice. GAs require the most mask designs, but their placement and routing are more flexible and less constrained, potentially allowing higher area utilization and performance than the SASICs and VPGAs. VPGAs have the greatest area and performance penalties. SASICs sacrifice some placement flexibility but maintain some routing flexibility and thus are placed between GAs and VPGAs in terms of area and performance. FPGAs are best for fast prototyping and low-volume applications, but they are relatively slow and use a lot of area and power. Their regularity and programmability may increase their usage in ultra-DSM technologies. Some research suggested that PLA-based design may compete with structured chip designs (SCs) in terms of area, performance, and power while offering improved manufacturability and predictability.

In terms of availability of design automation tools, GAs have been on the market for a long time, so EDA tools are readily available; Cadence and Synopsys both provide tools for various GA design stages. Academic VPGA tools are available [58]. Most major electronic design automation (EDA) companies as well as FPGA vendors themselves provide FPGA design tools.

7.6 SUMMARY

In this chapter, we have discussed structured digital design, including datapaths, PLAs, and memories. Regular chip structures like GAs, SASICs, and VPGAs were also discussed. Mainstream design automation approaches for these structures were outlined. The regularity of structured designs can offer high performance, compact layout, reduced mask cost, and improved manufacturability but requires additional design effort and knowledge. Many structure-specific algorithms have been developed to identify, maintain, and exploit regularity. In practice, many structured designs are still prepared with little automation. However, automated layout methods for structured designs, for example, the detection and pattern-based layouts for bit-slicing-based datapath components, are becoming more attractive solutions as both design complexity and design requirements increase. This area still remains critical for further research in the physical design automation domain and promises new commercial methodologies.

REFERENCES

1. International Technology Roadmap for Semiconductors (ITRS) reports, http://www.itrs.net/.
2. D. A. Patterson and J. L. Hennessy, *Computer Architecture, A Quantitative Approach*, Appendix A, Morgan Kaufmann, San Francisco, CA, 1996.
3. M. J. Flynn and S. F. Oberman, *Advanced Computer Arithmetic Design*, Wiley-Interscience, New York, 2001.
4. H. Xiang, M. Cho, H. Ren, M. Ziegler, and R. Puri, Network flow based datapath bit slicing, *Proceedings of the 2013 ACM International Symposium on International Symposium on Physical Design*, Santa Rosa, California, USA, 2013, pp. 139–146.
5. R. Zimmermann, Non-heuristic optimization and synthesis of parallel prefix adders, *International Workshop on Logic and Architecture Synthesis*, 1996, pp. 123–132.
6. P. M. Kogge and H. S. Stone, A parallel algorithm for the efficient solution of a general class recurrence equations, *IEEE Transactions on Computers*, C-22:786–793, 1973.
7. J. Sklansky, Conditional sum addition logic, *IRE Transactions on Electronic Computers*, 9:226–231, 1960.
8. R. P. Brent and H. T. Kung, A regular layout for parallel adders, *IEEE Transactions on Computers*, C-31:260–264, 1982.
9. R. E. Ladner and M. J. Fischer, Parallel prefix computation, *Journal of ACM*, 27(4):831–838, 1980.
10. C. Zhuo et al., 64-bit prefix adders: Power efficient topologies and design solutions, *Custom Integrated Circuits Conference, 2009 (CICC '09)*, San Jose, CA, 2009, pp. 179–182.
11. D. Harris, A taxonomy of parallel prefix networks, *Proceedings of Asilomar Conference on Signals, Systems and Computers*, Pacific Grove, California, USA, 2004.
12. J. Liu et al., Optimum prefix adders in a comprehensive area, timing and power design space, Asia and South Pacific Design Automation Conference (*ASPDAC*), Yokohama, Japan, 2007, pp. 609–615.
13. M. Snir, Depth-size trade-offs for parallel prefix computation, *Journal of Algorithms*, 7(2):185–201, 1986.
14. J. P. Fishburn, A depth decreasing heuristic for combinational logic; or How to convert a Ripple-Carry adder into a Carry-Lookahead adder or anything in-between, *Proceeding DAC '90 Proceedings of the 27th ACM/IEEE Design Automation Conference*, Orlando, FL, 1990, pp. 361–364.
15. J. Liu et. al., An algorithm approach for genetic parallel adders, *International Conference on Computer-Aided Design*, San Jose, California, USA, 2003, pp. 734–740.
16. S. Roy et al., Towards optimal performance-area trade-off in adders by synthesis of parallel prefix structures, *Design Automation Conference*, Austin, Texas, USA, 2013, pp. 48:1–48:8.
17. T. Matsunaga and Y. Matsunaga, Area minimization algorithm for parallel prefix adders under bitwise delay constraints, *Great Lakes Symposium on VLSI*, Stresa-Lago, Maggiore, Italy, 2007, pp. 435–440.
18. I. Neumann, D. Stoffel, M. Berkelaar, and W. Kunz, Layout-driven synthesis of datapath circuits using arithmetic reasoning, *International Workshop on Logic and Synthesis*, Laguna Beach, CA, 2003, pp. 1–6.
19. P. C. McGeer, R. K. Brayton, A. L. Sangiovanni-Vincentelli, and S. Sahni, Performance enhancement through the generalized bypass transform, *International Conference on Computer-Aided Design*, Santa Clara, California, USA, 1991, pp. 184–187.
20. M. Cho, H. Xiang, H. Ren, M. Ziegler, and R. Puri, LatchPlanner: Latch placement algorithm for datapath-oriented high-performance VLSI designs, *International Conference on Computer-Aided Design*, San Jose, California, USA, 2013.
21. A. Chowdhary, S. Kale, P. Saripella, N. Sehgal, and R. Gupta, A general approach for regularity extraction in datapath circuits, *International Conference on Computer Aided Design*, San Jose, CA, 1998, pp. 332–338.
22. T. Kutzschebauch, Efficient logic optimization using regularity extraction, *International Conference on Computer Design*, San Jose, California, USA, 2000, pp. 487–493.
23. S. R. Arikati and R. Varadarajan, A signature-based approach to regularity extraction, *International Conference on Computer-Aided Design*, San Jose, California, USA, 1997, pp. 542–545.

24. T. Cormen, C. Leiserson, R. Rivest, and C. Stein, *Introduction to Algorithms*, MIT Press, Cambridge, MA, 2001.

25. R. K. Ahuja, T. L. Magnanti, and J. B. Orlin, *Network Flows*, Prentice Hall, Englewood Cliffs, NJ, 1993.

26. H. Murata, K. Fujiyoshi, S. Nakatake, and Y. Kajitani, VLSI module placement based on rectangle packing by the sequence-pair, *IEEE Transactions on Computer Aided Design Integrated Circuits Systems*, 15:1518–1524, 1996.

27. D. Paik and S. Sahni, Optimal folding of bit sliced stacks, *IEEE Transactions on Computer Aided Design*, 12:1679–1685, 1993.

28. M. Moe and H. Schmit, Floorplanning of pipelined array modules using sequence pair, *International Symposium on Physical Design*, Del Mar, California, USA, 2002, pp. 143–150.

29. V. G. Moshnyaga et al., Layout-driven module selection for register-transfer synthesis of sub-micron ASICs, *International Conference on Computer-Aided Design*, Santa Clara, California, USA, 1993, pp. 100–103.

30. F. Mo and R. K. Brayton, A timing-driven module-based chip design flow, *Design Automation Conference*, San Diego, California, USA, 2004, pp. 67–70.

31. V. G. Moshnyaga and K. Tamaru, A floorplan-based methodology for data-path synthesis of sub-micron ASICs, *IEICE Transactions on Information System*, E79-D:1389–1396, 1996.

32. J. S. Yim and C. M. Kyung, Data path layout optimization using genetic algorithm and simulated annealing, *IEEE Proceedings of Computers and Digital Techniques*, 145(2):135–141, 1998.

33. T. T. Ye and G. De Micheli, Data path placement with regularity, *International Conference on Computer-Aided Design*, San Jose, California, USA, 2000, pp. 264–271.

34. S. Chou, M.-K. Hsu, and Y.-W. Chang, Structure-aware placement for datapath-intensive circuit design, *Design Automation Conference*, San Francisco, California, USA, 2012.

35. S. Ward, D. Ding, and D. Pan, PADE: A high-performance mixed-size placer with automatic datapath extraction and evaluation through high-dimensional data learning, *ACM/IEEE Design Automation Conference*, San Francisco, California, USA, 2012.

36. S. Ward, M.-C. Kim, N. Viswanathan, Z. Li, C. Alpert, E. Swartzlander, and D. Pan, Keep it straight: Teaching placement how to better handle designs with datapaths, *Proceedings of. ACM International Symposium on Physical Design (ISPD)*, Napa, California, USA, March 2012.

37. S. Raman, S. S. Sapatnekar, and C. J. Alpert, Datapath routing based on a decongestion metric, *International Symposium on Physical Design*, San Diego, California, USA, 2000, pp. 122–127.

38. R. Brayton, G. Hachtel, C. McMullen, and A. Sangiovanni-Vincentelli, *Logic Minimization Algorithms for VLSI Synthesis*, Kluwer Academic, Dordrecht, the Netherlands, 1984.

39. R. Brayton, G. Hachtel, and A. Sangiovanni-Vincentelli, Multi-level logic synthesis, *Proceedings of IEEE*, 78:264–300, 1990.

40. T. Chelcea and S. M. Nowick, Resynthesis and peephole transformations for the optimization of large-scale asynchronous systems, *Design Automation Conference*, New Orleans, Louisiana, USA, 2002, pp. 405–410.

41. A. J. Kim, C. Bamji, Y. Jiang, and S. Sapatnekar, Concurrent transistor sizing and buffer insertion by considering cost–delay tradeoffs, *International Symposium on Physical Design*, Napa, California, USA, 1997, pp. 130–135.

42. Y. B. Dhong and C. P. Tsang, High-speed CMOS POS PLA using predischarged OR array and charge sharing AND array, *IEEE Transactions of Circuits System II: Analog Digital Signal Process*, 39:557–564, 1992.

43. R. Rudell and A. Sangiovanni-Vincentelli, Multiple-valued minimization for PLA optimization, *IEEE Transactions on Computer Aided Design*, 6:727–750, 1987.

44. R. A. Wood, A high-density programmable logic array chip, *IEEE Transactions on Computer*, C-28:602–608, 1979.

45. N. A. Sherwani, *Algorithms for Physical Design Automation*, Kluwer Academic, Dordrecht, the Netherlands, 1993.

46. D. F. Wong, H. W. Leong, and C. L. Liu, PLA folding by simulated annealing, *IEEE Journal of Solid-State Circuits*, SC-22:208–215, 1987.

47. G. Hachtel, A. R. Newton, and A. Sangiovanni-Vincentelli, An algorithm for optimal PLA folding, *IEEE Transactions on Computer Aided Design Integrated Circuits Systems*, 1:63–77, 1982.

48. S. Khatri, R. Brayton, and A. Sangiovanni-Vincentelli, Cross-talk immune VLSI design using a network of PLAs embedded in a regular layout fabric, *International Conference on Computer-Aided Design*, San Jose, California, USA, 2000, pp. 412–418.

49. F. Mo and R. K. Brayton, Whirlpool PLAs: A regular logic structure and their synthesis, *International Conference on Computer-Aided Design*, San Jose, California, USA, 2002, pp. 543–550.

50. F. Mo and R. K. Brayton, River PLA: A regular circuit structure, *Design Automation Conference*, New Orleans, Louisiana, USA, 2002, pp. 201–206.

51. F. Mo and R. K. Brayton, Checkerboard: A regular structure and its synthesis, *International Workshop on Logic and Synthesis*, Laguna Beach, California, USA, 2003, pp. 7–13.

52. F. Mo and R. K. Brayton, PLA-based regular structures and their synthesis, *IEEE Transactions on Computer Aided Design Integrated Circuits Systems*, 22(6):723–729, 2003.

53. F. Mo and R. K. Brayton, *Regular Fabrics in Deep Sub-Micron Integrated-Circuit Design*, Kluwer Academic, Dordrecht, the Netherlands, 2004.

54. J. Tou, P. Gee, J. Duh, and R. Eesley, A submicrometer CMOS embedded SRAM compiler, *IEEE Journal of Solid-State Circuits*, 27:417–424, 1992.

55. H. Shinohara, N. Matsumoto, K. Fujimori, Y. Tsujihashi, H. Nakao, S. Kato, Y. Horiba, and A. Tada, A flexible multi-port RAM compiler for data path, *IEEE Journal of Solid-State Circuits*, 26:343–349, 1991.

56. A. Chandna, C. Kibler, R. Borwn, M. Roberts, and K. Sakallah, The Aurora RAM compiler, *Design Automation Conference*, San Francisco, CA, 1995, pp. 261–266.

57. T. Sasao, M. Matsuura, and Y. Iguchi, A cascade realization of multiple output function for reconfigurable hardware, *International Workshop on Logic Synthesis*, Lake Tahoe, CA, 2001, pp. 225–230.

58. T. Okamoto, T. Kimoto, and N. Maeda, Design methodology and tools for NEC Electronics' structured ASIC ISSP, *International Symposium on Physical Design*, Phoenix, Arizona, USA, 2004, pp. 90–96.

59. K.-C. Wu and Y.-W. Tsai, Structured ASIC, evolution or revolution?, *International Symposium on Physical Design*, Phoenix, Arizona, USA, 2004, pp. 103–106.

60. C. Patel, A. Cozzie, H. Schmit, and L. Pileggi, An architectural exploration of via patterned gate array, *International Symposium on Physical Design*, Monterey, California, USA, 2003, pp. 184–189.

61. L. Pileggi, H. Schmit, A. Strojwas, P. Gopalakrishnan, V. Kheterpal, A. Koorapaty, C. Patel, V. Rovner, and K. Y. Tong, Exploring regular fabrics to optimize the performance-cost trade-off, *Design Automation Conference*, Anaheim, CA, 2003, pp. 782–787.

62. Xilinx, http://www.xilinx.com/products.html.

63. Altera, https://www.altera.com/products.html.

Routing

8

Gustavo E. Téllez, Jin Hu, and Yaoguang Wei

CONTENTS

8.1 INTRODUCTION

Routing refers to the tasks, algorithms, and electronic design automation (EDA) tools that are used throughout the integrated circuit (IC) design process to create *interconnect* (*metallization*) shapes.* Every IC is composed of devices, including switching devices (*transistors*) and passive devices, that are fabricated by depositing materials layer by layer. The initial set of layers are used to fabricate devices; the latter set of layers, collectively called metallization, supply the power and provide the necessary connections between the devices (Figure 8.1). Routing is key in driving power, area, cost, and performance optimization of an IC, as it influences early technology decisions and design *floorplanning*, drives placement decisions, and is a principal part of physical synthesis tools. It is also the last major construction step of IC design flows and thus heavily impacts the productivity of a design team.

The total length of metal shapes in an IC is measured in kilometers (km) packed into an area of a few hundred squared centimeters (cm²); today's ICs can contain 4 km/cm² of interconnect. Physical characteristics of a metal layer (e.g., height, minimum spacing, and minimum width) are closely related to its electrical characteristics and are chosen to maximize performance and wire density. A high-performance IC in 2014 can have 15 layers of metal, and each layer can have its own height. The metal layer height increases from the bottom to the top of the IC (Figure 8.2a), with a modern microprocessor using six different metal heights. The choice of interconnect height varies depending on trade-offs between IC cost, electrical performance, and design effort. The 2012 International Technology Roadmap for Semiconductors expects IC wire density to grow by 27% every 2 years [1], with an additional metal layer added every 4 years. Routing automation addresses the technical challenges brought by these trends of metallization

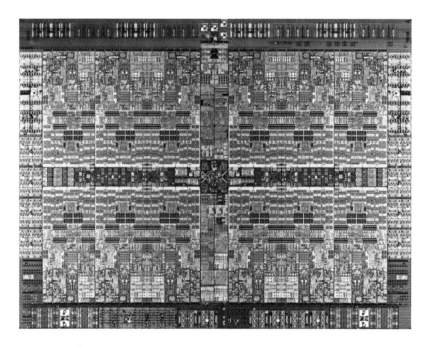

FIGURE 8.1   Image of the bottom metallization of the IBM Power8 processor.

---

* This chapter is a rewrite of the chapter "Routing" by Lou Scheffer in the first edition of the *Electronic Design Automation for Integrated Circuits Handbook*.

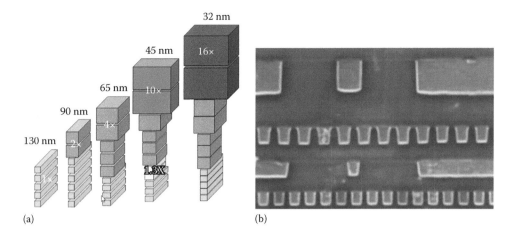

FIGURE 8.2 Illustrations of complex modern metal stacks. (a) Evolution of metal stacks, widths and heights for the 130nm through the 32nm nodes, and (b) four layer of metal cross-section micro-photograph including two metal heights.

complexity growth. Routers are also used by other algorithms to evaluate their quality of results (e.g., congestion analysis in a congestion-driven placement algorithm). Routers drive other algorithms (e.g., global routing for buffer insertion) and perform specialized tasks such as power, clock, and signal routing.

This chapter gives an overview of routing topics and algorithms, along with implementation improvements and applications to other areas in EDA. The following routing tasks will be discussed: (1) custom routing (Section 8.3), (2) digital signal routing (Sections 8.4 through 8.7), (3) routing timing models (Section 8.8), (4) clock routing (Section 8.9), (5) interconnect synthesis (Section 8.10), and (6) design for manufacturability (DFM) in routing (Section 8.11). The complementary task of power routing is discussed in detail in Chapter 23.

8.2 SHAPES, WIRES, VIAS, AND SHAPE CHECKING

IC routing creates a set of layered metallization shapes, where each shape describes where metals are deposited. Even though manufactured metallization shapes are really 3D polyhedra, each metal layer is designed with a uniform height. Therefore, the representation of each layer's metal shapes consists of 2D polygons. The metal layers of an IC are broken up into two classes: wiring layers and via layers. Wiring layers consist of long rectangles (or wires) designed with minimum width and minimum spacing requirements. Typically, each wiring layer has a horizontal or vertical preferred direction, which alternates successively from wiring layer to wiring layer (Figure 8.2b). Via layers consist of square shapes (or nearly square) that connect overlapping wiring segments of neighboring wiring layers.

Signal routing (Sections 8.4 through 8.7) and clock routing (Section 8.9) methods are applied to three concepts: *nets*, *terminals* or *pins*, and *junctions* or *Steiner points*. The main routing concept is the *net*. A net is a collection of wires and vias. A complete IC design is viewed, for the purposes of routing, as a set of nets or *netlist*. Nets are composed of a set of *pins* or *terminals* and *junctions* or *Steiner points*. It is expected that all parts of a net operate at the same voltage. To ensure this, terminals in a net are connected together by *wires* on a metal layer and by *vias* or *via cells*, which connect the metal layer below, the via or cut layer in the middle, and the metal layer above. Three or four wires meet to form a Steiner point (Section 8.5).

Terminals or pins are represented by *pin shapes*, which can be (1) real design shapes or (2) virtual connection points, representing real shapes. Pin shapes represent connection targets for a router. The router has completed connecting a net, if it has properly connected to all of its pins. When a pin contains multiple pin shapes, they may be considered (1) *strongly connected*, (2) *weakly connected*, or (3) *must-connect*. The router is allowed any number of pin shape connections in a strongly connected pin. The router is limited to connect to one pin shape on a

TABLE 8.1   Examples of Design Rules Affecting Routing That Have Emerged over the Last 10 Years

Tech. Node (nm)	Diff-Net Spacing Rule	Same-Net Rule
65	1D width table, via cluster, run-length-dependent via	Short edge, min area, 1D metal-via overlap
45	2D width–width table	Width-dependent metal-via enclosure
32	3D width–width–span table, 3 bodies	Wrong-way width, bar via, span-based via enclosure
22	Line end, line end to line side and interlayer via	Centerline enclosure, discrete widths, coarse edge grid
14	Double patterning metal and via, unidirectional routing	Inner vertex metal to via spacing

Detailed descriptions of these rules are found in [2]. The table includes design rules between shapes in different nets, or *diff-net* rules, and rules between shapes in the same net, or *same-net* rules.

weakly connected pin. A must-connect pin requires the router to connect to all of the pin shapes. For simplicity, the rest of this chapter assumes single-shaped pins. Routers must avoid *blockage shapes*, which represent areas where routing resources are not available for a given net. For example, the shapes that make other nets constitute blockages.

Routing follows minimum geometric requirements imposed by the manufacturing process in the form of *design rules*, which can be classified as either *diff-net* rules, that is, rules between shapes of different nets (e.g., minimum-spacing rules), or *same-net* rules, that is, rules between shapes of the same net (e.g., minimum-width rules). With the advent of subresolution lithography, many new rules impacting routing have emerged (Table 8.1). At every new technology node, new design rule types are added. Therefore, increasing design rule complexity is the principal challenge to the development of practical industrial routers.

Before the design is sent to the manufacturing facility, the design must be checked to verify that it meets these constraints. This verification process includes *design rule checking (DRC)*, a set of algorithms that verifies a given set of layered shapes will pass all geometric technology rules. A violation occurs if a DRC rule is not honored (e.g., the distance between two shapes is less than the minimum spacing). Routers must also ensure that the routes are electrically consistent with the original intentions of the design. When two shapes of different nets touch each other, they form a *short*; when a net is not completely connected, it has *opens* (Figure 8.17). The process of checking the design's correctness is called "layout versus schematic" (LVS) checking. In the context of routing, a subset of LVS checking, namely, *connectivity checking* between the terminals of a net, is sufficient, and this process is often also called LVS checking. The results of routing are expected to ultimately be LVS and DRC clean. The reader is referred to Chapter 20 for detailed discussions on DRC. Additional manufacturing requirements are typically considered as *DFM* constraints (Section 8.11). The reader is also referred to Chapter 22 for additional discussion on DFM.

8.3 CUSTOM ROUTING

Custom routing encompasses several routing problems that are traditionally solved with manual techniques using an interactive layout editor [3]. Custom routing is applied when automated routers leave LVS and DRC errors, when routing is extremely regular and packed, or when the electrical requirements on the routing are difficult to automate. In general, custom routing enables an expert to address those problems beyond the reach of automated routers. *Analog routing* addresses routing problems for nets that carry analog signals. These nets must not only meet connectivity but also satisfy special, additional electrical properties. *Leaf cell routing* refers to the routing problems that include connections to the transistor layers. Both analog routing and leaf cell routing are traditional custom routing tasks where successful research efforts have resulted in routing automation and will be discussed next.

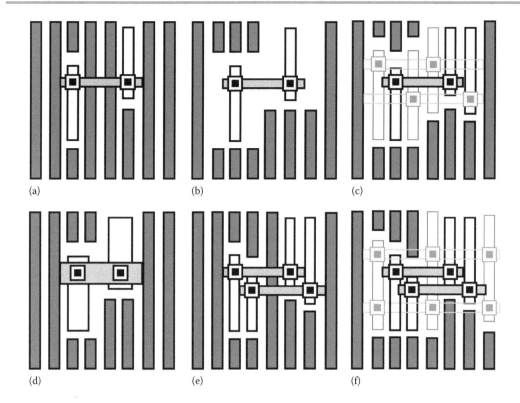

FIGURE 8.3   Illustrations of routings with different wire traits. Filled tracks represent available space and light-colored nonfilled shapes are shield wires. (a) Default, (b) isolated, (c) shielded, (d) nonminimum width, (e) differential pair, and (f) shielded diff. pair.

The electrical properties of wiring are controlled through the use of *wire traits* (Figure 8.3), which specify the physical properties of the wires including the layer width and spacing, a specific list of via cells, preferred routing layers, and *taper rules*, that is, special rules for pin access. The complex modeling of the complete requirements for routing in modern technologies is accomplished by several industry standard specification formats, the most popular standard for interchange of rules and routing data being the *LEF/DEF* standard [2].

Analog routing is distinguished by having a large number of electrically constrained nets. The solution of analog routing problems requires routers to strictly follow *special wire-trait* constraints. These wire traits include (1) nonminimum width routing (Figure 8.3d), (2) isolated or nonminimum space routing (Figure 8.3b), (3) shielded routing (Figure 8.3c), (4) differential pair routing (Figure 8.3e), (5) shielded differential pair routing (Figure 8.3f), and (6) matched or symmetric pair routing. The thinning of wires at input pins, known as *tapering*, and via-cell selection at output pins also play important roles in controlling electromigration and parasitics in analog routing. Analog routing techniques are also used in digital ICs. Section 8.10 provides further details.

Algorithms for analog routing [4] must support nonminimum width (Figure 8.3d) and isolated (Figure 8.3b) wire traits; the remaining more complex traits can be derived. As such, due to the nondiscrete nature of wiring pitches, *gridless routing* is most suited to handle analog routing requirements (Section 8.4).

Related to analog routing, *leaf cell* routing problems also exhibit specialized characteristics. The problem instances are small, on the order of tens of nets, allowing an expert to produce better results than an automated solution. These tasks emerge during automatic *standard-cell* generation, for example. Therefore, the challenge to an automatic leaf cell router is to produce results that are comparable or better than expert layouts, within a reasonable amount of time. Leaf cell routing also considers the creation of required space for pins during higher-level routing (Figure 8.4). Recent work on this problem can be found in [5].

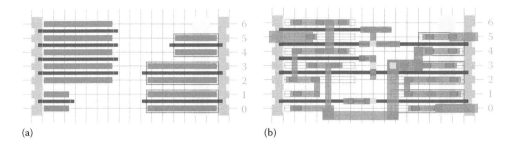

FIGURE 8.4   Transistor-level (a) placement and (b) routing, with routing shapes on Metal1 shown in gray.

8.4 SINGLE-NET POINT-TO-POINT ROUTING

The basis of most routing algorithms is the point-to-point path search that finds a valid least-cost path P from n_0 to n_1 on a graph $G(V, E)$, $n_0, n_1 \in V$, and $P \subset E$. Formally, let $G(V, E)$ denote the routing graph, with *vertices V (nodes)* and *edges E,* which connect pairs of nodes. The path cost, cost (P), is defined as the sum of edge costs, denoted cost (e) for all $e \in P$.

Algorithm 8.1 Pseudocode of Dijkstra's Algorithm
Input: graph $G(V,E)$, from node $n_0 \in V$ to node $n_1 \in V$
Output: least-cost path P from n_0 to n_1

```
P ← ∅; Q ← ∅;
for all v ∈ V do
  cost[v] ←∞ ; parent[v] ←NULL; visited[v] ←FALSE;
end for
cost[n₀] ← 0; v ← n₀;
while v ≠ n₁ and v ≠ NULL do
  visited[v] ←TRUE ;
  for all e = (w, v), e ∈ E, w ∈ V do
    k ← cost[v] + cost (e);
   if k < cost[w] then
     cost[w] ← k; parent[w] ← v; Q ← Q ∪ {w};
   end if
  end for
  Find v ∈ Q, with min cost[v], and visited[v] = FALSE;
 end while
if v = NULL then
 NO_PATH_FOUND;
else
 while v ≠ NULL do
  P ← P ∪ {v}; v ← parent[v];
 end while
end if
return P;
```

For unweighted graphs, where all edge costs are equal, a *breadth-first search* finds least-cost paths in $O(|V| + |E|)$ time, where $|V|$ is the number of nodes and $|E|$ is the number of edges in the routing graph. For a weighted graph, where edge costs are nonuniform and nonnegative, that is, cost $(e) \geq 0$, algorithms such as *Dijkstra's algorithm* [6] are employed to find paths in at most $O(|E| + |V|\log|V|)$ time. Dijkstra-based algorithms also find least-cost paths between a node and all other nodes in the graph. Scalable routing solutions rely both on clever graph construction methods and on the strategies chosen to manage edge costs. Algorithm 8.1 shows the pseudocode of Dijkstra's algorithm; Figure 8.5 illustrates a sample progression.

A standard implementation of Dijkstra's algorithm requires $O(|E| + |V|\log|V|)$ time. A quasi-linear time algorithm with $O(|E| + |V| \times f(|V|))$ time is possible when $0 \leq$ cost $(e) \leq |V|$,

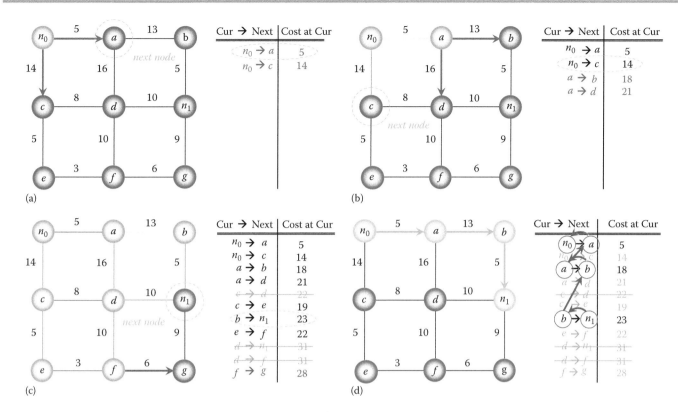

FIGURE 8.5   Illustration of Dijkstra's algorithm. (a) All paths from n_0 are considered. Node a has the least-cost node. (b) The algorithm calculates the cost from n_0 to all the neighbors of a. (c) Every subsequent iteration considers the node with the least cost that has not been visited. The process continues until all nodes have been visited or n_1 is the node with the least cost. (d) Starting from n_1, the algorithm backtraces through the parents of the nodes to determine the shortest path to n_0.

$\mathrm{cost}(e) \in \mathbb{Z}_{\geq 0}$. Here, $f(|V|)$ is dependent on the implementation of the priority-queue access time. The first such algorithm was proposed by Dial [7] with $f(|V|) = |V|$, but better algorithms exist [8]. Another algorithm for the single-source, shortest-paths problem with integer weights, though not based on Dijkstra's algorithm, achieves $O(|E|)$ runtime [9].

In practice, the dominant factor in the runtime of path-finding algorithms is the size of the search region $O(|E| + |V|)$, which can be prohibitive if $|E|$ or $|V|$ is large. Consequently, much of routing research has focused on methods to reduce search space. Bidirectional search starts and expands simultaneously from both n_0 and n_1 [10]. The *A*-search* [11], or goal-oriented path search, adds a lower-bound cost estimate of the unexplored portion of the path. This estimate is then used to guide the search along the most likely path to succeed. The *A**-search performs no worse than Dijkstra's algorithm but can speed up searches substantially in practice [12]. Bidirectional path searching can be combined with *A**-search algorithms to produce even faster path searches [13].

Several other methods for search-space reduction decrease the size of the search space by reducing its granularity. A straightforward method is to not search beyond the bounding box of n_0 and n_1. A complex method, known as "global routing," searches for paths on a coarse grid (Section 8.7.1). Once all nets in a netlist are routed on the coarse grid, the path of coarse grid nodes is converted into a set of contiguous areas for *detailed routing*, thus reducing the search space. The *interval-based path search* is a detailed routing algorithm that accumulates equivalent contiguous empty grid locations, creating regions (intervals) instead of a grid node [14]. A *gridless tile-based routing* performs the path search on a map of polygons or tiles that represent free space [15]. The gridless router returns a sequence of connected space polygons, which represent a point-to-point path. Space-based gridless routers result in a smaller space search but require very fast geometric area query and update algorithms and complex costing schemes. Another class of gridless path-search techniques, connection graph–based search, seeks to find algorithms whose complexity is a function of the number of routing obstacles in the search region. Connection graph–based searches lead to nonuniform grid graphs [16]. In practice, gridless routers produce high-quality routes, but their algorithmic overhead results in relatively high runtimes, relegating their role to smaller-scale, difficult custom routing problems and finishing tasks.

Routing algorithms can give up the optimality of the resulting path in favor of a smaller search time. As the traditional data structure upon which path searching is done is a 2D or a 3D grid graph (Figure 8.9), many algorithms leverage the uniformity of the search space. The *line-probe algorithm* by Hightower [17] is an early example. This algorithm expands from n_0 in one direction until an obstacle is found or until the route can no longer get closer to n_1. The algorithm then switches directions and repeats until a path is found or no further options are available. This algorithm neither guarantees that a path can be found if one exists nor does it ensure that it finds a least-cost path. Several very fast algorithms have improved upon this initial approach. *Pattern routing* algorithms recognize that the majority of point-to-point paths follow simple *shapes* (patterns) that represent high-quality paths, that is, those that have a small number of *bends*, such as *L*s, *U*s, and *Z*s (Figure 8.6). In these cases, the point-to-point router visits the path outlined by the pattern and determines if that path is valid, that is, exists. If the fixed path is invalid, however, there is no guarantee that any path exists. The complexity of a pattern search depends on the desired quality of the result. A full *L* plus *Z* pattern search requires $O(|V|)$ time, where $|V|$ is the number of nodes inside the least bounding box of n_0 and n_1. A generalization of pattern routing, *monotonic routing* [18], seeks a path that does not contain a *U* turn (Figure 8.6d). The monotonicity property of the algorithm allows the graph to be modeled as a *directed acyclic graph* (*DAG*) with nonnegative edge costs. The edge directions are determined by order in which the nodes are processed. The shortest path algorithm on a DAG, G, can be solved in $O(|E| + |V|)$ time. While monotonic routing subsumes *L* and *Z* pattern routing, that is, every *L* and *Z* route is a monotonic route, it cannot determine whether a path exists.

The path search challenge is further complicated by technology constraints. Diff-net rules are generally, but not always, handled with blockage strategies, and same-net rules (Table 8.1) have increased in complexity and quantity over the last 10 years. Industrial routers address these problems in several ways. The router can be constrained to use a subset of wires and vias, and these can be selected to minimize the chances of same-net errors. After the path search is completed, it is possible during the backtrace step to make local choices (such as via selection), which minimize, eliminate, and/or trade-off errors. During the path search, the router may not consider some path topologies at the cost of optimality, if same-net errors are detected. This latter solution has many undesirable effects. Some recent algorithms are capable of handling limited cases of these rules. The problem of finding shortest paths such that every wire in the path satisfies a minimum length constraint has been investigated in [19], resulting in a $O(|V|^4 \log|V|)$ algorithm. A similar problem is investigated in [20], where the edge length constraints are imposed by the spacing of neighboring shapes. A different approach is taken by the multilabeling path search algorithm [21], whereby using path history concepts (e.g., previous route topologies) to limit some local shape patterns that lead to same-net errors can find a path in polynomial time without loss of optimality.

The point-to-point routing algorithms form the basic fundamental kernel for clock and signal routing as well as for global and detailed routing. Some of the algorithms described earlier have found applications in global routing, while others are more suited to detailed routing.

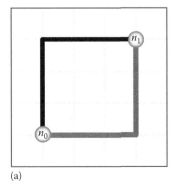 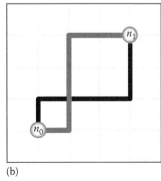 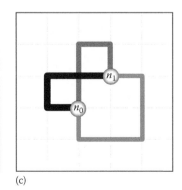 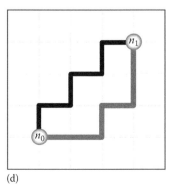

(a)   (b)   (c)   (d)

FIGURE 8.6   Different route patterns: (a) Ls, (b) Zs, (c) Cs or Us, and (d) monotonic.

8.5 SINGLE-NET MULTIPOINT ROUTING

Multipin or *multipoint routing* extends the concepts of point-to-point routing to nets with multiple pins. Nets have a *single driver* or *source pin* and at least one *sink pin*. Formally, a net N consists of a set of terminals $n_0, n_1,...,n_{|N|-1} \in N$, with a special source terminal n_0 and all other terminals sinks. A multipin routing of a net N results in a *routing tree* $T_N(V_N, E_N)$, where $n_i \in V_N$. Every vertex $n_i \in V_N$ has a corresponding coordinate, that is, $(x_i, y_i) = n_i$. The *Manhattan distance* between two vertices $n_i, n_j \in N$ is Dist $(n_i, n_j) = |x_i - x_j| + |y_i - y_j|$. An edge $e_{i,j} \in E_N$ connects vertices n_i, n_j in the tree, with length $l(e) = \text{Dist}(n_i, n_j)$. The principal metric for routing is *wirelength*, defined as $W(T_N) = \sum l(e)$. Wirelength in this chapter uses the Manhattan distance metric.

Multipin routing is done in three steps: (1) compute a multipin tree T_N for net N (with any $n \in N$ as the root), (2) decompose T_N into point-to-point subproblems, and (3) perform point-to-point routing on each subproblem (Section 8.4).

One common method to construct a spanning tree $T_N(V_N, E_N)$ with $V_N = N$ for a net N is the *rectilinear minimum spanning tree (RMST)* algorithm. This algorithm finds the least-cost tree T_N^{RMST} that connects all pins $n \in N$ without additional nodes. The simplest RMST algorithm starts with a *complete graph*, that is, every node has a direct connection to all other nodes, which puts a lower bound on the computation of $\Omega(|N|^2)$, and was proposed by Prim [22]. Algorithms using a sweep line of the pins can avoid the construction of the complete graph and enable RMST construction in $O(|N|\log|N|)$ time [23]. Figure 8.7b shows an example of different RMSTs for the same three-pin net.

The *minimum rectilinear Steiner tree (MRST)* problem emerges when additional, so-called Steiner, nodes can be added in the multipin tree $T_N^{MRST}(V_N, E_N)$, that is, $V_N \subset V$, where V is the set of all available nodes. The resulting *Steiner* tree consists of both the original terminals of N and possible additional Steiner nodes. The problem of computing an MRST is NP-complete but has been extensively studied in the literature [24]. The construction of such a tree introduces the *Hanan grid*, a minimal grid in which the tree must exist [25]. The Hanan grid is defined by extending a ray from every pin toward the borders of the net's bounding box (Figure 8.7a). The intersection points of the grid—locations that use the x coordinate of one pin and the y coordinate of another pin—are possible Steiner nodes to be added to the multipin net. The length of an RMST is an upper bound of the MRST and $W(T_N^{RMST})/W(T_N^{MRST}) \leq 1.5$. For this reason, the simplest multipin routing algorithms use minimum spanning trees. Robust algorithms exist for the optimal construction of Steiner trees for nets with fewer than 10 pins, using table lookup techniques [26]. These approaches are very practical, given that the majority of nets in an IC inherently have very few pins. Using divide-and-conquer techniques, that is, solving the subproblem by table looking and merging the results, produces very efficient Steiner trees for large nets. The rectilinear Steiner tree problem has been extended to handle obstacles in [27], where the problem is solved by combining obstacle avoidance graphs, RMSTs, and MRST computations.

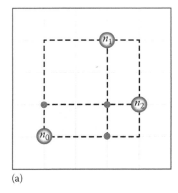
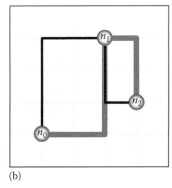
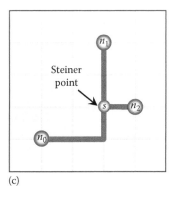

(a)　　　　　　　　　(b)　　　　　　　　　(c)

FIGURE 8.7 Examples of rectilinear minimum spanning tree (RMST) and minimum rectilinear Steiner tree (MRST). (a) A three-pin net with its Hanan grid (black lines) and potential Steiner point locations (red). (b) Two different RMSTs (red and black paths). (c) An MRST with a single Steiner point.

Multipin routing also addresses metrics beyond geometric considerations. For instance, in clock routing (Section 8.9), special consideration is given to *skew*. Timing-driven and buffering-driven routing (Section 8.10) also introduce several unique routing problems.

8.6 CLASSIC MULTINET TWO-LAYER ROUTING

At the dawn of semiconductor technology (in the 1970s and early 1980s), the number of available routing layers was confined to two, with additional (but limited) routing to a very highly resistive polysilicon (Poly) layer dedicated for transistor gates. Due to the restricted routing resources, algorithms were developed to trade off routing and area, giving rise to *two-layer routing* techniques. While further algorithmic development is no longer necessary, two-layer routing has several useful applications (e.g., avoiding higher metal layers during IP development and three-layer routing).

Traditional two-layer routing techniques are applied within rectangular routing regions known as *channels* and *switchboxes*. In channels, pins are located on opposite sides of the region, and in switchboxes, pins are located on all four sides of the regions (Figure 8.8). By convention, a channel is oriented horizontally, with pins located on the top and bottom of the region. These regions have only two routing layers—one vertical (Poly or Metal2) and one horizontal (Metal1); a route that spans across both layers is connected by vias. For more modern designs, *over-the-cell (OTC) routing* allows wires to be routed over standard cells on a third horizontal layer (e.g., Metal3). Channels are allowed to grow in one direction to complete routing. Switchboxes are allowed on all four sides of the switchbox to obtain full routing. OTC regions are not allowed area growth. The goal of all channel, switchbox, and OTC routing algorithms is to fully connect all multipin nets. If this is not possible, the goal is to connect as many nets as possible.

Channel routing methods include the left-edge algorithm originally developed in [28] and are based on the vertical constraint graph (VCG). To address cycles in the VCG, Deutsch [29] introduced the dog-leg enhancement to the left-edge algorithm by splitting n-pin nets in $(n-1)$ horizontal segments. Switchbox routing algorithms are largely extended from channel routing algorithms [30]. OTC routing algorithms were developed in conjunction to channel routing. OTC routers route a select set of nets outside the channels, leaving the remaining nets to be routed inside the channels [31].

Channel, switchbox, and OTC routing techniques laid the foundation for modern industrial routers. With the advent of multiple layers of metal, these methods gave way to routing in a *sea-of-gates* placement environment, where the area between rows is no longer sacrificed but can be used to alleviate congestion or improve routability.

8.7 MODERN MULTINET ROUTING

The modern approach toward multinet routing splits the routing flow into several steps, resulting in a divide-and-conquer routing paradigm. These steps include (1) global routing, (2) pin escape and local routing, (3) track routing, and (4) detailed or finish routing. The global router quickly provides a coarse-grained, approximate solution to the multinet routing problem. The pin escape

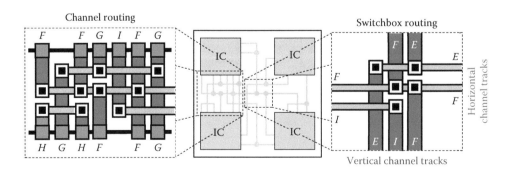

FIGURE 8.8   An example of channel routing and switchbox routing.

and local routing algorithms resolve pin access and reserve pin access space. Track routing then converts the global routing solution to a finer-grained routing track assignments. Detailed routing then creates the exact routes each net will use on each metal layer. The final solution is expected to be DRC and LVS clean with no opens or shorts and must also be compliant with all wire-trait constraints.

8.7.1 GLOBAL ROUTING MODELS

Global routing has undergone intense research in the last two decades. Global routers handle millions of nets and find several applications in a physical design flow: as estimators of placement and floorplanning quality by providing congestion and timing metrics, as drivers for algorithms like buffering (or repeater insertion), and as the global solvers for the detailed routers.

The global routing paradigm simplifies the routing problem by coarsening the routing grid and reducing the routing graph. The coarse grid point now becomes a rectangular region called a "gcell" or a "global tile." Global routing algorithms typically operate on a 2D or 3D grid graph $G(V, E)$ that is abstracted from the physically available metal routing layers (Figure 8.9). The vertices $v \in V$ represent the gcells, and edges $e \in E$ between the nodes represent the boundary between adjacent gcells. An edge e has a length $l(e)$, which is the distance between the centers of the nodes it connects. In a 3D grid graph, nodes belong to a layer, edges between nodes on the same layer represent wire paths, and edges between nodes in separate layers represent via locations. A 2D grid graph can be obtained by compressing the nodes at the same coordinate of a 3D graph into a single node (Figure 8.9).

A set of nets $\mathcal{N}$ or a *netlist* (Section 8.4) must be connected or routed over the grid graph G. The terminals of each net in the netlist are commonly modeled to be at the center of the gcells. In global routing, multiple nets may share the same edge. The *capacity* cap (e) represents the amount of available space across e, that is, how many wires fit through the side of a gcell. If routing blockages are present, then the edge capacity is reduced accordingly. The *flow* or *demand* usage (e) represents how many routing resources have been consumed across e. In a 3D grid graph, a net N that crosses an edge e requires width(N, e) width and space(N, e) distance from all other objects. Therefore, a net N has demand usage$(N, e) = $ width$(N, e) + $ space(N, e) on an edge e, which is a

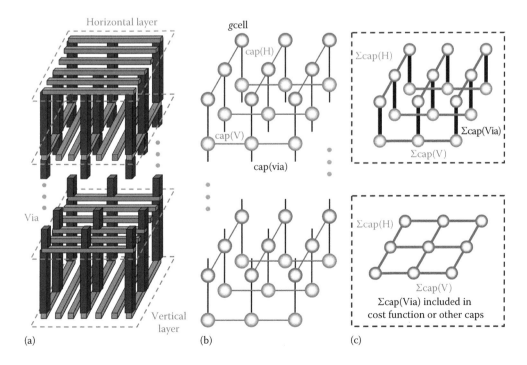

FIGURE 8.9 Modeling of global routing graphs from metal layers. (a) Metal layers, (b) 3D route modeling based on metal layers, and (c) 2D route modeling based on the 3D grid.

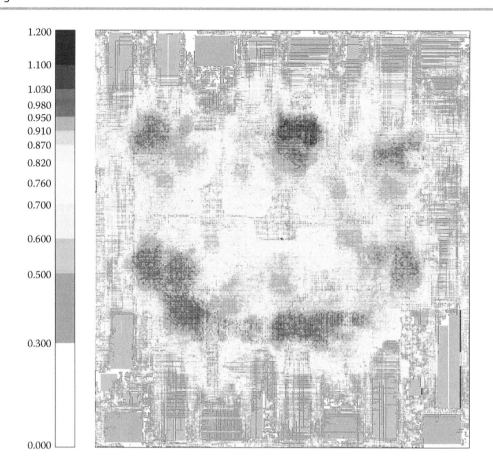

FIGURE 8.10 A congestion map plotting $\eta(e)$ (color scale on left) over a chip floorplan. Gray areas denote macro blocks within the floorplan.

function of the layers of e and the wire trait of N. Since layer information is not present in a 2D grid graph, net demand is simplified.

The quality of global routing solutions is measured by the relationship between supply and demand across all edges. One such metric is *congestion* or *density*, that is, the ratio of demand to supply over an edge, and is defined as $\eta(e) = usage(e)/cap(e)$. Another metric is *overflow*, which is defined as $\phi(e) = \max(usage(e) - cap(e), 0)$. Global routing seeks to route all nets $N \in \mathcal{N}$ on G such that $\phi(E) = 0$ while minimizing an objective, commonly total wirelength $\mathcal{W}(\mathcal{N}) = \sum W(T_N)$. A visual inspection of routing quality is often encapsulated in *congestion maps* (Figure 8.10).

8.7.2 GLOBAL ROUTING APPROACHES

Every global routing algorithm relies on successive calls to a single-net or path search algorithm that (1) finds a least-cost path P_N that connects a net N, where each edge $e_N \in P_N$ has cost $cost(e_N)$ and (2) updates usage (e_N) and $cap(e_N)$ accordingly after P_N has been determined. As such, the strategy for deriving $cost(e)$ holds fundamental importance. Early renditions of cost modeling sought to model congestion by incorporating wire supply and demand metrics [32]. Additional cost models include $cost(e) = \eta(e)$ and $cost(e) = \eta(e)^2$ that allow overcongestion of an edge and others that prohibit edge overcongestion [33]:

$$cost(e) = \begin{cases} \dfrac{usage(e)+1}{usage(e)-cap(e)} & : usage(e) < cap(e), \\ \infty & : usage(e) \geq cap(e). \end{cases}$$

TABLE 8.2   Global Routing Algorithms, Approaches, and Detailed Techniques

Approach	Technique
Sequential	Minimize max congestion using Steiner trees [34]
	Min-weighted Steiner trees [35]
Rip-up and Reroute	Optimal rip-up set [36]
	Simple iterative rip-up and reroute [37]
	3D global router with rip-up and reroute [38]
Hierarchical	Top down [39], bottom up [40]
	Min-cost MCF [41]
Formal	Maximum concurrent flow [42]
	Min–max resource sharing [43]

Table 8.2 summarizes the common global routing approaches. The global routing problem can be formally modeled as a *minimum-cost multicommodity flow (MCF)* problem [41], where the objective is to minimize total wirelength, as represented by the *transportation costs,* subject to *bundle-* or *edge-capacity* constraints. This initial iterative improvement algorithm finds a feasible solution (if one exists) but without runtime guarantees. The authors of [42] then applied, to global routing, an approximation algorithm for the *maximum concurrent flow problem* [44], with the objective to minimize the maximum edge *density.* This resulted in the first algorithm with an approximate optimality guarantee. In this approach, each net is initially routed by a convex combination of feasible Steiner *forests* (a set of Steiner trees), and the initial solution of the algorithm is a provably near-optimal fractional solution. A randomized-rounding algorithm [45] is then used to select a final solution, which with high probability increases congestion by a bounded amount. The final step in this approach is a *rip-up and reroute* stage to ensure that all nets are connected, where nets that violate capacity constraints are removed from the grid graph (ripped up) and reconnected (rerouted). An improvement to this framework was proposed in [46] by developing a simpler and faster maximum concurrent flow approximation algorithm based on [47]. This improvement enabled finding solutions with approximation guarantees on large industrial-sized designs.

A recent formal global routing approach, based on *Lagrangian relaxation,* finds solution for the *min–max resource sharing problem,* where the objective is to minimize the maximum resource consumption [43]. In this formulation, the problem of sharing a limited set of resources $\mathcal{R}$ among a finite set of consumers $\mathcal{C}$ is solved. The algorithm assumes that there is a *block $b_c \in \mathcal{B}_C$* or *convex set* of feasible solutions for each consumer $c \in \mathcal{C}$ and a nonnegative continuous convex function $g_c(b_c)$ specifying the resource consumption of each element b_c in the block. The computation of the vector of *Lagrangian multipliers* ω or *weights* depends on the existence of a *block solver* that is capable of finding a $(1 + \epsilon)$-approximate minimization of $\omega^T \times g_c(b_c)$.

In the implementation of [43], the netlist is defined as the set of consumers $(\mathcal{C} = \mathcal{N})$ and the resources include the edges in the constraint graph. A special resource is created for the objective and allows for multiple objectives. For example, the wirelength objective is modeled by $g_c^{\mathcal{W}} = \mathcal{W}(\mathcal{N}) / \mathcal{W}^{UB}$, where $\mathcal{W}^{UB}$ is an upper bound on the minimum wirelength. The block solver is a Dijkstra-based path search algorithm whose cost function for each net is derived from $\omega^T \times g_c(b_c)$. The path search cost is a weighted trade-off between congestion and wirelength (and any other objective):

$$\mathrm{cost}(c, e) = \omega(e) \cdot \mathrm{usage}(c,\ e) + \omega^{\mathcal{W}} \cdot l(e).$$

The Lagrangian multipliers are computed in successive iterations of refinement, with the edge congestion cost modeled exponentially. Similar to [42], the resulting solution is represented as a convex combination of Steiner forests and requires an intermediate rounding step and a final rip-up and reroute stage. The min–max resource sharing strategy for iteratively updating Lagrangian multipliers is guaranteed to converge to an approximate optimum in polylog time. The current best min–max resource sharing algorithm results in a practical global routing approximation algorithm, with a $(1+\epsilon)$-quality guarantee from the optimum, and polylog runtime with respect to $|\mathcal{N}|$, $|G|$, and the approximation parameter $1/\epsilon^2$. This framework (1) handles both convex constraints and convex optimization objectives as a function of the routing solution; (2) can be used to (simultaneously) optimize various objectives, for example, routability (Section 8.7.3), yield (Section 8.11), timing, and power; and (3) is scalable to solve industrial-sized designs. Timing is addressed by several techniques: (1) assignment of wire traits to nets, (2) net weighting or net prioritization, (3) Steiner length constraints, and (4) route topology controls. High-performance nets might also be constrained further by the specification of wire traits. Typically, these constraints are created externally and implicitly modeled by the global router. Similar techniques can be used to control capacitance, which combined with net-centric activity factors, can also be used to minimize the power.

The 2007 and 2008 ISPD contests on global routing created a surge of interest in fast global routing algorithms, resulting in very fast software packages. The authors of [48] surveyed several of these techniques, including the authors of [49], where their approach first routes nets on a 2D grid graph using techniques, such as (1) low-effort pattern routing on uncongested areas, (2) iterative tree-segment shifting, (3) monotonic routing, and (4) rip-up and reroute based on shortest paths, and then uses *layer assignment* to map 2D routes into 3D grid graph. The global convergence paradigm uses heuristically chosen edge costs that account for an edge's congestion *history* or how often an edge has been congested. An edge in greater demand at past iterations will have its cost increase so as to discourage demand. If the demand remains high, the cost is increased further, until the demand reacts. This negotiation is performed on multiple edge costs at the same time so that the routes choose the most appropriate edges. This intuition forms the basis of *negotiation congestion–based routing* (*NCR*) and can be formalized by Lagrangian relaxation. An implementation of NCR [50] introduces a history cost component $h(e)$ that increases every time (iteration) an edge is congested. In addition, the cost function accounts for congestion $p(e) = f(\eta(e))$ and the distance to the target $d(e)$:

$$\text{cost}(e) = h(e) + \alpha \cdot \text{p}(e) + \beta \cdot d(e),$$

where
    α is a scale factor that ensures $h(e)$ never exceeds $\eta(e)$
    β is a heuristically calculated constant

After 2D routing, the authors of [49] performed layer assignment by solving an ILP that minimizes the number of vias within a bounded region. During layer assignment, this bounded region is iteratively extended until the full grid is encompassed.

8.7.3 CONGESTION ANALYSIS

The goal of congestion analysis is to predict the *routability* or route distribution of a design by identifying routing *hot spots* or localized areas of high congestion. Congestion analysis typically evaluates the quality of other design flow operations and drives incremental optimizations. As such, runtime takes priority over routing quality. Congestion analysis approaches include: (i) non-constructive probabilistic estimation techniques [51], (ii) global routing solutions [52], and (iii) detailed routing solutions.

While an approach based on detailed routing estimates promises the greatest accuracy, it may require prohibitively long runtimes, especially during early design flow stages when detailed routes are not available. Careful experimentation shows that probabilistic methods are largely inaccurate and not robust enough to address the characteristics of modern designs (e.g., routing obstacles, multiple routing layers, and different layer spacings). The most attractive methods stem from global routing, as they can accurately and quickly provide a high-level analysis for any routing landscape. However, this approach (1) is usually unaware of local- or intra-gcell congestion, that is, congestion seen by the detailed router and (2) can provide false positives, that is, indicate a region that has high congestion, but where congestion can be mitigated if given further routing iterations. To this end, *short-net prerouting* and allowing nets greater flexibility in detouring greatly mitigate local congestion. Another approach [53] addresses these drawbacks by modeling local routing resources and applying smoothing techniques to reduce the number of noisy *hot spots*.

To perform congestion analysis, *congestion maps* (Figure 8.10) are used as a first-order (visual) evaluation of the design routability, where every edge is assigned a color based on its congestion. However, more detailed metrics are needed to effectively use routability information for optimization, including (1) overflow-based metrics, (2) net congestion–based metrics, and (3) edge congestion–based metrics.

Overflow-based metrics, including total overflow $\Sigma\phi(e)$ and maximum overflow $\max(\phi(e))$, measure the excess of routing usage versus routing capacity on global routing edges in a global routing graph (Section 8.7.1). However, these metrics can misrepresent the design routability, especially in the presence of *hot spots* (localized areas with extreme congestion).

Net congestion–based metrics [54] are based on *net congestion* $\Lambda(N)$, which is defined as the $\max(\eta(e_N))$, for all the grid graph edges e_N crossed by net N. These metrics include (1) *average net congestion* $ACN(\delta)$, which is the average of $\Lambda(N)$ for the top δ% congested nets, and (2) the worst congestion index $WCI(z)$, which is the number of nets with $\Lambda(N) \geq z$, $z \in \mathbb{R}_{>0}$. However, this type of metric cannot distinguish between nets that span a single congested edge and nets that span multiple congested edges.

Edge congestion–based metrics were developed to address the drawbacks of using net congestion–based metrics. The histogram approach [53] of the *average edge congestion* $ACE(\delta)$ computes the average $\eta(e)$ of the top δ% congested edges. When δ is small, $ACE(\delta)$ provides a focused evaluation. When δ is larger, $ACE(\delta)$ provides a more global routability evaluation, which is useful when comparing design points. A composite edge congestion metric *peak weighted congestion* $PWC = [ACE(0.5) + ACE(1) + ACE(2) + ACE(5)]/4$ was used in several placement contests [55].

8.7.4 ROUTABILITY-DRIVEN PLACEMENT

Routability-driven placement is one such application that heavily relies on congestion analysis, as the traditional half-perimeter wirelength metric is no longer sufficient to estimate routability for modern technology nodes [56]. Routability-driven placers most commonly use congestion maps to identify regions where routing will be difficult and guide movement decisions.

Once the placer has access to a congestion map, it converts the map into its own cost functions. It then applies optimizations during various stages of the placement methodology: (1) global placement, (2) modifying intermediate solutions, (3) legalization and detailed placement, and (4) postplacement processing step. Figure 8.11 illustrates the progression of the routability-driven placer SimPLR [57], which is representative of other techniques.

8.7.5 PIN ESCAPE AND TRACK ASSIGNMENT

When assigning routing tracks to signals, the nets' endpoints must correspond to cells' input and output ports. The *pin escape* problem determines each net's detailed pin locations by allocating or reserving essential pin access resources to the corresponding cells. This problem can be solved either (1) before global routing so as to provide the global router with accurate local congestion estimates or (2) after global routing so as to be consistent with the global routing solution. A typical standard cell architecture has tightly packed pins on Metal1, with no available routing space both inside the cell and on Metal1. Therefore, pin escape algorithms escape a Metal1 pin to a Metal3 track or a neighboring Metal1 pin connected to the same net. Pin escape methods are also applied at the boundary of macro blocks (e.g., register files or small memories) with a large number of tightly packed boundary pins on upper layers. Moreover, in high-performance routing, *over-the-macro* routing at a high layer of metal can easily become constrained when accessing macro pins on lower layers, due to the differences in the metal pitches and the requirements on low-resistance vias. Pin escape algorithms are explored in [58].

The *track assignment* problem maps global routes into detailed routes [59], where each global wire is assigned a *detailed track*, for example, physical location, such that *shorts* and *spacing*

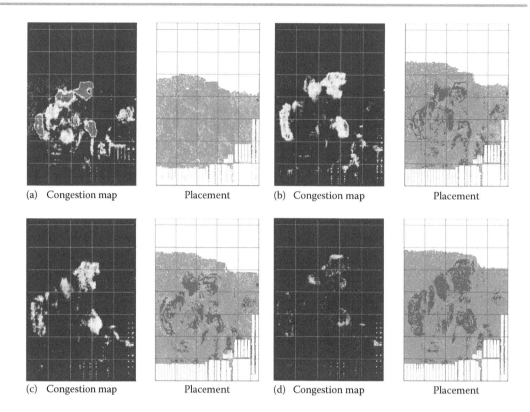

FIGURE 8.11 Progression of routability-driven placement. Red color indicates locations where cell utilization was temporarily decreased. (a) Initial placement (right) with areas of high congestion. (b) Intermediate placement where congestion is reduced. (c) Intermediate placement where congestion is further reduced. (d) Near-final placement where congestion is largely removed.

errors are minimized. Track assignment algorithms divide the global wires into *panels*, which comprise rows (or columns) of gcells. For each panel, the problem is divided into a minimal set of *cuts*. For each cut, a bipartite matching problem is solved, where the aim is to assign a set of wires to a set of tracks. Track assignment algorithms are also suited for producing solutions for mixed-width and gridless routing and can be used for noise avoidance algorithms [60]. However, as track assignment operates in 1D space, it cannot account for 2D effects, for example, shorts at vias and line ends. Furthermore, track assignment algorithms can result in incomplete connections using the given pin assignments. For these reasons, the last steps of a signal router rely on a rip-up and reroute phase from a *detailed router* (Section 8.7.6).

8.7.6 MULTINET SIGNAL ROUTING

Multinet signal routing consists of (1) track routing or (2) global routing–driven, region-bounded path searching, finished by (3) sequential rip-up and reroute. The global routing–driven, region-bounded path searching algorithm consists of two steps: converting the global routes into layer-wise search areas and then applying a multipoint routing algorithm within these areas. Due to the sequential and greedy nature of this solution, a path that is routed earlier may utilize resources that are needed for later routes. Consequently, if multinet routing cannot find resources for all nets but ends with routing violations, then another iteration (phase) of signal routing is required to resolve subsequent detailed routing conflicts.

Multinet signal routing has been formulated in several ways in the context of small routing instances (e.g., cell routing in Section 8.3). In [61], multinet signal routing is formulated as a *Boolean satisfiability (SAT)* problem, where a set of potential routes is enumerated for each net and a set of *constraint satisfiability formulae* place restrictions on which routes can be simultaneously used (e.g., lack of available resources on a routing edge). A valid SAT solution determines the route for each net. In [5], the problem is approached by formulating a Steiner tree packing problem, in combination with a flow formulation for finding Steiner trees, using mixed integer linear programming. Both of these approaches are capable of dealing with complex modern design rules. Unfortunately, these formal approaches are severely limited in the size of problem that they can solve using reasonable resources. As a result, large multinet signal routing problems are solved using rip-up and reroute heuristics.

Rip-up and reroute schemes apply to both global routing and signal routing and are classified in one of two general approaches. The *progressive rerouting* scheme considers one net at a time, removing considered nets from the problem when done, as proposed in [37]. As each net is considered, each of its component connections will be ripped up and rerouted. This approach may be repeated for several passes. The iterative improvement scheme, such as the one used in [36], is based on keeping track of the constraints violated in the design. At each pass, all connections containing errors are ripped up and rerouted. Constraint violations are updated after reroute, and decisions for additional passes are made depending on the progress of the algorithm based on the constraint violations. The iterative improvement schema works best with the general signal routing problem, so we will focus on this schema in the remainder of this section.

An iterative improvement rip-up and reroute algorithm, often employed in a *detailed router* or *search-and-repair router*, aims to take a partial routing with opens, shorts, and DRC violations and to repair these using path searching and other techniques. The end result of the detailed routing algorithms is a complete, violation-free set wires and vias for a given netlist. The key algorithm to all detailed routers is a path search engine, with variable controls over the cleanliness of its results and of its path costing structure. A second important element of these algorithms is the set of checks or rules the design must meet. A third element is an algorithm that determines the set of objects that will go through rip-up. By repeating the search, rip-up, and reroute operations within controlled path finding areas, with larger and larger rip-up sets and with adjusted costs, a detailed router can be made to converge to clean results, except for the hardest of congestion problems.

The hardest congestion problems require careful allocation of routing resources and are not solvable by conventional rip-up and reroute. In this case, multiple connections will be simultaneously selected for rip-up, and a heuristic algorithm will order and reroute the ripped-up connections to achieve a desired improvement. There are several rip-up and reroute techniques that attempt to solve these hard problems: *weak rip-up* techniques, *prioritized path searching* schemes, and *decongestion* schemes. Weak rip-up schemes involve solving routing problems using planar routing methods such as river routing, wire untangling, and wire push aside techniques. Prioritized path searching techniques involve storing a priority for connections, which may be updated as the router iterates, allowing a net to be routed while causing some shorts and then driving the reroute order of the shorted nets with the selected priorities. Decongestion techniques involve selecting connections that must be in an area, routing them cleanly, and then driving other connections to be routed outside of that area, finally finishing with cleanup rerouting. The most difficult challenge with all multinet rip-up and reroute methods is to avoid oscillations and to determine when and where the router must give up. A more formal description for rip-up and reroute using integer linear programming and Lagrangian relaxation can be found in [62].

8.7.7 PARALLEL ROUTING

Modern routing problems can come in very large sizes: signal routers must be able to route 10 million nets overnight. Due to the limitations of single-thread performance, routing algorithms must be capable of parallel operation. A few paradigms for router parallelization exist. Regional methods partition the routing problem geometrically, allowing parallel routing within each region. This technique is suitable for distributed and multithreaded computer architectures. Unfortunately, known partition-based routing algorithms scale poorly, due to boundary conditions and uneven resource use. Net-based parallelization techniques scale well but are best suited for multithreaded architectures, require careful coding, and suffer from convergence challenges. Fine-grained parallelization for core algorithms such as Dijkstra's algorithm on a grid graph has been proposed, even with implementations on specialized hardware. These methods have remained only of theoretical interest.

8.8 WIRING DELAY AND SLEW

Before discussing timing-aware routing algorithms, particularly clock routing and high-performance signal routing, it is important to understand wire delay models. The key metrics for the dynamic behavior of wiring are transition *delay* and *slew*. Delay measures the time that it takes for a signal transition to propagate through a wire. Slew measures the time that it takes for a signal to change state, such as from high to low (Figure 8.12). There are two basic models of wire delay: the *linear delay model* and the *Elmore resistance–capacitance (RC) delay model*. The linear delay model is useful as a first-order approximation for wire delay and offers a reasonably accurate approximation for the delay of a long-buffered tree [63]. The Elmore delay model forms the basis for a reasonable approximation for the delay of an *RC* tree, which models the parasitics of most practical wired networks. More accurate models exist but are significantly more time consuming and less appropriate for routing.

The linear delay model assumes that delay is proportionate to wirelength. For theoretical and first-order routing optimization, weighted wirelength can be used as a delay metric. A more sophisticated model, derived for the prediction of delays for buffered trees, includes a delay penalty component for branching β, can have a component to account for gate loads at sinks $\gamma(N)$ for a net N, and accounts for different time of flight properties of layers α. Formally, let $T_N = G(V_N, E_N)$ represent a rooted, wired tree with root n_0 and sinks $n_1, n_2, \ldots, n_{|N|-1}$ on net N. Let the set $P(n_i) = n_0; n_i$ be a source-to-sink path in T_N. Then a complete linear delay model to a sink $n_i \in N$ is

$$\tau_{\mathrm{LIN}}(n_i) = \gamma(n_i) + \sum_{e_{j,k} \in P(n_i)} \left[\alpha_{j,k} \cdot l(e_{j,k}) + \beta_k \right].$$

Wiring affects timing through the contribution of its parasitics. Important parasitics include resistance, capacitance, and, in very special cases such as long high-frequency interconnect (package-level and radio frequency ICs), inductance as well. Wire resistance is a function of wirelength L, wire width W, layer conductance ρ_L, and layer height T: $R_w = \rho_L(L/WT)$. For a given technology, layer heights are given, so typically a sheet resistance $R_L = (\rho_L/T)$ is known, and therefore, wire resistance is

$$R_W = R_L \frac{L}{W}.$$

Wire capacitance is much more complicated than wire resistance because it involves interactions of a given wire with its environment. Accurately calculating wire capacitances involves accounting for activity between every neighboring wire laterally, above and below, and between wires and the chip substrate, leading to the 3D capacitance extraction problem. For the purpose of optimizations, it is useful to have expressions for wire capacitance as a function of the same parameters as resistance. Such expressions capture the different components of wire capacitance, parallel plate capacitance C_{PP}, lateral or wire to wire capacitance C_{WW}, and fringe capacitance C_F, and therefore, wire capacitance is $C_W = C_{PP} + C_{WW} + C_F$. Examples of empirical expressions for

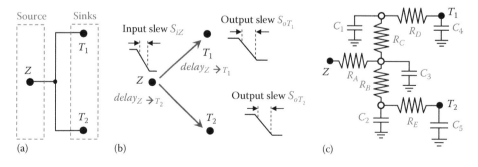

FIGURE 8.12 Example of an RC tree for a CMOS VLSI circuit. (a) Wire, (b) delay and slew models, and (c) parasitic RC model.

capacitance can be found in [64] and are a function of the dielectric constant of the insulator and a layer-to-layer distance. The equations of capacitance are commonly simplified by assuming that for a given layer, the height is given, the distance to the next layers is fixed, and there is an average distance to the nearest-neighbor wires:

$$C_W \simeq C_L L W.$$

Elmore delay is currently among the most important metrics in timing optimization [65] and was proposed in [66]. The basic time domain behavior of digital IC wiring is modeled using RC networks. Since most network topologies used are trees, an important and well-studied area in timing analysis is that of the modeling of RC trees.

Let $T_N(V_N, E_N)$ represent a rooted RC tree on net N. A vertex $v_i \in V_N$ represents a voltage node that contains a capacitance C_{w_i} and an edge $e_{i,j} \in E$ represents a resistance $R_{w_{i,j}}$. A subtree T_{n_i} of T_N, rooted at vertex $n_i \in N$, is the tree induced from T_N by including n_i and all vertices in N below n_i.

Two important quantities in wire delay computation are the *downstream capacitance* seen below a node v_i, Cap(v_i), and the total *upstream resistance* seen above a node v_i, Res(v_i), which is defined by

$$\mathrm{Cap}(v_i) = \sum_{v_j \in T_N(v_i)} C_{w_j} \quad \text{and} \quad \mathrm{Res}(v_i) = \sum_{e_{k,j} \in P(v_i)} R_{w_{k,j}}.$$

Given two vertices v_i and v_j, let $a(v_i, v_j)$ denote the first vertex above both vertices in $P(v_i) \cap P(v_j)$. Then the Elmore delay $\tau_{\mathrm{ELM}}(v_i)$ at a node v_i is defined as

$$\tau_{ELM}(v_i) = \sum_{v_k \in V} \mathrm{Res}\big(a(v_i, v_k)\big) \cdot C_{w_k} = \sum_{e_{j,k}, v_k \in P(v_i)} R_{w_{j,k}} \cdot \mathrm{Cap}(v_k).$$

There are therefore two equivalent methods for the computation of Elmore delay: a bottom-up RC summation where, along a sink-to-source path, $R_{w_{j,k}}\mathrm{Cap}(v_k)$ are added and a top-down method where in a source-to-sink path the RC sum $\mathrm{Res}(v_i)C_{w_i}$ is accumulated.

If the input waveforms are ramps, then the *PDF extension to ramp inputs* model can be used to calculate delay and output slew of a wire [67].

8.9 CLOCK ROUTING

Clock routing is a topic rich in research and in practice. There are several different approaches toward clock network design, with dramatically different requirements on routing. The basic timing requirement on clock design is *skew*. Skew is the maximum difference in the arrival times of the clock signal at the synchronizing elements. Skew is added directly to the minimum system clock period, and therefore, a primary objective in IC design is *zero skew*. Minimizing the total power of the clocking network is also a key objective in clock design.

At low clock frequencies, a clocking network can be viewed as a special routing problem on a high fanout tree, with the zero-skew requirement. As frequency increases, a clock tree must become a buffered clock tree, where the skew requirement falls primarily on the buffering of the tree, with no special routing requirements. As the frequency of a clock continues to rise and variability effects come into play, large specialized buffers for the upper levels of clock tree combined with zero-skew routing provide practical solutions. At the highest clock frequencies, typical of modern microprocessors, hybrid clock distributions centered around the clock mesh are necessary. This section will investigate several routing algorithms and approaches toward the problems in clock design.

The zero-skew clock routing problem is then to construct a minimum total wirelength clock tree such that the source-to-sink delay is the same for all sinks. The first clock tree construction algorithms focused on path length–balanced trees. Early clock tree constructions used H-trees that work well under limited circumstances. The first algorithm to address general distribution of pins is the method of means and medians [68], where a clock topology is created by a top-down recursive partitioning of the sinks. At each step, two equally sized subsets of sink pins are created by geometric median partitioning. Routing is performed top down, to the median points, and at the final partition to the sinks, resulting in an $O(|N|\log|N|)$ runtime. The next improvement in clock routing results in control over the total wirelength of the tree and is the minimum-weighted matching algorithm proposed in [69]. This algorithm uses a bottom-up recursive traversal. At each step, the algorithm computes a min-cost maximum cardinality matching of the sink positions (or an approximation), which chooses pairs of pins that get connected for minimum total wirelength. For each connection, a balance point that minimizes linear skew is computed and passed to the next level of recursion. Since the balance point is not always conveniently located, a heuristic algorithm, H-flipping, is added to improve the balance points. The runtime of this method is dependent on the matching algorithm, resulting in algorithms with complexities ranging from $O(|N|\log|N|)$ to $O(|N|^{2.5}\log|N|)$.

The next improvement in clock routing considers the Elmore delay model, leading to the exact zero-skew algorithm [70]. The exact zero-skew algorithm applies to any clock tree, traverses the clock tree bottoms up, computes Elmore delays, and creates a zero-skew subtree at each step. When zero-skew subtrees are merged, an added amount of wirelength is added to one of the trees to match the Elmore delays. Since the method accounts for wire parasitics and sink loads, it also compensates for uneven loads in the trees. The method runs in linear time on the size of the tree.

The clock routing algorithms do not directly address the problem of actually realizing (or embedding) the clock trees. This problem was addressed by several works, including [71]. The resulting algorithm is known as the deferred-merge embedding algorithm, and it can be applied to both problems with linear delay and Elmore delay models. The deferred-merge embedding algorithm takes in a given clock tree topology and realizes the topology in two passes of the clock tree. The first pass is a bottom-up phase that determines the regions for the placement of every balance point in the tree. The second top-down pass selects the exact balance points in

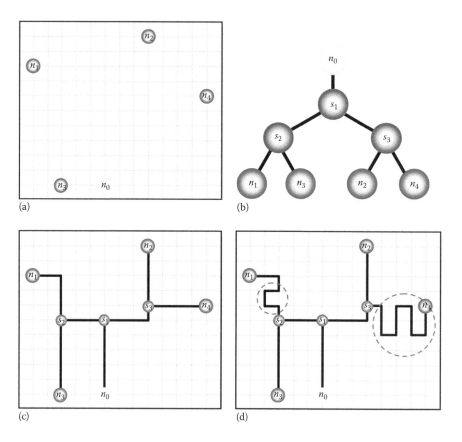

(a)

(b)

(c)

(d)

FIGURE 8.13 Illustration of clock tree routing. (a) Unrouted clock net, (b) clock net topology, (c) routed clock net by DME, and (d) clock net after wire snaking.

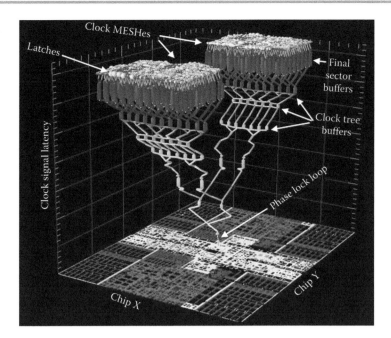

FIGURE 8.14   Illustration of a clock distribution network, from a PLL to latches, showing clock signal delay propagation in the vertical axis. The clock network is a hybrid clock mesh from an IBM Power6 microprocessor. (From Restle, P.J. et al., *IEEE J. Solid-State Circuits*, 36(5), 792, 2001.)

the tree and actually does the routing. Given a specific tree and for the linear delay model, Boese and Kahng [71] proved that this algorithm produces an optimal wirelength tree, with minimum source-to-sink distance. This algorithm provides good but not optimal results with the Elmore delay model. Figure 8.13 illustrates an example of a clock tree generated by the deferred-merge embedding algorithm and subsequent improvements to balance the local clock skew.

Pushing to the limits of achievable clock frequencies, nontree routing methodologies are used in clock designs. Modern microprocessors use clock meshes, which are hierarchical designs combining several different techniques (Figure 8.14). In clock meshes, the clock is globally distributed, not unlike a power distribution, by a mesh designed with the lowest resistance metallization available to the microprocessor. The mesh is driven at key points by a carefully balanced buffered clock tree network. Local clock buffers, or clock trees, then connect to the mesh directly and ultimately drive latches. In this style of clock design, each clocked component is given a delay budget from the mesh, and therefore, timing-driven routing algorithms with minimal wirelength become the ideal solution. Tight timing, electromigration, and reliability constraints lead to clustered latches driven by local clock buffers. In these latch clusters, the local clocks are routed by severely constrained routing solutions, such as spine clock trees.

8.10 INTERCONNECT SYNTHESIS

The importance of interconnect to IC performance has been recognized for two decades. In a seminal paper [73], Cong introduces the concept of *interconnect synthesis* that determines optimal or near-optimal interconnect topology, wire ordering, buffer/repeater locations and sizes, and wire traits to meet the performance and signal reliability requirements of all nets under the congestion constraint. Interconnect synthesis is invoked during physical synthesis, after placement stage, but before detailed routing stage. A typical interconnect synthesis stage may contain several steps: (1) timing-driven global routing with a linear delay model; (2) initial repeater insertion, optionally following the global routing paths; (3) repeater-tree optimization steps, for slew and slack; and (4) timing-driven global routing with the Elmore delay model. In this section, we examine algorithms in the area of interconnect synthesis.

Techniques in interconnect synthesis originated from research on wire sizing [74], which is primarily focused on optimization under the Elmore delay model. A negative result in this space was reported in [75], where it was shown that the ratio of the maximally attainable signal speed with optimal nonuniform wire sizing and buffering to that with optimal uniform wire sizing and buffering is ~1.03. Hence, the most effective strategy of timing-driven wiring optimization for high-performance designs is the selection of wire traits, combined with repeater or buffer insertion [76]. The resulting algorithm produces a set of solutions, driven by a discrete library of repeaters and a discrete set of wire traits. A single solution can then be selected to match the timing/area/power requirements of the network. This algorithm assumes a fixed routing tree topology and therefore requires a repeater-aware router.

Routing algorithms can optimize propagation delays by exploring different tree topologies. Algorithms optimizing topologies for the linear delay model are suitable for prebuffering stages, while those optimizing for Elmore delay are useful in the postbuffering stages. For the following discussion, the linear model with no vertex cost, that is, $\beta_k = 0$, and no sink component, that is, $\gamma(n_i) = 0$, is assumed. Timing is introduced into the routing problem by recognizing that some sinks are timing critical and by observing that their paths to the source must be delay optimized. In [77], the concept of bounded-radius minimum spanning trees is proposed. The radius of a tree is defined as a shortest path from the source to the farthest sink, that is, $R = \max(\tau_{\text{LIN}}(n_i))$. Figure 8.15 illustrates different topologies that trade off the radius and cost of the routing tree. The problem is to find a minimum-cost routing tree, with a radius bounded by $(1+\epsilon)R_M$, where $\epsilon \geq 0$ and R_M is the radius of the *shortest path tree* that is constructed by computing the shortest paths from the source pin to each sink pin independently. In [77], Prim's algorithm is altered to include a test that guarantees the radius constraint. When the test fails, a search is made for a connection that guarantees the radius. This algorithm guarantees the radius constraint but does not guarantee a tree of minimum cost at a runtime complexity of $O(|N|^2)$. A second solution starts with a minimum spanning tree; then during a depth first traversal of the initial tree, nodes are selected to become part of a source-rooted shortest path tree, based on the radius requirement. The resulting tree is a mix between the minimum spanning tree and the shortest path tree with a bounded radius, with a cost factor $1+(2/\epsilon)$ of the cost of the minimum spanning tree. An improvement on this method is provided by [78], where an algorithm with the same complexity as a Dijkstra shortest path tree search is designed to find a tree where every source-to-sink path has a length bounded by the minimum source-to-sink distance. This latest algorithm implements a Prim–Dijkstra trade-off, and given a parameter $0 \leq c \leq 1$, it guarantees that for every pin n_i, the length of the source-to-sink path is bounded by $c\,\tau_{\text{LIN}}(n_i) \leq \text{Dist}(n_0, n_i)$. When $c = 0$, the algorithm produces a minimum spanning tree, that is, equivalent to Prim's algorithm, and when $c = 1$, the algorithm constructs a shortest path tree and therefore is equivalent to Dijkstra's algorithm. Bounded-radius Steiner trees follow from the construction of the bounded-radius spanning trees.

Many repeater tree generation algorithms perform two steps: a routing step and then a repeater insertion step. The repeater insertion step in some algorithms also has the capability of performing layer assignment; for samples of such an algorithm, see [76]. Routing algorithms that are designed to create repeater trees have to address several unique challenges:

1. The blockages for repeater insertion and routing can be different (e.g., placement blockages and routing blockages can be different), and therefore, routing trees for repeater insertion must go through the areas where repeater can be inserted legally.
2. Since repeaters have a limited reach (due to speed or slew constraint), routing for repeaters is allowed to go over a blocked area but only for a limited distance on a given layer.
3. Repeater tree routing must also deal with sink polarity requirements.

In [79], the problems of Steiner trees with buffer blockages and buffer bays are considered. Recent work on repeater tree topology exploration with the linear delay model can be found in [63].

Steiner trees with all pins routed at a minimum radius are *rectilinear Steiner arborescences* (*RSAs*). Formally, an RSA is a minimum-cost Steiner tree where for every sink pin in the tree, $\tau_{\text{LIN}}(n_i) = \text{Dist}(n_0, n_i)$, with $\beta_i = 0$ (Figure 8.16). This problem is NP-hard [80]. The first $O(|N|\log|N|)$-time heuristic algorithm with a factor of 2 approximation bound for this problem was presented in [81]. A practical result comparing various timing-driven tree construction algorithms in [82] found that the pervasive use of arborescences results in 4% wirelength increases but with about 4%–5% reductions in cycle time.

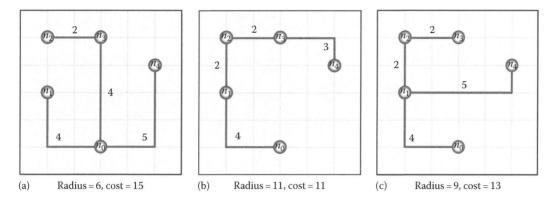

(a)   Radius = 6, cost = 15   (b)   Radius = 11, cost = 11   (c)   Radius = 9, cost = 13

FIGURE 8.15   Different topologies demonstrating the radius-cost trade-off. (a) Small radius, high cost; (b) high radius, small cost; and (c) medium radius, medium cost.

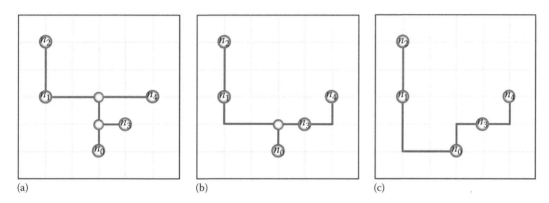

(a)   (b)   (c)

FIGURE 8.16   Example of a net routed with different trees. (a) Steiner arborescence, (b) MRST, and (c) RMST.

Topology exploration and multipin routing with the Elmore delay model are explored in [83]. In this work, it is established that Elmore delay exhibits fidelity with more accurate delay models. In addition, a greedy low delay tree heuristic with an $O(|N|^3)$ time complexity is proposed and compared with an optimal branch-and-bound algorithm.

A timing-driven global router, capable of utilizing the algorithms described in this section, is the central tool for interconnect synthesis. An example of global routing with timing constraints can be found in [84], where a general framework that can use any single-net routing algorithm and any delay model is provided.

In summary, a modern interconnect synthesis tool brings together global routing, buffer insertion, and wire-trait optimization to solve timing and electrical problems of circuits under the congestion constraints.

8.11 DESIGN FOR MANUFACTURABILITY AND ROUTING

Manufacturability has long been defined by the pass/fail criteria of *DRC* and *LVS* checking. *Yield*, which is the actual fraction of good manufactured parts, depends on many factors that can be optimized in a design. *DFM* seeks to improve the yield and the reliability of designs above and beyond the baseline achievable by following binary design rules (please refer to Chapter 22 for more details on DFM). Yield in manufacturing is driven by systematic and random defects, which can be addressed in routing. Defects in metallization serve as an important contribution to yield loss and reliability problems, and therefore, routing has a strong impact on yield (Figure 8.17).

Pioneering work on DFM started with the modeling of yield as a function of layouts. The critical area analysis method developed in [85] for the modeling of random defects provided the initial framework for DFM in routing. Random defects in interconnect are principally, but

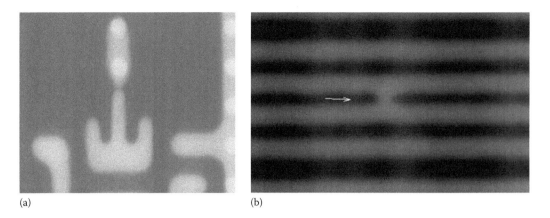

FIGURE 8.17   Metallization defects. (a) an open and (b) a short.

not exclusively, (1) *via opens*, (2) *wire shorts*, and (3) *wire opens*. Early research in DFM routing addressed these three defects primarily and will therefore be the focus in this section. Random-defect-driven DFM methods in routing improve the yield of wiring by minimizing the likelihood of random defects. The random-defect model, applied to a set of parallel wires on layer ℓ of length L, width W, and spacing S, predicts the average probability of faults including opens and shorts for parallel wires. The average probabilities of wire opens $\theta_{O,\ell}$ and wire shorts $\theta_{S,\ell}$, under a $1/r^3$ defect size distribution assumption (r is the defect size), are

$$(8.1) \qquad \theta_{O,\ell} \simeq K_O \cdot \frac{L}{W} \quad \text{and} \quad \theta_{S,\ell} \simeq K_S \cdot \frac{L}{S},$$

where K_O and K_S are technology-dependent constants for open and short defects, respectively. The average probability of faults measures the absolute likelihood of an individual type of failure. For an entire chip, the relative importance of different defect types is modeled per layer ℓ by a defect density $d_{i,\ell}$ for each failure type i. The average number of faults per layer is denoted as λ_ℓ. A chip with area A has an average number of faults λ. The yield, Y_P, can be estimated as follows:

$$(8.2) \qquad \lambda_\ell = \sum_i d_{i,\ell} \cdot \theta_{i,\ell}, \quad \lambda = A \cdot \sum_\ell \lambda_\ell, \quad Y_P = e^{-\lambda},$$

where $\theta_{i,\ell}$ is the average probability of failure for defect type i on layer ℓ. Shorts and opens are two such failure types, and their defect densities are denoted as $d_{S,\ell}$ and $d_{O,\ell}$.

According to Equation 8.1, to minimize opens and shorts, to the first order, wirelength minimization is an appropriate yield objective. Furthermore, minimizing opens requires increased wire width while minimizing shorts requires enlarged wire space. Consequently, open and short defect minimization leads to conflicting objectives, and a trade-off between space and width must be made with limited routing resources. Specifically, to minimize shorts, available space must be distributed evenly between wires, which is known as "wire spreading." This method was practical in aluminum technologies because $d_{S,\ell} \gg d_{O,\ell}$. Since the wirelength objective of routing results in packed wires (Figure 8.18a), wire spreading algorithms distribute spacing as evenly as possible to minimize shorts (Figure 8.18b). Wire spreading reduces wire coupling capacitance and thus improves timing. Postrouting wire spreading, based on linear programming techniques, was proposed in [86]. Methods for wire spreading during routing are described in [87]. Compared with minimization of shorts, minimization of opens is much more difficult, because it causes wire width to increase, which increases wire capacitance and may degrade timing. An algorithm simultaneously minimizing opens and shorts in track routing stage is proposed in [88]. An approach for yield optimization in global routing that can handle Equation 8.2 as an objective is presented in [43]. In summary, the full problem of minimizing wiring opens and shorts has to be solved by combining yield optimization and a timing-driven router with the wire spreading capability.

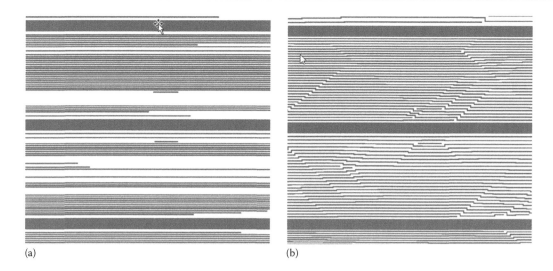

(a)   (b)

FIGURE 8.18   Single layer of routing (a) without wire spreading and (b) with wire spreading.

In deep-submicron technologies, many sources of systematic defects have emerged, which can be addressed by routing techniques. These sources of systematic defects include via yield and reliability, antenna effects, optical proximity correction (OPC) effects, and density effects. Unlike random defects, these effects have been systematically understood but have complex physical drivers, which are generally not sufficiently captured by DRC and LVS checks. Therefore, they have to be avoided with additional efforts made in the manufacturing process and EDA tools.

Via shapes are the smallest, most numerous, and most difficult shapes to manufacture in an IC. The first reported method for improving the manufacturability of vias was patented in [89]. Redundant via insertion can be implemented by an algorithm for postrouting insertion of vias that are more robust than the minimum size single vias. Examples of such vias can be seen in Figure 8.19. One kind of via (e.g., double and loop via) consists of two or more via shapes covered by one metal shape above and one metal shape below, and the redundant via shapes in the via improve its robustness. Vias need not be redundant to be robust: larger via shapes can also improve yield and reliability. Figure 8.19 shows the drawings of some of the multiple choices for robust via solutions in modern technologies. Given a set of vias to select from, the task of redundant via insertion algorithms consists of ranking these vias in terms of robustness, followed by attempts to replace vias with more robust variants in a DRC clean manner. Via yields can be further improved by the addition of nontree routing [90].

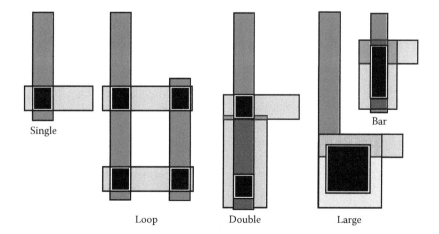

FIGURE 8.19   Choices of vias: single vias are smallest but have the worse yield, resistance, and reliability; loops and double vias are redundant solutions; bar and large vias are better in yield, reliability, and resistance than singles.

Another well-known systematic defect is the *antenna effect*. The antenna effect is a phenomenon of plasma-induced gate oxide degradation caused by charge accumulation on metal. During the fabrication process, the antenna effect leads to faults due to electrostatic discharge through the gate. A path for charge dissipation is formed when metal is connected to a diffusion, such as the source or drain of a transistor. The antenna effect occurs in a partially fabricated net when a sufficiently large amount of metal is connected to a gate but remains isolated from diffusion, as illustrated in Figure 8.20. There are three solutions for this problem: postrouting insertion of wire jumpers reducing the metal area connected to the gate [91] (Figure 8.21a and b), postrouting insertion of diode protect cells [92] (Figure 8.21c), and proactive antenna avoidance routing [93] (Figure 8.22).

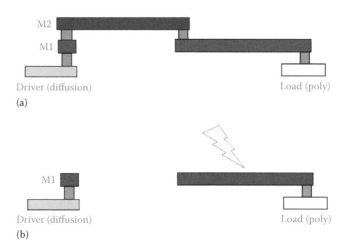

FIGURE 8.20 (a) A cross section of a net routing solution using two layers of metal. (b) The antenna effect problem on the net routing solution. During fabrication, before the addition of Metal2, the long Metal1 wire connected to the gate becomes charged and then destroys the gate.

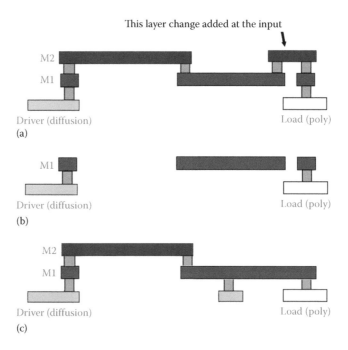

FIGURE 8.21 (a) The routing solution with addition of a wire jumper around the load side. (b) With the wire jumper, the Metal1 wire is disconnected from the gate during fabrication to avoid the antenna effect. (c) By adding an extra diode around the load side, the Metal1 wire will connect to the diffusion of the diode during fabrication, fixing the antenna effect.

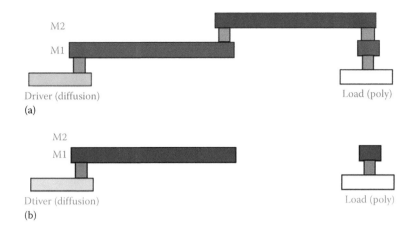

FIGURE 8.22   (a) The routing solution avoids connecting the long Metal1 wire to the gate of the load during fabrication. (b) The routing solution in (a) prevents the antenna effect.

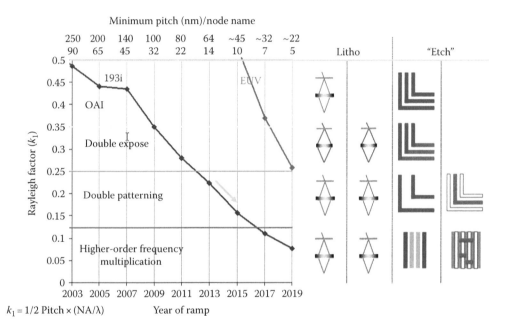

FIGURE 8.23   Overview of the dimensional scaling that is enabled by increasingly complex multiple patterning techniques. (From Liebmann, L. et al. Demonstrating production quality multiple exposure patterning aware routing for the 10 nm node, *Proceedings of SPIE*, 2014.)

OPC effects are driven by lithography. These effects are caused by the gap between the wavelength of the manufacturing light source and the minimum manufactured feature size. As seen in Figure 8.23, such subwavelength lithography has been the norm for IC fabrication since the 65 nm node, adopted by the industry around 2005. OPC methods, also called "resolution enhancement techniques" (RET), adjust and add subresolution features into layouts partially compensating for the resolution gap (Figure 8.24a). These compensations in turn lead to limits on printable features (Figure 8.24b). Consequently, fabricated shapes do not match drawn shapes, and several deformations can occur, including corner rounding, line end pullback, wire narrowing, pattern-dependent wire width variation, and ultimately forbidden pitches. The reader is referred to Chapter 21 for detailed discussions on RET.

There are several approaches in routing to address OPC effects. Design rules are created with the objective to leave enough space in the design for the OPC tools. The capabilities of the design system can be limited by imposing radical design restrictions (RDRs) [95]. Examples of RDRs

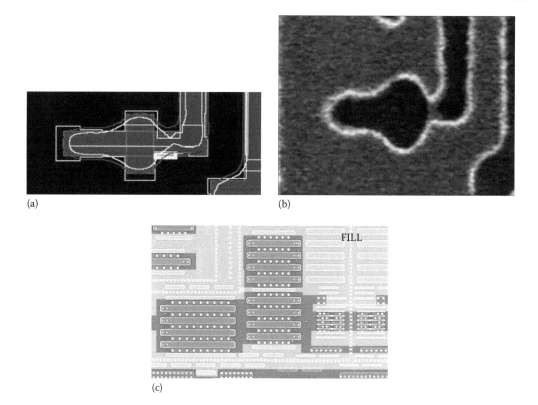

FIGURE 8.24   Optical proximity correction, (a) complex shape in blue with OPC shape in yellow, and predicted contours in red and light blue, (b) photomicrograph of manufactured complex shape, and (c) photomicrograph with fill shapes highlighted.

include prohibiting wrong-way routing and allowing only discrete widths. OPC effects can be addressed directly inside the routers. OPC-aware routing has been approached by adding avoidance techniques and through check and repair strategies. Fast detection of OPC problems for fix-up has been addressed by fast model–based engines and by using pattern recognition [96]. Search and repair techniques driven by several pattern-recognition schemes have been proposed in [97].

A class of systematic defects is driven by variations in the IC manufacturing process. Manufacturing variation can prevent a circuit from satisfying the specifications or, even worse, may lead to circuit failure. An important example of such manufacturing variation is that induced during *chemical mechanical polishing (CMP)*. CMP is used in the manufacturing process to polish the wafer whenever a planar surface is required. Planar surfaces are a key component in the manufacturing of ICs with multiple layers of metal, and therefore, CMP is a major source of variation and systematic defects in metallization. The variations in metal height caused by CMP are largely dependent on the wire density. Uniform wire density leads to smaller variations. A common way to reduce the CMP variation is to insert dummy metal shapes, also called dummy fills (Figure 8.24c). Fill shapes can be created during or after routing. However, dummy fills could increase coupling capacitance. Another way to reduce the CMP variations is to improve the uniformity of pattern density in routing. The work in [98] proposed a routing algorithm that directly targets to minimize the amount of dummy fills inserted for CMP requirements.

Multipatterning, starting with *double patterning*, is a technique used for resolution enhancement beyond OPC. Multipatterning techniques offer frequency doubling (or better) at the cost of design complexity. An introductory text on resolution enhancement is found in [99]. The limit for single patterning with OPC, with the current lithography solutions, is the 22 nm node. Later nodes starting with the 20 nm node will use double, triple, and perhaps even higher levels of multipatterning for their critical layers. The continued evolution of more complex multipatterning will be affected by the availability of economic alternatives such as *extreme ultraviolet lithography*. Currently, there are three technology nodes, 20, 14, and 10 nm, which will use double and/or triple patterning techniques.

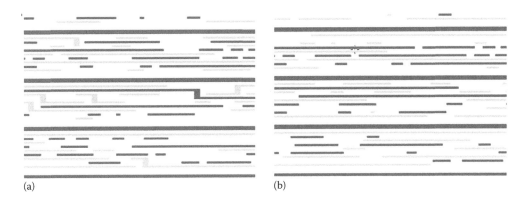

FIGURE 8.25   Single layer of colored routing for a double-patterning process: (a) double-patterning compliant, colored, bi-directional routing, and (b) double-patterning and RDR compliant unidirectional routing.

Multipatterning fabrication techniques impose significant new problems to routing and routing infrastructure. Double patterning is the first multipatterning technique in use, and it divides the shapes in a layer into two masks or patterns. In a layout, the modeling of these two masks is accomplished by a technique called coloring. New design rules are introduced to enforce the spacing limitations between the shapes based on their color, including a *same-color spacing* S_{SC} and a *diff-color spacing* S_{DC}. Two different color shapes may connect, creating a stitch. Given an uncolored layout of a single layer, the assignment of colors to shapes is solved by finding a two-coloring solution for a conflict graph, where the nodes represent the shapes and the edges represent shapes whose spacing is less than S_{SC}. For a k-patterning technology, the problem is then to find a k-coloring solution of the conflict graph. The process of assigning colors to shapes is also known as "decomposition." In the case of double patterning, where $k = 2$, polynomial time algorithms are known for finding a coloring. For triple patterning, where $k = 3$, the coloring problem is known to be NP-complete, but several practical approaches have been proposed [100]. The problem gets harder when stitching is allowed, and in that case, we must determine which shapes to be split and stitched and also the locations for stitches.

There are two ways to achieve a colored design. The first approach starts with a design in an uncolored state, and then through decomposition, DRC, and repair steps, the design will be colored correctly. In this approach, since the initial design is after routing, a *color-friendly* routing algorithm will minimize the chance of problems during decomposition. In [101], the path-search algorithm used in routing is altered to achieve this goal. The second approach toward a colored design is to color shapes by construction and therefore requires that designers and construction tools including routers produce colored designs at the beginning. This approach gives rise to the *color-aware* routing problems. A color-aware routing solution is shown in Figure 8.25a. Furthermore, combining RDRs with coloring, the routers can produce a practical solution for double patterning as shown in Figure 8.25b.

8.12 CONCLUSIONS

Routing plays an important role in many aspects of the EDA design flow, where it significantly affects performance, power, and area optimizations. As technology nodes change, the physical shapes of the routing tracks and pins exhibit different properties (Section 8.2) and have led to various routing methodologies. Custom routing (Section 8.3) techniques are applied for analog circuits and localized areas that have DRC and LVS errors. For signal nets, single-net routing algorithms include point-to-point routing (Section 8.4) and multipoint routing (Section 8.5). Multiple-net signal-net routing techniques span from the classic multinet two-layer routing (Section 8.6) to modern global routing methodologies (Section 8.7). Since the interconnect topology and layer RC parasitics can affect delay and slew greatly (Section 8.8), routing could have a big impact on the timing optimization. Because of the influences of routing on timing, specialized paradigms such as clock routing (Section 8.9) and interconnect synthesis (Section 8.10) have emerged as active research problems. In addition, routing also has big impact on manufacturability and yield, and therefore, various factors have to be considered in routing to improve the overall yield (Section 8.11).

Routing algorithms have been actively researched since the 1950s and continue to be an active area of study. As routing is the most technology-aware step in the EDA design flow, its many algorithms must be continually improved to suit emerging technologies. Since routers are the last construction tool used in IC design, their quality of results has important impacts on IC design teams. The history of routing algorithms is rich, and closely linked to the advancement of the EDA and semiconductor industries, and will continue to be so in the future.

ACKNOWLEDGMENTS

The authors would like to first thank the editors, Luciano Lavagno, Grant Martin, and especially Prof. Igor Markov, for giving them this opportunity and, of course, for their guidance. The authors would also like to thank Professors Jens Vygen, Dirk Müller, and Stephan Hougardy from Bonn University for their help and comments in reviewing this chapter. This chapter on routing was inspired by the original chapter in the first edition of *Electronic Design Automation for Integrated Circuits Handbook* by Louis K. Scheffer, formerly from Cadence Design Systems and now at the Howard Hughes Medical Institute.

REFERENCES

1. International Technology Roadmap for Semiconductors. ITRS Interconnect 2012 update. http://www.itrs.net/ITRS%201999-2014%20Mtgs,%20Presentations%20&%20Links/2013ITRS/Summary2013.htm. Accessed on December 8, 2015.
2. Si2. Lef/def exchange format ver 5.3 to 5.7. http://www.si2.org/openeda.si2.org/projects/lefdef. Accessed on December 8, 2015.
3. J. K. Ousterhout, G. T. Hamachi, R. N. Mayo, W. S. Scott, and G. S. Taylor. Magic: A VLSI layout system. In *Proceedings of the ACM/IEEE Design Automation Conference*, Albuquerque, NM, 1984, pp. 152–159.
4. J. M. Cohn, D. J. Garrod, R. A. Rutenbar, and L. R. Carley. KOAN/ANAGRAM II: New tools for device-level analog placement and routing. *IEEE Journal of Solid-State Circuits*, 26(3):330–342, 1991.
5. S. Hougardy, T. Nieberg, and J. Schneider. BonnCell: Automatic layout of leaf cells. In *Proceedings of the Asia-South Pacific Design Automation Conference*, Yokohama, Japan, 2013, pp. 453–460.
6. E. W. Dijkstra. A note on two problems in connexion with graphs. *Numerische Mathematik*, 1(1):269–271, 1959.
7. R. B. Dial. Algorithm 360: Shortest-path forest with topological ordering [H]. *Communications of the ACM*, 12(11):632–633, 1969.
8. R. K. Ajuha, K. Melhorn, J. B. Orlin, and R. E. Tarjan. Faster algorithms for the shortest path problem. *Journal of the ACM*, 37:213–233, 1990.
9. M. Thorup. Undirected single-source shortest paths with positive integer weights in linear time. *Journal of the ACM*, 46:362–394, 1999.
10. I. S. Pohl. *Bi-Directional and Heuristic Search in Path Problems*. PhD thesis, Stanford University, Stanford, CA, 1969.
11. P. E. Hart, N. J. Nilsson, and B. Raphael. A formal basis for the heuristic determination of minimum cost paths. *IEEE Transactions on Systems Science and Cybernetics*, 4(2):100–107, 1968.
12. S. Peyer, D. Rautenbach, and J. Vygen. A generalization of Dijkstra's shortest path algorithm with applications to VLSI routing. *Journal of Discrete Algorithms*, 7(4):377–390, 2009.
13. A. V. Goldberg, H. Kaplan, and R. F. Werneck. Reach for A*: Efficient point-to-point shortest path algorithms. In *Proceedings of the International Workshop on Algorithm Engineering and Experiments*, Miami, FL, 2006, pp. 129–143.
14. A. Hetzel. A sequential detailed router for huge grid graphs. In *Proceedings of Design, Automation and Test in Europe*, Paris, France, 1998, pp. 332–338.
15. D. Cross, E. Nequist, and L. Scheffer. A DFM aware, space based router. In *Proceedings of the ACM International Symposium on Physical Design*, Austin, TX, 2007, pp. 171–172.

16. J. Cong, J. Fang, and K. Khoo. DUNE: A multi-layer gridless routing system with wire planning. In *Proceedings of the ACM International Symposium on Physical Design*, San Diego, CA, 2000, pp. 12–18.

17. D. W. Hightower. A solution to line-routing problems on the continuous plane. In *Proceedings of the ACM/IEEE Design Automation Conference*, Las Vegas, NV, 1969, pp. 1–24.

18. M. Pan and C. Chu. FastRoute 2.0: A high-quality and efficient global router. In *Proceedings of the Asia-South Pacific Design Automation Conference*, Yokohama, Japan, 2007, pp. 250–255.

19. J. Maßberg and T. Nieberg. Rectilinear paths with minimum segment lengths. *Discrete Applied Mathematics*, 161(12):1769–1775, 2013.

20. F. Y. Chang, R. S. Tsay, W. K. Mak, and S. H. Chen. MANA: A shortest path maze algorithm, under separation and minimum length nanometer rules. *IEEE Transactions on Computer-Aided Design of Integrated Circuits and Systems*, 32(10):1557–1568, 2013.

21. F. Nohn. *Detailed Routing im VLSI-Design unter Berücksichtigung von Multiple-Patterning*. PhD thesis, University of Bonn, Bonn, Germany, 2012.

22. R. C. Prim. Shortest connection networks and some generalizations. *Bell System Technical Journal*, 36(6):1389–1401, 1957.

23. H. Zhou, N. Shenoy, and W. Nicholls. Efficient minimum spanning tree construction without Delaunay triangulation. In *Proceedings of the Asia-South Pacific Design Automation Conference*, Yokohama, Japan, 2001, pp. 192–197.

24. A. B. Kahng and G. Robins. *On Optimal Interconnections for VLSI*. Springer, New York, 1995.

25. M. Hanan. On Steiner's problem with rectilinear distance. *SIAM Journal on Applied Mathematics*, 14(2):255–265, 1966.

26. C. Chu and Y. C. Wong. FLUTE: Fast lookup table based rectilinear Steiner minimal tree algorithm for VLSI design. *IEEE Transactions on Computer-Aided Design of Integrated Circuits and Systems*, 27(1):70–83, 2008.

27. G. Ajwani, C. Chu, and W. K. Mak. FOARS: FLUTE based obstacle-avoiding rectilinear Steiner tree construction. *IEEE Transactions on Computer-Aided Design of Integrated Circuits and Systems*, 30(2):194–204, 2011.

28. A. Hashimoto and J. Stevens. Wire routing by optimizing channel assignment within large apertures. In *Proceedings of the Design Automation Workshop*, Atlantic City, NJ, 1971, pp. 155–169.

29. D. N. Deutsch. A "dogleg" channel router. In *Proceedings of the ACM/IEEE Design Automation Conference*, San Francisco, CA, 1976, pp. 425–433.

30. W. K. Luk. A greedy switchbox router. *Integration, the VLSI Journal*, 3(2):129–139, 1985.

31. D. Braun, J. L. Burns, F. Romeo, A. Sangiovanni-Vincentelli, K. Mayaram, S. Devadas, and H. K. T. Ma. Techniques for multilayer channel routing. *IEEE Transactions on Computer-Aided Design of Integrated Circuits and Systems*, 7(6):698–712, 1988.

32. R. Nair, S. J. Hong, S. Liles, and R. Villani. Global wiring on a wire routing machine. In *Proceedings of the ACM/IEEE Design Automation Conference*, Las Vegas, NV, 1982, pp. 224–231.

33. J. Hu and S. S. Sapatnekar. A survey on multi-net global routing for integrated circuits. *Integration, the VLSI Journal*, 31(1):1–49, 2001.

34. C. Chiang, M. Sarrafzadeh, and C. K. Wong. Global routing based on Steiner min-max trees. *IEEE Transactions on Computer-Aided Design of Integrated Circuits and Systems*, 9(12):1318–1325, 1990.

35. C. Chiang, C. K. Wong, and M. Sarrafzadeh. A weighted Steiner tree-based global router with simultaneous length and density minimization. *IEEE Transactions on Computer-Aided Design of Integrated Circuits and Systems*, 13(12):1461–1469, 1994.

36. B. Ting and B. N. Tien. Routing techniques for gate array. *IEEE Transactions on Computer-Aided Design of Integrated Circuits and Systems*, 2(4):301–312, 1983.

37. R. Nair. A simple yet effective technique for global wiring. *IEEE Transactions on Computer-Aided Design of Integrated Circuits and Systems*, 6(2):165–172, 1987.

38. L.-C. E. Liu and C. Sechen. Multi-layer chip-level global routing using an efficient graph-based steiner tree heuristic. In *Proceedings of the European Conference on Design and Test*, 1997, 311pp.

39. M. Burstein and R. Pelavin. Hierarchical wire routing. *IEEE Transactions on Computer-Aided Design of Integrated Circuits and Systems*, 2(4):223–234, 1983.

40. M. Marek-Sadowska. Global router for gate array. In *Proceedings of the IEEE International Conference on Computer Design*, Port Chester, NY, 1984, pp. 332–337.

41. E. Shragowitz and S. Keel. A global router based on a multicommodity flow model. *Integration, the VLSI Journal*, 5(1):3–16, 1987.

42. R. C. Carden IV, J. Li, and C.-K. Cheng. A global router with a theoretical bound on the optimal solution. *IEEE Transactions on Computer-Aided Design of Integrated Circuits and Systems*, 15(2):208–216, 2006.

43. D. Müller, K. Radke, and J. Vygen. Faster min–max resource sharing in theory and practice. *Mathematical Programming Computation*, 3(1):1–35, 2011.

44. F. Shahrokhi and D. W. Matula. The maximum concurrent flow problem. *Journal of the ACM*, 37(2):318–334, 1990.

45. P. Raghavan and C. D. Tompson. Randomized rounding: A technique for provably good algorithms and algorithmic proofs. *Combinatorica*, 7(4):365–374, 1987.

46. C. Albrecht. Global routing by new approximation algorithms for multicommodity flow. *IEEE Transactions on Computer-Aided Design of Integrated Circuits and Systems*, 20(5):622–632, 2006.

47. N. Garg and J. Koenemann. Faster and simpler algorithms for multicommodity flow and other fractional packing problems. *SIAM Journal on Computing*, 37(2):630–652, 2007.

48. M. D. Moffitt, J. A. Roy, and I. L. Markov. The coming of age of (academic) global routing. In *Proceedings of the ACM International Symposium on Physical Design*, Portland, OR, 2008, pp. 148–155.

49. Y. J. Chang, Y. T. Lee, and T. C. Wang. NTHU-Route 2.0: A fast and stable global router. In *Proceedings of the IEEE/ACM International Conference on Computer-Aided Design*, San Jose, CA, 2008, pp. 338–343.

50. M. Cho, K. Lu, K. Yuan, and D. Z. Pan. BoxRouter 2.0: Architecture and implementation of a hybrid and robust global router. In *Proceedings of the IEEE/ACM International Conference on Computer-Aided Design*, San Jose, CA, 2007, pp. 503–508.

51. J. Lou, S. Thakur, S. Krishnamoorthy, and H. S. Sheng. Estimating routing congestion using probabilistic analysis. *IEEE Transactions on Computer-Aided Design of Integrated Circuits and Systems*, 21(1):32–41, 2002.

52. M. Pan and C. Chu. IPR: An integrated placement and routing algorithm. In *Proceedings of the ACM/IEEE Design Automation Conference*, San Diego, CA, 2007, pp. 59–62.

53. Y. Wei, C. Sze, N. Viswanathan, Z. Li, C. J. Alpert, L. Reddy, A. D. Huber, G. E. Tellez, D. Keller, and S. S. Sapatnekar. Techniques for scalable and effective routability evaluation. *ACM Transactions on Design Automation of Electronic Systems*, 19(2):17:1–17:37, 2014.

54. C. J. Alpert and G. E. Tellez. The importance of routing congestion analysis. DAC Knowledge Center Online Article, 2010. https://www.researchgate.net/profile/Gustavo_Tellez/publication/242748131_The_Importance_of_Routing_Congestion_Analysis. Accessed on December 8, 2015.

55. N. Viswanathan, C. Alpert, C. Sze, Z. Li, and Y. Wei. The DAC 2012 routability-driven placement contest and benchmark suite. In *Proceedings of the ACM/EDAC/IEEE Design Automation Conference*, San Francisco, CA, 2012, pp. 774–782.

56. C. J. Alpert, Z. Li, M. M. Moffitt, G. J. Nam, J. A. Roy, and G. E. Tellez. What makes a design difficult to route. In *Proceedings of the ACM International Symposium on Physical Design*, San Francisco, CA, 2010, pp. 7–12.

57. M. C. Kim, J. Hu, D. J. Lee, and I. L. Markov. A SimPLR method for routability-driven placement. In *Proceedings of the IEEE/ACM International Conference on Computer-Aided Design*, 2011, pp. 67–73.

58. T. Nieberg. Gridless pin access in detailed routing. In *Proceedings of the ACM/IEEE Design Automation Conference*, 2011, pp. 170–175.

59. S. Batterywala, N. Shenoy, W. Nicholls, and H. Zhou. Track assignment: A desirable intermediate step between global routing and detailed routing. In *Proceedings of the IEEE/ACM International Conference on Computer-Aided Design*, 2002, pp. 59–66.

60. R. Kay and R. A. Rutenbar. Wire packing: A strong formulation of crosstalk-aware chip-level track/layer assignment with an efficient integer programming solution. In *Proceedings of the ACM International Symposium on Physical Design*, San Diego, CA, 2000, pp. 61–68.

61. N. Ryzhenko and S. Burns. Standard cell routing via Boolean satisfiability. In *Proceedings of the ACM/IEEE Design Automation Conference*, San Francisco, CA, 2012, pp. 603–612.

62. J. Salowe. Rip-up and reroute. In Charles J. Alpert, Dinesh P. Mehta, and Sachin S. Sapatnekar, Eds., *Handbook of Algorithms for Physical Design Automation*, CRC Press, Boca Raton, FL, 2008, pp. 615–626.

63. C. Bartoschek, S. Held, D. Rautenbach, and J. Vygen. Efficient generation of short and fast repeater tree topologies. In *Proceedings of the ACM International Symposium on Physical Design*, San Jose, CA, 2006, pp. 120–127.

64. T. Sakurai. Closed-form expressions for interconnection delay, coupling, and crosstalk in VLSIs. *IEEE Transactions on Electron Devices*, 40(1):118–124, 1993.

65. F. Liu and S. S. Sapatnekar. Metrics used in physical design. In Charles J. Alpert, Dinesh P. Mehta, and Sachin S. Sapatnekar, Eds., *Handbook of Algorithms for Physical Design Automation*, CRC Press, Boca Raton, FL, 2008, pp. 29–52.

66. W. C. Elmore. The transient response of damped linear networks with particular regard to wide-band amplifiers. *Journal of Applied Physics*, 19(1):55–63, 1948.

67. C. V. Kashyap, C. J. Alpert, F. Liu, and A. Devgan. Closed-form expressions for extending step delay and slew metrics to ramp inputs for RC trees. *IEEE Transactions on Computer-Aided Design of Integrated Circuits and Systems*, 23(4):509–516, 2004.

68. M. A. B. Jackson, A. Srinivasan, and E. S. Kuh. Clock routing for high-performance ICs. In *Proceedings of the ACM/IEEE Design Automation Conference*, San Jose, CA, 1990, pp. 573–579.

69. A. Kahng, J. Cong, and G. Robins. High-performance clock routing based on recursive geometric matching. In *Proceedings of the ACM/IEEE Design Automation Conference*, San Francisco, CA, 1991, pp. 322–327.

70. R. S. Tsay. An exact zero-skew clock routing algorithm. *IEEE Transactions on Computer-Aided Design of Integrated Circuits and Systems*, 12(2):242–249, 1993.

71. K. D. Boese and A. B. Kahng. Zero-skew clock routing trees with minimum wirelength. In *Proceedings of Annual IEEE International ASIC Conference and Exhibit*, Piscataway, NJ, 1992, pp. 17–21.

72. P. J. Restle, T. G. McNamara, D. A. Webber, P. J. Camporese, K. F. Eng, K. A. Jenkins, D. H. Allen et al. A clock distribution network for microprocessors. *IEEE Journal of Solid-State Circuits*, 36(5):792–799, 2001.

73. J. Cong. An interconnect-centric design flow for nanometer technologies. *Proceedings of the IEEE*, 89(4):505–528, 2001.

74. S. S. Sapatnekar. Wire sizing as a convex optimization problem: Exploring the area-delay tradeoff. *IEEE Transactions on Computer-Aided Design of Integrated Circuits and Systems*, 15(8):1001–1011, 1996.

75. C. J. Alpert, A. Devgan, J. P. Fishburn, and S. T. Quay. Interconnect synthesis without wire tapering. *IEEE Transactions on Computer-Aided Design of Integrated Circuits and Systems*, 20(1):90–104, 2001.

76. Z. Li, C. J. Alpert, S. Hu, T. Muhmud, S. T. Quay, and P. G. Villarrubia. Fast interconnect synthesis with layer assignment. In *Proceedings of the ACM International Symposium on Physical Design*, Portland, OR, 2008, pp. 71–77.

77. J. Cong, A. B. Kahng, G. Robins, M. Sarrafzadeh, and C. K. Wong. Provably good performance-driven global routing. *IEEE Transactions on Computer-Aided Design of Integrated Circuits and Systems*, 11(6):739–752, 1992.

78. C. J. Alpert, T. C. Hu, J. H. Huang, and A. B. Kahng. A direct combination of the Prim and Dijkstra constructions for improved performance-driven global routing. In *Proceedings of the IEEE International Symposium on Circuits and Systems*, Chicago, IL, 1993, pp. 1869–1872.

79. C. J. Alpert, G. Gandham, J. Hu, J. L. Neves, S. T. Quay, and S. S. Sapatnekar. Steiner tree optimization for buffers, blockages, and bays. *IEEE Transactions on Computer-Aided Design of Integrated Circuits and Systems*, 20(4):556–562, 2001.

80. W. Shi and C. Su. The rectilinear Steiner arborescence problem is NP-complete. In *Proceedings of ACM-SIAM Symposium on Discrete Algorithms*, San Francisco, CA, 2000, pp. 780–787.

81. S. K. Rao, P. Sadayappan, F. K. Hwang, and P. W. Shor. The rectilinear Steiner arborescence problem. *Algorithmica*, 7(1–6):277–288, 1992.

82. C. J. Alpert, A. B. Kahng, C. N. Sze, and Q. Wang. Timing-driven Steiner trees are (practically) free. In *Proceedings of the ACM/IEEE Design Automation Conference*, San Francisco, CA, 2006, pp. 389–392.

83. K. D. Boese, A. B. Kahng, B. A. McCoy, and G. Robins. Fidelity and near-optimality of Elmore-based routing constructions. In *Proceedings of the IEEE International Conference on Computer Design*, Cambridge, MA, 1993, pp. 81–84.

84. J. Hu and S. S. Sapatnekar. A timing-constrained algorithm for simultaneous global routing of multiple nets. In *Proceedings of the IEEE/ACM International Conference on Computer-Aided Design*, San Jose, CA, 2000, pp. 99–103.

85. C. H. Stapper. Modeling of defects in integrated circuit photolithographic patterns. *IBM Journal of Research and Development*, 28(4):461–475, 1984.

86. C. Bamji and E. Malavasi. Enhanced network flow algorithm for yield optimization. In *Proceedings of the ACM/IEEE Design Automation Conference*, Las Vegas, NV, 1996, pp. 746–751.

87. G. E. Tellez, G. Doyle, P. Honsinger, S. Lovejoy, C. Meiley, G. Starkey, and R. Wilcox Jr. Method for improving wiring related yield and capacitance properties of integrated circuits by maze-routing, US Patent 6,305,004, 2001.

88. M. Cho, H. Xiang, R. Puri, and D. Z. Pan. TROY: Track router with yield-driven wire planning. In *Proceedings of the ACM/IEEE Design Automation Conference*, San Diego, CA, 2007, pp. 55–58.

89. L. R. Darden, W. J. Livingstone, J. H. Panner, P. E. Perry, W. F. Pokorny, and P. S. Zuchowski. Redundant vias, US Patent 6,026,224, 2000.

90. M. T. Buehler, J. M. Cohn, D. J. Hathaway, J. D. Hibbeler, and J. Koehl. Use of redundant routes to increase the yield and reliability of a VLSI layout, US Patent 7,308,669, 2007.

91. B. Y. Su and Y. W. Chang. An exact jumper insertion algorithm for antenna effect avoidance/fixing. In *Proceedings of the ACM/IEEE Design Automation Conference*, San Diego, CA, 2005, pp. 325–328.

92. L. D. Huang, X. Tang, H. Xiang, D. F. Wong, and I. M. Liu. A polynomial time-optimal diode insertion/routing algorithm for fixing antenna problem [IC layout]. *IEEE Transactions on Computer-Aided Design of Integrated Circuits and Systems*, 23(1):141–147, 2004.

93. T. Y. Ho, Y. W. Chang, and S. J. Chen. Multilevel routing with antenna avoidance. In *Proceedings of the ACM International Symposium on Physical Design*, Phoenix, AZ, 2004, pp. 34–40.

94. L. Liebmann, V. Gerousis, P. Gutwin, M. Zhang, G. Hana, and B. Cline. Demonstrating production quality multiple exposure patterning aware routing for the 10nm node. In *Proceedings of SPIE*, 2014.

95. L. W. Liebmann. Layout impact of resolution enhancement techniques: impediment or opportunity? In *Proceedings of the ACM International Symposium on Physical Design*, Monterrey, CA, 2003.

96. J. Ghan, N. Ma, S. Mishra, C. Spanos, K. Poolla, N. Rodriguez, and L. Capodieci. Clustering and pattern matching for an automatic hotspot classification and detection system. In *Proceedings of SPIE*, 2009, pp. 727516.

97. D. Ding, J. R. Gao, K. Yuan, and D. Z. Pan. AENEID: A generic lithography-friendly detailed router based on post-RET data learning and hotspot detection. In *Proceedings of the ACM/IEEE Design Automation Conference*, San Diego, CA, 2011, pp. 795–800.

98. Y. Wei and S. S. Sapatnekar. Dummy fill optimization for enhanced manufacturability. In *Proceedings of the ACM International Symposium on Physical Design*, San Francisco, CA, 2010, pp. 97–104.

99. L. Liebmann and A. Torres. A designer's guide to sub-resolution lithography: Enabling the impossible to get to the 14 nm node. *IEEE Design & Test*, 30(3):70–92, 2013.

100. J. L. Gross, J. Yellen, and P. Zhang. *Handbook of Graph Theory*. Chapman and Hall/CRC Press, Boca Raton, FL, 2nd edn, 2013.

101. M. Cho, Y. Ban, and D. Z. Pan. Double patterning technology friendly detailed routing. In *Proceedings of the IEEE/ACM International Conference on Computer-Aided Design*, 2008, pp. 506–511.

Physical Design for 3D ICs

9

Sung-Kyu Lim

CONTENTS

9.1 INTRODUCTION

A major focus of the electronics industry today is to miniaturize ICs by exploiting advanced lithography technologies. This trend is expected to continue to the 7 nm node and, perhaps, beyond. However, due to the increasing power, performance, and financial bottlenecks beyond 7 nm, the semiconductor industry and academic researchers have begun to actively look for alternative solutions. This search has led to the current focus on thinned and stacked 3D ICs, initially by wire bond, later by flip chip and package on package (POP), most recently by through-silicon via (TSV) [16,40,41,48,82], and in the near future by monolithic 3D ICs [7,78,67,68].* A 3D IC provides the possibility of arranging and interconnecting digital and analog functional blocks across multiple dies at a very fine level of granularity, as illustrated in Figure 9.1. This shortens interconnect, which naturally translates into reduced delay and power consumption [6,32,48,63,77]. Historically, 3D IC technology was first adopted in real-time image sensing [42] and recently by DRAM [33,37] and flash memory [25]. Advances in 3D IC integration and packaging are gaining momentum and have become of critical interest to the semiconductor community. However, the lack of physical design tools that can handle TSVs and 3D die stacking—in addition to cost and yield—delays the mainstream acceptance of this technology.

This chapter focuses on TSV-based 3D ICs that are different from 2.5D ICs based on interposer technologies [71,75,76]. These 3D ICs are built by stacking bare dies and connecting them with TSVs. TSVs are deployed in dies with active devices, not just passive components as in 2.5D ICs. The design methods discussed in this chapter are *chip oriented* as opposed to *package oriented*, because our focus is on the quality of the chip, not its package. In a 3D IC built with TSVs, the individual dies are fabricated separately and later bonded and connected with TSVs. Another emerging technology for die stacking is monolithic 3D integration, where the dies are *grown* on top of each other in a sequential fashion. This chapter focuses on 3D ICs built with TSVs, not monolithic 3D integration, while we provide an overview of recent EDA work on the latter topic in Section 9.5. We review design challenges, algorithms, design methodologies, and technical details for key physical design steps targeting 3D ICs built with TSVs. We also examine related simulation results to evaluate the effectiveness of these techniques on real designs. The history of physical design tool development for TSV-based 3D ICs is short, and new research is reported at an increasing pace, as described in Section 9.5. Instead, we focus on the following practical topics that 3D IC designers face during 3D IC adoption:

---

* Monolithic 3D IC technology has been commercialized in late 2014 by Samsung for flash memory [25]. However, the application of this technology to logic is far behind: only a single or a small number of gates have been demonstrated [7,67,68,78].

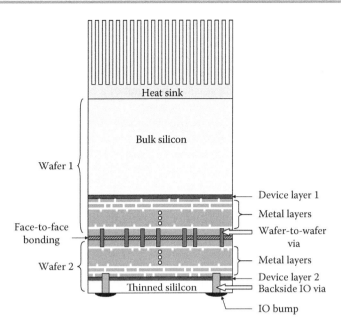

FIGURE 9.1   A two-tier 3D IC with face-to-face bonding. (Image from Black, B. et al., Die stacking (3D) microarchitecture, in *Proceedings of Annual International Symposium Microarchitecture*, 2006.)

- *TSV placement in block-level 3D ICs*: In 3D ICs, block-level designs offer various advantages over designs integrated at other levels of granularity, such as gate level, because they promote the reuse of IP blocks. One fundamental problem in this design style is the placement of TSVs, which have a profound impact not only on traditional physical design objectives such as area, wirelength, timing, and power [36,81] but also on reliability metrics including thermal and mechanical stress [30]. We study trade-offs among the various ways to place TSVs in block-level designs, where TSVs are placed between the blocks or form arrays at strategic locations.
- *Low-power 3D IC design with block folding*: Low power is a potential key benefit of 3D ICs, yet few thorough design studies explored it. We study several physical design methodologies to reduce power consumption in 3D ICs. We use a large-scale commercial-grade microprocessor (OpenSPARC T2) as a benchmark. Our specific focus is on functional module partitioning schemes that further reduce wirelength and buffer usage of individual modules used in a 3D IC design. We also study the needs for and the development of several new physical design tools to serve this purpose.

In this chapter, we present self-contained algorithms and methodologies that EDA engineers can implement from scratch, or in the form of pre- and postprocessing, or plug-ins to an existing tool—either 2D or 3D IC tool—and reproduce the related results. In addition, we describe how 3D IC designers can use various commercial tools built for 2D ICs to handle 3D IC physical design with power, performance, and area (PPA) optimization, as well as multi-physics simulation and reliability analysis. Our choices for particular tools used in this chapter are not due to the lack of features in other tools (in fact, our decision should not discourage the use of other tools). On the contrary, we hope to inspire the development of new EDA algorithms, software, and methodologies. We conclude each section by discussing related issues and implications that EDA engineers may face during tool development for 3D IC physical design.

9.2 WHAT IS A THROUGH-SILICON VIA?

A through-silicon via (TSV) is a micron scale via that vertically penetrates either the silicon substrate or the entire IC stack. TSV is currently the de facto standard for the die-to-die interconnect in 3D ICs. A TSV is typically made up of a copper pillar surrounded by a silicon dioxide liner.

In addition, tungsten pillars and benzocyclobutene (BCB) liners are used to mitigate TSV-induced mechanical stress issues such as delamination and cracks. As of early 2015, TSVs are usually a single-digit micron in diameter and two-digit micron in height [16,40,41,82]. The RC parasitics vary even more widely depending on the material, process, and geometric parameters. The TSV resistance is roughly an order of magnitude smaller than that of a back-end-of-line (BEOL) via, while the capacitance is one or two orders of magnitude larger. The *keep-out zone* (KOZ) is defined as a rectangular region surrounding a TSV, where the placement of devices is strictly forbidden. A KOZ is required to minimize adverse issues such as cracks and timing variations of nearby devices.

Depending on when the TSVs are fabricated, two major TSV types exist: via first and via last, as illustrated in Figure 9.2a. In the via-first case, TSVs are fabricated before CMOS or BEOL (back end of line) metallization. The dimensions of via-first TSVs are typically smaller (1–10 μm diameter), with aspect ratios (height:diameter) of 5:1 to 10:1. Via-last TSVs, on the other hand, are created after BEOL or bonding, essentially when the wafer is finished. In this case, the processing can be done at the foundry or the packaging house. The via-last TSV diameter is typically wider (10–50 μm), with aspect ratios of 2:1:5:1.

An important benefit of TSV-based 3D IC is wirelength reduction, which leads to power and delay savings [6,32,48,63,77]. However, the sheer size of TSVs is identified as a major impediment for the greater usage of TSVs. According to the 2013 ITRS [26], the TSV diameter is projected to range from 1.5 to 1.0 μm between 2009 and 2015. However, the area of a four-transistor NAND gate is projected to range from 0.82 to 0.20 μm² during the same period. This means that the area ratio between TSVs and logic gates is projected to increase from 2.74 (= 2.25/0.82) to 5 (= 1.0/0.20). This TSV-to-gate-size ratio becomes even larger if the KOZ is considered. This area overhead issue—which becomes the burden for physical design to minimize—directly influences the achievable performance-power trade-off curves of 3D ICs. Moreover, it was demonstrated in [37] that TSVs at the full-chip level can easily occupy 20%–30% of the die area even after careful partitioning. These concerns call for EDA tools that carefully consider the impact of TSV size during the partitioning, placement, routing, and clock tree synthesis stages of physical implementation. Tools are also needed to capture the RC parasitics of these large TSVs and their impact on circuit power and performance.

9.2.1 THROUGH-SILICON VIA (TSV) MANUFACTURING

There are two main technologies for manufacturing TSVs: dry etching (Bosch etching) and laser drilling [48]. Laser drilling is faster and cheaper but cannot produce TSVs with small diameter or high aspect ratio. Thus, laser drilling is preferred for via-last TSVs, while the Bosch etching is more frequently used for via-first TSVs [48]. Polysilicon, copper, and tungsten are the most popular materials for TSV fill. Silicon dioxide is a popular material for the liner that sits between the TSV and the silicon substrate for insulation purposes.

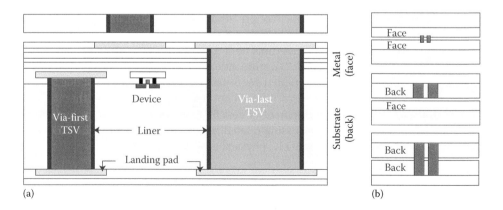

FIGURE 9.2 (a) TSV type and landing pads (shown with face-to-back bonding), (b) die bonding styles (shown with via-first TSV). (Image from Lim, S.K., SV-aware 3D physical design tool needs for faster mainstream acceptance of 3D ICs, in *ACM DAC Knowledge Center*, 2010.)

After the formation of the metal-filled holes, the chips are thinned and bonded together. Thinning is done by grinding, chemical mechanical planarization (CMP), or by a wet chemical process. A silicon or glass carrier is typically used here, where the wafer is turned upside down and temporarily bonded to the carrier. Depending on which sides of the dies are bonded together, there exist three types of bonding styles, namely, face-to-face, face-to-back, and back-to-back, as illustrated in Figure 9.2b. Some of the popular bonding technologies [41] include oxide fusion, metal-to-metal, copper-to-copper, micro-bumping, and polymer adhesive bonding. Note that face-to-face bonding does not utilize TSVs because the interconnect between the dies is established by using metal layers only.

From the perspective of physical design, via-first TSVs are less intrusive because they interfere only with the device, M1, and top layers, whereas via-last TSVs interfere with all layers in the die, as illustrated in Figure 9.2a. Via-first TSVs have their landing pads on M1 and the top metal layers, whereas via-last TSVs have their landing pads only on the top metal layers. A landing pad includes keep-out zones around it to reduce electrical coupling and mechanical damage to nearby devices and interconnects. The connections between via-first TSVs are made using local interconnect and vias between adjacent dies, whereas via-last TSVs are stacked on top of each other, as illustrated in Figure 9.2a. Therefore, via-first TSVs are normally used for signal and clock delivery, whereas power delivery networks utilize via-last TSVs.

9.2.2 3D IC TOOL AVAILABILITY AS OF EARLY 2015

Ansys [2] offers the following modeling and simulation tools for 3D IC:

- RedHawk: Simulator for simultaneous switching noise, decoupling capacitance, and on-chip and off-chip inductance for 3D IC.
- Sentinel-TI: Thermal simulation and mechanical stress integrity analysis platform for stacked die/3D IC designs.

Atrenta [5] offers SpyGlass Physical 2.5D/3D that provides early estimates of area, timing, and routability for RTL designers without the need for physical design expertise or tools for 2.5D/3D IC. It provides valuable physical reports and rules to identify area, congestion, and timing issues at the early stages of the 3D IC design.

Cadence [10] offers

- Encounter: 3D IC physical design tool (placement, optimization, routing) for custom and digital designs
- QRC Extraction: 3D IC verification and analysis tool
- Encounter DFT Architect: Design-for-test tool for 3D ICs
- Design IP for wide-I/O controller
- SiP (System-in-Package) Co-design: IC/Package co-design tool

Mentor Graphics [50] offers

- Calibre: 3D IC physical verification, extraction, LVS, and DFM for 3D IC products: SiP, silicon interposers, or stacked die with TSVs
- Tessent: Deterministic scan testing, embedded pattern compression, built-in self test, specialized embedded memory test and repair, and boundary scan tool for 2.5D and 3D ICs

Synopsys [72] offers

- DFTMAX Test Automation: Design-for-Test tool for stacked die and TSV
- DesignWare STAR Memory System IP: Integrated memory test, diagnostic and repair solution
- IC Compiler: 3D IC place-and-route tool, including TSV, microbump, silicon interposer redistribution layer (RDL) and signal routing, power mesh creation, and interconnect checks

- StarRC Ultra: Parasitic extraction tool with a support for TSV, microbump, interposer RDL, and signal routing metal
- HSPICE and CustomSim: Multi-die interconnect simulation and analysis tool.
- PrimeRail: IR drop and EM analysis tool for 3D IC
- IC Validator: DRC for microbumps and TSVs, LVS connectivity checking between stacked die
- Galaxy Custom Designer: Custom editor for silicon interposer RDL, signal routing, and power mesh
- Sentaurus Interconnect: Thermo-mechanical stress analyzer to evaluate the impact of TSVs and microbumps used in multi-die stacks

Xilinx offers 3D IC EDA tools to the customers of its 3D FPGA devices [81].

Among the other vendors, 3D IC layout editors are offered by Micro Magic [51] and R3 Logic [64], while 3DInCites [1] offers an up-to-date list of 3D EDA activities.

9.3 TSV PLACEMENT IN BLOCK-LEVEL 3D ICs

In general, block-level designs offer various advantages over designs done at lower levels of granularity, such as gate level, because they promote the reuse of existing hard IP blocks. The same philosophy applies to 3D ICs, where the IPs can be assembled into multiple tiers and connected with intra- and/or inter-tier vias and interconnects. In addition, chip-scale IPs such as a multi-core design or an entire L2 cache can be easily stacked and assembled in 3D ICs. A major physical design challenge in block-level 3D IC designs is TSV placement. In this section, we review trade-offs among various ways to place TSVs in block-level designs, where TSVs are placed between the blocks or form small and/or large arrays at strategic locations. We discuss three practical options studied in [4], namely, TSV farm, TSV distributed, and TSV whitespace. Other options published in the literature include [39,74]. Depending on the location of through-silicon vias (TSVs) in the bottom die, a redistribution layer (RDL) may become necessary on the backside of the bottom tier to connect the two dies, as shown in Figure 9.3.

Among several possible configurations, we focus on a two-tier 3D IC, where the bottom die has a larger footprint. A typical example of such a stacking includes a hybrid memory cube [73] (= 3D DRAM) stacked on top of a multi-core processor. Both dies are facing down so that the heat sink is located above the backside (= bulk) of the top die, and C4 bumps are below the frontside (= top metal layer) of the bottom die. This stacking allows for better power delivery and potentially better cooling if the top die consumes more power. *We further assume that the design of the top die is fixed so that we focus on the block-level design of the bottom die.*

9.3.1 TSV PLACEMENT STYLES

When face-to-back bonding is utilized between two dies with different die sizes, redistribution-layer (RDL) routing on the backside of the bottom die is required in some cases. If some TSVs in the bottom die are outside the footprint area of the top die, RDL routing is necessary to connect the TSVs to the bonding pads of the top die, as illustrated in Figure 9.3a. But if all TSVs inserted in

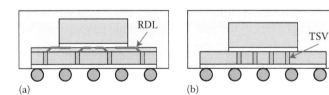

FIGURE 9.3 A side view of a 3D IC. (a) With a redistribution layer (RDL), (b) without an RDL. (Image from Athikulwongse, K. et al., Block-level designs of die-to-wafer bonded 3D ICs and their design quality tradeoffs, in *Proceedings of Asia and South Pacific Design Automation Conference*, 2013.)

the bottom die are inside the footprint area of the top die as shown in Figure 9.3b, the TSVs in the bottom die can be directly bonded to the bonding pads in the top die, without any RDL routing.

Although the RDL allows connections between TSV landing pads on the backside of the bottom die and the bonding pads in the top die, it causes several negative effects. First of all, typical wires on the RDL are wide, possibly as wide as the wires on the topmost metal layers. Thus, their parasitic capacitance is much higher than local metal wires and cause timing degradation and dynamic power overhead. In addition, the large minimum pitch between adjacent wires in the RDL limits the minimum TSV pitch in a TSV array. For example, if four TSVs are placed in a 2 × 2 array, they can be placed as close to each other as possible. However, if 25 TSVs are placed in a 5 × 5 array, the TSV in the center cannot be routed by an escape routing unless the TSV pitch is several times larger than the minimum pitch.

Our discussion distinguishes two options that are available for the design of 3D ICs with different die sizes: (1) insert all TSVs inside the footprint area of the top die so that RDL routing is not required or (2) insert TSVs wherever they are needed and perform RDL routing. The former limits TSV locations but does not require RDL wires. The latter provides a higher degree of freedom on TSV locations but requires RDL wires. In addition, different TSV insertion styles lead to very different layout qualities. We study three different design styles: TSV farm (without RDLs), TSV distributed (with RDLs and regularly placed TSVs), and TSV whitespace (with RDLs and irregularly placed TSVs), as shown in Figure 9.4.

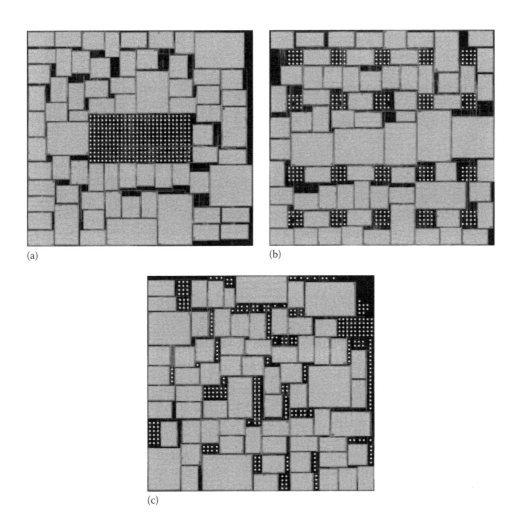

(a)

(b)

(c)

FIGURE 9.4 TSV placement styles in the block-level 3D IC. (a) TSV-farm, (b) TSV-distributed, (c) TSV-whitespace styles. TSVs are shown in white. (Image from Athikulwongse, K. et al., Block-level designs of die-to-wafer bonded 3D ICs and their design quality tradeoffs, in *Proceedings of Asia and South Pacific Design Automation Conference*, 2013.)

9.3.2 A BLOCK-LEVEL 3D IC DESIGN FLOW

In [4], TSV insertion and floorplanning are performed in the bottom die as follows: in the TSV-farm and the TSV-distributed styles, TSVs are pre-placed in arrays and treated as obstacles during floorplanning. In the TSV-farm style, an array of TSVs are placed in the middle of the bottom die. In the TSV-distributed style, on the other hand, TSVs are placed all over the bottom die. Therefore, some of the TSVs are placed outside the footprint area of the top die. After the TSVs are pre-placed, the floorplanning of the blocks is manually (or automatically using an obstacle-aware floorplanner) performed. The following factors are considered for each style:

- *TSV farm*: Since functional blocks and TSVs should not overlap, the blocks are placed around the TSV farm. Since the TSV farm area is usually large and occupies a prime location in the chip footprint, this style may cause significant wirelength overhead if the blocks are highly connected. On the other hand, if the inter-block connectivity is not high, the TSV farm in the center does not cause a significant wirelength overhead.
- *TSV distributed*: In this style, TSVs may not cause a significant wirelength overhead. This is because TSVs are grouped in small arrays unlike the one large array in the TSV-farm style. However, some large blocks may have very few locations available for their placement because they cannot be placed in the space between adjacent TSV arrays. This design restriction may degrade wirelength, timing, and power. However, the TSV-distributed style promotes low operating IC temperature and low TSV stress because of the even distribution of TSVs.
- *TSV whitespace*: In this style, a 3D floorplanner is used first to obtain TSV-whitespace style layouts. After floorplanning, TSVs are manually inserted into the whitespace between blocks. Therefore, TSVs are placed in irregular positions, unlike the other two styles. When there is not enough whitespace, the floorplan is perturbed by shifting blocks to create or expand whitespace. Since a 3D floorplanner is invoked without any restrictions imposed by TSVs, this style is expected to optimize the traditional objectives such as power, performance, and area (PPA) better than other design styles.

Another noteworthy work in TSV management for block-level 3D ICs is by [39]. The authors propose two styles, namely, *legacy 2D* and *TSV islands*. In the legacy 2D style, functional blocks are first floorplanned, and then TSVs are inserted in the whitespace for inter-block connection. This style resembles the TSV-whitespace case explained earlier, where the location of TSVs is irregular and TSVs do not tend to form groups. In the TSV island style, TSVs are grouped to form small islands. Unlike the TSV-distributed case earlier, however, the location of these islands is irregular. The authors presented a net clustering approach to group TSVs into islands while not degrading the initial 3D floorplan quality.

9.3.3 A 3D IC DESIGN EVALUATION METHODOLOGY

A timing and power analysis flow for 3D IC [34] is shown in Figure 9.5. First, parasitic resistance and capacitance of each die are extracted using, for example, Cadence QRC Extraction. Since the face-to-back die bonding style is assumed, the capacitive coupling between the bottom and the top dies is negligible.* The parasitic resistance and capacitance of the redistribution layer (RDL) are extracted next.

For 3D static timing analysis, the top and the bottom dies are represented as modules in a top-level Verilog file. A top-level SPEF file is also created. It includes not only the parasitic resistance and capacitance of both dies but also resistance and capacitance of TSVs and the RDL wires. For an accurate power analysis, the switching activity of all logic cells is obtained by a functional simulation of the whole chip. Synopsys PrimeTime is used to perform static timing and power analysis, using the combined SPEF file that contains parasitics in both dies.

---

* If we use face-to-face bonding, this inter-die coupling must be extracted. Currently, such a tool does not exist.

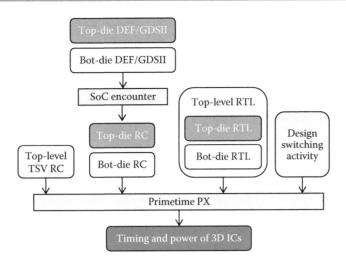

FIGURE 9.5   A timing and power analysis flow for 3D ICs. (Image from Athikulwongse, K. et al., Block-level designs of die-to-wafer bonded 3D ICs and their design quality tradeoffs, in *Proceedings of Asia and South Pacific Design Automation Conference*, 2013.)

The thermal analysis flow for 3D IC is shown in Figure 9.6. This flow is built based on a commercial tool, namely, Ansys FLUENT, and enhanced with custom plug-ins. First, a meshed structure is created, where each thermal tile contains material composition information, such as copper and dielectric density in the tile. This information is extracted from GDSII layout files, which include logic cells as well as TSVs. These files together with the power dissipation of each logic cell are supplied to the layout analyzer. The layout information of a tile consists of the total power dissipated in the tile, and thermal conductivity computed from the materials inside, such as poly-silicon used for transistor gates, tungsten used for vias, copper used for TSVs, and dielectric material. With a sufficiently small thermal tile size, the equivalent thermal conductivity can be computed based on a thermal resistive model [80]. Once the thermal equations are built, FLUENT solves them to obtain temperature values at all thermal tiles.

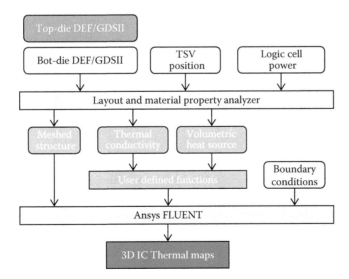

FIGURE 9.6   A GDSII layout-level thermal analysis flow for 3D ICs. (Image from Athikulwongse, K. et al., Block-level designs of die-to-wafer bonded 3D ICs and their design quality tradeoffs, in *Proceedings of Asia and South Pacific Design Automation Conference*, 2013.)

The mechanical stress of a 3D IC layout can be analyzed using the stress analyzer obtained from [29]. The inputs to the analyzer are die size, TSV diameter, TSV locations, simulation grid density, and pre-computed data of TSV stress tensor. The analyzer outputs a von Mises stress map [18], which is a widely used mechanical reliability diagnostic. The computation of stress at a point affected by multiple TSVs is based on the principle of linear superposition of stress tensors. With stress tensors obtained from finite element analysis (FEA) using a commercial tool such as ABAQUS FEA, we can perform a full-chip stress analysis.

9.3.4 SIMULATION RESULTS

In this simulation, we use a 45 nm technology [55]. An open-source hardware IP core [56] is synthesized using an open cell library [53]. We assume a high thermal conductivity molding compound [24]. The total numbers of gates in the benchmark design is 1,363,536. The total number of functional blocks, inter-block nets, and TSVs used are 69, 1853, and 312, respectively. The same partitioning and thus the same number of TSVs are used in all the three TSV placement styles for fair comparisons. The TSV size is 10 μm, and TSV pitch is 30 μm. The parasitic capacitance and resistance are 50 fF and 50 mΩ, respectively. RDL wire width and spacing of 0.4 μm is used in the experiments.

9.3.4.1 WIRELENGTH AND TIMING RESULTS

The silicon area, footprint, and block-to-block (B2B) and RDL wirelength of the three styles are shown in Table 9.1. The same area and footprint for all three styles are used: 3.979 and 2.766 mm², respectively. The TSV-farm style shows the shortest wirelength because all the TSVs occupy only one area in the middle of the die, confining the obstruction of an optimal block placement to a small area. The TSV-distributed style shows the longest block-to-block wirelength (27% longer than the TSV-farm style) because the TSV arrays distributed all over the die obstruct an optimal block placement. The TSV-whitespace style shows a slightly longer wirelength (2%) than the TSV-farm style because we start from optimal block placement and moves blocks only when it is necessary to insert TSVs in some positions. Most TSVs are inserted in the original whitespace and do not interfere with the placement of the blocks very much. In addition, the TSV-distributed and the TSV-whitespace styles require RDL routing, as shown in Figure 9.7.

The longest path delay (LPD), without and with timing optimization, are also shown in Table 9.1. The timing optimization proposed in [44] is used with the target delay of 1.25 ns. Without timing optimization, none of the designs meets the target delay; however, the TSV-farm style shows the shortest delay. With timing optimization, all designs almost met the target delay, and the delay of the TSV-farm style is still the shortest. The delay of the TSV-distributed and the TSV-whitespace styles is longer than that of the TSV-farm style by 10% and 15%, respectively. Because of the long wirelength, it is hard to optimize both the TSV-distributed and the TSV-whitespace styles. In addition, no buffer can be added along the RDL routing because the routing is on the backside of the bottom die.

TABLE 9.1   Comparison of Wirelength and Longest Path Delay (LPD) with or without Timing Optimization

Design Style	Wirelength (m)			LPD (ns)		
	Block-to-Block		**RDL**	**w/o**	**w/opt.**	
TSV farm	1.447	(100.00%)	—	3.136	1.293	(100.00%)
TSV distributed	1.842	(+27.30%)	0.170	4.252	1.425	(+10.20%)
TSV whitespace	1.483	(+2.46%)	0.176	4.568	1.492	(+15.38%)

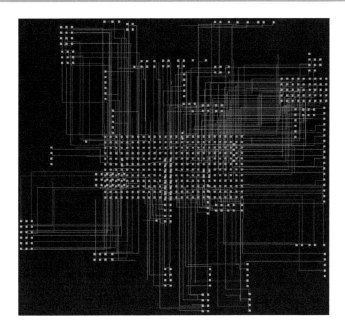

FIGURE 9.7 A redistribution layer (RDL) routing for the TSV-whitespace style, where the two dies are bonded using a wide-I/O interface. (Image from Athikulwongse, K. et al., Block-level designs of die-to-wafer bonded 3D ICs and their design quality tradeoffs, in *Proceedings of Asia and South Pacific Design Automation Conference*, 2013.)

9.3.4.2 POWER CONSUMPTION, THERMAL, AND STRESS RESULTS

We now compare power consumption, thermal, and mechanical stress among the three TSV placement styles when they are operating at their *maximum frequency*, as shown in Table 9.1.* First, the total power consumption is shown in Table 9.2. We observe that the TSV-distributed and the TSV-whitespace styles consume 6% and 10% less power than the TSV-farm style.

The maximum, minimum, and average temperatures are shown in Table 9.2. Although the minimum and average temperatures across all the three designs are close, the maximum temperature of the three designs is different. The TSV-distributed style shows that the lowest maximum temperature, not because it consumes less power—resulting from relatively low speed—but primarily because TSVs distributed all over the die, helps conduct heat. The TSV-farm style shows a high maximum temperature because TSVs in the center of the die cannot help conduct heat from high-power blocks far from them. The TSV-whitespace style also shows high maximum temperature although it consumes the least power for the same reason. The thermal profiles of the TSV-farm, TSV-distributed, and TSV-whitespace styles computed, based on the maximum operating speed, are shown in Figure 9.8. We see that TSVs help reduce temperature, and the local cool spots correspond to TSV array locations. The TSV-distributed style shows the lowest maximum temperature because TSVs are distributed across the die. The TSV-whitespace style exhibits the highest temperature because high-power blocks can be far from TSVs.

TABLE 9.2   Comparison of Power Consumption and Temperature

Design Style	P_{total} (mW)		T_{max} (°C)	T_{min} (°C)	T_{ave} (°C)
TSV farm	1183	(100.00%)	76.87	38.04	47.56
TSV distributed	1107	(−6.40%)	62.43	39.15	46.28
TSV whitespace	1065	(−9.99%)	77.04	38.65	46.19

* Note that it is also possible to conduct simulations under the same clock frequency and compare power and thermal qualities. This iso-performance comparison—although meaningful—is beyond the scope of this chapter.

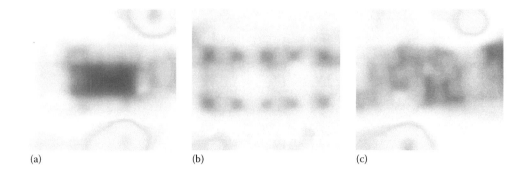

(a)   (b)   (c)

FIGURE 9.8 Temperature maps for (a) TSV-farm, (b) TSV-distributed, and (c) TSV-whitespace styles. The actual values are reported in Table 9.2. (Image from Athikulwongse, K. et al., Block-level designs of die-to-wafer bonded 3D ICs and their design quality tradeoffs, in *Proceedings of Asia and South Pacific Design Automation Conference*, 2013.)

The maximum and average stress values are shown in Table 9.3. The area with stress higher than 10 MPa (mega-pascal) is also shown in the table. Despite high TSV density, the TSV-farm style shows the lowest maximum stress among the designs. This is primarily due to the phenomenon called *destructive interference of stress*, where some vertical and horizontal stress vectors cancel each other in a TSV array [28]. We see that the maximum von Mises stress values reduce because of the interference. The average stresses above the 10-MPa threshold on the die, on the other hand, show the opposite trend. The TSV-farm style shows the highest average stress. When many TSVs are packed into a confined area, the impact of interference from non-neighboring TSVs becomes noticeable. This phenomenon may overwhelm the destructive interference of stress and accumulate high levels of overall stress. Therefore, the TSV-farm style shows higher average stress values compared with others. Last, the TSV-whitespace style shows the largest area of stress above the threshold and the TSV-farm style the smallest. It is mainly due to the fact that the area occupied by the TSV arrays and their keep-out zones is the smallest in the TSV-farm case. The stress profiles of three different styles are shown in Figure 9.9.

9.3.4.3 SUMMARY

We explored design trade-offs among three practical TSV placement styles. Because of the absence of RDL wiring, the TSV-farm style showed the best timing. The design in this style shows the highest average stress, but the area impacted by stress is the smallest. This means that a high level of stress is confined to a small area, and thus, the overall reliability could be worse. The TSV-distributed style showed the worst wirelength because TSV arrays interfere with block placement. However, it showed the lowest temperature because TSVs distributed across the die help reduce temperature.

Simulation results shown in this chapter are heavily design and technology dependent. The wirelength, timing, power, thermal, and stress results can vary significantly from one design to another based on the following factors: the total number of blocks and their intra- and inter-die connectivity (= TSV and RDL requirements), TSV and RDL dimension and keep-out-zone (KOZ) requirements, the device and interconnect technologies, material properties

TABLE 9.3   Comparison of TSV Mechanical Stress

Design Style	σ_{max} (MPa)		$\sigma_{ave,\sigma>10}$ (MPa)		Area$_{\sigma>10}$ (mm²)	
TSV farm	676.78	(100.0%)	150.20	(100.0%)	0.353	(100.00%)
TSV distributed	691.29	(+2.1%)	97.73	(−34.9%)	0.598	(+69.7%)
TSV whitespace	688.99	(+1.8%)	88.95	(−40.8%)	0.695	(+97.2%)

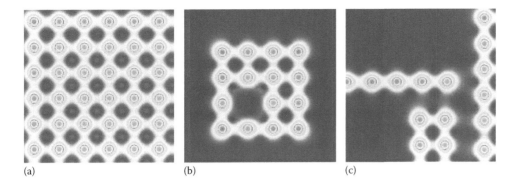

FIGURE 9.9 Stress maps for (a) TSV-farm, (b) TSV-distributed, and (c) TSV-whitespace styles. The actual values are reported in Table 9.3. (Image from Athikulwongse, K. et al., Block-level designs of die-to-wafer bonded 3D ICs and their design quality tradeoffs, in *Proceedings of Asia and South Pacific Design Automation Conference*, 2013.)

and their coefficient-of-thermal expansion (CTE) mismatch, etc. Thus, the TSV-farm versus TSV-distributed versus TSV-whitespace comparisons discussed in this chapter are to be treated as case studies. These comparisons illustrate that the EDA tools that handle the design, analysis, and optimization of 3D ICs under these requirements are the key in choosing the best possible options for a target application to be implemented in a 3D IC.

9.3.5 NEEDS FOR EDA TOOL DEVELOPMENT

We suggest the following requirements to the algorithm and tool developers of TSV placement tools for block-level 3D IC designs. First, it is crucial to understand the power, performance, area (PPA), and multi-physics (= electro-thermo-mechanical) reliability trade-offs among different design styles. While this chapter explores three styles, namely, TSV-farm, TSV-distributed, and TSV-whitespace, additional design styles are possible [39]: TSVs can be placed along the periphery or at some custom locations specified by the designers. The PPA and multi-physics reliability qualities can differ significantly among these options, and the tool needs to offer accurate assessment of these TSV placement solutions.

Second, block-level design for 3D IC, that is, 3D floorplanning, will still be performed manually for small- and medium-size designs. In this case, PPA and reliability analysis will be the only tool required, where accuracy and runtime will be the key objectives. In case an automatic floorplanner is desired to handle very large floorplanning problems, the tool must offer either (1) comparable quality solutions at a fraction of runtime or (2) better quality solutions while not requiring a prohibitive runtime, both compared with manual floorplanning. In either case, the key challenge is that the optimization engine used in the floorplanner, including Analytical [21,87], Simulated Annealing [14,17], Genetic Algorithm, Machine Learning, etc., needs to effectively and efficiently search the solution space and evaluate the PPA and reliability of each candidate solution quickly but accurately.

Third, the floorplanner must be able to handle various technological options available in 3D IC, including the TSV geometries (size, keep-out-zone [KOZ] requirement, etc.), bonding styles (face-to-face, face-to-back, back-to-back), and multi-physics properties of the chip elements. These parameters may be fixed in the early design stage, or the choice may be given to the designers to choose the best option. In case of the latter, the additional dimensions for optimization that must be explored, for example, 1 μm versus 2 μm KOZ, face-to-face versus face-to-back, silicon dioxide versus benzo-cyclo-butene liner for TSV, etc., will further complicate the overall floorplanning process. Algorithms and tools developed for this purpose will need to accurately capture the impact of these choices on full-chip PPA and reliability while searching for optimum solutions.

9.4 LOW-POWER 3D IC DESIGN WITH BLOCK FOLDING

We review the 3D block folding methods described in [32] that are developed to reduce power consumption in 3D ICs on top of the traditional 3D floorplanning. This study is based on the OpenSPARC T2 (an 8-core 64-bit SPARC SoC) design database and Synopsys 28 nm process design kit (PDK) with nine metal layers that are both available to the academic community.* We first discuss how to build, analyze, and optimize GDSII-level 2D and two-tier 3D layouts using industry EDA tools and enhancements. Based on this design environment, we study how to rearrange blocks into 3D to reduce power. Next, we explore block folding methods, that is, partitioning a block into two subblocks and bonding them to achieve power savings in the 3D design. We employ a mixed-size 3D placer for block folding. Last, we demonstrate system-level 3D power benefits by assembling folded blocks.

9.4.1 TARGET BENCHMARK AND 3D IC DESIGN FLOW

The OpenSPARC T2, an open-source commercial microprocessor from Sun Microsystems with 500 million transistors, consists of 53 blocks including eight SPARC cores (SPC), eight L2-cache data banks (L2D), eight L2-cache tags (L2T), eight L2-cache miss buffers (L2B), and a cache crossbar (CCX). Each block is synthesized with 28 nm cell and memory macro libraries. For the 2D design, we follow the original T2 floorplan [54] as much as possible, as shown in Figure 9.10. In addition, special care is taken to optimize both connectivity and data flow between blocks to reduce inter-block wirelength.

The RTL-to-GDSII tool chain for 3D IC design used here is based on commercial tools, and enhanced with in-house tools to handle TSVs and 3D stacking. With the initial design constraints, the entire 3D netlist is synthesized. The layout of each die is done separately based on the 3D floorplanning result. With a given target timing constraint, cells and

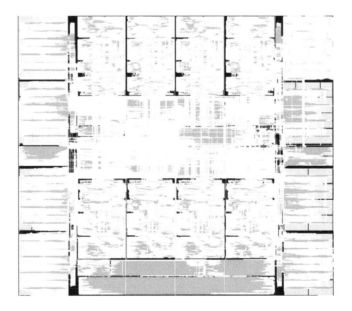

FIGURE 9.10   GDSII layouts of OpenSPARC T2 (full chip): 2D IC design (9 × 7.9 mm²). (Image from Jung, M. et al., On enhancing power benefits in 3D ICs: Block folding and bonding styles perspective, in *Proceedings of ACM Design Automation Conference*, 2014.)

---

* Synopsys has developed a 32/28 nm Interoperable PDK for its University Program members to use specifics of modern technologies. This PDK enables students to master design of digital, analog, and mixed-signal ICs, using the latest Synopsys Custom Implementation tools and utilizing IP-free technology, with parameters and peculiarities close to real processes.

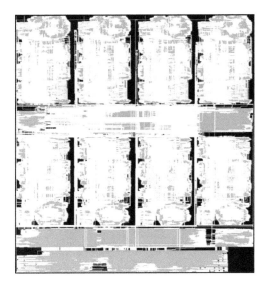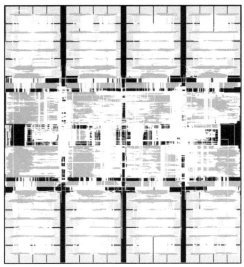

FIGURE 9.11   3D IC GDSII layouts of OpenSPARC T2 (full chip): core/cache stacking (6 × 6.4 mm², # TSV = 3263). (Image from Jung, M. et al., On enhancing power benefits in 3D ICs: Block folding and bonding styles perspective, in *Proceedings of ACM Design Automation Conference*, 2014.)

memory macros are placed in each block. Note that we only utilize regular-Vt (RVT) cells as a baseline. The netlists and the extracted parasitic files are used for 3D static timing analysis, using Synopsys PrimeTime to obtain new timing constraints for each block's I/O pins as well as die boundaries (= TSVs). In this section, we assume two-tier, face-to-back bonded 3D ICs. We use the following parameters for TSV: diameter 3 μm, height 18 μm, pitch 6 μm, resistance 0.043 Ω, and capacitance 8.4 fF.

With these new timing constraints, we perform block-level and chip-level timing optimizations (buffer insertion and gate sizing) as well as power optimizations (gate sizing) using Cadence Encounter. We improve the design quality through iterative optimization steps such as pre-CTS (clock tree synthesis), post-CTS, and post-route optimizations. We utilize all nine metal layers for the SPC design, which requires the most routing resources among all blocks. We use seven layers for all other blocks. Thus, the top two metal layers can be utilized for over-the-block routing in the chip-level design. Figure 9.11 shows a 3D IC design, where all cores are partitioned into one tier and all L2 cache in another. This is one of the most popular approaches to die partitioning. We use this design as another baseline—in addition to the 2D design shown in Figure 9.11—for comparison with a new partitioning scheme, named the *block folding* described in the next section.

9.4.2 A BLOCK FOLDING METHODOLOGY

So far in this chapter, a block-level design style is used in both 2D and 3D ICs. In this case, each block in a 3D IC design occupies a single tier, and TSVs are placed outside the blocks to connect them. In this section, we study *block folding*, where we take the tier-partitioning approach into a finer-grained level: we partition a single block into multiple tiers under the same footprint and connect them with TSVs that are placed inside the folded block.

9.4.2.1 BLOCK FOLDING CRITERIA

For the block folding to provide power saving, the following criteria need to be met in the target block to be folded:

■ The target block must consume a high enough portion of the total system power. Otherwise, the power saving from the block folding could be negligible at the system level. Blocks that consume over 1% of the total system power are listed in Table 9.4. Note that the total power portion of SPC, L2D, and L2T is the average of the eight

TABLE 9.4   2D IC Design Characteristics Used for Block Folding Candidate Selection

Block	Total Power Portion (%)	Net Power Portion (%)	# Long Wires (K)	Remark
SPC	5.8	55.1	27.7	CPU clock, 8×
RTX	3.6	44.4	27.5	I/O clock
CCX	2.8	57.6	12.4	CPU clock
L2D	2.1	29.2	6.5	8×
L2T	1.8	48.5	6.0	8×
RDP	1.7	48.9	5.2	I/O clock
TDS	1.3	43.1	4.8	I/O clock
DMU	1.1	40.7	5.4	I/O clock

Long wires are defined as the wires longer than 100× the standard cell height. The CPU clock runs at 500 MHz and the I/O clock at 250 MHz.

corresponding blocks. Thus, SPC, L2D, and L2T are outstanding target blocks. In addition, RTX and CCX consume high power as a single block and hence could provide nonnegligible power benefit if folded.

- The net power portion of the target block needs to be high. If the block is cell power dominated,* the wirelength reduction of the folded block may not reduce the total power noticeably. Therefore, SPC and CCX are attractive blocks to fold. L2D shows a relatively low net power portion compared with other blocks, as it is the memory dominated block that contains 512 kB (32·16 kB memory macros in our implementation).
- The target block must contain many long wires so that the wirelength decreases, and hence, the net power reduction in the folded block can be maximized. In this study, we define long wires as wires longer than 100× the standard cell height. We observe that SPC, RTX, and CCX have a large number of long wires.

In our study, we fold five blocks: SPC, CCX, L2D, L2T, and RTX. In the following sections, we discuss block folding methodologies for SPC, CCX, and L2D. Each block shows distinctive folding characteristics. Before this, we briefly explain the mixed-size 3D placer that is employed for block folding.

9.4.2.2 FOLDING EXAMPLES

In T2, eight cores use the cache crossbar (CCX) to exchange data stored in eight L2-cache banks. This crossbar is divided into two separate modules, the processor-to-cache crossbar (PCX) and the cache-to-processor crossbar (CPX). There are no signal connections between these two blocks except for the clock and a few test signals. The PCX occupies 48% of the block area and utilizes 48% of the CCX I/O pins, and the CPX uses the rest of them. Thus, the natural way to fold this crossbar is by placing the entire PCX block in one die and the CPX in another die, along with related I/O pins.

Sample 2D and 3D crossbar layouts are shown in Figure 9.12. Interestingly, in the 2D design, we see that the PCX and CPX blocks are separated into several groups. The PCX has eight sources (SPCs) and nine targets (eight L2-cache banks and the I/O bridge). The PCX I/O pin locations are determined based on the target core and L2-cache bank locations in the chip-level floorplan, which in turn attracts connected cells. Because of this, the cells of the PCX block tend to be placed far apart, which degrades cell-to-cell wirelength significantly. However, folding the crossbar eliminates this problem and hence cell-to-cell wirelength decreases by 31.7% compared with

---

* Cell power, also known as the *internal power*, is the power consumed within the boundary of a cell, including intra-cell switching power and short-circuit power.

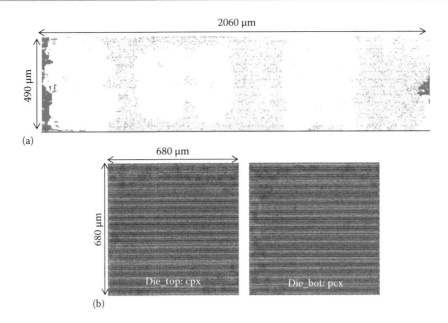

FIGURE 9.12   Cache crossbar (CCX) module 2D IC and 3D IC layouts. (a) A 2D IC design. The Cache-to-Processor (CPX) sub-module is highlighted with white color. (b) A 3D IC design (# TSV = 4). (Image from Jung, M. et al., On enhancing power benefits in 3D ICs: Block folding and bonding styles perspective, in *Proceedings of ACM Design Automation Conference*, 2014.)

the 2D. The folded crossbar leads to 54.6% reduced footprint, 28.8% shorter wirelength, 62.5% less buffer count, and 32.8% power reduction over the 2D counterpart.

Note that only four signal TSVs are used in this folded design, and this is due to the unique characteristics of the CCX module itself. However, we must consider the connections in and out of CCX to cores and cache blocks so that the overall TSV count is minimized in the full-chip 3D IC layout. Last, we examine whether different 3D partitions with more 3D connections can provide better power savings. However, as we increase the TSV count up to 6393, the 3D power benefit reduces down to 23.4%, largely due to the area overhead of TSVs (13.3%).

The single L2-cache data bank contains a 512 kB memory array. This L2D is further divided into four logical sub-banks. In our implementation, each sub-bank group is partitioned into eight blocks of size 16 kB each. The L2D is a memory macro dominated design, and hence, there are not many 3D partitioning options to balance area after folding. Thus, two sub-banks are placed in each die along with related logic cells. Although the buffer count and wirelength reduce by 33.5% and 6.4%, respectively, in the folded L2D, their impact on the total power saving is not significant (5.1% reduction over 2D), as shown in Table 9.5. This is because both cell and leakage power are dominated by memory macros, which 3D folding cannot improve unless these memory macros themselves are folded. Additionally, the net power portion is only about 29% of the total power in 2D, and hence, the small net power reduction in 3D does not lead to a noticeable total power reduction. Still, the footprint area reduction of 48.4% is nonnegligible, and this might affect chip-level design quality.

In case of the SPARC core (SPC), we employ the block folding strategy for one additional step: we fold the blocks *inside* SPC, which contains 14 blocks including two integer execution units (EXU), a floating point and graphics unit (FGU), five instruction fetch units (IFU), and a load/store unit (LSU). This SPC is the highest power consuming block in T2. We apply the same block folding criteria discussed in Section 9.4.2 and fold six blocks as shown in Figure 9.13. We call this second-level folding. With this second-level folding, we obtain 9.2% shorter wirelength, 10.8% less buffers, and 5.1% reduced power consumption than the SPC without second-level folding, that is, a block-level 3D design of the SPC. Additionally, this 3D SPC achieves 21.2% power saving over the 2D SPC.

TABLE 9.5   Comparison between 2D IC and 3D IC Level-2 Cache Data (L2D) Module Designs

L2D	2D	3D	Diff (%)
Footprint (mm²)	2.54	1.31	−48.4
Wirelength (m)	3.41	3.19	−6.4
# cells (×10⁶)	53.1	42.2	−20.5
# buffers (×10⁶)	38.1	25.3	−33.5
Total power (mW)	**172.9**	**164.0**	**−5.1**
Cell power (mW)	25.8	24.6	−4.7
Net power (mW)	50.5	44.5	−11.9
Leakage power (mW)	96.6	94.9	−1.8

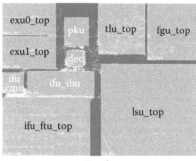
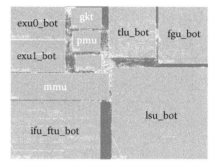

Top die   Bottom die

FIGURE 9.13   Second-level folding of a SPARC core. Six blocks inside the core shown in black text are folded. (Image from Jung, M. et al., On enhancing power benefits in 3D ICs: Block folding and bonding styles perspective, in *Proceedings of ACM Design Automation Conference*, 2014.)

9.4.3 FULL-CHIP ASSEMBLY AND SIMULATION RESULTS

Based on the criteria for block folding discussed in Section 9.4.2, SPC, CCX, L2D, L2T, and RTX have been folded in the full-chip T2 design. Unlike the other four blocks, RTX runs at I/O clock frequency (250 MHz). In addition, almost all signals to/from RTX are connected with MAC, TDS, and RDP that form a network interface unit (NIU) with RTX. Thus, the impact of RTX folding is limited to the RTX block and NIU. In this study, we use a T2 design with all five types of blocks folded.

For the F2B (face-to-back) bonding, the bottom die of folded blocks uses up to M7 (TSV landing pad at M1) as in unfolded blocks, while the top die utilizes up to M9 (TSV landing pad at M9). Thus, M8 and M9 can be used for over-the-block routing including folded blocks in the die bottom. The only exception is SPC that uses up to M9 for both dies, as this block requires the most routing resources. This is why SPCs are placed in the top and the bottom of the chip, as shown in Figure 9.14. Otherwise, these SPC blocks will act as inter-block routing blockages. We place CCX in the center. There are about 300 wires between CCX and each SPC (or L2T). Thus, in this implementation, wires between CCX and L2T are much shorter than those between CCX and SPC. All other control units (SIU, NCU, DMU, and MCU) are placed in the center row as well. Finally, NIU blocks are placed at the bottom-most part of the chip, as most connections are confined to the NIU.

Up to this point, both 2D and 3D designs utilize only regular-Vth (RVT) cells. However, the semiconductor industry has been using multi-Vth cells to further optimize power, especially for leakage power, at the cost of more complex power distribution network design. We employ high-Vth (HVT) cells to examine their impact on power consumption in 2D and 3D designs.

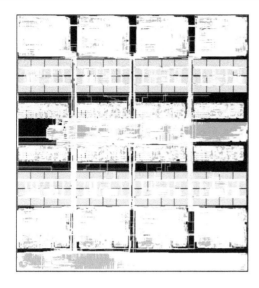

FIGURE 9.14 3D IC GDSII layouts of OpenSPARC T2 (full chip): block folding with TSVs (6 × 6.6 mm², #TSV = 69,091). (Image from Jung, M. et al., On enhancing power benefits in 3D ICs: Block folding and bonding styles perspective, in *Proceedings of ACM Design Automation Conference*, 2014.)

Each HVT cell is around 30% slower, yet has 50% lower leakage, and 5% smaller cell power consumption than the RVT counterpart.

We now compare three full-chip T2 designs: 2D IC, 3D IC without folding (core/cache stacking), and 3D IC with block folding (five types of blocks folded), all with a dual-Vth (DVT) cell library. Detailed comparisons are shown in Table 9.6. We first observe higher HVT cell usage in 3D designs, especially for the 3D with folding case (94.0% of cells are HVT). This is largely due to better timing in 3D designs and helps reduce power in 3D ICs further. We observe that 3D with folding case reduces the total power by 20.3% compared with the 2D and by 10.0% compared with the 3D without folding case. This clearly demonstrates the effectiveness of block folding in large-scale commercial-grade 3D designs for power reduction.

TABLE 9.6   A Full-Chip T2 Comparison among 2D IC, 3D IC without Block Folding (Core/Cache Stacking), and 3D IC with Block Folding (Five Types of Blocks Folded) Designs

	2D	3D w/o Folding	3D w/o Folding
Footprint (mm²)	71.1	38.4 (−46.0%)	40.8 (−42.6%)
Wirelength (m)	339.7	321.3 (−5.5%)	309.6 (−8.9%)
# Cells (×10⁶)	7.41	7.09 (−4.3%)	6.83 (−7.8%)
# Buffers (×10⁶)	2.89	2.37 (−17.9%)	2.23 (−22.8%)
# HVT cells (×10⁶)	6.50 (87.8%)	6.38 (90.0%)	6.42 (94.0%)
# TSV	0	3263	69,091
Total power (W)	**8.240**	**7.113 (−13.7%)**	**6.570 (−20.3%)**
Cell power (W)	1.770	1.394 (−21.2%)	1.175 (−33.6%)
Net power (W)	4.467	3.966 (−11.2%)	3.806 (−14.8%)
Leakage power (W)	2.003	1.753 (−12.4%)	1.589 (−24.2%)

The same dual-Vth design technique is applied to all cases. The numbers in parentheses indicate the difference against the 2D, except for the high-Vth (HVT) cell count reported as a % of the total cell count.

9.4.3.1 SUMMARY

We studied the power benefit of 3D ICs with an OpenSPARC T2 chip. To further enhance the 3D power benefit on top of the conventional 3D floorplanning method, the impact of block folding methodologies was explored. With the aforementioned methods, a total power saving of 20.3% was achieved against the 2D counterpart. Note that the 3D power benefit will increase with a faster clock. With better timing in 3D, the discrepancy in terms of cell size and HVT cell usage between 2D and 3D designs will increase, which in turn will enhance the 3D power savings. Ongoing efforts address thermal issues in various 3D design styles with different bonding styles, the impact of parasitics such as TSV-to-wire coupling capacitance on 3D power, and other sources of 3D power benefit loss. Interested readers are referred to [32] for more details.

9.4.4 NEEDS FOR EDA TOOL DEVELOPMENT

The first physical design tool that is needed to support block folding is a mixed-size 3D placer that can handle macros, gates, and TSVs, together for all dies in the stack. For a given block to be folded, the tool must partition the macros and the gates into multiple dies, while optimizing the number of connections (= TSVs) across the dies so that the PPA overhead of TSVs is minimized. The next step is to place the objects into multiple dies while optimizing the overall PPA and reliability. A TSV-based 3D placer based on a system of supply/demand of placement space was presented in [36] but lacks the capability to handle hard macros. This capability can be added by treating a hard macro as a large cell that demands some placement space. However, this leads to large whitespace regions, called halos, in the vicinity of hard macros. Spindler et al. [70] solved this issue by reducing the declared size of the hard macros. However, we observe that this tactic is insufficient for extremely large hard macros such as memory banks in the L2 cache, for which halos still exist.

An I/O pin partitioner for folded blocks is also needed because the inter-block routing quality is largely affected by block I/O pins. In the extreme case, when all I/O pins of folded blocks are placed in the die bottom, routing congestion and detour will be serious in this die. This phenomenon in turn increases coupling capacitance and thus net power consumption. Such a bad inter-block design quality can degrade intra-block design metrics as well. Therefore, I/O pins of folded blocks need to be partitioned so that the inter-block wirelength in both dies is balanced.

A 3D static timing analysis (STA) tool is a must for any 3D IC designer. The tool needs to handle multiple dies simultaneously and perform fast and accurate timing calculations. In addition, all of the parasitics including TSV and micro-bump related, as well as cross-die elements, must be extracted for a correct timing calculation. A *true 3D* buffer insertion tool is also required for effective timing closure in 3D IC. In the current tool flow, we first calculate timing constraints at die boundaries (= TSV connections), using a 3D STA tool. We then perform buffer insertion and gate sizing for each die separately, using those timing constraints. We chose this suboptimal approach simply because of the lack of a 3D IC buffer inserter. In an *all-dies-together* or true 3D approach, buffers are inserted and gates are sized while processing all dies at the same time instead of individual dies separately.

9.5 TO PROBE FURTHER

9.5.1 3D IC FLOORPLANNING

Cong et al. [14] use a combined bucket and 2D array (CBA) representation to better explore the solution space of the multi-tier module packing problem. They employ a fast but less accurate hybrid resistive model and another accurate but relatively slow resistive model selectively within their floorplanning to incorporate thermal awareness. Healy et al. [21] present a multi-objective micro-architectural floorplanning algorithm for high-performance processors implemented using 3D ICs. The floorplanner determines the dimension and placement locations of the functional modules, taking into consideration thermal reliability, area, wirelength, vertical overlap, and bonding-aware layer partitioning. This hybrid floorplanning approach combines linear programming and simulated annealing.

Zhou et al. [89] use a three-stage force-directed optimization flow combined with legalization techniques that eliminate block overlaps during multi-layer floorplanning. A temperature-dependent leakage model is used to permit optimization based on the feedback loop connecting thermal profile and leakage power consumption. Falkenstern et al. [17] focus on 3D floorplan and power/ground (P/G) co-synthesis, which builds the floorplan and the P/G network concurrently. Their tool integrates a 3D B*-tree floorplan representation, a resistive P/G mesh, and a simulated annealing (SA) engine to explore the 3D floorplan and P/G network.

Tsai et al. [74] propose a two-stage 3D fixed-outline floorplan algorithm, where stage one simultaneously plans hard macros and TSV blocks for wirelength reduction, while the next stage improves the wirelength by reassigning signal TSVs. In [39], the authors show how to integrate 2D IP blocks into 3D chips, without altering their layout. Their main idea is to optimize whitespace for TSV insertion. Experiments indicate that the overhead of the proposed integration is small, which can help accelerate the industry adoption of 3D integration.

9.5.1.1 3D IC PLACEMENT

In [19], thermal vias are assigned to specific areas of a 3D IC and used to adjust their effective thermal conductivities. Their method uses finite element analysis (FEA) to calculate temperatures quickly during each iteration, while making iterative adjustments to these thermal conductivities in order to achieve a desired thermal objective. Cong et al. [13] propose several techniques to obtain 3D IC layouts from 2D IC placement solutions through transformations. They present different types of folding transformations, where cutlines are drawn and the chip is folded, to obtain 3D IC placements. They provide a conflict-net graph based technique to reassign the cell layers to reduce wirelength further.

The work of Kim et al. [37] is the first one, which reveals the significant area overhead of TSVs in 3D IC placement. They present a force-directed 3D gate-level placement that efficiently handles TSVs. The authors developed a technique for irregular TSV placement, where cells and TSVs are placed together. In the case of regular TSV arrays, the cells are placed first, and next, they assign TSVs to nets to complete routing. Athikulwongse et al. [3] present techniques to exploit TSV stress to improve the timing of the design. They incorporate stress-aware mobility models into a force-directed placement framework to improve the overall timing of the design. They also provide techniques to handle (1) regular TSV arrays, where only the standard cells are moved, and (2) irregular TSV placements, where both standard cells and TSVs are placed together.

Hsu et al. [22] propose an algorithm that first performs 3D analytical global placement with a density optimization step. This reserves whitespace for TSVs that will be inserted later. Next, the authors perform TSV insertion into the layout and perform TSV-aware standard cell legalization. Finally, they perform tier-by-tier detailed placement to optimize the wirelength further. Cong et al. [12] propose a thermal-aware 3D placement method that considers both the thermal effect and the area impact of TSVs. They demonstrate that the minimum temperature can be achieved by making the TSV area in each placement bin proportional to the lumped power consumption in that bin, together with the bins in all the tiers directly above it.

9.5.1.2 3D IC ROUTING AND BUFFERING

Cong and Zhang [15] compare multiple methods for thermal via insertion. They propose a heuristic algorithm by planning thermal vias in vertical and horizontal directions. The authors claim a 200× speed-up compared with a nonlinear programming problem formulation with a thermal resistive model, with 1% difference in solution quality. Pathak and Lim [61] present an algorithm for 3D Steiner tree construction and TSV relocation. A constructive Steiner routing algorithm is used for the initial tree construction, while a linear programming method is subsequently used to refine TSV locations for thermal optimization. The proposed algorithm reduces delay and maximum temperature with comparable performance versus greedy methods.

Two buffering algorithms are presented in [43]: van Ginneken–based and slew-aware buffer insertion. The authors develop an efficient way to propagate slew information during the buffer insertion for the nets that contain TSVs. They demonstrate significant buffer count and negative slack (both the worst and the total) savings, compared with van Ginneken and another baseline that is based on Cadence Encounter. The authors in [57,62] present an algorithm for signal routing considering electro-migration (EM). Higher priority is given to the EM-critical nets and higher mean time to failure (MTTF) is achieved. The total number of nets with EM violations can also be reduced. Hsu et al. [23] use signal TSVs to improve heat dissipation in 3D ICs. The authors compare their results to [15] and claim 23% less TSVs used for thermal reduction. TSVs are initially placed and later relocated considering temperature reduction.

9.5.1.3 3D IC POWER DISTRIBUTION NETWORK ROUTING

Yu et al. [83] discuss how to place power/ground (P/G) vias, considering power and thermal issues simultaneously. The optimization is performed by calculating an RLC matrix for power distribution and a thermal resistance matrix for thermal optimization. The authors demonstrate up to 45.5% non-signal via count reduction by simultaneous optimization using dynamic programming. Zhou et al. [88] discuss decoupling-capacitance insertion with MIM (metal–insulator–metal) and CMOS capacitors. A sequence-of-linear-programs method is used, and the authors show that denser power grids help reduce voltage droops. A congestion model is built for evaluating area congestion in each design cell.

Healy and Lim [20] compare two different PG TSV placement strategies: TSV distributed versus TSV clustered. The TSV-distributed placement topology spreads the TSV locations evenly across the design, while TSV-clustered topology groups the TSVs within a small area. The authors demonstrate a 50% lower IR drop and 42% lower noise on power distribution network (PDN) with distributed TSV, compared with clustered TSV. Savidis et al. [66] present a test chip for 3D PDN noise measurement. TSV density and decoupling capacitance impact are experimentally described with measurement power grid noise extracted by a source follower sense circuit. A design without decoupling capacitance is measured to have much lower resonance frequency than those with decoupling capacitance. Song et al. [69] present a technique that utilizes TSVs as decoupling capacitors to reduce PDN noise. Using an extra stacked chip with TSVs as decoupling capacitors is shown to be more effective than on-chip or off-chip decoupling capacitors. They showed that with the extra tier, the parallel resonance frequency point is eliminated more effectively, compared with off-chip decoupling capacitors.

In [58], electro-migration for 3D IC PDN is modeled with a focus on multi-scale via structure, that is, TSVs and local vias used together for vertical power delivery. The work investigates the interplay of TSVs and conventional local vias in 3D ICs. The authors also study the impact of structure, material, and initial void size on EM-related lifetime of multi-scale via structures. In [86], a transient modeling of EM in TSV and TSV-to-wire interfaces in the power delivery network of 3D ICs is carried out. Atomic depletion and accumulation, effective resistance degradation, and full-chip-scale PDN lifetime degradation due to EM are captured in their model. The results show that voids and hillocks grow at various TSV-to-wire interfaces and degrade the effective resistance of TSVs significantly.

9.5.1.4 3D IC CLOCK ROUTING

Minz et al. [52] and Zhao et al. [85] are the first publications that show that multiple TSVs used in a clock tree for 3D ICs reduce wirelength and thus power consumption, compared with the single-TSV case. They present the 3D MMM (MMM = method of means and medians) algorithm that builds a 3D abstract tree for a given TSV bound under wirelength versus TSV congestion trade-off. This abstract tree is then embedded, buffered, and refined under the given nonuniform thermal profiles so that the temperature-dependent skews are minimized and balanced. Zhao et al. [84] provide a technique to construct pre-bond testable clock trees. If a 3D clock tree is built using multiple clock TSVs, each die—except for the die that contains the clock source—contains multiple subtrees that are not connected. These dies require multiple clock probe pads during pre-bond testing. The authors present a routing method to build a temporary tree that connects together these subtrees and use a single clock probe pad for low-cost and low-power testing. This work is extended by Kim and Kim [38] to consider co-optimization of TSV usage and power consumption.

In [35], the authors present techniques to build a clock tree for a 3D IC that consists of one GPU and a four-layer stacked DRAM connected on a 2.5D interposer. The clock generator is on the GPU chip and is fed through the interposer to clock pins on the 3D DRAM module. The authors provide an improved clock tree in the 3D DRAM module that reduces skew and jitter. The authors of [49] propose a fault-tolerant 3D clock tree that significantly reduces the overhead compared with redundant TSV for clock TSV. The authors propose a fault-tolerant TSV unit that makes use of the existing 2D redundant trees designed for pre-bond testing and thus has a minimum area overhead, while maintaining the same yield. Chae et al. [11] present a post-silicon tuning methodology, called tier adaptive body biasing (TABB), to reduce skew and data path variability in 3D clock trees. The proposed TABB uses specialized on-die sensors to independently detect the process corners of NMOS and PMOS devices and, accordingly, tune the body biases of NMOS/PMOS devices to reduce the clock skew variability.

9.5.1.5 PHYSICAL DESIGN FOR MONOLITHIC 3D IC

Monolithic 3D IC is a vertical integration technology that builds two or more tiers of devices sequentially rather than bonding two independently fabricated dies together using bumps and/or TSVs [7]. Compared with other existing 3D integration technologies (wire-bonding, interposer, through-silicon-via, etc.), monolithic 3D integration allows ultra fine-grained vertical integration of devices and interconnects, thanks to the extremely small size of inter-tier vias, typically local-via-sized (70 nm in diameter).

Bobba et al. [9] propose two different strategies of stacking standard cells in monolithic 3D that utilize 2D tools: intra-cell stacking and cell-on-cell stacking. Intra-cell stacking requires the modification of standard cell design but permits a direct reuse of 2D IC tools. In case of cell-on-cell stacking, the authors propose a placement tool based on commercial tools and an LP formulation for tier assignment. In [46], the authors present physical design techniques for transistor-level monolithic 3D ICs. They first build a cell library that consists of 3D gates and then model their timing and power characteristics. They perform iso-performance comparisons and demonstrate a significant power benefit over 2D. They also demonstrate that this benefit increases at future technology nodes.

Samal et al. [65] present a comprehensive study of thermal effects in monolithic 3D ICs. They develop a fast and accurate block-level thermal model using a nonlinear regression technique. They then incorporate this model into a simulated-annealing based floorplanner to reduce the maximum temperature of the monolithic 3D IC with minimal wirelength overhead. In [58], the authors present a power-performance study of block-level monolithic 3D ICs, under inter-tier performance differences that are caused by the manufacturing process. They first present an RTL-to-GDSII floorplanning framework that can handle soft blocks. Next, they model the inter-tier performance differences and present a floorplanning framework that can generate circuits immune to these differences. Panth et al. [60] present physical design techniques for gate-level monolithic 3D ICs. These techniques place the gates into half the footprint area of a 2D IC, using a commercial 2D engine. Next, the gates are partitioned into multiple tiers to give high-quality 3D solutions. The authors also present techniques to utilize the commercial tool for timing optimization, clock tree synthesis, and inter-tier via insertion.

9.5.1.6 3D IC MECHANICAL RELIABILITY ANALYSIS AND DESIGN OPTIMIZATION

Yang et al. [81] propose a TSV stress-aware timing analyzer and show how to optimize layout for a better performance. They show that TSV stress–induced timing variations can be as much as 10% for an individual cell and full-chip designs. Jung et al. [28] presented a full-chip TSV interfacial crack analysis flow and design optimization methodology to alleviate TSV interfacial crack problems in 3D ICs. First, they analyze TSV interfacial cracks at the TSV/dielectric liner interface caused by TSV-induced thermo-mechanical stress. Then, they explore the impact of TSV placement in conjunction with various associated structures, such as landing pad and dielectric liner on TSV interfacial cracks.

In [29], the authors show how TSV structures affect stress field and mechanical reliability in 3D ICs. They also present an accurate and fast full-chip stress and mechanical reliability analysis flow, which can be applicable to placement optimization for 3D ICs. Results show that KOZ size, TSV size, liner material/thickness, and TSV placement are the key design parameters to reduce the mechanical reliability problems in TSV-based 3D ICs. This work is extended in [31], where the authors show how package elements affect the stress field and the mechanical reliability on top of the TSV-induced stress in 3D ICs. This chapter shows that the mechanical reliability of TSVs in the bottom-most die in the stack is highly affected by packaging elements and that this effect decreases as it moves onto the upper dies.

9.5.1.7 3D IC LOW-POWER PHYSICAL DESIGN METHODOLOGIES

In [27], the power benefit of 3D ICs is demonstrated with an OpenSPARC T2 core. Four design techniques are explored to optimize power in 3D IC designs: 3D floorplanning, intra-block level metal layer usage control, dual-Vth design, and functional module folding. With these methods, total power savings of 21.2% were achieved.* Lee and Lim [45] demonstrated how the power consumption of the buses in GPUs can be reduced with 3D IC technologies. To maximize the power benefit of 3D ICs, the authors claim that finding a good partition and floorplan solution is critical. To further enhance the 3D power benefit versus the conventional 3D floorplanning method for GPU, block folding methodologies and bonding style impact were explored [32]. The authors also developed an efficient method to find face-to-face via locations for two-tier 3D ICs and showed more 3D power reduction in F2F bonding than in F2B.

REFERENCES

1. 3DInCites. http://www.3dincites.com/tag/3d-design-tools/, November, 2015.
2. Ansys 3D IC Tools. http://www2.apache-da.com/products/redhawk/, November, 2015.
3. K. Athikulwongse, A. Chakraborty, J.-S. Yang, D. Pan, and S. K. Lim. Stress-driven 3D-IC placement with TSV keep-out zone and regularity study. In *Proceedings of IEEE International Conference on Computer-Aided Design*, San Jose, CA, 2010.
4. K. Athikulwongse, D. H. Kim, M. Jung, and S. K. Lim. Block-level designs of die-to-wafer bonded 3D ICs and their design quality tradeoffs. In *Proceedings of Asia and South Pacific Design Automation Conference*, Yokohama, Japan, 2013.
5. Atrenta 3D IC Tools. http://www.atrenta.com/products/spyglass-physical.htm5, November, 2015.
6. K. Banerjee, S. Souri, P. Kapur, and K. Saraswat. 3-D ICs: A novel chip design for improving deep-submicrometer interconnect performance and systems-on-chip integration. *Proceedings of the IEEE*, 89(5):602–633, 2001.
7. P. Batude et al. Advances in 3D CMOS sequential integration. In *Proceedings of IEEE International Electron Devices Meeting*, Baltimore, MD, 2009.
8. B. Black et al. Die stacking (3D) microarchitecture. In *Proceedings of Annual International Symposium Microarchitecture*, Orlando, FL, 2006.
9. S. Bobba, A. Chakraborty, O. Thomas, P. Batude, T. Ernst, O. Faynot, D. Pan, and G. De Micheli. CELONCEL: Effective design technique for 3-D monolithic integration targeting high performance integrated circuits. In *Proceedings of Asia and South Pacific Design Automation Conference*, Yokohama, Japan, 2011.
10. Cadence 3D IC Tools. http://www.cadence.com/solutions/3dic/, November, 2015.
11. K. Chae, X. Zhao, S. K. Lim, and S. Mukhopadhyay. Tier adaptive body biasing: A post-silicon tuning method to minimize clock skew variations in 3-D ICs. *IEEE Transactions on Components, Packaging and Manufacturing Technology*, 3(10):1720–1730, 2013.
12. J. Cong, G. Luo, and Y. Shi. Thermal-aware cell and through-silicon-via co-placement for 3D ICs. In *Proceedings of ACM Design Automation Conference*, San Diego, CA, 2011.
13. J. Cong, G. Luo, J. Wei, and Y. Zhang. Thermal-aware 3D IC placement via transformation. In *Proceedings of Asia and South Pacific Design Automation Conference*, Yokohama, Japan, 2007.

---

* This work is presented in Section 9.4.

14. J. Cong, J. Wei, and Y. Zhang. A thermal-driven floorplanning algorithm for 3D ICs. In *Proceedings of IEEE Internatioanl Conference on Computer-Aided Design*, San Jose, CA, 2004.

15. J. Cong and Y. Zhang. Thermal via planning for 3-D ICs. In *Proceedings of IEEE International Conference on Computer-Aided Design*, San Jose, CA, 2005.

16. G. V. der Plas et al. Design issues and considerations for low-cost 3D TSV IC technology. In *IEEE International Solid-State Circuits Conference*, San Francisco, CA, 2010, pp. 148–149.

17. P. Falkenstern, Y. Xie, Y.-W. Chang, and Y. Wang. Three-dimensional integrated circuits (3D IC) floorplan and power/ground network co-synthesis. In *Proceedings of Asia and South Pacific Design Automation Conference*, Taipei, Taiwan, 2010.

18. S. Franssila. *Introduction to Microfabrication*. John Wiley & Sons, Chichester, U.K., 2004.

19. B. Goplen and S. Sapatnekar. Placement of thermal vias in 3-D ICs using various thermal objectives. *IEEE Transactions on Computer-Aided Design of Integrated Circuits and Systems*, 25(4):692–709, 2006.

20. M. Healy and S. K. Lim. Distributed TSV topology for 3-D power-supply networks. *IEEE Transactions on VLSI Systems*, 20(11):2066–2079, 2012.

21. M. Healy, M. Vittes, M. Ekpanyapong, C. Ballapuram, S. K. Lim, H.-H. Lee, and G. Loh. Multiobjective microarchitectural floorplanning for 2-D and 3-D ICs. *IEEE Transactions on Computer-Aided Design of Integrated Circuits and Systems*, 26(1):38–52, 2007.

22. M.-K. Hsu, Y.-W. Chang, and V. Balabanov. TSV-aware analytical placement for 3D IC designs. In *Proceedings of ACM Design Automation Conference*, June, 2011.

23. P.-Y. Hsu, H.-T. Chen, and T. Hwang. Stacking signal TSV for thermal dissipation in global routing for 3D IC. In *Proceedings of Asia and South Pacific Design Automation Conference*, Yokohama, Japan, 2013.

24. X. Hu et al. High thermal conductivity molding compound for flip-chip packages. US Patent 2009/0004317 A1, 2009.

25. J.-W. Im et al. A 128Gb 3b/cell V-NAND flash memory with 1Gb/s I/O rate. In *IEEE International Solid-State Circuits Conference*, 2015, pp. 1–3.

26. ITRS. The International Technology Roadmap for Semiconductors. http://www.itrs.net.

27. M. Jung et al. How to reduce power in 3D IC designs: A case study with OpenSPARC T2 core. In *Proceedings of IEEE Custom Integrated Circuits Conference*, September, 2013.

28. M. Jung, X. Liu, S. Sitaraman, D. Z. Pan, and S. K. Lim. Full-chip through-silicon-via interfacial crack analysis and optimization for 3D IC. In *Proceedings of IEEE International Conference on Computer-Aided Design*, San Jose, CA, 2011.

29. M. Jung, J. Mitra, D. Pan, and S. K. Lim. TSV stress-aware full-chip mechanical reliability analysis and optimization for 3D IC. In *Proceedings of ACM Design Automation Conference*, San Diego, CA, 2011.

30. M. Jung, J. Mitra, D. Pan, and S. K. Lim. TSV Stress-aware full-chip mechanical reliability analysis and optimization for 3D IC. *Communications of the ACM*, 57(1):107–115, 2014.

31. M. Jung, D. Pan, and S. K. Lim. Chip/package co-analysis of thermo-mechanical stress and reliability in TSV-based 3D ICs. In *Proceedings of ACM Design Automation Conference*, San Francisco, CA, 2012.

32. M. Jung, T. Song, Y. Wan, Y. Peng, and S. K. Lim. On enhancing power benefits in 3D ICs: Block folding and bonding styles perspective. In *Proceedings of ACM Design Automation Conference*, San Francisco, CA, 2014.

33. U. Kang et al. 8Gb 3D DDR3 DRAM using through-silicon-via technology. In *IEEE International Solid-State Circuits Conference*, San Francisco, CA, USA, 2009, pp. 130–131, 131a.

34. D. Kim et al. 3D-MAPS: 3D massively parallel processor with stacked memory. In *IEEE International Solid-State Circuits Conference*, San Francisco, CA, 2010.

35. D. Kim, J. Kim, J. Cho, J. S. Pak, J. Kim, H. Lee, J. Lee, and K. Park. Distributed multi TSV 3D clock distribution network in TSV-based 3D IC. In *Proceedings of IEEE Electrical Performance of Electronic Packaging*, San Jose, CA, 2011.

36. D. H. Kim, K. Athikulwongse, and S. K. Lim. A study of through-silicon-via impact on the 3D stacked IC layout. In *Proceedings of IEEE International Conference on Computer-Aided Design*, San Jose, CA, 2009.

37. J.-S. Kim et al. A 1.2V 12.8GB/s 2Gb mobile Wide-I/O DRAM with 4x128 I/Os using TSV-based stacking. In *IEEE International Solid-State Circuits Conference*, San Francisco, CA, 2011, pp. 496–498.

38. T.-Y. Kim and T. Kim. Clock tree synthesis with pre-bond testability for 3D stacked IC designs. In *Proceedings of ACM Design Automation Conference*, Anaheim, CA, 2010.

39. J. Knechtel, I. Markov, and J. Lienig. Assembling 2-D blocks into 3-D chips. *IEEE Transactions on Computer-Aided Design of Integrated Circuits and Systems*, 31(2):228–241, 2012.

40. J. Knickerbocker et al. Three-dimensional silicon integration. *IBM Journal of Research and Development*, 52(6):553–569, November 2008.

41. M. Koyanagi, T. Fukushima, and T. Tanaka. High-density through silicon vias for 3-D LSIs. *Proceedings of the IEEE*, 97(1):49–59, 2009.

42. M. Koyanagi, Y. Nakagawa, K.-W. Lee, T. Nakamura, Y. Yamada, K. Inamura, K.-T. Park, and H. Kurino. Neuromorphic vision chip fabricated using three-dimensional integration technology. In *IEEE International Solid-State Circuits Conference*, San Francisco, CA, pp. 270–271, 2001.

43. Y.-J. Lee, I. Hong, and S. K. Lim. Slew-aware buffer insertion for through-silicon-via-based 3D ICs. In *Proceedings of IEEE Custom Integrated Circuits Conference*, San Jose, CA, 2012.

44. Y.-J. Lee and S. K. Lim. Timing analysis and optimization for 3D stacked multi-core microprocessors. In *Proceedings of IEEE International 3D Systems Integration Conference*, Munich, Germany, 2010.

45. Y.-J. Lee and S. K. Lim. On GPU bus power reduction with 3D IC technologies. In *Proceedings of Design, Automation and Test in Europe*, Dresden, Germany, 2014.

46. Y.-J. Lee, D. Limbrick, and S. K. Lim. Power benefit study for ultra-high density transistor-level monolithic 3D ICs. In *Proceedings of ACM Design Automation Conference*, Austin, TX, 2013.

47. S. K. Lim. TSV-aware 3D physical design tool needs for faster mainstream acceptance of 3D ICs. In *ACM DAC Knowledge Center*, 2010.

48. J.-Q. Lu. 3-D Hyperintegration and packaging technologies for micro-nano systems. *Proceedings of the IEEE*, 97(1):18–30, 2009.

49. C.-L. Lung, Y.-S. Su, S.-H. Huang, Y. Shi, and S.-C. Chang. Fault-tolerant 3D clock network. In *Proceedings of ACM Design Automation Conference*, San Diego, CA, 2011.

50. Mentor Graphics 3D IC Tools. http://www.mentor.com/solutions/3d-ic-design/, November, 2015.

51. Micro Magic 3D IC Tools. http://www.micromagic.com/, November, 2015.

52. J. Minz, X. Zhao, and S. K. Lim. Buffered clock tree synthesis for 3D ICs under thermal variations. In *Proceedings of Asia and South Pacific Design Automation Conference*, Seoul, Korea, 2008.

53. NanGate Inc. NanGate 45nm Open Cell Library, 2009.

54. U. Nawathe et al. An 8-core 64-thread 64b power-efficient SPARC SoC. In *IEEE International Solid-State Circuits Conference*, 2007.

55. North Carolina State University. FreePDK45, 2009.

56. OpenCores.org. OpenCores. http://opencores.org, 2009.

57. J. Pak, S. K. Lim, and D. Pan. Electromigration-aware routing for 3D ICs with stress-aware EM modeling. In *Proceedings of IEEE International Conference on Computer-Aided Design*, San Jose, CA, 2012.

58. J. Pak, S. K. Lim, and D. Pan. Electromigration study for multi-scale power/ground vias in TSV-based 3D ICs. In *Proceedings of IEEE International Conference on Computer-Aided Design*, San Jose, CA, 2013.

59. S. Panth, K. Samadi, Y. Du, and S. K. Lim. Power-performance study of block-level monolithic 3D-ICs considering inter-tier performance variations. In *Proceedings of ACM Design Automation Conference*, San Francisco, CA, 2014.

60. S. A. Panth, K. Samadi, Y. Du, and S. K. Lim. Design and CAD methodologies for low power gate-level monolithic 3D ICs. In *Proceedings of International Symposium on Low Power Electronics and Design*, La Jolla, CA, 2014.

61. M. Pathak and S. K. Lim. Performance and thermal-aware steiner routing for 3-D stacked ICs. *IEEE Transactions on Computer-Aided Design of Integrated Circuits and Systems*, 28(9):1373–1386, 2009.

62. M. Pathak, J. Pak, D. Pan, and S. K. Lim. Electromigration modeling and full-chip reliability analysis for BEOL interconnect in TSV-based 3D ICs. In *Proceedings of IEEE International Conference on Computer-Aided Design*, San Jose, CA, 2011.

63. R. Patti. Three-dimensional integrated circuits and the future of system-on-chip designs. *Proceedings of the IEEE*, 94(6):1214–1224, 2006.

64. R3 Logic 3D IC Tools. http://www.r3logic.com/, November, 2015.

65. S. Samal, S. Panth, K. Samadi, M. Saedi, Y. Du, and S. K. Lim. Fast and accurate thermal modeling and optimization for monolithic 3D ICs. In *Proceedings of ACM Design Automation Conference*, San Francisco, CA, 2014.

66. I. Savidis, S. Kose, and E. Friedman. Power noise in TSV-based 3-D integrated circuits. *IEEE Journal of Solid-State Circuits*, 48(2):587–597, 2013.

67. C.-H. Shen et al. Heterogeneously integrated sub-40 nm low-power epi-like Ge/Si monolithic 3D-IC with stacked SiGeC ambient light harvester. In *Proceedings of IEEE International Electron Devices Meeting*, San Francisco, CA, 2014.

68. M. Shulaker, T. Wu, A. Pal, L. Zhao, Y. Nishi, K. Saraswat, H.-S. Wong, and S. Mitra. Monolithic 3D integration of logic and memory: Carbon nanotube FETs, resistive RAM, and silicon FETs. In *Proceedings of IEEE International Electron Devices Meeting*, San Francisco, CA, 2014, pp. 27.4.1–27.4.4.

69. E. Song, K. Koo, J. S. Pak, and J. Kim. Through-silicon-via-based decoupling capacitor stacked chip in 3-D-ICs. *IEEE Transactions on Components, Packaging and Manufacturing Technology*, 3(9):1467–1480, 2013.

70. P. Spindler, U. Schlichtmann, and F. M. Johannes. Kraftwerk2—A fast force-directed quadratic placement approach using an accurate net model. In *IEEE Transactions on Computer-Aided Design of Integrated Circuits and Systems*, 2008.

71. V. Sundaram et al. Low cost, high performance, and high reliability 2.5D silicon interposer. In *IEEE Electronic Components and Technology Conference*, Las Vegas, NV, 2013, pp. 342–347.

72. Synopsys 3D IC Tools. http://www.synopsys.com/solutions/endsolutions/3d-ic-solutions/, November, 2015.

73. The Hybrid Memory Cube. http://www.hybridmemorycube.org.

74. M.-C. Tsai, T.-C. Wang, and T. Hwang. Through-silicon via planning in 3-D floorplanning. *IEEE Transactions on VLSI Systems*, 19(8):1448–1457, 2011.

75. R. Tummala. Moore's law meets its match (system-on-package). *IEEE Spectrum*, 43(6):44–49, 2006.

76. R. Tummala and J. Laskar. Gigabit wireless: System-on-a-package technology. *Proceedings of the IEEE*, 92(2):376–387, February 2004.

77. Various Authors. Special issue on 3-D integration technologies. *Proceedings of the IEEE*, 97(1):1–175, 2009.

78. S. Wong, A. El-Gamal, P. Griffin, Y. Nishi, F. Pease, and J. Plummer. Monolithic 3D integrated circuits. In *Proceedings of International Symposium on VLSI Technology, Systems and Applications*, Hsinchu, Taiwan, 2007, pp. 1–4.

79. Xilinx 3D IC Tools. http://www.xilinx.com/products/silicon-devices/3dic.html, November, 2015.

80. C. Xu et al. Fast 3-D thermal analysis of complex interconnect structures using electrical modeling and simulation methodologies. In *Proceedings of IEEE International Conference on Computer-Aided Design*, San Jose, CA, 2009.

81. J.-S. Yang, K. Athikulwongse, Y.-J. Lee, S. K. Lim, and D. Pan. TSV stress aware timing analysis with applications to 3D-IC layout optimization. In *Proceedings of ACM Design Automation Conference*, Anaheim, CA, 2010.

82. K. B. Yeap et al. A critical review on multiscale material database requirement for accurate three-dimensional IC simulation input. *IEEE Transactions on Device and Materials Reliability*, 12(2):217–224, 2012.

83. H. Yu, J. Ho, and L. He. Allocating power ground vias in 3D ICs for simultaneous power and thermal integrity. *ACM Transactions on Design Automation of Electronics Systems*, 14(3):1–31, 2009.

84. X. Zhao, D. Lewis, H.-H. Lee, and S. K. Lim. Low-power clock tree design for pre-bond testing of 3-D stacked ICs. *IEEE Transactions on Computer-Aided Design of Integrated Circuits and Systems*, 30(5):732–745, 2011.

85. X. Zhao, J. Minz, and S. K. Lim. Low-power and reliable clock network design for through silicon via based 3D ICs. *IEEE Transactions on Components, Packaging and Manufacturing Technology*, 1(2):247–259, 2011.

86. X. Zhao, Y. Wan, M. Scheuermann, and S. K. Lim. Transient modeling of TSV-wire electromigration and lifetime analysis of power distribution network for 3D ICs. In *Proceedings of IEEE International Conference on Computer-Aided Design*, San Jose, CA, 2013.

87. P. Zhou, Y. Ma, Z. Li, R. Dick, L. Shang, H. Zhou, X. Hong, and Q. Zhou. 3D-STAF: Scalable temperature and leakage-aware floorplanning for 3D ICs. In *Proceedings of IEEE International Conference on Computer-Aided Design*, San Jose, CA, 2007.

88. P. Zhou, K. Sridharan, and S. Sapatnekar. Optimizing decoupling capacitors in 3D circuits for power grid integrity. *IEEE Design and Test of Computers*, 26(5):15–25, 2009.

Gate Sizing

Stephan Held and Jiang Hu

10

CONTENTS

10.1 INTRODUCTION

Gate sizing is a central step for achieving timing closure and minimizing the power consumption of integrated circuits. It refers to determining the widths of the transistors inside a logic gate, where wider transistors imply a faster charging and discharging of the output capacitance. As a drawback, larger transistors increase the input capacitances and thus the load, delays, and transition times at the upstream gates. An increased transition time slows down sibling gates driven by common predecessors.

The delay characteristics of a gate are also affected by its threshold voltage V_t. By varying the oxide thickness or materials in the gate, the threshold voltage can be lowered to speed it up. As a drawback, the leakage current I_{leak} grows exponentially with a falling threshold voltage:

$$I_{leak} = ae^{-bV_t}$$

for constants $a,b \in \mathbb{R}_+$ (see Section 3.1.3). As every alternative threshold voltage on a design requires a separate production step with additional masks, usually no more than four threshold voltages are available.

Further techniques for optimizing gate delays have a smaller impact on the end result compared to sizing and V_t-assignment and are, therefore, seldom used. First, the ratio of the p-type and n-type transistor sizes (β-*ratio*) is adjusted, trading off the rising and falling signal delay. Second, the sizes of serially connected transistors inside a gate are modified to balance the delays from different inputs to the output pin (*tapering*).

Most integrated circuits are designed using a *circuit library* $\mathcal{B}$ that contains a variety of predesigned physical layouts implementing different logic functions and are of varying sizes and threshold voltages. Instead of implementing each gate in the netlist individually, each gate is mapped to one of the library circuits. Let $\mathcal{G}$ denote the set of all gates in the netlist. The implementation of all gates is described by a vector $x \in \mathcal{B}^{\mathcal{G}}$, where $x_g \in \mathcal{B}$ specifies the physical implementation of $g \in \mathcal{G}$. Of course, the logical function of a gate restricts the set of implementations to which it can be assigned. By $X \subseteq \mathcal{B}^{\mathcal{G}}$ we denote the set of assignments, where each gate is assigned to a logically equivalent library circuit. In *discrete gate sizing*, $\mathcal{B}$ and thus X are finite sets. In *continuous gate sizing*, logic gates can take any size between a lower- and upper-size bound, extending the solution space compared to discrete sizing.

A key reason for using discrete gate libraries is that preprocessed timing models for each library gate significantly speed up timing analysis compared to the numerical simulation of continuously designed custom gates. Thus, they allow the processing of multimillion gate instances by physical design tools. More recently, FinFET transistors introduced additional discreteness in standard cell design. FinFETs are sized by adding multiple fins of unit size.

In microprocessor design, chips used to be partitioned into small custom circuits of a few thousand gates, where the layout was essentially defined manually. Here, numerical timing simulation is possible, and transistors can be sized continuously for optimum sizes, β-ratios, and taperings. For algorithms designed for *transistor sizing* of custom circuits, we refer the reader to the optimization methods in [7,42] that combine timing simulation with gradient descent or interior point methods. The benefit of the flexibility of transistor sizing is hampered by the restrictions imposed by partitioning the design into small blocks. Thus, transistor sizing on small blocks has mostly been replaced by discrete gate sizing on larger units, even in microprocessor design.

In this chapter, we first present theoretical results on discrete and continuous gate sizing in Section 10.2. Then in Section 10.3, we focus on practical algorithms for discrete gate sizing.

Readers that are mostly interested in practical algorithms can skip the theoretical part in Section 10.2. We begin with the basic concepts of gate sizing.

10.1.1 STATIC TIMING

Physical design optimization is usually based on static timing constraints, represented by a directed acyclic *timing graph* G (Chapter 6). G contains a vertex for each pin on the chip, and edges are inserted between source and sinks of a net and the inputs and outputs of logic gates. Starting at primary inputs and register outputs, the latest possible arrival time a_v is propagated to each pin $v \in V(G)$ along the topological order of G. Figure 10.1 shows a small netlist of a chip and its timing graph G.

Static timing analysis implies the following (simplified) mathematical constraints:

$$(10.1) \qquad \begin{aligned} a_v & \geq 0 & \forall v \in PI \cup RO, \\ a_v + \mathrm{delay}_{v,w}(x) & \leq a_w & \forall (v,w) \in E(G), \\ a_v & \leq T & \forall v \in PO \cup RI, \end{aligned}$$

where
  $E(G)$ is the edge set of the graph
  PI, PO are the sets of primary inputs and outputs
  RI, RO are register inputs and outputs
  T is the design cycle time

The solution vector $x \in X$ determines the gate sizes and threshold voltages. The edge delays $\mathrm{delay}_{v,w}(x)$ are functions of x, as we will describe in Section 10.1.2.

By introducing a super-vertex $v^\star$ representing the time zero ($a_{v^\star} = 0$) and edges with constant delay zero or $-T$, the three constraint types can be written in an uniform way:

$$(10.2) \qquad a_v + \mathrm{delay}_{v,w}(x) \leq a_w \quad \forall (v,w) \in E(G).$$

10.1.2 DELAY MODELS

Traditionally, the shape of a signal transition is approximated by (1) its arrival time, which is the point where the voltage is half the supply voltage V_{dd}, and (2) its slew, which is the time of the transition between 10% and 90% V_{dd}, as indicated in Figure 10.2. The 10/90 interval is chosen arbitrarily. It should reflect the region where the voltage changes almost linearly, and sometimes, the slew is defined by the 20/80 or 30/70 intervals.

The delay between an input and an output signal is the difference of their arrival times. For higher accuracy, the signal transition can also be approximated by piecewise linear functions with multiple pieces. The most accurate delay analysis is based on numerical circuit simulation using the SPICE software [26]. Due to the high running times of simulation, industry standard static timing sign-off tools, such as Synopsys PrimeTime or Cadence Tempus, use discrete gate

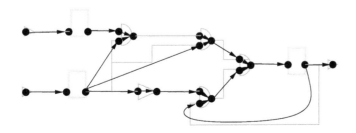

FIGURE 10.1   Example of a timing graph and its underlying netlist.

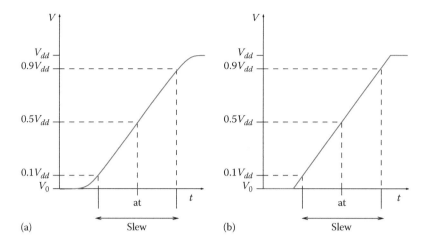

FIGURE 10.2   A (rising) signal transition (a) and its approximation (b).

libraries with preprocessed delay characteristics to analyze multi-million gate designs, providing SPICE simulation as an option for the accurate analysis of the most critical paths.

Traditionally, the gate delays were approximated by two nondecreasing functions: one for the delay and the other for the output slew. Both of these were taking the load capacitance and input slew as arguments and were computed by interpolated table lookup. Wire delays were computed as Elmore delays [10] or using asymptotic waveform emulation to account for resistive shielding [32]. Due to the growing importance of capacitive coupling between wires, the current source model has become the most popular delay model [8], usually analyzing stages from the input pins of a gate to the inputs of the downstream gates.

The delay rules are characterized only within some upper bounds for the load capacitance and input slews. Meaningful delays are only given if the capacitance and slew limits are met. Thus, capacitance and slew constraints have a higher priority than the delay constraints in Equation 10.2.

10.1.2.1   THE RC DELAY MODEL

A delay model that is often used in the literature is the RC delay model [4,11]. Here, the Elmore delay formula is also used for modeling gate delays. The delay through a gate is proportional to the product of its resistance and load capacitance. While the resistance of a gate is inversely proportional to the gate size, the input pin capacitance of a gate is proportional to the size. Let us assume that x_g reflects the size of a gate $g \in \mathcal{G}$, that is, $x_g \in \mathbb{R}$. Now, the RC delay through g coincides for all timing arcs to the output pin. It is defined as

$$\mathrm{delay}_g(x) = \frac{r_g}{x_g} \left(C_{wire} + \sum_{g' \in fanout(g)} c_{g'} x_{g'} + c_g^{intrinsic} \right),$$

where $C_{wire}, r_g, c_{g'}, c_g^{intrinsic} \in \mathbb{R}_+$ are constants for the wire capacitance of the output net, the base resistance, the base input pin capacitance, and the intrinsic capacitance, which depend on the gate type. Wire delays are also modeled as RC delays using the Elmore formula [10]. Under the RC delay model, efficient algorithms for gate sizing are possible, as we will discuss in Section 10.2.

A further simplification is the logical effort model [37], where the resistances and capacitances are normalized to a reference inverter, yielding a technology-independent delay model, where delays depend only on the topology of the pull-up and pull-down transistor networks in the gate. In this model, a minimum delay solution for a single path can be computed by uniformly distributing the so-called stage delay. From the stage delay, an optimum size can be derived locally at every gate. However, this approach does not easily extend to general netlists. Neither model reflects the nonlinear nature of the delays. In the RC delay model, the gate delay is a linear

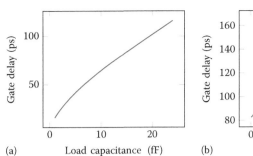
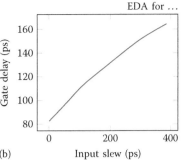

FIGURE 10.3   A typical gate delay as a (concave) function of load capacitance (a) and input slew (b), given a fixed input slew (left) and load capacitance (right).

function in the load capacitance, while in reality it is rather a concave function, as shown in Figure 10.3. The nonlinearity is further amplified by the slew dependency of the delay. Therefore, these two models are used at most to guess initial sizes before placement but not during subsequent optimization.

10.1.2.2 MODEL-INDEPENDENT DELAY CHARACTERISTICS

Delay and slew are increasing functions of load capacitance and input slew as in Figure 10.3. Changing the size of a gate influences the delays and slews at the gate itself as well as all its neighboring gates and all gates in their downstream cone. Assuming a constant input slew, an increasing size reduces the delay through the gate and the output slews and thereby decreases delays and slews at the downstream gates. On the other hand, it increases the load capacitance of the upstream gates, their delays and output slews, and thereby the delays of the other sink gates in the input nets as well as their successors. So, in fact, the input slew increases with the gate size. If the upstream gate is very load sensitive, a strongly increasing input slew may degrade the delay through the upsized gate instead of reducing it.

10.1.3 IMPACT ON POWER AND AREA

For details on power analysis, see Chapter 3 or [44]. Here, we emphasize the factors relevant for gate sizing. The power consumption $P(g)$ of a logic gate $g \in \mathcal{G}$ is composed of the dynamic power $P_{dynamic}(g)$ that occurs when a gate changes its logic state and static power $P_{static}(g)$ that occurs while the gate is in a steady state:

$$P(g) = P_{dynamic}(g) + P_{static}(g).$$

The dynamic power is, in turn, composed of the switching power and the short-circuit power:

$$P_{dynamic}(g) = P_{switching}(g) + P_{short}(g),$$

where the $P_{switching}$ measures the power for charging and discharging its input pin capacitances. Wire capacitances are usually considered constant during gate sizing as interconnects can stay in place after minor adjustments to ensure pin access. For the purpose of gate sizing, we assume that the dynamic power of a gate is more or less proportional to its size. Other factors that influence the dynamic power are the supply voltage that is not sensitive to gate sizing and the switching behavior, which is hard to predict and usually estimated by a switching frequency per net.

Modeling becomes more complicated for the short-circuit power, which is observed under simulation. It is dissipated at low-V_t gates during an input signal transition. If the threshold voltage is below 50% of the supply voltage, there is a period where both p-type and n-type transistors in the gate are turned on, forming a short circuit. The short-circuit current increases with a larger input slew, a lower threshold voltage, and a larger gate size. If a gate has steep input slews or a high threshold voltage, the short-circuit current is usually negligible. As low-V_t gates are used mostly on critical parts, small delays and slews are targeted anyway. But a designer or a tool should be aware that downsizing a gate to save power will increase the short-circuit power at downstream gates with a low threshold voltage. Often, the short-circuit power is not considered directly but mitigated by enforcing low slew limits for low-V_t gates.

While dynamic and short-circuit power dissipation occurs while a gate is switching, the leakage current is also present at a steady state. There is a permanent small current leaking through the transistor gates. This current is higher for lower threshold voltages and wider gates, with an exponential dependency on the threshold voltage. In particular, for mobile devices, it is critical to keep the static power small. Among many physical factors, the leakage power depends on the current state of the input signals. These state patterns are not affected by gate sizing. We assume that the leakage power of a gate is defined by its physical realization and not by the realization of other gates.

For simplicity, gate sizing algorithms estimate power by separable functions, where the contribution of one gate does not affect the contributions of other gates:

$$(10.3) \qquad P^{total} = P^{total}_{static} + P^{total}_{dynamic} = \sum_{g \in \mathcal{G}} \left(P_{static}(g) + P_{dynamic}(g) \right).$$

This is justifiable for the switching power and the leakage power. The short-circuit power can and usually is added as a small slew-independent constant to the dynamic power after imposing low upper bounds on the input slews of low-V_t gates.

10.1.3.1 AREA

The form factor of a gate is also an important aspect of gate sizing. If the chip gets locally overfilled, it can become difficult or even infeasible to legalize the gates. This can, of course, also be adjusted by changing the floorplan, but at some point in the design cycle, gate sizing should match local placement density targets to avoid disturbance by legalization. At the very end, it should avoid gate overlaps. In the literature, placement density is a rarely considered constraint for gate sizing. One approach is proposed in [6]. Here, a single global density violation is punished by a weighted quadratic penalty in the objective function during gate sizing. Gate sizing is repeated with an increasing penalty weight until the density constraint is met. Production tools, such as those used in [41], partition the chip into small placement bins and meet the placement density target inside each bin.

10.1.4 PROBLEM FORMULATIONS

Timing, power, and area can be considered as objectives or constraints in various combinations. A typical objective in gate sizing is to minimize the power consumption such that the timing constraints in Equation 10.2 are met:

$$(10.4) \qquad \begin{aligned} \min \quad & P^{total}(x) \\ s.t. \quad & a_v + \mathrm{delay}_{v,w}(x) \le a_w \qquad \forall (v,w) \in E(G) \\ & x \in X \end{aligned}$$

In practice, area constraints must be fulfilled as well. Early in the design cycle, when global gate sizing is performed, timing constraints are usually unsatisfiable. Therefore, minimizing the cycle time T or the absolute total negative slack (TNS) is usually the prevalent goal during most parts of the design cycle, where the TNS is the sum of negative slacks at the endpoints of signal paths. This can be done in linear combination with power and area or imposing area constraints. But one could also ask for the minimum area under timing and power constraints.

10.2 THEORETICAL BACKGROUND

The discrete and nonlinear nature of the problem makes gate sizing practically and mathematically difficult. Assuming well-behaved delay functions and continuous sizability, gate sizing becomes tractable to some extent. In this section, we give an overview on the most important theoretical results for discrete and continuous gate sizing.

10.2.1 DISCRETE SIZING

Even under mild assumptions on the delay model, discrete gate sizing is a hard problem, as the connection to the discrete time-cost trade-off problem shows. Here, we are given an acyclic directed graph $G = (V, E)$, and for each edge $e \in E$, there is a finite set $\mathcal{T}_e = \{t_e^1, \ldots, t_e^k\} \subset \mathbb{R}$ of edge delays and a nonincreasing cost function $c_e : \mathcal{T}_e \to \mathbb{R}_+$. In the *deadline problem*, we wish to choose a delay for each edge such that the length of each path stays within a given deadline D minimizing the total cost. This problem cannot be approximated within a factor of 1.36 unless $P = NP$, because it is as hard as the vertex cover problem [9,14]. Assuming the *unique games conjecture*, which is a stronger assumption than $P \neq NP$, there is even no constant-factor approximation algorithm [39]. Theoretically, polynomial time algorithms or approximation algorithms with better guarantees may exist for special delay functions. But in gate sizing, the delay functions appear rather more involved due to their nonlinearity and nonseparability. The nonseparability means that the delay of an edge in the timing graph depends on the sizes of multiple gates.

For netlists with a tree structure and certain delay models, pseudo-polynomial dynamic programming (DP) algorithms have been used successfully. One of the first was presented in [3]. They can be turned into a fully polynomial time approximation scheme as a consequence of the approximation scheme for buffering by [17]. For netlists with bounded width, that is, with a path-like structure, an approximation scheme for delay minimization obeying an upper bound on the power was developed in [22].

When nonlinear delay models with slew propagation come into play, the only known algorithms for general discrete gate sizing that provide a size guarantee while meeting the timing constraints are based on complete enumeration, potentially sped up by branch and bound [45]. Such techniques can be used to postoptimize small groups of gates [20].

10.2.2 CONTINUOUS SIZING

Mathematical tractability improves the plain gate sizing problem (without discrete V_t changes) if one assumes continuously sizable gates and posynomial delay functions, which are defined as follows. For each gate $g \in \mathcal{G}$, the realization x_g can be chosen continuously within a lower and an upper bound $l_g \leq x_g \leq u_g$, where $l_g, u_g \in \mathbb{R}$. Now the set X of feasible assignments consists of gate sizing solutions $x \in X = \{x' \in \mathbb{R}^{\mathcal{G}} : l_g \leq x_{g'} \leq u_g \text{ for all } g \in \mathcal{G}\}$.

A posynomial function $f : \mathbb{R}^n \to \mathbb{R}$ is of the form

$$(10.5) \qquad f(x) = \sum_{k=1}^{K} c_k x_1^{b_{1k}} x_2^{b_{2k}} \ldots x_n^{b_{nk}},$$

where $c_k > 0$ and $b_{ik} \in \mathbb{R}$. Posynomial delay functions occur, for instance, if gates are modeled as RC circuits with $b_{ij} \in \{-1, 0, 1\}$ [4,11].

The posynomial formulation for the circuit/transistor sizing problem was introduced by Fishburn and Dunlop [11]. A globally optimum solution to the transistor sizing problem based on *geometric programming* was first given in [35]. The nice property of geometric programs is that they can be turned into *convex programs* by the variable transformation $y_i = \log x_i$. Therefore, a local optimum is always a global optimum.

General-purpose geometric programming solvers mostly implement an interior point method for convex programs with polynomial running time. The drawback of interior point methods is that their iterations require a significant memory and running time. Solvable instance sizes are limited to approximately 100,000 gates according to [1,2], which also give a detailed tutorial on geometric programming.

10.2.3 LAGRANGIAN RELAXATION

A prominent practical approach to gate sizing is based on Lagrangian relaxation (LR), which was analyzed in detail in [4] and later in [43]. As this is also the motivation for many discrete gate sizing heuristics, we will describe this approach in detail.

In the LR approach, the timing constraints are moved to the objective weighted by Lagrange multipliers $\lambda \in \mathbb{R}_{\geq 0}^{E(G)}$ that penalize constraint violations. Applied to Equation 10.4, this leads to the following Lagrange function (called *Lagrangian*):

$$(10.6) \qquad \mathcal{L}(\lambda,x,a) = P^{total}(x) + \sum_{(v,w)\in E(G)} \lambda_{v,w}\big(a_v + \mathrm{delay}_{v,w}(x) - a_w\big).$$

For given $\lambda \geq 0$, the *Lagrangian dual function*

$$(10.7) \qquad L^{\star}(\lambda) := \inf\Big\{\mathcal{L}(\lambda,x,a) : x \in X, a \in \mathbb{R}^{V(G)}\Big\}$$

is a lower bound for the objective value in Equation 10.4. To get the best possible lower bound, we have to solve

$$(10.8) \qquad \sup_{\lambda \geq 0} L^{\star}(\lambda).$$

If the gate sizing problem in Equation 10.4 has a strictly feasible solution, Equations 10.4 and 10.8 have the same value and the Lagrangian duality gap is zero.

If no strictly feasible solutions exists, Equation 10.8 may not have a finite solution. Note that for a minimum cycle time T in Equation 10.1 the problem is not strictly feasible, because the constraints on the critical path can only be satisfied with equality. However, Wang et al. [43] specify regularity conditions for the objective function and the delay functions so that the duality gap between the gate sizing problems (Equation 10.4) and the Lagrangian dual problem (Equation 10.8) is zero. But the supremum in Equation 10.8 is not attained by a finite value, and instead, the multipliers of the tight constraints tend toward infinity.

The crucial observation in [4] is that the Lagrangian can be separated into two parts, one depending only on the size vector x and the other only on the arrival time vector $\mathbf{a}$:

$$(10.9) \qquad \mathcal{L}(\lambda,x,\mathbf{a}) = \mathcal{L}_1(\lambda,x) + \mathcal{L}_2(\lambda,\mathbf{a})$$

with

$$(10.10) \qquad \mathcal{L}_1(\lambda,x) = P^{total}(x) + \sum_{(v,w)\in E(G)} \lambda_{v,w}\mathrm{delay}_{v,w}(x)$$

and

$$\mathcal{L}_2(\lambda,\mathbf{a}) = \sum_{(v,w)\in E(G)} \lambda_{v,w}(\mathbf{a}_v - \mathbf{a}_w)$$

$$(10.11) \qquad = \frac{1}{2}\sum_{v\in V(G)} \mathbf{a}_v\left(\sum_{(v,w)\in E(G)} \lambda_{v,w} - \sum_{(u,v)\in E(G)} \lambda_{u,v}\right).$$

Now Equation 10.11 shows that the Lagrange multipliers must obey network *flow conservation* constraints:

$$\sum_{(v,w)\in E(G)} \lambda_{v,w} = \sum_{(u,v)\in E(G)} \lambda_{u,v}.$$

Otherwise, as the arrival time variables a_v ($v \in V(G)$) are unrestricted,

$$\min_a \mathcal{L}_2(\lambda,\mathbf{a}) = -\infty \quad \text{and thus as } X \text{ is compact} \quad \mathcal{L}^\star(\lambda) = -\infty.$$

Thus, to solve the Lagrangian dual problem in Equation 10.8, the Lagrange multipliers must be restricted to the space $\mathcal{N}$ of nonnegative network flows, as can also be derived from the Karush–Kuhn–Tucker optimality conditions [4]. In this case, $\mathcal{L}_2(\lambda,\mathbf{a}) = 0$ for all a and $\mathcal{L}(\lambda,x,\mathbf{a}) = \mathcal{L}_1(\lambda,x)$.

The Lagrangian dual problem (Equation 10.8) is usually solved by a subgradient method and involves solving the Lagrangian subproblem (Equation 10.7) in every iteration.

10.2.3.1 SOLVING THE LAGRANGIAN SUBPROBLEM

The Lagrangian subproblem (Equation 10.7) is again a geometric program with the box constraints $x \in X$. It can be solved by a simple descent algorithm. In [4], convergence was proven for a simple greedy algorithm if gates are modeled as RC circuits. In fact, it converges with a linear convergence rate [5,19,34]. These theoretical statements rely on a specific gate delay model. Empirically, linear running times were also observed for general delay models. They give reasons to apply LR in gate sizing [4,12,21,23,24,28,40].

10.2.3.2 SOLVING THE LAGRANGIAN DUAL PROBLEM

By the Lagrangian duality theory, $\mathcal{L}^\star$ is a concave function in the domain $\mathcal{N}$ of nonnegative network flows. It is continuous but not everywhere differentiable in $\mathcal{N}$. A prominent method for solving the Lagrangian dual problem (Equation 10.8) is the projected subgradient method [4].

Starting with some initial vector of multipliers $\lambda_0 \geq 0$, it determines in iteration $k \geq 0$, a subgradient s_k at $\mathcal{L}^\star(\lambda_k)$ in ascent direction and updates the multipliers by

(10.12) $$\lambda_{k+1} = \pi_{\mathcal{N}}(\lambda_k + \alpha_k s_k),$$

where
$\alpha_k > 0$ is the step length
$\pi_{\mathcal{N}}$ is the projection to the set $\mathcal{N}$ of nonnegative network flows

For the plain subgradient method without projection and Lipschitz continuous functions, convergence rates have been established. However, $L^\star$ is not Lipschitz continuous at the transitions between $\mathcal{N}$ and $\mathbb{R}_{\geq 0}^{E(G)} \setminus \mathcal{N}$. Polyak [29] showed that the solution sequence $(\lambda_k)_{k\in\mathbb{N}}$ in the projected subgradient method contains a subsequence that converges to the optimum, that is, the limit superior $\overline{\lim}_{k\to\infty} \mathcal{L}^\star(\lambda_k)$ converges. To prove this, the series α_k must fulfill $\alpha_k \xrightarrow{k\to\infty} 0$ and $\sum_{k=0}^{\infty} \alpha_k = \infty$. The projected subgradient steps are iterated until the duality gap $\left(P^{total}(x_k) - \mathcal{L}^\star(\lambda_k) \right)$ and the maximum or total delay constraint violation are within some positive error bound. Note that the optimum solution $x^\star$ will likely be located on the boundary of the feasible set and most iteration values x_k will not satisfy all delay constraints.

Computing the exact projection $\pi_{\mathcal{N}}$ is a *quadratic minimum-cost flow problem* that can be solved in polynomial time [25]. However, this would still be too time consuming in practice. Therefore, essentially, all papers employing the subgradient method for gate sizing resort to a heuristic [40] that maps the multipliers into $\mathcal{N}$, node by node in a backward topological order.

Convergence results similar to that in [29] are lacking for most of the employed heuristics. In [33], convergence was shown when projecting with the method of alternating projections. This implies that one can proceed as in [40] and replace their heuristic local mapping into $\mathcal{N}$ by a projection.

10.3 PRACTICAL ALGORITHMS

The gate sizing problem is so complicated and involves so many details that theoretically elegant approaches are inadequate. Many practical techniques have been developed and shown to be effective.

10.3.1 SENSITIVITY-BASED HEURISTICS

If one iteratively sizes a single gate at a time, it is natural to choose a gate whose change achieves the largest benefit at the lowest cost. The TILOS algorithm in [11] was the first to systematically use this idea. Since then, it became the rationale behind many sensitivity-based gate sizing heuristics. The idea is intuitive yet can be very effective if implemented well. One recent example is [15], which will be described in this section. An earlier work [38] will also be introduced to illustrate different flavors of this approach.

A sensitivity-based gate sizing heuristic usually consists of two phases—one for timing improvement and the other for power reduction. It typically starts with the timing phase followed by the power phase. Sometimes, the two phases can be interleaved and applied iteratively.

The sensitivity in the timing phase points to changes with the largest timing improvement and the least power overhead. In [15], it is defined by

$$\text{Sensitivity}_{timing} = \frac{\Delta TNS}{\Delta \text{power}^{power\_exponent}}$$ (10.13)

where
  TNS is the total negative slack
  Δ indicates change
  Sizing based on this sensitivity aims to minimize the absolute TNS. Since TNS is expensive to
    compute precisely, it is approximated by

$$\Delta TNS(m_i^k) = \sum_{g_j \in N_i} -\Delta \text{delay}_j^k \cdot \sqrt{\text{NPaths}_j}$$ (10.14)

where m_i^k means changing gate $g_i \in \mathcal{G}$ to implementation k. The set N_i is the neighbors of g_i and consists of gates that share fanin nodes with g_i. The delay change Δdelay_j^k of $g_j \in N_i$ is due to the size or V_t (threshold voltage) change at g_i. The notation $N\text{Paths}_j$ tells the total number of fanin and fanout gates of gate g_j that is affected by the delay change at g_j. The power exponent is between 0 and 3. Extensions to this approach including an integration with a complex yet relatively slow sign-off timer can be found in [20].

In [38], the timing phase sensitivity is defined as

$$\text{Sensitivity}_{timing} = \frac{1}{\Delta \text{power}} \sum_{arcs} \frac{-\Delta \text{delay}}{\text{slack}_{arc} - \text{slack}_{min} + K}$$ (10.15)

where
  Δpower is the power increase
  Δdelay is the delay change when upsizing a gate
  K is a constant

The arcs are (falling and rising) input arcs to a gate. In the timing recovery phase, one or several gates with the maximum sensitivity are selected to be sized up or assigned to a lower V_t level. This is repeated until the timing constraints are satisfied or no more improvement can be found. Upon each gate size or V_t change, an incremental timing analysis is performed to obtain timing updates.

A power reduction phase starts when the overall timing slack is nonnegative. In [15], the power-phase sensitivity is defined as

(10.16)
$$\text{Sensitivity}_{power} = \frac{-\Delta\text{power} \cdot \text{slack}}{\Delta\text{delay} \cdot \#\text{paths}}$$

where
  slack is at the output pin of the gate
  #paths is the total fanin and fanout of the gate

The power-phase sensitivity in [38] is defined by

(10.17)
$$\text{Sensitivity}_{power} = -\Delta\text{power} \sum_{arcs} \frac{\text{slack}_{arc}}{\Delta\text{delay}}$$

In the power reduction phase, gates with large sensitivity are selected for downsizing or for increasing the V_t level. This is repeated until the timing slack reaches minimal acceptable values.

Specialized heuristics for power minimization maintaining timing feasibility are given in [31]. Here, not only one gate is powered down at a time, but an independent set of gates is chosen based on its total sensitivity.

10.3.2 DEDICATED WORST-SLACK MAXIMIZATION

To improve the worst slack near the end of gate sizing, the focus of local search can be adjusted by changing the objectives in Equations 10.13 and 10.15: Instead of taking the sum of sensitivities among all neighbors or input arcs, the worst slack over all upstream driver gates is maximized.

Under this objective, a gate on the critical path is sized for minimum delay. Furthermore, the size of a gate with a more critical upstream gate cannot increase because this would increase the capacitance and worsen the slack at the upstream gate. Instead, its size rather decreases to reduce the load capacitance and delay of the upstream gate.

For the second reason, it is important to consider not only the gates on the critical path but also their less critical downstream gates in the local search [18,20] regardless of the local objective. In this setting, it can be beneficial to proceed in a forward topological order because an increased gate size on the critical path improves the input slews in the fanout and might allow a more aggressive downsizing. With such an extended gate selection and the local worst-slack objective explained earlier, close to optimum global worst slacks are achievable [18].

10.3.3 CONTINUOUS OPTIMIZATION WITH SELF-SNAPPING

With guidance from the function gradient, continuous optimization is usually more capable of handling large problems than discrete optimization. However, V_t levels are highly discrete, and its assignment is very difficult to be cast into a continuous optimization problem. The work by Shah et al. [36] overcomes this difficulty by implementing each gate with two parallel parts of different V_t levels (Figure 10.4) and by sizing the two parts continuously. Since gate sizes are available at a finer granularity than V_t levels, the discretization of a continuous sizing solutions is relatively manageable [16]. Interestingly, it is observed in [36] that solving this new sizing problem through convex programming usually results in integer V_t assignment solutions. More specifically, usually one part (either low V_t or high V_t) of a parallel gate has size 0. It is further mathematically proved that V_t assignment always self-snaps to integer solutions if there is no constraint on the gate sizes.

10.3.4 PRACTICAL GATE SIZING BY LAGRANGIAN RELAXATION

LR is a popular approach in solving gate sizing problems [4,12,21,23,24,28,40] mainly due to its theoretical advantage on the trade-off between conflicting objectives, such as performance and power. The foundation framework of LR-based gate sizing is introduced in the pioneer work [4] and summarized in Section 10.2.3.

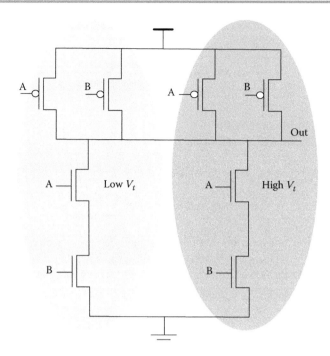

FIGURE 10.4   Implementing a NAND2 gate by a parallel connection of two logic gates with different V_t levels.

Later on, many practical techniques have been proposed to improve or extend this framework. The Lagrangian dual problem is usually solved by subgradient methods [4]. In [40], it is observed that the practical convergence of subgradient methods is very sensitive to the initial solution and step size. The convergence is faster if the initial solution satisfies timing constraints. The work by Tennakoon and Sechen [40] suggests invoking the iterative greedy algorithm [4] at the beginning so that a timing-feasible initial solution is obtained. It also shows how to estimate Lagrange multiplier values corresponding to the initial solution. With such an initial solution, the LR can focus more on area minimization subject to timing constraints rather than delay reduction. Tennakoon and Sechen [40] also propose the following Lagrange multiplier update method that does not rely on a constant step size:

$$(10.18) \qquad \lambda_{ji,new} = \begin{cases} \lambda_{ji}\dfrac{a_i}{A_0} & \text{if } i \in PO, j \in \text{fanin}(i) \\[2mm] \lambda_{ji}\dfrac{a_j}{a_i - D_i} & \text{if } j \in \text{fanin}(i) \\[2mm] \lambda_{ji}\dfrac{D_i}{a_i} & \text{if } i \in PI, j \in \text{fanin}(i) \end{cases}$$

where
 λ_{ji} is the multiplier for timing arc (j, i)
 a_i is the arrival time at node i
 A_0 is the required arrival time at the primary output (PO)
 D_i is the delay of gate i

Please note that the primary input (PI) nodes of a combinational logic circuit are driven by flip-flops and, therefore, they have fanin nodes.

Note that the multipliers are updated multiplicatively in Equation 10.18 and not in subgradient direction. In practice, multiplicative updates often lead to faster convergence than the subgradient update shown in Equation 10.12, as demonstrated by the results discussed in [12] on ISPD 2012 and 2013 gate sizing benchmarks.

LR is also combined with DP-like solution search [23] so that a discrete solution is obtained directly without the error-prone rounding procedure. DP has been quite successful for problems with tree topologies, such as buffer insertion. Since the topology of combinational logic circuits is a directed acyclic graph, the solution propagation of DP faces the challenge of history consistency constraint. When two candidate solutions are merged at a node, the choices on their common ancestor nodes should be identical. This constraint entails either large memory space or long computation time as a large volume of history information needs to be stored or traced. Moreover, it complicates pruning, which is a key component of the DP. The work of Liu and Hu [23] suggests an interesting idea to solve this challenge. It first runs DP with the history consistency constraint relaxed. Although the solution may be illegal, it provides a timing bound for each node. Next, it applies a greedy-like heuristic to obtain a history consistent solution. Since the heuristic utilizes the timing bound obtained from the DP, it is not myopic as a stand-alone greedy algorithm. Then, the algorithm continues solution search in a manner that is between DP and a greedy heuristic. Overall, it strives to gain the greatest possible benefit from DP, while avoiding excessive enumeration for resolving inconsistencies.

The algorithm in [28] is also based on LR and DP. Unlike most previous works that constrain only the longest path slack to be nonnegative, the problem formulation of [28] attempts to minimize the absolute TNS and includes it as a part of the objective function. For cases without timing-feasible solutions, the result from [28] is more usable. The algorithm of this work is built upon a graph model as illustrated in Figure 10.5. In this model, gate delay and power are captured in node and edge weight. Then, the Lagrangian subproblem, which optimizes a weighted sum of gate delay and power, is solved by DP. In order to avoid the troublesome history consistency problem, the entire circuit is partitioned into a set of disjoint trees where DP is conducted on each individual tree.

In [24], the Lagrangian subproblem is solved by a discrete greedy heuristic. The algorithm proceeds in circuit traversals. When a gate is encountered during traversal, its implementation options are enumerated and the one that minimizes the overall cost function is selected. This heuristic is very simple yet effective.

10.3.5 HEURISTICS BASED ON TIMING BUDGETING

Another approach to handle the trade-off between circuit timing and power is through timing budgeting [27]. First, a minimum delay sizing solution is obtained using a certain simple heuristic. Next, the timing slack is allocated to each gate. Last, each gate is sized down or changed to higher

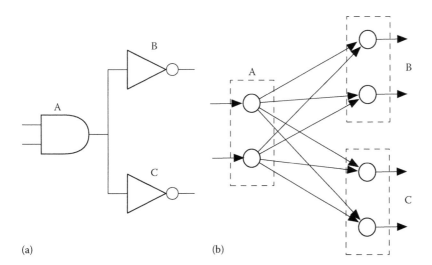

(a)   (b)

FIGURE 10.5   (a) Circuit and its (b) graph model assuming each gate has only two implementation options.

V_t level according to its slack budget. The key component of this approach is a systematic allocation of timing slack to each node, which can be formulated as follows:

(10.19)
$$\text{Max} \sum_{v_i \in V} \delta_i \cdot x_i$$

(10.20)
$$\text{s.t.} \quad a_i + d_j + x_j \le a_j \quad \forall v_i, v_j \in V, \quad v_i \in fanin(v_j)$$

(10.21)
$$a_j \le T \quad \forall v_j \in PO$$

(10.22)
$$0 \le x_i \le U_i \quad \forall v_i \in V$$

where
a_i, d_i, and x_i are the arrival time, delay, and slack budget for node v_i
T is the required arrival time at the PO
U_i is an upper bound for the budget at node v_i
δ_i is the power reduction per delay increase

The budgeting problem can be solved by linear programming [27] or network flow algorithms [13].

Noticing the correlation between gate delay and signal slew rate, the work of [18] proposes to size gates through slew budgeting instead of delay budgeting. The slew target for each gate is initialized with the design rules. Then, in a reverse topological order traversal of the circuit, a size is found for each gate to meet its slew target. When sizing a gate, its fanout gate sizes have already been updated, and therefore, the load capacitance is known. Its input slew can be estimated according to the slew target of its fanin gates. After one circuit traversal, slew targets are adjusted according to timing criticality. Timing critical gates are assigned with tighter slew constraints, and the slew-driven sizing is repeated. Since slew estimation is much faster than timing analysis, this approach can easily deal with very large circuits using a relatively slow sign-off timer.

10.3.6 GATE SIZING IN ELECTRONIC DESIGN AUTOMATION TOOLS

Some commercial tools use global gate sizing algorithms based on LR (Section 10.3.4) or timing budgeting (Section 10.3.5) early in the design flow to get good starting solutions. Such global solutions are improved by local search, where sensitivity-based heuristics (Section 10.3.1) are used predominantly.

Gate sizing is usually not performed as an isolated task but incrementally alternated or integrated with interconnect optimization, timing-driven detailed placement, or logic restructuring. Again sensitivity-based heuristics are mostly used in the interplay with other operations [30].

As LR does not guarantee fast convergence, there is always the danger of a timing degradation, when applied late in the flow. Again sensitivity-based heuristics that do not introduce new path delay violations are prevalent for timing final power reduction [31].

10.4 CONCLUSIONS AND OUTLOOK

Gate sizing is an important and challenging problem in very-large-scale integration design. The continuous gate sizing problem without V_t assignment can be solved in polynomial time for posynomial delay functions.

In practice, discrete gate libraries are used, and no polynomial time algorithm with a performance guarantee is known, except for trees or path-like netlists. On the other hand, there has been substantial practical improvement, in particular, after the ISPD gate sizing contests 2012 and 2013 [12,15,20,21,23,24]. It can be expected that several of these new heuristics will find their way into electronic design automation tools. Thereby, the major challenge is the interaction with sign-off timing engines that, due to the support of many features, tend to be much slower than the slim static timing implementations used in research papers. It is not clear how far the currently best solutions are away from the optimum. Even if LR is used, the Lagrangian subproblems are not solved to optimality. A fast computation of good lower bounds for the power consumption is an important open problem.

On the theoretical side, discrete gate sizing algorithms with good performance guarantees are lacking. As explained in Section 10.2.1, it aims high to achieve a constant-factor approximation algorithm. However, providing a polynomial-time algorithm with a logarithmic performance guarantee would be a significant result.

REFERENCES

1. S. Boyd, S.-J. Kim, L. Vandenberghe, and A. Hassibi. A tutorial on geometric programming. *Optimization and Engineering*, 8(1), 67–127, 2007.
2. S. P. Boyd, S.-J. Kim, D. D. Patil, and M. A. Horowitz. Digital circuit optimization via geometric programming. *Operations Research*, 53(6), 899–932, 2005.
3. P. K. Chan. Algorithms for library-specific sizing of combinational logic. *Proceedings of the 27th ACM/IEEE Design Automation Conference*, Orlando, FL, 1990, pp. 353–356.
4. C. P. Chen, C. C.-N. Chu, and D. F. Wong. Fast and exact simultaneous gate and wire sizing by Lagrangian relaxation. *IEEE Transactions on Computer-Aided Design*, 18(7), 1014–1025, 1999.
5. C. Chu and D. F. Wong. VLSI circuit performance optimization by geometric programming. *Annals of Operations Research*, 105, 37–60, 2001.
6. J. Cong, J. Lee, and G. Luo. A unified optimization framework for simultaneous gate sizing and placement under density constraints. *Proceedings of 2011 IEEE International Symposium on Circuits and Systems (ISCAS)*, Rio De Janeiro, Brazil, 2011, pp. 1207–1210.
7. A. R. Conn, I. M. Elfadel, W. W. Molzen, Jr., P. R. O'Brien, P. N. Strenski, C. Visweswariah, and C. B. Whan. Gradient-based optimization of custom circuits using a static-timing formulation. *Design Automation Conference (DAC)*, New Orleans, LA, 1999, pp. 452–459.
8. J. F. Croix and D. F. Wong. Blade and razor: Cell and interconnect delay analysis using current-based model. *Proceedings of 40th ACM/IEEE Design Automation Conference*, Anaheim, CA, 2003, pp. 386–389.
9. I. Dinur and S. Safra. On the hardness of approximating minimum vertex cover. *Annals of Mathematics*, 162, 439–485, 2005.
10. W. C. Elmore. The transient response of damped linear networks with particular regard to wide-band amplifiers. *Journal of Applied Physics*, 19(1), 55–64, 1948.
11. J. P. Fishburn and A. E. Dunlop. TILOS: A posynomial programming approach to transistor sizing. *Proceedings of the IEEE/ACM International Conference on Computer-Aided Design*, Santa Clara, CA, 1985, pp. 326–328.
12. G. Flach, T. Reimann, G. Posser, M. de O. Johann, R. Reis. Effective method for simultaneous gate sizing and Vth assignment using Lagrangian relaxation. *IEEE Transactions on Computer Aided Design of Integrated Circuits and Systems*, 33(4), 546–557, 2014.
13. S. Ghiasi, E. Bozorgzadeh, P.-K. Huang, R. Jafari, and M. Sarrafzadeh. A unified theory of timing budget management. *IEEE Transactions on Computer-Aided Design*, 25(11), 2364–2375, 2006.
14. A. Grigoriev and G. J. Woeginger. Project scheduling with irregular costs: Complexity, approximability, and algorithms. *Acta Informatica*, 41, 83–97, 2004.
15. J. Hu, A. B. Kahng, S. Kang, M.-C. Kim, and I. L. Markov. Sensitivity-guided metaheuristics for accurate discrete gate sizing. *Proceedings of the IEEE/ACM International Conference on Computer-Aided Design*, San Jose, CA, 2012, pp. 233–239.
16. S. Hu, M. Ketkar, and J. Hu. Gate sizing for cell-library-based designs. *IEEE Transactions on Computer-Aided Design*, 28(6), 818–825, 2009.
17. S. Hu, Z. Li, and C. J. Alpert. A fully polynomial time approximation scheme for timing driven minimum cost buffer insertion. *Proceedings of the 46th Annual Conference on Design Automation*, San Francisco, CA, 2009, pp. 424–429.
18. S. Held. Gate sizing for large cell-based designs. *Proceedings of Design, Automation and Test in Europe Conference*, Nice, France, 2009, pp. 827–832.
19. S. Joshi and S. Boyd. An efficient method for large-scale gate sizing. *IEEE Transactions on Circuits and Systems*, 55(9), 2760–2773, 2008.
20. A. B. Kahng, S. Kang, H. Lee, I. L. Markov, and P. Thapar. High-performance gate sizing with a signoff timer. *Proceedings of the International Conference on Computer-Aided Design*, San Jose, CA, 2013, pp. 450–457.
21. L. Li, P. Kang, Y. Lu, and H. Zhou. An efficient algorithm for library-based cell-type selection in high-performance low-power designs. *Proceedings of the IEEE/ACM International Conference on Computer-Aided Design*, San Jose, CA, 2012, pp. 226–232.

22. C. Liao and S. Hu. Approximation scheme for restricted discrete gate sizing targeting delay minimization. *Journal of Combinatorial Optimization*, 21(4), 497–510, 2011.

23. Y. Liu and J. Hu. A new algorithm for simultaneous gate sizing and threshold voltage assignment. *IEEE Transactions on Computer-Aided Design*, 29(2), 223–234, 2010.

24. V. S. Livramento, C. Guth, J. L. Güntzel, and M. de O. Johann. Fast and efficient Lagrangian relaxation-based discrete gate sizing. *ACM Transactions on Design Automation of Electronic Systems*, 19(4), Article No. 40, 2014.

25. M. Minoux. A polynomial algorithm for minimum quadratic cost flow problems. *European Journal of Operational Research*, 18(3), 377–387, 1984.

26. L. W. Nagel and D. O. Pederson. SPICE (Simulation Program with Integrated Circuit Emphasis). Technical Report M382, EECS Department, University of California, Berkeley, CA, 1973.

27. D. Nguyen, A. Davare, M. Orshansky, D. Chinnery, B. Thompson, and K. Keutzer. Minimization of dynamic and static power through joint assignment of threshold voltages and sizing optimization. *Proceedings of the ACM/IEEE International Symposium on Low Power Electronics and Design*, Seoul, Korea, 2003, pp. 158–163.

28. M. Ozdal, S. Burns, and J. Hu. Algorithms for gate sizing and device parameter selection for high-performance designs. *IEEE Transactions on Computer-Aided Design*, 31(10), 1558–1571, 2012.

29. B. T. Polyak. A general method for solving extemum problems. *Doklady Akademii Nauk SSSR*, 174(1), 33–36, 1967. Translation in *Soviet Mathematics Doklady*, 8(3), 593–597, 1967.

30. R. Puri, M. Choudhury, H. Qian, and M. Ziegler. Bridging high performance and low power in processor design. *Proceedings of the 2014 International Symposium on Low Power Electronics and Design*, La Jolla, CA, 2014, pp. 183–188.

31. H. Qian and E. Acar. Timing-aware power minimization via extended timing graph methods. *ASP Journal of Low Power Electronics*, 3(3), 318–326, 2007.

32. C. Ratzlaff and L. T. Pillage. RICE: Rapid interconnect circuit evaluation using asymptotic waveform evaluation. *IEEE Transactions on Computer-Aided Design*, 13(6), 763–776, 1994.

33. D. Rautenbach and C. Szegedy. A subgradient method using alternating projections. Technical Report 04940, Research Institute for Discrete Mathematics, University of Bonn, Bonn, Germany, 2004.

34. D. Rautenbach and C. Szegedy. A class of problems for which cyclic relaxation converges linearly. *Computational Optimization and Applications*, 41, 52–60, 2008.

35. S. S. Sapatnekar, V. B. Rao, P. M. Vaidya, and S.-M. Kang. An exact solution to the transistor sizing problem for CMOS circuits using convex programming. *IEEE Transactions on Computer-Aided Design*, 12(11), 1621–1634, 1993.

36. S. Shah, A. Srivastava, D. Sharma, D. Sylvester, D. Blaauw, and V. Zolotov. Discrete vt assignment and gate sizing using a self-snapping continuous formulation. *Proceedings of the IEEE/ACM International Conference on Computer-Aided Design*, 2005, pp. 705–712.

37. R. F. Sproull and I. E. Sutherland. Logical effort: Designing for speed on the back of an envelope. *IEEE Advanced Research in VLSI C*, Sequin, ed., MIT Press, Cambridge, MA, 1991, pp. 1–16.

38. A. Srivastava, D. Sylvester, and D. Blaauw. Power minimization using simultaneous gate sizing, minimization using simultaneous GatDual-Vdd and Dual-Vth assignment. *ACM/IEEE Design Automation Conference*, San Diego, CA, 2004, pp. 783–787.

39. O. Svensson. Hardness of vertex deletion and project scheduling. *Theory of Computing*, 9(24), 759–781, 2013.

40. H. Tennakoon and C. Sechen. Gate sizing using Lagrangian relaxation combined with a fast gradient-based pre-processing step. *Proceedings of the IEEE/ACM International Conference on Computer-Aided Design*, San Jose, CA, 2002, pp. 395–402.

41. L. Trevillyan, D. S. Kung, R. Puri, L. N. Reddy, and M. A. Kazda. An integrated environment for technology closure of deep-submicron IC designs. *IEEE Design & Test of Computers*, 21(1), 14–22, 2004.

42. A. Wächter, C. Visweswariah, and A. R. Conn. Large-scale nonlinear optimization in circuit tuning. *Future Generation Computer Systems*, 21(8), 1251–1262, 2005.

43. J. Wang, D. Das, and H. Zhou. Gate sizing by Lagrangian relaxation revisited. *IEEE Transactions on Computer-Aided Design of Integrated Circuits and Systems*, 28(7), 1071–1084, 2009.

44. N. H. E. Weste and D. Harris. *CMOS VLSI Design: A Circuits and Systems Perspective*, 4th edn., Addison-Wesley Publishing Company, Boston, MA, 2010.

45. T.-H. Wu and A. Davoodi PaRS: Parallel and near-optimal grid-based cell sizing for library-based design. *IEEE Transactions on Computer-Aided Design of Integrated Circuits and Systems*, 28(11), 1666–1678, 2009.

Clock Design and Synthesis

Matthew R. Guthaus

CONTENTS

11.1 INTRODUCTION

Clock design is the challenging task of distributing one or more clock signals throughout an entire chip while minimizing power, variation, skew, and jitter, and yet simultaneously being resource conscious [33,40]. The design problem is akin to analog design where trade-offs between multiple competing objectives depend on the methodology and experience with prior designs. Even clock signals are far from perfect square digital waveforms.

Because of these challenges, clock design is among the most misunderstood and inconsistent portions of any design methodology. There is no single "right way." While the primary goals are generally similar, the techniques to achieve these goals can be drastically different from one company to another and from one design methodology and toolset to another.

11.1.1 METRICS

A clock signal has a *period* that specifies the duration of a repeated high and low pattern. The period is inversely related to the integrated circuit frequency. The *duty cycle* of the clock is the ratio of high to low time in the period, usually 50%. The *insertion delay*, or just delay, is the time that takes the clock signal to propagate through the clock tree to the sinks. The *sinks* are the final receiving endpoints of the clock signal, which are either the clock pins of a sequential logic cells or the input pins of local clock buffers in hierarchical designs. Clock signal delays are usually measured at the point where the voltage is 50% of the supply voltage. Sometimes, this delay point is replaced with the switching threshold of an inverter in edge-triggered systems.

Global skew: If the underlying logic paths are unknown, the *global clock skew* is measured among all pairs of sinks, which implies the equivalent statement

(11.1)
$$\text{Skew} = \max_{i \in \text{Sinks}}(d_i) - \min_{j \in \text{Sinks}}(d_j).$$

This defines skew as the difference between the slowest and fastest clock sink delays, yet avoids enumerating all logical pairs or performing expensive static timing analysis.

Local skew: The difference in the delays of a pair of clock sinks, d_A and d_B, is the *local clock skew* (t_k) and is defined as the difference

(11.2)
$$t_k = d_A - d_B,$$

which can be positive or negative. The maximum local skew of a clock is defined as the maximum local clock skew between any logically connected pair of sinks

(11.3)
$$\text{Skew} = \max_{k \in \text{Paths}} t_k.$$

Skew can be global, local, or maximum local depending on context. Skew, preferably local, is used in static timing analysis to determine if the timing constraints of individual paths are satisfied. However, an astute reader who has already read the static timing analysis chapter may note that clock skew can also be leveraged to improve circuit performance through what is known as "useful skew" [31]. Useful skew purposely alters the skew between sending and receiving sequential cells on critical paths to satisfy setup or hold constraints. Useful skew ultimately relies on static timing analysis to determine if timing constraints are met. This method is useful to help timing sign-off but is often impractical if all timing paths are not yet finalized. In addition, the accurate control of clock delays is often difficult, which complicates practical application.

While skew measurements are the most common clock metrics, there are several other very important metrics including jitter, slew, power, robustness, and resource usage. The relative importance of each is debatable, as confirmed by the variety of clock design methodologies with different priorities.

Jitter: Similar to skew, the delay to a single sink may vary from period to period introducing clock *jitter.* Jitter in an on-chip clock distribution is usually caused by the clock source (oscillator or phase-locked loop), power supply noise, or interconnect cross-coupling. Jitter is equally important as clock skew since it affects timing paths in a similar way. In some designs, jitter can be of the same magnitude as skew or more.

Slew: The transition time between the high and low (or low and high) logic level is characterized by the clock *slew* rate. This slew is then defined as the time to transition from 10% to 90% or 90% to 10% of the supply voltage. Slew is important because it directly affects the setup time, hold time, and clock-to-output delay of sequential elements. While fast slew rates are desirable for performance, excessively fast slew rates can lead to cross-coupling noise in logic signals adjacent to the clock, and the increased power supply IR drop can then worsen the clock skew.

Power: There is a significant trade-off between performance and power in digital design, especially clock network design. Very low skew clock distributions can consume excessive portions of total chip power and have been cited as up to 70% of total chip power [5]. More commonly, the clock power budget is on the order of 20%–40% of total chip power, and so, power is often as important as performance. The primary reason for the large power is that the clock signal must be distributed throughout an entire chip and must drive a large number of sequential elements. Power can be minimized by reducing total capacitance, switching rate, frequency, and even voltage swing of the clock signal. Section 11.4 specifically focuses on power optimization.

Robustness: Robustness is also a major concern in clock design since it directly affects performance and can even render chips nonfunctional due to yield-related timing errors. There are numerous sources of variation [72] that can affect the robustness of a clock network including manufacturing process (P), supply voltage (V), on-chip temperature (T), and cross talk between the clock and signal wires (X). These are often referred to as PVTX variations or commonly just PVT variation. Furthermore, process variation can be caused by either *front-end* (device) parameters or *back-end* (interconnect) parameters. Variation that is observed from die-to-die (D2D) is called *inter-die variation,* while variation that occurs within a die (WID) is called *intra-die variation,* or occasionally on-chip variation (OCV). WID variation is uncorrelated or partially correlated, while D2D variation has full correlation across a single chip. Typically, clocks are made robust through redundancy; feedback with adaptive circuitry, or simply, improved sensitivity analysis to evaluate variation implications. Minimizing clock network insertion delay is often considered a proxy for variability, but it is only appropriate for correlated D2D variation, and the reality of variation is more complicated when WID is considered.

Resource usage: Resource usage is also an important metric that clock distributions must consider. While power minimization dictates the use of fewer buffer and interconnect resources in general, the physical allocation of routing, buffer and power supply decoupling resources must compete with logic modules throughout a chip. Specifically, global interconnect layers are limited and must be used in clocks, power supplies, and global busses. Similarly, active device area is used for clock buffers and dedicated clock power supply decoupling capacitances, but this must compete with the actual chip logic and its own power supply decoupling.

11.1.2 METHODOLOGIES

The differing importance of the previous metrics means that there are drastically different clock design methodologies. The most significant difference is whether a methodology is top-down or bottom-up [40]. Typically, aggressive designs will have a global clock distribution that is predesigned while lower modules are being completed [55,95]. Less aggressive designs, on the other hand, are created flat or bottom-up (if hierarchical). Because of this methodology difference, aggressive designs typically have problem instances with fewer clock endpoints and have physical constraints imposed by the hierarchical floorplan [92]. Less aggressive designs have problem instances with many clock endpoints but have more physical flexibility by interacting with floorplanning, placement, and routing.

In aggressive designs, circuit design is generally incomplete during system planning, with local clock distributions in individual modules/blocks merely estimated and being changed significantly during design completion and timing closure. To accommodate this, a system designer specifies the floorplan and coordinates the block/module designers. Using the floorplan, the global clock is predesigned in a top-down methodology, since it competes for shared routing resources and requires reservation of silicon area between modules for global clock buffers and decoupling capacitance. These resources are in contention with power distribution, local modules and system-wide connection between these same modules. These aggressive design methodologies utilize hierarchical design, which means clock design instances can be either global or local. At the global level, timing paths are typically not known until late in the design cycle, so the goal is zero skew. The local levels, however, typically use similar methodologies to less aggressive designs.

Less aggressive designs are often synthesized flat or, at the least, modules are much larger than aggressive designs. In these cases, a modification to a module results in a re-spin of the entire synthesis, placement and routing steps, which can perform drastic changes in the clock design to assist timing closure. Contrary to aggressive global clock distribution, these clocks do not target zero skew since timing paths within and between local modules are known. Local distributions in aggressive designs also consider per path skew much like these less aggressive designs.

The aggressive design approach consumes more power, due to redundancy in the global levels and a zero skew goal, but these allow independent optimization of local modules due to skew insensitivity, which helps high-performance timing closure. In addition, the redundancy in the global distribution better tolerates manufacturing variability that is necessary to achieve high performance. Such a hierarchical approach is often necessary due to the extreme size of the designs and insufficient CAD tool capacity or a very-high-frequency clock target. The less aggressive approaches, on the other hand, rely heavily on the scalability of automated algorithms and require a less manual design effort. This has the potential to perform cross-hierarchical optimization at the expense of increased tool and methodology complexity.

11.2 GLOBAL CLOCK DISTRIBUTION

Since global clock distributions span large areas of the chip, they are particularly sensitive to variations in metal interconnects, clock load capacitance, temperature, and the power supply. They are, however, not very sensitive to device variation since global clock buffers are typically large, which reduces common device-specific sources such as threshold variation.

11.2.1 SYMMETRIC TREES

Symmetric trees such as H-trees [6] as shown in Figure 11.1 are viable candidates for global distribution since they limit sensitivity to D2D variation but are still sensitive to WID variation. The main challenge with symmetric clock trees is that clock sinks are not symmetrically distributed, and so, electrically symmetric [87,89] and delay sensitivity symmetric [38] trees have been proposed. In global clocks, symmetric trees are often used to distribute the root clock signal to a multitude of drivers in a more robust global distribution network that can tolerate WID variation in addition to D2D variation.

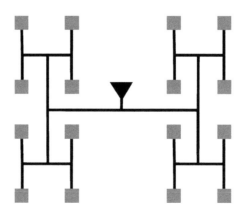

FIGURE 11.1   H-Tree clocks are robust to correlated D2D variation due to their physical symmetry.

11.2.2 CLOCK MESHES AND SPINES

The primary way of addressing variation is to compensate with redundant devices and wires. The amount and structure of the redundancy can have a broad range, depending on how much variation and power can be tolerated. Clock grids and spine networks are the preferred choice for high-end microprocessors or aggressive designs that push performance limits. On the other hand, redundant trees have had some attention in academia and are increasingly used in industry.

Clock spines, as shown in Figure 11.2, are a wide metal trunk with low resistance and therefore low skew between elements on the same spine. This enables more than one spine driver on the spine to compensate for incoming delay variation. Skew between spines due to uneven loading or incoming delay variation, however, can still be problematic. No published design automation algorithms have focused on clock spine distributions, but these are often used in some industry methodologies.

Clock meshes or grids, as shown in Figure 11.3, compensate for spine-to-spine variation with a 2D redundant structure. This additional redundancy ensures more paths between the mesh drivers and the clock sinks, which results in decreased variation. The additional redundancy, however, increases both the total capacitance and the size of the clock drivers. Clock mesh algorithms primarily use a minimum total wire length formulation based on meshworks [79] and produce very sparse regular meshes with minimum wire length. These can still be sensitive to

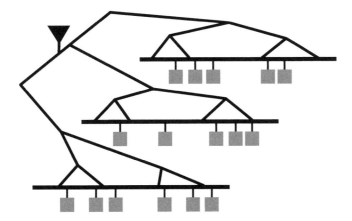

FIGURE 11.2 Clock spines enable low variation by attaching sequential cells to a low resistance interconnect trunk.

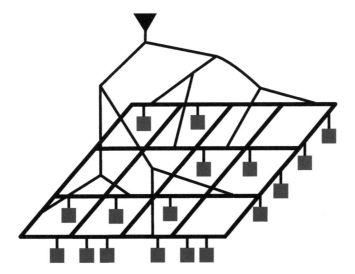

FIGURE 11.3 Clock meshes offer the most robustness to variation at the expense of high dynamic power and short-circuit power between clock grid drivers.

variation, so other approaches have taken the opposite approach by starting with a dense mesh and performing mesh wire removal to reduce power without sacrificing robustness [105]. More recently, several works have revisited some of the basic assumptions about clock grids, namely, the uniformity of grid wires and the placement of buffers at grid wire crossings (grid nodes). Specifically, Abdelhadi et al. [1] improved mesh reduction by considering local skew constraints, while Guthaus et al. [39,41] proposed a technique to customize grids for nonuniform sink distributions using a grid wire clustering algorithm. Another recent approach used binary programming to perform mesh reduction to trade-off skew and power [21]. On the other hand, Xiao et al. [111] explored methods of mixing grids and trees for power and skew trade-off, while considering blockages.

Besides mesh routing, mesh buffering is extremely important since a significant portion of mesh power consumption is due to short-circuit power between buffers. This is primarily due to skew among the inputs of the driving buffers, since the clock distribution that drives the mesh buffers is not redundant and is susceptible to variation. A sensitivity-based buffer removal was proposed by Guthaus et al. [39,41], which removed redundant or ineffective buffers using iterative sizing and removal. This leads to a dramatic reduction in power consumption due to decreased short-circuit power and reduced total buffer size. In another approach, Flach et al. [32] examined a mesh buffer displacement algorithm based on sensitivities to find skew optimal positions for grid buffers.

While mesh buffering and routing are important, these assume that an entire set of sinks are connected directly to the mesh. In reality, most meshes drive local clock buffers that distribute to the sequential elements using a local tree. While this is common in industry methodologies, Lu et al. [68] presented the first algorithm to do such an optimization, using register activity information.

The major complication with integrating automated mesh (or spine) synthesis approaches into high-performance designs is the contention for metal resources with power supply distribution. System designers frequently reserve particular global interconnect tracks specifically for power supply and clock distribution, and so, such algorithms may not have complete freedom in wire placement. Two works have specifically considered obstacle avoiding meshes [89,111] but have not considered this in coordination with system-level planning.

11.2.3 REDUNDANT TREES

To save power over meshes and spine methods while retaining robustness, redundant trees selectively insert redundancy to decrease variation between critical subtrees. This is the opposite of the clock mesh reduction schemes, by starting with a non-redundant solution and selectively adding redundancy. This redundancy was initially proposed as single cross-links, as shown in Figure 11.4, which attempt to connect nearby subtrees that suffer from skew variation to improve robustness [78,104].

Cross-link techniques present challenges of their own, however. Cyclic RC graph analysis with multiple drivers is complicated compared to the analysis of clock trees. Cross-links can lead to short-circuit power consumption, due to multiple drivers on a single net. In Figure 11.4, for example, if the left buffer switches before the right buffer, the two buffers will be driving different voltages onto the same net. This was already discussed as being prevalent in clock meshes, yet it becomes extremely problematic in very sparse redundant trees when cross-links are between skewed nets. It is desirable to insert cross-links between nets with high variation, and yet, this is exactly what causes additional short-circuit power. The most recent works have found that there is some benefit to inserting cross-links high up in clock trees, but, in general, the extreme sparsity of the cross-links and the physical distances between subtrees with variation limit the benefit at lower levels [71]. Another direction was proposed by Markov and Lee, which uses auxiliary trees instead of simple cross-links [59]. "Tree fusion," as the authors call it, finds critical sink pairs and their least common ancestor. An auxiliary tree is then synthesized and fused to the initial tree to provide redundant paths and reduce variation. This can still have difficulties with short-circuit power of multiple drivers but increases the feasibility of redundancy between distant nodes and increases robustness.

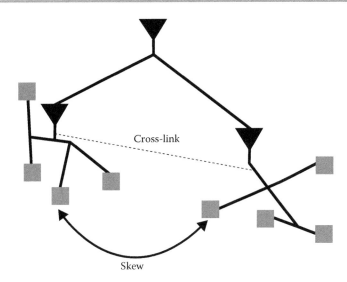

FIGURE 11.4   Clock trees with cross-links can add sparse redundancy to improve robustness.

11.2.4 POST-SILICON AND RUNTIME TUNING

Post-silicon clock tuning techniques can remove unwanted skew due to manufacturing and even runtime thermal and power supply variation. The clock tree itself can be repaired using digital- or analog-controlled adjustable delay buffers (ADBs). Post-silicon techniques are expensive in terms of routing, delay adjustment, and most importantly, skew detection. Therefore, it is usually only considered at the global level.

Geannopoulos and Dai proposed a hardware-based solution that deskews by measuring the phase difference between spines in a spine-based clock network [34]. Kapoor et al. took a similar approach for standard trees by adding return paths from all clock sinks [52]. The return paths are selectively enabled, depending on which paths are the most skewed. This method is very flexible but also requires significant overhead due to the additional routing, phase detection, and tri-state inverters for the return paths. Tsai et al. selectively insert ADBs with clock skew scheduling, considering variation [98] and even statistical timing analysis [101]. Specifically, they inserted ADBs on critical clock sinks, which allow the clock tree to be repaired during at-speed testing.

In addition to post-silicon tuning for manufacturing variation, several schemes have been proposed to use thermal sensor data to compensate for runtime on-chip thermal variation. Thermal variation has been directly considered by moving buffers and wires to compensate during design time [12,20,70], but thermal variation is difficult to model since it relies quite heavily on the workload. Optimizing for more than a single workload is difficult, in addition to the difficulty of determining a common thermal profile. Other approaches therefore use ADBs along with on-chip temperature data to detect potential skew problems and set ADBs to deskew the clock correspondingly [9,63].

While post-silicon techniques have the advantage that they do not rely on process models, they consume significant overhead in terms of the programmable delay elements and extra routing. At-speed testing is already a difficult problem without clock tuning. Similarly, thermal timing sign-off is difficult without having to verify that an adaptive thermal clock operates correctly.

11.3 LOCAL CLOCK DISTRIBUTION

Clock trees are preferred at the local level because they can distribute a clock to nonuniformly spread sinks while using few buffering and routing resources, and therefore, being more power efficient compared to the global clock distribution. The lower power, however, means that local clocks tend to be more susceptible to PVTX variation than their global counterparts. This section presents the details of several approaches to automated local clock tree synthesis.

The primary objective of clock tree synthesis is to create the buffered routed tree such that the skew and power are minimized. Formally, let $S = \{s_1, s_2, \ldots, s_n\} \subset \Re^2$ be the set of clock sink locations in a Manhattan plane. Each clock sink has associated with it a load capacitance, c_i, and optionally a desired delay offset, d_i, relative to the ideal clock for useful skew. The *abstract topology* of the set S, $T_t(S)$, is a rooted binary tree with the leaves corresponding to each sink in the set S. The *routing* of a clock tree is a mapping of the internal nodes of the abstract topology tree into the Manhattan plane to form a mapped topology, $T_m(S)$. Subsequently, all non-vertical and non-horizontal edges are decomposed, or *embedded* into rectilinear shapes, by adding internal nodes and mapping these nodes to locations in the Manhattan plane. The mapped embedded tree is referred to as the routed clock tree, $T_r(S)$. Buffers can be inserted before, during, or after routing to bound slew rates at the sinks while minimizing skew. Buffers can be inserted at internal nodes of a tree, or new nodes can be created by splitting an edge at a location in the Manhattan plane. The final sizes of the buffers and wires can be adjusted in a final *tuning* step, once the tree is mapped, routed, and buffered.

The first subsection discusses routing and topology generation algorithms. Each of these operations is done to minimize skew, or achieve zero skew, with the minimum amount of wire length. The next section provides an overview of buffer insertion algorithms for clock trees. The input to this phase is a routed clock tree, and the objective is to insert buffers at nodes or along wires such that slew rates are constrained and skew and power are minimized. The third subsection presents algorithms for tuning buffer and wire sizes. Finally, the last section presents some of the many combined methodologies of topology generation, routing, buffering, and tuning.

11.3.1 ROUTING AND TOPOLOGY GENERATION

The method of means and medians (MMM) [50] is the first top-down algorithm that heuristically minimizes clock skew, which attempts to balance the clock sinks into two partitions at each level of clock tree hierarchy, as shown in Figure 11.5. In Figure 11.5a, the top-level region is partitioned

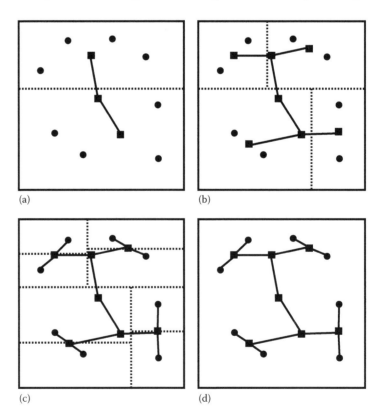

(a)   (b)

(c)   (d)

FIGURE 11.5   The Method of Means and Medians (MMM) performs top-down partitioning to balance subtrees and routes to the mean locations. (a) First level partition, (b) second level partition, (c) third level partition, and (d) final tree.

into two subregions of equal size, based on the median sink location in the y-dimension. Then the mean location of the original region is routed to the mean location of each subregion. The mean location is the center of mass of all the clock sinks. The procedure is repeated while altering between the horizontal and vertical dimensions in Figure 11.5b and c, until the final tree is constructed, as in Figure 11.5d. Clock skew is only minimized heuristically and the resulting tree may not have zero skew, but the trees use very low wire length and power. MMM has a worst-case complexity of $O(n \log n)$ for n clock sinks.

Soon after MMM, another approach to clock routing was proposed using a recursive, bottom-up geometric matching algorithm (GMA) [51], which is shown in Figure 11.6. At each level of the hierarchy in a bottom-up order, the algorithm constructs a set of $n/2$ segments connecting the n endpoints pairwise such that no two segments share an endpoint. The matching is selected to minimize the total wire length at the given level, using any bipartite matching algorithm. After the matching is determined, a tapping point on each segment is used to determine where the next higher level should connect. The tapping point is not necessarily the midpoint of the segment but is the point to minimize the skew between the current pairs. Assuming that wire jogs are added when needed, the GMA algorithm can attain zero skew. However, unlike MMM, the wire length and power of the resulting trees can be poor. GMA has a worst-case complexity of $O(n^2 \log n)$ for n clock sinks.

The MMM and GMA approaches are convenient to implement and are efficient for large clock trees and are still used in many commercial tools. However, neither is nonoptimal due to fundamental assumptions. The MMM does not know about the subtrees that have not been created yet, which can result in unwanted skew. The GMA does not know how trees will be merged further up in the hierarchy, which can result in wasted wire length and power. In the early to mid-1990s,

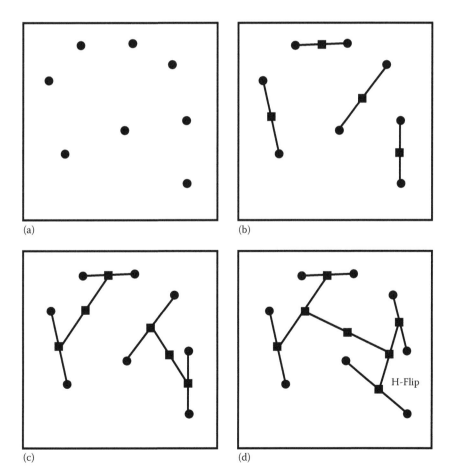

(a)

(b)

(c)

(d)

FIGURE 11.6 The Geometric Matching Algorithm (GMA) performs bottom-up matching to minimize wire length at any given level. (a) Initial locations, (b) first level matching, (c) second level matching, and (d) final tree with H-flip optimization.

three groups independently presented algorithms that address this suboptimality by performing a two-pass optimization [7,15,102] to obtain a zero-skew tree with minimal wire length. The same basic algorithm, Deferred Merge Embedding (DME), was later generalized to bounded skew trees [25,48]. The first pass performs a bottom-up merging of the sinks to find all potential zero-skew merging locations. The choice of the exact location of the merge, however, is deferred, until the parent root location is decided so that the location can be selected to minimize the overall wire length of the tree and save power while still obtaining zero skew. Most clock tree synthesis algorithms today are derived from a DME approach.

The bottom-up phase of DME finds a merging segment for each internal node of the abstract topology (Figure 11.7). A *merging segment* (MS) is the locus of points that define the area of the zero-skew merge. With rectilinear routing and equal resistance and capacitance on the layers, each MS is an intersection of two Manhattan circles (diamonds), as illustrated in Figure 11.8, and is always a line with slope of +1 or −1 or a point. The sizes of the Manhattan circles, d_1 and d_2, are selected to exactly balance the delay of the subtrees under the Elmore delay model [29]. In Figure 11.8, the closest distance (D) between two MS is used to connect MS1 and MS2 using minimum wire length. The problem of finding the size of the Manhattan circles can be simplified to dividing the total distance, $D = d_1 + d_2$, between d_1 and d_2 for the two subtrees rooted at MS1 and MS2, respectively. This can be illustrated electrically by Figure 11.9. Each subtree, MS1 and MS2, has a load capacitance, L_1 and L_2, and an insertion delay, D_1 and D_2. Now, suppose we want to solve for d_1 and call it, x. d_2 can be defined as the remainder of the distance, $d_2 = D − x$. Given a unit wire resistance, r, and capacitance, c, the resistance and capacitance of each wire segment are

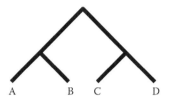

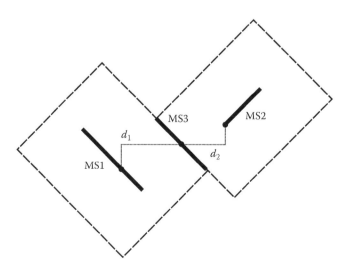

FIGURE 11.7   The abstract syntax tree determines the topology of the clock tree.

FIGURE 11.8   The intersection of two diamonds with radii d_1 and d_2 forms a Manhattan arc with slope ±1.

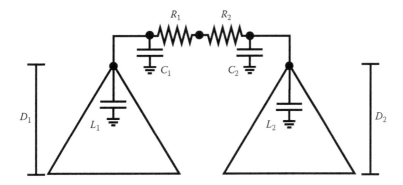

FIGURE 11.9 The merging distance is computed to equalize the Elmore delay of the current level plus any delay in a subtree.

$$
(11.4) \qquad
\begin{aligned}
R_1 &= rd_1 = rx & C_1 &= cd_1 = cx \\
R_2 &= rd_2 = r(D-x) & C_2 &= cd_2 = c(D-x),
\end{aligned}
$$

respectively. To find the allocation of the minimum distance, we can now equate the Elmore delay of the two trees to be merged with the new wire segment to connect them to get

$$
(11.5) \qquad D_1 + R_1\left(\frac{C_1}{2} + L_1\right) = D_2 + R_2\left(\frac{C_2}{2} + L_2\right).
$$

By substituting Equation 11.5 and solving the single quadratic equation for x, we get

$$
(11.6) \qquad x = \frac{D_2 - D_1 + rD\left(L_2 + \dfrac{cD}{2}\right)}{r\left(L_1 + L_2 + cD\right)}.
$$

If $x < 0$ or $x > D$, this means that the two delays cannot be equalized given the minimum separation distance. In this case, the MS with the larger insertion delay is selected as the merging location and extra wire is inserted to equalize the delay.

Figure 11.10 shows an example with four sinks (A–D) during the bottom-up (Figure 11.10a) and top-down (Figure 11.10b) phases of zero-skew DME using the abstract topology of Figure 11.7. First, sinks A and B are merged into MS1 and sinks C and D are merged into MS2. Then MS1 and MS2 are merged into MS3. After the bottom-up phase is completed, the root of the topology tree consists of a single MS, MS3. The root of the clock tree can be placed

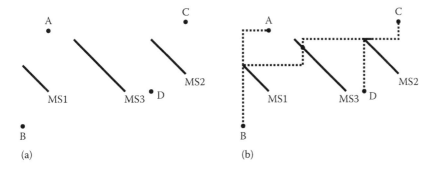

FIGURE 11.10 DME calculates all feasible merging points bottom-up while deferring the exact wire locations until the top-down embedding phase. (a) DME bottom-up merging determines the locus of all points for each node in the abstract topology tree that provides a feasible zero-skew routing. (b) DME top-down embedding greedily connects each MS to the next lower segments using minimum wire length.

anywhere on MS3, or the point on MS3 closest to the desired root location. Then, in the top-down phase, the location on the child MS that is closest to the parent location is selected. The actual wire route is embedded, and the procedure is repeated until the sinks are reached. The resulting tree has provably minimal wire length, given the fixed topology, and zero skew in nominal process conditions. The bottom-up and top-down DME algorithms are shown in Algorithms 11.1 and 11.2, respectively.

Algorithm 11.1 Bottom-Up DME Merge Algorithm
Require: Topology $T_t(S)$; set of sink locations (loc) S
Ensure: Merge segments $ms(v)$ for each node v in topology $T_t(S)$

```
 1: for each node v in T_t(S) (bottom-up order) do
 2:    if v is a sink node then
 3:       ms(v) ←{loc (v)}
 4:    else
 5:       Let l and r be the children of v
 6:       D ←Distance (ms(l),ms(r))
 7:       (d_l,d_r) = CalculateMergeLengths (l,r,D)
 8:       mr(l) ←CreateMergeRegion (ms(l),d_l)
 9:       mr(r) ←CreateMergeRegion (ms(r),d_r)
10:       ms(v) ← mr(l) ∩ mr(r)
11:    end if
12: end for
```

Algorithm 11.2 Top-Down DME Embedding Algorithm
Require: Merge segments $ms(v)$ for each node v in topology $T_t(S)$
Ensure: Zero-skew routed clock tree $T_r(S)$

```
 1: for each node v in T_t(S) (top-down order) do
 2:    if v is the root then
 3:       Choose any q ∈ ms(v)
 4:       loc (v) ← q
 5:    else
 6:       Let p be parent of v
 7:       d_p = Distance (p,ms(v))
 8:       mr(p) ←CreateMergeRegion (p,d_p)
 9:       Choose any q ∈ mr(p) ∩ ms(v)
10:       loc (v) ← q
11:    end if
12: end for
```

DME provides an optimal wire length solution with exactly zero skew, given a fixed abstract topology. It is able to do this with a worst-case complexity of $O(n)$ for n clock sinks, but this is because the input topology is fixed. DME algorithms can be combined with greedy clustering with local improvements [28], partitioning [50], or matching [51] algorithms to generate the topology on the fly during the bottom-up portion of DME. Recently, some improvements were made in iterative geometric matching using a dual-MST algorithm [66].

11.3.2 BUFFERING

Nearly, all non-clock, and some clock, buffering techniques are based on the seminal dynamic programming buffer insertion algorithm of van Ginneken, which produces optimal minimum delay buffered nets under the Elmore delay model [103]. The algorithm does this by traversing a fixed net routing in a bottom-up manner. The sinks of the buffered tree consist of a single solution. At each node during the traversal, the algorithm can merge any combination of the optimal minimum delay subtree bufferings and can add buffers to either one, both, or none of these subtree solutions. The key to the algorithm is that only a linear number of solutions are kept, and these solutions form the boundary of the delay–capacitance curve.

Unlike van Ginneken's algorithm that focuses on delay optimization, the primary goal of clock tree buffer insertion is to preserve signal quality at the clock tree sinks. In the past, simple clock trees used a monolithic multi-stage driver to drive the entire clock tree. This approach however is inadequate, as integration levels rise and process dimensions scale. Delay of long wires can grow quickly if unbuffered, and more importantly, slew rates degrade rapidly. Clock buffering has many challenges that are different from signal buffering. The most obvious is that skew, rather than delay, is the primary criterion. But power consumption and sink slew rates are also important. Clock networks are typically much larger than signal nets, and so, the runtime efficiency is extremely important and multi-objective buffering methods [43,58,81] used in signal wires are not practical.

The fundamental dynamic programming paradigm used in signal buffering [103] is inadequate for clock trees. By definition, a dynamic programming algorithm has three properties:

1. *Overlapping subproblems:* The problem must be broken down into subproblems that can be reused several times.
2. *Optimal substructure:* An optimal solution must consist of optimal subsolutions.
3. *Memoization:* Given the same arguments, the function will always return the same result.

For clock trees, the optimal substructure property does not hold, since an optimal solution does not necessarily have optimal subsolutions. Given two subtrees with unequal skews, for example, the subtree with the lesser skew wastes power and is not needed to have minimum skew. Despite this, at least one researcher has proposed to use a similar paradigm by enumerating large numbers of subtree candidates with differing delay and power criteria and picking the best solution [100]. The algorithm is not able to guarantee that a solution is feasible and requires significant runtime even for small benchmarks.

In contrast, many heuristic buffer methods have been proposed and are more commonly used in combination with a later tuning stage. Zeng et al. presented an algorithm that inserts a fixed number of buffers on each clock tree path and then greedily moves and combines buffers to improve the overall skew of the clock tree [114]. Xi and Dai presented a method to perform iso-radius buffering [110]. In this approach, the radius is the wire distance of a buffer from the root, and additional buffer levels are added until a skew constraint is satisfied. Both of these methods create solutions with the same number of buffers on each root to sink path, but unequal buffer loading can lead to skew problems.

Other algorithms have tried to use the number of buffers on a root to sink path as a heuristic for balancing clock tree skew, while optimizing other criteria such as slew rates and power. The *buffer skew* is the maximum difference in the number of buffers on any clock tree branches but does not directly translate to delay skew. Tellez and Sarrafzadeh proposed a greedy algorithm for zero buffer skew trees with the optimal minimum buffer solution [94]. To control slew rates, a limit on the capacitance of all buffers is used to guide the insertion process. The greedy algorithm, as with other approaches, works in a bottom-up manner and uses two simple rules, one on edges and one at nodes:

1. *Edge rule:* If the capacitance at the source of the edge is greater than the limit, split the edge and insert a node and buffer at the point on the edge nearest the root such that the downstream capacitance does not violate the limit.
2. *Node rule:* Add the necessary buffers to each subtree to equalize the number of buffer levels in each subtree. If the capacitance at the node after these buffers are inserted is greater than the limit, drive each subtree with an additional buffer.

Tellez and Sarrafzadeh go on to prove that the greedy algorithm inserts the minimum number of buffers such that the capacitance limit is not violated anywhere in the tree. They also propose a heuristic that allows an unequal but limited amount of buffer skew. The greedy bottom-up algorithm, however, was proven to be suboptimal with nonzero buffer skew bounds [3]. Albrecht et al. provided a dynamic programming algorithm that determines the optimal buffering given the nonzero buffer skew bound [3]. A similar algorithm to the previous two was proposed to reduce slew violations in large fanout, non-clock nets [4].

11.3.3 BUFFER AND WIRE SIZING

Since many of the computationally reasonable buffer insertion algorithms are heuristic, the final trees generally do not have minimum skew and/or power. Many researchers have proposed algorithms to minimize clock skew using buffer and/or wire sizing. These algorithms are often referred to as clock tree tuning, since the topology, wire locations, and buffer locations are usually fixed. Adjustments to the buffer sizes and wire widths enable fine skew tuning, yet many of the algorithms use simplified Elmore delay models initially to find a better global solution. Specifically, Chen et al. [17] proposed a dynamic programming algorithm, Kay and Pileggi [53] proposed convex optimization to obtain a bounded solution, and Saaied et al. [83] and Pullela et al. [77] proposed greedy local optimizations.

Most of the previous algorithms use simplified delay models for both interconnect and buffers due to runtime overhead. However, input slew rates can affect buffer delay, and resistive shielding is not present in an Elmore interconnect model. Other researchers have focused on using Spice-accurate models for final buffer and wire tuning to further minimize skew and power. Wang and Marek-Sadowska first proposed to use sequential linear programming (SLP) to optimize buffer sizes in a clock tree for minimum power given general skew constraints [107]. Guthaus et al. proposed a sequential quadratic programming technique for clock skew that uses the sum of local skews to perform a better global search [36]. This formulation includes variability through statistical static timing analysis (SSTA) that considers the impact of both buffer and wire variability on clock skew variability with both WID and D2D variation. Most recently, Rakai et al. formulated this as an iterative geometric programming problem [80] with improved results.

11.3.4 COMBINED APPROACHES

A significant amount of work has tried to combine the previous clock tree synthesis steps into a single monolithic optimization to obtain an "optimal" clock tree. DME has been combined with buffer insertion and wire sizing [60], buffer insertion/sizing and wire sizing have been combined [100], buffer sizing has been combined with useful skew budgeting [107], clock root gating has been combined with buffer insertion [30], etc. Shih et al. proposed a three-stage clock tree synthesis algorithm using topology generation, tapping-point determination, and routing [88]. A Dynamic Nearest-Neighbor Algorithm (DNNA) generates the tree topology by combining a dynamic number of pairs using a multi-objective cost. Interestingly, it does not consider skew or insertion delay, which is only considered in the next state during tapping-point determination that is based on the traditional DME algorithm, with a capacitance bound for slew constraints. Finally, the Walk-Segment Breadth First Search (WSBFS) routes the wires between buffers and tapping points. Clock optimization, however, often has more constraints than the primary skew objective. These can include insertion delay, slew rates, noise sensitivity, blockage avoidance, peak power minimization, predefined macro subtrees, etc. Each issue further complicates optimization. Some researchers have proposed frameworks that consist of small steps that are more focused and can handle the nuances of all the different constraints [97].

11.3.5 VARIATION-AWARE LOCAL CLOCKS

While global distributions primarily address variation through redundancy, a significant amount of work has been done on the analysis of local clock tree skew sensitivity to both WID and D2D variation [2,62,113]. However, most local clock optimization techniques have either not considered variation or have focused on worst-case bounded clock skew [23,56,64,99], which can be too conservative for meaningful results when all sources of variation are considered. Timing corner-based analysis can be potentially risky by missing the worst-case corner entirely [106]. At least two works have indirectly improved robustness by centering useful skew in setup/hold timing windows during clock skew scheduling [54,73], but these ignore interconnect variations. Guthaus et al. [37] were the first to consider wire embedding to adjust skew variation sensitivity to both D2D and WID sources. Since then, the ISPD 2009 clock synthesis contest considered global skew with only D2D variation, using a clock latency range (CLR) metric and two supply voltage corners [92]. Several researchers addressed CLR with multiple supply voltages [18,57,61,87]. In 2010, a continuation of the contest used Monte Carlo simulation of local skew, based not on logical path connectivity but sink proximity [91]. The results were evaluated with both supply and wire variation, but this did not consider physical models of variation as discussed in prior surveys on high-performance clock synthesis [40].

11.4 LOW-POWER TECHNIQUES

The majority of the clock power is dynamic power, but short-circuit power is often not negligible. Short-circuit currents are not very significant if signal slew rates are appropriately bounded, except when redundant cross-links or grids are used to compensate for variation. Most power reduction techniques focus on dynamic power of local clock distributions, since they are the majority of the power consumption. Dynamic switching power is

$$(11.7) \qquad P_{dynamic} \approx \alpha C V_{dd}^2 f,$$

where

    α is the switching rate
    C is the total capacitance
    V_{dd} is the supply voltage
    f is the frequency

11.4.1 ELECTRICAL AND PHYSICAL OPTIMIZATION

The clock tree topology, routing, buffering, and tuning algorithms all minimize the total wire length and total buffer size in order to reduce the total capacitance. This capacitance reduction directly reduces dynamic power consumption but leaves the other variables such as V_{dd}, f, and α in Equation 11.7 untouched.

Another potential lever to reduce power consumption of clocks is with dynamic voltage and frequency scaling (DVFS). Dynamic power is proportional to fV_{dd}^2, which provides a cubic power reduction. Since a decreased supply voltage means slower circuits, both V_{dd} and f can be reduced simultaneously to adapt to a dynamic workload requirement [8,74]. This allows reduction of both the clock network power and the logic circuit itself. Some work has addressed clock skew with multiple supply voltages [18,57,61,87]. Specifically, the clock latency range (CLR) proposed in the 2009 ISPD clock synthesis contest [92] measured clock skew over multiple supply voltage corners.

Other approaches have proposed to use an ultra-low supply voltage of the clock swing alone [115]. These low-voltage swing clocks are often differentiated to improve robustness to supply variation and noise [86]. Another work examined the extreme case of clock skew in subthreshold circuits that are very sensitive to threshold voltage variations, and not significantly affected by interconnect resistances [96]. Most recently, a new current-mode-sensing flip-flop promises to sense very small current fluctuations in interconnect rather than full voltage swings [49].

Short-circuit power in clock grids can be very significant, since multiple buffers simultaneously drive the grid to reduce variability. Wilke et al. proposed a new grid buffer design that uses a small high-impedance state to reduce short-circuit power [108]. An iterative sensitivity-based grid buffer removal approach has been shown to remove ineffective buffers that consumed a significant amount of this short-circuit power [39,41].

11.4.2 CLOCK GATING

The most frequently used industry technique to reduce power is clock gating. This is available in nearly every commercial tool. Clock gating reduces the switching activity α in Equation 11.7. High-performance designs typically have multiple levels of clock gating, including architectural root gating for entire modules, and local clock gating. Local clock gating uses the enable information on sequential elements to logically cluster them so that portions of a clock tree can be disabled with a clock gater when not in use. *A clock gater* is a logic gate with an enable signal so that clocks may only be enabled when the underlying logic is active. Many techniques have been proposed for local gating [16,30,75], but only root gating is feasible on clock grids or when cross-links are present. Farrahi et al. provided both an exact algorithm based on dynamic programming and an approximation algorithm based on recursive geometric matching [30]. Oh and Pedram extended a similar concept to include the power consumption of the control logic needed to perform the clock root gating [75]. All gating algorithms must consider the extra overhead from the control logic and clock gater along with any wire length increase to adjust the topology for improved activity reduction. Placing gaters higher in a tree allows a more efficient use of their overhead but may decrease the portion of time that gating can be done. It is also important to consider variability in clock gating, as clock gaters can introduce both nominal skew and skew variation [10,11,14]. Lu et al. were the first to consider simultaneous buffering and clock gater insertion in clock trees while consider clock slew constraints [65]. Lu et al. were also the first to consider register placement and clustering in a clock mesh with local clock gaters [69]. While this is commonly used in industry using custom scripts and methodologies, no publications have previously discussed such a methodology in practice.

11.4.3 RESONANT CLOCKING

One additional method to reduce dynamic power is to use resonance in the clock network. Resonance is the property that the imaginary parts of impedance or admittance cancel out at a particular frequency known as the resonant frequency. This has been used extensively in the analog/RF fields to build high-quality passive filter networks, but there are three distinctly different approaches to utilize this for low-power "resonant" clocks. These include standing wave (salphasic) [19,76], traveling wave (rotary) [109], and LC tank resonant [13,22,26,84,116] clock distributions. Standing and traveling wave clocks assume that on-chip inductance forms transmission lines, whereas the LC tank resonant clocks assume lumped inductors. LC tank resonant clocks can further be distinguished as monolithic or distributed. Monolithic LC tank clock distributions use a single tank circuit, whereas distributed LC tank clock addresses the distributed parasitic interconnect in large modern chips.

A standing wave is a non-traveling vibration formed by the interference of two harmonic waves of the same frequency and amplitude. In standing wave clocks, the phase of all points is equal, but the amplitude varies with position. The loss of the transmission lines attenuates the amplitude of the waves and hence introduces a residual traveling wave that leads to some clock skew. This was initially used for board-level clock distribution [19] but has also been demonstrated for on-chip clock distribution [76]. Transconductors can be used to address the varying amplitude and transmission line loss [76].

A rotary clock is a closed parallel loop formed by an inner and outer transmission line [109]. Back-to-back inverters are connected to this parallel loop to compensate for the energy loss. Unlike standing wave clocks, the amplitude of all positions in a rotary clock loop is the same while the phase varies with position. The differences between sink phases bring about extra work in clock timing and synchronization, since clock sinks can be attached to different positions on the loop. Automated methods for routing resonant clocks have addressed many of these issues [93,67] including zero clock skew operation [42], and prototype designs have been demonstrated [112]. At least one company has a commercial product to support rotary clocking [44].

The LC tank resonant clocks are similar to conventional clocks by providing a clock signal with constant phase and magnitude; thus, it is easier to integrate in a traditional IC flow than standing and traveling wave clocks. However, the LC tank resonant clocks require additional passive components that resonate at a particular frequency to reduce power consumption. LC tank resonant clocks modify the clock impedance by inserting inductor–capacitor (LC) "tank" circuits to cancel the aforementioned reactive portion of the clock impedance. In a parallel LC circuit, this ideally leaves infinite electrical impedance at the resonant frequency, while in a series LC tank this ideally leaves zero electrical impedance. Increasing the clock network impedance allows clock driver transistor sizes to be dramatically reduced, which saves both power and silicon area. In addition, the oscillatory behavior of an LC tank circuit recovers some of the energy as it is transferred back and forth between the clock/decoupling capacitance and the inductor. Early adoption of these techniques resulted in monolithic LC tank clocks [22,116,27], but this does not consider parasitic resistance in clock distribution networks that decrease the overall efficiency. Distributed LC tank resonant clocks have been shown to save 30%–80% of dynamic power compared to buffered clocks [13,47,84,85]. Several methodologies have been proposed to synthesize global H-tree resonant clocks [82], asymmetric global resonant clock trees [35], hierarchical local resonant clocks [24], and resonant clock grids, using custom inductor sizing [46,47] and inductor-library-based sizing [45]. At least one company has a product to support LC tank resonant clocking [26] in commercial designs [84,85]. Recently, monolithic LC tanks have been simulated using additional harmonics to improve the sinusoidal slew rates using higher-order harmonics [90].

11.5 CONCLUSIONS

Clock distribution is and will likely remain one of the most controversial steps in integrated circuit design. Both integrated circuit and EDA companies have proprietary methodologies with different preferences of evaluation metrics and how to achieve these goals. This can result in vastly different clock architectures. In addition, clock synthesis is difficult to research, since ignoring any single metric can result in unfair comparisons. In this chapter, we explained the importance of each metric and surveyed how many of these metrics have been addressed in both global and local clock synthesis. In particular, the importance of clock robustness to a multitude of sources of variability has been a primary theme. However, this robustness is often at the expense of power consumption, which is also summarized. There are still many open problems for future research not only in clock synthesis methodologies but also by incorporating new low-power clocking circuits and technologies.

REFERENCES

1. A. Abdelhadi, R. Ginosar, A. Kolodny, and E. G. Friedman. Timing-driven variation-aware non-uniform clock mesh synthesis. In *ACM Great Lakes Symposium on VLSI (GLSVLSI)*, Providence, RI, 2010, pp. 15–20.
2. A. Agarwal, D. Blaauw, and V. Zolotov. Statistical clock skew analysis considering intra-die process variations. In *IEEE/ACM International Conference on Computer-Aided Design (ICCAD)*, San Jose, CA, 2003, pp. 914–920.
3. C. Albrecht, A. B. Kahng, B. Liu, I. Mandoiu, and A. Zelikovsky. On the skew-bounded minimum buffer routing tree problem. *IEEE Transactions on Computer-Aided Design of Integrated Circuits and Systems (TCAD)*, 22(7), 2003, pp. 937–945.
4. C. Alpert, A. B. Kahng, B. Liu, I. Mandoiu, and A. Zelikovsky. Minimum-buffered routing of non-critical nets for slew rate and reliability control. In *IEEE/ACM International Conference on Computer-Aided Design (ICCAD)*, San Jose, CA, 2001, pp. 408–415.
5. C. J. Anderson, J. Petrovick et al. Physical design of a fourth-generation POWER GHz microprocessor. In *International Solid-State Circuits Conference (ISSCC)*, San Francisco, CA, 2001, pp. 232–233.
6. H. Bakoglu, J. T. Walker, and J. D. Meindl. A symmetric clock-distribution tree and optimized high-speed interconnections for reduced clock skew in ULSI and WSI circuits. In *International Conference on Computer Design (ICCD)*, 1986, pp. 118–122.
7. K. Boese and A. Kahng. Zero-skew clock routing trees with minimum wirelength. In *International ASIC Conference and Exhibit*, 1992, pp. 17–21.
8. T. Burd and R. Broderson. Design issues for dynamic voltage scaling. In *IEEE International Symposium on Low Power Electronics and Design (ISLPED)*, Rapallo, Italy, 2000, pp. 9–14.

9. A. Chakraborty, K. Duraisami, A. Sathanur, P. Sithambaram, L. Benini, A. Macii, E. Macii, and M. Poncino. Dynamic thermal clock skew compensation using tunable delay buffers. In *IEEE International Symposium on Low Power Electronics and Design (ISLPED)*, Tegernsee, Germany, 2006, pp. 162–167.

10. A. Chakraborty, G. Ganesan, A. Rajaram, and D. Z. Pan. Analysis and optimization of NBTI induced clock skew in gated clock trees. In *IEEE Design, Automation and Test in Europe (DATE)*, Nice, France, 2009, pp. 296–299.

11. A. Chakraborty and D. Z. Pan. Skew management of NBTI impacted gated clock trees. In *IEEE International Symposium on Physical Design (ISPD)*, 2010, pp. 127–133.

12. A. Chakraborty, P. Sithambaram, K. Duraisami, A. Macii, E. Macii, and M. Poncino. Thermal resilient bounded-skew clock tree optimization methodology. In *IEEE Design, Automation and Test in Europe (DATE)*, Munich, Germany, 2006, pp. 832–837.

13. S. Chan, P. Restle, K. Shepard, N. James, and R. Franch. A 4.6 GHz resonant global clock distribution network. In *IEEE International Solid-State Circuits Conference (ISSCC)*, San Francisco, CA, 2004, pp. 342–343.

14. C.-M. Chang, S.-H. Huang, Y.-K. Ho, J.-Z. Lin, H.-P. Wang, and Y.-S. Lu. Type-matching clock tree for zero skew clock gating. In *ACM/IEEE Design Automation Conference (DAC)*, Anaheim, CA, 2008, pp. 714–719.

15. T.-H. Chao, Y.-C. Hsu, and J. M. Ho. Zero skew clock net routing. In *ACM/IEEE Design Automation Conference (DAC)*, Anaheim, CA, 1992, pp. 518–523.

16. W.-C. Chao and W.-K. Mak. Low-power gated and buffered clock network construction. *ACM Transactions on Design Automation of Electronic Systems (TODAES)*, 13(1):1–20, 2008.

17. C.-P. Chen, Y.-P. Chen, and D. F. Wong. Optimal wire-sizing formula under the Elmore delay model. In *ACM/IEEE Design Automation Conference (DAC)*, Las Vegas, NV, 1996, pp. 487–490.

18. Y.-Y. Chen, C. Dong, and D. Chen. Clock tree synthesis under aggressive buffer insertion. In *ACM/IEEE Design Automation Conference (DAC)*, Anaheim, CA, 2010, pp. 86–89.

19. V. Chi. Salphasic distribution of clock signals for synchronous systems. *IEEE Transactions on Computers*, 43(5):597–602, May 1994.

20. M. Cho, S. Ahmedtt, and D. Z. Pan. TACO: Temperature aware clock-tree optimization. In *IEEE/ACM International Conference on Computer-Aided Design (ICCAD)*, San Jose, CA, 2005, pp. 582–587.

21. M. Cho, D. Z. Pan, and R. Puri. Novel binary linear programming for high performance clock mesh synthesis. In *IEEE/ACM International Conference on Computer-Aided Design (ICCAD)*, San Jose, CA, 2010, pp. 438–443.

22. J.-Y. Chueh, M Papaefthymiou, and C Ziesler. Two-phase resonant clock distribution. In *IEEE International Symposium on Very Large Scale Integration (ISVLSI)*, Tampa, FL, May 2005, pp. 65–70.

23. J. Chung and C.-K. Cheng. Skew sensitivity minimization of buffered clock tree. In *IEEE/ACM International Conference on Computer-Aided Design (ICCAD)*, San Jose, CA, 1994, pp. 280–283.

24. W. Condley, X. Hu, and M. R. Guthaus. A methodology for local resonant clock synthesis using lc-assisted local clock buffers. In *IEEE/ACM International Conference on Computer-Aided Design (ICCAD)*, San Jose, CA, 2011, pp. 503–506.

25. J. Cong and C.-K. Koh. Minimum-cost bounded-skew clock routing. In *IEEE International Symposium on Circuits and Systems (ISCAS)*, Seattle, WA, 1995, pp. 215–218.

26. Cyclos Semiconductor. Cyclify: Resonant clock ASIC flow. http://www.cyclos-semi.com/.

27. A. Drake, K. Nowka, T. Nguyen, J. Burns, and R. Brown. Resonant clocking using distributed parasitic capacitance. *IEEE Journal of Solid-State Circuits (JSSC)*, 39(9):1520–1528, 2004.

28. M. Edahiro. A clustering-based optimization algorithm in zero-skew routings. In *ACM/IEEE Design Automation Conference (DAC)*, Dallas, TX, 1993, pp. 612–616.

29. W. C. Elmore. The transient response of damped linear networks. *Journal of Applied Physics*, 19:55–63, January 1948.

30. A. H. Farrahi, C. Chen, A. Srivastava, G. Tellez, and M. Sarrafzadeh. Activity-driven clock design. *IEEE Transactions on Computer-Aided Design of Integrated Circuits and Systems (TCAD)*, 20(6):705–714, 2001.

31. J. P. Fishburn. Clock skew optimization. *IEEE Transactions on Computers*, 39(7):945–951, 1990.

32. G. Flach, G. Wilke, M. Johann, and R. Reis. A mesh-buffer displacement optimization strategy. In *IEEE International Symposium on Very Large Scale Integration (ISVLSI)*, Lixouri, Greece, 2010.

33. E. G. Friedman. Clock distribution networks in synchronous digital integrated circuits. In *Proceedings of the IEEE*, 2001, pp. 665–692.

34. G. Geannopoulos and X. Dai. An adaptive digital deskewing circuit for distribution networks. *IEEE International Solid-State Circuits Conference (ISSCC)*, San Francisco, CA, 1998.

35. M. R. Guthaus. Distributed LC resonant clock tree synthesis. In *IEEE International Symposium on Circuits and Systems (ISCAS)*, Rio, Brazil, 2011, pp. 1215–1218.

36. M. R. Guthaus, D. Sylvester, and R. B. Brown. Clock buffer and wire sizing using sequential quadratic programming. In *ACM/IEEE Design Automation Conference (DAC)*, San Francisco, CA, 2006, pp. 1041–1046.

37. M. R. Guthaus, D. Sylvester, and R. B. Brown. Process-induced skew reduction in deterministic zero-skew clock trees. In *IEEE Asia and South Pacific Design Automation Conference (ASP-DAC)*, Yokohama, Japan, 2006, pp. 84–89.

38. M. R. Guthaus, D. Sylvester, and R. B. Brown. Clock tree synthesis with data-path sensitivity matching. In *IEEE Asia and South Pacific Design Automation Conference (ASP-DAC)*, Seoul, Korea, 2008, pp. 498–503.

39. M. R. Guthaus, G. Wilke, and R. Reis. Non-uniform clock mesh optimization with linear programming buffer insertion. In *ACM/IEEE Design Automation Conference (DAC)*, Anaheim, CA, 2010, pp. 74–79.

40. M. R. Guthaus, G. Wilke, and R. Reis. Revisiting automated physical synthesis of high-performance clock networks. *ACM Transactions on Design Automation of Electronic Systems (TODAES)*, 18(2):31:1–31:27, April 2013.

41. M. R. Guthaus, X. Hu, G. Wilke, G. Flache, and R. Reis. High-performance clock mesh optimization. *ACM Transactions on Design Automation of Electronic Systems (TODAES)*, 18(3):33:1–33:17, 2013.

42. V. Honkote and B. Taskin. Skew-aware capacitive load balancing for low-power zero clock skew rotary oscillatory array. In *International Conference on Computer Design (ICCD)*, Amsterdam, Netherlands, 2010, pp. 209–214.

43. M. Hrkic and J. Lillis. Buffer tree synthesis with consideration of temporal locality, sink polarity requirements, solution cost, congestion, and blockages. *IEEE Transactions on Computer-Aided Design of Integrated Circuits and Systems (TCAD)*, 22(4):481–491, 2003.

44. http://www.multigig.com.

45. X. Hu, W. Condley, and M. R. Guthaus. Library-aware resonant clock synthesis (LARCS). In *ACM/IEEE Design Automation Conference (DAC)*, San Francisco, CA, June 2012, pp. 145–150.

46. X. Hu and M. R. Guthaus. Distributed resonant clock grid synthesis (ROCKS). In *ACM/IEEE Design Automation Conference (DAC)*, San Diego, CA, 2011, pp. 516–521.

47. X. Hu and M. R. Guthaus. Distributed LC resonant clock grid synthesis. *IEEE Transactions on Circuits and Systems I (TCAS-I)*, 2012.

48. D. J.-H. Huang, A. B. Kahng, and C.-W. A. Tsao. On the bounded-skew clock and steiner routing problems. In *ACM/IEEE Design Automation Conference (DAC)*, San Francisco, CA, 1995, pp. 508–513.

49. R. Islam and M. R. Guthaus. Current-mode clock distribution. In *IEEE International Symposium on Circuits and Systems (ISCAS)*, Melbourne, Australia, 2014, pp. 1203–1206.

50. M. A. B. Jackson, A. Srinivasan, and E. S. Kuh. Clock routing for high performance ICs. In *ACM/IEEE Design Automation Conference (DAC)*, Orlando, FL, 1990, pp. 573–579.

51. A. B. Kahng, J. Cong, and G. Robins. High-performance clock routing based on recursive geometric matching. In *ACM/IEEE Design Automation Conference (DAC)*, San Francisco, CA, 1991, pp. 322–327.

52. A. Kapoor, N. Jayakumar, and S. P. Khatri. A novel clock distribution and dynamic de-skewing methodology. In *IEEE/ACM International Conference on Computer-Aided Design (ICCAD)*, San Jose, CA, 2004, pp. 626–631.

53. R. Kay and L. T. Pileggi. EWA: Efficient wiring-sizing algorithm for signal nets and clock nets. *IEEE Transactions on Computer-Aided Design of Integrated Circuits and Systems (TCAD)*, 17(1):40–49, 1998.

54. I. S. Kourtev and E. G. Friedman. A quadratic programming approach to clock skew scheduling for reduced sensitivity to process parameter variations. In *International ASIC/SOC Conference*, Washington, DC, 1999, pp. 210–215.

55. N. A. Kurd, J. S. Barkatullah, R. O. Dizon, T. D. Fletcher, and P. D. Madland. Multi-GHz clocking scheme for Intel Pentium 4 microprocessor. In *IEEE International Solid-State Circuits Conference (ISSCC)*, San Francisco, CA, 2001, pp. 404–405.

56. W.-C. D. Lam and C.-K. Koh. Process variation robust clock tree routing. In *IEEE Asia and South Pacific Design Automation Conference (ASP-DAC)*, Shanghai, China, 2005, pp. 606–611.

57. D. Lee and I. L. Markov. Contango: Integrated optimization of SoC clock networks. In *Design, Automation Test in Europe Conference Exhibition (DATE)*, March 8–12, Dresden, Germany, 2010, pp. 1468–1473.

58. R. Li, D. Zhou, J. Liu, and X. Zeng. Power-optimal simultaneous buffer insertion/sizing and wire sizing. In *IEEE/ACM International Conference on Computer-Aided Design (ICCAD)*, San Jose, CA, 581pp., 2003.

59. D.-J. Lin and I. L. Markov. Multilevel tree fusion for robust clock networks. In *IEEE/ACM International Conference on Computer-Aided Design (ICCAD)*, San Jose, CA, 2011, pp. 632–639.

60. I.-M. Liu, T.-L. Chou, A. Aziz, and D. F. Wong. Zero-skew clock tree construction by simultaneous routing, wire sizing and buffer insertion. In *IEEE International Symposium on Physical Design (ISPD)*, 2000, pp. 33–38.

61. W.-H. Liu, Y.-L. Li, and H.-C. Chen. Minimizing clock latency range in robust clock tree synthesis. In *IEEE Asia and South Pacific Design Automation Conference (ASP-DAC)*, Taipei, Taiwan, 2010, pp. 389–394.

62. Y. Liu, S. R. Nassif, L. T. Pileggi, and A. J. Strojwas. Impact of interconnect variations on the clock skew of a gigahertz microprocessor. In *ACM/IEEE Design Automation Conference (DAC)*, Los Angeles, CA, 2000, pp. 168–171.

63. J. Long, J. C. Ku, S. O. Memik, and Y. Ismail. A self-adjusting clock tree architecture to cope with temperature variations. In *International Conference on Computer-Aided Design (ICCAD)*, San Jose, CA, 2007, pp. 75–82.

64. B. Lu, J. Hu, G. Ellis, and H. Su. Process variation aware clock tree routing. In *IEEE International Symposium on Physical Design (ISPD)*, 2003, pp. 174–181.

65. J. Lu, W.-K. Chow, and C.-W. Sham. Fast power- and slew-aware gated clock tree synthesis. *IEEE Transactions on Very Large Scale Integration (VLSI) Systems*, 20(11):2094–2103, 2012.

66. J. Lu, W.-K. Chow, C.-W. Sham, and E. F.-Y. Young. A dual-MST approach for clock network synthesis. In *IEEE Asia and South Pacific Design Automation Conference (ASP-DAC)*, Taipei, Taiwan, 2010, pp. 467–473.

67. J. Lu, V. Honkote, X. Chen, and B. Taskin. Steiner tree based rotary clock routing with bounded skew and capacitive load balancing. In *IEEE Design, Automation and Test in Europe (DATE)*, Grenoble, France, 2011, pp. 455–460.

68. J. Lu, X. Mao, and B. Taskin. Clock mesh synthesis with gated local trees and activity driven register clustering. In *IEEE/ACM International Conference on Computer-Aided Design (ICCAD)*, San Jose, CA, 2012, pp. 691–697.

69. J. Lu, X. Mao, and B. Taskin. Integrated clock mesh synthesis with incremental register placement. *IEEE Transactions on Computer-Aided Design of Integrated Circuits and Systems (TCAD)*, 31(2):217–227, 2012.

70. J. Minz, X. Zhao, and S. K. Lim. Buffered clock tree synthesis for 3D ICs under thermal variations. In *IEEE Asia and South Pacific Design Automation Conference (ASP-DAC)*, Seoul, Korea, 2008, pp. 504–509.

71. T. Mittal and C.-K. Koh. Cross link insertion for improving tolerance to variations in clock network synthesis. In *IEEE International Symposium on Physical Design (ISPD)*, 2011, pp. 29–36.

72. S. R. Nassif. Modeling and forecasting of manufacturing variations (embedded tutorial). In *IEEE Asia and South Pacific Design Automation Conference (ASP-DAC)*, Yokohama, Japan, 2001, pp. 145–150.

73. J. L. Neves and E. G. Friedman. Optimal clock skew scheduling tolerant to process variations. In *ACM/IEEE Design Automation Conference (DAC)*, Las Vegas, NV, 1996, pp. 623–629.

74. K. J. Nowka et al. A 32-bit powerPC system-on-a-chip with support for dynamic voltage scaling and dynamic frequency scaling. *IEEE Journal of Solid-State Circuits (JSSC)*, 37(11):1441–1447, 2002.

75. J. Oh and M. Pedram. Gated clock routing minimizing the switched capacitance. In *IEEE Design, Automation and Test in Europe (DATE)*, Paris, France, 1998, pp. 692–697.

76. F. O'Mahony, C. Yue, M. Horowitz, and S. Wong. Design of a 10GHz clock distribution network using coupled standing-wave oscillators. In *ACM/IEEE Design Automation Conference (DAC)*, Anaheim, CA, June 2003, pp. 682–687.

77. S. Pullela, N. Menezes, and L. T. Pillage. Reliable non-zero skew clock trees using wire width optimization. In *ACM/IEEE Design Automation Conference (DAC)*, Dallas, TX, 1993, pp. 165–170.

78. A. Rajaram, J. Hu, and R. Mahapatra. Reducing clock skew variability via cross links. In *ACM/IEEE Design Automation Conference (DAC)*, San Diego, CA, 2004, pp. 18–23.

79. A. Rajaram and D. Z. Pan. Meshworks: An efficient framework for planning, synthesis and optimization of clock mesh networks. In *IEEE Asia and South Pacific Design Automation Conference (ASP-DAC)*, Seoul, Korea, 2008, pp. 250–257.

80. L. Rakai, A. Farshidi, L. Behjat, and D. Westwick. Buffer sizing for clock networks using robust geometric programming considering variations in buffer sizes. In *IEEE International Symposium on Physical Design (ISPD)*, 2013, pp. 154–161.

81. R. R. Rao, D. Blaauw, D. Sylvester, C. J. Alpert, and S. Nassif. An efficient surface-based low-power buffer insertion algorithm. In *IEEE International Symposium on Physical Design (ISPD)*, 2005, pp. 86–93.

82. J. Rosenfeld and E. Friedman. Design methodology for global resonant H-tree clock distribution networks. *IEEE Transactions on Very Large Scale Integration (VLSI) Systems*, 15(2):135–148, February 2007.

83. H. Saaied, D. Al-Khalili, A. J. Al-halili, and M. Nekili. Simultaneous adaptive wire adjustment and local topology modification for tuning a bounded-skew clock tree. *IEEE Transactions on Computer-Aided Design of Integrated Circuits and Systems (TCAD)*, 24(10):1637–1643, 2005.

84. V. Sathe, S. Arekapudi, C. Ouyang, M. Papaefthymiou, A. Ishii, and S. Naffziger. Resonant clock design for a power-efficient high-volume x86–64 microprocessors. In *IEEE International Solid-State Circuits Conference (ISSCC)*, San Francisco, CA, February 2012, pp. 68–70.

85. V. S. Sathe, S. Arekapudi, A. Ishii, C. Ouyang, M. C. Papaefthymiou, and S. Naffziger. Resonant-clock design for a power-efficient, high-volume x86–64 microprocessor. *IEEE Journal of Solid-State Circuits (JSSC)*, 48(1):140–149, 2013.

86. D. C. Sekar. Clock trees: Differential or single ended? In *International Symposium on Quality Electronic Design (ISQED)*, Santa Clara, CA, 2005, pp. 548–553.

87. X.-W. Shih and Y.-W. Chang. Fast timing-model independent buffered clock-tree synthesis. In *ACM/IEEE Design Automation Conference (DAC)*, Anaheim, CA, 2010, pp. 80–85.

88. X.-W. Shih, C.-C. Cheng, Y.-K. Ho, and Y.-W. Chang. Blockage-avoiding buffered clock-tree synthesis for clock latency-range and skew minimization. In *IEEE Asia and South Pacific Design Automation Conference (ASP-DAC)*, Taipei, Taiwan, 2010.

89. H. Skinner, X. Hu, and M. R. Guthaus. Harmonic resonant clocking. In *IFIP/IEEE International Conference on Very Large Scale Integration (VLSI-SoC)*, Santa Cruz, CA, 2012, pp. 59–64.

90. C. N. Sze. ISPD 2010 high performance clock network synthesis contest. In *IEEE International Symposium on Physical Design (ISPD)*, 2010.

91. C. N. Sze, P. Restle, G.-J. Nam, and C. J. Alpert. Clocking and the ISPD'09 clock synthesis contest. In *IEEE International Symposium on Physical Design (ISPD)*, 2009, pp. 149–150.

92. B. Taskin, J. Demaio, O. Farell, M. Hazeltine, and R. Ketner. Custom topology rotary clock router with tree subnetworks. *ACM Transactions on Design Automation of Electronic Systems (TODAES)*, 14(3):1–14, 2009.

93. G. E. Tellez and M. Sarrafzadeh. Minimal buffer insertion in clock trees with skew and slew rate constraints. *IEEE Transactions on Computer-Aided Design of Integrated Circuits and Systems (TCAD)*, 16(4):333–342, 1997.

94. M. G.-R. Thomson, P. J. Restle, and N. K. James. A 5GHz duty-cycle correcting clock distribution network for the POWER6 microprocessor. In *IEEE International Solid-State Circuits Conference (ISSCC)*, San Francisco, CA, 2006, pp. 1522–1529.

95. J. Tolbert, X. Zhao, S.-K. Lim, and S Mukhopadhyay. Slew-aware clock tree design for reliable sub-threshold circuits. *IEEE International Symposium on Low Power Electronics and Design (ISLPED)*, San Francisco, CA, 2009, pp. 15–20.

96. L. Trevillyan, D. Kung, R. Puri, L. N. Reddy, and M. A. Kazda. An integrated environment for technology closure of deep-submicron IC designs. *IEEE Design and Test of Computers*, 21(1):14–22, 2004.

97. J.-L. Tsai, D.-H. Baik, C. C. P. Chen, and K. K. Saluja. A yield improvement methodology using pre- and post-silicon statistical clock scheduling. In *IEEE/ACM International Conference on Computer-Aided Design (ICCAD)*, San Jose, CA, 2004, pp. 611–618.

98. J.-L. Tsai and C. C.-P. Chen. Process variation robust and low-power zero-skew buffered clock-tree synthesis using projected scan-line sampling. In *IEEE Asia and South Pacific Design Automation Conference (ASP-DAC)*, Shanghai, China, 2005, pp. 1168–1171.

99. J.-L. Tsai, T.-H. Chen, and C. C. P. Chen. Zero skew clock-tree optimization with buffer insertion/sizing and wire sizing. *IEEE Transactions on Computer-Aided Design of Integrated Circuits and Systems (TCAD)*, 23(4):565–573, 2004.

100. J.-L. Tsai, L. Zhang, and C. C.-P. Chen. Statistical timing analysis driven post-silicon tunable clock-tree synthesis. In *IEEE/ACM International Conference on Computer-Aided Design (ICCAD)*, San Jose, CA, 2005, pp. 575–581.

101. R.-S. Tsay. Exact zero skew. In *IEEE/ACM International Conference on Computer-Aided Design (ICCAD)*, San Jose, CA, 1991, pp. 336–339.

102. L. P. P. P. van Ginneken. Buffer placement in distributed RC-tree networks for minimal Elmore delay. In *IEEE International Symposium on Circuits and Systems (ISCAS)*, New Orleans, LA, 1990, pp. 865–868.

103. G. Venkataraman, N. Jayakumar, J. Hu, P. Li, S. Khatri, A. Rajaram, P. McGuinness, and C. J. Alpert. Practical techniques to reduce skew and its variations in buffered clock networks. In *IEEE/ACM International Conference on Computer-Aided Design (ICCAD)*, San Jose, CA, 2005, pp. 592–596.

104. G. Venkataraman, F. Zhuo, J. Hu, and P. Li. Combinatorial algorithms for fast clock mesh optimization. In *IEEE/ACM International Conference on Computer-Aided Design (ICCAD)*, San Jose, CA, 2006, pp. 563–567.

105. C. Visweswariah. Death, taxes and failing chips. In *ACM/IEEE Design Automation Conference (DAC)*, San Francisco, CA, 2003, pp. 343–347.

106. K. Wang and M. Marek-Sadowska. Buffer sizing for clock power minimization subject to general skew constraints. In *ACM/IEEE Design Automation Conference (DAC)*, San Diego, CA, 2004, pp. 159–164.

107. G. Wilke, R. Fonseca, C. Mezzomo, and R. Reis. A novel scheme to reduce short-circuit power in mesh-based clock architectures. In *Symposium on Integrated Circuits and System Design (SBCCI)*, Gramado, Brazil, 2008, pp. 117–122.

108. J. Wood, T. C. Edwards, and S. Lipa. Rotary traveling-wave oscillator arrays: A new clock technology. *IEEE Journal of Solid-State Circuits (JSSC)*, 36(11):1654–1664, 2001.

109. J. G. Xi and W. W.-M. Dai. Buffer insertion and sizing under process variation for low power clock distribution. In *ACM/IEEE Design Automation Conference (DAC)*, San Francisco, CA, 1995, pp. 491–496.

110. L. Xiao, Z. Xiao, Z. Qian, Y. Jiang, T. Huang, H. Tian, and E. F. Y. Young. Local clock skew minimization using blockage-aware mixed tree-mesh clock network. In *IEEE/ACM International Conference on Computer-Aided Design (ICCAD)*, San Jose, CA, 2010, pp. 458–462.

111. Z. Yu and X. Liu. Implementing multiphase resonant clocking on a finite-impulse response filter. *IEEE Transactions on Very Large Scale Integration (VLSI) Systems*, 17(11):1593–1601, 2009.

112. S. Zanella, A. Nardi, A. Neviani, M. Quarantelli, S. Saxena, and C. Guardiani. Analysis of the impact of process variations on clock skew. *IEEE Transactions on Semiconductor Manufacturing*, 13(4):401–407, 2000.

113. X. Zeng, D. Zhou, and W. Li. Buffer insertion for clock delay and skew minimization. In *IEEE International Symposium on Physical Design (ISPD)*, 1999, pp. 36–41.

114. Q. K. Zhu and M. Zhang. Low-voltage swing clock distribution schemes. In *IEEE International Symposium on Circuits and Systems (ISCAS)*, Sydney, Australia, 2001, pp. 418–421.

115. C. Ziesler, S. Kim, and M. Papaefthymiou. A resonant clock generator for single-phase adiabatic systems. In *IEEE International Symposium on Low Power Electronics and Design (ISLPED)*, Huntington Beach, CA, 2001, pp. 159–164.

Exploring Challenges of Libraries for Electronic Design

James Hogan, Scott T. Becker, and Neal Carney

12.1   INTRODUCTION

Explaining how to best design libraries is usually a difficult task. We all know that a library is a collection of design behavior models at specific points in the design process, but in order to fully understand what it means to design libraries, we have to explore the intricacies and challenges of designing libraries. This includes examining what it means to design libraries, understanding the background, exploring the design process, and, perhaps, even analyzing the business models for libraries.

12.2   WHAT DOES IT MEAN TO DESIGN LIBRARIES?

Good designers must optimize constraints to achieve market requirements in terms of a finite number of cost functions. For example, they have to consider costs, performance, features, power consumption, quality, and reliability. These considerations are pretty universal for any design—even if you were designing a car, cell phone, or toaster.

On top of the traditional design constraints, the dramatic shortening of product life cycles also impacts design engineers. Often, this concern results from the dominance of consumer applications in the marketplace. The components of this trend are as follows:

- Hardware continues to be commoditized.
- Spiraling design costs lead to an increasing use of design platforms.
- Original device manufacturers (ODMs) build private label hardware (e.g., Wal-Mart).
- Original equipment manufacturers (OEMs) differentiate software and build brand value (e.g., Dell).
- Value is created through algorithms, system architecture, and software.
- Partitioning hardware and software has become the key decision.
- Reusing design platforms provides market leverage over multiple product cycles.

Thus, there are two opposing issues you must consider when designing libraries: time to market and costs. Industry economics (related partially to the complex nature of manufacturing small geometry silicon and short product lives driven by consumers) have little room for political or technical arguments. Instead, the answer for many system providers is reusable design platforms. Some key advantages of design platforms are as follows:

- Allowing a provider to capture multiple market segments by amortizing large and growing design costs and reducing time to market.
- Reducing the number of core processor architectures, while allowing more differentiation at the software application level.
- Increasing the percentage of mixed-signal designs as high-speed and mobile applications are integrated into a single silicon system-on-chip (this does not need to be a SoC it could be a multi-die solution or even a small form factor by board implementation).
- Providing flexibility of outsourcing and integrating pre-verified intellectual property (IP) functional blocks.
- Increasing hardware and software programmability. For the system architects, the trade-off is hardware or software. Software offers flexibility but costs silicon real estate, degraded system performance, and increased power consumption.

■ Enabling special emphasis or "special sauce" to be captured in custom blocks or in software.

■ Including retargeted IP. Previously used IP substantially reduces functional risk.

12.3 HOW DID WE GET HERE, ANYWAY?

Electronic system design has evolved in the last 30 years into a hierarchical process, which generally can be separated into three groups: system design, hardware and software implementation, and manufacturing and test. Although each group has its own area of optimization, they each must maintain the design intent originally specified in the system requirements. Each level of the design hierarchy must preserve the design intent of the preceding level (Figure 12.1).

As the design progresses through the hierarchy, the details of design intent become more specific. Design intent at the system design level guarantees that the system performs the desired function under certain specifications, such as power and speed. At the hardware implementation level, design intent is preserved by block- and instance-level specifications. Finally, at the manufacturing level, design intent is preserved by the lowest-level primitives, such as transistors and metallization.

As an example, we can analyze the design of a cell phone.

The system supplier (e.g., Nokia) is typically the company that is familiar to the consumer. With an ever-increasing complexity in electronic systems, consumers have come to rely on a brand to make their purchase decisions. If all brands have equal qualities, the market becomes commoditized, and the only factor then is cost. For example, brands that command a premium price are Sony and Apple Computers. To avoid commoditization, system suppliers must continue to increase their systems' features and performance. Their tasks include market definition, product specification, brand identity, and distribution. The system designer in a company like Nokia works from a specification and describes the design in terms of behavior. The behavior can be expressed in a high-level language such as C, System C, or C++.

At hardware and software implementation, the implementation company (e.g., Texas Instruments) will create a logical and physical description. Brands in this area are less of a factor with the consumer (the notable exception being the lingo "Intel Inside," which managed to commoditize everyone else on the motherboard). The main concerns are design closure for performance, cost, and power. In the implementation phase, there must be a convergence of the design or closure. The litmus test for closure is whether the design has met the performance specifications in physical implementation.

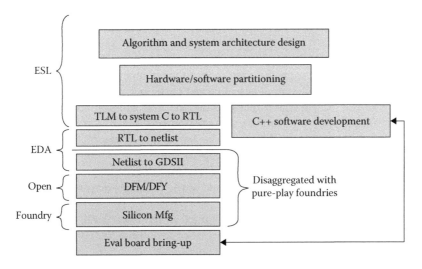

FIGURE 12.1 Design hierarchy.

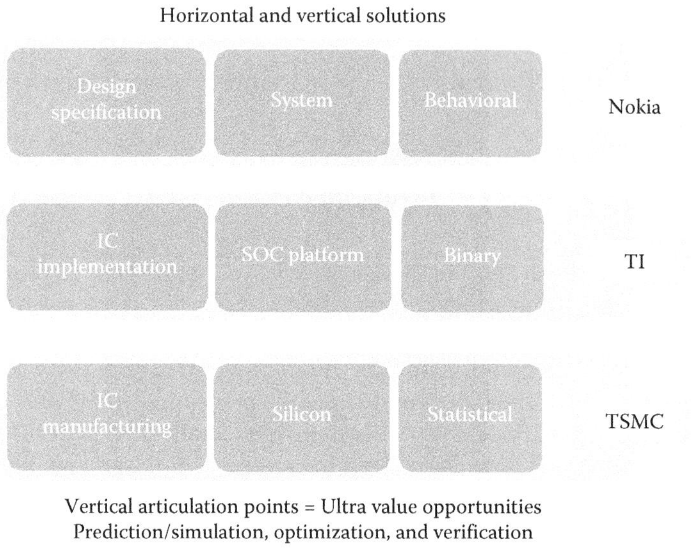

Horizontal and vertical solutions

Vertical articulation points = Ultra value opportunities
Prediction/simulation, optimization, and verification

FIGURE 12.2   Horizontal and vertical design solutions for cell phone example.

At the manufacturing level, the manufacturing company (e.g., TSMC or Flextronics) provides manufacturing integration and test, process control, supply logistics, and capital utilization. In electronics design manufacturing, the measure of success is different from implementation. At the implementation phase, there is a binary decision (i.e., were the specifications met or not?). In manufacturing, it is statistical. Does the finished good fall in the statistical standard deviation that defines a good product?

In an increasingly disaggregated design chain, there must be efficient and accurate methods to communicate both vertically and horizontally to ensure the integrity of design intent. Figure 12.2 shows how horizontal and vertical solutions may appear for our cell phone example.

12.3.1 WHAT DEFINES HORIZONTAL DESIGN SPACE?

At any given step in the vertical hierarchy of the design process, there is a need to explore the design space. Engineers at each level must guarantee that the design intent from preceding levels is preserved. This entails understanding the design constraints, optimization of the design, and validation of the implementation at the current level.

Generally, the progression horizontally is the following:

- *Measure*: Using instruments automatic test equipment (ATE) or a computer program (timing analyzer) to measure the attributes of the current design.
- *Model*: A parameterized mathematical model of the behavior found through measurement, the accuracy of which is directly related to the level of statistical control. Typically, there is a trade-off between model accuracy and complexity. Often, developers look to develop the perfect model. This, in turn, drives more computation time and subsequently reduces the number of possible experiments due to limited resources. Generally speaking, there is more value in exploring a larger design space.
- *Analyze*: Tools to view and edit the results of the designer's experiments in optimizing the design against constraints.
- *Simulate*: An engine using the models defined in the lower levels of the design hierarchy to explore "what if" scenarios. This can expose more widely the opportunities to optimize.
- *Verify*: Often, the same engine is used in simulation, but with assertions on the design, to verify that design intent was not changed during design optimization.

- *Optimize*: A process that uses weighted design constraints to perform experiments to best fit the design. The optimization process can be performed using an automated or manual approach, depending upon the complexity of optimization and the number of experiments required. The drive is toward automation, but this is not always feasible.

Libraries in the horizontal solution will contain models, pre-verified IP, and a set of rules (test benches, design rules, or lower-level models) for verifying the design. A test bench implements an automated method for generating stimulus to apply to the model of the design, along with the means of comparing the expected results with the simulated results.

In terms of economic value, in the history of electronic design automation (EDA), more value has been awarded to simulation and optimization products that follow analysis in the horizontal flow. While modeling is a very difficult but necessary step, it has not been able to extract the same value.

12.3.2 WHAT IS A VERTICAL DESIGN SOLUTION?

In the vertical axis of the design chain, the critical step is verification of the design as it moves through the vertical design chain flow.

An excellent example is at the physical verification step, where you verify whether an IC design has met the manufacturer's rules. The best execution of verification is to use the same physical verification engine and rules that the manufacturing semiconductor foundry will use to verify the design for manufacturing. In the past, verification tests were based solely on geometric rules, but they now rely increasingly on a model-based approach to ensure manufacturability. It has become an extremely difficult problem to ensure that design intent is maintained, as the design moves from a binary to a statistical world. Several companies (e.g., ARM, Virage) have established an excellent business of developing standard physical library elements for pure-play foundries (e.g., UMC, TSMC) that already take into account the manufacturing variability bridging graphic data system II (GDSII) to silicon.

In another example, at the system level of the design chain, the software developer relies on hardware behavior models to optimize software application. In this area, the software engineer has no concept of a clock, as a hardware design would. The processor model only needs to be "cycle" accurate to ensure that the application can be verified effectively. Several companies (e.g., Virtutech, VAST, and Carbon) have offered their approach to modeling the processor behavior.

12.4 COMMERCIAL EFFORTS

These examples have illustrated that there is a need for IP or libraries in both the horizontal and vertical design chain. This need was recognized far back in the early 1970s. Early efforts such as in [1,2] concentrated on supplying the physical design of standard cells. Then new forms of modeling were added, typically behavioral and timing [3]. Integrated design and manufacturing houses explicitly built libraries to ease vertical integration [4], and vertically integrated systems implicitly included libraries [5,6]. As processes became more complex, libraries were also characterized for faults [7], signal integrity [8], and yield [9]. There are now companies such as ARM [10] and websites such as Design and Reuse [11], which are completely devoted to IP and libraries.

Economically, the commercial efforts that have done well have developed a *de facto* standard for communicating design intent across the design chain. The design chain adopts them as the standard protocol to communicate design data. They have enjoyed a proprietary position through a combination of business models and technologies, enabling them to establish broader product and service portfolios. Examples are Synopsys with their DesignWare and timing libraries, Rambus with memory controllers, Artisan with standard cells and memories, and Meta Software with SPICE models.

It is interesting that the market capitalization for the major IP companies has grown significantly (e.g., ARM, RAMBUS, Tessera, and Synopsys with DesignWare). These commercial IP companies have been able to capture significant value from the design chain, as it has become disaggregated. What is different with regard to the traditional EDA companies is that once IP is established, there is room for only one dominant supplier. As the design chain disaggregates, the requirements of preserving design intent create high-value opportunities (i.e., pre-verified IP, EDA tools, or semiconductor manufacturing). In the physical IP space, the use of the IP can insulate the designer from the changing physics of manufacturing, so the designer can focus on creating value in the design, and not on whether manufacturing can deliver the design intent. The details are embedded in the physical IP.

IP will continue to increase in value, as the design chain continues to disaggregate, and the difficulty of communicating design intent increases. For example, the complexities of manufacturing, both in the physics (lithography, etch, and CMP) and in logistics (global supply chains), require efficient and accurate virtual use of information by teams that know no political or time boundaries. The winners will solve not only the modeling problem technically but also the business challenges of addressing this 24/7/365 world.

There will be growth in the IP and libraries in the horizontal domain as current EDA models continue to mature. We can think of it in terms of an "evolution" that will identify more parameters (e.g., 48 term BSIM4 SPICE models that now include power), new analysis technologies (software and hardware) for dealing with the increased compute and data volumes, and new assertion cases for verification.

12.5 WHAT MAKES THE EFFORT EASIER?

IP and library standardization of information would be a great step forward. Standardization is usually the instrument of market expansion. Common interfaces between vertical domains allow the free and unencumbered exchange of information.

12.6 ENEMIES OF PROGRESS

The design solutions that made winners in the past will not work with the speed of the market today. The following is a list of issues that have caused problems in the past, especially in IP and libraries. These should be avoided if possible:

- Warring tribes that miss the "Big Picture" of consumer demands (UNIX vs. DEC/VMS; VHSIC Hardware Description Language vs. Verilog; System Verilog vs. SystemC, etc.).
- Unclear benefits such as the supplier controlling the agenda.
- Companies holding onto proprietary formats, until it becomes detrimental to their growth. Some enlightened suppliers have seen the future and have opened their standards.
- Political motivation that limits competition (DivX and IBM Microchannel).
- Customer edict without sufficient vendor input (e.g., OLA, another logical format [ALF], VHDL, Wal-Mart, and radio frequency identification [RFID]).
- Difficult to adopt and no mapping of existing infrastructure. No one has the time or money to do something that does not have a clear and efficient cost/benefit over existing solutions.
- The installed base is incompatible, and the cost to adopt is prohibitive.
- Standards that are ahead of their time, and solutions looking for problems.
- Perfection can be the enemy of a good solution. Good enough will win the day. Do not postpone for perfection.
- Abstraction without effective automation. This is generally a gap between domains that are not interconnected. As a result, the standard becomes useless.

12.7 ENVIRONMENTS THAT DRIVE PROGRESS

There are circumstances that make IP less of a limiting factor and more of an enabler. In order to increase system complexity and continue manufacturing that enables differing versions of IP, you need to consider the following attributes:

- Design is increasingly driven by cost, efficiency, and functionality over raw performance. Does the world need any more MIPS?
- The effective communication of design intent and the use of models to the surrounding levels in the design chain.
- Cost pressures will cause industry consolidation around new market aggregation layers (IP and library companies have and will continue to capture value). Other companies like eSilicon and OpenASIC are successfully using design and logistics information to develop new business models.
- Reliance on standards will amortize the cost by members of standard conforming groups.
- Open IP standards are allowing for a new road map of requirements and the development of commercial offerings (PCI-X, DDR1/2). Systems companies will innovate in the applications not in the protocols.
- Standards (products, architectures, interfaces, and abstractions) enable economies of scale and the reuse of platforms over multiple product cycles.

12.8 LIBRARIES AND WHAT THEY CONTAIN

The following lists are examples of libraries at various levels in the vertical design hierarchy (from low to high) as well as the methods for communication:

12.8.1 LOW-LEVEL PHYSICAL IP

- These include standard cells, memory, analog/mixed signal IP, and interface IP, such as I/O cells and PHYs. (PHY is a generic term referring to a special electronic circuit or functional block of a circuit, which provides physical access to a digital interconnect cable.)
- Examples: ARM Holdings, Virage, internal design groups, etc.
- Typical libraries enable variations to optimize area, speed power, and increase the yield. These are implemented via configuration or cell variants.
- Outputs to higher levels in the design hierarchy: cell layout (typically GDS-II), routing model (typically LEF), functional models (typically Verilog), timing models (typically. lib), circuit models (typically SPICE), and yield models (often PDF solutions format).
- Inputs from lower levels in the design hierarchy (manufacturing): geometric rules, transistor device models, and yield trade-off rules.

12.8.2 HIGH-LEVEL PHYSICAL IP (CORES)

- Examples: ARM, MIPS, and OAK.
- Typical libraries offer time to market advantages with a functional and performance guarantee. Offer no aspect ratio control or functional flexibility.
- Outputs to higher levels in the design hierarchy: cell layout, routing model (abstract), functional model, timing model, and yield model.
- Inputs from lower levels in the design hierarchy: process models, lower-level IP models, and system specification.

12.8.3 HIGH-LEVEL SOFT IP

- Examples: serial bus controllers, processor cores.
- Advantages: flexible aspect ratio and usable with many processes.
- Disadvantages: may be difficult to communicate design intent to the physical implementer, complex soft IP may diminish due to complexities introduced in the manufacturing process. It is hard to provide a performance guarantee, since detailed implementation is unknown at design time.
- Inputs from higher levels in the design hierarchy: block-level specifications for performance and functionality.
- Outputs to lower levels in the design hierarchy: logic net list, performance parameters, and placement information.

12.8.4 SYSTEM DESIGN/IMPLEMENTATION

- Inputs from higher levels in the design hierarchy: high-level C code and system specifications.
- Inputs from lower levels in the design hierarchy: low-level models for function, timing, power, and area.
- Outputs to lower levels in the design hierarchy: block-level specifications and overall system specifications for performance and cost.

12.8.5 SYSTEM ARCHITECTURE

- System specification: market-driven specification starts the design process.
- Outputs to lower levels in the design hierarchy: system functional code communicates system functionality along with the system performance requirements to the system design phase.

12.9 UPDATE: LITHOGRAPHY LIMITS, NEW TRANSISTORS, AND IMPLICATIONS FOR LIBRARIES

At technology nodes with feature sizes below 40 nm, library design necessitates a much more intimate interaction between the manufacturing process engineers and the design engineers. This is due to the fact that advances in lithography have plateaued in terms of resolution with 193 nm light sources in immersion-based scanners. This has imposed new constraints on library designers, particularly in the physical implementation and layout of library elements.

In addition, new transistor structures like FinFETs will have a significant impact on library architectures, circuit design, and modeling.

12.9.1 193 NM IMMERSION LITHOGRAPHY

The minimum feature size of an optical system is directly proportional to the wavelength of light and inversely proportional to the numerical aperture (NA) of the optical system. Light sources have reached a limit of 193 nm (deep ultraviolet excimer laser). Light sources of 193 nm were first widely deployed at the 65 nm node. The next advance in shorter wavelength light sources has been focused on extreme ultraviolet (EUV) with wavelengths of 13.5 nm. EUV is still in development, facing a number of technical challenges to see wide-scale deployment.

With wavelengths being limited to 193 nm, improvements in numerical aperture have been made using water instead of air as the interface. Water's refractive index allows an increase to a NA of 1.35 from 0.93 in today's immersion lithography systems. Immersion lithography was first deployed at 45 nm and will be the workhorse system down through at least the 14 nm node.

12.9.2 DESIGN PROCESS CO-OPTIMIZATION

With both wavelength and NA having plateaued, another variable that can help improve the resolution of smaller feature sizes is controlling the complexity of the patterns that are required to be printed. Complex patterns with bends and jogs in close proximity to each other are difficult to resolve and set up interference patterns that are unpredictable. In more advanced technologies, standard cell designers have to deal with proximity effects. This entails placing a cell surrounded by representative neighboring cells and extracting the parametric variations due to these neighboring cells. These variations are then factored into the electrical characterization of the cells. In addition, design margins for variability on a broader scale across the entire chip must be factored in with OCV (on chip variability) factors.

The easiest pattern for an optical system to resolve is a series of straight lines on a fixed pitch. Since transistor variation is a critical concern, manufacturers have begun requiring restrictions at the 32/28 nm node on the gate-level layout that essentially results in the gate layer being a series of straight-line patterns. This restriction forces cell library designers to develop alternative cell architectures and physical implementation methodologies.

In advanced technology nodes, the limits of optical lithography in terms of process variability and design rule restrictions are driving a much closer collaboration between library designers and the process development community. This collaboration must not only be close but must start much earlier in the design of a new node, as trade-offs between the manufacturing process and its impact on design must be rationalized early in the development cycle. From an industry standpoint, this has led manufactures, including "pure-play" foundries, to invest in library designs teams while at the same time forging closer alliances with customer library teams as well as third-party library entities.

12.9.3 BEYOND THE RESOLUTION LIMIT

Even straight-line patterns run into resolution issues at feature sizes associated with the 20 nm node. In the absence of shorter wavelength lithography like EUV, the techniques associated with double patterning or multiple patterning are the only solution available. A simplistic illustration of double patterning for a series of straight lines would be creating two lower resolution masks, with alternate lines on each mask. This shifts the lithography challenge from resolution to alignment accuracy, as the second pass line patterns must be accurately interleaved with the first pattern in order to achieve accurate spacing between all the lines. There are many techniques in the industry to accomplish this type of resolution improvement.

Some involve multiple optical exposures, and others use a single exposure and multiple process steps to implement a technique referred to as pitch division to increase resolution. In general, the impact on cell library designers involves restrictions on layout rules to enable these multi-patterning approaches to be implemented.

The combination of layout restrictions for variability and multi-patterning is driving the EDA (Electronic Design Automation) industry to also collaborate more closely, and early, in the process development phase. The complexity of dealing with these restrictions and their interactions must be automated, in order to be feasible to implement the billion transistor designs that 28 and 20 nm nodes enable. Additional complexities introduced by new transistor structures will put additional burdens on EDA companies to help automate and model the interactions of these new technologies.

12.9.4 NEW TRANSISTORS

At 20 nm and below, companies are beginning to deploy new transistor structures, in particular FinFET or 3D Tri-Gate, that have very fundamental design differences and implications for cell library design. These transistor structures have dramatically improved power-performance characteristics and are hence attractive for all applications, in particular for the proliferation of battery-operated mobile devices [12].

The discrete size of the fin in these devices leads to a quantization in terms of transistor sizing. The width of the transistor is no longer a continuum but a multiple of the number of fins. This is a significant change that standard cell library designers must deal with, in architecting the overall library or family of libraries, in terms of area, performance, and power. Another key issue for FinFETs is the 3D nature of the structure and the complex modeling implications of this [13]. As these devices are new, many of the design rules and models remain proprietary and in some cases are still preliminary depending upon the manufacturer. Cell library innovation using FinFETs is currently an active industry topic and opportunity for differentiation in the industry.

12.10 SUMMARY

Standard cell-based designs remain the dominant approach for the implementation of SoC, ASICs being deployed in the vast majority of today's consumer, communications, and computer products. Libraries must be deployed in advance of the time that leading edge processes are brought on line, in order for designers to have products taping out to fill the massive fabs brought online in support of these high-volume consumer applications. In parallel, the EDA industry is grappling with many simultaneous technical challenges. They include

- Increasing product complexity, the drive to improve design productivity and shorten design cycles
- Increasing the complexity of manufacturing processes, forcing the knowledge of manufacturing physics up the design hierarchy to cause multiple non-convergent optimization problems (Litho and DFM/DFY)
- Critically verifying and solving the equivalence problem of design intent
- Increasing need for a higher level of design abstraction/handoff to foster more design starts
- Optimizing multiple simultaneous design objectives, including power, timing, signal integrity, etc.

The market has matured—the EDA standards process must also rise to a new level. This will include IP as well as tools in their design solutions. IP and library products that address the stated challenges by bridging the horizontal and vertical design chain articulation points are and will be more valuable. New process technologies require library designers to have not only an in-depth knowledge of circuit design but also a broad knowledge of adjacent technology considerations in terms of device design and process development. Close collaboration among companies and entities in manufacturing, library design, chip design, and EDA is essential for success in advanced technology nodes.

REFERENCES

1. R.L. Mattison, A high quality, low cost router for MOS/LSI, *Proceedings of Ninth Design Automation Workshop*, 1972, pp. 94–103.
2. A. Feller, Automatic layout of low-cost quick-turnaround random-logic custom LSI devices, *Proceedings 13th Design Automation Conference*, 1976, pp. 79–85.
3. I. Jones, Characterization of standard cell libraries, *Proceedings of CICC*, 1985, pp. 438–441.
4. A. Martinez, S. Dholakia, and S. Bush, Compilation of standard cell libraries, *IEEE J. Solid-State Circ.*, SC-22, 190–197, 1987.
5. J. Rosenberg, D. Boyer, J. Dallen, S. Daniel, C. Poirier, J. Poulton, D. Rogers, and N. Weste, A vertically integrated VLSI design environment, *Proceedings of the 20th Conference on Design Automation*, 1983.
6. A.F. Hutchings, R.J. Bonneau, and W.M. Fisher, Integrated VLSI CAD systems at digital equipment corporation, *Proceedings of the 22nd ACM/IEEE Conference on Design Automation*, 1985.
7. J. Khare, W. Maly, and N. Tiday, Fault characterization of standard cell libraries using inductive contamination analysis (ICA), *Proceedings of 14th VLSI Test Symposium*, Princeton, NJ, 1996, pp. 405–413.

8. Artisan inks deals with Cadence, Synopsys, *Electronic News*, November 19, 2002, http://www.reed--electronics.com/electronicnews/article/CA260087.html.
9. P. Clarke, PDF solutions, Virage to add DFM to cell libraries, *EETimes*, October 6, 2004.
10. ARM, see http://www.arm.com.
11. Design and ReUse, http://www.us.design-reuse.com.
12. L. Collins, How to design with FinFETs, May 29, 2013, http://www.techdesignforums.com/practice/technique/how-to-design-with-finfets/.
13. Synopsys, FinFET: The promises and the challenges, 2012, https://www.synopsys.com/COMPANY/PUBLICATIONS/SYNOPSYSINSIGHT/Pages/Art2-finfet-challenges-ip-IssQ3-12.aspx.

Design Closure

Peter J. Osler, John M. Cohn, and David Chinnery

13

CONTENTS

13.1 INTRODUCTION

Design closure is the process by which a VLSI design is modified from its initial description to meet a growing list of design constraints and objectives. This chapter describes the common constraints in VLSI design and how they are enforced through the steps of a design flow.

Every chip starts off as someone's idea of a good thing: "If we can make a part that performs function X, we will all be rich!" Once a concept is established, someone from marketing says, "in order to make this chip and sell it profitably, it needs to cost $\$C$ and run at frequency F." Someone from manufacturing says, "in order to make this chip's targets, it must have a yield of $Y\%$." Someone from packaging says, "it has to fit in the P package and dissipate no more power than W watts." Eventually, the team generates an extensive list of all the constraints and objectives that need to be met in order to manufacture a product that can be sold profitably. The management then forms a design team, consisting of chip architects, logic designers, functional verification engineers, physical designers, and timing engineers, and tasks them to create the chip to these specifications. Other chapters in this book have dealt with the details of each specific step in this design process (static timing analysis, placement, routing, etc.). This chapter looks at the overall *design closure* process, which takes a chip from its initial design state to the final form in which all of its design constraints are met.

We begin this chapter by briefly introducing a reference design flow. We then discuss the nature and evolution of design constraints. This is followed by a high-level overview of the dominant design closure constraints that currently face a VLSI designer. With this background, we present a step-by-step walk-through of a typical design flow for *application-specific integrated circuits* (ASICs) and discuss the ways in which design constraints and objectives are handled at each stage. We will conclude with some thoughts on future design closure issues.

13.1.1 EVOLUTION OF THE DESIGN CLOSURE FLOW

Designing a chip used to be a much simpler task. In the early days of VLSI, a chip consisted of a few thousand logic circuits that performed a simple function at speeds of a few MHz. Design closure at this point was simple: if all of the necessary circuits and wires "fit," the chip would perform the desired function. Since that time, the problem of design closure has grown to orders of magnitude more complex. Modern logic chips can have millions to billions of logic elements switching at speeds of up to several GHz. This improvement has been driven by Moore's law of scaling of technology, which has introduced a number of new design considerations. As a result, a modern VLSI designer must simultaneously consider the performance of the chip against a list of dozens of design constraints and objectives, including performance, power, signal integrity, reliability, and yield. We will discuss each of these design constraints in more detail in Section 13.1.2.

In this chapter, the distinction between constraints and objectives is not that important and we will use the words interchangeably. A *constraint* is a design target that must be met in order for the design to be considered successful. For example, a chip may be required to run at a specific frequency in order to interface with other components in a system. In contrast, an *objective* is a design target where more (or less) is better. For example, yield is generally an objective that is maximized to lower manufacturing cost.

13.1.1.1 ASIC DESIGN FLOW

In response to the growing list of constraints, the design closure flow has evolved from a simple linear list of tasks to a very complex highly iterative flow such as the following simplified ASIC design flow:

1. *Concept phase*: The functional objectives and architecture of a chip are developed.
2. *RTL logic design*: The architecture is implemented in a register-transfer-level (RTL) language and then simulated to verify that it correctly performs the desired functions.
3. *Floorplanning*: The top-level RTL modules of the chip are assigned to large regions of the chip, input/output (I/O) pins are assigned, and large macro objects (e.g., memory arrays) are placed.
4. *Synthesis*: The RTL is mapped into a gate-level netlist in the target technology of the chip.
5. *Placement*: The gates in the netlist are assigned to nonoverlapping locations on the chip.
6. *Pre-CTS placement optimization*: Iterative logical and placement transformations to close performance and power constraints. Idealized clock skew constraints are used up to this point in the design flow.
7. *Clock tree synthesis* (CTS): Buffered clock trees are introduced into the design, and clock gaters are cloned to drive their loads.
8. *Post-CTS optimization*: Further logic and placement optimization, as clock skews are known, including optimization to fix hold time violations. Register positions may be fixed to avoid further perturbation to the clock trees.
9. *Routing*: The wires that connect the gates in the netlist are added.
10. *Post-route optimization*: Remaining performance, noise, and yield violations are removed. Power or area optimization is performed, downsizing and swapping to low-power cells as wire loads are known accurately. Final checks are done.

We will use this reference flow throughout this chapter to illustrate points about design closure. The purpose of the flow is to take a design from concept phase to working chip.

13.1.1.2 EVOLUTION OF DESIGN CONSTRAINTS

The complexity of the design flow is a direct result of the addition and evolution of the list of design closure constraints. To understand this evolution, it is important to understand the *life cycle* of a design constraint. The influence of design constraints on the flow typically evolves in the following manner:

1. *Early warnings*: Before chip issues begin occurring, academics and industry visionaries make dire predictions about the future impact of some new technological issue.
2. *Hardware problems*: Sporadic hardware failures start showing up in the field due to the new effect. Post-manufacturing redesign and hardware respins are required to get the chip to function.

3. *Trial and error*: Constraints on the effect are formulated and used to drive post-design checking. Violations of the constraints are fixed manually with *engineering change orders* (ECOs), minimizing the scope of the changes and avoiding a respin through the design flow where possible.

4. *Find and repair*: The large number of violations of the constraints drives the creation of automatic post-route analysis and repair flows.

5. *Predict and prevent*: Constraint checking moves earlier in the flow, using predictive estimations of the effect. These drive optimizations to prevent violations of the constraints.

A good example of this evolution can be found in the coupling noise constraint. In the mid-1990s (180 nm node), industry visionaries described the impending dangers of coupling noise, long before chips failed [1]. By the mid-late 1990s, noise problems cropped up in advanced microprocessor designs. By 2000, automated noise analysis tools were available and were used to guide manual fix-up [2]. The total number of noise problems identified by the analysis tools quickly became too many to correct manually. In response, EDA companies developed the noise avoidance flows that are currently in use in the industry [3]: early in the flow, critical wires (e.g., clock nets) may be assigned non-default rules (NDR) to have additional spacing and/or shielding to reduce cross-coupling, and smaller drive strength cells may not be allowed for long wires; estimates of signal integrity (SI) noise allow post-CTS optimization to try to address the issue; and SI noise victims will be upsized and aggressors downsized, in post-route optimization to resolve the majority of the remaining SI violations.

At any point in time, the constraints in the design flow are at different stages of their life cycle. For example, at the time of this writing, performance optimization is the most mature and well into the fifth phase with the widespread use of *timing-driven* design flows. Power and defect-oriented yield optimization are well into the fourth phase. Power supply integrity, a type of noise constraint, is in the third phase. Circuit-limited yield optimization is in the second phase. A list of the first-phase impending constraint crises can be found in the *International Technology Roadmap for Semiconductors* (ITRS) 15-year-outlook technology road maps [4].

13.1.1.3 ADDRESSING CONSTRAINTS IN THE DESIGN FLOW

As a constraint matures in the design flow, it tends to work its way from the end of the flow to the beginning. As it does this, it also tends to increase in complexity and in the degree that it contends with other constraints. Constraints move up in the flow due to one of the basic paradoxes of design: *accuracy versus influence*. Specifically, the earlier in a design flow a constraint is addressed, the more flexibility there is to address the constraint. Ironically, the earlier one is in a design flow, the more difficult it is to predict compliance. For example, an architectural decision to pipeline a logic function can have a far greater impact on total chip performance than any amount of post-route fix-up. At the same time, accurately predicting the performance impact of such a change before the RTL is synthesized, let alone placed or routed, is very difficult.

This paradox has shaped the evolution of the design closure flow in several ways. First, it requires that the design flow is no longer composed of a linear set of discrete steps. In the early stages of VLSI, it was sufficient to break the design into discrete stages, that is, first do logic synthesis, then do placement, and then do routing. As the number and complexity of design closure constraints have increased, the linear design flow has broken down. In the past, if there were too many timing constraint violations left after routing, it was necessary to loop back, modify the tool settings slightly, and re-execute the previous placement steps. If the constraints were still not met, it was necessary to reach further back in the flow, modify the RTL, and repeat the synthesis and placement steps. This type of looping is both time consuming and unable to guarantee convergence. It is possible to loop back in the flow to correct one constraint violation, only to find that the correction induced another unrelated violation.

To minimize the requirement for frequent iteration, the design flow has evolved to use estimates of downstream attributes. As an example, wire length and gate area are used to drive upstream optimization of timing and power dissipation. As the design evolves, these approximations are refined and used to drive more precise optimizations. In the limit, the lines that separate two sequential steps such as synthesis and placement can be removed to further improve downstream constraint estimates. For example, *physical synthesis* now includes global placement within the synthesis step in the EDA flow, and this placement can be used as a starting point for pre-CTS optimization (instead of an unplaced gate-level netlist) to improve timing correlation and results. In today's most advanced design closure flows, the lines between once-discrete steps have been blurred to the point that steps such as synthesis, placement, and routing can be simultaneously co-optimized [5,6]. The evolution from discrete stand-alone design steps to integrated co-optimizations has had a profound influence on the software architecture of EDA flows.

Modern design closure suites are composed of three major components: a central database; a set of optimization engines, that is, logic optimization, placement, and routing; and a set of analysis engines, for example, *static timing analysis*, *power analysis*, and *noise analysis*. The central database manages the evolving state of the design. The optimization engines modify the database directly, while the analysis engines track incremental changes in the database and report back on the results. By allowing a certain degree of independence between optimization engines and analysis tools, this *data-driven* architecture greatly simplifies the addition of new design constraints or the refinement of existing ones. More detail on data-driven EDA software architecture can be found in Chapter 1.

13.1.1.4 EVOLUTION OF CHIP TIMING CONSTRAINTS

The evolution of chip timing constraints provides a good overall illustration of a constraint's movement up the flow.

In the first integrated circuits, analyzing performance consisted of summing the number of levels of logic in each path. After a couple of technology generations, hardware measurements began to show that this simple calculation was becoming less accurate. In response, performance analysis tools were extended to take into consideration that not all gates have the same delay. This worked for a time, until measured hardware performance again began to deviate from predictions. It was clear that gate output loading was becoming a factor. The performance analysis flow was modified to determine the delay of a gate, based on the total input capacitances of the gates it drove.

As process technologies scaled smaller, wire capacitance became a more significant portion of the total load capacitance. Wire length measures were used to calculate total wire capacitive load, which was converted into a delay component. As performance increased further, wire resistance became a factor in all interconnects, so delays were approximated using a simple, single-pole *Elmore RC delay* model. In the early 1990s, it was observed that the Elmore approximation was a poor predictor of wire delay for multi-sink wires, and so more complex *moment matching* [7] methods were introduced. As timing analysis matured further, it became increasingly difficult to correct all of the timing constraint problems found after routing was complete. As the flow matured from *find and repair* to *predict and prevent*, it became increasingly important to accurately predict the total wire delay as part of a *timing-driven* design flow. Crude models of wire delay were added to placement, in order to preferentially shorten wires on timing critical paths.

The first placement-based wire delay models used gross estimates of total wire length, such as calculating the bounding box of all pins on a wire. As interconnect delay grew in importance, such nonphysical wire length estimates proved to be poor predictors of actual wire delay. Delay estimates based on Steiner tree approximation of routed wire topologies were introduced to predict delay better. As interconnect delay increased further, it became necessary to push wire length calculation even further back into pre-placement logic synthesis. Initially, this was done using simple *wire load models* characterized by placed block size, which assigned average wire-length and wire capacitance based on fanout count. Eventually, this too proved too inaccurate for high-performance logic.

By the late-1990s, placement-driven *physical synthesis* [5,8] was introduced to bias logic synthesis, based on predictions of critical wire length. As the handling of timing constraints matured, it also began to contend more with other constraints. During the early days of design closure, chip timing problems could be mitigated by increasing the drive strength of all circuits on a slow path. However, doing this increases active power and also increases the chance of coupling noise onto adjacent wires.

We will now shift our focus to learn more about the specific design constraints that are facing chip designers. Armed with this, we will then begin a detailed walk-through of a typical design closure flow.

13.1.2 INTRODUCTION OF DESIGN CONSTRAINTS

Chip designers face an ever-growing list of design constraints that must be met for the design to be successful. The purpose of the design closure flow is to move the design from concept to completion, while eliminating all the constraint violations. This section will briefly introduce the current taxonomy of design constraints, objectives, and considerations and discuss the impact and future trends for each. For the sake of this discussion, we will divide the constraint types into economic, realizability, performance, power, signal integrity, reliability, and yield.

13.1.2.1 ECONOMIC CONSTRAINTS

While not explicitly technical, the economic constraints governing design closure are perhaps the most important constraints of all. Economic constraints pertain to the overall affordability and marketability of a chip. Economic constraints include *design cost, time to market*, and *unit cost*.

- *Design cost constraints* govern the *non-recurring expense* (NRE) associated with completing a design. This includes the cost of the skilled resources needed to run the design flow, of any test hardware, and of additional design passes that occur due to errors. These costs also include the amortized cost of facilities, computer resources, design software licenses, etc., which the team needs to complete the task. Design cost can be traded off against other design constraints such as performance and power, because increased effort and skill generally will yield better optimization. Missing a design cost estimate can be a very serious problem, as cost overruns generally come out of projected profit.
- *Time-to-market constraints* govern the schedule of a design project. This includes the time required for development, manufacturing, and any additional hardware respins necessary to yield a functional and manufacturable part in sufficient volumes to meet the customers' requirements. Time-to-market cost can be traded off against other design constraints such as performance and power because, like increased design effort, an increased design time generally yields better optimization, albeit at the expense of design cost. Missing a time-to-market constraint can imply missing a customer deadline or market window. In competitive market segments, being late to market can mean the difference between huge profits and huge losses.
- *Unit cost constraints* govern the cost of each manufactured chip. This includes the cost of the chip itself accounting for any yield loss, the package, the cost of the module assembly, and the cost of all testing and reliability screens. Unit cost is a strong function of chip die size, chip yield, and package cost. This can be traded off against design cost and time to market, by allowing additional effort to optimize density, power, and yield. Missing a unit cost constraint can make a chip noncompetitive in the marketplace.

13.1.2.2 REALIZABILITY CONSTRAINTS

The most basic constraint of VLSI design is "does the chip fit and does it work?" These "realizability" constraints pertain to the basic logical correctness of the chip. Realizability constraints include *area constraints, routability constraints*, and *logical correctness constraints*.

- *Area constraints* are one of the most basic constraints of any VLSI design, that is, does the chip "fit" in the desired *die size* and *package*? To fit, the total area required by the sum of the chip's circuitry, plus additional area required for routing, must be less than the total useable area of the die. The die size and package combination must also support the type and number of I/O pins required by the design. Because silicon area and package complexity are major components of chip cost, minimizing the die size and the package complexity are major goals in cost-sensitive designs. The impact of mispredicting capacity can be great. If the required area is overestimated, die utilization is low and the chip costs more than it should. If the required area is underestimated, the design must move to a larger die size or more complex package, increasing cost and adding design time.

■ *Routability constraints* ensure that the resulting placement can be completely connected legally by a router. The impact of mispredicting routability can impact average wire length, which can affect timing. If routability is compromised enough, the designers must manually route the overflows that could not be routed automatically. In the worst case, mispredicting routability can require that the chip be bumped up to a larger die size at great impact to cost and schedule. The challenges of this problem have been growing due to a number of factors, such as increasing gate count; increased complexity of metallurgy, including complex via stack rules and multiple wire widths and heights; manufacturability constraints; double patterning; and increased interrelations between performance and routing. However, the addition of extra routing layers and improvements to routing technology have alleviated this problem to some extent. Routability problems can be mitigated by adding additional area for routing, which contends with chip area constraints.

■ *Logical correctness constraints* ensure that the design remains logically correct through all manipulations used by design closure. Modern design flows make significant use of local logical transformations such as buffering, inversion "bubble-pushing" with De Morgan's laws, logic cloning and de-cloning, and retiming to meet performance constraints. Today's optimization tools can perform logic resynthesis to speed up timing critical paths that are exposed only once the design is place and routed. As these logical transformations have become more complex, there is an increased possibility of introducing errors. To ensure that logical correctness is maintained, modern flows use equivalence checking to verify that the design function remains unchanged after each design step. Furthermore, many designs now have several operating and sleep modes, requiring careful verification of turnoff and turnon behavior to avoid floating signals from power-gated off logic driving "awake" logic, and ensure correct initialization of logic woken up from a sleep state. Obviously, the impact of failing logic correctness is that the chip no longer functions as intended.

13.1.2.3 PERFORMANCE CONSTRAINTS/OBJECTIVES

Once we ensure that the design will fit and is logically correct, the primary function of design closure is to ensure that the chip performance targets are met. Traditionally, chip timing has received the most focus as a design objective. Performance constraints can be divided into *setup* and *hold time constraints.*

■ *Setup* timing analysis establishes the longest delay, the *critical path* among all paths, which sets the maximum speed at which a chip can run. Setup timing is also known as late-mode timing analysis. Setup timing considerations can either be a constraint, that is, the chip must run at least as fast as x; or they can be a design objective, that is, the faster this chip, the better it runs. Setup timing constraints are enforced by comparing late-mode static timing results against desired performance targets and clock cycle time. Setup timing constraint violations are removed by decreasing the gate or interconnect delay along critical paths. Gate delays can be reduced by decreasing logic depth, increasing gate drive strength, increasing supply voltage, or substituting gates with *low transistor threshold voltage* (low V_t) logic. Interconnect delay can be reduced by decreasing wire length, adding buffers, widening wires to reduce their resistance at the cost of increased capacitance, or increasing spacing between wires to reduce cross-coupling capacitance. Setup timing constraint violations can also be resolved by allowing more time for a path to evaluate through the addition of *useful clock skew.*

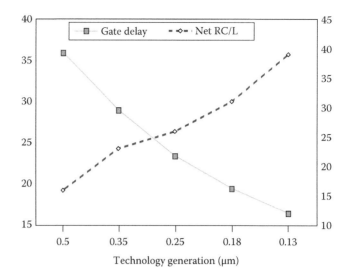

FIGURE 13.1   Interconnect and gate delay scaling trends.

Meeting the setup timing constraint for all paths is becoming increasingly difficult due to the combination of increasing chip complexity and clock speeds, combined with an increasingly pronounced "roll-off" of technology performance scaling. Both device performance and interconnect performance are failing to keep pace with the decades-long Moore's law improvement rate. More worrisome is that interconnect performance is scaling even more slowly than gate performance, as illustrated in Figure 13.1. In fact, increases in wiring resistance are beginning to cause *reverse scaling*, in which the relative interconnect delay actually increases with each new technology node. This implies more design closure effort in additional buffering and multicycle pipelining of signals that need to go a long way across the chip. Optimizations for setup timing constraints on one path may contend with setup timing constraints on other paths. For example, increasing the gate size to reduce delay on one critical path may increase gate load on another critical path. Setup timing optimizations also contend with power optimization. Most techniques used to optimize the setup time increase the total chip power—for example, gate upsizing, buffering, and low threshold logic substitution.

The impact of missing a setup timing constraint is that the chip is slower than required. In most designs, missing the setup timing constraints implies increasing the cycle time of a circuit. This may imply that the system specification must be renegotiated, or at worst case, redesigned at great cost of money and time. In some rare cases however, the final timing objectives may be negotiable, for example, one may be able to sell a slower microprocessor for less money.

■ *Hold* timing constraints are designed to prevent a path from being too fast to hold the signal steady while it is captured at the receiving register. Hold timing is also known as early-mode timing analysis. Unlike setup timing constraints, hold timing is always a constraint and never an objective. A single hold violation means that the chip will not function at any speed. Hold timing issues are difficult to estimate early in the design flow. They generally must be analyzed and corrected after final clock routing is added to the design. Then, static timing analysis for hold verifies that the final logic state is stable before the capture clock fires. This testing must account for possible clock skew induced by manufacturing variations. It is necessary to run static timing over multiple *process corners*, for example, fast wiring with slow logic and fast logic with slow wiring, to predict these cases correctly. Poor control of clock overlap can also give rise to a large number of hold time violations. Hold timing violations are removed by adding delay elements such as buffers to fast paths. By adding circuitry, hold constraints contend with area, power, and setup timing constraints.

13.1.2.4 POWER CONSTRAINTS/OBJECTIVES

Power dissipation can either be a constraint, that is, the chip must consume no more than *x* Watts; or it can be a design objective, that is, the less power this chip uses, the longer the battery will last. As chip geometries have scaled, the total chip power has become an increasingly important design closure constraint.

Power can be divided into two types: *dynamic* power and *leakage* power, which can contribute different portions to the total power consumption of a circuit, depending on its operation mode. *Active power* is used when the chip is on, and *standby power* is dissipated when the chip is off. Some ICs have a variety of different shallows through deep sleep states, where increasingly more of the design is power-gated off, with only a small portion of the circuit active waiting for input. Likewise, active operation modes may range in use from low power, low frequency, low voltage, (even near-threshold voltage operation) to turbo-mode peak operation with a frequency above that permitted by thermal design power limits, until the operating temperature exceeds predefined bounds, which triggers frequency reduction to avoid damaging the circuit.

- *Dynamic power* is dissipated through the charging and discharging of the capacitance of the switching nodes. Active power is proportional to the sum of $\frac{1}{2}FSCV^2$ of all switching signals on a chip, where *F* is the clock frequency, *S* the *switching factor*, that is, the average fraction of clock cycles in which the signal switches, *C* the total capacitive load presented by logic fanout and interconnect, and *V* the supply voltage. Active power increases due to higher chip logic densities, increased switching speeds, and a slowdown in voltage scaling due to limits on scaling gate oxide thickness. Active power can be lowered by decreasing gate size or wire load, and reducing switching frequency via clock gating. Reducing gate size to reduce active power also reduces the drive strength of the gate and thus contends directly with performance optimization. This gives rise to the essential power/performance trade-off.
- *Leakage power* is due to current that leaks through a device channel or transistor gate, even when it is turned off. Leakage power is increasing relative to dynamic power in MOSFET transistors, due to the use of smaller active transistor gate geometries, thinner transistor gate oxides that increase gate tunneling leakage, and lower threshold voltages used to offset the increase in gate delay with lower supply voltage. Subthreshold leakage power is a function of supply voltage, temperature, device threshold voltage, and logical state. Figure 13.2 shows the active and leakage power density trends by technology node.

There are several common approaches to reduce leakage power. Leakage power can be reduced by swapping faster transistors to slower transistors with *longer channel length* or *high threshold*

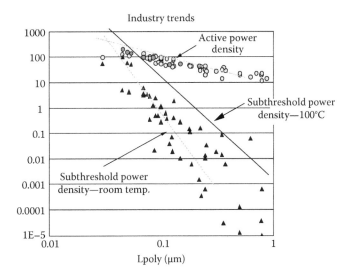

FIGURE 13.2 Active and leakage power trends.

voltage (high V_t) logic, which increases fabrication costs due to the additional implant masks. *Power gating* can disable the power supply to inactive portions of the chip at the cost of additional area for the sleep header or footer transistors for the power gates, which when on have some additional voltage drop that also reduces performance. The supply voltage may also be lowered or reverse-body bias applied to reduce leakage. These methods come at the price of reduced circuit performance, except where the cells impacted are only on paths that are not timing critical. FinFET transistors have been introduced in 22 nm and smaller process technologies, significantly reducing leakage power, but these new process technologies are more expensive, increasing unit cost.

In high-performance applications such as server microprocessors, power constraints are generally imposed by the amount of heat that a particular chip, system, and package combination can remove, before the chip temperature rises too high to allow proper function. In low-performance applications such as consumer products, factors such as battery life, packaging, and cooling expense are the major limits. Missing either an active or standby power constraint can necessitate costly redesign of chip, package, or system. In the worst case, it can render a chip unusable in its intended application.

13.1.2.5 SIGNAL INTEGRITY CONSTRAINTS

Signal integrity constraints prevent chip function from being disrupted by electrical *noise*. Signal integrity constraints include *power integrity constraints* and *coupling constraints*.

- *Power integrity constraints* are used to ensure that the chip power supply is robust enough to limit unacceptable supply voltage variations. As chip logic elements switch, current is sourced through the chip power routing. The current must either come from off-chip or from charge stored in the on-chip capacitance. This reserve capacity on-chip is in the diffusion structures, routing, and any *decoupling capacitors* (decaps) connected to the power bus. The decaps are added to provide small reservoirs of charge to smooth out switching transients. If it cannot be supplied by local decaps, a switching event with net current I induces a voltage drop $\Delta V = IR + LdI/dt$, as it flows through the resistance R and the inductance L of the power supply routing, where dI/dt is the rate of change of the current with time. These components are depicted in Figure 13.3.

The voltage drop breaks down roughly into three components: *IR drop*, which is affected by average current flow I and power bus resistance R; *steady-state AC voltage drop*, which is affected primarily by intra-cycle current variation and local decoupling capacitance; and *switching response*, which is affected mainly by the switching current variation and package inductance. When added together, the voltage drops induced by each switching event lead to potentially large variations in supply voltage, both in space and in time. Figure 13.4 shows a map of the spatial distribution of the average or

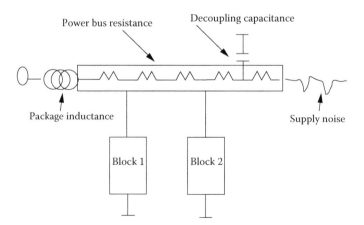

FIGURE 13.3 Power bus voltage drop components.

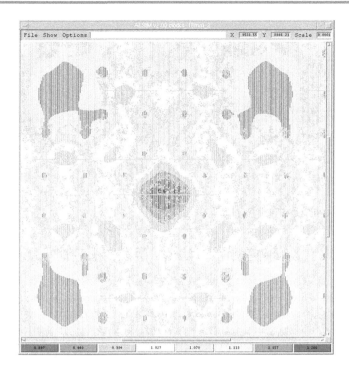

FIGURE 13.4   Power supply voltage drop map.

steady-state voltage drop across a large ASIC, as calculated by IBM's ALSIM tool. The resulting voltage fluctuations cause time-varying delays across a chip. If large enough, these delay variations lead to setup or hold timing violations. Power voltage transients can also introduce disruptive electrical noise into sensitive analog circuitry. The relative magnitude of power voltage variations has increased, as total chip power and operating frequency have increased, and supply voltages have decreased.

Power supply integrity can be improved by increasing the wire width, package power/ground pin count, and via count used for power supply routing. Transient power integrity can be further improved by the judicious use of decaps. These optimizations contend with capacity and routability constraints.

■ *Coupling constraints* are used to ensure that inter-signal coupling noise does not disrupt chip timing or logical function. Noise is always a design constraint as even a single violation is sufficient to render a chip inoperable. Coupling noise occurs when a voltage transition on a noisy *aggressor* wire causes current to be injected into an adjacent sensitive *victim* wire through the mutual capacitance of the two wires. The injected current ΔI is proportional to CdV/dt, where C is the mutual capacitance, and dV/dt is the rate of change of the voltage transition with time, that is, the *signal slew rate*. If the coupled signal exceeds the logic threshold on the victim wire, an incorrect logic value or *glitch* is induced in the victim circuit, as shown in Figure 13.5. If the victim wire is transitioning during the coupling event, its delay is affected. If not modeled correctly, this variation in delay can cause unexpected variations in chip performance. If these delay variations are large enough, they can lead to improper operation.

Coupling has increased markedly due to increased switching speeds; less noise margin due to decreased supply voltage; and increased cross-coupling capacitance with decreased inter-wire spacing and increased wire aspect ratios. Cross-coupling noise is most easily addressed by downsizing aggressors to reduce the signal slew rate, and/or upsizing victims to reduce their susceptibility, at the cost of increased delay and increased area, respectively. Cross-coupling capacitance can be reduced by segregating noisy and sensitive wiring with increased inter-wire spacing, and in extreme cases, by adding shielding wires—these optimizations contend with the routability constraint.

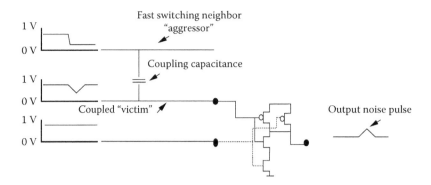

FIGURE 13.5 Coupling-induced logic glitch.

13.1.2.6 RELIABILITY CONSTRAINTS

Once the basic performance targets are met, we need to ensure that the chip will function properly through its required working life. Reliability concerns are related to processes that may allow a chip to function correctly immediately after manufacture but may cause the chip to malfunction at some later point in time. In the best case, this type of unexpected chip failure may present a costly inconvenience to the customer. In mission-critical functions such as automotive, avionics, or security, the result of a malfunction can be a risk to life. Many reliability factors can affect the long-term operation of a chip. Most fit into one of the three categories: *device wear-out constraints*, *interconnect wear-out constraints*, and *transient disruption constraints*.

- *Device wear-out constraints* relate to mechanisms that can cause a gradual shift of transistors' electrical characteristics over time. Two common concerns are hot carrier injection (HCI) and negative bias threshold instability (NBTI). HCI occurs when electrons in the channel of a transistor are accelerated by the high electric field found near the drain of devices, which are on, or switching. These highly energetic electrons are injected into the gate oxide where they create electron or hole traps. Over time, these traps lead to charge build-up in the gate, which changes the threshold voltage. For fast input slew rates, for example, below 100 ps in 45 nm technology, HCI increases as slew rate decreases, and as the number of fanouts increase [9]. HCI is accelerated for devices that have high applied gate voltages, high switching activity, and drive large loads. NBTI is similar in effect but does not require high electric fields. It affects both switching and non-switching devices. NBTI is accelerated by high operating temperatures and can be induced during burn-in test. The effect of these threshold shifting mechanisms is increasing over time due to the use of thinner gate oxides and lower threshold voltages required by technology scaling. Shifting the threshold of devices has a direct effect on the delay through a device. As it ages, delay can either increase or decrease, depending on the nature of the injected charge and the type of device. In time, the performance shift may be large enough to cause the chip to malfunction. Device wear-out can be minimized by reducing the current, by using the lowest supply voltage necessary, and by reducing load capacitance on the outputs of fast slew rate signals, both of which must be traded off against performance.
- *Interconnect wear-out constraints* relate to mechanisms that can cause a gradual shift in the electrical characteristics of chip interconnect with time. The principal mechanism is *electro-migration* (EM). EM occurs when ballistic collisions between energetic electrons and the metal atoms in the interconnect cause the interconnect atoms to creep away from their original position. This movement of metal can thin wires to the point that their resistance increases or they fail completely. Like device wear-out, EM-induced wire wear-out can affect delay to the point that the chip begins to malfunction. In the limit, EM can cause wires to completely open, which clearly changes the function of the chip. Electromigration is accelerated by increased current densities.

Additionally, the new low-permittivity dielectric materials introduced to help performance have inferior thermal characteristics. The result is an increase in wire self-heating, which further accelerates EM. EM problems can be mitigated by widening high-current wires and lowering the loads on high-duty cycle signals. These mitigations contend with both routability and performance constraints.

- *Transient disruption constraints* relate to mechanisms that cause a sudden failure of devices or interconnect. The two most common mechanisms are *electro-static discharge* (ESD) and *soft error upset* (SEU). ESD occurs when unacceptably high voltage is inadvertently presented to chip structures. This can occur either due to *induced charge build-up* during manufacturing or *a post-manufacture ESD event.*

Induced charge build-up can be caused by certain manufacturing processes that involve high electric fields that can induce charge on electrically isolated structures, such as transistor gates. If too much charge is induced, the resultant electric field can damage sensitive gate oxides. Induced charge ESD is mitigated by insuring that all gate inputs are tied to at least one diffusion connection. This allows a leakage path to ground, which prevents build-up of dangerously high fields. In some cases, this requires the addition of special *floating gate contacts.*

Post-manufacture ESD events occur when high voltages are inadvertently applied to the chip I/O by improper handling, installation, or grounding of equipment and cables. When an ESD event occurs, the gate of any device connected to the transient high voltage is destroyed due to the high field induced in its gate oxide. In some cases, the wiring that connects the pin to the device may also be destroyed due to the induced transient currents. Post-manufacture ESD events can be minimized by proper chip handling and by the addition of *ESD protection diodes* on all chip I/Os. These diodes protect the chip I/Os by shunting high-voltage transients to ground.

Soft error upsets are recoverable events caused by high-energy charged particles that either originate from outer space or from nearby radioactive materials. The carriers induced by the charged particle as it travels through the silicon substrate of the chip can disrupt the logic state of sensitive storage elements. The amount of charge required to upset a logic gate is dependent on its *critical charge.* As device structures shrink, the amount of charge required to cause a logic upset is decreasing. SEUs are a concern for circuits operating at high altitude and in space, for example, avionics, as well as for memory arrays and dynamic logic [10]. SEUs are best addressed at the architecture level through the addition of logical redundancy or error-correction logic, which is not a design closure step, per se.

13.1.2.7 YIELD CONSTRAINTS

For all its sophistication, semiconductor manufacturing remains an inexact science. Random defects and variations in device characteristics can be introduced at almost any step of manufacturing, which can cause a chip to not function as intended. The greater the number of chips affected by manufacturing errors, the more the chips must be manufactured to guarantee a sufficient number of working chips. Mispredicting yield can require costly additional manufacturing, expensive delays, and possible product supply problems. In this way, yield may be considered a cost-oriented design constraint. In many cases, maximizing yield is considered an economic design objective.

Foundries focus much of their effort on yield improvement, particularly on the introduction of a new technology for which initial yields may be only a few percent and insufficient for mass production. Early adopters of a new technology run the risk of being delayed to market, due to insufficient foundry capacity because of low yield [11]. Test chips are fabricated to validate a new process and identify device problems to mitigate this risk. Foundries will specify somewhat conservative design constraints to fabless design companies to try to ensure reasonable yield.

Design companies that own their fabrication plants, such as Intel, can optimize their fabrication process for high-volume designs (e.g., change V_t implant dosage to improve speed), and trade-off yield versus design performance and other constraints. Some additional design margin may be adopted to ensure that those process parameters can be tweaked safely, for example, more conservative hold time margins.

Today, most IC design companies are fabless, fabricating their ICs with pure-play foundries that do not themselves design ICs. Only those with high volumes and high price per part can negotiate with the foundries for significant process tweaks, such as additional minimum width metal layers that require double patterning, to achieve higher circuit density.

At the cost of additional testing, which is prohibitive for low-cost parts, fabricated ICs can be binned into high-speed and low-speed parts. It is also more common nowadays for parts to be binned between high power and low power, which have a higher selling point for the mobile sector. Binning is another way to trade-off yield versus design cost, recovering some value for less desirable ICs, providing that there is sufficient market for them.

The two types of yield constraints that must be considered during design closure are *defect-limited yield* and *circuit-limited yield*:

- *Defect-limited yield constraints* relate to a product that is rendered faulty during manufacturing due to *foreign material defects* or *printability defects*. Foreign material defects result when small bits of material accidentally fall on the chip surface or the mask reticule and interfere with the proper creation of a chip structure. Printability defects result when local geometry, chip topography, optical interference, or other manufacturing processes prevent correct printing of a desired width or spacing. These defects may take the form of unintended shorts between adjacent structures, unexpected holes in insulating materials such as device gates, or unintended opens in a conductor. Defect-limited yield can be improved by decreasing the amount of *critical area*, that is, the inter-geometry spacing that is less than or equal to the size of likely defects. Critical area can be minimized by using *relaxed* or *recommended design rules* rather than *minimum design rules,* wherever density will allow. This additional spacing contends with capacity and routability constraints.
- *Circuit-limited yield constraints* relate to yield loss due to the effect of manufacturing variations on chip performance. Despite advances in every phase of processing, there remain uncontrollable variations in the properties of the dimensions and materials of the finished product. These small variations cause identically designed transistors or wires to have significantly different electrical characteristics. The most-studied variation is *across chip line-width variation* (ACLV), which creates perturbations in the critical gate length of transistors. ACLV has many causes including uneven etching due to local shape density variations as well as lithographic distortions caused by optical interactions between shape regions. As interconnect dimensions have shrunk, variations in wire delay are becoming equally significant. For example, wire width variation of ±20% causes wire delay variation in minimum width wires of −7.5% to +15% in 32 nm technology [12] and −20% to +44% in 7 nm technology [13]. A large portion of the wire delay variation is due to lithographic distortion and issues related to etch rate variations in processing steps, such as *chemical mechanical polishing* (CMP).

Many trends contribute to the growing concern about parametric yield, including increasingly deep *sub-resolution lithography* and increasingly complex manufacturing processes. In addition, decreased device and interconnect dimensions contribute to the relative impact of variation. For example, decreased device channel dimensions give rise to *micro-implant* dopant variation, in which the distribution of the relatively small number of implanted dopant ions creates small differences in transistor threshold voltages. Similarly, gate oxide thickness approaches only five or so atomic layers. In such small configurations, a change of just one atomic layer can induce a quantized threshold voltage shift of nearly 10%, as shown in Figure 13.6.

These parametric variations in turn give rise to statistical variations in design performance characteristics, such as delay and leakage power. Figure 13.7 shows a typical manufacturing distribution of a large sample of *performance screen ring oscillator* (PSRO) circuits used to characterize the performance of a microprocessor. In the example, the PSRO circuits in the

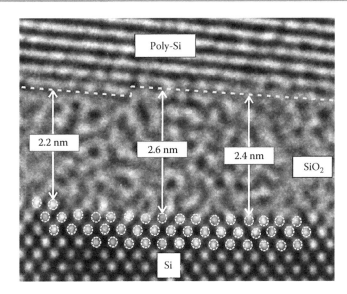

FIGURE 13.6   Gate oxide thickness variation.

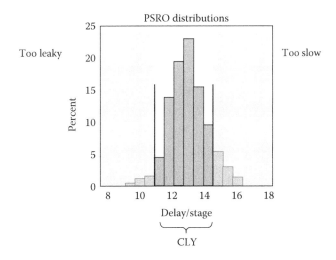

FIGURE 13.7   Performance screen ring oscillator (PSRO) delay distribution and circuit-limited yield (CLY).

left-most tail of the distribution represent the fastest circuits, but these same circuits violate the leakage power constraint and must be discarded. Similarly, the PSRO circuits on the right tail have low leakage, but they are too slow and fail to meet the setup timing constraint and must also be scrapped. The portion of the distribution between these two bounds is the circuit-limited yield.

These variations may be observable both when measuring the same device on different chips (*inter-die*) or between identical devices on the same chip (*intra-die*). Both inter-die and intra-die variations cause the actual performance of a given chip to deviate from its intended value. For example, Figure 13.8 illustrates inter-die variations, as measured using identical ring oscillators placed on each die on a 200 mm silicon wafer. The areas of identical shading have identical frequency measurements. The total range of variation is 30%.

Parametric yield is emerging as a design closure constraint for structured logic. We will discuss how this will affect the design closure flow in our discussion of the future of design closure in Section 13.3.

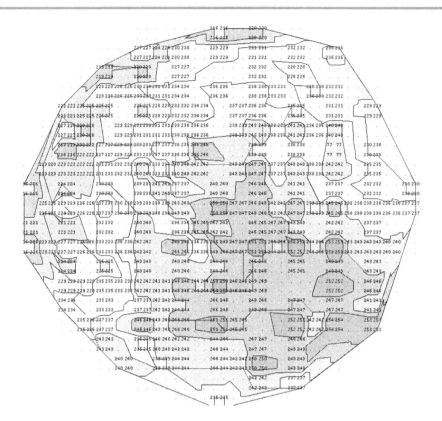

FIGURE 13.8   Wafer map of inter-die parametric variations.

13.2 CURRENT PRACTICE

In this section, we will examine the design closure implications of each phase of the ASIC design flow, shown in Figure 13.9. At each phase, we will examine the design constraints that are addressed, estimations and approximations that are made, and the trade-offs that can be made between constraints. We will also explore the interactions—both forward and backward—between the phases.

13.2.1 CONCEPT PHASE

The design concept phase is to set the overall scope of a chip project. During the concept phase, many aspects of the design are estimated: gate count; cell area utilization; area increase to alleviate routing congestion; operating frequencies; supply voltages; power dissipation; I/O count; yield percentages; fabrication process technology; and requirements for special processing, such as embedded dynamic RAM and analog circuits.

There are a number of choices to be made for process technology: technology generation; high performance or low power; transistor threshold voltages and channel lengths; low-k dielectric insulator to reduce cross-coupling; standard cell height; metal widths and number of metal layers that will require more expensive immersion lithography and double patterning. For example, shorter standard cell height increases area density, reducing wire length and power consumption, at the cost of reduced performance for smaller drive strengths [14] and reduced routing porosity as there are fewer routing tracks.

All of these factors are estimated and combined into a business case: the chip cost to manufacture; its yield; its design time; its design cost; and its market price. Incorrectly estimating any of these factors can significantly impact the business case, sometimes to the point of jeopardizing the entire project.

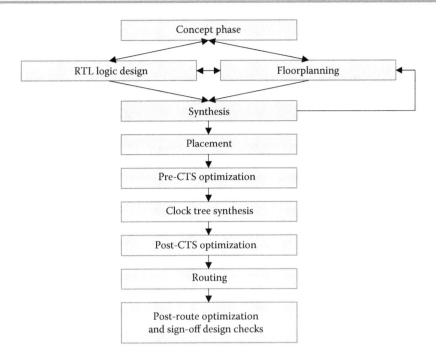

FIGURE 13.9   The design closure flow.

Many trade-offs are made during this phase. The design is still at an abstract state, so changing significant aspects is easy to do: changing die sizes; making the power design more robust; adding or deleting RAMs, functional units, or changing the number of processor cores, to trade-off power for speed, speed for area, and area for design time. The challenge is to estimate characteristics of the design as accurately as possible, with only an inexact notion of what the design will eventually look like. The aspects of the design that need to be estimated are as follows:

- *Amount of logic*: This is estimated, based on a number of factors, including the size requirements of large reused components, for example, memory, embedded processors, I/O circuitry, arithmetic functions, and other data path functions. The amount of small gate-level *glue logic* is estimated by comparing with past designs, experience, and technology insights. We can accurately calculate the required active logic area from the logic count.
- *Die size*: This is derived from the active logic area by adding extra space for routing. First, a target placement density is chosen. Then an area increase factor based on empirical rules of *routability* is applied to the chosen placement density. The area increase factor may be adjusted upward, based on the type of design. A densely interconnected structure, such as a cross-bar switch, will require additional routing area. The density can be adjusted up or down by trading off design time. Higher densities can be achieved with extra effort in manual placement. Finally, the limiting factor in die size may be the number of I/Os. For example, the die size of an extremely high pin-count design with simple logic functionality will be defined by the I/O count.
- *Defect-limited yield*: This factor is calculated from logic count and the die size. Yield prediction is based on empirical tables, which take into consideration logic density, die size, and technology information, as shown in Figure 13.10.
- *Performance*: This is estimated by examining factors such as maximum logical path length. In the absence of an actual design, the critical path length is derived by experience with similar designs, technology parameters, *a priori* knowledge of the performance of embedded IPs, required I/O rates, and estimated interconnect delays based on projected die size, voltage, and power limitations. Performance is a strong function of voltage, so the voltage is set to achieve performance targets defined by system or marketing constraints.

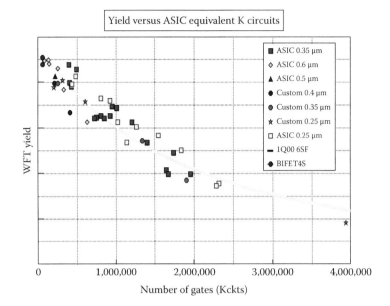

FIGURE 13.10 Yield versus number of gates.

- *Power*: Dynamic power is estimated based on performance, supply voltage, technology parameters, and empirical information about switching factors. Leakage power is estimated based on logic count, technology parameters, supply voltage, threshold voltages, and channel lengths. There are significant architectural levers that can affect the power. For example, supply voltage or frequency scaling can be specified, and significant subsystems can be identified as candidates for clock gating and power gating.
- *Package*: Once the power is known, the package can be determined. Designers will choose the cheapest package that supports the frequency of operation, power dissipation, I/O requirements, and mechanical constraints.
- *Unit cost*: This is derived from yield, die size, and package cost. Projected volumes from marketing are also a factor in unit cost calculations.

In reality, the calculations are more complicated than this list implies. There are many trade-offs that are made. For example, architectural accommodations can favor performance at the expense of power and die size, such as adding additional arithmetic units to improve throughput. There are trade-offs balancing area for yield by adding redundancy, such as extra word lines on array structures. Likewise, adding error-correction logic trades off area for increased reliability. Design effort is balanced against unit cost, time to market, performance, and power. The calculations outlined earlier are made and revisited, as different trade-off decisions are made.

Platform-based design [15] allows for a much higher degree of accuracy of these early estimates. This is where a design is made up of one or more reused large components including processors, integer and floating-point arithmetic units, digital signal processors, memories, etc.

After all the trade-offs have been made, the product of the concept phase is an architectural specification of the chip and a set of design constraints. These are fed forward to the floorplan and logic design phases.

13.2.2 RTL LOGIC DESIGN

The logic design phase involves implementing the register transfer logic (RTL) description of the chip based on the concept phase architectural specification. First, the specification is mapped into a hierarchical RTL structure in as clear and concise a fashion as possible and the details are filled in. Then the RTL is simulated against a set of test cases to ensure compliance with the specifications. Because logic design and floorplanning are so intimately linked, the two steps generally proceed in parallel with each other.

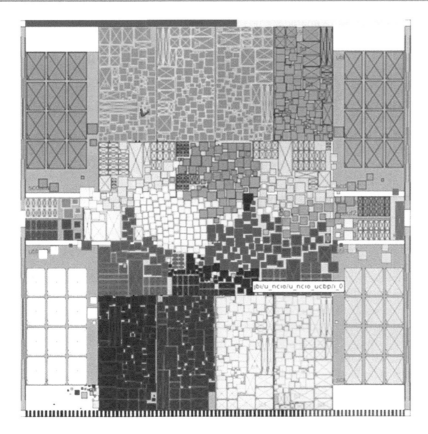

jbi/u_ncio/u_ncio_ucbp/i_0

FIGURE 13.11 Floorplan generated from RTL in Mentor Graphic's RealTime Designer™. (From Mentor Graphics, RTL synthesis for advanced nodes: Realtime designer, white paper, 2014. http://www.mentor.com/products/ic_nanometer_design/techpubs/request/rtl-synthesis-for-advanced-nodes-realtime-designer-87068.)

There are three main design closure aspects that are dealt with during the logic design phase: performance, power, and routability. Because the design is so easy to change at this phase, mitigations of problems in these areas are easy to implement. It is difficult to measure any of these parameters directly from the RTL, which has not yet been mapped to gates, and is not yet placed or routed. However, there are now tools that can provide a more accurate analysis with a quick pass through floorplanning (see Figure 13.11), synthesis, global placement, and global routing [16]. The RTL logic implementation is improved in further design iterations by feedback on the performance, power consumption, and routability from subsequent design phases, such as logic synthesis and placement.

The design performance is improved at the RTL by identifying a set of critical paths and modifying the RTL to reduce the logic depth. Reducing the amount of logic in a path is done by analyzing the logic to identify portions that are not required, or which can be refactored to reduce the logic depth. Owing to the hierarchical nature of design, there are often logical redundancies that can be removed by restructuring the RTL hierarchy. Performance can also be improved by *path balancing* or *retiming*; moving logic from one side of a long path's source or capture *latches* or *flip-flops* to the other.

There is also significant leverage for the mitigation of power issues at the RTL. The RTL can be modified to use *clock gating* to save dynamic power by shutting off the clock switching to idle portions of logic and *power gating* to save both active and standby power by switching off the power to unused portions of the design. Both clock and power gating require careful attention to ensure that gating signals are calculated correctly and arrive in time to allow logic to stabilize as it comes out of its idle state.

The RTL can be simulated across a suite of typical usage scenarios to automatically identify additional clock gating opportunities to reduce power consumption. This helped to achieve a 25% reduction in power consumption on AMD's Jaguar core versus the previous design generation Bobcat, which had already been heavily clock gated [17]. Designers can better architect the

RTL to specify clock gating enable signals that are not timing critical and can clock-gate more logic. Synthesis tools can infer clock gating logic, but it is more fine-grained, providing less power reduction, and may be limited due to timing criticality. In particular, the clock gating enable signal must be stable when the clock is high, so the enable signal is captured by a negative phase latch to prevent glitches (e.g., by using an integrated clock gating cell), and thus, the enable signal must be generated within half a clock-cycle. An RTL designer can generate the enable signal from logic on the previous clock cycle, avoiding timing problems.

In addition to these logical transformations, power/performance trade-offs can be made by using *frequency scaling* or *voltage scaling*. In frequency scaling, portions of the design that can run more slowly are segregated into more power-efficient lower-frequency clock domains, while more performance-critical logic is assigned to higher-frequency, and therefore, higher-power domains. Voltage islands allow a similar power/performance trade-off by assigning less performance-critical logic to a lower voltage "island." Using lower supply voltage reduces both active and standby power at the cost of additional delay. Voltage islands also require *level-shifting logic* that must be added to allow logic-level translation between circuits running at different voltages. The granularity of voltage islands needs to be chosen carefully to ensure that the benefits of their implementation outweigh their performance, area, and power overhead. Some designs also utilize power gating with dynamic voltage and frequency scaling, adjusting the design performance and cores that are operating based on the system load [18].

Routability can also be optimized during the logic design phase by identifying congested regions. The RTL hierarchy can be restructured so that interconnected logic in a congested region is in the same portion of hierarchy to allow better logical optimization and placement. In extreme cases, this allows designers to hand-instantiate gate-level logic for the RTL and to manually place it, which is particularly useful in regular dataflow or "bit-sliced" logic.

Owing to the enormous amount of simulation and verification time to ensure logical correctness, logic design is the most time-consuming phase of the design closure flow and happens in parallel with the rest of the flow. At regular intervals, the RTL is brought into a consistent state (e.g., matching sets of I/O pins between logic blocks) to iterate through the rest of the design flow. These trial runs give engineers responsible for subsequent steps, opportunities to tune their recipes and timing budgets, and provide feedback to the logic designers and floorplanners about factors such as timing critical paths, particularly those between logic blocks, and poor structure.

The product of this phase is a hierarchical RTL design and amended constraints. These are fed forward to the final stages of floorplanning and the logic synthesis phase.

13.2.3 FLOORPLANNING

The floorplanning phase prepares a design for the placement of the standard cell logic and other larger objects. Figure 13.12 shows the floorplan of a large ASIC design. Owing to their tight linkage, floorplanning generally proceeds in parallel with the logic design phase. From the floorplan, rough load capacitance values are calculated for global nets, which can be used to guide logic synthesis.

Design work in the floorplanning phase includes placing large *macro* objects; creating power/ ground grids; routing upper metal layers for clock distribution; and creating rows in the remaining areas for the placement of the standard cell logic. Large objects consist of memory macros such as register arrays, Random Access Memory (RAMs), and Content Addressable Memory (CAMs); I/O macros with interface logic and wiring between the I/O cells and pads; large clock buffers to drive the clock grid or clock trees; decaps; analog circuits such as phase lock loops; and any hand-designed *custom macro* function blocks. In a hierarchical design, pin locations of subhierarchy blocks are assigned as part of the floorplan, as are restricted placement areas. White space may be reserved in the floorplan for different purposes: to instantiate repeater buffers on global nets; to clone clock gaters and enable latches for leaf-level clock mesh distribution, aligning with upper routing levels of the clock mesh; and for the placement of large sleep header or footer cells for power gating, along with always-on logic and state retention flip-flops to preserve state, when other logic enters sleep mode.

By establishing macro and I/O placement, floorplanning has a large impact on the interconnect delay on critical paths. As interconnect scaling continues to worsen, the importance of

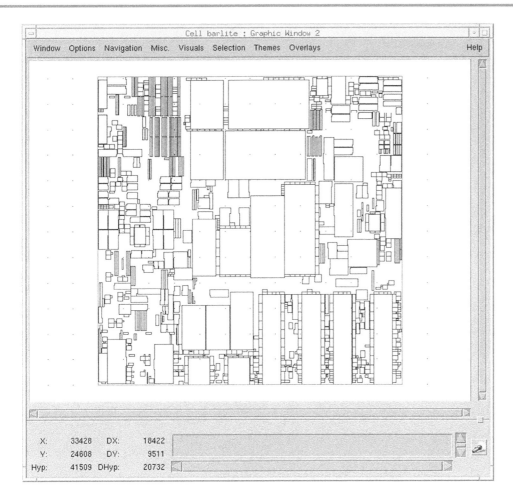

FIGURE 13.12   ASIC floorplan showing large objects.

floorplanning is increasing. The major design closure aspects that are treated in this phase are routability, performance, power supply integrity, and power.

The initial macro placement is guided by insights about their interconnectivity, expected participation in critical paths, and their impact on routing congestion. A common method is to place macros around the periphery of a design to reduce congestion due to these large obstructions. Narrow whitespace "channels" between macros, or between macros and the floorplan boundary, can be troublesome for both routing and placement. For logic placed in such a narrow channel, placement optimization may run out of whitespace to upsize cells or place buffers to speed up a critical path. A keepout halo may be specified around a macro to avoid cross-coupling noise from wires parallel to the macro [19] and to avoid logic being placed in the narrow channels. The floorplan is refined based on feedback from subsequent steps, such as insight into critical performance and congestion issues from post-placement optimization, global wiring, and timing analysis runs.

In this context, global wiring refers to routing of the longer wires across the chip, typically across other blocks, rather than within a block or between neighboring blocks. If global wires are too long to meet timing constraints, the floorplan may need to be changed to move those blocks closer together, registers and storage buffers may need to be inserted, or other similar solutions.

Routability is the primary design closure consideration dealt with during the floorplanning phase. The placement of the macros significantly affects the eventual congestion of the design. Macros with a high degree of interconnection are placed close together, with central areas of reduced placement density defined to accommodate the high wiring load—the classic example of this is a cross-bar switch. There are a number of floorplanning tools that can provide assistance, ranging from the simplest that give an abstract view of the macros, the glue logic, and their interconnectivity as in Figure 13.13 to virtual prototyping tools that do quick low-accuracy synthesis,

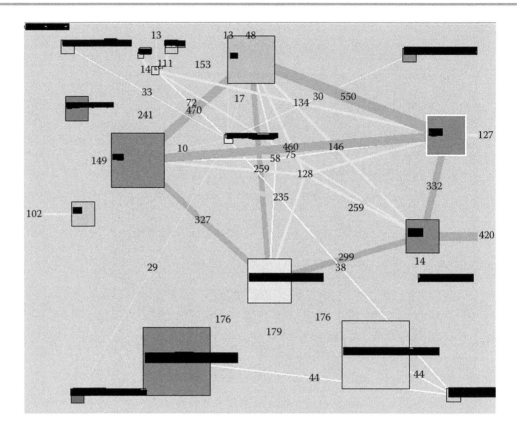

FIGURE 13.13   Interconnectivity and logic size visualization tool.

placement, and routing and give almost real-time feedback. Later, feedback from the actual place-ment and routing steps is used to adjust the location of the macros to reduce congestion.

Performance problems are also addressed during the floorplanning phase. The key action is to place macros that have timing-critical connections close together. Initially, this is done with *a priori* knowledge of the timing paths in the design. The floorplan is adjusted based on either low-accuracy timing feedback from virtual prototyping tools or higher accuracy feedback from timing runs, performed after placement and global wiring become available. Floorplanning insights from these sources are used to drive restructuring of the RTL in the logic design phase to keep clock domains and critical logic closer together.

Power supply integrity can also be addressed during the floorplanning phase. Decoupling capacitors are placed to provide power supply isolation to reduce voltage drop, and decaps are also used with clock meshes that have insufficient capacitance in the mesh itself to avoid voltage drop due to significant simultaneous currents when the clock switches. Decaps may specifically be placed around particularly noisy elements, such as CAMs and large clock drivers. The initial power supply distribution is based on factors such as the location of chip power pins, power requirements of large fixed objects, expected standard cell utilization densities, and locations and sizes of voltage islands. Post-placement feedback is used to augment the power supply design if necessary. If power density is estimated to be too high in certain areas, the placement density in these areas may be reduced, or supply and ground rails in those areas may be widened, to meet electromigration constraints.

Power is addressed indirectly during floorplanning. The most important power optimization at this stage is the planning of the voltage islands introduced in the logic design phase. Each voltage island requires the design and analysis of its own separate power supply and power pin routing.

During the floorplanning phase, there are a number of trade-offs that are made. Addition of decaps or lowered placement density regions can cause extra wire length, which impacts per-formance. Overdesign in the power distribution can significantly affect routability, which again can impact performance. Finding the right design points among all these factors can often take a number of iterations between floorplanning, logic design, and the subsequent design flow steps.

The hand-off from the floorplanning phase is a detailed floorplan, including macro placements, power grids, move bounds, pin assignments, circuit rows, I/O wiring, and amended constraints. This floorplan is used both in physical logic synthesis and in placement.

13.2.4 LOGIC SYNTHESIS

The purpose of the logic synthesis step is to map the RTL description of the design to a netlist rendered in library elements of the chosen technology, meeting the performance targets with the fewest number of gates, as detailed in Chapter 2. The main design closure issues addressed in the phase are performance, power, and area. First, the RTL is compiled into a technology-independent netlist format. Then, this netlist is subjected to logical analysis to identify and remove redundancies and balance cones of logic. Next, the optimized technology-independent netlist is mapped into technology-dependent gates. Finally, the technology-dependent netlist is timed, and critical timing paths are optimized to try to meet the delay constraints.

Because of the relative ease of running logic synthesis with different optimization targets, designers use it to explore the design space to make the best performance, power, and area trade-offs for their design. Logic synthesis applies a number of techniques to the technology-mapped netlist to achieve the area, power, and performance goals. In order to do this, these factors need to be measured: logic area is measured by adding up the sizes of the various gates; power is assumed to be a function of gate size—reducing gate sizes reduces power. Measuring performance is more complicated, and logic synthesis is the first phase to rely heavily on static timing analysis. As the clock distribution circuitry has yet to be added, idealized clock arrival times are applied to launch and capture clocks at registers. The logic synthesis step must also anticipate the impact of buffer insertion that will be performed during and after placement to prevent interconnect delay from being over estimated.

Since the standard cell logic gates have yet to be placed, *wire load models* were historically used in the synthesis step to estimate the gate loads and wire delays to provide a reasonable sizing and buffering solution in gate-level optimization after technology mapping. Wire load models estimate the wire capacitance and resistance versus the number of fanouts. Wire load models were based on chip size for a given technology and could be characterized for a particular design, once routed, to improve accuracy when next iterating through the design flow. There are subtle interactions between the wire load–based parasitic estimation of the logic synthesis phase and subsequent placement and pre-CTS placement optimization phases. If the wire load models overestimate loading, then the power levels in the gates in the resulting design passed to placement will be excessively large, with no real way to recover the overdesign. However, if the parasitic estimation is optimistic, then optimization in this phase will not focus on the correct problems. In general, erring on the side of optimism produces better results, especially with the advent of truly effective timing-driven placement flows. Chips were synthesized with zero wire load models for local signals, and load estimates for global signals were derived from the floorplanning step.

As wire loads and delays have became more significant, *physical synthesis* improved the accuracy by including global placement of the gates, with wire load capacitance and wire RC delay estimated from a Steiner tree for a wire's connectivity. Newer approaches to logic synthesis interleave quick floorplanning, global placement, and global routing, with resynthesis from RTL to rapidly achieve production quality synthesis results, as in Mentor Graphics' RealTime Designer™ [16]. RTL optimization can remap the logic to different structures, providing wider opportunities to close timing than can be achieved in gate-level optimization. Whether from RTL or at a gate level, resynthesis can fix timing critical paths or severe routing congestion that may only become apparent after placement or routing steps—these can be fixed, respectively, by reducing logic depth and by remapping to more complex cells to reduce pin count.

There are a number of environmental factors that need to be set in synthesis to guide static timing analysis, such as voltage and temperature, based on information from the concept phase, and guard banding for manufacturing variation and reliability factors such as HCI and NBTI. Performance is measured using incremental static timing analysis, and transforms that trade off area, performance, and power are applied.

	Frequency	Cumulative	%	Cumulative%
−1.20		0	0	0.00	0.00
−1.00	*	2	2	0.02	0.02
−0.80		0	2	0.00	0.02
−0.60	*	4	6	0.04	0.05
−0.40	*	4	10	0.04	0.09
−0.20		0	10	0.00	0.09
0.00		0	10	0.00	0.09
0.20	*****	479	489	4.30	4.39
0.40	******	607	1096	5.45	9.85
0.60	*********	814	1910	7.31	17.16
0.80	***	292	2202	2.62	19.78

```
        ----+---+----+
        400  800  1000
```

Worst slack = −0.98
Number of negative slack = 10
Total tests = 11,132

FIGURE 13.14 Timing-slack histogram.

The basic approach to reducing power consumption and minimizing standard cell area is reducing gate sizes in a sub-step in the design flow, where all the gate sizes in the design are resized. This includes reassignment of transistor threshold voltages and channel lengths where permissible to trade off leakage power versus performance. In addition to gate-level optimization in the synthesis step, area minimization may be used at other steps in the design flow, including for very similar objectives such as leakage power, dynamic power, and maximum utilization.

Critical paths are individually timing corrected using a slack-take-down approach, sometimes referred to as *WNS optimization*. Critical paths are ordered from the worst negative slack (WNS) to the least. The worst timing path has one or more timing optimization transforms applied to it, improving it "toward the good" on the critical path list (see Figure 13.14). The process loops back to create a new ordered list of bad paths for further optimization to progressively improve WNS. Timing paths may be grouped for WNS optimization to attack the worst path in each group, rather than just focusing on the single worst WNS path across the design.

This slack-take-down is repeated until all paths meet the performance constraint, or there are no more optimizations that can be applied to the top critical paths. Further timing optimization may be performed on some of the worse timing violations to reduce total negative slack (TNS). The WNS and TNS optimization approach is used extensively in the subsequent design closure phases. A short list of some of these optimizations used to improve timing on the critical path includes the following:

- *Retiming* of logic across register boundaries to balance the amount of logic between registers.
- *Rewiring* where timing critical signals are moved "later" further toward the sinks in a cone of logic to reduce the overall path delay.
- *Refactoring* where a group of logic is mapped back into technology independent form and then resynthesized to preferentially shorten the logic depth of a critical path.
- *Cloning* where a section of logic is replicated to allow logical fanout to be divided over more drivers.
- *Repowering* where a logic function is replaced by a similar function with higher drive strength. (Repowering used the other way, to reduce gate sizes, is the work horse transform for both area reduction and power mitigation in the logic synthesis phase.)

The first time that accurate gate count and timing are available in the logic synthesis phase is after technology mapping. Previous phases have all relied on gate count and timing estimates. Surprises encountered when these numbers become available cause looping back to the floorplanning phase to reallocate space due to excess logic, or to reposition floorplanned elements for performance reasons, or to go back to the logic design phase for various area and performance mitigations available in that phase.

The output of the logic synthesis phase is an area, performance, and power optimized technology-mapped netlist. This is combined with the floorplan and the most recent updated list of constraints and passed onto the placement phase. Physical synthesis tools can now also provide the global placement therein, as a seed to the placement step to improve design convergence.

13.2.5 PLACEMENT

The purpose of the placement phase is to assign cell locations that shorten timing-critical wires and minimize wiring congestion, as detailed in Chapter 5. Figure 13.15 shows the placement of a small portion of a larger ASIC design.

Until recently, placers solved congestion and performance problems by creating a minimum wire length placement—both min-cut or quadrisection placement techniques provide good results. As interconnect delay has become more dominant, it has become necessary to make the placement flow more timing driven.

An effective timing-driven placement flow involves two placement passes: *global placement* and *detailed placement*. Before the global placement, the design is preprocessed to remove artifacts such as buffers on large-fanout nets, repeaters on long nets, and scan and clock connections, which may bias the placement improperly. Then, a timing-independent congestion–mitigation placement is run. Next, a series of timing optimizations including gate sizing, buffer tree insertion, and long wire buffering are performed. These optimizations require the analysis of only slew and capacitance violations. Resized gates and new buffers and repeaters are added to the design, without regard to legal placement constraints.

The design is then timed, using ideal clocks and interconnect delays calculated from Steiner estimates of wire topology. The timing problems that are identified are the "hard" problems that need to be fixed by the timing-driven detailed placement. A set of attraction factors, or *net weights*, are then calculated, which bias the placement engine to move timing-critical objects closer together. These net weights are used to guide an incremental detailed placement of the optimized design. Finally, scan chains are reconnected and reordered (see Figure 13.16), and well taps are added.

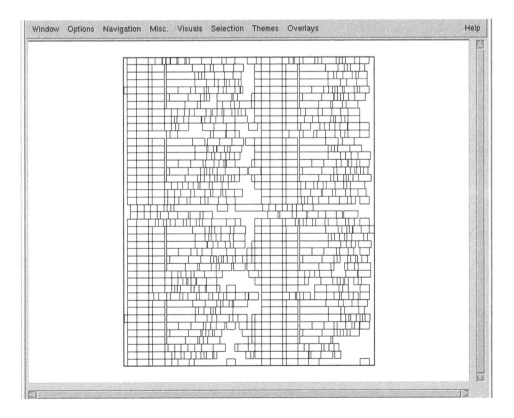

FIGURE 13.15 Logic placement.

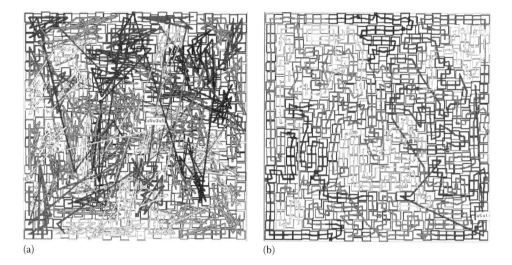

FIGURE 13.16   Register scan chain connectivity from scan stitching in synthesis (a) and with scan reordering after placement (b), as shown in Mentor Graphic's RealTime Designer™. (From Mentor Graphics, RTL synthesis for advanced nodes: Realtime designer, white paper, 2014. http://www.mentor.com/products/ic_nanometer_design/techpubs/request/rtl-synthesis-for-advanced-nodes-realtime-designer-87068.)

A key point in this timing-driven flow is the stability of the placement algorithm. A small change in the input to the placement engine, such as adding attractions between a small percentage of the objects being placed, must generate a small change in the result. If this is not the case, although the timing-driven placement will have pulled the timing-critical objects from the first placement closer together, a completely new set of timing problems will manifest themselves. This same stability can help limit the disruption caused by late *engineering change orders* (ECOs) to chip logic.

Another key consideration in placement is design *hierarchy*. All of the previous design steps—concept, logic design, floorplanning, and logic synthesis—rely on hierarchy to limit problem complexity. At the placement stage, it is possible to either keep the logic design hierarchy or *flatten* it, whether partially or fully.

Let us pause to quickly overview what is typical in a logic design hierarchy and the limitations that hierarchy imposes. Top-level RTL for a design of millions of gates typically comprises quite a few blocks that may range in size from 50,000 to 500,000 gates, such as a floating point unit. These large modules may comprise various subblocks for functional units, control logic, etc. Some subblocks may be further subdivided and may include small hand-crafted RTL modules for specific logic, register banks, and so forth. Logic optimizations permitted across hierarchy are typically limited to buffering, inversion, and port-punching (adding new I/Os in intermediate levels of hierarchy) for scan and buffering. Inversion and port-punching may or may not be allowed, depending on any intended reuse of the sub-module. Refactoring and other powerful optimization techniques are not permitted across logic hierarchy, which is also a necessary limitation for formal verification versus RTL at the module hierarchy level.

Voltage regions and power-gated portions of the design usually correspond to logic hierarchies that have particular voltage requirements to achieve a given performance, or to reduce power, or can go into standby or sleep states at a given time. Coarse clock gating is also generally done at a logical hierarchy block level. As there are particular power supply and clock distribution requirements for these design portions, it is common to maintain some intermediate level of hierarchy for them.

If the hierarchy is kept, each hierarchical unit is placed separately, and then the units are combined to form the chip. Retaining all or some of the hierarchy allows better parallelization of design effort, easier reuse of design, and faster incorporation of design changes. Certain designs blocks, for example, shader blockers in a graphics processor, may be replicated across the top-level chip to minimize design effort, maintaining the same cell placement, sizes, wiring et al. through to post-route optimization. Hierarchy is a natural way to keep the highest-frequency portions of a design physically compact, which can help reduce clock skew therein. For better area usage and to reduce wire length between cross-hierarchy interconnected logic, placement tools may be permitted to ignore logical hierarchy when placing cells, but this prevents the reuse of the physical design.

In flat design, the borders between some or all logical hierarchies are dissolved, and the flattened logic is placed together. Generally, flattening a design allows a significantly better optimization of performance and power during placement and subsequent design steps and requires less manual effort. It should be noted that some flat placement flows provide some form of *move bounds* mechanism to manage the proximity of the most performance-critical circuits. The debate on advantages and disadvantages of flat versus hierarchy continues in the industry and is worthy of its own chapter.

The output from the placement phase is a placed and sized netlist—including buffers, repeaters, and well taps—which is passed to the pre-CTS placement optimization phase.

13.2.6 PRE-CTS PLACEMENT OPTIMIZATION

The purpose of the pre-CTS placement optimization phase is to apply placement, timing, and congestion-aware transformations to the logic to fix critical setup timing paths and power problems left over from the placement phase. The timing-driven flow outlined in the placement phase is excellent at localizing large numbers of gates to solve general performance and routability problems. However, after the placement phase, there remain performance-critical paths that need to be fixed individually. In addition, local power and routability issues are considered while making these optimizations. This phase has two steps. First, electrical problems such as maximum input slew violations at signal sink pins and maximum load capacitance violations at output pins are corrected. These problems are fixed by gate sizing, buffering of large fanout nets, and repeater insertion. Second, the design is timed and optimized, using the slack-take-down WNS and TNS optimization approach.

The timing environment includes using ideal clocks, worst-case setup timing rules, and wiring parasitics extracted from Steiner wires or global routes. This is the first place in the design closure flow where useful information about spatially dependent power-supply *IR* voltage drop is available—this information is applied to the timing model where it is used to adjust individual gate delays. Feedback from this analysis can be used to guide redesign of the power supply routing.

The logic optimizations used in the second step are similar to the WNS and TNS timing optimization transforms, mentioned in the logic synthesis phase. They are augmented in placement optimization to also consider assignment of locations to changed gates, impact to placement density, and wiring congestion. When logic is modified at this step, its placement must be *legalized* to a legitimate nonoverlapping placement site. This generally involves moving nearby logic as well, which can induce new timing or congestion constraint violations. These new violations are queued to be optimized. The step is complete when no more resolvable violations exist. An important part of this phase is the tool infrastructure that allows for the incremental analysis of timing and congestion, caused by simultaneous changes to the logical netlist and the placement (see Section 1.4.3 for more details). As the placement changes, previously calculated interconnect delays are invalidated, and when new timing values are requested, these values are recalculated using Steiner tree or global routing estimates of the new routing topology. To measure congestion, an incrementally maintained global routing congestion map is used.

Routability can be affected in a number of different ways in this phase. Routing congestion information is used to assign lower placement densities in overly congested regions. As transforms are applied and logic moves, the regions with a lower density limit are avoided, mitigating congestion. This can impact timing, as gates that cannot be placed in their optimal locations, due to placement density constraints, are placed further away. Similarly, upsizing and buffering of timing critical paths to speed them up can also be limited by the available area nearby. Another method of congestion mitigation is via congestion avoidance buffer placement, guiding the routes that long nets take by placing the repeaters along those nets in less congested regions. Figure 13.17 shows a post-placement chip routability map.

Power and area can be recovered at this stage with the application of gate resizing. Despite the fact that the bulk of placement is already done, reducing the gate sizes wherever possible provides additional space for subsequent changes and additions and reduces power. Also, at this point, there is enough accuracy in the timing to optimize the choice of the cells' *transistor threshold*

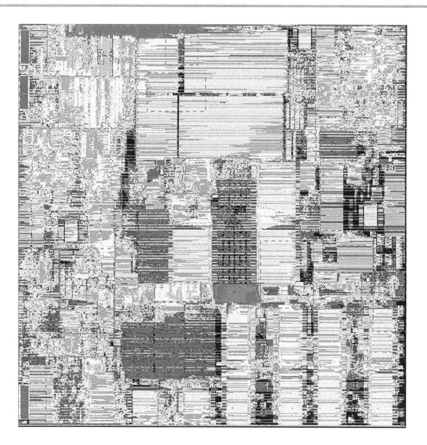

FIGURE 13.17 Chip routability map.

voltages. A gate may be swapped to another cell with a different threshold voltage that has the same logical function and footprint of its standard threshold voltage counterparts. High threshold voltage logic can be used to reduce leakage power at the cost of increased delay—off-critical path standard threshold voltage gates can be replaced with their high threshold voltage equivalents. Particularly timing critical paths can be swapped to low threshold voltage logic at the cost of higher leakage power. Figure 13.18 shows a delay comparison between low and standard threshold logic. These substitutions are done generally by sequentially substituting logic on the most critical paths, until all timing violations are resolved, or until the amount of low threshold logic exceeds some leakage power-limited threshold. Low-V_t substitutions can be done at this phase, or after clocking, when more timing accuracy is available. The main cost of additional

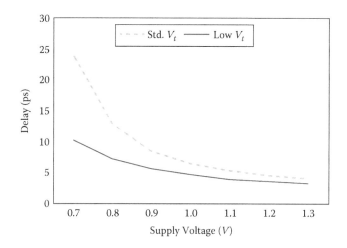

FIGURE 13.18 Effect of low threshold voltage logic on delay versus supply voltage.

transistor threshold voltages is the additional mask per threshold voltage; an alternative is to use longer channel length transistors at the cost of somewhat higher transistor capacitances and correspondingly somewhat higher dynamic power. It is not uncommon for today's designs to use a mix of two or three transistor threshold voltages and two or three channel lengths for fine-grained delay versus power trade-offs.

Clock power can be decreased by reducing the clock wire lengths by placing nearby the registers that have the same clock enable. ASICs commonly use flip-flops for registers, though latches may also be used instead, to reduce the clock load and clock power at the expense of substantially increased hold time constraints, or requiring a pulsed clock to turn the latches off quickly. The clock power can be further reduced by swapping those registers with the same enable to multi-bit latches or flip-flops that can share common clock circuitry. The cost for moving the registers to physically cluster them is the increase in the delay on the data signals, into and out of the registers, and displacement of other logic from good locations.

The product of the pre-CTS placement optimization phase is a timing-closed legally placed layout, which is passed to the clock tree synthesis design phase.

13.2.7 CLOCK TREE SYNTHESIS (CTS)

This phase inserts clock buffers and wires to implement the clock distribution logic. Figure 13.19 shows the clock distribution logic of a large ASIC design. By this phase, the amount of design optimization that can be accomplished is more limited. The main design closure issues dealt with in this phase are performance, signal integrity, manufacturability, and power supply integrity.

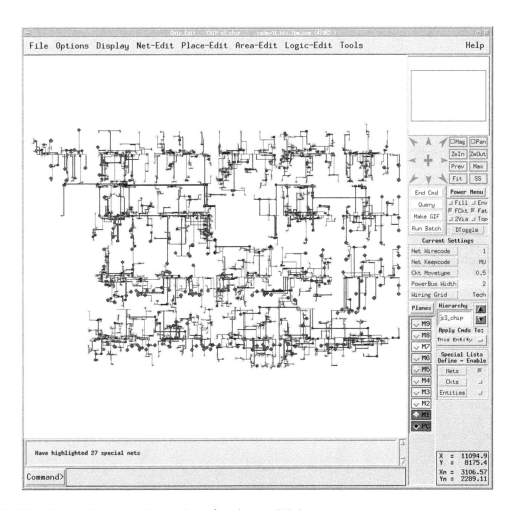

FIGURE 13.19 Clock distribution logic for a large ASIC design.

This step in an ASIC design flow is typically *clock tree synthesis* (CTS), where a deep tree of clock buffers is constructed to drive the clock from a single source to the registers, or to the enable latches and clock gaters that drive the registers. A simple clock gater is an AND2 gate with clock and enable inputs. Negative phase enable latches are required to prevent glitching of the enable signal to the clock gaters while the clock is high. ASICs usually use *integrated clock gaters* (ICGs), where the AND2 gate and enable latch are combined into a single cell, which are well supported by commercial EDA tools.

It can be advantageous to separate the enable latches and clock gaters. Compared to using ICGs that have one enable latch for each clock gater, fewer enable latches are needed. An enable latch can directly drive several clock gaters, and the enable signal from the latch can be buffered to drive more gaters. As fewer enable latches are needed, this reduces the clock load and the clock power [20]. The CTS step includes cloning of the ICGs or cloning of both the enable latches and gaters, with enable buffering to the cloned enable latches and from the enable latches to the clock gaters where needed.

Multi-source CTS (MSCTS) uses multiple clock sources to reduce the depth of the clock trees. This reduces the clock skew due to process variability, and due to variation in delay at each clock level, with differences in load capacitance and RC delay. The clock sources are typically placed in a grid across the design, where they are needed. These MSCTS clock sources may be driven by an upper level clock mesh or by an H-tree or X-tree topology of clock buffers [21]. Alternatively, a full clock mesh may be used, with a fixed number of levels of clock buffers and gaters that need cloning, typically one, two, or three levels—trading off clock skew versus the additional power to distribute the clock signal on the mesh [20]. Here is a rough comparison of the clock skew trade-offs in 32 nm technology: CTS has typical clock skews of 100 ps or more to as low as 50 ps, with significant additional manual work; MSCTS can achieve 30–50 ps clock skew [21]; and a clock mesh topology can reduce skew to between 10 and 30 ps, depending on the number of clock buffer or gater levels, from the mesh to the register [20].

Whether CTS, MSCTS, or a clock mesh is used, the first step in this design phase is the clustering of registers, with cloning and placement of the final stage of clock buffers, gaters, or ICGs, to drive those register clusters. Then, the final stage buffers et al., are clustered and the process repeats recursively in reverse topological order back to the clock source(s). In CTS and MSCTS, the clock buffers are placed with priority in the standard cell logic regions, which causes other logic to be moved out from under the clock buffers. In contrast, clock mesh synthesis may reserve whitespace to place the clock buffers et al., directly below the clock mesh to minimize additional wire RC delay contributing to clock skew and to reduce wire load on the clock mesh. After the clock buffers have been placed, the clock wires are inserted into the design.

Managing skew in the clock trees is critically important. Any unplanned skew is deducted directly from the cycle time. The buffers and wires in the clock distribution logic are carefully designed to provide as little unplanned skew as possible. A number of different clock topologies can be used to achieve low clock skew, as discussed earlier. For many years, *zero skew* was the optimization goal of clock design. In an ideal zero-skew clock, the clock signal arrives at every register in a clock domain at precisely the same moment.

Recently, flows have begun using *useful skew* to further improve performance. By advancing or delaying clocks to registers on the critical path where there is a significant difference between the slack of the data input and data output signal, the cycle time of the design can be improved. Network flow algorithms are used to recalculate the useful skew targets for all the critical registers in the design. Implementation of a *skew schedule* is done by clustering registers with similar skew targets, and delaying clock arrival times to them, by the addition of delay buffers or extra wires. Calculating the skew schedule and then implementing it in this phase can yield a significant cycle time improvement. Useful skew can also help reduce voltage drop, as registers will not be switching at the same time and so the simultaneous current draw is less. Care is needed with useful skew to avoid introduction of too many hold violations.

If the useful skew schedule is fed back to the logic synthesis or pre-CTS placement optimization phases, a further performance improvement and area reduction can be realized. The register positions and any assignment to multi-bit registers are typically fixed during CTS. However, a few iterations of optimization, permitting the registers and other cells to move and be resized et al, with interleaved incremental CTS and useful skew, can significantly improve performance, reduce clock power, and improve design convergence [22].

Minimizing the clock power is important because a significant portion of a chip's active power is dissipated in the clock logic and routing. Clock buffers are tuned to ensure that the minimum gate size is used to meet the clock slew and latency requirements. Since most clock power is dissipated in the final stage or *leaves* of the clock tree, particular care is applied to minimize leaf-level clock wire length. Reducing the leaf-level wire and pin load capacitance allows further power savings by downsizing or removing buffers in the clock distribution logic, for example, enabling a 1.5× overall dynamic power reduction, for a clock mesh topology [20].

There are also interesting power trade-offs to be made, around the way clock gating is handled in a clock tree. Moving clock gaters as close as possible to the root of the clock trees allows for the greatest portion of the clock tree to be turned off. However, this requires a logical OR of the enable signals to the gaters that are merged up the clock tree, reducing the fraction of cycles for which that clock signal may be off; or separate clock trees are required for the sets of registers with different clock enables, increasing the total capacitance of the clock trees. The power savings are limited due to these issues with either approach. Coarse and fine-grained clock gating may be needed to best reduce the clock activity, but there are also power overheads to generate the additional fine-grained enable signals, and there must be a reasonable number of registers that can share the common enable to justify the power overhead of generating that enable signal [20].

Transitions on clock wires have sharp edge rates and as a result can cause coupling noise to adjacent wires. In addition, sequential elements can be susceptible to noise on their clock inputs. As a result, clock wires are sometimes shielded with parallel wires or given non-minimum spacing from adjacent wires to protect against coupling noise and to reduce signal dependent *clock jitter*. Wider spacing of clock wires also reduces the wire capacitance and reduces the wire RC delay contribution to clock skew at the cost of increased routing congestion. The switching of high-power clock buffers can also introduce significant power supply noise. To combat this, it is a good practice to surround the main clock buffers with a large number of decaps; though the need for this is reduced by the wire capacitances in a clock mesh topology, which further helps reduce clock skew in clock mesh distribution.

As manufacturing variation has become worse, it has become increasingly important to do *variation-aware clocking*. Variation-aware clocking techniques include the use of matched clock buffers, wider metal lines, and a preferential use of thicker wiring layers. In sensitive cases, it is necessary to perfectly match the order of the layer and via usage of all paths to skew-sensitive registers. It is also helpful to group registers that share critical timing paths onto the same branches of a clock tree, as they will not suffer clock skew between them for the portion of the clock tree that is in common. Timing can then use *common path pessimism removal* (CPPR) to account for the improved skew tolerance derived from the shared portion of the clock tree [23]. CPPR is also known as clock reconvergence pessimism removal (CRPR).

The clock distribution is detail routed in this design step to avoid circuitous routing to minimize the clock skew. This is before most other nets are routed, though designers may have manually inserted pre-routes for certain highly critical nets. The register positions are usually fixed beyond this point in the design flow to avoid rerouting the detailed clock routes. Fixing the register positions also improves design convergence, by minimizing perturbations to the clock trees that will impact clock skew for multiple timing start and end points.

The output of this design phase is a fully placed, presumably routable layout, with fully instantiated and detail-routed clock trees. This is passed to the post-CTS optimization phase.

13.2.8 POST-CTS OPTIMIZATION

The purpose of this phase is to clean up the performance problems caused by the introduction of the clocks. This is usually the first step at which hold time violations are directly addressed, though useful skew may be used earlier in the flow to reduce hold violations.

Two factors degrade performance at this design stage. First, the clock distribution logic can now be timed fully using 2.5D or 3D capacitance and resistance extraction of the real clock wires. Up until this point, all timing has been done with idealized clocks. If the idealized clocks were assigned properly with sufficient guard bands for skew, the introduction of the real clocks is fairly painless—only a few problems surface. However, estimating skew has become increasingly difficult with greater impact of on-chip variation (OCV) on newer technologies, which contributes significant skew to clock paths that are not in common, where CPPR does not apply [22]. OCV analysis applies conservative timing deratings to cells on the clock and data paths to account for the possible process variation at a given process corner [24]. The second factor that degrades the performance is the replacement of the logic under the clock buffers—when the clock buffers are placed in the design, other logic is moved out from underneath them. The performance problems introduced by these two factors are fixed by applying the same timing and placement-aware transforms that are applied in the pre-CTS placement optimization phase.

The added accuracy of timing the real clock distribution logic in this phase allows for the correction of hold time problems. The workhorse hold time fix is the introduction of delay gates between the launch and capture latches. Sometimes, the launch and capture clocks can be de-overlapped by useful skew to eliminate hold time violations. Useful skew optimization in the post-CTS step is limited, typically adding a buffer or tapping off a different point in the clock tree, to minimize modifications to the clock distribution logic which can impact timing to many registers. Useful skew is best used at this step only when a large number of setup or hold time violations can be eliminated with a single change.

To properly analyze timing problems at this stage, and to ensure that any fixes inserted do not break other timing paths, timing analysis must be run on multiple timing "corners" simultaneously. There corners are combinations of different process and operating conditions (supply voltage and temperature). Gate and wire delays are characterized, based on specific parametric assumptions: for example, *worst*, *nominal*, and *best* case. Four standard corners were typically used for timing analysis to ensure correct circuit operation:

1. Slow process, worst-case voltage and temperature
2. Slow process, best-case voltage and temperature
3. Fast process, worst-case voltage and temperature
4. Fast process, best-case voltage and temperature

As variation effects become more pronounced, these set of corners must be extended to include the possibility of *process mistracking* between device types and interconnect layer characteristics. This requirement for multiple simultaneous timing runs further complicates the infrastructure requirements of the design closure tool flow.

Finally, in preparation for inevitable engineering changes, empty spaces in the standard cell regions of the chip are filled with metal programmable "spare" gates. These cells provide unused transistors, which can be configured into logic gates by customizing one or more metal layers. This technique can be used to implement emergency fixes for design problems found late in the flow, or after hardware has been built. It is often possible to provide a patch to a design by rebuilding only one or two mask levels.

The output of the post-CTS optimization phase is a timing-closed clock instantiated presumably routable netlist. This is passed to the routing phase.

13.2.9 ROUTING

The purpose of the routing phase is to add wires to the design, as detailed in Chapter 8. By this phase, all of the main timing objectives of the design should be met. It is very difficult for timing problems to be fixed in routing; rather, it is routing's job to deliver on the "promises" made by the wire length and congestion estimates used in synthesis, placement, and so on. There are three major design closure issues dealt with in this phase: performance, signal integrity, and manufacturability.

Routing is generally broken up into two steps: *global* routing and *detailed* routing. Global routing determines routes between a coarse grid of global routing cells (g-cells), abstracting the permissible locations to connect to each gate's pins. Detailed routing specifies the actual routes that will be taken, the wires on each metal layer, where the wires connect to pins, and vias in between layers, subject to design rules and routing blockages.

As discussed in prior sections, global routing may be used earlier in the flow in placement or even physical synthesis to improve the accuracy of wire resistance and capacitance and routing congestion estimates. The clock tree synthesis step is where the first additional detailed routes are added for the design for clock network distribution. There may also be designer-specified pre-routes in the floorplan, along with the clock grid and power/ground grids. The remaining detailed routing of data paths is left until after post-CTS optimization, after which most gates will remain in the same position, subject to a more limited post-route optimization.

To improve design convergence, it is important that detailed routes match moderately well with Steiner routes or global routes used in earlier design flow steps. The Steiner or global routing capacitance and resistance estimates must correlate without too much inaccuracy versus the detailed routes, and correlation is improved by the actual detailed wire routes using similar paths and layer assignments to those used in the earlier estimates.

The goal of global routing is to localize all wires on the chip in such a way that all the routing capacity and demands of the chip are roughly balanced. To do this, the chip is partitioned into horizontal and vertical wiring tracks, where each track has some fixed capacity. Steiner wires are generated for all the nets, and these are laid into the tracks. Routability is optimized during global routing by permuting the track assignments to flow excess capacity out of highly utilized tracks. At this step, the wiring *porosity* of fixed objects is also analyzed to ensure a correct calculation of routing capacity. Based on routing density, estimates of wire spacing given the number of available routing tracks and how inter-wire spacing impacts etching the metals, wire resistance and capacitance are estimated from the global routes. As interconnect scaling has worsened, it is important to assign layer usage as well as track assignments during global routing. Color assignment for double or multi-patterned layers is also necessary to improve accuracy of resistance and capacitance estimates. Overestimating track and layer capacity can have large impacts on the actual routed net length.

The important performance-related design closure mitigation available during global routing is the preferential routing of timing-critical nets, where pre-identified timing-critical nets are routed first and to the extent possible will follow a direct path with the minimum number of jogs between the source and sinks. Coupling noise can also be addressed at this phase by forcing the global router to segregate noisy and sensitive nets into different global routing tracks [25]. Subject to routing congestion, additional wire spacing may be used to reduce cross-coupling capacitance, which both help speed up timing critical paths and reduce SI noise.

The detailed routing step maps the global routes into the real wire resources on the chip. In this phase, the real wire topology is determined and all wire widths, layers, vias, and contact choices are made. As wires are added, wiring congestion increases and average wire length goes up. To ensure that timing constraints are met, timing-critical signals are wired first. This gives them access to the shortest wire length and preferred wiring layers. In addition, certain signals may be assigned to be wired on specific layers or specific widths for electrical reasons. For example, long timing-critical wires may be designated as wide wires on *thick* upper wiring layers to minimize resistance. Particularly, skew-sensitive situations such as buses or differential signals may require *balanced routing* in which the lengths, topology, and layer usage of two or more skew-sensitive wires are matched. Wide busses are often prewired to ensure that they have balanced delay.

In addition to managing performance, wiring must optimize for coupling noise. The detail router can be guided to segregate noisy and sensitive nets [25] and to insert additional empty space between them to further reduce coupling if needed. In some cases, it may be necessary to route adjacent grounded shielding nets to protect particularly sensitive signals or to prevent interference by a particularly noisy signal. Coupling issues on busses can be improved by using *random Z-shaped routing* [26] to prevent overly long parallel wires. Inductance effects in large busses can be mitigated somewhat by the addition of interspaced power or ground wires, which serve as low-resistance current return paths.

Manufacturability can also be optimized during the routing phase. Routing can add redundant vias and contacts to improve both reliability and yield. As interconnect variability becomes more important, routing can also add wide wires and matched vias and layers to reduce

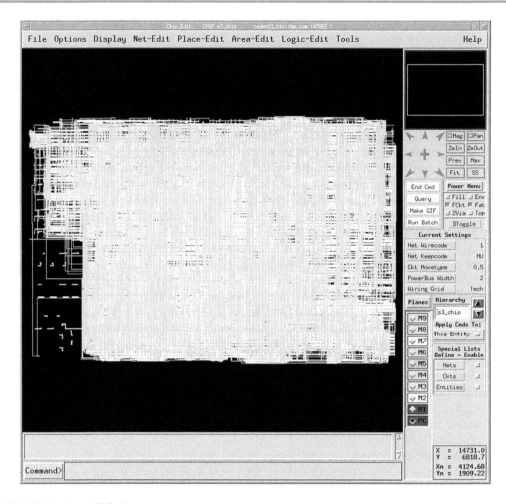

FIGURE 13.20   ASIC chip routing.

back-end-of-the-line variability on sensitive wires. Overall, manufacturability can be further improved by adding extra *metal fill shapes* during routing to ensure more uniform density, which improves dimensional control during lithography and CMP.

If congestion estimates were inaccurate or other constraints such as coupling decreased routability, detailed routing may not be able to complete all the wires, leaving *opens* between wires or design rule violations. The remaining wiring violations may be manually fixed if there are only a few hundred of them at most. Otherwise, the designers must return to earlier flow steps to mitigate the severe routing congestion—for example, specifying lower area density limits at routing hot spots during placement or modifying the floorplan. Figure 13.20 shows a section of a fully routed ASIC.

The output of the routing phase is a wired and legally placed netlist. This is passed to the post-route optimization phase.

13.2.10 POST-ROUTE OPTIMIZATION AND FINAL SIGN-OFF

This phase deals with optimizing out the last problems using the most accurate analyses. The main design closure issues dealt with in this phase are performance, power consumption, signal integrity, yield, and reliability. At this point, since the design is nearly finished, any changes that are made must be localized.

Timing is now analyzed using fully extracted wires with real clocks. Any timing problems that occur at this point are fixed with gate sizing, buffering, automatic or manual rerouting, and wire widening. Where there is additional timing slack due to overestimating the wiring resistance and capacitance prior to routing, gates may be downsized and buffers may be removed to reduce power consumption. For example on a 32 nm CPU core, clock gaters were oversized due to inaccurate Steiner tree wire load estimates in CTS; in post-route optimization, 60% of the gaters could be downsized [20].

Footprint-compatible cell swaps provide the least disruption to the existing detailed routes at this point in the flow. The most common footprint swaps are changing to a faster cell with worse leakage, lower transistor threshold voltage, and/or shorter channel length; or a slower cell with less leakage, higher threshold voltage, and/or longer channel length. However since 32 nm process technology, transistor well isolation constraints have limited opportunities for footprint V_t-swaps. Another alternative is to swap to a lower- or higher-power cell variant that is footprint compatible. Rather than minimizing the cell area for all library cells, cells can be designed to have the same footprint—for example, to allow swapping from faster, higher-power flip-flops to low-power flip-flops that gate the internal slave latch clock, when the data input has not changed [27]. Similarly, a good technique to minimize clock skew in post-route optimization is to provide "shadow" clock gaters in the cell library to swap to, allowing clock skew to be closely match based on the actual detail-routed wire load.

In modern high-performance designs, footprint swaps seldom suffice to close a design. Resizing a cell at its current position may be possible, provided there is sufficient whitespace. However, even downsizing a cell can sometimes introduce minimum spacing violations due to different edge-type restrictions with neighbors, or minimum spacing constraints for metal fill, whereas the original cell may have abutted the neighboring cells and not had the same issue. Thus, it may be necessary to perform post-route *whitespace* optimization, placing a resized cell or added buffer in the nearest available whitespace. Post-route whitespace optimization usually requires an automated detailed routing ECO to fix the wiring connectivity.

If there is insufficient whitespace nearby, detailed placement *legalization* is needed, which can be quite disruptive and require more extensive rerouting. If too much rerouting is required, automated ECO routing will fail, and the design must be fully rerouted, which is quite problematic for design convergence, as the wiring parasitics used for post-route optimization will now be inaccurate. At that point, it may be better to respin the design through the flow, making changes at earlier design steps to address the problems that appeared in post route—for example, *derating* wire loads to be more conservative based on the post-route extracted parasitics, applying non-default routing rules to particularly problematic timing critical nets, and so forth.

Now that real wires are available, mutual capacitances can be extracted to drive noise analysis. Noise analysis uses a simple model of coupling, combined with the analysis of possible logic switching windows derived from timing, to determine if adjacent wire switching will create timing violations or logic glitches. If errors are found, they may be fixed by reducing common run length, spreading wires, adding shields, adding buffers, resizing gates to modify slew rates, and rerouting to segregate wires. Yield issues may also be analyzed and corrected in a similar manner. Wire-limited yield can be addressed at this stage by increasing wire width, increasing inter-wire spacing, or decreasing common run length between wires. Yield can also be optimized via wire spreading (see Figure 13.21), though this processing can complicate optical proximity processing and can interfere with wire uniformity.

Any timing problems, coupling or yield constraint violations, found at this point can be fixed manually or by automatic routing-based optimization, which uses the same logic, placement, and timing optimization framework described in pre-CTS placement optimization. In such an automated flow, wires with constraint violations are deleted and queued for rerouting. Final closure is performed using transformations that adjust gate sizes, do minimal logic modification, and minimally adjust placement density on the offending wire's logical and physical neighbors.

In addition to these final optimizations, there are a number of final checks and optimizations that are run on the design. One of these is electro-static discharge *checking and optimization. Floating gate* detection, and the addition of any necessary floating *gate contacts*, ensures that all gates in the design electrically connect to at least one diffusion. The second is *wiring antenna* detection and correction to remove any unterminated wiring segments that might be left in the design by previous editing. Antennas can also cause ESD problems. Reliability checks are performed to ensure that no wire violates its electromigration limit and

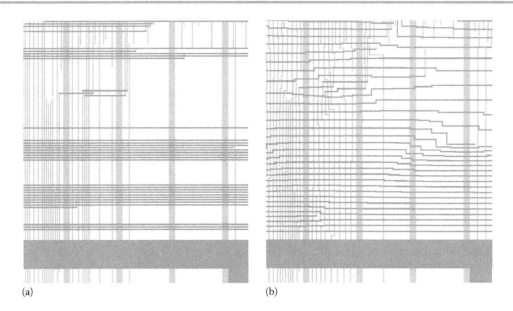

FIGURE 13.21   Before (a) and after (b) wire spreading.

any wire that does is modified via wire-widening and load adjustment. Finally, design rule checking and logical-to-physical verification are performed.

Final timing sign-off is performed using *variation-aware timing* by checking the design against an exhaustive set of process corners. This exhaustive checking tests the design for robustness to possible mistracking between layers, device strengths, multiple supply voltages, etc. This guarantees conservatively that the design will be manufacturable over the entire process window.

The product of this phase is a fully placed, routed, and manufacturable design, ready to be sent to the mask house and fabrication.

13.3 THE FUTURE OF DESIGN CLOSURE

The next big challenges in design closure will be dealing with the increased importance of *power-limited performance optimization* and the increased need to do *design for variability*.

13.3.1 POWER-LIMITED PERFORMANCE OPTIMIZATION

Today's design flows still typically treat performance as the primary optimization objective, with constraints such as power and area, handled as secondary concerns. As the significance of both dynamic and leakage power continues to increase, the design flow will have to be modified to treat the power/performance trade-off as the primary optimization objective. Because power is best addressed early in the flow, the most leverage will come from advances in the concept and logic design phases. The most important design closure advances in this area will need to be the creation of power-oriented design exploration and optimization flows. To enable these flows, the industry will need to develop more accurate early power prediction techniques.

In the logic synthesis, placement, and routing stages, the flow will have to be modified to manage the power impact of every optimization, rather than sequentially optimizing for performance and then assessing the power impact. The flow will also have to be modified for more aggressive power management design techniques, including a wider use of dynamic voltage scaling and power gating. Design closure flows will also need to be extended to handle emerging circuit-oriented power management techniques. These techniques are likely to include the use of new circuit families, and techniques such as *dynamic body bias. Reverse body bias* increases the threshold voltage, reducing leakage at the cost of reduced gate drive strength and hence lower performance. *Forward body bias* improves performance at the cost of higher leakage. Body bias can be used with a wide voltage range in fully depleted silicon-on-insulator (FDSOI) process technology from 28 nm today to 10 nm, which is currently in development [28].

13.3.2 DESIGN FOR VARIABILITY

In Section 13.1.2.7, we mentioned the trend of increasing the impact of parametric variability for both devices and wires, as chip geometry continues to shrink. As this has occurred, the relative amount of performance guardbanding has had to increase. This forced conservatism effectively reduces the amount of performance gain extractable from new technology nodes. As we move into 14 nm design and below, we are now entering a phase in design closure where management of parametric variability is a first-order optimization objective. Design closure for variability can be addressed in two distinct ways. The first is *design for manufacturability* (DFM) and the second is by doing *statistically driven optimization.*

Design for manufacturability concerns has been slowly working their way into the design closure flow. DFM optimizations involve constructively modifying the design to minimize sources of variability, in order to make the design more robust to manufacturing variations. There are many constructive DFM techniques that have been or are currently being automated as part of the design flow. These include the following:

Matching devices: Identical circuits tend to track each other better than dissimilar circuits in the face of manufacturing variation. Design closure flows can exploit this by using identical buffering in skew-sensitive circuits such as clocks. For example, clock skew can be reduced by using clock buffers and gaters that have the same transistor threshold voltage, rather than allowing alternate threshold voltage cells that will have different process variations. Furthermore, to help minimize skew with a clock mesh at a leaf level where clock gaters are used, the few unconditional clock buffers might be implemented as clock gaters with the enable tied high to minimize variation by using the same cell type.

Variation-aware routing: Automatic routers are being modified to create more variation-tolerant routing. Optimizations include the use of wide wires on variation-sensitive routing; the use of geometrically balanced (i.e., matching topology, layer, and via usage) routing for critically matched signals; and the design of maximum common subtrees to reduce process-induced delay variation in skew-sensitive routes such as clocks.

Geometric regularity: Designs with a high degree of device and wiring regularity tend to have lower levels of dimensional variation. This is because both optical lithography and etch processes such as *CMP* respond better with uniform shapes densities. Regularity can be imposed at many levels of the design. At the cell level, regularity can be achieved by placing all critical geometry such as transistor polysilicon gates on a uniform grid, with additional edge-spacing restrictions between cells to add a poly fill for regularity where necessary. At a placement level, the flow can ensure that all gates see essentially uniform density on critical polysilicon and contact layers. At a routing level, the flow can enforce a fixed routing grid and uniform routing density through careful route planning and selective addition of routing fill. Some foundries now require strict assignments of wire routes to pre-colored routing tracks for double patterning and disallow wrong-way routing for self-aligned double patterning and wire widths must be a fixed integer multiple of the minimum width available from a limited number of wire widths.

Adaptive circuit techniques: Chips will include a greater number of *adaptive variability management* circuits. These circuits will either use automatic feedback or digital controls to "dial out" variability. For example, adaptive control could be used to cancel a manufacturing-induced offset in a differential receiver or could be used to adjust out delay variation in a critical signal's timing.

Statistical static timing analysis (SSTA): This analysis provides much of the benefit of exhaustive corner analysis [29]. Statistical timing tools characterize delay, using statistical delay distributions derived from measured hardware. Rather than calculating path delay as the sum of all the worst-case (or best-case) delays, statistical timing calculates a probability distribution for delays. This distribution indicates the likelihood of a given path having a specific delay. By calculating this measure for all paths, one can determine the percentage of manufactured chips that will run at a given speed. Calculating delay in this way avoids the compounding conservatism of using the worst-case delay of all elements for the circuit.

With SSTA, a design can be optimized to an acceptable circuit-limited yield value, reducing the performance lost to overdesign, rather than designing to an exhaustive set of worst-/best-case parameters. However, SSTA is used only in a couple of design companies like IBM and Intel, and SSTA has not been widely adopted elsewhere. Instead, other approaches have been used such as advanced OCV (AOCV), parametric OCV (POCV), and statistical OCV (SOCV) [30]. POCV and SOCV use path-based analysis, which is less pessimistic than AOCV and less computationally expensive than SSTA.

13.4 CONCLUSION

We have defined the design closure problem and provided an overview of the design constraints of major interest in modern process technologies. We detailed how design flows develop to better consider design constraints and objectives, to improve design convergence, and achieve design closure. How design constraints are handled in the flow evolves, as the constraints become more significant and techniques are developed to analyze and automatically fix them, then estimate their impact and mitigate such issues earlier in the design flow.

New constraints continue to evolve, and the design closure problem grows more complex and more interesting, with some quite difficult issues to address in today's designs and process technologies. This drives significant, further development of EDA tools and design techniques.

ACKNOWLEDGMENTS

The authors acknowledge Dr. Leon Stok, Dr. Ruchir Puri, Dr. Juergen Koehl, and David Hathaway for their helpful input and feedback on this chapter.

REFERENCES

1. Shepard, K. and Narayanan, V., Noise in deep submicron digital design, *International Conference on Computer-Aided Design*, San Jose, CA, 1996, pp. 524–531.
2. Shepard, K., Narayanan, V., and Rose R., Harmony: Static noise analysis of deep submicron digital integrated circuits, *IEEE Transactions on Computer-Aided Design*, 18(8), 1132–1150, August 1999.
3. Shepard, K. and Narayanan, V., Conquering noise in deep-submicron digital ICs, *IEEE Design and Test of Computers*, 15(1), 51–62, 1998.
4. International Technology Roadmap for Semiconductors, 2013 Edition. http://www.itrs.net/ Links/2013ITRS/Home2013.htm. Accessed on May 2015.
5. Trevillyan, L., Kung, D., Puri, R., Reddy, L., and Kazda, M., An integrated design environment for technology closure of deep-submicron IC designs, *IEEE Design and Test of Computers*, 21(1), 14–22, 2004.
6. Shenoy, N., Iyer, M., Damiano, R., Harer, K., Ma, H.-K., and Thilking, P., A robust solution to the timing convergence problem in high-performance design, *International Conference on Computer Design*, Austin, TX, 1999, pp. 250–254.
7. Pillage, L.T. and Rohrer, R.A., Asymptotic waveform evaluation for timing analysis, *IEEE Transactions on Computer-Aided Design*, 9(4), 352–366, April 1990.
8. Alpert, C., Chu, C., and Villarrubia, P., The coming of age of physical synthesis, *International Conference on Computer-Aided Design*, San Jose, CA, 2007, pp. 246–249.
9. Huard, V. et al., CMOS device design-in reliability approach in advanced nodes, *International Reliability Physics Symposium*, Montreal, Quebec 2009, pp. 624–633.
10. Slayman, C., Soft error trends and mitigation techniques in memory devices, *Reliability and Maintainability Symposium*, Lake Buena Vista, FL, 2011, pp. 1–5.
11. Moammer, K., AMD and Nvidia 20 nm and 16 nm GPUs slightly delayed, WCCF Tech, December 2014. http://wccftech.com/amd-nvidia-20nm-16nm-delayed/. Accessed on May 2015.
12. Nassif, S., Nam, G., and Banerjee, S., Wire delay variability in nanoscale technology and its impact on physical design, *International Symposium on Quality Electronic Design*, Santa Clara, CA, 2013, pp. 591–596.

13. Ceyhan, A. and Naeemi, A., Cu/Low-k interconnect technology design and benchmarking for future technology nodes, *IEEE Transactions on Electron Devices*, 60(12), 4041–4047, November 2013.
14. Ueno, K. et al., A design methodology realizing an over GHz synthesizable streaming processing unit, *IEEE Symposium on VLSI Circuits*, Kyoto, Japan, 48–49, 2007.
15. Sangiovanni-Vincentelli, A., Defining platform-based design. *EETimes*, February 2002. http://www.eetimes.com/document.asp?doc_id=1204965. Accessed on May 2015.
16. Mentor Graphics, RTL synthesis for advanced nodes: Realtime designer, White Paper, 2014. http://www.mentor.com/products/ic_nanometer_design/techpubs/request/rtl-synthesis-for-advanced-nodes-realtime-designer-87068. Accessed on May 2015.
17. Kommrusch, S., Implementing an efficient RTL clock gating analysis flow at AMD, White Paper, 2014. http://calypto.com/en/blog/2014/04/15/white-paper-implementing-an-efficient-rtl-clock-gating-analysis-flow-at-amd/. Accessed on May 2015.
18. Kosonocky, S., Practical power gating and dynamic voltage/frequency scaling issues, *Hot Chips*, Stanford, CA, 2011.
19. Rodgers, R., Knapp, K., and Smith, C., Floorplanning principles, SNUG Silicon Valley, San Jose, CA, 2005.
20. Chinnery, D. et al. Gater expansion with a fixed number of levels to minimize skew, SNUG Silicon Valley, Santa Clara, CA, 2012.
21. Foley D. et al., A low-power integrated x86-64 and graphics processor for mobile computing devices, *Journal of Solid-State Circuits*, 47(1), 220–231, January 2012.
22. Cunningham, P., Swinnen, M., and Wilcox, S., Clock concurrent optimization, *Electronic Design Processes Workshop*, Monterey, CA, 2009.
23. Bhardwaj, S., Rahmat, K., and Kucukcakar, K., Clock-reconvergence pessimism removal in hierarchical static timing analysis, US Patent 20120278778, 2012.
24. Jarrar, A. and Taylor, K., On-chip variation and timing closure, *EDN Magazine*, pp. 61–64, June 2006.
25. Stohr, T., Alt, H., Hetzel, A., and Koehl, J., Analysis, reduction and avoidance of cross talk on VLSI chips, *International Symposium on Physical Design*, 1998, pp. 211–218.
26. Kastner, R., Bozorgzadeh, E., and Sarrafzadeh, M., Pattern routing: Use and theory for increasing predictability and avoiding coupling, *IEEE Transactions on Computer-Aided Design*, 21(7), 777–790, November 2002.
27. McIntyre, H. et al., Design of the two-core x86-64 AMD 'Bulldozer' module in 32 nm SOI CMOS, *Journal of Solid-State Circuits*, 47(1), 1–13, 2012.
28. Flatresse, P., UTBB FD-SOI: The technology for extreme power efficient SOCs, keynote at the *Telecommunications Circuits Laboratory*, Lausanne, Switzerland, January 2014.
29. Visweswariah, C., Death, taxes and failing chips, *IEEE/ACM Design Automation Conference*, Anaheim, CA, 2003, pp. 343–347.
30. Bautz, B. and Lokanadham, S., A slew/load-dependent approach to single-variable statistical delay modeling, *International Workshop on Timing Issues in the Specification and Synthesis of Digital Systems*, Santa Cruz, CA, 2014.

Tools for Chip-Package Co-Design

14

Paul D. Franzon and Madhavan Swaminathan

CONTENTS

14.1 INTRODUCTION

Chip-package co-design refers to design scenarios in which the design of the chip impacts the package design or vice versa. Computer-aided tools are needed for co-design in situations where simple bookkeeping is insufficient. The most classical and the most used chip-package co-design tools are the I/O buffer interface standard (IBIS) macromodeling tools for the conversion of integrated circuit (IC) and I/O buffer information into a format suited for rapid co-simulation. Tool issues of IBIS will be discussed toward the end of this chapter.

However, the need for more sophisticated tools for chip-package co-design is rapidly appearing. High-frequency designs require a more accurate modeling of the chip and package components. Reduced order macromodeling is often needed, especially to speed up the IC portion of a simulation. Chip-package co-synthesis is starting to emerge as a new area, especially for simultaneous floorplanning, pin assignment, and routability analysis for high pin-count packages and 3D ICs.

14.2 DRIVERS FOR CHIP-PACKAGE CO-DESIGN

Until recently, there was relatively little need for chip-package co-design. The package, and subsequently the board, could be designed after the fact, mainly with the sole focus of obtaining connectivity. The only information that needed to be communicated between the chip and the package design was pin functionality.

As operating frequencies and system density increased, this "over the wall" sequential design strategy no longer worked. Table 14.1 summarizes the issues that have become important over the

TABLE 14.1   Summary of Some of the Main issues, Solutions and Open Challenges in Chip-Package Co-design

Type of System	System Issue	Technological Solutions	Design Solutions	Open Challenges
Digital	Board/package level timing and noise		Simulation with accurate timing and accurate I/O macromodels	
	SSN Management	Package- Embedded Decoupling Capacitors	Chip-Package Cosimulation	SSN-accurate Macromodels for core and I/O
	High Speed Transceiver Design	Flip-chip solder bump; low-loss materials	Equalization	Continued scaling beyond 10 Gbps; Increased high-speed pin counts; Cost-effectiveness
	High pin counts	high-density laminates; 3D ICs	System floorplanning and pin assignment; Coverification	Effective floorplanning and pin assignment tools
	Thermal dissipation and stress management	Heat spreaders, advanced air cooling, spray cooling, etc.	Thermal design.	Thermal macromodeling.
Mixed Signal	Cost-effective passive integration (Ls, Cs, baluns, antennae, etc.) for miniaturization and cost reduction	SOP technologies (embedded passives; 3D-IC)	Cosimulation and Co-extraction	Simulation models
	Noise management e.g. for VCOs and LNAs		Cosimulation and Co-extraction; Noise reduction	Accurate SSN prediction

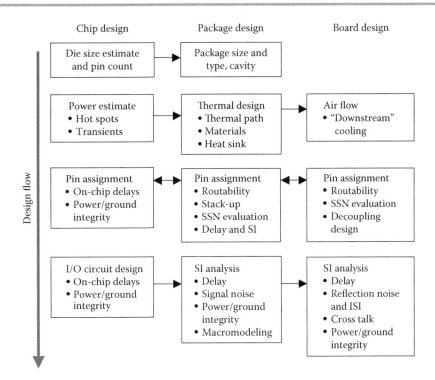

FIGURE 14.1 Major flows and design activities used in chip-package co-design.

last decade, along with the solutions available and challenges still open. The next two sections discuss this table in detail, first looking at digital co-design and then mixed-signal co-design. The following section provides a brief overview of the most successful co-design tool to date, the IBIS macromodeling language before presenting the conclusions, and an annotated bibliography.

Figure 14.1 presents a summary flow of the major design activities involved in chip-package co-design. Early planning requires determining the design size plan, anticipated I/O counts, and selection of the basic package type. Often, package selection might require a detailed evaluation of some of the later activities in the flow. Although thermal design is shown it is usually conducted concurrently with the other activities. For more power hungry chips, it has become important to pay attention to "hot spots" and thermal transients when specific chip functions turn on or off.

Pin assignment and I/O electrical designs are major co-design activities that, especially for higher pin counts and mixed-signal systems, require a simultaneous evaluation of chip, package, and board. For example, poor pin assignment will lead to larger packages and higher levels of simultaneous switching noise (SSN). Delay planning must include budgets for chip, package, and board. The SSN management will require macromodeling of on-chip drivers and other sources of power/ground transients (e.g., clock distribution circuits) and a detailed simulation of the package, including any embedded capacitors or the board. For a large digital system, SSN modeling and simulation are often the most complex task in this flow, because in mixed-signal systems, RF, and analog circuits must be carefully separated from digital circuits, since excess noise from the latter can prevent a phase-locked loop (PLL) from even locking.

14.3 DIGITAL SYSTEM CO-DESIGN ISSUES

Table 14.1 lists digital issues roughly in their chronological emergence as co-design drivers. When system clock speeds started exceeding a few tens of megahertz, it became necessary to design the traces on a printed circuit board (PCB) to meet timing (delay) and reflection noise requirements. Two pieces of information had to be passed from the system/IC designer to the board designer. The first was a list of timing slacks or what board delays would be acceptable. Timing slacks can easily be conveyed via a spreadsheet or similar simple mechanisms. The second information was a reasonably accurate simulatable model of the chip I/O, including package parasitics.

A transistor-level SPICE (or other circuit-level simulator—see Chapter 17) model is rarely a desirable solution, mainly because it conveys proprietary data about the chip design and fabrication process and also because chip transistor models are often proprietary to a specific simulator. Instead, a portable macromodel is needed. Since the mid-1990s, the highly successful IBIS macromodel has fulfilled this need, but new, more accurate techniques are emerging (see Section 14.5).

As IC transistor counts increase, the amount of SSN increases. Simultaneous switching noise occurs when on-chip transients lead to current pulses (di/dt spikes) in the power and ground system. Since any practical power and ground system have inductance associated with it, noise is induced (as $V = L\,di/dt$) on the power and ground rails. High noise levels can lead to transient errors in the chip. While most of the di/dt pulsing is caused by the output switching, on-chip logic blocks also contribute to it. Techniques to control SSN include the provision of decoupling capacitors between power and ground and also design methods to reduce L and di/dt. Accurate SSN simulation requires a combination of fairly detailed macromodels for on-chip di/dt sources, along with high fidelity models of the package and the PCB. There is no widely accepted macromodeling procedure for the di/dt models, and thus, this is an important open issue. While there are existing proofs of accurate SSN co-simulation, these are generally confined to sophisticated teams working in vertically integrated companies. No general methodology is available, although one is sorely needed.

The SSN management has started permeating into co-simulation and co-design issues. An example would be co-design of on-chip and off-chip decoupling capacitors, together with optimized pin assignment, so as to minimize the effective inductance. While sophisticated teams have performed such a co-design, the methods for doing it effectively is when the chip design team and system design team are vendors, and the customers are yet to be established. Many signal integrity textbooks document the impact of pin assignment on SSN. In general, it is important to intersperse power and ground pins among signal pins in order to control SSN. The ratio and degree of interspersion depend on the pin bit rate and overall SSN issues.

With increased inter-chip bandwidths also comes the requirement for multi-Gbps I/O and high pin counts. True co-design is needed to successfully build multi-Gbps I/O. Since the channel characteristics are strongly frequency dependent, on-chip signal processing or other circuit techniques are used to compensate (equalize) for the poor channel frequency response. Macromodels of IBIS are not used here due to the use of circuit and DSP techniques, and the high degree of modeling fidelity required at these speeds. Transistor-level design is required along with accurate PCB and package characteristics. However, some emerging macromodeling techniques are under investigation.

Chips with over 1000 pins are available today, and multi-thousand pin packages are anticipated in the future. With such high I/O pin counts, the details of chip I/O location are important. Doing a poor job at pin assignment could greatly increase the layer count and cost of the package and PCB or even make the system unbuildable! Quality I/O assignment requires simultaneous consideration of chip floorplan and package/PCB routability. Relatively, little work has been done in this area of integrated co-synthesis of chip and package. Anecdotal evidence shows that good I/O pin assignment can reduce the amount of package-level wiring by more than 10%.

Three-dimensional ICs, in which chips are stacked and integrated with internal vias, will likely exhibit very high pin counts. Upcoming via technologies will be able to support tens of thousands of "pins" between adjacent chips in the 3D stack. Given the high pin count, automatic co-design will be needed in any but the most regular system. At the least, a 3D floor planner with interlayer pin assignment will be needed. Chapter 9 deals with some of these issues.

After design, co-verification is important for these high pin-count systems, but it is fortunately more straightforward. It is relatively easy to import the package design into the IC design tools (Figure 14.2) and conduct verification tasks such as extraction and comparison of the extracted view with the initial schematic.

The final issue in Table 14.1 is thermal dissipation. Today, thermal chip-package co-design is not practiced, since the package can be safely designed after the chip. However, in the future, the spatial and temporal rates of change of heat flux will increase. Chip-package co-design will be needed to match these rates of change to the capability of the package. For example, a high spatial variation of heat flux could lead to unacceptable temperature gradients and hot spots, and might best be fixed through an adjustment in the chip floorplan. Similarly, the temporal rate of change might have to be reduced for the same reason. This issue is likely to be particularly important in 3D chip stacks, wherein heat removal is particularly difficult.

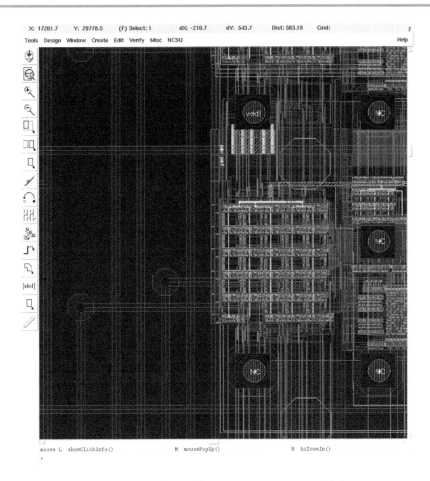

FIGURE 14.2   Result of a flip-chip package after importation into an IC design environment for the purposes of coverification.

14.4 MIXED-SIGNAL CO-DESIGN ISSUES

For mixed-signal systems, the combination of increasing operating frequencies, together with a strong drive toward miniaturization, leads to an increasing need for co-verification and co-design. These steps are very important for the correct execution of a System-on-Package (SOP) design, in which digital and analog RF IC portions are separately, or together, combined with on-package passives, such as inductors, capacitors, baluns, and antennae. This combination allows separate optimization of the analog/RF and digital IC processes, tightly integrated with passives that perform better than their on-chip alternatives, and are smaller than discrete components.

For example, an inductor placed on a package structure can be built with a higher Q factor (lower losses) than its on-chip equivalent. It has been shown that SOP-based filters and antennae can have lower insertion losses than their discrete equivalents. A chip package with co-implemented radio is likely to have lower power consumption and a smaller size, compared to its conventional equivalent. One drawback of the SOP approach is the increasing stress it places on the need for good parametric models to enable co-simulation of the transistors and SOP passives.

Accurate analog transistor models are needed together with accurate models for the passives integrated onto the package. Such models must include the impact of processing variations. Collecting such models is a tedious and time-consuming task, especially for the first run on a new technology. Good parametric models are needed to size the passives correctly, while sizing the transistors at the same time. When such models are lacking, multiple design-build test cycles are often needed to achieve a working design.

In some circumstances, the use of an SOP might increase the modeling and design effort required. For example, if an RF chip is placed just above a package-integrated inductor or antenna, the designer should model and simulate the current induced in the on-chip circuits due to the integrated passive component and the impact of the chip substrate on the performance of the passive component. For the former, it might be necessary to move a highly sensitive circuit, such as a low-noise amplifier (LNA), away from the peak fields induced by the passive component. If shield layers are used, it is important to model the amount of digital SSN introduced onto the "grounds," especially floating grounds. Remember, there is no such thing as a perfect voltage reference—common mode noise is always a concern.

Even if integrated passives are not used, accurate co-simulation is important to achieve a quality design. Often, the most critical issue is the impact of the digitally sourced SSN on the jitter of VCOs and noise figure in LNAs. Phase noise and noise figure are generally hard to simulate accurately. However, to come close, the power and ground noise need to be included. To predict accurately the power and ground noise requires good di/dt modeling and the inclusion of accurate package and PCB parameters for the power and ground system. True co-simulation is also needed. (See also Chapter 26, where very similar problems are discussed.)

14.5 I/O BUFFER INTERFACE STANDARD AND OTHER MACROMODELS

The IBIS has been a highly successful technique to macromodel I/O buffers. Although an accurate count of the number of users is not available, a web search shows over 800 companies using IBIS, and over 1000 users have downloaded the SPICE to IBIS tool. The reasons for its success are as follows:

- No proprietary information is conveyed in an IBIS model.
- It is sufficiently accurate to lead to design success for timing and noise prediction in board-level designs operating at clock rates of around or just above 100 MHz.
- It can be automatically produced from a SPICE netlist, using the SPICE to IBIS utility.
- IBIS-compatible simulators can be very fast, as IBIS IV curves are continuous and analytic (continuous in the first derivative).

The circuit elements in an IBIS-compatible output model are shown in Figure 14.3. The pull-up and pull-down components are used in the output models only. They include an I-V data table and ramp rates (to give accurate rise and fall times). The power_clamp and gnd_clamp curves are used in input models and output models to enable outputs. They are intended to model the clamp diodes that turn on when the voltage goes too far outside the Gnd–Vcc range. They are also captured as I-V tables. All I-V tables are interpolated during simulation. L_pkg, R_pkg, and C_pkg are a simple attempt to capture package parasitics. Note that there is a lot of ambiguity here, as to how to account for mutual inductance and capacitance between neighboring leads in the package. It is up to the IBIS vendor whether to leave them out or to simply add them in; neither is entirely accurate. C_comp is used to capture the on-chip pin capacitances, including the pad capacitance and the drain and sources capacitances of the final drive transistors.

While highly successful, IBIS does have a number of limitations. It tends to overpredict SSN (Figure 14.4). The main reason for this overprediction is that the pull-up and pull-down components do not lose current drive capability during the power ground noise collapse event, as they do in

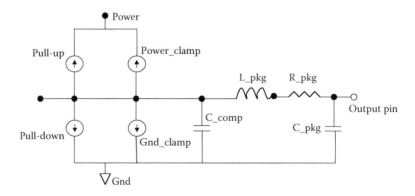

FIGURE 14.3   IBIS output behavioral model.

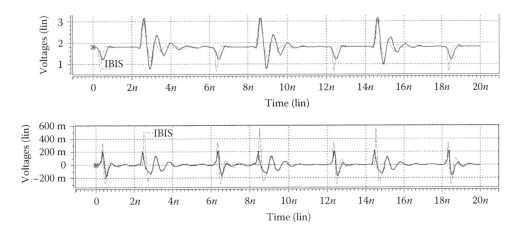

FIGURE 14.4   Power (top) and ground (bottom) signals for IBIS (broken line) and SPICE (solid line).

the full fidelity simulation (i.e., normally i_out and di/dt is reduced during the voltage collapse and this is not captured in IBIS).

The simplicity of IBIS models also leads to limitations. The model can be incorrect if used in a different loading environment than it was produced for. In particular, the ramp times for the pull-up and pull-down are very load specific. If package parasitics or the characteristics of the transmission line being driven change, then the I-V tables should be produced for the new load and not just pulled from a previous model. Also, the IBIS model is not sufficiently accurate for multi-Gbps simulations.

Because of these limitations, later versions of the IBIS standard include the ability to support a wide range of AMS (Analog and Mixed Signal) models, which are essentially equations captured in a language that can convey any I/O relationship. However, the production of AMS models is up to the user, and the Spice to IBIS converters do not support an automatic production of a higher fidelity model. But several methods have been established, which permit the capture of higher-level models, including "template" modeling, "black box" modeling and "surrogate" modeling (see the references below).

14.6 PATHFINDING

The trend in the semiconductor and packaging industry is toward system integration, where more functionality can be embedded in a smaller volume. This leads to ICs assembled side by side on a conventional package (2D), ICs assembled on a high-density package called the interposer (2.5D), and ICs placed on top of each other assembled on an interposer (3D). In both the 2.5D and 3D embodiments, the vertical interconnections on a tight pitch provide for a large number of I/O terminals that need to be connected through high-density wiring in the interposer. Meanwhile, other forms of 3D integration such as Package on Package (POP), which have lower I/O counts, are also being pursued due to cost reasons. As the die-to-die connections (I/O terminals) increase, the dimensions of these interconnections are decreasing, leading to ICs coming closer together with lower parasitics for the wires. This is lowering the energy expended per word (8 bits), from around 4.8 nJ/word for 2D integration to less than 7 pJ/word for 3D integration. In Figure 14.5, DDR3 (Double Data Rate), LPDDR (Low-Power Double Data Rate), and Wide I/O are the applications driving such technologies. With technologies rapidly changing, a combination of such technologies is necessary to meet both the cost and performance targets. Hence, systems of the future will contain a mixture of technologies, where as an example, ICs will be assembled onto an interposer using a combination of wire bond and micro-bump technology. The ICs will be interconnected using high-density redistribution layers and through silicon vias in the interposer which provide the necessary conduit for communication between the ICs. A host of other materials and structures in the interposer, such as embedded passives, will provide for additional functionality. With such high levels of integration, the interface between the chip and the package will become transparent, and hence, chip-package co-design issues will become even more important.

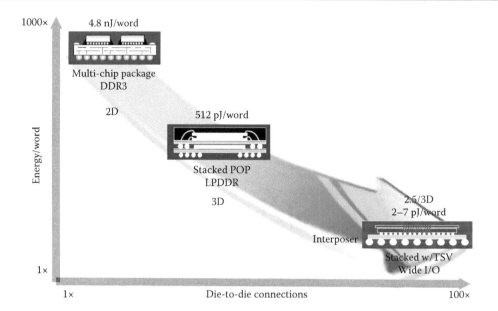

FIGURE 14.5   Chip-package integration trend.

The integration of such disparate technologies for a system architect or designer can be very challenging, since doubts often exist as to whether the combination of such technologies can meet the performance and cost targets required. These doubts can continue even after a combination of technologies is chosen, since the structures used in the design need careful evaluation as to whether they will meet the performance targets. In addition, process variations and other electrical interactions can make the technologies difficult to implement. A new suite of tools is therefore emerging for chip-package co-design especially targeted toward 2.5D and 3D integration that enables early analysis of the designs prior to implementation. This can help minimize expensive modifications to either the technologies chosen or the structures being implemented. These tools are being classified as PATHFINDING tools. Pathfinding includes electrical, mechanical, and thermal analysis at an exploratory stage of the design process that can help reduce the overall cost of the design cycle. Cost modeling and floorplanning represent important aspects of pathfinding as well. Figure 14.6 shows a design cycle where a typical EDA flow consisting of front-end design and back-end verification tools is augmented with the pathfinding tools. The exploratory phase of the flow using the pathfinding tools is very important since it can have the largest impact on design in terms of both performance and cost.

An example of pathfinding is shown in Figure 14.7, where the objective is to evaluate the coupling between through silicon vias (TSV) referenced to a power/ground (P/G) grid for a specific

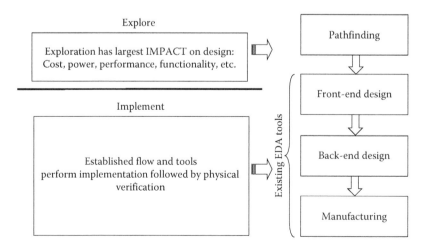

FIGURE 14.6   Role of pathfinding.

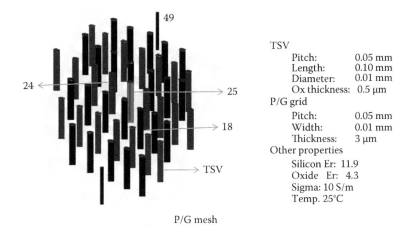

TSV
Pitch: 0.05 mm
Length: 0.10 mm
Diameter: 0.01 mm
Ox thickness: 0.5 µm
P/G grid
Pitch: 0.05 mm
Width: 0.01 mm
Thickness: 3 µm
Other properties
Silicon Er: 11.9
Oxide   Er: 4.3
Sigma: 10 S/m
Temp. 25°C

FIGURE 14.7   Through silicon vias over power grid.

process technology, with details as shown in the figure. Since cross talk between TSVs is a major concern due to tight via pitch, the outcome of this exercise is a modification in the physical dimensions of the TSV array (pitch, oxide thickness…) such that cross talk is maintained within the budgeted level. For this example, a pathfinding tool should be capable of creating the TSV array referenced to a P/G grid with ease, where the physical dimensions are parameterized and can be modified by the user. The near-end cross talk between the center via 25 and vias 24 and 18, are shown in Figure 14.8 at two temperatures, namely, 25°C and 100°C. Since thermal hot spots can occur when ICs are stacked on each other, the designer needs to gauge the impact of temperature on cross talk. The temperature in the interposer can be determined using a thermal pathfinding tool which, when combined with electrical pathfinding, can be used to estimate the level of cross talk expected. In Figure 14.8, a temperature rise from 25°C to 100°C in the interposer, estimated using the thermal pathfinding tool, decreases cross talk by 2.5 dB at 5 GHz due to a reduction in the silicon conductivity. Processes are never perfect and often at times lead to via tapers. Evaluating the effect of via tapers on cross talk is an important aspect of the exploratory design process. In Figure 14.9, a via taper from 10 to 8 µm has a 0.5 dB effect on cross talk at 5 GHz, which provides valuable information to the designer for prioritizing the effect of parameters such as temperature and via taper on cross talk. The results from this analysis are an updated list of dimensions and rules that are manufacturable and can then be used for physical implementation.

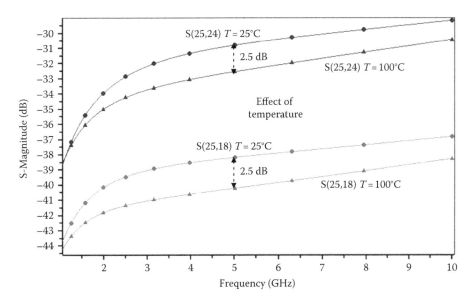

FIGURE 14.8   Near-end cross talk between through silicon vias—effect of temperature.

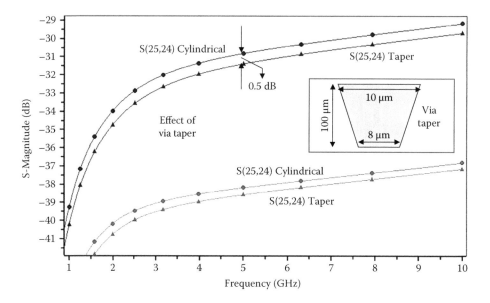

FIGURE 14.9 Near-end cross talk between through silicon vias—effect of via taper.

14.7 MULTI-SCALE AND MULTI-PHYSICS ISSUES

Energy efficiency is becoming a major driver for electronic systems. The optimum efficiency for transistors is typically achieved at low power supply voltages that are close to the threshold voltage of the transistor. At these supply voltages, the effect of temperature variations on the performance of the transistors can be very large. Hence, predicting the voltage drops and temperature gradients in an IC becomes very important. With the trend toward integration, thermal hot spots are created in the system, which in turn can increase voltage drops. Similarly, the current flowing through the interconnections can generate heat, which in turn can affect the thermal gradients. The interaction between the electrical and thermal domains occurs through Joule heating. This effect is becoming especially pronounced for 3D integration where heat cannot escape easily. Since the power supply and cooling solutions are outside of the IC, the chip-package interactions play a very large role in determining both the voltage drops and thermal gradients in a system. Hence, for back-end verification, an entire system needs to be analyzed, which poses a unique problem due to the difference in length scales between the IC and the package. Therefore, a big challenge for 3D integration is electrical–thermal co-analysis (multi-physics) at the system's level (multi-scale).

As an example of multi-scale and multi-physics analysis, consider the 3D system shown in Figure 14.10, where TSVs are used to connect dies (bare ICs prior to packaging) to each other using micro-bump technology (not shown). The center die contains a clock distribution network (CDN), distributed as an H-tree that feeds the top and bottom IC. The stacked IC is assembled on an interposer containing TSVs using micro bumps, which is then mounted on a printed circuit board (PCB) using solder balls. A power supply and clock generator on the PCB are used to supply power to the dies and provide the clock signal to the CDN, respectively. A heat sink attached to the top die along with fans and thermal interface materials enables the removal of heat from the dies. As the circuits switch, the center die (CDN) draws 10 W of power, while die1 and die2 each draw 20 W. A scenario, as in Figure 14.10, when analyzed, can lead to temperature gradients in the center die as shown in Figure 14.11, due to the nonuniformity of the power maps within each die. When the CDN is overlaid on the temperature profile, as in Figure 14.11, the hot spots affect both interconnect and buffer delay leading to varying delays as the clock signal traverses the CDN. The delay is also affected by the voltage drops in the power distribution network. This results in the clock arriving at different times at the end of the CDN (End 1, 4, 13, and 16) causing large skews, leading to an eye diagram with excessive jitter, as shown in Figure 14.12. This example highlights some of the issues that need to be tackled for 3D integration, where chip-package co-design tools are required for analysis.

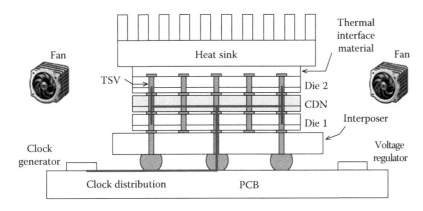

FIGURE 14.10   3D Stacked ICs on interposer and PCB.

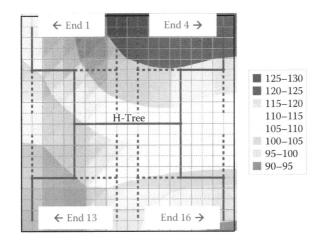

FIGURE 14.11   Temperature gradient and clock distribution.

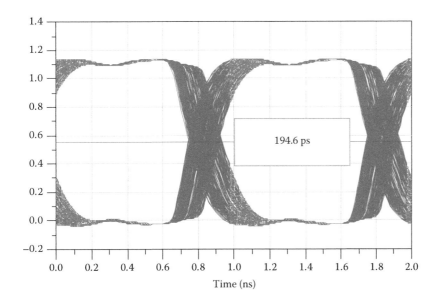

FIGURE 14.12   Eye diagram at the edge of the clock tree.

14.8 CONCLUSIONS

Traditionally, when one refers to EDA tools for chip-package co-design, modeling tools are usually implied. The most classic and successful tool examples are the reduced complexity macromodeling tools available to support IBIS I/O models. With the trend toward 3D integration, tools that support pathfinding and multi-scale/multi-physics analysis are emerging. However, co-design is not just about modeling. In the future, physical co-synthesis will be needed to support the design of high pin-count systems.

ANNOTATED BIBLIOGRAPHY

SYSTEM ON PACKAGE TECHNOLOGIES AS IT IMPACTS CO-DESIGN:

Tummala, R., M. Swaminathan, M.M. Tentzeris, J. Laskar, Gee-Kung Chang, S. Sitaraman, D. Keezer et al., The SOP for miniaturized, mixed-signal computing, communication, and consumer systems of the next decade, *IEEE Trans. Adv. Pack.*, 27, 250–267, 2004.

DIGITAL CO-DESIGN AND CO-VERIFICATION:

Schaffer, J.T., A. Glaser, S. Lipa, and P. Franzon, Chip package codesign of a triple DES processor, *IEEE Trans. Adv. Pack.*, 27, 194–202, 2004.

Shen, M., L. Zheng, and H. Etenhunen, Robustness enhancement through chip-package co-design for high-speed electronics, *Proceedings of the Fifth International Symposium on Quality Electronic Design*, 2004, Vol. 3, pp. 184–189, CA.

Varma, A.K., A.W. Glaser, and P.D. Franzon, CAD flows for chip-package coverification, *IEEE Trans. Adv. Pack.*, 28, 96–101, 2005.

Wang, J., K.K. Muchherla, and J.G. Kumar, A clustering based area I/O planning for flip-chip technology, *Proceedings of the Fifth International Symposium on Quality Electronic Design*, 2004, Vol. 2, pp. 196–201, CA.

MIXED-SIGNAL CO-DESIGN EXAMPLES:

Brebels, S., J. Ryckaert, B. Come, S. Donnay, W. De Raedt, E. Beyne, and R. Mertens, SOP integration and codesign of antennas, *IEEE Trans. Adv. Pack.*, 27, 341–351, 2004.

Donnay, S., P. Pieters, K. Vasen, W. Diels, P. Wambacq, W. De Raedt, E. Beyne, M. Engles, and I. Bolsens, Chip-package codesign of a low-power 5-GHz RF front end, *Proc. IEEE*, 88, 1583–1597, 2000.

Nayak, G., and P.R. Mukund, Chip package codesign of a heterogeneously integrated 2.45 GHz, CMOS VCO using embedded passives in a silicon package, *Proceedings of the 17th International Conference on VLSI Design*, 2004, Vol. 4, pp. 627–630, Mumbai, India.

IBIS AND OTHER MACROMODELING:

Mutnury, B., M. Swaminathan, and J. Libous, Macromodeling of nonlinear I/O drivers using Spline functions and finite time difference approximation, *Twelfth Tropical Meeting on Electrical Performance of Electronic Packaging (EPEP 2003)*, Princeton, NJ, 2003.

Stievano, I., I. Maio, F. Canavero, and C. Siviero, Parametric macromodels of differential drivers and receivers, in Advanced packaging, *IEEE Trans. Adv. Pack.*, 28, 189–196, 2005.

Varma, A., A. Glaser, S. Lipa, M. Steer, and P. Franzon, The development of a macromodeling tool to develop IBIS models, *Proceedings of IEEE Electrical Performance of Electronic Packaging*, Princeton, NJ, October 2003, pp. 177–280.

Zhu, T., M.B. Steer, and P.D. Franzon, Accurate and scalable IO buffer macromodle based on surrogate modeling, *IEEE Trans. CPMT*, 1(8), 1240–1249, 2011.

3D-ICS:

Chan, V., P. Chan, and M. Chan, Three-dimensional CMOS SOI integrated circuit using high-temperature metal-induced lateral crystallization, *Elect. Dev. IEEE Trans.*, 48, 1394–1399, 2001.
Cong, J., J. Wei, and Y. Zhang, A thermal-driven floorplanning algorithm for 3D ICs, *ICCAD* 2004, pp. 306–313, CA.

THERMAL CO-DESIGN:

Shadron, K. The importance of computer architecture in microprocessor thermal design, in thermal and thermomechanical phenomena in electronic systems, 2004, *ITHERM 2004*, *The Ninth Intersociety Conference*, 2004, Vol. 2, pp. 729–730.

PATH FINDING:

Martin, B., K. Han, and M. Swaminathan, A path finding based SI design methodology for 3D integration, *Electronics Technology and Components Conference (ECTC)*, Orlando, FL, June 2015.
Priyadarshi, S., W.R. Davis, M.B. Steer, and P.D. Franzon, Thermal pathfinding for 3-D ICs, components, packaging and manufacturing technology, *IEEE Trans.*, 4(7), 1159–1168, July 2014.
Swaminathan, M. and K. Han, Chapter 1: Design and modeling for 3D ICs and interposers, World Scientific Publishing Company, 2013.

MULTI-SCALE AND MULTI-PHYSICS:

Park, S.J., N. Natu, M. Swaminathan, B. Lee, S.M. Lee, W. Ryu, and K. Kim, Timing analysis for thermally robust clock distribution network design for 3D ICs, *Electrical Performance of Electronic Packaging and Systems Conference*, 2013, pp. 69–70, CA.
Swaminathan, M. and K. Han, Design and modeling for 3D ICs and interposers, World Scientific Publishing Company, 2013.

Design Databases

Mark Bales

CONTENTS

15.1 INTRODUCTION

When Chevrolet introduced C5 Corvette in 1997, its chassis was 4½ times sturdier than the chassis in the C4. This was the main reason that the C5 Corvette runs 0.99G on the skid pad versus 0.84G of the C4 and improved its handling to the point where it is competitive with even Porsche sports cars. The new C6 and C7 Corvette improve this to 1.09G and 1.11G, respectively. I can state with a high degree of certainty that *not a single person* buys a Corvette because of the chassis. It is a necessary feature but always takes a backseat to the flash.

So is the case with EDA design databases. The design database is at the core of any EDA system. It is expected to perform flawlessly with a very high level of performance and be as miserly with memory as possible. The end users tend to ignore the database, unless something goes wrong. A file is corrupted, the design gets too large to fit in memory, or a design has 10,000,000 cells all superimposed at 0, which performs poorly. Of course, in every one of these cases, the database is at fault. Application developers have their own requirements on design databases. The API must be so intuitive that it can be used without a manual; it must be completely tolerant of any bad data given to it and fail in a graceful way; it must maintain a high level of performance even when the database is used for purposes for which it was never designed. Nevertheless, a design database *is* the heart of the system. While it is possible to build a bad EDA tool or flow on *any* database, it is *impossible* to build a good EDA tool or flow on a *bad* database.

This chapter describes the place of a design database in an integrated system. It documents some historical databases and design systems by way of reference, for those who are interested in how we got to the state of the art. We then focus on the current best practices in the EDA design database area, starting with a generic conceptual schema and building in complexity on that. Real-world examples of commercially available design databases are presented, with special emphasis placed on the OpenAccess (OA) Coalition and their community-open-sourced

EDA database. Finally, a series of historic and current references are given as a resource for those who would like to learn more. As there is too much information to present in these pages, readers are encouraged to seek out the current references.

15.2 HISTORY

15.2.1 DESIGN DATABASES AS PART OF AN INTEGRATED SYSTEM

In examining EDA design databases, it is useful to look at a hypothetical tool architecture. It lets us determine which parts are considered part of the design database and which are the application levels. In Figure 15.1, the portion within the dotted line represents the design database components. On the left is the language system (which, although not directly part of the database, will be used by *parameterized designs*, as described below in Section 15.2.5.1). Built on top of the database are the algorithmic engines within the tool (such as timing, floor planning, place and route, or simulation engines), with the highest level representing the applications built from these component blocks. Note that the scope of the design database includes the actual design, library information, technology information, and the set of translators to and from external formats. All of these components will be discussed in this chapter.

15.2.2 MATURE DESIGN DATABASES

In many instances, of mature design databases exist in the EDA industry as a basis for both commercial EDA tools and proprietary EDA tools developed by the CAD groups of major electronics companies. Systems from IBM [Bushroe97], Hewlett-Packard [Infante78], [Wilmore79], SDA Systems (Cadence), ECAD (Cadence), High-Level Design Systems [HLD95], and many other companies have been developed over the last 30 years and continue to be the basis of integrated circuit (IC) design systems today. Many of these systems took ideas from university research programs such as those by [Keller82], [Crawford84], [Ousterhout84], [Harrison86], and successfully productized them.

Most of the mature design databases have evolved to the point where they can represent netlist data, layout data, and the ties between the two. They are hierarchical so as to allow for reuse and application in smaller designs. They can support styles of layout from digital through pure analog and many styles of mixed-signal designs. Most of these systems have a performance and/or capacity problems as designs grow larger and larger. Although it is possible to increase capacity by using 64-bit versions of these systems, their performance falls off even more. A new generation of databases that have higher capacity while maintaining good performance is needed. These databases had been developed over the last decade.

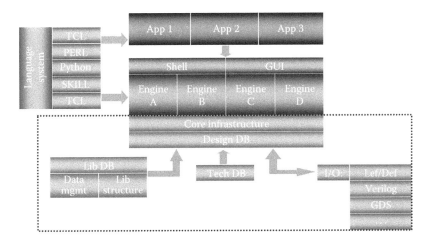

FIGURE 15.1 High-level architecture of an EDA system.

15.2.3 MODERN DATABASE EXAMPLES

15.2.3.1 OPENACCESS DESIGN DATABASE

Given the importance of a common design database in the EDA industry, the OA Coalition was formed to develop, deploy, and support an open-sourced EDA design database with shared control [OAC03]. The data model presented in the OA database (DB) provides a unified model that currently extends from structural RTL through GDSII-level mask data and now into the reticle and wafer space. It provides a capability rich enough to support digital, analog, and mixed-signal design data. It provides technology data that can express foundry process design rules through 10 nm and contains the definitions of the layers and purposes used in the design, definitions of VIAs and routing rules, definitions of operating points used for analysis, and so on. OA makes extensive use of IC-specific data compression techniques to reduce the memory footprint and to address the size, capacity, and performance problems of previous DBs.

The details of the data model and the OA DB implementation have been described previously [Santos03]. As of 2015, OA is the only modern IC database where the implementation is publicly available, so the examples in this chapter primarily refer to the OA implementation.

15.2.3.2 OPEN MILKYWAY AND THE IC COMPILER II DATA MODEL

Perhaps in reaction to the OA Coalition, in early 2003, Synopsys decided to open up their Milkyway physical design database and started the MAP-In (Milkyway Access Program). This has been described in [Brevard03]. The similarities to OA outweigh the differences. Milkyway is a more mature database and has many more chips taped out using it. At the same time, its data model is a subset of what is available in OA, with about 70% of the OA model available in Milkyway. It is written in C and was never built to be an open-sourced or a standard model. As a result, its API is not as clean as OA. Its internal implementation is not available even for MAP-In members, so no comments may be made about the implementation, other than to stress that this is a tried-and-true production-level EDA database. One observation that can be made is that the capacity of the Milkyway database seems to be similar to that of other databases of the same age (e.g., DF-II from Cadence). In that regard, it is probable that OA is as much as an order of magnitude smaller than the Milkyway data.

Starting in 2010, the MAP-In program has been ramped down and Milkyway is no longer freely available. In 2014, Synopsys Users Group conference, a new data model that is a follow-on to Milkyway, was presented as part of the introduction of IC Compiler II. It is a full reimplementation in C++ of a model that is similar to OA but is tailored for synthesis and optimization. It is not available outside of its use in IC Compiler II, and hence, little is known about it publicly. Its memory is typically 3–5× smaller than Milkyway and up to 10× faster, so it is similar in performance to OA [SNPS14].

15.2.3.3 MAGMA, MENTOR, AND OTHERS

Other significant design databases have been built by Mentor Graphics (their Falcon database, one of the first in the industry written in C++) and Magma Design Automation.* Neither of these databases is available publicly, so little can be said about their features or performance relative to other industry standards. Like Milkyway is for Synopsys, Falcon seems to be a stable and mature platform for Mentor's IC products. Falcon seems to be similar in capability to Milkyway. Magma's Talus database is not just a disk format with an API but an entire system built around their DB as a central data structure. This is very advanced and is more advanced than the levels of integration currently taking place using OA. Again, since the details of the system are not publicly available, a direct comparison of features or performance is not possible. Looking at the capabilities of the Magma tools, it would indicate that this DB has a similar functionality to OA and may be capable of representing behavioral (synthesis input) information.

---

* Magma Design Automation is now part of Synopsys, Inc.

15.2.4 FUNDAMENTAL FEATURES

Figure 15.2 shows the general-purpose physical/logical database. In this diagram, a shaded bubble represents a database object. Unshaded bubbles represent classes that are base classes for some of the objects. The dashed lines represent the class hierarchy relationships. The solid lines represent relationships among the objects, and their endpoint arrows represent the ordinality of the relationship. A single solid arrow represents a 1-to-1 relationship. An example of this is the Term to Net relationship. A Term may not exist without a corresponding Net. A double hollow arrow represents a 1 to 0-or-more relationship. An example of this is a Net to Term relationship. A net may have one or more associated terminals, but it does not necessarily need to have even one.

15.2.4.1 THE VIEW AS THE BASIC UNIT

The fundamental basic unit of a design block is called a "Design." It has also been called a "cell," "view," "block," "cellview," "representation," and other names in other databases. In essence, a Design is a container that holds the contents of what a designer thinks of as a component within the design. Most designs are hierarchical, and they are composed of many Design objects. In a flat design, there is only one Design object. It then contains every piece of physical geometry, all of the connectivity, all of the timing and place and route (P&R) objects, and in general everything about the design with the exception of the technology information and the directory information for the storage of the Design in a library. Each of these classes of information is kept elsewhere.

The Design can be viewed as being the file descriptor for the information in this design component. In addition to containing general information about the component, it is the starting point from which all of the other information contained in this Design may be found.

15.2.4.2 SHAPES AND PHYSICAL GEOMETRY

The simplest and longest-lived (from a historical viewpoint) information in a Design is *shapes*. These represent the physical shapes used to construct the mask patterns for the implementation of the IC masks. While some shapes are used purely for annotations or other forms of graphical display, most shapes are actually part of the design and become part of the ultimate masks that are built at the tail end of the IC design flow. Shapes include the *Dot*, the *Path*, the *Rect*, the *Polygon*, the *Ellipse*, *Text*, the *VIA*, and the *Route*. They are most often assigned to a *Layer*, which is an object that links together all of the shapes in a given Design on the same mask layer.

The *Dot* (also called a "Point" or "Keystone" in other databases) represents a 0D object, which just has a position in the X–Y plane. Some implementations may assign a "size" to the dot, which is used for display purposes. Some implementations may assign a layer to the dot as well. Dots are used to represent the center of a pin to represent an anchor point for some sort of constraint or annotation and possibly to represent the origin of a cell. Since they are a 0D object, their use is semantic, and they do not become part of the mask, even if they are on a mask layer.

The *Path* (also called a "Line" or "Segment" in other databases) represents a 1D object, which is defined by two (or more) points in the X–Y plane. Since it has a width that may be greater than zero, it is in reality a 2D object, but it is more useful to think of it as a line plus a width. Note that some databases consider a zero-width path and a path with nonzero width as different objects. Some implementations restrict a Path to a single segment (defined by its two endpoints). Most implementations allow a Path to have a series of points. Paths are assigned to a layer. If their width is greater than zero and they are assigned to a mask layer, they become part of the ultimate mask for that layer. Paths may be kept as *Manhattan*, meaning their segments are always parallel to one of the X or Y coordinate axes. They may be *Manhattan plus 45*, meaning their angle is one of 0, 45, 90, 135, or 180, or they may be *All-angle*, meaning there is no restriction on the slope of the path segments. The type of path is defined by its least-restrictive segment. In other words, if every segment but one in a path is Manhattan and the last is All-angle, then the entire path is considered All-angle. Paths have an end style, which describes a possible extension of the path beyond its endpoints by an amount in a direction given by the slope of the end segment. The path end style may be *flush* (with no extension), *half width* (where the extension is half the width

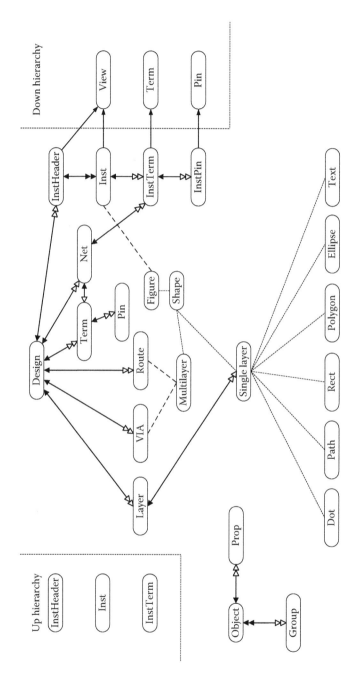

FIGURE 15.2 General EDA logical/physical database schema diagram.

of the wire), *variable* (where the amount of extension is explicitly specified), *octagonal* (where the end forms an octagon of the same width as the path centered around the endpoint of the path segment), or (in rare cases today) *round* (with the diameter of the half-circle end matching the width of the segment).

The *Rect* (sometimes called a "Box" or by its full name of "Rectangle") is the simplest and historically most used shape. It is a simple rectangle whose sides are parallel to the X–Y coordinate axes. It can be represented by two points, typically the *lower left* and *upper right* (although some systems use *upper-left* and *lower-right* points, still others use a single point plus a *width* and *height*). Each Rect is assigned to a layer and becomes part of the mask for that layer.

The *Polygon* (sometimes called a "Poly") is a shape defined by a closed set of points in the X–Y plane. Similar to a Path, it may be Manhattan, Manhattan plus 45, or All-angle. The type of polygon is defined by the least restrictive of the segments that make up its boundary. Most systems assume that the closing edge of the polygon goes from the last point to the first point, although some databases represent the first and last points explicitly as the same point. Some databases allow holes to be explicitly represented in polygons, but most databases have the hole split at some point, so the polygon touches itself but does not overlap. Polygons are not allowed to be self-intersecting in any modern system. As with the other shapes to this point, polygons are assigned to a layer and become part of the mask.

The *Ellipse* (sometimes simplified as a *Circle*) may sometimes be used in analog IC design or when the database is used to represent a printed circuit board. Their greatest use is in representing connection points, both in IC layout and mainly in schematic diagrams. The ellipse has axes that are parallel with the coordinate axes. An ellipse where the defining points are coincident forms a circle—some databases only represent circles and do not allow the full freedom of the ellipse. Unless used in analog design, the ellipses are almost never part of the final IC mask. They are most often used as annotation shapes in nonmask layers.

The *Text* (sometimes called a "String" or "Message") is a nonmask shape that is used for labels and other forms of annotation. While there is often a need for "mask text," this is usually done with polygons. Modern text objects may allow for different fonts, styles, and sizes. Their size may be represented using a word processing "point" value, or it may be given in units that match the ones being used in the Design. Text strings are assigned to a layer, which controls their display color.

The *VIA* is the first shape introduced that represents a multilayer object. As such, it is not assigned to a layer but is rather assigned directly to the Design or carried along with a *Route* (described next). It is defined by two conducting layers and a contact-cut layer that will join the two. VIAs may be simple, including only a single cut, or they may be quite complex, having an array of cuts, possibly with some cuts missing. Their definition is usually described in the technology database in modern systems. In older databases, they may be represented as anything from a single contact cut (with the connecting metal layers implicitly derived from the cut), a series of shapes joined together (which can be horrendously inefficient), an instance of a separate cell (again, very inefficient), or some form of parameterized cell (which is defined later).

The *Route* (shown in Figure 15.3) is a grouping mechanism that represents a multilayer (and in this case multiobject) shape. It is like a Path, in that it goes from one point to another, may include a single segment or many segments, and may be Manhattan, Manhattan plus 45, or All-angle. It differs from the Path in that it may change width, layer (through a VIA), or both during

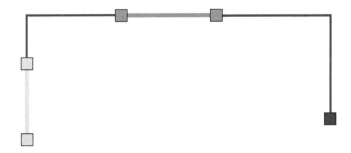

FIGURE 15.3   An example of a route.

the course of its travels. It also has an explicit record of what its endpoints are connected to. It is a grouping of many paths and VIAs, all in a single object. Its use is prevalent in constructive algorithms (routers) and, when present, can be leveraged for very-high-performance incremental routing and rerouting as well as physical shape tracing.

15.2.4.3 HIERARCHY

Most IC designs are hierarchical. Even for designs that are small enough to be implemented as a single Design, the leaf library cells are kept separate from the Design for reasons of capacity, update, and just sense. This means that a database should be hierarchical, and every database of the modern era is. Hierarchy is established through an *instance/master* relationship, in which a Design may be *instantiated* or placed within another design. In the simplest case described earlier, library leaf cells are placed and then connected using additional shapes to implement the circuit under construction. The Inst object refers to a Design (called the master). The Inst has a placement location and an orientation used for placement. The location is just an offset of the origin (the 0, 0 point) of the Design being instantiated from the origin of the higher-level Design containing the Inst. The orientation is most generally one of the eight Manhattan-based orientations that are achieved by giving a rotation of 0°, 90°, 180°, or 270° and then being able to mirror invert each of these rotated instantiations. Some databases support fewer of these eight possible orientations, and a few databases support all-angle orientations so that a Design may be, for example, rotated by 38°. Most modern databases are limited to the eight Manhattan-based orientations. Note that the hierarchy is most often *folded*, in that each Inst of the same master Design points to that single master Design (see Figure 15.4).

The introduction of hierarchy and physical instantiation prompts a discussion of two additional concepts, those of units and coordinate spaces. In every physical design database, there is some notion of units. Most databases keep coordinate data as integers, because rounding errors that might occur in floating-point numbers would be unacceptable if they resulted in layout shapes not touching, for example. There is then a mapping from these integer database units (often abbreviated as *DBU*s for Database Units) into physical dimensions. OA represents this mapping in the technology data and uses a *single* such mapping for every Design in the entire IC implementation. This makes for a simpler, more robust system as it becomes impossible to have a unit mismatch between Designs across the instance/master boundary. Other systems have used a single mapping throughout, for example, considering each DBU as a nanometer. Other systems allow a different mapping per each Design. This can make for an error-prone system; since the databases present coordinates as the raw integers, it becomes the responsibility of each application to understand when hierarchy boundaries have been crossed and to make the necessary adjustments in units. A single application in a flow that does not properly treat differing units can

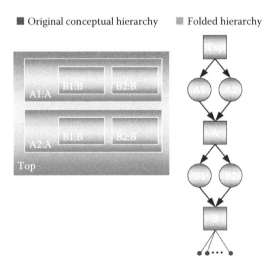

FIGURE 15.4 An example of a folded hierarchy.

corrupt data or generate incorrect analyses. From experience, using different units for different parts of a design is confusing to both man and machine and should be avoided wherever possible, even if the database supports it.

Each Design has its own coordinate space. In a hierarchical physical design, the coordinates of shapes within an instantiated master are different inside the master and from the point of view of the Inst. The coordinates are transformed by the translation and orientation of the Inst in the instantiating Design. In a multi-level hierarchy, these transformations must be concatenated. Most modern databases provide support for this so that an ultimate flattening of the physical design data that must be done in order to create the IC masks is simple.

15.2.4.4 CONNECTIVITY AND HIERARCHICAL CONNECTIVITY

One of the advantages of design databases of the modern era is the unification of physical layout with connectivity. Being able to represent simultaneously a netlist along with the shapes that implement the design and being able to relate specific shapes to specific nets within the netlist provided a quantum step in capability within the EDA tools. The typical connectivity structures include a *Net* that represents a logical or physical node within the netlist, a *Term* that represents an available connection point on a Design, and an *InstTerm* that represents a connection between a Net and the Term on the master of a given Inst. These constructs form the basic scalar connectivity. In addition, these constructs can be combined into a *Bus*, which is an aggregation of nets that are all related and have the same root name, or a *Bundle*, which is an aggregation of nets that are not related and have different root names. This is true for both the *Net* and *Term* (and in some databases the *InstTerm*), so we have the *BusNet* and *BusTerm* (and sometimes the *BusInstTerm*).

The *Net* is the core construct for connectivity. It represents a logical or physical node within the netlist. It has a name. It provides a linkage of zero (or more) *Term*s that "export" the net so that it becomes visible when the containing Design is instantiated. It provides a linkage of zero (or more) *InstTerm*s that represent connections to Terms within Designs that have Insts at this level of hierarchy. The InstTerm provides the linkage of connectivity across the hierarchy.

The *Term* is a net that is exported or made visible outside the Design to which it belongs. In some databases, it may have a name that is different from the Net to which it is connected, while other systems have a single name for the net and terminal. The Term usually contains a linkage of the *Pin* shapes that define the physical connection points and layers for the physical implementation. Most modern databases allow multiple Terms for a single Net (sometimes called "shorted terminals"), while some restrict the relationship to a single Term per Net. These shorted terminals have become necessary in the modern era mostly due to post-synthesis optimizations that remove one or more gates, having the side effect of shorting two or more terminals together. Systems that cannot support shorted terminals have problems with modern-day netlists that come from synthesis tools.

15.2.4.5 GENERAL CONSTRUCTS

Two fundamental constructs that can be used with any object in a modern design database are the *Prop* and the *Group*. Props (also called "Property") form a simple extension mechanism that is intended for use in adding data that are sparsely populated. For example, if you need to add a property "GridShape" that identifies shapes that are part of a power grid, you can do this with a property and *not* incur the overhead of storing this information on every shape, even though the properties are used only on a small fraction of the overall shapes. Props are of simple basic types, such as Boolean, integer, float, and string. Some databases extend these basic types with specialized variations, such as filename and mapped name. Some databases provide for default values for the properties; some restrict the legal set of property names per object type (to control unwarranted proliferation and abuse of properties); and some add range values and/or enumerations to indicate legal values for the properties.

The *Group* is used to extend the database relationships among objects. Most databases allow groups to be either a collection or a set (where each member must be unique). Many implementations allow groups to be declared as having a unique name or a common name. In this way, groups can be used to establish specific relationships among objects or a class of relationship in the nonspecific name case. An example of the former is to have a group called "GridShapes" that represents the set of shapes that are used in a power grid. An example of the latter would be a series of groups called "macroArray" where each such group represents a set of macro cells that should be placed in a physical array structure.

Note that the use of both properties and groups is discouraged in general. They are very expensive in terms of space and moderately expensive in terms of performance. In a typical implementation, if the percentage of objects in a given class that can have a specific property or group grows above about 15%, it is better to implement the extension as a "native" object within the database for superior space and speed performance. OA provides an extension Attribute mechanism that makes this possible without change to the core database, and this preferred mechanism is described in a subsequent section.

15.2.4.6 API FORMS

Along with the general evolution of API forms in computer science, the form of design database APIs has evolved. Most modern database implementations use C++ classes to represent the objects in the database. In most implementations, the constructors and destructors for the objects are special methods, and the default constructors and destructors are disabled. This is because the memory management of the objects must be closely controlled to allocate the objects within their associated Design and the techniques that are used in the internal database implementation to represent the data in an efficient manner require the locality of reference that these customized constructors supply. In addition, it is a good idea for the API that the developers use to make it perfectly clear when they are creating and deleting objects.

In most modern implementations, the conceptual schema represented by the public classes is very different from the internal implementation of the database. This is done to allow ongoing trade-offs between speed and space performance as well as to allow implementation-level improvements to be made without needing to revise the interface of the public API. In this way, the lifetime of the public APIs may remain unchanged through several different implementations of that API.

Because of this separation, most implementations provide access to all members through methods rather than a direct form of access. This has the added advantage of making it possible to provide a very robust form of *observer* (also called a callback, trigger, notifications, or daemon), which can be used to detect any change of interest in the data. The methods are generally broken up into "set()" and "get()" methods to make the code more readable and less error prone.

In addition to the data that are accessed through member functions, special functions called "Iterators" (also called traversal or generation methods) are created to allow traversal of the relationships among the data. For example, an Iterator is created from a Design that will allow iteration through all nets in the Design. Some older implementations use macros to implement these traversal functions as a kind of loop, but most modern implementations create an object that contains the state of the iteration *and* that is used in a pure control loop within the application program. Iterators require special care to deal with situations where the design is modified during iteration.

15.2.4.7 NOTIFICATIONS

Notifications were mentioned in the previous section. They provide a mechanism where interested application-level code can register interest in specific changes to specific kinds of objects. This keeps code dependencies clean, as the low-level data model does not need to be aware of or depend on higher-level application code. It can form the basis for a system that uses *lazy evaluation* to construct high-performance incremental application engines. For example, a timer in a flow can register interest in changed nets and use those data to invalidate nets within the timer. An incremental router can change a small fraction of nets in the design. The timer (and presumably an extractor with its own observer clients) notes the changes to the nets. The next timing update performs an incremental extraction followed by an incremental timing update. This can make these operations run an order of magnitude faster than the full updates. Proper use of notifications can make incremental systems reliable and efficient.

Notifications are very powerful, and with this power comes responsibility. It is possible for a poorly written observer client to cause slowdown of an entire EDA system, sometimes to the point of making it unusable. If an observer client does substantially more work than the call that initiated the notification, if the interaction between data model and observer client cause expensive recomputation to be done in what should be a simple client, or if the observer clients are active at inappropriate times or on objects or designs where they are not needed, then they will work to the detriment of the system. It is important for client writers to keep these potential problems in mind as they write observer client code.

15.2.4.8 UTILITY LAYER

The set of classes and methods that would allow a database to be fully created, modified, queried, traversed, and deleted constitutes the core set of database functions. There are many common functions that are done on top of this core set, and it makes sense to unify them into a "utility layer" of classes and methods that come along with the code database. Most modern databases implement some such functions. Here are descriptions of a few of the possible utility areas.

15.2.4.8.1 Region Query   One operation that is almost universal throughout physical design tools is the ability to quickly search for shapes contained in an area of interest. Most modern databases provide a capability of doing this efficiently, and this capability is used by everything, from graphical display to design-rule checking (DRC) utilities to P&R utilities. Most often, some form of tree-based organization (K-D tree, quad-tree, etc. [Rosenberg85]) is used. Some implementations consider the region-query structures to be part of the database and persistently store them. Others recompute these data when they are needed and focus on high-performance computation to make this practical.

15.2.4.8.2 Error Handling   There are two main styles of error handling used in modern design databases, *return codes* and *exceptions.* In the return-code style, each method returns an error code when it fails, or it might return a known bad value and *then* provide a way of determining the error code. Exception-based methods will "throw" an exception when an error occurs, and it must be "caught" by the application-level code. This area is a somewhat controversial one. People who prefer the return-code style note that uncaught exceptions can make for system crashes that can enrage end users. People who prefer the exception style note that unchecked return codes can make for programs that fail in obscure ways and possibly corrupt data. OA uses the exception style of error handling, and this method seems to result in code that is more robust over time.

15.2.4.8.3 Boolean Mask Operations   Another often used utility is the ability to perform Boolean mask operations (AND, OR, NOT, NAND, XOR, GROW) on mask shapes. These can be used as the basis of a DRC capability, to create blockages, to cut power-grid structures, and for many other physical purposes. The algorithms used for this capability are complex and often have a high value placed on them (see Chapters 20 and 22 for more details). Most publicly supported design databases either charge extra for this capability, or, as is the case in OA, provide a way to hook in customer-supplied Boolean mask ops. This provides for a standard API while giving the end user the flexibility of applying a mask op tool set of appropriate performance, capacity, and price.

15.2.4.8.4 Checkpointing   A very handy utility layer is one that can implement check-pointing. This is the ability to save a design, perhaps along with a small or large amount of tool state, in a form such that it can be restored at a future time. This can take the form of a simple saving and bundling of data into a checkpoint file or may be as complex as a system that saves all of virtual memory into a file. In any case, the goal is to allow a user (or sometimes an application) to save the tool state in the middle of a long-running flow or application engine.

15.2.5 ADVANCED FEATURES

15.2.5.1 PARAMETERIZED DESIGNS

Most of the *Design*s that exist are just aggregations of shapes and/or connectivity that represent a particular design. In many cases, however, a library or technology requires many cells that share a common design but have different sizes. The simplest example of this is a VIA, which can easily be imagined as a more complex object with a set of parameters. In fact, this is what the VIA object is. There is the notion of a *Parameterized Design* that abstracts this concept, allowing a user to write a small piece of code, which, when coupled with a set of the parameters, will evaluate a Design with layout and/or connectivity. This requires that the database support one or more extension languages. Historically, the first extension languages were proprietary and specific to the database, but recent efforts use conventional programming languages and APIs. For example, OA supports C++ and now TCL. Unofficial additions to OA include Python and PERL, and the proprietary language SKILL is supported by Cadence. In order to be successful, the system must be architected so that any user of the given database can at least *read* the Parameterized Designs. OA provides the capability of shipping an interpreter along with the parameterized library designs, making the use of this capability portable. OA also provides for the management of *variants* (a Design evaluated with a specific set of parameters) in order to make the system as efficient as possible. Several other systems support this capability but usually in a single-language-specific way.

15.2.5.2 NAMESPACES AND NAME MAPPING

A concept that every design database must deal with at some point is *name mapping*. Every database, every file format, and every hardware description language (such as Verilog or VHDL) have a legal name syntax and may (or may not) have some characters that convey semantic meaning. Examples of this are hierarchy delimiters, bus-name constructs, and path-terminator separating constructs. If a design database supports characters with semantic meaning, there needs to be some form of escaping to be able to treat these characters in names *without* their semantic meaning.

In addition to this complexity is the fact that of the many formats that can drive a third-party tool, each one has a slightly different version of its own name mapping and namespace characteristics. The problem is simple. When you input from one format and then output to the same format, the names need to remain as they originally were. When names are input in one format and then output in another, the transformations that occur must be consistent, so another tool that uses these files can be presented with a consistent group of files. As an example, consider a third-party tool that takes the library data as an LEF file, the netlist as a Verilog file, and the parasitic data as an SPEF file. Each of these file formats has a different namespace, and it is tricky to represent names in any design database so that they will be consistent when they meet up in a single tool.

The most modern approach is to use some variation of algorithmic conversion of names so that the transformations are invertible. This was the method used in the Pillar database [Pillar95] and is the method used in OA. A namespace is defined for each different format, and a native namespace is used within the design database. Other systems can deal with the problem by building maps to map names that are not representable in both namespaces. This can be problematic, as if the map is lost, or if the names are changed out of context of the map, then it may no longer be possible to correctly map the names.

15.2.5.3 UNDO/REDO

Every user of the most basic utility program is familiar with the notion of *undo*. Almost but not quite as familiar is the companion notion of *redo*. Most users of EDA tools want these capabilities in the system, and this presents special issues for the database designer. First off, an undo/redo system must be an integral part of a design database if it is to be reliable. If there are two ways to modify a data model, one that goes through an undo/redo layer and another one that does not, you can guarantee that some application will call the wrong layer and a resulting mixture of undoable and nonundoable modifications will spell disaster. This means that there cannot be an access layer in a data model that supports undo/redo that bypasses the undo/redo system.

On the other hand, the performance and overhead of the undo/redo system are of equal importance. If users move a single rectangle in an interactive command, they can reasonably expect that it would be possible to undo that operation. Likewise, if a very large selected set of objects is deleted and immediately discovered to be an error, most users would expect to be able to undo the deletion and restore the objects. In these cases, the overhead of the undo/redo system is a welcome price to pay for the safety for interactive operations.

For large-scale commands, however, it may not pay to suffer the undo/redo overhead. For example, in a router, it seems more likely the user will preserve the ability to *undo* the router by saving separately the input and output designs rather than using a fundamental undo/redo system. Therefore, it must be possible to have commands that modify the data model and yet *do not support* undo/redo. Some systems may leave this choice in the hands of the user, classifying commands as interactive or batch and giving user-level control over which make use of the undo system and which do not. This let the end user choose the performance overhead they are willing to pay for and let them build the highest-performance flows that they can.

15.2.5.4 POWER FORMATS AND VOLTAGE AREAS

With the increasing popularity of multi-voltage design, it became necessary for a means of describing *power intent* for a design from the very early stages, perhaps even before logic synthesis has been performed. This might mean that no specific gates yet exist that could be connected to specific power and ground voltages. In addition, it is simpler for the early stages of power design if higher-level blocks can have their power supplies designated rather than requiring explicit wiring of PG pins for all leaf-level cells. This allows timing analysis and estimation to be more accurate without requiring full mapping of logic elements or the explicit PG connections. Two popular formats currently exist. The Unified Power Format (UPF) is a standard (IEEE 1801) and is based on a donation from the Accellera organization, supported by Mentor Graphics and Synopsys [Karmann13], [Hazra10], [Gourisetty13]. The alternative Common Power Format (CPF) is a standard administered by the Silicon Integration Initiative (Si2) [Carver12], [Lakshmi11]. It was donated to Si2 by Cadence Design Systems. Each of these standards allows an evolving association of power supplies to the components of an IC. Early on in the design, the associations can be symbolic. As the design progresses and actual placement and routing of physical leaf cells must take place, the associations become more concrete. Ultimately, power supplies must be wired to each cell in the design, so this is a natural evolution within the design flow.

UPF and CPF represent a symbolic or logical form of the connectivity [Pangrle11].

For the physical constraints that accompany the power formats, a construct called a Voltage Area is used (see Figure 15.5). This construct has one or more regions and represents locations on a chip or subblock where a specific supply association is in effect. This construct is used in floorplanning to show the different areas where different power domains are located. The placers, routers, and optimizers use these data to properly place logic in an area with compatible power and can also use this information to properly place level shifting logic, retention logic (for voltage areas that are switched off), and so on. Generally, there is a one-to-one correspondence between a voltage area and a power domain described in the UPF or CPF.

15.2.5.5 PLACE-AND-ROUTE CONSTRUCTS

Many place-and-route-specific constructs may be added to an EDA design database. (See Chapters 5 and 8 for a more detailed description of these operations.) *Blockages* and *Halos* or *Keepouts* describe areas where operations such as placement or routing are prohibited. A *Cluster* or *Bound* defines a group of instances that are clustered together and possibly an area where they should be placed. A *GCell* (for *Global routing Cell*) divides the chip area up into a grid and is used to store global routing and congestion information. A *Steiner* point is a virtual object that acts as a junction object for more-than-two-way route junctions. A *Track* is an object that defines routing tracks for a region within the design. It is generally used to limit routing in some areas by limiting

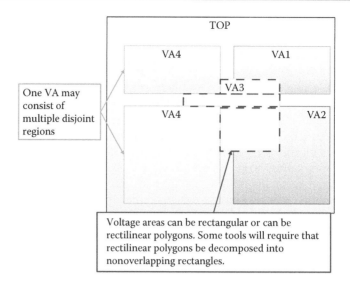

FIGURE 15.5 Voltage areas.

the available layers and to specify the available tracks, which might change near large subblocks or macrocells. A *Site* is similar to a *Track*, but instead of defining locations for routing, it defines locations for placement. Site definitions are used to create *Site Rows* in regions of the chip. A leaf cell has an associated Site that indicates where it must go. Site Rows can overlap, and this is often done to allow single-, double- and even triple-height cells to inhabit the same region.

These objects may or may not be implemented in some design databases. OA implements all of these objects and you can find a detailed description of these items in [Mallis03].

Other place-and-route constructs can include *Route Guides*, which are the inverse of blockages and indicate areas where routes are specifically allowed to go on specific layers. A similar construct is a *Route Corridor*, which further limits the allowed routes to specific nets in a given area. These constructs are not found in OA at this time but are used in other systems.

Many constraints can be placed on nets, so they are routed in specific ways. *Shielding* surrounds a sensitive net with shield routes that are connected to ground (or perhaps power) nets. *Balanced Routes* relate a pair of nets and ensure that the overall parasitics for the nets are kept the same within a defined tolerance. *Cable Routing* takes a bus or bundle of nets and routes them together as a larger entity. The number and types of these constraints are growing over time. Sometimes, nets can have multiple constraints applied at the same time. For instance, Shielding is often used with Cable Routing between the individual nets within a bus or bundle to minimize coupling while ensuring they are routed together. This allows the shields to be shared between adjacent signals and makes for a more efficient implementation overall.

15.2.5.6 TEST CONSTRUCTS

Most modern digital IC designs include some logic for testing of the chip. Level-sensitive scan design is one such method in which special "scan flops" are linked together with an alternate serial interface. This allows a circuit to be stopped and to have its data readout from the scan flops or to have data written into the flops to set a particular state to test the circuit's reaction when proceeding from that state. Managing the complexity of this testing resource is a challenge. At a bare minimum, it is necessary to identify the scan ports and scan nets in a design, so their connectivity is not used to bias placement. Otherwise, the normal operation of the chip could be compromised for nets that are most often used at slower-than-normal speeds and that should not be considered in circuit performance. Often, it is desirable to give the user control over the ability of placers and routers to reorder *Scan Chain* elements. In some cases, it may be desirable to maintain a fixed order or proximity of specific scan chain elements in a *Scan Group*, while the ordering of other elements or groups in the overall scan chain is not fixed. Physical hierarchy complicates the implementation of scan chains, especially if the geometric layout for the chip makes it most efficient to have the scan chain for the design cross subblocks multiple times. This will have the side effect of adding multiple scan ports to the subblocks, and these must be managed properly to maintain a port signature for the block that is as close to the original (user-defined) port signature as is possible. This last requirement is a general one and is necessary to reduce problems when trying to use the physical blocks within a simulation test harness that was created for the original logical subblock definition.

15.2.5.7 TIMING AND PARASITIC CONSTRUCTS

Modern-day design databases contain constructs for dealing with timing-driven design and analysis. *Parasitic* elements represent the results of parasitic extraction, storing a network representing the full parasitic model, including the resistor, capacitor, and inductor elements. It is possible to store a reduced form of the network, either as a pole-residue model or as a simpler *Elmore* model. It is necessary that each net is able to be represented at a different level of abstraction, which reflects on the fact that the routing is fully complete for some nets but only partially complete for others. *Timing* models for Designs must be possible and can serve both as the de facto model for leaf cells and as an abstraction for an intermediate-level Design. Constraints must be stored both in the technology area (as definitions of operating conditions) and within the Design (as design-specific constraints). Timing arcs should be able to be stored to allow for long-term lazy evaluation of timing information. Few modern databases contain much of this information. The Magma database contains most of it. OA contains everything but timing arcs and specific timing constraints, through a publicly available extension, include these objects [Xiu05].

15.2.5.8 VIRTUAL GLOBAL NETS

A "net" that spans repeater (buffer and inverter) logic is called a virtual global net or *supernet*. When optimizing designs that include repeaters, it is convenient to be able to ignore the repeaters. In the past, some optimization systems would start by removing all repeaters and then (re)optimizing a net after understanding its full extent and all its endpoint connections.

It is not necessary to remove actual repeaters to get this information. A (generally nonpersistent) object can be built that ignores repeaters and presents a single net that "sees through" the repeaters. This gives the same effect as if all the repeaters had been removed but does so without modifying the design at all. Once a net is optimized, the information about the already-existing repeaters on the virtual global net can be used in conjunction with the current optimization to better leverage the existing repeaters. This can reduce the need for incremental placement, routing, extraction, and timing, as a greater percentage of existing repeaters are reused.

15.2.5.9 OCCURRENCE MODELS AND LOGICAL/PHYSICAL MAPPING

An advanced feature beyond the scope of this chapter is the addition of Occurrence Models to the database. Remember that most modern design databases use *folded* hierarchy, in that every instance of a given master Design points to the same Design. This has the advantage of making it easy to fix a problem in a cell and have it reflected everywhere instantly. However, there is no place in a folded database to attach data that will be specific to each unique *occurrence* of a master within a chip design (see Figure 15.6). Most systems today deal with this problem by using auxiliary data structures to represent the *unfolded* (occurrence) hierarchy, or they proceed to make copies of every intermediate-level cell that is used more than once, a process called "uniquification."

OA contains an Occurrence Model called EMH for Embedded Module Hierarchy. It includes a logical view (called the "Module" view), a physical view (called the "Block" view), and an unfolded occurrence view (called the "Occurrence" view) that relates the two. Figure 15.7 shows a logical hierarchy with its corresponding flat physical hierarchy, and Figure 15.8 shows the EMH representation for this conceptual design, illustrating the Module, Occurrence, and Block hierarchies side by side. This is very useful as you can refer to an object by using its logical name or its

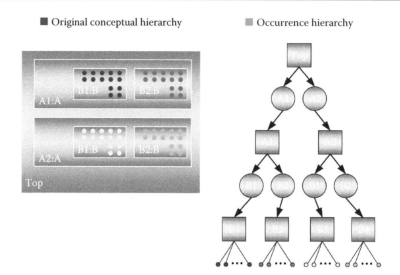

FIGURE 15.6 Different occurrences of the same master.

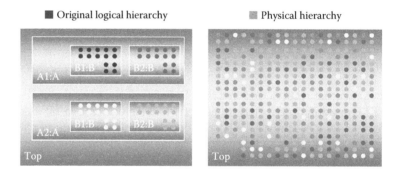

FIGURE 15.7 A logical hierarchy with its corresponding flat physical hierarchy.

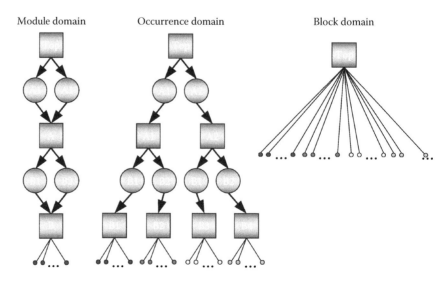

FIGURE 15.8 Three components of the Embedded Module Hierarchy of OA.

physical name, and this correlation persists even through some repartitioning. One limitation of the OA model is that it is not permitted to move a logical Module past a physical Block boundary. This means that some common forms of repartitioning are not yet possible with the current version of OA. See the OA manual [Mallis03] for more details.

15.2.5.10 EXTENSIBILITY

One key feature of a design database is extensibility. We have already discussed Props and Groups and know that while they are suitable for simple extensions, they are too costly to use for anything substantial. A modern design database must have some way of quickly adding new objects and relationships. If the database is proprietary, the only issue with changing the database is the effects on customer migration. Most customers will not move to a new version of EDA tools that require a database change. They will wait until the current chip is finished. This can lead to excessive time being spent supporting two or more versions of the database and hence the tools. If, on the other hand, a more portable extension mechanism is available, the core information may remain unchanged while the extensions are added.

This is the approach taken by OA. Extension mechanisms to add attributes to existing objects, create new objects, and establish new relationships among new and/or existing objects are available in OA. These are like Props that have their definition separate from the data, which makes them much more efficient in storage space. While they take some work to set up, they are nearly identical in performance and capacity to similar additions that would be added in a "native" way.

Another mechanism that is used stores data in named groupings. An older tool may not understand one section of data but can skip over those data. A newer tool can read all the data that might be present. This provides both extensibility and backward *and forward* compatibility. It is a broader mechanism than the attributes described earlier and has greatest applicability for larger-scale changes.

As a last comment, interoperability among a set of tools that originate from many different sources is greatly improved by using an extensibility approach. As long as the core data are understood by all tools, the additions may be kept and communicated only when necessary. This makes for more robust multi-vendor tool flows.

15.2.6 TECHNOLOGY DATA

Any design database must be able to represent process-level design rules, such as spacing and width requirements, as well as rules for the construction of VIAs. The rules must be able to represent technology nodes down to at least 10 nm. This means that, for example, apposition- and width-based spacing rules are possible. As feature sizes shrink ever further, antenna rules come into play as well [Mercier03]. In addition, there should be a way of recording data to be used for *constructing* routing as well as for *analyzing* shapes. OA uses the *Constraint* and *ConstraintGroup* objects to implement these capabilities.

Most processing rules, especially for the smaller process nodes, are proprietary and subject to very strict security. As a result, they cannot be discussed here in any meaningful way.

15.2.7 ENGINEERING CHANGE ORDER SUPPORT

A very common occurrence is late-stage changes to the logic in a design because of full testing or even last-minute design changes. The engineering change order process needs support from a data model to allow for efficient and incremental update of the design. A useful (and almost mandatory) utility takes an existing and new version of a netlist and identifies differences. This utility can be quite complex as the N−1 version of a netlist has likely gone through substantial optimization, buffering, and other modifications to meet performance requirements. It can be very difficult to distinguish differences due to changes between the N−1 and N versions of a netlist versus the changes due to the tool flow operating on the version N−1 netlist.

Some form of Change Management is useful in a design system. Such a utility layer keeps track of changes that are made during optimization and keeps that information for later use when doing the aforementioned diff. This can make it simple to perform a three-way diff and hence easily distinguish netlist version changes from flow changes.

15.2.8 LIBRARY DATA AND STRUCTURES: DESIGN DATA MANAGEMENT

15.2.8.1 LIBRARY ORGANIZATION: FROM DESIGNS TO DISK FILES

Because most IC implementation teams are consisted of many members nowadays, and sometimes as many as a hundred members, the design database must allow for the individual Design databases that comprise a chip hierarchy to be worked on simultaneously. Each Design may be worked on by at most a single person. A finer-grained level of concurrent access is not warranted. This makes it natural to divide the data so that each individual Design is contained in one or more files that may be put under design data management control (it also makes it possible to use the native operating system file locking and access permissions, which reduces implementation effort and may already be familiar to the designer). In general, a library is a directory and in it contained a number of Designs, each of which is identified by a unique Cell and View name. Beyond that, the only requirements are to store things in such a way as to make the performance of searching for and reading Designs as high as is humanly possible. Most end users would like a library organization scheme that makes it easy to locate a file in the file system given a library directory and the cell and view names. Historically, this has been done by using directories for cells (and in some cases views). While this worked well in the past, modern designs are seeing an explosion in the number of unique Designs, and directory searches are very poor when the number of Cells in a given library might reach 50,000 or more. In addition, in such cases, the actual data might be relatively small, and the overhead for a directory might be as large as or larger than the Design file. This means that the traditional lib/cell/view structures found on disk will have a poor speed and space performance.

OA provides two different mechanisms for mapping Designs to disk files. The first is an upward-compatible system that represents the data in the traditional file system-based way. The second is a server-based method, which has significantly higher performance than the older method. The effects of this are only visible in designs with more than about 5000 Cellviews.

15.2.8.2 DESIGN DATA MANAGEMENT

Design data management is a controversial area in which it is virtually impossible to achieve a consensus. The best that can be hoped for is to build a flexible enough system that it will allow incorporation of any design-data management (DDM) system, either commercial or homegrown. OA has taken this approach starting with its 2.2 release. Instances of interfacing this to RCS, ClearCase, Synchronicity, and IC Manage exist, along with many OA users who have adapted their own proprietary DDM systems.

15.2.9 INTEROPERABILITY MODELS

It is useful to look at what interoperability requirements are from an application developer's point of view. We can look at data interoperability levels ranging from file or interchange format through a common in-memory model and discuss which mechanism is appropriate in which circumstances. In addition, we can look at the control-interface level for components within a part of an interoperable system. Given an underlying framework that supports the interoperability required at the data level, we will discuss what is necessary at the algorithmic and application levels to insure interoperability. File-level interoperability is the simplest kind.

It has the benefit that it decouples the tools involved from each other and probably from the exact version of the interchange format used. It provides a readable format (in most cases) that can be used for purposes of debugging and testing. Using files presents many challenges and issues, however. The time needed to parse a readable format is usually much greater than that required to read a specific database file. The size of the readable files is usually much greater than the design data in a database format. Interchange formats act as "lossy filters" when compared to design databases. If a format does not contain a particular type of information, then a "round-trip" through that format will lose possibly critical information. For example, the Verilog format does not contain physical data such as cell locations and routing data. The DEF format does not contain much of the information about intermediate levels of logic hierarchy (which would be found in the Verilog representation) and does not even represent the full spectrum of physical data that is found in a data model such as OA. This means that using either of these file formats for design transfer (or even both in conjunction) will lose data. File-level interchange is most useful for applications that create a different form of output and are not responsible for round-trip fidelity of the data. For example, using the GDS format to drive a physical DRC program loses much of the original data, but this is not important because the physical data are not brought back using GDS. Rather, a report or a file of annotations for DRC errors is created and can be annotated back on the original data.

Using a database, but with a different in-memory model, treats it almost as an interchange format but without several of the problems. It shares the file-format advantage of allowing the tools to remain uncoupled. It does not directly provide a readable version of the data, but databases often provide a readable archiving or versioning format that can be used if the need arises. Databases mitigate almost all of the problems found in file-based formats. They require little or no parsing and are hence much faster to read. Applications only need to be concerned with the relevant data and the database system manages the rest of the data. In this sense, they are not "lossy" like the file-based formats. Round-trip fidelity is achieved by updating the relevant information in the database system. The problems unique to database use arise when significant mapping must be done between the database information model and the in-memory model used by the application. This can make opening and saving a design too slow and can even cause problems in the fidelity of the data mapping in the worst case. In addition, it is very difficult to share components among applications that use the same database but have different in-memory models. If this is a requirement, you are much better off developing a common in-memory model.

Once you have a common in-memory model, you have the best of the worlds.

If you take the additional step of making the system components incremental and based upon the core in-memory model wherever possible, then you can collect an entire set of application components: technology and algorithmic engines that can be combined and reused in many different products. This lends consistency to the different tool suites within a design flow, reduces development and maintenance costs over time, makes for a modular system that can remain vital by easily upgrading modules one at a time, and provides a rich base for more rapid development of new products and technologies.

Some would think (with more than a bit of justification) that having a common in-memory model can slow down progress. It is true that any sweeping change must be well planned and deployed. It is less true that additions must be well planned; they often benefit from a developmental deployment where an initial design is tested with actual products and refined before being more broadly deployed. The need for rapid progress can be met through the use of the extension mechanisms mentioned in the previous section. This allows rapid prototyping, the ability to keep proprietary extensions private, and makes sure that the progress of the standard does not limit the ability of the tools to keep up with changes in technology. See [Xiu05] for an example of the use of this mechanism.

Even though the common in-memory model provides the greatest gains, it is important to choose an integration method that is well matched to the task. We continue to dismiss file-based formats for the problems previously mentioned. However, using a database as an exchange format is appropriate if

- The subsystem is not incremental.
- The I/O time as a fraction of the total run time is small.
- The components within the subsystem are built on a different in-memory model and switching would be expensive or would cause degradation in performance.

It is possible to build a highly productive design flow that uses tools plugged into a common database as a backplane that interfaces to each tool in the flow in the most appropriate manner. When done in a careful and elegant way, the integration seems seamless.

One advantage the common in-memory model gives over the use of the database as an interchange format is the ability to reuse system components. The biggest impact this makes on the overall design flow is the level of consistency and convergence it brings. Timing analysis gives answers right after synthesis that correlate properly with the answers given after physical implementation and extraction, even if the analysis is done in different tools.

15.3 CONCLUSIONS

Modern EDA databases must support designs heading toward 1 billion components. They must be thread safe and support almost linear scalability in application algorithms. They have to provide interactive levels of performance, even with the largest designs. They must have physical constructs and technology rules that support the most advanced nodes at 10 nm and below. They must be written in a language like C++ that supports object-oriented design while maintaining the highest levels of performance necessary to support the ever-increasing design sizes and the ever-increasing levels of sophistication in applications that use the data model. They must have a modern software architecture, being modular enough to support their use in fully integrated EDA tool suites as well as in stand-alone utilities.

Many such examples exist today in commercial tools. The OA system remains the only publicly available database. It contains a very large number of data constructs that has grown over the years of its use. The proprietary nature of advanced process nodes at 20 nm and below works against the open nature of OA and has shifted the advantage back to the proprietary databases supported by the major EDA companies. Whether the industry will shift back toward openness or continue down the current path of proprietary solutions is still an open question. In any event, the increasing complexity of ICs will continue to drive EDA database technology in the foreseeable future.

REFERENCES

The references listed represent many of the important papers on design databases. They are broken up into two sections—Historical and Current References. For many of the papers (especially the historical ones), the main thrust of the paper is not on the design database, so you will need to read through these papers to find the description of the data structures, models, and actual database technology used.

HISTORICAL REFERENCES

[Bales03] Bales M., Facilitating EDA flow interoperability with the OpenAccess design database, *Electronic Design Processes 2003 Workshop*, Monterey, CA, April 2003.

[Barnes92] Barnes T., Harrison D., Newton A.R., Spickelmier R., *Electronic CAD Frameworks*, Kluwer Academic Publishers, Boston, MA, 1992.

[Bingley92] Bingley P., ten Bosch O., van der Wolf P., Incorporating design flow management in a framework based CAD system, *Proceedings of the 1992 IEEE/ACM International Conference on Computer-Aided Design*, Santa Clara, CA, November 1992.

[Blanchard03] Blanchard T., Assessment of the OpenAccess standard: Insights on the new EDA industry standard from Hewlett-Packard, a beta partner and contributing developer, *International Symposium on Quality Electronic Design*, San Jose, CA, March 2003.

[Brevard03] Brevard L., Introduction to Milkyway, *Electronic Design Processes Workshop 2003*, Monterey, CA, April 2003.

[Bushroe97] Bushroe R.G., DasGupta S., Dengi A., Fisher P., Grout S., Ledenbach G., Nagaraj N.S, Steele R., Chip hierarchical design system (CHDS): A foundation for timing-driven physical design into the 21st century, *International Symposium on Physical Design*, Napa Valley, CA, April 1997.

[Chan98] Chan F.L., Spiller M.D., Newton A.R., WELD—An environment for web-based electronic design, *Proceedings of the 35th Conference on Design Automation*, San Francisco, CA, June 1998.

[Chin91] Chin G., Dietrich W., Jr., Boning D., Wong A., Neureuther A., Dutton R., Linking TCAD to EDA—Benefits and issues, *Proceedings of the 28th Conference on Design Automation*, San Francisco, CA, June 1991.

[Cottrell03] Cottrell D., Grebinski T., Interoperability beyond design: Sharing knowledge between design and manufacturing, *International Symposium on Quality Electronic Design*, San Jose, CA, March 2003.

[Crawford84] Crawford J., An electronic design interchange format, *Proceedings of the 21st Conference on Design Automation*, Albuquerque, NM, June 1984.

[Darringer03] Darringer J., Morrell J., Design systems evolution and the need for a standard data model, *Electronic Design Processes 2003 Workshop*, Monterey, CA, April 2003.

[Eurich86] Eurich J., A tutorial introduction to the electronic design interchange format, *Proceedings of the 23rd Conference on Design Automation*, Las Vegas, NV, June 1986.

[Filer94] Filer N., Brown M., Moosa Z., Integrating CAD tools into a framework environment using a flexible and adaptable procedural interface, *Proceedings of the Conference on European Design Automation*, Grenoble, France, September 1994.

[Grobman01-1] Grobman W., Boone R., Philbin C., Jarvis B., Reticle enhancement technology trends: Resource and manufacturability implications for the implementation of physical designs, *International Symposium on Physical Design*, Sonoma County, CA, April 2001.

[Grobman01-2] Grobman W., Thompson M., Wang R., Yuan C., Tian R., Demircan E., Reticle enhancement technology: Implications and challenges for physical design, *Proceedings of the 38th Conference on Design Automation*, Las Vegas, NV, June 2001.

[Harrison86] Harrison D., Moore P., Spickelmier R., Newton A.R., Data management and graphics editing in the Berkeley design environment, *Proceedings of the 1986 IEEE/ACM International Conference on Computer-Aided Design*, Santa Clara, CA, November 1986.

[Infante78] Infante B., Bracken D., McCalla B., Yamakoshi S., Cohen E., An interactive graphics system for the design of integrated circuits, *Proceedings of the 15th Conference on Design Automation*, Las Vegas, NV, June 1978.

[Keller82] Keller K.H., Newton A.R., Ellis S., A symbolic design system for integrated circuits, *Proceedings of the 19th Conference on Design Automation*, Las Vegas, NV, June 1982.

[Khang03] Khang A., Markov I., Impact of interoperability on CADIP reuse: An academic viewpoint, *International Symposium on Quality Electronic Design*, San Jose, CA, March 2003.

[Liu90] Liu L.-C., Wu P.-C., Wu C.-H., Design data management in a CAD framework environment, *Proceedings of the 27th Conference on Design Automation*, Orland, FL, June 1990.

[Mallis03] Mallis D., Leavitt E., *OpenAccess: The Standard API for Rapid EDA Tool Integration*, Silicon Integration Initiative, Austin, TX, 2003.

[Markov02] Adya S.N., Yildiz M.C., Markov I.L., Villarrubia P.G., Parakh P.N. Madden P.H., Benchmarking for large-scale placement and beyond, *International Symposium on Physical Design*, San Diego, CA, April 2002.

[Mitsuhashi80] Mitsuhashi T., Chiba T., Takashima M., Yoshida K., Integrated mask artwork analysis system, *Proceedings of the 17th Conference on Design Automation*, Minneapolis, MN, June 1980.

[Nash78] Nash D., Topics in design automation data bases, *Proceedings of the 15th Conference on Design Automation*, Las Vegas, NV, June 1978.

[OAC03] Silicon Integration Initiative, *OpenAccess: Goals and Status*, https://www.si2.org/events_dir/2003/date/oaintro.ppt, June 2003.

[Otten03] Otten R.H.J.M., Camposano R., Groeneveld P.R., Design automation for deepsubmicron: Present and future, *Proceedings of the 2002 Design, Automation and Test in Europe Conference and Exhibition*, Paris, France, March 2002.

[Ousterhout84] Ousterhout J., Hamachi G., Mayo R., Scott W., Taylor G., Magic: A VLSI layout system, *Proceedings of the 21st Conference on Design Automation*, Albuquerque, NM, June 1984.

[Pillar95] High Level Design Systems, *Pillar Language Manual*, Version 3.2, Personal Communication, September 1995.

[Riepe03] Riepe M.A., Interoperability, datamodels, and databases, *Electronic Design Processes 2003 Workshop*, Monterey, CA, April 2003.

[Roberts81] Roberts K., Baker T., Jerome D., A vertically organized computer-aided design database, *Proceedings of the 18th Conference on Design Automation*, Nashville, TN, June 1981.

[Rodman02] Rodman P., Collett R., Lev L., Groeneveld P., Lev L., Nettleton N., van den Hoven L., Tools or users: Which is the bigger bottleneck?, *Proceedings of the 39th Conference on Design Automation*, New Orleans, LA, June 2002.

[Rosenberg85] Rosenberg J.B., Geographical data structures compared: A study of data structures supporting region queries, *IEEE Transactions on Computer-Aided Design*, 4(1), 53–67, January 1985.

[Santos02] Santos J., OpenAccess architecture and design philosophy, *OpenAccess 2002 Conference*, San Jose, CA, April 2002.

[Santos03] Santos J., Overview of OpenAccess: The next-generation database for IC design, *OpenAccess 2003 Conference*, San Jose, CA, April 2003.

[Schurmann97] Schürmann B., Altmeyer J., Modeling design tasks and tools—The link between product and flow model, *Proceedings of the 34th Conference on Design Automation*, Anaheim, CA, June 1997.

[Silva89] Silva M., Gedye D., Katz R., Newton R., Protection and versioning for OCT, *Proceedings of the 26th Conference on Design Automation*, Las Vegas, NV, June 1989.

[Shenoy02] Shenoy N.V., Nicholls W., An efficient routing database, *Proceedings of the 39th Conference on Design Automation*, New Orleans, LA, June 2002.

[Smith89] Smith W.D., Duff D., Dragomirecky M., Caldwell J., Hartman M., Jasica J., d'Abreu M.A., FACE core environment: The model and its application in CAE/CAD tool development, *Proceedings of the 26th Conference on Design Automation*, Las Vegas, NV, June 1989.

[Soukup90] Soukup J., Organized C: A unified method of handling data in CAD algorithms and databases, *Proceedings of the 27th Conference on Design Automation*, Orlando, FL, June 1990.

[Spiller97] Spiller M., Newton A.R., EDA and the network, *Proceedings of the 1997 IEEE/ACM International Conference on Computer-Aided Design*, San Jose, CA, November 1997.

[VanDerWolf88] van der Wolf P., van Leuken T.G.R., Object type oriented data modeling for VLSI data management, *Proceedings of the 25th Conference on Design Automation*, Anaheim, CA, June 1988.

[Wilmore79] Wilmore J., The design of an efficient data base to support an interactive LSI layout system, *Proceedings of the 16th Conference on Design Automation*, San Diego, CA, June 1979.

[Xiu05] Xiu Z., Papa D., Chong P., Albrecht C., Kuehlmann A., Rutenbar R.A., Markov I.L., Early research experience with OpenAccess gear: An open source development environment for physical design, *ACM International Symposium on Physical Design*, San Francisco, CA, *2005 (ISPD05)*, pp. 94–100.

CURRENT REFERENCES

[Bertolino07] Bertolino A., Jinghua G., Marchetti E., Polini A., Automatic test data generation for XML schema-based partition testing, *Second International Workshop on Automation of Software Test*, Minneapolis, MN, 2007.

[Blanchard03] Blanchard T., Assessment of the OpenAccess standard: Insights on the new EDA industry standard from Hewlett-Packard, a beta partner and contributing developer, *Proceedings of the Fourth International Symposium on Quality Electronic Design*, San Jose, CA, 2003.

[Carver12] Carver S., Mathur A., Sharma L., Subbarao P., Urish S., Qi W., *Low-Power Design Using the Si_2 Common Power Format*, Design & Test of Computers, IEEE, 2012.

[Cirstea07] Cirstea M., VHDL for industrial electronic systems integrated development, IEEE *International Symposium on Industrial Electronics*, Montreal, Quebec, Canada, 2006.

[Cottrell03] Cottrell D.R., Grebinski T.J., Interoperability beyond design: Sharing knowledge between design and manufacturing, *Proceedings of the Fourth International Symposium on Quality Electronic Design*, San Jose, CA, 2003.

[Ferguson03] Ferguson J., The glue in a confident SoC flow, *Proceedings of the Third IEEE International Workshop on System-on-Chip for Real-Time Applications*, Calgary, Alberta, Canada, 2003.

[Gibson10] Gibson P., Ziyang L., Pikus F., Srinivasan S., A framework for logic-aware layout analysis, *11th International Symposium on Quality Electronic Design (ISQED)*, San Jose, CA, 2010.

[Gourisetty13] Gourisetty V., Mahmoodi H., Melikyan V., Babayan E., Goldman R., Holcomb K., Wood T., Low power design flow based on Unified Power Format and Synopsys tool chain, *Third Interdisciplinary Engineering Design Education Conference (IEDEC)*, Santa Clara, CA, 2013.

[Guiney06] Guiney M., Leavitt E., An introduction to openaccess an open source data model and API for IC design, *Asia and South Pacific Conference on Design Automation*, Yokohama, Japan, 2006.

[Hazra10] Hazra A., Mitra S., Dasgupta P., Pal A., Debabrata B., Guha K., Leveraging UPF-extracted assertions for modeling and formal verification of architectural power intent, *47th ACM/IEEE Design Automation Conference (DAC)*, Anaheim, CA, 2010.

[Karmann13] Karmann J., Ecker W., The semantic of the power intent format UPF: Consistent power modeling from system level to implementation, *23rd International Workshop on Power and Timing Modeling, Optimization and Simulation (PATMOS)*, Karlsuhe, Germany, 2013.

[Karunaratne04] Karunaratne M., Sagahyroon A., Weerakkody A., An advanced library format for ASIC design, *Canadian Conference on Electrical and Computer Engineering*, Niagara Falls, Ontario, Canada, 2004.

[Kazmierski03] Kazmierski T., Yang X.Q., A secure web-based framework for electronic system level design, *Design, Automation and Test in Europe Conference and Exhibition*, Munich, Germany, 2003.

[Lakshmi11] Lakshmi M.S., Vaya P., Venkataramanan S., Power management in SoC using CPF, *Third International Conference on Electronics Computer Technology (ICECT)*, Kanyakumari, India, 2011.

[Li09] Li Y., Wang Y.F., Xu X.F., Chen G., Chen Y., Guo D.H., Implementation of web-based platform for IC design and project management, *Third International Conference on Anti-counterfeiting, Security, and Identification in Communication*, Hong Kong, China, 2009.

[Madec13] Madec M., Pecheux F., Gendrault Y., Bauer L., Haiech J., Lallement C., EDA inspired open-source framework for synthetic biology, *IEEE Biomedical Circuits and Systems Conference (BioCAS)*, Rotterdam, the Netherlands, 2013.

[Mercier03] Mercier J., Dao T., Flechner H., Jean B., Oscar D.B., Aum P.K., Process induced damages from various integrated circuit interconnection designs—Limitations of antenna rule under practical integrated circuit layout practice, *Eighth International Symposium Plasma- and Process-Induced Damage*, Corbeil-Essonnes, France, 2003.

[Ofek06] Ofek H., EDA vendor adoption, *Asia and South Pacific Conference on Design Automation*, Yokohama, Japan, 2006.

[Pangrle11] Pangrle B., Biggs J., Clavel C., Domerego O., Just K., Beyond UPF & CPF: Low-power design and verification, *Design, Automation & Test in Europe Conference & Exhibition (DATE)*, Grenoble, France, 2011.

[Papa06] Papa D.A., Markov I.L., Chong P., Utility of the OpenAccess database in academic research, *Asia and South Pacific Conference on Design Automation*, Yokohama, Japan, 2006.

[Seng05] Seng W.K., Palaniappan S., Yahaya N.A., A framework for collaborative graphical based design environments, *Proceedings of the 16th International Workshop on Database and Expert Systems Applications*, Copenhagen, Denmark, 2005.

[SNPS14] IC Compiler II: Building a scalable platform to enable the next 10X in physical design, Synopsys White Paper, https://www.synopsys.com/cgi-bin/imp/pdfdla/pdfr1.cgi?file=iccii_infrastructure_wp.pdf.

[Tong06] Tong K., Bian J., Wang H., Universal data model platform: The data-centric evolution for system level codesign, *10th International Conference on Computer Supported Cooperative Work in Design*, Nanjing, China, 2006.

[Yu10] Hua Y., Wen-Quan W., Zhong L., General toolkit for discrete estimation of distribution algorithms, *International Conference of Information Science and Management Engineering (ISME)*, Xi'an, China, 2010.

FPGA Synthesis and Physical Design

Mike Hutton, Vaughn Betz, and Jason Anderson

CONTENTS

16

16.1 INTRODUCTION

Since their introduction in the early 1980s, Field-Programmable Gate Arrays (*FPGAs*) have evolved from implementing small glue-logic designs to implementing large complete systems. Programmable logic devices range from lower-capacity nonvolatile devices such as Altera MAX™, Xilinx CoolRunner™, and MicroSemi ProASIC™ to very-high-density static RAM (SRAM)-programmed devices with significant components of hard logic (ASIC). The latter are commonly called FPGAs. Most of the interesting CAD problems apply to these larger devices, which are dominated by Xilinx (Virtex™, Kintex™, Artix™ families) and Altera (Stratix™, Arria™, Cyclone™ families) [1]. All of these are based on a tiled arrangement of lookup table (LUT) cells, embedded memory blocks, digital signal processing (DSP) blocks, and I/O tiles.

FPGAs have been historically used for communications infrastructure, including wireless base stations, wireline packet processing, traffic management, protocol bridging, video processing, and military radar. The growth domains for FPGAs include industrial control, automotive, high-performance computing, and datacenter compute acceleration.

The increasing use of FPGAs across this wide range of applications, combined with the growth in logic density, has resulted in significant research in tools targeting programmable logic. Flows for High-Level Synthesis (Vivado HLS™) and OpenCL™ programming of FPGAs have recently emerged to target the new application domains. Xilinx Zynq™ and Altera Cyclone/Arria SoC™ contain embedded processors and introduce new CAD directions.

There are two branches of FPGA CAD tool research: one is concerned with developing algorithms for a given FPGA and another is a parallel branch that deals with developing the tools required to design FPGA architectures. This emphasizes the interdependence between CAD and architecture: unlike ASIC flows where CAD is an implementation of a design in silicon, the CAD flow for FPGAs is an embedding of the design into a device architecture with fixed cells and routing. Some algorithms for FPGAs continue to overlap with ASIC-targeted tools, notably language and technology-independent synthesis. However, technology mapping, routing, and aspects of placement are notably different.

Emerging tools for FPGAs now concentrate on power and timing optimization and the growing areas of embedded and system-level design. Going forward, power optimizations will be a combination of semiconductor industry–wide techniques and ideas targeting the programmable nature of FPGAs, closely tuned to the evolving FPGA architectures. System-level design tools will attempt to exploit two of the key benefits of FPGAs—fast design and verification time combined with a high degree of programmable flexibility.

FPGA tools differ from ASIC tools in that the core portions of the tool flow are owned by the silicon vendors. Third-party EDA tools exist for synthesis and verification, but with very few exceptions, physical design tools are supplied by the FPGA vendor, supporting only that vendor's products. The two largest FPGA vendors (Altera and Xilinx) offer complete CAD flows from language extraction, synthesis, placement, and routing, along with power and timing analysis. These complete design tools are Quartus™ for Altera and Vivado™ for Xilinx.

After some introductory description of FPGA architectures and CAD flows, we will describe research in the key areas of CAD for FPGAs roughly in flow order: high-level synthesis (HLS), register-transfer level (RTL) and logic synthesis, technology mapping, placement, and routing.

16.1.1 ARCHITECTURE OF FPGAs

To appreciate CAD algorithms for FPGAs, it is necessary to have some understanding of FPGA devices.

Figure 16.1 shows a high-level block diagram of a modern FPGA, showing block resources. This is not any specific commercial device but an abstraction of typical features. FPGAs contain high-speed serial I/O or transceivers, usually including the Physical Coding Sublayer processing for a variety of different protocols, and sometimes an ASIC IP (embedded intellectual property or macro) block for PCI Express or Ethernet. All FPGAs have a general-purpose parallel I/O, which can be programmed for different parallel standards, and sometimes have dedicated external memory interface controllers for SRAM or DDR. Designer logic will target embedded memory blocks (e.g., 10 kB or 20 kB block RAM with programmable organization), DSP blocks (multiply-and-accumulate blocks, built with ASIC standard cell methodology), and logic blocks (Altera Logic Array Block [LAB], Xilinx Configurable Logic Block [CLB]) comprising the LUTs. The device may have an embedded processor, shown here as a dual-core CPU with memories and a set of peripheral devices.

Figure 16.2 illustrates the routing interface. A LAB or CLB cluster is composed of a set of *logic elements* (LEs) and local routing to interconnect them. Each LE is composed of an LUT, a register, and a dedicated circuitry for arithmetic functions. Figure 16.2a shows a four-input LUT (4 LUT). The 16 bits of a programmable LUT mask specify the truth table for a four-input function, which is then controlled by the A, B, C, and D inputs to the LUT. Figure 16.2b shows an example of an LE (from Stratix I), comprising a 4 LUT and a D flip-flop (DFF) with selectable control signals. The arithmetic in this LE and a number of other devices is accomplished

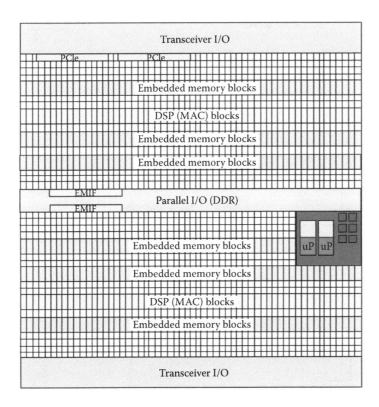

FIGURE 16.1 FPGA high-level block diagram.

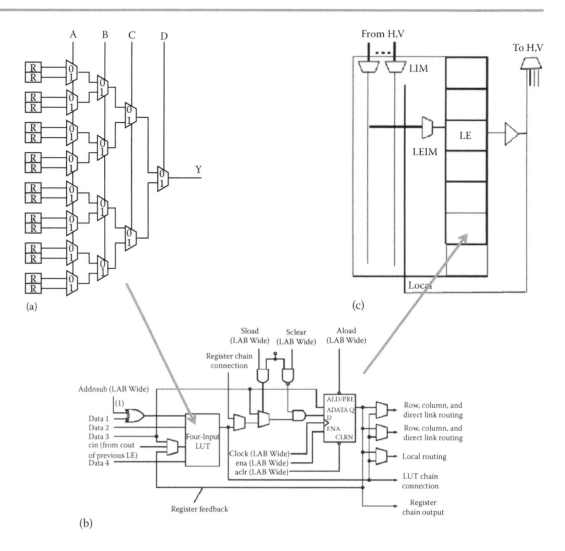

FIGURE 16.2   Inside an FPGA logic block. (a) 4 LUT showing LUT mask, (b) 4 LUT Stratix Logic element, and (c) LAB cluster showing routing interface to inter-logic-block H and V routing wires.

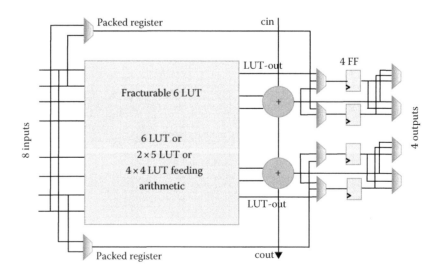

FIGURE 16.3   Fracturable LUT/FF (Stratix V ALM).

by using the two 3 LUTs independently to generate sum and ripple-carry functions. The carry-out becomes a dedicated fifth connection into the neighboring LE and optionally replaces the data3 or C input. By *chaining* the carry-out of several LEs together, a fast multibit adder can be formed. Figure 16.2c shows how several LEs are grouped together to form a LAB. A subset of the global horizontal (H) and vertical (V) lines that form the FPGA interconnect can drive into a number of *LAB input multiplexers* (LIMs). The signals generated by the LIMs (along with feedback from other LEs in the same LAB) can drive the inputs to the LE through *logic-element input multiplexers* (LEIMs). On the output side of the LE, *routing multiplexers* allow signals from one or more LEs to drive H and V routing wires in order to reach other LABs. These routing multiplexers also allow H and V routing wires to connect to each other in order to form longer routes that can reach across the chip. Each of the LAB input, LE input, and routing multiplexers is controlled by SRAM configuration bits. Thus, the total SRAM bits programming the LUT, DFF, modes of the LE and the various multiplexers comprise the configuration of the FPGA to perform a specific logical function.

There is extensive literature on FPGA architecture exploration; the reader is referred to Betz [2] for a description of early FPGA research. Depending on the architectural choices, some of the blocks in Figure 16.1 may not be present, for example, the device could have no DSP blocks, the routing could be organized differently, and control signals on the DFF could be added or removed. More modern devices do not have the simplified 4 LUT/DFF structure shown in Figure 16.2. To reduce circuit depth and address common six-input functions such as a 4:1 multiplexer, they are now more often 6 LUT based and this 6 LUT usually has additional circuitry that allows it to be *fractured* to compute two different functions of less than 6 inputs each. Lewis describes one such architecture in [3], which is shown in Figure 16.3.

The Versatile Place and Route (VPR) toolset [4] introduced what is now the standard paradigm for empirical architecture evaluation. An architecture specification file controls a parameterized CAD flow capable of targeting a wide variety of FPGA architectures, and individual architecture parameters are swept across different values to determine their impact on FPGA area, delay, and recently power, as shown in Figure 16.4. Some example parameters include LUT and cluster sizes, resource counts, logic-element characteristics (e.g., arithmetic and secondary signal structures), lengths, and number of wires. These are used both to generate the architecture and to modify the behavior of the CAD tools to target it. There have been about 100 research papers published that use VPR for either architecture exploration or evaluating alternative CAD algorithms. The open-source VPR has recently been updated [5] to include a wide range of new architectural enhancements, notably Verilog RTL synthesis, memory, and arithmetic support, in combination with new benchmarks for architecture research (and renamed VTR for Verilog-to-Routing).

The interaction of CAD and architecture is highlighted by Yan [6], who illustrated the sensitivity of architecture results to the CAD algorithms and settings used in experiments and how this could dramatically alter design conclusions. There have been a number of studies related to power optimization and other architectural parameters. For example, Li [7] expanded the VPR toolset to evaluate voltage islands on FPGA architectures, and Wilton [8] evaluated CAD and architecture targeting embedded memory blocks in FPGAs.

16.1.2 CAD FLOW FOR FPGAS

The core RTL CAD flow seen by an FPGA user is shown in Figure 16.5. After design entry, the design is elaborated into operators (e.g., adders, multiplexers, multipliers), state machines, and memory blocks. Gate-level logic synthesis follows, then technology mapping to LUTs and registers. Clustering groups of LEs into logic blocks (LABs) and placement determines a fixed location for each logic or other block. Routing selects which programmable switches to turn on in order to connect all the terminals of each signal net, essentially determining the configuration settings for each of the LIM, LEIM, and routing multiplexers described in Section 16.1.1. Physical resynthesis, shown between placement and routing in this flow, is an optional step that resynthesizes the netlist to improve area or delay now that preliminary timing is known. Timing analysis [9] and power analysis [10] for FPGAs are approximated at many points in the flow to guide the

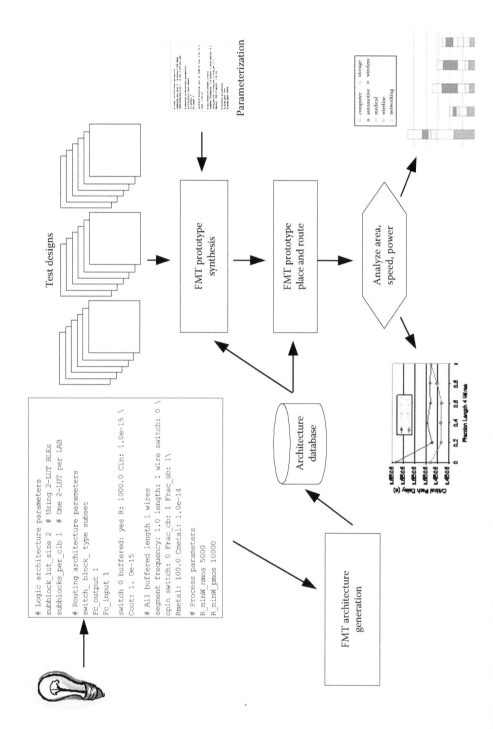

FIGURE 16.4 FPGA modeling toolkit flow based on Versatile Place and Route.

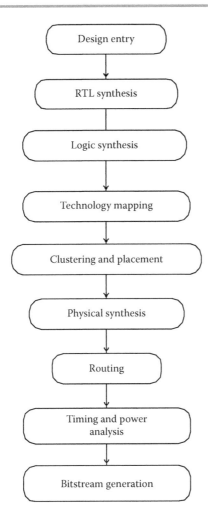

FIGURE 16.5   FPGA CAD flow.

optimization tools but are performed for final analysis and reporting at the end of the flow. The last step is bit-stream generation, which determines the sequence of 1's and 0's that will be serially loaded to configure the device. Notice that some ASIC CAD steps are absent: sizing and compaction, buffer insertion, clock tree, and power grid synthesis are not relevant to a prefabricated architecture.

Prior to the design entry phase, a number of different HLS flows can be used to generate high-level design language (HDL) code—domain-specific languages for DSP or other application domains and HLS from C or OpenCL are several examples that will be discussed in Section 16.2.

Research on FPGA algorithms can take place at all parts of this flow. However, the academic literature has generally been dominated by synthesis (particularly LUT technology mapping) at the front end and place-and-route algorithms at the back end, with most other aspects left to commercial tools and the FPGA vendors. The literature on FPGA architecture and CAD can be found in the major FPGA conferences: *the ACM International Symposium on FPGAs (FPGA), the International Conference on Field-Programmable Logic and Applications (FPL), the International Conference on Field-Programmable Technology (FPT), the International Symposium on Field-Programmable Custom-Computing Machines (FCCM)* [11], as well as the general CAD conferences.

In this chapter, we will survey algorithms for both the end-user and architecture development FPGA CAD flows. The references we have chosen to include are representative of a large body of research in FPGAs, but the list cannot be comprehensive. For a more complete discussion, the reader is referred to [12,13] for the overall FPGA CAD flow and [14,15,16] for FPGA-specific synthesis.

16.2 HIGH-LEVEL SYNTHESIS

Though the majority of current FPGA designs are entered directly in VHDL or Verilog, there have been a number of attempts to raise the level of abstraction to the behavioral or block-integration level with higher-level compilation tools. These tools then generate RTL HDL, which is shown in Figure 16.5. This section briefly describes HLS tools for domain-specific tasks and then the emerging commercial HLS. It closes with research challenges in the field.

16.2.1 DOMAIN-SPECIFIC SYNTHESIS

Berkeley Design Technologies Inc. [17] found that FPGAs have better price/performance than DSP processors for many DSP applications. However, HDL-based flows are not natural for most DSP designers, so higher-level DSP design flows are an important area of research. Altera DSP Builder™ and Xilinx System Generator™ link the MATLAB® and Simulink® algorithm exploration and simulation environments popular with DSP designers to VHDL and Verilog descriptions targeting FPGAs. See Chapter 8 of [12] for a case study on the use of Xilinx System Generator™.

For embedded processing and connecting complex systems, the FPGA vendors provide a number of system-level interconnect tools such as Altera's QSYS™ and Xilinx IP Integrator™ to enable the stitching of design blocks across programmable bus interconnect and to the embedded processors.

The Field-Programmable Port Extender (FPX) system of Lockwood [18] developed system-level and modular design methodologies for the network processing domain. The modular nature of FPX also allows for the exploration of hardware and software implementations. Kulkarni [19] proposed a methodology for mapping networking-specific functions into FPGAs.

Maxeler Technologies [20] offers a vertical solution for streaming dataflow applications that include software tools and FPGA hardware. The user describes the application's dataflow graph in a variant of the Java language—including the portion of the application to run on the FPGA and a portion to run on a connected x86 host processor. The company provides integrated desktop and rack-scale x86/FPGA hardware platforms comprising Intel processors connected with Altera/Xilinx FPGAs.

16.2.2 HLS

HLS refers to the automated compilation of a *software* program into a *hardware* circuit described at the RTL in VHDL or Verilog. For some HLS tools, the software program is expressed in a standard language, such as C or C++, whereas other tools rely on extended versions of standard languages [21–23] or even entirely new languages as in the case of BlueSpec [24]. Most HLS tools provide a mechanism for the user to influence the hardware produced by the tool via (1) constraints in a side file provided to the HLS tool, (2) pragmas in the source code, or (3) constructs available in the input language. HLS tools exist for both ASIC and FPGA technologies, with the main difference being that the FPGA tools have models for the speed, area, and power of the resources available on the target FPGA, including soft logic and hard IP blocks. Regardless of the specific input language, constraints, and target IC media, we consider the defining feature of HLS to be the translation of an untimed clockless behavioral input specification into an RTL hardware circuit. The RTL is then subsequently synthesized by FPGA vendor synthesis. HLS thus eases hardware design by raising the level of abstraction an engineer uses, namely, by permitting the use of software design methodologies.

HLS is attractive to two different types of users: (1) hardware engineers who wish to shorten/ease the design process, perhaps just for certain modules in the system, and (2) software engineers who lack hardware design skills but wish to glean some of the energy and speed benefits associated with implementing computations in hardware versus software. To be sure, recent studies have shown that FPGA hardware implementations can outperform software implementations by orders of magnitude in speed and energy [25]. Traditionally, FPGA circuit design has required hardware skills; however, software engineers outnumber hardware engineers by 10:1 [26]. FPGA vendors are keenly interested in broadening access to their technology to include software developers who can use FPGAs as computing platforms, for example, to implement accelerators that work alongside traditional processors. Indeed, both Altera and Xilinx have invested heavily in HLS in recent years, and both companies have released commercial solutions: Altera's OpenCL SDK performs HLS for an OpenCL program, while Xilinx's AutoESL-based HLS accepts a C program as input. Both are overviewed in further detail in the remainder of this section, along with recent academic FPGA HLS research.

Figure 16.6 shows the HLS flow. The program is first parsed/optimized by a front-end compiler. Then, the allocation step determines the specifications of the hardware to be synthesized, for example, the number and types of the functional units (e.g., number of divider units). Speed constraints and characterization models of the target FPGA device are also provided to the allocation step (e.g., the speed of an 8-bit addition, a 16-bit addition). Following allocation, the scheduling step assigns the computations in the (untimed) program to specific hardware clock cycles, defining a finite-state machine (FSM). The subsequent binding step assigns (binds) the computations from the C to specific hardware units, while adhering to the scheduling results. For example, a program may contain 10 division operations, and the HLS-generated hardware chosen in the allocation step may contain 3 hardware dividers. The binding step assigns each of the 10 division operations to one of the dividers. Memory operations, for example, loads/stores, are also bound to specific ports on specific memories. The final RTL generation step produces the RTL to be passed to back-end vendor tools to complete the implementation steps, which will be described later in Sections 16.3 and 16.4.

Often, there is a gap in quality—area, speed, and power—between HLS auto-generated hardware and that designed by a human hardware expert, especially for applications where there exists a particular spatial hardware layout, such as in the implementation of a fast Fourier transform. The impact of that gap can be higher in custom ASIC technologies than in FPGAs, where the entire purpose of a custom implementation is to achieve the best-possible power/speed in the lowest silicon area. With FPGAs, as long as the (possibly bloated) synthesized implementation fits in the target device, additional area is usually not a problem. For these reasons, it appears that FPGAs may be the IC medium through which HLS will enter the mainstream of hardware design.

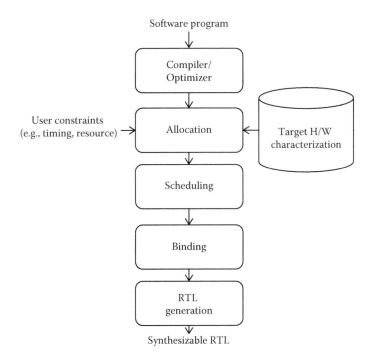

FIGURE 16.6 High-level synthesis design flow.

16.2.2.1 COMMERCIAL FPGA HLS TOOLS

In 2011, Xilinx purchased AutoESL, an HLS vendor spawned from research at the University of California, Los Angeles [27]. AutoESL has since become VivadoHLS—Xilinx's commercial HLS solution [28], which supports the synthesis of programs written in C, C++, and SystemC. By using pragmas in the code, the user can control the hardware produced by VivadoHLS, for example, by directing the tool to perform *loop pipelining*. Loop pipelining is a key performance concept in HLS that permits a loop iteration to commence before its prior iteration has completed—essentially implementing loop-level parallelism. Loop pipelining is illustrated in Figure 16.7, where the left side of the figure shows a C code snippet having three addition operations in the loop body, and the right side of the figure shows the loop pipelined schedule. In this example, it is assumed that an addition operation takes one cycle. In cycle #1, the first addition of the 0th loop iteration is executed. In cycle #2, the second addition of the 0th loop iteration and the first addition of the 1st loop iteration are executed. Observe that by cycle #3, three iterations of the loop are in flight at once, utilizing three adder functional units—referred to as the steady state of the loop pipeline. The entire loop execution is concluded after N+2 clock cycles. The ability to perform loop pipelining depends on the data dependencies between loop iterations (loop-carried dependencies) and amount of hardware available. In VivadoHLS, one would insert the pragma, #pragma AP pipeline II=1, just prior to a loop to direct the tool to pipeline the loop with an *initiation interval* (II) of 1, meaning that the tool should start a loop iteration every single cycle. Additional pragmas permit the control of hardware latency, function inlining, loop unrolling, and so on. A recent study showed that for certain benchmarks, VivadoHLS produces hardware of comparable quality to human-crafted RTL [29].

Altera takes a different approach to HLS versus Xilinx by using OpenCL [30,31] as the input language. OpenCL is a C-like language that originated in the graphics computing domain but has since been used as an input language for varied computing platforms. In OpenCL, parallelism is expressed *explicitly* by the programmer—a feature aligned nicely with an FPGA's spatial parallelism. An OpenCL program has two parts: a host program that runs on a standard processor and one or more *kernels*, which are C-like functions that execute computations on OpenCL devices—in this case, an Altera FPGA connected to the host x86 processor via a PCIe interface. Altera's OpenCL SDK performs HLS on the kernels to produce deeply pipelined implementations that aim to keep the underlying hardware as busy as possible by accepting a new thread into the pipeline every cycle, where possible. The HLS-generated kernel implementations connect to the host processor via a PCIe interface. A recent demonstration [32] for a fractal video compression application showed the OpenCL HLS providing 3× performance improvement over a GPU and two orders of magnitude improvement over a CPU.

16.2.2.2 ACADEMIC FPGA HLS RESEARCH

HLS has also been the focus of recent academic research with several different research frameworks under active development: GAUT is an HLS tool from the Université de Bretagne Sud that is specifically designed for DSP applications [32]. Shang is a generic (application-agnostic) HLS tool under development at the Advanced Digital Sciences Center in Singapore [33]. Riverside optimizing

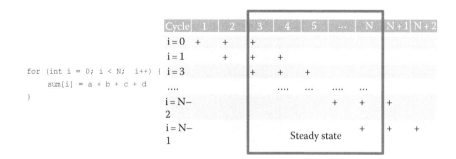

FIGURE 16.7 Polyhedral loop model.

compiler for configurable computing (ROCCC) [34] is a framework developed at UC Riverside that is specifically aimed at streaming applications. Bambu [35] and Dwarv [36] are generic HLS tools being developed at Politecnico di Milano and Delft University of Technology, respectively. Of these tools, Bambu and Sheng have been made open source. A binary is available for GAUT and ROCCC. Dwarv is not available publicly as it is built within a proprietary compiler framework.

LegUp is an open-source FPGA HLS tool from the University of Toronto, which was first released in 2011 and is presently on its third release [37]. LegUp accepts C as input and synthesizes the program entirely to hardware or, alternately, to a hybrid system containing a MIPS soft processor and one or more HLS-generated accelerators. It specifically targets Altera FPGAs, with the processor and accelerators connecting to one another using a memory-mapped on-chip bus interface. LegUp is implemented as back-end passes of the open-source low-level virtual machine (LLVM) compiler framework [38], and therefore, it leverages the parser and optimizations available in LLVM. Scheduling in LegUp is formulated mathematically as a linear program [39], and binding is implemented using a weighted bipartite matching approach [40]—operations from the C are matched to hardware units, balancing the number of operations assigned to each unit. Figure 16.8 shows the LegUp design flow. Beginning at the top left, a C program is compiled and run on a self-profiling processor, which gathers data as the program executes. The profiling data are used to select portions of the program (at the function level of granularity) to implement as accelerators. The selected functions are passed through HLS, and the original program is modified, with the functions replaced by wrappers that invoke and communicate with the accelerators. Ultimately, a complete FPGA-based processor/accelerator system is produced.

Beyond the traditional HLS steps, several FPGA-specific HLS studies have been conducted using LegUp. Resource sharing is an HLS area reduction technique that shares a hardware functional unit (e.g., a divider) among several operations in the input program (e.g., multiple division operations in the source). Resource sharing can be applied whenever the operations are scheduled in different clock cycles, and it requires adding multiplexers to the inputs of the functional unit to steer the correct input signals into the unit, depending on which operation is executing. Hadjis [41] studied resource sharing in the FPGA context, where, because multiplexers are costly to implement with LUTs, the authors showed there to be little area savings to be had by resource sharing unless the target resource is large. Dividers and repeated patterns of interconnected computational operators were deemed worth sharing, and the authors also showed the benefits of sharing to depend on whether the target FPGA architecture contained 4 LUTs or 6 LUTs, with

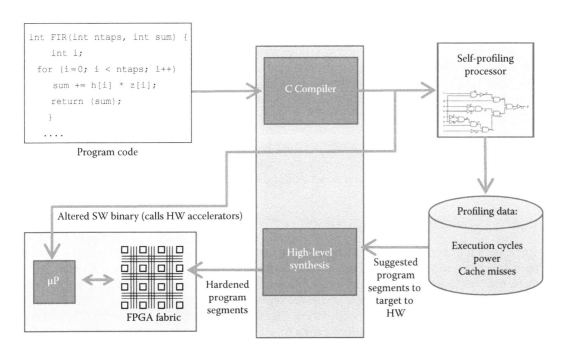

FIGURE 16.8 Loop pipelining.

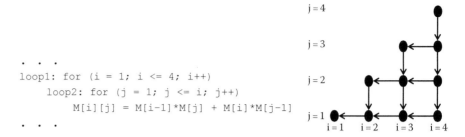

```
   .   .   .
loop1: for (i = 1; i <= 4; i++)
   loop2: for (j = 1; j <= i; j++)
       M[i][j] = M[i-1]*M[j] + M[i]*M[j-1]
   .   .   .
```

FIGURE 16.9   LegUp design flow.

sharing being more profitable in 6-LUT-based architectures owing to the steering multiplexers being covered within the same LUTs as the shared logic. In another FPGA-centric study, Canis [42] takes advantage of the high-speed DSP blocks in commercial FPGAs, which usually can operate considerably faster than the surrounding system (with LUTs and programmable interconnect). Their idea was to multipump the DSP blocks, operating them at 2× the clock frequency of the system, thereby allowing 2 multiply operations per system cycle.

There has also been considerable recent HLS research that, while not unique to FPGAs, nevertheless uses FPGAs as their test vehicle. Huang [43] studied the 50+ compiler optimization passes distributed with LLVM (e.g., constant propagation, loop rotation, and common subexpression elimination) and assessed which passes were beneficial to LegUp HLS-generated hardware. The authors also considered customized recipes of passes specifically tailored to an application and showed that it is possible to improve hardware performance by 18% versus using the blanket "-O3" optimization level. Choi [44] added support for pthreads and OpenMP to LegUp: two widely used software parallelization paradigms. The authors synthesize parallel software threads into parallel operating hardware accelerators, thereby providing a relatively straightforward way for a software engineer to realize spatial parallelism in an FPGA. Zheng et al. studied multicycling combinational paths in HLS—a technique applicable to combinational paths with cycle slack to lower the minimum clock period [45] to raise performance. There has also been work on supporting other input languages, such as CUDA, where, like OpenCL, parallelism is expressed explicitly [46].

Another popular HLS topic in recent years has been on the use of the *polyhedral model* to analyze and optimize loops in ways that improve the HLS hardware (Figure 16.9). The iteration space of the nested loops on the left side of the figure is shown as black points on the right side of the figure. Observe that the iteration space resembles a geometrical triangle—a polyhedron. The arrows in the figure illustrate the dependencies between loop iterations. Polyhedral loop analysis and optimization [47] represent loop iteration spaces mathematically as polyhedra and can be applied to generate alternative implementations of a loop where the order of computations, and thereby the dependencies between iterations, is changed. A straightforward example of such a manipulation would be to interchange the inner and outer loops with one another, and polyhedral loop optimizers also consider optimizations that are considerably more complex. Optimization via the polyhedral model may permit loop nests to be pipelined with a lower initiation interval to improve hardware performance [48] or to reduce the amount of hardware resources required to meet a given throughput [49] for the synthesis of partitioned memory architectures [50] and for the optimization of off-chip memory bandwidth [51].

16.2.2.3 RESEARCH CHALLENGES FOR HLS

There are a number of future challenges to the widespread adoption of HLS. First is the need for debugging and visualization tools for HLS-generated hardware. Presently, with many HLS tools, the user must resort to logic simulation and waveform inspection to resolve bugs in the hardware or its integration with the surrounding system—a methodology that is unacceptable and incomprehensible to software engineers. Likewise, the machine-generated RTL produced by HLS tools is often extremely difficult for a human reader to follow, making it difficult for the engineer to understand the hardware produced and determine how to optimize it. A second key challenge is the need to raise the quality of HLS hardware from the power, performance, and area perspectives, to narrow the gap between HLS and human-crafted hardware. Within this challenge, a present issue is that the quality of the hardware produced depends on the specific style of the input program. For example, some HLS tools may only support a subset of the input language, or the output may depend strongly on the input coding style. Another issue on this front is that, presently, it is difficult for the HLS tool to be able to leverage the parallelism available in the target fabric, given that the input to the HLS tool is typically a sequential specification. The extraction of parallelism is mitigated somewhat in some cases by the use of parallel input languages, like OpenCL™, where more of the parallelism is specified by the designer.

16.3 LOGIC SYNTHESIS

This chapter covers the application of traditional RTL and logic synthesis to FPGA design and then overviews technology mapping algorithms specific to the LUT-covering problem of FPGA synthesis.

16.3.1 RTL SYNTHESIS

RTL synthesis includes the inference and processing of high-level structures—adders, multipliers, multiplexers, buses, shifters, crossbars, RAMs, shift registers, and FSMs—prior to decomposition into generic gate-level logic. In commercial tools, 20%–30% of logic elements eventually seen by placement are mapped directly from RTL into the device-specific features rather than being processed by generic gate-level synthesis. For example, arithmetic is synthesized into carry-select adders in some FPGAs and ripple carry in others. An 8:1 multiplexor will be synthesized differently for 4 LUT or 6 LUT architectures and for devices with dedicated multiplexor hardware.

Though RTL synthesis is an important area of work for FPGA and CAD tool vendors, it has very little attention in the published literature. One reason for this is that the implementation of operators can be architecture specific. A more pedantic issue is simply that there are historically very few public-domain VHDL/Verilog front-end tools or high-level designs with arithmetic and other features, and both are necessary for research in the area. There are some promising improvements to research infrastructure, however. The VTR toolset [5] provides an open-source tool containing a full Verilog analysis and elaboration front end to VPR (referring to the physical design portion) and enables academic research for the first time to address the RTL synthesis flow and to examine CAD for binding and mapping.

In commercial FPGA architectures, dedicated hardware is provided for multipliers, adders, clock enables, clear, preset, RAM, and shift registers. Arithmetic was discussed briefly earlier; all major FPGAs either convert 4 LUTs into 3-LUT-based sum and carry computations or provide dedicated arithmetic hardware separate from the LUT. The most important effect of arithmetic is the restriction it imposes on placement, since cells in a carry chain must be placed in fixed relative positions.

One of the goals of RTL synthesis is to take better advantage of the hardware provided. This will often result in different synthesis than that which would take place in an ASIC flow because the goal is to minimize LEs rather than gates. For example, a 4:1 mux can be implemented optimally in two 4 LUTs, as shown in Figure 16.10. RTL synthesis typically produces these premapped cells and then protects them from processing by logic synthesis, particularly when they occur in a bus and there is a timing advantage from a symmetric implementation. RTL synthesis will recognize barrel shifters ("`y <= x >> s`" in Verilog) and convert these into shifting networks as shown in Figure 16.11, again protecting them from gate-level manipulation.

Recognition of register control signals such as clock enable, clear/preset, and synchronous/asynchronous load signals adds additional complications to FPGA synthesis. Since these preexist in the logic cell hardware (see Figure 16.2), there is a strong incentive to synthesize to them even when it would not make sense in ASIC synthesis. For example, a 4:1 mux with one constant input does not fit in a 4 LUT, but when it occurs in a datapath, it can be synthesized for most commercial LEs by using the LAB-wide (i.e., shared by all LEs in a LAB) synchronous load signal as a fifth input. Similarly, a clock enable already exists in the hardware, so register feedback can be

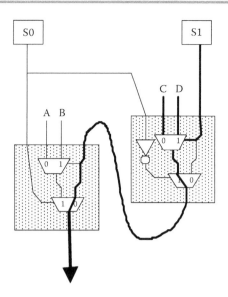

FIGURE 16.10   Implementing a 4:1 mux in two 4 LUTs.

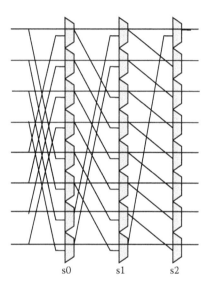

FIGURE 16.11   An 8-bit barrel-shifter network.

converted to an alternative structure with a clock enable to hold the current value but no routed register feedback. For example, if $f = z$ in Figure 16.12, we can express the cone of logic with a clock enable signal $CE = c_5 \cdot c_3 \cdot c_1'$ added to DFF z, and MUX(f, g) replaced simply by g—this is a win for bus widths of two or more. This transformation can be computed with a binary decision diagram (BDD) or other functional techniques. Tinmaung [52] applies BDD-based techniques for this and other power optimizations.

Several algorithms have addressed the RTL synthesis area, nearly all from FPGA vendors. Metzgen and Nancekievill [53,54] showed algorithms for the optimization of multiplexer-based buses, which would otherwise be inefficiently decomposed into gates. Most modern FPGA architectures do not provide on-chip tri-state buses, so multiplexers are the only choice for buses and are heavily used in designs. Multiplexers are very interesting structures for FPGAs because the LUT cell yields a relatively inefficient implementation of a mux, and hence, special-purpose hardware for handling multiplexers is common. Figure 16.12 shows an example taken from [53], which restructures buses of multiplexers for better technology mapping (covering) into 4 LUTs. The structure on the left requires 3 LUTs per bit to implement in 4 LUTs, while the structure on the right requires only 2 LUTs per bit.

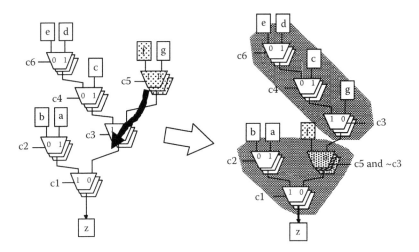

FIGURE 16.12 Multiplexer bus restructuring for LUT packing.

Also to address multiplexers, newer FPGA architectures have added clever hardware to aid in the synthesis of multiplexer structures such as crossbars and barrel shifters. Virtex devices provide additional stitching multiplexers for adjacent LEs, which can be combined to efficiently build efficient and larger multiplexers; an abstraction of this composable LUT is shown in Figure 16.13a. These are also used for stitching RAM bits together when the LUT is used as a 16-bit RAM (discussed earlier). Altera's adaptive logic module [55], shown abstractly in Figure 16.13b, allows a 6 LUT to be fractured to implement a 6 LUT, two independent 4 LUTs, two 5 LUTs that share two input signals, and also two 6 LUTs that have 4 common signals and two different signals (a total of 8). This latter feature allows two 4:1 mux with common data and different select signals to be implemented in a single LE, which means crossbars and barrel shifters built out of 4:1 mux can use half the area they would otherwise require.

At the RTL, a number of FPGA-specific hardware optimizations can be made for addressing the programmable hardware. For example, Tessier [56] examines the alternate methods for producing logical memories out of the hard-embedded memory blocks in the FPGA: a logical 16K word by 16b word memory can be constructed either by joining sixteen 16Kx1 physical memories with shared address or by joining sixteen 1Kx16 memories with an external mux built of logic. These come with trade-offs on performance and area versus power consumption, and Tessier

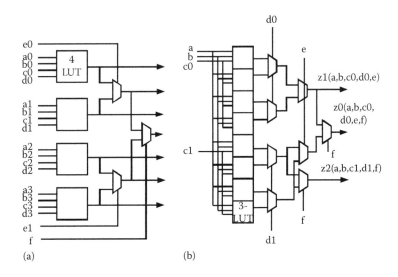

(a) (b)

FIGURE 16.13 Composable and fracturable logic elements: (a) Composable LUT and (b) fracturable LUT.

shows up to 26% dynamic power reduction (on memory blocks) by synthesizing the min-power configurations versus the min-area versions.

16.3.2 LOGIC OPTIMIZATION

Technology-independent logic synthesis for FPGAs is similar to ASIC synthesis. Berkeley system for interactive synthesis (SIS) [57] and more recently Mischenko's AIG-based [58] synthesis implemented in the ABC system [59] are in widespread use. Nearly, all FPGA-based synthesis research is now based on ABC as a base tool in the way that VPR is the standard for physical design. The general topic of logic synthesis is described in [14,15]. Synthesis tools for FPGAs contain basically the same two-level minimization algorithms and algebraic and Boolean algorithms for multilevel synthesis. Here, we will generally restrict our discussion to the differences from ASIC synthesis and refer the reader to the chapter on logic synthesis in this book [14] for the shared elements.

One major difference between standard and FPGA synthesis is in cost metrics. The target technology in a standard cell ASIC library is a more finely grained cell (e.g., a two-input NAND gate), while a typical FPGA cell is a generic k-input LUT. A 4 LUT is a 16-bit SRAM LUT mask driving a 4-level tree of 2:1 mux controlled by the inputs A, B, C, D (Figure 16.2b). Thus, A + B + C + D and AB + CD + AB'D' + A'B'C' have identical costs in LUTs, even though the former has 4 literals and the latter 10—the completeness of LUTs makes input counts more important than literals. In general, the count of two-input gates correlates much better to 4 LUT implementation cost than the literal-count cost often used in ASIC synthesis algorithms, but this is not always the case, as described for the 4:1 mux in the preceding section. A related difference is that inverters are free in FPGAs because (1) the LUT mask can always be reprogrammed to remove an inverter feeding or fed by an LUT and (2) programmable inversion at the inputs to RAM, IO, and DSP blocks is available in most FPGA architectures. In general, registers are also free because all LEs have a built-in DFF. This changes cost functions for retiming and state-machine encoding as well as designer preference for pipelining.

Subfactor extraction algorithms are much more important for FPGA synthesis than those commonly reported in the academic literature where ASIC gates are assumed. It is not clear whether this arises from the much larger and more datapath-oriented designs seen in industrial flows (versus open-source gate-level circuits), from the more structured synthesis from a complete HDL to gates flow, or due to the larger cell granularity. In contrast, algorithms in the class of speed _ up in SIS do not have significant effects on circuit performance for commercial FPGA designs. Again, this can be due either to the flow and reference circuits or to differing area/depth trade-offs. Commercial tools perform a careful balancing of area and depth during multilevel synthesis.

The synthesis of arithmetic functions is typically performed separately in commercial FPGA tools (see Section 16.3.1). Prior to VTR and some more recent versions of ABC, public tools like SIS synthesize arithmetic into LUTs, which distort their behavior in benchmarking results. Though reasonable for gate-level synthesis targeting ASICs, this can result in a dramatic shift in critical paths and area metrics when targeting FPGAs—an LUT–LUT delay may be as much as 10× the delay of a dedicated cin–cout path visible in Figure 16.2b. Typical industrial designs contain 10%–25% of logic cells in arithmetic mode, in which the dedicated carry circuitry is used with or instead of the LUT [60].

Retiming algorithms from general logic synthesis [61] have been adapted specifically for FPGAs [62], taking into account practical restrictions such as metastability (retiming should not be applied to synchronization registers), I/O versus core timing trade-offs (similar asynchronous transfers), power-up conditions (which can be provably unsatisfiable), and the abundance of registers (unique to FPGAs). Retiming is also used as part of physical resynthesis algorithms, as will be discussed later.

There are a number of resynthesis algorithms that are of particular interest to FPGAs, specifically structural decomposition, functional decomposition, and postoptimization using *set of pairs of functions to be distinguished* (SPFD)-based rewiring. SPFDs exploit the inherent flexibility in LUT-based netlists. SPFDs were proposed by Yamashita [63]. SPFDs are a generalization of *observability don't-care* functions. The on/off/dc set of functions is represented abstractly as a bipartite graph denoting distinguishing edges between minterms. A coloring of the graph thus gives an alternative implementation of the function. An inherent flexibility of LUTs in FPGAs is that they do not need to represent inverters, because these can always be absorbed by changing the destination node's LUT mask. By storing distinctions rather than functions, SPFDs generalize this to allow for more efficient expressions of logic.

Cong [64,65] applied SPFD calculations to the problem of rewiring a previously technology-mapped netlist. The algorithm consists of precomputing the SPFDs for each node in the network, identifying a target wire (e.g., a delay-critical input of an LUT after technology mapping), and then trying to replace that wire with another LUT output that satisfies its SPFD. The don't-care sets in the SPFDs occur in the internal nodes of the network where flexibility exists in the LUT implementation after synthesis and technology mapping. Rewiring was shown to have benefits for both delay and area.

An alternative more FPGA-specific approach to synthesis was taken by Vemuri's BDS system [66,67] building on BDD-based decomposition [68]. These authors argued that the separation of technology-independent synthesis from technology mapping disadvantaged FPGAs, which need to optimize LUTs rather than literals due to their greater flexibility and larger granularity (many SIS algorithms target literal count). The BDS system integrated technology-independent optimization using BDDs with LUT-based logic restructuring and used functional decomposition to target decompositions into LUT mappings. The standard sweep, eliminate, decomposition, and factoring algorithms from SIS were implemented in a BDD framework. The end result uses a technology mapping step but on a netlist more amenable to LUT mapping. Comparisons between SIS and BDS-pga using the same technology mapper showed area and delay benefits attributable to the BDD-based algorithms.

16.3.3 TECHNOLOGY MAPPING

Technology mapping for FPGAs is the process of turning a network of primitive gates into a network of LUTs of size at most k. The constant k is historically 4, though many recent commercial architectures have used fracturable logic cells with $k = 6$: Altera starting with Stratix II and Xilinx with Virtex V. LUT-based technology mapping is best seen as a covering problem, since it is both common and necessary to cover some gates by multiple LUTs for an efficient solution. Figure 16.14, taken from the survey in [69], illustrates this concept in steps from the original netlist (a), covering (b), and final result (c). Technology mapping aims for the least unit depth combined with the least number of cells in the mapped network.

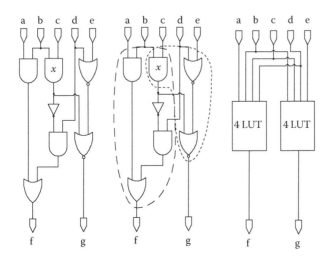

FIGURE 16.14 Technology mapping as a covering problem. (From Ling, A. et al., FPGA technology mapping: A study of optimality, in *Proceedings of the Design Automation Conference*, 2005.)

FPGA technology mapping differs significantly from library-based mapping for cell-based ASICs and uses different techniques. The most successful attempts divide into two paradigms: dynamic programming approaches derived from Chortle [70] and cut-based approaches branching from FlowMap [71]. Technology mapping is usually preceded by the decomposition of the netlist into two-input gates using an algorithm such as DOGMA [72].

In the first FPGA-specific technology mapping algorithm, Chortle [70], the netlist is decomposed into two-input gates and then divided into a forest of trees. Chortle computes an optimum set of *k-feasible* mappings for the current node. A *k*-feasible mapping is a subcircuit comprising the node and (some of) its predecessors such that the number of inputs is no more than *k* and only the mapped node has an output edge. Chortle combines solutions for children within reach of one LUT implemented at the current node, following the dynamic programming paradigm. Improvements on Chortle considered not trees but maximum fanout-free cones (MFFCs), which allowed for mapping with duplication. Area mapping with no duplication was later shown to be optimally solvable in polynomial time for MFFCs [73]. But, perhaps contrary to intuition, duplication is important in improving results for LUTs because it allows nodes with fanout greater than the one to be implemented as internal nodes of the cover; this is required to obtain improved delay and can also contribute to improved area. Figure 16.14b shows a mapping to illustrate this point.

A breakthrough in technology mapping research came with the introduction of FlowMap [71] proposed by Cong and Ding. In that work, the authors consider the combinational subcircuit rooted at each node of a Boolean network. A cut of that subcircuit divides it into two parts: one part containing the root node, referred to as A′, and a second part containing the rest of the subcircuit, referred to as A. In mapping to LUTs, one cares *only* that the number of signals to every LUT does not exceed *k*, the number of LUT inputs. Consequently, the portion of the circuit, A′, can be covered by an LUT as long as the number of signals crossing from A to A′ does not exceed *k*. FlowMap uses network flow techniques to find cuts of the circuit network in a manner that provably produces a depth-optimal mapping (for a given fixed decomposition into two-input gates).

All state-of-the-art approaches to FPGA technology mapping use the notion of *k*-feasible cuts. However, later work has shown that network flow methods are not needed to find such cuts. In fact, it is possible to find all *k*-feasible cuts for every node in the network. To achieve this, the network is traversed in topological order from primary inputs to outputs. The set of *k*-feasible cuts for any given node is generated by combining cuts from its fanin nodes and discarding those cuts that are not *k*-feasible. Figure 16.15 gives an example, where it is assumed we are at the point of computing the cuts for node *z*, in which case, the cuts for *x* and *c* have already been computed (owing to the topological traversal order). In this example, C_x is a cut for node *x*, and C_c is a cut for node *c*. We can find a cut for node *z* by combining C_x and C_c, producing cut C_z in the figure.

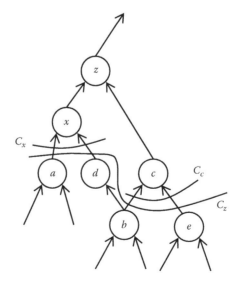

FIGURE 16.15   Example of cut generation.

Taking all such pairs yields the complete *k*-feasible cut set for node *z*. Schlag [74] proved that this cut generation approach is comprehensive—it does not miss out on finding any cuts. Technology mapping with cuts then starts (1) finding all cuts for each node in the network, (2) selecting a best cut for each node, and (3) constructing a mapping solution in reverse topological order using the selected best cuts.

CutMap by Cong and Hwang [75] improved the area of mapping solutions, while maintaining the property of optimal delay. The key concept in CutMap is to first compute a depth-optimal mapping and then, based on the slack in that mapping, remap portions of the network to minimize area. That is, for portions of the input network that are on the critical path, the cuts seen as best are those that minimize delay. Conversely, for portions of the network that are not on the critical path, the cuts seen as best are those that minimize area. Empirical results for CutMap show 15% better area than FlowMap with the same unit delay at the cost of longer runtime.

DAOmap [76] is a more recent work that generates *k*-feasible cones for nodes as in CutMap and then iteratively traverses the graph forward and backward to choose implementation cuts balancing area and depth. The forward traversal identifies covering cones for each node (depth optimal for critical and area optimal for noncritical nodes), and the backward traversal then selects the covering set and updates the heights of remaining nodes. The benefit of iteration is to relax the need for delay-optimal implementation once updated heights mark nodes as no longer critical, allowing greater area improvement, while maintaining depth optimality. DAOmap additionally considers the potential node duplications during the cut enumeration procedure, with look-ahead pruning. It is worth mentioning that while it is possible to compute a depth-optimal mapping for a given input network in polynomial time, computing an area-optimal mapping was shown to be NP-hard by Farrahi [77]. However, clever heuristics have been devised on the area front, such as the concept of *area flow* [78], which, like DAOmap, iterates over the network to find good mappings for multifanout nodes. Specifically, area flow–based mapping makes smart decisions regarding whether a multifanout node should be replicated in multiple LUTs or should be the fanout node of a subcircuit covered by an LUT in the final mapping.

The ABC logic synthesis framework [59] also incorporates technology mapping to LUTs. It mirrors prior work in that it computes and costs cuts for each node in the network. However, rather than storing *all* cuts for each node in the network (which can scale exponentially in the worst case), ABC only stores a limited set of *priority* cuts—a set whose size can be set by the user [79]. Experimental results demonstrated little quality loss, despite considering a reduced number of cuts. A circuit having a certain function can be represented in a myriad of ways, and more recent work in ABC has considered the structure of the input network on the technology mapping results and proposed ways to perform technology mapping on several different networks (representing the same circuit), choosing the best results from each [80,81].

16.3.4 POWER-AWARE SYNTHESIS

Recently, synthesis and technology mapping algorithms have begun to address power [82]. Anderson and Najm [83] proposed a modification to technology mapping algorithms to minimize node duplication and thereby minimize the number of wires between LUTs (as a proxy for dynamic power required to charge up the inter-LUT routing wires). EMAP by Lamoreaux [84] modifies CutMAP with an additional cost function component to favor cuts that reuse nodes already cut and those that have a low-activity factor (using probabilistic activity factors) to trap high-activity nodes inside clusters with lower capacitance. Chen [85] extends this type of mapping (proposed in the paper) to a heterogeneous FPGA architecture with dual voltage supplies, where high-activity nodes additionally need to be routed to low-V_{DD} (low power) LEs and critical nodes to high-V_{DD} (high-performance) LEs.

Anderson [86] proposes some interesting ideas on modifying the LUT mask during synthesis to place LUTs into a state that will reduce leakage power in the FPGA routing. Commercial tools synthesize clock enable circuitry to reduce dynamic power consumption within blocks. These are likely the beginning of many future treatments for power management in FPGAs. In a different work, Anderson [87] shows a technique for adding guarded evaluation of functions to reduce dynamic power consumption—this is a technique unique to FPGAs, as it utilizes the leftover circuitry in the FPGA that would otherwise be unused. The technique adds a new cost function to technology mapping that looks for unobservable logic (i.e., don't-care cones) and gates the activity to the subdesign, trading off additional area versus power savings. Hwang [88] and Kumthekar [89] also applied SPFD-based techniques for power reduction.

16.4 PHYSICAL DESIGN

The physical design flow for FPGAs consists of clustering, placement, physical resynthesis, and routing; we discuss each of these phases in the following subsections.

Commercial tools have additional preprocessing steps to allocate clock and reset signals to special low-skew *global clock networks*, to place phase-locked loops, and to place transceiver blocks and I/O pins to meet the many electrical restrictions imposed on them by the FPGA device and the package. With the exception of some work on placing FPGA I/Os to respect electrical restrictions [90,91], however, these preprocessing steps are typically not seen in any literature.

FPGA physical design can broadly be divided into *routability-driven* and *timing-driven* algorithms. Routability-driven algorithms seek primarily to find a legal placement and routing of the design by optimizing for reduced routing demand. In addition to optimizing for routability, timing-driven algorithms also use timing analysis to identify critical paths and/or connections and attempt to optimize the delay of those connections. Since the majority of delay in an FPGA is contributed by the programmable interconnect, timing-driven placement and routing can achieve a large circuit speedup versus routability-driven approaches. For example, a Xilinx commercial CAD system achieves an average of 50% higher design performance with full effort timing-driven placement and routing versus routability-only placement and routing at the cost of 5× runtime [92].

In addition to optimizing timing and routability, some recent FPGA physical design algorithms also implement circuits such that power is minimized.

16.4.1 PLACEMENT AND CLUSTERING

Since nearly all FPGAs have clustering (into CLB or LAB structures), physical design usually consists of a clustering phase followed by direct placement or a two-step (global and detailed) placement.

16.4.1.1 PROBLEM FORMULATION

The placement problem for FPGAs differs from the placement problem for ASICs in several important ways. First, placement for FPGAs is a slot assignment problem—each circuit element in the technology-mapped netlist must be assigned to a discrete location, or slot, on the FPGA device of a type that can accommodate it. Figure 16.1 shows the floorplan of a typical modern FPGA. An LE, for example, must be assigned to a location on the FPGA where an LE has been fabricated, while an input/output (I/O) block or RAM block must each be placed in a location where the appropriate resource exists on the FPGA. Second, there are usually a large number of constraints that must be satisfied by a legal FPGA placement. For example, groups of LEs that are placed in the same logic block have limits on the maximum number of distinct input signals and the number of distinct clocks they can use, and cells in carry chains must be placed together as a macro. Finally, all routing in FPGAs consists of prefabricated wires and transistor-based switches to interconnect them. Hence, the amount of routing required to connect two circuit elements, and the delay between them, is a function not just of the distance between the circuit elements but also of the FPGA routing architecture. The amount of (prefabricated) routing is also strictly limited, and a placement that requires more routing in some region of the FPGA than which exists there cannot be routed.

16.4.1.2 CLUSTERING

A common adjunct to FPGA placement algorithms is a bottom-up clustering step that runs before the main placement algorithm in order to group related circuit elements together into clusters (LABs, RAM, and DSP blocks in Figure 16.1). Clustering reduces the number of elements to place, improving the runtime of the main placement algorithm. In addition, the clustering algorithm usually deals with many of the complex FPGA legality constraints by grouping primitives (such as LEs in Figure 16.2) into legal function blocks (e.g., LABs in Figure 16.2), simplifying legality checking for the main placement algorithm.

Many FPGA clustering algorithms are variants of the VPack algorithm [93]. VPack clusters LEs into logic blocks by choosing a *seed* LE for a new cluster and then greedily packing the LE with the highest *attraction* to the current cluster until no further LEs can be legally added to the cluster. The attraction function is the number of nets in common between an LE and the current cluster. The T-VPack algorithm by Marquardt [94] is a timing-driven enhancement of VPack, where the attraction function for an LE, L, to cluster C becomes

$$(16.1) \qquad \text{attraction}(L) = .75 \cdot \text{criticality}(L,C) + .25 \frac{\left| \text{Nets}(L) \cap \text{Nets}(C) \right|}{\text{MaxNets}}$$

The first term gives higher attraction to LEs that are connected to the current cluster by timing-critical connections, while the second term is taken from VPack and favors grouping LEs with many common signals together. Somewhat surprisingly, T-VPack improves not only circuit speed versus VPack but also routability, by absorbing more connections within clusters. The iRAC [95] clustering algorithm achieves further reductions in the amount of routing necessary to interconnect the logic blocks by using attraction functions that favor the absorption of small nets within a cluster and by sometimes leaving empty space in clusters. The study of [96] showed that the Quartus II commercial CAD tool also significantly reduces routing demand by not packing clusters to capacity when it requires grouping unrelated logic in a single cluster.

Feng introduced an alternative clustering approach with PPack2 [97]. Instead of greedily forming clusters one by one with an attraction function, PPack2 recursively partitions a circuit into smaller and smaller partitions until each partition fits or nearly fits into a single cluster. A rebalancing step after partitioning moves LEs from overfull clusters to clusters with some spare room and prefers moves between clusters that are close in the partitioning hierarchy. This approach minimizes the packing of unrelated logic into a single cluster, creating a clustered netlist for the placement step that has better locality. PPack2 is able to reduce wiring by 35% and circuit delay by 11% versus T-VPack, without increasing the cluster count when clusters have no legality constraints other than a limit on the number of logic cells they can contain. When clusters have limits on the number and type of signals they can accommodate, PPack2 requires a postprocessing step to produce legal clusters and this can moderately increase the cluster count versus T-VPack.

The AAPack algorithm of [98] uses the T-VPack approach to clustering but adds much more complex legality checking to ensure the clusters created are legal and routable when the connectivity within a cluster is limited by the architecture. Using these complex legality checkers, AAPack can cluster not only logic cells into logic blocks but also other circuit primitives such as RAM slices and basic multipliers into RAM and DSP blocks, respectively. AAPack uses a variety of heuristics to check if a group of primitives are routable in a cluster; usually, fast heuristics provide an answer, but if they fail, a slower routing algorithm checks the cluster legality.

Lamoureaux [84] developed a power-aware modification of T-VPack that adds a term to the attraction function of Equation 16.1 such that LEs connected to the current cluster by connections with a high rate of switching have a larger attraction to the cluster. This favors the absorption of nets that frequently switch logic states, resulting in lower capacitance for these nets and lower dynamic power.

16.4.1.3 PLACEMENT

Simulated annealing is the most widely used placement algorithm for FPGAs due to its ability to adapt to different FPGA architectures and optimization goals. However, the growth in FPGA design size has outpaced the improvement in CPU speeds in recent years, and this has created a need to speed up placement by using multiple CPUs in parallel, incorporating new heuristics within an annealing framework, or by using other algorithms to create a *coarse* or starting placement that is usually then refined by an annealer.

```
P = InitialPlacement ();
T = InitialTemperature ();
while (ExitCriterion () == False) {
  while (InnerLoopCriterion () == False) {/* "Inner Loop" */
    P_new = PerturbPlacementViaMove (P);
    ΔCost = Cost (P_new) - Cost (P);
    r = random (0,1);
    if (r < e^(-ΔCost/T)) {
      P = P_new; /* Move Accepted */
    }
  } /* End "Inner Loop" */
  T = UpdateTemp (T);
}
```

FIGURE 16.16   Pseudocode of a generic simulated annealing placement algorithm.

Figure 16.16 shows the basic flow of simulated annealing. An initial placement is generated, and a placement perturbation is proposed by a *move generator*, generally by moving a small number of circuit elements to new locations. A *cost function* is used to evaluate the impact of each proposed move. Moves that reduce cost are always accepted or applied to the placement, while those that increase cost are accepted with probability $e^{-(\Delta\text{cost}/\text{T})}$, where T is the current *temperature*. Temperature starts at a high level and gradually decreases throughout the anneal, according to the *annealing schedule*. The annealing schedule also controls how many moves are performed between temperature updates and when the *ExitCriterion* that terminates the anneal is met.

There are two key strengths of simulated annealing that many other approaches lack:

1. It is possible to enforce all the legality constraints imposed by the FPGA architecture in a fairly direct manner. The two basic techniques are either to forbid the creation of illegal placements in the move generator or to add a penalty cost to illegal placements.
2. By creating an appropriate cost function, it is possible to directly model the impact of the FPGA routing architecture on circuit delay and routing congestion.

VPR [1,93,94] contains a timing-driven simulated annealing placement algorithm as well as timing-driven routing. The VPR placement algorithm is usually used in conjunction with T-VPack or AAPack, which preclusters the LEs into legal logic blocks. The placement annealing schedule is based on the monitoring statistics generated during the anneal, such as the fraction of proposed moves that are accepted. This adaptive annealing schedule lets VPR automatically adjust to different FPGA architectures. VPR's cost function also automatically adapts to different FPGA architectures [94]:

$$(16.2) \quad \text{Cost} = (1-\lambda) \sum_{i \in \text{AllNets}} q(i) \left[\frac{bb_x(i)}{C_{av,x}(i)} + \frac{bb_y(i)}{C_{av,y}(i)} \right] + \lambda \sum_{j \in \text{AllConnections}} \text{criticality}(j) \cdot \text{delay}(j)$$

The first term in Equation 16.2 causes the placement algorithm to optimize an estimate of the routed wirelength, normalized to the average wiring supply in each region of the FPGA; see Figure 16.17 for an example computation. The wirelength needed to route each net i is estimated as the sum of the x- and y-directed (bb_x and bb_y) span of the bounding box that just encloses all the terminals of the net, multiplied by a fanout-based correction factor, $q(i)$. Figure 16.17b shows that for higher-fanout nets, the half perimeter of the net bounding box underestimates wiring; multiplying by $q(i)$ helps correct this bias. Figure 16.17b also shows that Equation 16.2 tends to underestimate wiring when an FPGA contains longer wiring segments, as some of the wiring segment may extend beyond what is needed to route a net. So long as an FPGA contains at least some short-wiring segments, however, Equation 16.2 is usually sufficiently accurate to guide the placement algorithm. In FPGAs with differing amounts of routing available in different regions or directions, it is beneficial to move wiring demand to the more routing-rich areas, so the estimated wiring required is divided by the average routing capacity over the bounding box in the appropriate direction ($C_{av,x}$ and $C_{av,y}$).

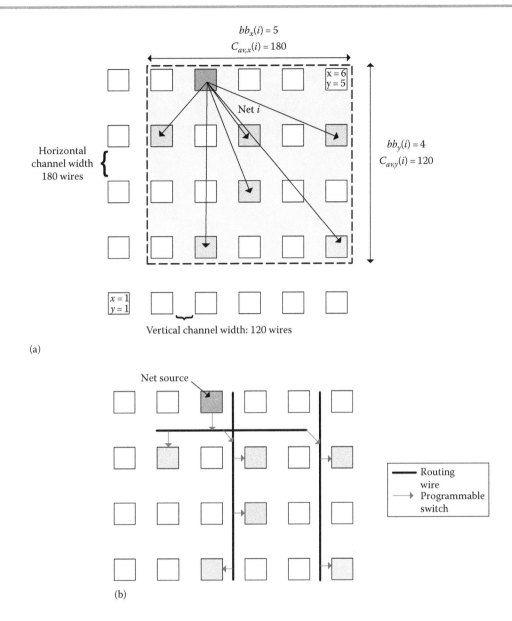

FIGURE 16.17 (a) Estimating the wiring cost of a net by half-perimeter bounding box and (b) best-case routing of the same net on an FPGA with all length four wires.

The second term in Equation 16.2 optimizes timing by favoring placements in which timing-critical connections have the potential to be routed with low delay. To evaluate the second term quickly, VPR needs to be able to quickly estimate the *delay* of a connection. To accomplish this, VPR assumes that the delay is a function only of the difference in the coordinates of a connection's endpoints, $(\Delta x, \Delta y)$, and invokes the VPR router with each possible $(\Delta x, \Delta y)$ to determine a table of delays versus $(\Delta x, \Delta y)$ for the current FPGA architecture before the simulated annealing algorithm begins. The *criticality* of each connection in the design is determined via periodic timing analysis using delays computed from the current placement.

Many enhancements have been made to the original VPR algorithm. The PATH algorithm from Kong [99] uses a new timing criticality formulation in which the timing criticality of a connection is a function of the slacks of all paths passing through it, rather than just a function of the worst-case (smallest) slack of any path through that connection. This technique significantly improves timing optimization and results in circuits with 15% smaller critical path delay on average. Lin [100] models unknown placement and routing delays as statistical process variation to make decisions during placement that maximize the probability of improving the circuit speed.

The SCPlace algorithm [101] enhances VPR so that a portion of the moves are *fragment moves* in which a single logic cell is moved instead of an entire logic block. This allows the placement algorithm to modify the initial clustering, and it improves both circuit timing and wirelength. Lamoureaux [84] modified VPR's cost function by adding a third term, *PowerCost*, to Equation 16.2:

$$(16.3) \qquad \text{PowerCost} = \sum_{i \in \text{AllNets}} q(i) \Big[bb_x(i) + bb_y(i) \Big] \cdot \text{activity}(i)$$

where activity(i) represents the average number of times net i transitions per second. This additional cost function term reduces circuit power, although the gains are less than those obtained by power-aware clustering.

Independence [102] is an FPGA placement tool that can effectively target a very wide variety of FPGA routing architectures by directly evaluating (rather than heuristically estimating) the routability of each placement generated during the anneal. It is purely routability driven, and its cost function monitors both the amount of wiring used by the placement and the routing congestion:

$$(16.4) \qquad \text{Cost} = \sum_{i \in \text{Nets}} \text{RoutingResources(i)} + \lambda \cdot \sum_{k \in \text{RoutingResources}} \max(\text{occupancy}(k) - \text{capacity}(k), 0)$$

The λ parameter in Equation 16.4 is a heuristic weighting factor. Independence uses the Pathfinder routing algorithm ([127], discussed in detail in Section 16.4.3) [128] to find new routes for all affected nets after each move and allow wire congestion by routing two nets on the same routing resource. Such a routing is not legal. However, by summing the overuse of all the routing resources in the FPGA, Independence can directly monitor the amount of routing congestion implicit in the current placement. The Independence cost function monitors not

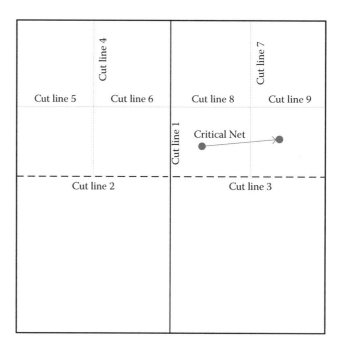

FIGURE 16.18 Typical recursive partitioning sequence for placement. The critical net will force the right terminal to be in the bottom partition when the design is partitioned along cut line 9.

only routing congestion but also the total number of *routing resources* (wires and block inputs/outputs) used by the router to create a smoother cost function that is easier for the annealer to optimize. Independence produces high-quality results on a wide variety of FPGA architectures but requires very high CPU time.

Sankar and Rose [103] seek the opposite trade-off of reduced quality for extremely low run-times. They create a hierarchical annealer that clusters the logic blocks twice to reduce the size of the placement problem, as shown in Figure 16.18. They first group logic blocks into level-one clusters of approximately 64 logic blocks and then cluster four of these level-one clusters into each level-two cluster. The level-two clusters are placed with a greedy (temperature = 0) anneal seeded by a fast constructive initial placement. Next, each level-one cluster is initially placed within the boundary of the level-two cluster that contained it, and another temperature = 0 anneal is performed. Finally, the placement of each logic block is refined with a low-starting-temperature anneal. For very fast CPU times, this algorithm significantly outperforms VPR in terms of achieved wirelength, while for longer permissible CPU times, it lags VPR.

Maidee [104] also seeks reduced placement runtime but does so by creating a *coarse placement* using recursive bipartitioning of a circuit netlist into smaller squares of the physical FPGA, as shown in Figure 16.18, followed by a low-starting-temperature anneal to refine the placement. This algorithm also includes modifications to the partitioning algorithm to improve circuit speed. As recursive partitioning proceeds, the algorithm records the minimum length each net could achieve, given the current number of partitioning boundaries it crosses and the FPGA routing architecture. Timing-critical connections to terminals outside of the region being partitioned act as anchor points during each partitioning. This forces the other end of the connection to be allocated to the partition that allows the critical connection to be made short, as shown in Figure 16.19. Once partitioning has proceeded to the point that each region contains only a few cells, the placement is fine-tuned by a low-temperature anneal with VPR. This step allows blocks to move anywhere in the device, so early placement decisions made by the partitioner, when little information about the critical paths or the final wirelength of each net was available, can be reversed. The technique achieves wirelength and speed comparable to VPR, with significantly reduced CPU time.

Analytic techniques are another approach to create a coarse placement. Analytic algorithms are based on creating a continuous and differentiable function of a placement that approximates routed wirelength. Efficient numerical techniques are used to find the global minimum of this function, and if the function approximates wirelength well, this solution is a placement with good wirelength. However, the global minimum is usually an illegal placement with overlapping blocks, so constraints and heuristics must be applied to guide the algorithm to a legal solution. Analytic approaches have been very popular for ASIC placement but have been less widely used for FPGAs, likely due to the more difficult legality constraints and the fact that delay is a function

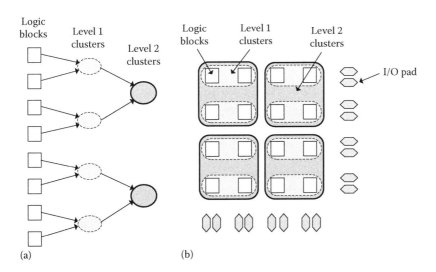

FIGURE 16.19   Hierarchical annealing placement algorithm. (a) Multilevel clustering, (b) place large clusters, uncluster and refine placement.

of not just wirelength but also the FPGA routing architecture. Analytic approaches scale well to very large problems, however, and this has resulted in increased interest in their use for FPGAs in recent years.

Gort and Anderson [105] develop the HeAP algorithm by adapting the SimPL [106] analytic placement algorithm to target heterogeneous FPGAs that contain RAM, DSP, and logic blocks. HeAP approximates the half perimeter of the bounding box enclosing each net with a smooth function and minimizes the sum of these wirelengths by solving a matrix equation to determine the (x,y) location of each block. This solution places more blocks of a certain type in some regions than the chip can accommodate. HeAP spreads the blocks out while maintaining their relative positions as much as possible, adds new terms to the matrix equation to reflect the desired spreading, and solves again. This solve/spread procedure iterates many times until the placement converges, and the authors of HeAP found that controlling which blocks are movable in each iteration is very important in heterogeneous FPGAs. Allowing all blocks to be placed by the solver and then spread simultaneously can result in blocks of different types (e.g., RAM and logic) that were *solved* to be close together moving in different directions during spreading. Better performance is achieved when some iterations of solve/spread place only one type of block (e.g., RAM blocks only) with other blocks being kept in their prior locations. HeAP provides better results at low CPU times than the commercial Quartus II placer, while the simulated annealing–based Quartus II placer produces higher quality results when longer CPU times are permitted.

HeAP uses an iterative and greedy swap algorithm to refine the final placement; essentially, this is a temperature = 0 anneal. The Synopsys commercial CAD tools also combine analytic placement with annealing for final refinement [107] but still allow some hill climbing during the anneal.

Another approach to speeding up placement is to leverage multiple CPUs working in parallel. The commercial Quartus II algorithm speeds up annealing in two ways [108]: first by using *directed moves* that explore the search space more productively than purely random moves and second by evaluating multiple moves in parallel. Evaluating moves in parallel yields a speedup of 2.4× versus a serial algorithm, without compromising quality. This parallel placer also maintains *determinism* (obtains the same results for every run of a certain input problem) by detecting when two moves would access the same blocks or nets (termed a "collision") and aborting and later retrying the move that would have occurred later in a serial program. An et al. [109] parallelized the simpler VPR annealer by prechecking if multiple moves would interact and evaluating them in parallel when they would not—they achieved speedups of 5× while maintaining determinism and 34× with a nondeterministic algorithm, with negligible impact on result quality. Goeders [110] took an alternative approach of changing the moves and cost function to guarantee that multiple CPUs could each optimize a different region in parallel without ever making conflicting changes to the placement. Their approach is deterministic and achieves 51× speedup over serial VPR, but at a cost of a 10% wirelength increase and a 5% circuit slowdown.

16.4.2 PHYSICAL RESYNTHESIS OPTIMIZATIONS

Timing visibility can be poor during FPGA synthesis. What appears to be a noncritical path can turn out to be critical after placement and routing. This is true for ASICs too, but the problem is especially acute for FPGAs. Unlike an ASIC implementation, FPGAs have a predefined logic and routing fabric and hence cannot use drive-strength selection, wire sizing, or buffer insertion to increase the speed of long routes.

Recently, physical (re)synthesis techniques have arisen both in the literature and in commercial tools. Physical resynthesis techniques for FPGAs generally refer either to the resynthesis of the netlist once some approximate placement has occurred, and thus, some visibility of timing exists, or local modifications to the netlist during placement itself. Figure 16.20 highlights the difference between the two styles of physical resynthesis flow. The *iterative* flow of Figure 16.20a iterates between synthesis and physical design. The advantage of this flow is that the synthesis tool is free to make large-scale changes to the circuit implementation, while its disadvantage is that the placement and routing of this new design may not match the synthesis tool expectations, and hence, the loop may not converge well. The *incremental* flow of Figure 16.20b instead makes

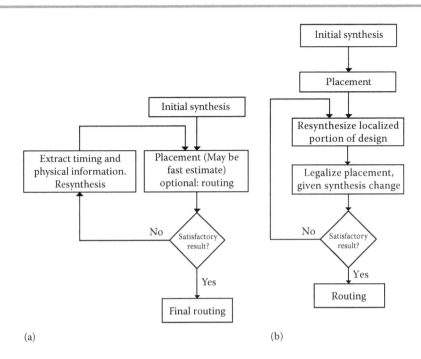

FIGURE 16.20   Physical resynthesis flows: Example of (a) "iterative" and (b) "incremental" physical synthesis flow.

only more localized changes to the circuit netlist such that it can integrate these changes into the current placement with only minor perturbations. This flow has the advantage that convergence is easier, since a legal or near-legal placement is maintained at all times, but it has the disadvantage that it is more difficult to make large-scale changes to the circuit structure.

Commercial tools from Synopsys (Syncplicity) [1] follow the *iterative* flow. They resynthesize a netlist given output from the FPGA vendor place and route tool and provide constraints to the place and route tool in subsequent iterations to assist convergence. Lin [111] described a similar academic flow in which remapping is performed either after a placement estimate or after the actual placement delays are known. Suaris [112] used timing budgets for resynthesis, where the budget is calculated using a quick layout of the design. This work also makes modifications to the netlist to facilitate retiming in the resynthesis step. In a later improvement [113], the flow was altered to incrementally modify the placement after each netlist transform, assisting convergence.

There are commercial and academic examples of the *incremental* physical resynthesis flow as well. Schabas [114] used logic duplication as a postprocessing step at the end of placement, with an algorithm that simultaneously duplicates logic and finds legal and optimized locations for the duplicates. Logic duplication, particularly on high-fanout registers, allows significant relaxation on placement critical paths because it is common for a multifanout register to be pulled in multiple directions by its fanouts, as shown in Figure 16.21. Chen [115] integrated duplication throughout a simulated annealing–based placement algorithm. Before each temperature update, logic duplicates are created and placed if deemed beneficial to timing, and previously duplicated logic may be *unduplicated* if the duplicates are no longer necessary.

Manoharajah [116,117] performed local restructuring of timing-critical logic to shift delay from the critical path to less critical paths. An incremental placement algorithm then integrates any changed or added LUTs into a legal placement. Ding [118] gave an algorithm for postplacement pin permutation in LUTs. This algorithm reorders LUT inputs to take advantage of the fact that each input typically has a different delay. Also, this algorithm also swaps inputs among several LUTs that form a logic cone in which inputs can be legally swapped, such as an AND-tree or EXOR-tree. An advantage of this algorithm is that no placement change is required, since only the ordering of inputs is affected and no new LUTs are created.

Singh and Brown [119] present a postplacement retiming algorithm. This algorithm initially places added registers and duplicated logic at the same location as the original logic and then

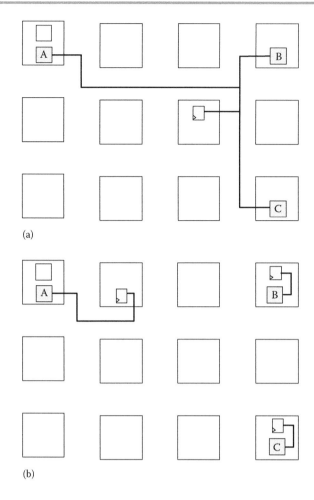

FIGURE 16.21   Duplicating registers to optimize timing in physical resynthesis: (a) Register with three time-critical output connections. (b) Three register duplicates created and legally placed to optimize timing.

invokes an incremental placement algorithm to legalize the placement. Chen and Singh [120] describe an improved incremental placement algorithm that reduces runtime.

In an alternative to retiming, Singh and Brown [121] employed unused PLLs and global clocking networks to create several shifted versions of a clock and developed a postplacement algorithm that selects a time-shifted clock for each register to improve speed. This approach is similar to retiming after placement but involves shifting clock edges at registers rather than moving registers across combinational logic. Chao-Yang and Marek-Sadowska [122] extended this beneficial clock-skew timing optimization to a proposed FPGA architecture, where clocks can be delayed via programmable delay elements on the global clock distribution networks.

16.4.3 ROUTING

Routing for FPGAs is unique in that it is a purely discrete problem, in that the wires already exist as part of the underlying architecture. This section describes historical development and state-of-the-art FPGA routing algorithms.

16.4.3.1 PROBLEM FORMULATION

All FPGA routing consists of prefabricated metal wires and programmable switches to connect the wires to each other and to the circuit element input and output pins. Figure 16.22 shows an example of FPGA routing architecture. In this example, each routing channel contains four wires

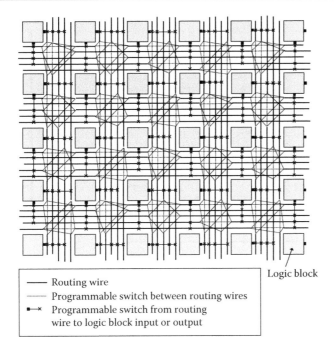

FIGURE 16.22   Example of FPGA routing architecture.

of length 4—wires that span four logic blocks before terminating—and one wire of length 1. In the example shown in Figure 16.22, the programmable switches allow wires to connect only at their endpoints, but many FPGA architectures also allow programmable connections from interior points of long wires as well.

Usually, the wires and the circuit element input and output pins are represented as nodes in a *routing-resource graph*, while programmable switches that allow connections to be made between the wires and pins become directed edges. Programmable switches can be fabricated as pass transistors, tri-state buffers, or multiplexers. Multiplexers are the dominant form of programmable interconnect in recent FPGAs due to a superior area-delay product [123]. Figure 23 shows how a small portion of an FPGA's routing is transformed into a routing-resource graph. This graph can also efficiently store information on which pins are logically equivalent and hence may be swapped by the router, by including *source* and *sink* nodes that connect to all the pins that can perform a desired function. It is common to have many logically equivalent pins in commercial FPGAs—for example, all the inputs to an LUT are logically equivalent and may be swapped by the router. A legal routing of a design consists of a tree of routing-resource nodes for each net in the design such that (1) each tree electrically connects the net source to all the net sinks and (2) no two trees contain the same node, as that would imply a short between two signal nets.

Since the number of routing wires in an FPGA is limited and the limited number of programmable switches also creates many constraints on which wires can be connected to each other, congestion detection and avoidance is a key feature of FPGA routers. Also since most delay in FPGAs is due to the programmable routing, timing-driven routing is important to obtain the best speed.

16.4.3.2 TWO-STEP ROUTING

Some FPGA routers operate in two sequential phases, as shown in Figure 16.24. First, a global route for each net in the design is determined, using channeled global routing algorithms that are essentially the same as those for ASICs. The output of this stage is the series of channel segments through which each connection should pass. Next, a detailed router is invoked to determine exactly which wire segment should be used within each channel segment. The SEGA [124] algorithm finds detailed routes by employing different levels of effort in searching the routing graph.

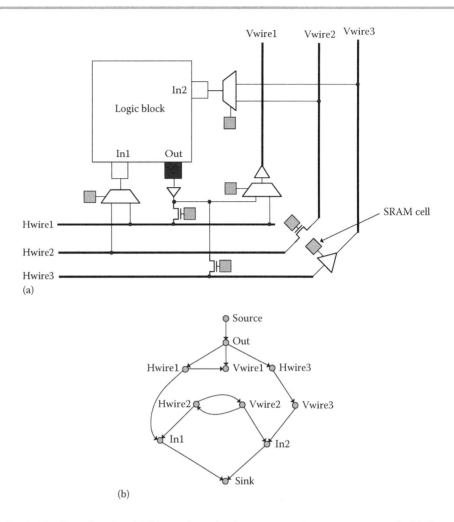

FIGURE 16.23 Transforming FPGA routing circuitry to a routing-resource graph: (a) Example of FPGA routing circuitry. (b) Equivalent routing-resource graph.

A search of only a few routing options is conducted first in order to quickly find detailed routes for nets that are in uncongested regions, while a more exhaustive search is employed for nets experiencing routing difficulty. An alternative approach by Nam formulates the FPGA detailed routing problem as a Boolean satisfiability problem [125]. This approach guarantees that a legal detailed routing (which obeys the current global routing) will be found if one exists but can take high CPU time.

The divide-and-conquer approach of two-step routing reduces the problem space for both the global and detailed routers, helping to keep their CPU times down. However, the flexibility loss of dividing the routing problem into two phases in this way can result in significantly reduced result quality. The global router optimizes only the wirelength of each route and attempts to control congestion by trying to keep the number of nets assigned to a channel segment comfortably below the number of routing wires in that channel segment. However, the fact that FPGA wiring is prefabricated and can be interconnected only in limited patterns makes the global router's view of both wirelength and congestion inaccurate. For example, a global route one logic block long may require the detailed router to use a wire that is four logic blocks long to actually complete the connection, thereby wasting wire and increasing delay. Figure 16.25 highlights this behavior; the global route requires 9 units of wire, but the final wires used in the detailed routing of the net are 13 wiring units long in total. Similarly, a global route where the number of nets assigned to each channel segment is well below the capacity of each segment may still fail detailed routing because the wiring patterns (i.e., the limited connectivity between wires) may not permit this pattern of global routes.

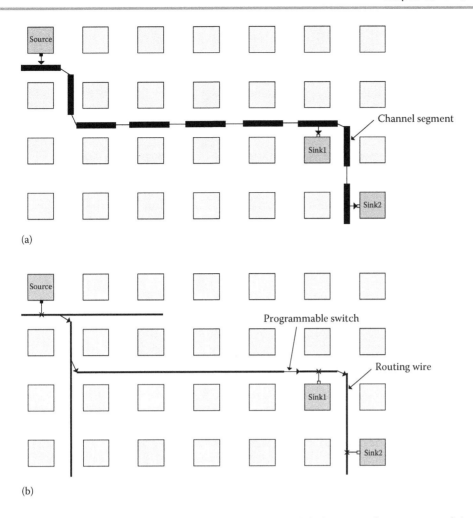

(a)

(b)

FIGURE 16.24   Two-step FPGA routing flow: (a) Step one, global routing chooses a set of channel segments for a net. (b) Step two, detailed routing wires within each channel segment and switches to connect them.

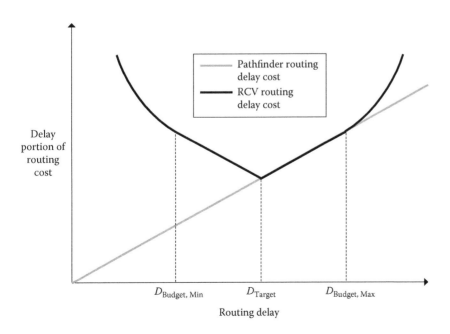

FIGURE 16.25   Routing Cost Valleys routing delay cost compared to Pathfinder routing delay cost.

16.4.3.3 SINGLE-STEP ROUTERS

Most modern FPGA routers are *single-step* routers that find routing paths through the routing-resource graph in a single unified search algorithm. Most such routers use some variant of a maze router [126] as their inner loop—a maze router uses Dijkstra's algorithm to search through the routing-resource graph and find a low-cost path to connect two terminals of a net. Single-step FPGA routers differ primarily in their costing of various routing alternatives and their congestion resolution techniques.

The Pathfinder algorithm by McMurchie and Ebeling [127] introduced the concept of *negotiated congestion routing*, which now underlies many academic and commercial FPGA routers. In a negotiated congestion router, each connection is initially routed to minimize some metric, such as delay or wirelength, with little regard to congestion, or overuse of routing resources. After each *routing iteration*, in which every net in the circuit is ripped up and rerouted, the cost of congestion is increased such that it is less likely that overused nodes will occur in the next routing iteration. Over the course of many routing iterations, the increasing cost of congestion gradually forces some nets to accept suboptimal routing in order to resolve congestion and achieve a legal routing.

The congestion cost of a node is

$$(16.5) \qquad \text{CongestionCost}(n) = \big[b(n) + h(n) \big] \cdot p(n)$$

where
    $b(n)$ is the base cost of the node
    $p(n)$ is the present congestion of the node
    $h(n)$ is the historical cost of the node

The base cost of a node could be its intrinsic delay, its length, or simply 1 for all nodes. The present congestion cost of a node is a function of the overuse of the node and the routing iteration. For nodes that are not currently overused, $p(n)$ is one. In early routing iterations, $p(n)$ will be only slightly higher than 1 for nodes that are overused, while in later routing iterations, to ensure congestion is resolved, $p(n)$ becomes very large for overused nodes. $h(n)$ maintains a congestion history for each node. $h(n)$ is initially 0 for all nodes but is increased by the amount of overuse on node n at the end of each routing iteration. The incorporation of not just the present congestion but also the entire history of congestion of a node, into the cost of that node, is a key innovation of negotiated congestion. Historical congestion ensures that nets that are trapped in a situation where all their routing choices have present congestion can see which choices have been overused the most in the past. Exploring the least historically congested choices ensures new portions of the solution space are being explored and resolves many cases of congestion that the present congestion cost term alone cannot resolve.

In the Pathfinder algorithm, the complete cost of using a routing-resource node n in the routing of a connection c is

$$(16.6) \qquad \text{Cost}(n) = \big[1 - \text{Crit}(c) \big] \text{CongestionCost}(n) + \text{Crit}(c)\text{Delay}(n)$$

The criticality is the ratio of the connection slack to the longest delay in the circuit:

$$(16.7) \qquad \text{Crit}(c) = \frac{\text{Slack}(c)}{D\max}$$

The total cost of a routing-resource node is therefore a weighted sum of its congestion cost and its delay, with the weighting being determined by the timing criticality of the connection being routed. This formulation results in the most timing-critical connections receiving delay-optimized routes, with non-timing-critical connections using routes optimized for minimal wirelength and congestion. Since timing-critical connections see less cost from congestion, these connections are also less likely to be forced off their optimal routing paths due to congestion—instead, non-timing-critical connections will be moved out of the way of timing-critical connections.

The VPR router [1,93] is based on the Pathfinder algorithm but introduces several enhancements. The most significant enhancement is that instead of using a breadth-first search or an A^* search through the routing-resource graph to determine good routes, VPR uses a more aggressive directed search technique. This directed search sorts each routing-resource node, n, found during graph search toward a sink, j, by a total cost given by

$$(16.8) \qquad \text{TotalCost}(n) = \text{PathCost}(n) + \alpha \cdot \text{ExpectedCost}(n, j)$$

Here, $\text{PathCost}(n)$ is the known cost of the routing path from the connection source to node n, while $\text{ExpectedCost}(n, j)$ is a prediction of the remaining cost that will be incurred in completing the route from node n to the target sink. The *directedness* of the search is controlled by α. An α of 0 results in a breadth-first search of the graph, while α larger than 1 makes the search more efficient but may result in suboptimal routes. An α of 1.2 leads to improved CPU time without a noticeable reduction in result quality.

An FPGA router based on negotiated congestion, but designed for very low CPU times, is presented by Swartz [129]. This router achieves very fast runtimes through the use of an aggressive directed search during routing graph exploration and by using a *binning* technique to speed the routing of high-fanout nets. When routing the kth terminal of a net, most algorithms begin the route toward terminal k by considering every routing-resource node used in routing the previous $k - 1$ terminals. For a k-terminal net, this results in an $O(k^2)$ algorithm, which becomes slow for large k. By examining only the portions of the routing of the previous terminals that lie within a geographical *bin* near the sink for connection k, the algorithm achieves a significant CPU reduction.

Wilton developed a cross talk–aware FPGA routing algorithm [130]. This algorithm enhances the VPR router by adding an additional term to the routing cost function that penalizes routes in proportion to the amount of delay they will add to neighboring routes due to cross talk, weighted by the timing criticality of those neighboring routes. Hence, this router achieves a circuit speedup by leaving routing tracks near those used by critical connections vacant.

Lamoureaux [84] enhanced the VPR router to optimize power by adding a term to the routing node cost, Equation 16.6, which includes the capacitance of a routing node multiplied by the switching activity of the net being routed. This drives the router to achieve low-energy routes for rapidly toggling nets.

The Routing Cost Valleys (RCV) algorithm [131] combines negotiated congestion with a new routing cost function and an enhanced slack allocation algorithm. RCV is the first FPGA routing algorithm that not only optimizes long-path timing constraints, which specify that the delay on a path must be less than some value, but also addresses the increasing importance of short-path timing constraints, which specify that the delay on a path must be greater than some value. Short-path timing constraints arise in FPGA designs as a consequence of hold time constraints within the FPGA or of system-level hold time constraints on FPGA input pins and system-level minimum clock-to-output constraints on FPGA output pins. As with ASIC design, increasing process variation in clock trees can result in the need to address hold time in addition to setup. To meet short-path timing constraints, RCV will intentionally use slow or circuitous routes to increase the delay of a connection and guarantee minimum delays when required.

RCV allocates both short-path and long-path slack to determine a pair of delay budgets, $D_{\text{Budget,Min}}(c)$ and $D_{\text{Budget,Max}}(c)$, for each connection, c, in the circuit. A routing of the circuit in which every connection has a delay between $D_{\text{Budget,Min}}(c)$ and $D_{\text{Budget,Max}}(c)$ will satisfy all the long-path and short-path timing constraints. Such a routing may not exist for all connections however, so, where possible, it is desirable if connection delays lie at the middle of the delay window defined by $D_{\text{Budget,Min}}(c)$ and $D_{\text{Budget,Max}}(c)$, which is referred to as $D_{\text{Target}}(c)$. The extra timing margin achieved by connection c may allow another connection on the same path to have a delay outside its delay budget window, without violating any of the path-based timing constraints. Figure 16.25 shows the form of the RCV routing cost function compared to that of the original Pathfinder algorithm. RCV strongly penalizes routes that have delays outside the delay budget window and weakly guides routes to achieve D_{Target}. RCV achieves superior results on short-path timing constraints and also outperforms Pathfinder in optimizing traditional long-path timing constraints.

Routing is a time-consuming portion of the FPGA CAD flow, motivating work by Gort and Anderson [132] to reduce the CPU time of the negotiated congestion routing. They first find that in rerouting every net, each routing iteration is wasteful. By rerouting only those nets that are illegally routed (use some congested resources), they achieve a 3× speedup versus VPR. The authors then achieve a further 2.3× speedup by routing multiple nets in parallel when the bounding boxes of those nets do not overlap, and hence, routings constrained to lie within the net terminal bounding boxes will not interact. This algorithm remains deterministic and achieves the same result quality as VPR.

16.5 CAD FOR EMERGING ARCHITECTURE FEATURES

As FPGA architectures incorporate new hardware features, new CAD flows are needed to support them. Two such areas that have attracted significant research interest are new power management hardware and 2.5D or 3D FPGA systems that are built from multiple silicon dice.

16.5.1 POWER MANAGEMENT

As the previous sections of this chapter have described, power can be saved at every stage of the FPGA CAD flow by making power-aware decisions concerning a design's implementation. However, while an ASIC CAD tool can change the underlying hardware to save power—by making a low-voltage island or using high-V_T transistors, for example—an FPGA tool can only work with the hardware that exists in the FPGA. This has led to several proposals to augment FPGA hardware with features that enable more advanced power management.

Li et al. investigated several different FPGA architectures where the logic blocks and routing switches could select from either a high- or low-voltage supply [133]. They found that hardwiring the choice of which logic blocks used high V_{dd} and which used low V_{dd} led to poor results, while adding extra transistors to allow a programmable selection per design worked well. They augmented the FPGA CAD flow to incorporate a voltage assignment step after placement and routing and found that they could reduce FPGA power by 48%, at a cost of 18% delay increase due to the voltage selection switches.

The commercial Stratix III FPGA [134] added programmable back-bias at the *tile* granularity to reduce leakage power. A tile is a pair of logic blocks along with all their adjacent routing, and the nMOS transistors in a tile can use either the conventional body voltage of 0 V for maximum speed or a negative voltage for lower leakage. After placement and routing, a new CAD step chooses the body voltage for each tile and seeks to back-bias as many tiles as possible without violating any timing constraints. Typically, this approach is able to set 80% of the tiles to the low-leakage state, reducing the FPGA static power by approximately half.

Huda et al. [135] proposed a small change to FPGA routing switch design that enables unused routing wires to become free-floating capacitors. They then modify the VPR router to prefer routing rapidly switching signals on wires that are adjacent to *free-floating* routing wires to reduce their effective coupling capacitance and hence dynamic power. They see a routing power reduction of 10%–15% with only a 1% delay increase.

16.5.2 MORE-THAN-2D INTEGRATION

While FPGA capacity has grown for many years in line with Moore's law, there is an increasing interest in even larger FPGA systems that combine multiple silicon dice. Xilinx's Virtex-7 series FPGAs contain up to four 28 nm FPGA dice connected via microbumps to a 65 nm silicon interposer that creates approximately 10,000 connections between adjacent dice [136], as shown in Figure 16.26. This system enables very large *2.5D* FPGAs with over 2 million logic cells, and the Xilinx Vivado CAD system allows designers to target it as if it was a single very large FPGA. Hahn and Betz investigate such systems in [137] and show that with suitable modification to the placement algorithm to understand that wires crossing the interposer are relatively scarce and slow,

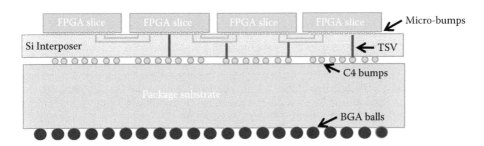

FIGURE 16.26 2.5D Virtex-7 FPGA built with silicon interposer technology.

the system does indeed behave like a single FPGA. So long as the interposer crossing provides at least 30% of the vertical wiring within an FPGA, this 2.5D device remains quite routable. Even more ambitious systems that stack multiple FPGA dice on top of each other and connect them with through-silicon vias are investigated in [138]. The CAD system of [138] first employs partitioning to assign blocks to different silicon layers and then places blocks within a single die/layer.

16.6 LOOKING FORWARD

In this chapter, we have surveyed the current algorithms for FPGA synthesis, placement, and routing. Some of the more recent publications in this area point to the growth areas in CAD tools for FPGAs, specifically HLS [30,33].

Power modeling and optimization algorithms are likely to continue to increase in importance as power constraints become ever more stringent with shrinking process nodes. Placement and routing for multi-VDD and other power-reduction modes have had several publications [83,85–87], but many have yet to be fully integrated into commercial tools. A workshop at the FPGA 2012 conference [139] discussed these and other emerging topics such as high-performance computing acceleration on FPGAs.

Timing modeling for FPGA interconnect will need to take into account variation, multicorner, cross talk, and other physical effects that have been largely ignored to date and incorporate these into the optimization algorithms. Though this work parallels the development for ASIC (e.g., PrimeTime-SI™), FPGAs have some unique problems in that the design is not known at fabrication.

Producing ever-larger designs without lengthening design cycles will require tools with reduced runtime and will likely also further accelerate the adoption of HLS flows. As FPGAs incorporate processors, fast memory interfaces, and other system-level features, tool flows that enable complete embedded system development will also become ever more important.

REFERENCES

INTRODUCTION

1. See www.<companyname>.com for commercial tools and architecture information.
2. Betz, V., Rose, J., and Marquardt, A. *Architecture and CAD for Deep-Submicron FPGAs*, Kluwer, February 1999.
3. Lewis, D. et al., Architectural enhancements in Stratix V, in *Proceedings of the 21st ACM International Symposium on FPGAs*, 2013, pp. 147–156.
4. Betz, V. and Rose, J., Automatic generation of FPGA routing architectures from high-level descriptions, in *Proceedings Seventh International Symposium on FPGAs*, 2000, pp. 175–184.
5. Luu, J. et al., VTR 7.0: Next generation architecture and CAD system for FPGAs, *ACM Transactions Reconfigurable Technology and Systems*, 7(2), 6:1–6:30, June 2014.
6. Yan, A., Cheng, R., and Wilton, S., On the sensitivity of FPGA architectural conclusions to experimental assumptions, tools and techniques, in *Proceedings of the 10th International Symposium on FPGAs*, 2003, pp. 147–156.

7. Li, F., Chen, D., He, L., and Cong, J., Architecture evaluation for power-efficient FPGAs, in *Proceedings of the 11th International Symposium on FPGAs*, 2003, pp. 175–184.
8. Wilton, S., SMAP: Heterogeneous technology mapping for FPGAs with embedded memory arrays, in *Proceedings of the Eighth International Symposium on FPGAs*, 1998, pp. 171–178.
9. Hutton, M., Karchmer, D., Archell, B., and Govig, J., Efficient static timing analysis and applications using edge masks, in *Proceedings of the 13th International Symposium FPGAs*, 2005, pp. 174–183.
10. Poon, K., Yan, A., and Wilton, S.J.E., A flexible power model for FPGAs, *ACM Transactions on Design Automation of Digital Systems*, 10(2), 279–302, April 2005.
11. Conference websites: www.isfpga.org, www.fccm.org, www.fpl.org, www.icfpt.org.
12. DeHon, A. and Hauck, S., *Reconfigurable Computing*, Morgan Kaufman, San Francisco, CA, 2007.
13. Chen, D., Cong, J., and Pan, P., FPGA design automation: A survey, *Foundations and Trends in Electronic Design Automation*, 1(3), 195–330, October 2006.
14. Khatri, S., Shenoy, N., Khouja, A., and Giomi, J.C., Logic synthesis, *Electronic Design Automation for IC Implementation, Circuit Design, and Process Technology*, L. Lavagno, I.L. Markov, G.E. Martin, and L.K. Scheffer, eds., Taylor & Francis Group, Boca Raton, FL, 2016.
15. De Micheli, G., *Synthesis and Optimization of Digital Circuits*, McGraw Hill, New York, 1994.
16. Murgai, R., Brayton, R., and Sangiovanni-Vincentelli, A. *Logic Synthesis for Field-Programmable Gate Arrays*, Kluwer, Norwell, MA, 2000.

SYNTHESIS: HLS

17. Berkeley Design Technology Inc., Evaluating FPGAs for communication infrastructure applications, in *Proceedings of the Communications Design Conference*, 2003.
18. Lockwood, J., Naufel, N., Turner, J., and Taylor, D., Reprogrammable network packet processing on the Field-Programmable Port Extender (FPX), in *Proceedings of the Ninth International Symposium FPGAs*, 2001, pp. 87–93.
19. Kulkarni, C., Brebner, G., and Schelle, G., Mapping a domain-specific language to a platform FPGA, in *Proceedings of the Design Automation Conference*, 2004.
20. Maxeler Technologies, https://www.maxeler.com/.
21. Impulse Accelerated Technologies, http://www.impulseaccelerated.com/.
22. Auerbach, J., Bacon, D., Cheng, P., and Rabbah, R., Lime: A Java-compatible and synthesizable language for heterogeneous architectures, in *Object-Oriented Programming, Systems, Languages and Applications*, 2010, pp. 89–108.
23. Grotker, T., Liau, S., and Martin, G., *System Design with SystemC*, Kluwer, Norwell, MA, 2010.
24. Bluespec Inc., http://www.bluespec.com/.
25. Cong, J. and Zou, Y., FPGA-based hardware acceleration of lithographic aerial image simulation, *ACM Transactions on Reconfigurable Technology and Systems*, 2(3), 17.1–17.29, 2009.
26. United States Bureau of Labor Statistics. *Occupational Outlook Handbook 2010–2011 Edition*, 2010.
27. Cong, J., Fan, Y., Han, G., Jiang, W., and Zhang, Z., Platform-based behavior-level and system-level synthesis, in *Proceedings of IEEE International SOC Conference*, Austin, TX, 2006, pp. 199–202.
28. Xilinx Vivado High-Level Synthesis, http://www.xilinx.com/products/design-tools/vivado/integration/esl-design/, 2014.
29. Cong, J., Liu, C., Neuendorffer, S., Noguera, J., Vissers, K., and Zhang, Z. High-level synthesis for FPGAs: From prototyping to deployment, *IEEE Transactions on Computer-Aided Design of Integrated Circuits and Systems*, 30(4), 473–491, 2011.
30. Chen, D. and Singh, D., Fractal video compression in OpenCL: An evaluation of CPUs, GPUs, and FPGAs as acceleration platforms, in *IEEE/ACM Asia and South Pacific Design Automation Conference*, 2013, pp. 297–304.
31. Altera SDK for OpenCL, http://www.altera.com/products/software/opencl/opencl-index.html, 2014.
32. Coussy, P., Lhairech-Lebreton, G., Heller, D., and Martin, E., GAUT—A free and open source high-level synthesis tool, in *Proceedings of the IEEE/ACM Design Automation and Test in Europe*, University Booth, 2010.
33. Zheng, H., Shang high-level synthesis, https://github.com/OpenEDA/Shang/.
34. Villarreal, J., Park, A., Najjar, W., and Halstead, R. Designing modular hardware accelerators in C with ROCCC 2.0, in *IEEE International Symposium on Field-Programmable Custom Computing Machines*, 2010, pp. 127–134.
35. Pilato, C. and Ferrandi, F., Bambu: A free framework for the high-level synthesis of complex applications, in *ACM/IEEE Design Automation and Test in Europe*, University Booth, 2012.
36. Nane, R., Sima, V., Olivier, B., Meeuws, R., Yankova, Y., and Bertels, K., DWARV 2.0: A CoSy-based C-to-VHDL hardware compiler, in *International Conference on Field-Programmable Logic and Applications*, 2012, pp. 619–622.

37. Canis, A., Choi, J., Aldham, M., Zhang, V., Kammoona, A., Czajkowski, T., Brown, S., and Anderson, J. LegUp: Open source high-level synthesis for FPGA-based processor/accelerator systems, *ACM Transactions on Embedded Computing Systems*, 13(2), 1:1–1:25, 2013.
38. Lattner, C. and Adve, V., LLVM: A compilation framework for lifelong program analysis & transformation, in *International Symposium on Code Generation and Optimization*, 2004, pp. 75–88.
39. Cong, J. and Zhang, Z., An efficient and versatile scheduling algorithm based on SDC formulation, in *IEEE/ACM Design Automation Conference*, 2006, pp. 433–438.
40. Huang, C., Che, Y., Lin, Y., and Hsu, Y., Data path allocation based on bipartite weighted matching, in *IEEE/ACM Design Automation Conference*, 1990, pp. 499–504.
41. Hadjis, S., Canis, A., Anderson, J., Choi, J., Nam, K., Brown, S., and Czajkowski, T. Impact of FPGA architecture on resource sharing in high-level synthesis, in *ACM/SIGDA International Symposium on Field Programmable Gate Arrays*, 2012, pp. 111–114.
42. Canis, A., Anderson, J., and Brown, S., Multi-pumping for resource reduction in FPGA high-level synthesis, in *IEEE/ACM Design Automation and Test in Europe Conference*, 2013, pp. 194–197.
43. Huang, Q., Lian, R., Canis, A., Choi, J., Xi, R., Brown, S., and Anderson, J., The effect of compiler optimizations on high-level synthesis for FPGAs, in *IEEE International Symposium on Field-Programmable Custom Computing Machines*, 2013, pp. 89–96.
44. Choi, J., Anderson, J., and Brown, S., From software threads to parallel hardware in FPGA high-level synthesis, in *IEEE International Conference on Field-Programmable Technology (FPT)*, 2013, pp. 270–279.
45. Zheng, H., Swathi T., Gurumani, S., Yang, L., Chen, D., and Rupnow, K., High-level synthesis with behavioral level multi-cycle path analysis, in *International Conference on Field-Programmable Logic and Applications*, 2013.
46. Papakonstantinou, A., Gururaj, K., Stratton, J., Chen, D., Cong, J., and Hwu, W., FCUDA: Enabling efficient compilation of CUDA kernels onto FPGAs, in *Symposium on Application-Specific Processors*, 2009, pp. 35–42.
47. Zuo, W., Liang, Y., Li, P., Rupnow, K., Chen, D., and Cong, J., Improving high-level synthesis optimization opportunity through polyhedral transformations, in *ACM/SIGDA International Symposium on Field-Programmable Gate Arrays*, 2013, pp. 9–18.
48. Bastoul, C., Cohen, A., Girbal, S., Sharma, S., and Temam, O, Putting polyhedral loop transformations to work, in *International Workshop on Languages and Compilers for Parallel Computing*, College Station, TX, 2003, pp. 209–225.
49. Cong, J., Huang, M., and Zhang, P., Combining computation and communication optimizations in system synthesis for streaming applications, in *ACM/SIGDA International Symposium on Field Programmable Gate Arrays*, 2014, pp. 213–222.
50. Wang, Y., Li, P., and Cong, J., Theory and algorithm for generalized memory partitioning in high-level synthesis, in *ACM/SIGDA International Symposium on Field Programmable Gate Arrays*, 2014, pp. 199–208.
51. Bayliss, S. and Constantinides, G., Optimizing SDRAM bandwidth for custom FPGA loop accelerators, in *ACM/SIGDA International Symposium on Field Programmable Gate Arrays*, 2012, pp. 195–204.

SYNTHESIS: LOGIC OPTIMIZATION

52. Tinmaung, K. and Tessier, R., Power-aware FPGA logic synthesis using binary decision diagrams, in *Proceedings of the 15th International Symposium on FPGAs*, 2007.
53. Metzgen, P. and Nancekievill, D., Multiplexor restructuring for FPGA implementation cost reduction, in *Proceedings of the Design Automation Conference*, 2005.
54. Nancekievill, D. and Metzgen, P., Factorizing multiplexors in the datapath to reduce cost in FPGAs, in *Proceedings of the International Workshop on Logic Synthesis*, 2005.
55. Hutton, M. et al., Improving FPGA performance and area using an adaptive logic module, in *Proceedings of the 14th International Symposium Field-Programmable Logic*, 2004, pp. 134–144.
56. Tessier, R., Betz, V., Neto, D., Egier, A., and Gopalsamy, T., Power-efficient RAM-mapping algorithms for FPGA embedded memory blocks, *IEEE Transactions on Computer-Aided Design of Integrated Circuits and Systems*, 26(2), 278–290, February 2007.
57. Sentovich, E. et al., SIS: A system for sequential circuit analysis, Tech Report No. UCB/ERL M92/41, UC Berkeley, Berkeley, CA, 1992.
58. Mischenko, A., Brayton, R., Jiang, J., and Jang, S., DAG-aware AIG re-writing: A fresh look at combinational logic synthesis, in *Proceedings of the ACM DAC*, 2006, pp. 532–536.
59. Mischenko, A. et al., ABC—A system for sequential synthesis and verification, http://www.eecs.berkeley.edu/~alanmi/abc/, 2009.

60. Luu, J. et al., On hard adders and carry chains in FPGAs, in *International Symposium on Field-Programmable Custom Computing Machines (FCCM)*, 2014.

61. Shenoy, N. and Rudell, R., Efficient implementation of retiming, in *Proceedings of the International Conference on CAD (ICCAD)*, 1994, pp. 226–233.

62. van Antwerpen, B., Hutton, M., Baeckler, G., and Yuan, R., A safe and complete gate-level register retiming algorithm, in *Proceedings of the IWLS*, 2003.

63. Yamashita, S., Sawada, H., and Nagoya, A., A new method to express functional permissibilities for LUT-based FPGAs and its applications, in *Proceedings of the International Conference on CAD (ICCAD)*, 1996, pp. 254–261.

64. Cong, J., Lin, Y., and Long, W., SPFD-based global re-wiring, in *Proceedings of the 10th International Symposium on FPGAs*, 2002, pp. 77–84.

65. Cong, J., Lin, Y., and Long, W., A new enhanced SPFD rewiring algorithm, in *Proceedings of the International Conference on CAD (ICCAD)*, 2002, pp. 672–678.

66. Vemuri, N., Kalla, P., and Tessier, R., BDD-based logic synthesis for LUT-based FPGAs, *ACM Transactions on Design Automation of Electronic Systems*, 7(4), 501–525, 2002.

67. Yang, C., Ciesielski, M., and Singhal, V., BDS A BDD-based logic optimization system, in *Proceedings of the Design Automation Conference*, 2000, pp. 92–97.

68. Lai, Y., Pedram, M., and Vrudhala, S., BDD-based decomposition of logic functions with application to FPGA synthesis, in *Proceedings of the Design Automation Conference*, 1992, pp. 448–451.

TECHNOLOGY MAPPING FOR FPGAS

69. Ling, A., Singh, D.P., and Brown, S.D., FPGA technology mapping: A study of optimality, in *Proceedings of the Design Automation Conference*, 2005.

70. Francis, R.J., Rose, J., and Chung, K., Chortle: A technology mapping program for lookup table-based field-programmable gate arrays, in *Proceedings of the Design Automation Conference*, 1990, pp. 613–619.

71. Cong, J. and Ding, E., An optimal technology mapping algorithm for delay optimization in lookup table based FPGA designs, *IEEE Transactions on CAD*, 13(1), 1–12, 1994.

72. Cong, J. and Hwang, Y. Structural gate decomposition for depth-optimal technology mapping in LUT-based FPGA designs, *ACM Transactions on Design Automation of Digital Systems*, 5(2), 193–225, 2000.

73. Cong, J. and Ding, Y., On area/depth trade-off in LUT-based FPGA technology mapping, *IEEE Transactions on VLSI*, 2(2), 137–148, 1994.

74. Schlag, M., Kong, J., and Chan, P.K., Routability-driven technology mapping for lookup table-based FPGA's, *IEEE Transactions on Computer-Aided Design of Integrated Circuits and Systems*, 13(1), 13–26, 1994.

75. Cong, J. and Hwang, Y., Simultaneous depth and area minimization in LUT-based FPGA mapping, in *Proceedings of the fourth International Symposium FPGAs*, 1995, pp. 68–74.

76. Chen, D. and Cong, D., DAOmap: A depth-optimal area optimization mapping algorithm for FPGA designs, in *Proceedings of the International Conference on CAD (ICCAD)*, November 2004.

77. Farrahi, A. and Sarrafzadeh, M., Complexity of the lookup-table minimization problem for FPGA technology mapping, *IEEE Transactions on Computer-Aided Design of Integrated Circuits and Systems*, 13(11), 1319–1332, 2006.

78. Manohararajah, V., Brown, S.D., and Vranesic, Z.G., Heuristics for area minimization in LUT-based FPGA technology mapping, *IEEE Transactions on Computer-Aided Design of Integrated Circuits and Systems*, 25(11), 2331–2340, 2006.

79. Mischenko, A. et al., Combinational and sequential mapping with priority cuts, in *Proceedings of the IEEE International Conference on CAD (ICCAD)*, 2007, pp. 354–361.

80. Mischenko, A., Chatterjee, S., and Brayton, R., Improvements to technology mapping for LUT-based FPGAs, in *Proceedings of the 14th International Symposium on FPGAs*, 2006.

81. Mischenko, A., Brayton, R., and Jang, S., Global delay optimization using structural choices, in *Proceedings of the 18th International Symposium on FPGAs*, 2010, pp. 181–184.

POWER-AWARE SYNTHESIS

82. Farrahi, A.H. and Sarrafzadeh, M., FPGA technology mapping for power minimization, in *Proceedings of the International Workshop on Field-Programmable Logic and Applications*, 1994.

83. Anderson, J. and Najm, F.N., Power-aware technology mapping for LUT-based FPGAs, in *Proceedings of the International Conference on Field-Programmable Technology*, 2002.

84. Lamoreaux, J. and Wilton, S.J.E., On the interaction between power-aware CAD algorithms for FPGAs, in *Proceedings of the International Conference on CAD (ICCAD)*, 2003.

85. Chen, D., Cong, J., Li, F., and He., Low-power technology mapping for FPGA architectures with dual supply voltages, in *Proceedings of the 12th International Symposium on FPGAs*, 2004, pp. 109–117.

86. Anderson, J., Najm, F., and Tuan, T., Active leakage power estimation for FPGAs, in *Proceedings of the 12th International Symposium on FPGAs*, 2004, pp. 33–41.

87. Anderson, J. and Ravishankar, C., FPGA power reduction by guarded evaluation, in *Proceedings of the 18th International Symposium on FPGAs*, 2010, pp. 157–166.

88. Hwang, J.M., Chiang, F.Y., and Hwang, T.T., A re-engineering approach to low power FPGA design using SPFD, in *Proceedings of the 35th ACM/IEEE Design Automation Conference*, 1998, pp. 722–725.

89. Kumthekar, B. and Somenzi, F., Power and delay reduction via simultaneous logic and placement optimization in FPGAs, in *Proceedings of the Design and Test in Europe (DATE)*, 2000, pp. 202–207.

PHYSICAL DESIGN

90. Anderson, J., Saunders, J., Nag, S., Madabhushi, C., and Jayarman, R., A placement algorithm for FPGA designs with multiple I/O standards, in *Proceedings of the International Conference on Field Programmable Logic and Applications*, 2000, pp. 211–220.

91. Mak, W., I/O placement for FPGA with multiple I/O standards, in *Proceedings of the 11th International Symposium on FPGAs*, 2003, pp. 51–57.

92. Anderson, J., Nag, S., Chaudhary, K., Kalman, S., Madabhushi, C., and Cheng, P., Run-time conscious automatic timing-driven FPGA layout synthesis, in *Proceedings of the 14th International Conference on Field-Programmable Logic and Applications*, 2004, pp. 168–178.

CLUSTERING

93. Betz, V. and Rose, J., VPR: A new packing, placement and routing tool for FPGA research, in *Proceedings of the Seventh International Conference on Field-Programmable Logic and Applications*, 1997, pp. 213–222.

94. Marquardt, A., Betz, V., and Rose, J., Timing-driven placement for FPGAs, in *Proceedings of the International Symposium on FPGAs*, 2000, pp. 203–213.

95. Singh, A. and Marek-Sadowska, M., Efficient circuit clustering for area and power reduction in FPGAs, in *Proceedings of the International Symposium on FPGAs*, 2002, pp. 59–66.

96. Murray, K., Whitty, S., Luu, J., Liu, S., and Betz, V., Titan: Enabling large and realistic benchmarks for FPGAs, in *Proceedings of the International Conference on Field-Programmable Logic and Applications*, 2013, pp. 1–8.

97. Feng, W., Greene, J., Vorwerk, K., Pevzner, V., and Kundu, A., Rent's rule based FPGA packing for routability optimization, in *Proceedings of the International Symposium on FPGAs*, 2014, pp. 31–34.

98. Luu, J., Rose, J., and Anderson, J., Towards interconnect-adaptive packing for FPGAs, in *Proceedings of the International Symposium on FPGAs*, 2014, pp. 21–30.

PLACEMENT

99. Kong, T., A novel net weighting algorithm for timing-driven placement, in *Proceedings of the International Conference on CAD (ICCAD)*, 2002, pp. 172–176.

100. Lin, Y., He, L., and Hutton, M., Stochastic physical synthesis considering pre-routing interconnect uncertainty and process variation for FPGAs, *IEEE Transactions on VLSI Systems*, 16(2), 124–133, 2008.

101. Chen, G. and Cong, J., Simultaneous timing driven clustering and placement for FPGAs, in *Proceedings of the International Conference on Field Programmable Logic and Applications*, 2004, pp. 158–167.

102. Sharma, A., Ebeling, C., and Hauck, S., Architecture adaptive routability-driven placement for FPGAs, in *Proceedings of the International Conference on Field-Programmable Logic and Applications*, 2005, pp. 95–100.

103. Sankar, Y. and Rose, J., Trading quality for compile time: Ultra-fast placement for FPGAs, in *Proceedings of the International Symposium on FPGAs*, 1999, pp. 157–166.

104. Maidee, M., Ababei, C., and Bazargan, K., Fast timing-driven partitioning-based placement for Island style FPGAs, in *Proceedings of the Design Automation Conference (DAC)*, 2003, pp. 598–603.

105. Gort, M. and Anderson, J., Analytical placement for heterogeneous FPGAs, in *Proceedings of the International Conference on Field-Programmable Logic and Applications*, 2012, pp. 143–150.

106. Kim, M.-C., Lee, D., and Markov, I., SimPL: An effective placement algorithm, *IEEE Transactions on CAD*, 31(1), 50–60, 2012.

107. Wu, K. and McElvain, K., A fast discrete placement algorithm for FPGAs, in *Proceedings of the International Symposium on FPGAs*, 2012, pp. 115–119.

108. Ludwin, A. and Betz, V., Efficient and deterministic parallel placement for FPGAs, *ACM Transactions on Design Automation of Electronic Systems*, 16(3), 22:1–22:23, June 2011.

109. An, M., Steffan, G., and Betz, V., Speeding up FPGA placement: Parallel algorithms and methods, in *Proceedings of the International Symposium on Field-Configurable Custom Computing Machines*, 2014, pp. 178–185.

110. Goeders, J., Lemieux, G., and Wilton, S., Deterministic timing-driven parallel placement by simulated annealing using half-box window decomposition, in *Proceedings of the International Conference on Reconfigurable Computing and FPGAs*, 2011, pp. 41–48.

PHYSICAL RESYNTHESIS

111. Lin, J., Jagannathan, A., and Cong, J. Placement-driven technology mapping for LUT-based FPGAs, in *Proceedings of the 11th International Symposium on FPGAs*, 2003, pp. 121–126.

112. Suaris, P., Wang, D., and Chou, N., Smart move: A placement-aware retiming and replication method for field-programmable gate arrays, in *Proceedings of the Fifth International Conference on ASICs*, 2003.

113. Suaris, P., Liu, L., Ding, Y., and Chou, N., Incremental physical re-synthesis for timing optimization, in *Proceedings of the 12th International Symposium on FPGAs*, 2004, pp. 99–108.

114. Schabas, K. and Brown, S., Using logic duplication to improve performance in FPGAs, in *Proceedings of the 11th International Symposium on FPGAs*, 2003, pp. 136–142.

115. Chen, G. and Cong, J., Simultaneous timing-driven placement and duplication, in *Proceedings of the 13th International Symposium on FPGAs*, 2005, pp. 51–61.

116. Manohararajah, V., Singh, D, Brown, S., and Vranesic, Z., Post-placement functional decomposition for FPGAs, in *Proceedings of the International Workshop on Logic Synthesis*, 2004, pp. 114–118.

117. Manohararajah, V., Singh, D.P., and Brown, S., Timing-driven functional decomposition for FPGAs, in *Proceedings of the International Workshop on Logic and Synthesis*, 2005.

118. Ding, Y., Suaris, P., and Chou, N., The effect of post-layout pin permutation on timing, in *Proceedings of the 13th International Symposium on FPGAs*, 2005, pp. 41–50.

119. Singh, D. and Brown, S., Integrated retiming and placement for FPGAs, in *Proceedings of the 10th International Symposium on FPGAs*, 2002, pp. 67–76.

120. Chen, D. and Singh, D., Line-level incremental resynthesis techniques for FPGAs, in *Proceedings of the 19th International Symposium on FPGAs*, 2011, pp. 133–142.

121. Singh, D. and Brown, S., Constrained clock shifting for field programmable gate arrays, in *Proceedings of the 10th International Symposium on FPGAs*, 2002, pp. 121–126.

122. Chao-Yang, Y. and Marek-Sadowska, M., Skew-programmable clock design for FPGA and skew-aware placement, in *Proceedings of the International Symposium on FPGAs*, 2005, pp. 33–40.

ROUTING

123. Lemieux, G. and Lewis, D., *Design of Interconnection Networks for Programmable Logic*, Kluwer, Norwell, MA, 2004.

124. Lemieux, G. and Brown, S., A detailed router for allocating wire segments in FPGAs, in *Proceedings of the Physical Design Workshop*, 1993, pp. 215–226.

125. Nam, G.-J., Aloul, F., Sakallah, K., and Rutenbar, R., A comparative study of two Boolean formulations of FPGA detailed routing constraints, in *Proceedings of the International Symposium on Physical Design*, 2001, pp. 222–227.

126. Lee, C.Y., An algorithm for path connections and applications, *IRE Transactions on Electronic Computers*, EC-10(2) 346–365, 1961.

127. McMurchie, L. and Ebeling, C., PathFinder: A negotiation-based performance-driven router for FPGAs, in *Proceedings of the Fifth International Symposium on FPGAs*, 1995, pp. 111–117.

128. Youssef, H. and Shragowitz, E., Timing constraints for correct performance, in *Proceedings of the International Conference on CAD*, 1990, pp. 24–27.

129. Swartz, J., Betz, V., and Rose, J., A fast routability-driven router for FPGAs, in *Proceedings of the Sixth International Symposium on FPGAs*, 1998, pp. 140–151.

130. Wilton, S., A crosstalk-aware timing-driven router for FPGAs, in *Proceedings of the Ninth ACM International Symposium on FPGAs*, 2001, pp. 21–28.
131. Fung, R., Betz, V., and Chow, W., Slack allocation and routing to improve FPGA timing while repairing short-path violations, *IEEE Transactions on CAD*, 27:4, pp. 686–697, April 2008.
132. Gort, M. and Anderson, J., Accelerating FPGA routing through parallelization and engineering enhancements, *IEEE Transactions on CAD*, 31:1, pp. 61–74, January 2012.

CAD FOR EMERGING ARCHITECTURE FEATURES

133. Li, F., Lin, Y., and He, L., Field programmability of supply voltages for FPGA power reduction, *IEEE Transactions on CAD*, 13:9, pp. 752–764, April 2007.
134. Lewis, D., et al., Architectural enhancements in Stratix-III and Stratix-IV, in *Proceedings of the International Symposium on FPGAs*, 2009, pp. 33–41.
135. Huda, S., Anderson, J., and Tamura, H., Optimizing effective interconnect capacitance for FPGA power reduction, in *Proceedings of the International Symposium on FPGAs*, pp. 11–19, 2014.
136. Chaware, R., Nagarajan, K., and Ramalingam, S., Assembly and reliability challenges in 3D integration of 28 nm FPGA die on a large high density 65 nm passive interposer, in *Proceedings of the IEEE Electronic Components and Technology Conference*, 2012, pp. 279–283.
137. Hahn Pereira, A. and Betz, V., CAD and architecture for interposer-based multi-FPGA systems, in *Proceedings of the International Symposium on FPGAs*, 2014, pp. 75–84.
138. Ababei, C., Mogal, H., and Bazargan, K., Three-dimensional place and route for FPGAs, *IEEE Transactions on CAD*, 25:6, pp. 1132–1140, June 2006.
139. "FPGAs in 2032", *Workshop at the 2012 ACM/IEEE International Symposium on FPGAs* (slides available at www.tcfpga.org).

Analog and Mixed-Signal Design

Simulation of Analog and RF Circuits and Systems

17

Jaijeet Roychowdhury and Alan Mantooth

CONTENTS

17.1 INTRODUCTION

Circuit simulation has always been a crucial component of analog system design, even more so today. Ever-shrinking DSM technologies result in both analog and digital designs, proving to be less ideal. Indeed, the distinction between digital and analog design is blurred; from the digital standpoint, analog effects, often undesired, are becoming pervasive. Integrated radio frequency (RF) and communication system design involving both analog and digital components on the same substrate, now constitutes an important part of the semiconductor industry's growth, while traditionally digital circuits (such as microprocessors) are now critically limited by analog effects such as delays, the need to synchronize internal busses, and so on. In short, advances in analog, RF, digital, and mixed-signal design over the last few years, combined with the effects of shrinking technologies, have spurred a renaissance in simulation. Old simulation problems have assumed renewed significance and new simulation challenges—in some cases already addressed by novel and elegant algorithmic solutions—have arisen.

Every circuit designer is familiar with the program SPICE, the original circuit simulation program released into the public domain in the mid-1970s by the University of California at Berkeley. The basic algorithms and capabilities within SPICE engendered an entire industry of circuit simulation tools, with noted commercial offerings such as HSPICE, PSPICE, and SPECTRE enjoying widespread adoption. Several of these programs have evolved from originally providing roughly the same capabilities as Berkeley SPICE, into tools with significantly more advanced simulation capabilities. Furthermore, the aforementioned resurgence of interest and research in advanced circuit simulation algorithms promises more powerful capabilities, which will enable designers to make higher-performance circuits and systems faster, better, and with fewer mistakes.

In this chapter, we provide a quick tour of modern circuit simulation. The style we adopt attempts to combine mathematical rigor with physical intuition, in order to provide a simple, clean exposition that links the various facets of the subject logically. We start with showing how one can write circuit equations as nonlinear differential equations and comment on why it is a good idea to do so. Next, we delve into "device models"—i.e., the equations that describe current—voltage and charge—voltage relationships of the semiconductor devices that pervade most modern circuits. We then show how the basic circuit analyses can be simply derived as specific ways of solving the circuit's differential equations. We touch on an important feature typical of circuit equations—sparsity—that makes these analyses computationally feasible for large circuits, and hence practically useful.

Next, motivated by recent (as of 2006) interest in RF circuit simulation, we review more advanced analysis techniques for the circuit's differential equations, starting with the computation of periodic steady states. We explain two alternative methods in the frequency and time domains—harmonic balance (HB) and shooting—for computing periodic steady states that have complementary numerical properties. We show how the equations involved in these analyses lose sparsity, and hence can be much more difficult to solve computationally than those for the basic analyses above. We then explain how methods based on preconditioned iterative solution of matrix equations can be used to alleviate the computational problem to a large extent.

A recurring theme in circuit simulation is the simultaneous presence of fast and slow rates of time variation, which presents challenges for basic analyses such as transient simulation. We show how the use of multiple artificial time scales can be used to separate fast/slow behavior in a fundamental manner at the circuit equation level, leading to a special circuit equation form called multitime partial differential equations (MPDEs). We then outline how MPDEs can be solved numerically using different techniques to solve fast/slow problems efficiently. We touch on special MPDE forms for oscillators and also outline how MPDEs can be used as a link to enable automatic macromodeling of time-varying systems.

Finally, we touch on the important issue of (statistical) noise analysis of circuits. We first outline the fundamental concepts of basic stationary noise analysis and explain the kinds of circuits it applies to. Then we show how stationary noise analysis is insufficient for noise analysis of RF, switching, and other nonlinear circuits. We explain the fundamental concepts of cyclostationary noise and outline the use of these concepts for the noise analysis of RF circuitry, including computational aspects. Finally, we touch on the important problem of noise analysis of oscillators—in particular, calculating phase noise and jitter in oscillators.

17.2 DIFFERENTIAL-ALGEBRAIC EQUATIONS FOR CIRCUITS VIA MODIFIED NODAL ANALYSIS

The fundamental premise behind virtually all types of computer simulation is to first abstract a physical system's behavior using appropriate mathematical representations and then to use algorithms to solve these representations by computer. For most kinds of circuits, the appropriate mathematical abstraction turns out to be systems of differential equations, or more precisely, systems of nonlinear, differential-algebraic equations (DAEs). In the circuit context, the following special DAE form can be shown to be appropriate:

$$q(x) + f(x) = b(t).$$ (17.1)

In Equation 17.1, all quantities (except t, the time variable) are vectors of size n, which are (roughly) the size of the circuit represented by the DAE. The nonlinear vector functions $q(x)$ and $f(x)$ represent the charge/flux and resistive parts of the circuit respectively, while $b(t)$ is a forcing term representing external inputs from independent current and voltage sources. This DAE form is obtained by writing out the fundamental charge and potential conservation equations of the circuit—Kirchoff's current laws (KCL) and Kirchoff's voltage law (KVL)—together with the branch constitutive relationships (BCR) of the individual elements. We refer the reader to e.g., [1–3], for a systematic exposition of how these equations can be obtained for a circuit topology. It is not difficult, however, to understand the basic procedure of defining the quantities in Equation 17.1 and to apply it to any circuit of interest, "by hand" or via a computer procedure. Consider the simple circuit shown in Figure 17.1, consisting of a voltage source driving a parallel-RC load through a nonlinear diode. The electrical quantities of interest in this circuit are the two node voltages $v_1(t)$ and $v_2(t)$, as well as the current through the voltage source, $i(t)$.

The resistor's action on the circuit is defined by its current—voltage relationship (Ohm's law), $i_R(t) = v_2(t)/R$; similarly, the capacitor's current–voltage relationship is given by $i_C(t) = (d/dt)(Cv_2(t))$. The current—voltage relationship of the diode element, which is nonlinear, is given by (see Section 17.3.2)

$$i = d(v_1 - v_2) = I_s\left(e^{\frac{v_1 - v_2}{v_T}} - 1\right)$$ (17.2)

where I_s and V_t are constant numbers. Finally, the voltage source is captured simply by its setting the voltage of node 1 to its value: $v_1(t) = E$. These four relationships, one for each of the four circuit elements, are the BCRs relevant to this circuit.

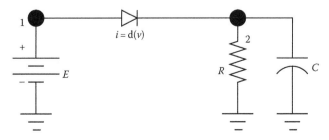

FIGURE 17.1 Simple diode–resistor–capacitor circuit.

By summing currents at each node of interest, the circuit equations are obtained. At node 1, we obtain

(17.3)
$$i(t) + d(v_1 - v_2) = 0$$

while at node 2, we have

(17.4)
$$\frac{d}{dt}(Cv_2) + \frac{v_2}{R} - d(v_1 - v_2) = 0$$

Finally, we also have (for the voltage source)

(17.5)
$$v_1 = E.$$

These are known as the modified nodal analysis equations (MNA equations) of the circuit.

By writing Equations 17.3 through 17.5 in vector form, the form of Equation 17.1 can be obtained. First, the voltage/current unknowns are written in vector form as

(17.6)
$$x = \begin{bmatrix} v_1(t) \\ v_2(t) \\ i(t) \end{bmatrix}.$$

Define the vector function $q(x)$ in Equation 17.1 to be

(17.7)
$$q(x) \equiv \begin{bmatrix} 0 & & \\ & C & \\ & & 0 \end{bmatrix} x = \begin{bmatrix} 0 \\ Cv_2(t) \\ 0 \end{bmatrix}.$$

$f(x)$ to be

(17.8)
$$f(x) \equiv \begin{bmatrix} i(t) + d(v_1 - v_2) \\ \frac{v_2}{R} - d(v_1 - v_2) \\ v_1 \end{bmatrix},$$

and the vector $b(t)$ to be

(17.9)
$$b(t) = \begin{bmatrix} 0 \\ 0 \\ E(t) \end{bmatrix}.$$

With these definitions, Equation 17.1 simply represents the circuit Equations 17.3 through 17.5, but in vector DAE form. The utility of writing circuit equations in this canonical form is that a variety of mathematical and computational techniques may be brought to bear on Equation 17.1, without having to worry about the details of $f(x)$, $q(x)$, and $b(t)$, so long as they can be realistically assumed to have some simple properties like continuity, smoothness, etc. Indeed, the DAE form Equation 17.1 is not restricted to circuits; it subsumes virtually all physical systems, including those from mechanical, optical, chemical, etc., applications.

It will be readily understood, of course, that DAEs for typical industrial circuits are usually much larger and more complex than the one for the simple example of Figure 17.1. Much of the complexity of these equations stems from the complicated equations describing the *semiconductor devices*, such as metal oxide semiconductor (MOS) and bipolar transistors, that populate circuits in large numbers. These equations, called *device models*, are described further in Section 17.3.

17.3 DEVICE MODELS

Many books that describe the details of various device models available in circuit simulators [4–8] have been written. In these few pages, it is not possible to delve into the derivation of device physics that represent any specific device models. Rather, the focus of this section will be on the role the models play and constraints placed on models by simulators. This section first describes models in the context of traditional circuit simulation and then touches on two key device models, the diode and MOSFET, which are illustrative of issues relevant to this chapter. In each case, some remarks are made with respect to versions of these models suitable for power electronic applications as well.

17.3.1 ROLE OF MODELS IN CIRCUIT SIMULATION

In the context of this chapter, models are the representation of the properties of devices or a group of interconnected devices by means of mathematical equations, circuit representations (i.e., macromodels), or tables. Care has been taken to avoid defining models strictly as electrical devices. In the era of hardware description languages (HDLs), circuit simulation and modeling have expanded beyond the boundaries of passive and active electrical devices [9–12]. However, for the purpose of this section, our focus will remain on electrical device models.

Semiconductor device models such as the MOSFET involve many complicated equations. Often these equations are defined piecewise according to the different regions of operation observed in the device. Timing studies performed on circuit simulations indicate that most of the computational effort in network analysis is spent on evaluation of these complicated model equations. This is not hard to imagine when "compact" model implementations of advanced MOSFET technologies are over 15,000 lines of C code in SPICE3 [13]. While much of this code involves data structures in the simulator, several thousand lines of code are devoted to describing model behavior alone. As such, models and modeling techniques are as important to addressing simulation efficiency and accuracy as the solution algorithms employed. In fact, the fidelity of simulation results will be dictated by the weakest link between the two, so both are vital to the simulation process.

Figure 17.2 below depicts a very high level view of the simulation process, which specifically pulls out the role the models play in the solution of the system. This flow diagram begins with the processing of the input netlist to the simulator and proceeds to the formulation of the system of equations. The system formulation step involves both the interconnectivity of the specific models in the circuit as well as the matrix stamps of the models. The interconnectivity defines what models contribute current to which nodes and thus leads to the matrix statement of KCL. The matrix stamps of the models, which will be described further in Section 17.4, indicate the specific contributions a model makes to various rows and columns of the matrix. Once the system formulation is complete, the process proceeds to the numerical solution based on the user-specified analysis. In any event, one thing that is common to all analyses (e.g., DC, AC, transient) and

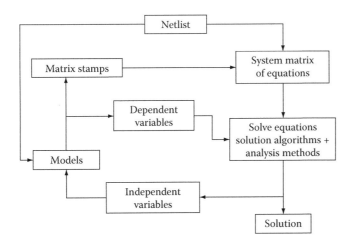

FIGURE 17.2 Depiction of the role of models in the traditional circuit simulation process.

solution methods (e.g., Newton–Raphson [NR]) is the need to evaluate the models' constitutive relationships at each iteration, subject to any bypass techniques that may be employed in a given algorithm. This is the model evaluation time referred to above. The independent variables indicated in Figure 17.2 are those variables being solved for by the simulator and are thus given to the models as known values for an iteration. Then, the model is evaluated and provides the dependent variable values back to the simulator for the purpose of determining if the simulator's guess for the independent variable values was the correct solution to the system.

17.3.2 DIODE MODEL

For the model of a basic p–n junction diode, the constitutive relations are derived in [14] and also shown in [4,15]. Figure 17.3 shows the basic large-signal model of the diode. The series resistance is a straightforward application of Ohm's law. The current source is represented by the BCR

(17.10)
$$I_d = I_S\left(e^{\frac{V_d}{nV_T}} - 1\right),$$

where
I_S is the saturation current
V_T the thermal voltage
V_d the diode junction voltage
n the emission coefficient of the diode

The saturation current is a function of the geometry of the device and is given by

(17.11)
$$I_s = Aqn_i^2\left(\frac{D_p}{L_pN_D} + \frac{D_n}{L_nN_A}\right),$$

where
A is the cross-sectional area of the junction
q the electronic charge
n_i the intrinsic carrier concentration
N_A and N_D the acceptor and donor concentrations
$L_n(D_n)$ and $L_p(D_p)$ are the diffusion lengths (diffusion constants) for electrons in p-type material and holes in n-type material, respectively

The thermal voltage is equal to

(17.12)
$$V_T = \frac{kT}{q},$$

where
$k = 1.3806 \times 10^{-23}$ J/K is Boltzmann's constant
T is the absolute temperature in K
$q = 1.6022 \times 10^{-19}$ C the electronic charge

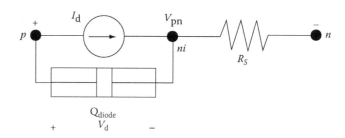

FIGURE 17.3   Large-signal model topology of a p–n junction diode.

The well-known relationships in Equations 17.10 through 17.12 can be used to illustrate some of the options that a model developer has when creating a model for circuit simulation. In this case, a single continuous (and continuously differentiable) equation has been used to describe the BCR. This is the most desirable scenario for the solution algorithms in traditional circuit simulators. Most solvers require continuity in the first derivative of the BCRs in order to avoid convergence problems associated with the Jacobian. Saber [16,17] is an exception to this requirement, but even in that case models that possess continuous first derivatives simulate faster and more robustly than those with only continuous BCRs. From a pragmatic perspective, fewer convergence problems are introduced with BCRs that are not piecewise defined [8]. A second option that faces the model developer is the degree of physics to represent in the model. Most diode models simply specify the saturation current as a parameter and avoid calculating it from Equation 17.11. This allows diodes to be effectively modeled without the need to know doping concentrations and other low-level process information about how the device was constructed. It still remains possible to make the saturation current scalable with area by simply providing a scale factor for I_S in the model equations. Then, since I_S varies so significantly with temperature, primarily due to the n_i^2 factor in Equation 17.11, an empirical relationship is commonly used to model the variation of I_S with temperature, as observed from data sheets or measurements.

The governing equation for the charge in the diode in Figure 17.3 is

$$(17.13) \qquad Q_{\text{diode}} = Q_{\text{diff}} + Q_{\text{depl}}$$

where

Q_{diode} is the total charge
Q_{diff} is the charge stored due to diffusion in the diode
Q_{depl} is the charge stored in the depletion region of the junction

The relationships [4] for each of these components of charge are given by

$$(17.14) \qquad Q_{\text{diff}} = \tau_T I_d$$

where τ_T is known as the mean transit time of the diode and is typically a model parameter. This charge-storage effect is only significant in forward bias and thus a piecewise definition for the charge is typically created at this point by setting $Q_{\text{diff}} = 0$ for reverse bias voltages. This charge equation is continuous as is its derivative.

The depletion charge is also defined piecewise as follows:

$$(17.15)$$

$$Q_{\text{depl}} = \begin{cases} V_{j0}C_{j0} \dfrac{\left(1 - \dfrac{V_d}{V_{j0}}\right)^{1-m}}{m-1} \\[2em] V_{j0}C_{j0} \dfrac{\left(1-f_c\right)^{1-m}}{m-1} + \dfrac{C_{j0}}{\left(1-f_c\right)^{1+m}}\left[1 - f_c\left(1+m\right)\left(V_d - f_c V_{j0}\right) + m\dfrac{V_d^2 - \left(f_c v_{j0}\right)^2}{2V_{j0}}\right] \end{cases} \begin{matrix} V_d < f_c V_{j0}, \\[1em] V_d \geq f_c V_{j0.} \end{matrix}$$

where

f_c is a fitting parameter (typically = 0.5)
m is the grading coefficient of the junction (= 0.5 for a step junction)
C_{j0} is the zero-bias junction capacitance
V_{j0} is the built-in junction potential

In this case, a great deal of care is taken to produce a continuous charge and a continuous capacitance (i.e., first derivative) for Equation 17.15. The diode model possesses nonlinear behavior in the current—voltage relationships as well as in the charge relationships. Details of how such behavior is "stamped" into the circuit matrices and thus used in the flow of Figure 17.2 will be covered in the next section.

The modeling of power diodes is conceptually the same as that for low-voltage diodes. The primary differences arise from the different conditions the device is subjected to and the physical structure of the device. These differences lead to a different set of assumptions for deriving the model. In the end, this does have an impact on the nature of the equations that must be implemented. In the case of power diode models, the levels of injection are significantly higher, requiring the model developer to account for ambipolar diffusion effects and pay particular attention to conductivity modulation within lightly doped regions. Some power diodes are manufactured with a lightly doped "base" region between the p and n regions, making a p−ν−n or a p−π−n diode. Such diodes are generically referred to as p−*i*−n diodes, where the *i* refers to an almost intrinsic region. Figure 17.4 is a cross-sectional diagram of a p−*i*−n diode. In these devices, holes are injected from the p region and electrons are injected from the n region into the base. The normally high-resistance region of the base becomes saturated with carriers and the on-state resistance of the region becomes quite low. When the device is switched off, the carriers must be removed from the base region before the device can effectively turn off.

Through mechanisms such as recombination and carrier sweep-out effects, the device has a significant reverse recovery where the current through the device changes from positive to negative and then recovers back to near zero (Figure 17.5). The depletion region that forms during this recovery process changes width, producing a nonquasi-static effect that must also be accounted for to model reverse recovery accurately. The diode model in [18] possesses all of this behavior which typical SPICE diode models do not.

Figure 17.6 shows the large-signal topology of the power diode model in [18]. The DC characteristics are far more complicated than those described above for the low-voltage diode. Due to the injection levels, there are four separate exponential relationships that are used to model low-level recombination, normal injection, high-level injection, and emitter (i.e., end region) recombination effects. All of this DC behavior is represented by the current I_d represented by the diode symbol in Figure 17.6. An additional series resistance effect to R_S is R_{MOD}, which represents the conductivity modulation of the base region. The nonlinear charge effects are represented by four different charges in parallel with the diode current in Figure 17.6. The capacitance C_j is the voltage-dependent diode

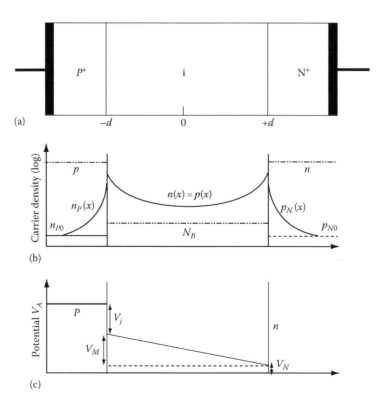

FIGURE 17.4   (a) Cross-sectional diagram of a p−*i*−n power diode; (b) corresponding carrier distribution profile; and (c) potential distribution for high-level injection conditions.

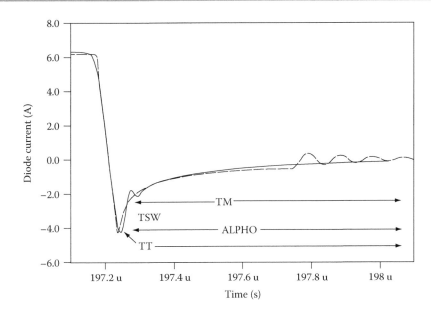

FIGURE 17.5 Reverse recovery waveform for a silicon carbide p–*i*–n power diode. (For further reading on SiC diode modeling, please see McNutt, T. R. et al., *IEEE Trans. Power Electron*, 19, 573, 2004.)

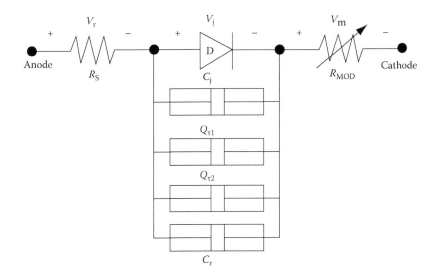

FIGURE 17.6 Large-signal model topology of power diode.

junction capacitance, Q_{SW} and Q_R represent the two-time constant response of the base charge, and C_r is the moving boundary redistribution capacitance. The total diode voltage v_d is the sum of the contact voltage v_r, the junction voltage v_j, and the midregion voltage v_m.

17.3.3 MOSFET MODELS

The most commonly used device model in circuit simulation is the MOSFET model, given the volume of integrated circuits that are designed using this workhorse device. MOS technology has undergone rapid advances that have enabled the semiconductor industry to keep pace with Moore's Law. As a result of these advances, MOSFET models have been an ongoing subject of research for the past two decades. Five versions of the BSIM model have been developed by a research team from UC Berkeley [19]. Many semiconductor companies have made impressive MOSFET modeling efforts including Texas Instruments, Motorola, Philips, and IBM to name a few.

Other research efforts including the EKV model [20] and the latest efforts on the surface potential (SP) model [21] have made impacts on device modeling. It is well beyond the scope of this section to delve into any significant details of these or any other leading-edge MOSFET model. Our intent here is merely to provide another example of how linear and nonlinear BCRs arise in device models and segue into how such nonlinearities are dealt with in simulators. For this purpose, we will illustrate a simple MOSFET model with nonlinear drain–source current and charges inspired by the work in [22].

The MOSFET equations for the three regions of drain–source current i_{DS} are given below:

$$(17.16) \quad i_{DS} = \begin{cases} 0, & v_{GS} - V_t \leq 0 \quad (\text{cut-off}) \\ \dfrac{\beta}{2}(v_{GS} - V_t)^2, & v_{DS} \geq V_{GS} - V_t > 0 \quad (\text{saturation}) \\ \beta\left[(v_{GS} - V_t)v_{DS} - \dfrac{1}{2}v_{DS}^2\right], & v_{GS} - V_t > V_{DS} \text{ and } v_{GS} - V_t > 0 \quad (\text{triode}) \end{cases}$$

where
  V_t is the effective threshold voltage
  $\beta = k'(W/L)(1 + \lambda v_{DS})$.

The governing equations for the bulk–drain and bulk–source junctions are similar to those provided in the previous section, but typically do not possess series resistance effects. The governing equations for the intrinsic MOSFET charge are taken from [22] over four regions of operation as shown below:

Below flatband; $V_{GB} \leq V_{fb}$:

$$Q_{gate} = C_{ox}(V_{GB} - V_{fb})$$

$$Q_{bulk} = -Q_{gate}$$

(17.17)

$$Q_{drain} = 0$$

$$Q_{source} = 0$$

Below threshold; $V_{GDt} \leq V_{GSt} \leq 0$:

$$Q_{gate} = C_{ox}\gamma\left(\sqrt{\dfrac{\gamma^2}{2} + v_{GB} - v_{fb}} - \dfrac{\gamma}{2}\right)$$

(17.18)

$$Q_{bulk} = -Q_{gate}$$

$$Q_{drain} = 0$$

$$Q_{source} = 0$$

Saturation; $V_{GDt} \leq 0$ and $V_{GSt} > 0$:

$$Q_{gate} = C_{ox}\left(\dfrac{2}{3}v_{GSt} + (V_t - V_{bi})\right)$$

$$Q_{bulk} = -C_{ox}(V_t - V_{bi})$$

(17.19)

$$Q_{drain} = -\dfrac{4}{15}C_{ox}\,V_{GSt}$$

$$Q_{source} = -\dfrac{2}{5}C_{ox}v_{GSt}$$

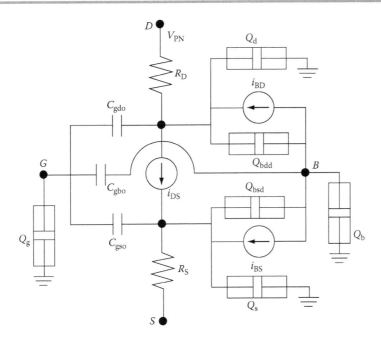

FIGURE 17.7   Large-signal model topology of a MOSFET.

Triode; $V_{GDt} \geq 0$ and $V_{GSt} > V_{GDt}$:

(17.20)

$$Q_{gate} = C_{ox}\left[V_t - V_{bi} + \frac{2}{3}\left(v_{GSDt} + v_{GSt} - \frac{v_{GSt}v_{GDt}}{v_{GSt} + v_{GDt}} \right) \right]$$

$$Q_{bulk} = -C_{ox}(V_t - V_{bi})$$

$$Q_{drain} = -\frac{C_{ox}}{3}\left[\frac{1}{5}v_{GDt} + \frac{4}{5}v_{GSt} + v_{GDt}\left(\frac{v_{GDt}}{v_{GSDt}} \right) + \frac{1}{5}\frac{v_{GSt}v_{GDt}(v_{GDt} - v_{GSt})}{v_{GSDt}^2} \right]$$

$$Q_{source} = -\frac{C_{ox}}{3}\left[\frac{1}{5}v_{GSt} + \frac{4}{5}v_{GDt} + v_{GSt}\left(\frac{v_{GSt}}{v_{GSDt}} \right) + \frac{1}{5}\frac{v_{GSt}v_{GDt}(v_{GSt} - v_{GDt})}{v_{GSDt}^2} \right]$$

In the above, $v_{GDt} = v_{GD} - V_t$, $v_{GSDt} = v_{GSt} - v_{GDt}$, $V_{bi} = V_{fb} + 2\phi_f$, and $C_{ox} = \varepsilon_{Si}WL/t_{ox}$.

The large-signal model topology of this simple MOSFET model is given in Figure 17.7. In addition to the drain-source current, body diodes, and nonlinear charges, some linear capacitances are shown as well as series resistances on the drain and source.

As in the case of the power diode, several additional effects must be taken into account when dealing with power MOSFET models. The device structure is very different in that power MOSFETs are typically vertical conduction devices, where the source and gate are on the top surface and the drain is on the back surface of the die or wafer. Many of the same issues arise as they did in the case of the power diode. Higher levels of injection, conductivity modulation of a drift layer, and moving-boundary conditions are just some of the major effects that must be modeled.

17.4 BASIC CIRCUIT SIMULATION: DC ANALYSIS

In Section 17.2, we noted how any circuit could be written as a set of nonlinear differential equations Equation 17.1, while in Section 17.3, we looked into the details of some of the more complex equations involved, notably those from semiconductor models. We now look at how to solve Equation 17.1 numerically for a basic design problem: finding the DC operating point of the circuit, i.e., solving the circuit when all voltages/currents within it remain constant with time.

Finding the DC operating point is a fundamental and basic requirement in the design of virtually every circuit. The voltages and currents in a circuit at its DC operating point provide important information about correct functioning of the circuit. The DC operating point is essential not only as a first basic check of circuit operation, but is also a prerequisite for further analyses. Small-signal AC analysis, transient analysis, noise analysis, etc., all rely on a prior DC operating point having been calculated; furthermore, the operating point is also useful for steady-state and envelope analyses.

When no quantity in Equation 17.1 is changing with time, the input $b(t)$ must be a constant vector b_{DC}; and the circuit variables $x(t)$ must similarly be unchanging, i.e., $x(t) \equiv x_{DC}$ When nothing changes with time, all derivatives become zero, hence Equation 17.1 becomes

$$(17.21) \qquad f(x_{DC}) = b_{DC}, \quad \text{or} \quad g(x_{DC}) \equiv f(x_{DC}) - b_{DC} = 0.$$

This is a *nonlinear equation system*, where the unknown to be solved for is the vector x_{DC}; solving such a system of equations is generally a nontrivial numerical task. In circuit simulation applications, a numerical method known as the Newton–Raphson algorithm is most often used to solve this system.

17.4.1 THE NEWTON–RAPHSON ALGORITHM FOR NONLINEAR EQUATION SOLUTION

The Newton–Raphson (NR) algorithm (e.g., [23]) is a technique that is widely used in many disciplines for finding solutions to nonlinear equations. It is an *iterative* method—that is to say, it starts from a guess regarding the solution x_0 of Equation 17.21, and keeps refining the guess until it finds a solution. It is important to note that there is no guarantee that NR will succeed in finding a solution, although it often does so in practice. Therefore, the algorithm is typically implemented to announce failure after some number of iterations (e.g., after 100 tries).

NR relies on having available to it two functionalities related to the nonlinear function $g(x)$ in Equation 17.21:

1. It needs a means to evaluate $g(x)$, given any x
2. It needs a means to evaluate the derivative of $g(x)$, i.e., $J(x) = dg(x)/dx$, given any x. Note that since both x and $g(x)$ are vectors of size n, J is an $n \times n$ matrix. The matrix J is known as the *Jacobian matrix* of $g(x)$.

The following is an outline of the basic NR algorithm, in MATLAB-like pseudo-code:
```
1. function solution = NR (xguess)
2. x = xguess;
3. finished = 0;
4. epsilon = 1e-10;
5. while (~finished) % iteration loop
6.    gx = g(x); % evaluate g(x)
7.    if (norm(gx) < epsilon)
8.        finished = 1; % solution to accuracy epsilon found
9.        solution = x;
10.       break;
11.   end % of if
12.   Jx = dgdx(x); % find the Jacobian matrix J(x)
13.   delta_x = -inverse(Jx)*gx; % calculate update delta_x
14.   %to current guess by solving J(x) delta_x = -g(x)
15.   x = x + delta_x; % update current guess; re-do loop
16. end % of while
```

Observe that the key step is line 13, where the current guess for the solution is updated by an amount Δx, calculated by solving

$$(17.22) \qquad J(x)\Delta x = -g(x) \quad \text{or} \quad \Delta x = -J^{-1}(x)g(x).$$

The above involves solving a *linear matrix equation*, with the Jacobian matrix $J(x)$, at every iteration of the NR algorithm. The computational ease with which this can be done depends on the structure of the Jacobian matrix, which we will discuss in more detail below.

17.4.2 DERIVATIVE ("JACOBIAN") MATRICES AND DEVICE "STAMPS"

To illustrate the structure of the Jacobian matrix, consider again the example of Figure 17.1, together with its $f(x)$ defined in Equation 17.8. The Jacobian matrix of $f(x)$ in Equation 17.8 is

$$(17.23) \qquad J(x) = \begin{pmatrix} d'(v_1 - v_2) & -d'(v_1 - v_2) & 1 \\ -d'(v_1 - v_2) & \dfrac{1}{R} + d'(v_1 - v_2) & 0 \\ 1 & 0 & 0 \end{pmatrix}$$

where $d'(\cdot)$ is the derivative of the diode's current–voltage relationship, defined in Equation 17.2. Observe, first, that each element in the circuit has a characteristic pattern of entries that it contributes to the Jacobian matrix. For example, the resistor R from node 2 to ground contributes the pattern

$$(17.24) \qquad \begin{pmatrix} \cdot & \cdot & \cdot \\ \cdot & 1/R & \cdot \\ \cdot & \cdot & \cdot \end{pmatrix}$$

while the diode between nodes 1 and 2 contributes the pattern

$$(17.25) \qquad \begin{pmatrix} d'(v_1 - v_2) & d'(v_1 - v_2) & \cdot \\ -d'(v_1 - v_2) & -d'(v_1 - v_2) & \cdot \\ \cdot & \cdot & \cdot \end{pmatrix}$$

The pattern of Jacobian entries contributed by each element is called its *matrix stamp*. The Jacobian matrix $J(x)$ thus consists of the addition of the stamps of all the elements in the circuit.

17.4.3 JACOBIAN SPARSITY AND ITS IMPORTANCE

Observe also that several entries in the Jacobian matrix in Equation 17.23 are zero. As the size of the circuit grows larger, the number of zero entries in the matrix predominates. The reason for this is easy to see—if the circuit has roughly n nodes and each node is connected to only 2 or 3 circuit elements (as is typical), the total number of nonzero matrix stamps is of the order of $2n$ or $3n$. As a result, the remaining entries of the matrix (which has a total of n^2 entries) must be zero or "unstamped". Such matrices, where there are relatively few nonzero entries and many zero entries, are called *sparse matrices*. When a matrix has no zero entries, it is called *dense*. We have just noted that because each node in typical circuits is connected only to a few elements, Jacobian matrices of circuits tend to be sparse.

The fact that circuit matrices are usually sparse is of enormous importance for circuit simulation. The reason is that, while the Newton update step in Equation 17.22 is in general very computationally expensive for arbitrary matrices J (especially if J is dense), efficient techniques, collectively dubbed "sparse matrix technology," exist for solving for Δx in Equation 17.22 when J is sparse. More specifically, the computational complexity of general linear equation solution (e.g., when J is dense) is $O(n^3)$, while that for typical sparse J is $O(n)$. When the circuit size n reaches the thousands or tens of thousands, this difference in computational complexity is extremely significant from a practical standpoint. Thus, sparsity of circuit matrices is critical for enabling efficient and practically effective linear circuit solutions; so far, we have noted the importance of this linear solution only for DC analysis (as part of the NR loop in line 13), but we will see that exploiting sparsity is critical for virtually all other analyses related to circuits.

17.5 STEADY-STATE ANALYSIS

It is often important in RF design to find the periodic steady state of a circuit driven by one or more periodic inputs. For example, a power amplifier driven to saturation by a large single-tone input is operating in a periodic steady state. A variant is the quasiperiodic steady state, i.e., when the circuit is driven by more than one signal tone; for example, an amplifier driven by two closely spaced sinusoidal tones at 1 GHz and 990 MHz. Such excitations are closer approximations of real-life signals than pure tones and are useful for estimating intermodulation distortion.

The workhorse of analog verification, SPICE (and its derivatives), can of course be applied to find the (quasi) periodic steady state of a circuit, simply by performing a time-stepping integration of the circuit's differential equations ("transient simulation") long enough for the transients to subside and the circuit's response to become (quasi)periodic. This approach has several disadvantages, however. In typical RF circuits, the transients take thousands of periods to die out, and hence the procedure can be very inefficient. Further, harmonics are typically orders of magnitude smaller than the fundamental, hence long transient simulations are not well suited for their accurate capture, because numerical errors from time-stepping integration can mask them. These issues are exacerbated in the presence of quasiperiodic excitations, because simulations need to be much longer—e.g., for excitations of 1 GHz and 990 MHz, the system needs to be simulated for thousands of multiples of the common period, 1/10 MHz, yet the simulation time-steps must be much smaller than 1 ns, the period of the fastest tone. For these reasons, more efficient and accurate specialized techniques have been developed. We will focus on two different methods with complementary properties, Harmonic Balance (HB) and shooting.

17.5.1 HARMONIC BALANCE AND SHOOTING

In the well-known method of HB (e.g., [24–32]), $x(t)$ and $b(t)$ of Equation 17.1 are expanded in a Fourier series. The Fourier series can be one-tone for periodic excitations (e.g., 1 GHz and its harmonics) or multitone in the case of quasiperiodic excitations (e.g., 1 GHz and 990 MHz, their harmonics, and intermodulation mixes). The DAE is rewritten directly in terms of the Fourier coefficients of $x(t)$ (which are unknown) and of $b(t)$; the resulting system of nonlinear equations is larger by a factor of the number of harmonic/mix components used, but are algebraic (i.e., there are no differential components). Hence they can be solved numerically using, e.g., the well-known NR method [23].

Example 17.5.1

We illustrate the one-tone procedure with the following scalar DAE:

$$(17.26) \qquad \dot{x} + x - \varepsilon x^2 - \cos(2\pi 1000t) = 0$$

First, we expand all time variations in a Fourier series of (say) $M = 3$ terms, i.e., the DC component, fundamental, and second harmonic components:

$$x(t) = \sum_{i=-2}^{2} X_i e^{j2\pi i 10^3 t}$$

Here, X_i are the unknown Fourier coefficients of $x(t)$. For notational convenience, we express them as the vector $X = [X_2, \ldots, X_{-2}]^T$.

Similarly, $x^2(t)$ is also expanded in a Fourier series in t; the Fourier coefficients of this expansion are functions of the elements of X, which we denote by $F_i(X)$.

$$x^2(t) = \sum_{i=-2}^{2} \sum_{k=-2}^{2} X_i X_k e^{j2\pi(i+k)10^3 t} = \sum_{i=-2}^{2} F_i(X) e^{j2\pi i 10^3 t} + \text{higher terms}$$

In this case, where the nonlinearity is a simple quadratic, F_i can be obtained analytically; but in general, numerical techniques like those used in HB need to be employed for computing these functions. For convenience, we write $F(X) = [F_2(X), \ldots, F_{-2}(X)]^T$

We also write the Fourier coefficients of the excitation $\cos(2\pi 1000 t)$ as a vector $B = [0, 1/2, 0, 1/2, 0]^T$. Finally, we write the differential term x also as a vector of Fourier coefficients. Because the differentiation operator is diagonal in the Fourier basis, this becomes simply ΩX, where $\Omega = j2\pi 1000 \ \text{diag}(2,1,0,-1-2)$ is the diagonal frequency-domain differentiation matrix.

Invoking the orthogonality of the Fourier basis, we now obtain the HB equations for our DAE:

$$H(X) \equiv \Omega X + X - \varepsilon F(X) - B = 0$$

This is a set of nonlinear algebraic equations in five unknowns, and can be solved by numerical techniques such as NR.

The above example illustrates that the size of the HB equations is larger than that of the underlying DAE, by a factor of the number of harmonic/mix components used for the analysis. In fact, the HB equations are not only larger in size than the DAE, but also considerably more difficult to solve using standard numerical techniques. The reason for this is the dense structure of the derivative, or Jacobian matrix, of the HB equations. If the size of the DAE is n, and a total of N harmonics and mix components are used for the HB analysis, the Jacobian matrix has $Nn \times Nn$. Just the storage for the nonzero entries can become prohibitive for relatively moderate values of n and N; for example, $n = 1000$ (for a medium-sized circuit) and $N = 100$ (e.g., for a two-tone problem with about 10 harmonics each) require 10 GB of storage for the matrix alone. Further, inverting the matrix, or solving linear systems with it, requires $O(N^3 n^3)$ operations, which is usually infeasible for moderate to large-sized problems. Such linear solutions are typically required as steps in solving the HB equations, for example by the NR method. Despite this disadvantage, HB is a useful tool for small circuits and few harmonics, especially for microwave circuit design. Moreover, as we will see later, new (as of 2006) algorithms have been developed for HB that make it much faster for larger problems.

Another technique for finding periodic solutions is the shooting method (e.g., [33–36]). Shooting works by finding an initial condition for the DAE that also satisfies the periodicity constraint. A guess is made for the initial condition, the system is simulated for one period of the excitation using time—stepping DAE solution methods, and the error from periodicity used to update the initial condition guess, often using an NR scheme.

More precisely, shooting computes the *state transition function* $\Phi(t, x_0)$ of Equation 17.1. $\Phi(t, x_0)$ represents the solution of the system at time t, given initial condition x_0 at time 0. Shooting finds an initial condition x^* that leads to the same state after one period T of the excitation $b(t)$; in other words, shooting solves the equation $H(x) \equiv \Phi(t, x) - x = 0$.

The shooting equation is typically solved numerically using the NR method, which requires evaluations of $H(x)$ and its derivative (or Jacobian) matrix. Evaluation of $H(x)$ is straightforward using time-stepping, i.e., transient simulation, of Equation 17.1. However, evaluating its Jacobian is more involved. The Jacobian matrix is of the same size n as the number of circuit equations, but it is dense, hence storage of its elements and linear solutions with it is prohibitive in cost for large problems. In this respect, shooting suffers from size limitations similar to HB. In other respects though, shooting has properties complementary to HB. The following list compares and contrasts the main properties of HB and shooting:

- *Problem size*: The problem size is limited for both HB and shooting, due to the density of their Jacobian matrices. However, since the HB system is larger by a factor of the number of harmonics used, shooting can handle somewhat larger problems given the same resources. Roughly speaking, sizes of about 40 circuit elements for HB and 400 for shooting represent practical limits.
- *Accuracy/dynamic range*: Because HB uses orthogonal Fourier bases to represent the waveform, it is capable of very high dynamic range—a good implementation can deliver 120 dB of overall numerical accuracy. Shooting, being based on time-stepping solution of the DAE with time-steps of different sizes, is considerably poorer in this regard.
- *Handling of nonlinearities:* HB is not well suited for problems that contain strongly non-linear elements. The main reason for this is that strong nonlinearities (e.g., clipping elements) generate sharp waveforms that do not represent compactly in a Fourier series basis. Hence many harmonics/ mix components need to be considered for an accurate simulation, which raises the overall problem size.

   Shooting, on the other hand, is well suited for strong nonlinearities. By approaching the problem as a series of initial value problems for which it uses time-stepping DAE methods, shooting is able to handle the sharp waveform features caused by strong non-linearities quite effectively.
- *Multitone problems:* A big attraction of HB is its ability to handle multitone or quasiperiodic problems as a straightforward extension of the one-tone case, by using multitone Fourier bases to represent quasiperiodic signals. Shooting, on the other hand, is limited in this regard. Since it uses time-stepping DAE solution, shooting requires an excessive number of timepoints when the waveforms involved have widely separated rates of change; hence it is not well suited for such problems.

17.5.2 FAST METHODS

A disadvantage of both HB and shooting is their limitation to circuits of relatively small size. This was not a serious problem as long as microwave/RF circuits contained only a few nonlinear devices. Since the mid-1990s however, economic and technological developments have changed this situation. The market for cheap, portable wireless communication devices has expanded greatly, leading to increased competition and consequent cost pressures. This has spurred on-chip integration of RF communication circuits and the reduction of discrete (off-chip) components. On-chip design techniques favor the use of many integrated nonlinear transistors over even a few linear external components. Hence the need has arisen to apply HB and shooting to large circuits in practical times.

To address this issue, so-called *fast* algorithms have arisen to enable both HB and shooting to handle large circuits. The key property of these methods is that computation/memory usage grows approximately linearly with problem size. The enabling idea behind the improved speed is to express the dense Jacobian matrices as sums and products of simpler matrices that are either sparse, or have very regular structure and so can be applied/inverted efficiently. Using these expansions for the Jacobian, special solution algorithms called *preconditioned iterative linear solvers* are applied to solve linear equations involving the Jacobian, without forming it explicitly.

A detailed description of fast techniques is beyond the scope of this chapter; the interested reader is referred to [29,31,32,36,37] for further information. Here we outline the main ideas behind these methods in a simplified form using Example 17.5.1 for illustration, and summarize their main properties.

From Example 17.5.1, the Jacobian matrix of the HB system is

$$J = \frac{\partial H(X)}{\partial X} = \Omega + I - \varepsilon \frac{\partial F(X)}{\partial X}$$

Now, $F(X)$ in this case represents the vector of Fourier coefficients of the nonlinear term $f(x) = x^2$. One way in which these can be computed numerically is (1) use the inverse Fast Fourier Transform (FFT) to convert the Fourier coefficients X into samples of the time-domain waveform $x(t)$, then (2) evaluate the nonlinear function $f(x) = x^2(t)$ at each of these samples in the time domain, and finally, (3) use the FFT to reconvert the time-domain samples of $f(x)$ back to the frequency domain, to obtain $F(X)$. The derivative of these operations can be expressed as

$$\frac{\partial F}{\partial X} = DGD^*$$

where

 D is a block-diagonal matrix with each block equal to the discrete Fourier transform (DFT) matrix
 D^* is its inverse
 G is a diagonal matrix with entries $f'()$ evaluated at the time-domain samples of $x(t)$

Hence the overall Jacobian matrix can be represented as

$$J = \Omega + I - \varepsilon DGD^*$$

Observe that each of the matrices in this expansion is either sparse or consists of DFT matrices. Hence multiplication of J with a vector is efficient, since the sparse matrices can be applied in approximately linear time, and the DFT matrix and its inverse can be applied in $N \log N$ time using the FFT, where N is the number of harmonics. It is this key property, that multiplications of J with a vector can be performed in almost-linear computation despite its dense structure, that enables the use of preconditioned iterative linear methods for this problem.

Preconditioned iterative linear methods (e.g., [38–40]) are a set of numerical techniques for solving linear systems of the form $Jc = d$. Modern iterative solvers like QMR [39] and GMRES [38] use Krylov-subspace techniques for superior performance. The key feature of these solvers is that the only way in which J is used is in matrix–vector products Jz. This constrasts with traditional methods for linear solution that use Gaussian elimination or variants like LU factorizations directly on elements of J. Due to this property of iterative linear solvers, it is not necessary to even form J explicitly in order to solve linear systems with it, so long as a means is available for computing matrix–vector products with it.

As we have seen above, products with the HB Jacobian can be conveniently computed in almost-linear time without having to build the matrix explicitly. Hence preconditioned linear iterative techniques are well suited to solving the linear systems that arise when solving the nonlinear HB equations using the NR method. If the iterative linear methods use only a few matrix–vector products with the Jacobian to compute the linear system's solution, and the NR is well behaved, the overall cost of solving the HB equations remains almost linear in problem size.

An important issue with preconditioned iterative linear solvers, especially those based on Krylov subspace methods, is that they require a good *preconditioner* to converge reliably in a few iterations. The convergence of the iterative linear method is accelerated by applying a preconditioner $\tilde{J}$, replacing the original system $Jc = d$ with *the preconditioned* system $\tilde{J}^{-1}Jc = \tilde{J}^{-1}d$, which has the same solution. For robust and efficient convergence, the preconditioner matrix $\tilde{J}$ should be in some sense a good approximation of J, and also "easy" to invert, usually with a direct method such as LU factorization. Finding good preconditioners that work well for a wide variety of circuits is a challenging task, especially when the nonlinearities become strong.

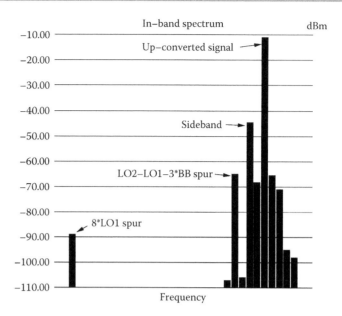

FIGURE 17.8   Quadrature modulator spectrum.

The ideas behind the fast techniques outlined above are applicable not just to HB but also to shooting [36]. Jacobian matrices in to-from shooting can be decomposed into products and sums of the sparse circuit Jacobian matrices. Preconditioned linear iterative techniques can then be applied to invert the Jacobian efficiently.

As an example of the application of the fast methods, consider the HB simulation of an RFIC quadrature modulator reported in [32]. The circuit of about 9500 devices was simulated by fast HB: a three-tone excitation, a baseband signal at 80 kHz, and local oscillators at 178 MHz and 1.62 GHz. The size of the circuit's DAE was $n = 4800$; the three tones, their harmonics, and mixes totalled $N = 4320$ components. Simulating a circuit with these specifications is completely infeasible using traditional HB techniques.

Using fast HB, the simulation required only 350 MB of memory, and took 5 days of computation on an SGI 150 MHz R4400 machine. The results of the simulation are shown in Figure 17.8.

17.6 MULTITIME ANALYSIS

In the previous section, we noted that HB and shooting have complementary strengths and weaknesses, stemming from their use of Fourier and time-domain bases respectively. While HB is best suited for multitone problems that are only mildly nonlinear, shooting is best for single-tone problems that can be strongly nonlinear. Neither method is suitable for circuits that have both multitone signals (i.e., widely separated time scales of variation) *and* strongly nonlinear components. With greater RF integration, tools that can analyze precisely this combination of circuit characteristics effectively are required. In this section, we review a promising family of techniques, based on partial differential equations (PDEs) using multiple artificial time scales [30,41–45].

Consider the waveform $y(t)$ shown in Figure 17.9, a simple two-tone signal given by

(17.27)
$$y(t) = \sin\left(\frac{2\pi}{T_1}t\right)\sin\left(\frac{2\pi}{T_2}t\right), T_1 = 0.02 \text{ s}, T_2 = 1 \text{ s}$$

The two tones are at frequencies $f_1 = 1/T_1 = 50$ Hz and $f_2 = 1/T_2 = 1$ Hz, i.e., there are 50 fast-varying cycles of period $T_1 = 0.02$ s modulated by a slowly varying sinusoid of period $T_2 = 1$ s. If each fast cycle is sampled at n points, the total number of time steps needed for one period of the slow modulation is nT_2/T_1 To generate Figure 17.9, 15 points were used per cycle, hence the total

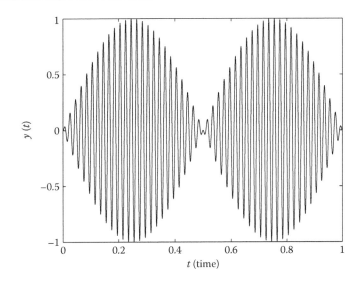

FIGURE 17.9 Example of a two-tone quasiperiodic signal $y(t)$.

number of samples was 750. This number can be much larger in applications where the rates are more widely separated, e.g., separation factors of 1000 or more are common in electronic circuits. Now consider a multivariate representation of $y(t)$ using two artificial timescales, as follows: for the "fast-varying" parts of $y(t)$, t is replaced by a new variable t_1; for the "slowly varying" parts, by t_2. The resulting function of two variables is denoted by

(17.28)
$$\hat{y}(t_1,t_2) = \sin\left(\frac{2\pi}{T_1}t_1\right)\sin\left(\frac{2\pi}{T_2}t_2\right)$$

The plot of $\hat{y}(t_1,t_2)$ on the rectangle $0 \le t_1 \le T_1$, $0 \le t_2 \le T_2$ shown in Figure 17.10. Observe that $\hat{y}(t_1,t_2)$ does not have many undulations, unlike $y(t)$ in Figure 17.9. *Hence it can be represented by relatively few points, which, moreover, do not depend on the relative values of T_1 and T_2.* Figure 17.10 was plotted with 225 samples on a uniform 15×15 grid—three times fewer than for Figure 17.9. This saving increases with increasing separation of the periods T_1 and T_2.

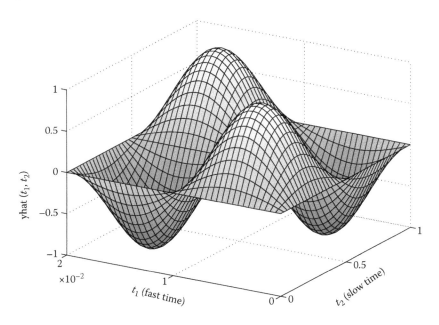

FIGURE 17.10 Corresponding two-periodic bivariate form $\hat{y}(t_1, t_2)$.

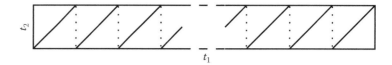

FIGURE 17.11 Path in the t_1–t_2 plane.

Further, note that $\hat{y}(t_1,t_2)$ is periodic with respect to both t_1 and t_2, i.e., $\hat{y}(t_1 + T_1, t_2 + T_2) = \hat{y}(t_1,t_2)$. This makes it easy to recover $y(t)$ from $\hat{y}(t_1,t_2)$, simply by setting $t_1 = t_2 = t$, and using the fact that $\hat{y}$ is biperiodic. It is easy, from direct inspection of the three-dimensional plot of $\hat{y}(t_1,t_2)$, to visualize what $y(t)$ looks like. As t increases from 0, the path given by $\{t_i = t \bmod T_i\}$ traces the sawtooth path shown in Figure 17.11. By noting how $\hat{y}$ changes as this path is traced in the t_1–t_2 plane, $y(t)$ can be traced.

When the time scales are widely separated, inspection of the bivariate waveform directly provides information about the slow and fast variations of $y(t)$ more naturally and conveniently than $y(t)$ itself.

We observe that the bivariate form may require far fewer points to represent numerically than the original quasiperiodic signal, yet it contains all the information needed to recover the original signal completely. This observation is the basis of the partial differential formulation to be introduced shortly. The waveforms in a circuit are represented in their bivariate forms (or multivariate forms if there are more than two timescales). The key to efficiency is to solve for these waveforms directly, without involving the numerically inefficient one-dimensional forms at any point. To do this, it is necessary to first describe the circuit's equations using the multivariate functions. If the circuit is described by the differential Equation 17.1, then it can be shown that if $\hat{x}(t_1,t_2)$ and $\hat{b}(t_1,t_2)$ denote the bivariate forms of the circuit unknowns and excitations, then the following MPDE is the correct generalization of Equation 17.1 to the bivariate case:

$$(17.29) \qquad \frac{\partial q(\hat{x})}{\partial t_1} + \frac{\partial q(\hat{x})}{\partial t_2} + f(\hat{x}) = \hat{b}(t_1,t_2)$$

More precisely, if $\hat{b}$ is chosen to satisfy $b(t) = \hat{b}(t,t)$, and x satisfies Equation 17.29, then it can be shown that $x(t) = \hat{x}(t,t)$ satisfies Equation 17.1. Also, if Equation 17.1 has a quasiperiodic solution, then Equation 17.29 can be shown to have a corresponding bivariate solution.

By solving the MPDE numerically in the time domain, strong nonlinearities can be handled efficiently. Several numerical methods are possible, including discretization of the MPDE on a grid in the t_1–t_2 plane, or using a mixed time–frequency method in which the variation along one of the timescales is expressed in a short Fourier series. Quasiperiodic and envelope solutions can both be generated, by appropriate selection of boundary conditions for the MPDE. Sparse matrix and iterative linear methods are used to keep the numerical algorithms efficient even for large systems.

As an example, Figure 17.12 depicts the output voltage of a switched-capacitor integrator block, obtained from a multitime simulation based on the above concepts. The cross-section parallel to the signal timescale represents the envelope of the signal riding on the switching variations. By moving these cross-sections to different points along the clock timescale, the signal envelope at different points of the clock waveform can be seen.

17.6.1 AUTONOMOUS SYSTEMS: THE WARPED MPDE

When the DAEs under consideration are oscillatory, frequency modulation (FM) can be generated. Unfortunately, FM cannot be represented compactly using multiple timescales as easily as the waveform in Figure 17.10. We illustrate the difficulty with an example. Consider the following prototypical FM signal:

$$(17.30) \qquad x(t) = \cos(2\pi f_0 t + k\cos(2\pi f_2 t)), \qquad f_0 \gg f_2$$

with instantaneous frequency

$$(17.31) \qquad f(t) = f_0 - k f_2 \sin(2\pi f_2 t)$$

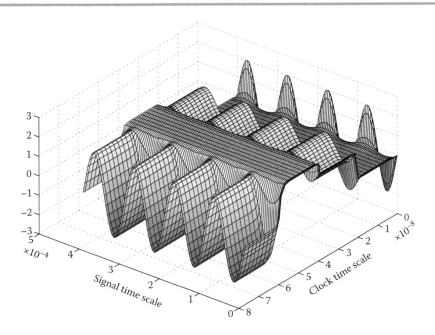

FIGURE 17.12   Multitime output waveform of SC integrator.

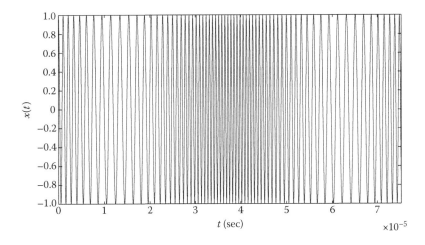

FIGURE 17.13   FM signal.

$x(t)$ is plotted in Figure 17.13 for $f_0 = 1$ MHz, $f_2 = 20$ kHz, and modulation index $k = 8\pi$. Following the same approach as for Equation 17.27, a bivariate form can be defined to be

$$(17.32) \qquad \hat{x}(t_1, t_2) = \cos(2\pi f_0 t_1) + k\cos(2\pi f_2 t_2)), \qquad \text{with } x(t) = \hat{x}_1(t, t).$$

Note that $\hat{x}_1$ is periodic in t_1 and t_2, hence $x(t)$ is quasiperiodic with frequencies f_0 and f_2. Unfortunately, $\hat{x}(t_1, t_2)$, illustrated in Figure 17.14, is not a simple surface with only a few undulations like in Figure 17.10. When $k \gg 2\pi$, i.e., $k \approx 2\pi m$ for some large integer m, then $\hat{x}(t_1, t_2)$ will undergo about m oscillations as a function of t_2 over one period T_2. In practice, k is often of the order of $f_0/f_2 \gg 2\pi$, and hence this number of undulations can be very large. Therefore, it becomes difficult to represent $\hat{x}_1$ efficiently by sampling on a two-dimensional grid. It turns out that resolving this problem requires the stretching, or warping of one of the timescales. We illustrate this by returning to Equation 17.30. Consider the following new multivariate representation:

$$(17.33) \qquad \hat{x}_2(\tau_1, \tau_2) = \cos(2\pi\tau_1),$$

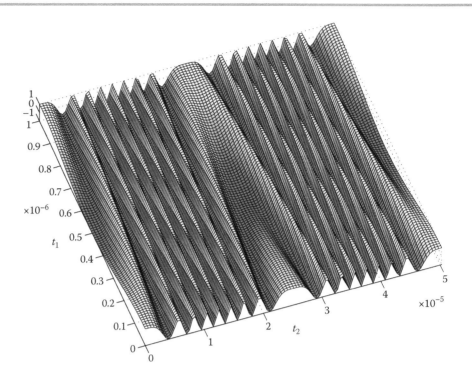

FIGURE 17.14 $\hat{x}_1$ unwarped bivariate representation of FM signal.

together with the warping function

$$(17.34) \qquad \phi(\tau_2) = f_0\tau_2 + \frac{k}{2\pi}\cos(2\pi f_2\tau_2).$$

We now retrieve our one-dimensional FM signal i.e., Equation 17.30 as

$$(17.35) \qquad x(t) = \hat{x}_2(\phi(t), t).$$

Note that both $\hat{x}_2$ and ϕ, given in Equations 17.33 and 17.34, can easily be represented with relatively few samples, unlike $\hat{x}_1$ in Equation 17.32. What we have achieved with Equation 17.34 is simply a stretching of the time axis differently at different times, to even out the period of the fast undulations in Figure 17.13. The extent of the stretching or the derivative of $\phi(\tau_2)$ at a given point is simply the local frequency $\omega(\tau_2)$, which modifies the original MPDE to result in the warped multirate partial differential equation (WaMPDE)

$$(17.36) \qquad \omega(\tau_2)\frac{\partial q(\hat{x})}{\partial \tau_1} + \frac{\partial q(\hat{x})}{\partial \tau_2} + f(\hat{x}(\tau_1, \tau_2)) = b(\tau_2).$$

The usefulness of Equation 17.36 lies in that specifying

$$(17.37) \qquad x(t) = \hat{x}(\phi(t), t), \phi(t) = \int_0^t \omega(t_2)\,\mathrm{d}\tau_2$$

results in $x(t)$ being a solution to Equation 17.1. Furthermore, when Equation 17.36 is solved numerically, the local frequency $\omega(\tau_2)$ is also obtained, which is desirable for applications such as VCOs and also difficult to obtain by any other means.

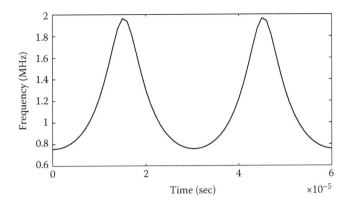

FIGURE 17.15 VCO: frequency modulation.

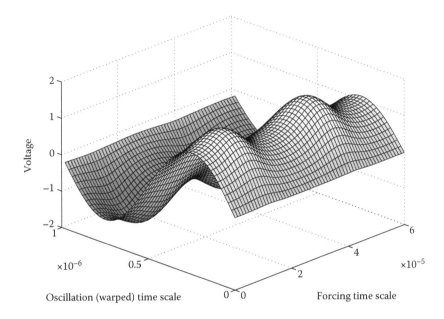

FIGURE 17.16 VCO: bivariate representation of capacitor voltage.

As an example, Figure 17.15 shows the changing local frequency in an LC tank VCO simulated with WaMPDE-based numerical techniques. The controlling input to the VCO was about 30 times slower than its nominal frequency. Figure 17.16 depicts the bivariate waveform of the capacitor voltage. It is seen that the controlling voltage changes not only the local frequency, but also the amplitude and shape of the oscillator waveform.

The circuit was also simulated by traditional numerical ODE methods ("transient simulation"). The waveform from this simulation, together with the one-dimensional waveform obtained by applying Equation 17.37 to Figure 17.16, are shown in Figure 17.17. Frequency modulation can be observed in the varying density of the undulations.

17.6.2 MACROMODELING TIME-VARYING SYSTEMS

Another useful application of multiple time scales is in macromodeling linear time-varying (LTV) systems equation [46,47]. These approximations are adequate for many apparently nonlinear systems, like mixers and switched-capacitor filters, where the signal path is designed to be linear, even though other inputs (e.g., local oscillators, clocks) cause "nonlinear" parametric changes to the system. LTV approximations of large systems with few inputs and outputs are particularly useful, because it is possible to automatically generate *macromodels* or *reduced-order*

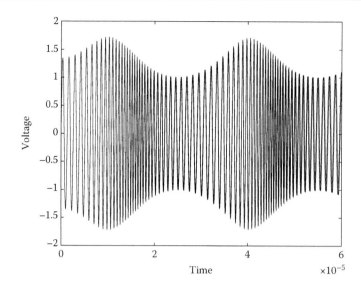

FIGURE 17.17   VCO: WaMPDE vs. transient simulation.

models of such systems. The macromodels are much smaller dynamical systems than the originals, but retain similar input–output behavior within a given accuracy. Such macromodels are useful in verifying systems hierarchically at different levels of abstraction, an important task in communication system design.

While mature techniques are available for the simpler task of reduced-order modeling of linear time-invariant (LTI) systems (e.g., [48–54]), a difficulty in extending them to handle LTV systems has been the interference of the time variations of the system and the input. By separating the two with artificial time variables, the MPDE provides an elegant solution to this problem.

The time-varying small-signal equations obtained by linearizing Equation (17.1) around a steady-state solution are given by

(17.38)
$$C(t)\dot{x}(t) + G(t)x(t) = ru(t)$$
$$y(t) = d^{\mathrm{T}}x(t)$$

In Equation 17.38, the input to the system is the scalar $u(t)$ while the output is $y(t)$. If the above equation is Laplace-transformed (following the LTI procedure), the system time variation in $C(t)$ and $G(t)$ interferes with the I/O time variation through a convolution. The LTV transfer function $H(t,s)$ is therefore hard to obtain; this is the difficulty alluded to earlier. The problem can be avoided by casting Equation 17.38 as an MPDE:

(17.39)
$$C(t_1)\left[\frac{\partial \hat{x}}{\partial t_1}(t_1,t_2) + \frac{\partial \hat{x}}{\partial t_2}(t_1,t_2)\right] + G(t_1)\hat{x}(t_1,t_2) = ru(t_2)$$
$$\hat{y}(t_1,t_2) = d^{\mathrm{T}}\hat{x}(t_1,t_2), \quad y(t) = \hat{y}(t,t)$$

Notice that the input and system time variables are now separated. By taking Laplace transforms in t_2 and eliminating x, the time-varying transfer function $H(t_1,s)$ is obtained:

(17.40)
$$Y(t_1,s) = \underbrace{\left\{d^{\mathrm{T}}\left[C(t_1)\left\{\frac{\partial}{\partial t_1} + s\right\} + G(t_1)\right]^{-1}[r]\right\}}_{H(t_1,s)}U(s)$$

Observe that $H(t_1, s)$ in Equation 17.40 is periodic in t_1; hence, discretizing the t_1 axis, it can also be represented as *several* time-invariant transfer functions $H_i(s) = H(t_{1i},s)$. Or, a frequency-domain

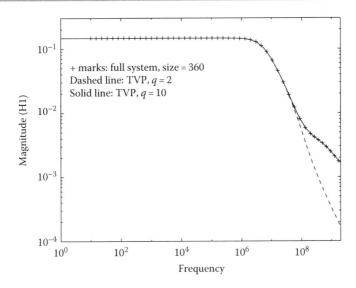

FIGURE 17.18 I-channel mixer $H_1(s)$: reduced vs. full system.

discretization using harmonics of the t_1 variation can be used. Once an equivalent system of LTI transfer functions has been obtained, existing reduced-order modeling techniques for LTI systems can be used to find a smaller system of equations, in the same form as Equation 17.39, that has the same input–output relationship to within a given accuracy.

The reduced-order modeling technique (dubbed *time-varying Padé*, or TVP) was run on an RFIC I-channel mixer circuit of size about $n = 360$ nodes, excited by a local oscillator at 178 Mhz [47]. A frequency-domain discretization of the t_1 axis in Equation 17.39 was employed in the model reduction process. Figure 17.18 shows frequency plots of $H_1(s)$, the up-conversion transfer function (the first harmonic w.r.t t_1 of $H(t_1, s)$). The points marked "1" were obtained by direct computation of the full system, while the lines were computed using the reduced models of size $q = 2$ and 10, respectively.* Even with $q = 2$, a size reduction of two orders of magnitude, the reduced model provides a good match up to the LO frequency. When the order of approximation is increased to 10, the reduced model is identical up to well beyond the LO frequency. The reduced models were more than three orders of magnitude faster to evaluate than the original system, and hence are useful for system-level verification.

The poles of the reduced models for $H_1(s)$, easily calculated on account of their small size, are shown in Table 17.1. These are useful in design because they constitute an excellent approximations of the full system's poles, which are difficult to determine otherwise.

TABLE 17.1   Poles of $H_1(s)$ for the I-Channel Buffer/Mixer

TVP, $q = 2$	TVP, $q = 10$
−5.3951 e + 06	−5.3951 e + 06
−6.9196e + 07 − j3.0085e + 05	−9.4175e + 06
	−1.5588e + 07 − j2.5296e + 07
	−1.5588e + 07 + j2.5296e + 07
	−6.2659e + 08 − j1.6898e + 06
	−1.0741e + 09 − j2.2011e + 09
	−1.0856e + 09 + j2.3771e + 09
	−7.5073e + 07 − j1.4271e + 04
	−5.0365e + 07 + j1.8329e + 02
	−5.2000e + 07 + j7.8679e + 05

---

* The order q of the reduced model is the number of state variables in its differential equation description.

17.7 NOISE IN RF DESIGN

Predicting noise correctly in order to minimize its impact is central to RF design. Traditional circuit noise analysis is based on three assumptions: that noise sources and their effects are *small enough* not to change the operating point; that all noise sources are *stationary*; and that the small-signal linearization of the circuit is *time invariant.* These assumptions break down when there are large signal variations, as is typical in RF circuits. Because of changing operating points, small-signal linearizations do not remain constant but become *time-varying.* In addition, noise sources that depend on operating point parameters (such as shot noise and flicker noise) also vary with time and no longer remain stationary. Finally, even though noise sources remain small, their impact upon circuit operation may or may not. In nonautonomous (driven) circuits, circuit effects of small noise remain small, allowing the use of linearized *mixing noise* analysis. In autonomous circuits (oscillators), however, noise creates frequency changes that lead to large deviations in waveforms over time—this phenomenon is called *phase noise.* Because of this, analysis based on linearization is not correct, and nonlinear analysis is required.

Figure 17.19 illustrates mixing noise. A periodic noiseless waveform in a circuit is shown as a function of time. The presence of small noise corrupts the waveform, as indicated. The extent of corruption at any given time remains small, as shown by the third trace, which depicts the difference between the noiseless and noisy waveforms. The noise power can, however, vary depending on the large signal swing, as indicated by the roughly periodic appearance of the difference trace—depicting *cyclostationary* noise, where the statistics of the noise are periodic.

Figure 17.20 illustrates oscillator phase noise. Note that the noisy waveform's frequency is now slightly different from that of that of the noise-free one leading to increasing deviations between the two with the progress of time. As a result, the difference between the two does not remain small, but reaches magnitudes of the order of the large signal itself. Small additive corruptions remain here also just as in the mixing noise case, but the main distinguishing characteristic of oscillator noise is the frequency deviation.

The difference between mixing and phase noise is also apparent in the frequency domain, shown in Figure 17.21. Noise-free periodic waveforms appear as the impulse in the upper graph. If this is corrupted by small mixing noise, the impulse is not modified, but a small possibly broadband noise floor appears. In the case of free-running oscillators, the impulse disappears in the presence of any noise, no matter how small. It is replaced by a continuous spectrum that peaks at the oscillation frequency, and retains the power of the noise-free signal. The width and shape of this *phase noise spectrum* (i.e., the spread of power over neighboring frequencies) is related to the amount and nature of noise in the circuit.

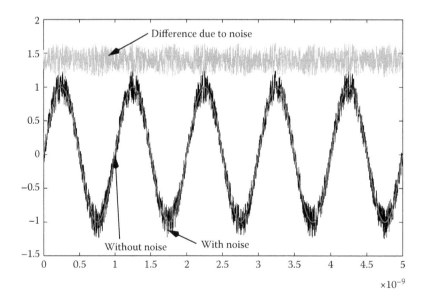

FIGURE 17.19   Time-domain view of mixing noise.

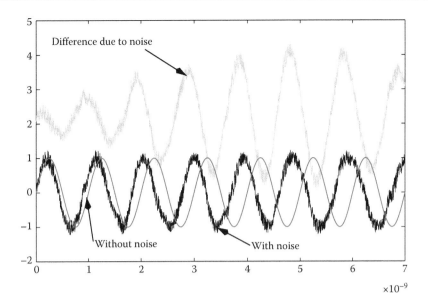

FIGURE 17.20 Time-domain view of phase noise.

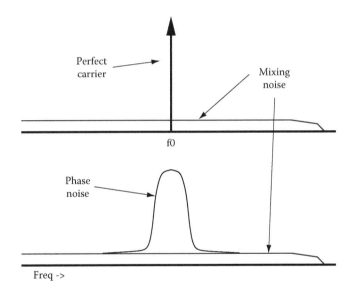

FIGURE 17.21 Frequency-domain view of mixing and phase noise.

17.7.1 MIXING NOISE

Correct calculation of noise in nonlinear circuits with large signal swings (e.g., mixers and gain—compressed amplifiers) requires a sufficiently powerful stochastic process model. In the following, we use cyclostationary time-domain processes (e.g., [55–58]), although a different but equivalent formulation, i.e., that of correlated processes in the frequency domain (e.g., [27,59]), is often used. The statistics of cyclostationary processes (in particular, the second-order statistics) are periodic or quasiperiodic, and hence can be expressed in Fourier series. The coefficients of the Fourier series, termed *cyclostationary components*, capture the variations of noise power over time. The DC term of the Fourier series, or the *stationary* component, is typically the most relevant for design, since it captures the average noise power over a long time. It is important to realize, though, that calculating the correct value of the stationary component of noise over a circuit does require *all* the Fourier components to be properly accounted for. Basing calculations only on the stationary component at each node or branch current in the circuit will, in general,

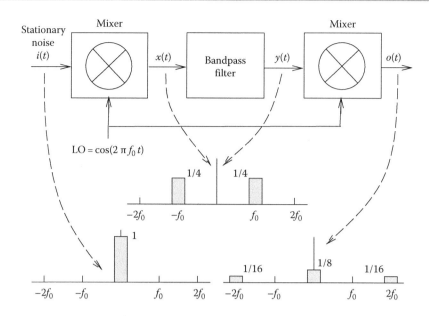

FIGURE 17.22   Mixer–filter–mixer circuit: naïve analysis.

produce wrong results. This is analogous to computing the DC term of the product of two sinusoidal waveforms by simply multiplying the DC terms of each.

We highlight the need for cyclostationary analysis with an example. The circuit of Figure 17.22 consists of a mixer, followed by a bandpass filter, followed by another mixer. This is a simplification of, for example, the bias-dependent noise generation mechanism in semiconductor devices [60]. Both mixers multiply their inputs by a local oscillator of frequency f_0, i.e., by $\cos(2\pi f_0 t)$. The bandpass filter is centered around f_0 and has a bandwidth of $B \ll f_0$. The circuit is noiseless, but the input to the first mixer is stationary band-limited noise with two-sided bandwidth B.

A naive attempt to determine the output noise power would consist of the following analysis, illustrated in Figure 17.22. The first mixer shifts the input noise spectrum by $\pm f_0$ and scales it by 1/4. The resulting spectrum is multiplied by the squared magnitude of the filter's transfer function. Since this spectrum falls within the pass-band of the filter, it is not modified. Finally, the second mixer shifts the spectrum again by $\pm f_0$ and scales it by 1/4, resulting in the spectrum with three components shown in the figure. The total noise power at the output, i.e., the area under the spectrum, is 1/4th that at the input.

This common but simplistic analysis is inconsistent with the following alternative argument. Note that the bandpass filter, which does not modify the spectrum of its input, can be ignored. The input then passes through only the two successive mixers, resulting in the output noise voltage $o(t) = i(t)\cos^2(2\pi f_0 t)$. The output power is

$$o^2(t) = i^2(t)\left[\frac{3}{8} + \frac{\cos(2\pi 2 f_0 t) + \cos(2\pi 4 f_0 t)}{2}\right]$$

The average output power consists of only the 3/8 $i^2(t)$ term, since the cosine terms time-average to zero. Hence the average output power is 3/8th of the input power, 50% more than that predicted by the previous naive analysis. This is, however, the correct result.

The contradiction between the arguments above underscores the need for cyclostationary analysis. The autocorrelation function of any cyclostationary process $z(t)$ (defined as $R_{zz}(t,\tau) = E[z(t)z(t + \tau)]$, $E[\cdot]$ denoting expectation) can be expanded in a Fourier series in t:

(17.41)
$$R_{zz}(t,\tau) = \sum_{i=-\infty}^{\infty} R_{Zi}(\tau)e^{ji2\pi f_0 t}$$

$R_{zi}(\tau)$ are termed *harmonic autocorrelation functions.* The periodical time-varying power of $z(t)$ is its autocorrelation function evaluated at $\tau = 0$, i.e., $R_{zz}(t,0)$. The quantities $R_{zi}(0)$ represent the harmonic components of the periodically varying power. The average power is simply the value of the DC or *stationary component, $R_{z0}(0)$*.* The frequency-domain representation of the harmonic autocorrelations are termed *harmonic power spectral densities* (HPSDs) $S_{zi}(f)$ of $z(t)$, defined as the Fourier transforms

$$(17.42) \qquad S_{Zi}(f) = \int_{-\infty}^{\infty} R_{Zi}(\tau)e^{-j\pi f\tau}d\tau$$

Equations can be derived that relate the HPSDs at the inputs and outputs of various circuit blocks. By solving these equations, any HPSD in the circuit can be determined.

Consider, for example, the circuit in Figure 17.22. The input and output HPSDs of a perfect cosine mixer with unit amplitude can be shown [61] to be related by

$$(17.43) \qquad S_{vk}(f) = \frac{S_{u_{k-2}}(f - f_0)}{4} + \frac{S_{u_k}(f - f_0) + S_{u_k}(f + f_0)}{4} + \frac{S_{u_{k+2}}(f + f_0)}{2}$$

where u and v denote the input and output, respectively. The HPSD relation for a filter with transfer function $H(f)$ is [61]

$$(17.44) \qquad S_{v_k}(f) = H(-f)H(f + kf_0)S_{u_k}(f)$$

The HPSDs of the circuit are illustrated in Figure 17.23. Since the input noise $i(t)$ is stationary, its only nonzero HPSD is the stationary component $S_{i_0}(f)$, assumed to be unity in the frequency band $[-B/2, B/2]$, as shown. From Equation 17.43 applied to the first mixer, *three* nonzero HPSDs (S_{x_0}, S_{x_2}, and $S_{x_{-2}}$, shown in the figure) are obtained for $x(t)$. These are generated by shifting the input PSD by $\pm f_0$ and scaling by 1/4; in contrast to the naive analysis, the stationary HPSD is not the only spectrum used to describe the up-converted noise. From Equation 17.44, it is seen that the ideal bandpass filter propagates the three HPSDs of $x(t)$ unchanged to $y(t)$.

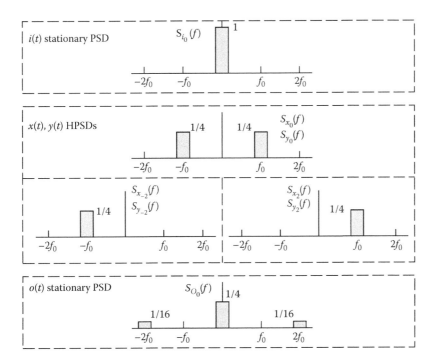

FIGURE 17.23 HPSDs of mixer–filter–mixer circuit.

---

* Stationary processes are a special case of cyclostationary processes, where the autocorrelation function (hence the power) is independent of the time t; it follows that $R_{zi}(\tau) \equiv 0$ if $i \neq 0$.

Through Equation 17.43, the second mixer generates five nonzero HPSDs, of which only the stationary component $S_{00}(f)$ is shown in the figure. This is obtained by scaling and shifting not only the stationary HPSD of $y(t)$, but also the cyclostationary HPSDs, which in fact contribute an extra 1/4 to the lobe centered at zero. The average output noise (the shaded area under $S_{00}(f)$) equals 3/8 of the input noise.

We now sketch the general procedure for analyzing mixing noise in circuits. The noise sources within a circuit can be represented by a small additive term $Au(t)$ to Equation 17.1, where $u(t)$ is a vector of noise sources, and A an incidence matrix capturing their connections to the circuit. Equation 17.1 is first solved for a (quasi)periodic steady state in the absence of noise, and then linearized as in Equation 17.38, to obtain

$$(17.45) \qquad C(t)\dot{x} + G(t)x + Au(t) = 0$$

where $x(t)$ represents the small-signal deviations due to noise. Equation 17.45 describes a linear periodically time-varying (LPTV) system with input $u(t)$ and output $x(t)$. The system can be characterized by its time-varying transfer function $H(t,f)$. $H(t,f)$ is periodic in t and can be expanded in a Fourier series similar to Equation 17.41. Denote the Fourier components (*harmonic transfer functions*) by $H_i(f)$.

Since $u(t)$ and $x(t)$ are vectors, their autocorrelation functions are *matrices* $R_{zz}(t,\tau) = E[z(t) z^T(t + \tau)]$, consisting of auto-and cross-correlations. Similarly, the HPSDs $S_{zi}(f)$ are also matrices. It can be shown [58] that the HPSD matrices of y and u are related by:

$$(17.46) \qquad S_{xx}(f) = \mathcal{H}(f)S_{uu}(f)\mathcal{H}^*(f)$$

$\mathcal{H}(f)$ (the *conversion matrix*) is the following block-structured matrix (f^k denotes $f+kf_0$):

$$(17.47) \qquad \mathcal{H}(f) = \begin{pmatrix} & \vdots & \vdots & \vdots & \\ \cdots H_0(f^1) & H_1(f^0) & H_2(f^{-1}) \cdots \\ \cdots H_{-1}(f^1) & H_0(f^0) & H_1(f^{-1}) \cdots \\ \cdots H_{-2}(f^1) & H_{-1}(f^0) & H_0(f^{-1}) \cdots \\ & \vdots & \vdots & \vdots & \end{pmatrix}$$

$S_{uu}(f)$ and $S_{xx}(f)$ are similar to $\mathcal{H}(f)$: their transposes $S_{zz}^T(f)$ have the same structure, but with $H_i(f^k)$ replaced by $S_{zi}^T(f^k)$.

Equation 17.46 expresses the output HPSDs contained in $S_{xx}(f)$, in terms of the input HPSDs contained in $S_{uu}(f)$, and the harmonic transfer functions of the circuit contained in $\mathcal{H}(f)$. The HPSDs of a single output variable $x_p(t)$ (both auto- and cross-terms with all other output variables) are available in the *p*th column of the central block-column of $S_{xx}^T(f)$. To pick this column, $S_{xx}^T(f)$ is applied to a unit vector E_{0p}, as follows ($\bar{}$ denotes the conjugate):

$$(17.48) \qquad S_{xx}^T(f)E_{0p} = \bar{H}(f)S_{uu}^T(f)H^T(f)E_{0p}$$

Evaluating Equation 17.48 involves two kinds of matrix–vector products, $\mathcal{H}(f)z$ and $S_{uu}(f)\,z$ for some vectors z. Consider the latter product first. If the inputs $u(t)$ are stationary, as can be assumed without loss of generality [61], then $S_{uu}(f)$ is block-diagonal. In practical circuits, the inputs $u(t)$ are either uncorrelated or sparsely correlated. This results in each diagonal block of $S_{uu}(f)$ being either diagonal or sparse. In both cases, the matrix–vector product can be performed efficiently.

The product with $\mathcal{H}(f)$ can also be performed efficiently by exploiting the relation $\mathcal{H}(f) = J^{-1}(f) \mathcal{A}$ [26]. $\mathcal{A}$ is a sparse incidence matrix of the device noise generators, hence its product with a vector can be computed efficiently. $J(0)$ is the HB Jacobian matrix [31] at the large-signal solution $x^*(t)$. $J(f)$ is obtained by replacing kf_0 by $kf_0 + f$ in the expression for the Jacobian. The product $J^{-1}z$ can therefore be computed efficiently using the fast techniques outlined in Section 17.5.2. As a

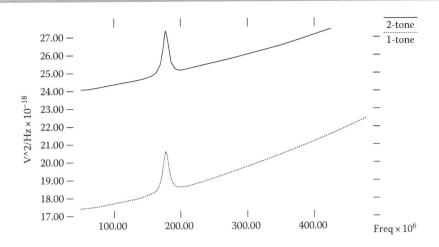

FIGURE 17.24   Stationary PSDs for the I–Q mixer/buffer circuit.

result, Equation 17.48 can be computed efficiently for large circuits to provide the auto- and cross-HPSDs of any output of interest.

For example, a portion of the Lucent W2013 RFIC, consisting of an I-channel buffer feeding a mixer, was simulated using Equation 17.48. The circuit consisted of about 360 nodes, and was excited by two tones—a local oscillator at 178 MHz driving the mixer, and a strong RF signal tone at 80 kHz feeding into the I-channel buffer. Two noise analyses were performed. The first analysis included both LO and RF tones (sometimes called a three-tone noise analysis). The circuit was also analyzed with only the LO tone to determine if the RF signal affects the noise significantly. The two-tone noise simulation, using a total of 525 large-signal mix components, required 300 MB of memory, and for each frequency point, took 40 min on an SGI machine (200 MHz, R10000 CPU). The one-tone noise simulation, using 45 harmonics, needed 70 MB of memory and took 2 minutes per point.

The stationary PSDs of the mixer output noise for the two simulations are shown in Figure 17.24. It can be seen that the presence of the large RF signal increases the noise by about 1/3. This is due to *noise folding*, the result of devices being driven into nonlinear regions by the strong RF input tone. The peaks in the two waveforms located at the LO frequency are due to up and down conversion of noise from other frequencies.

17.7.2 PHASE NOISE

Even small noise in an oscillator leads to dramatic changes in its frequency spectrum and timing properties, i.e., to phase noise. This effect can lead to interchannel interference and increased bit-error rates (BER) in RF communication systems. Another manifestation of the same phenomenon, jitter, is important in clocked and sampled-data systems: uncertainties in switching instants caused by noise can affect synchronization.

Although a large body of literature is available on phase noise,* treatments of the phenomenon from the design perspective have typically been phenomenological, e.g., the well-known treatment of Leeson [62]. Most analyses have been based on linear time-invariant or time-varying approaches, which though providing useful design guidelines, contain qualitative inaccuracies— e.g., they can predict infinite noise power. Recently (as of 2006), however, the work of Kärtner [63] and Demir et al. [64] have provided a more correct understanding of phase noise. Here, we sketch the approach in [64].

The starting point for phase noise analysis is Equation 17.1, reproduced here for oscillators with no external forcing:

$$(17.49) \qquad\qquad \dot{q}(x) + f(x) = 0$$

---

* BSIM homepage [19] contains a list of references.

We assume Equation 17.49 to be the equation for an oscillator with an orbitally stable,* nontrivial periodic solution, i.e., an oscillation waveform $x_s(t)$. With small noise generators in the circuit, possibly dependent on circuit state, the equation becomes

$$(17.50) \qquad \dot{q}(x) + f(x) = B(x)b(t)$$

where $b(t)$ now represents small perturbations.

When $b(t)$ is small, it can be shown [66] that the originally periodic oscillation $x_s(t)$ changes to

$$(17.51) \qquad x(t) = x_s(t + \alpha(t)) + y(t)$$

where $y(t)$ remains small, but $\alpha(t)$ (a time/phase deviation) can grow unboundedly with time, no matter how small the perturbation $b(t)$ is (see Figure 17.25). For driven circuits (the mixing noise case) $\alpha(t)$ remains bounded and small and its effects can therefore be lumped into the $y(t)$ term. This is the difference illustrated in Figures 17.19 through 17.21. The underlying reason for this difference is that oscillators by their very definition are phase unstable, and hence phase errors build up indefinitely.

Furthermore, it can be shown that $\alpha(t)$ is given by a nonlinear scalar differential equation

$$(17.52) \qquad \dot{\alpha} = v_1^{\mathrm{T}}(t + \alpha(t))B(x_s(t + \alpha(t)))b(t)$$

where $v_1(t)$ is a periodic vector function dubbed the perturbation projection vector (PPV). The PPV, which is characteristic of an oscillator in steady state and does not depend on noise parameters, is an important quantity for phase noise calculation. Roughly speaking, it is a "transfer function" that relates perturbations to resulting time or phase jitter in the oscillator. The PPV can be found only through a linear time-varying analysis of the oscillator around its oscillatory solution, and simple techniques to calculate it using HB or shooting are available [67].

In general, Equation 17.52 can be difficult to solve analytically. When the perturbation $b(t)$ is white noise however, it can be shown that $\alpha(t)$ becomes a Gaussian random walk process with linearly increasing variance ct, where c is a scalar constant given by

$$(17.53) \qquad c = \frac{1}{T}\int_0^T v_1^{\mathrm{T}}(t)B(x_s(t))B^{\mathrm{T}}(x_s(t))v_1(t)\,\mathrm{d}t$$

with T being the period of the unperturbed oscillation.

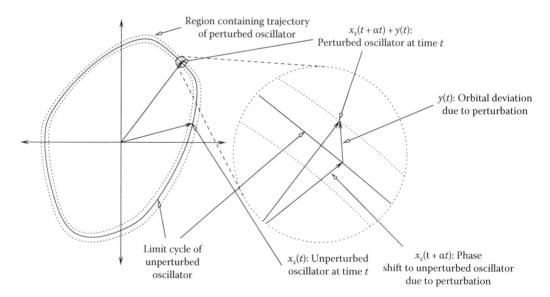

FIGURE 17.25 Oscillator trajectories.

---

* See, e.g., [65] for a precise definition; roughly speaking, an orbitally stable oscillator is one that eventually reaches a unique, periodic waveform with a definite magnitude.

This random-walk stochastic characterization of the phase error $\alpha(t)$ implies that:

1. The average spread of the jitter (mean-square jitter) increases *linearly* with time, with cT being the jitter per cycle.
2. The spectrum of the oscillator's output, i.e., the power spectrum of $x_s(t + \alpha(t))$, is *Lorenzian** about each harmonic. For example, around the fundamental (with angular frequency $\omega_0 = 2\pi/T$ and power P_{fund}), the spectrum is

$$(17.54) \qquad S_p(f) = P_{fund} \frac{\omega_0^2 c}{w_0^4 c^2/4 + (2\pi f - \omega_0)^2}$$

This means that the spectrum decays as $1/f^2$ beyond a certain knee distance away from the original oscillation frequency and its harmonics, as is well known for white noise in oscillators [62]. The $1/f^2$ dependence does not, however, continue as $f \rightarrow 0$, i.e., close to and at the oscillation frequency; instead, the spectrum reaches a finite maximum value.
3. The oscillator's output is a *stationary* stochastic process.

The Lorenzian shape of the spectrum also implies that the power spectral density at the carrier frequency and its harmonics have a finite value, and that the total carrier power is preserved despite spectral spreading due to noise. Equation 17.52 can also be solved for colored noise perturbations $b(t)$ [66], and it can be shown that if $S(f)$ is the spectrum of the colored noise, then the phase noise spectrum generated falls as $S(f)/(f - f_0)^2$, away from f_0.

Numerical methods based on the above insights are available to calculate phase noise. The main effort is calculating the PPV; once it is known, c can be calculated easily using Equation 17.53 and the spectrum obtained directly from Equation 17.54. The PPV can be found from the time-varying linearization of the oscillator around its steady state. Two numerical methods can be used to find the PPV. The first calculates the time-domain monodromy (or state-transition) matrix of the linearized oscillator explicitly, and obtains the PPV by eigendecomposing this matrix [64]. A more recent (as of 2006) method [67] relies on simple postprocessing of internal matrices generated during the solution of the operator's steady state using HB or shooting, and as such, can take advantage of the fast techniques of Section 17.5.2. The separate contributions of noise sources, and the sensitivity of phase noise to individual circuit devices and nodes, can be obtained easily.

As an example, the oscillator in Figure 17.26 consists of a Tow–Thomas second-order bandpass filter and a comparator [68]. If the OpAmps are considered to be ideal, it can be shown that this oscillator is equivalent (in the sense of the differential equations that describe it) to a parallel RLC circuit in parallel with a nonlinear voltage-controlled current source (or equivalently, a series RLC

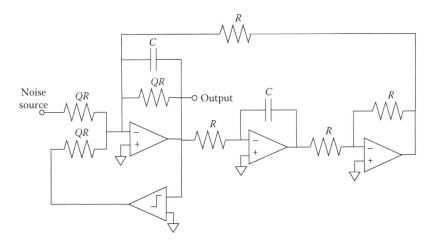

FIGURE 17.26   Oscillator with a band-pass filter and a comparator. (For further information on such oscillators, please see [68].)

---

* A Lorenzian is the shape of the squared magnitude of a one-pole lowpass filter transfer function.

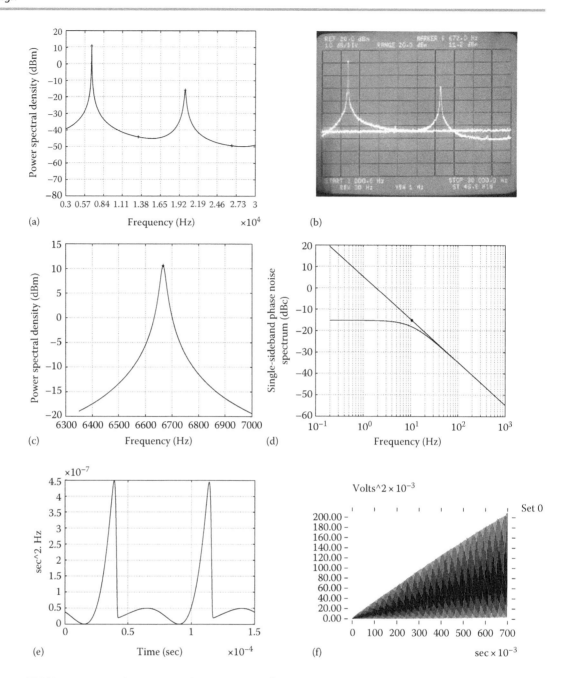

FIGURE 17.27   Phase noise characterisation for the oscillator in Figure 17.26. (a) PSD of oscillator output, (b) spectrum analyzer measurement, (c) PSD around first harmonic, (d) single-sideband phase noise spectrum, (e) squared magnitude of PPV, and (f) mean-square difference between perturbed and unperturbed systems.

circuit in series with a nonlinear current-controlled voltage source). In [68], authors breadboarded this circuit with an external white noise source (intensity of which was chosen such that its effect is much larger than the other internal noise sources), and measured the PSD of the output with a spectrum analyzer. For $Q = 1$ and $f_o = 6.66$ kHz, a phase noise characterization of this oscillator was performed to yield the periodic oscillation waveform $x_s(t)$ for the output and $c = 7.56 \times 10^{-8} \text{sec}^2 \text{Hz}$. Figure 17.27a shows the PSD of the oscillator output and Figure 17.27b shows the spectrum analyzer measurement.* Figure 17.27c shows a blown-up version of the PSD around the first harmonic. The single-sideband phase noise spectrum is shown in Figure 17.27d. The oscillator model that was

---

* The PSDs are plotted in units of dBm.

simulated has two state variables and a single stationary noise source. Figure 17.27e shows a plot of the periodic nonnegative scalar (essentially the squared magnitude of the PPV)

$$v_1^{\mathrm{T}}(t)B(x_s(t))B^{\mathrm{T}}(x_s(t))v_1(t) = (v_1^{\mathrm{T}}(t)B)^2$$

where B is independent of t since the noise source is stationary. Recall that c is the time average of this scalar that is periodic in time.

c can also be obtained relatively accurately in this case using Monte Carlo analysis (in general, however, Monte Carlo based on transient simulations can be extremely time-consuming and also inaccurate, and should be avoided except as a sanity check). The circuit was simulated with 10,000 random excitations and the results averaged to obtain the mean-square difference between the perturbed and unperturbed systems as a function of time. Figure 17.27f illustrates the result, in which the slope of the envelope determines c. The Monte Carlo simulations required small time-steps to produce accurate results, since numerical integration methods easily lose accuracy for autonomous circuits.

17.8 CONCLUSIONS

In this chapter, we have taken the reader through a quick tour of both basic and advanced topics in analog simulation—from writing circuit equations, to solving them numerically for simple analyses (like DC), to steady-state, envelope, multitime, and noise simulation. We have also looked into special challenges for oscillatory circuits. We hope that this tour has served to provide the reader an appreciation of concepts in simulation, in some depth as well as breadth.

ACKNOWLEDGMENTS

Roychowdhury takes pleasure in acknowledging Alper Demir and Amit Mehrotra for joint work on which much of Section 17.7.2 is based. He would also like to acknowledge his other colleagues in the Design Principles Research Department of Bell Laboratories during the period 1995–2000, notably Peter Feldmann, David Long, Bob Melville, and Al Dunlop. Discussions with these colleagues have influenced the style and content of much of this chapter, in particular Sections 17.2 through 17.7.

REFERENCES

1. L.O. Chua and P-M. Lin, *Computer-Aided Analysis of Electronic Circuits: Algorithms and Computational Techniques*, Prentice-Hall, Englewood Cliffs, NJ, 1975.
2. K.S. Kundert and A. Sangiovanni-Vincentelli, Simulation of nonlinear circuits in the frequency domain, *IEEE Transactions on Computer-Aided Design of Integrated Circuits System*, CAD-5, 521–535, October 1986.
3. K.S. Kundert, G.B. Sorkin, and A. Sangiovanni-Vincentelli, Applying harmonic balance to almost-periodic circuits, *IEEE Transactions on Microwave Theory and Techniques*, MTT-36, 366–378, 1988.
4. P. Antognetti and G. Massobrio, *Semiconductor Device Modeling with SPICE*, McGraw-Hill, New York, 1988, Chapter 1.
5. W. Liu, *MOSFET Models for SPICE Simulation Including BSIM3v3 and BSIM4*, Wiley, New York, 2001.
6. Y.P. Tsividis, *Operation and Modeling of the MOS Transistor*, McGraw-Hill, New York, 1987.
7. I.E. Getreu, *Modeling the Bipolar Transistor*, Tektronix, Inc., Beaverton, OR, 1976.
8. D. Foty, *MOSFET Modeling with SPICE: Principles and Practice*, Prentice Hall, Upper Saddle River, NJ, 1997.
9. K. Kundert and O. Zinke, *The Designer's Guide to Verilog-AMS*, Kluwer Academic Publishers, Norwell, MA, 2004.
10. D. Fitzpatrick and I. Miller, *Analog Behavioral Modeling with the Verilog-A Language*, Kluwer Academic Publishers, Norwell, MA, 1998.
11. H.A. Mantooth and M. Fiegenbaum, *Modeling with an Analog Hardware Description Language*, Kluwer Academic Publishers, Norwell, MA, 1995.

12. P. Ashenden, G. Peterson and D. Teegarden, *System Designer's Guide to VHDL-AMS*, Morgan Kaufman, San Francisco, CA, 2003.

13. P. Su, S.K.H. Fung, S. Tang, F. Assaderaghi and C. Hu, BSIMPD: A partial-depletion SOI MOSFET model for deep-submicron CMOS designs, *IEEE Proceedings of Custom Integrated Circuits Conference*, Orlando, FL, 2000, pp. 197–200.

14. B.G. Streetman, *Solid State Electronic Devices*, Prentice-Hall, New York, 1980, pp. 172–173.

15. A.S. Sedra and K.C. Smith, *Microelectronic Circuits*, 5th edn., Oxford University Publishing, New York, 2004, Chapter 3.

16. H.A. Mantooth and M. Vlach, Beyond SPICE with Saber and MAST, *IEEE Proceedings of International Symposium on Circuits System*, San Diego, CA, Vol. 1, May 1992, pp. 77–80.

17. T.R. McNutt, A.R. Hefner, H.A. Mantooth, J.L. Duliere, D. Berning, and R. Singh, Silicon carbide PiN and merged PiN Schottky power diode models implemented in the Saber circuit simulator, *IEEE Transactions on Power Electronics*, 19, 573–581, 2004.

18. H.A. Mantooth and J.L. Duliere, A unified diode model for circuit simulation, *IEEE Transactions on Power Electronics*, 12, 816–823, 1997.

19. BSIM Homepage, http://www-device.eecs.berkeley.edu/bsim3/ Accessed on December 3, 2015.

20. C. Enz, F. Krummenacher, and E. Vittoz, An analytical MOS transistor model valid in all regions of operation and dedicated to low-voltage and low-current applications, *Journal of Analog Integrated Circuits Signal Process*, 8, 83–114, 1995.

21. G. Gildenblat, H. Wang, T.-L Chen, X. Gu, and X. Cai, SP: An advanced surface-potential-based compact MOSFET model, *IEEE Journal of Solid-State Circuits*, 39, 1394–1406, 2004.

22. K.A. Sakallah, Y.T. Yen, and S.S. Greenberg, First-order charge conserving MOS capacitance model, *IEEE Transactions of Computer Aided Design*, 9, 99–108, 1990.

23. W.H. Press, S.A. Teukolsky, W.T. Vetterling, and B.P. Flannery, *Numerical Recipes—The Art of Scientific Computing*. Cambridge University Press, Cambridge, U.K., 1989.

24. K.S. Kundert, J.K. White, and A. Sangiovanni-Vincentelli, *Steady-State Methods for Simulating Analog and Microwave Circuits*, Kluwer Academic Publishers, Norwell, MA, 1990.

25. M.S. Nakhla and J. Vlach. A piecewise harmonic balance technique for determination of periodic responses of nonlinear systems, *IEEE Transactions of Circuit System*, CAS-23, 85, 1976.

26. S.A. Haas, *Nonlinear Microwave Circuits*, Artech House, Norwood, MA, 1988.

27. V. Rizzoli and A. Neri, State of the art and present trends in nonlinear microwave CAD techniques, *IEEE Transactions on MTT*, 36, 343–365, 1988.

28. R.J. Gilmore and M.B. Steer, Nonlinear circuit analysis using the method of harmonic balance—A review of the art. Part I. Introductory concepts. *International Journal of Microwave Millimeter Wave CAE*, 1, 22–37, 1991.

29. M. Rösch, Schnell simulation des stationären Verhaltens nichtlinearer Schaltungen. PhD thesis, Technischen Universität München, München, Germany, 1992.

30. R. Mickens, *Oscillations in Planar Dynamic Systems*, World Scientific, Singapore, 1995.

31. R.C. Melville, P. Feldmann, and J. Roychowdhury, Efficient multi-tone distortion analysis of analog integrated circuits, *Proceedings of IEEE CICC*, Santa Clara, CA, May 1995, pp. 241–244.

32. D. Long, R.C. Melville, K. Ashby, and B. Horton, Full chip harmonic balance, *Proceedings of IEEE CICC*, Santa Clara, CA, May 1997, pp. 379–382.

33. T.J. Aprille and T.N. Trick, Steady-state analysis of nonlinear circuits with periodic inputs, *Proceedings of IEEE*, 60, 108–114, 1972.

34. S. Skelboe, Computation of the periodic steady-state response of nonlinear networks by extrapolation methods, *IEEE Transactions on Circuit System*, CAS-27, 161–175, 1980.

35. A. Nayfeh and B. Balachandran, *Applied Nonlinear Dynamics*, Wiley, New York, 1995.

36. R. Telichevesky, K. Kundert, and J. White, Efficient steady-state analysis based on matrix-free Krylov subspace methods, *Proceedings of the IEEE DAC*, San Francisco, CA, 1995, pp. 480–484.

37. P. Feldmann, R.C. Melville, and D. Long, Efficient frequency domain analysis of large nonlinear analog circuits, *Proceedings of IEEE CICC*, San Diego, CA, May 1996, pp. 461–464.

38. Y. Saad, *Iterative Methods for Sparse Linear Systems*, PWS, Boston, MA, 1996.

39. R.W. Freund, Reduced-order modeling techniques based on Krylov subspaces and their use in circuit simulation, Technical Report 11273-980217-02TM, Bell Laboratories, 1998.

40. R.W. Freund, Reduced-order modeling techniques based on Krylov subspaces and their use in circuit simulation, *Application Computer Control Signal Circuits*, 1, 435–498, 1999.

41. J. Kevorkian and J.D. Cole, *Perturbation Methods in Applied Mathematics*, Springer, Berlin, Germany, 1981.

42. E. Ngoya and R. Larcheveque, Envelop transient analysis: A new method for the transient and steady state analysis of microwave communication circuits and systems, *Proceedings of the IEEE MTT Symposium*, San Francisco, CA, 1996.

43. H.G. Brachtendorf, G. Welsch, R. Laur, and A. Bunse-Gerstner, Numerical steady-state analysis of electronic circuits driven by multi-tone signals, *Electrical Engineering*, 79, 103–112,1996.

44. J. Roychowdhury, Efficient methods for simulating highly nonlinear multi-rate circuits, *Proceedings of the IEEE DAC*, Anaheim, CA, 1997.

45. O. Narayan and J. Roychowdhury, Multi-time simulation of voltage-controlled oscillators, *Proceedings of the IEEE DAC*, New Orleans, LA, June 1999.

46. J. Phillips, Model reduction of time-varying linear systems using approximate multipoint Krylov-subspace projectors, *Proceedings of the ICCAD*, San Jose, CA, November 1998.

47. J. Roychowdhury, Reduced-order modelling of time-varying systems, *IEEE Transactions on Circuit System II Signal Process*, 46, 1273–1288, November 1999.

48. L.T. Pillage and R.A. Rohrer, Asymptotic waveform evaluation for timing analysis, *IEEE Transactions on CAD*, 9, 352–366, 1990.

49. X. Huang, V. Raghavan, and R.A. Rohrer, AWEsim: A program for the efficient analysis of linear(ized) circuits, *Proceedings of the ICCAD*, Santa Clara, CA, November 1990, pp. 534–537.

50. E. Chiprout and M.S. Nakhla, *Asymptotic Waveform Evaluation*, Kluwer Academic Publishers, Norwell, MA, 1994.

51. P. Feldmann and R.W. Freund, Efficient linear circuit analysis by Padé approximation via the Lanczos process, *IEEE Transactions on CAD*, 14, 639–649, 1995.

52. P. Feldmann and R.W. Freund, Reduced-order modeling of large linear subcircuits via a block Lanczos algorithm, *Proceedings of the IEEE DAC*, San Francisco, CA, 1995, pp. 474–479.

53. P. Feldmann and R.W. Freund, Circuit noise evaluation by Padé approximation based model-reduction techniques, *Proceedings of the ICCAD*, San Jose, CA, November 1997, pp. 132–138.

54. A. Odabasioglu, M. Celik, and L.T. Pileggi, PRIMA: Passive reduced-order interconnect macromodelling algorithm, *Proceedings of the ICCAD*, San Jose, CA, November 1997, pp. 58–65.

55. W. Gardner, *Introduction to Random Processes*, McGraw-Hill, New York, 1986.

56. T. Ström and S. Signell, Analysis of periodically switched linear circuits. *IEEE Transactions on Circuits System*, CAS-24, 531–541, 1977.

57. M. Okumura, H. Tanimoto, T. Itakura, and T. Sugawara, Numerical noise analysis for nonlinear circuits with a periodic large signal excitation including cyclostationary noise sources, *IEEE Transactions on Circuits System I Fundamental Theory Application*, 40, 581–590, 1993.

58. J. Roychowdhury, D. Long, and P. Feldmann, Cyclostationary noise analysis of large RF circuits with multitone excitations, *IEEE Journal of Solid-State Circuits*, 33, 324–336, 1998.

59. V. Rizzoli, F. Mastri, and D. Masotti, General noise analysis of nonlinear microwave circuits by the piecewise harmonic balance technique, *IEEE Transactions on MTT*, 42, 807–819, 1994.

60. A.R. Kerr, Noise and loss in balanced and subharmonically pumped mixers: Part 1—Theory, *IEEE Transactions on MTT*, MTT-27, 938–943, 1979.

61. J. Roychowdhury and P. Feldmann, A new linear-time harmonic balance algorithm for cyclostationary noise analysis in RF circuits, *Proceedings of the ASP-DAC*, Chiba, Japan, 1997, pp. 483–492.

62. D.B. Leeson, A simple model of feedback oscillator noise spectrum, *Proceedings of IEEE*, 54, 329, 1966.

63. F. Kärtner, Analysis of white and f^a noise in oscillators, *International Journal of Circuit Theory Applications*, 18, 485–519, 1990.

64. A. Demir, A. Mehrotra, and J. Roychowdhury, Phase noise in oscillators: A unifying theory and numerical methods for characterization, *IEEE Transactions on Circuits System I Fundamental Theory Application*, 47, 655–674, 2000.

65. M. Farkas, *Periodic Motions*, Springer, Berlin, Germany, 1994.

66. A. Demir, Phase noise in oscillators: DAEs and colored noise sources, *Proceedings of the ICCAD*, San Jose, CA, 1998, pp. 170–177.

67. A. Demir, D. Long, and J. Roychowdhury, Computing phase noise eigenfunctions directly from steady-state Jacobian matrices, *Proceedings of the ICCAD*, San Jose, CA, November 2000 pp. 283–288.

68. A. Dec, L. Toth, and K. Suyama, Noise analysis of a class of oscillators, *IEEE Transactions of Circuits System*, 45, 757–760, 1998.

Simulation and Modeling for Analog and Mixed-Signal Integrated Circuits

18

Georges G.E. Gielen and Joel R. Phillips

CONTENTS

18.1 INTRODUCTION

This chapter presents an overview of the modeling and simulation methods that are needed to design and embed analog and RF blocks in mixed-signal integrated systems (ASICs, SoCs, and Systems in Package). The design of these integrated systems is characterized by growing design complexities and tight time-to-market constraints. Handling these requires mixed-signal design methodologies and flows that include system-level architectural explorations and hierarchical design refinements with behavioral models in the top-down design path, along with detailed behavioral model extraction and efficient mixed-signal behavioral simulation in the bottom-up verification path. Mixed-signal simulation methods at different hierarchical levels are reviewed, and techniques to generate analog behavioral models, including regression-based methods as well as model-order reduction techniques, are described in detail. This chapter also describes the generation of performance models for analog circuit synthesis and of symbolic models that provide designers with insight into the relationships governing the performance behavior of a circuit.

With the evolution toward ultra-deep-submicron and nanometer CMOS technologies [1], the design of complex integrated systems (ASICs, SoCs, and SiPs), is emerging not only in consumer-market applications such as telecom and multimedia, but also in more traditional application domains like automotive and instrumentation. Driven by cost reduction, these markets demand low-cost, optimized and highly integrated solutions with very challenging performance specifications. These integrated-systems are increasingly mixed-signal designs, embedding on a single die high-performance analog or mixed-signal blocks and possibly sensitive RF front-ends, together with complex digital circuitry-(multiple processors, some logic blocks, and several large memory blocks) that form the core of most electronic systems today. In addition to the technical challenges related to the increasing design complexity and the problems posed by analog–digital integration, shortening time-to-market constraints put pressure on the design methodology and tools used to design these systems.

Hence, the design of today's integrated systems calls for mixed-signal design methodologies and flows that include system-level architectural explorations and hierarchical design refinements with behavioral models in the top-down design path to reduce the chance of design iterations and to improve the overall optimality of the design solution [2]. In addition, to avoid design errors before tape-out, detailed behavioral model extraction and efficient mixed-signal behavioral simulation are needed in the bottom-up verification path. This chapter presents an overview of the modeling and simulation methods used in this context.

The chapter is organized as follows. Section 18.2 addresses mixed-signal design methodologies and describes techniques and examples for architectural exploration and top-down hierarchical design refinement. Section 18.3 discusses mixed-signal simulation techniques. Section 18.4 describes analog and mixed-signal behavioral simulation and the corresponding hardware description languages. It also gives an overview of techniques to automatically generate analog behavioral models, including regression-based methods as well as model-order reduction techniques. Also the generation of performance models for analog circuit synthesis is described. Section 18.5 then presents methods to generate symbolic models that provide designers with insight into the relationships governing the performance behavior of a circuit. Conclusions are drawn in Section 18.6, followed by an extensive list of references.

18.2 TOP-DOWN MIXED-SIGNAL DESIGN METHODOLOGY

The growing complexity of the systems that can be integrated on a single die today, in combination with the tightening time-to-market constraints, results in a growing design productivity gap. That is why new (as of 2006) design methodologies are being developed that allow designers to shift to a higher level of abstraction, such as the use of platform-based design, object-oriented system-level hierarchical design refinement flows, hardware–software co-design, and IP reuse, on top of the already established use of CAD tools for logic synthesis and digital place and route. However, these flows have to be extended to incorporate embedded analog/RF blocks.

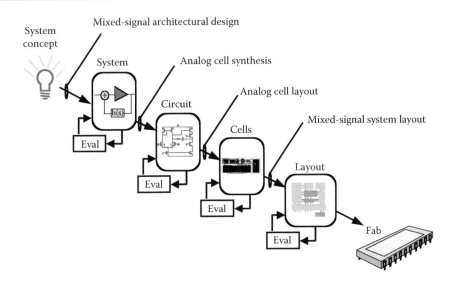

FIGURE 18.1   Top-down view of the mixed-signal IC design process.

A typical top-down design flow for mixed-signal integrated systems may appear as shown in Figure 18.1, where the following distinct phases can be identified: system specification, architectural design, cell design, cell layout, and system-layout assembly [2,3]. The advantages of adopting a top-down design methodology are:

- The possibility to perform system-architectural exploration and a better overall system optimization (e.g., finding an architecture that consumes less power) at a high level before starting detailed circuit implementations;
- The elimination of problems that often cause overall design iterations, such as the anticipation of problems related to interfacing different blocks;
- The possibility to do early test development in parallel to the actual block design, etc.

The ultimate advantage of top-down design therefore is to catch problems early in the design flow, and as a result have a higher chance of first-time success with fewer or no overall design iterations, hence shortening design time, while at the same time obtaining a better overall system design. A top-down design example will be presented later on. The methodology, however, does not come for free, and requires some investment from the design team, especially in terms of high-level modeling and setting up a sufficient model library for the targeted application. Even then, there remains the risk that at higher levels in the design hierarchy also, low-level details (e.g., matching limitations, circuit nonidealities, and layout effects) may be important to determine the feasibility or optimality of an analog solution. The high-level models used must therefore include such effects to the extent possible, but it remains difficult in practice to anticipate or model everything accurately at higher levels. Besides the models, efficient simulation methods are also needed at the architectural level in order to allow efficient interactive explorations. The subjects of system exploration and simulation as well as behavioral modeling will now be discussed in more detail.

18.2.1 SYSTEM-LEVEL ARCHITECTURAL EXPLORATION

The general objective of analog architectural system exploration is twofold [4,5]. First, a proper (and preferably optimal) architecture for the system has to be decided upon. Second, architectural blocks must be specified, so that the overall system meets its requirements at minimum implementation cost (chip area, etc.). The aim of a system-exploration environment is to provide the system designer with the platform and the supporting tool-set to explore different architectural alternatives in a short time and to take the above decisions based on quantified rather than heuristic information.

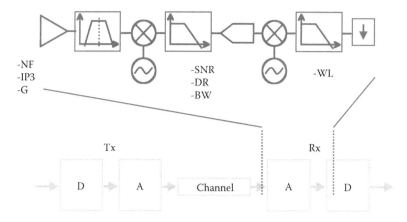

FIGURE 18.2 Digital telecommunication link, indicating a possible receiver front-end architecture with some building block specifications to be determined during front-end architectural exploration.

Consider, for instance, the digital telecommunication link of Figure 18.2. It is clear that digital bits are going into the link to be transmitted over the channel, and that the received signals are being converted again into digital bits. One of the major considerations in digital telecom system design is the bit error rate, which characterizes the reliability of the link. This bit error rate is not only impacted by the characteristics of the transmission channel itself, but also by the architecture chosen for the transmitter and receiver front-end, and by the performances achieved and the non-idealities exhibited by the analog/RF blocks in this front-end. For example, the noise figure and nonlinear distortion of the input low-noise amplifier (LNA) are key parameters. Similarly, the resolution and sampling speed of the analog-to-digital converter (ADC) used may have a large influence on the bit error rate, but they also determine the requirements for the other analog sub-blocks: a higher ADC resolution may relax the filtering requirements in the transceiver, resulting in simpler filter structures, though it will also consume more power and chip area than a lower-resolution converter. At the same time, the best trade-off solution, i.e., the minimum required ADC resolution and therefore also the minimum power and area, depends on the architecture chosen for the transceiver front-end.

Clearly, there is a large interaction between system-level architectural decisions and the performance requirements for the different subblocks, which in turn are bounded by technological limits that shift with every new technology process being employed. Hence, it is important to offer designers an exploration environment where they can define different front-end architectures and analyze and compare their performance quantitatively and derive the necessary building block specifications. Today, the alternative architectures explored are still provided by the system designer, but future tools might also derive or synthesize these architectures automatically from a high-level language description [6].

The important ingredients that are needed to set up such an architectural exploration environment are [4,5]:

- A fast high-level simulation method that allows the evaluation of the performance (e.g., SNR or BER) of the front-end;
- A library of high-level (behavioral) models for the building blocks used in the targeted application domain, including a correct modeling of the important building block non-idealities (offset, noise, distortion, mirror signals, phase noise, etc.);
- Power and area estimation models that, starting from the block specifications, allow estimation of the power consumption and chip area that would be consumed by a real implementation of the block, without really designing the block.

The above ingredients allow a system designer to explore interactively front-end architectures. Combining this with an optimization engine would additionally allow optimization of the selected front-end architecture in determining the optimal building block requirements so as to meet the system requirements at minimum implementation cost (power/area). Repeating this optimization for different architectures then makes a quantitative comparison between these architectures possible before they are implemented down to the transistor level. In addition, the high-level exploration environment would also help in deciding on other important system-level decisions, such as determining the optimal partitioning between analog and digital implementations in a mixed-signal system [7], or deciding on the frequency planning of the system, all based on quantitative data rather than ad hoc heuristics or past experiences.

To some extent, this approach can be implemented in existing commercial tools such as COSSAP, PTOLEMY, Matlab/Simulink, ADS, and SPW. However, not all desired aspects for system-level exploration are readily available in the present commercial system-level simulators, asking for more effective and more efficient solutions to be developed. To make system-level exploration really fast and interactive, dedicated algorithms can be developed that speed up the calculations by maximally exploiting the properties of the system under investigation and using proper approximations where possible. ORCA, for instance, is targeted toward telecom applications and uses dedicated signal spectral manipulations to gain efficiency [8]. A more recent (as of 2006) development is the FAST tool which performs a time-domain dataflow type of simulation without iterations [9], and which easily allows dataflow co-simulation with digital blocks. Compared to commercial simulators such as COSSAP, PTOLEMY, or SPW, this simulator is more efficient because it uses block processing instead of point-by-point calculations for the different time points in circuits without feedback. In addition, the signals are represented as complex equivalent baseband signals with multiple carriers. The signal representation is local and fully optimized, as the signal at each node in the circuit can have a set of multiple carriers and each corresponding equivalent baseband component can be sampled with a different time-step depending on its bandwidth. Large feedback loops, especially when they contain nonlinearities, are however more difficult to handle with this approach. A method to simulate efficiently bit error rates with this simulator has been presented in Ref. [10].

Example 18.1

As an example [4,5], consider a front-end for a cable TV modem receiver, based on the MCNS standard. The MCNS frequency band for upstream communication on the CATV network is from 5 to 42 MHz (extended subsplit band). Two architectures are shown in Figure 18.3: (a) an all-digital architecture where both the channel selection and the down-conversion are done in the digital domain, and (b), the classical architecture where the channel selection is performed in the analog domain.

A typical input spectrum is shown in Figure 18.4. For this example we have used 12 QAM-16 channels with a 3 MHz bandwidth. We assume a signal variation of the different channels of maximally ±5 dB around the average level. The average channel noise is 30 dB below this level. Figures 18.5 and 18.6 show the spectrum simulated by ORCA [8] for the all-digital architecture of Figure 18.3a. Figure 18.5 shows the spectrum of all eight channels after initial filtering at the output of the ADC, whereas Figure 18.6 shows the spectrum of the desired channel at the receiver output after digital channel selection and quadrature down-conversion. The desired channel signal and the effects of the channel noise, the ADC

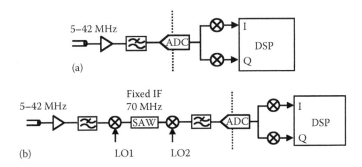

FIGURE 18.3 Two possible architectures for a cable TV application: (a) all-digital architecture, (b) classical architecture.

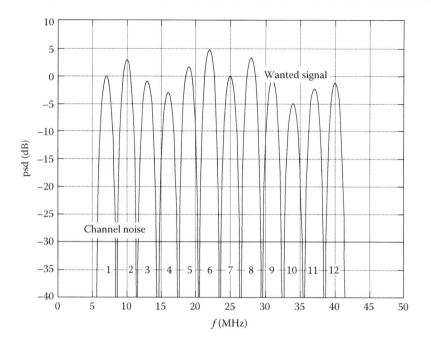

FIGURE 18.4 Typical input spectrum for a CATV front-end architecture using 12 QAM-16 channels.

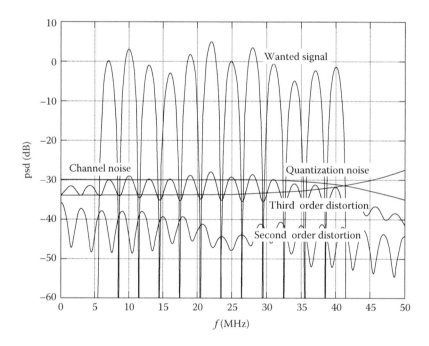

FIGURE 18.5 Simulated spectrum of the eight input channels at the output of the ADC in the all-digital CATV architecture.

quantization noise, and the second- and third-order distortion are generated separately, providing useful feedback to the system designer. The resulting SNDR is equal to 22.7 dB in this case, which corresponds to a symbol error rate of less than 10^{-10} for QAM-16.

By performing the same analysis for different architectures, and by linking the required subblock specifications to the power or chip area required to implement the subblocks, a quantitative comparison of different alternative architectures becomes possible with respect to (1) their suitability to implement the system specifications, and (2) the corresponding implementation cost in power consumption or silicon real estate. To assess the

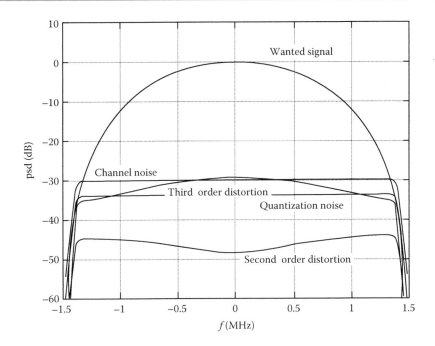

FIGURE 18.6   Simulated spectrum for the desired channel at the receiver output in the all-digital CATV architecture, indicating both the signal as well as noise and distortion added in the receiver.

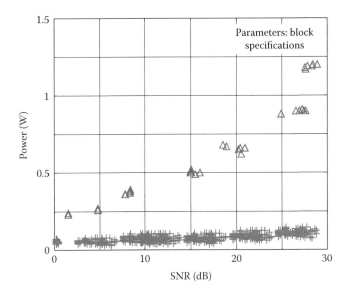

FIGURE 18.7   Power consumption comparison between the all-digital CATV architecture (triangles) and the classical architecture (crosses), as a function of the required SNR. (From Wambacq, P. et al., High-level simulation and power modeling of mixed-signal front-ends for digital telecommunications, *Proceedings of the International Conference on Electronics, Circuits and Systems* (*ICECS*), September 1999, pp. 525–528. With permission.)

latter, high-level power or area estimators must be used to quantify the implementation cost. In this way, the system designer can choose the most promising architecture for the application at hand.

Figure 18.7 shows a comparison between the estimated total power consumption required by the all-digital and by the classical CATV receiver architectures of Figure 18.3, as a function of the required SNR [11]. These results were obtained with the simulator

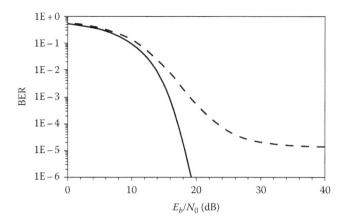

FIGURE 18.8   Simulated BER analysis result for a 5 GHz 802.11 WLAN architecture with (dashed), and without (straight), nonlinear distortion included. (From Vandersteen, G. et al., Efficient bit-error-rate estimation of multicarrier transceivers, *Proceedings of the Conference on Design, Automation and Test in Europe* (*DATE*), 2001, pp. 164–168. With permission.)

FAST [9]. Clearly, for the technology used in the experiment, the classical architecture still required much less power than the all-digital solution.

Finally, Figure 18.8 shows the result of a BER simulation with the FAST tool, for a 5 GHz 802.11 WLAN architecture [9]. The straight curve shows the result without taking into account nonlinear distortion caused by the building blocks; the dashed curve takes this distortion into account. Clearly, the BER considerably worsens in the presence of nonlinear distortion. Note that the whole BER analysis was performed in a simulation time that is two orders of magnitude faster than traditional Monte Carlo analysis performed on a large number of OFDM symbols.

18.2.2 EXAMPLE OF TOP-DOWN DESIGN

Top-down design is already heavily used in industry today for the design of complex analog blocks, such as Delta–Sigma converters or phase-locked loops (PLL). In these cases, first a high-level design of the block is done with the block represented as an architecture of subblocks, each modeled with a behavioral model that includes the major non-idealities as parameters, rather than a transistor schematic. This step is often done using Matlab/Simulink and it allows the determination of the optimal architecture of the block at this level, together with the minimum requirements for the subblocks (e.g., integrators, quantizers, VCO, etc.), so that the entire block meets its requirements in some optimal sense. This is then followed by a detailed device-level (SPICE) design step for each of the chosen architecture's subblocks, targeted to the derived subblock specifications. This is now illustrated for a PLL.

Example 18.2

The basic block diagram of a PLL is shown in Figure 18.9. If all subblocks like the phase-frequency detector or the voltage-controlled oscillator (VCO) are represented by behavioral models instead of device-level circuits, then enormous time savings with regard to simulation time can be obtained during the design and verification phase of the PLL. For example, for requirements arising from a GSM-1800 design example (frequency range around 1.8 GHz, phase noise −121 dB/Hz at 600 kHz frequency offset, settling time of the loop for channel frequency changes below 1 ms within 1×10^{-6} accuracy), the following characteristics can be derived for the PLL subblocks using behavioral simulations with generic behavioral models for the subblocks [12]: $A_{LPF} = 1$, $K_{VCO} = 1 \times 10^6$ Hz/V, $N_{div} = 64$, $f_{LPF} = 100$ kHz. These specifications are then the starting point for the device-level design of each of the subblocks.

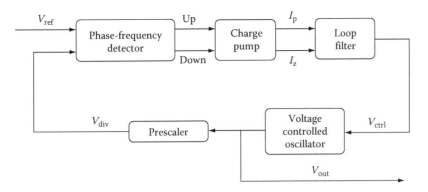

FIGURE 18.9   Basic block diagram of a phase-locked loop analog block.

18.3 MIXED-SIGNAL AND BEHAVIORAL SIMULATION

We will now describe the state of the art in mixed-signal and behavioral simulation and modeling.

18.3.1 ANALOG AND MIXED-SIGNAL SIMULATION

Circuit simulation began with the early development of the SPICE simulator [13,14] which even today remains the cornerstone of many circuit designs. This simulator numerically solves the system of nonlinear differential-algebraic equations (DAE) that characterize the circuit by using traditional techniques of numerical analysis. For example, to compute the evolution of the circuit's response in time, the time-derivative operator is first discretized by an implicit integration scheme such as the backward-Euler or trapezoidal rule. At each time point, the resulting set of nonlinear algebraic equations is solved iteratively via the Newton–Raphson method, with the matrix solution required to update the approximate solution computed using sparse matrix techniques. For details, refer to the companion chapter on circuit simulation in this volume [15]. Advances in mathematics and the evolutionary improvement of the core numerical algorithms (e.g., adaptive time-step control, sparse matrix factorization, and improved convergence) have over the years contributed to a vast number of commercial CAD tools. Many descendents of the SPICE simulator are now marketed by a number of CAD vendors and many IC manufacturers have in-house versions of the SPICE simulator that have been adapted to their own proprietary processes and designs. A few examples of the many commercial SPICE-class simulators are HSPICE (Synopsys), Spectre (Cadence Design Systems), and Eldo (Mentor Graphics). SPICE or its many derivatives have evolved into an established designer utility that is being used both during the design phase (often in a designer-guided trial-and-error fashion) and for extensive postlayout design verification.

Although SPICE is a general-purpose circuit simulator, its adaptive time-stepping approach is very slow for circuits with widely separated time constants. This is why for certain circuit classes, more dedicated, faster solutions have been developed. For instance, simulators for switched-capacitor [16] and switched-current circuits, as well as discrete-time DS modulators [17] take advantage of the switched timing of the circuits to cut down on simulation time. Another important domain is RF simulation, where shooting and harmonic balance techniques have been developed to directly simulate the steady-state behavior of these circuits without having to follow slow initial transients or many carrier cycles in a slow signal envelope [18].

With the explosion of mixed-signal designs, the need has also arisen for simulation tools that allow not only simulation of analog or digital circuits separately, but also simulation of truly mixed analog–digital designs [19]. Simulating the large digital parts with full SPICE accuracy results in very long overall simulation times, whereas efficient event-driven techniques exist to simulate digital circuits at higher abstraction levels than the transistor level. Therefore, mixed-mode simulators were developed that glue together an accurate SPICE-like analog simulator to an efficient digital simulator. The general scheme of these so-called glued mixed-signal simulators is shown in Figure 18.10. These simulators partition the netlist into digital and analog parts that are each simulated by a separate

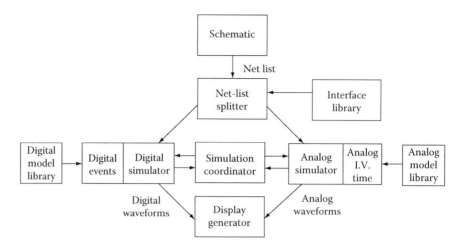

FIGURE 18.10   General scheme of a mixed-signal simulator, gluing an analog and a digital simulation kernel together.

kernel that interfaces through a simulation "backplane": an analog kernel with adaptive time-step control, and a digital kernel with its event-driven mechanism. The integration of two disparate simulation technologies requires reconciliation of the different computation models. Conversion of the signals between analog and digital signal representations and the appropriate driving/loading effects are addressed by inserting interface elements at the boundaries between the analog and digital circuitry. Effective treatment of the synchronization between the analog simulation kernel with its non-uniform time-steps due to its adaptive time-step control and the digital simulation kernel with its event-driven mechanism determines the efficiency of the overall simulation. Such synchronization is needed at each time point when there is a signal event at the boundary between the analog and digital parts of the system: whenever an analog signal at a boundary node passes the logic's thresholds, an event is created on the digital simulator side, and alternatively a digital event at a boundary node is translated into the corresponding analog signal transition on the analog simulator side.

Problems arising due to the weak integration mechanism in these glued simulators, such as poor simulation performance, difficulty in describing modules with mixed semantics, and difficulty in debugging across the mixed-signal boundary, have recently (as of 2006) led to more integrated simulators that combine the analog and digital capabilities in a single kernel. Such simulators that feature the ability to seamlessly mix Verilog-AMS, VHDL-AMS (see below for details), and SPICE descriptions are today available in the commercial marketplace. As representative examples of modern mixed-signal simulators, we would mention the AdvanceMS (Mentor Graphics) and AMS Designer tools (Cadence Design Systems).

Besides the eternal problem of the limited accuracy of the device models used in SPICE, the main problems with the standard SPICE simulator are that it is essentially a structural circuit simulator, and that its CPU time increases rapidly with the size of the circuit, making the simulation of really large designs infeasible. This is why in the past years the need has arisen for higher levels of abstraction to describe and simulate analog circuits more efficiently, at the expense of a (little) loss in accuracy.

18.3.2 ANALOG BEHAVIORAL SIMULATION

There are (at least) four reasons for using higher-level analog modeling (functional, behavioral, or macromodeling) for describing and simulating mixed-signal systems [2]:

1. The simulation time of circuits with widely spaced time constants (e.g., oversampling converters, PLLs, etc.) is quite large since the time-step control mechanism of the analog solver follows the fastest signals in the circuit. Use of higher-level modeling for the blocks will accelerate the simulation of these systems, particularly if the "fast" time-scale behavior can be "abstracted away", e.g., by replacing transistor-level descriptions of RF blocks by baseband-equivalent behavioral models.

2. In a top-down design methodology based on hierarchical design refinement (as in Figure 18.1 and the example of Section 18.2.2) at higher levels of the design hierarchy, there is a need for higher-level models describing the pin-to-pin behavior of the circuits in a mathematical format, rather than representing it as an internal structural netlist of components. This is unavoidable during top-down design, since at higher levels in the design hierarchy, the details of the underlying circuit implementation are simply not yet known and hence only generic mathematical models can be used.

3. A third use of behavioral models is during bottom-up system verification when these models are needed to reduce the CPU time required to simulate the block as part of a larger system. The difference is that in this case the underlying implementation is known in detail, and that peculiarities of the block's implementation can be incorporated in the model as far as possible without slowing down the simulation too much.

4. Fourthly, when providing or using analog IP macrocells in a system-on-a-chip context, the virtual component has to be accompanied by an executable model that efficiently models the pin-to-pin behavior of the virtual component. This model can then be used in system-level design and verification by the SoC integrating company, even without knowing the detailed circuit implementation of the macrocell [20].

For all these reasons, analog/mixed-signal (AMS) behavioral simulation models are needed, that describe analog circuits at a higher level than the circuit level, i.e., that describe the input–output behavior of the circuit in a mathematical model rather than as a structural network of basic components. These higher-level models must describe the desired behavior of the block (e.g., amplification, filtering, mixing, or quantization) and simulate efficiently, while still including the major non-idealities of real implementations with sufficient accuracy.

In addition, high-level models also form a key part of several analog circuit synthesis and optimization systems, both at the circuit level as well as for the hierarchical synthesis of more complex blocks. Most of the basic techniques in both circuit and layout synthesis today rely on powerful numerical-optimization engines coupled to "evaluation engines" that qualify the merit of some evolving analog circuit or layout candidate [2]. This basic scheme of optimization-based analog circuit sizing is shown in Figure 18.11. The most general but also by far the slowest solution is to call the transistor-level simulator as an evaluation engine during each iteration of the optimization of a circuit. The CPU time can be reduced significantly by replacing the simulations by model evaluations. These models can be behavioral simulation models as described above, or they can be what are termed *performance models* [21]. Rather than modeling the input–output behavior, performance models directly relate the achievable performances of a circuit (e.g., gain, bandwidth, or slew rate) to the design variables (e.g., device sizes and biasing). In the synthesis procedure, calls to the transistor-level simulation are then replaced by performance model evaluations, resulting in substantial speed-ups of the overall synthesis, once the performance models have been created

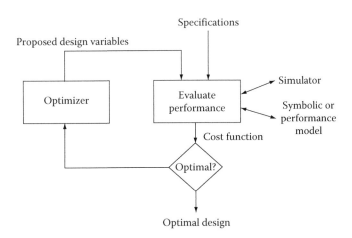

FIGURE 18.11   Basic flow of optimization-based analog circuit sizing.

and calibrated. For the synthesis of more complex analog blocks, a hierarchical approach is needed, in which higher-level models are indispensable to bridge between the different levels [22,17].

Example 18.3

For example, the realistic dynamic behavior of a (current-steering) digital-to-analog converter shows finite settling and ringing (glitch), as shown in Figure 18.12. Such behavior can mathematically be approximated by superposition of an exponentially damped sine for the glitch and a hyperbolic tangent [23]:

$$
\begin{aligned}
i_{\text{out}} = A_{\text{gl}}\sin\left(\frac{2\pi}{t_{\text{gl}}}(t-t_0)\right)\exp\left(-\sin(t-t_0)\frac{2\pi}{t_{\text{gl}}}(t-t_0)\right) \\
+\frac{\text{level}_{i+1}-\text{level}_i}{2}\tanh\left(\frac{2\pi}{t_{\text{gl}}}(t-t_0)\right)+\frac{\text{level}_{i+1}-\text{level}_i}{2}
\end{aligned}
$$

(18.1)

where level_i and level_{i+1} are the DAC output levels before and after the considered transition, and where the parameters such as A_{gl}, t_0, and t_{gl} need to be determined, for instance, by regression fitting to simulation results of a real circuit. Figure 18.13 compares the response of the behavioral model (with parameter values extracted from SPICE simulations) with SPICE simulation results of the original circuit. The speedup in CPU time is almost three orders of magnitude, while the error is below 1% [23].

Table 18.1 gives an overview of the different analog hardware description levels considered in design practice today, and the implications of those abstraction levels on the modeling [24]. The *circuit level* is the traditional level where a circuit is simulated as a network of physical components. In a *macromodel*, an equivalent but computationally cheaper circuit representation is used that has approximately the same behavior as the original circuit. Equivalent sources combine the effect of several other elements that are eliminated from the netlist. The simulation speedup is roughly proportional to the number of nonlinear devices that can be eliminated. In a *behavioral or functional model*, a purely mathematical description of the input–output behavior of the block is used. This typically will be in the form of a set of DAE or transfer functions. At the behavioral level, conservation laws still

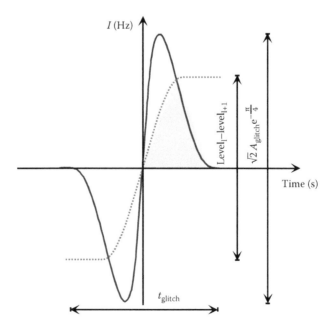

FIGURE 18.12   Typical dynamic behavior of a current-steering digital-to-analog converter output showing both finite settling and glitch behavior when switching the digital input code.

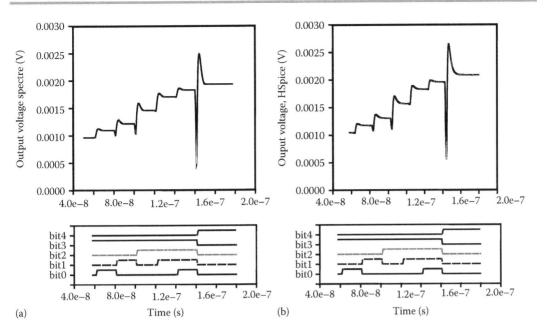

FIGURE 18.13 Comparison between the device-level simulation results (b) and the response of the behavioral model (a). (From Vandenbussche, J. et al., Systematic design of high-accuracy current-steering D/A converter macrocells for integrated VLSI systems, *IEEE Trans. Circ. Syst.*, Part II, 48, 300, 2001. With permission.)

TABLE 18.1   Different Analog Hardware Description Levels to Describe Analog Circuits

Level	Modeling Primitives	Implications
Functional	Mathematical signal flow description per block, connected in signal flow diagram	No internal block structure; conservation laws need not be satisfied on pins
Behavioral	Mathematical description (equations, procedures) per block	No internal block structure; conservation laws must be satisfied on pins
Macromodel	Simplified circuit with controlled sources	Spatially unrelated to actual circuit; conservation laws must be satisfied
Circuit	Connection of SPICE primitives	Spatially one-to-one related to actual circuit; conservation laws must be satisfied

have to be satisfied; at the functional level this is no longer the case, and the simulated system turns into a kind of signal-flow diagram.

The industrial use of analog higher-level (functional, behavioral, macro-) modeling is today enabled by the availability of standardized mixed-signal hardware description languages such as VHDL-AMS [25,26] and Verilog-AMS [27,28], both of which are extensions of the corresponding digital hardware description languages, and both of which are supported by commercial simulators today. These languages allow description and simulation of separate analog circuits, separate digital circuits, and mixed analog–digital circuits, at the different abstraction levels mentioned above. In general, they also allow description and simulation of both electrical and nonelectrical systems, as long as they can be modeled by a set of (nonlinear) DAE. Note that while originally restricted to low-to-medium frequencies with circuits having lumped elements only, efforts to standardize the extension of these languages, e.g., VDHL-AMS, toward RF and microwave circuits with distributed elements, have been started in recent (as of 2006) years. Likewise, effort is being made to leverage language-based modeling to modernize descriptions of the transistor models themselves by means of language extensions that support automatic compilation of semiconductor device compact models [29,30].

Example 18.4

Below is an illustrative example of a VHDL-AMS model for a circuit combining a VCO and a frequency divider (DIVN):

```
entity vcodivn is
generic (vco_gain: real:5 1.0e6;
                f0: real:5 1.0e9;
         rin: real:5 1.0e6;
         propdelay: time: 5 5.0 ns;
         pn: integer:5 64);
port (terminal input, ref: electrical;
      signal sdived: out std_ulogic);
end entity vcodivn;
architecture behavioral of vcodivn is
quantity phase: real:5 0.0;
quantity freq: real:5 0.0;
quantity vin across iin through input to ref;
begin
    break phase 5. 0.0 when
       phase'above(real(pn/2));
    freq 55 f0 1 vco_gain*vin;
    phase 55 freq'integ;
    iin 55 vin/rin;
    triggerOutput: process is
          variable ostate: std_ulogic:5 '0';
          begin
          wait on phase'above(real(pn/2));
          ostate:5 not ostate;
          sdived,5 ostate after propdelay;
    end process triggerOutput;
end architecture behavioral;
```

The model contains an entity part, describing the circuit as a box, i.e., the parameters and the pins of the model, and an architecture part, describing the behavior of the circuit. In the example, the circuit has five parameters (after the keyword "generic") and two electrical terminals and one signal port (after the keyword "port"). The architecture description first defines the analog variables ("quantities") that are being used and then describes the equations that govern the behavior of the circuit. Algebraic equations and an integral equation (freq'integ means applying the integral operator to the quantity freq) are used. The model is a mixed analog–digital model as it also contains a standard VHDL process description as part of the model.

18.3.3 ENTRY AND ANALYSIS ENVIRONMENTS

While transistor- and behavioral-level simulation tools provide the core analysis capability for analog design, practical usability of simulation tools for design and verification of complex circuits has been enabled by advances in user interface technology: schematic entry systems, graphical interfaces, waveform displays, and scripting languages. The bulk of analog design at present takes place within such integrated design environments that tie analysis capabilities together with physical layout, verification, and parasitic extraction tools. While emerging analog synthesis tools [2] are beginning to appear in commercial use, the more traditional environments are centered around the capture–simulate–analyze task loop. In this, loop circuit topology and properties are described via a graphical schematic entry tool. From the database description thus created, a circuit netlist or other elaborated structural circuit description is generated, which is read and processed by the simulator to produce simulation results data (i.e., waveforms). Simulation waveforms are either processed in-line to obtain circuit performance measurements (delay, rise time, distortion, etc.) or stored in a waveform database for later analysis by a waveform display and calculation tool. As commercial examples of such environments, we would mention the Virtuoso Analog Design Environment (aka "Artist") from Cadence Design Systems, and the Cosmos system marketed by Synopsys.

18.4 ANALOG BEHAVIORAL AND POWER MODEL GENERATION TECHNIQUES

One of the biggest problems today is the lack of systematic methods to create good analog behavioral or performance models—a skill not yet mastered by the majority of analog designers—as well as the lack of any tools to automate this process. Fortunately, in recent years (as of 2006), research has started to develop methods that can automatically create models for analog circuits, both behavioral models for behavioral simulation and performance models for circuit sizing. Techniques used here can roughly be divided into fitting or regression approaches, constructive approaches, and model-order reduction methods.

18.4.1 FITTING OR REGRESSION METHODS

In the *fitting or regression approaches,* a parameterized mathematical model is proposed or constructed by the model developer, and the values of the parameters p are selected as to best approximate the known (simulated) circuit behavior, as schematically depicted in Figure 18.14. A systematic approach to regression-based model construction consists of several steps:

1. Selection of an appropriate model structure or template, often done ad hoc by the designer based on knowledge of the circuit behavior. The possible choices of model are vast. Some of the more common include polynomials, rational functions [31], and artificial neural networks. Recently (as of 2006), EDA researchers have begun to develop more constructive methods to build models from underlying circuits by utilizing results from statistical inference [32] and data mining [33], and we expect to see regression tree, k-nearest neighbor, and kernel forms such as support vector machines [34–36] to become more prominent in the future. Posynomial forms have attracted particular interest for optimization applications [37,21], as optimization problems involving models of this form can be recast as convex programs, leading to very efficient sizing of analog circuits.
2. Creation or selection of the simulation data to which to fit the model via an appropriate design-of-experiments scheme [21].
3. Selection of a model fidelity criterion. For example, the model can be fit by a least-squares error optimization where the model response matches the simulated (or measured) time-domain response of the real circuit as closely as possible in an average sense [25]. This is schematically depicted in Figure 18.14. The error could, for instance, be calculated as

(18.2)
$$\text{Error} = \int_0^T \left\| v_{\text{out,real}}(t) - v_{\text{out,model}}(t) \right\|^2 dt$$

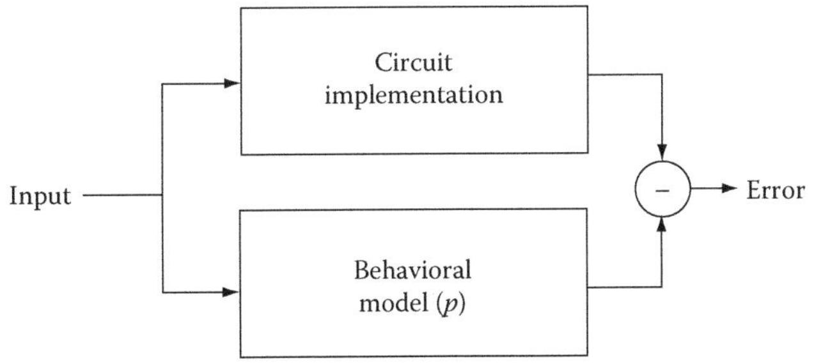

FIGURE 18.14 Basic flow of fitting or regression approach for analog behavioral model generation.

4. Selection of the optimization procedure to select the parameters (in some cases this step and the previous one are combined), such as by gradient descent or other gradient-based optimization, "boosting" [38], or stochastic optimization such as Markov-chain Monte Carlo or simulated annealing.
5. Validation of the final model. Without specific attention paid to model validation, it is quite common to find "overfit" models. Such models may have small error for the simulation data on which they were "trained," but very poor accuracy when slightly different circuit excitations are introduced when the model is put into use. Regularization may be introduced in step 3 to attempt to suppress such behavior, e.g., by modifying the model fidelity criterion to penalize large coefficients in a least-squares fit.

It should be clear that these fitting approaches can, in principle, be very generic as they consider the block as a blackbox and only look at the input–output behavior of the block which can easily be simulated (or measured). Once the model is generated, it becomes an implicit model of the circuit. However, hiding in each of the steps outlined above are daunting practical challenges. Chief among these comes the first step: for any hope of success, first a good model template must be proposed, which is not always trivial to do in an accurate way without knowing the details of the circuit. Even when good choices are possible, it may happen that the resulting model is specific for one particular implementation of the circuit. Likewise, the training set must exercise all possible operating modes of the circuit, but these can be hard to predict in advance.

To address these challenges, progress made in other areas such as in research on time-series prediction (e.g., support vector machines [35,36]) and data-mining techniques [33] are being pursued.

18.4.2 SYMBOLIC MODEL GENERATION METHODS

The second class of methods, the *constructive approaches*, try to generate or build a model from the underlying circuit description. Inherently, these are therefore white-box methods, as the resulting model is specific for the particular circuit, but there is a higher guarantee than with the fitting methods that it tracks the real circuit behavior well in a wider range. One approach, for instance, uses symbolic analysis techniques to generate first the exact set of algebraic/differential equations describing the circuit, which are then simplified within a given error bound of the exact response using both global and local simplifications [39]. The resulting simplified set of equations then constitutes the behavioral model of the circuit and tracks the behavior of the circuit nicely. The biggest drawback, however, is that the error estimation is difficult, and for nonlinear circuits heavily depends on the targeted response. Until now, the gains in CPU time obtained in this way are not high enough for practical circuits. More research in this area is definitely needed.

18.4.3 MODEL-ORDER REDUCTION METHODS

The third group of methods, the *model-order reduction methods*, are mathematical techniques that generate a model with lower order for a given circuit by direct analysis and manipulation of its detailed, low-level description; for example the nonlinear differential equations in a SPICE simulator, or the resistor–capacitor model describing extracted interconnect. Classical model-order reduction algorithms take as input a linear, time-invariant set of differential equations describing a state-space model, for example,

$$(18.3) \qquad \frac{dx}{dt} = Ax + Bu, \quad y = Cx + Du$$

where
   x represents the circuit state
   u are the circuit inputs
   y are the circuit outputs
   the matrices A, B, C, and D determine the circuit properties

As output, model-order reduction methods produce a similar state-space model $\tilde{A}$, $\tilde{B}$, $\tilde{C}$, $\tilde{D}$, but with a state vector $\tilde{x}$ (thus matrix description) of lower dimensionality, i.e., of lower order:

$$(18.4) \qquad \frac{d\tilde{x}}{dt} = \tilde{A}\tilde{x} + \tilde{B}u, \quad \tilde{y} = \tilde{C}\tilde{x} + \tilde{D}u$$

These reduced-order models simulate much more efficiently, while closely approximating the exact response; for example matching the original model closely up to some specified frequency.

Originally developed to reduce the complexity of interconnect networks for timing analysis, techniques such as asymptotic waveform evaluation (AWE) [40] used Padé approximation to generate a lower-order model for the response of the linear interconnect network. The early AWE efforts used explicit moment-matching techniques which were not numerically stable, and thus could not produce higher-order models that were needed to model circuits more complicated than resistor–capacitor networks, and Padé approximations often generate unstable or nonpassive reduced-order models. Subsequent developments using Krylov-subspace-based methods [41,42] resulted in methods like PVL that overcome many of the deficiencies of the earlier AWE efforts, and passive model construction is now guaranteed via projection-via-congruence such as used in PRIMA [43].

In recent years (as of 2006), similar techniques have also been extended in an effort to create reduced-order macromodels for analog/RF circuits. Techniques have been developed for time-varying models, particularly periodically time-varying circuits [44,45], and for weakly nonlinear circuits via polynomial-type methods that have a strong relation to Volterra series [45–47]. Current research (as of 2006) focuses on methods to model more strongly nonlinear circuits (e.g., using trajectory piecewise-linear [48] or piecewise-polynomial approximations [49]), and is starting to overlap with the construction of performance models, through the mutual connection to the regression and data-mining ideas [33,35,36].

Despite the progress made so far, more research in the area of automatic or systematic behavioral model generation or model-order reduction is certainly needed.

18.4.4 POWER/AREA ESTIMATION METHODS

Besides behavioral models, other crucial elements to compare different architectural alternatives and to explore trade-offs during system-level exploration and optimization are accurate and efficient power and area estimators [50]. They allow the assessment and comparison of the optimality of different design alternatives. Such estimators are functions that predict the power or area that is going to be consumed by a circuit implementation of an analog block (e.g., an ADC) with given specification values (e.g., resolution and speed). Since the implementation of the block is not yet known during high-level system design, and considering the large number of different possible implementations for a block, it is very difficult to generate these estimators with high absolute accuracy. However, for the purpose of comparing different design alternatives, the tracking accuracy of estimators with varying block specifications is of much more importance.

Such functions can be obtained in two ways. A first possibility is the derivation of analytic functions or procedures that given the block's specifications, return the power or area estimate. An example of a general yet relatively accurate power estimator that was derived based on the underlying operating principles for the whole class of CMOS high-speed Nyquist-rate ADCs (such as flash, two-step, pipelined, etc., architectures) is given by [50]

$$(18.5) \qquad \text{Power} = \frac{V_{dd}^2 L_{min}(F_{sample} + F_{signal})}{10^{(-0.1525 \cdot \text{ENOB} + 4.8381)}}$$

where

F_{sample} and F_{signal} are the clock and signal frequency respectively

ENOB is the effective number of bits at the signal frequency

The estimator is technology-scalable (V_{dd} and L_{min} are parameters of the model), and has been fitted with published data of real converters, and for more than 85% of the designs checked, the estimator has an accuracy better than 2.2×. Similar functions are developed for other blocks, but of course often a more elaborate procedure than a simple formula is needed. For example, for the case of high-speed continuous-time filters [50], a crude filter synthesis procedure in combination with operational transconductor amplifier behavioral models had to be developed to generate accurate results. The reason is that the filter implementation details, and hence the power and chip area, vary quite significantly with the specifications, requiring a rough filter synthesis to be performed to gather sufficient implementation detail knowledge to allow reliable estimates of power and area.

A second possibility to develop power/area estimators is to extract them from a whole set of data samples from available or generated designs through interpolation or fitting of a predefined function or an implicit function, e.g., a neural network. As these methods do not rely on underlying operating principles, extrapolations of the models have no guaranteed accuracy.

In addition to power and area estimators, *feasibility functions* are also needed that limit the high-level optimization to realizable values of the building block specifications. These can be implemented under the form of functions (e.g., a trained neural network or a support vector machine [51]) that return whether a block is feasible or not, or of the geometrically calculated feasible performance space of a circuit (e.g., using polytopes [52] or using radial base functions [53]).

18.5 SYMBOLIC ANALYSIS OF ANALOG CIRCUITS

Analog design is a very complex and knowledge-intensive process, which heavily relies on circuit understanding and related design heuristics. Symbolic circuit analysis techniques have been developed to help designers gain a better understanding of a circuit's behavior. A symbolic simulator is a computer tool that takes as input an ordinary (SPICE type) netlist, and returns as output (simplified) analytic expressions for the requested circuit network functions in terms of the symbolic representations of the frequency variable and (some or all of) the circuit elements [54,55]. These simulators perform the same function that designers traditionally do by hand analysis (even the simplification). The difference is that the analysis is now done by the computer, which is much faster, can handle more complex circuits, and does not make as many errors. An example of a complicated BiCMOS opamp is shown in Figure 18.15. The (simplified) analytic expression for the differential small-signal gain of this opamp in terms of the symbolic small-signal parameters of the opamp's devices, has been analyzed with the SYMBA tool [56], and is shown below:

$$(18.6) \qquad A_{V0} = \frac{g_{m,M2}}{g_{m,M1}} \frac{g_{m,M4}}{(g_{o,M4}g_{o,M5} / g_{m,M5} + g_{mb,M5}) + (G_a + g_{o,M9} + g_{o,Q2} / \beta_{Q2})}$$

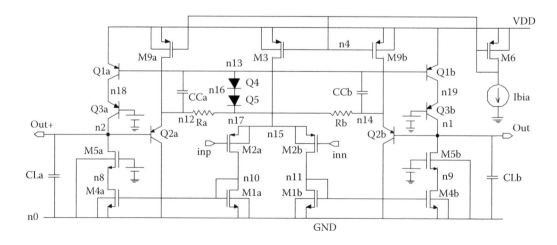

FIGURE 18.15 BiCMOS operational amplifier to illustrate symbolic analysis.

The symbolic expression gives a better insight into which small-signal circuit parameters predominantly determine the gain in this opamp and how the user has to design the circuit to meet a certain gain constraint. In this way, symbolic circuit analysis is complementary to numerical (SPICE) circuit simulation that was described in the previous section. Symbolic analysis provides a different perspective that is more suited for obtaining insight into a circuit's behavior and for circuit explorations, whereas numerical simulation is more appropriate for detailed design validation once a design point has been decided upon. In addition, the generated symbolic design equations also constitute a model of the circuit's behavior that can be used in CAD tasks, such as analog synthesis, statistical analysis, behavioral model generation, or formal verification [54].

At this moment, only symbolic analysis of linear or small-signal linearized circuits in the frequency domain is possible, both for continuous- and discrete-time (switched) analog circuits [54,55,57]. In this way, symbolic expressions can be generated for transfer functions, impedances, noise functions, etc. In addition to understanding the first-order functional behavior of an analog circuit, a good understanding of the second-order effects in a circuit is equally important for the correct functioning of the design in its system application later on. Typical examples are the Power Supply Rejection Ratio (PSRR) and Common Mode Rejection Ratio (CMRR) of a circuit, which are limited by the mismatches between circuit elements. These mismatches are represented symbolically in the formulas. Another example is the distortion or intermodulation behavior, which is critical in telecom applications. To this end, the technique of symbolic simulation has been extended to the symbolic analysis of distortion and intermodulation in weakly nonlinear analog circuits where the nonlinearity coefficients of the device small-signal elements appear in the expressions [46].

Exact symbolic solutions for network functions are however too complex for linear(ized) circuits of practical size, and even impossible to calculate for many nonlinear effects. Even rather small circuits lead to an astronomically high number of terms in the expressions, which can neither be handled by the computer nor interpreted by the circuit designer. Therefore, since the late 1980s, and in principle similar to what designers do during hand calculations, dedicated symbolic analysis tools have been developed that use heuristic simplification and pruning algorithms based on the relative importance of the different circuit elements to reduce the complexity of the resulting expressions and retain only the dominant contributions within user-controlled error tolerances. Examples of such tools are ISAAC [57], SYNAP [58], and ASAP [59] among many others. Although successful for relatively small circuits, the fast increase of the CPU time with the circuit size restricted their applicability to circuits between 10 and 15 transistors only, which was too small for many practical applications.

In the past years, however, an algorithmic breakthrough in the field of symbolic circuit analysis has been realized. The techniques of simplification before and during the symbolic expression generation, as implemented in tools such as SYMBA [56] and RAINIER [60], highly reduce the computation time and therefore enable the symbolic analysis of large analog circuits of practical size (like the entire 741 opamp or the example of Figure 18.15). In simplification before generation (SBG), the circuit schematic or some associated matrix or graph(s), are simplified before the symbolic analysis starts [61,62]. In simplification during generation (SDG), instead of generating the exact symbolic expression followed by pruning the unimportant contributions, the desired simplified expression is built up directly by generating the contributing dominant terms one by one in decreasing order of magnitude, until the expression has been generated with the desired accuracy [56,60]. In addition, the technique of determinant decision diagrams (DDDs) has been developed as a very efficient canonical representation of symbolic determinants in a compact nested format [63]. The advantage is that all operations on these DDDs are linear with the size of the DDD, but the DDD itself is not always linear with the size of the circuit. Very efficient methods have been developed using these DDDs [63,64].

All these techniques however, still result in large, expanded expressions, which restricts their usefulness for larger circuits. Therefore, for really large circuits, the technique of hierarchical decomposition has been developed [65,66]. The circuit is recursively decomposed into loosely connected subcircuits. The lowest-level subcircuits are analyzed separately, and the resulting symbolic expressions are combined according to the decomposition hierarchy. This results in the global nested expression for the complete circuit, which is much more compact than the expanded expression. The CPU time increases about linearly with the circuit size, provided that the coupling between the different subcircuits is not too strong. Also the DDD technique has been combined with hierarchical analysis in Ref. [67].

Another recent (as of 2006) extension is toward the symbolic analysis of linear periodically time-varying circuits, such as mixers [68]. The approach generalizes the concept of transfer functions to harmonic transfer matrices, generating symbolic expressions for the transfer function from any frequency band from the circuit's input signal to any frequency band from the circuit's output signal.

In addition, recently (as of 2006), approaches have also started to appear that generate symbolic expressions for large-signal and transient characteristics, for instance, using piecewise-linear approximations or using regression methods that fit simulation data to predefined symbolic templates. A recent (as of 2006) example of such a fitting approach is the automatic generation of symbolic posynomial performance models for analog circuits, which are created by fitting a pre-assumed posynomial equation template to simulation data created according to some design-of-experiments scheme [21]. These kinds of methods are very promising, since they are no longer limited to simple device models nor to small-signal characteristics only, but they still need some further research.

Based on the many research results in this area over the last decade, it can be expected that symbolic analysis techniques might soon be part of the standard tool suite of every analog designer, as an add-on to numerical simulation.

18.6 CONCLUSIONS

The last few years (as of 2006) have seen significant advances in both design methodology and CAD tool support for analog, mixed-signal, and RF designs. Mixed-signal simulators allow simulation of combined analog and digital circuits. The emergence of commercial AMS simulators supporting multiple analog abstraction levels (functional, behavioral, macromodel, and circuit level) enables top-down design flows in many industrial scenarios. In addition, there is increasing progress in system-level modeling and analysis allowing architectural exploration of entire systems, as well as in mixed-signal verification to anticipate problems related to embedding the analog blocks in a digital environment. A crucial element to enable this is the development of techniques to generate efficient behavioral models. An overview of model generation techniques has been given, including regression-based methods as well as model-order reduction techniques. Also the generation of performance models for analog circuit synthesis and of symbolic models that provide designers with insight into the relationships governing the performance behavior of a circuit have been described. Finally, the progress in symbolic analysis of both linear(ized) and weakly nonlinear analog circuits has been presented.

Despite this enormous progress in research and commercial EDA offerings that today enable the efficient design of analog blocks for embedding in mixed-signal integrated systems (ASICs, SoCs or SiPs), several problems still remain to be solved. Especially, behavioral model generation remains a difficult art that needs more research work toward automatic model generation techniques. Also, more advanced mixed-signal verification tools for signal integrity analysis (crosstalk, electromagnetic interference, etc.) need to be developed.

ACKNOWLEDGMENTS

The authors acknowledge the PhD students who have contributed to the reported results, as well as the IEEE and IEE for their kind permission to allow to reuse some parts of earlier publications in their overview chapter.

REFERENCES

1. International Technology Roadmap for Semiconductors, 2004, http://public.itrs.net. Accessed on July 2015.
2. G. Gielen and R. Rutenbar, Computer-aided design of analog and mixed-signal integrated circuits, *Proc. IEEE*, 88, 1825–1854, 2000.

3. H. Chang, E. Charbon, U. Choudhury, A. Demir, E. Felt, E. Liu, E. Malavasi, A. Sangiovanni-Vincentelli, and I. Vassiliou, *A Top-Down Constraint-Driven Design Methodology for Analog Integrated Circuits*, Kluwer Academic Publishers, Dordrecht, the Netherlands, 1997.

4. G. Gielen, Modeling and analysis techniques for system-level architectural design of telecom frontends, *IEEE Trans. Microw. Theory Tech.*, 50, 360–368, 2002.

5. G. Gielen, Top-down design of mixed-mode systems: Challenges and solutions, in *Advances in Analog Circuit Design*, Chapter 18, J. Huijsing, W. Sansen, R. Van der Plassche, Eds., Kluwer Academic Publishers, Dordrecht, the Netherlands, 1999.

6. The MEDEA+ Design Automation Roadmap, 2003, http://www.medea.org. Accessed on July 2015.

7. S. Donnay, G. Gielen, and W. Sansen, High-level analog/digital partitioning in low-power signal processing applications, *Proceedings of the Seventh International Workshop on Power and Timing Modeling, Optimization and Simulation (PATMOS)*, Louvain-la-Neuve, Belgium, 1997, pp. 47–56.

8. J. Crols, S. Donnay, M. Steyaert, and G. Gielen, A high-level design and optimization tool for analog RF receiver front-ends, *Proceedings of the International Conference on Computer-Aided Design (ICCAD)*, San Jose, CA, 1995, pp. 550–553.

9. P. Wambacq, G. Vandersteen, Y. Rolain, P. Dobrovolny, M. Goffioul, and S. Donnay, Dataflow simulation of mixed-signal communication circuits using a local multirate, multicarrier signal representation, *IEEE Trans. Circ. Syst., Part I*, 49, 1554–1562, 2002.

10. G. Vandersteen et al., Efficient bit-error-rate estimation of multicarrier transceivers, *Proceedings of the Conference on Design, Automation and Test in Europe (DATE)*, Munich, Germany, 2001, pp. 164–168.

11. P. Wambacq, G. Vandersteen, S. Donnay, M. Engels, I. Bolsens, E. Lauwers, P. Vanassche, and G. Gielen, High-level simulation and power modeling of mixed-signal front-ends for digital telecommunications, *Proceedings of the International Conference on Electronics, Circuits and Systems (ICECS)*, Pafos, Cyprus, September 1999, pp. 525–528.

12. B. De Smedt and G. Gielen, Models for systematic design and verification of frequency synthesizers, *IEEE Trans. Circ. Syst., Part II: Analog Digit. Signal Process.*, 46, 1301–1308, 1999.

13. L. Nagle and R. Rohrer, Computer analysis of nonlinear circuits, excluding radiation (CANCER), *IEEE J. Solid-State Circ.*, 6, 166–182, 1971.

14. A. Vladimirescu, *The SPICE Book*, Wiley, New York, 1994.

15. J. Roychowdhury and A. Mantooth, Simulation of analog and RF circuits and systems, Chapter 17, *Electronic Design Automation for IC Implementation, Circuit Design, and Process Technology*. L. Lavagno, I.L. Markov, G.E. Martin, and L.K. Scheffer (eds.), CRC Press, Boca Raton, FL, 2016, pp. 417–453.

16. S. Fang, Y. Tsividis, and O. Wing, SWITCAP: A switched-capacitor network analysis program, *IEEE Circ. Syst. Mag.*, 5, 4–10, 1983.

17. K. Francken and G. Gielen, A high-level simulation and synthesis environment for delta–sigma modulators, *IEEE Trans. Comput. Aided Des.*, 22, 1049–1061, 2003.

18. K. Kundert, Introduction to RF simulation and its applications, *IEEE J. Solid-State Circ.*, 34, 1298–1319, 1999.

19. R. Saleh, B. Antao, and J. Singh, Multi-level and mixed-domain simulation of analog circuits and systems, *IEEE Trans. Comput. Aided Des.*, 15, 68–82, 1996.

20. Virtual Socket Interface Alliance, Several documents including VSIA Architecture Document and Analog/Mixed-Signal Extension Document, http://www.vsia.org.

21. W. Daems, G. Gielen, and W. Sansen, Simulation-based generation of posynomial performance models for the sizing of analog integrated circuits, *IEEE Trans. Comput. Aided Des.*, 22, 517–534, 2003.

22. R. Phelps, M. Krasnicki, R. Rutenbar, L.R. Carley, and J. Hellums, A case study of synthesis for industrial-scale analog IP: Redesign of the equalizer/filter frontend for an ADSL CODEC, *Proceedings of the ACM/IEEE Design Automation Conference (DAC)*, Los Angeles, CA, 2000, pp. 1–6.

23. J. Vandenbussche et al., Systematic design of high-accuracy current-steering D/A converter macrocells for integrated VLSI systems, *IEEE Trans. Circ. Syst.*, Part II, 48, 300–309, 2001.

24. A. Vachoux, J.-M. Bergé, O. Levia, and J. Rouillard, Eds., *Analog and Mixed-Signal Hardware Description Languages*, Kluwer Academic Publishers, Dordrecht, the Netherlands, 1997.

25. E. Christen and K. Bakalar, VHDL-AMS—A hardware description language for analog and mixed-signal applications, *IEEE Trans. Circ. Syst., Part II: Analog Digit. Signal Process.*, 46, 1999, 1263–1272.

26. IEEE 1076.1 Working Group, *Analog and Mixed-Signal Extensions to VHDL*, http://www.eda.org/vhdl-ams/.

27. K. Kundert and O. Zinke, *The Designer's Guide to Verilog-AMS*, Kluwer Academic Publishers, Dordrecht, the Netherlands, 2004.

28. *Verilog-AMS Language Reference Manual*, version 2.2, November 2004, http://www.eda.org/verilog-ams/.

29. B. Wan, B.P. Hu, L. Zhou, and C.-J.R. Shi, MCAST: An abstract-syntax-tree based model compiler for circuit simulation, *Proceedings of the Custom Integrated Circuits Conference (CICC)*, San Jose, CA, September 2003, pp. 249–252.

30. L. Lemaitre, G. Coram, C. McAndrew, and K. Kundert, Extensions to Verilog-A to support compact device modeling, *Proceedings of the Workshop on Behavioral Modeling and Simulation (BMAS)*, San Jose, CA, October 2003, pp. 134–138.

31. B. Antao and F. El-Turky, Automatic analog model generation for behavioral simulation, *Proceedings of the IEEE Custom Integrated Circuits Conference (CICC)*, Boston, MA, May 1992, pp. 12.2.1–12.2.4.

32. T. Hastie, R. Tibshirani, and J. Friedman, *The Elements of Statistical Learning*, Springer, Heidelberg, Germany, 2001.

33. H. Liu, A. Singhee, R. Rutenbar, and L.R. Carley, Remembrance of circuits past: Macromodeling by data mining in large analog design spaces, *Proceedings of the Design Automation Conference (DAC)*, New Orleans, LA, June 2002, pp. 437–442.

34. B. Scholkopf and A. Smola, *Learning with Kernels*, MIT Press, Cambridge, MA, 2001.

35. J. Phillips, J. Afonso, A. Oliveira, and L.M. Silveira, Analog macromodeling using kernel methods, *Proceedings of the IEEE/ACM International Conference on Computer-Aided Design (ICCAD)*, San Jose, CA, November 2003, pp. 446–453.

36. Th. Kiely and G. Gielen, Performance modeling of analog integrated circuits using least-squares support vector machines, *Proceedings of the Design, Automation and Test in Europe (DATE) Conference*, Paris, France, February 2004, pp. 448–453.

37. M. Hershenson, S. Boyd, and T. Lee, Optimal design of a CMOS op-amp via geometric programming, *IEEE Trans. Comput. Aided Des. Int. Circ. Syst.*, 20, 1–21, 2001.

38. Y. Freund and R. Schapire, A short introduction to boosting, *J. Jpn. Soc. Artif. Intell.*, 14, 771–780, 1999.

39. C. Borchers, L. Hedrich, and E. Barke, Equation-based behavioral model generation for nonlinear analog circuits, *Proceedings of the IEEE/ACM Design Automation Conference (DAC)*, Las Vegas, NV, 1996, pp. 236–239.

40. L. Pillage and R. Rohrer, Asymptotic waveform evaluation for timing analysis, *IEEE Trans. Comput. Aided Des.*, 9, 352–366, 1990.

41. L.M. Silveira et al., A coordinate-transformed Arnoldi algorithm for generating guaranteed stable reduced-order models of RLC circuits, *Proceedings of the IEEE/ACM International Conference on Computer-Aided Design (ICCAD)*, San Jose, CA, 1996, pp. 288–294.

42. P. Feldmann and R. Freund, Efficient linear circuit analysis by Padé approximation via the Lanczos process, *IEEE Trans. Comput. Aided Des.*, 14, 639–649, 1995.

43. A. Odabasioglu, M. Celik, and L. Pileggi, PRIMA: Passive reduced-order interconnect macromodeling algorithm, *IEEE Trans. Comput. Aided Des.*, 17, 645–654, 1998.

44. J. Roychowdhury, Reduced-order modeling of time-varying systems, *IEEE Trans. Circ. Syst., Part II*, 46, 1273–1288, 1999.

45. J. Phillips, Projection-based approaches for model reduction of weakly nonlinear, time-varying systems, *IEEE Trans. Comput. Aided Des.*, 22, 171–187, 2003.

46. P. Wambacq, G. Gielen, P. Kinget, and W. Sansen, High-frequency distortion analysis of analog integrated circuits, *IEEE Trans. Circ. Syst., Part II*, 46, 335–345, 1999.

47. L. Peng and L. Pileggi, NORM: compact model order reduction of weakly nonlinear systems, *Proceedings of the Design Automation Conference*, Anaheim, CA, June 2003, pp. 472–477.

48. M. Rewienski and J. White, A trajectory piecewise-linear approach to model order reduction and fast simulation of nonlinear circuits and micromachined devices, *IEEE Trans. Comput. Aided Des.*, 22, 155–170, 2003.

49. N. Dong and J. Roychowdhury, Piecewise polynomial nonlinear model reduction, *Proceedings of the Design Automation Conference*, Anaheim, CA, June 2003, pp. 484–489.

50. E. Lauwers and G. Gielen, Power estimation methods for analog circuits for architectural exploration of integrated systems, *IEEE Trans. Very Large Scale Integr. (VLSI) Syst.*, 10, 155–162, 2002.

51. F. De Bernardinis, M. Jordan, and A. Sangiovanni-Vincentelli, Support vector machines for analog circuit performance representation, *Proceedings of the Design Automation Conference (DAC)*, Anaheim, CA, June 2003, pp. 964–969.

52. P. Veselinovic et al., A flexible topology selection program as part of an analog synthesis system, *Proceedings of the IEEE European Design & Test Conference (ED&TC)*, Paris, France, 1995, pp. 119–123.

53. R. Harjani and J. Shao, Feasibility and performance region modeling of analog and digital circuits, *Kluwer Int. J. Analog Integr. Circ. Signal Process.*, 10, 23–43, 1996.

54. G. Gielen, P. Wambacq, and W. Sansen, Symbolic analysis methods and applications for analog circuits: a tutorial overview, *Proc. IEEE*, 82, 287–304, 1994.

55. F. Fernández, A. Rodríguez-Vázquez, J. Huertas, and G. Gielen, *Symbolic Analysis Techniques—Applications to Analog Design Automation*, IEEE Press, Washington, DC, 1998.

56. P. Wambacq, F. Fernández, G. Gielen, W. Sansen, and A. Rodríguez-Vázquez, Efficient symbolic computation of approximated small-signal characteristics, *IEEE J. Solid-State Circ.*, 30, 327–330, 1995.

57. G. Gielen, H. Walscharts, and W. Sansen, ISAAC: A symbolic simulator for analog integrated circuits, *IEEE J. Solid-State Circ.*, 24, 1587–1596, 1989.

58. S. Seda, M. Degrauwe, and W. Fichtner, A symbolic analysis tool for analog circuit design automation, *Proceedings of the IEEE/ACM International Conference on Computer-Aided Design (ICCAD)*, Santa Clara, CA, 1988, pp. 488–491.

59. F. Fernández, A. Rodríguez-Vázquez, and J. Huertas, Interactive AC modeling and characterization of analog circuits via symbolic analysis, *Kluwer Int. J. Analog Integr. Circ. Signal Process.*, 1, 183–208, 1991.

60. Q. Yu and C. Sechen, A unified approach to the approximate symbolic analysis of large analog integrated circuits, *IEEE Trans. Circ. Syst., Part I*, 43, 656–669, 1996.

61. W. Daems, G. Gielen, and W. Sansen, Circuit simplification for the symbolic analysis of analog integrated circuits, *IEEE Trans. Comput. Aided Des.*, 21, 395–407, 2002.

62. J. Hsu and C. Sechen, DC small signal symbolic analysis of large analog integrated circuits, *IEEE Trans. Circ. Syst., Part I*, 41 (12), 817–828, 1994.

63. C.-J. Shi and X.-D. Tan, Canonical symbolic analysis of large analog circuits with determinant decision diagrams, *IEEE Trans. Comput. Aided Des.*, 19, 1–18, 2000.

64. W. Verhaegen and G. Gielen, Efficient DDD-based symbolic analysis of linear analog circuits, *IEEE Trans. Circ. Syst., Part II: Analog Digit. Signal Process.*, 49, 474–487, 2002.

65. J. Starzyk and A. Konczykowska, Flowgraph analysis of large electronic networks, *IEEE Trans. Circ. Syst.*, 33, 302–315, 1986.

66. O. Guerra, E. Roca, F. Fernández, and A. Rodríguez-Vázquez, A hierarchical approach for the symbolic analysis of large analog integrated circuits, *Proceedings of the IEEE Design Automation and Test in Europe Conference (DATE)*, Paris, France, 2000, pp. 48–52.

67. X.-D. Tan and C.-J. Shi, Hierarchical symbolic analysis of analog integrated circuits via determinant decision diagrams, *IEEE Trans. Comput. Aided Des.*, 19, 401–412, 2000.

68. P. Vanassche, G. Gielen, and W. Sansen, Symbolic modeling of periodically time-varying systems using harmonic transfer matrices, *IEEE Trans. Comput. Aided Des.*, 21, 1011–1024, 2002.

Layout Tools for Analog Integrated Circuits and Mixed-Signal Systems-on-Chip

A Survey

Rob A. Rutenbar, John M. Cohn, Mark Po-Hung Lin, and Faik Baskaya

19

CONTENTS

Abstract

Layout for analog circuits has historically been a manual, time-consuming, and trial-and-error task. The problem is not so much the size (in terms of the number of active devices) of these designs but rather the plethora of possible circuit and device interactions: from the chip substrate, from the devices and interconnects themselves, and from the chip package. In this short survey we briefly enumerate the basic problems faced by those who need to do layout for analog and mixed-signal designs and look at the evolution of the design tools and geometric/electrical optimization algorithms directed at these problems.

19.1 INTRODUCTION

Layout for digital integrated circuits (ICs) is usually regarded as a difficult task because of the *scale* of the problem: millions of gates, kilometers of routed wires, complex delay, and timing interactions. Analog designs and the analog portions of mixed-signal systems-on-chips (SoCs) are usually much smaller—up to 100 devices in a cell, usually less than 20,000 devices in a complete subsystem—and yet they are nothing if not *more* difficult to lay out. Why is this? The answer is that the complexity of analog circuits is not so much due to the number of devices, as to the complex *interactions* among the devices, the various continuous-valued performance specifications, the fabrication process, and the operating environment.

This would be less of a problem if analog circuits and subsystems were rare or exotic commodities, or if they were sufficiently generic that a few stable designs could be easily retargeted to each new application. Unfortunately, neither is true. The markets for application-specific ICs (ASICs), application-specific standard parts (ASSPs), and high-volume commodity ICs are characterized by an increasing level of integration. In recent years, complete systems that previously occupied separate chips are now being integrated on a single chip. Examples of such "systems on a chip" include telecommunications ICs such as modems, wireless designs such as components in radio frequency (RF) receivers and transmitters, and networking interfaces such as local area network ICs. Although most functions in such integrated systems are implemented with digital (or especially digital signal processing) circuitry, the analog circuits needed at the interface between the electronic system and the "real" world are now being integrated on the same die for reasons of cost and performance.

The booming market share of mixed-signal ASICs in complex systems for telecommunications, consumer, computing, biomedical, automotive, and Internet of Things applications is one direct result of this. But along with this increase in achievable complexity has come a significant increase in design complexity. A the same time, many present ASIC application markets are characterized by shortening product life cycles and time-to-market constraints. This has put severe pressure on the designers of these analog circuits, especially on those who lay out these designs. If cell-based library methodologies were workable for analog (as they are for semi-custom digital designs), layout issues would be greatly mitigated. Unfortunately, such methodologies fare poorly here. Most analog circuits are one of a kind or at best few of a kind on any given IC. Today, they are usually designed and laid out by hand. The problem recurs at the system level: the discipline of row-based layout as the basis for large function blocks, which is so successful in digital designs, is not (yet) as rigorously applied in analog designs.

Despite these problems, there is a thriving research community working to make custom analog layout tools a practical reality. This brief survey attempts to describe the history and evolution of these tools and extends two earlier reviews [1,2]. Our survey is organized as follows. We begin with a brief taxonomy of problems and strategies for analog layout in Section 19.2. Then, in Section 19.3, we review attacks on the cell-level layout problem. In Section 19.4, we review mixed-signal system-level layout problems. In Section 19.5, we review recent work on field-programmable analog arrays (FPAAs). Section 19.6 offers some concluding remarks. We end with an extensive and annotated bibliography.

19.2 ANALOG LAYOUT PROBLEMS AND APPROACHES

Before trying to categorize the various geometric strategies that have been proposed for analog layout, it is essential to first understand what the electrical problems that affect analog design are. We enumerate here briefly the salient effects that layout can have on circuit performance. References [27,110,138] together comprise a fairly complete treatment of these issues. There are really three core problems, which we first describe. We then briefly survey solution strategies here.

19.2.1 LOADING PROBLEMS

The nonideal nature of inter-device wiring introduces capacitive and resistive effects, which can degrade circuit performance. At sufficiently high frequencies, inductive effects arise as well. There are also parasitic RLC elements associated with the geometry of the devices themselves, for example, the various capacitances associated with MOS diffusions. All these effects are remarkably sensitive to a detailed (polygon-level) layout. For example, in MOS circuits, layout designers have a useful degree of freedom in that they can fold large devices (i.e., devices with large width-to-length ratios) and alter both their overall geometric shape and detailed parasitics. Folding transforms a large device (large channel width) into a parallel connection of smaller devices. The smaller devices are *merged*: parallel side-by-side alignment allows a single source or drain diffusion to be shared between adjacent gate regions, thus minimizing overall capacitance. Every diffused structure also has an associated parasitic resistance, which varies with its shape. These resistances can be reduced by minimizing the aspect ratio of all diffusions (reducing the width of the device), merging diffusions when possible, and strapping diffusion with low-resistance layers, such as metal where possible. Of course, this "strapping" may interfere with signal routing. Cohn et al. [37] offer a careful treatment of the layout issues here for MOS devices.

One of the distinguishing characteristics of analog layout, in comparison to digital layout, is the amount of effort needed to create correct layouts for atomic devices. MOS devices with large width-to-length ratios, bipolar devices with large emitter areas, etc., are common in analog designs and require careful attention to detail. Passive components that implement resistors, capacitors, or inductors are also more frequent in analog design and require a careful layout (see, e.g., Bruce et al. [16] as an example of procedural generation of complex MOS devices). Extremely low-level geometric details of the layout of individual devices and passives can have a significant circuit-level impact.

19.2.2 COUPLING PROBLEMS

Layout can also introduce unexpected signal coupling into a circuit, which may inject an unwanted electrical noise or even destroy its stability through unintended feedback. At lower frequencies, coupling may be introduced by a combination of capacitive, resistive, or thermal effects. At higher frequencies, inductive coupling becomes an issue. Especially, in the modern deep submicron digital processes in which analog systems are being integrated, coupling is an increasing problem. Metal conductors couple capacitively when two metal surfaces are sufficiently close, for example, if wires run in parallel on a single layer or cross on adjacent layers. If a parallel run between incompatible signals is unavoidable, a neutral wire such as a ground or reference line can be placed between them as a coupling shield.

Current flowing through a conductor also gives rise to a fluctuation in the voltage drop across the conductor's finite resistance. This fluctuation is then coupled into all the devices attached to the conductor. This effect is particularly problematic in power supply routing for analog cells on digital ICs. Sensitive analog performance often depends on an assumption of moderate stability in the power rails—which may not be true if the power distribution network is improperly laid out (see Refs. [136,138] for a detailed treatment of the issues in mixed-signal power distribution).

Signals can also be coupled through the silicon substrate or bulk, either through capacitive, resistive, or thermal effects. Because all devices share the same substrate, noise injected into the substrate is capacitively or resistively coupled into every node of the circuit. This is particularly problematic when analog circuits must share a substrate with an inherently noisy high-speed digital logic. On mixed-signal ICs, conventional solutions focus on *isolation*, either by locating the sensitive analog far away from the noise-injecting source or by surrounding the noise-sensitive circuits with a low-impedance diffusion guard ring to reduce the substrate noise level in a particular area. Unfortunately, the structure of the substrate and details of the layout of the power supply network that bias the substrate greatly affect even the qualitative behavior of this coupling. For example, epitaxial substrates have such a low resistivity that injected noise can "reappear" far from its origin on the chip surface. Simple isolation schemes do not always work well here. Bulk substrates are somewhat more resistive and noise may remain more local. Evolving silicon-on-insulator (SOI) processes may dramatically reduce this problem, but they are not yet in widespread use for commodity mixed-signal designs. References [102,119,138] are a good starting point for an analysis of the substrate-coupling problem in mixed-signal design.

In addition, since silicon is an excellent conductor of heat, local temperature variations due to current changes in a device can also cause signal coupling in thermally sensitive devices nearby. This phenomenon is most prevalent in bipolar or Bi–CMOS processes. By placing thermally sensitive devices far away from high-power thermally dissipating devices, this effect can be reduced. Placing matching devices symmetrically about thermally "noisy" sources can also be effective in reducing the effects of thermal coupling.

19.2.3 MATCHING PROBLEMS

Unavoidable variations that are present in all fabrication processes lead to small mismatches in the electrical characteristics of identical devices. If these mismatches are large enough, they can affect circuit performance by introducing electrical problems, such as offsets. Four major layout factors can affect the matching of identical devices: *area*, *shape*, *orientation*, and *separation*.

Device area is a factor because semiconductor processing introduces unavoidable distortions in the geometry, which make up the devices. Creating devices using identical geometry (identical shape) improves matching by ensuring that both devices are subject to the same (or at least *similar*) geometric distortions. Similarly, since the proportional effect of these variations tends to decrease as the size of the device increases, matching devices are usually made as large as the circuit performance and area constraints will allow. Since many processing effects, for example, ion-implantation, introduce anisotropic geometric differences, devices that must match should also be placed in the same orientation. Finally, spatial variations in process parameters tend to degrade the matching characteristics of devices, as their separation increases. This is largely due to process-induced gradients in parameters such as mobility or oxide thickness. Sensitivity to these effects can be reduced by placing devices, which must match well in close proximity. Devices that must be *extremely* well matched may be spatially interdigitated in an attempt to cancel out the effects of global process gradients.

Device matching, particularly of bipolar devices, may also be degraded by thermal gradients. Two identical devices at different points on a thermal gradient will have slight differences in VBE for a given collector current. To combat this, it is common practice to arrange thermally sensitive matching devices symmetrically around thermally generating noise sources. The parasitic capacitive and resistive components of interconnect can also introduce problems of matching in differential circuits, that is, those circuits that comprise two matching halves. A mismatch in the parasitic capacitance and resistance between the two matching halves of the circuit can give rise to offsets and other electrical problems. The most powerful technique used to improve interconnect parasitic matching is layout symmetry, in which the placement and wiring of matching circuits are forced to be identical, or as in the case of differential circuits, mirror symmetric.

19.2.4 LAYOUT SOLUTION STRATEGIES

We mentioned a variety of solution techniques in the aforementioned enumeration of layout problems: careful attention to atomic device layout, MOS merging, substrate noise isolation, symmetric layout, etc. However, these are really low-level *tactics* for dealing with specific problems for specific circuits in specific operating or fabrication environments. The more general question that we wish to address next is how the analog portion of a large mixed-signal IC is laid out—and what are the overall geometric *strategies* here?

We note first that like digital designs, analog designs are attacked hierarchically. However, analog systems are usually significantly smaller than their digital counterparts: 10,000–20,000 analog devices versus 100,000–1,000,000 digital gates. Thus, analog layout hierarchies are usually not as deep as their digital counterparts. The need for low-level attention to coupling and interaction issues is another force that tends to flatten these hierarchies. The typical analog layout hierarchy comprises three fundamentally different types of layout tasks:

1. *Cell-level layout*: The focus here is really device-level layout, placement, and routing of individual active and passive devices. At this level, many of the low-level matching, symmetry, merging, reshaping, and proximity management tactics for polygon-level optimization are applied. The goal is to create cells that are suitably insulated not only from fluctuations in fabrication and operating environment but from probable coupling with neighboring pieces of layout.
2. *System-level layout*: The focus here is on cell composition, arranging, and interconnecting the individual cells to complete an analog subsystem. At this level, isolation is one major concern: from nearby noisy digital circuits coupled in via substrate, power grid, or package. The other concern is signal integrity. Some of the digital signals from neighboring digital blocks need to penetrate into any analog regions, and these signals may be fundamentally incompatible with some sensitive analog signals.
3. *Programmable layout*: The focus here is on applying field-programmable gate array (FPGA; see, e.g., Rose et al. [151]) ideas to analog designs. The idea is to completely bypass the need to do custom cells. Rather, a set of programmable active and passive elements is connected with programmable wiring to achieve low-performance analog functions. This is a relatively recent implementation strategy, but we expect to see it grow.

We visit the ideas behind each of these layout strategies in the following three sections.

19.3 ANALOG CELL LAYOUT STRATEGIES

For our purposes, a "cell" is a small analog circuit, usually comprising not more than about 100 active and passive devices, which is designed and laid out as a single atomic unit. Common examples include operational amplifiers, comparators, voltage references, analog switches, oscillators, and mixers.

19.3.1 BASIC STRATEGIES

The earliest approaches to custom analog cell layout relied on procedural module generation. These approaches are a workable strategy when the analog cells to be laid out are relatively static, that is, necessary changes in device sizing or biasing result in little need for global alterations in device layout, orientation, reshaping, etc. Procedural generation schemes usually start with a basic geometric template (sometimes called a *topology* for the circuit), which specifies all necessary device-to-device and device-to-wiring spatial relationships. The generation completes the template by correctly sizing the devices and wires, respacing them as necessary. References [7,12] are examples dedicated mainly to opamps. The mechanics for capturing the basic design specifications can often be as familiar as a common spreadsheet interface (see, e.g., [13]). Owen et al.'s work [15] is a more recent example focused on opamp generation,

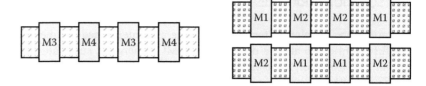

FIGURE 19.1 Device stacking with single-row inter-digitized or multi-row common-centroid placement.

and describes both languages for specifying these layouts and several optimized layout results. The system at Philips [14] is another good example of practical application of these ideas on complex circuits. Bruce et al. [16] show an example of module generation useful for atomic MOS devices (Figure 19.1).

Often however, changes in circuit design require a full custom layout, which can be handled with a *macrocell-style* strategy. The terminology is borrowed from digital floorplanning algorithms, which manipulate flexible layout blocks, arrange them topologically, and then route them. For analog cells, we regard the flexible blocks as devices to be reshaped and reoriented as necessary. Module generation techniques are used to generate the layouts of the individual devices. A placer then arranges these devices, and a router interconnects them—all while attending to the numerous parasitics, and couplings to which analog circuits are sensitive.

The earliest techniques used a mix of knowledge-based and constructive techniques for placement and routers usually adapted from common semicustom digital applications [20,22–24]. For example, a common constructive placement heuristic is to use the spatial relationships in a drawn circuit schematic as an initial basis from mask-level device placement [25]. Unfortunately, these techniques tended to be rather narrow in terms of which circuits could be laid out effectively.

The ILAC tool from CSEM was an important early attempt in this style [19,24]. It borrowed heavily from the best ideas in digital layout: efficient slicing tree floorplanning with flexible blocks, global routing via maze routing, detailed routing via channel routing, and area optimization by compaction. The problem with this approach was that it was difficult to extend these primarily digital algorithms to handle all the low-level geometric optimizations that characterize expert manual design. Instead, ILAC relied on a large, very sophisticated library of device generators.

ANAGRAM and its successor KOAN/ANAGRAM II from CMU kept the macrocell style but reinvented the necessary algorithms from the bottom-up, incorporating many manual design optimizations [21,27,29]. For example, the device placer KOAN relied on a very small library of device generators and migrated important layout optimizations into the placer itself. KOAN could dynamically fold, merge, and abut MOS devices and thus discover desirable optimizations to minimize parasitic capacitance during placement. KOAN was based on an efficient simulated annealing algorithm [3]. (Recent placers have also extended ideas from many topological module-packing representations, such as sequence pair [5], to handle analog layout tasks.) KOAN's companion, ANAGRAM II, was a maze style detailed area router, capable of supporting several forms of symmetric differential routing, mechanisms for tagging compatible and incompatible classes of wires (e.g., noisy and sensitive wires), parasitic cross talk avoidance, and over-the-device routing. Other device placers and routers operating in the *macrocell-style* have appeared (see, e.g., Refs. [28,33–36]), confirming its utility.

In the next generation of cell-level tools, the focus shifted to quantitative optimization of performance goals. For example, KOAN maximized MOS drain-source merging during layout, and ANAGRAM II minimized cross talk but without any specific quantitative performance targets. The routers ROAD [31] and ANAGRAM III [32] use improved cost-based schemes that route instead to minimize the deviation from acceptable parasitic bounds derived from designers or sensitivity analysis. The router in Ref. [40] can manage not just parasitic sensitivities but also basic yield and testability concerns. Similarly, the placers in Refs. [30,38,39] augment a KOAN-style model with sensitivity analysis so that performance degradations due to layout parasitics can be accurately controlled. Other tools in this style include Ref. [41].

In the fourth generation of CMOS analog cell research, the device placement task has been separated into two distinct phases: device *stacking*, followed by *stack placement*. By rendering the circuit as an appropriate graph of connected drains and sources, it is possible to identify natural clusters of MOS devices that ought to be merged—called *stacks*—to minimize parasitic capacitance. Malavasi et al. [76,77] and Malavasi and Pandini [80] give an exact algorithm to extract all the optimal stacks, and the placer in Refs. [78,79] extends a KOAN-style algorithm to choose dynamically the right stacking and the right placement of each stack. Basaran and Rutenbar [81] offer another variant of this idea. Instead of extracting all the stacks (which can be time-consuming since the underlying algorithm is exponential), this technique extracts one optimal set of stacks very fast (in linear time). The technique is useful in either the inner loop of a layout algorithm (to evaluate quickly a merging opportunity) or an interactive layout editor (to stack quickly a set of devices optimally) (Figure 19.2).

In the most recent generation of analog cell-level layout design tools, device matching during stack extraction and generation is further considered. A stack contains only a set of MOS devices whose device size and electrical properties are either identical or of specific ratios. To generate the layout of a stack with device matching, the devices usually form a single-row inter-digitized placement or a multi-row common-centroid placement within the device stack (Figure 19.1). Lin et al. [18] introduced a pattern-based approach and a layout automation flow, which integrates device stacking and stack placement with the consideration of device matching within each stack. The circuit patterns, placement patterns, and routing patterns of a stack can be customized and incrementally added to the pattern database of their tool. Based on the predefined patterns, their tool can automatically extract circuit patterns for device stacking and generate various layouts of each stack with device matching (Figure 19.2).

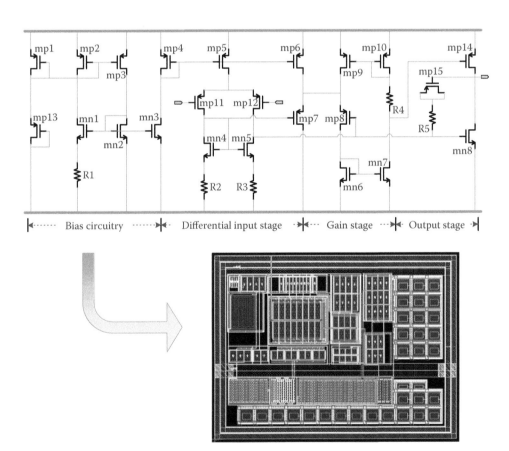

FIGURE 19.2 Example of matching-based extraction, placement, and routing for a two-stage CMOS amplifier using Lin et al.'s approach. (From Lin, P.-H. et al., A matching-based placement and routing system for analog design, *Proceedings of International Symposium on IEEE VLSI-DAT*, Hsinchu, Taiwan, 2007, pp. 16–19; Courtesy of SpringSoft, Inc.)

The notion of using sensitivity analysis to quantify the impact on final circuit performance of low-level layout decisions (e.g., device merging, symmetric placement/routing, parasitic coupling due to specific proximities, etc.) has emerged as a strategy to link the various approaches being taken for cell-level layout and system assembly. Several systems from U.C. Berkeley are notable here. The influential early formulation of the sensitivity analysis problem was by Choudhury and Sangiovanni-Vincentelli [92], who not only quantified layout impacts on circuit performance but also showed how to use nonlinear programming techniques to map these sensitivities into constraints on various portions of the layout task. Charbon et al. [94] extended these ideas to handle constraints involving nondetermininistic parasitics of the type that arise from statistical fluctuations in the fabrication process. In related work, Charbon et al. [93] also showed how to extract critical constraints on symmetry and match directly from a device schematic.

One final problem in the macrocell style is the separation of the placement and routing steps. In manual cell layout, there is no effective difference between a rectangle representing a wire and one representing part of a device. They can each be manipulated simultaneously. In a place-then-route strategy, one problem is estimating how much space to leave around each device for the wires. One solution strategy is *analog compaction* (see, e.g., [82,84,85]), in which we leave extra space during device placement and then compact. A more radical alternative is a *simultaneous device place and route*. An experimental version of KOAN [94] supported this by iteratively perturbing both the wires and the devices, with the goal of optimally planning polygon-level device and wire layout interactions.

As wireless and mobile design has proliferated, more RF designs have appeared. These offer yet another set of challenges at the cell level. RF circuits and higher frequency microwave circuits have unique properties, which make their automated layout impossible with the techniques developed for lower-frequency analog cells. Because every geometric property of the layout of an individual wire—its length, bends, and proximity to other wires or devices—may play a key role in the electrical performance of the overall circuit, most RF layouts are optimized for performance first and density second.

Most layout work targeting RF circuits comprise interactive tools that aid the designer to speed manual design iterations [87,88]. Other work in the area includes semiautomated approaches, which rely on the knowledge of the relative position of all cells [86]. However, these template-based approaches with predefined cells do limit the design alternatives possible. Charbon et al. [89] introduced a performance-driven router for RF circuits. In their approach, sensitivity analysis is employed to compute upper bounds for critical parasitics in the circuit, which the router then attempts to respect. Aktuna et al. [90,91] introduced the idea of a device-level floorplanning for these circuits; using a genetic algorithm, the tool simultaneously evolves placement and detailed routing under constraints on length, bends, phase, proximity, and planarity (Figure 19.3).

FIGURE 19.3 Microwave frequency (~60 GHz) device-level floorplan shows devices, planar wiring on a single layer, and several instances of "detours," which are inserted to match lengths and electrical properties of wires in this small design. (From Aktuna, M. et al., *IEEE Trans. CAD*, 18, 375, 1999. With permission.)

19.3.2 MANIPULATION OF LAYOUT CONSTRAINTS

Modern analog layout strategies solve the loading, coupling, and matching problems by incorporating various topological and geometrical layout design constraints. Common analog layout constraints include symmetry or symmetry island, boundary, proximity, regularity, common centroid, current/signal path, and current density (see Table 19.1).

The layout design constraints, which are usually associated with two or more devices/nets, can be either manually assigned or automatically extracted from a circuit netlist. The algorithms presented in [54,93,94,95] can detect symmetry constraints. In addition, Long et al. extracted current and signal paths/flows from an analog circuit. Eick et al. [95] further extracted common-centroid, proximity, alignment, and other hierarchical placement constraints based on building block recognition and structural analysis of a circuit netlist. According to their experimental results, the more constraints are extracted and identified, the better the placement quality achieved by a modern analog placer, such as Plantage [45].

In order to generate a DRC-free device-level placement and to satisfy all the aforementioned topological layout constraints, most of the modern analog placement approaches applied topological representations instead of absolute ones, as the topological representations can effectively and efficiently represent and generate various nonoverlapped placements. The main weakness of the absolute method lies in the fact that it may generate an infeasible placement with overlapped modules. Therefore, a post-processing step must be performed to eliminate this condition, which implies a longer computation time.

The topological representations, which have been adopted by modern analog placement approaches, include slicing tree, O-tree, B*-tree, sequence pair (SP), transitive closure graphs (TCG), and corner block list (CBL). The feasibilities of various topological layout constraints based on different topological representations were also explored during the past decade. Table 19.1 summarizes the topological layout constraints handled by modern analog placement approaches and the corresponding topological representations they applied.

Symmetry is the most common and important topological analog layout constraints, which has been extensively studied in the literature. Symmetry constraints can be formulated in terms of symmetry types, symmetry groups, symmetry pairs, and self-symmetric devices. In an analog layout design, a symmetry group may contain some symmetry pairs and self-symmetric devices with respect to a certain symmetry type. A symmetry type may correspond to a symmetry axis, in either horizontal or vertical direction. Balasa et al. derived the symmetric-feasible conditions for several popular floorplan representations, including O-tree [60], B*-tree [65], and sequence pair [62]. To efficiently explore the solution space in the symmetric-feasible binary trees, they augmented B*-trees using various data structures, including segment trees [66], red–black trees [67], and deterministic skip lists [73]. Three more recent works [48,72,74] took further advantage of the symmetric-feasible condition of the sequence pair [62]. Koda et al. [74] proposed a linear programming–based method, and Tam et al. [48] introduced a dummy node and additional constraint edges for each symmetry group after obtaining a symmetric-feasible sequence pair.

TABLE 19.1   Topological Layout Constraints Handled by Modern Analog Placement Approaches and the Corresponding Topological Representations They Applied

	Slicing Tree	O-Tree	B*-Tree	SP	TCG	CBL
Boundary	[49]		[46,47]	[48]		
Common centroid			[45]	[50]		
Proximity	[52]		[4,47,51]	[48]		
Signal/Current path	[55]		[75]			
Regularity			[57]	[58]		
Symmetry	[55,63]	[60]	[44–47,51,57,59,65–69, 73,75]	[48,50,58,62,70, 71,72]	[53,64]	[61]
Symmetry island	[55]		[44,46,47,51,59,75]			

Source: Wu, P.-H., *IEEE Trans. Comput. Aided Des. Integr. Circ. Syst.*, 33(6), 879, June 2014.

TABLE 19.2   Perturbation and Packing Time of Different Topological Representations for Analog Placement Considering Symmetry Constraints

Topological Representations for Analog Placement Considering Symmetry Constraints	Perturbation Time	Packing Time
B*-tree [65]	$O(\lg n)$	$O(n^2)$
B*-tree w/Seg. Tree [66]	$O(\lg n)$	$O(n \lg n)$
B*-tree w/RB tree [67]	$O(\lg n)$	$O(n \lg n)$
B*-tree w/Skip list [73]	$O(\lg n)$	$O(n \lg n)$
O-tree [60]	$O(\lg n)$	$O(n^2)$
SP [62]	$O(1)$	$O(n^2)$
SP w/LP [74]	$O(1)$	$O(n^2)$
SP w/Dummy [48]	$O(1)$	$O(n^2)$
SP w/Priority Queue [72]	$O(1)$	$O(mn \lg \lg n)$
TCG-S [64]	$O(n^2)$	$O(n^2)$
TCG [53]	$O(n)$	$O(n^2)$
Symmetry island–feasible B*-tree [44]	$O(\lg n)$	$O(n)$
Symmetry island–feasible slicing tree [55]	$O(\lg n)$	$O(n^2)$

Krishnamoorthy et al. [72] proposed an $O(mn \lg \lg n)$ packing-time algorithm by employing the priority queue, where m is the number of symmetry groups and n is the number of devices. Lin et al. [64] introduced a basic set of symmetric-feasible conditions in the transitive closure graphs. Zhang et al. [53] further presented a complete set of symmetric-feasible conditions in the transitive closure graphs and improved the perturbation scheme for the symmetric-feasible transitive closure graphs.

As the difference of an electrical parameter between two identical devices is proportional to the square of the distance between the two devices [6], it is of significant importance for the symmetric devices of a symmetry group to be placed in close proximity. To model a symmetry group in close proximity, Lin et al. [44] introduce the concept of symmetry islands, which is a placement of a symmetry group in which each device in the group abuts at least one of the other devices in the same group, with all devices in the symmetry group forming a connected placement. They further explored the feasibility of generating symmetry-island placements based on B*-trees and presented a linear-time-packing algorithm for the placement with symmetry constraints. In addition to the symmetry island–feasible conditions in B*-trees, Wu et al. [55] also presented the symmetry island–feasible conditions in slicing trees. Table 19.2 summarizes the symmetry constraints handled by modern analog placement approaches based on different topological representations that they applied.

There are several open problems in cell-level layout. For example, the optimal way to couple the various phases of cell layout—stacking, placement, routing, compaction—to each other and back to circuit design (or redesign) remains a challenging problem. However, there is now a maturing base of workable transistor-level layout techniques to build on. We note that commercial tools offering device-level analog layout synthesis have emerged and have now been applied in a range of industrial applications in production use. Figure 19.4, an industrial analog cell produced by a commercial analog layout synthesis tool, is one such automatic layout example [42,43].

19.4 MIXED-SIGNAL SYSTEM LAYOUT

A mixed-signal *system* is a set of custom analog and digital functional blocks. At the system level, the problem is really an *assembly* problem [2]. Assembly means block floorplanning, placement, global, and detailed routing (including the power grid). Apart from parasitic sensitivities, two new problems at the chip level are *coupling* between noisy signals and sensitive analog signals and *isolation* from digital switching noise that couples through the substrate, power grid, or package.

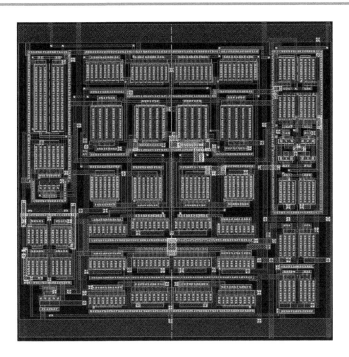

FIGURE 19.4   Commercial amplifier circuit, with custom device generators, custom wells, symmetric placement and routing, all synthesized automatically. (Courtesy of Cadence Design Systems , San Jose, CA.)

Like at the cell level, procedural generation remains a viable alternative for well-understood designs with substantial regularity. Many signal processing applications have the necessary highly stylized and regular layout structure. Procedural generation has been successful for many switched capacitor filters and data converters [14,96,100], especially regular, array-style blocks [17].

More generally though, work has focused on custom placement and routing at the block level, with layout optimization aimed at not only area but also signal coupling and isolation. Trnka et al. [124] offered one very early attempt aimed at bipolar array-style (i.e., fixed device image) layout by adapting semi-custom digital layout tools to this analog layout image. For row-based analog standard cell layout, an early elegant solution to the coupling problem was the *segregated channels* idea of authors in Refs. [121,122] to alternate noisy digital and sensitive analog wiring channels in a row-based cell layout. The strategy constrains digital and analog signals never to be in the same channel and remains a practical solution when the size of the layout is not too large. However, in modern multilevel interconnect technologies, this rigorous segregation can be overly expensive in area.

For large designs, analog channel routers were developed. In Gyurscik and Jeen [125], it was observed that a well-known digital channel routing algorithm [123] could be easily extended to handle critical analog problems that involve varying wire widths and wire separations needed to isolate interacting signals. Work at Berkeley substantially extended this strategy to handle complex analog symmetries and the insertion of shields between incompatible signals [126,127,129]. This work also introduced the idea of *constraint mapping*, which begins with parasitic sensitivities available from the analysis of the system (or the cell) to be laid out and transforms them into hard bounds on the allowable parasitics of each wire in each channel. The mapping process is itself a nonlinear programming problem, in this case, a quadratic programming formulation. These tools are particularly effective for stylized row-based layouts such as switched capacitor filters, where complex routing symmetries are necessary to balance subtle parasitics, and adjoint simulation methods can yield the necessary sensitivities.

The WREN [128] and WRIGHT [130,131] systems from CMU generalized these ideas to the case of arbitrary layouts of mixed functional blocks. WREN comprises both a mixed-signal global router and a channel router. WREN introduced the notion of *SNR-style* (signal-to-noise ratio) constraints for incompatible signals, and both the global and detailed routers strive to comply with designer-specified noise rejection limits on critical signals. WREN incorporates a constraint mapper (influenced by Choudhury and Sangiovanni-Vincentelli [92]) that transforms input noise rejection constraints from across-the-whole-chip form used by the global router into the per-channel per-segment form necessary for the channel router (as in [129]). WRIGHT uses a KOAN-style annealer to floorplan the blocks, but with a fast substrate noise-coupling evaluator, so that a simplified view of the substrate noise influences the floorplan. WRIGHT used a coarse-resistive mesh model with numerical pruning to capture substrate coupling; the approach in Charbon et al. [117] uses semi-analytical substrate modeling techniques, which allow fast update when blocks move, and can also support efficient noise sensitivity analysis (Figure 19.5). Based on the substrate noise model in [120], Cho and Pan [132,133] introduced the block preference directed graph, which is constructed based on the inherent noise characteristics of analog and digital blocks, to capture the preferred relative locations for substrate noise minimization during floorplanning with topological representations, such as B*-tree and sequence pair.

The substrate coupling problem is an increasingly difficult one, as more and faster digital logic is placed side by side with sensitive analog parts. One avenue of relevant work here seeks to model the substrate accurately, efficiently extract tractable electrical models of its conduction of overall chip noise, and understand qualitatively how various isolation mechanisms (e.g., separation, guard rings) will work. Of late, this has been an active area. References [101,103,105,107,114] address basic computational electromagnetic attacks on modeling and analysis of substrate coupling. The approaches vary in their discretization of the substrate, their numerical technique to solve for the point-to-point resistance between two devices in the substrate, and their model-order reduction techniques to reduce potentially large extracted circuit-level substrate models to smaller, more efficient circuit models. Su et al. [102] offer experimental data from test circuits on the mechanisms of substrate noise conduction for CMOS mixed-signal designs in epitaxial substrates. Charbon et al. [117] and Miliozzi et al. [116] address substrate coupling in the context of linking substrate modeling with the generation of constraints on allowable noise in the synthesis and layout process. Mitra et al. [106] and Miliozzi et al. [115] address the problem of estimating substrate current injection; Mitra et al. [106] use a circuit-level switching model and circuit simulation and transform simulation results into an equivalent single-tone sinusoid

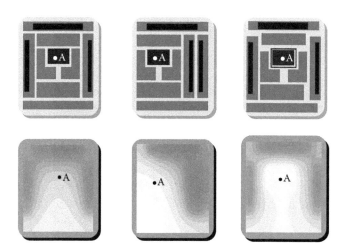

FIGURE 19.5 Example of a small, synthetic floorplan optimized by WRIGHT [130,131] under substrate noise constraints. Dark bars show noise sources; the module labeled "A" is sensitive to the overall noise level. From left to right, we see the result of floorplans generated under increasingly tight constraints on the allowed noise seen at "A." Bottom figures show iso-voltage contours in the substrate. At the far right, we see "A" has had a guard ring added to meet the tight noise spec. (From Mitra, S. et al., Substrate-aware mixed-signal macrocell-placement in WRIGHT, *Proceedings of IEEE Custom IC Conference*, San Diego, CA, 1994, pp. 529–532; Mitra, S. et al., *IEEE JSSC*, 30, 269, 1995.)

with the same total energy as the original random switching waveform. Miliozzi et al. [115] use a digital simulator to capture simple digital switching waveforms that are then combined with precharacterized circuit-level injection models to estimate block-level injection. Tsukada and Makie-Fukuda [118] suggest an active guard ring structure to mitigate substrate noise, based on some of these modeling ideas. Verghese and Allstot [119] offer a survey of substrate modeling, extraction, reduction, and injection work, along with a review of how substrate issues are dealt with in current mixed-signal design methodologies.

The thermal issue becomes more and more important in analog and mixed-signal layouts, especially when a system integrates high-power devices and other thermally sensitive devices into the same chip. The RF system is one of the examples, which contains power devices in power amplifier and thermally sensitive matched devices in other functional blocks. The heat coming from power devices may degrade electrical matching of thermally sensitive matched devices. References [59,61] addressed the thermal-driven and thermal-aware analog or mixed-signal placement. Liu et al. [61] proposed to place thermally sensitive matched devices under a nonuniform thermal profile, while Lin et al. [59] introduced a placement approach to achieve a uniform thermal profile for easily placing thermally sensitive matched devices.

Another important task in mixed-signal system layout is power grid design. Digital power grid layout schemes usually focus on connectivity, pad-to-pin ohmic drop, and electromigration effects. But these are only a small subset of the problems in high-performance mixed-signal chips, which feature fast-switching digital systems next to sensitive analog parts. The need to mitigate unwanted substrate interactions, handle arbitrary (nontree) grid topologies, and to design for transient effects such as current spikes, are serious problems in mixed-signal power grids. The RAIL system [134,138] addresses these concerns by casting mixed-signal power grid synthesis as a routing problem that uses fast AWE-based [4] linear system evaluation to model electrically the entire power grid, package, and substrate during layout. By allowing changes in both grid topology (where segments, power pins, and substrate contacts are located) and grid segment sizing, the tool can find power grid layouts to optimize AC, DC, and transient noise constraints. Techniques such as [106,115] are useful to estimate the digital switching currents needed here for power grid optimization. Chen and Ling [139] discuss a similar power distribution formulation applied to digital circuits (Figure 19.6).

Most of these system layout tools often rely on mature core algorithms from similar digital layout problems. Many have been prototyped both successfully and quickly. Several full, top-to-bottom prototypes have emerged (see, e.g., [117,140,143]).

There are many open problems here. Signal coupling and substrate/package isolation are overall still addressed via rather *ad hoc* means. There is still much work to be done to enhance existing constraint-mapping strategies and constraint-based layout tools to handle the full range of industrial concerns and to be practical for practicing designers.

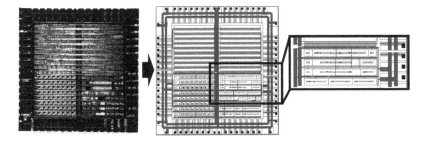

FIGURE 19.6   Example of a mixed-signal power grid from a commercial IBM ASIC, redesigned automatically by the RAIL tool [134,138] to meet strict AC, DC, and transient specifications. (From Stanisic, B.R. et al., Power distribution synthesis for analog and mixed signal ASICs in RAIL, *Proceedings of IEEE Custom IC Conference*, San Diego, CA, 1993, pp. 17.4.1–17.4.5; Stanisic, B.R. et al., *Synthesis of Power Distribution to Manage Signal Integrity in Mixed-Signal ICs*, Kluwer Academic Publishers, Norwell, MA, 1996.)

19.5 FIELD-PROGRAMMABLE ANALOG ARRAYS

Finally, we mention an analog layout style, which is radically different from those mentioned earlier. In the digital realm, FPGAs have revolutionized digital prototyping and rapid time-to-market designs [151]. Integrating programmable logic elements and programmable interconnect, these devices allow rapid customization with no fabrication steps. An obvious question is: can a similar technology be adapted for analog and mixed-signal designs? The apparent answer is a qualified "yes."

Early work, such as [144,146], directly mimicked the FPGA style of small primitive functions connectable by programmable interconnect. However, the loading on the signal path, which is already problematic in digital designs, proved even more deleterious here. Later designs such as [147,150] moved up to higher-level building blocks (e.g., OPAMPs, switches, capacitor arrays), and also focused a new energy on sensitive analog design of the programmable interconnect. These early efforts mostly used switched-capacitor techniques as programmable interconnects for minimizing the effects of routing parasitics on the signals. The first commercially available FPAA products were offered by the Anadigm Company in 2002, who were using switched-capacitor techniques for routing [152].

Switched-capacitor techniques, on the other hand, had serious bandwidth limitations on the signals to be processed. To overcome this limitation, continuous-time switches were used; but to minimise the impact of these switches on the signal integrity, architectures that offer special routing techniques had to be developed. In 2002, a single floating-gate MOSFET, which is also the enabling technology for flash memories, was proposed as the switch to be used in the routing fabric of the reconfigurable analog signal processor (RASP) architecture [153]. Several updates were made to the RASP family in the following years, eventually achieving a maximum bandwidth of 57 MHz through the switch matrix, and 5 MHz for a first-order low-pass filter implementation on the IC [154]. In addition, a flexible placement and routing tool was also developed for the different versions of RASP family as well as other similar FPAA topologies [155]. Another FPAA topology uses only a hexagonal chain of Operational Transconductance Amplifiers (OTA) that can be turned on or off to establish a signal path simultaneously with the desired gain elements [156].

FPAA layout consists of a rather small basic Configurable Analog Block (CAB) layout that is replicated over the entire IC, in addition to a digital circuit that can be used for programming and configuration of the device. So far, this task has not been very difficult; and so, layout is not really a problem. However, it will be interesting to see if these designs become more complicated (e.g., larger digital blocks with smaller analog blocks) in the future, and if, automatic layout becomes a requirement.

19.6 CONCLUSIONS

There has been substantial progress on tools for custom analog and mixed-signal circuit layout. Cast mostly in the form of numerical and combinatorial optimization tasks, linked by various forms of sensitivity analysis and constraint mapping, and leveraged by ever faster workstations, these tools are beginning to have a practical—even commercial—application. There remain many open problems here, along with some newly created ones, as analog circuits are increasingly embedded in unfriendly digital deep submicron designs. Given the demand for mixed-signal ICs, we expect no reduction in the interest in various layout tools to speed the design of these important circuits.

ACKNOWLEDGMENTS

We are grateful to our colleagues at Carnegie Mellon for their assistance in the preparation of this chapter. Rick Carley, Mehmet Aktuna, and Brian Bernberg in particular helped with early drafts. Sachin Sapatnekar of the University of Minnesota also offered helpful comments on the final version of the paper. The writing of this survey was supported in part by the NSF under contract 9901164, by the SRC, and by the Ministry of Science and Technology, Taiwan, under grant number NSC 102-2221-E-194-065-MY2.

REFERENCES

For clarity, we have grouped the references by topic, in roughly the order in which each topic area is visited in our review.

GENERAL REFERENCES

1. R.A. Rutenbar, Analog design automation: Where are we? Where are we going? *Proceedings of IEEE Custom IC Conference*, San Diego, CA, 1993, pp. 13.1.1–13.1.7.
2. L.R. Carley, G.G.E. Gielen, R.A. Rutenbar, and W.M.C. Sansen, Synthesis tools for mixed-signal ICs: Progress on frontend and backend strategies, *Proceedings of the ACM/IEEE Design Automation Conference*, Las Vegas, NV, 1996, pp. 298–303.
3. S. Kirkpatrick, C.D. Gelatt, and M.P. Vecchi, Optimization by simulated annealing, *Science*, 220, 671–680, 1983.
4. L.T. Pillage and R.A. Rohrer, Asymptotic waveform evaluation for timing analysis, *IEEE Trans. CAD*, 9, 352–366, 1990.
5. H. Murata, K. Fujiyoshi, S. Nakatake, and Y. Kajitani, VLSI module placement based on rectangle-packing by the sequence-pair method, *IEEE Trans. CAD*, 15, 1518–1524, 1996.
6. M. Pelgrom, A. Duinmaijer, and A. Welbers, Matching properties of MOS transistors, *IEEE J. Solid-State Circ.*, 24(5), 1433–1439, October 1989.

CELL-LEVEL MODULE GENERATORS

7. J. Kuhn, Analog module generators for silicon compilation, *VLSI Syst. Des.*, 75–80, 1987.
8. E. Berkcan, M. d'Abreu, and W. Laughton, Analog compilation based on successive decompositions, *Proceedings of ACM/IEEE Design Automation Conference*, Atlantic City, NJ, 1988, pp. 369–375.
9. H. Koh, C. Séquin, and P. Gray, OPASYN: A compiler for CMOS operational amplifiers, *IEEE Trans. CAD*, 9, 113–125, 1990.
10. H. Onodera et al., Operational amplifier compilation with performance optimization, *IEEE JSSC*, SC-25, 460–473, 1990.
11. J.D. Conway and G.G. Schrooten, An automatic layout generator for analog circuits, *Proceedings of EDAC*, Brussels, Belgium, 1992, pp. 466–473.
12. J.P. Harvey et al., STAIC: An interactive framework for synthesizing CMOS and BiCMOS analog circuits, *IEEE Trans. CAD*, 1402–1416, 1992.
13. R. Henderson et al., A spreadsheet interface for analog design knowledge capture and re-use, *Proceedings of IEEE CICC*, San Diego, CA, 1993, pp. 13.3, 13.1.1–13.3.4.
14. G. Beenker, J. Conway, G. Schrooten, and A. Slenter, Analog CAD for consumer ICs, Chapter 15, in *Analog Circuit Design*, J. Huijsing, R. van der Plassche, and W. Sansen, Eds., Kluwer Academic Publishers, Dordrecht, the Netherlands, 1993, pp. 347–367.
15. B.R. Owen, R. Duncan, S. Jantzi, C. Ouslis, S. Rezania, and K. Martin, BALLISTIC: An analog layout language, *Proceedings of IEEE Custom IC Conference*, Santa Clara, CA, 1995, pp. 41–44.
16. J.D. Bruce, H.W. Li, M.J. Dallabetta, and R.J. Baker, Analog layout using ALAS!, *IEEE JSSC*, 31, 271–274, 1996.
17. G. van der Plas, J. Vandenbussche, G. Gielen, and W. Sansen, Mondriaan: A tool for automated layout of array-type analog blocks, *Proceedings of IEEE Custom IC Conference*, Santa Clara, CA, 1998, pp. 485–488.
18. P.-H. Lin, H.-C. Yu, T.-H. Tsai, and S.-C. Lin, A matching-based placement and routing system for analog design, *Proceedings of IEEE International Symposium on VLSI Design Automation and Test*, Hsinchu, Taiwan, 2007, pp. 16–19.

DEVICE-LEVEL PLACEMENT AND ROUTING

19. J. Rijmenants, T.R. Schwarz, J.B. Litsios, and R. Zinszner, ILAC: An automated layout tool for CMOS circuits, *Proceedings of the IEEE Custom IC Conference*, Rochester, NY, 1988, pp. 7.6/1–7.6/4.
20. M. Kayal, S. Piguet, M. Declerq, and B. Hochet, SALIM: A layout generation tool for analog ICs, *Proceedings of the IEEE Custom IC Conference*, Rochester, NY, 1988, pp. 7.5/1–7.5/4.
21. D.J. Garrod, R.A. Rutenbar, and L.R. Carley, Automatic layout of custom analog cells in ANANGRAM, *Proceedings of ICCAD*, Santa Clara, CA, 1988, pp. 544–547.

22. M. Kayal, S. Piguet, M. Declerq, and B. Hochet, An interactive layout generation tool for CMOS ICs, *Proceedings of IEEE ISCAS*, Espoo, Finland, 3, 2431–2434, 1988.

23. M. Mogaki et al., LADIES: An automatic layout system for analog LSI's, *Proceedings of ACM/IEEE ICCAD*, Santa Clara, CA, 1989, pp. 450–453.

24. J. Rijmenants, J.B. Litsios, T.R. Schwarz, and M.G.R. Degrauwe, ILAC: An automated layout tool for analog CMOS circuits, *IEEE JSSC*, 24, 417–425, 1989.

25. S.W. Mehranfar, STAT—A schematic to artwork translator for custom analog cells, *Proceedings of IEEE Custom IC Conference*, Boston, MA, 1990, pp. 30.2/1–30.2/4.

26. S. Piguet, F. Rahali, M. Kayal, E. Zysman, and M. Declerq, A new routing method for full-custom analog ICs, *Proceedings of IEEE Custom IC Conference*, Boston, MA, 1990, pp. 27.7/1–27.7/4.

27. J.M. Cohn, D.J. Garrod, R.A. Rutenbar, and L.R. Carley, New algorithms for placement and routing of custom analog cells in ACACIA, *Proceedings of IEEE Custom IC Conference*, Boston, MA, 1990, pp. 27.6/1–27.6/4.

28. E. Malavasi, U. Choudhury, and A. Sangiovanni-Vincentelli, A routing methodology for analog integrated circuits, *Proceedings of ACM/IEEE ICCAD*, Santa Clara, CA, 1990, pp. 202–205.

29. J.M. Cohn, D.J. Garrod, R.A. Rutenbar, and L.R. Carley, KOAN/ANAGRAM II: New tools for device-level analog placement and routing, *IEEE JSSC*, 26, 330–342, 1991.

30. E. Charbon, E. Malavasi, U. Choudhury, A. Casotto, and A. Sangiovanni-Vincentelli, A constraint-driven placement methodology for analog integrated circuits, *Proceedings of IEEE CICC*, Boston, MA, 1992, pp. 28.2.1–28.2.4.

31. E. Malavasi and A. Sangiovanni-Vincentelli, Area routing for analog layout, *IEEE Trans. CAD*, 12, 1186–1197, 1993.

32. B. Basaran, R.A. Rutenbar, and L.R. Carley, Latchup-aware placement and parasitic-bounded routing of custom analog cells, *Proceedings of ACM/IEEE ICCAD*, Santa Clara, CA, 1993, pp. 415–421.

33. M. Pillan and D. Sciuto, Constraint generation and placement for automatic layout design of analog integrated circuits, *Proceedings of IEEE ISCAS*, London, U.K., 1994, pp. 355–358.

34. G.J. Gad El Karim, R.S. Gyurcsik, and G.L. Bilbro, Sensitivity driven placement of analog modules, *Proceedings of IEEE ISCAS*, London, U.K., 1994, pp. 363–366.

35. J.A. Prieto, J.M. Quintana, A. Rueda, and J.L. Huertas, An algorithm for the place-and-route problem in the layout of analog circuits, *Proceedings of IEEE ISCAS*, London, U.K., 1994, pp. 491–494.

36. E. Malavasi, J.L. Ganley, and E. Charbon, Quick placement with geometric constraints, *Proceedings of IEEE Custom IC Conference*, Santa Clara, CA, 1997, pp. 561–564.

37. J.M Cohn, D.J. Garrod, R.A. Rutenbar, and L.R. Carley, *Analog Device-Level Layout Automation*, Kluwer, Dordrecht, the Netherlands, 1994.

38. K. Lampaert, G. Gielen, and W. Sansen, Direct performance-driven placement of mismatch-sensitive analog circuits, *Proceedings of the European Design and Test Conference*, Paris, France, 1995, p. 597.

39. K. Lampaert, G. Gielen, and W.M. Sansen, A performance-driven placement tool for analog integrated circuits, *IEEE JSSC*, 30, 773–780, 1995.

40. K. Lampaert, G. Gielen, and W. Sansen, Analog routing for manufacturability, *Proceedings of IEEE CICC*, San Diego, CA, 1996, pp. 175–178.

41. C. Brandolese, M. Pillan, F. Salice, and D. Sciuto, Analog circuits placement: A constraint driven methodology, *Proceedings of IEEE ISCAS*, Atlanta, GA, 1996, Vol. 4, pp. 635–638.

42. O. Stephan, Electronica: Cell-builder tool anticipates analog synthesis, *EE Times*, 1998, p. 9.

43. A.H. Shah, S. Dugalleix, and F. Lemery, High-performance CMOS-amplifier design uses front-to-back analog flow, *EDN Magazine*, Reed Electronics Group, 2002, pp. 83–91.

TOPOLOGICAL PLACEMENT AND CONSTRAINT HANDLING

44. P.-H. Lin, Y.-W. Chang, and S.-C. Lin, Analog placement based on symmetry-island formulation, *IEEE Trans. Comput. Aided Des. Integr. Circuits Syst.*, 28, 791–804, June 2009.

45. M. Strasser, M. Eick, H. Graeb, U. Schlichtmann, and F.M. Johannes, Deterministic analog circuit placement using hierarchically bounded enumeration and enhanced shape functions, *Proceedings of IEEE/ACM International Conference on Computer-Aided Design*, San Jose, CA, 2008, pp. 306–313.

46. C.-W. Lin, J.-M. Lin, C.-P. Huang, and S.-J. Chang, Performance-driven analog placement considering boundary constraint, *Proceedings of ACM/IEEE Design Automation Conference*, Anaheim, CA, 2010, pp. 292–297.

47. H.-F. Tsao, P.-Y. Chou, S.-L. Huang, Y.-W. Chang, M.P.-H. Lin, D.-P. Chen, and D. Liu, A corner stitching compliant b*-tree representation and its applications to analog placement, *Proceedings of IEEE/ACM International Conference on Computer-Aided Design*, San Jose, CA, 2011, pp. 507–511.

48. Q. Ma, L. Xiao, Y.-C. Tam, and E.F.Y. Young, Simultaneous handling of symmetry, common centroid, and general placement constraints, *IEEE Trans. Comput. Aided Des. Integr. Circ. Syst.*, 30, 85–95, January 2011.

49. F.Y. Young, D.F. Wong, and H.H. Yang, Slicing floorplans with boundary constraints, *IEEE Trans. Comput. Aided Des. Integr. Circ. Syst.*, 1, 1385–1389, September 1999.

50. L. Xiao and E.F.Y. Young, Analog placement with common centroid and 1-D symmetry constraints, *Proceedings of IEEE/ACM Asia and South Pacific Design Automation Conference*, Yokohama, Japan, 2009, pp. 353–360.

51. P.-H. Lin and S.-C. Lin, Analog placement based on hierarchical module clustering, *Proceedings of ACM/IEEE Design Automation Conference*, Anaheim, CA, pp. 50–55, 2008.

52. F.Y. Young, D.F. Wong, and H.H. Yang, Slicing floorplans with range constraint, *IEEE Trans. Comput. Aided Des. Integr. Circ. Syst.*, 19, 272–278, February 2000.

53. L. Zhang, C.-J.R. Shi, and Y. Jiang, Symmetry-aware placement with transitive closure graphs for analog layout design, *Proceedings of IEEE/ACM Asia South Pacific Design Automation Conference*, Seoul, Korea, March 2008, pp. 180–185.

54. D. Long, X. Hong, and S. Dong, Signal-path driven partition and placement for analog circuit, *Proceedings of IEEE/ACM Asia and South Pacific Design Automation Conference*, Yokohama, Japan, 2006, pp. 694–699.

55. P.-H. Wu, M.P.-H. Lin, T.-C. Chen, C.-F. Yeh, T.-Y. Ho, and B.-D. Liu, Exploring feasibilities of symmetry islands and monotonic current paths in slicing trees for analog placement, *IEEE Trans. Comput. Aided Des. Integr. Circ. Syst.*, 33(6), 879–892, June 2014.

56. F.Y. Young and D.F. Wong, Slicing floorplans with pre-placed modules, *Proceedings of IEEE/ACM International Conference on Computer-Aided Design*, San Jose, CA, 1998, pp. 252–258.

57. P.-Y. Chou, H.-C. Ou, and Y.-W. Chang, Heterogeneous B*-trees for analog placement with symmetry and regularity considerations, *Proceedings of IEEE/ACM International Conference on Computer-Aided Design*, San Jose, CA, 2011, pp. 512–516.

58. S. Nakatake, M. Kawakita, T. Ito, M. Kojima, K. Izumi, and T. Habasaki, Regularity-oriented analog placement with diffusion sharing and well island generation, *Proceedings of IEEE/ACM Asia and South Pacific Design Automation Conference*, Taipei, Taiwan, 2010, pp. 305–311.

59. M.P.-H. Lin, H. Zhang, M.D.F. Wong, and Y.-W. Chang, Thermal-driven analog placement considering device matching, *IEEE Trans. Comput. Aided Des. Integr. Circ. Syst.*, 30, 325–336, March 2011.

60. Y. Pang, F. Balasa, K. Lampaert, and C.-K. Cheng, Block placement with symmetry constraints based on the O-tree non-slicing representation, *Proceedings of ACM/IEEE Design Automation Conference*, 2000, pp. 464–467.

61. J. Liu, S. Dong, Y. Ma, D. Long, and X. Hong, Thermal-driven symmetry constraint for analog layout with CBL representation, *Proceedings of IEEE/ACM Asia South Pacific Design Automation Conference*, Yokohama, Japan, January 2007, pp. 191–196.

62. F. Balasa and K. Lampaert, Symmetry within the sequence-pair representation in the context of placement for analog design, *IEEE Trans. Comput. Aided Des. Integr. Circ. Syst.*, 19, 721–731, July 2000.

63. M.P.-H. Lin, B.-H. Chiang, J.-C. Chang, Y.-C. Wu, R.-G. Chang, and S.-Y. Lee, Augmenting slicing trees for analog placement, *Proceedings of IEEE International Conference on Synthesis, Modeling, Analysis and Simulation Methods and Applications*, Seville, Span, 2012, pp. 57–60.

64. J.-M. Lin, G.-M. Wu, Y.-W. Chang, and J.-H. Chuang, Placement with symmetry constraints for analog layout design using TCG-S, *Proceedings of IEEE/ACM Asia and South Pacific Design Automation Conference*, Shanghai, China, 2005, pp. 1135–1138.

65. F. Balasa, Modeling non-slicing floorplans with binary trees, *Proceedings of IEEE/ACM International Conference on Computer-Aided Design*, San Jose, CA, 2000, pp. 13–16.

66. F. Balasa, S.C. Maruvada, and K. Krishnamoorthy, Efficient solution space exploration based on segment trees in analog placement with symmetry constraints, *Proceedings of IEEE/ACM International Conference on Computer-Aided Design*, San Jose, CA, 2002, pp. 497–502.

67. F. Balasa, S.C. Maruvada, and K. Krishnamoorthy, Using red-black interval trees in device-level analog placement with symmetry constraints, *Proceedings of IEEE/ACM Asia and South Pacific Design Automation Conference*, Kitakyushu, Japan, 2003, pp. 777–782.

68. F. Balasa, S.C. Maruvada, and K. Krishnamoorthy, On the exploration of the solution space in analog placement with symmetry constraints, *IEEE Trans. Comput. Aided Des. Integr. Circ. Syst.*, 23, 177–191, February 2004.

69. S.C. Maruvada, K. Krishnamoorthy, S. Annojvala, and F. Balasa, Placement with symmetry constraints for analog layout using red-black trees, *Proceedings of IEEE International Symposium on Circuits and Systems*, Bangkok, Thailand, 2003, pp. 489–492.

70. F. Balasa and S.C. Maruvada, Using non-slicing topological representations for analog placement, *IEICE Trans. Fund. Electr.*, E84-A, 2785–2792, November 2001.

71. F. Balasa and K. Lampaert, Module placement for analog layout using the sequence-pair representation, *Proceedings of ACM/IEEE Design Automation Conference*, New Orleans, LA, 1999, pp. 274–279.

72. K. Krishnamoorthy, S.C. Maruvada, and F. Balasa, Topological placement with multiple symmetry groups of devices for analog layout design, *Proceedings of IEEE International Symposium on Circuits and Systems*, New Orleans, LA, 2007, pp. 2032–2035.
73. S. Maruvada, A. Berkman, K. Krishnamoorthy, and F. Balasa, Deterministic skip lists in analog topological placement, *Proceedings of IEEE International Conference of ASIC*, October 2005, Vol. 2, pp. 834–837.
74. S. Koda, C. Kodama, and K. Fujiyoshi, Linear programming-based cell placement with symmetry constraints for analog IC layout, *IEEE Trans. Comput. Aided Des. Integr. Circ. Syst.*, 26(4), 659–668, April 2007.
75. H.-C. Oh, S.-C. Chang-Chien, and Y.-W. Chang, Simultaneous analog placement and routing with current flow and current density considerations, *Proceedings of ACM/IEEE Design Automation Conference*, Austin, TX, June 2013.

OPTIMAL MOS DEVICE STACKING

76. E. Malavasi, D. Pandini, and V. Liberali, Optimum stacked layout for analog CMOS ICs, *Proceedings of IEEE Custom IC Conference*, San Diego, CA, 1993, p. 17.1.1.
77. V. Liberali, E. Malavasi, and D. Pandini, Automatic generation of transistor stacks for CMOS analog layout, *Proceedings of IEEE ISCAS*, Chicago, IL, 1993, pp. 2098–2101.
78. E. Charbon, E. Malavasi, D. Pandini, and A. Sangiovanni-Vincentelli, Simultaneous placement and module optimization of analog ICs, *Proceedings of ACM/IEEE Design Automation Conference*, San Diego, CA, 1994, pp. 31–35.
79. E. Charbon, E. Malavasi, D. Pandini, and A. Sangiovanni-Vincentelli, Imposing tight specifications on analog ICs through simultaneous placement and module optimization, *Proceedings of the IEEE Custom Integrated Circuits Conference* (CICC), San Diego, CA, 1994, pp. 525–528.
80. E. Malavasi and D. Pandini, Optimum CMOS stack generation with analog constraints, *IEEE Trans. CAD*, 14, 107–112, 1995.
81. B. Basaran and R.A. Rutenbar, An O(n) algorithm for transistor stacking with performance constraints, *Proceedings of ACM/IEEE DAC*, Las Vegas, NV, 1996, pp. 221–226.

DEVICE-LEVEL COMPACTION AND LAYOUT OPTIMIZATION

82. R. Okuda, T. Sato, H. Onodera, and K. Tamuru, An efficient algorithm for layout compaction-problem with symmetry constraints, *Proceedings of IEEE ICCAD*, Santa Clara, CA, 1989, pp. 148–151.
83. J. Cohn, D. Garrod, R. Rutenbar, and L.R. Carley, Techniques for simultaneous placement and routing of custom analog cells in KOAN/ANAGRAM II, *Proceedings of ACM/IEEE ICCAD*, Santa Clara, CA, 1991, pp. 394–397.
84. E. Felt, E. Malavasi, E. Charbon, R. Totaro, and A. Sangiovanni-Vincentelli, Performance-driven compaction for analog integrated circuits, *Proceedings of IEEE Custom IC Conference*, San Diego, CA, 1993, pp. 17.3.1–17.3.5.
85. E. Malavasi, E. Felt, E. Charbon, and A. Sangiovanni-Vincentelli, Symbolic compaction with analog constraints, *Int. J. Circuit Theory Appl.*, 23, 433–452, 1995.

RADIO FREQUENCY CELL LAYOUT

86. J.F. Zurcher, MICROS 3—A CAD/CAM program for fast realization of microstrip masks, *1985 IEEE MTT-S International Microwave Symposium Digest*, St. Louis, MO, June 1985.
87. R.H. Jansen, LINMIC: A CAD package for the layout-oriented design of single- and multi-layer MICs/MMICs up to mm-wave frequencies, *Microwave J.*, 29, 151–161, 1986.
88. R.H. Jansen, R.G. Arnonld, and I.G. Eddison, A comprehensive CAD approach to design of MMICs up to MM-wave frequencies, *IEEE J. MTT-T*, 36, 208–219, 1988.
89. E. Charbon, G. Holmlund, B. Donecker, and A. Sangiovanni-Vincentelli, A performance-driven router for RF and microwave analog circuit design, *Proceedings of IEEE Custom Integrated Circuits Conference*, Santa Clara, CA, 1995, pp. 383–386.
90. M. Aktuna, R.A. Rutenbar, and L.R. Carley, Device level early floorplanning for RF circuits, *Proceedings of ACM International Symposium on Physical Design*, Monterey, CA, 1998, pp. 57–64.
91. M. Aktuna, R.A. Rutenbar, and L.R. Carley, Device level early floorplanning for RF circuits, *IEEE Trans. CAD*, 18, 375–388, 1999.

CONSTRAINT GENERATION AND MAPPING TO PHYSICAL DESIGN

92. U. Choudhury and A. Sangiovanni-Vincentelli, Automatic generation of parasitic constraints for performance-constrained physical design of analog circuits, *IEEE Trans. CAD*, 12, 208–224, 1993.
93. E. Charbon, E. Malavasi, and A. Sangiovanni-Vincentelli, Generalized constraint generation for analog circuit design, *Proceedings of IEEE/ACM ICCAD*, Santa Clara, CA, 1993, pp. 408–414.
94. E. Charbon, P. Miliozzi, E. Malavasi, and A. Sangiovanni-Vincentelli, Generalized constraint generation in the presence of non-deterministic parasitics, *Proceedings of ACM/IEEE International Conference on CAD*, Santa Clara, CA, 1996, pp. 187–192.
95. M. Eick, M. Strasser, K. Lu, U. Schlichtmann, and H.E. Graeb, Comprehensive generation of hierarchical placement rules for analog integrated circuits, *IEEE Trans. CAD Integr. Circ. Syst.*, 30(2), 180–193, 2011.

SYSTEM-LEVEL MIXED-SIGNAL MODULE GENERATORS

96. W.J. Helms and K.C. Russel, Switched capacitor filter compiler, *Proceedings of IEEE CICC*, Rochester, NY, 1986, pp. 125–128.
97. H. Yaghutiel, A. Sangiovanni-Vincentelli, and P.R. Gray, A methodology for automated layout of switched capacitor filters, *Proceedings of ACM/IEEE ICCAD*, Santa Clara, CA, 1986, pp. 444–447.
98. G. Jusef, P.R. Gray, and A. Sangiovanni-Vincentelli, CADICS—Cyclic analog-to-digital converter synthesis, *Proceedings of IEEE ICCAD*, Santa Clara, CA, 1990, pp. 286–289.
99. H. Chang, A. Sangiovanni-Vincentelli, F. Balarin, E. Charbon, U. Choudhury, G. Jusuf, E. Liu, E. Malavasi, R. Neff, and P. Gray, A top-down, constraint-driven methodology for analog integrated circuits, *Proceedings of IEEE Custom IC Conference*, Boston, MA, 1992, pp. 8.4.1–8.4.6.
100. R. Neff, P. Gray, and A. Sangiovanni-Vincentelli, A module generator for high speed CMOS current output digital/analog converters, *Proceedings of IEEE Custom IC Conference*, Santa Clara, CA, 1995, pp. 481–484.

SUBSTRATE MODELING, EXTRACTION, AND COUPLING ANALYSIS

101. T.A. Johnson, R.W. Knepper, V. Marcello, and W. Wang, Chip substrate resistance modeling technique for integrated circuit design, *IEEE Trans. CAD*, CAD-3, 126–134, 1984.
102. D.K. Su, M. Loinaz, S. Masui, and B. Wooley, Experimental results and modeling techniques for substrate noise in mixed-signal integrated circuits, *IEEE JSSC*, 28, 420–430. 1993.
103. N. Verghese, D. Allstot, and S. Masui, Rapid simulation of substrate coupling effects in mixed-mode ICs, *Proceedings of IEEE Custom IC Conference*, San Diego, CA, 1993, pp. 18.3.1–18.3.4.
104. F. Clement, E. Zysman, M. Kayal, and M. Declerq, LAYIN: Toward a global solution for parasitic coupling modeling and visualization, *Proceedings of IEEE Custom IC Conference*, San Diego, CA, 1994, pp. 537–540.
105. R. Gharpurey and R.G. Meyer, Modeling and analysis of substrate coupling in ICs, *Proceedings of IEEE Custom IC Conference*, Santa Clara, CA, 1995, pp. 125–128.
106. S. Mitra, R.A. Rutenbar, L.R. Carley, and D.J. Allstot, A methodology for rapid estimation of substrate-coupled switching noise, *Proceedings of IEEE Custom IC Conference*, Santa Clara, CA, 1995, pp. 129–132.
107. N.K. Verghese, D.J. Allstot, and M.A. Wolfe, Fast parasitic extraction for substrate coupling in mixed-signal ICs, *Proceedings of IEEE Custom IC Conference*, Santa Clara, CA, 1995, pp. 121–124.
108. I.L. Wemple and A.T. Yang, Mixed signal switching noise analysis using Voronoi-tessellated substrate macromodels, *Proceedings of ACM/IEEE Design Automation Conference*, San Francisco, CA, 1995, pp. 439–444.
109. N. Verghese and D. Allstot, SUBTRACT: A program for efficient evaluation of substrate parasitics in integrated circuits, *Proceedings of ACM/IEEE ICCAD*, Santa Clara, CA, 1995, pp. 194–198.
110. N. Verghese, T. Schmerbeck, and D. Allstot, *Simulation Techniques and Solutions for Mixed-Signal Coupling in Integrated Circuits*, Kluwer Academic Publishers, Norwell, MA, 1995.
111. T. Smedes, N.P. van der Meijs, and A.J. van Genderen, Extraction of circuit models for substrate crosstalk, *Proceedings of ACM/IEEE ICCAD*, Santa Clara, CA, 1995, pp. 199–206.
112. K.J. Kerns, I.L. Wemple, and A.T. Yang, Stable and efficient reduction of substrate model networks using congruence transforms, *Proceedings of ACM/IEEE ICCAD*, Santa Clara, CA, 1995, pp. 207–214.
113. N.K. Verghese, D.J. Allstot, and M.A. Wolfe, Verification techniques for substrate coupling and their application to mixed signal IC design, *IEEE JSSC*, 31, 354–365, 1996.

114. R. Gharpurey and R.G. Meyer, Modeling and analysis of substrate coupling in integrated circuits, *IEEE JSSC*, 31, 344–352, 1996.

115. P. Miliozzi, L. Carloni, E. Charbon, and A.L. Sangiovanni-Vincentelli, SUBWAVE: A methodology for modeling digital substrate noise injection in mixed-signal ICs, *Proceedings of IEEE Custom IC Conference*, San Diego, CA, 1996, pp. 385–388.

116. P. Miliozzi, I. Vassiliou, E. Charbon, E. Malavasi, and A. Sangiovanni-Vincentelli, Use of sensitivities and generalized substrate models in mixed-signal IC design, *Proceedings of ACM/IEEE Design Automation Conference*, Las Vegas, NV, 1996, pp. 227–232.

117. E. Charbon, R. Gharpurey, R.G. Meyer, and A. Sangiovanni-Vincentelli, Semi-analytical techniques for substrate characterization in the design of mixed-signal ICs, *Proceedings of ACM/IEEE ICCAD*, Santa Clara, CA, 1996, pp. 455–462.

118. T. Tsukada and K.M. Makie-Fukuda, Approaches to reducing digital noise coupling in CMOS mixed signal LSIs, *IEICE Trans. Fundam. Electron. Commun. Comput. Sci.*, E80-A(2), 263–275, 1997.

119. N.K. Verghese and D.J. Allstot, Verification of RF and mixed-signal integrated circuits for substrate coupling effects, *Proceedings of IEEE Custom IC Conference*, Santa Clara, CA, 1997, pp. 363–370.

120. B.E. Owens, S. Adluri, P. Birrer, R. Shreeve, S.K. Arunachalam, and K. Mayaram, Simulation and measurement of supply and substrate noise in mixed-signal ICs, *IEEE J. Solid-State Circ.*, 40(2), 382–391, February 2005.

SYSTEM-LEVEL MIXED-SIGNAL PLACEMENT AND ROUTING

121. C.D. Kimble, A.E. Dunlop, G.F. Gross, V.L. Hein, M.Y. Luong, K.J. Stern, and E.J. Swanson, Autorouted analog VLSI, *Proceedings of IEEE Custom Integrated Circuits Conference*, Portland, OR, 1985, pp. 72–78.

122. A.E. Dunlop, G.F. Gross, C.D. Kimble, M.Y. Luong, K.J. Stern, and E.J. Swanson, Features in the LTX2 for analog layout, *Proceedings of IEEE ISCAS*, Kyoto, Japan, 1985, pp. 21–23.

123. H.H. Chen and E. Kuh, Glitter: A gridless variable width channel router, *IEEE Trans. CAD*, CAD-5, 459–465, 1986.

124. J. Trnka, R. Hedman, G. Koehler, and K. Lading, A device level auto place and wire methodology for analog and digital masterslices, *IEEE ISSCC Digest of Technical Papers*, San Francisco, CA, 1988, pp. 260–261.

125. R.S. Gyurcsik and J.C. Jeen, A generalized approach to routing mixed analog and digital signal nets in a channel, *IEEE JSSC*, 24, 436–442, 1989.

126. U. Choudhury and A. Sangiovanni-Vincentelli, Use of performance sensitivities in routing analog circuits, *Proceedings of IEEE ISCAS*, New Orleans, LA, 1990, pp. 348–351.

127. U. Choudhury and A. Sangiovanni-Vincentelli, Constraint generation for routing analog circuits, *Proceedings of ACM/IEEE DAC*, Orlando, FL, 1990, pp. 561–566.

128. S. Mitra, S. Nag, R.A. Rutenbar, and L.R. Carley, System-level routing of mixed-signal ASICs in WREN, *Proceedings of ACM/IEEE ICCAD*, Santa Clara, CA, 1992, pp. 394–399.

129. U. Choudhury and A. Sangiovanni-Vincentelli, Constraint-based channel routing for analog and mixed analog/digital circuits, *IEEE Trans. CAD*, 12, 497–510, 1993.

130. S. Mitra, R.A. Rutenbar, L.R. Carley, and D.J. Allstot, Substrate-aware mixed-signal macrocell -placement in WRIGHT, *Proceedings of IEEE Custom IC Conference*, San Diego, CA, 1994, pp. 529–532.

131. S. Mitra, R.A. Rutenbar, L.R. Carley, and D.J. Allstot, Substrate-aware mixed-signal macrocell-placement in WRIGHT, *IEEE JSSC*, 30, 269–278, 1995.

132. M. Cho and D.Z. Pan, Fast substrate noise-aware floorplanning with preference directed graph for mixed-signal SOCs, *Proceedings of IEEE/ACM Asia and South Pacific Design Automation Conference*, Yokohama, Japan, 2006, pp. 765–770.

133. M. Cho and D.Z. Pan, Fast substrate noise aware floorplanning for mixed signal SOC designs, *IEEE TVLSI*, 16(12), 1713–1717, December 2008.

MIXED-SIGNAL POWER DISTRIBUTION LAYOUT

134. B.R. Stanisic, R.A. Rutenbar, and L.R. Carley, Power distribution synthesis for analog and mixed signal ASICs in RAIL, *Proceedings of IEEE Custom IC Conference*, San Diego, CA, 1993, pp. 17.4.1–17.4.5.

135. B.R. Stanisic, R.A. Rutenbar, and L.R. Carley, Mixed-signal noise decoupling via simultaneous power distribution design and cell customization in RAIL, *Proceedings of IEEE Custom IC Conference*, San Diego, CA, 1994, pp. 533–536.

136. B.R. Stanisic, N.K. Verghese, R.A. Rutenbar, L.R. Carley, and D.J. Allstot, Addressing substrate coupling in mixed-mode IC's: Simulation and power distribution synthesis, *IEEE JSSC*, 29, 226–238, 1994.

137. B.R. Stanisic, R.A. Rutenbar, and L.R. Carley, Addressing noise decoupling in mixed-signal ICs: Power distribution design and cell customization, *IEEE JSSC*, 30, 321–326, 1995.

138. B.R. Stanisic, R.A. Rutenbar, and L.R. Carley, *Synthesis of Power Distribution to Manage Signal Integrity in Mixed-Signal ICs*, Kluwer Academic Publishers, Norwell, MA, 1996.

139. H.H. Chen and D.D. Ling, Power supply noise analysis methodology for deep submicron VLSI chip design, *Proceedings of ACM/IEEE Design Automation Conference*, Anaheim, CA, 1997, pp. 638–643.

EXAMPLES OF COMPLETE ANALOG LAYOUT FLOWS

140. R. Rutenbar et al., Synthesis and layout for mixed-signal ICs in the ACACIA system, in *Analog Circuit Design*, J.H. Huijsing, R.J. van de Plassche, and W.M.C Sansen, Eds., Kluwer Academic Publishers, Norwell, MA, 1996, pp. 127–146.

141. I. Vassiliou, H. Chang, A. Demir, E. Charbon, P. Miliozzi, and A. Sangiovanni-Vincentelli, A video driver system designed using a top-down constraint-driven methodology, *Proceedings of ACM/IEEE ICCAD*, San Jose, CA, 1996, pp. 463–468.

142. E. Malavasi, E. Felt, E. Charbon, and A. Sangiovanni-Vincentelli, Automation of IC layout with analog constraints, *IEEE Trans. CAD*, 15, 923–942, 1996.

143. H. Chang, E. Charbon, U. Choudhury, A. Demir, E. Felt, E. Liu, E. Malavasi, A. Sangiovanni-Vincentelli, and I. Vassiliou, *A Top-Down, Constraint-Driven Design Methodology for Analog Integrated Circuits*, Kluwer Academic Publishers, Norwell, MA, 1997.

FIELD PROGRAMMABLE ANALOG ARRAYS

144. M. Sivilotti, A dynamically configurable architecture for prototyping analog circuits, *Proceedings of the Fifth MIT Conference on Advanced Research in VLSI*, Cambridge, MA, 1988, pp. 237–258.

145. E.K.F. Lee and P.G. Gulak, A field programmable analog array based on MOSFET transconductors, *Electron. Lett.*, 28, 28–29, 1992.

146. E.K.F. Lee and P.G. Gulak, A transconductor-based field programmable analog array, *IEEE ISSCC Digest of Technical Papers*, San Francisco, CA, 1995, pp. 198–199.

147. H.W. Klein, Circuit development using EPAC technology: An analog FPGA, *Proceedings of the SPIE, International Society for Optical Engineering*, Philadelphia, PA, 2607, 1995, 136–144.

148. A. Bratt and I. Macbeth, Design and implementation of a field programmable analogue array, *Proceedings of ACM International Symposium on FPGAs*, Monterey, CA, 1996, pp. 88–93.

149. P. Chow and P.G. Gulak, A field programmable mixed analog digital array, *Proceedings of ACM International Symposium on FPGAs*, Monterey, CA, 1995, pp. 104–109.

150. C.A. Looby and C. Lyden, A CMOS continuous time field programmable analog array, *Proceedings of ACM International Symposium on FPGAs*, Monterey, CA, 1997, pp. 137–141.

151. S.D. Brown, R.J. Francis, J. Rose, and Z.G. Vranesic, *Field Programmable Gate Arrays*, Kluwer Academic Publishers, Norwell, MA, 1992.

152. Anadigm: Supplier of dynamically programmable analog signal processors (dpASP), USA. http://www.anadigm.com.

153. T.S. Hall, P. Hasler, and D.V. Anderson, Field programmable analog arrays: A floating-gate approach, in *FPL'02: Proceedings of the Reconfigurable Computing Is Going Mainstream, 12th International Conference on Field Programmable Logic and Applications*, Springer-Verlag, London, U.K., 2002, pp. 424–433.

154. A. Basu, S. Brink, C. Schlottmann, S. Ramakrishnan, C. Petre, S. Koziol, F. Baskaya, C.M. Twigg, and P. Hasler, A floating-gate-based field-programmable analog array, *IEEE J. Solid-State Circ.*, 45(9), 1781–1794, 2010.

155. F. Baskaya, D.V. Anderson, P. Hasler, and S.K. Lim, A generic reconfigurable array specification and programming environment (GRASPER), *Proceedings of ECCTD*, Antalya, Turkey, August 2009, pp. 619–622.

156. J. Becker, F. Henrici, S. Trendelenburg, M. Ortmanns, and Y. Manoli, A continuous-time hexagonal field-programmable analog array in 0.13 μm CMOS with 186MHz GBW, *Digest of Technical Papers in IEEE Solid-State Circuits Conference*, February 2008, pp. 70–596.

Physical Verification

Design Rule Checking

20

Robert Todd, Laurence Grodd, Jimmy Tomblin,
Katherine Fetty, and Daniel Liddell

CONTENTS

20.1 INTRODUCTION

After the physical mask layout for a circuit is created using a specific design process, it is evaluated by a set of geometric constraints, or rules, for that process. The main objective of design rule checking (DRC) is to achieve a high overall yield and reliability for the design. To meet this goal of improving die yields, DRC has evolved from simple measurement and Boolean checks to more involved rules that modify existing features, insert new features, and check the entire design for process limitations such as layer density. While design rule checks do not validate the design's logical functionality, they verify that the structure meets the manufacturing constraints for a given design type and process technology.

A completed layout consists not only of the geometric representation of the design but also data that provide support for the manufacture of the design. With each new technology advance, DRC includes additional manufacturing-related elements, and EDA vendors work with the manufacturing companies to develop tools to help manage design verification for these elements. Three such elements are design for manufacturing (DFM), pattern matching (PM), and multi-patterning technology (MPT).

Before discussing how DRC verification works, an understanding of DRC concepts is useful.

20.2 CONCEPTS

The physical mask layout consists of shapes on drawn layers that are grouped into one or more cells. A cell may contain a placement of another cell. If the entire design is represented in one cell, it is a flat design; otherwise, it is a hierarchical design. Figure 20.1 shows an example of a hierarchical design with a top cell that contains other instances of cells and primitive objects. In the lower left, a flat view of one cell is magnified to show its content.

In early design verification stages, the layout database may contain text objects, properties, or other attributes that further define the purpose of an object in the layout or provide information that ties an object or set of objects to the logical representation of the design.

Geometries in a verification flow are grouped into layers. Most verification systems provide unlimited numbers of layers of the following types:

- Drawn layers represent the original layout data. They are merged on input to the verification system to remove any overlap or abutment of geometry on the same layer.
- Polygon layers are the output of a layer creation operation such as a Boolean operation, a topological polygon operation, or a geometric measurement function. Here are pseudocode examples:

```
Gate = POLY Boolean AND DIFF
Floating_met1 = MET1 having no shared CONTACT area
Big_island = area of ACTIVE > 100 square units
```

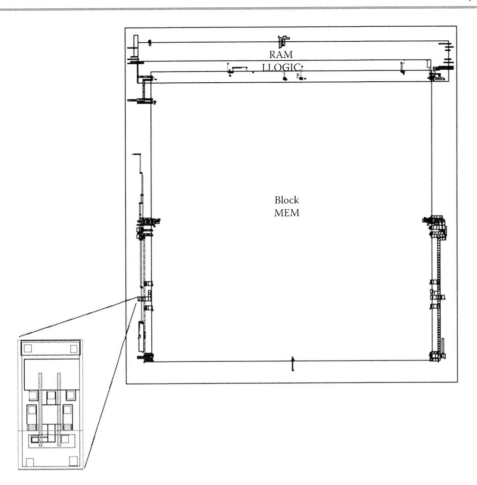

FIGURE 20.1 Mask data design hierarchy.

- Edge layers represent the edges of merged polygons as categorized by length, angle, or other attributes. Examples:

```
Long_met1_edge = length of MET1 edge > 10 units
Poly_gate_edges = POLY edges inside of DIFF
```

- Error layers contain clusters of one to four edges from a DRC spatial measurement for use in graphical result representation. Example:

```
enclosure of CONTACT by MET1 < 0.6 units
```

Figure 20.2 illustrates these layer types.

20.2.1 DRC OPERATIONS

To perform DRC, one must be able to select particular data, perform a rich set of operations on the selected data, and choose the output format. DRC operation output is either reported as an error or provided as an input to another DRC operation. Some operations have more information to return than a polygon or edge alone may convey, and so, they create a separate report with additional details.

The following design rule pseudocode combines two DRC operations and returns edge clusters representing the errors:

```
X = CP edges enclosed by ME between 2.24 and 2.26 units
Rule1 = output enclosure of X by POLY < 1.25 units
```

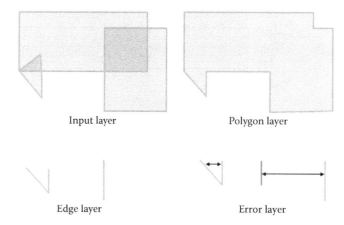

Input layer   Polygon layer

Edge layer   Error layer

FIGURE 20.2   Layer types.

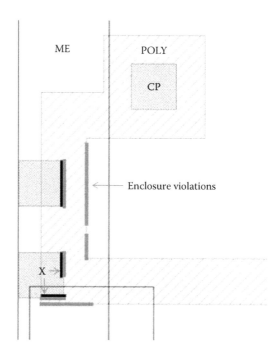

FIGURE 20.3   Error-based layer results.

Figure 20.3 shows the initial layers analyzed by Rule1 in a small region. The first DRC operation creates layer X to represent all the CP edges that are enclosed by the ME layer between 2.24 and 2.26 units. The three violations of the second enclosure operation between X and POLY are represented as edge clusters.

20.2.2 LANGUAGE-BASED DRCS

A complete DRC system provides a common language to describe all phases of layout verification. The language allows the same rules to be applied to checks for flat, cell/block based, and hierarchical full-chip modes. In addition to its primary role of specifying design rules, this DRC language may also derive layers for a process technology, define devices, and specify parasitic parameters.

The following pseudocode shows how DRC operations in a language-based system may be combined to modify a layout, by adding slots on wide metal lines and adding fill shapes to low-density regions of the design.

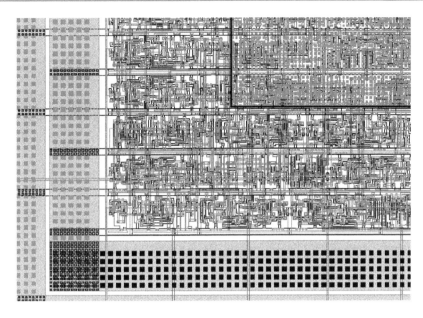

FIGURE 20.4   Metal slot and fill insertion.

```
MET1 is layer 16
// smooth any small metal jogs
resize_met1 = shrink MET1 by 3 units then expand by 3 units
// shrink the wide metal for proper slot enclosure
fat_met1 = shrink resize_met1 by 0.6 units
// metal1 square slots for Cu process
Rule met1_slots = output 0.5 x 0.5 unit squares spaced 0.6 units
inside of fat_met1
met1_density = area ratio of MET1 to 100 x 100 unit
grid box < 0.1
met1_fill = 0.3 x 0.3 unit squares spaced 0.25 units
inside of met1_density
met1_exp = expand MET1 by 0.25 units
Rule met1_mask = output met1_fill having no shared met1_exp area
```

First, the MET1 layer is resized in two steps to smooth the contour of the metal route and to select only metal regions that may validly contain a slot. The met1_slots rule creates squares of the specified dimensions inside the regions of fat_met1 derived previously. These squares are output as met1_slots results. The second part of this example calculates the ratio of MET1 area to a 10,000 square unit region of the layout, selecting regions containing an unacceptably low area (density) of MET1. The second met1_fill layer contains fill shapes in these low-density regions. The generated squares are placed a minimum distance away from the original MET1 shapes.

The results of this pseudocode are shown in Figure 20.4. The dark squares are metal slots along the bottom of the frame, and the metal fill is in the upper right.

20.3 DESIGN FOR MANUFACTURING

Traditional physical verification is dominated by simple design rule checks that identify sensitive layout features known to fail during manufacturing. As the technology for creating smaller device features advances, checking that these new devices are properly structured also becomes more complex. At advanced nodes, the increasing sensitivity of manufacturing to a combination of design features leads to the need for a rating system [1]. The rating system is employed by rules intended to improve manufacturing yield and robustness of the fabricated circuit. Different aspects of the design may be close to failure, and choices for improving the layout can be made by a designer or by an automated tool based upon the rating system.

```
sd_prop = LAYER_PROPERTY sd contact
   [sd_cnt = contact count in sd
    ratio = if sd_cnt > 0 then
               set (area of sd)/(area of contact)
            else
               set -1]
```

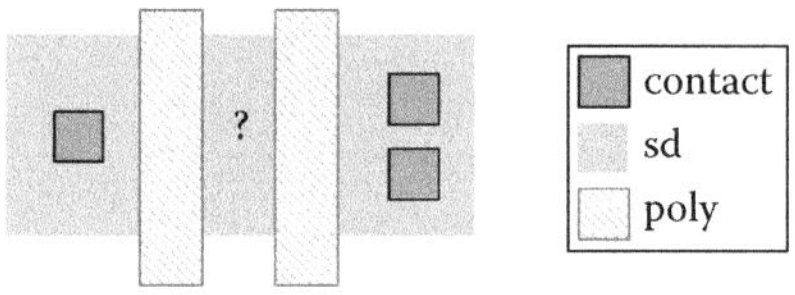

FIGURE 20.5   Layer with calculated properties.

One mechanism for recording abstract information about a single object or a collection of objects on one or more layers is a property. A property is an attribute containing numeric or string values (possibly both) that permit the classification of objects. Properties can be read from the layout database or be generated and attached to geometric objects during a verification run.

20.3.1 OPERATIONS AND PROPERTY ASSIGNMENT

Equation-based DRC checks allow a user to filter one or several layers with a combination of relationship equations and to assign each object on the layer a property result. The property assigned to a shape, a layer, or a cell definition is interpreted by a separate DRC operation and can be stored as part of the results data. The rule writer defines the equation and provides meaning to the property.

Figure 20.5 illustrates a LAYER_PROPERTY operation to count the contacts in a source-drain (sd) layer. The resulting sd_prop layer has a property assigned to each sd object to indicate the contact count and the ratio of sd area to contact area. The LAYER_PROPERTY equation accounts for arithmetic errors that might occur, such as when no contacts are in the sd region. The sd_prop layer can then be used later in the flow to classify sd polygons by the property values and report them.

20.4 PATTERN MATCHING

Pattern matching (PM) is the process of locating specific areas in a design that are identical or nearly identical to a defined set of layer geometries. A PM tool scans a design for specific predefined patterns of shapes. The patterns may be specified as containing one or more layers inside a rectilinear extent for the tool to match in the design layout. PM can find targeted patterns for special DRC checks, identify known lithographic problem areas, and create layers for classifying design features for other processing.

The major capabilities of a PM tool include the following:

- Enable the user to build a pattern, including elements that are spatially fixed or movable, relative to other elements in the pattern.
- Store a pattern library of the individual patterns.
- Select a subset of patterns from the pattern library.
- Search the design and choose regions that match patterns.
- Present results.

Figure 20.6 shows the results from running the PM tool with a library of two patterns. The layout has a highlighted box in the center for an area matching Pattern_1. Pattern_2 has a match in the upper right. Although an area in the lower right looks very similar to Pattern_2, it is not identical,

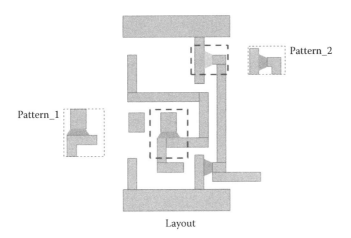

Layout

FIGURE 20.6 Pattern matching in a layout.

and so not highlighted. Once areas of the design that match a pattern are selected, they can be used for further analysis.

The PM tool can find matches of exact or *fuzzy* patterns. As shown in Figure 20.6, exact patterns permit no variability of edge placement relative to other edges in the pattern and obtain a match, although rotation and reflection of the pattern are permitted. Fuzzy patterns permit variability in edge placements within defined limits to obtain a match.

20.4.1 CAPTURING A PATTERN AND BUILDING A PATTERN LIBRARY

Individual patterns are created in an editing tool that marks specific areas of a design. For each area, the user marks in the design, a separate window displays a copy of visible layers, and the user identifies the edges or shapes for creating the pattern template. This pattern-capture capability allows the user to identify variable edges of the pattern. Each variable edge also has a range of allowed displacement.

The patterns are aggregated into libraries for the PM tool to match against a design. There may be over 100,000 patterns in a complete pattern library. Some foundry rule deck providers may build sample pattern libraries that are available as part of the process technology.

20.4.2 RUNNING THE PM TOOL

Once the pattern library is created, it is included as part of the rule deck for running against the layout. The library is applied like any other design rule check, and designers might simply use an existing library without having created it. Figure 20.7 shows the user's flow for using a pattern matching tool as part of a physical verification flow.

A match from the PM tool can be an error to report or an input to another processing step for a design rule check.

20.5 MULTI-PATTERNING

Technology advances allow layout features to be placed closely together such that it becomes difficult for them to produce accurately if placed on the die in a single manufacturing step. The drawn layer in the design layout represents the desired physical output, but one mask may be insufficient to produce that layer on the die.

When the 45 nm half-pitch processes were introduced, the lithographic solutions to create the physical mask for small objects required double exposure or double patterning [2]. This meant that a single drawn layer was split in to two. Initially, the configuration of the layers targeted for

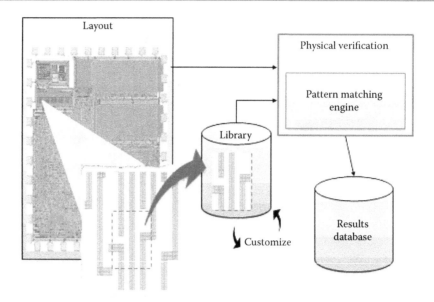

FIGURE 20.7 Pattern matching in the physical verification flow.

double patterning was primarily rectangles, so simple DRC spacing rules sufficed. As the requirement for double patterning began to include metal routing layers, other geometric configurations were included.

The physical verification steps for DRC include the ability to check that a drawn layer can be split into two (or more) output mask layers and meet a set of relationship constraints for overlap of and spacing between objects. This capability is known as multi-patterning technology (MPT).

An MPT function examines the design to verify that each mask layer can be divided into a *split layer* set given the manufacturing constraints. Figure 20.8 shows an example of a mask layer initially represented as a simple original layer. Objects on the mask layer that are too close together for reliable manufacture must be split into separate layers. The composite of the split layers represents the original mask layer.

Future advances will require splitting a mask layer into three or more layers.

Shapes on split layers may still be too close together for reliable manufacture. The DRC tool reports this situation to the designer using conflict rings or conflict paths, as shown in Figure 20.9.

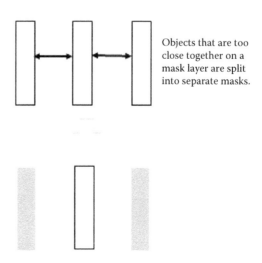

Objects that are too close together on a mask layer are split into separate masks.

FIGURE 20.8 Splitting of a mask layer in double patterning.

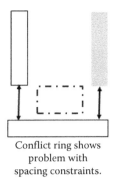

Conflict ring shows
problem with
spacing constraints.

FIGURE 20.9   Conflict ring.

When a conflict ring is reported, the designer needs to change the spacing for the affected shapes. More than one split layer may need to be modified to maintain the intent of the design. All the conflict rings must be resolved before the design can pass physical verification.

20.5.1 USING STITCHES TO RESOLVE CONFLICTS

Some flows allow for using a *stitch* to resolve a conflict ring or conflict path. A stitch is a region of the mask layer that is output on two split layers into which the original mask layer is divided. A stitch may be an original layer, an overlap of the split layers, or an object generated during the physical verification run. Stitches must also pass specific design rule checks.

Some systems have a stitch generation function. This function has a complete set of requirements for a valid stitch for the mask layer. The function scans the mask layer and creates a collection of valid stitch candidates. These stitch candidates are passed into the verification tool and are used to help automatically resolve conflicts in the layout.

20.5.2 DENSITY BALANCING BETWEEN SPLIT MASK LAYER SETS

Ideally, the manufacturing process for each of the split masks should be the same, implying that the split layers together should have roughly the same density across the extent of the original mask layer. Density measurement functions analyze the split layers, and the designer resolves any discrepancies. The design flow may allow stitches to improve the relative density of the split layers.

20.6 WHEN TO PERFORM DRC

Initially, DRC was considered primarily at the cell creation and block assembly levels of design, and physical layout was done by hand [3]. As the complexity in the layout increased, DRC became a requirement at more stages in the manufacturing process.

Interactive checks verify small cells or cell libraries, often from within the layout editor. As blocks are assembled, the integrated blocks also need to be verified. Areas between blocks and transitions from the block to the routing areas are verified for layout accuracy. After the full chip is assembled, DRCs may also create fill objects, insert slots in wide metal routes, create and analyze features for Optical Proximity Correction, or insert Sub-Resolution Assist Features (SRAFs, or scatter bars). System-on-chip (SoC) designs require checks at both the interactive and batch phases of the design. At the full-chip phase, DRC is used as a ready-to-manufacture design certification tool.

20.7 FLAT DRC

In older verification systems, the cell-based hierarchical input for a design was flattened, and the entire design was internally represented as one large cell. Overlaps between shapes on a layer were removed, or merged, and design rule checks were performed on the merged data.

Errors in the design were reported from the top-level view of the design. Since the input hierarchy no longer existed, any output from the system was represented as flat data.

20.8 HIERARCHICAL DRC

As designs become more complex, their verification in flat mode rapidly becomes impractical, consuming too many compute and storage resources and taking too long to complete. Modern verification systems take advantage of the input design hierarchy and other repetitions found in a physical layout design to identify blocks of the design that are analyzed once and then reused to significantly reduce verification time. The DRC results are reported at the lowest practical level of the design hierarchy.

Graphical output from a hierarchical verification tool retains or enhances the original design hierarchy. The tool can optimize the hierarchy using various processes, like interconnect layer cell recognition, automatic via recognition, selective placement flattening for billions of simple cells (like vias), or dense overlaps of large placements, and expanding certain types of array placements and duplicate placement removal.

20.9 GEOMETRIC ALGORITHMS FOR PHYSICAL VERIFICATION

All design rule checking programs, whether hierarchical or flat, require low-level algorithms that analyze geometric relationships between primitive data objects such as polygons and edges [4]. Many computational geometry algorithms are applied to perform this analysis, which seek to minimize the time and space resources required when the number of objects is large.

20.9.1 SCAN LINE–BASED ANALYSIS

Scan line–based sweep algorithms [5,6] have become the predominant form of low-level geometric analysis. A scan line sweep analyzes relationships between objects that intersect a virtual line, either vertical or horizontal, as that line is swept across the layout extent. Figure 20.10 shows a

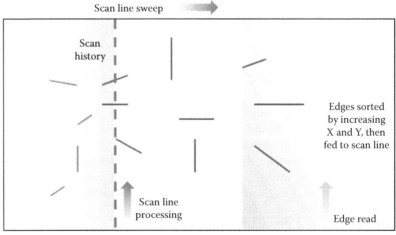

FIGURE 20.10 Scan line sweep analysis of a layout.

scan line moving across the extent of the layout, analyzing geometric data represented as edges. The edges provided as input to the scan line are ordered in increasing X and increasing Y.

20.9.1.1 TIME COMPLEXITY OF THE SCAN LINE ALGORITHM

In practice, the number of objects intersecting the scan line is $O(\sqrt{n})$. This has advantages in both space and time. For space, only $O(\sqrt{n})$ objects need to be in active memory for analysis. For time, there are $\sqrt{n}$ scan line sweep points with $\sqrt{n}$ objects at each sweep point, so the time complexity is approximately

$$O(\sqrt{n} \times \sqrt{n}) = O(n).$$

In fact, the $O(n)$ assumption is slightly optimistic and most implementations are between $O(n \log n)$ and $O((\sqrt{n})^3)$. Any type of object can be placed in a scan line sweep and, as a result, this approach is very flexible.

Typically, either edges or trapezoids are used for most implementations [7]. Edge representations have an advantage of directly representing the geometries being analyzed. Trapezoids, which are formed by fracturing input polygons along either axis, provide several performance optimizations, but require additional bookkeeping complexity due to the false boundaries created from fracturing.

By expanding the scan line to have some width using a history band, separation or distance, relationships can be readily handled.

Another issue that must be addressed by the low-level algorithms that support all-angle geometries arises from the fact that all layout objects have vertices that lie on a finite x–y grid. Orthogonal geometries intersect each other only at grid points. Non-orthogonal geometry can intersect off-grid, and the resulting perturbations must be addressed to provide robust implementations. One such method enhances the scan line to include advanced geometric snap-rounding algorithms [8].

20.10 HIERARCHICAL DATA STRUCTURES

Introducing design hierarchy to the DRC process requires additional structures for processing data. Hierarchical DRC operations determine the subset of data that can be acted upon on a per-cell basis. Each cell is processed once, and the results are applied to multiple areas of the original layout. Data outside of the subset are promoted up the hierarchy to the lowest point at which they can be accurately acted upon by the DRC operation on a cell-specific basis. Hierarchical layout verification incorporates these concepts:

Intrinsic geometries: Given a layer L and a cell C, an intrinsic geometry of C on L is a shape (geometry) on layer L common to every instance of cell C.
Interaction geometries: Given a layer L and a cell C, an interaction geometry of C on layer L is a region where intrinsic geometries on layer L intersect C at some point higher up the hierarchy.
Promotion: The process of moving intrinsic geometries up the hierarchy, as necessary, in order to achieve sufficient cell-specific context to accurately execute an operation.

Intrinsic geometries are promoted based on their proximity to the interaction geometries in the cell on related layers. Promotion is also dependent on the algorithmic intricacies of the type of DRC operation being executed. When promotion ceases, the intrinsic geometry may normally be analyzed or manipulated on a per-cell basis.

20.10.1 AREA INTERACTION AND PROMOTION

To show how the concepts of intrinsic geometries, interaction geometries, and object promotion work in a hierarchical DRC operation, consider the following operation.

```
Z = X Boolean AND Y
```

In Figure 20.11, cell B is placed in cell Top. Object 1 is an intrinsic geometry in cell Top, and objects 2, 3, and 4 are intrinsic geometries in cell B. Object 5 is an interaction geometry showing the overlap of object 1 in Top with cell B.

The AND operation is first performed in cell B. Objects 3 and 4 are sufficiently remote from 5 and are processed by the AND operation in B. The intersection of objects 2 and 5 is promoted into cell Top. The AND operation is then completed in cell Top because no further promotion is required. Figure 20.12 shows the result.

Layer Z contains two intrinsic geometries, one in cell Top and one in cell B. A layer Z interaction geometry is also created in cell B to show the overlap of the intrinsic Z shape in Top. In Figure 20.12, this interaction geometry would be coincident with the intersection of objects 2 and 5.

20.10.2 CONNECTIVITY MODELS

Another useful data structure for verification is a connectivity model, which encapsulates geometric interactions within a layer or between several layers, into a logical structure associated with the geometries.

Connectivity models in flat verification can be easily implemented by encapsulating the interaction sets as a single unique number. Hierarchical connectivity models require a more complex encapsulation, using the concepts of pins and nets.

Within any given cell, a net organizes the geometric interactions with regard to connectivity. A net may also be an external pin, an internal pin, or both. An internal pin forms a connection to a net within the hierarchical subtree of the cell, while an external pin forms a connection to a

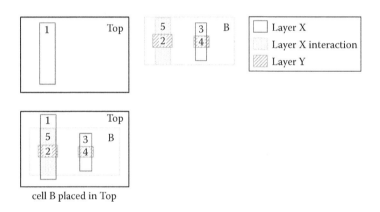

FIGURE 20.11   Cell-based geometry instantiation.

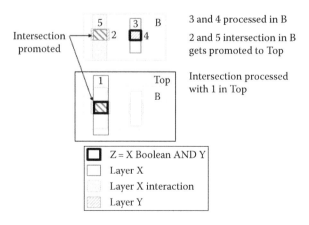

FIGURE 20.12   Hierarchical layer operation.

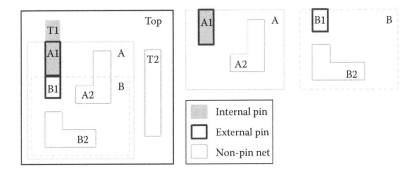

FIGURE 20.13   Internal and external pins.

net outside the hierarchical subtree of the cell. Pins allow hierarchical algorithms to traverse a net in the hierarchy both up, via an external pin, and down, via an internal pin. A net is considered complete at the top-most cell, in which that net is not an external pin.

Examples of internal and external pins are shown in Figure 20.13. In the design, net B1 is an external pin of cell B. When cell B is placed in cell A, B1 connects to A1. Net A1 is an external pin of cell A. Net A1 is also an internal pin to cell A when cell B is placed in A. When cell A is placed in cell Top, net A1 connects to net T1—an internal pin to Top. The net is then complete in cell Top. Nets B2, A2, and T2 make no connections outside their parent cells, and so, they are not pins.

Hierarchical algorithms using connectivity models must work hand in hand with the geometrical promotion techniques described earlier. Logical promotion, via dynamic creation of new logical pins, must accompany geometric promotion.

20.10.2.1 POLYGONAL CONNECTIVITY

The polygon connectivity model determines which geometries interact with each other on a particular layer. The polygon connectivity model is useful for those topological operations that require information about polygonal relationships.

Consider the operation

```
Z = X overlaps Y
```

The operation selects all layer X polygons that overlap a Y polygon or have a coincident edge with a layer Y polygon. For flat verification, this operation is comparatively simple since full polygons in both X and Y are present at the same (single) hierarchical level. In hierarchical verification, a single polygon may be broken across hierarchical boundaries and exist at multiple locations in the hierarchy. Flattening data in order to get full polygons at a single hierarchical level is undesirable because flattening causes explosive growth of the data set and degrades the hierarchy for future operations involving Z. Fortunately, selective geometric and logical promotion, along with careful traversal of the nets and pins of the polygon connectivity models in X and Y, can produce hierarchical output in Z without excessive flattening.

20.10.2.2 NODAL CONNECTIVITY

The nodal connectivity model, if specified, exists for the entire design and determines the geometries that interact with each other on layers predefined as having electrical connectivity. This electrical connectivity is defined by the user with an operation between two or more layers. The complete connectivity sequence includes all the interconnect layers, as suggested by this pseudocode:

```
connect M1 and POLY through CONTACT
connect M1 and M2 through VIA1
connect M2 and M3 through VIA2
. . .
```

This connectivity is essential for connectivity-based DRC checks, device recognition, and circuit topology extraction.

The complete electrical network can be determined in a single pass, or it can be built up incrementally layer by layer. Incremental connectivity allows specialized DRC checks to be performed after a layer has its connectivity determined and before a subsequent layer is added to the network. Incremental connectivity is particularly needed for operations that compute electrostatic charge accumulation by calculating area ratios between layers on the same electrical node. If incremental connectivity were not used, many copies of the interconnect layers would be necessary in order to partition the connectivity for all the appropriate area ratio calculations to occur. The memory required to do this is impractical, and so, incremental connectivity is employed.

20.11 PARALLEL COMPUTING

Due to the enormous data and rule volume at current nodes, it is essential for a physical verification tool to support parallel processing [9]. The number of separate operations in a DRC flow at 20 nm, for example, is approaching 50,000. The number of geometries is typically 250–500 billion just on drawn layers and many orders of magnitude greater on derived layers through the entire flow.

The DRC tool must support both SMP and distributed architectures, as well as a combination of the two, and should scale well to 200 or more processors. Most users of any DRC application expect 24 hour, and even overnight, turnaround times and are willing to utilize whatever hardware resources are required to achieve this goal. If the DRC tool cannot take full advantage of the customer's hardware resource then it is not a viable product.

How is parallelism achieved? There are two obvious avenues called cell-based parallelism and operation-based parallelism.

Cell-based parallelism takes advantage of the inherent parallelism of the hierarchy itself [10]. Let $C = \{C_1,...,C_n\}$ be the set of all cells in the design. Given any operation, such as a Boolean or a DRC spacing measurement, the idea is that there are subsets C' of C, for which the work required to generate the operation can be done in parallel over all cells in C'. However, there are two limiting factors in this approach.

The first limiting factor is that a promotion-based hierarchical algorithm requires the work for all cells placed in any given cell A throughout its sub-hierarchy to be complete or partially complete before work can begin on A itself. This impedes scalability near the top of the graph defining the hierarchy.

The second limiting factor in cell-based parallelism is that the design may not be very hierarchical to begin with. For example, a modern SoC design may have macros with hundreds of thousands of standard cell placements, each of which execute practically instantaneously. However, promoted data at the macro level can be computationally intensive, again limiting scalability. This, however, is not as much of a problem as first appears and can be mitigated with clever injection strategies that introduce extra levels of hierarchy into the macro via artificially created cells called bins [11].

Operation-based parallelism takes advantage of the parallelism built into the rule deck itself. A simple example is that a spacing check on METAL1 has no dependencies on a VIA3/METAL4 enclosure check, and they can obviously be executed in parallel. The analysis of a typical sub-20 nm flow shows that out of 50,000 or more separate operations, there are often hundreds at any given point in the flow that await immediate execution (the input layers have been generated) and large numbers of these may be executed independently. This approach, by itself, also has certain limitations.

First, the operation graph is isomorphic to the directed acyclic graph underlying the hierarchical design and has the same dependency restrictions—in this case, an operation may not be executed before its input data set (products of other operations) has been generated.

Next, there are often a few long-duration operations, such as in multi-patterning, which are extremely computationally intensive and may cause what is referred to as a tail if they end up executing by themselves near the end of the flow. Another issue is the difficulty in managing connectivity models, and the fact that the nodal model is shared between all affected operations in the rule deck. This means that the execution of connectivity-based operations effectively needs to be serialized. For example, a flow that only checks voltage-based rules may have minimal opportunities for operation parallelism. Current research avenues include injection strategies for operation parallelism similar to injection for cell-based parallelism.

Optimum scalability can be achieved by combining both cell-based and operation-based parallelism [12]. Operations are executed in parallel to the greatest extent possible, and each separate operation is generated using cell-based parallelism. The complexity in this approach manages the multiple levels of mutex requirements to protect the hierarchical database model that is being simultaneously manipulated, read, and added to by N operations. This complexity may be mitigated somewhat by strategies such as duplicating the portion of the database model required only by the operation and mapping that model (and virtually the operation itself) to a separate process.

20.12 PROSPECTS

Design size and complexity in sub-28 nm processes have resulted in substantial increases in chip production costs as well as a lag in technology node advances versus Moore's law projections [13]. Successful DRC applications will continue to adapt to the challenges posed by increasingly sophisticated design and manufacturing requirements.

REFERENCES

1. C. Gérald, M. Gary, F. Gay et al. A high-level design rule library addressing CMOS and heterogeneous technologies. *IEEE International Conference on IC Design & Technology*, Austin, TX, 2014.
2. W. Yayi and R.L. Brainard. *Advanced Processes for 193-nm Immersion Lithography*. SPIE *Press* Bellingham, WA, (2009), pp. 215–218.
3. M. Carver and L. Conway. *Introduction to VLSI Systems* (Addison-Wesley, Reading, MA, 1980), pp. 91–111.
4. M. de Berg, O. Cheong, M. van Kreveld et al. *Computational Geometry: Algorithms and Applications*, 3rd edn. (Springer-Verlag, Berlin, Germany, 2008).
5. M. I. Shamos and D. J. Hoey. Geometric intersection problems. *Proceedings 17th Annual Conference on Foundations of Computer Science* Houston, TX, (1976), pp. 208–215.
6. J. L. Bently and T. A. Ottmann. Algorithms for reporting and counting geometric intersections. *IEEE Transactions on Computers* C-28(9) (1979): 643–647.
7. K.-W. Chiang, S. Nahar, and C.-Y. Lo. Time-efficient VLSI artwork analysis algorithms in GOALIE2. *IEEE Transactions on Computer-Aided Design* 8(6) (1989): 640–647.
8. W. Barry and A. Michael. *Parallel Programming: Techniques and Applications Using Networked Workstations and Parallel Computers* (Prentice Hall, Upper Saddle River, NJ, 1998).
9. H. John Stable snap rounding. *Computational Geometry: Theory and Applications* 46(4) (2013): 403–416.
10. Z. Bozkus and L. Grodd. Cell based parallel verification on an integrated circuit design. US Patent 6,397,372 (1999).
11. L. Grodd Placement based design cells injection into an integrated circuit. US Patent 6,381,731 (2002).
12. L. Grodd, R. Todd, and J. Tomblin. Distribution of parallel operations. Japan Patent 5,496,986 (2014).
13. Z. Or-Bach. Moore's lag shifts paradigm of semi industry. *EE Times* (September 3, 2014). http://www.eetimes.com/author.asp?section_id=36&doc_id=1323755 (accessed October 10, 2014).

Resolution Enhancement Techniques and Mask Data Preparation

21

Franklin M. Schellenberg

CONTENTS

21.1 INTRODUCTION

Traditionally, an IC became ready for fabrication after an IC design was converted into a physical layout, the timing verified, and the polygons certified to be DRC-clean. The data files representing the various layers were shipped to a mask shop, which used mask-writing equipment to convert each data layer into a corresponding mask, and the masks were then shipped to the fab where they were used to repeatedly manufacture the designs in silicon.

In the past, the creation of the layout was the end of EDA's involvement in this flow. However, as Moore's Law has driven features to ever-smaller dimensions, new physical effects that could be ignored in the past now have an impact on the features that are formed on the silicon wafer. So, even though the final layout may represent what is desired in silicon, the layout can still undergo dramatic alteration through several EDA tools before the masks are fabricated and shipped.

These alterations are required, not to make any change in the device as designed, but to simply allow the manufacturing equipment, often purchased and optimized for making ICs one or two generations behind, to deliver the new devices. The intent of these alterations is to precompensate for known manufacturing distortions that are inherent in the manufacturing process. These distortions can arise in almost any processing step: lithography, etching, planarization, and deposition, all of which introduce distortions of some kind. Fortunately, when these distortions are measured and quantified, an algorithm for their compensation can also be determined.

The first part of this chapter is concerned with these compensation schemes, particularly for the compensations required for lithographic processes. These lithographic compensations are usually grouped under the heading resolution enhancement techniques (RET), and this chapter will describe these techniques and the consequences of their implementation on the IC design in some detail. They are also sometimes categorized under the more general category of "design for manufacturability" (DFM), but we will not attempt to give an exhaustive treatment to all the possible manufacturing effects that may be treated in this same way, but instead will direct the reader to the references for further information.

The second part of this chapter is concerned with the final step, that of converting the final compensated layout into a mask-writer data format. Although previously a simple translation task, the changes required for RET can create huge data volume problems for the mask writer, and so some care must be applied to mitigate these negative effects.

21.2 LITHOGRAPHIC EFFECTS

Although various processes [1,2] are used to create the physical structures of circuit elements in various layers of an IC, the physical dimensions of those structures are defined using lithography [3–5]. In a lithographic process, the wafer is coated with a sensitized resist material, which is exposed and processed to selectively remove the resist (see Figure 21.1). The subsequent

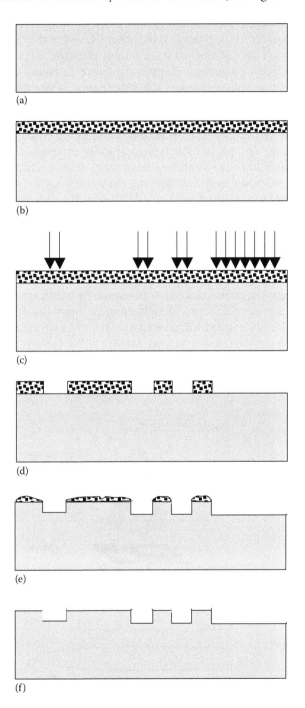

FIGURE 21.1   The photolithography process flow: exposure, development, and etching. (a) Begin with a silicon wafer, (b) coat silicon wafer with photoresist, (c) expose photoresist to a pattern of radiation, (d) dissolve exposed resist, (e) etch wafer, and (f) remove photoresist. Areas protected by photoresist remain unetched; only etched material remains.

manufacturing step, whether it is implantation, etching, deposition, or some other step, occurs only where the resist is not protecting the surface.

The most common process for patterning wafers is optical lithography using ultraviolet (UV) light as the source. In this case a photomask (also called a reticle), with the layout pattern for a given layer mapped on it as a pattern of transparent or absorbing regions, is created for each layer [6,7]. A reduced image of the photomask is formed on the resist-coated wafer using a highly specialized exposure tool, called a stepper or scanner. This process therefore resembles a traditional photographic printing process, with the photomask corresponding to the negative, the stepper to the enlarger, and the wafer to the photographic paper. However, the tolerances on mechanical motion and lens quality are far tighter than those for any photographic system.

The elements of a typical lithographic stepper are shown in Figure 21.2. A UV light source (usually an excimer laser) is collimated, and the illumination is shaped to illuminate the photomask from behind. The light passing through the photomask is collected and focused as a reduced image (typically 4× smaller) on the resist-coated wafer, by using a complex lens designed to be as aberration-free as possible. The resolution of the image is the primary figure of merit for a lithographic system. The photomask image typically corresponds to one field, typically 26–32 mm in size, so on a 300-mm-diameter silicon wafer, there is space for hundreds of fields. Therefore, once a field is exposed by the photomask, the wafer stage steps to the next field, where the exposure is repeated (hence the term "stepper"). Clearly, the alignment of the photomask to the pre-existing layers on the wafer must be exact, making overlay the second critical figure of merit for lithography.

For some recent tool configurations, only a part of the photomask is illuminated at one time, and the mask and wafer are scanned to expose the entire field. These systems are called "scanners" or "step-and-scan" systems [4,5].

When light illuminates the photomask, the transmitted light diffracts, with light from regions with higher spatial frequencies diffracting at higher angles. The relationship governing the resolution in imaging can be understood by recognizing that the mathematical description of the diffraction of light in an imaging system is equivalent to a 2-D Fourier transform operation [8], with the wavefront at the lens pupil plane corresponding to the transform. This transform pattern

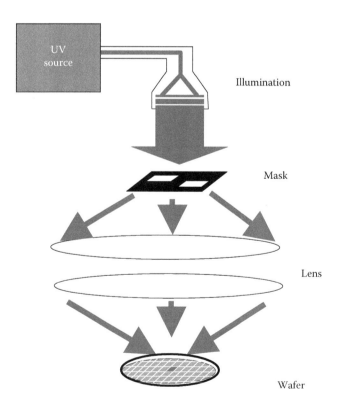

FIGURE 21.2 The elements of a typical UV stepper.

then diffracts and is recollected by the remaining lens elements to form the image. The formation of the image is mathematically equivalent to taking the inverse transform of the pupil plane [8,9].

Because the stepper lens pupil has a finite aperture, the light diffracted at the highest angles (corresponding to the finest gratings) is truncated at the pupil. The imaging system, therefore, effectively acts as a low-pass filter, removing the highest spatial frequencies from the 2-D layout pattern. The measure of this aperture, defined as the sine of the largest angle as light falls on the wafer, is called the numerical aperture (NA) of the system. This is one of the key factors limiting resolution [10].

The other factor limiting resolution is the wavelength of the light source itself. With smaller wavelengths, the diffraction angles are smaller, and more light can be collected and passed through the pupil. These factors come together in what is often called the "Rayleigh equation" relating the smallest linewidth (or technically, the smallest half-pitch [HP] dimension for equal lines and spaces) with the wavelength λ, refractive index n of the imaging medium, and lens NA [11–13]:

$$L_w = k_1 \frac{\lambda}{nNA}$$

The factor k_1 represents the degree of difficulty of a process. The traditional resolution limit of an optical system would give $k_1 = 0.61$, anything larger can be easily resolved. The absolute limit for single-exposure imaging is $k_1 = 0.25$; below this value the modulation transfer function of any lens system is effectively zero. It is the mid-range of $0.25 < k_1 < 0.61$ that represents the challenge for optical lithography.

Therefore, there are two obvious paths to improving the capabilities of optical lithography. Historically, mercury lamps with emission lines at 436 and 365 nm were readily available, and so improvements in resolution came from better lens design with larger values of NA, increasing from $NA = 0.3$ in 1982 to $NA = 0.93$ in 2008. However, NA is the sine of an angle, and has an absolute maximum of 1.0. Design of lenses with NA larger than 0.93, becomes increasingly difficult and expensive for very little improvement in resolution, indicating that this is not a path open for development in the future.

"Effective" values of $NA > 1$, sometimes called "hyper-NA" lithography, can be achieved if n is increased [14,15]. For exposure in air, $n \approx 1$, but for immersion in water, $n \approx 1.44$. For a water immersion system designed with $\lambda = 193$ nm and $NA = 0.935$, this is equivalent to having an $NA = 1.35$ (or $\lambda = 134$ nm) in an $n = 1$ system. Recent models of lithographic tools have now been introduced in which the lens–wafer interface is filled with flowing de-gassed water, effectively exploiting this effect [16,17].

The second obvious path to improvement is to reduce the wavelength λ. This is shown in Table 21.1 [3–5,14–28]. Lithographic exposure has followed a steady progression from the use of

TABLE 21.1   Lithography Sources

Source	Wavelength (nm)	Tool NA	Comments	Reference
Hg g-line	436	0.3		[3–5]
Hg i-line	365	0.45–0.62		[3–5]
KrF excimer	248	0.63–0.70		[4,5,18]
ArF excimer	193	0.63–0.95		[4,14–18]
ArF excimer + immersion	193	0.93	Water immersion Effective NA = 1.35	[16,17]
Xe or Sn plasma	13.4	0.3	EUV project—under development	[19]
Synchrotron	1.2	Contact/proximity printing	X-ray lithography project discontinued	[20]
E-beam direct write	0.017	Beam technology	5 keV electrons 13,000 parallel beams	[21–24]
Extended E-Beam Source	0.0039	Beam technology	100 keV electrons REBL (Reflective DPG)	[26–28]

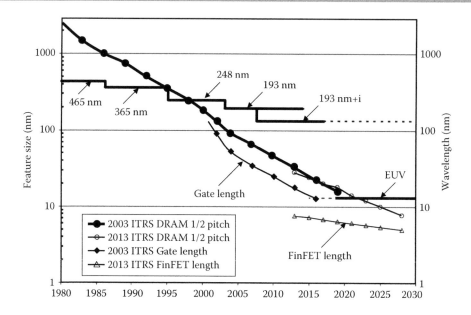

FIGURE 21.3   Ever-shrinking linewidths in accordance with Moore's Law became sub-wavelength in 1998. 193nm+i represents 193 nm water immersion lithography. EUV represents Extreme UV Lithography at 13.4 nm. (Data adapted from *The International Technology Roadmap for Semiconductors (ITRS)*, 2003 and 2013 Editions.)

the mercury lamp atomic lines at 436–365 nm (near-UV), and more recently excimer lasers have been used at 248 and 193 nm in the deep UV (DUV) [18].

However, as shown in Figure 21.3, even with this decrease in wavelength, the ever-shrinking linewidths, required to stay on the cost-per-function curve of Moore's Law [29–32], dictate that the feature sizes are now much smaller than the wavelength of the light. In current production fabs, circuit half-pitches of 25 nm are now being routinely manufactured [32], with FinFET gate dimensions [32–35] as small as 18 nm (smaller than $\lambda/10$) [32].

Other lithographic approaches exist. These are also shown in Table 21.1. Extreme ultraviolet (EUV) sources at $\lambda = 13.4$ nm are being developed as a possible alternative [19], and are forecast to be introduced to production fabs around 2017. However, questions remain about whether the technology can achieve its goals with an acceptable number of defects and a reasonable cost of ownership. X-ray lithography using synchrotron radiation to generate photons with $\lambda = 1.2$ nm have been built and demonstrated [20], but again, the cost of ownership for these systems proved to be impractical.

High-energy electron beams ($\lambda = 0.0055$ nm for 50 keV electrons) have been successfully deployed in beam systems to directly pattern features on the order of a few nm [36], or even smaller in scale [37], but these systems address the pattern to be written serially, and are consequently extremely slow when used to write patterns over large areas [21]. Interactions between charged electrons in transit to the wafer, as well as additional exposure due to secondary electrons scattered within the photoresist, can create electron proximity effects that also distort the image being written [38–40]. Electron beam tools are, therefore only commonly used to write the master photomasks to be used for lithography [6,7], where high fidelity and resolution are more important than rapid processing.

E-beam systems with large numbers of parallel beams that write different sections of the wafer are under development [22]. One such system designed by MAPPER Lithography uses 13,000 independent 5 keV e-beams that are individually directed and controlled [23,24]. Projection electron beam systems using masks have been introduced [25], but these have additional problems compensating for the interactions of the charged particles, and so have been discontinued. Another approach to electron projection lithography is the reflective Electron Beam Lithography (REBL) system [26–28]. This replaces the static physical mask with a dynamically programmable electron beam digital pattern generator (DPG) [28], to "reflect" a large area pattern of 100 keV electron beams (analogous to the digital light projector [DLP] devices used for digital projection). Demonstrations of these systems have been made, but commercial tools have not yet become commercially viable.

The push to stay on Moore's Law, given the limitations of NA and wavelength, has dictated the development of low-k_1 lithography solutions to perform lithography in this sub-wavelength regime. The general category these come under is RET [41]. We now turn our attention to these.

21.3 RET FOR SMALLER k1

An electromagnetic wave has four independent variables that define it: an amplitude, a phase, the direction of propagation, and the direction of the electric field (polarization). The first three variables have been exploited to provide resolution enhancement, while polarization is currently an active topic for exploration. In addition to these physical effects, RET approaches that utilize multiple exposures, as well as materials which "grow" circuit-like features through self-assembly are also paths to enhanced resolution. This section discusses each of these phenomena in turn.

21.3.1 AMPLITUDE

The photomasks typically used in lithography are coated with an opaque mixture of chrome and chrome oxide compounds, and are patterned by removing the opaque layer. This makes the masks effectively binary—transmitting ($T = 1.0$) or opaque ($T = 0$). It would therefore seem that there is little to control for amplitude.

However, although the high resolution of the electron beam writing systems used to make photomasks can produce sharp corners and well-resolved structures, the resolution limits of the stepper lens make the lens act effectively as a low-pass filter for the various spatial frequencies in the 2-D layout. This manifests itself in three distinct effects:

1. A bias between isolated and dense structures (iso-dense bias)
2. A pullback of line-ends from their desired position (line-end pullback)
3. Corner rounding

These come under the general heading of "optical proximity effects" [41–44] and are illustrated in Figure 21.4 [44].

To compensate for these effects, the layout can be adjusted so that the image matches the desired pattern better. The actions, described in more detail below, generally serve to add higher spatial frequency content to the layout, to adjust for the spatial frequency components that are attenuated by the lens system. This is illustrated in Figure 21.5. These corrections have traditionally been given the name "optical proximity correction" (OPC). As other effects besides these proximity effects have been introduced to the solutions, the meaning of the acronym has broadened to "optical and process correction."

21.3.2 PHASE

Interference phenomena produce fringes of dark and light, that can be exploited to enhance the contrast of an image. A single-phase transition from 0° to 180° becomes an interference fringe which, in imaging, becomes a very thin dark line, as illustrated in Figure 21.6. With special exposure conditions, extremely thin individual lines can be imaged [11,12].

The use of phase requires that light passing through different regions has different path lengths at various refractive indices. This is achieved by etching the photomask to different depths in different regions. The etch depth is given by [46]:

$$d_{etch} = \frac{1}{2}\frac{\lambda}{(n-1)}$$

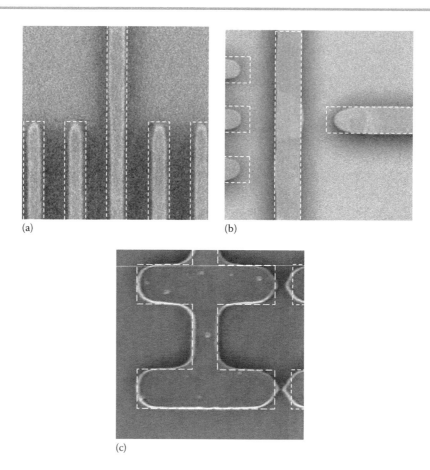

FIGURE 21.4 Typical image fidelity problems that can be corrected through OPC: (a) Iso-dense bias; (b) line-end pullback; and (c) corner rounding. (From Schellenberg, F.M. et al., SEMATECH J111 Project: OPC validation, *Optical Microlithography XI, Proceedings of SPIE*, Vol. 3334, 1998, pp. 892–911. With permission.)

A mask with such phase-shifting structures is typically called a phase-shifting mask (PSM). For quartz at $\lambda = 193$ nm and $n = 1.56$, the etch depth becomes 172 nm.

Various names exist for certain types of PSMs [45–55]. The most straightforward version, in which the phase of alternating apertures is phase-shifted, is called "alternating PSM" (or altPSM) [45–48]. Features with phase-shifted sub-resolution assist features (SRAF), sometimes nicknamed "outriggers," have been explored [48], as have features with edges augmented with phase-shifters, also called "rim-shifters." [49] Use of a single-phase edge to form a pattern is a phase-edge PSM, also sometimes called "chromeless PSM." [50,51] Another "chromeless" technique merges the dark fringes of two phase edges to form a larger dark line [52,53]. This is commonly used in combination with off-axis illumination (OAI), in a technique called "phase-shifted chromeless and off-axis" (PCO) or "chromeless phase lithography" (CPL) [54,55].

21.3.3 AMPLITUDE AND PHASE

The previous techniques produce clear regions somewhere on the mask that shift the phase of the light 180°. A variation of phase-shifting also shown in Table 21.2, called "attenuated PSM," is constructed with a choice of material composition and thickness, such that the transmission is weak (typically 6%–9%, well below the exposure threshold of the photo-resist), and 180° out of phase with the completely transparent regions. This has the effect of increasing contrast at the edges, which arises from the zero crossing for the electric field at the transition from dark to light [56]. This is illustrated in Figure 21.7.

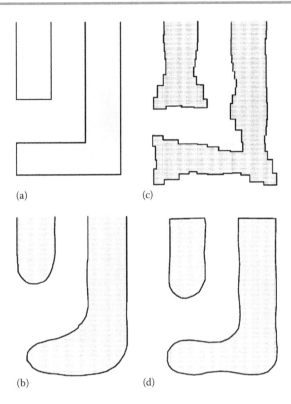

(a)

(b)

(c)

(d)

FIGURE 21.5   (a) Original drawn layout; (b) corresponding image of the layout of (a) on a wafer showing rounding and line-end pullback; (c) layout modified with OPC (original layout is shown faintly underneath); and (d) corresponding image of the layout of (c) showing that the image of the modified layout is much closer to the designer's intent (i.e., the original layout).

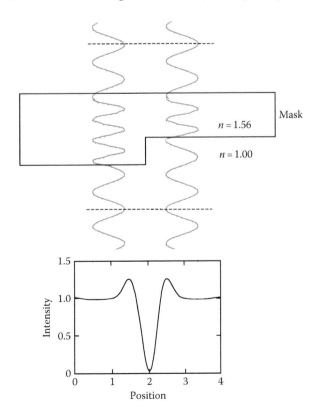

FIGURE 21.6   Cross-section of a phase-shifting mask, and an intensity plot of the interference fringe formed in the image of the phase edge.

TABLE 21.2

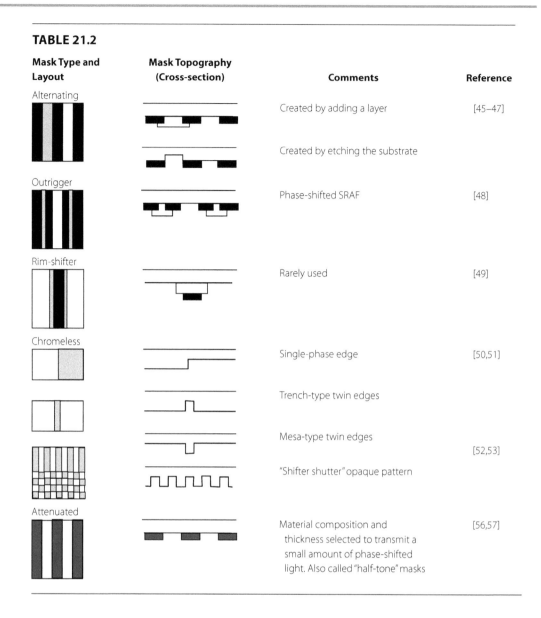

Mask Type and Layout	Mask Topography (Cross-section)	Comments	Reference
Alternating		Created by adding a layer	[45–47]
		Created by etching the substrate	
Outrigger		Phase-shifted SRAF	[48]
Rim-shifter		Rarely used	[49]
Chromeless		Single-phase edge	[50,51]
		Trench-type twin edges	
		Mesa-type twin edges	[52,53]
		"Shifter shutter" opaque pattern	
Attenuated		Material composition and thickness selected to transmit a small amount of phase-shifted light. Also called "half-tone" masks	[56,57]

Although the overall layout of an altPSM mask resembles a binary mask, the additional contrast given by having the phase shift can significantly improve the image quality. OPC can also be applied, assuming that the effects of the phase are also anticipated in the correction algorithms. These photomasks are fabricated from a mixture of material, usually a combination of molybdenum (Mo) and silicon (Si), deposited such that the film has the desired transmission and 180° phase shift. For more information on attenuated materials, see Ref. [57].

21.3.4 OFF-AXIS ILLUMINATION (OAI)

A traditional lithography system uses an uniform illumination falling perpendicular onto the photomask. However, if light falling at an angle is used, the diffraction from certain high spatial frequencies can be enhanced, effectively increasing the resolution [58]. This is illustrated in Figure 21.8 [42,59–66].

Various patterns of OAI have been demonstrated. These are shown in Figure 21.9 [42,59–66]. Annular illumination is the most common [59], but does not offer the greatest possible benefit. Quadrupole illumination can emphasize certain pitches very well and can work for layouts with Manhattan geometries, but diagonal lines will fail to print using such a system [61,62,64,65]. With dipole illumination, the greatest emphasis of certain specific pitches is achieved, but only

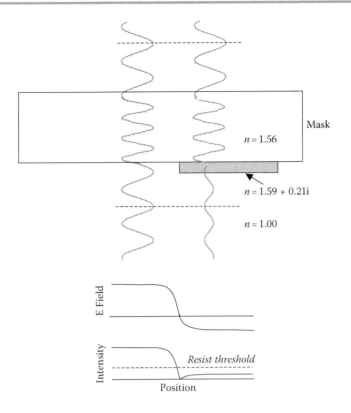

FIGURE 21.7 Cross-section of an attenuated phase-shifting mask, and plots of the corresponding image amplitude (**E** field) and intensity.

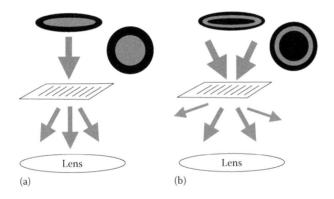

FIGURE 21.8 (a) Conventional circular illumination; and (b) off-axis (annular) illumination.

for features aligned with the dipole [66]. Features of the orthogonal variation do not print. To print a Manhattan layout, a double exposure with horizontal and vertical dipoles must be done. For an arbitrary layout, more exposures may be necessary to produce the desired features.

Since only the light of the same polarization can form interference patterns, the polarization properties of the light can have an important impact on the image formed, especially when PSMs are used.

21.3.5 POLARIZATION

The fourth independent variable of the electromagnetic wave is polarization. The illumination in a typical lithography system is unpolarized, or more accurately, randomly polarized. Until

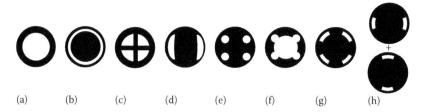

(a)   (b)   (c)   (d)   (e)   (f)   (g)   (h)

FIGURE 21.9 Off-axis pupil maps: (a) conventional illumination, (b) annular illumination [59], (c) "four-fold" source [42], (d) "separated" source [60], (e) quadrupole illumination [62, 63], (f) "CQUEST" illumination [61], (g) QUASAR illumination [64,65], and (h) horizontal and vertical dipole illumination [66]. ((b) From Fehrs, D.L. et al., Illuminator modification of an optical aligner, *Proceedings of KTI Microelectronics Seminar INTERFACE'89*, 1989, pp. 217–230; (c) Rosenbluth, A. et al., A critical examination of submicron optical lithography using simulated projection images, *J. Vac. Sci. Technol. B*, 1, 1190–1195, 1983; (d) Asai, S. et al., *J. Vac. Sci. Technol. B*, 10, 3023–3026, 1992; (e) Shiraishi, N. et al., New imaging technique for 64M-DRAM, *Optical/Laser Microlithography V, Proceedings of SPIE*, Vol. 1674, 1992, pp. 741–752; Ogawa, T. et al., The effective light source optimization with the modified beam for the depth-of-focus enhancements, *Optical/Laser Microlithography VII, Proceedings of SPIE*, Vol. 2197, 1994, pp. 19–30; (f) Noguchi, M. et al., Subhalf micron lithography system with phaseshifting effect, *Optical/Laser Microlithography V, Proceedings of SPIE*, Vol. 1674, 1992, pp. 92–104; (g) Socha, R.J. et al., Forbidden pitches for 130 nm lithography and below, *Optical Microlithography XIII, Proceedings of SPIE*, Vol. 4000, 2000, pp. 1140–1155; Schellenberg, F.M. and Capodieci, L., Impact of RET on physical layouts, in *Proceedings of the 2001 ISPD*, ACM, New York, 2001, pp. 52–55; (h) Eurlings, M. et al., 0.11-μm imaging in KrF lithography using dipole illumination, *Lithography for Semiconductor Manufacturing II, Proceedings of SPIE*, Vol. 4404, 2001, pp. 266–278.)

now, there has been relatively little need to explore this further. However, in the sub-wavelength domain, this may be impossible to ignore.

As feature sizes on the photomask approach the wavelength of light, they begin to act like wire grid polarizing structures [67], depending on the conductivity of the opaque material on the mask at optical frequencies. They may therefore begin to polarize actively the transmitted light, but only in certain geometry-dependent regions [68]. Surface plasmons may also be excited at the surface of the photomask, changing the local transmission factors [69]. The impact on the software flow for a layout compensating for these effects will be minimal. The impact instead will be felt through the accuracy of the models available to the RET software tools, with inaccurate models degrading the correction quality. We will therefore not discuss further these polarization and plasmon models here.

21.3.6 DOUBLE PATTERNING

Isolated features in silicon that are much thinner than the half pitch can always be achieved through various processing techniques, such as over-etching. However, with the limit for single exposure half pitch using scanners having water immersion being about 35 nm (corresponding to a $k_1 = 0.25$), and with EUV tools having a wavelength of 13 nm not yet ready for production, other solutions have been invented to allow the progression of Moore's Law to continue into the 22 nm regime.

The most straightforward approach is one of double patterning. In this approach, the layout to be fabricated on a single IC layer is split into two mask patterns, each with features having a pitch at least twice as large as the final desired pitch. The final wafer layer is then created using two aligned exposures with the complementary masks.

Several methods have been developed for multiple exposures [70–82]. In some, the cells of a layout may be decomposed into two complementary structures in which the pitch in each exposure is above the pitch limit of the lithography process. A double exposure, especially when using dark field masks or subtractive imaging processes with material deposition, creates the sum of the two exposures. This is illustrated in Figure 21.10, adapted from an example by Liebmann et al. [71]. For this approach, however, an overlay between the two exposures can be critical to the device performance.

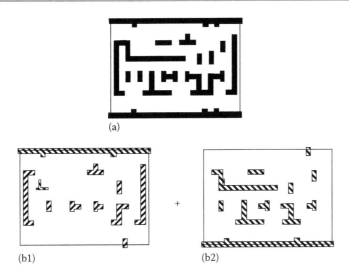

FIGURE 21.10 (a) Layout for a representative cell and (b1) and (b2): decomposition of the cell from (a) into two layouts for double patterning. (Adapted from Liebmann L. et al, Decomposition-aware standard cell design flows to enable double-patterning technology, *Design for Manufacturability through Design-Process Integration V, Proceedings of SPIE*, Vol. 7974, 2011. With permission.)

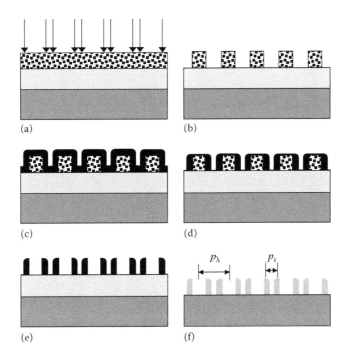

FIGURE 21.11 Self-aligned "Spacer" fabrication process for creating features at a pitch p_s from lithographic patterns at the much larger pitch p (a) Exposure of resist on polysilicon on silicon (b) processing of photoresist (c) deposit hard mask (SiO$_2$) (d) etchback of SiO$_2$ (e) remove resist (f) final etch into polysilicon. (Data adapted from Yaegshi, H. et al., *J. Photopolym. Sci. Technol.*, 24, 491, 2011.)

A different process that is more overlay-independent is a "spacer" process [72–75]. This allows a particular large pitch to be fabricated in a first exposure, and the wafer is then processed so that the edges of the original exposure pattern define intermediate masking features. With various subsequent processing steps, this creates features defined at the edges of the initial printed pattern, cutting the pitch by half. This is illustrated in Figure 21.11 [74]. Although somewhat complicated, this regular pattern of parallel lines at twice the density of the original pattern can be manufactured reliably and repeatably.

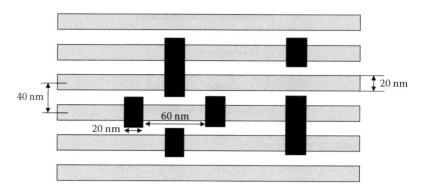

FIGURE 21.12   Complementary lithography using a first periodic pattern with a 40 nm pitch and a second pattern of "cut-lines" using a second exposure step. (Drawing adapted from illustrations in Borodovsky, Y., *Complementary Lithography at Insertion and Beyond*, Semicon West 2012, June 11, 2012, San Francisco, CA.)

Once this regular periodic pattern has been created, it can be used for layouts that are indeed entirely periodic. It can also be adapted to form distinct, individual structures on a periodic pitch, by cutting certain features using a second, "complementary" exposure in order to define individual circuit features [76–78]. This is illustrated in Figure 21.12 [77]. This complementary exposure can be carried out using a conventional optical lithography tool, or using an electron beam tool if higher resolution is needed for the cut patterns [78].

With the success of double patterning techniques to extend below the single exposure $k_1 = 0.25$, triple [79,80], quadruple [81], or even sextuple [82] patterning processes and design approaches are being investigated to extend the technique for future IC generations.

21.3.7 DIRECTED SELF ASSEMBLY (DSA)

Once the concept of circuit fabrication using large areas of regular periodic patterns was proven as workable, the application of a recent development in polymer chemistry to create these periodic patterns has received increasing interest [83–93].

Co-polymers are polymers that have two distinct types of polymer, such as polystyrene-*b*-poly(methyl methacrylate) (PS-*b*-PMMA) attached to each other, as illustrated in Figure 21.13 [83–86]. Individually each of these polymers may have very different properties, such as resistance to etching and solubility. When a solution of co-polymers is allowed to reach thermal equilibrium, the polymer components of the co-polymer tend to segregate to the degree that they can, and, depending on the relative molecular weight of the two components, can form parallel lamella, cylindrical columns, or isolated spheres of one polymer embedded in the other [83–86]. If the polymers have, for example, different solubilities, a solvent may then dissolve one of the polymer components while leaving the other behind. Dissolution of the PMMA portion while leaving PS behind can have the effect of forming periodic lines and spaces of PS on a substrate that can serve as a resist for subsequent etching.

The normal segregation of these components, called "self-assembly," may have a somewhat random appearance, as shown in Figure 21.14a. However, with a suitable chemical or mechanical texturing of the substrate on which the polymer is coated, the orientation of the structures can be "directed," forming the patterns in Figure 21.14b. For the lamella as shown in Figure 21.14b for PS-*b*-PMMA, the "natural" half-pitch dictated by using relatively equal molecular weights of ~30,000 g/mol is on the order of 10–20 nm [90,93], making this an inexpensive, optics-free method for forming periodic resist structures.

Several co-polymers suitable for the formation of parallel lines, contact holes, and other circuit-like structures using "directed self-assembly" (DSA) have been the subject of research in recent years [83–93]. Although the ability to control exactly where a particular line may connect to the circuit elements defined by more conventional lithography is still an issue, DSA processes for forming features on a 10 nm scale are so inexpensive that rapid adoption into manufacturing is expected.

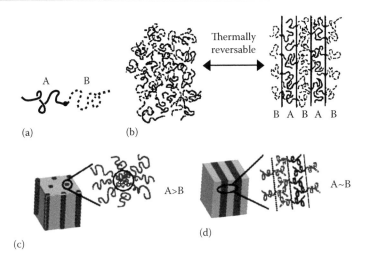

FIGURE 21.13 (a) A single PS-*b*-PMMA co-polymer strand; (b) polymer phase separation for PS-*b*-PMMA. When the molecular weights of the two components are (c) unequal, cylinders form; and (d) approximately equal, lamella form. (Adapted from Schellenberg, F.M. and Torres, J.A.R., Contemporary design paradigms applied to novel lithography, SEMATECH 2006 Litho Forum Poster IM06, Slides 3 & 4.)

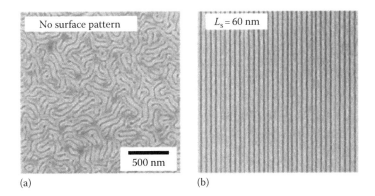

FIGURE 21.14 (a) Planar phase separation for PS-*b*-PMMA on an untreated surface, and (b) Planar phase separation for PS-*b*-PMMA on a surface treated with a directional coating. (Adapted from Stoykovich, M. et al., Directed assembly of block copolymer blends into nonregular device-oriented structures, *Science*, 308, 1442–1446, 2005. Reprinted with permission from AAAS.)

The design paradigms that need to change when large areas of the circuit layout may simply be large empty spaces in which circuit structures are expected to "grow themselves" will be addressed in Section 21.4.5.2.

21.4 SOFTWARE IMPLEMENTATIONS OF RET SOLUTIONS

The main RET solutions can all be implemented in software that alters the layouts of an integrated circuit. As such, it belongs to a part of an EDA design flow.

The correct insertion point for RET solutions has been a matter of some debate. A simplified flow diagram [94] of the process steps in the design of a typical IC, without explicitly accommodating RET, is shown in Figure 21.15. Given this flow, insertion of RET was initially carried out after the layout had been generated, as an augmentation of the "data massaging" step after the layout has been completed and verified. However, insertion after verification runs the risk that the layout changes may introduce some unforeseen effect. An alternative insertion point for RET can be found as part of the layout creation/verification steps. This is illustrated in

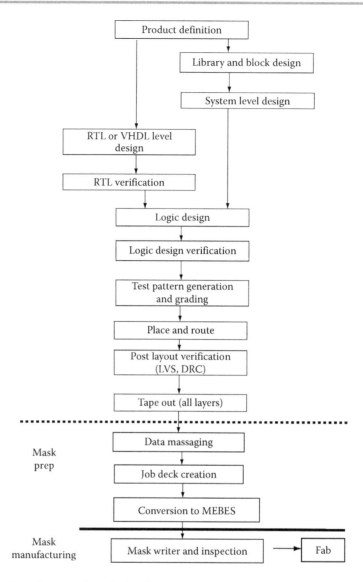

FIGURE 21.15 Data flows as identified in the SEMATECH Litho/Design Workshops (a) raw scattering bars (b) default cleanup (c) prioritize by layer. (From Schellenberg, F.M., Design for manufacturing in the semiconductor industry: The litho/design workshops, *Proceedings of the 12th International Conference on VSLI Design*, IEEE Computer Society Press, Los Alamitos, CA, 1999, pp. 111–119. With permission.)

Figure 21.16 [95]. Insertion in earlier parts of the physical verification flow means that, in principle, the results can be verified again after corrections have been made, reducing the risk to the functionality of the overall product. Modification of the design rules, using what are sometimes called "restrictive design rules (RDRs)," has been proposed as a way of accomplishing this insertion [96,97].

Although it is clear that the insertion of RET solutions before there is even a layout would be counterproductive, there are certain places (for example, in a place-and-route tool) where some awareness of the lithographic limitations might help. After we review the implementations of the various RET solutions, we will return to this question in Section 21.4.6.

21.4.1 OPTICAL PROXIMITY CORRECTION (OPC)

The simple manipulation of layouts to achieve better circuit performance is not a new concept. However, several approaches to the alteration of layout polygons have been developed so that all come under the general heading of "optical proximity correction" (OPC).

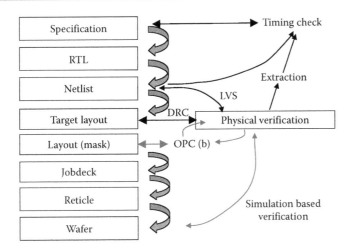

FIGURE 21.16   Simplified flow showing the insertion of OPC integrated with the steps of physical verification. (From Schellenberg, F.M., Advanced data preparation and design automation, *Photomask and Next-Generation Lithography Mask Technology IX*, Proceedings of SPIE, Vol. 4754, 2002, pp. 54–65. With permission.)

21.4.1.1 RULES-BASED OPC

The geometric manipulations needed to implement OPC might appear to be fairly straightforward, and initial attempts to implement OPC followed what is now called *rules-based OPC*.

In rules-based OPC, the proximity effects, as discussed in Section 21.3, are characterized, and specific solutions devised for specific geometric configurations. The layout is then searched using a DRC tool or a similar geometry-based software engine to find these geometric configurations. Once they are found, the problem areas are then replaced by the specific solution. The first OPC software products introduced were rules-based software systems [98].

For iso-dense biasing, the "rules" simply represent a set of biases that are added to geometries, with different biases used when the nearest neighbors have certain distances. This can be easy to conceptualize, and can be implemented in a compact look-up table format.

For line-end pullback, a solution that extends a line with, for example, a hammerhead structure can be substituted for every line end. When printed, the extended line-end falls much closer to the desired location, instead of being pulled back. Look-up tables can also be used to implement these solutions, with different sizes and shapes for line ends inserted for different original linewidths and nearest-neighbor conditions.

Corner rounding is generally addressed using a serif (or antiserif in the case of an inside corner). Again, the size and shape of the serif can be predetermined, and a look-up table for various feature sizes and nearest-neighbor conditions can be created. An example of a rule-based implementation using Calibre SVRF script [99] is shown below. The rules can be encompassed in only a few lines of code. The results of running this script are shown in Figure 21.17.

```
HAMMERHEAD = OPCLINEEND MET1
      WIDTH <= 0.3 HEIGHT > 0.1 END 0.02 SERIF 0.02 0.01
   // OPCLINEEND to explained elsewhere
MET1_EDGE1 = MET1 OUTSIDE EDGE HAMMERHEAD
CORNER_OUT = INT [MET1_EDGE1] <= 0.08 ABUT==90 INTERSECTING ONLY
CORNER_PAD = (OPCBIAS CORNER_OUT MET1
      SPACE > 0.3 <= 0.4 MOVE 0.01
      SPACE > 0.4 MOVE 0.02) NOT MET1
MET1_EDGE2 = MET1_EDGE1 NOT COINCIDENT EDGE CORNER_OUT
OPC { OPCBIAS MET1_EDGE2 MET1
 xxxxxSPACE > 0.3 <= 0.4 WIDTH <= 0.3 OPPOSITE EXTENDED 0.3 MOVE
0.01
}
```

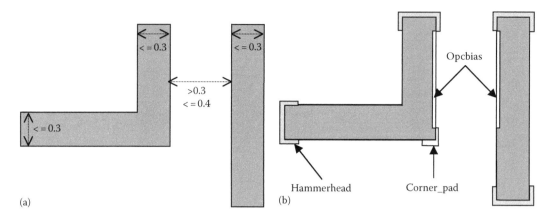

(a)

(b)

FIGURE 21.17   (a) Original layout (target layout); (b) layout modified with a rule-based OPC script using biasing to correct proximity effects, hammerheads for line-end shortening, and a corner serif for corner rounding.

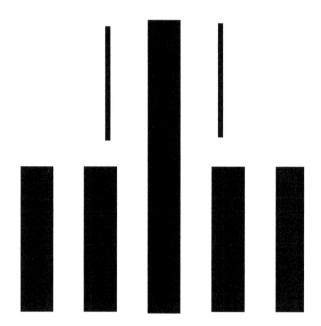

FIGURE 21.18   Layout with simple SRAF to compensate for iso-dense bias.

21.4.1.2 SUB-RESOLUTION ASSIST FEATURES (SRAF)

A special case of rules-based OPC involves the insertion of SRAF, sometimes also known as "scattering bars." [100,101] These are typically additional opaque features, introduced to address the iso-dense bias problem, that are themselves too small to be resolved by the imaging system. A simple case of SRAF insertion is shown in Figure 21.18. When SRAF are inserted into the layout, they provide a dense-like environment for the isolated feature. The isolated features therefore print more like the dense features, reducing the problem. There are certainly special cases for introducing SRAF, which depend on nearest-neighbor proximity. The size and placement of the SRAF are also open to some optimization: SRAF that are too large may print, but those that are too small will not have the desired effect. Placement can also be important, depending on nearest-neighbor distance.

Additional SRAF can also be placed to further enhance the effect. This is illustrated in Figure 21.19. However, as is clear from the illustration, adding more SRAF to the layout can depend highly on nearest-neighbor geometry, and also depends on the size of the isolated feature, and the room available in the immediate neighborhood. This again can be made much easier by the preparation of look-up tables based on feature size and nearest-neighbor spacing.

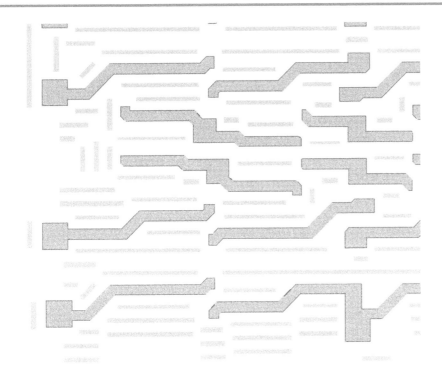

FIGURE 21.19 Layout with printing features (dark) and SRAF (light grey).

SRAF are typically added along 1-D structures (e.g., lines). Conflicts can arise, however, when SRAF are added for two orthogonal features that intersect. This is illustrated in Figure 21.20. For cases like this, prioritization schemes need to be developed that allow certain SRAF to be retained or extended while others are truncated or eliminated.

21.4.1.3 MODEL-BASED OPC

Rules-based approaches work well for simple cases. However, for complex layouts, the number of feature sizes and geometric environments can be huge, and it is not possible to encompass the behavior in a manageable set of rules. For this, *model-based OPC* has been developed.

The original proposal for model-based OPC was made in the early 1980s as an academic exercise in image processing [102–105]. In that work, small layouts were digitized into individual

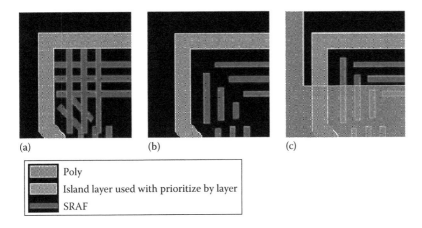

FIGURE 21.20 SRAF prioritization for a poly layer. (a) Raw output after the insertion of SRAF (b) layout after initial SRAF cleanup and (c) final layout, showing an extension (priority) given to SRAFs that overlap the island layer (also shown).

pixels, and the lithography system modeled as a low-pass filter followed by a high-contrast square law photosensor (the photo-resist). Iterative processing of the layout with several proposed correction functions was carried out and results compared. The imaging system was modeled rather simply, using a fast Fourier transform (FFT), low-pass filter, and inverse FFT to create the image. Although a simple model, this actually mimics what a real imaging system ideally does in controlling the wavefront with refractive surfaces. The intensity and phase of the image at the ideal pupil plane of the lens system will be mathematically equivalent to the Fourier transform of the object itself. Since the pupil has a finite extent, the low-pass filter has a physical meaning, as characterized by the NA of the imaging system, which forms the image by taking the inverse transform of the pupil plane.

After the image is computed, the transfer of the gray-scale image into resist is modeled as a simple constant-threshold function to define the photo-resist boundaries. For image pixels that exceeded the desired tolerance, several correction functions were introduced, and the computation run again. As many as 500 iterations were necessary to reach convergence in such a system.

The results, as illustrated in Figure 21.21, are quite different from the results a rule-based approach would generate [103]. Here, additional "ripples" are generated in the edges of the lines, and irregular SRAF appear naturally out of the algorithm, and not at predetermined sizes and distances, as a rule-based implementation would suggest. Although producing counterintuitive mask layouts, the corresponding wafer images are far closer to the desired patterns.

A problem with such a pixel-based approach was that large amounts of computation are used on regions where computation is not necessary. For example, in a large dark region, an FFT-based approach must still compute the transform and inverse transform for all the dark pixels, including those in the center that will certainly print as dark in any circumstance.

To streamline the process, contemporary model-based OPC differs from the pixel-based approach in two ways. First, most contemporary model-based OPC software systems are edge-based. Here, the polygons of the layout are divided into edge fragments. Traditional GDSII layouts already contain these edge-based definitions, in that the polygons are defined by vertex points defining polygon edges. For model-based OPC, additional edge vertices may be introduced to allow the desired fine motion of edge fragments [106]. This is illustrated in Figure 21.22. Once these edge fragments are defined, only certain points on or near the edges that are expected to require adjustment are selected in advance for simulation. This approach reduces the total amount of computation and eliminates computation of image values in regions where there is no ambiguity in the outcome.

Second, the image simulation itself is based on the Hopkins method, not on FFTs [107]. The Hopkins simulation integrates the image over all the source points, then adds the various contributions and produces an aerial image of the intensity of the electric field at the wafer.

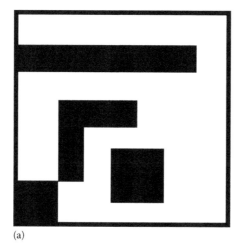
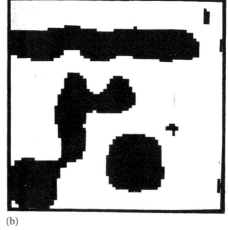

(a)   (b)

FIGURE 21.21 (a) Original layout (target) and (b) results of Saleh's model-based OPC calculation to achieve this target. (From Nashold, K.M. and Saleh, B.E.A., *J. Opt. Soc. Am. A*, 2, 635, 1985. With Permission.)

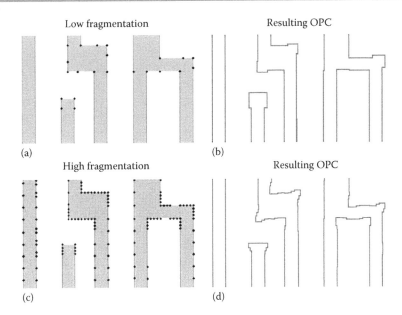

FIGURE 21.22   Effect of fragmentation settings on model-based OPC results. (a) Layout with fragmentation points selected to be far apart, (b) the OPC layout generated using (a), (c) Layout with fragmentation points selected to be denser, and (d) the OPC layout generated using (c).

This can be significantly more efficient computationally, especially if the series expansion of the computation is reasonably truncated after a certain number of kernels are used. However, most lithographic patterns are further distorted by the exposure in the photo-resist and by the subsequent processing. This means that the most accurate models will not be a constant-threshold model, as was used in the original Saleh work, but will require a more complex evaluation of the threshold conditions to be used.

Models, such as those of Rieger and Stirniman [108] or the variable-threshold resist models as presented by Cobb [109], typically require a test pattern to be fabricated on a photomask and printed using the wafer process in question. A set of measurements from that test pattern as printed on the wafer characterizes the signature of the effects of the process, including the proximity effects mentioned above (iso-dense bias, line-end pullback, and corner rounding). An example of a pitch curve from such a test pattern is shown in Figure 21.23 [109]. From a comparison of the actual line placement with the predicted line placement, variable-threshold models for edge placement can be generated. These empirical adjustments to the aerial image model can replace a more detailed process simulator for the various steps (e.g., post-exposure baking for the photo-resist, diffusion phenomena, and plasma-etch biasing).

Using such a model-based approach, convergence for the final edge placement is generally achieved in a few iterations.

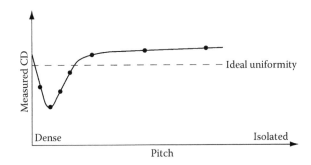

FIGURE 21.23   Pitch curve characterizing 1-D iso-dense bias for model-based OPC calibration. (After Cobb, N.B., Fast optical and process proximity correction algorithms for integrated circuit manufacturing, Ph.D. Dissertation, University of California at Berkeley, 1998.)

21.4.1.4 INVERSE OPC

The approaches to OPC described earlier have tended to take a simple approach to a layout: a particular graphical representation of portions of the layout is loaded, the image for that portion is simulated, and the relative positions of the line segments in the portion are iteratively adjusted to minimize a set of edge placement errors corresponding to predetermined edge placement sites.

A more complex generalization of the problem is to break the entire region of the layout into a grid of pixels, with a final target layout matching the desired circuit patterns, having a 0 or 1 value mapped to the presence or absence of a circuit feature. The input layout (or mask pattern) may also begin as identical to this desired layout, but a transformation of this matrix representation to compensate for the distortions of the lithography system will produce an output matrix (the image) that differs significantly from the desired features—as described above, corners will be rounded, lines will fall short, sub-resolution features will disappear, etc.

To correct the input function, an error function representing the difference between the actual image and the desired image can then be determined by various methods, and the input layout may then be systematically adjusted according to various optimization algorithms, such as systematic pixel-flipping, to minimize this error function. In some regards, this reflects the approach originally outlined with Saleh's original proposal [102–105], but now enabled for billions of polygons using contemporary computing power and matrix computation techniques [110].

For layouts at generations smaller than 65 nm, the density of the edge-based approach begins to approach the density of the pixels in the layout, and switching to a pixel-based OPC approach may actually become more computationally efficient for smaller features [111].

Such approaches may go by the name of Matrix OPC [112], pixel-based OPC [113], or Inverse OPC [114–122]. The layouts so produced may appear to be counter-intuitive, as illustrated in Figure 21.24 [117], and boundary conditions to restrict the inverse mask patterns to be manufacturable by conventional mask fabrication processes may be important [122].

21.4.2 PHASE-SHIFTING MASK (PSM)

The geometric operations needed to implement phase-shifting in a photomask layout may seem to be straightforward, especially if alternating aperture/dark-field masks are used (see Table 21.2). However, certain topological structures, such as "T"-shaped junctions, may lead to phase conflicts that cannot be easily solved [123,124]. An example of conflict structures is shown in Figure 21.25.

21.4.2.1 DARK-FIELD PHASE SHIFTING

When fine pitches are required in a dark-field mask, for example, for the lower metal layers of an IC, a novel approach to solve the problem has been presented by Ooi et al. [125,126]. Here, the

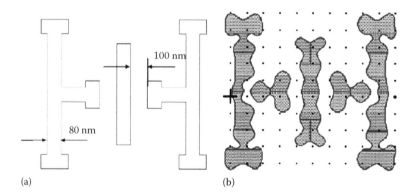

(a)   (b)

FIGURE 21.24 (a) Sample poly layout pattern, and (b) the corresponding mask layout computed using inverse lithography. (From Liu, Y. et al., Inverse lithography technology principles in practice: Unintuitive patterns, *25th Annual BACUS Symposium on Photomask Technology Proceedings of SPIE*, Vol. 5992, 2005, p. 599231. With permission.)

FIGURE 21.25   Layout structures that represent topologies leading to "phase conflicts."

phases are assigned prior to the finalization of the layout. Then, using compactor software, the final layout is only created at the fine pitch using phase-shifting where there are no phase conflicts. In other situations, the feature sizes and spacings are kept larger so that phase-shifting is not needed and no conflict occurs.

Fabrication of the mask now becomes relatively straightforward, in that, two writing steps are required, one to clear the chrome and the other to etch the phase-shifted structures.

21.4.2.2 BRIGHT-FIELD PHASE-SHIFTING

When dark fringes are required for thin lines, such as for very small transistor gates, a bright-field approach to phase-shifting is called for. This assigns 0° and 180° on the opposite sides of the features desired to be thin and has been proven extremely effective at shrinking transistor gates [127]. This is illustrated in Figure 21.26.

However, since both sides of the gate are not infinitely large, the phase polygons must somehow be closed. This represents the main problem of bright-field phase-shifting.

One solution to the problem is to create "partial shifters" of either 90° or 60°–120° combinations [128]. These partial phase steps scatter some light but do not create the strong phase interference that a 0°–180° transition does, and the dim fringe they create ideally never goes below the exposure threshold of the resist. This is illustrated in Figure 21.27.

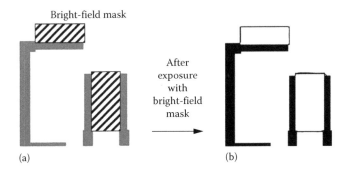

FIGURE 21.26   Bright-field shifiting. (a) Original layout pattern, with clear 0° areas (white), chrome opaque areas (dark), and clear regions with a 180° phase shift. (b) The image on the wafer after this mask is used for exposure. Not only do the opaque features print, but the phase edges print as well.

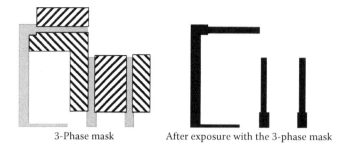

3-Phase mask   After exposure with the 3-phase mask

FIGURE 21.27 Partial-shifter approach. Left: original layout pattern, with clear 0° areas (white), chrome opaque areas (dark), and clear regions with a −90° phase shift or a +90° phase shift. Right: the image on the wafer after this mask is used for exposure. Regions bounded by both partial shifters have a relative 180° phase shift and print as dark fringes, while those with only a net 90° phase shift do not.

Using simple routing algorithms or, in some cases, simple design rules allows the boundaries of these partial shifters to be defined [129]. Although easily able to make fine lines and even dense groups of fine lines when alternating apertures are used, the drawback to this relatively straightforward technique is that the additional partial shifters use up space that could otherwise be dedicated to additional circuit elements, decreasing the overall possible device density. In addition, partial shifter masks are expensive as four distinct writing steps are now required to fabricate the different etch depths of the photomask, each requiring precise alignment. Defects in such masks are often impossible to repair, and only about 1 in 10 masks are typically defect-free.

An additional solution to this problem is to leave the boundaries of the phase-shifters on the PSM, and to use a second exposure to "trim" or remove these unwanted features [130]. This is illustrated in Figure 21.28. The advantage here is that the benefits of the phase-edge fringe can be achieved using masks that are cheaper than the multi-write defect-prone partial-shifter masks, but the disadvantage is that the multiple exposures reduce the wafer stepper throughput.

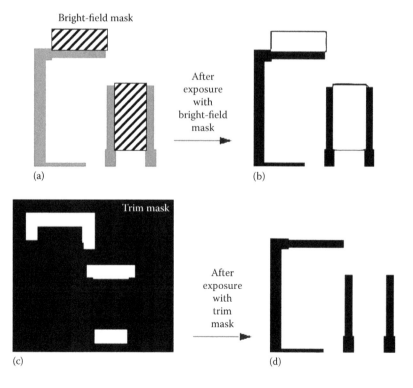

FIGURE 21.28 Trim-mask approach. After initial exposure to the mask in (a), the image shown in (b) still has unwanted bridging structures. With the additional exposure using a second trim mask (c), the unwanted features are also exposed and the final double-exposed image (d) is correct.

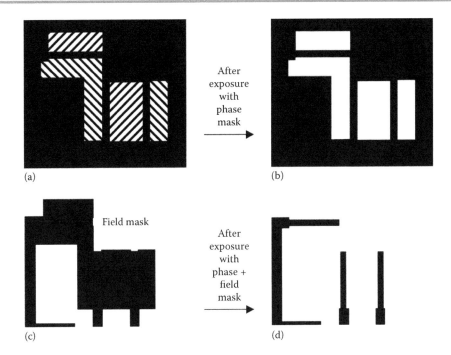

FIGURE 21.29 Dark-field phase-shifting trim-mask approach. After initial exposure to the mask in (a), the image shown in (b) has printed the fine features that benefit from phase-shifting, but everything else remains unexposed. Using a second exposure with a "field mask" shown in (c), the remaining larger dimensions of the layout are defined, leading to the desired image in (d).

An interesting variation on this approach is one in which the phase-shifters are designed on the dark-field second-exposure mask, leaving the other circuit features and larger "protection" features on the first exposure [131]. This is illustrated in Figure 21.29. This allows the polygons of other layers, e.g., the active layer when phase-shifting the gate layer, to serve as the starting point for the phase-mask polygons, eliminating the need for routing algorithms in the definition of the phase features. Furthermore, by limiting the area that is cleared in the phase mask, this reduces the susceptibility to defects, potentially making masks less expensive. However, the throughput disadvantages of the "trim"-mask approach are not mitigated in this approach. This means that these approaches to phase-shifting will only find application in large-volume products such as microprocessors, in which the higher mask and processing costs can be recovered.

We should also note that the problems OPC has been designed to fix are not eliminated with the adoption of phase-shifting. Although contrast may be improved, isolated lines can still behave differently than dense lines, line ends can still pull back, and corners can still be rounded. For this reason, OPC is still typically required after conversion of a layout to phase-shifting has been completed [132].

21.4.2.3 PHASE ASSIGNMENT ALGORITHMS

The decomposition of a layout for phase assignment has been studied extensively [133]. As discussed above, layouts that are originally created that conform to phase-compliant design rules, are the best to reduce potential phase conflicts and to make phase-shifting masks manufacturable.

Automatic algorithms that evaluate the phase compatibility of existing layouts, have also been developed [134–136]. One such approach using a conflict graph has been presented by Kahng et al. and is illustrated in Figure 21.30 [135]. In this approach, a conflict graph G corresponding to a layout is created, with G having a vertex assigned to each layout feature and edges between two vertices, if and only if, there is a phase conflict between two neighboring features. The problem of phase assignment then becomes a coloring problem, in which no odd cycles in the conflict graph may occur.

This is achieved by constructing a dual graph D for this conflict graph G, as also shown in Figure 21.30 [135], in which the nodes of the dual graph correspond to the planar faces of the

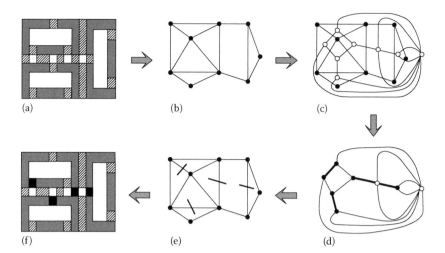

FIGURE 21.30   From the conflicts between features (a), the conflict graph is derived (b). The dual graph (c) is constructed. The vertices of odd degree are matched using paths in the dual graph (d), and the corresponding conflict edges are determined (e). Finally, the minimum set of conflicts to be deleted is determined (f). (With kind permission from Springer Science+Business Media: Berman, P., Kahng, A.B., Vidhani, D., and Zelikovsky, A., The T-join problem in sparse graphs: Applications to phase assignment problem in VLSI mask layout, *Proceedings of the Sixth International Workshop, WADS'99*, Lecture Notes in Computer Science 1663, Springer Verlag, London, U.K., 1999, pp. 25–36, Figure 1.)

conflict graph G. This identifies the locations of the potential conflicts, and algorithms that suggest their elimination can then be applied.

Implementation of these algorithms in a coloring schemes to "convert" established layouts or cell designs to phase-shifting have been developed [133], but in recent years, the trend has evolved more to creating phase-aware layouts or cell libraries at the original design phase, eliminating the need to actually convert a layout to phase-shifting.

21.4.3 OFF-AXIS ILLUMINATION (OAI)

The third approach to RET is OAI. Although one would think that the choice of illumination angle would not be a factor in the creation of IC layouts, there are in fact significant ramifications to the layout that must be discussed.

As mentioned above, OAI functions by emphasizing the diffraction from certain pitches at the expense of others. This leads to the concept of "forbidden pitches": lines with certain spacings that are far more difficult to print (have smaller focus and exposure latitudes) than lines of other spacings [64,65]. Although related to the angle of the illumination and the NA of the lens, the exact pitches where performance degrades can only be found by simulation using the desired OAI pattern. The results can be surprising: dense pitches may print very well under a particular kind of OAI, while contrast is very poor for certain larger pitches that would traditionally print very well. This is illustrated in Figure 21.31 [137].

To mitigate the problems caused by forbidden pitches, combinations of RET are used. This is where SRAF are most commonly applied. The insertion of SRAF transforms a region with a "forbidden" pitch into one where the pitch between polygons is smaller and no longer "forbidden." This is also illustrated in Figure 21.31.

The orientation of the pitches is also a factor. For symmetric annular illumination, all orientations are treated equally, but for quadrupole illumination (see Figure 21.9), only vertical and horizontal features diffract light well. Diagonal structures will fail to print, allowing only Manhattan layouts to be used with this illumination system (see Figure 21.32 [137]).

Likewise, dipole illumination allows exceptionally high contrast for dense features oriented orthogonally to the dipole, but features parallel to the dipole will not print. For 2-D layouts, dipole illumination therefore requires a double exposure. Since not all features are either horizontal or

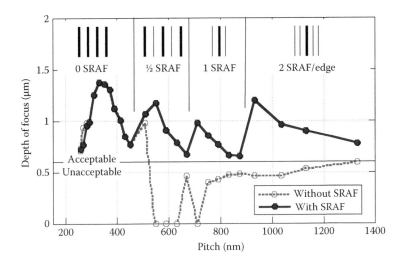

FIGURE 21.31 Off-axis illumination (OAI) and forbidden pitches, mitigated using SRAF. (From Schellenberg, F.M. et al., Adoption of OPC and the impact on design and layout, *Proceedings of the 38th Design Automation Conference*, ACM, New York, 2001, pp. 89–92. With permission.)

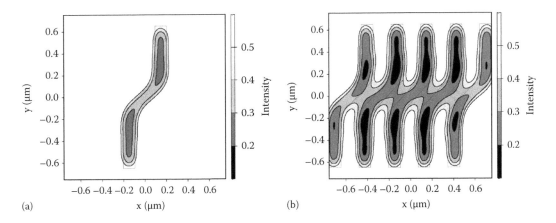

FIGURE 21.32 (a) Simulated image of an isolated line, and (b) dense lines with 45 nm portions under QUASAR illumination. The diagonal portions are significantly brighter, and will not be faithfully reproduced. (From Schellenberg, F.M. and Capodieci, L., Impact of RET on physical layouts, *Proceedings of the 2001 ISPD*, ACM, New York, 2001. With permission.)

vertical, the parsing of data into these two masks and the correct treatment of the points of intersection of horizontal and vertical lines can be complicated. A number of algorithms have been explored that allow the automatic conversion of these layouts.

An area of active development to improve simulation accuracy for OAI relies on collecting actual pupil maps of the illumination source and using these in the simulation algorithms. For generations at 90 nm and below, this can make a significant difference in the accuracy of the model. An ideal source map for QUASAR illumination and the experimental results of a measurement of the source map for a particular stepper is shown in Figure 21.33 [138]. The use of the actual map in the simulation software engine for the computation of OPC removed a 19 nm error in the model that had remained when the idealized source map had been used. For a 90-nm process, 19 nm represents a significant error.

Another area of improvement is in the customization of the illumination pattern for the particular layout. This can be very effective for ICs that will see large volume production, such as DRAMs [139–142]. Two examples of layouts, along with the source map that produced the optimal lithographic performance for each are shown in Figure 21.34 [140]. It is clear that these are far from the simple patterns for illuminators shown in Figure 21.9. However, using customized

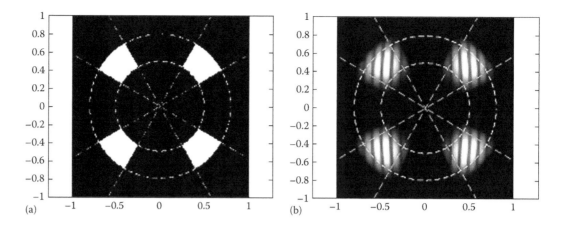

FIGURE 21.33 Pupil map of (a) an ideal QUASAR source, and (b) measured intensity map of an actual stepper illumination pattern. (From Granik, Y. and Cobb, N., New process models for OPC at sub-90nm nodes, *Optical Microlithography XVI*, *Proceedings of SPIE*, Vol. 5040, 2003, pp. 1166–1175. With permission.)

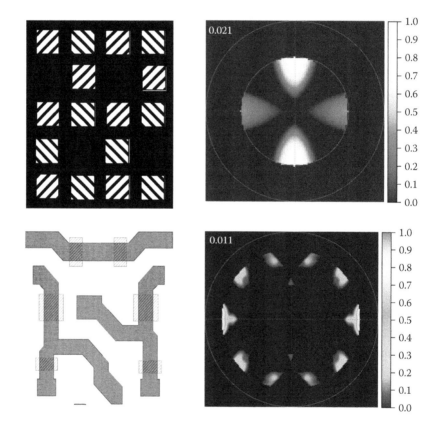

FIGURE 21.34 Mask/source optimization. On the left are representative layout patterns for (upper) a contact layer with alternating phase shifts, and (lower) the gate layer of an SRAM cell without phase-shifting. On the right are the corresponding optimized source illumination patterns that optimize lithographic performance. (From Granik, Y., *J. Microlith. Microfab. Microsys.*, 3, 509, 2004. With permission.)

diffractive optical elements, these can be produced and inserted into a stepper, further improving the potential lithographic performance.

This source-mask optimization (SMO) is especially enhanced by the introduction of dynamic illuminators, such as ASML's FlexRay [143–144]. These programmable reflective optical elements use adaptive optics techniques and devices (originally developed for telescopes to compensate for atmospheric distortions [145]) to allow programmable off-axis illumination to be used. This has eliminated the need to make a special dedicated OAI reflector (cost effective only for large-volume manufacturing) and allows SMO to be used with any layout for any volume of production.

21.4.4 POLARIZATION

As discussed above, polarization is an emerging variable that may need to be considered more rigorously as IC dimensions continue to shrink. Certain polarization effects causing contrast changes as SRAF become approximations to wire grid polarizers have been observed. Modeling these phenomena requires a full electromagnetic simulation of the fields around the photomask and knowledge of the actual optical properties n and k for the mask film as fabricated.

This can be complicated. Although nominally n and k, the real and imaginary parts of the complex refractive index of the materials of a photomask can be easily estimated assuming that the opaque mask material is chrome, masks in fact are not just chrome, but a mixture of chrome, chrome oxides, and other compounds, designed to make the mask opaque and antireflective. Table 21.3 shows the values for n and k at optical and lithography wavelengths [146,147]. The difference can be significant with materials that behave optically like metals (forming polarizers, etc.), becoming less metallic at shorter wavelengths, while the material properties of others become more metallic.

This is important because the various polarization effects (and also certain plasmon effects, which can lead to additional transmission through sub-wavelength apertures) depend on the metallic properties of the material. Without an accurate way to determine these material properties, there is no way to determine the effect they will have on imaging.

Simulation tools such as TEMPEST have been developed to carry out a finite-difference time domain (FDTD) simulation of the electromagnetic field at the mask and can predict these effects [148]. However, the amount of time that is needed to compute these for each aperture and polygon for an IC with millions of components would be prohibitive.

Techniques have been developed to form a compromise between accuracy and speed, by pre-computing certain images with a full FDTD solution, storing these primitive image components, and then adding these precomputed images as required for specific image layouts. This "domain decomposition method" (DDM) has been proven for masks where the topography effects must be precomputed with a 3-D solver and are currently showing promise when used for polarization and other E-M effects [149].

TABLE 21.3 **Optical Properties for the Complex Refractive Index n and k at Three Different Wavelengths for Metals Silver and Gold, Materials Commonly Found in the Photomask Industry, and an Insulator (Glass)**

	564 nm		248 nm		193 nm	
	n	k	n	k	n	k
Silver (Ag)[a]	0.120	3.45	1.298	1.35	1.028	1.18
Gold (Au)[a]	0.306	2.88	1.484	1.636	1.426	1.156
Chromium (Cr)[a]	3.215	4.40	0.85	2.01	0.84	1.65
CrN[b]	2.466	0.041	1.600	1.007	1.292	0.899
Cr_2O_3[b]	2.185	0.229	1.863	0.708	1.616	0.021
CrO_3[b]	1.874	0.070	1.722	0.392	1.634	0.6504
Silicon (Si)[a]	4.042	0.032	1.730	3.222	0.883	2.78
Glass (SiO_2)[a]	1.459	0($<10^{-6}$)	1.508	0($<10^{-6}$)	1.563	0($<10^{-6}$)

Note: Electrically conductive behavior is typically indicated by small values of n and large values of k; an insulator by extremely small values of k.

Sources:

[a] Palik, E.D., Ed., *Handbook of Optical Constants of Solids*, volume I, II, and III, Academic Press, San Diego, CA, 1998.

[b] http://www.rit.edu/kgcoe/microsystems/lithography/research/thinfilms.html.

21.4.5 RET INTEGRATION

Each of the topics above has been discussed in isolation, as a physical effect occurring in lithography that can be countered by changes in the IC layout. However, the more recent trend in the field of Resolution Enhancement Technologies is to take a "holistic" approach to layout optimization, and to apply several RET solutions simultaneously. Some examples of these combined solutions are phase-shifting masks that also apply OPC [124,133,137], layouts using SRAF that are also tuned to a particular off-axis illumination configuration [64–66], and source-mask optimization using polarized illumination [150].

To address the combined RET problem, EDA suites such as Calibre® from Mentor Graphics [151], Proteus products from Synopsys [152], and Brion software offered by ASML [153] have been brought to market, and there have been calls for tools that incorporate even more physical effects (such as etch effects) in their computations to create a holistic approach to RET [154]. This field has generally come to be known as "computational Lithography." [155]

21.4.5.1 DESIGN FOR MULTI-PATTERNING

Multiple patterning lithography as discussed in Section 21.3.6 requires that the circuit layout be parsed into multiple masks, each of which must obey the design rules for the selected lithography system and processing recipe [156–165]. Fortunately, much of the work originally developed for the double exposure of phase-shifting masks, discussed above, can be applied to the data parsing for multiple patterning [157].

Creating layouts for designs where processes that add features, such as the spacer patterning processes that double the feature density, may be straightforward, especially when a regular grating along with a cut-and-trim approach to modify a grating pattern formed by the spacers is employed [164,165]. However, this will be effective only if the process control and overlay for the manufacturing process is repeatable enough so the doubled patterns form reliably, allowing interconnects and other circuit features to form functioning circuits using the features that appear through the double patterning process.

21.4.5.2 DESIGN FOR SELF-ASSEMBLY

The "leap of faith" required when using process-generated features in technologies such as spacer patterning becomes even more acute when applying directed self-assembly (DSA) processes.

As discussed in Section 21.3.7 above, in a DSA process, a large blank region is patterned by conventional lithography techniques, and within the blank region, the polymers that will define small circuit features will self-assemble into line/space pairs or contact holes [83–93]. The first proposal for design modification for use with line/space pairs created by DSA was an adaptation of PSM phase assignment algorithms [166]. Other approaches have since been developed as more materials have been explored [167–170].

This is illustrated for an experimental design using contact holes in Figure 21.35 [169]. Shown in Figure 21.35a is the original layout of desired contact holes (the "target" pattern) converted into a "guiding pattern" (GP), shown in Figure 21.35 (b), according to certain DSA process rules. (This GP has been nicknamed a "seahorse" pattern by the authors [169].) This GP is then transformed into a mask layout, shown in Figure 21.35 (c), to be used in a lithography scanner in conjunction with a customized source map, also shown in Figure 21.35 (c). The printed patterns shown in Figure 21.35 (d), leave a large region in which the DSA contact patterns are to self-assemble.

Clearly, faith in the process to deliver billions of contacts in a single IC with statistically uniform dimensions will require a degree of process control and material uniformity that has yet to be achieved. However, given that the materials to carry out DSA are so inexpensive, and the alternative approaches that directly pattern features are so expensive, there is a great incentive to solve any DSA problems with uniformity and defects. It is expected that these materials will be part of routine IC production by the end of the decade [91].

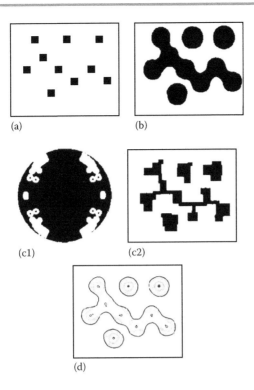

FIGURE 21.35   Contact patterning with directed self-assembly (DSA). (a) A desired contact layout, along with (b) a corresponding "guide pattern" to be created by lithography on the wafer; (c) the corresponding mask layout and source map, and (d) the image to be printed using the source of (c1) with the mask of (c2). (Adapted from Lai, K. et al., Computational aspects of optical lithography extension by directed self-assembly, *Optical Microlithography XXVI*, *Proceedings of SPIE*, Vol. 8683, 2013.)

21.4.5.3 DESIGN FOR DIRECT WRITE E-BEAM AND EUV LITHOGRAPHY

Although data conversions for e-beam writers (see Section 21.5 below) have been used for decades for photomask writing, special problems arise when trying to parse layout data into thousands of multiple e-beams for the direct-write systems as mentioned above [21–28]. In particular, local charging and beam deflection in transit from neighboring electron beams may add additional proximity effects to those already present in a single beam exposure [38–40] and create field stitching problems where the regions exposed by distinct beams intersect [171]. However, given that the physics of the charged particle interactions are well known, computation remains straightforward, and performance with a 10% tolerance has been demonstrated [171].

The introduction of lithographic scanners using extreme ultraviolet (EUV) wavelengths of $\lambda = 13$ nm has been promised for years [19], and yet problems with generating enough EUV source power, defect-free mask blanks, and sensitive photoresists have delayed the introduction of commercial systems. It is now forecast that EUV systems may be ready for production of the ICs with a half pitch between 13 and 20 nm [32] in 2017.

With the ongoing progress in RET lithography allowing conventional lithography systems to produce the features required for the 22 nm node, and even smaller, one of the arguments in favor of the introduction of higher k_1 EUV system (that expensive RET data processing could be avoided) no longer applies—with the smaller NA ($NA = 0.3$) of these systems, introduction in 2017 would entail the use of k_1 between 0.29 and 0.45, a domain for which RET has been necessary for conventional lithography.

It is therefore no surprise that extensive work has been done recently on applying all of the above described RET techniques to EUV systems, including OPC [172], sub-resolution assist features (SRAF) [173], and source-mask optimization (SMO) [174].

A unique problem that arises with EUV is that of flare [175–177]. With EUV masks and optical elements fabricated from multilayer stacks of molybdenum and silicon, scattering of photons will be different in densely patterned areas than in sparsely patterned areas. The additional scattered light can change the local exposure levels and therefore the patterned linewidths. Several EDA-based solutions to flare-based distortions have been proposed, including approaches that expand hierarchy for corrections derived from regional feature density functions [177] as well as local density [178].

What is clear is that the introduction of EUV, if it indeed ever happens, will be done using all the developed technology of RET applied to the technology from the first day of production.

21.4.6 RESTRICTIVE DESIGN RULES (RDR)

Low k_1 lithography, multi-patterning and DSA all succeed by creating patterns as close to regular, periodic gratings as possible. This is because gratings are the simplest structure that can be formed. A systematic approach to grating-based patterning is the subject of a recent DARPA project [179].

Luigi Capodieci has called this increasing use of 1D critical features a trend towards "design regularity." [180] Several articles and experiments have appeared promoting the benefits for manufacturing with the adoption of Restrictive Design Rules (RDR) to achieve these regular patterns [181–191]. The restrictions are as follows:

1. A restriction that all lines in a layer extend in only one dimension.
2. A defined regular minimum pitch and/or spacing distance to another feature (dictated by the minimum patterning resolution).
3. All spacings between features are required to occur at some multiple of this minimum pitch.

The adoption of RDR does not have to mean a decrease in circuit performance. Replacement of jogs that designers may be accustomed to using with regular landing pads need not degrade performance at all, and circuits (especially individual cell designs) may be produced that are both optimized for circuit performance and process reliability [180].

Geometrical regularity is now being standardized by the introduction of the concepts of "regular fabrics" and "regular bricks" [185–189] for cell libraries. An example of a "brick" is illustrated in Figure 21.36 [187]. These configurable fixed-size circuit components are characterized

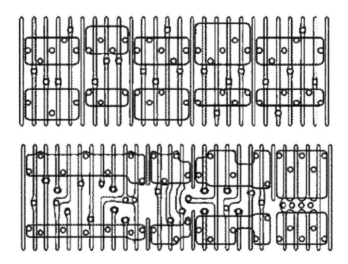

FIGURE 21.36 Diffusion, poly, and contact lithography variability-bands for pdBrix layout (top) and conventional layout (bottom). (From Liebmann, L. et al., Simplify to survive: Prescriptive layouts ensure profitable scaling to 32nm and beyond, *Design for Manufacturability through Design-Process Integration III, Proceedings of SPIE*, Vol. 7275, 2009. With permission.)

by extreme regularity at the layout level (and thus improved manufacturability), both individually and as assemblies. For some of these "bricks," the layout may be slightly larger than a less regular circuit, but the loss in area may be compensated by the increased manufacturing reliability.

Standard cells with extreme regularity for several levels (gate, contact, metal, etc.) and a number of electrical advantages have been created by the company Tela Innovations [190]. Tela has presented scalable cells that follow a gridded design strategy using a pre-determined set of layout sub-structures, with a fixed number of allowable shape placement grids, shape dimensions, and edge-to-edge relationships, and result in improved device and interconnect formation, which in turn reduces sub-threshold leakage and electrical variability [191].

21.5 MASK DATA PREPARATION

Once the mask data layout has been created and modified to accommodate various RET algorithms, the final photomask still needs to be written. Although in principle a straightforward process, there are several practical issues that require further manipulation of the layout data. Usually, the data must be flattened to some degree, and the polygons must be reduced to a simple set of structures (typically rectangles and trapezoids) that the machine can use to write the patterns directly. This process of data conversion is called *fracturing* [192].

21.5.1 MASK WRITERS: PHYSICS

Mask writers are machines that modulate an exposure beam to expose photoresist coated onto a mask blank in selected locations. Typically, either electron beams or laser beams are used. In one common architecture, the beam constantly scans in a predetermined pattern (a raster scan) and is turned on and off to write the pattern, as illustrated in Figure 21.37. This has the advantage of making the motion of the stage quite predictable and allows the task of writing to be divided into several exposure beams that handle different portions of the total exposure field. However, the tool also scans (with the beam off) over areas that require no exposure, reducing throughput for sparse layouts. As many as 24 beams simultaneously are used to write masks in some tools.

Other tools write with a "variable-shaped beam" (VSB), as illustrated in Figure 21.38. Here, a larger beam is shaped by an aperture into a primitive shape (usually a rectangle or a trapezoid), and the image of the aperture projected in individual "flashes" at appropriate locations. With VSB tools, many small and large features can be written using only single flashes, while no flashes are used for large areas that require no exposure. This can be more efficient for many types of layouts, particularly sparse ones such as contact hole layers. Additionally, some "vector scan" tools combine aspects of VSB tools (vectoring only to regions requiring exposure) and raster scanning

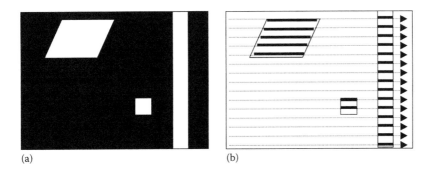

(a)   (b)

FIGURE 21.37 Exposure of (a) a representative layout geometry using (b) a raster scan. The beam coordinates for exposure are systematically swept over the area to be exposed, and the beam is turned on only as it passes over the areas to be cleared on the mask.

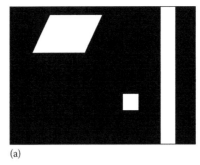 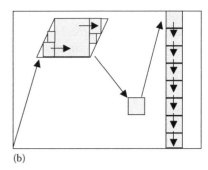

(a)   (b)

FIGURE 21.38   Exposure of (a) a representative layout geometry using (b) a variable shaped beam (VSB). The beam is "vectored" to particular locations to be exposed, and the beam size and shape are adjusted to expose the areas to be cleared on the mask.

(using raster scans to expose those regions once there). More information on different exposure strategies can be found in Refs. [6,7].

21.5.2 MASK DATA FRACTURING

Clearly, the exact machine instructions that will be required to write a mask will be different for these two tool architectures. For raster scan machines, the data are converted into primitive shapes (rectangles and trapezoids), and then converted by the tool itself into a pattern of scan lines to be used with each of the exposing beams. A common format used with raster scan machines is MEBES [193]. This format represents all the figures as composites of either rectangles or trapezoids and partitions the data into individual stripes. The dimensions of the stripes are a multiple of the minimum address unit specified in the particular data file. The mask is written stripe by stripe. The MEBES format has no hierarchy—all polygons occur uniquely within their stripe location. However, MEBES also uses a jobdeck language that controls the placement and writing sequence of MEBES files. In this regard, the jobdeck can be regarded as a second layer of hierarchy for an otherwise flat format.

VSB machines require that the layout be fractured into shots of acceptable size, and the appropriate stage motion instructions be generated to create the pattern. Examples of a layout broken down into shots are shown in Figure 21.39. Because VSB format data is a collection of shot sizes and locations unrelated to a specific writing stripe on the mask, they can have a limited hierarchy structure. How this is utilized is a proprietary function of the different mask-writing tools themselves.

The introduction of RET can significantly change the mask-writing characteristics of an IC layout, especially with VSB. A layout after edges have been moved with the application of OPC along with the corresponding shot layout, is shown in Figure 21.30. Clearly, the number of shots has increased dramatically, and the writing time would be expected to correspondingly increase. For the worst cases of aggressive OPC, mask-writing times approaching 24 h have been reported, as illustrated in Figure 21.40 [95].

With a judicious choice of OPC parameters, this can be significantly improved. An example of this is shown in Figures 21.41 through 21.43. Here, OPC has been run on a gate layer. The typical result, as shown in Figure 21.42, is a significant increase in shot count and hence writing time. However, the portions of the gate layout that are not overlapping the underlying active area can be interpreted as poly interconnects, which can be fabricated with a more relaxed tolerance than the more critical gate regions themselves. Performing a logical AND operation of the gate layer and the underlying active area allows the fragmentation for the OPC run to be set dynamically, with finer fragmentation in the gate regions over active and more relaxed fragmentation in the other areas. The resulting VSB layout, with significantly smaller shot count, is shown in Figure 21.43.

In the long run, as long as feature sizes continue to shrink while IC density increases, the volume of data required to describe an IC will continue to grow exponentially. To mitigate this,

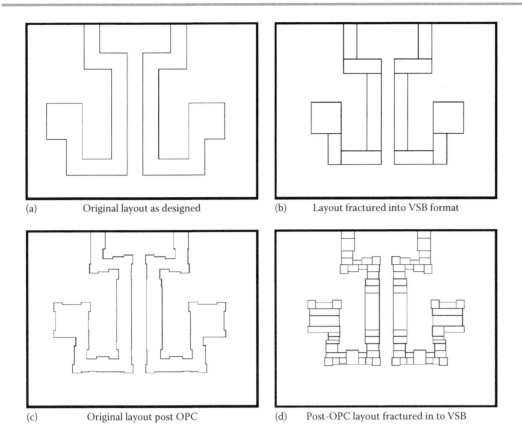

(a)   Original layout as designed

(b)   Layout fractured into VSB format

(c)   Original layout post OPC

(d)   Post-OPC layout fractured in to VSB

FIGURE 21.39 Visualization of the impact of the application OPC on the fractured result in VSB11 format. The more complex polygons require a more detailed trapezoiding, which results in a much larger number of trapezoids in the fractured output.

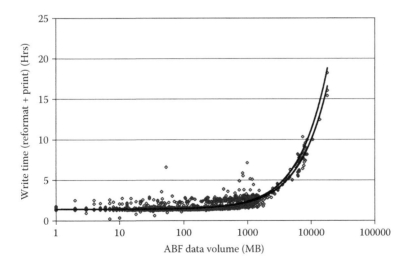

FIGURE 21.40 Measured writing times for an ALTA 3500 raster scan mask writer. Above 1 GB, write times increase in a dramatic and nonlinear manner. (Data courtesy Dupont Photomask. From Schellenberg, F.M., Advanced data preparation and design automation, *Photomask and Next–Generation Lithography Mask Technology IX, Proceedings of* SPIE, Vol. 4754, 2002, pp. 54–65. With permission.)

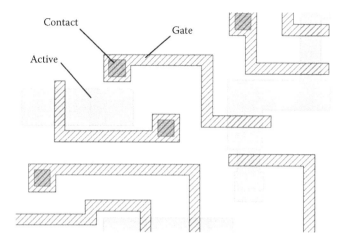

FIGURE 21.41 Representative portion of a microprocessor layout.

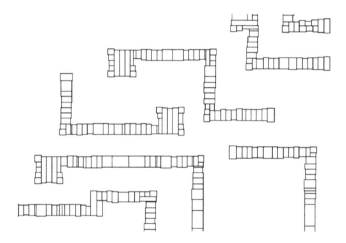

FIGURE 21.42 Fractured layout of the gate layer from Figure 21.41 in VSB11 format with standard fragmentation-uniform aggressiveness for the entire gate layer.

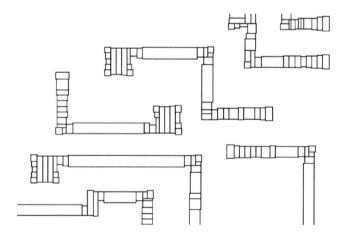

FIGURE 21.43 Fractured layout of the gate layer from Figure 21.41 in VSB11 format with interlayer-aware fragmentation. The layout is separated into the critical and noncritical regions. The aggressiveness of the fragmentation is retained for the critical areas while it is reduced for the noncritical areas. This significantly decreases the total data volume with no loss in performance.

in June 2003, a new data format called Open Artwork System Interchange Standard (OASIS) was approved by SEMI after 2 years of discussion and testing [194]. The goals achieved by this standard were

- Greater than 10× file size reduction for the same data when compared to GDSII
- Removal of 16- and 32-bit restrictions, allowing integers to extend to 64 bits and beyond
- Efficient handling of flat geometric data
- Improvement in the "richness" (e.g., annotation) that can be communicated along with the formatted files
- No significant increase in operation/fracture run times

OASIS achieves its primary goal by defining the coordinates of polygons by difference functions rather than absolute coordinates for all vertices. In other words, once the initial coordinates for the first vertex have been set, the following coordinates can be expressed in length differences Δx and Δy from the initial coordinate. Since the distance between vertices is usually very small (especially after refragmentation for OPC has occurred), a single byte is often enough to represent the additional vertex. The success of the format is represented by the comparisons for the files of Table 21.4 [195].

The adoption of OASIS in production flows is progressing, with more advanced nodes of IC manufacturing adopting OASIS as a matter of course. At the 65 nm node, more than 50% of the post-OPC data was generated in OASIS, and smaller nodes show progressively higher ratios, with the 32 nm node having over 80% of the OPC files and almost 70% of the fracture files as well represented in OASIS [196]. This trend is expected to continue as the nodes become progressively smaller, and data volumes continue to increase.

21.5.3 DATA PREPARATION WITH MASK PROCESS CORRECTION (MPC)

With all the attention that OPC has received in recent years for improving lithographic fidelity, it should be no surprise that the pattern fidelity for the production of photomasks has also come under some examination. Although the photomask features are typically used in a 4× reduction system, and the feature dimensions are therefore 4× larger than on the wafer, there is still a need to accurately fabricate SRAF and other OPC jogs and structures that are significantly smaller.

The introduction of mask processing correction (MPC) computations has added an additional level of complexity for photomask data processing [197,198]. In addition to normal charged

TABLE 21.4   File Size Comparison for GDSII and OASIS Representation of Design Data

Test Case	GDSII	OASIS	Ratio Case (GDSII/OASIS)
1	5316341760	26438663	201.1
2	6433212416	417658458	15.4
3	8750811136	515677720	17.0
4	995145728	19078143	52.2
5	2111688704	122104830	17.3
6	2344345600	109435610	21.4
7	8191951330	749401369	10.9
8	1077782528	38230183	28.2
9	2329663888	136189555	17.1
10	1150431232	92419711	12.4
11	2379108352	91049913	26.1
12	2046351360	111240392	18.4

Note: The test cases represent a variety of design styles and device types.

Source: Schulze, S., LaCour, P. and Grodd, L., OASIS-based data preparation flows: Progress report on containing data size explosion, *Design and Process Integration for Microelectronic Manufacturing II, Proceedings of SPIE,* Vol. 5379, 2004, pp. 149–157.

particle proximity corrections [21,38–40], distortions caused by mask processing effects (line pullback, corner rounding, etc.) are also included in the models used to adjust the electron beam exposure dose. These have now been built into the various software products for MDP as an option. Specific models applying inverse lithography to the e-beam exposure problem are also being explored [199].

21.6 SUMMARY

The final layout verification step in IC design used to be a simple matter of checking design rules. With more complex process distortions and manufacturing limitations being placed on designs by lithographic processes, RET has grown to be an important part of the tape-out process. Various insertion points have been proposed for RET, but in general the adoption of RET has been migrating up the design flow, finding its most common application integrated with layout generation and verification.

The various specific techniques involved with the layout changes for any of the RET approaches, whether they be OPC, phase-shifting, OAI, or new directed self-assembly (DSA) techniques, all require that efficient models be provided that can efficiently describe the expected wafer behavior. Once the prescriptive action has been determined, polygon manipulation itself is fairly straightforward and can be automated efficiently with new formats, such as OASIS. However, with the new physical processes required for 22 and 16 nm IC generations, the ability to calibrate models and predict wafer behavior will determine whether the full promise of RET can be achieved.

REFERENCES

1. M. Madou, *Fundamentals of Microfabrication: The Science of Miniaturization*, 2nd edn., CRC Press, Boca Raton, FL, 2002.
2. Y. Nishi and R. Doering, Eds., *Handbook of Semiconductor Manufacturing Technology*, Marcel Dekker, New York, 2000.
3. H. Levinson, *Principles of Lithography*, 2nd edn., SPIE Press, Bellingham, WA, 2005.
4. L. Thompson, C.G. Willson, and M. Bowden, Eds., *Introduction to Microlithography*, 2nd edn., American Chemical Society, Washington, DC, 1994.
5 J.R. Sheats and B. Smith, Eds., *Microlithography: Science and Technology*, Marcel Dekker, New York, 1998.
6. S. Rizvi, Ed., *Handbook of Photomask Manufacturing Technology*, CRC Press, Boca Raton, FL, 2005.
7. B. Eynon, Jr. and B. Wu, *Photomask Fabrication Technology*, McGraw Hill, New York, 2005.
8. J.W. Goodman, *Introduction to Fourier Optics*, McGraw-Hill, San Francisco, CA, 1968.
9. M. Bowden, The lithographic process: The physics, chap. 2 of Ref. [4], and B. Smith, Optics for photolithography, chap. 3 of Ref. [5].
10. F.M. Schellenberg, Ed., *Selected Papers on Resolution Enhancement Techniques in Optical Lithography*, SPIE Press, Bellingham, WA, 2004.
11. E. Abbe, Beiträge zur Theorie des Mikroskops und der mikroskopischen Wahrnehmung, *Archiv Mikroskopische Anatomie*, 9, 413–468, 1873, also reproduced in Ref. [10].
12. L. Rayleigh, On the theory of optical instruments, with special reference to the microscope, *Philos. Mag.*, 42, 167–195, 1896, also reproduced in Ref. [10].
13. B.J. Lin, Where is the lost resolution? *Optical Microlithography V, Proceedings of SPIE*, Santa Clara, CA, Vol. 633, 1986, pp. 44–50, also reproduced in Ref. [10].
14. B. Smith, A. Bourov, H. Kang, F. Cropanese, Y. Fan, N. Lafferty, and L. Zavyalova, Water immersion optical lithography at 193 nm, *J. Microlith. Microfab. Microsys.*, 3, 44–51, 2004.
15. B.J. Lin, Immersion lithography and its impact on semiconductor manufacturing, *J. Microlith. Microfab. Microsys.*, 3, 377–395, 2004.
16. S. Owa, H. Nagasaka, Y. Ishii, O. Hirakawa, and T. Yamamoto, Advantage and feasibility of immersion lithography, *J. Microlith. Microfab. Microsys.*, 3, 97–103, 2004.
17. Y. Wei and R.L. Brainard, *Advanced Processes for 193-nm Immersion Lithography*, SPIE Press, Bellingham, WA, 2009.
18. K. Jain, *Excimer Laser Lithography*, SPIE Press, Bellingham, WA, 1990.
19. C.W. Gwyn, R. Stulen, D. Sweeney, and D. Attwood, Extreme ultraviolet lithography, *J. Vac. Sci. Technol. B*, 16, 3142–3149, 1998.

20. F. Cerrina, X-ray lithography, Chapter 3, in *Handbook of Microlithography, Micromachining and Microfabrication, Vol. 1: Microlithography*, SPIE Press, Bellingham, WA, 1997.

21. M. McCord and M. Rooks, Electron beam lithography, Chapter 2, in *Handbook of Microlithography, Micromachining and Microfabrication, Vol. 1: Microlithography*, SPIE Press, Bellingham, WA, 1997.

22. B.J. Lin, Future of multiple-e-beam direct-write systems, *J. Micro/Nanolithogr. MEMS, MOEMS*, 11, 033011, 2012.

23. E. Slot, M.J. Wieland, G. de Boer, P. Kruit, G.F. ten Berge et al., MAPPER: High throughput maskless lithography, *Emerging Lithographic Technologies XII, Proceedings of SPIE*, San Jose, CA, Vol. 6921, 69211P, 2008.

24. M.J. Wieland, G. de Boer, G.F. ten Berge, M. van Kervinck, R. Jager, J.J.M. Peijster, E. Slot, S.W.H.K. Steenbrink, T.F. Teepen, and B.J. Kampherbeek, MAPPER: High-throughput maskless lithography, *Alternative Lithographic Technologies II, Proceedings of SPIE*, San Jose, CA, Vol. 7637, 76370F, 2010.

25. K. Suzuki, T. Fujiwara, K. Hada, N. Hirayanagi, S. Kawata, K. Morita, K. Okamoto, T. Okino, S. Shimizu, and T. Yahiro, Nikon EB Stepper: System concept and countermeasures for critical issues, *Emerging Lithographic Technologies IV, Proceedings of SPIE*, Santa Clara, CA, Vol. 3997, 2000, pp. 214–224.

26. M.A. McCord, P. Petric, U. Ummethala, A. Carroll, S. Kojima, L. Grella, S. Shriyan, C.T. Rettner, and C.F. Bevis, REBL: Design progress toward 16nm half-pitch electron-beam lithography, *Alternative Lithographic Technologies IV, Proceedings of SPIE*, San Jose, CA, Vol. 8323, 832811, 2012.

27. T. Gubiotti, J.F. Sun, R. Freed, F. Kidwingira, J. Yang, C. Bevis, A. Carroll et al., Reflective electron beam lithography: Lithography results using CMOS controlled digital pattern generator chip, *Alternative Lithographic Technologies V, Proceedings of SPIE*, San Jose, CA, Vol. 8680, 86800H, 2013.

28. A. Carroll, L. Grella, K. Murray, M.A. McCord, P. Petric, W.M. Tong, C.F. Bevis et al., The REBL DPG: Recent innovations and remaining challenges, *Alternative Lithographic Technologies VI, Proceedings of SPIE*, San Jose, CA, Vol. 9049, 904917, 2014.

29. G. Moore, Cramming more components onto integrated circuits, *Electronics*, 38, 114–117, 1965, also reproduced in Ref. [10].

30. G.E. Moore, Progress in digital integrated electronics, *IEDM 1975 Technical Digest*, IEEE, Piscataway, NJ, 1975, also reproduced in Ref. [10].

31. *International Technology Roadmap for Semiconductors, 2003 Edition*, Semiconductor Industry Association, Washington, DC, 2003, http://www.itrs.net/Links/2003ITRS/Home2003.htm. Accessed on October 13 2014.

32. *International Technology Roadmap for Semiconductors, 2013 Edition*, Semiconductor Industry Association, Washington, DC, 2013, http://www.itrs.net/Links/2013ITRS/Summary2013.htm. Accessed on October 13 2014.

33. X. Huang, W.-C. Lee, C. Kuo, D. Hisamoto, L. Chang, J. Kedzierski, E. Anderson et al., Sub 50-nm FinFET: PMOS, *IEDM '99 Technical Digest*, IEEE, Piscataway, NJ, 1999, pp. 67–70.

34. D. Hisamoto, W.-C. Lee, J. Kedzierski, H. Takeuchi, K. Asano, C. Kuo, E. Anderson, T.-J. King, J. Bokor, and C. Hu, FinFET-a self-aligned double-gate MOSFET scalable to 20 nm, *IEEE Trans. Electron Dev.*, 47, 2320–2325, 2000.

35. IntelPR, Intel 22nm 3-D Tri-Gate Transistor Technology, http://newsroom.intel.com/docs/DOC-2032. Accessed on December 1 2015.

36. D.R.S. Cumming, S. Thoms, J.M.R. Weaver, and S.P. Beaumont, 3 nm NiCr wires made using electron beam lithography and PMMA resist, *Microelectron. Eng.*, 30, 423–425, 1996.

37. W.F. van Dorp, X. Zhang, B.L. Feringa, T.W. Hansen, J.B. Wagner, and J.T. De Hosson, Molecule-by-molecule writing using a focused electron beam, *ACS Nano*, 6, 10076–10081, 2012.

38. T.H.P. Chang, Proximity effect in electron-beam lithography, *J. Vac. Sci. Technol.*, 12, 1271–1275, 1975.

39. G. Owen, Methods for proximity effect correction in electron lithography, *J. Vac. Sci. Technol. B*, 8, 1889–1892, 1990.

40. G. Owen and P. Rissman, Proximity effect correction for electron beam lithography by equalization of background dose, *J. Appl. Phys.*, 54, 3573–3581, 1983.

41. A.K.K. Wong, *Resolution Enhancement Techniques in Optical Lithography*, SPIE Press, Bellingham, WA, 2001.

42. A. Rosenbluth, D. Goodman, and B.J. Lin, A critical examination of submicron optical lithography using simulated projection images, *J. Vac. Sci. Technol. B*, 1, 1190–1195, 1983.

43. P. Chien and M. Chen, Proximity effects in submicron optical lithography, *Optical Microlithography VI, Proceedings of SPIE*, Santa Clara, CA, Vol. 772, 1987, pp. 35–40.

44. F.M. Schellenberg, H. Zhang, and J. Morrow, SEMATECH J111 Project: OPC validation, *Optical Microlithography XI, Proceedings of SPIE*, Santa Clara, CA, Vol. 3334, 1998, pp. 892–911.

45. Masato Shibuya, 透過照明用被投影原板 [Projection master for use with transmitted illumination], 公開特許公報 (A) 昭 57–62052, 特許公報 (B) 昭 62–50811 [Japan Patent Office Laid-Open Patent Publication (A) Showa 57-62052, Patent Publication (B) Showa 62– 50811] (filed Sept. 30, 1980, published April 14, 1982, issued October 27, 1987).

46. M.D. Levenson, N.S. Viswanathan, and R.A. Simpson, Improving resolution in photolithography with a phase-shifting mask, *IEEE Trans. Electron Dev.*, ED-29, 1828–1836, 1982.

47. M.D. Levenson, D.S. Goodman, S. Lindsey, P.W. Bayer, and H.A.E. Santini, The phase-shifting mask II: Imaging simulations and submicrometer resist exposures, *IEEE Trans. Electron Dev.*, ED-31, 753–763, 1984.

48. T. Terasawa, N. Hasegawa, T. Kurosaki, and T. Tanaka, 0.3-micron optical lithography using a phase-shifting mask, *Optical/Laser Microlithography II, Proceedings of SPIE*, San Jose, CA, Vol. 1088, 1989, pp. 25–33.

49. A. Nitayama, T. Sato, K. Hashimoto, F. Shigemitsu, and M. Nakase, New phase shifting mask with selfaligned phase shifters for a quarter micron lithography, *IEDM 1989 Technical Digest*, IEEE, Piscataway, NJ, 1989, pp. 57–60.

50. T. Tanaka, S. Uchino, N. Nasegawa, T. Yamanaka, T. Terasawa, and S. Okazaki, A novel optical lithography technique using the phase-shifter fringe, *Jpn. J. Appl. Phys.*, 30, 1131–1136, 1991.

51. H. Jinbo and Y. Yamashita, 0.2 mm or less i-line lithography by phase-shifting-mask technology, *IEDM 1990 Technical Digest*, 825–828, 1990.

52. K.K.H. Toh, G. Dao, R. Singh, and H. Gaw, Chromeless phase-shifted masks: A new approach to phase-shifting masks, *10th Annual Symposium on Microlithography, Proceedings of SPIE*, Sunnyvale, CA, Vol. 1496, 1990, pp. 27–53.

53. H. Watanabe, Y. Todokoro, Y. Hirai, and M. Inoue, Transparent phase shifting mask with multistage phase shifter and comb-shaped shifter, *Optical/Laser Microlithography IV, Proceedings of SPIE*, San Jose, CA, Vol. 1463, 1991, pp. 101–110.

54. J.F. Chen, J.S. Petersen, R. Socha, T. Laidig, K.E. Wampler, K. Nakagawa, G. Hughes, S. MacDonald, and W. Ng, Binary halftone chromeless PSM technology for λ/4 optical lithography, *Optical Microlithography XIV, Proceedings of SPIE*, Santa Clara, CA, Vol. 4346, 2001, pp. 515–533.

55. D.J. Van Den Broeke, J.F. Chen, T.L. Laidig, S. Hsu, K.E. Wampler, R.J. Socha, and J.S. Petersen, Complex two dimensional pattern lithography using chromeless phase lithography (CPL), *J. Microlith. Microfab. Microsys.*, 1, 229–242, 2002.

56. Y.-C. Ku, E.H. Anderson, M.L. Schattenburg, and H.I. Smith, Use of a pi-phase shifting x-ray mask to increase the intensity slope at feature edges, *J. Vac. Sci. Technol.*, B, 6, 150–153, 1988.

57. F.D. Kalk, R.H. French, H.U. Alpay, and G. Hughes, Attenuated phase shifting photomasks fabricated from Cr-based embedded shifter blanks, *Photomask and X-Ray Mask Technology, Proceedings of SPIE*, Kawasaki City, Kanagawa, Japan, Vol. 2254, 1994, pp. 64–70.

58. Ernst Abbe's lecture notes on off-axis resolution for microscopy are mentioned in Ref. [10], and also are presented by O. Lummer and F. Reiche, *Die Lehre von der Bildentstehung im Mikroskop von Ernst Abbe*, Vieweg und Sohn, Braunschweig, Germany, 1910.

59. D.L. Fehrs, H.B. Lovering, and R.T. Scruton, Illuminator modification of an optical aligner, *Proceedings of KTI Microelectronics Seminar INTERFACE'89*, San Diego, CA, 1989, pp. 217–230.

60. S. Asai, I. Hanyu, and K. Hikosaka, High performance optical lithography using a separated light source, *J. Vac. Sci. Technol. B*, 10, 3023–3026, 1992.

61. M. Noguchi, M. Muraki, Y. Iwasaki, and A. Suzuki, Subhalf micron lithography system with phaseshifting effect, *Optical/Laser Microlithography V, Proceedings of SPIE*, San Jose, CA, Vol. 1674, 1992, pp. 92–104.

62. N. Shiraishi, S. Hirukawa, Y. Takeuchi, and N. Magome, New imaging technique for 64M-DRAM, *Optical/Laser Microlithography V, Proceedings of SPIE*, San Jose, CA, Vol. 1674, 1992, pp. 741–752.

63. T. Ogawa, M. Uematsu, T. Ishimaru, M. Kimura, and T. Tsumori, The effective light source optimization with the modified beam for the depth-of-focus enhancements, *Optical/Laser Microlithography VII, Proceedings of SPIE*, San Jose, CA, Vol. 2197, 1994, pp. 19–30.

64. R.J. Socha, M.V. Dusa, L. Capodieci, J. Finders, J.F. Chen, D.G. Flagello, and K.D. Cummings, Forbidden pitches for 130 nm lithography and below, in *Optical Microlithography XIII, Proceedings of SPIE*, Santa Clara, CA, Vol. 4000, 2000, pp. 1140–1155.

65. F.M. Schellenberg and L. Capodieci, Impact of RET on physical layouts, *Proceedings of the 2001 ISPD*, ACM, New York, 2001, pp. 52–55.

66. M. Eurlings, E. van Setten, J.A. Torres, M.V. Dusa, R.J. Socha, L. Capodieci, and J. Finders, 0.11-μm imaging in KrF lithography using dipole illumination, *Lithography for Semiconductor Manufacturing II, Proceedings of SPIE*, Edinburgh, United Kingdom, Vol. 4404, 2001, pp. 266–278.

67. Lord Rayleigh, On the remarkable case of diffraction spectra described by Prof. Wood, *Philos. Mag.*, 14, 60–65, 1907.

68. A. Estroff, Y. Fan, A. Bourov, F.C. Cropanese, N.V. Lafferty, L.V. Zavyalova, and B.W. Smith, Maskinduced polarization, *Optical Microlithography XVII, Proceedings of SPIE*, Santa Clara, CA, Vol. 5377, 2004, pp. 1069–1080.

69. T.W. Ebbesen, H.J. Lezec, H.F. Ghaemi, T. Thio, and P.A. Wolff, Extraordinary optical transmission through sub-wavelength hole arrays, *Nature*, 391, 667–669, 1998.

70. C.-S. Koay, S. Holmes, K. Petrillo, M. Colburn, S. Burns et al., Evaluation of double-patterning techniques for advanced logic nodes, *Optical Microlithography XXIII, Proceedings of SPIE*, San Jose, CA, Vol. 7640, 764009, 2010.

71. L. Liebmann, D. Pietromonaco, and M. Graf, Decomposition-aware standard cell design flows to enable double-patterning technology, *Design for Manufacturability through Design-Process Integration V, Proceedings of SPIE*, San Jose, CA, Vol. 7974, 79740K, 2011.

72. C. Bencher, Y. Chen, H. Dai, W. Montgomery, and L. Huli, 22nm half-pitch patterning by CVD spacer self alignment double patterning (SADP), *Optical Microlithography XXI, Proceedings of SPIE*, San Jose, CA, Vol. 6924, 69244E, 2008.

73. K. Oyama, E. Nishimura, M. Kushibiki, K. Hasebe, S. Nakajima, H. Murakami, A. Hara et al., The important challenge to extend spacer DP process towards 22nm and beyond, *Advances in Resist Materials and Processing Technology XXVII Proceedings of SPIE*, San Jose, CA, Vol. 7639, 763907, 2010.

74. H. Yaegashi, Important challenge for the extension of Spacer DP process, *2010 International Symposium on Lithography Extensions*, Kobe, Japan, October 21, 2010, http://www.sematech.org/meetings/archives/litho/8940/pres/DP1_03_Hidetami Yaegashi.pdf. Accessed on April 27 2015.

75. H. Yaegshi, K. Oyama, K. Yabe, S. Yamauchi, A. Hara, and S. Natori, Important challenge for optical lithography extension utilizing double patterning process, *J. Photopolym. Sci. Technol.*, 24, 491–495, 2011.

76. Y. Borodovsky, *MP Processing for MP Processors*, SEMATECH Maskless Lithography and Multibeam Mask Writer Workshop, (New York, NY, May 2010).

77. Y. Borodovsky, *Complementary Lithography at Insertion and Beyond*, Semicon West 2012, San Francisco, CA, June 11, 2012.

78. D.K. Lam, E.D. Liu, M.C. Smayling, and T. Prescop, E-beam to complement optical lithography for 1D layouts, *Alternative Lithographic Technologies III Proceedings of SPIE*, San Jose, CA, Vol. 7970, 797011, 2011.

79. Rick Merritt, Intel Opens Door on 7nm, Foundry: EUV not needed at 10, 7nm, *EE Times* 9/11/2014, http://www.eetimes.com/document.asp?doc_id=1323865. Accessed on April 27 2015.

80. H. Tian, H. Zhang, Q. Ma, and M.D.F. Wong, Evaluation of cost-driven triple patterning lithography decomposition, *Design for Manufacturability through Design-Process Integration VII, Proceedings of SPIE*, San Jose, CA, Vol. 8684, 868407, 2013.

81. K. Nakayama, C. Kodama, T. Kotani, S. Nojima, S. Mimotogi, and S. Miyamoto, Self-aligned double and quadruple patterning layout principle, *Design for Manufacturability through Design-Process Integration VI, Proceedings of SPIE*, San Jose, CA, Vol. 8327, 83270V, 2012.

82. W. Kang and Y. Chen, Process characteristics and layout decomposition of self-aligned sextuple patterning, *Design for Manufacturability through Design-Process Integration VII, Proceedings of SPIE*, San Jose, CA, Vol. 8684, 86840F, 2013.

83. S.O. Kim, H. Solak, M. Stoykovich, N. Ferrier, J. de Pablo, and P. Nealey, Epitaxial self-assembly of block copolymers on lithographically defined nanopatterned substrates, *Nature*, 424, 411–414, July24, 2003.

84. M. Stoykovich, M. Müller, S.O. Kim, H. Solak, E. Edwards, J. de Pablo, and P. Nealey, Directed assembly of block copolymer blends into nonregular device-oriented structures, *Science*, 308, 1442–1446, June 3, 2005.

85. E. Edwards, M. Stoykovich, M. Müller, H. Solak, J. de Pablo, and P. Nealey, Mechanism and kinetics of ordering in diblock copolymer thin films on chemically nanopatterned substrates, *J. Polym. Sci. Part B: Polym. Phys.*, 43, 3444–3459, 2005.

86. F.M. Schellenberg and J.A.R. Torres, Contemporary design paradigms applied to novel lithography, SEMATECH 2006 Litho Forum, Poster IM06.

87. C.T. Black and O. Bezencenet, Nanometer-scale pattern registration and alignment by directed diblock copolymer self-assembly, *IEEE Trans. Nanotechnol.*, 3, 412–415, 2004.

88. C.T. Black, Self-aligned self assembly of multi-nanowire silicon field effect transistors, *Appl. Phys. Lett.*, 87, 163116, 2005.

89. C.T. Black, Integration of self assembly for semiconductor microelectronics, *Proceedings of the IEEE 2005 Custom Integrated Circuits Conference*, IEEE Piscataway, NJ, 2005, pp. 87–91.

90. J. Bang, U. Jeong, D.Y. Ryu, T.P. Russell, and C.J. Hawker, Block copolymer nanolithography: Translation of molecular level control to nanoscale patterns, *Adv. Mater.* 21,4769–4792, 2009.

91. D.J.C. Herr, Directed block copolymer self-assembly for nanoelectronics fabrication, *J. Mater. Res.* 26, 122–139, 2011.

92. K. Galatsis, K.L. Wang, M. Ozkan, C.S. Ozkan, Y. Huang, J.P. Chang, H.G. Monbouquette, Y. Chen, P.F. Nealey, and Y. Botros, Patterning and templating for nanoelectronics, *Adv. Mater.*, 22, 769–778, 2010.

93. H. Kim, H.-C. Kim, J. Cheng, C. Rettner, O.-H. Park, R. Miller, M. Hart, L. Sundström, and Y. Zhang, Self-aligned, self-assembled organosilicate line patterns of ~20nm half-pitch from block co-polymer mediated self-assembly, *Advances in Resist Materials and Processing Technology XXIV, Proceedings of SPIE*, San Jose, CA, Vol. 6519, 65191H, 2007.

94. F.M. Schellenberg, Design for manufacturing in the semiconductor industry: The litho/design workshops, *Proceedings of the 12th International Conference on VLSI Design*, IEEE Computer Society Press, Los Alamitos, CA, 1999, pp. 111–119.

95. F.M. Schellenberg, Advanced data preparation and design automation, *Photomask and Next-Generation Lithography Mask Technology IX, Proceedings of SPIE*, Yokohama, Japan, Vol. 4754, 2002, pp. 54–65.

96. L.W. Liebmann, G.A. Northrop, J. Culp, L. Sigal, A. Barish, and C.A. Fonseca, Layout optimization at the pinnacle of optical lithography, *Design and Process Integration for Microelectronic Manufacturing, Proceedings of SPIE*, Santa Clara, CA, Vol. 5042, 2003, pp. 1–14.

97. L. Capodieci, P. Gupta, A.B. Kahng, D. Sylvester, and J. Yang, Toward a methodology for a manufacturability-driven design rule exploration, *Proceedings of the 41st Design Automation Conference*, ACM, New York, 2004, pp. 311–316.

98. O.W. Otto, J.G. Garofalo, K.K. Low, C.-M. Yuan, R.C. Henderson, C. Pierrat, R.L. Kostelak, S. Vaidya, and P.K. Vasudev, Automated optical proximity correction: A rules-based approach, *Optical/Laser Microlithography VII, Proceedings of SPIE*, San Jose, CA, Vol. 2197, 1994, pp. 278–293.

99. SVRF (Standard Verification Rule Format) is the scripting language for Mentor Graphics Calibre® products.

100. J. Garofalo, C. Biddick, R.L. Kostelak, and S. Vaidya, Mask assisted off-axis illumination technique for random logic, *J. Vac. Sci. Technol. B*, 11, 2651–2658, 1993.

101. J.F. Chen and J.A. Matthews, Mask for photolithography, U.S. Patent 5,242,770, issued September 7, 1993.

102. B.E.A. Saleh and S.I. Sayegh, Reduction of errors of microphotographic reproductions by optimal correction of original masks, *Opt. Eng.*, 20, 781–784, 1981.

103. K.M. Nashold and B.E.A. Saleh, Image construction through diffraction-limited high-contrast imaging systems: An iterative approach, *J. Opt. Soc. Am. A*, 2, 635–643, 1985.

104. B.E.A. Saleh and K. Nashold, Image construction: Optimum amplitude and phase masks in photolithography, *Appl. Opt.*, 24, 1432–1437, 1985.

105. B.E.A. Saleh, Image synthesis: Discovery instead of recovery, Chapter 12, in *Image Recovery: Theory and Application*, H. Stark, Ed., Academic Press, Orlando, FL, 1987, pp. 463–498.

106. N.B. Cobb and A. Zakhor, Fast sparse aerial-image calculation for OPC, *15th Annual BACUS Symposium on Photomask Technology and Management, Proceedings of SPIE*, Santa Clara, CA, Vol. 2621, 1995, pp. 534–545.

107. H.H. Hopkins, On the diffraction theory of optical images, *Proc. R. Soc. London Ser. A*, 217, 408–432, 1953.

108. M. Rieger and J. Stirniman, Using behavior modeling for proximity correction, *Optical/Laser Microlithography VII, Proceedins of SPIE*, San Jose, CA, vol. 2197, 1994, pp. 371–376.

109. N.B. Cobb, Fast optical and process proximity correction algorithms for integrated circuit manufacturing, Ph.D. Dissertation, University of California at Berkeley, 1998.

110. X. Ma and G.R. Arce, *Computational Lithography*, John Wiley & Sons, Hoboken, NJ, 2010.

111. N.B. Cobb and Y. Granik, Dense OPC for 65nm and below, *25th Annual BACUS Symposium on Photomask Technology, Proceedings of SPIE*, Monterey, CA, Vol. 5992, 599259, 2005.

112. Y. Granik and N. Cobb, Matrix Optical Process Correction, US Patents 6,928,634 issued August 9, 2005; and 7,237,221, issued June 26, 2007.

113. Y. Borodovsky, W.-H. Cheng, R. Schenker, and V. Singh, Pixelated phase mask as novel lithography RET, *Optical Microlithography XXI, Proceedings of SPIE*, San Jose, CA, Vol. 6924, 69240E, 2008.

114. Y. Granik, Solving inverse problems of optical microlithography, *Optical Microlithography XVIII, Proceedings of SPIE*, San Jose, CA, Vol. 5754, 2005, pp. 506–526.

115. A. Poonawala and P. Milanfar, Mask design for optical microlithography: An inverse imaging problem, *IEEE Trans. Image Process.*, 16, 774–788, 2007.

116. A. Poonawala and P. Milanfar, A pixel-based regularization approach to inverse lithography, *Microelectron. Eng.*, 84, 2837–2852, 2007.

117. Y. Liu, D. Abrams, L. Pang, and A. Moore, Inverse lithography technology principles in practice: Unintuitive patterns, *25th Annual BACUS Symposium on Photomask Technology Proceedings of SPIE*, Monterey, CA, Vol. 5992, 599231, 2005.

118. L. Pang, Y. Liu, and D. Abrams, Inverse lithography technology (ILT): A natural solution for model-based SRAF at 45nm and 32nm, *Photomask and Next-Generation Lithography Mask Technology XIV, Proceedings of SPIE*, Yokohama, Japan, Vol. 6607, 660739, 2007.

119. L. Pang, Y. Liu, and D. Abrams, Inverse lithography technology (ILT) for advanced semiconductor manufacturing, *J. Exp. Mech.*, 22, 295–305, 2007.

120. A. Poonawala, Y. Borodovsky, and P. Milanfar, ILT for double exposure lithography with conventional and novel materials, *Optical Microlithography XX, Proceedings of SPIE*, San Jose, CA, Vol. 6520, 65202Q, 2007.

121. J. Zhang, W. Xiong, Y. Wang, and Z. Yu, A highly efficient optimization algorithm for pixel manipulation in inverse lithography technique, *IEEE/ACM International Conference on Computer-Aided Design, 2008. ICCAD 2008*, IEEE, Piscataway, NJ, 2008, pp. 480–487.

122. B.-G. Kim, S. S. Suh, S.G. Woo, and HanKu Cho, G. Xiao et al., Inverse lithography (ILT) mask manufacturability for full-chip device, *Photomask Technology 2009 Proceedings of SPIE*, Monterey, CA, Vol. 7488, 748812, 2009.

123. L. Liebmann, J. Lund, F.L. Heng, and I. Graur, Enabling alternating phase shifted mask designs for a full logic gate level: Design rules and design rule checking, *Proceedings of the 38th Design Automation Conference*, ACM, New York, 2001, pp. 79–84.

124. L. Liebmann, J. Lund, and F.L. Heng, Enabling alternating phase shifted mask designs for a full logic gate level, *J. Microlith. Microfab. Microsys.*, 1, 31–42, 2002.

125. K. Ooi, S. Hara, and K. Kojima, Computer aided design software for designing phase-shifting masks, *Jpn. J. Appl. Phys.*, 32, 5887–5891, 1993.

126. K. Ooi, K. Koyama, and M. Kiryu, Method of designing phase-shifting masks utilizing a compactor, *Jpn. J. Appl. Phys.*, 33, 6774–6778, 1994.

127. M. Fritze, J.M. Burns, P.W. Wyatt, D.K. Astolfi, T. Forte et al., Application of chromeless phase-shift masks to sub-100-nm SOI CMOS transistor fabrication, *Optical Microlithography XIII, Proceedings of SPIE*, Santa Clara, CA, Vol. 4000, 2000, pp. 388–407.

128. T. Terasawa, N. Hasegawa, H. Fukuda, and S. Katagiri, Imaging characteristics of multi-phase-shifting and halftone phase-shifting masks, *Jpn. J. Appl. Phys.*, 30, 2991–2997, 1991.

129. G. Galan, F. Lalanne, P. Schiavone, and J.M. Temerson, Application of alternating type phase shift mask to polysilicon level for random logic circuits, *Jpn. J. Appl. Phys.*, 33, 6779–6784, 1994.

130. H. Jinbo and Y. Yamashita, Improvement of phase-shifter edge line mask method, *Jpn. J. Appl. Phys.*, 30, 2998–3003, 1991.

131. H.-Y. Liu, L. Karklin, Y.-T. Wang, and Y.C. Pati, Application of alternating phase-shifting masks to 140-nm gate patterning: II. Mask design and manufacturing tolerances, *Optical Microlithography XI, Proceedings of SPIE*, Santa Clara, CA, Vol. 3334, 1998, pp. 2–14.

132. C. Spence, M. Plat, E. Sahouria, N. Cobb, and F. Schellenberg, Integration of optical proximity correction strategies in strong phase shifter design for poly-gate layer, *19th Annual Symposium on Photomask Technology, Proceedings of SPIE*, Monterey, CA, Vol. 3873, 1999, pp. 277–287.

133. F.-L. Heng, L. Liebmann, and J. Lund, Application of automated design migration to alternating phase shift mask design, *Proceedings of the 2001 ISPD*, ACM, New York, 2001, pp. 38–43.

134. A.B. Kahng, H. Wang, and A. Zelikovsky, Automated layout and phase assignment techniques for dark field alternating PSM, *SPIE 11th Annual BACUS Symposium on Photomask Technology, Proceedings of SPIE*, Sunnyvale, CA, Vol. 1604, 1998, pp. 222–231.

135. P. Berman, A. B. Kahng, D. Vidhani, and A. Zelikovsky, The T-join problem in sparse graphs: Applications to phase assignment problem in VLSI mask layout, *Proceedings of the 6th International Workshop, WADS'99, Lecture Notes in Computer Science* 1663, Springer Verlag, London, U.K., 1999, pp. 25–36.

136. P. Berman, A.B. Kahng, D. Vidhani, H. Wang, and A. Zelikovsky, Optimal phase conflict removal for layout of dark field alternating phase shifting masks, *IEEE Trans. Comput. Aided Des. Integr. Circuits Syst.*, 19, 175–187, 2000.

137. F.M. Schellenberg, L. Capodieci, and R. Socha, Adoption of OPC and the impact on design and layout, *Proceedings of the 38th Design Automation Conference*, ACM, New York, 2001, pp. 89–92.

138. Y. Granik and N. Cobb, New process models for OPC at sub-90nm nodes, *Optical Microlithography XVI, Proceedings of SPIE*, Santa Clara, CA, Vol. 5040, 2003, pp. 1166–1175.

139. A.E. Rosenbluth, S. Bukofsky, M. Hibbs, K. Lai, R.N. Singh, and A.K. Wong, Optimum mask and source patterns for printing a given shape, *J. Microlith. Microfab. Microsys.*, 1, 13–30, 2002.

140. Y. Granik, Source optimization for image fidelity and throughput, *J. Microlith. Microfab. Microsys.*, 3, 509–522, 2004.

141. R. Socha, X. Shi, and D. LeHoty. Simultaneous source mask optimization (SMO). *Photomask and Next-Generation Lithography Mask Technology XII, Proceedings of SPIE*, Yokohama, Japan, Vol. 5853, 2005, 180–193.

142. J. Bekaert, B. Laenens, S. Verhaegen, L. Van Look, D. Trivkovic et al., Experimental verification of source-mask optimization and freeform illumination for 22-nm node static random access memory cells, *J. Micro/Nanolithogr. MEMS MOEMS*, 10, 013008, 2011.

143. M. Mulder, A. Engelen, O. Noordman, G. Streutker, B. van Drieenhuizen et al., Performance of FlexRay: A fully programmable illumination system for generation of freeform sources on high NA immersion systems, *Optical Microlithography XXIII Proceedings of SPIE*, San Jose, CA, Vol. 7640, 76401P, 2010.

144. R. Socha, T. Jhaveri, M. Dusa, X. Liu, L. Chen, S. Hsu, Z. Li, and A.J. Strojwas, Design compliant source mask optimization (SMO), *Photomask and Next-Generation Lithography Mask Technology XVII Proceedings of SPIE*, Yokohama, Japan, Vol. 7748, 77480T, 2010.

145. J.M. Beckers, Adaptive optics for astronomy: Principles, performance, and applications, *Annu. Rev. Astron. Astrophys.*, 31, 13–62, 1993.

146. E.D. Palik, Ed., *Handbook of Optical Constants of Solids*, Vol. I, II, and III, Academic Press, San Diego, CA, 1998.

147. B. Smith at Rochester Institute of Technology (RIT) maintains a website that allows the computation of optical properties for semiconductor materials at, http://www.rit.edu/kgcoe/microsystems/lithography/research/thinfilms.html. Accessed on December 1 2015.

148. A.K.K. Wong and A.R. Neureuther, Rigorous three dimensional time-domain finite difference electromagnetic simulation, *IEEE Trans. Semicond. Manuf.*, 8, 419–431, 1995.

149. K. Adam and A.R. Neureuther, Domain decomposition methods for the rapid electromagnetic simulation of photomask scattering, *J. Microlith. Microfab. Microsys.*, 1, 253–269, 2002.

150. S.G. Hansen, Source mask polarization optimization, *J. Micro/Nanolithogr. MEMS MOEMS*, 10, 033003, 2011.

151. Mentor Graphics for more on Mentor Graphics Calibre software, *Design for Manufacturing with Calibre*, see http://www.mentor.com/products/ic_nanometer_design/design-for-manufacturing/. Accessed on December 1 2015.

152. 2015 Synopsys, Inc., For more on Synopsys DFM products, *Mask Synthesis and Data Prep*, see http://www.synopsys.com/Tools/Manufacturing/MaskSynthesis/Pages/default.aspx. Accessed on December 1 2015.

153. For more on ASML Brion software, see http://www.brion.com/.

154. A.B. Kahng, Lithography and design in partnership: A new roadmap, *Photomask Technology 2008, Proceedings of SPIE*, Monterey, CA, Vol. 7122, 712202, 2008.

155. X. Ma and G.R. Arce, *Computational Lithography*, John Wiley & Sons, New York, 2010.

156. G.E. Bailey, A. Tritchkov, J.-W. Park, L. Hong, V. Wiaux, E. Hendrickx, S. Verhaegen, P. Xie, and J. Versluijs, Double pattern EDA solutions for 32 nm HP and beyond, *Design for Manufacturability through Design-Process Integration, Proceedings of SPIE*, San Jose, CA, Vol. 6521, 65211K, 2007.

157. L. Liebmann, Z. Baum, I. Graur, and D. Samuels, DfM lessons learned from altPSM design, *Design for Manufacturability through Design-Process Integration II, Proceedings of SPIE*, San Jose, CA, Vol. 6925, 69250C, 2008.

158. L.W. Liebmann, J. Kye, B.-S. Kim, L. Yuan, and J.-P. Geronimi, Taming the final frontier of optical lithography: Design for sub-resolution patterning, *Design for Manufacturability through Design-Process Integration IV Proceedings of SPIE*, San Jose, CA, Vol. 7641, 764105, 2010.

159. L.T.-N. Wang, V. Dai, and L. Capodieci, Pattern matching for identifying and resolving non-decomposition-friendly designs for double patterning technology (DPT), *Design for Manufacturability through Design-Process Integration VII Proceedings of SPIE*, San Jose, CA, Vol. 8684, 868409, 2013.

160. L. Liebmann, V. Gerousis, P. Gutwin, M. Zhang, G. Han, and Brian Cline, Demonstrating production quality multiple exposure patterning aware routing for the 10NM node, *Design-Process-Technology Co-optimization for Manufacturability VIII, Proceedings of SPIE*, San Jose, CA, Vol. 9053, 905309, 2014.

161. Q. Ma, H. Zhang, and M.D.F. Wong, Triple patterning aware routing and its comparison with double patterning aware routing in 14nm technology, *Proceedings of the 49th Annual Design Automation Conference*, ACM, New York, 2012, pp. 591–596.

162. S.-Y. Fang, Y.-W. Chang, and W.-Y. Chen, A novel layout decomposition algorithm for triple patterning lithography, *IEEE Trans. Comput. Aided Des. Integr. Circuits Syst.*, 33, 397–408, 2014.

163. F. Nakajima, C. Kodama, H. Ichikawa, K. Nakayama, S. Nojima, and T. Kotani, Self-aligned quadruple patterning-aware routing, *Design-Process-Technology Co-optimization for Manufacturability VIII, Proceedings of SPIE*, San Jose, CA, Vol. 9053, 90530C, 2014.

164. Y. Kohira, Y. Yokoyama, C. Kodama, A. Takahashi, S. Nojima, and S. Tanaka, Yield-aware decomposition for LELE double patterning, *Design-Process-Technology Co-optimization for Manufacturability VIII Proceedings of SPIE*, San Jose, CA, Vol. 9053, 90530T, 2014.

165. Y.A. Badr, A.G. Wassal, and S. Hammouda, Split-it!: From litho etch litho etch to self-aligned double patterning decomposition, *Photomask Technology 2012, Proceedings of SPIE*, Monterey, CA, Vol. 8522, 852225, 2012.

166. F.M. Schellenberg and J.A.R. Torres, Using phase mask algorithms to direct self assembly, *Emerging Lithographic Technologies X Proceedings of SPIE*, San Jose, CA, Vol. 6151, 61513L, 2006.

167. J.-B. Chang, H.K. Choi, A.F. Hannon, A. Alexander-Katz, C.A. Ross, and K.K. Berggren, Design rules for self-assembled block copolymer patterns using tiled templates, *Nat. Commun.*, 5, 3305, 2013.

168. J. Qin, G.S. Khaira, Y. Su, G.P. Garner, M. Miskin, H.M. Jaeger, and J.J. de Pablo, Evolutionary pattern design for copolymer directed self-assembly, *Soft Matter*, 9, 11467–11472, 2013.

169. K. Lai, C.-c. Liu, J. Pitera, D.J. Dechene, A. Schepis et al., Computational aspects of optical lithography extension by directed self-assembly, *Optical Microlithography XXVI, Proceedings of SPIE*, San Jose, CA, Vol. 8683, 868304, 2013.

170. K. Lai, M. Ozlem, J.W. Pitera, C.-c.Liu, A. Schepis et al., Computational lithography platform for 193i-guided directed self-assembly, *Optical Microlithography XXVII Proceedings of SPIE*, San Jose, CA, Vol. 9052, 90521A, 2014.

171. P. Brandt, J. Belledent, C. Tranquillin, T. Figueiro, S. Meunier, S. Bayle, A. Fay, M. Milléquant, B. Icard, and M. Wieland, Demonstration of EDA flow for massively parallel e-beam lithography, *Alternative Lithographic Technologies VI, Proceedings of SPIE*, San Jose, CA, Vol. 9049, 904915, 2014.

172. C. Zuniga, M. Habib, J. Word, G.F. Lorusso, E. Hendrickx, B. Baylav, R. Chalasani, and M. Lam, EUV flare and proximity modeling and model-based correction, *Extreme Ultraviolet (EUV) Lithography II, Proceedings of SPIE*, San Jose, CA, Vol. 7969, 79690T, 2011.

173. M. Burkhardt, G. McIntyre, R. Schlief, and L. Sun, Clear sub-resolution assist features for EUV, *Extreme Ultraviolet (EUV) Lithography V Proceedings of SPIE*, San Jose, CA, Vol. 9048, 904838, 2014.

174. X. Liu, R. Howell, S. Hsu, K. Yang, K. Gronlund et al., EUV source-mask optimization for 7nm node and beyond, *Extreme Ultraviolet (EUV) Lithography V, Proceedings of SPIE*, San Jose, CA, Vol. 9048, 90480Q, 2014.

175. C. Krautschik, M. Ito, I. Nishiyama, and S. Okazaki, Impact of EUV light scatter on CD control as a result of mask density changes, *Emerging Lithographic Technologies VI, Proceedings of SPIE*, Santa Clara, CA, Vol. 4688, 2002, pp. 289–301.

176. M. Chandhok, S.H. Lee, and T. Bacuita, Effects of flare in extreme ultraviolet lithography: Learning from the engineering test stand, *J. Vac. Sci. Technol. B*, 22, 2966–2969, 2004.

177. F.M. Schellenberg, J. Word, and O. Toublan, Layout compensation for EUV flare, *Emerging Lithographic Technologies IX, Proceedings of SPIE*, San Jose, CA, Vol. 5751, 2005, pp. 320–329.

178. C. Maloney, J. Word, G.L. Fenger, A. Niroomand, G.F. Lorusso, R. Jonckheere, E. Hendrickx, and B.W. Smith, Feasibility of compensating for EUV field edge effects through OPC, *Extreme Ultraviolet (EUV) Lithography V, Proceedings of SPIE*, San Jose, CA, Vol. 9048, 90480T, 2014.

179. M.C. Fritze, Gratings of regular arrays and trim exposures (GRATE); DARPA Solicitation Number: DARPA-BAA-10-12, https://www.fbo.gov/index?s=opportunity&mode=form&id=9d4f67fa73e424bd 7dc1ac9eabdafc28&tab=core&_cview=1 (Accessed on December 1 2015.)

180. L. Capodieci, From optical proximity correction to lithography-driven physical design (1996-2006): 10 years of resolution enhancement technology and the roadmap enablers for the next decade, *Optical Microlithography XIX, Proceedings of SPIE*, San Jose, CA, Vol. 6154, 615401, 2006.

181. L.W. Liebman, Layout impact of resolution enhancement techniques: Impediment or opportunity?, *Proceedings of the 2003 International Symposium on Physical Design ISPD '03*, ACM, New York, 2003, pp. 110–117.

182. M. Lavin, F.L. Heng, and G. Northrop, Backend CAD flows for restrictive design rules *IEEE/ACM International Conference on Computer Aided Design, ICCAD 2004*, IEEE Piscataway, NJ, 2004, pp. 739–746.

183. L. Liebmann, D. Maynard, K. McCullen, N. Seong, E. Buturla, M. Lavin, and J. Hibbeler, Integrating DfM components into a cohesive design-to-silicon solution, *Design and Process Integration for Microelectronic Manufacturing III, Proceedings of SPIE*, San Jose, CA, Vol. 5756, 2005, pp. 1–12.

184. L. Stok, Design rules: From restriction to prescription, *IEEE 2007 EDP Workshop* (April 2007). http://www.eetimes.com/showArticle.jhtml?articleID=199000715. Accessed on April 27 2015.

185. L. Pileggi, H. Schmit, A. Strojwas, P. Gopalakrishnan, V. Kheterpal, A. Koorapaty, C. Patel, V. Rovner, and K. Tong, Exploring regular fabrics to optimize the performance-cost trade-off, *Proceedings of the 40th Annual Design Automation Conference*, ACM, New York, 2003, pp. 782–787.

186. V. Kheterpal, V. Rovner, T.G. Hersan, D. Motiani, Y. Takegawa, A.J. Strojwas, and L. Pileggi, Design methodology for IC manufacturability based on regular logic-bricks, *Proceedings of the 42nd Annual Design Automation Conference*, ACM, New York, 2005, pp. 353–358.

187. L. Liebmann, L. Pileggi, J. Hibbeler, V. Rovner, T. Jhaveri, and G. Northrop, Simplify to survive: Prescriptive layouts ensure profitable scaling to 32nm and beyond, *Design for Manufacturability through Design-Process Integration III, Proceedings of SPIE*, San Jose, CA, Vol. 7275, 72750A, 2009.

188. T. Jhaveri, V. Rovner, L. Pileggi, A.J. Strojwas, D. Motiani, V. Kheterpal, K.Y. Tong, T. Hersan, and D. Pandini, Maximization of layout printability/manufacturability by extreme layout regularity, *J. Micro/Nanolithogr. MEMS MOEMS*, 6, 031011, 2007.

189. K. Vaidyanathan, R. Liu, L. Liebmann, K. Lai, A.J. Strojwas et al., Design implications of extremely restricted patterning, *J. Micro/Nanolithogr. MEMS MOEMS*, 13, 031309, 2014.

190. Tela Innovations, Los Gatos, CA, http://www.tela-inc.com/. Accessed on December 1 2015.

191. S.P. Kornachuk and M.C. Smayling, New strategies for gridded physical design for 32nm technologies and beyond, *Proceedings of the 2009 International Symposium on Physical Design, ISPD'09*, ACM, New York, 2009, pp. 61–62.

192. P. van Adrichem and C. Kalus, Data preparation, chap. 2 of Ref. [6]; and P. DePesa, D. Kay, and G. Meyers, Data preparation, chap. 2 of Ref. [7].

193. MEBES is a format developed by ETEC Systems, and is now owned by Applied Materials (http://www.appliedmaterials.com/). Accessed on December 1 2015.

194. *SEMI P39-0304E2—OASIS—Open Artwork System Interchange Standard* can be downloaded from the SEMI Standards Organization, www.semi.org/. Accessed on April 27 2015.

195. S. Schulze, P. LaCour, and L. Grodd, OASIS-based data preparation flows: Progress report on containing data size explosion, *Design and Process Integration for Microelectronic Manufacturing II, Proceedings of SPIE*, Santa Clara, CA, Vol. 5379, 2004, pp. 149–157.

196. J.C. Davis, S. Schulze, S. Fu, and Y. Tong, Deployment of OASIS in the semiconductor industry—Status, dependencies and outlook, *26th European Mask and Lithography Conference, Proceedings of SPIE*, Grenoble, France, Vol. 7545, 75450B, 2010.

197. T. Lin, T. Donnelly, and S. Schulze, Model based mask process correction and verification for advanced process nodes, *Optical Microlithography XXII, Proceedings of SPIE*, San Jose, CA, Vol. 7274, 72742A, 2009.

198. J. Sturtevant, E. Tejnil, T. Lin, S. Schulze, P. Buck, F. Kalk, K. Nakagawa, G. Ning, P. Ackmann, F. Gans, and C. Buergel, Impact of 14-nm photomask uncertainties on computational lithography solutions, *J. Micro/Nanolithogr MEMS MOEMS*, 13, 011004, 2013.

199. J. Choi, J.S. Park, In K. Shin, and C.-Uk Jeon, Inverse e-beam lithography on photomask for computational lithography, *J. Micro/Nanolithogr MEMS MOEMS*, 13(1), 011003, 2013.

Design for Manufacturability in the Nanometer Era

22

Nicola Dragone, Carlo Guardiani, and Andrzej J. Strojwas

CONTENTS

22.1 INTRODUCTION

Achieving high-yielding designs in the state-of-the-art, ultralarge-scale integration (ULSI) technology has become an extremely challenging task due to the miniaturization and the complexity of leading-edge products. The design methodology called *design for manufacturing* (DFM) includes a set of techniques to modify the design of ICs in order to make them more manufacturable, that is, to improve their functional yield, parametric yield, or their reliability. Traditionally, in the pre-nanometer era, DFM consisted of a set of different methodologies trying to enforce some soft (recommended) design rules (DRs) regarding the shapes and polygons of the physical layout of an IC product. These DFM methodologies worked primarily at the full chip level. Additionally, worst-case simulations at different levels of abstraction were applied to minimize the impact of process variations on performance and other types of parametric yield loss. All these different types of worst-case simulations were essentially based on a set of worst-case (or corner) SPICE device parameter files that were intended to represent the variability of transistor performance over the full range of variation in a fabrication process.

With the advent of nanometer technologies, namely, at 130 nm and below, process and design systematic *yield loss mechanisms* (YLMs) have started to cross over random effects in the yield Pareto [1] (see Figure 22.1). Random YLMs are caused by random contaminants or defects (e.g., particles) and are therefore characterized by the absence of strong spatial, temporal, or other kinds of correlations. On the contrary, process-systematic YLMs are usually spatially correlated (e.g., their failure rate shows strong radial wafer patterns) or temporally correlated (e.g., lot-to-lot variations as a function of equipment).

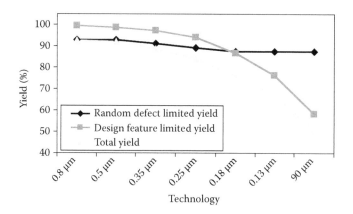

FIGURE 22.1   Yield limiters by technology node.

Design-systematic YLMs are strongly correlated to specific local physical layout patterns, and therefore their failure rate may vary by orders of magnitude as a function of the layout characteristics.

The same physical design feature (e.g., a contact or a via) may yield completely differently depending upon neighboring features (short correlation distance effects) such as the presence or lack of certain shapes (e.g., the presence of a minimum distance, small area, metal island, and via) next to it . Other types of design-systematic YLMs are characterized by medium or even relatively large correlation distances, for example, YLM due to CMP effects. Any design-systematic YLM is, however, characterized by an intrachip correlation pattern. In other words, they depend on local physical design properties. This design locality property of design-systematic YLMs considerably complicates the effectiveness of the traditional DR approach because DRs are applied globally to every physical design feature in a chip.

In order to cope with the design locality property of design-systematic YLMs, the total number of DRs exploded and the rules have become increasingly complex. The DRs for subnanometer technologies have been split into two sets: a fundamental set of DRs, i.e., rules that if violated, would cause a product's yield to drop to zero; and a set of "recommended DRs" (RDRs), i.e., rules that the design tools should try to implement as much as possible based upon specific layout patterns in the design. Unfortunately, it would be extremely impractical to include every such complex DR in the design tools to generate maximally manufacturable designs by construction.

For this reason, a large number of rule-driven reactive DFM methodologies have been introduced recently, which have the drawback of modifying the design patterns after timing closure and verification have been completed, hence introducing additional risk and delay in the design process completion. Moreover, several of the RDRs create some conflicts among other rules. For example, adding via redundancy may create a large number of small dielectric regions that are difficult to manufacture and may cause systematic failures. Model-based, proactive DFM philosophy consists of the development of accurate, silicon-verified YLM models that can evaluate the relative impact of each YLM and assess trade-offs. These models are then integrated in the design tool's cost function along with other design objectives such as speed, power, and signal integrity functions. Design tools can thus exploit the design locality property of design-systematic YLMs as well as global random YLMs to optimize manufacturability subject to the actual IC design specifications.

22.2 TAXONOMY OF YIELD LOSS MECHANISMS

The most important YLMs for VLSI ICs can be classified into several categories based on their nature. Functional yield loss remains the dominant factor and is caused by mechanisms such as misprocessing (e.g., equipment-related problems), systematic effects such as printability or planarization problems, and purely random defects. High-performance products may exhibit parametric design marginalities caused by either process fluctuations or environmental factors (such as supply voltage or temperature). The test-related yield losses, which are caused by incorrect testing, can also play a significant role. In this section, we present a detailed description of how YLMs have evolved in nanometer technologies, and some of the standard yield modeling approaches.

22.2.1 RANDOM YIELD LOSS MODELING

Random defects and contaminants are possibly the best-known and studied YLMs [1]. Typical random failure mechanisms involve active, poly, and metal shorts and opens due to particle defects, as well as contact and via opens due to formation defectivity. Some examples of open/short failures caused by random particles are shown in Figure 22.2.

Before deep submicron technologies, the yield loss was dominated by random defects, and yield was typically modeled as a function of the defect density and size distribution (Figure 22.3)

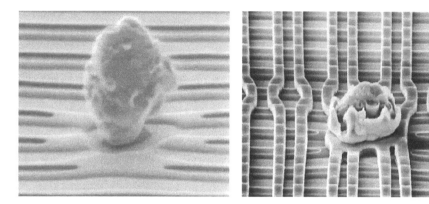

FIGURE 22.2 Random defects causing metal opens and shorts.

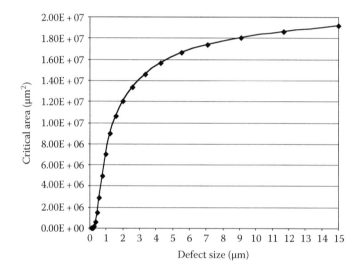

FIGURE 22.3 Critical area as a function of defect size distribution.

and of the amount of the chip area that is susceptible to defects (also known as critical area). The concept of critical area is shown graphically in Figure 22.4.

A number of different models [1] have been developed over the years starting from the simple Poisson to negative binomial to more complicated critical-area-based models, also taking into account the defect size distribution (DSD) as shown in Figure 22.5.

22.2.2 DESIGN-SYSTEMATIC YIELD LOSS

The contribution of systematic YLMs at the time when critical-area-based models were applied was typically less than 5% of the total, and often they were totally negligible for relatively mature processes. Critical-area-based yield models are not sufficient for state-of-the-art technologies that are affected by significant systematic YLMs. For example, it is often the case that two products characterized by similar critical area values have very different yields. A more comprehensive set of models had to be developed that included a parameterization of the yield loss of individual design components as a function of their design attributes in order to predict yield adequately, especially in the early production stages. Typical design-systematic failure mechanisms are due to mask misalignment, line-end and border effects, oxide stress, and more generally due to different DR marginalities sensitized by the impact of micro- and macro-loading. Figure 22.6 demonstrates the drastic profile difference between sparse and dense vias. Sparse vias or stacks are

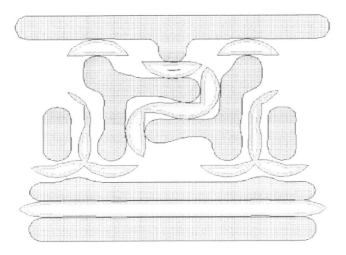

FIGURE 22.4   Graphical representation of critical area.

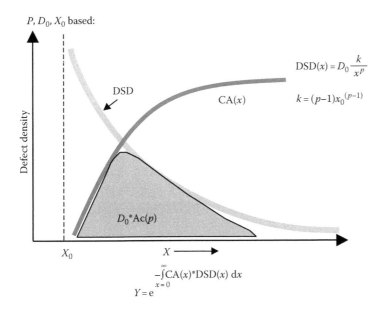

FIGURE 22.5   Random defect limited yield model.

much more likely to exhibit the Cu pull-up-type failures. These profile differences result in much higher fail rates for sparse vias than for vias placed in a dense environment.

22.2.3 PRINTABILITY YIELD LOSS

Printability failure mechanisms are often caused by errors in the optical proximity correction (OPC) algorithm and limited process window. Although it is sometimes difficult to separate potential flaws of the OPC algorithm from intrinsically "hard to print" features, DFM should only address the latter, as OPC/RET errors are, by definition, independent of the drawn layout features, and hence of the design. Therefore, in a DFM methodology, we are interested in identifying, characterizing, and fixing process window issues in the design features that are overly sensitive to lithography parameters variation. Common failures of this type are due to metal pull-back and consequent poor via coverage, metal necking (open), and metal bridging (short). Printability yield loss modeling must be based on physically accurate simulation of the printed shapes for the entire process window. Figure 22.7 illustrates printability simulation results of a layout pattern with high sensitivity to contact coverage.

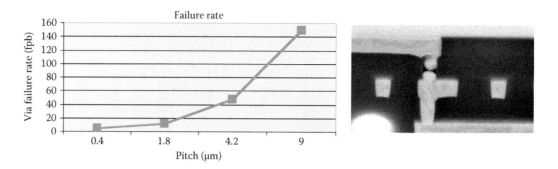

FIGURE 22.6 Pitch dependence of via failure rate in a sparse environment.

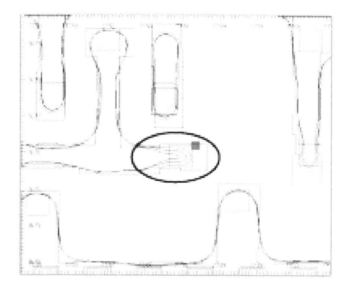

FIGURE 22.7 Simulation of contact coverage for different process corners.

22.2.4 PARAMETRIC YIELD LOSS

The electrical performances of IC devices are subject to natural random variation. Because of this, every IC performance parameter can be described as a random variable with a specific associated probability distribution. The tails of the probability distribution may fall outside the acceptable device specs and cause parametric yield loss. Parametric yield loss is characterized by inter- and intra-die level correlations, with the latter also known as device mismatch. There are two major sources of process variations that affect the electrical behavior of digital or analog circuits: electrical parameter variations, for example related to dopant fluctuations or oxide thickness variations; and lithography parameter variations, for example related to gate length or width variations. One more source of discrepancy between predicted and actual circuit performance is due to layout pattern-dependent device characteristics, such as stress-related drain extension-dependent I_{DSS} in MOSFET devices.

Figure 22.8 illustrates the dispersion of the I_{OFF} vs. I_{ON} data caused by poly CD variations for three different layout environments.

22.3 LOGIC DESIGN FOR MANUFACTURING

In the previous sections we illustrated the major YLMs and how they relate to the design attributes of devices and interconnections. We have shown that the design-dependent and printability systematic YLMs are becoming increasingly critical and that they dominate the design-dependent

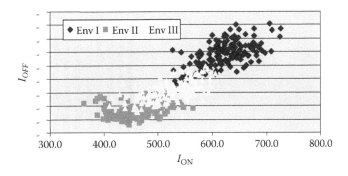

FIGURE 22.8   Effects of environment-dependent poly-CD variations on I_{ON}/I_{OFF}.

yield loss, i.e., the fraction of yield loss that is recoverable by adopting different design configurations, which ultimately defines the scope of DFM changes. The process sensitivity to such systematic YLMs is hard to predict and varies considerably as a function of the particular process flow implementation. Therefore, although with some experience it is possible to predict the list of the potential YLMs including systematic effects that may impact a given technology, it is often hard, if not impossible, to know which of these potential YLMs are the most critical, and the Pareto of the dominant effects. As a consequence, the key enabler for a DFM methodology is an accurate characterization and modeling of the process technology to determine process sensitivities of the leading YLMs and to determine design trade-offs. In the following section, we will focus on bulk and SOI (silicon on insulator) CMOS technologies facing the integration challenges of new materials, advanced next generation lithography, and yield limitations above and beyond random defectivity.

Generation of the process–design mechanism database requires several major components:

1. Development and implementation of the required characterization structures to span all the relevant process–design interaction effects and to generate the required physical data for accurate yield modeling.
2. A test and analysis infrastructure with sufficient throughput, including both cost effectiveness and time efficiency.
3. Development and calibration of yield models as a function of design attributes and based on the data from test structures.
4. Time-based scaling of process maturity to enable predictability of the yield impact of YLMs.

22.3.1 CHARACTERIZATION STRUCTURES

The ability to exploit fully the advantages of a DFM methodology requires a set of comprehensive characterization test structures. A comprehensive set of structures includes those required to characterize fully random, systematic, and parametric effects on the front-end-of-line (FEOL) and back-end-of-line (BEOL) process module yields. In addition, the structures must incorporate design of experiments (DOE) and response surface methodology [2,3] to enable product layout attribute dependence to take into account local and global loading, proximity, and density effects—as well as to optimize the amount of silicon area required to characterize each effect within the required confidence interval [1].

A large number of structures must be designed to capture defect densities on the order of parts per billion (ppb) and, at the same time, make efficient use of the available silicon area. Among those we can mention are harp, nest/comb, checkerboard, chains, and passive multipexers [4,5]. Figure 22.9 illustrates several structures used to characterize the DSD and open/short defectivity.

A rich DOE must be implemented in test structure design to identify process sensitivities and process–design interactions. Typical experiments include failure dependency on material density, both local and global, associated with particular layout patterns. Figure 22.10 illustrates a via experiment with variable chain and hole pitch, and metal density.

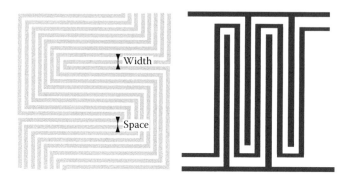

FIGURE 22.9   Nest and snake/comb structures for random defectivity characterization.

Characterization vehicles must include both calibration and verification structures for printability. Calibration structures are needed for tuning litho simulators for hotspot detection. Verification structures must be included to assess the quality of calibration and characterize process robustness against the variation in the parameters. Typical litho parameters that could play a critical role are exposure level, mask error, defocus, misalignment, and resist. Figure 22.11 illustrates structures designed to verify the quality of printability in environments, which can be considered *hard-to-print* for any OPC algorithm.

A heterogeneous set of structures should also cover the most relevant effects that have an impact on performance variability and performance predictability. Typical effects are CD line variations, poly and active corner rounding, and STI stress. Figure 22.12 illustrates a couple of structures that can be implemented to characterize poly CD variations with different local and global density levels, and the impact of layout styles on transistor performance.

22.3.2 TEST AND ANALYSIS INFRASTRUCTURE

The feasibility of fully utilizing a characterization test structure set and associated DOE, such as those presented in the previous section, must take into account the overall throughput capability as a function of testing and analysis. Novel structures that optimize the statistical sample size per

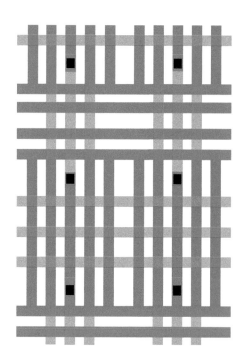

FIGURE 22.10   Via chain structure for characterizing failure rate dependency on the environment.

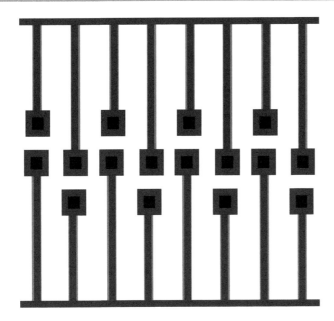

FIGURE 22.11 Litho verification structure with difficult-to-print OPC patterns.

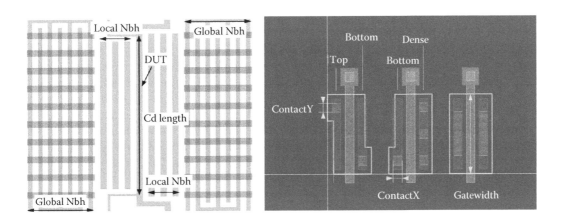

FIGURE 22.12 Structures for Idrive characterization: poly-CD variation and layout styles.

test structure are available and reduce the test time and silicon area required for accurate characterization [3]. Nonetheless, the amount of data generated by a comprehensive test structure set remains massive.

Reducing the time and cost of test as a function of automated test equipment (ATE)-related costs [6,7] should be complemented with the reduction of analysis time and costs when time-to-market pressures commonly demand early design tape-outs for test sample debugging, engineering samples for customers, product ramp in nonmature processes, and finally, long-term volume manufacturing. Therefore, the desired solution integrates and optimizes the ATE and analysis infrastructure [8] as much as possible to enable fast turnaround time such that the process characterization is used for calibrating yield models that can be utilized early in the product design cycle.

22.3.3 YIELD IMPACT MATRIX

Any type of DFM solution applied to an arbitrary circuit must consider two basic concepts. The first is the likelihood that a certain layout configuration will cause a fail. The second is the information on the number of occurrences of the layout configuration in the product design. The former

TABLE 22.1 Yield Impact Matrix

YIMP Matrix			Full Chip (%)	Limited Yield	
	Failure Mode			**SRAM (%)**	**LOGIC (%)**
Random defects	Active	Random	98	99	99
		Pattern dependent	99	99	100
		Total	97	98	99
	Poly	Random	97	98	99
		Pattern dependent	94	95	99
		Total	91	95	96
	Metal	Random	97	98	99
		Pattern dependent	97	98	99
		Total	94	95	99
	Holes		97	98	99
	Total		81	82	99
Systematic	Metal islands		87	89	90
	Pattern density		91	93	94
	Narrow space wide neighbor		97	99	100
	Via induced metal shorts		77	78	79
	Total		62	64	64

must be derived from process characterization, whereas the latter must be computed through product analysis and extraction of design attributes. By combining these two pieces of information, we obtain yield models, as presented in previous section. A very effective way to summarize the yield numbers associated with a particular design is represented by the yield impact matrix (YIMP) [9,10], which identifies the most critical contributors to yield loss in terms of circuit components and process modules. Table 22.1 shows a very simple YIMP in which columns represent design blocks and rows represent process modules. The relevance of such a representation is twofold: on the one hand it allows the identification of critical blocks and, on the other, it makes it possible to plan adequate DFM or process fixes to alleviate yield losses.

22.3.4 IP OPTIMIZATION

22.3.4.1 STANDARD CELL OPTIMIZATION

A substantial fraction of IC yield loss occurs in standard cell logic. Logic-limited yield associated with standard cells may become the most critical factor in designs such as graphic applications where logic content is much larger than the analog or memory content.

Composition of standard cell libraries, both in terms of their logic functionality and capability of steering current, often referred to as cell driving strength, is commonly being recognized as one of the most critical aspects to achieve high-quality synthesis [11,12]. A number of actions can be taken to improve standard cell manufacturability, such as adding feature redundancy, over-sizing, spacing, increasing uniformity, or density. However, the application of such techniques often conflicts with the optimization of other design parameters such as area, speed, and power. Therefore, the ability to quantify the contribution of different effects, and to rank them in order of priority is of utmost importance in a yield-aware design flow [13]. In addition to the above considerations, general-purpose standard cell libraries should also support differentiated portfolios of IC products that are characterized by different performance specifications and volume production requirements. Such requirements can be translated into design objectives for the standard cell library as follows:

- Maximize average yield, which is typically the most important manufacturing objective for high-volume parts.
- Minimize lot-to-lot, wafer-to-wafer, and within-wafer yield excursions that are often a major concern for less mature processes and are a key objective for relatively

low-volume products and obviously for engineering samples, where an excursion may mean no parts to test.

■ Minimize performance variability, hence provide tighter speed bins and minimize parametric yield loss, which is obviously a main objective for high-performance products such as micro processors.

These high-level manufacturability optimization objectives are in general characterized by different and often conflicting process–layout interaction mechanisms. A manufacturing-aware standard cell library should support all these objectives; therefore, it should contain cell modules that are optimized for each of the key process–design interaction mechanisms that are relevant for a given class. A nonexhaustive example set of effects that are addressed in each class can consist of the following:

■ Maximize average yield: hole (contact and via) opens, critical area for opens, shorts, etc.
■ Minimize yield variability: hole coverage by the upper/lower layer, metal islands, field transitions, active leakage paths, etc.
■ Minimize performance variability: poly flaring, STI stress, etc.

Figure 22.13 shows the high-level architecture of the proposed DFM standard cell library. A core library including a rich function set of high-density cells with all their required drive strength variants is complemented by additional cells that address one or more manufacturing objectives. Within each class and for each logic function/drive strength combination represented in the class, multiple yield strength variants are designed that implement different levels of yield vs. cell size and other design costs trade-off optimizations.

Some example statistics based on the application of a high-yield standard cell library design for a 130 nm process are reported in Table 22.2. The Limited Yield (LY) improvements have been obtained on two very similar 130 nm processes but at a different stage of maturity. On a million-gate design, the average functional yield improvement is 2% and 5%, respectively.

Figure 22.14 illustrates the distribution of performance degradation for the same library cells. Most of the cells show performance degradation between 1% and 10%, but we also have standard cells with worst-case timing arcs exceeding 20% degradation.

Figure 22.15 shows the high-yield cell area increase distribution. The average standard cell area increase for this implementation is 15%, but also reaches peaks of up to 50%. This clearly

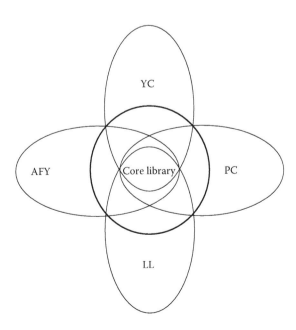

FIGURE 22.13 DFM standard cell library modules.

TABLE 22.2 Yield Improvements for High-Yield Library

	Equivalent Gates (Million)	LY Improvement (%)
Process 1 (mature)	1	2.08
	2	4.03
	3	5.84
Process 2 (ramp)	1	4.96
	2	9.27
	3	12.99

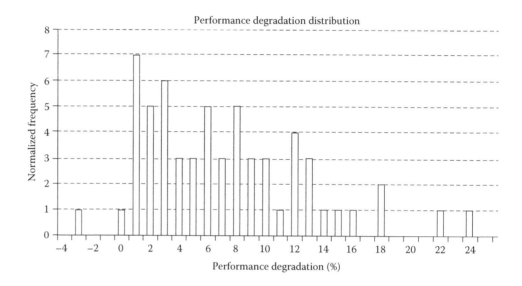

FIGURE 22.14 DFM performance degradation of a DFM library.

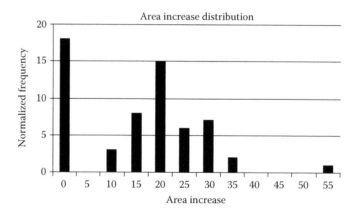

FIGURE 22.15 Area degradation of a DFM library.

demonstrates the need of accurate characterization of the yield benefit and of the separate contribution of all the important effects.

22.3.5 MEMORY OPTIMIZATION

Along with logic, memories represent another significant source of yield loss for a large class of ICs, and systems-on-chip (SoCs) in particular. Yield loss in memories can be due to failures occurring in the core or the periphery, with the former being usually most relevant except for very small arrays.

The memory bit-cell still plays a critical role during technology development because it is often used to tune and optimize the process flow and OPC. The need for larger embedded memories drives the implementation of very high-density bit-cell layouts and of aggressive ad hoc DRs along with custom manually optimized OPC.

The challenge of designing highly manufacturable bit-cells is therefore often overwhelming, given the fact that multiple issues need to be addressed at the same time. As already mentioned, the first goal is to design bit-cell layouts that are as compact as possible to drive down bit-cell size and ultimately memory size, speed, and power consumption. Dense, sub-DR bit-cells often pose severe printability and lithography process window problems. The integration between process and design is exacerbated in memories where stress can have severe impact on the electrical behavior of the circuitry. On top of this, performance, power, and leakage obviously must all be optimized. All these issues must be addressed during technology development and they often require a significant amount of resources, silicon, and time in order to achieve satisfactory results.

Embedded as well as commodity memories, such as SRAMs, ROMs, FLASH, and DRAMs, make use of different fault recovery methods in order to achieve reasonable yields. The simplest and most popular method is memory redundancy that provides spare memory elements such as spare rows or columns of bit-cells, which are fuse-selectable at electrical test or by embedded self-test logic.

Many semiconductor companies started developing multiple bit-cell variants to more flexibly address different application needs and to improve manufacturability. It is in fact often the case that multiple bit-cells are employed within a single SoC product; smaller bit-cells are used for large memory arrays that usually include a significant fraction of spare resources (as the overhead in terms of block size is still tolerable), whereas relaxed or design-rule compliant bit-cells are used in smaller blocks with little or often no redundancy. A critical DFM decision has thus to occur at the memory partitioning and allocation level, where yield should be included along with speed and power considerations when allocating logic memory units to physical arrays as this may have a significant impact on the floorplan design, on the chip performance, and on its yield. Of course, this requires a somewhat accurate model of the memory yield, as well as of its repairability as a function of the available redundancy. The effort required to design and validate multiple bit-cell memory variants is relatively small if planned properly, especially when taking into account that this effort is already dominated by the high-density bit-cell designs that push DRs the hardest.

For a sound DFM strategy, it is therefore recommended that high-density memory bit-cells are complemented by at least two or three variants (depending on critical YLMs) of high-yield layouts; for example, a design-rule compliant bit-cell and an intermediate density variant with feature redundancy and litho-friendly layout. The area penalty for alternative bit-cell variants may range from 30% to 70%. The timing penalty depends on the design styles and the memory array architecture. The availability of high-yield, fast bit-cells for small embedded memories is critical to achieve better SoC yields. In fact unrepairable memory errors are often a significant factor in SoC yield loss.

As discussed previously, yield modeling is a critical DFM factor as it enables both bit-cell selection and optimized redundancy allocation.

A key enabler methodology for memory yield modeling is based on a technique called micro-event analysis [1]. The main goal of micro-event analysis is to propagate the probability of failure from the microscopic or individual feature level (such as the probability of contact opens or critical area for metal shorts) to the electrical level (such as the probability of a bit-line short event or single vs. paired bit fail). By feeding process characterization data into a micro-event probability model, it is possible to compute the frequency of certain faults and thus predict the effectiveness of any given type and amount of memory redundancy, and ultimately its yield. For example, a micro-event associated with M2 shorts may generate a paired bit-line (columns) shorts signature, which can be repaired if, and only if, at least two redundant columns per physical memory bank are available. On the other hand, an open on the access transistor gate will produce a single bit failure, which is often a repairable event up to high fail rate densities.

As discussed previously, functional failures in the logic of the memory periphery are usually a much less likely event unless there are also some significant design-systematic effects such as increased fail rates in the transition region from the dense array to the surrounding logic.

Parametric faults in the memory sensing circuitry are often more significant. The DFM methods for analysis and optimization of parametric yield loss are described in the next section.

22.4 PARAMETRIC DESIGN FOR MANUFACTURING METHODOLOGIES

As discussed in the previous sections, the occurrence of random and systematic defects that modify the functionality of a circuit may cause functional yield losses. A defect-free circuit that has been tested for correct functionality can still fail to meet its performance specifications, such as speed, power consumption, and leakage. Yield loss caused by performance specs test failure is called parametric yield loss.

Analog circuits do not normally involve very large area or transistor counts, and they often adopt relaxed DRs, and hence their functional yield is often very high. However, both technology and voltage scaling increase the sensitivity of analog circuit performance to manufacturing variations, and thus high-performance, high-precision analog components are sensitive to die-to-die variation and mismatch. The design for manufacturability of analog and mixed-signal IP involves the verification and optimization of parametric variations over the measured or expected range of fabrication process perturbations.

Parametric yield loss in digital circuits often involves speed test failures—in other words the circuit meets its functional specs at low clock frequencies, but it starts failing once the internal clock speed is increased above a certain frequency value. Typically, this can happen because some of the logic gates on a critical path become too slow, thus causing an incorrect signal to be latched at the register outputs. On the contrary, it may also happen that the internal circuit synchronization is lost when some gates become too fast, thus causing hold-time violations or excessive clock skew.

Parametric/speed yield optimization of logic circuits involves accurate verification of critical path delays over all process corners by using static timing analysis (STA) or statistical static timing analysis (SSTA).

A simple way to detect parametric yield loss in products is to compare the product yield learning curve (equivalent product D_0 expressed in DEF/cm^2) to the learning curve of random defect monitors resulting, for example, from periodic measurements of random defect limited yield (RLY) structures of a characterization vehicle [14] as shown in Figure 22.16. Large deviations (slow down) from the ideal product yield learning curve with respect to RLY are an indication of the presence of parametric yield loss.

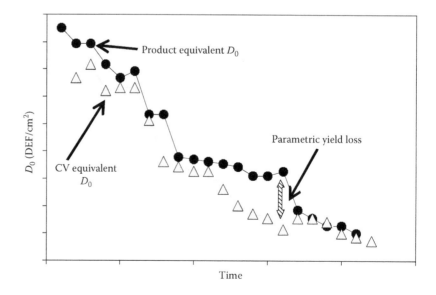

FIGURE 22.16 Identification of parametric yield loss by comparison with defect yield limited characterization vehicles.

Correlation of product test bins with process control monitor (PCMs) measurements in scribe lines is also often used in product engineering to track the presence of parametric yield loss and to take suitable corrective actions.

22.4.1 SOURCE OF PARAMETRIC YIELD LOSS

In general, parametric yield loss can be caused by both random and systematic effects. Random sources of parametric yield loss are primarily represented by random variation of the electrical properties of the layers forming active and passive devices. For example, random fluctuations of channel dopants, oxide thickness, and effective length in MOSFETS cause I_{DSS} variation across identically designed transistors. Random variation of electrical device parameters can be extremely local, and can cause two identically drawn devices placed close to each other to have different electrical behavior. This phenomenon is called device mismatch, or intra-die variation, as opposed to inter-die variation, i.e., such that device parameters are quasi-constant for devices on the same die but change from one die to another.

Systematic sources of parametric yield loss are generally caused by electrical device parameter variation that is a function of the device layout attributes. Examples of systematic sources of parametric yield loss are:

- Device layout pattern distortion caused by printability effects
- Context-dependent poly/active flaring causing W_{EFF}/L_{EFF} variation in MOSFETs
- Stress-related mobility variation in the MOSFET channel
- Contact position and contact resistance
- Narrow salicide formation issues causing output device impedance to vary as a function of the contact to gate spacing

Systematic sources of parametric yield loss, such as random sources, can have intra-die as well as inter-die components.

22.4.2 CORNER ANALYSIS

A typical approach to account for process variation is to simulate circuit performance at the so-called worst-case corners. These corners are estimated from the expected worst-case variation in key device parameters, such as I_{DSAT} and I_{OFF} of MOSFET transistors. The corners of the expected statistical distribution of the target performance parameters are translated into worst-case corner SPICE models, i.e., one or more combinations of SPICE model parameters, extracted from electrical device test data, or e-tests. The procedure of extracting corner model parameters from electrical test data often involves the identification of a set of first-order independent model parameters that can be uniquely extracted from specific regions of the $I–V$ characteristics of certain transistors (such as TOX and NCH from I_{DSAT} and V_{TSAT} of long-channel MOSFETs). Availability of physical device models, such as BSIM4 [15], helps to determine and extract (i.e., analytically inverting the model to obtain parameter values corresponding to specific e-test data) a suitable set of such first-order physical parameters.

The set of device model parameters that corresponds to a predefined percentile point (e.g., $\pm 3\sigma$) of the e-test data statistical distribution is then used to define the SPICE models that are used for circuit corner analysis.

Corner analysis is powerful because it allows for capture of the impact of process variation on circuit performance with a small number of electrical or timing simulations; however, it presents a number of significant limitations, such as:

- Special attention is required to account properly for model parameter correlation, including different parameters of the same device (e.g., VT0 and TOX) as well as the same parameters of different devices and device mismatch. Lack of proper handling of device parameter correlation leads to unphysical corner models and often to unnecessary pessimism.

■ Corner models are defined as $\pm 3\sigma$ (or $\pm N\sigma$) percentiles of certain key device parameters that are assumed to map to the corresponding percentile of the target circuit performance parameters. Owing to the nonlinear mapping between device and circuit performance, this assumption is often inaccurate and also leads to excessive pessimism.

■ Worst-case corners are typically derived by considering combinations of model parameter values that will cause slow or fast digital switching events. These do not necessarily result in the worst-case performance of analog circuits, where a different combination of parameter values could potentially be the worst case for many performances of interest (e.g., gain, offset voltage, and unity-gain bandwidth).

Statistical circuit simulation can be used to address these limitations.

22.4.3 STATISTICAL SPICE MODELS

The active and passive SPICE device parameters can be modeled statistically in order to capture the effect of the electrical variation of the device parameters and ultimately to be able to simulate parametric yield loss.

As mentioned in the previous section, statistical circuit simulation represents a more accurate alternative to the worst-case corner approach. Statistical circuit simulation requires statistical and mismatch SPICE models. These models represent the change in SPICE model parameters caused by manufacturing variation. Statistical SPICE models are derived by estimating the joint probability density function (JPDF) of a certain number of model parameters. By carefully maintaining the correlation between model parameters, these models capture the correlated variation in device characteristics. For analog design, statistical and mismatch SPICE models can be used for parametric yield assessment and robust design [16]. For digital designs, these models can be utilized for estimating realistic and application-specific worst-case corners that take into account the observed device characteristics [17].

The problem of extracting statistical SPICE models is more challenging than it may seem. In fact, the most popular SPICE models such as BSIM4 employ a large number of different parameters, many of which are actually empirical fitting coefficients [15]. These parameters are practically obtained by using nonlinear, multiobjective, numeric optimization methods that are often trapped in local minima.

Therefore, the approach of estimating model parameter statistics by fitting a separate model for each individual sample in a large collection of measurement data sets (I–V and C–V curves), besides being impractical, is also subject to noise and variance inflation caused by numerical optimization noise.

Statistical modeling methods that address these issues generally fall into one of the following two categories:

1. *Monte Carlo device and process simulation (technology CAD or TCAD)*: These methods have the advantage that physical model parameters, such as TOX or LEFF, can be directly obtained from the TCAD simulators' outputs, thus reducing the need for aggressive numerical fitting [18,19]. Although TCAD-based Monte Carlo methods are very useful during the early manufacturing phase, when a large and stable set of statistical device measurements is not available, the main drawback of these approaches is the time and effort required to calibrate TCAD tools adequately.

2. *Direct extraction*: These approaches apply direct inversion to estimate the value of a reduced subset (core or first-order independent set of physical parameters) of the device model parameters [20]. Direct extraction methods are definitely more practical although they often require availability of a large number of different types of measurements, which are often difficult to obtain in a production environment.

22.4.4 STATISTICAL CIRCUIT SIMULATION

22.4.4.1 MONTE CARLO ANALYSIS

Monte Carlo analysis is the most typical and straightforward method of statistical circuit simulation. A random number generator is used to sample the value of the parameters of a set of statistical SPICE models from their probability density function (PDF). For each random sample, the circuit is simulated and the output circuit performance statistics of interest are collected.

The random Monte Carlo samples need to maintain the same statistical correlation of the original population of device parameters that are used in the target circuit. There are different ways to do this. Correlated random Monte Carlo samples can be extracted from the JPDF of the original population if this is known. Alternatively, if an estimated value of the covariance or correlation coefficient matrix is known, a linear transformation known as principal component analysis (PCA) or principal factor analysis (PFA) can be applied to the original device parameters. In this case, SPICE device parameters are defined as linear combinations of a set of new variables, which can be normalized to be independent $N(0,1)$ random variables.

The typical output of Monte Carlo simulations is the sample distribution and statistics (such as sample mean and variance) of the circuit performance specifications. An example of such sample distribution for the noise figure (NF) of an RF LNA circuit is shown in Figure 22.17.

The main drawback of Monte Carlo analysis is that a number of circuit simulations are required in order to estimate the circuit output parameter statistics with reasonable accuracy.

In fact, the Monte Carlo output statistics, such as mean and standard deviation of circuit parameters, are sample estimates of the statistical parameters of the underlying population. The variance of sample estimates is inversely proportional to the sample size, i.e., to the number of Monte Carlo circuit simulations. Therefore, in order to estimate the true underlying population distribution parameters with a reasonable confidence level (e.g., >90%), the number of simulations required is easily on the order of a few thousands, for typical circuit parameter variability.

The plots in Figure 22.18 show an example of the confidence intervals at 95% and 99% confidence levels as a function of the sample size, showing that several hundreds of simulations are required in order to get within ±5% of the true average value.

22.4.4.2 RESPONSE SURFACE MODELING

The method known as response surface modeling (RSM) is a statistical technique that can be applied to statistical circuit analysis in order to circumvent the runtime complexity of Monte Carlo

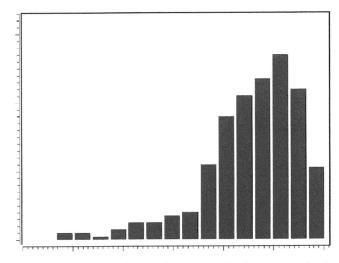

FIGURE 22.17   Example of Monte Carlo simulations results: frequency distribution of LNA noise figure.

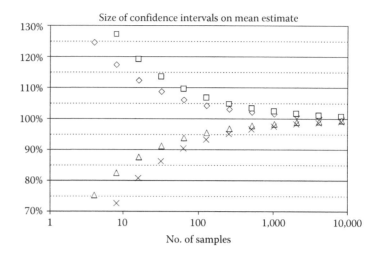

FIGURE 22.18   Confidence intervals vs. sample size.

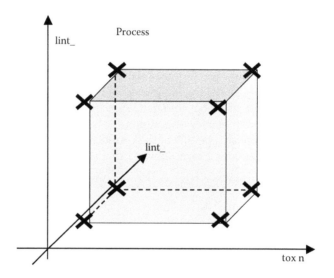

FIGURE 22.19   Full factorial DOE example in three variables.

simulations [21,22]. The principle of RSM is to determine an analytical regression model of the circuit response (e.g., VCO phase noise, amplifier BW, PM and GM, LNA gain, and NF and IP3) to SPICE device parameters variation. This regression model is typically obtained as the least square model fit of a set of circuit simulation results obtained by perturbing the SPICE device parameters according to a particular DOE plan. Figure 22.19 shows an example of a full factorial DOE in three variables.

The type of DOE plan is determined based on the regression model, the number of parameters, and the model accuracy requirements. The typical number of simulations that are required to build a quadratic regression model is less than a hundred, with 10–12 input statistical device parameters, which is typical for the state-of-the-art MOSFET technology.

After the circuit output response parameter has been modeled by RSM, the input device parameters can be perturbed according to a Monte Carlo sampling plan, and statistical circuit behavior can be estimated using the RSM model instead of circuit simulations.

In this way very tight error bounds on the statistical parameters of the output circuit response distribution can be achieved with a much smaller computational effort.

The RSM methodology has been integrated in a commercial EDA tool [22] and is routinely used in IC industry.

22.4.4.3 MISMATCH SIMULATION

Not only analog circuits and SRAM design, but also the clocking network of digital designs, are particularly sensitive to intra-die variation or mismatch. The reason is that in order to reduce the effect of die-to-die variability as well as to compensate for global environment parameter variation (e.g., temperature or supply voltage), circuit designers use differential circuit configurations that rely upon good matching properties of the semiconductor devices. Because of this reason, the problem of characterizing and modeling the intra-die component of device variability has been extensively studied by researchers in both academia and industry [23,24].

Nonetheless, including device mismatch in statistical circuit simulations is challenging because the number of statistical device parameters to be considered increases by a factor that is proportional to the number of transistors in the circuits. This inflation of parameters is a serious problem for RSM-based tools, as the number of DOE experiments (hence the number of simulations) grows exponentially with the number of input parameters.

This apparent roadblock can be circumvented by assuming that the parameters of identically drawn devices placed in close proximity, although not perfectly matched, are strongly correlated. Therefore, by decomposing the covariance matrix of a system of correlated random variables that describes the set of device parameters of a large circuit, it is possible to identify a small number of independent factors (linear combinations of the original correlated random variables) that can properly explain most of the system variance and correlation. This method [25], based on a two-step eigenvalue decomposition of the system covariance matrix, has been successfully applied to both analog and digital statistical circuit analysis [26,27].

22.5 DESIGN FOR MANUFACTURING INTEGRATION IN THE DESIGN FLOW: YIELD–AWARE PHYSICAL SYNTHESIS

22.5.1 EARLY YEARS: REACTIVE DESIGN FOR MANUFACTURING

In the early years, DFM optimizations were applied just before final layout finishing and mask data preparation (MDP). In fact, some of the early types of DFM can be seen as a special kind of layout finishing or MDP operation. For example, the introduction of feature redundancy, such as contact and via redundancy, wire spreading for critical area reduction, line biasing (over- or under-sizing), cheesing and dummy fill insertion, and to some extent OPC can be considered as the most basic form of DFM, and were all introduced after final tape-out of the mask layout—in other words after timing closure, and physical and signal integrity verification.

A typical design flow, which consists of these types of DFM operations, will be referred to as "reactive DFM" (for reasons that will become clear in the following discussion) is shown in Figure 22.20.

Despite its apparent simplicity and widespread popularity, reactive DFM presents some significant drawbacks. In fact, because reactive DFM manipulations occur after main timing closure and physical verification loops, they are supposed to leave the circuit's electrical (such as Rs and Cs) timing, and connectivity properties unchanged. This is of course not strictly possible, therefore, the typical approach is to limit the scope and location of DFM changes to transformations and places in the layout that have minimum impact on the circuit characteristics. Examples of such opportunistic optimizations are shown in Figure 22.21.

Although most of these simple modifications have minimum impact on the electrical characteristics, it is impossible to guarantee with absolute confidence that the impact is completely negligible, so reactive DFM methods may expose the designers to some risk and uncertainty. For example, redundant vias and contacts with their larger landing pads have a small but nonzero impact on the node resistance and capacitance; similarly slotting/cheesing (i.e., creating "holes" in fat Cu lines to improve CMP and achieve better planarity), and dummy fill insertion may change the interconnect coupling and grounded capacitance values and thus potentially invalidate timing and signal integrity simulation results.

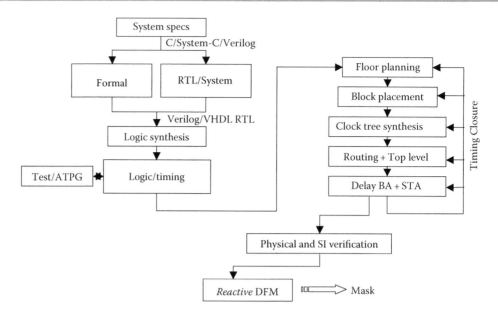

FIGURE 22.20 "Reactive DFM" design flow.

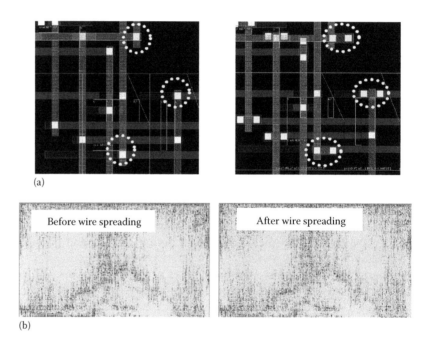

FIGURE 22.21 Opportunistic optimization examples: via redundancy (a) and wire spreading for CA optimization (b).

Another drawback of reactive DFM techniques is the additional lag time that they introduce between design tape-out and mask making. In fact, most of these techniques can only be effective at full chip level and may create some additional hazards when used hierarchically, as some of the connectivity checks may fail to catch issues at the interface between blocks.

For large VLSI designs with the final flat GDS database size of an order of several Gbytes, the pure processing times of any of the above-described DFM steps may take a couple of days. Assuming a few days of physical verification, and even without taking into account the potential time to fix eventual problems and reloop, the total time for reactive DFM may easily total from a few weeks up to more than a month.

22.5.2 PROACTIVE DESIGN FOR MANUFACTURING

The limitations of reactive DFM techniques stem from the fact that they are applied at the end of the design cycle instead of being integrated into the design flow.

In order to address the limitation of reactive DFM, it is thus necessary to develop DFM methods that are integrated upstream in the design flow and that are capable of synthesizing DFM-friendly physical layouts, while meeting all remaining design constraints (such as timing, power, and signal integrity) before timing closure and physical verification.

A key enabler of DFM integration in the design flow is the capability of precisely quantifying the yield and manufacturability impact of every different design choice. In fact, in this way it is possible to expand the cost function of the various design optimization steps in the design flow to include a yield metric. Design flow tools are thus made aware of the manufacturability cost of each different choice that they take and trade-off yield for existing slacks in the remaining design dimensions.

An example of such a proactive approach of DFM has been presented in [28], where the cost function of logic synthesis has been modified in order to take into account different types of yield metrics. Other proactive DFM approaches have been developed introducing yield cost and metrics at different stages in the design flow such as floorplanning, physical synthesis, and routing [29,30].

The key feature of any of these methods is the silicon characterization of accurate yield models of the circuit IP blocks, such as standard cells, interconnects, memories, and hard-macros, and the integration of such models into the cost function of the design tools.

The implementation described in [29], for example, makes use of accurate standard cell yield models to optimize the tech-mapping and placement steps of logic synthesis by trading off some of the available timing slack of logic cells placed off the critical path for increased design manufacturability.

Planning for manufacturability at earlier stages of the design flow, such as floorplanning and global routing, leaves increased margins for DFM optimizations when compared to a methodology that performs DFM optimizations after detailed routing and timing ECOs, when the physical design hierarchy is basically frozen. This allows substantial gains in terms of design manufacturability which may translate into 10% or more yield and GDPW (good die per wafer) improvement as per reported, for example in [30].

22.6 SUMMARY

Until the deep-submicron era, worst-case simulation and DR compliance were the only methods that design engineers had to worry about in order to ensure consistent product yields. However, with each new generation of process technology, DRs become more complicated trying to provide tolerance to increasingly complex layout systematic effects, while intra-die variability and pattern dependency of MOSFET parameters make worst-case simulation less and less manageable. In this chapter, we have focused on the DFM requirements for technologies in the nanometer range where the spectrum of physical phenomena that may impact manufacturability of products is mind-boggling. We have shown that the systematic characterization of these phenomena and their impact on IC yield and performance is of crucial importance for true DFM.

We have stated that the most accurate way of accounting for these effects is to provide accurate physical models and then simulate the actual IC layout to estimate the impact on yield and performance. Such a simulator requires calibration to the actual manufacturing process by specially designed test structures that cover all the possible layout patterns found in real products. To provide observability in the range of a few failures per billion as a function of layout attributes, such test chips must contain many specially designed layout patterns that require large die size up to the full reticle area. In return, the necessary DFM characterization is achievable with just a few wafers, which actually reduces the cost and (equally importantly) the turnaround time.

We have shown that such a modeling-based approach can then serve as the basis for generating guaranteed-to-yield IP blocks and as a yield sign-off tool for the entire IC layout.

Moreover, such a simulation-based model can also be used to generate guidelines for the physical-design tools (for place and route), without creating any extra effort for IC designers or drastically changing the design flow. We have shown that such an approach enables truly proactive DFM in which all the design modifications take place before verification and after tape-out.

As process technology moves into the sub-50 nm era, new approaches are being developed to maximize layout regularity that will guarantee manufacturability of giga-scale ICs designed with these technologies. The extreme version of such regularity will be required where guaranteed-to-yield layout patterns are created in this DFM environment and layout synthesis becomes equivalent to pattern assembly [31].

REFERENCES

1. W. Maly, Computer-aided design for VLSI circuit manufacturability, *Proc. IEEE*, 78, 356–390, 1990.
2. R. Myers and D. Montgomery, *Response Surface Methodology Process and Product Optimization Using Designed Experiments*, Wiley, New York, 2002.
3. S. Saxena et al., Tests structures and analysis techniques for estimation of the impact of layout on MOSFET performance and variability, *ICMTS 2004*, Awaji, Japan, 2004.
4. C. Hess, D. Stashower, B. Stine, G. Verma, and L. Weiland, Fast extraction of killer defect density and size distribution using a single layer short flow NEST structure, *IEEE 2000 International Conference on Microelectronic Test Structures*, Monterey, CA, March 2000, Vol. 13.
5. C. Hess, B.E. Stine, L.H. Weiland, T. Mitchell, M. Karnett, and K. Gardner, Passive multiplexer test structure for fast and accurate contact and via fail rate evaluation, Cork, Ireland, *ICMTS 2002*, 2002.
6. J. Horgan, TEST & ATE—Cost of test, Private Communication, http://www.edacafe.com/magazine/.
7. D. Hamling, ATE solutions for the global semiconductor industry, *Fabless Forum*, Vol. 11, Fabless Semiconductor Association, March 2004.
8. J. Orbon et al., Integrated electrical and SEM based defect characterization for rapid yield ramp, *SPIE 2004*, 2004.
9. D. Ciplickas, A. Joshi, S. Lee, and A.J. Strojwas, *Designing for High Product Yield*, Semiconductor International, October 2002.
10. PDF Solutions, Inc., *Yield Ramp Simulator User Manual v3.1*, Copyright 2001–2003.
11. K. Keutzer, K. Kolwicz, and M. Lega, Impact of library size on the quality of automated synthesis, *Proceedings of ICCAD*, Santa Clara, CA, 1987, pp. 120–123.
12. K. Scott and K. Keutzer, Improving cell libraries for synthesis, *Proceedings of CICC*, San Diego, CA, 1994, pp. 128–131.
13. N. Dragone et al., High yield standard cell libraries: Optimization and modeling, Santorini, Greece, *PATMOS 2004*, 2004.
14. K. Miyamoto, K. Inoue, I. Tamura, N. Kondo, H. Inoto, I. Ito, K. Kasahara, and Y. Oshikiri, Yield management for SoC vertical yield ramp, San Francisco, CA, *IEDM 2000*, 2000.
15. BSIM4v4 Manual, Department of Electrical and Computer Science, University of California, Berkeley, CA, 2004, http://www-device.eecs.berkeley.edu/~bsim/Files/BSIM4/BSIM480/BSIM480_Manual.pdf. Accessed on December 16 2015.
16. C. Guardiani, S. Saxena, P. Schumaker, P. McNamara, and D. Coder, An asymptotically constant, linearly bounded methodology for the statistical simulation of analog circuits including component mismatch effects, *Proceedings IEEE/ACM Design Automation Conference*, New Orleans, LA, 1999, pp. 15–18.
17. A.D. Fabbro, F. Franzini, L. Corce, and C. Guardiani, An assigned probability technique to derive realistic worst-case timing models of digital standard-cells, *Proceedings IEEE/ACM Design Automation Conference*, San Francisco, CA, 1995, pp. 702–706.
18. pdfab Manual, PDF Solutions, San Jose, CA, https://www.pdf.com/Home. Accessed on December 16 2015.
19. D.A. Hanson, R.J.G. Gossens, M. Redford, J. McGinty, J.K. Kibarian, and K.W. Micheals, Analysis of mixed-signal manufacturability with statistical TCAD, *IEEE Transactions on Semiconductor Manufacturing*, 1996, pp. 478–488.
20. J.C. Chen, C. Hu, C.-P. Wan, P. Benedix, and A. Kapoor, E-T based statistical modeling and compact statistical circuit simulation methodologies, *Proceedings of International Electron Device Meeting*, San Francisco, CA, 1996, pp. 635–638.
21. G. Nicollini and C. Guardiani, A 3.3-V 800-nV^{rms} noise, gain programmable CMOS microphone preamplifier design using yield modeling technique, *IEEE J. Solid State Circuits*, 28, 915–921, 1993.
22. Circuit Surfer Manual, PDF Solutions, San Jose, CA, https://www.pdf.com/Home. Accessed on December 16 2015.

23. C. Guardiani, A. Tomasini, J. Benkoski, M. Quarantelli, and P. Gubian, Applying a submicron mismatch model to practical IC design, *Proceedings of IEEE Custom Integrated Circuits Conference*, San Diego, CA, May 1994.

24. M. Quarantelli et al., Characterization and modeling of MOSFET mismatch of a deep submicron technology, *International Conference on Microelectronic Test Structures*, Monterey, CA, March 17–20, 2003, pp. 238–243.

25. C. Guardiani, S. Saxena, P. McNamara, P. Schumaker, and D. Coder, An asymptotically constant, linearly bounded methodology for the statistical simulation of analog circuits including component mismatch effects, *Proceedings of IEEE/ACM 37th Design Automation Conference*, Los Angeles, CA, June 2000.

26. P. McNamara, S. Saxena, C. Guardiani, H. Taguchi, E. Yoshida, N. Takahashi, K. Miyamoto, K. Sugawara, and T. Matsunaga, Design for manufacturability characterization and optimization of mixed-signal IP, *Proceedings of IEEE Custom Integrated Circuits Conference*, San Diego, CA, May 2001.

27. E. Malavasi, S. Zanella, J. Uschersohn, M. Misheloff, M. Cao, and C. Guardiani, Impact analysis of process variability on digital circuits with performance limited yield, *Proceedings IEEE International Workshop on Statistical Metrology*, Kyoto, Japan, June 2001.

28. A. Nardi and A.L. Sangiovanni-Vincentelli, Synthesis for manufacturability: A sanity check, *Proceedings of the Design, Automation and Test in Europe Conference and Exhibition*, Paris, France, February 16–20, 2004, Vol. 2, pp. 796–801.

29. C. Guardiani, N. Dragone, and P. McNamara, Proactive design for manufacturing (DFM) for nanometer SoC designs, *Custom Integrated Circuits Conference, Proceedings of IEEE 2004*, San Diego, CA, 2004, pp. 309–316.

30. J. Kibarian, C. Guardiani, and A.J. Strojwas, Design for manufacturability in nanometer era: System implementation and silicon results, *International Solid State Circuits Conference (ISSCC), Proceedings of the IEEE*, San Diego, CA, February 6–10, 2005, pp. 18–19.

31. V. Kheterpal, V. Rovner, T.G. Hersan, D. Motiani, Y. Takegawa, A.J. Strojwas, and L. Pileggi, Design methodology for IC manufacturability based on regular logic-bricks, *DAC*, Anaheim, CA, USA, 2005.

32. D. Ciplickas, S.F. Lee, and A.I. Strojwas, A new paradigm for evaluating IC yield loss, *Solid State Technol.*, 44(10), 47–52, October 2001.

33. D.J. Ciplickas, X.Li, and A.J. Strojwas, Predictive yield modeling of VLSIC's, *2000 Symposium on VLSI Technology, Statistical Metrology Workshop*, Honolulu, HI, June 2000.

34. C. Hess and L. Weiland, Determination of defect size distributions based on electrical measurements at a novel harp test structure, *Proceedings of IEEE 1997 International Conference on Microelectronic Test Structures*, Monterey, CA, March 1997, Vol. 10.

35. M. Quarantelli et al., Characterization and modeling of MOSFET mismatch of a deep submicron technology, *Proceedings of IEEE International Conference on Microelectronic Test Structures*, Monterey, CA, 2003.

36. E.P. Box and N.R. Draper, *Empirical Model-Building and Response Surfaces*, Wiley, New York, 1987.

Design and Analysis of Power Supply Networks

Rajendran Panda, Sanjay Pant, David Blaauw, and Rajat Chaudhry

23

CONTENTS

23.1 INTRODUCTION

Device scaling over several process technology generations has permitted the integration of an extremely large number of transistors on a single die. Combined with faster switching speeds achievable with advanced lithographic and fabrication technologies, the power dissipation of chips has grown at a very rapid pace. The operating voltage has been scaled down to reduce power dissipation, thus shrinking the voltage drop allowed on the chip power and ground distribution for the correct operation of the circuit. Unless the overall power distribution network is designed adequately for the power demands of the chip, the entire design will be vulnerable to many issues as discussed in this section. These issues and the related design challenges are obvious for high-performance processors that need to deliver a very large amount of power and accommodate more severe power transients. In reality, the problem is no less important or less challenging for low-power processors such as mobile application processors, as their power distribution network design has to consider many different operating modes resulting from clock gating, power gating, multiple power domains, dynamic voltage, and frequency-scaled operation and also has to be implemented using fewer routing layers and cheaper packages due to cost consideration.

The power supply network encompasses all devices and wiring involved in the generation of DC power supply voltages for the chip's circuits. It also includes all wiring (and decoupling capacitances) distributed across the die and package of the chip, and the discrete components (and their wiring) for generation and regulation of power supply voltages on the circuit boards in the system. The current return (ground) network is also a key element of the power supply distribution. In the remainder of this chapter, we will use the popular acronym PDN (for power distribution network) to refer to both the on-chip and off-chip power supply systems.

23.1.1 POWER SUPPLY NOISE

Due to resistance in the PDN, a voltage difference, commonly referred to as the *IR drop*, is observed between the supply origin and loads. For an instantaneous current $i(t)$ flowing through a wire segment of resistance R, a voltage of $i*R$ is dropped across the segment in the direction of current flow. The voltage drop seen at any point in the network is the sum of voltage drops of segments of any path from a supply source to that point.

The package brings power and ground supplies to the die either by means of package leads in wire-bonded chips or through controlled collapse chip connections, known as C4 bumps [1], in flip-chip technology. While the resistance of the package PDN itself is quite small, the inductance, contributed mainly by the package-to-die interconnections (leads or bumps), is significant. The inductance responds sluggishly to fast changes in the delivered current by causing a dynamic change in local supply voltage that is proportional to the amount of inductance (L) and the rate of change of current (di/dt). This component is referred to as the *L*di/dt drop* (although it may be a negative or positive contribution to the supply voltage). The sum of resistive and inductive losses, namely, $i*R + L*di/dt$, is the power supply noise seen by the chip's circuits and can be problematic if it is excessive.

Capacitors between power and ground distribution networks, referred to as the decoupling capacitors (or *decaps* for short), act as local charge storage and help in mitigating the *L*di/dt* voltage drop at loads. Parasitic capacitances between power and ground wires, diffusion and gate capacitances of devices, and capacitances between wells and substrate provide a certain amount of decoupling intrinsically. Unfortunately, these parasitic capacitors are not enough to smooth out the noise to within safe limits, and hence additional dedicated decoupling capacitance structures need to be placed on the die at strategic locations [2–4]. These added devices are not free; they increase the die area and the leakage power dissipation. The distributed resistance, decoupling capacitance, and inductance in the power network form a complex resistance, inductance, and capacitance (RLC) network with resonant nodes [5,6]. During the operation of the chip, if power transients occur at frequencies close to these resonance nodes, large voltage drops can develop in the grid.

23.1.2 PERFORMANCE AND POWER ISSUES

Voltage droops at the gates of a logic signal transition path slow the signal propagation along that path, and so the maximum speed at which a circuit can be operated safely is limited by the worst-case voltage noise in the PDN [7,8]. In clock generation circuits, the power supply noise contributes to clock jitter [9], which reduces the useful part of the clock cycle time available for logic computation.

The impact of PDN noise on total dissipated power and the preservation of operating margins of a design are other concerns. As the dynamic power dissipation of a chip is proportional to the square of the supply voltage, raising the supply voltage by, say, 15% to compensate for the worst-case noise in the chip will result in a 30% increase in power dissipation. Moreover, the voltage regulator (VR) designed to handle a wider noise band needs a bulkier low-pass filter and is less efficient.

Timing margins are impacted adversely by power supply noise either directly (e.g., additional setup and hold margins at clocked sequential circuits for supply variation) or indirectly (e.g., greater margins for higher clock arrival skew [10] and jitter [9]). Noise from the power supply also propagates as noise in both analog and digital signals. As a result, the functional (glitch) noise will have to be margined appropriately. PDN noise thus creates a need for higher design efforts to tackle functional noise issues.

23.1.3 RELIABILITY ISSUES

A poorly designed power network contributes to chip reliability problems such as electromigration failure in the wires and gate oxide failure in the devices.

The flow of unidirectional current in a metal wire segment for an extended time period causes transport of some metal mass itself in the direction opposite to that of current flow. The metal ions move by gaining momentum from the conducting electrons, creating gaps and hillocks at certain locations of the conductor over a period of time, and may eventually lead to open or short circuit failures. The mean time to failure of the conductor is dependent (superlinearly) on the inverse of the current density in the conductor [11]. Even before a catastrophic open/short failure, the resistance of the conductors can increase substantially, causing the IR drop to be elevated. To guarantee a certain lifetime of the reliable operation of a chip, the current densities through the PDN wires will have to be limited to levels appropriate for the chip and the fabrication process technology. The power network design should therefore ensure that conductors are sized adequately and the current distribution has no current density hot spots.

Reversing the direction of current in a conductor segment contributes to some recovery and healing of the electromigration mechanism. However, this recovery is only partial. Therefore, a conservative check for current densities could use time-averaged currents in a single direction (ignoring the recovery due to the flow in the other direction) or, if using time-averaged bidirectional currents, rely on conservative margins in the determination of load currents or the duty factor. The structure of a typical multilayered PDN, with supply connections on the highest metal layer and loads connected to the lower layers, causes the current in wire segments to exhibit unidirectional flow. This is due to the gradual decrease in pitch from higher to lower layers, and as a result, the current in a segment of upper tiers is the aggregate of currents in several segments of lower tiers. Current in segments of lower layers may switch direction more often due to spatial shifting of switching activity from one region of the chip to another and thus may show weaker directionality. However, higher layer segments show strong directionality and are thus more prone to electromigration failure.

Gate oxide failures may be triggered and accelerated by repeated overstressing of the gate terminals of devices with overshooting input voltages resulting from a poorly designed PDN. Even when there is no failure, the excess stress will contribute to increased gate leakage.

23.1.4 ELECTROMAGNETIC COMPATIBILITY ISSUES

PDN design has to consider electromagnetic compatibility issues in certain products as well, for example, microcontrollers for automotive applications [12]. High frequency noise in long wire traces of the power supply system (typically in the package and board traces) is a significant source of electromagnetic radiation. For systems where radiation is regulated very stringently, the power delivery design has to filter noise with adequate decoupling capacitance on the die and ensure that the energy in the high frequency band of the noise spectrum in power traces is within limits.

23.2 IMPEDANCE AND RESONANCE

Figure 23.1a shows a simplified model of the chip power distribution system consisting of the system board, package, and the die. A VR placed off-chip is shown as a variable DC source with a series resistance. Later in this section, we will discuss the integration of on-chip VRs. The electrical models for the board and package are shown as series–parallel RLC. The die is represented by a single independent current source element, which is the aggregate of the currents drawn by all the active devices, in series with an effective resistance of the on-die PDN and in parallel to a capacitance, which is the sum of intrinsic and intentionally added decoupling capacitors.

Figure 23.1b shows the characteristic of the *looking-in* impedance Z in the frequency domain. The impedance graph shows a peak at the natural or resonant frequency of the overall power system [6]. Since the R, L, and C components are in reality distributed, the system may have more than one resonant node [5]. Current variations occurring at different frequencies each contribute to noise in the power supply voltage in the amount of $I(f)*Z(f)$. If a power system is to have not more than $x\%$ noise on the node voltages, we need to satisfy that

$$\left| \int_0^\infty Z(f)^* I(f) df \right| \leq x V_{dd} / 100$$

The power supply noise problem is so acute in today's designs that it becomes necessary to focus on both the variables on the left side of this equation, with very significant consequences for the needed design effort. Reducing $Z(f)$ is largely the task involved in the power distribution design itself, while controlling the magnitude and frequency of currents (characterized by the $I(f)$ spectrum) requires design efforts that span the architecture, circuit design, and physical implementation teams. We will defer the discussion of controlling currents and current transients to a later section and focus here instead on controlling the impedance factor.

Figure 23.2a shows the impedance of an initial network and Figures 23.2b and c illustrate how this impedance is reduced and reshaped during power network design. The resistive component R (which equates to $Z(0)$ and is also called the active component) is a highly effective control

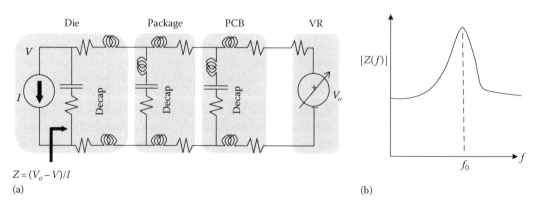

$Z = (V_o - V)/I$

(a)

(b)

FIGURE 23.1   (a) Simple model of a power distribution network (PDN) with off-chip VR. (b) Looking-in impedance Z of the PDN.

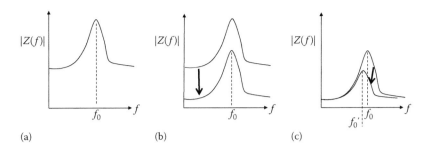

(a)   (b)   (c)

FIGURE 23.2 (a) Frequency dependence of impedance of a power distribution network. (b) Reduction of impedance by reducing the resistance. (c) Reduction of peak impedance.

variable as its effect pervades the entire frequency spectrum. Reducing wiring resistances has the effect of pushing the entire impedance curve down while approximately preserving the shape. This is illustrated in Figure 23.2b. This task requires significant routing resources and design efforts. Common measures include

- Widening of power and ground wires
- Adopting denser wiring pitches
- Strapping wires with more vias
- In some extreme cases, adding another layer to the metal stack

The reshaping of impedance shown in Figure 23.2c is done by altering its reactive component. The impedance peaks are reduced by reducing the inductive parasitic elements and increasing the decoupling capacitance in the PDN. Measures to reduce the peaks include

- Increasing the number of power/ground connections at the board/package and package/die interfaces
- Trimming inductances of traces and vias in the board and package
- Adding more decoupling capacitors on the die, package, and the circuit board

The resonance frequency is shown to move down (from f_0 to f_0') in Figure 23.2c, as is typical when on-die decaps are added. It may, however, be noted that the resonance frequency may shift in the other direction if package inductance is reduced or is shielded by addition of embedded decaps in the package or by moving the package decaps closer to the die. In either case, the impedance peak will be reduced.

Attention should be paid to the magnitude of current variations at or close to the resonance frequencies since their contribution is weighted heavily by the impedance peaks. It should be noted that reducing the resistance of the power network has a side effect of rendering the resonance peaks to be quite sharp. A simple LRC model shown in Figure 23.3 is helpful in understanding the noise and resonance behaviors.

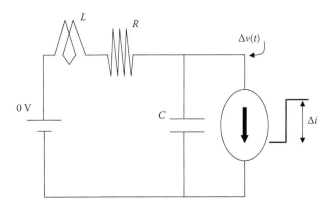

FIGURE 23.3 A simplified model of the power distribution network.

The capacitance in this model is predominantly the decoupling capacitance on the die and the inductance is primarily from the package. The response $\Delta V(t)$ to a step input of ΔI is the familiar waveform given by

$$\Delta V(t) = \Delta I * R + \Delta I * \sqrt{\frac{L}{C}}\, e^{-\frac{R}{2L}t} \sin(\omega_r t - \theta)$$

where

$$\omega_r = \frac{1}{\sqrt{LC}}$$

is the chip-package resonance frequency.

The noise voltage is a damped sinusoidal waveform superimposed on a $\Delta I*R$ base. The noise is damped by the $R/2L$ factor and oscillates at the resonance frequency, ω_r. The maximum noise is determined by ΔI, L, and C. As inductance is governed by physical dimensions [13,14], it cannot be easily reduced beyond a limit. Decoupling capacitors help to reduce the noise, but they have a diminishing return owing to the $1/\sqrt{C}$ relationship. In view of this behavior, controlling sudden power variations (or fast ramps) is the most direct and effective method available to reduce the PDN noise.

The chip-package resonance frequencies of current technologies are typically much smaller than their core clock frequencies. This has certain consequences. On-chip decoupling capacitance effectively suppresses the high frequency noise (due to power variations within a clock cycle). However, low frequency noise due to power fluctuations averaged over many cycles, especially noise at frequencies close to the resonance frequency, sustains oscillations over many clock cycles, and the capacitors on the die cannot supply charge for that many cycles. Adding more capacitance further decreases the chip-package resonance frequency, making the decoupling less effective. Thus, controlling power fluctuations becomes a very important design objective and will be discussed in Section 23.4.

On-chip voltage regulation is an attractive alternative solution to the transient noise issues. Figure 23.4 shows a power system with an on-chip VR. Linear, as well as switched, regulators have been used for on-chip integration [15,16]. The benefits of on-chip integration of VR are multifold:

- The VR on the die is closer to the loads and intercepts the inductance and resistance of the package and the printed circuit board (PCB), reducing the mid-frequency resonance depicted in Figure 23.2 significantly. The result is a faster response time to power transients.
- Multiple regulators can be distributed on the chip near different load centers, such as the core-cache clusters. The distributed scheme allows individual regulators to respond appropriately to correct the local noise they sense and thus facilitate operating all the regulators more efficiently.
- For a chip design with multiple discrete power domains, a distributed VR scheme yields better shielding of noise between the power domains.
- Real estate on the PCB occupied by off-chip VR components is freed up, resulting in a better form factor of the system and potentially an overall cost reduction.
- On-chip switched regulators can achieve faster switching speeds and so will need smaller filtering elements (L and C). This makes the trade-off between response time and efficiency more attractive for the on-chip implementation.

However, on-chip voltage regulation also has many challenges for design and integration. A low dropout (LDO) linear regulator designed to eliminate an external capacitor [15] is an easy integration option, and it provides a faster response time though at lower efficiency. Power considerations, especially for high-performance systems, favor the use of switched regulators and LC filtering (even for LDO). Capacitance is more easily integrated on the die, whereas integration of inductance is challenging and costly. Air core inductors placed in the package or mounted directly on the die have been considered. Air core inductors do not saturate, but the efficiency is poor due to their relatively high DC resistance (DCR). Inductors based on ferromagnetic materials, on

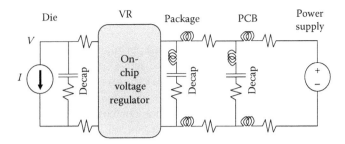

FIGURE 23.4 Power distribution network with on-chip voltage regulator.

the other hand, have a lower DCR, but suffer from saturation at higher load currents. Regulator design with multiple switch and inductor pairs connected in parallel, but switching in a multiphase staggered manner, has been proposed as it inherently reduces the ripples to be filtered and thus will require smaller filtering components to be integrated. It can thus be seen that one of the challenges in designing on-chip voltage regulation is achieving the best trade-off between efficiency, response time, and ease of integration. Another challenge lies in ensuring that the design has no stability issues from the complex interactions between the multiple regulators and load domains [17]. Since the PDN is very large and also includes the active devices from regulators, sensors, and controllers, circuit simulation and stability analysis required to guide the design become extremely challenging.

The R, L, and C of the PDN with on-chip or off-chip VR need to be accurately modeled to analyze and modify its dynamic behavior. Most components can be extracted and modeled quite accurately using the extraction tools for PCB, package, and chip layouts. The modeling of intrinsic (i.e., parasitic) decoupling capacitance in circuits, however, is not straightforward. The gate, diffusion, well, and wiring capacitance in circuit blocks that are coupled from internal nodes to power/ground provide a nontrivial amount of decoupling action when the internal nodes are stationary (i.e., not switching). Ignoring this contribution leads to overestimation of decoupling capacitance cells to be added to the design. A simple and effective way to model the intrinsic decoupling capacitors is to characterize the basic circuit structures as capacitance (with series resistance) when these circuits are not switching and instantiate the characterized RC values after discounting C by a factor equal to the switching factor of that circuit instance [18]. Figure 23.5 illustrates this characterization idea. The circuit is put in random but valid quiet states. The V_{ac} variation (about 10% of V_{dd}) around the V_{dd} bias mimics the supply noise. The capacitance is measured and averaged over multiple quiet states and over the V_{ac} frequency range of about 0.1X–10X of the clock frequency.

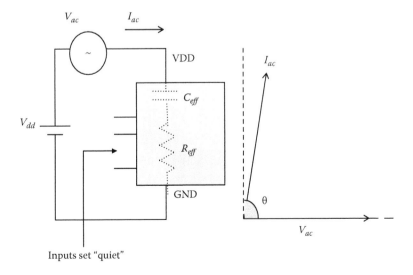

FIGURE 23.5 Characterization of intrinsic decoupling capacitance of circuits.

23.3 POWER NETWORK ANALYSIS

Power network verification is a very challenging task, primarily due to the very large size of the electrical circuits that need to be simulated. All the active devices on a chip, several millions to billions in count, are connected to only a handful of power nets. Since the parasitic elements and devices on a power net form a tightly connected system, meaning the voltage response in every part of the network is affected by the excitation current in some part, the entire net has to be simulated as a whole for best accuracy. However, Spice-like simulation of a circuit with such a large number of active and passive elements is impractical. So this giant task is split into two simpler decoupled tasks as defined here:

1. Transistor-, gate-, or RTL-level power estimation simulations for the circuit blocks, assuming an ideal power supply
2. Simulation of the power nets with the passive parasitic elements and currents obtained from the first task attached to the PDN at the appropriate nodes

Although currents drawn by active circuits and supply voltages at the locations of these circuits are dependent on each other, this circular dependency is ignored by treating the currents as independent of the voltages. As a second order effect, the error due to this is small (typically about 0.5% of V_{dd}) and conservative. This simplification is necessary to make PDN analysis practical, but this still requires a very large passive network with 10^6–10^9 circuit nodes to be solved.

23.3.1 STATIC AND DYNAMIC ANALYSES

Power grid simulation involves solving a system of differential equations using circuit simulation approaches, such as the *nodal or modified nodal analysis* [19]:

$$(23.1) \qquad Gx(t) + Cx'(t) = b(t)$$

where
  G is the conductance matrix
  C is the matrix resulting from capacitive and inductive elements
  $b(t)$ is the vector of ideal voltage sources and time-varying current sources
  $x(t)$ is the vector of node voltages, inductor currents, and currents drawn from voltage sources
  $x'(t)$ is the time derivative of $x(t)$

This differential system is very efficiently solved in the time domain by reducing it to a linear algebraic system:

$$(23.2) \qquad \left(G + \frac{C}{h}\right)x(t) = b(t) + \frac{C}{h}x(t-h)$$

using the backward Euler technique with a small fixed time step, h. As the new conductance matrix on the left-hand side, $(G + C/h)$, does not change during simulation, it can be preprocessed or factored for a one-time cost and efficiently reused to solve the system at successive time points in a transient simulation.

In static analysis, only resistances in the power network are considered and constant (DC) currents are applied as loads to drive the analysis, so that the results show the IR component of the noise. Equation 23.2 in this case reduces to

$$(23.3) \qquad G \cdot V = I$$

where

　　G is the conductance matrix
　　V is the vector of node voltages (unknowns)
　　I is the vector of load currents

Power supply sources to the network are modeled as Norton current sources that contribute elements to the diagonal of the G matrix and to the vector I on the right hand side.

Solution of sparse linear systems is a well-studied area and numerous *direct, iterative, and hybrid* techniques are available to solve Equation 23.2 [20,21]. However, the extremely large size of PDN systems has demanded a continued research into solution methods. Direct methods decompose the left-hand side matrix for a one-time cost as a product of invertible upper and lower triangular matrices, known as the L and U factors, and then solve the system at every time step of a transient simulation in an inexpensive backward and forward substitution procedure using these factors. When Equation 23.2 is formulated to have only node voltages as variables, the conductance matrix can be shown to be positive definite and can be factored more efficiently into LL^T. If sufficient memory is available, factorization of the matrix can be done for multiple time steps, for instance, h, 2* h, 4* h, and 8* h, so that the transient simulation can be carried out more efficiently by dynamically choosing the time step, depending on input activity, output response, and error behavior.

The single biggest problem with direct methods is the need for large memory to store the factors. Memory management relies on very high sparsity of the factored matrix and the resultant factors so that only their nonzero entries need be stored. However, as the fills (i.e., nonzeroes in the factors) grow superlinearly if not in a quadratic fashion, the memory requirement of factoring methods is very sensitive to the sparseness of the matrix. The amount of fills also depends on the ordering of the matrix. Fortunately, the power network matrix is very sparse since each node in the network is connected only to a very few nodes. The high sparseness, combined with matrix reordering heuristics, achieves a very slow growth of fills during the factoring process and thus makes it a viable approach even for very large size systems. Where a matrix is not very sparse or the memory resource in the computer is limited, iterative methods become more advantageous than direct methods.

Iterative methods start with a guess value for the solution and progressively refine it by reducing a norm of the error vector with computations involving chiefly matrix-vector multiplication. Thus, their memory demand does not grow significantly. However, this comes at the price of longer solution times of iterations whose convergence is slowed by the ill-conditioning of the matrix. Improving the condition number of the matrix by preconditioning [20–22] makes the iterative methods competitive with direct methods. The preconditioned conjugate gradient (PCG) method is widely used in the industry for PDN simulation. Successful use of other iterative methods, such as the multigrid [23,24] and random walk [25] methods, has also been reported. Since faster convergence is critical in achieving practical and competitive run times for PDN simulation, much attention has been given to designing very effective preconditioners [26]. Preconditioners based on incomplete Cholesky factors, fast transforms [27], multigrid reduction [24], and random walk [25,28] have been used quite successfully in conjunction with the PCG method.

23.3.2 HIERARCHICAL ANALYSIS

As a strongly coupled system, the power network needs to be simulated as one system for correct results. The work reported in [29] demonstrates strong locality of currents in regions bounded by the C4 bump power supply connections to the die. This work proposed a method to divide the network into partitions and simulate each partition independently to get an approximate solution that is very close to the exact solution when the locality is strong. The authors of [30] formulated a hierarchical analysis approach that gives the same (exact) solution as the full flat simulation, with partitions formed arbitrarily or defined naturally by the design hierarchy. Considering a two-level design hierarchy for simplicity of description, the network in each hierarchical block is modeled independently as a local partition. The global partition includes the interface nodes between the global and the hierarchical blocks. Mathematical abstracts of the partitions, which are the current–voltage relations at the interface of the partitions (referred to as the *macromodels*), are generated. A global simulation using only the macromodels is done to determine the interface node voltages, and then individual partitions are solved using the interface voltages. This idea is extendable to hierarchies deeper than two levels.

Figure 23.6 illustrates the idea of macromodeling. Assuming the network is partitioned into m partitions and all the interface nodes are in partition 1 (called the global partition), then the linear system in Equation 23.2 or 23.3 can be written in the following form:

(23.4)
$$
\begin{bmatrix}
g_{11} & g_{12} & g_{13} & \cdots & g_{1m} \\
g_{21} & g_{22} & 0 & \cdots & 0 \\
g_{31} & 0 & g_{33} & & \vdots \\
\vdots & \vdots & & \ddots & \vdots \\
g_{m1} & 0 & \cdots & \cdots & g_{mm}
\end{bmatrix}
\begin{bmatrix}
v_1 \\ v_2 \\ v_3 \\ \vdots \\ v_m
\end{bmatrix}
=
\begin{bmatrix}
i_1 \\ i_2 \\ i_3 \\ \vdots \\ i_m
\end{bmatrix}
$$

By eliminating the nodes of partitions 2 through m, one can reduce this equation to a system containing only the global partition nodes:

(23.5)
$$ Av_1 = i_1 - S $$

Matrix A is the Schur complement of initial matrix with respect to g_{11}, and S is a vector of currents at the interface nodes whose effect is the same as all the internal currents. A and S are obtained from

(23.6)
$$ A = g_{11} - \sum_{k=2}^{m} g_{1k}\, g_{kk}^{-1}\, g_{k1} $$

(23.7)
$$ S = \sum_{k=1}^{m} g_{1k}\, g_{kk}^{-1} i_k $$

The hierarchical approach presents both memory and performance advantages due to parallelism. Factoring of the g_{kk} matrices and computation of contributions to A and S from the partitions can be done in parallel. After solving the global system (Equation 23.5), the voltages of nodes in the partitions can also be back-solved in parallel. The reduced matrix A is very dense, so in situations when it is also very large, an iterative solver is more useful to solve the global partition (Equation 23.5), while other partitions can be solved with a direct solver, thus utilizing a hybrid solution approach overall. Needless to say, a good partitioning strategy minimizes the number of interface nodes between partitions. Hypergraph partitioning utilities such as *hMetis* [31] can be used to help achieve that goal.

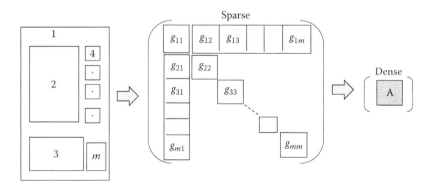

FIGURE 23.6   Hierarchical analysis and matrix reduction.

23.3.3 VECTORLESS ANALYSIS

The number of operational modes for which a chip must be verified is increasing due to wider adoption of power gating, clock gating, dynamic voltage, frequency scaling, and other design features. The complexity of power grid analysis is exacerbated by the need to carry out long simulations with many current traces from functional vectors such that a chip's various operating modes and input vector patterns are adequately covered [32]. However, simulation is never exhaustive, and even limited simulation demands lots of resources. Moreover, vector-based simulation is not appropriate for an incomplete and evolving design, where only partial information about currents is available.

Techniques to determine maximum currents in a pattern-independent manner [33] or to generate simulation vectors for maximum instantaneous currents [34] have been proposed. One difficulty with vectors that maximize currents is that they do not necessarily maximize IR drop or transient noise in the PDN. Artificial instruction sequences (known as *power viruses* or *smoke vectors*) composed by a chip's architects to target various high power modes and create severe power transients may be the best inputs a power grid designer has for verification. Yet, these may be excessively pessimistic leading to overdesign and may still miss some worst-case power grid conditions. In view of these difficulties, providing guarantees as to the performance of a power grid using vector-based simulation is nearly impossible.

The authors of [35] advanced the idea of vectorless PDN verification and proposed an approach for determining the maximum static voltage drop at nodes of interest in a power network without requiring exact current distributions, using only a set of constraints on the currents. This vectorless approach is extended in [36,37] to determine worst-case transient power noise.

For the static case, when the local and global currents are constrained, the problem is solved [35] as an optimization problem using linear programming. In practice, it is common to have current constraints (viz., lower and upper bounds) on each circuit instance as a whole, whether the instance is a leaf instance or a nonleaf instance. This is illustrated in Figure 23.7. In such circumstances, the worst-case analysis can be very efficiently carried out using sensitivities of node voltages to instance currents (known as *adjoint sensitivities*) and allocating the maximum total current to individual instances in a systematic manner as per the sensitivities [5]:

$$k_1 \leq I_A \leq k_2$$

$$k_3 \leq I_{B1}, \quad I_{B2} \leq k_4$$

$$k_5 \leq I_{B3} \leq k_6$$

$$k_7 \leq I_{C1}, \quad I_{C2} \leq k_8$$

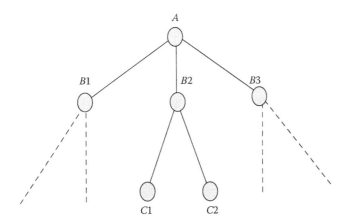

FIGURE 23.7   An example of hierarchical current constraints.

For the system $GV = I$, the (adjoint) sensitivity of voltage at node k, V_k, to the current drawn from node j, I_j, can be computed by solving the system with a unit current injected at node j and no other currents and measuring the voltage V_k. It is easy to see that the sensitivities of V_k to all the currents are really the entries in the kth column of the G^{-1} matrix and all dV_k/dI_j are therefore obtained by solving the system once as

$$\left[\frac{dV_k}{dI_1} \quad \frac{dV_k}{dI_2} \quad \cdots \quad \cdots \quad \frac{dV_k}{dI_n}\right]^T = G^{-1}e_k^T$$

In this, e_k is a vector of zeroes except the kth entry, which is 1. The voltage drop at node k is maximized by maximizing $\sum_j \frac{dV_k}{dI_j} I_j$ through allocation of node currents based on sensitivities, subject to the current constraints. It is convenient to derive and use voltage-to-current sensitivities between collections of nodes. Using the additive property of the node sensitivity, the sensitivity of a node voltage to the total current of a block, b_j, can be expressed as

$$\frac{dV_k}{dI_{b_j}} = \sum_i k_{ji} \frac{dV_k}{dI_i}$$

where k_{ji} is the fraction of current of block j contributed to current at node i.

Likewise, if X is a collection of nodes (such as a collection of representative nodes from one or more regions), the voltage sensitivity of that collection to a node current is the sum of sensitivities of the nodes in the collection with appropriate weighting:

$$\frac{dV_X}{dI_j} = \sum_i \propto_i \frac{dV_i}{dI_j}$$

Using these two collective sensitivities, worst-case analysis can be carried out targeting regions or nodes of interest and such analysis would require only as many number of solutions of the linear system as the number of nodes of interest.

Figure 23.8 illustrates the application of the worst-case analysis. It shows IR drop plots from nine worst-case analyses of a PDN. Each analysis targeted one of the nine regions (marked by the red square boxes in their respective graphs) that together cover the entire PDN. Approximately 50 nodes in each region, specified by the designer, were taken to represent each region. The voltage drops at these nodes were maximized with equal weights subject to the specified maximum total current for the chip and the hierarchical current constraints as shown in Figure 23.7. In the graphs, the voltage drops at the loads are plotted on a color scale spanning from the dark green color (signifying the lowest drop) to the dark red color (signifying the highest drop).

23.3.4 INCREMENTAL ANALYSIS

In the course of a chip design, many corrections made to a power network may involve only localized modifications while the bulk of the network is unchanged. Fast incremental analysis techniques that can reuse much of the computations done in a previous full scale analysis will help to reduce the design-analysis-design iteration time. They are useful for *what-if* studies also.

When the changes affect very few circuit components, a *large scale sensitivity* method [38] can be used to quickly compute the new solution. For N changes, this requires N forward–backward substitutions using the original matrix factors, as well as solution of an $N \times N$ system. However, N can be quite large (e.g., few millions) even for a modest change to the network. So, highly efficient incremental methods are needed. The hierarchical analysis discussed in Section 23.3.3 inherently has some incremental capability. When modifications are made in one of the partitions, the new analysis can reuse the macromodels of the unmodified partitions. The *random walk* method for power network analysis [25] is another method suited for incremental local updates. A hybrid

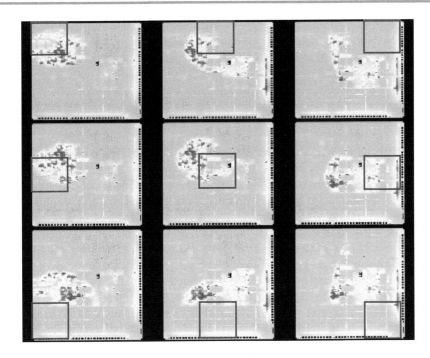

FIGURE 23.8   Application of worst-case analysis.

(direct and iterative) approach proposed in [39], which is based on fictitious domains, provides a very competitive method that can handle very large PDN modifications in an incremental manner. Figure 23.9 illustrates the idea behind the method.

In the figure, R_1 is the unmodified network, R_2 is the region in the original network to be modified, and R_2^* is the modified R_2. The method solves three simpler problems $P1$, $P2$, and $P3$ iteratively. $P1$ solves the original network, but injecting at the interface between R_1 and R_2 a fictitious current ΔI. The new voltages V at the interface are used to compute the currents entering R_2^* (in $P2$) and R_2 (in $P3$). The difference in these currents, ΔI, is used in the next iteration of $P1$. When V and ΔI do not change appreciably, the iterations have converged and the voltages in $P1$ give the solution to the modified network. As only the currents change in $P1$, solving $P1$

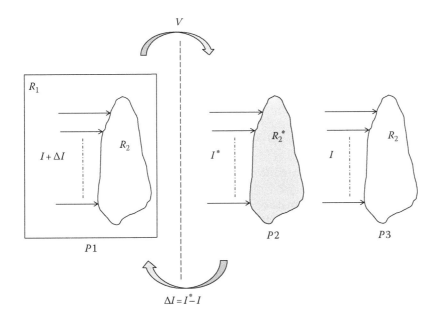

FIGURE 23.9   Subproblems and iterative solution based on fictitious domain decomposition.

is accomplished through a forward and backward substitution. *P2* and *P3* solve only a fraction of the original network and their solutions can be accelerated through a one-time factorization. The convergence is improved by enlarging the modified region slightly to include a thin layer of the unmodified region.

23.4 POWER NETWORK DESIGN

Invariably, all high-performance microprocessors (and most other chip designs) use a mesh grid topology using multiple layers of metal interconnects for power distribution on the die. Some high-performance designs have also used dense mesh power planes on some layers [40]. The mesh topology is characterized by a very uniform footprint of power and ground wires that run horizontally and vertically in alternate layers, with vias at their intersections. A mesh provides advantages of very low resistance, a high degree of redundancy for failure, a highly uniform current distribution, and ease of design and construction. A mesh can be specified by the layers it spans and the width and spacing of wires in each layer. A mesh topology is used for both the global (chip-level) distribution, as well as for local distributions within large circuit blocks, though with different footprint parameters depending on the local loading.

Since many different power mesh configurations meet the specified IR drop criteria, it is valuable to recognize the inherent strengths of certain configurations over others. At nanometer-level scaling, the resistivity of metal wires is strongly influenced by the wire cross section. Very slim wires suffer 30%–50% higher resistivity, due to scattering, dishing, and cladding, than wide wires [41]. However, this consideration needs to be weighed appropriately with choices for wiring pitches, especially for the lower-level routing layers, because densely pitched power/ground grid wiring provides shielding or semishielding to a larger number of signal routes than a sparsely pitched configuration.

The challenge in PDN design, besides its complexity, is that there are many unknowns until the very end of the design cycle. Nevertheless, decisions about design parameters (viz., the number of layers, widths, and pitches of wires), collateral inputs for package design (viz., the number and placement of bumps or bond wires), and on-chip decoupling capacitance requirements have to be made at very early stages, when a large part of the chip design has not even begun. Power grid problems revealed at very late stages are usually very difficult and expensive to fix. A methodology was presented in [42], which helps to design an initial power grid and refine it progressively at various design stages. Needless to say, such a methodology will need a very flexible analysis framework, providing tools and techniques for performing sign-off quality analysis, as well as early and mid design stage analysis with approximate models of R, L, C and currents constructed from incomplete design data.

Considering the risk that a bad power distribution poses to the success of a design project, early power grid design decisions are made, understandably, in a very conservative manner in the presence of many unknowns. When a previous generation of the design is available, many parameters of the new design are derived by scaling the old ones for the change in process technology, operating voltage, frequency, and the (scaled) size of circuits in functional blocks. Layerwise, wiring resources to be allocated for the PDN and the grid foot prints may be derived in this manner. Alternatively, a uniform trial power mesh, down to the lowest metal layer, may be constructed, iteratively refined through analysis, and used as the basis. For very early estimates, one could make quick calculations using closed form expressions [43] assuming a uniform grid topology and a uniform loading of the network. However, if nonuniformities need to be considered, a static analysis can provide accurate estimates.

For the most part of the design cycle, static (DC) analysis in one of the modes (hierarchical, vectorless, or incremental) serves as the workhorse for constructing and debugging PDN issues based solely on IR drop targets, which are often padded to provide for transient noise, uncertainties in current estimates, etc. The effort needed to run transient analysis (viz., extracting and preparing the R, L, and C models for the on-chip and off-chip components, as well as the current excitation traces) is very high and the run times for analysis are longer. In view of this, transient analysis is often carried out with simplified models for the die network or the package, to assist with package evaluation and decap placement.

During early design phases, simpler 1-D models for resistances and area-based DC currents suffice for static analysis. Likewise, per-pin or per-bump inductance estimates and area-based decoupling estimates suffice for transient analysis. These models are to be refined and extracted more accurately as the design progresses and reliable collateral data from completed parts of the design become available. For valid results from analysis carried out on incomplete designs, a designer will have to provide approximate models for the incomplete parts, and so it is very important that the tool and flow is flexible enough to allow such usage models.

As noted in Section 23.2, many of the measures to guarantee reliable operation of the PDN during power transients go beyond PDN design and span circuit, microarchitectural, power management, and even software design. Circuit designs for voltage droop monitoring and adaptive clocking provide droop-compensated clock periods to enable operating the chip at lower timing margin and lower nominal V_{dd} [40,44]. Load lining is another technique where the VR module raises the supply voltage during low processor activity to provide voltage headroom for potential supply droops. The transient noise issues are best addressed at the source by exercising control over operational events causing abrupt changes in the magnitude of current drawn. Clock gating and fine-grained DVFS induce high frequency noise that, to a certain extent, can be effectively handled by on-chip decoupling capacitors. Excessive noise can be avoided by fine-tuning gating actions using module decay counters or avoiding simultaneous gating using queue-based control mechanisms [45]. Voltage sensors can be used to detect noise emergencies and initiate appropriate control actions such as power throttling. Analysis of software routines has been proposed to identify and correct code sequences and loops causing voltage emergencies [46]. In essence, power saving and power management actions need to be cognizant of the di/dt that will result from these actions [47].

Power gating is another potential source for severe di/dt issues if the design is not properly done and verified. Fine-grained power gating [48] (shown in Figure 23.10) partitions a large circuit block into multiple clusters of logic gates and turns on and off these clusters individually. Due to the finer control, this scheme has higher leakage power saving potential and also lends to better control of di/dt.

Coarse-grained power gating (shown in Figure 23.11), on the other hand, turns on and off an entire circuit block or a core, saves the state retention elements, and so is vulnerable to

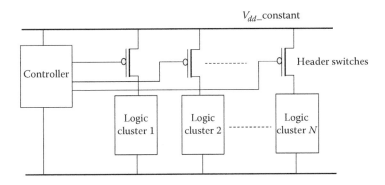

FIGURE 23.10   Fine-grained power gating.

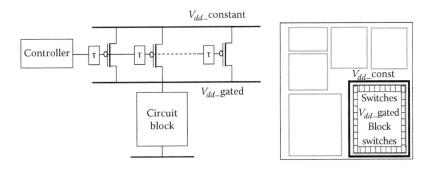

FIGURE 23.11   Coarse-grained power gating.

severe *di/dt* issues. Modular switches (typically PMOS devices) are arranged in arrays and controlled by delay chains so that the block can be powered up or down gradually. The number of switch modules needed is determined by the IR drop caused by the switches when they are ON and the speed at which the module needs to be powered on. The delay elements in the chain and the driver strength of the controller determine the operating point on the trade-off between power-on speed and in-rush current (or transient noise). Simulation for power-up or power-down analysis needs to include nonlinear active devices of the sensing and controller circuits, and a simplified RC model of the powered network is helpful in reducing the simulation run time [5].

23.5 CONCLUSIONS

Design and verification of the power supply and distribution systems are critical tasks in the design of integrated chip systems since power supply integrity issues have dire consequences. We presented various analysis techniques for verifying and optimizing the PDN at different design stages. A linear system solution, which forms the core of power grid simulation, was discussed, highlighting the complexity of the problem and presenting popular approaches for different types of analysis. It was emphasized that while static voltage drop issues can be addressed with adequate layout resources, dynamic noise issues arising from power transient events demand much engineering effort. All stages of chip design, from architecture to circuit and physical implementation, need to be cognizant of these issues in order to avoid them or to put in place adequate mechanisms to control them. With further reduction in the supply voltage due to process scaling and increasing architectural complexity of processors, PDN noise issues will continue to demand significant design and verification effort.

ACKNOWLEDGMENTS

Thanks to Oracle's Tom Dillinger, Bob Masleid, and Alexander Korobkov for reviewing this chapter and making several suggestions to improve the contents and the presentation.

REFERENCES

1. L. Miller, Controlled collapse reflow chip joining, *IBM J. Res. Dev.*, 13(3), 239–250, 1969.
2. H. Chen and D. Ling, Power supply noise analysis methodology for deep-submicron VLSI chip design, in *34th ACM/IEEE Design Automation Conference*, Anaheim, CA, 1997.
3. H. Su, S. Sapatnekar, and S. Nassif, Optimal decoupling capacitor sizing and placement for standard-cell layout designs, *IEEE Trans. Comput. Aided Des. Integr. Circuits Syst.*, 22, 428–436, 2003.
4. S. Zhao, K. Roy, and C.-K. Koh, Decoupling capacitance allocation and its application to power-supply noise-aware floorplanning, *IEEE Trans. Comput. Aided Des. Integr. Circuits Syst.*, 21, 81–92, 2002.
5. E. Chiprout, R. Panda, F. Najm, and B. Krauter, Tutorial: Design and analysis of high-performance package and die power delivery networks, in *Design Automation Conference*, San Diego, CA, 2007.
6. P. Larsson, Resonance and damping in CMOS circuits with on-chip decoupling capacitance, *IEEE Trans. Circuits Syst. I, Fundam. Theory Appl.*, 45, 849–858, 1998.
7. S. Vemuru, Effects of simultaneous switching noise on the tapered buffer design, *IEEE Trans. Very Large Scale Integr. VLSI Syst.*, 3, 290–300, 1997.
8. L. Chen, M. Marek-Sadowska, and F. Brewer, Buffer delay change in the presence of power and ground noise, *IEEE Trans. Very Large Scale Integr. VLSI Syst.*, 11, 461–473, 2003.
9. P. Heydari and M. Pedram, Analysis of jitter due to power-supply noise in phase-locked loops, in *Proceedings of IEEE Custom Integrated Circuits Conference*, Orlando, FL, 2000.
10. R. Saleh, S. Hussain, S. Rochel, and D. Overhauser, Clock skew verification in the presence of IR-drop in the power distribution network, *IEEE Trans. Comput. Aided Des. Integr. Circuits Syst.*, 19, 635–644, 2000.
11. J. R. Black, Electromigration failure modes in aluminum metallization for semiconductor devices, *Proc. IEEE*, 57, 1587–1594, 1969.

12. S. Dobrasevic, EMC guidelines for MPC5500-based systems, Freescale Semiconductor Application Note, AN2706 Rev. 0.1, 01/2005.
13. F. Grover, *Inductance Calculations*, New York: Dover, 1954.
14. A. Ruehli, Inductance calculations in a complex integrated circuit environment, *IBM J. Res. Dev.*, 16, 470–481, 1972.
15. R. J. Milliken, J. Silva-Martínez, and E. Sánchez-Sinencio, Full on-chip CMOS low-dropout voltage regulator, *IEEE Trans. Circuits Syst. I, Reg. Papers*, 54(9), 1879–1890, 2007.
16. W. Kim, M. S. Gupta, G.-Y. Wei, and D. Brooks, Enabling on-chip switching regulators for multi-core processors using current staggering, in *Workshop on Architectural Support for Gigascale Integration*, San Diego, CA, 2007.
17. S. Lai, B. Yan, and P. Li, Stability assurance and design optimization of large power delivery networks with multiple on-chip voltage regulators, in *International Conference on Computer-Aided Design*, San Jose, CA, 2012.
18. R. Panda, D. Blaauw, R. Chaudhry, V. Zolotov, B. Young, and R. Ramaraju, Model and analysis for combined package and on-chip power grid simulation, in *Proceedings of the International Symposium on Low-Power Electronics and Design*, Rapallo, Italy, 2000.
19. C. Ho, A. Ruehli, and P. Brennan, The modified nodal approach to network analysis, *IEEE Trans. Comput. Aided Des. Integr. Circuits Syst.*, CAS-22, 504–509, 1975.
20. G. Golub and C. V. Loan, *Matrix Computations*, Baltimore, MD: The John Hopkins University Press, 1989.
21. L. Pillage, R. Rohrer, and C. Visweswariah, *Electronic Circuit and System Simulation Methods*, New York: McGraw-Hill, 1994.
22. T. Chen and C. Chen, Efficient large-scale power grid analysis based on preconditioned Krylov-subspace iterative methods, in *Proceedings of the Design Automation Conference*, Las Vegas, NV, 2001.
23. W. Briggs, *A Multigrid Tutorial*, Philadelphia, PA: SIAM, 1987.
24. J. Kozhaya, S. Nassif, and F. Najm, A multigrid-like technique for power grid analysis, *IEEE Trans. Comput. Aided Des. Integr. Circuits Syst.*, 21, 1148–1160, 2002.
25. H. Qian, S. Nassif, and S. Sapatnekar, Power grid analysis using random walks, *IEEE Trans. Comput. Aided Des. Integr. Circuits Syst.*, 24, 1204–1224, 2005.
26. D. Chen and S. Toledo, Implementation and evaluation of Vaidya's preconditioners, *Preconditioning*, 2001.
27. K. Daloukas, N. Evmorfopoulos, G. Drasidis, M. Tsiampas, P. Tsompanopoulou, and G. I. Stamoulis, Fast transform-based preconditioners for large-scale power grid analysis on massively parallel architectures, in *International Conference on Computer-Aided Design*, San Jose, CA, 2012.
28. J. Wang, Deterministic RW preconditioning for PG analysis, in *International Conference on Computer-Aided Design*, San Jose, CA, 2012.
29. E. Chiprout, Fast flip-chip power grid analysis via locality and grid shells, in *International Conference on Computer-Aided Design*, San Jose, CA, 2004.
30. M. Zhao, R. Panda, S. Sapatnekar, and D. Blaauw, Hierarchical analysis of power distribution networks, *IEEE Trans. Comput. Aided Des. Integr. Circuits Syst.*, 21, 159–168, 2002.
31. G. Karypis, R. Aggarwal, V. Kumar, and S. Shekhar, Multilevel hypergraph partitioning: Applications in VLSI domain, *IEEE Trans. VLSI Syst.*, 7(1), 69–79, 1999.
32. H. Qian, S. Nassif, and S. Sapatnekar, Early-stage power grid analysis for uncertain working modes, *IEEE Trans. Comput. Aided Des. Integr. Circuits Syst.*, 24, 676–682, 2005.
33. H. Kriplani, F. Najm, and I. Hajj, Pattern independent maximum current estimation in power and ground buses of CMOS VLSI circuits: Algorithms, signal correlations and their resolution, *IEEE Trans. Comput. Aided Des. Integr. Circuits Syst.*, 14, 1998.
34. A. Krstic and K. Cheng, Vector generation for maximum instantaneous current through supply lines for CMOS circuits, in *Proceedings of the 34th ACM/IEEE Design Automation Conference*, Anaheim, CA, 1997.
35. D. Kuoroussis and F. Najm, A static pattern-independent technique for power grid voltage verification, in *Proceedings of the ACM/IEEE Design Automation Conference*, Anaheim, CA, 2003.
36. I. A. Ferzli, F. N. Najm, and L. Kruse, A geometric approach for early power grid verification using current constraints, in *International Conference on Computer-Aided Design*, San Jose, CA, 2007.
37. I. A. Ferzli, E. Chiprout, and F. N. Najm, Verification and codesign of the package and die power delivery system using wavelets, *IEEE Trans. Comput. Aided Des. Integr. Circuits Syst.*, 29(1), 92–102, 2010.
38. J. Vlach and K. Singhal, *Computer Methods for Circuit Analysis and Design*, 2nd edn., Springer Science & Business Media, 1993.
39. Y. Fu, R. Panda, B. Reschke, S. Sundareswaran, and M. Zhao, A novel technique for incremental analysis of on-chip power distribution networks, in *International Conference on Computer-Aided Design*, San Jose, CA, 2007.

40. J. Hart, S. Butler, H. Cho, and Y. Ge, 3.6 GHz 16-core SPARC SoC processor in 28 nm, in *IEEE International Solid-State Circuits Conference*, San Francisco, CA, 2013.

41. S. X. Shi and D. Z. Pan, Wire sizing with scattering effect for nanoscale interconnection, in *ASP-DAC*, 2006.

42. A. Dharchoudhury, R. Panda, D. Blaauw, R. Vaidyanathan, B. Tutuianu, and D. Bearden, Design and analysis of power distribution networks in PowerPC microprocessors, in *Proceedings of the 35th ACM/IEEE Design Automation Conference*, San Francisco, CA, 1998.

43. K. Shakeri and J. D. Meindl, Compact physical IR-drop models for chip/packager co-design of gigas-cale integration, *IEEE Trans. Electron Devices*, 52(6), 1087–1096, 2005.

44. A. Grenat, S. Pant, R. Rachala, and S. Nafziger, Adaptive clocking system for improved power efficiency in a 28 nm x86−64 microprocessor, in *IEEE International Solid-State Circuits Conference*, San Francisco, CA, 2014.

45. F. Mohamood, M. Healy, S. K. Lim, and H.-H. Lee, A floorplan-aware dynamic inductive noise controller for reliable processor design, in *39th Annual International Symposium on Microarchitecture*, Orlando, FL, 2006.

46. M. Gupta, K. Rangan, M. Smith, and G.-Y. Wei, Towards a software approach to mitigate voltage emergencies, in *International Symposium on Low Power Electronics and Design*, Portland, OR, 2007.

47. R. Joseph, D. Brooks, and M. Martonosi, Control techniques to eliminate voltage emergencies in high performance processors, in *HPCA*, 2003.

48. J. Kao, S. Narendra, and A. Chandrakasan, MTCMOS hierarchical sizing based on mutual exclusive discharge patterns, in *Design Automation Conference*, San Francisco, CA, 1998.

Noise in Digital ICs

Igor Keller and Vinod Kariat

CONTENTS

24.1 INTRODUCTION

The growing size, complexity, and more stringent time-to-market requirements of modern designs necessitate highly automated RTL-to-GDS flows [1]. Such flows describe designs at multiple levels of abstraction, starting from high-level languages during synthesis and ending with electrical circuits and polygons at the timing and electrical sign-off steps. Meeting a set of targets at the end of the design cycle typically requires multiple design-flow iterations, often repeating several steps. The adjacent steps of a flow must use compatible models and algorithms; otherwise, physical characteristics of the design may abruptly change and no progress will be made. Design flows are carefully assembled to produce quality results with fewer iterations. For instance, synthesis incorporates physical modeling in order to better capture wiring resources and predict routing congestion. Likewise, physical optimization often benefits from near-sign-off accuracy of delay calculations in order to pass a design to final sign-off timing analysis with as little disruption as possible.

An efficient design flow benefits from a shared data representation across steps, thus avoiding slow file-based exchange. Greater integration of design-flow steps is also motivated by the wealth of complex models describing more sophisticated physical phenomena arising in advanced nodes. Analysis engines utilizing these models require costly multiyear development efforts and must be reused in different steps of the flow. For instance, modern delay calculation in the presence of noise and process variations is so complex that either the entire engine or its main components are reused in various steps of the flow. The modeling of noise effects, which was earlier performed by mostly stand-alone analysis tools, is now becoming an integral part of nearly all steps of the flow, starting from physical synthesis and preroute optimization.

The trend toward greater integration of various steps benefits single-vendor flows, where a single EDA vendor develops and supports models, engines, infrastructure, and user interface. While reducing flexibility of design houses in their choices of tools and vendors, single-vendor flows are much more efficient, have better convergence properties, and provide better performance, power, and area than custom flows integrating various point tools from different vendors. To that end, as of this chapter revision (2015), all foundries collaborate with major EDA vendors to build the so-called reference flows (e.g., SMIC, TSMC, and Global Foundries).

As we will see later in this chapter, noise affects various metrics of design, such as functional correctness, timing, power, area, and reliability. The first two characteristics have a lot in common and are normally the main objectives of optimization. Consequently, in this chapter, we mostly focus on functional noise and noise-on-delay effects and leave other aspects out of the scope.

In the early days of *static timing analysis* (STA) (in the early 1990s), delay calculation was simple and based on a 2-D table lookup, using transition time and lumped load as parameters, with wire delays and cross talk effects mostly ignored.

Further scaling to the 250 nm process node introduced several physical effects, which could not be ignored anymore: wire delay and unintended interactions between signals (*noise* or cross talk). While initial methods were simple, inclusion of wire delay and cross talk in STA signified a drastic change in the methodology of delay calculation [2–4]. Around the same time (in the late 1990s and early 2000s) functional noise became a consideration for digital design, paving the way to the formulation of the static noise analysis (as described below in Section 24.5) [5] and the emergence of the first commercial noise analyzers. From the 130 nm process node and below, performance and correctness of a design could not be assured without considering noise effects. Yet they were considered in a simplistic way as perturbations of noiseless timing: incremental signal integrity (SI) delays were computed by a separate SI analyzer and then added on top of the so-called base delays—a delay in the absence of noise.

As design technology continues its migration toward nanoscale nodes (28 nm and below), more physical phenomena become significant in timing and noise analyses: increased process variability, high sensitivity of results to waveform details, nonlinear pin capacitance, wire-dominated delay, and so on. These effects triggered multiple innovations in delay calculation and noise analysis, the most noticeable of which are

- ■ Merger of base delay calculation and SI delay into a single analysis
- ■ Migration from mostly table lookup–based analysis to mostly simulation-based analysis
- ■ Use of detailed waveforms in analysis as opposed to simple abstraction of waveforms
- ■ Tighter integration variability support though either efficient multimode multicorner analysis or statistical analysis

A significant increase in analysis complexity due to the new effects on the one hand and tighter margins for allowed inaccuracy on the other led to a much greater simulation cost of modern delay and SI analyzers. This, in turn, requires better algorithms, capable of utilizing multicore and distributed computing infrastructure.

Noise can have many drastic consequences for digital designs: (1) it can make the design slower, (2) it can make the design fail completely, and (3) it can create yield problems. For a chip designer, the cost of such a failure is very high, and it includes mask costs, engineering costs, and opportunity cost due to delayed product introduction. For a high-volume chip, a 3-month product launch delay can lead to a $500 million loss in product revenues [6].

The rest of this chapter is organized as follows. We first discuss the technology and design factors that have made noise a significant problem at nanometer technology nodes. We then discuss how noise affects the operation of a digital circuit. We review noise analysis techniques and algorithms in some detail, and we finally discuss how noise can be handled during the design process.

24.2 WHAT IS DIGITAL NOISE?

In analog circuits, noise arises from fundamental physical sources, such as thermal noise (caused by random motion of carriers), flicker noise (related to material defects), and shot noise (caused by gate leakage current). These noise sources on the one hand present a lower limit to the smallest signal that can be amplified and on the other hand define an upper limit to useful amplification. In digital circuits, in contrast, noise arises from the operation of the circuit itself, primarily the switching of other signals or propagation of noise from other parts of the circuit. This so-called digital noise affects the timing and functionality of a digital circuit and is the main focus of this chapter.

Noise can generate different effects in a digital design. They are usually classified into three main *noise effects*:

1. It can cause a signal to exhibit the wrong value, leading to a functional failure.
2. It can cause a signal to arrive too late, thus causing a so-called setup-timing failure. With this type of failure the chip can still work, albeit at a lower frequency than intended.
3. It can cause a signal to arrive too early, thus causing a so-called hold timing failure. This type of failure is most often fatal, as the chip cannot work even at a lower frequency.

Before we study these effects in detail, let us introduce some terminology, which is widely used in the noise analysis domain.[*] A signal net that causes noise is usually called an "aggressor net," and the signal net that is affected by an aggressor is usually called the "victim net."[†] Note that the same net can be a victim in one situation and an aggressor in another.

There are several ways in which an aggressor can cause noise on a victim; we call them "noise types." The most well-known noise type is "coupling noise," often also referred to as "capacitive cross talk": this is when an aggressor is coupled to a victim through capacitance, often called "coupling capacitance." If an aggressor is connected to a victim through one or more transistors, it can cause other types of noise: *propagated noise, charge-sharing noise, Miller noise,* and other types of noise.[‡] As we will see in later sections, these types of noise can contribute to all three main noise effects, often in a nontrivial way. In the following, we explain the different noise types in detail.

24.2.1 COUPLING NOISE

Cross talk, or coupling noise, is the most common type of noise in modern digital systems. It happens when an aggressor, coupled to a victim via a coupling capacitance, undergoes a rising or a falling transition. When coupling noise happens on a quiet victim, it usually has the shape of a spike, or a glitch, which in this chapter we will refer to as "glitch noise." As we will show later, if glitch noise is sufficiently high, it can contribute to a functional failure, that is, to noise effect (I) defined earlier. If the victim signal is also transitioning, coupling noise may slow down or speed up the transition, which may lead to timing failures defined earlier as noise effects (II) and (III), respectively.

To demonstrate how coupling noise leads to glitch noise, consider the simple circuit shown in Figure 24.1a. The circuit consists of a quiet victim *Vic* coupled to an aggressor *Agg*. In this example, the victim's driver is an inverter whose input is tied to V_{dd} and the aggressor is rising.

In order to obtain a quantitative insight into how different factors affect glitch magnitude, let us simplify the circuit in Figure 24.1a further:

- Ignore the resistance of the aggressor and victim nets and represent them only as lump capacitances rather than as distributed RC networks.
- Approximate the impedance of the driver holding the victim net at its steady state by a resistance *r*.
- Model the aggressor using an ideal voltage with the shape of a saturated-ramp waveform $u(t)$.

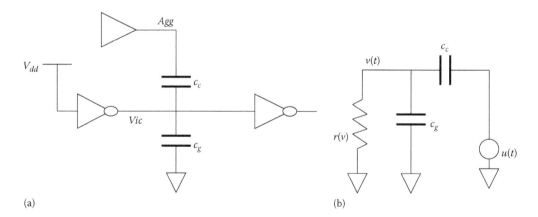

(a) (b)

FIGURE 24.1 Simple circuit for demonstrating glitch noise characterizing: (a) original and (b) simplified circuit.

---

[*] Various publications on noise analysis may use a different terminology.
[†] In this chapter we often call them simply as "aggressor" and "victim," respectively.
[‡] The term "aggressor" is often defined only for coupling noise; in this chapter we use a more general definition.

The simplified circuit, shown in Figure 24.1b, is simple enough to solve analytically for voltage response at the *Vic* node. Using the Kirchhoff current law, we obtain the following differential equation:

$$c_{tot}\frac{dv}{dt}+\frac{v}{r}=c_c\frac{du(t)}{dt} \qquad (24.1)$$

where
  $c_{tot} = c_g + c_c$ and $v(t)$ are the total capacitance and response on the victim, respectively
  $u(t)$ denotes the voltage source modeling the aggressor transition

Using the initial condition $v|_{t=0} = 0$, we find a solution in the following form:

$$v(t)=\frac{c_c}{c_{tot}}\int_0^t\frac{du(t')}{dt'}e^{-t'/\tau}dt' \qquad (24.2)$$

Here, $\tau = rc_{tot}$, often called the "victim time constant," describes how fast a glitch decays after an aggressor finishes its transition. The closed-form solution given by Equation 24.2 allows us to derive several useful characteristics of the response. First, note that for any shape of aggressor transition $u(t)$ that is constant before and after the transition, the response $v(t)$ has the shape of a glitch, that is, it has zero value before the aggressor's transition and it eventually attains zero value after the transition.

Now consider an important case where the characteristic time of the aggressor transition T is large compared to the victim time constant: $T \gg \tau$. In this case the solution can be written as

$$v(t)=c_cr\frac{du(t)}{dt}+o\left(\frac{\tau}{T}\right) \qquad (24.3)$$

As we can see from Equation 24.3, the magnitude of the response is directly proportional to the coupling capacitance c_c, the holding resistance r, and the time derivative of the aggressor transition $u(t)$, which is $\sim T^{-1}$. Since $c_c < c_{tot}$, the magnitude is asymptotically bounded by τ/T, meaning that the glitch is small when the aggressor transition is slow.

In another interesting case, $T \ll \tau$, the following is the solution (24.2) for times $t \ll \tau$:

$$v(t)=\frac{c_c}{c_{tot}}u(t)+o\left(\frac{T}{\tau}\right) \qquad (24.4)$$

This means that the initial part of the victim response has the shape of the aggressor transition, with the attenuation factor c_c/c_{tot}, and the magnitude of the glitch is bounded by $v_{dd}c_c/c_{tot}$.

The most important characteristic of noise glitch is its maximum, or peak, v_{peak}. Let us derive v_{peak} as a function of the circuit's parameters in the case where the aggressor transition has the shape of a rising saturated linear ramp:

$$u(t)=\begin{cases} 0, & t<0 \\ V_{dd}\dfrac{t}{T}, & 0\le t\le T \\ V_{dd}, & t\ge T \end{cases}$$

The response can be written in the following piecewise form:

$$v(t)=\begin{cases} 0, & t<0 \\ \dfrac{rc_cV_{dd}}{T}(1-e^{-t/\tau}), & 0\le t\le T \\ v_{peak}e^{-t/\tau}, & t\ge T \end{cases}$$

Here, v_{peak} is the maximum value of the glitch noise reached at $t = T$ and can be expressed as a function of the circuit's parameters as

$$v_{peak} = \left[\frac{c_c V_{dd}}{c_{tot}} \right] f\left(\frac{T}{\tau} \right) \tag{24.5}$$

Here, the first factor combines the electrical properties of the circuit, and the second factor is the following nonlinear function of ratio of two characteristic times of the circuit

$$f(x) = \frac{1 - e^{-x}}{x} \tag{24.6}$$

As can be seen from these equations, the glitch peak is directly proportional to the victim driver holding resistance r and V_{dd} and to the victim's ratio c_c/c_{tot}, whereas it is inversely proportional to the aggressor's transition time T. This gives us a good indication of where we will see cross talk problems. Nets with higher driver holding resistance (i.e., lower drive strength) and higher coupling capacitance are most susceptible to cross talk. If the aggressors switch faster (i.e., have smaller transition times), the cross talk will be worse. Equations 24.5 and 24.6 are often used in cross talk analysis programs as a simple and efficient filter to eliminate low-risk nets from further analysis.

Let us now look at the effects that coupling noise can have on delay. Figure 24.2a shows a similar circuit as before, but this time the victim also undergoes a transition. First, consider the case where the victim and aggressor transitions, $v(t)$ and $u(t)$, respectively, have the same shape and directions: $v(t) = u(t)$. In this case, voltages at both terminals of the coupling capacitors c_c in Figure 24.2a change in unison, and the current injected into the victim from the aggressor is zero. Now, consider the case where victim and aggressor transitions have opposite directions: $v(t) = V_{dd} - u(t)$. In this case, the current through the coupling capacitor is $2c_c \, dv/dt$. In both cases, the impact of the aggressor on the victim can be modeled using the simple circuit shown in Figure 24.2b, where the aggressor is *decoupled* from the victim, meaning that an equivalent coupling capacitor, often called Miller capacitor factor [4], is connected to ground with a multiplicative factor K_{MCF}. The resulting ground capacitance of the victim is adjusted to account for coupling noise effects as

$$\hat{c}_g = c_g + K_{MCF} c_c \tag{24.7}$$

In the two cases considered earlier, where aggressor and victim transitions have the same shape and switch in the same or opposite directions, the factor takes values 0 and 2, respectively. Equation 24.7 can also be used to approximate coupling noise impact on delay with multiple aggressors and arbitrary aggressor and victim waveforms. In these cases, K_{MCF} may be computed

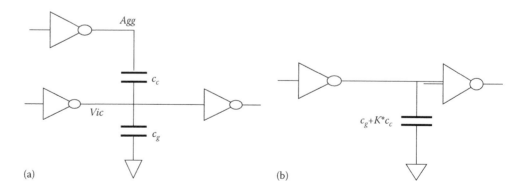

(a)   (b)

FIGURE 24.2 Simple circuit for demonstrating delay noise: (a) original and (b) simplified circuit.

as a function of circuit parameters, including transition times and timing windows (TWs)* of victim and aggressor [4]. Similar to the case of glitch noise, the simple estimation formula for coupling noise effects on delay given by Equation 24.7 can be used to determine the nets that are susceptible to noise problems in STA and filter a large number of nets from a more costly analysis.

24.2.2 PROPAGATED NOISE

As explained in the previous section, capacitive interaction between signals can create unwanted cross-talk glitches on a signal line. The presence of a cross talk glitch at the input of a gate can create a few different effects:

- It can weaken the driving power of the gate; this effect is called "driver weakening."
- Under the right conditions, the noise glitch can actually appear at the output of the gate; see Figure 24.3. It is called "propagated noise."

This propagated noise can combine with the noise on the output net and continue downstream to other gates. Propagated noise must be considered along with cross-talk glitches to understand whether the circuit will function correctly in the presence of noise, since each gate is affected by the total noise that it sees.

24.2.3 CHARGE-SHARING NOISE

Standard-cell-based designs typically have to deal only with capacitive coupling and propagated noise. However, full-custom designs often implement dynamic circuit design techniques such as domino logic. Although the use of dynamic circuits has reduced significantly since first edition of this book in 2006, they are still occasionally used in high-performance processors. These dynamic circuits are sensitive to an effect known as "charge sharing" [5,7]. Here, a noise glitch can be induced on an otherwise quiet signal due to two nets coupled through a transistor, rather than through interconnect capacitance.

Figure 24.4a shows a domino circuit, and some waveforms under different charge-sharing conditions are shown in Figure 24.4b. The circuit is first precharged by setting the clock signal (clk) to 0. This turns off the n-transistor at the bottom of the stack and turns on the p-transistor at the top of the stack. This will store charge on the output node of the circuit. This is called the precharge phase. When the clock signal changes its value to 1, the circuit is said to be in its evaluation phase. At this time, the inputs from the previous stages become valid. Consider the case where inputs A1, A2, A3, and A4 switch from 0 to 1, while inputs B1, B2, B3, and B4 remain at 0. This should not change the value of the output node. However, the output waveforms for this case are shown in Figure 24.4b. The solid line represents the case where there is no *keeper* transistor

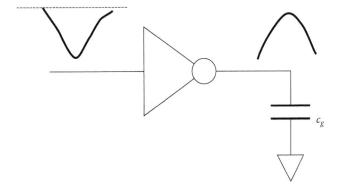

FIGURE 24.3   Propagated noise.

* A timing window represents the interval between the earliest and the latest times when a particular signal can switch.

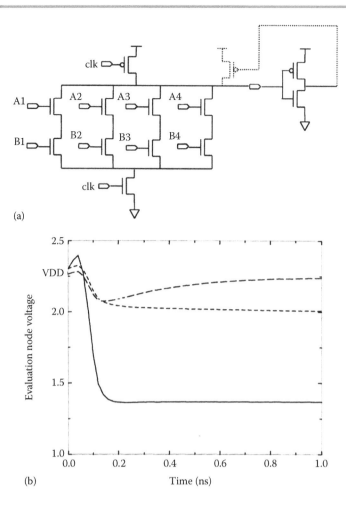

FIGURE 24.4   Charge-sharing noise: (a) Circuit and (b) waveforms. (Copyright 1998 IEEE.)

(solid p-transistor shown dotted on the upper right side of the circuit). The dotted waveforms represent the output node voltage with different strengths for the keeper transistor. As can be seen from the figure, the redistribution of charge within different nodes of this dynamic circuit creates a significant noise problem for this particular circuit.

24.2.4 MILLER NOISE

This type of noise until recently drew little attention even though it is as widespread as the coupling noise. The reasons include the relatively low magnitude of Miller noise in technology nodes above 28 nm and the high complexity of modeling it. In advanced process nodes, 28 nm and below, the coupling capacitance between the gate and source terminals of transistors increases relative to the total capacitance of wires. At the same time, wire resistance increases relative to gate resistance. As a result, the Miller noise becomes a prominent type of noise in advanced nodes.

The Miller noise becomes noticeable if a transition at the output of a receiver (or at the first stage of a multistage cell) is fast relative to its input transitions and if the receiver's pin capacitance is a significant fraction of the total capacitance of the previous net.

The Miller noise is demonstrated in Figure 24.5 where input *a* of a NOR gate is held at "0." The other input *b* transitions from "1" to "0" that causes output *y* to transition from "0" to "1." As a result, the coupling capacitance connecting output *y* and input *a* will inject a positive charge on the net connected to input *a* causing a distortion of signal "0" there.

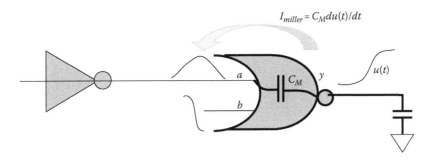

FIGURE 24.5   Miller noise is by coupling capacitance between input and output nodes of receiving devices.

As seen from the Figure 24.5, the nature of Miller is very similar to that of coupling noise—both are due to a current through a capacitor, which connects victim net to a receiver output net undergoing a transition. However, modeling of Miller noise is much more complex due to the following factors:

■ The coupling capacitance C_M between input and output (or internal) nodes of the receiving cell, often called the Miller capacitance, is nonlinear and needs to be calculated as part of cell characterization.
■ The Miller current $I_M = C_M du(t)/dt$ depends on the output transition waveform $u(t)$ of receiver's first stage, which in turn is a function of the input transition. This cyclic dependency may require several iterations to resolve.
■ One can break the mentioned cyclic dependency and model receivers with a current source. However, computation of voltage response due to the current sources attached at the far end of interconnect requires a more expensive RC reduction procedure, where *taps* (observation points) are modeled as *ports* (excitation points). Since the cost of creation of a reduced-order model (ROM) of interconnect is a strong function of the number of ports, converting taps into ports for Miller calculation will lead to a higher cost of ROM generation, to its larger size, and to an eventually higher cost of computing responses using a larger ROM.

The Miller noise impacts both glitches and propagation delay. While it is too weak to cause significant glitch noise by itself, it may combine with other noise sources on a victim to cause glitch failures in the downstream logic. It is usually not a dominant source of glitch noise problems in modern designs.

The effect of Miller noise on delay is more profound: it is one of the top physical effects influencing delay in advanced process nodes even in absence of other sources of noise, because it happens on most nets of a design. It causes distortion of a later part of input transition, which in turn increases the delay of receiving gates.

To enable adequate modeling of Miller noise in delay calculation, timing cell models have also been enhanced to include the necessary data [8].

24.2.5 LEAKAGE NOISE

Leakage noise becomes important in advanced process nodes, starting from 20 nm [9], especially in designs using low V_{th} to make circuits switch faster. Figure 24.6 explains the nature of leakage noise. An inverter shown in the figure has "1" at its input, which logically turns off the p-transistor and turns on the n-transistor of the inverter. However, the p-transistor may still conduct (*leak*) a subthreshold current, often referred to as "leakage current." This current is normally very low, but sometimes it is strong enough to raise voltage at the output by several millivolts. Even though leakage noise by itself is normally negligible, it can have measurable effects on glitch and delay noise in low-voltage circuits [10].

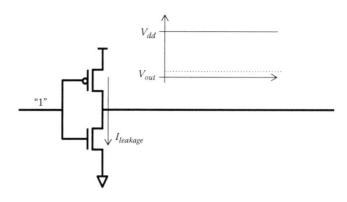

FIGURE 24.6 Leakage noise is caused by subthreshold current that shifts baseline voltage.

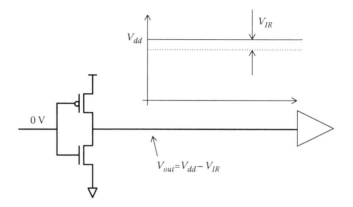

FIGURE 24.7 Supply noise.

24.2.6 SUPPLY NOISE

This type of noise has been known since the emergence of the first static noise analyzers [5]. The origin of supply noise is the variation of voltage in power and ground rails also known as IR drop. Rail analysis became mandatory in advanced process nodes due to the increased impact of IR drop on delay and noise. Figure 24.7 demonstrates supply noise on a simple circuit consisting of an inverter having 0 V at the input. As a result, the p-transistor connects the output node with the power rail, where due to circuit activity the nominal voltage may go down by V_{IR}, causing a reduction of V_{dd} by the same amount at the input of the receiver, as shown in the figure. This reduction of voltage, called "supply noise," can cause glitch noise and delay noise on the receiver, either by itself or along with other sources of noise.

24.3 NOISE EFFECTS IN DIGITAL DESIGNS

Earlier, we briefly introduced the three main noise effects. In this section, we consider them in more detail, and specifically we explain how they affect digital designs and how different noise types contribute to them. In the first subsection, which explains glitch noise in detail, we show how it can lead to the first of the three noise effects introduced earlier: functional failure. The second subsection covers the other two noise effects, specifically speedup and slowdown of transitions, which may lead to timing failures.

24.3.1 GLITCH NOISE

As discussed earlier, when a victim net is at a quiet state, the noise induced on it usually has the shape of a voltage glitch or pulse. This is because before the noise occurs, the victim's driver holds

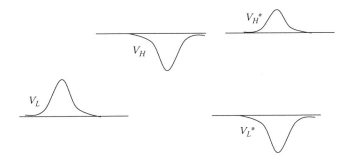

FIGURE 24.8   Four types of glitch noise.

the net at a stable voltage level corresponding to either logic "0" or "1," and it then returns the net to its original stage after the noise.* This noise effect is therefore called "glitch noise".

It is useful to define four different types of glitch noise, as shown in Figure 24.8. If the glitch is directed upward from the *low* voltage level V_{gnd} (usually 0 V), we call it a V_L glitch [5]. Similarly, if a glitch is directed downward from the *high* voltage level V_{dd}, we call it a V_H glitch. Although these two glitch types are most prevalent in noise analysis, it is sometimes important to consider a glitch that starts at V_{dd} and is directed upward (sometimes called overshoot glitch) and, similarly, a glitch starting at V_{gnd} and directed downward (undershoot glitch); these are called V_H^* and V_L^* noise glitches, respectively.

In Section 24.2, we introduced the various noise types and explained each type using simple circuits. It is helpful to characterize the noise types in terms of their importance and the frequency of occurrence. Since the early days, when designers just started considering noise in validation of their circuits, the coupling noise has been the primary type of noise. After decades of evolution of process technology, this kind of noise remains the strongest and the most prevalent type of noise in modern designs. The propagated noise, even though very important as it ultimately can lead to functional failure in the case of glitch noise, happens less frequently, since for it to appear at a gate output, a significant glitch noise needs to happen at its input. The supply noise and Miller noise types occur almost as often as the coupling noise. However, they are relatively weak and are often ignored in analysis. The charge-sharing noise, which happens primarily in dynamic circuits, is not common anymore and is only important in full-custom designs. Finally, the leakage noise has become more significant at 20 nm and below, although standard-cell designers usually prevent it by construction.

Practical circuits may experience multiple sources of noise, either the same or of different types. For example, if a victim has two coupling aggressors, we can say that there are *two* noise sources of coupling noise type. In another example, if a victim has propagated noise, supply noise, and coupling noise from two aggressors, we can say that there are *four* sources of noise.

Multiple sources of noise possibly of different types can combine together and lead to a stronger glitch noise. One such scenario is depicted in Figure 24.9, where a victim net $V1$ has several noise sources, including two coupling aggressors (shown in the figure) and supply noise (not shown). Combined, these sources lead to a significant V_H glitch noise, which propagates to the next net $V2$ to become a propagated glitch of type V_L. There it combines with coupling noise to produce a large V_L glitch at the input D of the latch.

It is important to note that glitch noise is a nonlinear phenomenon. As we saw earlier, glitch noise magnitude for most noise types is a nonlinear function of the circuit parameters, often exhibiting amplification, like in the case of propagated noise.

---

* This is however not always the case. For some gates like latches, which include *keepers* connected directly to the output, a glitch at the output net may flip the value from one logical level to another. When this happens, the downstream logic will see an incorrect signal, which constitutes functional failure of the circuit. That is why it is standard practice to protect the gate's internal node with an inverter; see Section 24.6 on the discussion about noise analysis in implementation and sign-off flows.

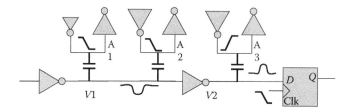

FIGURE 24.9 Multiple noise sources cause large glitch at the input of a sequential element.

In addition, several sources of noise, of the same or different types, interact in a nonlinear fashion, often resulting in combined glitch noise being stronger than a linear sum of glitch noises computed for each source at a time. As an example of such nonlinear interaction, consider a scenario where coupling noise occurs at both input and output of a cell and the two noise glitches overlap in time. As we discussed in Section 24.2.2, a sufficiently strong input noise will cause a propagated glitch at the output of the cell, where it will combine with coupling noise to produce a stronger combined noise. In addition, the input glitch will also have the driver weakening effect on the cell, which will reduce its ability to resist to the coupling noise at its output. As a result, the coupling noise will become stronger, causing the combined noise to be stronger than a linear sum of propagated and coupling noise glitches.

Let us now see how glitch noise can create a functional failure in a design. Consider again the example in Figure 24.9 with a strong V_L glitch noise at the pin D of a latch. The schematic of a simple latch is shown in Figure 24.10a. When the clock rises, it opens the latch for the signal to propagate to the internal node, where it will be stored when the clock falls.

If the clock closes the latch around the time when the internal node is perturbed by the V_L glitch propagating to it, the keeper that is supposed to hold the signal at the intended stage may inadvertently flip the voltage to the opposite logic value if the glitch noise is sufficiently strong. Figure 24.10b shows the voltages of the clock and of the internal node signals, as obtained from

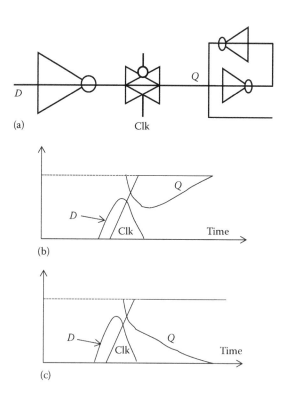

FIGURE 24.10 Schematics (a) and SPICE simulation showing internal node of a latch disturbed by glitch noise: (b) Noise is small enough to corrupt the value, and (c) noise is high enough to change the value.

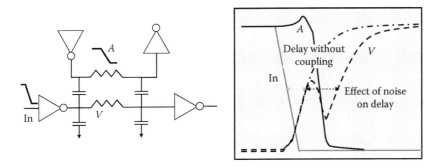

FIGURE 24.11 Effect of noise on delay.

a SPICE* simulation. As seen from the figure, the keeper is able to return the signal to its original (correct) value. Figure 24.10c shows the case where noise at the input D is too strong for the keeper to sustain the correct value—after a certain time when the internal node hovers near the metastable state, it is attracted toward the wrong value.

This example shows how a glitch noise can cause an incorrect value to be stored inside a sequential element and ultimately lead to a functional failure of a circuit. Note that even a small change in input glitch noise may lead to a dramatic change in the circuit behavior.

This strong nonlinearity is a distinguished characteristic of the glitch noise effect, which requires a high-accuracy analysis for adequate modeling and reliable prediction of circuit behavior. In the following discussion, we will describe ways to efficiently and accurately compute glitch noise.

24.3.2 DELAY NOISE

In addition to functional problems, noise can also create timing problems. Consider the case where a given victim net V is transitioning from 0 to V_{DD}. Now, if the neighboring net A switches at the same time from V_{DD} to 0, this will slow down the transition of the victim net A. This situation is shown in Figure 24.11. We call this phenomenon "delay slowdown." If the neighboring net V were to switch in the same direction as the victim net (i.e., from 0 to V_{DD}), then this will result in speeding up the transition of the victim. This is called "delay speedup." Both of these effects are influenced by the same factors as noise glitch, namely, coupling capacitance, transition time of the aggressor, and the drive strength of the victim's driver. In this case, the victim parameter of interest is its switching waveform, which is distorted by the noise effect. Collectively, we call delay speedup and delay slowdown as *noise-on-delay* effects.

24.4 WHY HAS NOISE BECOME A PROBLEM FOR DIGITAL CHIPS?

Today no design is taped-out without comprehensive noise analysis. Due to the significant similarity between glitch noise and noise-on-delay effects, a modern noise analyzer is tightly integrated with an STA tool, where delay noise is analyzed at the same time as delays are calculated. Since the delay calculation topic is beyond the scope of this chapter, the rest of this chapter will focus on glitch noise effects and its modeling in modern analyzers, unless noted otherwise. For details see Chapter 6.

The key reason why CMOS is the most prevailing technology in digital designs is its ability to suppress noise and yet amplify useful signals. In fact, this noise immunity of CMOS technology is what makes digital design truly "digital." The noise immunity of a CMOS circuit is easily explained through a *DC noise margin* [11]. For simplicity, consider the DC transfer characteristic of an inverter as shown in Figure 24.12. As is seen from the figure, when the input voltage is between 0 and V_{IL}, the output remains near V_{DD}, or logic 1. The output changes to logic 0 when

---

* We generically refer to all circuit simulators as SPICE for simplicity.

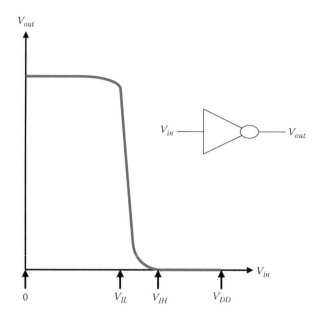

FIGURE 24.12 Inverter transfer characteristic and DC noise margins.

the input voltage reaches V_{IH}. The CMOS gate is, therefore, immune to voltage fluctuations from 0 to V_{IL} when the input is at logic 0 and from V_{DD} to V_{IH} when the input is at logic 1.

As manufacturing technology scales down toward 20/16 nm and below, the following factors coalesce to degrade the noise immunity of digital designs:

- Transistors became faster, leading to faster transition times, which contribute to higher coupling noise as explained in Section 24.2.1.
- Changes in wire geometries cause an increase of the side wall capacitance at the expense of ground capacitance. Wires have become taller, narrower, and much closer (Figure 24.13). As a result, a much higher percentage of the total capacitance of a net comes from coupling with its neighbors, thus increasing the magnitude of coupling noise.
- Another side effect of narrower wires is their higher resistivity. For a victim net, wire resistance reduces the driver ability to cope with cross talk induced at the receiver side of the wire. For an aggressor, the higher resistance shields some of the capacitance, leading to a lower effective load seen by the aggressor driver and again to faster transition times.
- Threshold voltages became lower, which makes victim drivers more susceptible to propagated glitch noise and aggressor drivers faster to switch. On the other hand, due to the rapid voltage scaling driven by mobile applications, the ratio of V_{th}/V_{dd} increases, which reduces the effective holding strength of drivers.
- Although transition times become smaller in general, they scale down slower than cell delays. This leads to higher overlap of input and output transitions of a cell, which makes the cell more prone to noise-on-delay effects.
- Transition times also scale down slower than the clock cycle, making waveform effects a more acute problem.

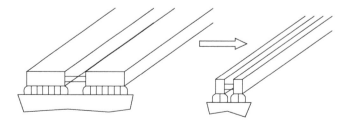

FIGURE 24.13 Wire cross section scaling.

These effects have increased the interactions between signals and decreased the noise immunity of digital CMOS circuits. This has led to noise being a significant problem for digital ICs, which must be considered by every digital chip designer prior to tape-out. Section 24.5 discusses how noise considerations affect the design methodology. A fair question to consider is whether noise problems can be completely eliminated by design methodology constraints. For example, can we completely eliminate the need to analyze a design for noise problems by restricting the number of neighbors any given net is allowed to have and by inserting additional buffers often? In practice, if we restrict the amount of neighboring wiring to completely safe levels, there is a significant area increase and associated cost. Similarly, adding sufficient buffers to achieve safe levels leads to significant degradation in performance. Furthermore, as we will see in later sections, the same amount of noise may be harmful at one place in the design and not harmful at another place in the design. Hence, a designer must analyze a design for noise issues and fix noise problems that can lead to chip failures while allowing other less significant noise effects to still exist in the design.

24.5 MODERN STATIC NOISE ANALYSIS

As one can see from the previous section, multiple noise effects need to be analyzed in a digital IC. Prior to the introduction of static noise analysis [5], digital designers dealt with noise using a variety of ad hoc techniques based on circuit simulation. This was practical only for full-custom design styles, where a designer works on a very small circuit block—and still is very cumbersome even for that case. *Static noise analysis* is a systematic way to deal with all the noise effects in a digital design.

In order to explain the motivation and roots of static noise analysis, let us first briefly review how a circuit designer might go about analyzing a design for noise effects using circuit simulation. Typically, a designer would take the following steps to conduct the verification:

- Create a list of potential noise sources, which may include different types, such as coupling and charge sharing.
- For each noise source, develop a way to excite the circuit so that the expected noise source will occur.
- Create a SPICE netlist that represents the desired excitation; run SPICE and record the results.
- Analyze the simulation results and decide whether any redesign is required.

Although easy to describe, this procedure is error prone and unwieldy in practice. Some of the difficulties are as follows: the number of noise sources can explode, creating excitations either requires tracing logic or cutting the SPICE netlist to insert appropriate voltage sources at the right places, it is difficult to determine the right transition and delay values, and there is no clear criteria to decide whether a noise problem requires a redesign.

Static noise analysis overcomes these difficulties through a systematic procedure. A high-level procedural description of static noise analysis is shown in Table 24.1. The steps outlined there are general enough to apply to both the analysis of functional noise effects and the effect of noise on timing;

TABLE 24.1   High-Level Description of Static Noise Analysis

- Divide the circuit into small individually analyzable partitions.
- Sort the partitions in order from source nodes to sink notes (topological ordering).
- In topological sort order, for each partition:
  - Create noise sources.
  - Create simulation condition for each source.
  - Simulate each noise source and calculate amount of noise.
  - Determine worst-case combination of noise sources and relative timing of noise source.
  - Calculate worst-case combined noise.
- Propagate worst-case noise to output of partition and perform appropriate tests.

---

they are also general enough to apply to both full-custom and standard-cell designs. In the rest of this section, we will discuss each of those steps in more detail and describe how the steps can be implemented differently for different analysis scenarios.

24.5.1 CIRCUIT PARTITIONING AND ORDERING

For ease and speed of analysis, each partition must be as small as possible. However, it is important to partition in a way that preserves circuit behavior; this can be accomplished by keeping some CMOS behavior in mind.

Let us first consider the case where we are dealing with a full-custom, transistor-level netlist. We want to make sure that all transistors that are strongly connected are analyzed together. For CMOS technology, this suggests that the most appropriate cut point is the gate terminal of a transistor. Hence, we create each partition with all the transistors that are connected through source or drain terminal connections, that is, through the channel. Such a partition is called a "channel-connected component" (CCC) [18]. In addition to the transistors, each CCC will contain all the nets and pins associated with the transistors in the set.

When we analyze a CCC, we also need to include all the nets that have significant coupling capacitances to any of the nets in the CCC (some nets that have insignificant coupling can be eliminated through *simple electrical filtering*). These nets are the aggressors to the nets that belong to the CCC and may be modeled in a simpler form than the nets within the CCC itself. For example, one could simply model them as being driven by ideal voltage sources and not their real, highly nonlinear drivers. We define a *net complex* as a net with its parasitic elements and all other nets that have significant coupling to it, including their parasitic elements. For each CCC, we need to include the net complexes of the nets within the CCC. In general, most CCCs consist of only a few transistors and hence lead to quick analysis. However, certain types of circuits, like barrel shifters and multipliers, can lead to very large CCCs. In these cases, rigorous analysis of a CCC can become intractable and special heuristics may be required.

We can now extend this concept easily to standard-cell designs as well. Since the CCCs are contained within a cell, we can replace the aforementioned CCC concept by the idea of a *stage*. A stage is a cell driving a net; we include the appropriate electrical model for the driving cell, the driven net and its parasitic elements, and the net complex associated with that net. The receiving cell can be modeled in a simple fashion, for example, as a single fixed capacitance, or using a more sophisticated model based on a nonlinear capacitance or a current source.

Once the circuit has been partitioned, we need to analyze the partitions in the right order. Before we can analyze a partition, we must have all the analysis results from any partitions that precede it and provide input values to it. Let us use some graph terminology to describe this. Let each partition be a node n in a circuit graph G. Also let there be an edge E from any node $n1$ to a node $n2$ if the output of node corresponding to a gate $n1$ connects to an input of a gate corresponding to node $n2$. In other words, the edges represent the connectivity represented by each net. We now need to do a *topological sort* [14] on the graph G to decide the order in which the partitions are analyzed. If the circuit was fully combinational and contained no loops, this would be sufficient (i.e., if the circuit graph was a directed acyclic graph). A real circuit requires some modifications. First of all, the graph can be cut at latch boundaries. Also, combinational loops must be cut. In the simplest case, a combinational loop can be cut at any arbitrary net. However, more sophisticated algorithms are often employed for preserving important information about the loop, but these are beyond the scope of this chapter.

In general, static noise analysis may be categorized as a *breadth-first search* of the circuit under analysis, and it is very similar to the approach taken in STA [16].

24.5.2 ANALYZING PARTITIONS

Once the circuit has been partitioned appropriately and the partitions have been ordered for analysis, we can analyze each partition in detail. The fundamental analysis approach is similar

---

TABLE 24.2   Steps to Analyze Each Partition

- Identify *noise sources* to analyze.
- For each noise source:
  - Create sensitization condition.
  - Create electrical analysis view of the noise source.
  - Simulate and calculate response.
- Pick combination of noise source and their relative alignments.
- Calculate the combined response.

---

for both full-custom and cell-level designs, although the actual engines employed are different. Table 24.2 sketches the different steps involved in analyzing a partition at a high level. We will describe each of these steps in more detail here:

1. *Identify noise sources*: In the case of coupling noise analysis, this includes deciding which aggressors must be included. As discussed later in this chapter, both *electrical filtering* and TW *filtering* can be employed to reduce the set of relevant aggressors or noise sources in general.

2. *Analyze noise sources*: Once all the noise sources are identified, we need to analyze each noise source. In order to do this, we need to perform the following steps:

   ▪ *Create an electrical view of the noise source*: Conceptually, this step is equivalent to what designers would do in the ad hoc technique described earlier, where they create a SPICE netlist for simulation. The main difference is that this will be done in an automated way and some of the devices will be replaced with simplified models. For instance, an aggressor will be represented with voltage sources and receivers with simple capacitors. The interconnect parasitics are often represented by a reduced RC or RLC or RLCK model. We will discuss some of the models in a Section 24.5.6; for a detailed model classification, see [8].

   ▪ *Sensitize the circuit of the partition*: In order to select the right models, we also need to create an appropriate *sensitization condition* for the driver. For example, consider the case where the driver is a NAND gate and the output is being held at a logic value of "1." For this case, we need to have both inputs at "0." We need to choose an appropriate model for this input condition: in the case of a transistor model, it means setting up the input vector for the SPICE simulation to have the right values, while for a cell model it may mean selecting the appropriate holding resistance or the set of $I–V$ characteristics from a set of precharacterized library models. The end result of this step is an electrical view that can be simulated or *solved* for the resulting noise response on the output pins of the partition.

   ▪ *Simulate a noise source*: During this step, the electrical view created in the previous step is evaluated to calculate the noise contribution of this particular source. Typically, the noise glitch contribution due to this source is calculated, and the effect of noise on delay is calculated in two separate steps. Depending on the type of electrical model, different types of engines are used to solve the electrical view. The simplest one to explain and understand is a SPICE model. In this case, usually a SPICE engine with an application programming interface (API) is employed. The API is used to create the netlist of circuit elements and to instruct the SPICE engine to run for a given amount of time. The appropriate output nodes will be instrumented to measure the voltages—either the entire waveform or the peak voltage. An example of the required electrical model for a cross talk noise source is shown in Figure 24.14. In this case, the driver is represented using transistor models, the interconnect is represented using resistors and capacitors, the aggressor is represented using a saturated-ramp voltage source, and the receivers are represented by equivalent passive capacitive loads.

3. *Select combination of noise sources and relative alignment*: This is a crucial step where many pragmatic trade-offs must be considered. The objective of this step is to select a conservative yet realistic subset of all possible noise sources that can occur together to

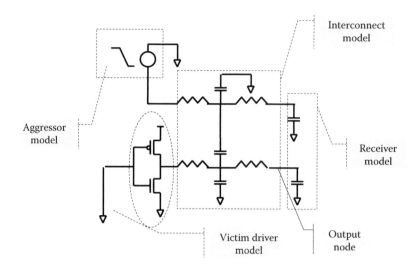

FIGURE 24.14   SPICE electrical model for simulation of a coupling noise source.

create the worst possible noise on an output node of this partition. Some of the factors to consider are the following:

■ Which combinations of noise sources can occur together? There will be several different sets of noise sources that can occur together based on the TWs of the different signals. When considering effect of noise on delay, we must also look at the TW of the victim signals.

■ For a given set of noise sources, what is the relative timing of each of the aggressors? Different relative timings can produce very different combined noise on the output of the partition. Finding the worst-case relative timing of the different aggressors, or their relative *alignment*, is critical for accuracy and is discussed in detail in [21].

■ Should you consider noise sources, which are only partially overlapping? For example, the latest time a particular signal can switch may be 10 ps before the earliest time another signal can switch, but the effects of these two signals can overlap at the receiver.

■ Given a set of signals, which have overlapping TWs, what logical relationships can you consider to eliminate some combinations? How do you consider logical relationships with timing information [17]? Can you consider a small subset of logical relationships without exploding the complexity of analysis?

Each commercial noise analysis tool makes slightly different choices; there is no single definitive answer for each of the aforementioned questions.

4. *Create combined noise response*: Once step 3 is completed, this step is relatively straightforward. If the electrical model used in steps 2 and 3 is linear (i.e., it contains no nonlinear circuit models like transistors), then linear superposition principles can be applied to calculate the combined noise response of the circuit. In the case of glitch noise, the glitches can be added together to create the combined glitch. For calculating the effect of noise on delay, the combined glitch waveform can be added to the victim transition waveform to create the new (noisy) waveform, from which the delay and slew need to be measured. However, in most cases the underlying circuit is strongly nonlinear, and hence the linear superposition principle may not be accurate enough. In these cases, when nonlinear models are used, a final *combined simulation* is usually required to get the combined response. In this final simulation, all the aggressor noise sources are instantiated with appropriate relative alignments, and the output noise effect is measured from the response of such simulation. For noise-on-delay simulation, the victim's driver input signal is also set up to switch at the appropriate time.

In case multiple sets of noise sources may be active simultaneously, we need to find which of them will cause the strongest noise effect. One way of doing it is to estimate the noise effect for each set and choose the one where the estimate is the strongest. Sometimes, such an estimate cannot be made accurate enough to guarantee the worst set is chosen. Then the described steps of analyzing the set of noise sources need to be applied to more than one set. After that, the calculated noise effects are compared to produce the worst-case one.

Once the worst-case set is determined and the corresponding noise effect is calculated, the analysis of the stage (or partition) is complete. However, before the analyzer moves to the next stage, we need to determine whether the noise effect on the current stage is strong enough to violate user-imposed limits on the amount of noise allowed per stage.

For example, in glitch noise analysis the peak of a glitch at the input of a receiving stage is often used for this purpose. In this case, the current stage is declared *failing* and the stage needs to be revisited in the implementation or optimization flow for fixing the glitch failure. Normally the strong noise effect causing failure on a given stage is not propagated forward, to prevent massive failures of the subsequent stages. Instead, the strong noise is discarded or limited to a certain magnitude to allow a meaningful analysis of its fanout logic, assuming that the failing stage is going to be fixed later.

In case the noise level at a given stage does not cause a failure, the combined noise effect is saved and then propagated to the next stage, where it can contribute to a larger noise or even a failure. Sometimes it is impossible to predict where a noise effect is too strong. For instance, in delay calculation it is not known how much noise should be considered as too strong until arrival times are propagated to the final flip-flop and timing checks are done. Likewise, one may allow even large glitch noise to occur and propagate forward and only check for failure at the storage cells such as latches and flip-flops.

Such methodology is less conservative but can be costly in terms of memory: the saved result depends on the type of analysis and can be bulky, often including detailed noisy waveforms.

Determination of noise failure is based on the so-called noise metrics. We will describe various noise metrics in the next section. The choice of noise metrics varies from tool to tool, and it is often configured by designers.

24.5.3 NOISE METRICS

In the previous section, we have seen how to calculate the noise effect at the output of a stage or partition. In order to determine whether the calculated noise will cause incorrect operation of the circuit, some metrics are required; different metrics are required for glitch noise and noise-on-delay effects.

First, let us consider glitch noise; the following metrics are typically employed for determining the severity of a glitch noise problem. The different metrics for glitch noise are shown in Figure 24.15.

- *Glitch peak*: This noise metric is the simplest. We merely define an upper bound for any allowed noise glitch peak. Typically, glitch peak limits are defined with respect to V_{dd}. For instance, if we set the glitch peak limit to 40% of V_{dd}, then with a power supply of 1.2 V, only glitches below 0.48 V will be allowed and any higher noise glitches will be labeled as noise failures. This metric is easy to calculate but it can be inaccurate as it ignores the shape of the noise pulse. In reality, the shape of the noise pulse impacts significantly the noise in downstream logic. A sharp, narrow pulse is much less dangerous than a wide pulse of the same height. Often, DC noise margins [11] are used as the glitch peak limits, which make this metric very conservative. A DC noise margin is illustrated pictorially in Figure 24.15a. This metric can have a single value for the entire design, in which case the DC noise margin of the most noise-sensitive gate in the entire design must be applied throughout the design. In order to reduce pessimism, cell-specific DC noise margins can be applied.

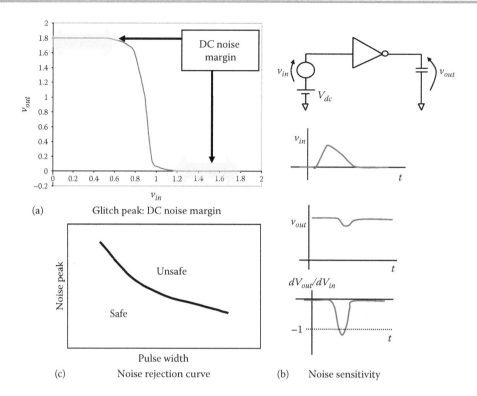

FIGURE 24.15 Glitch noise metrics: (a) Glitch peak, DC noise margin. (b) Noise sensitivity. (c) Noise rejection curve.

- *Noise rejection curve*: Noise rejection curves modify the glitch peak metric to also take the glitch shape into account. The glitch width/glitch peak space is divided into safe and unsafe regions. For a given glitch peak, a glitch with a smaller width would be safe, but one with a larger width would be unsafe. This is pictorially shown in Figure 24.15c.
- *Receiver output sensitivity*: The idea behind this metric is very simple: if a noise is strong enough to push a receiver into its amplification region, then it is a problem. Otherwise, the noise is being suppressed by the receiver and will not have a significant impact on downstream logic. Receiver output sensitivity is shown pictorially in Figure 24.15b. In practice, this metric can be implemented via a time-domain sensitivity calculation in the simulation engine.
- *Receiver output peak*: This is a modification of receiver output sensitivity. Instead of checking whether the receiver enters its amplification region, we check the propagated noise peak at the output of the receiver. Given the transfer characteristics of CMOS gates, usually there will be no propagated noise until the gate is well into its amplification region. This metric has gained wide popularity in modern analyzers, as fewer noise failures will be reported using this metric than with the metrics described earlier.
- *Noise propagation to latches*: All of the previous metrics are stage based. They evaluate the noise at the output of a stage and then evaluate its effect relative to its receiver. However, a glitch noise is not a problem unless it propagates all the way to a storage element. Noise propagation moves the tests away from all the combinational stages to the storage element stages. This is the most aggressive and least pessimistic noise metric for glitch noise.

Now, let us consider the effect of noise on delay. The metric of significance for a delay change is of course its effect on timing. However, we need to consider where the delay change is measured, and there are two approaches to measuring the delay change:

- *Receiver input measurement*: In this case, the delay change is measured at the input of the receiver. First, a nominal delay for the stage is calculated without any coupling effects. Then a "noisy delay" is calculated, which includes coupling, and the difference

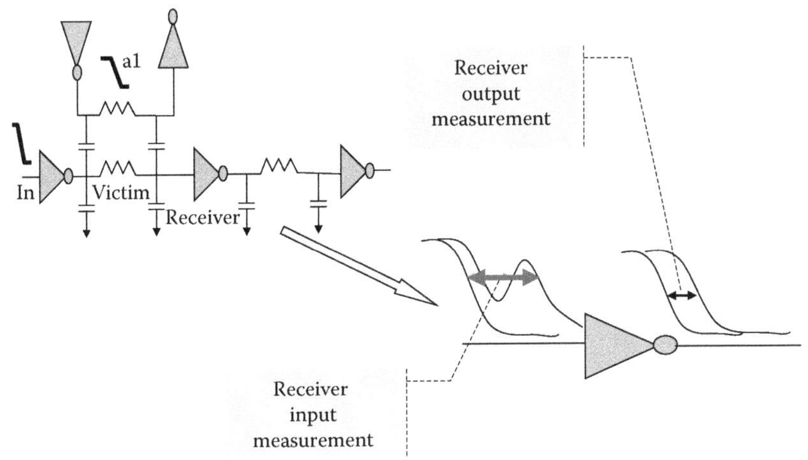

FIGURE 24.16   Delay measurement at receiver input and output.

between the two is the effect of noise on delay. All of these measurements are made at the input of the receiver. This change in delay at the input of each stage is essentially added to the path delay.

■ *Receiver output measurement*: In this case, the delay change is measured at the output of the receiver. As in the case of glitch noise, this method is widely used today thanks to a lower pessimism compared to the previous method. Because CMOS gates often behave as low-pass filters for narrow spikes at their input, the receiver is able to absorb the change in the input waveform with only minor change in the output waveform.

Both of the aforementioned approaches are shown in Figure 24.16. Note that unlike glitch noise, the noise-on-delay effect is not used to flag a noise failure on intermediate stages of a timing path. An exception to this rule is the so-called double-clocking failure, where a bumpy transition on a clock net is often flagged as a violation.

24.5.4 MULTIPLE AGGRESSOR CONSIDERATIONS

As we mentioned earlier, multiple noise sources may induce noise on a victim net. A typical example is where a wire is routed close to several other wires. In this case, the victim will have several coupling aggressors and an analysis needs to calculate the combined effect from all the aggressors. The worst-case scenario is to assume that all the neighbors will contribute to the noise on the victim. However, this is usually not the case in reality; several factors can make the situation less severe:

■ Different nets are expected to switch at different times; we say that each net has a *TW*. We can reduce the number of active aggressors by considering the overlap of the TWs of different aggressors. We call this "TW filtering."

■ Different nets may always switch in opposite directions; or there might be logical relationships between different nets that prevent a net from switching when another net switches. Using this information, one can reduce the number of active aggressors; we call this "logic filtering."

In practice, TW information is very easy to obtain from STA [14,15], which already tracks the earliest and latest arrival times at every pin. The transition times of aggressors and victims can also be obtained from STA. All commercially available noise analysis tools provide TW filtering capabilities, and these capabilities are applied in the noise analysis of the vast majority of digital ICs.

Logic filtering has also been implemented within the framework of noise analysis tools [17]. However, it has proven much more elusive to implement in practice, since it is computationally very expensive and the improvements in analysis so far do not justify the cost. While the general case is difficult to handle, some special cases can be handled with low complexity. For example, many noise analyzers can recognize the fact that two aggressors are connected via an inverter and therefore will always switch in opposite directions.

24.5.5 TIMING WINDOW CONSIDERATIONS

TW filtering provides significant reductions in the number of noise problems that must be considered and analyzed, and hence noise analysis tools pay careful attention to how noise windows are computed and used. Figure 24.17 illustrates how TWs are used to filter glitch noise sources. The main factors to consider in computing TWs are (1) the relationship between timing and noise, (2) the order of computing TWs, and (3) the determination of TW overlap. We will discuss each of these topics in this section.

First, let us look at the interaction between TW computation and noise-on-delay analysis. Essentially, for a given victim net, we look at the earliest and latest arrival times of each potential aggressor (its TW) to decide whether the aggressor can switch during the time that the victim can switch. After selecting the set of aggressors, we compute noise-induced slowdown and speedup for the given net. These values change the delay of the current net and hence affect the TW of the net itself, as well as of the downstream nets. Since the victim–aggressor relationship is reciprocal, some of the aggressors' TWs depend upon the victim's TW. To solve this interdependency, we need to perform multiple rounds of TW and noise-on-delay computations. This process is usually called "TW iteration" and is discussed extensively in [12], including the convergence criteria for the iterations. The necessary number of iterations required for convergence is often debated, and it is usually controlled by user settings.

For the first pass of analysis, the initial TWs can be either nominal (computed without noise) or infinite. In the former case, the first pass of analysis assumes no noise-induced delay changes, and then the TWs expand as the early and late arrival times are modified for noise effects. In the latter case the noise-induced delay changes are maximized during the first pass because no aggressors are eliminated due to TW filtering, while as iterations progress, the TWs are successively narrowed. Even though the former method has runtime advantages, starting the iterations with infinite TWs is a preferred method nowadays, because TWs remain conservative throughout the iterations.

Early and late arrival times are usually measured based on the switching waveform crossing a specific threshold voltage, usually set to 50% of V_{DD}. Consider the case where the latest arrival time of a signal is slightly before the earliest arrival time of another signal. The waveforms resulting from this case are shown in Figure 24.18. The TWs do not overlap, yet the waveforms partially overlap and hence can affect each other. To avoid potential optimism, TWs are usually padded by half of the slew for each signal.

Even though a TW constructed by using the earliest and latest arrival times covers the entire range of time for which a signal can switch, there are usually times during this interval when the signal will not transition; these can be considered as *holes* in the TW [13]. With some extra overhead in timing analysis, it is possible to keep an approximate track of TW holes. Some noise analysis tools use this approach to add an extra level of filtering.

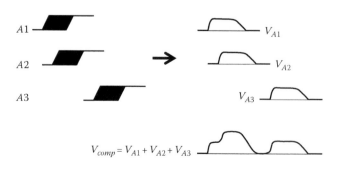

FIGURE 24.17 Timing window filtering.

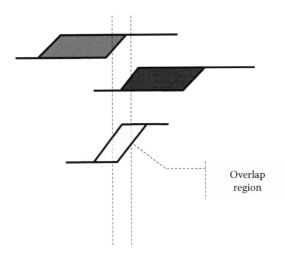

FIGURE 24.18 Timing window padding.

Another type of TW information may be used to add further filtering to cross talk glitch analysis. A noise glitch can create a functional failure only if it arrives at a latch during the time that the latch is sensitive to the value of the data input; this is a narrow window of time around a clock edge. We can calculate a possible TW for the glitch (a *noise window*) based on the timing of the aggressors and calculate the propagation delay for the glitch from the victim net to the latch. We can eliminate many glitches as harmless if they arrive at the latch during a time when the noise glitch cannot affect the state of the latch. This type of noise window filtering is treated in detail in [22].

24.5.6 ELECTRICAL MODELS AND ANALYSIS

In this section, we will delve deeper into the models and solvers that are used in noise analysis. The speed and accuracy of any noise analysis tool depend on the models and solvers used for the analysis. Furthermore, there is no perfect model or solver, which delivers both the best accuracy and the best performance. Every model and every solver express a trade-off between accuracy and performance.

Different models are required to cover different aspects of the circuit in question [8]. The most important models to consider are the following:

- *Driver models*: These models are used to represent the drivers for each stage or partition and are probably the most important model for noise analysis. Similar but slightly different models are employed for glitch analysis and delay analysis.
- *Interconnect models*: These are usually ROMs used to represent the interconnect parasitics in a compact form, so that it can be analyzed much faster.
- *Receiver models*: These models represent some characteristics of the receiving gate of the stage being analyzed.
- *Aggressor models*: These models represent drivers of aggressors.

Depending on the type of model used and the computational requirements, different electrical solvers may be used for computing the response of the electrical network of a partition. The solvers range from closed-form analytical equations to fully SPICE-like solvers. In the rest of this section, we will discuss specific models and solvers in some more detail.

24.5.6.1 DRIVER MODELS

The early static noise analysis tools were implemented at the transistor level [23], and the driver model was obviously composed of transistors. It consisted of a CCC driving a coupled interconnect model;

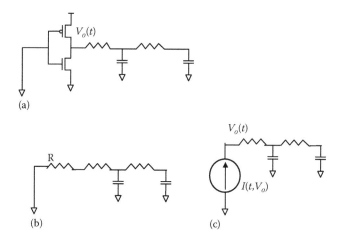

FIGURE 24.19   Driver models: (a) Transistor driver model. (b) Resistor driver model. (c) *I–V* driver model.

often, the inactive (i.e., OFF) transistors of the CCC were dropped. An example of a transistor-based driver model is shown in Figure 24.19a. A transistor-based driver model preserves the nonlinear behavior completely and hence is very flexible. However, it requires a SPICE simulation to evaluate the model, which can be very expensive.

At the other extreme, early cell-based noise analysis tools [19] used a single resistor *R* for representing the driver; this is shown in Figure 24.19b. This model is easy to characterize and also very easy to simulate efficiently by a linear network simulator. This model can provide reasonable accuracy for glitch peak, but cannot accurately model the noise waveform. Also, since the actual behavior of the driver is nonlinear, a single resistance value can only accurately represent the behavior of the driver in some linear subregion. The model must be characterized in the subregion where the driver has the highest resistance, and then that value must be used to represent the driver across its entire region of operation. This will overestimate the noise in all other subregions. Furthermore, the resistance used to model the driver for glitch noise (when the driver is holding its state) is different from the resistance used to model the driver for noise on delay, since the switching resistance of a cell is different (usually higher) than the holding resistance.

A refinement of the resistor-based model is to use different resistance values for different regions. The selection of the right resistance value for a particular victim driver is still a difficult problem, since it depends on the amount of noise—we can iterate on the selection of the resistance and the calculation of the noise, but this will be computationally expensive and take away from the simplicity and efficiency of the model.

A compromise between transistor models and resistor models is a specialized noise model based on storing the current–voltage (*I–V*) characteristics of the driver cell [20], as shown in Figure 24.19c. Instead of modeling individual transistors in the driving cell, we model the nonlinear electrical behavior of the entire cell through its *I–V* characteristics. These models may be based on DC (or static) *I–V* characteristics or on transient *I–V* characteristics. One significant advantage of the *I–V* model over the resistor model is that it can effectively model the behavior of the driver in both glitch noise and delay noise analyses. More details of this model are beyond the scope of this chapter, and the interested reader is referred to [20].

Most commercial tools use a combination of various driver models. Typically, they may use a fixed resistance for electrical filtering, an *I–V* characteristic–based model for the bulk of the computation, and transistor models selectively in cases where the highest accuracy is required.

24.5.6.2 RECEIVER MODELS

In most cases, receiver models tend to be simpler than driver models and are used for fewer calculations. Typically, when calculating a noise response on a stage or partition, a very simple receiver model is used: a simple capacitor. In the simplest case, the same capacitance is used for all calculations: base delay (without noise), glitch noise, and delay noise. A slightly more sophisticated approach is to use different capacitances for each of these cases; there may even be a separate capacitance for each switching direction.

More information is required for propagating cross-talk glitches. The most accurate model for this purpose is again a transistor-based model. Since many glitches can be filtered from further propagation by using thresholds, it is possible to limit the number of places where noise must be propagated, so it is quite feasible to use transistor models for noise propagation. Another alternative is to build noise propagation tables, where the height and width of the output noise pulse are characterized and stored based on the input noise pulse width, input noise, pulse height and the capacitance seen by the gate. This characterization tends to be expensive and cannot easily capture the effect of the shape of the noise waveform. For noise on delay, a different model is required in the case of receiver output measurement. This can be accomplished by an *I–V* characteristic–based cell model [20].

24.5.6.3 INTERCONNECT MODELS

In the simplest case, the interconnect can be modeled through a lumped RC network. A lumped model is shown in Figure 24.20b. Here, the entire RC network is represented by a few lumped parasitic elements. In the model shown, all the resistance along the victim is summed into a single resistor, and all the capacitance to ground is lumped into a single capacitance, while the entire coupling capacitance to the aggressor is lumped into a single coupling capacitance. Refinements of this scheme may include resistance for the aggressor or a PI model for the victim, which consists of two grounded capacitors separated by a resistor.

A more sophisticated interconnect modeling scheme is to use an RC reduction technique [27], as shown in Figure 24.20c. The objective of RC reduction is to replace the detailed RC network

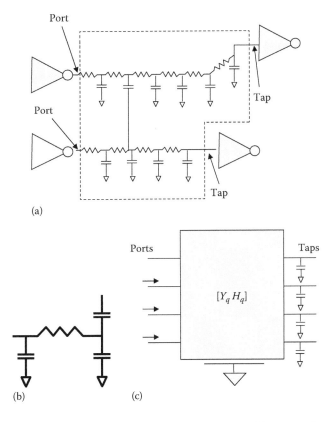

FIGURE 24.20   Interconnect models: (a) Original interconnect. (b) Lumped model. (c) Reduced-order model.

with an equivalent mathematical model, which provides similar response at the *ports* and *taps** of the RC interconnect network to that of the original unreduced network. A reduction in size is achieved because electrical characteristics of the intermediate nodes in the RC network are not preserved.

Different RC reduction techniques are available with differing accuracy and performance characteristics and features. Some RC reduction techniques can generate a reduced model, which can be represented in terms of positive R and C values (realizable models), while other techniques may create a reduced interconnect model represented by a set of transfer functions. Some reduction techniques can handle mutual and self-inductances (and the reduction process is then called RLCK reduction). A detailed mathematical treatment of interconnect reduction is beyond the scope of this chapter. Interested readers can consult Refs. [24–32] for more details.

24.5.6.4 AGGRESSOR MODELS

Aggressor models are usually quite simple as well. In the past, the most commonly used aggressor model was a *linear saturated-ramp* voltage source. This is a three-piece voltage waveform as shown in Figure 24.21a. The waveform starts with an initial value and then switches to a final value with a strictly linear shape with a given transition time. The advantage of this model is that it is very simple to model and to treat analytically. In reality, the aggressor waveform is much smoother. This can be modeled to some degree by adding a driver resistance in front of the saturated-ramp waveform to create a softer waveform as shown in Figure 24.21b. Modern analyzers use a piecewise linear waveform that better represents the different parts of the aggressor waveform. When utmost accuracy is required, the aggressor driver can be represented using a nonlinear I–V model or even transistors; the latter is computationally very expensive and is usually avoided.

24.5.7 SOLVER CONSIDERATIONS

In order to achieve any given accuracy and performance requirements, a noise analysis tool must use both the right models and the right solvers. While simpler models can be handled by a more complex solver, albeit at a lower performance, a given solver will limit the type of models that can be handled. The simplest solver is based on an analytic closed-form solution of the type we derived in Section 24.2.1. It can usually be applied only with a lumped interconnect model and a resistor-based driver cell model, and it is very useful for electrical filtering. A more complex analytic solver, similar to those discussed in [18], may be employed when distributed RC interconnect models are employed along with resistor-based cell models.

At the other end of the spectrum, the most complex and flexible solver used in noise analysis is a SPICE engine. Typically, this SPICE engine is accessed through an API. SPICE is able to

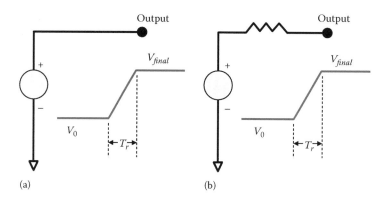

(a)   (b)

FIGURE 24.21   Attacker models: (a) Saturated-ramp model. (b) Modified saturated-ramp model.

---

* A *port* is an input terminal, while a *tap* is an output terminal. Figure 24.22c shows the ports and taps of the network.

handle all the different types of driver models; special device models may only need to be created for handling *I–V* cell models, and some modifications are necessary for handling reduced RC networks. Since noise analysis involves many small simulations rather than one large simulation, the SPICE engine may be tuned in different ways to improve its performance. For more details on simulation algorithms, please see the chapter on circuit simulation.

A special-purpose numerical nonlinear solver is usually used for many of the simulations during the detailed analysis. Such a solver is able to handle the specialized cell *I–V* models, but not the transistors. These solvers will make several simplifications to improve their performance. An example of such a simplification is to fix the time steps used in simulation in a SPICE-like engine. Often, a hybrid linear/nonlinear approach is employed. In that case, a nonlinear driver model is connected to a simple interconnect model, often preserving only few top moments of the driving-point impedance, and this circuit is solved using a nonlinear numerical solver. The computed response at the driver output node is then used as an excitation in a linear simulator to compute responses at the receiver ends of the interconnect. Efficient implementations of such techniques can be found in [13,20].

24.6 NOISE ANALYSIS IN IMPLEMENTATION AND SIGN-OFF FLOWS

Ruling out noise problems is insufficient to know that noise problems exist—we need to be able to create a design that works correctly. In this section, we will consider some prevention and correction techniques that may be applied in the design process. However, a caveat is in order: many of these techniques are heuristics employed in the design methodology or in particular design tools and are hence difficult to discuss exhaustively. The effectiveness of any methodology or tool heuristics is in the quality of the design that is produced, and hence the discussion in this section should be considered only providing examples of possible approaches.

First of all, in order to understand noise prevention and correction, consider again Equations 24.5 and 24.6 and the quantitative characteristics of coupling glitch noise derived from them. There, we saw that the noise induced on a victim by a specific aggressor is

1. Directly proportional to the ratio of the coupling capacitance to the total capacitance
2. Inversely proportional to the transition time of the aggressor
3. Directly proportional to the holding (or drive) resistance of the victim driver

These relationships tell us of the parameters that can control noise in the design. By making different design changes, we can alter one or more of these parameters to either reduce potential noise problems or to fix a particular noise problem. First, let us see what correction steps can be taken to fix a noise problem, and then we will see how to extend them to noise prevention. The typical correction steps, often called "transforms," are driver upsizing, aggressor downsizing, buffer insertion, and routing changes. We will discuss each of these transforms in more detail:

■ *Driver upsizing*: The victim driver cell is made stronger by upsizing. This will reduce the victim holding resistance, hence leading to a smaller noise. The victim, however, may also act as an aggressor to other nets. We, therefore, must be careful to ensure that the upsizing does not create noise problems elsewhere, by analyzing the parts of the design possibly affected by this transform.

■ *Buffer insertion*: In this approach, instead of upsizing the victim driver, a buffer is inserted at an appropriate point in the victim net. This will either require a complete rerouting of the victim net or a partial rerouting to connect the inserted buffer. This helps reduce noise in multiple ways. For the part of the original net that is in front of the inserted buffer, the coupling capacitance is decreased, and hence the effective strength of the driving cell is increased. In general, this transform also improves timing, since the RC delay will be reduced even if the gate delay is increased. For the part of the net that is after the inserted buffer, a better driver is provided. Multiple buffers may be inserted on the same net; inverters can be inserted if care is taken to ensure that they are inserted in pairs.

- *Aggressor downsizing*: This works by increasing the transition time of the attacking net by reducing the strength of its driver. This transform can cause the aggressor to become more susceptible to noise, or even not to meet timing.
- *Routing changes*: Routing changes can be very effective for fixing noise problems. The main cause of cross talk noise is coupling between wires, and routing affects both the overall coupling capacitance and its ratio to total capacitance. Many different routing heuristics can be employed to reduce the total amount of noise. The simplest technique is to add extra space around critical nets by leaving vacant tracks; however, this is rather expensive. Often, the router can selectively add extra space whenever it is able to do so without sacrificing its ability to route the design; this technique is usually called "soft spacing." It is also possible to change the order of nets and routing layers to prevent nets from running parallel to each other for a long distance. In addition, TW information may be used to decide which nets can run parallel to each other. Commercial routers typically use a combination of these techniques to achieve good results.

After this overview of the basic correction techniques, let us see how to put them together into a flow; this is shown in Figure 24.22. After the initial routing is completed, a full analysis of the block (or chip) is performed. If no errors are found, the process can stop. If there are noise errors or noise-induced timing errors, then the list of errors is usually fed to a correction program, which heuristically determines the changes to be made. These may include both cell changes and routing changes; even if only cell changes are made, corrective routing is typically required to hook up the inserted buffers. The correction program typically uses quick noise estimates to ensure that its corrections are effective, but the final effectiveness of the changes is verified by another round of noise analysis. The process iterates until there are no errors left; in practice, only a few automatic iterations are permitted before the designer evaluates the results and makes some changes to the settings and tolerances.

While it is good to have a correction flow, it is still important to practice noise prevention strategies throughout the design flow. However, there are certain design implementation steps where critical decisions are made, and it is important to focus on these design steps. (The various steps are discussed in Chapters 1 and 13.)

Particular attention must be paid to buses and other very dense wiring structures during the floor-planning step, especially if there are large inflexible IP blocks in the design. It is possible to create a situation as in Figure 24.23. Here, the random logic made up of standard cells is separated at the two ends of the chip, with many wires running between them. However, because of the placement of the two hard IP blocks, all the wires have to travel in a narrow area between the two blocks.

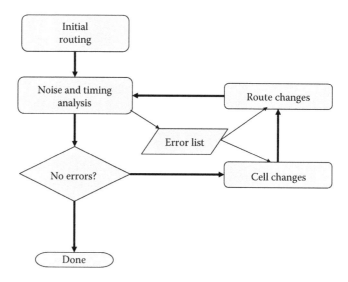

FIGURE 24.22 Noise correction flow.

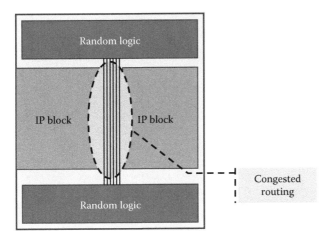

FIGURE 24.23 Floor-plan-generated noise issues.

This will cause many wires to run in parallel for long lengths, leading to cross talk problems. Given the narrow area of the channel, it will not be possible to insert buffers to solve this problem.

Another important prevention step is to ensure that good transition time limits are set during the physical optimization step. If some nets have long transition times, they might be susceptible to cross talk noise. Nets that are not timing critical can become so due to cross talk effects. It is also very important to have a noise avoidance strategy in the initial routing. This can be in the form of limiting parallel lengths of wires whenever possible. However, more sophisticated routers usually employ simple heuristic noise calculators within the inner loop of the router to predict which nets are sensitive to noise. These nets can then be routed with special consideration to reduce coupling.

24.7 SUMMARY AND CONCLUSIONS

Noise has become a design metric of significance equal to timing and area; it is not feasible to finish a digital design without considering noise. Noise must be considered throughout the design flow, and a comprehensive noise analysis and correction strategy must be applied in the design implementation tool suite. In this chapter, we have discussed the basic noise considerations, noise analysis strategies, and some techniques for noise prevention. The references provide more in-depth treatment for any reader wishing to further pursue this subject.

ACKNOWLEDGMENTS

For the many valuable discussions, I thank our colleagues and professional acquaintances that contributed significantly to the material in this chapter. In particular we acknowledge Venkat Thanvantri, Joel Phillips, Ram Iyer, Manuj Verma, and Ratnakar Goyal for their valuable suggestions and useful discussions.

Thanks to Grant Martin, Igor Markov, and Luciano Lavagno for reviewing the initial drafts of this chapter.

REFERENCES

1. D. Chinnery, D. Hathaway, L. Stok, and K. Keutzer, Design flows, *Electronic Design Automation for IC Implementation, Circuit Design, and Process Technology*, L. Lavagno, I.L. Markov, G.E. Martin, and L.K. Scheffer, eds., Taylor & Francis Group, Boca Raton, FL, 2015.
2. J. Qian, S. Pullela, and L.T. Pillage, Modeling the "effective capacitance" of RC interconnect, *IEEE Transactions on Computer-Aided Design*, 13, 1526–1535, December 1994.
3. F. Dartu and L.T. Pileggi, Calculating worst-case gate delays due to dominant capacitance coupling, *ITC*, 809–818, October 1997.

4. P. Chen, D.A. Kirkpatrick, and K. Keutzer, Miller factor for gate-level coupling delay calculation, *Proceedings of ICCAD*, San Jose, CA, November 2000, pp. 68–74.

5. K.L. Shepard, V. Narayanan, and R. Rose, Harmony: Static noise analysis of deep submicron digital integrated circuits, *IEEE Transactions on Computer-Aided Design of Integrated Circuits and Systems*, 18(8), 1132–1150, August 1999.

6. A. Khan, Recent developments in high-performance system-on-chip IC design, *International Conference on IC Design and Technology*, Austin, TX, 2004.

7. J. Pretorius, A. Shubat, and C. Salama, Charge redistribution and noise margins in domino CMOS logic, *Circuits Systems, IEEE Transactions*, 33, 786–793, 1986.

8. I. Keller, K.H. Tam, and V. Kariat, Challenges in gate level modeling for delay and SI at 65 nm and below, *Proceedings of Design Automation Conference*, Anaheim, CA, June 8–13, 2008, pp. 468–473.

9. Process Integration, Devices, and Structures. ITRS Roadmap, 2013 Edition, http://public.itrs.net.

10. A. Tajalliand and Y. Leblebici, Design trade-offs in ultra-low-power digital nanoscale CMOS, *IEEE Transactions on Circuits and Systems I: Regular Papers*, 58(9), 2189–2200, September 2011.

11. J.P. Uyemura, *Circuit Design for CMOS VLSI*, Kluwer Academic Publishers, Dordrecht, the Netherlands, 1992, pp. 80–83.

12. H. Zhou, N. Shenoy, and W. Necholls, Timing analysis with crosstalk as fixpoints on complete lattice, *Proceedings of DAC*, Las Vegas, NV, 2001, pp. 714–719.

13. J.M. Wang, P. Chen, and O. Hafiz, A new continuous switching window computation with crosstalk noise, *Proceedings of 16th Symposium on Integrated Circuits and Systems Design*, Arizona University, Tucson, AZ, 2003, pp. 261–266.

14. R.B. Hitchcock Sr., G.L. Smith, and D.D. Cheng, Timing analysis of computer hardware, *IBM Journal of Research and Development*, 26(1), 100–105, 1982.

15. K.A. Sakallah, T.N. Mudge, and O.A. Olukotun, CheckTc and minTc: Timing verification and optimal clocking of synchronous digital circuits, Digest of Technical Papers, *IEEE International Conference on Computer-Aided Design, ICCAD-90*, Santa Clara, CA, November 11–15, 1990, pp. 552–555.

16. T.H. Cormen, C.E. Leiserson, and R.L. Rivest, *Introduction to Algorithms*, MIT Press, Cambridge, MA, 1996, pp. 484–487.

17. D. Chai, A. Kondratyev, Y. Ran, K.H. Tseng, Y. Watanabe, and M. Marek-Sadowska, Temporo-functional crosstalk noise analysis, *Proceedings of Design Automation Conference*, Anaheim, CA, June 2–6, 2003, pp. 860–863.

18. R.E. Bryant, Boolean analysis of MOS circuits, *IEEE Transactions on Computer Aided Design*, CAD-6(4), 634–649, July 1987.

19. L. Scheffer, A roadmap of cad tool changes for sub-micron interconnect problems, *Proceedings of ISPD-97*, Napa Valley, CA, 1997, pp. 104–109.

20. I. Keller, K. Tseng, and N. Verghese, A robust cell-level crosstalk delay change analysis, *Proceedings of ICCAD*, San Jose, CA, 2004, pp. 147–154.

21. L.H. Chen and M. Marek-Sadowska, Aggressor alignment for worst case noise, *IEEE Transactions on CAD*, 20(5), 612–620, May 2001.

22. K. Tseng and V. Kariat, Static noise analysis with noise windows, *Design Automation Conference*, Anaheim, CA, 2003, pp. 864–868.

23. K.L. Shepard and V. Narayanan, Noise in deep submicron design, *Proceedings of ICCAD-96*, San Jose, CA, 1996, pp. 524–531.

24. L.T. Pillage and R.A. Rohrer, Asymptotic waveform evaluation for timing analysis, *IEEE Transactions on Computer Aided Design*, 9, 352–366, April 1990.

25. S.C. Chan and K.L. Shepard, Practical considerations in RLCK crosstalk analysis for digital integrated circuits, *Proceedings of ICCAD*, San Jose, CA, 2001, pp. 598–604.

26. A. Odabasioglu, M. Celik, and L. Pilleggi, PRIMA: Passive reduced-order interconnect macromodeling algorithm, *IEEE Transactions on Computer-Aided Design*, 17, 645–654, August 1998.

27. K.J. Kerns and A.T. Yang, Stable and efficient reduction of large multiport RC networks by pole analysis via congruence transformation, *IEEE Transactions on Computer-Aided Design*, 16, 734–744, July 1997.

28. H. Levy, W. Scott, D. MacMillen, and J. White, A rank-one update method for efficient processing of interconnect parasitics in timing analysis, *Proceedings of the Design Automation Conference*, Los Angeles, CA, June 2000, pp. 75–79.

29. C. Ratzlaff, N. Gopal, and L.T. Pillage, RICE: Rapid interconnect evaluator, *Proceedings of 28th DAC*, San Francisco, CA, June 1991, pp. 555–560.

30. D. Anastasakis, N. Gopal, S.Y. Kim, and L.T. Pillage, On the stability of approximations in asymptotic waveform evaluation, *Proceedings of 29th DAC*, Anaheim, CA, June 1992.

31. L.M. Silveira, M. Kamon, and L. White, EDTC, Efficient reduced-order modelling of frequency-dependent coupling inductances associated with 3-D interconnect structures, *Proceedings of EDTC*, Paris, France, March 1995, pp. 534—538.

32. J.R. Phillips and L.M. Silveira, Poor Man's TBR: A simple model reduction scheme, *IEEE Transactions on CAD*, 24(1), 43–55, January 2005.

Layout Extraction

25

William Kao, Chi-Yuan Lo, Mark Basel, Raminderpal Singh, Peter Spink, and Louis K. Scheffer

CONTENTS

25.1 INTRODUCTION

Layout extraction is the translation of the topological layout back into the electrical circuit it is intended to represent. This extracted circuit is needed for various purposes, including simulation, timing analysis, and logic to layout comparison (see the Chapters on Digital and Analog Simulation, Timing Analysis, and Formal Verification). Each of these functions requires a slightly different representation of the circuit, resulting in the need for multiple layout extractions. In addition, there may be a postprocessing step of converting the device-level circuit into a gate-level circuit [97–101], but this is not considered part of the extraction process.

The detailed functionality of an extraction process will depend on its system environment. The simplest form of extracted circuit may be in the form of a netlist, which is formatted for a particular simulator or analysis program. A more complex extraction may involve writing the extracted circuit back into the original database containing the physical layout and the logic diagram. In this case, by associating the extracted circuit with the layout and the logic network, the user can cross-reference any point in the circuit to its equivalent points in the logic and layout (cross-probing). For simulation or analysis, various formats of netlist can then be generated using programs that read the database and generate the appropriate text information.

In this chapter, we will make an informal distinction between *designed devices*, which are devices that are deliberately created by the designer, and *parasitic devices*, which were not explicitly intended by the designer but are inherent in the layout of the circuit.

Primarily, there are three different parts to the extraction process. These are designed device extraction, interconnect extraction, and parasitic device extraction. These parts are inter related since various device extractions can change the connectivity of the circuit, for example, resistors (whether designed or parasitic) convert single nets into multiple nodes. Usually one level of interconnect extraction is used with designed device extraction to provide a circuit for simulation or gate-level reduction, and a second level of interconnect extraction is used with parasitic device extraction to provide a circuit for timing analysis.

25.2 EARLY HISTORY

Early integrated circuits (ICs) were small, and designers manually determined the electrical model corresponding to the geometry, including devices, their interconnections, and estimates of parasitics. However, even in a small layout it was easy to miss a width or spacing error, so design rule checking (DRC) programs emerged to find these errors (see Chapter 20). DRC programs require a machine-readable layout, and designers soon realized such a representation could be used to derive automatically a circuit representation from the geometry. Extraction was first implemented as an afterthought to these early DRC programs—for example, the first mention of a computer program that processes layouts into netlists is a single paragraph at the end of a DRC paper [83]. As chips grew larger, and processes more sophisticated, extraction became an important task in its own right, and emerged as a specific task distinct from DRC, and soon there were programs (and papers) concentrating entirely on extraction [87]. Despite this divergence, the two tasks share considerable basic technology. The now-universal scan line approach, used in both DRC and extraction, appears in [84], and is improved to near current form in [88]. Early consideration of the use of hierarchy can be found in [85].

25.3 PROBLEM ANALYSIS

For layout extraction, there are eight main areas to be evaluated and analyzed. These are overall system capabilities, designed device extraction, connectivity extraction, parasitic in-line device (resistance) extraction, parasitic cross-coupled device (capacitance and inductance) extraction, and network reduction. These issues are summarized briefly here and covered in detail in later sections.

1. *Overall system.* What is the intended application—analog, digital, radio frequency (RF), or other? What are the input and output formats and databases? Is extraction flat or hierarchical? Batch or incremental?
2. *Creating the as-built silicon geometries from the drawn geometries.* The designer draws an idealized version of the overhead view of the layout. Extraction needs the as-built geometries of the devices and interconnect, including thicknesses, and this must be derived.
3. *Designed device extraction.* Find the designed devices and determine their parameters. Remove them from the underlying artwork, if needed, so that what remains is interconnect.
4. *Connectivity extraction.* Find which combinations of the remaining geometry form nets. Assign a name to each net. Determine which nets are connected to which pins of the devices, and which are externally visible.
5. *Resistance extraction.* Optionally, each net may be divided into one or more sub-nodes to account for the parasitic resistance of each net. This is a 2D problem. Along similar lines, the substrate may need to be subdivided as well. This is a 3D problem over a large area, and different techniques are used.
6. *Capacitance and inductance estimation.* Each piece of each net must have its capacitance and inductance estimated. Both capacitance and inductance may require both self and mutual terms.
7. *Reduction.* Straightforward extraction gives netlists that are far too big for most practical uses. Reduction generates smaller netlists with very similar properties.
8. *Process variation.* Because the fabrication process can vary significantly, even from place to place on the same die, the user may be interested in not only the nominal component values, but also in how they change with changes in physical dimensions.

We examine each of these steps in turn.

25.4 SYSTEM CAPABILITIES

The overall design of the extraction system is dictated by the design of the overall system in which it resides. Usually layout extraction is a single component in a much larger system involving a complex database, layout editor, simulator, design rule checker, etc. The two major impacts this has on the layout extraction are in the areas of hierarchical processing and results processing.

There are three main possibilities for hierarchical processing:

- First, the complete circuit being processed can be flattened to a single level of hierarchy. This is not usually done since it generates extremely large data sets and it eliminates the optimization of the cell-level repetitions. It does, however, facilitate the extraction of all designed and parasitic devices without imposing any design or layout restrictions.
- Second, the layout can be processed on a cell-by-cell basis, ignoring the hierarchical relationships between cells other than for the determination of inter-cell connectivity. This is a popular processing methodology since layout cells are normally designed to match the logic cells, which are normally complete mini-circuits in themselves. This method does, however, require some form of hierarchical processing to form connections

between cells, and it cannot handle intra-cell parasitic capacitances or devices formed across the hierarchy.

■ Third, a true hierarchical system can be designed which allows for the extraction of inter-cell parasitic devices and permits optimizations beyond the cell-level repetitions. This theoretically allows both fast processing and unrestricted layouts, but involves a complex special-purpose database and complex algorithms. Also, even if the extraction program allows it, it is often a good idea to prohibit designed device formation between cells as they violate the hierarchy and eliminate repetition optimizations.

For the purposes of defining extraction techniques, this document will look only at the flat- or cell-level extraction, without reference to the hierarchy. Extension to fully hierarchical extraction is a fascinating problem but beyond the scope of this chapter.

The ultimate result of layout extraction will be a circuit description in a form that can be used by a simulator or analysis program. This can be (and originally was) achieved by writing the extracted results out directly as a text file in the simulator language. Today however, it is more likely that the extraction program is part of a larger system. The overall system will enable such features as viewing the extracted circuit on top of the layout; cross-probing the extracted circuit relative to the layout and the logic diagram; and translating the extracted circuit into different text formats for different analysis programs. Each of these functions will require different information from the extraction process stored in the database. What the content and format of this output is will depend on the overall system and database design, but it will probably include physical shapes representing the extracted circuit and devices, in addition to circuit interconnect information to represent the logical extracted circuit.

For the purposes of defining extraction techniques, these detailed output requirements will be ignored. The focus will be on the detection of the circuit elements as logical entities.

A large system for the design of ICs will already have many tools available for the manipulation of layout data (polygons). Many of these can be utilized in the development of an extraction program to save development time and to minimize new buggy code. These tools may be low-level tools such as scan lines, sort programs, hash tables, etc., or they may be higher-level tools such as programs to generate logical operations like AND and OR between polygons. Choosing to use these tools will push the developer in specific directions when deciding on the techniques to use.

25.5 CONVERTING DRAWN GEOMETRIES TO ACTUAL GEOMETRIES

The designer of an IC draws an idealized overhead view of all the layers to be created. Accurate parasitic extraction, however, depends on accurate knowledge of the dimensions of structures as they are actually built on the wafer. For many years this difference was small and could be ignored, and extraction was based on the drawn geometries. However, with sub-wavelength lithography (see Chapter 21) and modern planarization techniques (such as chemical mechanical polishing [CMP]) this difference is no longer small, and must be accounted for. There are two main effects. First, the size and shape of the geometries may be different than drawn, primarily because of limitations in mask making, exposure, resist, and etching processes. Second, the thickness of the various layers is not specified at all by the designer, and must be determined (at least implicitly) by the extraction program.

The first step is to take into account the limitations of mask making, lithography, exposure, and etching. The designers, of course, may draw whatever they wish, but the as-built IC geometries will be limited by the resolution of light and lenses. Specialized software, often called optical proximity correction (OPC) or resolution enhancement technology (RET) is used to preprocess the user-defined geometries before making masks, so that the result on the wafer will best match the designer's intent. This process is not perfect, however. In some cases full correction is simply not possible (the physics of exposure prevents a truly square corner, for example) and in other cases a very accurate correction is deemed too expensive. Therefore, to best match the silicon result, the layout may need to be preprocessed to accurately represent what will really be built. For optical effects, the radius is small (perhaps a micron or so), but the edge positions can depend in complex ways upon the surrounding geometries and the OPC process.

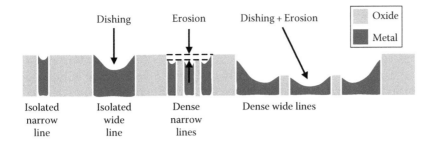

FIGURE 25.1   Layout dependent variation of metal thickness depending on local layout properties. This effect is caused by the chemical–mechanical polishing (CMP) process, which grinds down the metal faster than the oxide, since the metal is softer.

Next, the sizes of geometries in the vertical direction (perpendicular to the wafer) are not specified at all by the designer, and are determined entirely by the process. Originally, these dimensions were either specified by a technology file, or assumed implicitly by the extraction coefficients (amount of capacitance per unit area, for example) and were assumed to be the same for all geometries on a given layer. However, with modern processes, the vertical dimensions of shapes are a function of the surrounding neighborhood. For example, CMP makes it possible to stack many metal layers by grinding each layer flat as it is processed. However, metal and oxide grind at different rates, leading to phenomena such as "erosion" and dishing, as shown in Figure 25.1. These must be accounted for during the extraction process. In general, these effects depend both upon the details of features (such as the width and spacing to neighbors) and the overall density over the surrounding area (typically averaged over a few hundred microns) [89–91].

There are tools and techniques that can analyze these effects accurately, but most are simulation based and too slow for extraction of large circuits. Therefore, extractors usually use simple empirical models. For OPC, the most common model is an "edge bias," or displacement of the edge from the drawn position, which depends on the width and spacing to the neighbors. For CMP, the extractor may define the layer thickness as a function of wire width, spacing to neighbors, and local density integrated over some region.

25.6 DESIGNED DEVICE EXTRACTION

A designed device is any device required in the circuit that is explicitly created in the layout. These devices can be created by the inter-relationship of the active circuit layers as in the case of a metal-oxide-semiconductor (MOS) transistor or capacitor occurring anywhere when polysilicon overlaps diffusion (Figure 25.2a), or by the application of specific device definition layers as in the

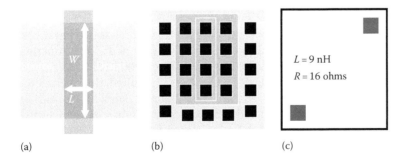

(a)   (b)   (c)

FIGURE 25.2   Three styles of defined devices. (a) Typical for MOS design—the transistor is defined by poly overlap diffusion, and the extractor should measure *W*, *L*, and the area of source and drain. (b) Typical bipolar design—transistor is defined by emitter inside base inside collector. Extractor should measure number of contacts, length of perimeter, and area of each polygon. (c) Device inside a cell commonly used for RF or analog design. Extractor should just make connections, but take device values from cell name or properties.

case of a designed bipolar transistor (Figure 25.2b), or by inclusion of specific cells implementing a device (Figure 25.2c), as is typical in RF design.

There are number of steps involved in extracting a device. Firstly, the existence of the device must be recognized. The exact form of the device must be extracted (and verified for correctness) and an instance of the device added to the circuit database. Each device must be given a unique identification and its terminal connections identified. If the existence of a device impacts the flow of connectivity in an interconnect layer, like an MOS transistor that breaks the continuity of diffusion, then that device area must be removed from the connectivity layer. Finally, if required, the parameters of the device must be measured.

The recognition of devices is linked directly to the circuit connectivity extraction. The connectivity extraction cannot be done until the device areas are removed from the interconnect layers, yet the devices cannot be fully recognized until the interconnect nets are extracted so the device terminal connections can be identified. (For example, the device identity is sometimes different depending on the number of distinct nets connected.) This means the two functions of device recognition and interconnect extraction must be done in parallel or interwoven.

There are five stages involved in designed device extraction:

1. Firstly, before any interconnect analysis is performed the device must be recognized, a device definition shape created, and the device given a unique identifier. Devices must be recognized by the specific combination of design layers, by the presence of a specially created device definition layer, or by the presence of designated cells (refer again to Figure 25.2). For each device that is recognized, a series of polygonal logical operations must be performed to create a single shape (polygon) that represents the physical location and extent of that device.

2. Second, if the existence of the device breaks the continuity of an interconnect layer, the area of that device must be removed from the interconnect layer prior to interconnect analysis being performed.

3. After interconnect analysis, the device definition shape must be related back to the associated interconnect layer to determine the existence and extent of terminals. Each terminal found must be identified as a physical and a nodal connection to the device. The layer involved in each connection will sometimes define the type of terminal (e.g., for MOS transistors, diffusion terminals will form sources and drains, and polysilicon will form gates).

4. Next, given the terminals and nodal connections to the device, it must be validated. Each device must have the correct number of terminals of each type. A different number may mean the device is badly formed, or it may mean it is a different device type. In addition, the nodes connecting to the device may affect the validity or type of device. For example, a MOS transistor having three physical source drains may be a badly formed device unless that is a valid configuration for the technology involved. However, if two of those terminals are connected to the same electrical node, the device has only two logical source drains and that may be acceptable. If the transistor has two physical source-drain terminals which are connected to the same electrical node, then functionally it is an MOS capacitor and may be redefined as that device type.

5. Finally, if the measurement of device parameters is required, this operation must be performed. The measurement process itself is fairly simple. For a MOS transistor where the parameters of gate length, gate width, and source and drain area are required, measuring the appropriate edge lengths of the gate polygon and the areas of the source and drain diffusions gives the required answers. For bipolar devices, the area and perimeter of each of the constituent polygons is required. Add to these the association of the correct terminal node number to each measurement and the parameters are complete. When and how these measurements are made will depend on the algorithms being used for general polygon manipulation and device recognition. It may be possible to make these measurements during the process of recognizing and extracting the device recognition shapes (gates) or it may be necessary to perform secondary measurements. For devices contained in explicit cells, the device parameters can be obtained from the name or the

properties of the master cell instance (or the occurrence, in the case of programmable cells, or pcells). For unusual devices, user-defined code may be needed to establish the device parameters. (See Chapter 15, for more information about pcells and extension languages, which are used for these tasks.)

A very real problem for device parameter measurement is the definition of the required parameters and their relationship to interconnect parasitic measurements. The layout is inherently a distributed system, but the desired output is almost always in terms of lumped elements. This conversion is inherently ambiguous. Consider Figure 25.3, and the questions it raises:

- If transistor source and drain area are required as parameters, how should the area of diffusion between gates A and B be considered? Is the total area applied to each gate or is it divided between them?
- If the source and drain area are used to define the inherent capacitance in a device, should the areas measured for device parameters be excluded from the interconnect parasitic capacitance measurement?
- For the extension of the diffusion to the right of gate B, how much should be allotted to the source/drain area and how much is just an interconnect path to the next device?
- The resistive polysilicon path between points C and D has a single logical connection point to gate B so the polysilicon area over the gate is still part of the interconnect resistance. However, the gate area itself is part of the device and can be included as part of the device characteristics or measurements. How should this be resolved?
- Not all devices are constructed as simple rectangles, as shown by transistor E, but SPICE (for example) wants a single value for the width W and the length L. A transistor gate can be shaped like an L, U, or Z, or even a more complex shape with any number of bends. Some technologies even allow Y shaped or matrix transistors. How is W and L for these devices defined?

Only the electrical engineer responsible for the technology can answer these questions and a major component of designing a system will be the interaction with that engineer to completely clarify what is required. The parameter definitions provided by the electrical engineer must cover all device types allowed in the technology for which the program is being written, and the measurement techniques developed for their extraction must be adapted to suit.

All the extracted information (device shapes and location, terminal locations and types, and nodal connections and parameters) must be stored back into a database accessible by other functions of the extraction process and ultimately by the programs that will generate the netlists required for the simulation and analysis programs.

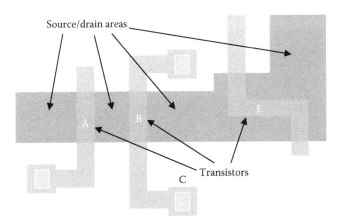

FIGURE 25.3   Diagram illustrating the inherent ambiguity of converting a distributed layout model into a lumped model.

25.7 CONNECTIVITY EXTRACTION

A circuit consists of devices, device terminals, I/O terminals and the network that interconnects them. The network is made up of conductive elements and the contacts that join them together. For the circuit to be useful in subsequent analysis programs, certain terminals or nodes in the circuit must be identified by their function name.

For clarification, in this context, a net is a physical collection of shapes in a layout that connect together to form a single logical entity, which ideally shares a single voltage value. A node is a point in the physical net that forms a terminal or branch. It can also be used to define the single logical entity in the circuit that is represented by a net.

For connectivity extraction the following issues must be addressed:

■ In-line designed device identification and removal. The first step in interconnect extraction must be the removal of any designed device that changes the connectivity of the circuit. This is addressed in Section 25.3.3. Figure 25.4 shows the interconnect layers for the circuit from Figure 25.2 after the designed devices have been removed.

■ Interconnect continuity recognition through contacts and vias. There are two main forms of connection in IC layouts. Layer to layer connections are defined by contacts and vias. Connections within the same layer are formed by abutment or overlap. Within a cell, connection by abutment is not needed, since two shapes on the same layer that touch are in general made part of the same polygon. Connection by abutment or overlap is more commonly used to connect cells together, or to connect interconnect to cells.

■ Circuit terminal identification. There are three forms of terminals to be considered during extraction. When a designed device is identified, the body of the device may form a single terminal as in the case of the gate of MOS transistor. This terminal can be recognized by the overlap of the device body and the terminal layer.

The body of a designed device may form a terminal by butting up against the edge of a shape on the interconnect layer, as in the case of the source and drain of an MOS transistor.

The input and output terminals of a circuit must also be identified, and this is normally done through a terminal definition layer such as a bonding pad with the addition of a

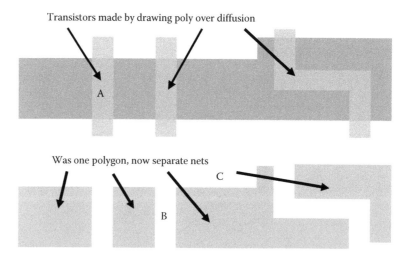

Transistors made by drawing poly over diffusion

A

Was one polygon, now separate nets

C

B

FIGURE 25.4 Why designed devices (transistors) must be removed from drawn layer diffusion before interconnect extraction. The upper panel shows a portion of a layout as drawn; the bottom panel shows diffusion as used for extraction. Note that poly gates do not need to be removed from the poly layer, but we need to make sure the gate capacitance is not double counted.

text label, or in some cases by the use of a text label alone without the identification layer. Additional issues are:

- Terminal naming and net labeling. In addition to the identification of I/O terminals, internal nets in the circuit may need specific identification to help the application programs that are going to be applied to the extracted circuit. These nets are normally identified simply with text labels.
- Internal net numbering. Every net extracted must be identified in some way. Some will be defined by net and terminal labels. All others must be given a unique identification, usually in the form of a number.
- Net subdivision through parasitic resistance identification. When parasitic resistance is required, the nets must be extracted twice. First the nonresistive interconnect must be extracted. Then for each physical net that contains shapes on resistive layers, a resistive sub-network must be extracted. The issues of resistance extraction will be dealt with later, but for the actual network extraction there is one additional issue that must be addressed.

For nodes in the circuit that form inputs and outputs, the exact point of the I/O terminal must be identified. For normal connectivity extraction, a label anywhere on the net suffices to identify the function of the net; however, if a resistive network is to be extracted, that is not sufficient. The resistance network must form a path between the I/O and the device terminals. Either an identifier shape must be used to exactly position the I/O terminal, or a text label must be placed at the exact location of the terminal.

25.8 PARASITIC RESISTANCE EXTRACTION

Once the connectivity extraction is complete, nets with resistive interconnect layers can be converted into parasitic resistance networks. This is best done one resistive layer at a time. Multiple layers in one net can be extracted separately but the connection between the layers, whether it is vias or sections of a nonresistive layer, must be identified as additional nodes in the network. There are a number of steps involved:

- Terminal location. Polygons on each resistive net must be related to contacts, I/O terminal definitions, and to device terminals to determine the exact physical position of the resistive network terminals.
- Breaking the nets into shapes. The basic idea is to break a complex shape into a collection of simpler shapes. The R of each simple shape can be estimated, then the results combined to create a resistive network representing the net. How the net is broken up determines the accuracy and efficiency of the process (see [23] for details).
- Resistance calculation. For each shape extracted from the layout, a resistance value must be calculated using the dimensions of the shape, its function (linear resistor, bend, junction, etc.), and a resistance coefficient value. In addition, the resistance values of the contacts must be calculated if these are required.

In modern copper processes, several additional effects complicate resistance extraction. Unlike aluminum interconnect, the entire conductor cross-section is not one homogeneous material. Instead, there is a thin layer of cladding, intended to keep the core of copper in place. The cladding is a much poorer conductor than the copper, and this must be taken into account, particularly for microwave frequencies where the skin effect concentrates current in this part of the wire. Also, for thin wires at deep sub-micron dimensions, a significant fraction of electrons scatter off the walls of the conductor. This leads to an increase in resistance of narrow lines that depend on the surface roughness of the conductor.

The resistance of contacts can be significant, and for accurate extractions it must be considered in more detail. There is the vertical resistance of the contact material itself which for a given technology is inversely proportional to the contact area, and there is the contact edge resistance (where the resistive material thins out as it dips over the edge of a via) that is inversely proportional to the length of periphery of a contact. Complicating the issue even further is the inclusion in the layout of contact arrays where the array itself forms a very complex resistive network or where the multiple resistive layers are in parallel. Two other issues are associated with resistance extraction:

■ Network reduction. In general, an accurate R reduction requires splitting a net into far more pieces than the user desires. Finding a smaller network with the same response is the problem of "reduction." The reduction that can be achieved depends strongly on the application and the need for accurate capacitance and inductance. This will be covered in more detail in Section 25.3.8.

■ Substrate coupling. Extracting the equivalent circuit for the substrate is similar to resistance calculations in that it requires dividing the network (here the substrate) into smaller pieces, then extracting each piece and computing a reduced model. The coupling mechanisms are shown in Figure 25.5, and the desired result is shown in Figure 25.6, where the entire substrate is condensed into a relatively simple electrical model. Substrate extraction differs from interconnect extraction since the problem is 3D, not 2D, and the networks are much larger. Therefore specialized techniques have been developed for both extraction and reduction of substrate models—(see Chapter 23). A wide variety of methods, including finite element, finite difference, and boundary element methods have been used for this task.

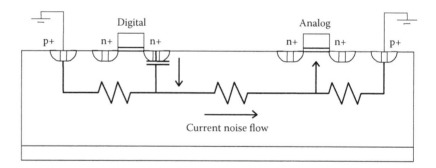

FIGURE 25.5   Current is injected into the substrate and flows to other parts of the circuit.

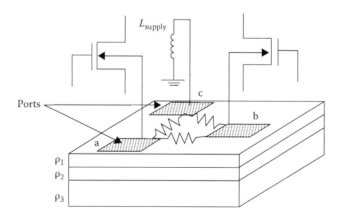

FIGURE 25.6   Substrate ports (nodes) created to connect extracted substrate resistive network to the circuit.

25.9 CAPACITANCE EXTRACTION TECHNIQUES

Capacitance and resistance are the dominant interconnect parasitics in the very deep sub-micron regime. Even though the recent processing technology advancements of copper interconnect reduces the effect of R, and the low k (<3) dielectric material reduces the effect of C, the continued scaling down of the feature size keeps the parasitic effects dominant, and makes accurate capacitance extraction a necessity. Capacitance extraction has advanced from 1D, 2D, 2.5D to 3D effects to meet the required accuracy. Here, the extraction definitions of 1D, 2D,..., etc. are given roughly to mark the evolutions.

In 1D capacitance extraction, the area and perimeter parameters of interconnect geometries are obtained. A fine-tuned set of area and perimeter weights per routing layer can be used to calculate capacitance values as an inner product.

$$Cap = (area\_amount, perimeter\_amount) \cdot (area\_weight, perimeter\_weight)$$

Such area and perimeter weights can be obtained by pre-characterization of an "average" environment of a wire. The area can be single layer or a combination of layer overlaps.

In 2D capacitance extraction, the parameter sets are extended to closely account for the same layer lateral capacitive effect. In other words, not only is the conductor's overlap of interest, but also the capacitive effect across emptiness (free space). In 2D mode, all capacitance effect are modeled as long routing structures with per unit length values. The pure 2D capacitance model does not address the effect of 3D structures (such as two wires in layers metal 2 (m2) and metal 3 (m3) crossing).

An extension to address 3D effects is the 2.5D method [39], where the 3D effect is modeled as a combination of two orthogonal 2D structures. This is illustrated in Figure 25.7.

In Figure 25.7, an m2 wire crosses an m1 wire. Along a vertical cut line (looking down), a 2D cross-sectional view is shown in the middle. Along a horizontal cut line (looking left-right), the other 2D cross-section view is shown to the right. By carefully composing a 3D solution from the two orthogonal 2D ones, most 3D effects are captured.

Obviously, true 3D extractors would go straight to a 3D pattern for a solution in the above crossing structure. However, there are numerous variations in 3D structures. Three dimensional capacitance extractors are not trivial extensions of 2D ones.

25.9.1 HOW MODERN 3D CAPACITANCE EXTRACTORS WORK

Modern capacitance extraction tools divide the full-chip extraction tasks into three major steps. The first step is called "technology pre-characterization." Given a description of the process cross-sections, tens of thousands of test structures are enumerated and simulated with 2D or 3D field solvers [40–43]. The resulting data are collected either to fit empirical formulas or to build look-up tables (call either one a pattern library). We cite [44] as a reference for model fitting using analytical equations. A good fit would require fewer simulation points. The number of patterns can be reduced by pattern reduction techniques.

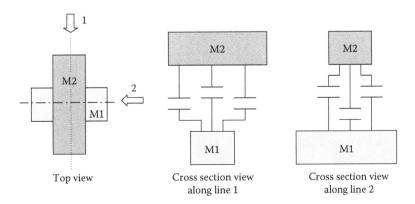

Top view   Cross section view   Cross section view
                along line 1   along line 2

FIGURE 25.7 Three-dimensional effect modeling using 2D structures.

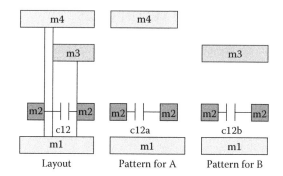

FIGURE 25.8 Layout pattern and two precharacterization patterns.

We cite [45] as a reference for pattern compression technique. Pattern compression reduces the total number of pre-characterization patterns. Specifically, consider the following layout pattern and two precharacterization patterns in Figure 25.8. In the layout pattern, c12 (m2 to m2 lateral capacitance) is to be obtained over some fixed separation distance. The bottom is fully covered by m1 plate while the top is partially covered by m3 plate and partially covered by m4 plate. A vertically disjoint partition is made to yield two volumes, A and B. There are two corresponding precharacterization patterns A and B, each producing a distinct c12 value at the fixed spacing for m2 to m2 lateral cap. The actual c12 can be composed from c12a and c12b.

Capacitance field solvers employ different numerical algorithms (see Chapter 29). Even though they often give similar answers for dense layout structures, they may give very different answers for sparse layout structures, depending on the problem setup and boundary conditions. Therefore, the precharacterization software should have the flexibility to incorporate any third party field solvers. The first step needs to be performed only once per process technology. The challenge in this area includes the handling of ever-more-complex processing technology: low k, air bubble dielectric, nonrectangular conductor cross-sections, conformal dielectrics, shallow trench isolations, etc. An effort of standardization on the process file format is sponsored by SII (Silicon Integration Initiative). (For more details please visit the website www.si2.org.)

The second major step in capacitance extraction is the geometric parameter extraction. The geometric parameters are an integral part of precharacterization. If a geometric pattern requires ten parameters to describe, each with five possible values, there is a corresponding pre-characterization of 5^{10} (10,000,000) patterns to simulate. This is assuming that all sample points are taken in each of the 10 parameters, resulting in a 10-dimensional table of the above size. This is clearly not feasible. On the other hand, if a geometric pattern can be described by very few parameters, then it is difficult for it to be accurate. In a full-chip situation, the run time of geometric parameter extraction can be very time/space consuming, with millions of interconnect polygons to analyze. Time- and space-efficient geometric processing algorithms are very important.

A geometric parameter reduction technique can be found in [46], where geometric parameters can be dramatically reduced by taking advantage of the shielding effect. Conductors two layers away from the main conductor of interest do not require a precise description. This is particularly useful for the very deep sub-micron geometry, where a far away conductor mesh behaves like a big plate.

The last step in capacitance extraction is to calculate capacitance from geometric parameters. Here, the geometric parameters are matched to entries in the pattern library. One major source of error is called the "pattern mismatch," where extracted geometry parameters do not have an exact match in the pattern library. At this time, there are two remedies to perform the capacitance calculations. One method is to enhance the pattern library by running field solvers at the full-chip extraction time. The other method is to employ heuristics to synthesize a solution from closely matched precharacterization patterns. Even if all the geometric patterns match the library completely, there could still be discontinuities in the layout pattern decomposition, which is another source of error. If the sources of errors can be identified and characterized, it is possible to estimate the error committed due to each error source. One such error bound can be obtained by the "empty" and "full" boundary conditions [47,50].

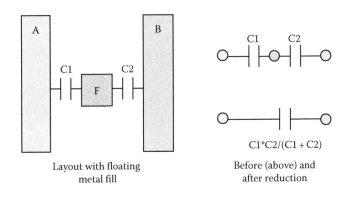

FIGURE 25.9   Area-fill pattern becomes a floating net.

25.9.2 CRITICAL NET EXTRACTION

Given that full-chip pattern matching capacitance extractors are subject to errors, they should allow a third party field solver to be plugged in to reextract critical nets. A SEMATECH critical net application program interface, or API, is a possible standard. In this API, coupling capacitances are supported. The standardization effort of this API is headed by SII.

25.9.3 AREA FILL

Modern chemical mechanical polishing (CMP) process requires area-fill patterns to help planarization. The area-fill patterns becomes a floating net in a chip. The presence of floating nets can make traditional delay calculation or signal integrity verification unreliable, or cause to fail completely. A typical solution is to ground the floating nets (either physically by wire routing or artificially by the extraction tool). In either approach, the resulting cap values are larger (more pessimistic) than reality. In Figure 25.9, a simple demonstration with three nets is shown. Net f is the floating net, which does not connect to anything.

Let us assume c153ff and c254ff. Using a grounding method, net a would have 3ff to ground, while net b, 4ff to ground; there is no coupling between a and b. Using the well-known series capacitor merging formula, the equivalent capacitance between net a and b becomes 1.7ff, which is much smaller than either one. Once the equivalent cap is obtained, both net a and net b will have 1.7ff to ground, and net a–b coupling is also 1.7ff.

In certain 3D capacitance field solvers (e.g., BEM), floating nets can be modeled with a zero total charge boundary condition on them (with equal electric potential on their surfaces). With such modeling, the same results can be obtained as above. However, the 3D solvers have limited capacity and cannot be used in the full-chip environment.

In reality, the area-fill patterns form an array of polygons, which means many floating nets need to be eliminated. A sparse capacitance matrix solver is required to eliminate them for aggressive designs.

25.10 INDUCTANCE EXTRACTION TECHNIQUES

For the idealized case of a lossless, homogeneous dielectric with an array of conductors, the inductance matrix $[L]$ can be derived directly from the capacitance matrix $[C]$ by

(25.1)
$$[L] = \frac{1}{v_0^2}[C]^{-1}$$

where v_0 is the phase velocity of the medium [51]. Unfortunately, in the IC domain, these assumptions do not hold and more complicated methods are necessary.

Inductance is much more complicated to extract than resistance or capacitance, because of the loop current definition of inductance. In other words, the calculation of inductance for a particular structure relies on knowledge of the return current paths, in addition to the current flow through the wire itself.

25.10.1 THE RETURN CURRENT PROBLEM

To complicate things further, not all of the return currents follow a simple DC path. Instead, they follow the path of least impedance, so the return path may change with frequency or even with the circuit state. Some return currents are in the form of displacement currents through the interconnect capacitances (Figure 25.10) or even the substrate. If these displacement currents are not taken into account and only the DC path is used, the loop inductance will be overestimated.

Capacitance is a function of the electric field density (i.e., flux) between conductors. By and large electric field lines conveniently terminate on adjacent conductors, physically limiting the scope of the problem. Furthermore, it is possible to limit artificially the scope of the problem by only taking into account nearest-neighbors. Doing so will only affect the accuracy of the capacitance calculation but would not result in an unstable RC interconnection model. This is not the case for inductance.

Artificially restricting the physical problem during inductance extraction by using a nearest-neighbor approximation leads not only to obvious inaccuracies but more insidiously, it can result in unstable interconnect models [52].

The magnetic field lines due to a current in a wire do not terminate nicely on adjacent conductors but will spread out to encompass as many wires as necessary to provide a sufficient return current path. Since arbitrarily eliminating some of these wires from consideration could result in ill-behaved models, the inductance matrix produced for a given structure will be dense (i.e., no wire interactions can be eliminated and therefore all off-diagonal elements will be nonzero).

It is therefore highly desirable to use some form of circuit reduction technique when including inductance in the interconnect models. Otherwise the size of the interconnect models produced could overwhelm the analysis tools. Note that not all circuit reduction methods can handle inductance, especially mutual inductance.

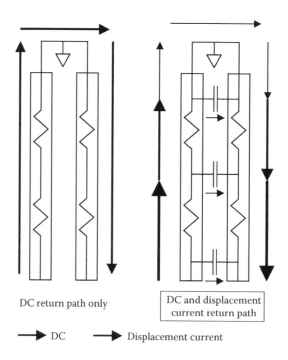

DC return path only

DC and displacement current return path

→ DC   → Displacement current

FIGURE 25.10 Inductance return current loop with and without considering displacement currents.

Since a simplistic nearest-neighbor strategy cannot be applied to the inductance problem, the simple pattern matching (i.e., rule based) methods used in capacitance extraction would not work (at least not for the general case of arbitrarily shaped and placed conductors).

25.10.2 NUMERICAL TECHNIQUES

Most inductance extraction methods in use today are numerically intensive and involve breaking the 3-D conductors into filaments or panels (Figure 25.11) [53–55]. Unfortunately, this method requires that the return current path be known, and may require meshing adjacent conductors, including the substrate. The rapid growth in computational resources required by such methods limit them to small problems on the order of a few dozen shapes. In reality, even a few lines above a nonideal substrate may overwhelm purely numerical methods.

The partial inductance [56,57] method avoids the problem of determining return current paths by assuming that the loops are closed at infinity.

As seen in Figure 25.12, the return loop of line 2 is closed at infinity. The partial inductance of a segment is defined by the flux due to the current in another segment captured inside this

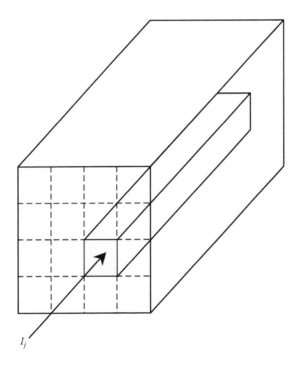

FIGURE 25.11 Example showing one of the current filaments created as part of the discretization of a rectangular conductor.

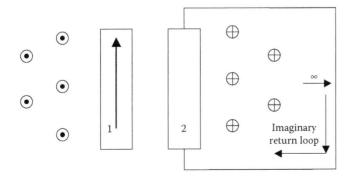

FIGURE 25.12 Computation of partial inductance using an imaginary return path at infinity.

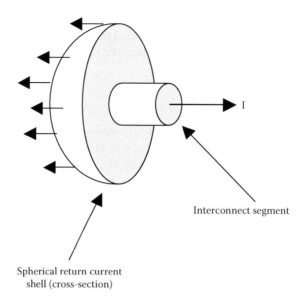

Interconnect segment

Spherical return current
shell (cross-section)

FIGURE 25.13 Example of an equipotential shell of return currents used to decrease the density of the partial inductance matrix.

semi-infinite loop. The total loop inductance is defined as the sum of the partial self- and mutual inductances along the closed path.

Since the area of each partial inductance loop is much larger than it needs to be, their values tend to be much larger than the actual inductance. The trade-off then in not having to know that the return path is both deviations (errors) from the actual values and potentially ill-conditioned matrices in the interconnect model due to these overly large values.

Various methods have been proposed [52] to sparsify a dense partial inductance matrix in order to reduce the complexity and simulation time of the interconnect model. These turn out to be impractical for a number of reasons. First, they do not reduce the computationally intensive extraction time significantly (at least compared to a rule based capacitance extraction tool). Secondly, the window around the target conductor for sub-micron designs is still fairly large, requiring the inclusion of a large number of wires in order to maintain stability.

Another method to produce a sparse inductance matrix is by use of equipotential shells [58,59]. Rather than defining the return current path at infinity, an equipotential shell is placed around the conductor segment. Instead of an electrical charge as one would use in capacitance calculations, this shell has a uniform current distribution (Figure 25.13). This shell then negates the magnetic vector potential created by the current in the segment for regions outside of the shell. This forces the mutual partial inductance to segments outside the shell to 0.

As long as the shell is a sphere or other equipotential surface, the only effect on the partial inductances inside the shell is to shift them by a constant. It turns out that in many cases this does not produce a corresponding change in the circuit solution [59].

Unfortunately, there are a couple of problems with this method. Firstly, creating the shells can be nontrivial when dealing with complex interconnect structures. If only part of a segment lies within another's shell, the partial inductance calculations become complex. Secondly, even though this technique may greatly enhance the sparsity of the resulting matrix, it is still computationally expensive and not well suited to large-scale (particularly full chip) problems.

25.10.3 RULE-BASED METHODS

Due to the return current problem, it is difficult to create a rule-based scheme such as that used in capacitance and resistance extraction. For the special case of long parallel lines, Krauter and Mehrotra [60] have proposed a quasi-rule-based scheme that uses closed-form expressions instead of numerically intensive iterative methods. By exploiting the fact that inductive issues are a greater problem in the long, wide lines found in upper-level routings, they were able to reduce the scope of the problem. In general, these upper-level routes are often composed of long stretches of parallel lines resembling a traditional microwave coplanar waveguide [51]. (A related approach [71] assumes return currents flow in the nearest power supply line.) This assumption allows the use of well-established, closed-form expressions for the inductance of such structures [56]. Unfortunately, there are closed-form expressions only for the simplest of structures.

25.10.3.1 ONLY EXTRACT INDUCTANCE WHEN YOU REALLY NEED TO

Since inductance is difficult to extract and reduce, it makes sense to extract it only when necessary. Several approaches have been proposed to determine exactly where inductance extraction is necessary.

One technique looks at the bandwidth of the signal to be transmitted. The faster the edge rate of a signal, the greater the bandwidth requirements for the interconnect structures. Insufficient bandwidth will result in severe distortion of the signal, affecting timing, power consumption, and other performance parameters. This need for high bandwidth implies that more complex interconnect models are necessary. Simple RC models are insufficient to simulate accurately the interconnection wiring in extreme conditions. A multistage RLC model would be more appropriate.

Modeling the interconnect as a true frequency-dependent transmission line captures the behavior of the line more accurately by including phenomenon such as retardation (i.e., time of flight). Note that it is impossible to generate a constant signal delay at all frequencies (due to the finite speed of light) with models using only discrete passive elements.

The critical length of a line (the length above which inductance must be included) can be determined from knowledge of the desired edge rate of the signal, along with the speed of propagation of the interconnect structure. Several rules-of-thumb can be used to determine what this critical length is. In general, an interconnect structure should be considered as a transmission line when its physical length approaches 1/4 and 1/10 the wavelength of the highest frequency component.

For the case of a simple microstrip line, it can be shown [64] that the wavelength at a given frequency is

$$(25.2) \qquad \lambda_{\mathrm{g}} = \frac{300}{F\sqrt{\varepsilon_{\mathrm{eef}}}}\,\mathrm{mm}$$

where $\varepsilon_{\mathrm{eef}}$ is the effective dielectric constant given by

$$(25.3) \qquad \varepsilon_{\mathrm{eef}} = \frac{1}{2}(\varepsilon_{\mathrm{r}} + 1)$$

and F is the frequency in GHz.

Consider for example, the case of a 500 MHz signal (square wave). Assuming that the highest frequency component we wish to pass is the sixth harmonic, then the line should be able to pass a signal at 3 GHz. Also assuming a relative dielectric constant of 4, the wavelength for the highest component is then 63.25 mm. Assuming the 1/4 l rule, the line length for which transmission line properties become important is about 16 mm. In other words, any line longer than about 16 mm should probably be considered for inductance modeling.

The choice of using an RC, RLC, or transmission line model is also dependent on what analysis will be performed with it. For short lines (a few mm), it may be possible to get away with a distributed RC model if timing is the sole criteria. Even though the waveform shapes will be significantly different than those obtained with an RLC model, the timing at the 50% points would not be drastically different.

The case is different for medium to long lines, however. An RC-only representation can underestimate delay and cross-talk values by upwards of 20% or more [65,66].

Another approach takes into account the relatively high resistance of IC interconnect. In a theoretical study using a distributed RLC transmission line with a CMOS driver, Ismail et al. [67]

have proposed metrics for determining what line lengths may experience inductive effects. For the case where the driver and line impedances match, it can be shown that line lengths that fall within the inequalities given in Equation 25.4 have significant inductive effects.

(25.4)
$$\frac{t_r}{2\sqrt{LC}} < l < \frac{2}{R}\sqrt{\frac{L}{C}}$$

where
   R, L, and C are the per unit length parameters of the line
   t_r is the rise time of the signal

If the line is shorter than the lower limit, then inductance is not important because the rise time is longer than the propagation time, so the line behaves as an equipotential. If the line is greater than the upper limit, inductance is not important because the resistive component is so large it overwhelms any inductive contribution. For narrow lines on most modern IC processes, these two regions overlap, and there is no length for which inductance is important. It may need to be considered only for wide wires on the upper metal layers.

25.10.3.2 SIMPLIFYING THE INDUCTANCE PROBLEM BY USING SPECIFIC DESIGN METHODOLOGIES

Another way to reduce the inductance extraction problem is to simplify the determination of the return current path by imposing a strict interconnection structure for all global routing. In other words, follow the path already established by high-speed printed circuit board (PCB) technologies.

For future large-scale, high-speed, DSM designs to meet performance goals, interconnect lines must have predictable delay and noise characteristics.

As shown in several studies [61–63], the most predictable and well-behaved interconnect lines are those with either ground planes above or below them, or neighboring ground/power lines (i.e., coplanar waveguides), as shown in Figure 25.14. As mentioned previously, the key to the behavior of an interconnect wire is its electrical length, not its physical dimensions.

In today's random interconnect environment, it is impractical to perform inductance extraction on even a modest scale except for a few special cases (e.g., coplanar structures). If sufficient care is not exercised when determining the return current paths, not only can significant errors arise but also the resulting models can be unstable. Restricting the interconnect structures to those with well-defined return paths (either power and ground planes or coplanar lines) will simplify the extraction of not only inductance but capacitance and resistance as well [61,68].

While resorting to these techniques will limit the density of the signal lines, the result will be simpler and faster CAD tools along with more predictable interconnect wiring. On the other hand the need for accuracy will require the use of 3D inductance extraction methods, such as explained in Chapter 29, and Refs. [69,70,72].

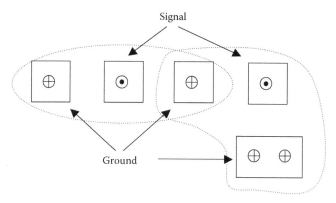

FIGURE 25.14 Example of a coplanar interconnect structure with alternating signal and ground return lines.

25.11 NETWORK REDUCTION

The techniques of the previous sections can generate very large netlists. This is particularly true if accurate resistances are attempted, if substrate analysis is included, or if mutual inductances are wanted. Each of these can generate huge numbers of internal nodes (for R and substrate) or mutual components (for inductance). This causes several problems—the storage alone is prohibitive, the downstream tools often run slowly (or fail entirely) when faced with huge networks, and the wide range of time constants in the raw networks causes numerical and stability problems.

In most cases, the intended use of the netlist does not strictly require all these internal nodes and components. In a digital circuit the user is primarily interested in the delay from an output to the inputs on the net. In an analog circuit, the user may be interested in the frequency response, or the transient response, at one particular output.

This defines the job of reduction—take a (usually large) netlist, and generate a smaller netlist (or a mathematical expression) that accurately represents the desired response. There are at least three variants in common use:

1. Take an $RC(L)$ netlist, and generate a smaller $RC(L)$ netlist. Reducers like these are generally built into extraction tools. The methods tend to be ad hoc, though they try to preserve basic properties like total C and total R [81,82].
2. Take an $RC(L)$ netlist, and generate an approximation to the desired result directly. The desired result is typically a delay, or a waveform from which delay and slope information can be derived. Elmore delay [12], and asymptoptic waveform evaluation [13–16] are the best known examples here. However, because of the increasing interconnect series resistance, these methods are no longer sufficiently accurate, thus requiring the use of higher order moment matching [17,18] or Krylov subspace methods [19–21].
3. Take an $RC(L)$ netlist, and generate a mathematical model (poles and zeros, a state space model, or rational polynomial expressions). This is most commonly used in RF designs, and there is a large and specialized literature on this topic. (See Chapter 18, for more details.)

25.12 PROCESS VARIATION

Unfortunately, even if CMP and OPC/RET are taken into account, the final result of fabrication is not always what was intended, due to variations both systematic and random. For systematic effects, as in the previous section, it is best to model them if possible. If they cannot be modeled, or if they are in fact truly random, then we are in the realm of *statistical analysis.* This is a complex area due to the wide variety of correlations that may exist and the difficulty of treating them accurately. For example, if the metal is thicker at a given point, it is not clear without detailed analysis whether the delay goes up (due to increased C), down (due to decreased R), or stays the same (due to cancellation of these effects).

Since the extractor does not know how the extracted data will be used, it cannot solve the correlation problem. What it *can* do is express how the extracted values will vary with changes in the physical structure. For example, if the local change in metal thickness is $\Delta t1$, the local change of width is $\Delta w1$, and the local change in interconnect thickness is $\Delta i1$, then the R of a wire might be expressed as

$$R = R_0 + k_1 \Delta t1 + k_2 \Delta w1$$

The capacitance C of the same wire might be

$$C = C_0 + k_3 \Delta t1 + k_4 \Delta w1 + k_5 \Delta i1$$

In this case, we would expect k_1, k_2, and k_5 to be negative, and k_3 and k_4 to be positive. Note that delays, slacks, moments, and other circuit metrics can be expressed in the same form, which is very useful during statistical timing and optimization [92,93].

The coefficients k_N are called "sensitivities," and the expression of values in this form is called an "affine representation." As a linear form it is not completely accurate, but for interconnect modeling in particular the accuracy is quite good [94]. The big advantage of this form is that it preserves the correlation information. See Ref. [95] and Chapter 6, for a more detailed discussion of this. A practical advantage of this form is that a single extraction run can be used instead of multiple interconnect corner run. Of course, such parasitic data needs to be stored somehow, but this can be done by extensions to existing databases (see Chapter 15).

25.13 CONCLUSIONS

Extracting a circuit from a physical description is conceptually simple, but quite complex in practice. The requirements, the required capacities, and extraction programs themselves, keep changing as new technologies are introduced. The best way to keep abreast of developments is to monitor the conferences, ISPD, DAC, ICCAD, and DATE, the IEEE journals, and the commercial trade publications that cover IC design.

REFERENCES

These include articles, not mentioned in the text, which the authors have found helpful.

1. M. Kamon, S. McCormick, and K. Shepard, Interconnect parasitic extraction in the digital IC design methodology, *Proceedings of the ICCAD*, 1999, pp. 223–230.
2. W.H. Kao, L. Chi-Yuan, M. Basel, and R. Singh, Parasitic extraction: Current state of the art and future trends, *IEEE Proceedings*, pp. 729–739.
3. D. Bailey and B. Benschneider, Clocking design and analysis for a 600 MHz alpha microprocessor, *IEEE J. Solid State Circ.*, 33, 1627–1633, 1998.
4. D. Edelstein et al., Full copper wiring in a sub-0.25 um CMOS ULSI technology, *Proceedings of the IEDM*, 1997, pp. 773–776.

FINITE DIFFERENCE METHODS

5. E. Dengi and R. Rohrer, Hierarchical 2-D field solution for capacitance extraction for VLSI interconnect modeling, *Proceedings of the 34th ACM/IEEE DAC*, Anaheim, CA, June 1997, pp. 127–132.

FINITE **ELEMENT METHODS**

6. G. Costache, Finite element method applied to skin effect problems in strip transmission lines, *IEEE Trans. Microw. Theory Tech.*, MTT-35, 1009–1013, 1987.

METHOD **OF MOMENTS**

7. W. Cao, R. Harrington, J. Mantz, and T. Sarkar, Multiconductor transmission lines in multilayered dielectric media, *IEEE Trans. Microw. Theory Tech.*, MTT-32, 439–450, 1984.

BOUNDARY **ELEMENT METHODS**

8. S. Rao, T. Sarkar, and R. Harrington, The electrostatic field of conducting bodies in multiple dielectric media, *IEEE Trans. Microw. Theory Tech.*, 32, 1441–1448, 1984.

9. E. Dengi and R. Rohrer, Boundary element method macromodels for 2-D hierarchical capacitance extraction, *Proceedings of the 35th ACM/IEE DAC*, June 1998, pp. 218–223.

3D INTERCONNECTS

10. W. Sun, W.M. Dai, and W. Hong, Fast parameters extraction of general 3-D interconnects using geometry independent measured equation of invariance, *Proceedings of the 33rd ACM/IEEE DAC*, 1996, pp. 105–110.
11. M. Kamon, N. Marques, and J. White, FastPep: A fast parasitic extraction program for complex three dimensional geometries, *Proceedings of the ICCAD*, San Jose, CA, 1997, pp. 456–460.

ELMORE DELAY, ASYMPTOPTIC WAVEFORM EVALUATION

12. W.C. Elmore, The transient response of damped linear networks with particular regard to wideband amplifiers, *J. Appl. Phys.*, 19, 55–63, 1948.
13. J. Rubinstein, J.P. Penfield, and M.A. Horowitz, Signal delay in RC tree networks, *IEEE Trans. CAD*, CAD-2, 202–211, 1983.
14. L.T. Pillage and R.A Rohrer, Asymptotic waveform evaluation for timing analysis, *IEEE Trans. CAD*, 9, 352–366, 1990.
15. C. Ratzlaff and L.T. Pillage, RICE: Rapid interconnect circuit evaluator using asymptotic waveform evaluator, *IEEE Trans. CAD*, 13(6), 763–776, June 1994.
16. J.E. Bracken, V. Raghavan, and R. Rohrer, Interconnect simulation with asymptotic waveform evaluation, *IEEE Trans. Circ. Syst.*, 39(11), 869–878, November 1992.

MOMENT MATCHING METHODS

17. E. Chiprout and M. Nakhla, Generalized moment-matching methods for transient analysis of interconnect networks, *Proceedings of the 29th ACM/IEEE DAC*, Anaheim, CA, June 1992, pp. 201–206.
18. J.R. Phillips, E. Chiprout, and D. Ling, Efficient full-wave electromagnetic analysis via model order reduction of fast integral transforms, *Proceedings of the 33rd ACM/IEEE DAC*, Las Vegas, NV, June 1996, pp. 377–382.

KRYLOV SUBSPACE METHODS

19. P. Feldmann and R. Freund, Efficient linear circuit analysis by Pade approximation via the Lanczos process, *IEEE Trans. CAD*, 14, 639–649, 1995.
20. K. Gallivan, E. Grimme, and P. Van Dooren, Asymptotic waveform evaluation via a Lanczos method, *Appl. Math. Lett.*, 7, 75–80, 1994.
21. A. Odabasioglu, M. Celik, and L. Pileggi, PRIMA: Passive reduced-order interconnect macromodeling algorithm, *Proceedings of the ICCAD*, San Jose, CA, August 1997, pp. 58–65.

RESISTANCE EXTRACTION

22. R. Singh, A review of substrate coupling issues and modeling strategies, *Proceedings of the IEEE Custom Integrated Circuits Conference*, 1999, pp. 491–499.
23. M. Horowitz and R.W. Dutton, Resistance extraction from mask layout data, *IEEE Trans. Comput. Aided Des. Integr. Circ. Syst.*, 2, 1983, 145–150.
24. D. Su, M.J. Loinaz, S. Masui, and B.A. Wooley, Experimental results and modeling techniques for substrate noise and mixed-signal integrated circuits, *IEEE J. Solid-State Circ.*, 28, 420–430, 1993.
25. N. Verghese, T. Schmerbeck, and D. Allstot, *Simulation Techniques and Solutions for Mixed-Signal Coupling in Integrated Circuits*, 2nd edn., Kluwer Academic, Dordrecht, the Netherlands, 1995.
26. X. Aragones, J. Luis Gonzalez, and A. Rubio, *Analysis and Solutions for Switching Noise Coupling in Mixed Signal ICs*, Kluwer Academic Publishers, Dordrecht, the Netherlands, 1999.
27. F.J. Clement, E. Zysman, M. Kayal, and M. Declercq, LAYIN: Toward a global solution for parasitic coupling modeling and visualization, *Proceedings of the IEEE Custom Integrated Circuits Conference*, 1994, pp. 537–540.

28. K. Nabors and J. White, Multipole-accelerated 3-D capacitance extraction algorithms for structures with conformal dielectrics, *Proceedings of the 29th ACM/IEEE Design Automation Conference*, 1992, pp. 710–715.

29. A.E. Ruehli and P.A. Brennan, Efficient capacitance calculations for three dimensional multiconductor systems, *IEEE Trans. Microw. Theory Tech.*, MTT-21, 76–82, February 1973.

30. N.K. Verghese and D. Allstot, SUBTRACT: A program for the efficient evaluation of substrate parasitics in integrated circuits, *Proceedings in IEEE/ACM International Conference on Computer-Aided Design*, 1995, pp. 194–198.

31. N. Verghese, D. Allstot, and M. Wolfe, Verification techniques for substrate coupling and their application to mixed-signal IC design, *IEEE J. Solid-State Circ.*, 31, 354–365, 1996.

32. R. Gharpurey and R. Meyer, Modelling and analysis of substrate coupling in integrated circuits, *IEEE J. Solid-State Circ.*, 31, 344–353, 1996.

33. T. Smedes, N.P. van der Meijs, and A.J. van Genderen, Boundary element methods for capacitance and substrate resistance calculations in a VLSI layout verification package, *Proceedings of the ELECTROSOFT'93*, July 1993, pp. 337–344.

34. R. Gharpurey, Hosur transform domain techniques for efficient extraction of substrate parasitics, *Proceedings of the IEEE/ACM International Conference on Computer-Aided Design*, 1997, pp. 461–467.

35. M. Chou and J. White, Multilevel integral equation methods for the extraction of substrate coupling parameters in mixed-signal IC's, *Proceedings of the 35th Design Automation Conference*, 1998, pp. 20–25.

36. J. Zhao, W.M. Dai, S. Kadur, and D.E. Long, Efficient three-dimensional extraction based on static and full-wave layered Green's functions, *Proceedings of the 35th Design Automation Conference*, 1998, pp. 224–229.

37. S. Kapur and D.E. Long, IES3: A fast integral equation solver for efficient 3-dimensional extraction, *Proceedings of the IEEE/ACM International Conference on Computer-Aided Design*, November 1997, pp. 448–455.

38. J. Zhao, W. Dai., R.C. Frye, and K.L. Tai, Green function via moment matching for rapid and accurate substrate parasitics evaluation, *Proceedings of the IEEE Custom Integrated Circuits Conference*, 1997, pp. 371–374.

CAPACITANCE EXTRACTION

39. S. Napper, Technical white paper on RC extraction, Technical Report, *EPIC Design Technology*, 1995.

40. K. Nabors and J. White, FastCap: A multipole-accelerated 3-D capacitance extraction program, *IEEE Trans. CAD*, 10, 1447–1459, 1991.

41. K. Nabors and J. White, Multipole-accelerated 3-D capacitance extraction algorithms for structures with conformal dielectrics, *Proceedings of the 29th ACM/IEEE DAC*, June 1993, pp. 710–715.

42. W. Hong, W.K. Sun, Z.H. Zhu, H. Ji, B. Song, and W. Dai, A novel dimension reduction technique for the capacitance extraction of 3-D VLSI interconnects, *MTT*, 46, 1037–1044, 1996.

43. W. Shi, J. Liu, N. Kakani, and T. Yu, A fast hierarchical algorithm for 3-D capacitance extraction, *Proceedings of the 35th ACM/IEEE DAC*, San Francisco, CA, June 1998, pp. 212–217.

44. U. Choudhury and A. Sangiovanni-Vicentelli, Automatic generation of analytical models for interconnect capacitances, *IEEE Trans. Comput. Aided Des. Integr. Circ. Syst.*, 14, 470–480, 1995.

45. N.D. Arora and K.V. Roal, R. Schumann, and L.M. Richardson, Modeling and extraction of interconnect capacitances for multilayer VLSI circuits, *IEEE Trans. Comput. Aided Des. Integr. Circ. Syst.*, 15, 58–67, 1996.

46. P.A. Habitz and I.L. Wemple, A simpler, faster method of parasitic capacitance extraction, *Electr. J.* October 11–15, 1997.

47. E. Dengi, A parasitic capacitance extraction method for VLSI interconnect modeling, Research Report: CMUCAD-97-29, Carnegie Mellon University, Pittsburgh, PA.

48. E. You, S. Yew Choe, C. Kim, L. Varadadesikan, K. Aingaran, and J. MacDonald, Parasitic extraction for multimillion-transistor integrated circuits: Methodology and design experience, *Proceedings of the IEEE 2000 Custom Integrated Circuits Conference, CICC*, May 21–24, 2000, pp. 491–494 (Hierarchical).

49. A. Kurokawa, T. Kanamoto, A. Kasebe, Y. Inoue, and H. Masuda, Efficient capacitance extraction method for interconnects with dummy fills, *Proceedings of the IEEE 2004 Custom Integrated Circuits Conference*, October 3–6, 2004, pp. 485–488.

50. M.W. Beattie and L.T. Pileggi, Error bounds for capacitance extraction via window techniques, *IEEE Trans. CAD*, 18, 311–321, 1999.

INDUCTANCE EXTRACTION

51. K.C. Gupta, R. Garg, and R. Chadha, *Computer Aided Design of Microwave Circuits*, Artech House, Bedham, MA, 1981.
52. M. Beattie and L. Pileggi, IC analyses including extracted inductance models, *Proceedings of 36th International Conference on Design Automation*, June 1999, pp. 915–920.
53. J. Wang, J. Tausch, and J. White, A wide frequency range surface integral formulation for 3-D inductance and resistance extraction, *Technical Proceedings of the 1999 International Conference on Modeling and Simulation of Microsystems*, 1999.
54. M. Kamon, N. Marques, Y. Massoud, L. Silveira, and J. White, Interconnect analysis: From 3-D structures to circuit models, *Proceedings of 36th International Conference on Design Automation*, June 1999, pp. 910–914.
55. Z. He, M. Celik, and L. Pileggi, SPIE: Sparse partial inductance extraction, *Proceedings of 34th International Conference on Design Automation*, June 1997, pp. 137–140.
56. E. Rosa, The self and mutual inductance of linear conductors, *Bull. Nat. Bur. Stand.*, 4, 301–344, 1908.
57. A. Ruehli, Inductance calculations in a complex integrated circuit environment, *IBM J. Res. Dev.*, 16, 470–481, 1972.
58. B. Krauter and L. Pileggi, Generating sparse partial inductance matrices with guaranteed stability, *International Conference on Computer Aided Design*, 1995, pp. 45–52.
59. M. Beattie, L. Alatan, and L. Pileggi, Equipotential shells for efficient partial inductance extraction, *Proceedings of the 1998 IEDM*, December 1998.
60. B. Krauter and S. Mehrotra, Layout based frequency dependent inductance and resistance extraction for on-chip interconnect and timing analysis, *Proceedings of 35th International Conference on Design Automation*, June 1998, pp. 303–308.
61. P. Restle, A. Ruehli, and S. Walker, Dealing with inductance in high-speed chip design, *Proceedings of 36th International Conference on Design Automation*, June 1999, pp. 904–910.
62. S. Morton, On-chip inductance issues in multiconductor systems, *Proceedings of 36th International Conference on Design Automation*, 1999, pp. 921–926.
63. A. Deutsch, G.V. Kopcsay, C.W. Surovic, B.J. Rubin, L.M. Terman, R.P. Dunne, T.A. Gallo, and R.H. Dennard, Modeling and characterization of long on-chip interconnections for high-performance microprocessors, *IBM J. Res. Dev.*, 39, 547–567, 1995.
64. T.C. Edwards, *Foundations for Microstrip Circuit Design*, Wiley, New York, 1981.
65. A. Deutsch, G. Kopcsay, P. Restle, H. Smith, G. Katopis, et al., When are transmission line effects important for on-chip interconnections? *IEEE Trans. Microw. Theory Tech.*, 45, 1836–1846, 1997.
66. A. Deutsch, H. Smith, C. Surovic et al., Frequency dependent crosstalk simulation for on-chip interconnections, *IEEE Trans. Adv. Pack.*, 22, 292–308, 1999.
67. Y. Ismail, E. Friedman, and J. Neves, Figures of merit to characterize the importance of on-chip inductance, *Proceedings of 35th International Conference on Design Automation*, June 1998, pp. 560–565.
68. N. van der Meijs and T. Smedes, Accurate interconnect modeling: Towards multi-million transistor chips as microwave circuits, *International Conference on Computer Aided Design*, 1996, pp. 244–251.
69. M. Kamon, M. Tsuk, C. Smithhisler, and J. White, Efficient techniques for inductance extraction of complex 3-D geometries, *Proceedings of the ICCAD*, Santa Clara, CA, 1992, pp. 438–442.
70. M. Kamon, M.J. Tsuk, and J. White, FASTHENRY: A multipole-accelerated 3-D inductance program, *IEEE Trans Microw. Theory Tech.*, 42, 1750–1758, 1994.
71. K.L. Shepard and Z. Tian, Return-limited inductances: A practical approach to on-chip inductance extraction, *IEEE Trans. CAD*, 19, 425–436, 2000.
72. K.L. Shepard, D. Sitaram, and Y. Zheng, Full-chip, three-dimensional, shapes-based RLC extraction, *Proceedings of the International Conference on Computer-Aided Design*, 2000, pp. 142–149.
73. M. Beattie and L. Pileggi, Efficient inductance extraction via windowing, *Proceedings of the Design Automation and Test in Europe (DATE)*, March 2001.

INTERCONNECT MODELING, MODEL ORDER REDUCTION, AND DELAY CALCULATION

74. B. Tutuianu, F. Dartu, and L. Pileggi, An explicit RC-circuit delay approximation based on the first three moments of the impulse response, *Proceedings of the 33rd ACM/IEEE DAC*, June 1996, pp. 611–616.
75. Y. Ismail, E. Friedman, and J. Neves, Equivalent Elmore delay for RLC trees, *IEEE Trans. CAD*, 19(1), 83–97, January 2000.

76. L.M. Miguel Silveira, M. Kamon, I. Elfaldel, and J. White, A coordinate-transformed Arnoldi algorithm for generating guaranteed stable reduced-order models of RLC circuits, *Proceedings of the International Conference on CAD*, San Jose, CA, November 1996, pp. 288–294.

77. I.M. Elfaldel and D. Ling, A block rational Arnoldi algorithms for multipoint passive model-order reduction of multiport RLC networks, *Proceedings of the ICCAD*, San Jose, CA, November 1997, pp. 66–71.

78. Y. Liu, L. Pileggi, and A. Strojwas, Model order reduction of RCL interconnect including variational analysis, *Proceedings of the 36th ACM/IEEE DAC*, June 1999, pp. 201–206.

79. Y. Ismail, E. Friedman, and J. Neves, Repeater insertion in tree structured inductive interconnect, *Proceedings of the 36th DAC*, June 1999, pp. 420–424.

80. J. Kanapka, J. Phillips, and J. White, Fast methods of extraction and sparsification of substrate coupling, *Proceedings of 37th Design Automation Conference*, June 2000, pp. 738–743.

81. B.N. Sheehan, TICER: Realizable reduction of extracted RC circuits, *ICCAD'99: Proceedings of the 1999 IEEE/ACM International Conference on Computer-Aided Design*, San Jose, CA, 1999, pp. 200–203.

82. V.B. Rao, J.P. Soreff, R. Ledalla, and F.L. Yang, Aggressive crunching of extracted RC netlists, *TAU'02: Proceedings of the Eighth ACM/IEEE International Workshop on Timing Issues in the Specification and Synthesis of Digital Systems*, Monterey, CA, 2002, pp. 70–77.

HISTORICAL

83. Yamin, M., XYTOLR—A computer program for integrated circuit mask design checkout, *Bell Syst. Tech. J.*, 51, 1581–1593, 1972.

84. H.S. Baird, Fast algorithms for LSI artwork analysis, *DAC'77: Proceedings of the 14th Conference on Design Automation*, 1977, pp. 303–311.

85. P. Losleben, Design validation in hierarchical systems, *DAC'75: Proceedings of the 12th Conference on Design Automation*, 1975, pp. 431–438.

86. R.M. Allgair and D.S. Evans, A comprehensive approach to a connectivity audit, or a fruitful comparison of apples and oranges, *DAC'77: Proceedings of the 14th Conference on Design Automation*, 1977, pp. 312–321.

87. B.T. Preas, B.W. Lindsay, and C.W. Gwyn, Automatic circuit analysis based on mask information, *DAC'76: Proceedings of the 13th Conference on Design Automation*, 1976, pp. 309–317.

88. U. Lauther, An O (N log N) algorithm for Boolean mask operations, *DAC'81: Proceedings of the 18th Conference on Design Automation*, 1981, pp. 555–562.

OPC, RET, AND CMP MODELING (CMP MODELING ALONE IS A HUGE AREA. THIS IS JUST AN INTRODUCTION)

89. C. Heitzinger, A. Sheikholeslami, F. Badrieh, H. Puchner, and S. Selberherr, Feature-scale process Simulation and accurate capacitance extraction for the backend of a 100-nm aluminum/TEOS process, *IEEE Trans. Electron Dev.*, 51, 1129–1134, 2004.

90. D.O. Ouma, D.S. Boning, J.E. Chung, W.G. Easter, V. Saxena, S. Misra, and A. Crevasse, Characterization and modeling of oxide chemical—Mechanical polishing using planarization length and pattern density concepts, *IEEE Trans. Semiconduct. Manuf.*, 15, 232–244, 2002.

91. T.E. Tugbawa, T.H. Park, and D.S. Boning, integrated chip-scale simulation of pattern dependencies in copper electroplating and copper chemical mechanical polishing processes. *Proceedings of the IEEE 2002 International Interconnect Technology Conference*, June 3–5, 2002, pp. 167–169.

STATISTICAL EXTRACTION

The paper by Scheffer has some extraction experiments showing the linear form is quite accurate. The papers by Sapatnekar and Visweswariah show how to use results in this form in timing analysis. The paper by Rutenbar is about the mechanics of manipulating the affine forms. The paper by Liu and Pillagi [78], referenced in Section 25.3.8, discusses how this form can be reduced.

92. H. Chang and S.S. Sapatnekar, Statistical timing analysis considering spatial correlations using a single pert-like traversal, *ICCAD'03: Proceedings of theIEEE/ACM International Conference on Computer-Aided Design*, 2003, p. 621.

93. C. Visweswariah, K. Ravindran, K. Kalafala, S.G. Walker, and S. Narayan, First-order incremental block-based statistical timing analysis, *DAC'4: Proceedings of the 41st Annual Conference on Design Automation*, 2004, pp. 331–336.

94. L. Scheffer, Explicit computation of performance as a function of process variation, *TAU'02: Proceedings of the Eighth ACM/IEEE International Workshop on Timing Issues in the Specification and Synthesis of Digital Systems*, 2002, pp. 1–8.

95. J.D. Ma and R.A. Rutenbar, Interval-valued reduced order statistical interconnect modeling, *Proceedings of the IEEE/ACM International Conference on Computer Aided Design, ICCAD-2004*, November 7–11, 2004, pp. 460–467.

96. A. Labun, Rapid method to account for process variation in full-chip capacitance extraction, *IEEE Trans. CAD*, 23, 941–951, 2004.

CONVERTING TRANSISTOR LEVEL TO GATE LEVEL

97. G. Ditlow, W. Donath, and A. Ruehli, Logic equations for MOSFET circuits, *Proceedings of the IEEE International Symposium Circuits and System*, May 1983, pp. 752–755.

98. Z. Barzilai, L. Huisman, G.M. Silberman, D.T. Tang, and L.S. Woo, Simulating pass transistor circuits using logic simulation machines, *Design Automation Conference*, 1983, pp. 157–163.

99. R.E. Bryant, Boolean analysis of MOS circuits, *IEEE Trans. CAD*, 6, 634–649, 1987.

100. D.T. Blaauw, D.G. Saab, P. Banerjee, and J. Abraham, Functional abstraction of logic gates for switch level simulation, *European Conference on Design Automation*, 1991, pp. 329–333.

101. R.E. Bryant, Extraction of gate level models from transistor circuits by four valued symbolic analysis, *International Conference in Computer-Aided Design*, 1991, pp. 350–353.

Mixed-Signal Noise Coupling in System-on-Chip Design

Modeling, Analysis, and Validation

Nishath Verghese and Makoto Nagata

CONTENTS

Abstract

The impact of noise coupling in mixed-signal integrated circuits is described and the techniques to model, analyze, and validate it are reviewed. The physical phenomena responsible for the creation of noise and its transmission mechanisms and media are described, and the parameters affecting the strength of the noise coupling are discussed. The different modeling approaches and computer simulation methods used in quantifying noise coupling phenomena are presented. Measurement techniques for substrate noise and subsequent validation of noise analysis results are also presented. Application of substrate noise analysis to placement and power distribution synthesis are also reviewed.

26.1 INTRODUCTION

The push for reduced cost, more compact circuit boards, and added customer features has provided incentives for the inclusion of analog functions on primarily digital MOS integrated circuits (ICs) forming mixed-signal ICs. In these systems, the speed of digital circuits is constantly increasing, chips are becoming more densely packed, interconnect layers are added, and analog resolution is increased. In addition, recent increase in wireless applications and its growing market are introducing a new set of aggressive design goals for realizing mixed-signal systems. Here, the designer integrates radio-frequency (RF) analog and base band digital circuitry on a single chip. The goal is to make single-chip radio frequency integrated circuits (RFICs) on silicon, where all the blocks are fabricated on the same chip. One of the advantages of this integration is low power dissipation for portability due to a reduction in the number of package pins and associated bond wire capacitance. Another reason that an integrated solution offers lower power consumption is that routing high-frequency signals off-chip often requires a 50 W impedance match, which can result in higher power dissipation. Other advantages include improved high-frequency performance due to reduced package interconnect parasitics, higher system reliability, smaller package count, smaller package interconnect parasitics, and higher integration of RF components with VLSI-compatible digital circuits. In fact, the single-chip transceiver is now a reality [1].

The design of such systems, however, is a complicated task. There are two main challenges in realizing mixed-signal ICs. The first challenging task, specific to RFICs, is to fabricate good on-chip passive elements such as high-Q inductors [2]. The second challenging task, applicable to any mixed-signal IC and the subject of this chapter, is to minimize noise coupling between various parts of the system to avoid any malfunctioning- of the system [3,4]. In other words, for successful system-on-chip integration of mixed-signal systems, the noise coupling caused by nonideal isolation must be minimized so that sensitive analog circuits and noisy digital circuits can effectively coexist, and the system operates correctly. To elaborate, note that in mixed—signal circuits, both sensitive analog circuits and high-swing high-frequency noise injector digital circuits may be present on the same chip, leading to undesired signal coupling between these two types of circuit via the conductive substrate. The reduced distance between these circuits, which is the result of constant technology scaling, exacerbates the coupling. The problem is severe, since signals of

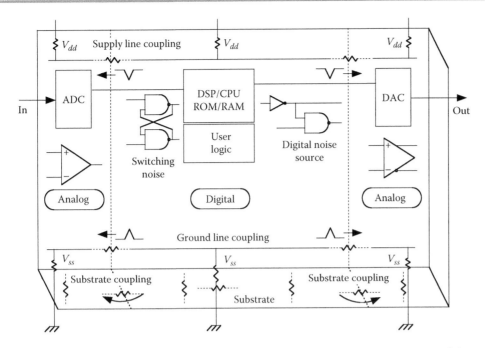

FIGURE 26.1   Digital noise coupling in a mixed-signal LSI circuit. For simplicity, some of the parasitic elements causing the coupling are not shown. (From Tsukada, T. and Makie-Fukuda, K., *IEICE-T. Fundam. Electr.*, E80-A, 263, 1997. With permission.)

different nature and strength interfere, thus affecting the overall performance, which demands higher clock rates and greater analog precisions.

The primary mixed-signal noise coupling problem from fast-changing digital signals, coupling to sensitive analog nodes, is shown schematically in Figure 26.1. Another significant cause of undesired signal coupling is the cross talk between analog nodes themselves owing to the high-frequency/high-power analog signals. One of the media through which mixed-signal noise coupling occurs is the substrate. Digital operations cause fluctuations in the underlying substrate voltage, which spreads through the common substrate causing variations in the substrate potential of sensitive devices in the analog section. Similarly, in the case of cross talk between analog nodes, a signal can couple from one node to another via the substrate. This phenomenon is referred to as "substrate coupling" or "substrate noise coupling."

In this chapter, we discuss the mixed-signal coupling problem and review various techniques to model, analyze, and validate it. In Section 26.2, the random noise inherent to electronic devices and the deterministic noise generated by circuits are differentiated. Then, we discuss the physical phenomena responsible for the creation of undesired signals in a digital circuit and the mechanisms of their transport to other parts of the system, mainly via the substrate. Various modeling approaches and simulation techniques of digital noise generation as well as substrate impedance network determining noise coupling are reviewed in Section 26.3. Section 26.4 describes an application of the analysis techniques to placement and power distribution synthesis.

26.2 MECHANISMS AND EFFECTS OF MIXED-SIGNAL NOISE COUPLING

26.2.1 DIFFERENTIATION BETWEEN RANDOM NOISE AND DETERMINISTIC NOISE

All undesired phenomena, behaviors, or influences that degrade performance are regarded as noise [5]. Noise in a mixed-signal IC can be classified into two types. One is the intrinsic noise of active and passive devices in the circuit and the other is the undesired signal (e.g., switching noise of digital circuits) coupled from other parts of the circuit. Intrinsic noise originates from various physical phenomena within the device itself; some examples are thermal noise, shot noise, and flicker noise. This type of noise, which has a nondeterministic nature, is an important consideration in the design of sensitive analog circuits such as receiver RF circuitry where signal levels could be very small. The level of this type of noise represents a minimum level of noise in a system, and its control is accomplished through optimal circuit design, topology selection, bandwidth limiting of signals, and semiconductor process control. The parameters that are used to quantify this type of noise are noise figure (NF) and signal-to-noise ratio (SNR). Having the circuit topology of a block as well as its circuit component's noise characteristics, one can determine the NF or SNR for the block. The noise is usually represented by an input referred noise source and has its own frequency spectrum.

Contrary to the first noise type, the second type, which is an undesired signal, has a deterministic nature and, theoretically, can be quantified both in the frequency domain and in the time domain. An example of the second type of noise is digital switching noise, which is the major source of undesired signals in mixed-signal ICs [6] and can be very destructive because it can be broadcast over great distances, acting on all transistors by modulating threshold voltage and gain, and directly coupling to the signal node. It can also increase the average delay of digital blocks [7]. In addition to digital switching noise, any high-frequency signal of analog circuits can behave as the source of an undesired signal for other circuits. Therefore, not only digital circuits but also analog circuits could play the role of aggressor blocks for other parts of the system. However, we will intensively focus on deterministic noise from digital ICs in the rest of this chapter.

26.2.2 COUPLING FROM DIGITAL INTEGRATED CIRCUITS

Noise can couple to the substrate through various mechanisms during the operation of digital ICs, for instance, through the impact-ionization phenomenon in device-level operation, capacitive coupling from parasitic passive components in primitive circuit-level operation, and resistive coupling with ground wirings in an entire IC operation. Designers need to recognize the relative importance of these possible noise-generation mechanisms to tackle undesired coupling during design. In recent years, substrate noise measurements have provided unvarnished waveforms and distinctive physical properties of substrate noise within a chip, which are quite helpful for intuitive understanding of the noise injection mechanisms.

CMOS digital circuits can inject dynamic noise into a substrate through junction capacitances at source and drain terminals of MOSFETs. Also, every well and interconnect on an IC couples capacitively to the substrate through a reverse biased bulk/well and an overlap capacitance to substrate, respectively. Only high-frequency components of a signal can couple to substrate via capacitive coupling. This capacitively coupled substrate current is important in mixed-signal circuits due to the presence of both a large number of switching digital nodes and high-impedance analog nodes. With decreasing technology feature sizes, the interconnect capacitances to substrate are becoming increasingly important.

In contrast to the parasitic resistive (ohmic) and capacitive coupling described above, noise current in a substrate can be induced by the impactionization phenomenon, where electron–hole pairs are generated and cause a current flow to the substrate when the electric field in the depleted region of the drain bulk of a MOS transistor becomes large. This can be transported to other parts of the system via the substrate, even under DC operating conditions. Although the growth of impactionization current cannot be ignored with the increasing speed of operation and the decreasing feature sizes of technology [8,9], the leakage of ground bounce and the size of capacitive coupling also significantly increases for larger scale of integration. Therefore, it can be said that the contribution of impactionization phenomenon to substrate noise is insignificant from a macroscopic view of the entire circuit operation, which Briaire has pointed out in [10,11], with detailed comparisons made by device and circuit simulations.

26.2.3 EFFECTS OF POWER/GROUND GRID, PACKAGE/BONDWIRE PARASITICS ON NOISE GENERATION

Since substrate noise is dominated by the leakage of power supply and ground bounce as discussed in Section 26.2.2, the interaction of power supply and transient currents with parasitic RLC impedance networks strongly affect the noise.

Interconnect layers are used to set the voltage of different terminals of a circuit and to transport the signals from one point to another point of the circuit. Interconnect layers have distributed parasitics such as inductance and resistance on each layer, and capacitance and mutual inductance between two interconnect layers and between an interconnect layer and the substrate. These parasitic elements can create or couple noise. When a digital gate switches, current transients pass through power supply interconnect layers and lead to transient voltage drops across the parasitic inductance and resistance of the power supply/ground (V_{dd}/G_{nd}) interconnect lines [12]. The current spikes can be proportional to the load capacitance. The voltage drops in the power grid cause power supply/ground line bounce, known as "V_{dd} bounce" or "*ground bounce*", and can be obtained from the following expression:

$$(26.1) \qquad V_{drop} = Ri + L\frac{di}{dt}$$

where R and L are the resistance and inductance parasitic to the interconnect layers and more dominantly to the package and bondwires. The second term on the right-hand side of Equation 26.1 is also referred to as "inductive noise," "Ldi/dt noise," or "delta-I noise." This equation implies that the ground and supply voltages bounce as a function of the switching current and its derivative. The former determines the resistive voltage drop, while the latter represents the inductive voltage drop across the supply lines. Another undesired effect of di/dt term is that it induces ringing due to parasitic reactance, which consists mainly of capacitance at pads and inductance of bonding [13,14]. Given the above discussion, a reference ground set by these layers may not be at real zero potential as it should be, and the same is true for the reference supply voltage. Through interconnect layers, the noise is distributed to other parts of the circuit, which are tied to the same power/ground network. In addition to direct coupling of the G_{nd}/V_{dd} bounce, if the same power lines are used for substrate bias, voltage changes in the power interconnect layers propagate to the substrate directly through substrate contacts or by means of the depletion capacitances associated with source and drain regions. This leakage is the primary cause of substrate noise as was observed in Figure 26.2, where ringing dominated in the substrate noise waveform when large di/dt was induced for the shorter interval (T_s) of inverter switching.

In reality, as in the equivalent model of a CMOS inverter of Figure 26.3, external ports are connected to lumped equivalent inductances and resistances representing links (e.g., bond wires) to the board. Although not shown, equivalent capacitances due to package cavities also exist. The nonzero package/bondwire parasitic elements cause signal coupling between the blocks directly

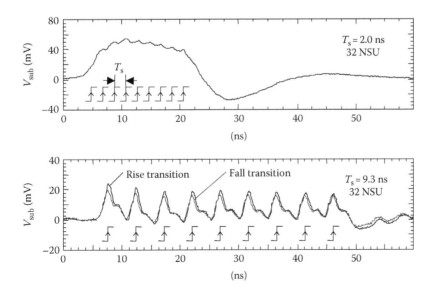

FIGURE 26.2 Measured substrate noise waveforms from transition-controlled noise source units (NSU) with stage delay T_s = 2.0 ns (upper) and 9.3 ns (lower).

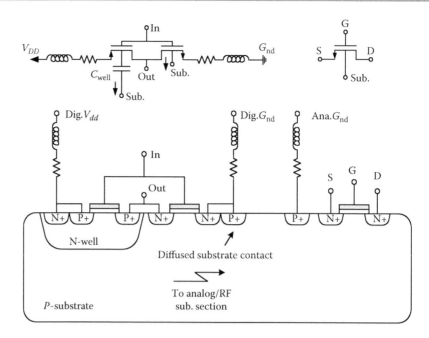

FIGURE 26.3   Noise coupling from the switching portion of the IC (CMOS-not gate) to the RF/analog portion via substrate. Some of the parasitic elements involved in the undesired signal coupling are shown. (From Verghese, N.K. and Allstot, D.J., *IEEE J. Solid-State Circ.*, 33, 314, 1998. With permission.)

(pin to pin and bondwire to bondwire mutual inductance and capacitance), or indirectly by initially coupling to the substrate and from the substrate to other devices. They can have a major impact on the amount of coupled switching noise in an IC design [15]. This is due to the fact that many cheaper packages have large bondwires and package pin inductances associated with the supplies carrying fast current variations, caused by digital gates and nondifferential analog circuits, resulting in sizeable voltage drops [16]. One should also note that since package/bondwire parasitics along with substrate contacts can behave as RLC circuits, switching currents induced by logic circuits cause ringing in the power supply rails and in the output drive circuitry [6,17,18]. The oscillations (ringing), which modify the substrate voltage with a peak voltage and a settling time, are attributed to this ringing.

On the one hand, a major contribution of parasitic capacitance, C, is found in the determination of noise amplitude. Figure 26.4 shows the reduction of peak substrate noise amplitude, V_p, vs. the volume of inactive logic cells (gray area) on the same cell row as the active logic cells fixedly located at one end (black area), where an on-chip noise probing technique is applied [19]. The noise decreases as the volume increases, since an inactive logic cell serves as a local charge reservoir, namely, decoupling capacitance. On the other hand, the effect of noise reduction deteriorates as the inactive logic cells are located away from the active logic cells due to screening by resistive impedance on power supply and ground rails.

26.2.4 COUPLING TO ANALOG INTEGRATED CIRCUITS

The chip substrate can act as a collector, integrator, and distributor of switching currents, causing noise coupling to locations across the chip [15,20]. The injection of currents into the substrate causes substrate biases to vary, which in turn cause variations in MOS threshold voltages, depletion capacitances, and other circuit bias and performance quantities. The MOS threshold voltage modulation is due to the substrate bias fluctuation, which is called the body effect. The MOS drain current I_D is a function of both the gate voltage V_{GS} and the substrate voltage V_{BS}, where dependencies are modeled by two transconductances, g_m and g_{mb} respectively. While the former results in normal MOS operation, the latter increases the body effect, which in turn increases the circuit noise in analog circuits. This is because g_{mb} is a function of

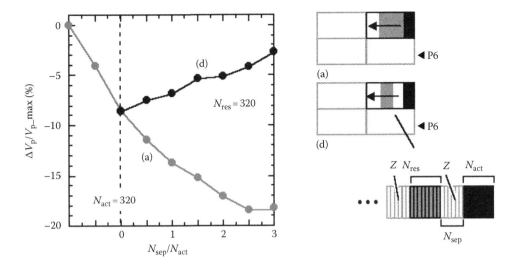

FIGURE 26.4   Effects of parasitic capacitances in the determination of noise amplitude. Number of inactive logic cells increase in curve (a), while the position of fixed number of inactive logic cells moves away from active logic cells in curve (d). (From Nagata, M. et al., Effects of power-supply parasitic components on substrate noise generation in large-scale digital circuits, *Symposium on VLSI Circuits Digest of Technical Papers*, Kyoto, Japan, June 14–16, 2001, pp. 159–162. With permission.)

the substrate voltage and changing the substrate voltage due to any undesired current leads to the fluctuation of the drain current. Hence, a substrate voltage disturbance affects the neighboring transistors. Note that substrate coupling is a problem both in pure analog circuits and in mixed-signal circuits [15]. Although the mechanism of coupling in the substrate is identical in both cases, the effect of the parasitic cross talk tends to be slightly different in these two classes of circuits. In purely analog circuits, the substrate acts as a signal feedback path and the substrate coupling can lead to changes in small signal performance functions like amplifier gain and bandwidth due to feedback properties. In mixed-signal circuits, the switching noise injected into the substrate is picked up by sensitive analog devices on the same substrate, through both their junction capacitances to substrate and through the MOS device body effect as stated above. This results in induced spikes of noise in both device currents and node voltages. It can be shown that the signal coupling via body effect becomes less important relative to depletion capacitance coupling as the substrate doping is increased [21]. It is noteworthy to mention that in addition to transistors, diffused resistors and capacitors may be affected capacitively by substrate noise [9].

26.3 MODELING OF MIXED-SIGNAL NOISE COUPLING

A complete analysis of substrate noise requires both extraction of the substrate as well as simulation. Any substrate noise analysis technique has to include some form of circuit simulation to assess the impact of substrate noise on a particular parameter of interest for a specific analog function. Extraction is the process by which the electrical equivalent model of the substrate, possibly including resistance, capacitance, or inductance, is determined. To extract a substrate accurately, the complex geometries of wells, contacts, well taps, diffusions, trenches, etc. must be extracted. Once extraction has been completed, simulation can be performed on a circuit including the extracted RC network for the substrate. To predict simulation of substrate noise requires some knowledge of the equivalent extracted network, as well as the nature and location of noise injectors that cause the noise. If a SPICE simulation is performed with devices and substrate parasitics present, the computational time required explodes very quickly; hence this approach is tractable only for analyzing small components of the order of a few hundred devices. Alternatives to circuit simulation of substrate noise are discussed below.

26.3.1 MODELING COUPLING TO SUBSTRATE

26.3.1.1 CAPTURING NOISE CURRENT

To model the effect of coupling, there are expressions or equivalent circuit elements that must be incorporated in a circuit simulator to calculate the current injected into the substrate and hence the signal received by sensitive analog circuits. Among them, device capacitances are included in the transistor models for SPICE-like circuit simulators. In the case of interconnect to substrate capacitances, layout extraction tools can easily extract these parasitic elements. Since these values are layout-dependent, they are extracted after the layout of the circuit is completed and are typically incorporated into the circuit simulator for postlayout simulation. A straightforward approach to capture noise current on power supply and ground wirings in an entire large scale digital circuit is to apply SPICE-like enhanced simulators to the extracted netlist. Those simulators provide the capability of hundred-thousands of components, through the use of macromodel devices based on a look-up table of current–voltage characteristics as well as hierarchical circuit macromodels relying on the similarity of circuit operation among primitive circuits with the same set of input signals. The contribution of impact-ionization phenomenon can be additionally included to the noise current, where analytical models representing I–V characteristics of the impact ionization current are built-in with the transistor model employed by SPICE. In one of these models, for example, the hot-electron-induced substrate current can be expressed in semianalytical form as [22]

$$(26.2) \qquad I_{sub} = C_1(V_{ds} - V_{dsat})I_d \exp\left(-\frac{C_2 t_{ox}^{1/3} \cdot x_j^{1/2}}{V_{ds} - V_{dsat}}\right)$$

where

C_1 and C_2 are process-related, empirically determined parameters
t_{ox} is the oxide thickness
x_j is the junction depth
V_{ds} is the drain–to–source voltage
V_{dsat} is the saturation voltage

Using the results obtained from device simulations or measurements, it is possible to determine the empirical coefficients C_1 and C_2, and incorporate impact ionization induced substrate currents into existing device models for circuit simulation. This straightforward approach can provide good accuracy in the noise current; however, one of the major drawbacks is the limitation of circuit scale to analyze due to the explosion of memory usage and CPU time for larger digital circuits.

26.3.1.2 NOISE CURRENT MACROMODELS

The use of simulators with higher levels of abstraction than transistor-level descriptions in SPICE can accelerate as well as improve the capacity of capturing noise current. There are many possible ways of macromodeling large scale digital operations.

A traditional idea of macromodeling assumes that a logic gate draws current with triangular waveshape when it toggles; the height and slope of the triangle depend on input–output logic function as well as fan-out load capacitance and are characterized in advance. Then, power supply current of a digital circuit with a given vector can be approximated by way of superimposing the triangular currents. Noise analyses use the captured noise current directly in terms of electromagnetic radiation such as EMI [23] and ground bounce where it interacts with impedance networks parasitic to power supply and ground wirings as well as to a substrate. This allows the substitution of gate-level logic simulation for transistor-level circuit simulation, which greatly improves the capability of noise analysis.

Another approach to modeling switching current sources utilizes the concept of substrate noise signature for each digital gate [24,25]. This methodology exploits the fact that any given logic gate injects a particular signal into the substrate through capacitive coupling and impact ionization. Such a signal, known as substrate injection pattern, is a unique fingerprint of the gate, input transition, circuit implementation, and technology. It can be accurately calculated using standard device

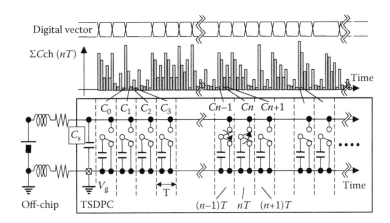

FIGURE 26.5 Time series divided parasitic capacitance (TSDPC) model. (From Nagata, M. et al., Modeling substrate noise generation in CMOS digital integrated circuits, *Proceedings of IEEE Custom Integrated Circuits Conference*, Orlando, FL, May 12–15, 2002, pp. 501–504.)

modeling and circuit simulation, and is better than the simplest signature of triangular waveform approximation described above. The substrate noise signature of the entire circuit is then evaluated using the substrate injection patterns and a precise analysis of the switching activity in the circuit's internal nodes. Switching activities are computed from user-specified input vector sequences by a gate-level simulation. Since the input vector is not known *a priori*, the user should simulate a realistic load or perform a worst/bestcase analysis. Further improvements applied to this approach are discussed in [26], where the power-supply current flowing from V_{dd} to G_{nd}, and noise current directly injected to a substrate, are separately characterized for every digital gate with every possible input combination and stored in a standard cell library as auxiliary information. The pairs of currents are combined in synthesizing substrate noise current accordingly to the switching activities.

In contrast to the previous signature-based techniques that approximate a noise current by the superposition of predetermined waveforms, time series divided parasitic capacitance (TSDPC) substitutes a single capacitor charging process for a mass of logic toggles that occur in a time interval [27]. A train of switched capacitors, such as shown in Figure 26.5, represents a large-scale digital circuit for simulating power-supply currents. The size of capacitance can be derived from toggle records by gate-level logic simulation. A circuit-level simulator solves power-supply current through real charge transfer process within digital circuits along with the parasitic impedances. This improves the accuracy of the noise current, and the simultaneous ground-bounce waveform that couples to the substrate.

26.3.2 MODELING SUBSTRATE PARASITICS

Due to its distributed nature, the substrate cannot be translated into a compact analytical model accounting for the entire chip area whose global effects are felt everywhere in the chip. In general, models for substrate coupling can be derived by one of three methods: from the full 3-D numerical simulation (Maxwell's equation), or using suitable discretization of a simplified form of Maxwell's equation, or using lumped element models. Common techniques to model the substrate include the finite difference method (FDM) followed by network reduction of the resulting mesh, or the boundary element method (BEM) followed by fast BEM matrix solution techniques.

26.3.2.1 BOX INTEGRATION FORMULATION

To obtain a distributed RC network, the box integration technique may be utilized [28]. Using this technique, a 3-D rectangular RC mesh network as the equivalent circuit representation of the modeled substrate is constructed. The mesh topology could be correlated to the circuit's physical design by distributing grid points according to the layout features on relevant fabrication photomasks. The substrate is treated as a 3-D mesh where each mesh edge is a parallel combination of a resistor and a capacitor. In this approach, the edge surfaces (boundaries) are assumed to be Neumann (reflective) boundaries for voltages, while the diffusion/active areas and contact areas are treated as Dirichlet (fixed) boundaries for voltages (equipotential regions) in the resulting 3-D RC mesh. These areas are represented as ports in the multiport network and connected to corresponding nodes in the electrical circuit [29]. Outside the diffusion/active areas which are called ports here, the substrate can be approximated as layers of uniformly doped semiconductor of varying doping density. In these regions, a box integration method [30] can be applied to spatially discretize simplified Maxwell's equations. Ignoring magnetic fields and using the identity $\nabla \cdot (\nabla \times A) = 0$, Maxwell's equations can be written as

$$\nabla \cdot J + \nabla \cdot \frac{\partial D}{\partial t} = 0 \tag{26.3}$$

where

J is the current density
D is the displacement vector
t is the time

Using the relations, $D = \varepsilon E$ and $J = \sigma E$, this reduces to

$$\sigma \nabla \cdot E + \varepsilon \frac{\partial}{\partial t}(\nabla \cdot E) = 0 \tag{26.4}$$

where

E is the electric field
σ is the conductivity
ε is the permitivity

The above equation can be discretized on the substrate volume either in differential form using FDM, or in integral form using the BEM, which is explained later in this section.

In the FDM technique, the substrate is expressed as a collection of square cubes as shown in Figure 26.6 [18]. An electric field normal to a contact plane of two adjacent cubes (i, j) with

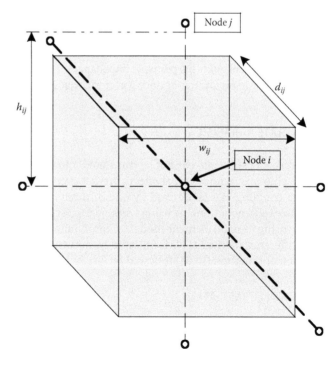

FIGURE 26.6   Substrate is expressed as a collection of square cubes. (From Verghese, N.K. et al., *Simulation Techniques and Solutions for Mixed-Signal Coupling in Integrated Circuits*, Kluwer, Boston, MA, 1995. With permission.)

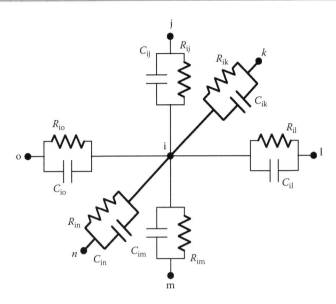

FIGURE 26.7 Resistances and capacitances around a mesh node in the electrical substrate mesh. (From Stanisic, B.R. et al., *IEEE J. Solid-State Circ.*, 29, 226, 1994. With permission.)

distance h_{ij} is given in Equation 26.5, and thus Equation 26.4 is rewritten as Equation 26.6 by Gauss' law under the assumption of uniform impurity concentration in the cube and by applying the box integration method [30,31]:

$$(26.5) \qquad E_{ij} = \frac{V_i - V_j}{h_{ij}}$$

$$(26.6) \qquad \sum_j \left[G_{ij}(V_i - V_j) + C_{ij}\left(\frac{\partial}{\partial t}V_i - \frac{\partial}{\partial t}V_j \right) \right] = 0$$

where

$$(26.7) \qquad G_{ij} = \sigma\left(\frac{W_{ij} \times d_{ij}}{h_{ij}} \right)$$

and

$$(26.8) \qquad C_{ij} = \varepsilon\left(\frac{W_{ij} \times d_{ij}}{h_{ij}} \right)$$

as shown in Figure 26.7.

This approach is not efficient because large 3-D meshes are created, which would be prohibitive to simulate using variable time-step trapezoidal integration techniques. For instance, consider the case of a lightly doped substrate where a lot of empty space (between ports) must be discretized to get an accurate estimate of port-to-port substrate admittance [20].

26.3.2.2 NETWORK REDUCTION

To address this problem, the generated linear RC network should be approximated by a smaller circuit that exhibits similar electrical properties. A small percentage of the network nodes, called ports, are physically connected to the external circuit (at the top surface of the substrate). In theory, an "equivalent" multi-port network (similar to a Thevenin equivalent circuit for a one-port network) can be formulated by eliminating a substantial fraction of the internal nodes. This technique is generally referred to as *"network reduction."* For network reduction using congruence transformations [28], full-network conductance and susceptance matrices are transformed to reduced equivalents, which can be directly realized with resistors and capacitors. This algorithm utilizes the well-conditioned symmetric Lanczos process, which exploits the specialized structure of the extracted substrate networks to formulate Pade approximations of the network port admittance. Congruence transformations are employed to ensure stability and to create reduced networks that are easily realizable with RC elements. The approximated networks are guaranteed to be passive, and are thus well behaved in subsequent simulations. The reduced models retain the accuracy of original models, but contain orders of magnitude of fewer circuit nodes.

The network reduction problem can be simplified if the capacitances in the RC mesh can be ignored. Neglecting the intrinsic substrate capacitance is a reasonable assumption for operating speeds of up to a few GHz and switching times of the order of 0.1 ns. This is due to the fact that the relaxation time of the substrate (outside of active areas and well diffusions) which is given by $\tau = \rho\varepsilon$, is of the order of 15 ps (with $\rho = 15$ Ω-cm and $\varepsilon_r = 11.9$). The capacitive behavior of the substrate outside the active area is negligible for frequencies below one tenth of τ. As we see here, if the capacitance to substrate, which is introduced by the depletion regions of well diffusions and interconnect overlying field oxide, can be accurately modeled as lumped circuit elements outside the mesh located near the chip surface, the substrate can be modeled as a purely resistive mesh. If the substrate is multilayered, then the network may not be approximated by a resistive network even at lower frequencies [15]. Note that in circuit simulators, such as SPICE, junction capacitances of active devices are modeled outside the mesh as lumped capacitances.

The substrate may be modeled as a resistive grid, which is determined through box integration or the Delaunay tessellation (for greater accuracy in and around the more interesting regions of the substrate) [12]. Using a simple DC macromodeling approach, the 3-D resistive mesh can be reduced to an equivalent set of $n(n + 1)/2$ resistances interconnecting the n ports [31,32]. In this approach, wells are considered ports and connected to lumped capacitances outside the mesh. Since the substrate is modeled as a purely resistive mesh, the macromodel consists of only the steady-state/DC values of the admittance parameters, with the higher-order mesh moment being zero. The computational complexity of the DC macromodel is much lower than that of the congruence transform-based method, and the resulting macromodel is more compact.

26.3.2.3 BOUNDARY ELEMENT METHOD USING GREEN'S FUNCTION

An alternative approach to solving the simplified Maxwell's equations is the BEM, which can be used for parasitic and substrate extraction [33]. In this method only the ports that connect the substrate to the devices/contacts/wells are discretized. BEM is more appealing compared to FDM, since in this method instead of discretizing the whole structure, only the relevant boundary features, the 2-D substrate contacts (port areas) are discretized, and hence the resulting matrix to be inverted in the network reduction process is much smaller albeit fully dense [34]. The extraction can be combined with a model-reduction technique to obtain simpler models for cross talk [35]. Another major advantage of the BEM is that it is not very discretization-dependent (unlike FDM) [20]. For example, by discretizing a port into a single panel, i.e., assuming a constant current density across the port, the results are within 10% of the actual answer. By a proper choice of the Green's function of the BEM, only those parts of the substrate boundary (called "contacts") have to be discretized that directly interact with the designed circuit.

The Green's function is used to determine point-to-point impedance between each pair of discretized ports as shown schematically in Figure 26.8. The resulting impedance matrix is then inverted to yield the required substrate admittances [36]. The Green's function is the potential at any point in a medium due to a current injected at any point in the medium, and can be determined for the substrate in quasianalytical form. The areas of the substrate that connect to the external world (device/contact areas) are discretized into a collection of n panels, and the contribution to the potential at each panel, due to currents injected at every panel is stenciled into an $n \times n$ matrix of impedances, which is then inverted to determine the substrate admittances. This technique effectively reduces a 3-D problem into a 2-D one.

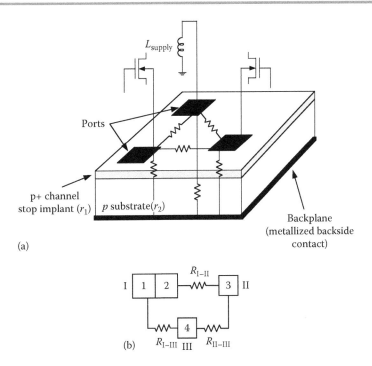

FIGURE 26.8   (a) Ports on a substrate resistively connected to one another; (b) Determining resistive coupling between ports using the BEM. (From Verghese, N.K. and Allstot, D.J., *IEEE J. Solid-State Circ.*, 33, 314, 1998. With permission.)

26.3.2.4 GREEN'S FUNCTION FORMULATION

For the resistive substrate case ($\varepsilon_r = 0$), the Maxwell's equations reduce to the well-known Laplace's equation [20]

$$\nabla^2\phi = 0 \qquad (26.10)$$

where ϕ is the electrostatic potential. Applying Green's theorem to the above equation gives the electrostatic potential at an observation point, r, due to a unit current injected at a source point, r', as

$$\phi(r) = \int_v J(r')G(r,r')\,\mathrm{d}^3r \qquad (26.11)$$

where
   $G(r,r')$ is the substrate Green's function satisfying the boundary conditions of the substrate
   $J(r')$ is the source current density

Since all the sources and observation points are at the defined ports on the substrate and these are planar and practically 2-D, the above volume integral reduces to a surface integral.

$$\phi(r) = \int_s J(r')G(r,r')\mathrm{d}a \qquad (26.12)$$

Essentially, this reduces a 3-D problem into a 2-D problem. In addition, since the Green's function implicitly takes the substrate boundaries into account, there is no need to consider them explicitly when solving the above equation where only the port areas (that actually connect to the substrate) need to be discretized to solve the equation.

   The Green's function of the substrate can be determined analytically using classical mathematical techniques [36,37] and has been reported in the literature [33,38]. The substrate Green's function $G(x,x',y,y')$, with (x,y) and (x',y') being the coordinate locations of the observation and source points on the substrate surface is:

$$(26.13) \qquad G(x,x',y,y') = \sum_{m=0}^{M} \sum_{n=0}^{N} f_{mn} \cos\left(\frac{m\pi x}{a}\right) \cos\left(\frac{m\pi x'}{a}\right) \sin\left(\frac{n\pi y}{b}\right) \sin\left(\frac{n\pi y'}{b}\right)$$

where f_{mn} for a homogeneously doped substrate is given by

$$(26.14) \qquad f_{mn} = \frac{C_{mn}}{ab\sigma} \tanh\left(\sqrt{\frac{m^2\pi^2}{a^2} + \frac{n^2\pi^2}{b^2}} \cdot c\right)$$

Here, C_{mn} is a constant, s the substrate conductivity, and (a, b, c) are the (X, Y, Z) substrate dimensions. For a multilayered substrate profile (of uniform sheet resistivities) a more complicated expression is obtained for f_{mn}. Once the Green's function is determined, Equation 26.12 remains to be solved [20]. The solution is obtained using a suitable discretization technique which discretizes each port on the substrate into a set of panels. A system of equations can be formulated that relate the currents and potentials at all panels in the system. In matrix form this is represented as

$$(26.15) \qquad \phi = Z_i$$

where each entry in the impedance matrix, z_{ij}, is given as

$$(26.16) \qquad z_{ij} = \int\int_{S_i \, S_j} J(r')G(r,r') \, da' da$$

In this equation, S_i and S_j are the surface areas of panel i and j, respectively. The impedance, given by Equation 26.16, can be analytically determined for rectangular panels once the Green's function is derived. The matrix, Z, is then inverted to obtain the substrate admittance matrix, Y. We can determine the substrate resistance between any two ports as the reciprocal of the sum of corresponding admittance matrix entries.

From a computational point of view, the direct evaluation of the quasianalytical Green's function given above involves several million floating-point multiplications and additions and since it must be repeated for every pair of panels, the formulation of the impedance matrix becomes an expensive task for large problems [36]. As an alternative technique, after discretizing the entire substrate surface into a uniform grid of panels, a 2-D discrete cosine transformation (DCT) can be utilized to pre-compute all the panel-to-panel impedances on the substrate in O($N \log N$) time [38]. Another problem with the BEM approach in general is that inversion of the dense $n \times n$ matrix is a cumbersome task. Direct LU factorization requires O(n^3) operations, which is clearly infeasible for a reasonably sized problem.

26.3.2.5 FAST SOLUTION OF BEM MATRICES

To improve the efficiency of the BEM, multilevel (multigrid) methods, which are efficient iterative techniques can be developed for first-kind integral equations defined over complicated geometries [35]. Based on the method, a multigrid iterative solver integrated with sparsification algorithms specially tuned to account accurately for substrate edge effect allows for almost an order of magnitude improvement in the speed of the BEM solution process.

Alternatively, a fast eigendecomposition technique that accelerates operator application in BEMs and avoids the dense matrix storage while taking all of the substrate boundary effects into account has been presented explicitly [34,39]. For efficient extraction of the substrate coupling model in a BEM formulation, the authors use the eigendecomposition-based technique in a

Krylov subspace solver. The model can be incorporated into a circuit simulator such as SPICE to perform coupled circuit–substrate simulation. To speed up the model-computation process at the cost of a slight decrease in accuracy, the use of pre-corrected-DCT (PcDCT) algorithm was also proposed [40,41]. The main idea behind the PcDCT algorithm is to realize that the effect of an injected current in a panel on the potential of another far-away panel can be considered the same for small variations in the distance between panels. A heavily doped bulk substrate was used in the experiment that the authors reported. The results indicated better accuracy as well a speed up of around 180 times compared to the vanilla Green's function method, and 12 times speed up compared to the eigendecomposition method. The memory requirements for the PcDCT algorithm were seen to be considerably smaller (20 times) than those of the vanilla Green's function method. In the example, some savings in memory (three times) were also obtained with respect to the unaccelerated eigendecomposition algorithm.

Several other problems plague the BEM-based substrate modeling and subsequent simulation problem [42]. First, the density of the extracted coupling matrix makes later circuit simulation prohibitively costly, because the now dense circuit matrix must be factored hundreds or thousands of times in each simulation. Second, most methods of obtaining the n columns of the coupling matrix require n matrix solutions, which is computationally quite costly, making it impractical to solve problems with n larger than a few hundred. To address these problems on a multiscale, wavelet-like basis for fast integral equation solutions has been proposed [42]. The wavelet basis efficiently represents coarse grain information of the IC geometry. Using such a basis, many entries would become small and could simply be dropped with only a small loss of accuracy. The wavelet basis has a multiresolution property. It was shown, for example, that reducing the number of non-zero elements by 90% led to 1% accuracy loss. The results presented showed that for a problem with a few thousand contacts, this method was almost ten times faster in constructing the matrix.

26.3.2.6 CHIP-LEVEL SUBSTRATE NETWORK EXTRACTION

SeismIC™ performs a mixed-signal noise simulation without the presence of devices (using equivalent noise sources) in order to compute the time- or frequency-domain substrate noise waveforms at the bulk nodes of interest and can be utilized to analyze chips with 1 million or more devices [43]. After the substrate noise waveforms have been computed, a circuit simulation with devices and noise sources attached can be performed to assess the impact of the noise on the subcircuits of interest. A typical flow for substrate noise analysis for verification using SeismIC is shown in Figure 26.9.

In order to efficiently model and analyze the substrate of large designs, an adaptive substrate modeling approach is used. This is achieved through the use of sensitivity analysis to determine which areas of the chip need high model accuracy and where the model accuracy can be relaxed without impacting the accuracy of the overall analysis. Noise sensitivity analysis is also used to measure the impact on substrate noise of a change in any given parameter. By calculating the sensitivity to various layout, process, and package parameters, the appropriate measures to minimize substrate noise are determined.

26.3.3 CHIP-LEVEL MIXED-SIGNAL SUBSTRATE COUPLING ANALYSIS

Once an accurate substrate extraction has been performed, the location and magnitude of noise injectors need to be determined to facilitate simulation of the substrate noise waveforms. The location of noise injectors can be determined from the layout and schematic netlist information. To determine the magnitude and phase of injected currents, some form of simulation input is required under assumed switching activity. Once this has been ascertained, the problem is reduced to solving a very large RC network with active current sources, as shown in Figure 26.10. The number of current sources can be extremely large, for example, a million transistor mixed-signal design may have a million current sources. To see how a large RC network driven by active

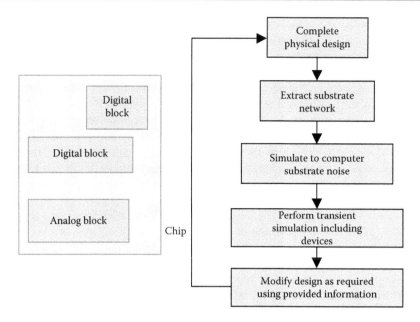

FIGURE 26.9 Verification flow for full chip level substrate noise analysis.

current sources is analyzed, consider that the voltage response at a bulk node of interest, vb, is desired. The voltage response can be written as follows:

$$(26.17) \qquad vb(s) = z1(s) \cdot i1(s) + z2(s) \cdot i2(s) + z3(s) \cdot i3(s) + \cdots$$

where
$\quad i1, i2, i3,\ldots$ are the current sources at various locations on the substrate
$\quad z1, z2, z3,\ldots$ are their corresponding impedances to the bulk node of interest

The current source values, $i1, i2, i3,\ldots$ can be determined from a simulation of the original circuit (without parasitics) by observing the currents flowing in the power/ground nodes and the device bulk terminals. This can be accomplished either with a transistor-level circuit simulator or a gate-level event driven simulator in conjunction with precharacterized cell libraries, as discussed in Section 26.3.1. The currents can be either time-domain waveforms or a composition of spectral values at every frequency ($s = j\omega$) of interest. The impedances, $z1, z2,\ldots$ can be obtained by inverting the admittance matrix formed by the RC substrate network and package inductances (Figure 26.10)

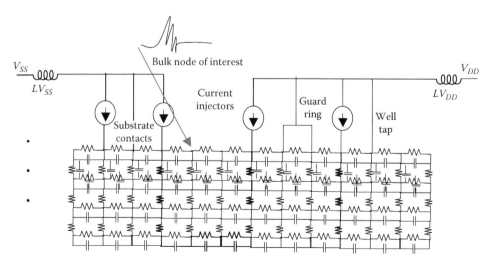

FIGURE 26.10 Simulation model for full chip substrate with a large number of noise injectors.

at every frequency ($s = j\omega$) of interest. The frequency-domain response of vb can be obtained by solving Equation 26.17 at every frequency of interest. Applying the inverse Laplace transform to this response results in the corresponding time-domain waveform.

One advantage of using Equation 26.17 to calculate the noise response of a bulk node of interest is that each individual noise contributor can be calculated independently. Hence, from Equation 26.17, the noise contribution from injector 1 at the bulk node of interest is $z1(s)i1(s)$. Similarly, $z2(s)i2(s)$ is the contribution from injector 2, $z3(s)i3(s)$ from injector 3 and so on. Thus, the most significant noise contributors can be identified and appropriate measures can be taken to minimize their impact.

26.3.3.1 MACROSCOPIC SUBSTRATE NOISE ANALYSIS USING HIGH-LEVEL SIMULATION

A macroscopic substrate noise model that expresses coupling noise as a function of logic state transition frequencies among digital blocks has been proposed [13,44]. The coupled noise is defined as one of the state variables in behavioral description of victim circuits such as analog circuits. The noise can be typically expressed by the superimposition of voltage changes arising from digital state transitions in unit time; thus a function of state transition frequencies is evolved in the digital block, which can be easily extracted from digital logic simulation. This results in the introduction of behavioral noise modeling using a hardware description language (HDL)-based system-level design [44–46]. A simulation system based on the model was implemented in a mixed-signal simulation environment as shown in Figure 26.11, where performance degradation of a second order Delta Sigma Analog-to-Digital Converter (ADC) coupled to digital noise sources was simulated. The computation of the noise proceeded according to a noise waveform function $F(f_{eff}, t)$ in parallel with a transient analysis of the mixed-signal circuit under design. The calculated noise waveform was injected into the analog circuit. Here, $f_{eff}(n)$ is the effective transition frequency, a global state transition count per unit time, T at the nth sampling interval defined by

$$(26.18) \qquad f_{eff}(n) = \frac{1}{T}\sum_{i=1}^{m}\left[W_i^+ N_i^+ + W_i^- N_i^-\right]$$

where
  $N_i(n)$ is the local state transition count of the ith digital sub-block
  W_i is the weight coefficient corresponding to a substrate coupling intensity of the sub-block to the sensitive analog circuit, and is a relative quantity among the digital sub-blocks

Superscripts "$+$" and "$-$" stand for rising and falling transitions, respectively. T is the noise sampling period introduced to discretize the noise-generation process and $F_{nclk}(= 1/T)$ is

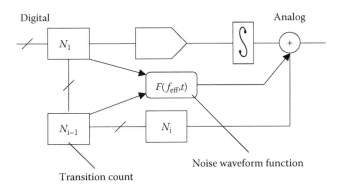

FIGURE 26.11 Proposed macroscopic substrate noise model. (From Nagata, M. and Iwata, A., *IEICE Trans. Fundam. Electr.*, E82-A, 271, 1999.)

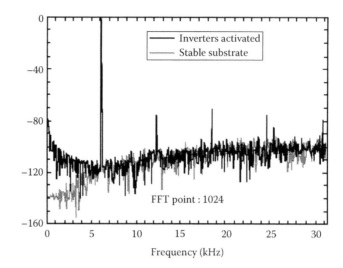

FIGURE 26.12 Simulated in-band power spectrum of second order delta sigma modulator coupled to substrate noise generator. (From Nagata, M. and Iwata, A., *IEICE Trans. Fundam. Electr.*, E82-A, 271, 1999. With permission.)

synchronous to the system clock (F_{sclk}). $F(f_{\text{eff}}, t)$ must be continuous and reflect the nature of its transient behavior. The author adopted a successive function system $\{\phi_n(t)\}$ of Equation 26.19 as $F(f_{\text{eff}}, t)$, where $\phi_n(t)$ is defined in $t \; \varepsilon \; [0,T]$ of the nth sampling period, and α, β are model parameters.

(26.19)
$$\phi_n(t) = \phi_{n-1}(T) + \alpha[f_{\text{eff}}(n) - f_{\text{eff}}(n-1)]\left[1 - \exp\left(-\frac{t}{\beta}\right)\right]$$

The weight coefficient W_i models the attenuation of the noise amplitude by distance and guard-banding, and the ratio of the noise amplitude for rising to falling transitions. To determine the coefficients beforehand, evaluating the substrate noise transmission by circuit simulation is required. Hence, one of the modeling methods for substrate equivalent circuits explained in Section 26.3.2 should be used. One of the shortcomings of this technique is that the coefficients are functions of technology, circuit, layout, etc., and therefore the circuit simulator must be run each time any of these parameters change. In addition, α and β are parameters that determine the amplitudes and the widths of the generated noise waveforms, respectively. These are dominated by the substrate structure and thus should be evaluated from experimental results with dedicated test chips consisting of simple noise sources like inverter arrays and wide bandwidth substrate noise sensors. However, the macro-modeling approach is appreciated for capturing the sensitivity, and measuring the response of mixed analog and digital circuits to coupling noise in terms of performance metrics such as SNR and BER, for which the designer cannot apply circuit-level simulation. For instance, the observed degradation of THD performance shown in Figure 26.12 agrees with the reported experiments [47], and therefore the model successfully expresses the interaction of delta-sigma modulation loop dynamics with transient voltage noises injected into analog signal paths mainly through integrators.

26.3.3.2 PERIODIC ANALYSIS OF MIXED-SIGNAL NOISE IN RADIO FREQUENCY CIRCUITS

For RF circuits, specialized simulation techniques for the analysis of periodic circuits can be used to calculate quickly the response of such circuits to mixed-signal noise [6,48]. Following a periodic steady-state operating point analysis, a transfer function analysis computes the transfer functions to the RF circuit output at a single frequency and from every noise source in the circuit at every input frequency (i.e., output frequency and all frequencies offset from it by a harmonic of the periodic signal). Using this approach, it is possible to compute transfer functions from the bulk node of every device in the (RF) circuit to a specific output at a given frequency of interest. Note that for a given frequency of interest, this results in a set of transfer functions for each bulk node. Once the transfer functions have been computed for the (RF) circuits, it no longer needs to be represented at the transistor level. A transient simulation can be performed on the digital (and analog) circuits to obtain the transient substrate noise signals. A postprocessing of the signal (Fourier transform) determines the equivalent noise spectra at the device bulk nodes. These noise frequency components multiplied with the transfer functions obtained above calculate the coupling of substrate noise to the RF circuit output. To calculate the periodic transfer functions, efficient matrix-free iterative methods for the periodic analysis of RF circuits are used [49–52]. The methodology was reported to have been applied to the verification of the transmit section of a portable radio front-end IC [6,48]. Measured results on the fabricated IC indicated an RF spur at the output of an up-conversion mixer (modulator) in the transmit section of the circuit, which was adequately predicted after analyzing the circuit using this method. With roughly 1900 devices, 717 nodes, and 3234 equations to solve, a transient analysis of the modulator and reference frequency generator would have required 2 days of computation to simulate 20 periods which are required for the modulator to attain a steady state. Using periodic analysis instead, a macromodel of the modulator (containing 982 devices, 438 nodes, and 1445 equations) was obtained in less than an hour of CPU time. Another hour was required to simulate the transient noise coupling from the reference generator.

26.4 MIXED-SIGNAL NOISE MEASUREMENT AND VALIDATION

Nagata et al. [53] proposed a direct sampling technique for substrate noise measurement, where a detector named SFLC consists of a P-channel source follower (SF) with an input probe located around P1 area on the surface of a P-type substrate and a latch comparator (LC) connecting to the SF output, as shown in Figure 26.13. The SF picks up substrate potential around the probe and the LC discretizes the SF's level-shifted output voltage through successive comparisons, with stepwise reference voltage externally provided and with sampling occurring at every latch operation. The SF provides good linearity with the input voltage range of the order of 1 V, along with gain of slightly less than unity and bandwidth of a few GHz, even when followed by the LC. The authors demonstrated waveform-accurate substrate noise measurements with voltage and time resolution of 100 μV and 100 ps, respectively, and also showed the consistency between substrate noise waveforms acquired by direct measurements and comparator-based indirect measurements described in the previous section [53]. Here, only the former can

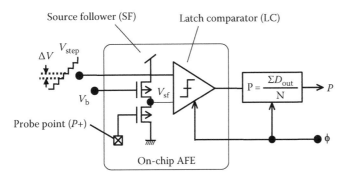

FIGURE 26.13 On-chip noise probing circuit consisting of source follower (SF) and latched comparator (LC). (From Nagata, M. et al., *IEEE Trans. Comput. Aid. Des.*, 19, 671, 2000. With permission.)

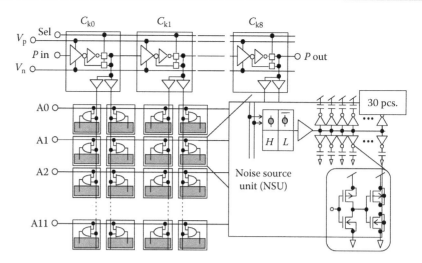

FIGURE 26.14   Transition controllable noise source (TCNS) circuit. (From Nagata, M. et al., *IEEE Trans. Comput. Aid. Des.,* 19, 671, 2000. With permission.)

achieve absolute-voltage quantitative evaluation of substrate noise while the latter is restricted to relative evaluation.

A transition-controllable noise source (TCNS) shown in Figure 26.14 includes a multiphase clock ($C_k[0:8]$) generator consisting of nine delay elements and a matrix of noise source unit (NSU) in the form of 9 rows × 12 columns, where the number of NSUs to activate by each edge of C_k can be set from 0 to 12. The delay element has bias voltages V_n and V_p for regulating rise and fall delay, respectively. In addition, inverse or noninverse transitions among adjacent noise source blocks are selected by the signal "Sel". The NSU has 30 inverters operating in parallel, where each inverter has a 50 fF load capacitor to the substrate that corresponds to typical parasitic capacitance of 2 fanout gates and local wiring. Minimum gate length is used and widths are chosen to have a switching time of roughly 200 ps for both rise and fall transitions when driving the load capacitors. The TCNS can generate substrate noise with controlled transitions in size, inter-stage delay, and direction [53].

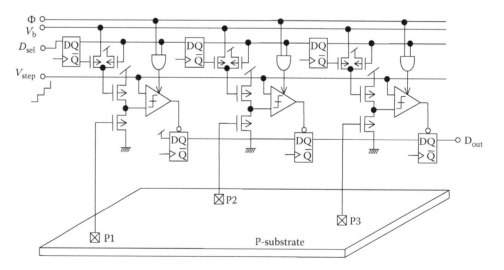

FIGURE 26.15   Arrayed SFLC detectors for multiple-point substrate noise measurement. (From Nagata, M. et al., Effects of power-supply parasitic components on substrate noise generation in large-scale digital circuits, *Symposium on VLSI Circuits Digest of Technical Papers*, Kyoto, Japan, June 14–16, 2001, pp. 159–162. With permission.)

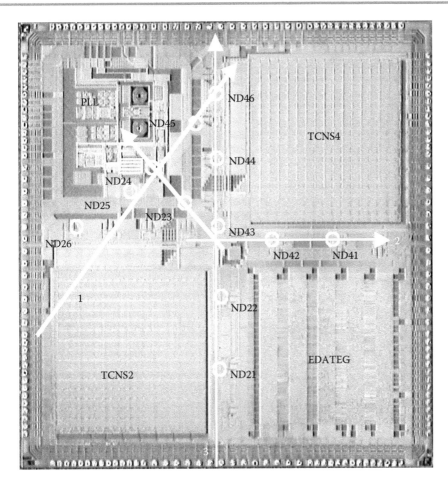

FIGURE 26.16   Micrograph of test chip carrying two TCNS circuits and twelve SF1LC detectors. (From Chu, W.K. et al., A substrate noise analysis methodology for large-scale mixed-signal ICs, *Proceedings of IEEE Custom Integrated Circuits Conference*, San Jose, CA, September 21–24, 2003, pp. 369–372. With permission.)

Multiple-point measurements on a single substrate were achieved by arraying the SFLC detectors shown in Figure 26.15 [19]. A combination of TCNS and arrayed SF + LC can be a reference structure for assessing substrate noise generation and substrate coupling for a given CMOS mixed-signal technology. An example of such a test chip, which was fabricated in a commercial 0.3 μm 3.3 V CMOS process with P-type bulk substrate is shown in Figure 26.16 [54]. The chip includes two TCNS blocks on the top-right and bottom-left quadrants, a victim PLL circuit in the top-left quadrant, and 12 SF + LC placed along four different axes at the periphery of the noise source and inside the PLL.

The substrate noise waveforms shown in Figure 26.2 were measured by SF + LC for TCNS running with the shortest delay among $C_k[0:8]$ (T_s) on the top of the figure, and with a larger delay in the bottom of the figure. The shortest delay causes single large peak noise due to large di/dt coupling to inductance parasitic to assembly, while positive peaks corresponding to each edge of C_k with rise and fall transitions appear for the larger T_s. The observed difference in substrate noise waveform from the identical digital noise circuit also results from the fact that the major cause of substrate noise is the leakage of ground bounce.

The location dependence of peak-to-peak substrate noise amplitude obtained by measurements with arrayed SF1LC detectors and by simulation with chip-level substrate network extraction and noise analysis is shown in Figure 26.17. The distance of each detector from the first one on the axis is listed on the x-axis of each graph. The average absolute error between these simulated and measured results is 4.5 dBV [54].

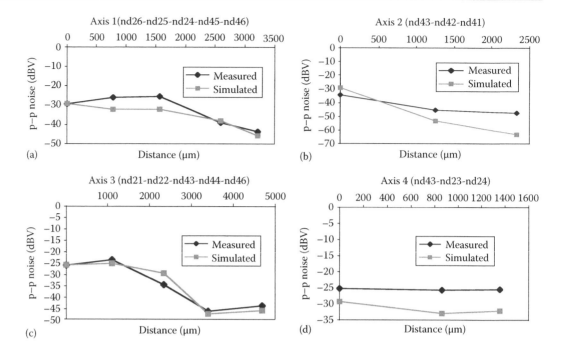

FIGURE 26.17 Simulated and measured peak-to-peak noise amplitude at various points along the 4 axes of detection shown in Figure 26.16. (From Chu, W.K. et al., A substrate noise analysis methodology for large-scale mixed-signal ICs, *Proceedings of IEEE Custom Integrated Circuits Conference*, San Jose, CA, September 21–24, 2003, pp. 369–372. With permission.)

26.5 APPLICATION TO PLACEMENT AND POWER DISTRIBUTION SYNTHESIS

Simulation of mixed-signal switching noise has been integrated into several automatic layout tools including a power distribution synthesis program (RAIL) that automates the design of the power distribution network [31,55] and a substrate aware placement tool (WRIGHT) [56,57]. In RAIL, the topology of the power grid, the sizing of individual segments, and the choice of I/O pad number and location are simultaneously optimized. The optimization, which employs combinational optimization techniques, is performed under tight dc, ac, and transient electrical constraints arising from the interaction of the power grid with the rest of the IC—notably via substrate coupling. Coupling effects are included in the cost function of a simulated annealing (SA) based power distribution synthesis system. In this work, linear macromodels for the digital switching logic circuits are created, and the capacitive and resistive coupling to the power rails are modeled as shown in Figure 26.18. Each logic circuit is replaced by its linear macromodel. To design a power grid, the tool begins with an initial "state" for the power bus geometry and power I/O pad configuration and then this geometry is perturbed to create a new candidate power grid or pad configuration and update the electrical models for the buses and I/O pads. In the next step, these models are combined with designer-supplied circuit macro-models for blocks being supplied by the power grid, and for the substrate. With this complete electrical model—power grid, blocks, pads, and substrate—the resulting electrical performance is evaluated and compared against designer constraints. For example, one might evaluate the coupled noise waveform at a sensitive node against a designer—supplied peak-to-peak noise amplitude constraint. Finally, the optimizer accepts or rejects the perturbation based on the result. The iterative improvement loop is continued until the optimizer determines no further improvement is possible. The main objective is to ensure that the power distribution as a whole (buses, power I/O cell assignment, and internal cell decoupling) is designed to meet dc voltage drop and current density constraints, while keeping transient voltage below user-specified targets.

In placement tools, traditionally area and the wire length have been the most important concerns, but other factors that deal with the interaction between analog and digital sections through

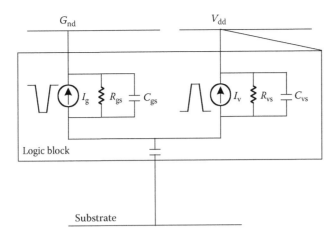

FIGURE 26.18 Simple linear macromodel of logic block for switching noise in power distribution synthesis. (From Stanisic, B.R. et al., *Proc. IEEE J. Solid-State Circ.*, 321, 1995. With permission.)

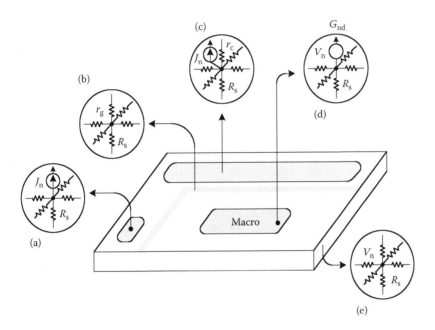

FIGURE 26.19 Substrate-resistive macromodel for placement. (a) Noise source, (b) guard ring, (c) contact, (d) noise detector, and (e) substrate. (From Mitra, S. et al., Substrate-aware mixed-signal macro-cell placement in WRIGHT, *Proceedings of IEEE Custom Integrated Circuits Conference*, San Diego, CA, May 1994, pp. 529–532. With permission.)

the common substrate have added a new dimension to this problem. A set of algorithms for handling substrate-coupled switching noise in an iterative placement framework implemented in a substrate-aware mixed-signal placement tool called WRIGHT has been described [56,57]. The focus here is on physical design, in particular chip-level macro-cell placement and the approach incorporates simplified switching noise estimation into an SA placement algorithm. A coarse-resistive grid method analyzing the coupling of digital switching noise into the analog macros on the chip is used as shown in Figure 26.19. The tool includes models for the chip substrate, noise sources, and receivers on the macrocells. In addition, mitigation measures such as guard rings were incorporated into the inner loop of the placer by low impedance ties from the substrate to the reference potential. The accuracy available in design tools to analyze the substrate noise is not needed and is unaffordable within a placement framework since such a tool must visit thousands of candidate placement solutions.

26.6 SUMMARY

To understand and address the problem of noise coupling in mixed-signal ICs many modeling methods and computer simulation techniques for mixed-signal noise coupling have been proposed. These efforts have been reviewed in this chapter. The physical phenomena responsible for the generation of the undesired signal have been discussed, and the media transporting the signal from the source to the destination described. In addition, different approaches for modeling the source and coupling media, and subsequent computer methods to simulate the coupling have been discussed. Measurement techniques for substrate noise and validation of computer simulation approaches have also been discussed. Finally, an application of substrate noise analysis to placement and power distribution synthesis has been reviewed.

REFERENCES

1. A. Rofougaran, G. Chang, J.J. Rael, J.Y.-C. Chang, M. Rofougaran, P.J. Chang, M. Djafari, M.-K. Ku, E.W. Roth, A.A. Abidi, and H. Samueli, A single-chip 900-MHz spread-spectrum wireless transceiver in 1-mm CMOS—Part I Architecture and transmitter design, *IEEE J. Solid-State Circ.*, 33, 515–534, 1998.
2. D.R. Pehlke, A. Burstein, and M.F. Chang, Extremely high-Q tunable inductor for Si-based RF integrated circuit applications, *IEEE International Electron Devices Meeting Digest*, Washington, DC, 1997, pp. 63–66.
3. J.A. Olmstead and S. Vulih, Noise problems in mixed analog-digital integrated circuits, *Proceedings of IEEE Custom Integrated Circuit Conference*, Portland, OR, May 4–7, 1987, pp. 659–662.
4. S. Masui, Simulation of substrate-coupling in mixed-signal MOS circuits, *Proceedings of VLSI Circuit Symposium*, Seattle, WA, June 4–6, 1992, pp. 42–43.
5. T. Tsukada and K. Makie-Fukuda, Approaches to reducing digital-noise coupling in CMOS mixed-signal LSI's, *IEICE-T. Fundam. Electr.*, E80-A, 263–275, 1997.
6. N.K. Verghese and D.J. Allstot, Computer-aided design considerations for mixed-signal coupling in RF integrated circuits, *IEEE J. Solid-State Circ.*, 33, 314–323, 1998.
7. E. Charbon, R. Gharpurey, R.G. Meyer, and A. Sangiovanni-Vincentelli, Substrate optimization based on semi-analytical techniques, *IEEE Trans. Comput. Aid. Des.*, 18, 172–190, 1999.
8. R.B. Merrill, W.M. Young, and K. Brehmer, Effect of substrate material on crosstalk in mixed analog/digital integrated circuit, *International Electron Devices Meeting Technical Digest*, San Francisco, CA, December 11–14, 1994, pp. 433–436.
9. X. Aragones and A. Rubio, Experimental comparison of substrate noise coupling using different wafer types, *IEEE J. Solid-State Circ.*, 34, 1405–1409, 1999.
10. J. Briaire and K.S. Krisch, Substrate injection and crosstalk in CMOS circuits, *Proceedings of IEEE Custom Integrated Circuits Conference*, San Diego, CA, May 16–19, 1999, pp. 483–486.
11. J. Briaire and K.S. Krisch, Principles of substrate crosstalk generation in CMOS circuits, *IEEE Trans. Comput. Aid. Des.*, 19, 645–653, 2000.
12. S. Mitra, R.A. Rutenbar, L.R. Carley, and D.J. Allstot, A methodology for rapid estimation of substrate-coupled switching noise, *Proceedings of IEEE Custom Integrated Circuits Conference*, Santa Clara, CA, May 1–4, 1995, pp. 129–132.
13. M. Nagata and A. Iwata, Substrate noise simulation techniques for analog-digital mixed LSI design, *IEICE Trans. Fundam. Electr.*, E82-A, 271–278, 1999.
14. M. Nagata and A. Iwata, Substrate crosstalk analysis in mixed signal CMOS integrated circuits, *Proceedings of IEEE Asia and South Pacific Design Automation Conference 2000 with EDA TechnoFair*, Yokohama, Japan, January 25–28, 2000, pp. 623–629.
15. S. Kiaei, D.J. Allstot, K. Hansen, and N.K. Verghese, Noise consideration for mixed-signal RF IC transceivers, *Wirel. Netw.*, 4, 41–53, 1998.
16. A.J. Rainal, Eliminating inductive noise of external chip interconnections, *IEEE J. Solid-State Circ.*, 29, 126–129, 1994.
17. D.K. Su, M.J. Loinaz, S.Masui, and B.A. Wooley, Experimental results and modeling techniques for substrate noise in mixed-signal integrated circuits, *IEEE J. Solid-State Circ.*, 28, 420–430, 1993.
18. N.K. Verghese, T.J. Schmerbeck, and D.J. Allstot, *Simulation Techniques and Solutions for Mixed-Signal Coupling in Integrated Circuits*, Kluwer Academic Publishers, Boston, MA, 1995.
19. M. Nagata, T. Ohmoto, Y. Murasaka, T. Morie, and A. Iwata, Effects of power-supply parasitic components on substrate noise generation in large-scale digital circuits, *Symposium on VLSI Circuits Digest of Technical Papers*, Kyoto, Japan, June 14–16, 2001, pp. 159–162.

20. N.K. Verghese, D.J. Allstot, and M.A. Wolfe, Verification techniques for substrate coupling and their application to mixed-signal IC design, *IEEE J. Solid-State Circ.*, 31, 354–365, 1996.

21. J.M. Casalta, X. Aragones, and A. Rubio, Substrate coupling evaluation in BiCMOS technology, *IEEE J. Solid-State Circ.*, 32, 598–603, 1997.

22. C. Hu, *VLSI Electronics: Microstructure Science*, Vol. 18, Academic Press, New York, 1981.

23. K. Shimazaki, H. Tsujikawa, S. Kojima, and S. Hirano, LEMINGS: LSI's EMI-noise analysis with gate level simulator, *Proceedings of IEEE International Symposium on Quality Electronic Design*, San Jose, CA, March 24–26, 2003, pp. 129–136.

24. E. Charbon, P. Miliozzi, L. Carloni, A. Ferrari, and A. Sangiovanni-Vincentelli, Modeling digital substrate noise injection in mixed-signal IC's, *IEEE Trans. Comput. Aid. Des.*, 18, 301–310, 1999.

25. P. Miliozzi, L. Carloni, E. Charbon, and A. Sangiovanni-Vincentelli, SUBWAVE: A methodology for modeling digital substrate noise injection in mixed-signal IC's, *Proceedings on IEEE Custom Integrated Circuits Conference*, San Diego, CA, May 5–8, 1996, pp. 385–388.

26. M. van Heijningen, M. Badaroglu, S. Donnay, M. Engels, and I. Bolsens, High-level simulation of substrate noise generation including power-supply noise coupling, *Proceedings of 37th Design Automation Conference*, Los Angeles, CA, June 5–9, 2000, pp. 446–451.

27. M. Nagata, T. Morie, and A. Iwata, Modeling substrate noise generation in CMOS digital integrated circuits, *Proceedings of IEEE Custom Integrated Circuits Conference*, Orlando, FL, May 12–15, 2002, pp. 501–504.

28. K.J. Kerns, I.L. Wemple, and A.T. Yang, Stable and efficient reduction of substrate model networks using congruence transforms, *Proceedings of IEEE/ACM International Conference on Computer-Aided Design*, San Jose, CA, November 5–9, 1995, pp. 207–214.

29. N.K. Verghese and D.J. Allstot, Rapid simulation of substrate coupling effects in mixed-mode IC's, *Proceedings of IEEE Custom Integrated Circuits Conference*, San Diego, CA, May 9–12, 1993, pp. 422–426.

30. S. Kumasiro, R.A. Rohrer, and A. Strowas, A new efficient method for transient simulation of three-dimensional interconnect structures, *Digest of the IEEE International Electron Devices Meeting*, San Francisco, CA, December 9–12, 1990, pp. 193–196.

31. B.R. Stanisic, N.K. Verghese, R.A. Rutenbar, L.R. Carley, and D.J. Allstot, Addressing substrate coupling in mixed-mode ICs: Simulation and power distribution systems, *IEEE J. Solid-State Circ.*, 29, 226–237, 1994.

32. F.J.R. Clement, E. Zysman, M. Kayal, and M. Declercq, LAYIN: Toward a global solution for parasitic coupling modeling and visualization, *Proceedings of IEEE Custom Integrated Circuits Conference*, San Diego, CA, May 1–4, 1994, pp. 537–540.

33. T. Smedes, Substrate resistance extraction for physics-based layout verification, *IEEE/PRORISC Workshop on Circuits Systems and Signal Processing*, Mierlo, the Netherlands, 1993, pp. 101–106.

34. J.P. Costa, M. Chou, and L.M. Silveria, Efficient techniques for accurate modeling and simulation of substrate coupling in mixed-signal IC's, *IEEE Trans. Comput. Aid. Des.*, 18, 597–607, 1999.

35. M. Chou and J. White, Multilevel integral equation methods for the extraction of substrate coupling parameters in mixed-signal IC's, *Proceedings on ACM/IEEE Design Automation Conference*, San Francisco, CA, June 15–19, 1998, pp. 20–25.

36. N.K. Verghese, D.J. Allstot, and M.A. Wolfe, Fast parasitic extraction for substrate coupling in mixed-signal ICs, *Proceedings of IEEE Custom Integrated Circuits Conference*, Santa Clara, CA, May 1–4, 1995, pp. 121–124.

37. J.D. Jackson, *Classical Electrodynamics*, Wiley, New York, 1962.

38. R. Gharpurey and R.G. Meyer, Modeling and analysis of substrate coupling in integrated circuits, *Proceedings of IEEE Custom Integrated Circuits Conference*, Santa Clara, CA, May 1–4, 1995, pp. 125–128.

39. J.P. Costa, M. Chou, and K.M. Silveira, Efficient techniques for accurate modeling and simulation of substrate coupling in mixed-signal ICs, *Proceedings of Design, Automation and Test in Europe*, Paris, February 23–26, 1998, pp. 892–898.

40. J.P. Costa, M. Chou, and K.M. Silveira, Precorrected-DCT techniques for modeling and simulation of substrate coupling in mixed-signal IC's, *Proceedings of IEEE International Symposium on Circuits and Systems, 1998; ISCAS 1998*, Monterey, CA, Vol. 6, May 31–June 3, 1998, pp. 358–362.

41. J.P. Costa, M. Chou, and K.M. Silveira, Efficient techniques for accurate extraction and modeling of substrate coupling in mixed-signal IC's, *Proceedings of Design, Automation and Test in Europe, 1999; Conference and Exhibition*, Munich, Germany, March 9–12, 1999, pp. 396–400.

42. J. Kanapka, J. Phillips, and J. White, Fast methods for extraction and sparsification of substrate coupling, *Proceedings of ACM/IEEE Design Automation Conference*, San Francisco, CA, 2000, pp. 738–743.

43. S. Ponnapalli, N. Verghese, W. Chu, and G. Coram, Preventing a noisequake, *IEEE Circ. Dev.*, 17, 19–28, 2001.

44. M. Nagata and A. Iwata, A macroscopic substrate noise model for full chip mixed-signal verification, *Symposium on VLSI Circuits Digest Technical Papers*, Kyoto, Japan, June 12–14, 1997, pp. 37–38.
45. M.K. Mayes and S.W. Chin, All Verilog mixed-signal simulator with analog behavioral and noise models, *Symposium on VLSI Circuits Digest Technical Papers*, Honolulu, HI, June 13–15, 1996, pp. 186–187.
46. M.K. Mayes and S.W. Chin, All Verilog mixed-signal simulator with analog behavioral and noise models, *IEEE-CAS Region 8 Workshop on Analog and Mixed IC Design Proceedings*, 1996, pp. 50–54.
47. T. Blalack and B.A. Wooley, The effects of switching noise on an oversampling A/D converter, *IEEE International Solid-State Circuits Conference on Digest of Technical Papers*, San Francisco, CA, 1995, pp. 200–201.
48. N.K. Verghese and D.J. Allstot, Verification of RF and mixed-signal integrated circuits for substrate coupling effects, *Proceedings of IEEE Custom Integrated Circuits Conf*erence, Santa Clara, CA, May 5–8, 1997, pp. 363–370.
49. R. Telichevesky, K. Kundert, I. Elfadel, and J. White, Fast simulation algorithms for RF circuits, *Proceedings of IEEE Custom Integrated Circuits Conference*, San Diego, CA, May 5–8, 1996, pp. 437–444.
50. R. Telichevesky, K. Kundert, and J. White, Receiver characterization using periodic small-signal analysis, *Proceedings of IEEE Custom Integrated Circuits Conference*, San Diego, CA, May 5–8, 1996, pp. 449–452.
51. R. Telichevesky, K. Kundert, and J. White, Efficient steady-state analysis based on matrix-free Krylov-subspace methods, *Proceedings on ACM/IEEE Design Automation Conf*erence, San Francisco, CA, June 12–16, 1995, pp. 480–484.
52. K.H. Kwan, I.L. Wemple, and A.T. Yang, Simulation and analysis of substrate coupling in realistically-large mixed-A/D circuits, *Symposium on VLSI Circuits Digest of Technical Papers*, Honolulu, HI, June 13–15, 1996, pp. 184–185.
53. M. Nagata, J. Nagai, T. Morie, and A. Iwata, Measurements and analyses of substrate noise waveform in mixed-signal IC environment, *IEEE Trans. Comput. Aid. Des.*, 19, 671–678, 2000.
54. W.K. Chu, N. Verghese, H.J. Cho, K. Shimazaki, H. Tsujikawa, S. Hirano, S. Doushoh, M. Nagata, A. Iwata, and T. Ohmoto, A substrate noise analysis methodology for large-scale mixed-signal ICs, *Proceedings of IEEE Custom Integrated Circuits Conference*, San Jose, CA, September 21–24, 2003, pp. 369–372.
55. B.R. Stanisic, R.A. Rutenbar, and L.R. Carley, Mixed-signal noise decoupling via simultaneous power distribution and cell customization in RAIL, *Proceedings of IEEE Custom Integrated Circuits Conference*, San Diego, CA, May, 1994, pp. 533–536.
56. S. Mitra, R.A. Rutenbar, L.R. Carley, and D.J. Allstot, Substrate-aware mixed-signal macro-cell placement in WRIGHT, *Proceedings of IEEE Custom Integrated Circuits Conference*, San Diego, CA, May 1994, pp. 529–532.
57. S. Mitra, R.A. Rutenbar, L.R. Carley, and D.J. Allstot, Substrate-aware mixed-signal macro-cell placement in WRIGHT, *IEEE J. Solid-State Circ.*, 30, 269–278, 1995.

Technology Computer-Aided Design

Process Simulation

Mark D. Johnson

CONTENTS

27.1 INTRODUCTION

Process simulation is the modeling of the fabrication of semiconductor devices such as transistors. The ultimate goal of process simulation is an accurate prediction of the active dopant distribution, the stress distribution, and the device geometry. Process simulation is typically used as an input for device simulation, the modeling of device electrical characteristics. Collectively, process and device simulation form the core tools for the design phase known as technology computer-aided design (TCAD). Considering the design process as a series of steps with decreasing levels of abstraction, synthesis would be at the highest level and TCAD, being closest to fabrication, would be the phase with the least amount of abstraction. Because of the detailed physical modeling involved, process simulation is almost exclusively used to aid in the development of single devices—whether discrete or as a part of an integrated circuit.

The fabrication of integrated circuit devices requires a series of processing steps called a process flow. Process simulation involves modeling all essential steps in the process flow in order to obtain dopant and stress profiles and, to a lesser extent, device geometry. The input for process simulation is the process flow and a layout. The layout is selected as a linear cut in a full layout for a 2-D simulation or a rectangular cut from the layout for a 3-D simulation.

Technology computer-aided design has traditionally focused mainly on the transistor fabrication part of the process flow, ending with the formation of source and drain contacts—also known as front-end of line manufacturing. The back-end of line manufacturing, for example interconnect and dielectric layers are not considered. One reason for delineation is the availability of powerful analysis tools such as electron microscopy techniques, scanning electron microscopy (SEM) and transmission electron microscopy (TEM), which allow for accurate measurement of device geometry. There are no similar tools available for accurate high-resolution measurement of dopant or stress profiles.

Nevertheless, there is a growing interest to investigate the interaction between front-end and back-end manufacturing steps. For example, back-end manufacturing may cause stress in the transistor region, changing device performance. These interactions will stimulate the need for better interfaces to back-end simulation tools or lead to integration of some of those capabilities into TCAD tools.

In addition to the recent expanding scope of process simulation, there has always been a desire to have more accurate simulations. However, simplified physical models have been most commonly used in order to minimize computation time. But shrinking device dimensions put increasing demands on the accuracy of dopant and stress profiles, so new process models are added for each generation of devices to match new accuracy demands. Many of the models were conceived by researchers long before they were needed, but sometimes new effects are only recognized and understood once process engineers discover a problem and experiments are performed. In any case, the trend of adding more physical models and considering more detailed physical effects will continue and may accelerate.

The history of commercial process simulators began with the development of the Stanford University Process Modeling (SUPREM) program.[1] Building upon this beginning with improved models, SUPREM II and SUPREM III were developed. Technology Modeling Associates, Inc. (TMA), which was formed in 1979, was the first company to commercialize SUPREM III. Later, Silvaco also commercialized SUPREM and named the product ATHENA. TMA commercialized

SUPREM-IV (2-D version) and called it TSUPREM4. In 1992, Integrated Systems Engineering (ISE) came out with the 1-D process simulator TESIM and the 2-D process simulator DIOS. At about the same time, development of a new 3-D process and device simulator began at TMA, and after TMA was acquired by Avant!, Corp. The product was released in 1998 as Taurus. Around 1994, the first version of the Florida Object Oriented Process Simulator (FLOOPS) was completed. FLOOPS was later commercialized by ISE in 2002. Another process simulator PROPHET was created around 1994 at Bell Labs, which later became Agere, but has not been sold commercially. In 2002, Synopsys acquired Avant!, Corp. and in 2004, it acquired ISE. Synopsys released Sentaurus Process in 2005 which it says combines the best features of FLOOPS, TSuprem4 and Taurus using FLOOPS as a platform. Besides these simulators, there are numerous other university and commercial simulators such as PROMIS, PREDICT, PROSIM, ICECREAM, DADOS, TITAN, MicroTec, DOPDEES, and ALAMODE.

This chapter describes the fabrication steps most often modeled with process simulation tools including both the important physical effects, and models and techniques used to simulate them. The process steps most often associated with process simulation are ion implantation, annealing (diffusion and dopant activation), etch, deposition, oxidation, and epitaxy. Other common steps include CMP, silicidation, and reflow. In addition to the current state of the art, some attempt will be made to look forward to what new physical effects will be required to develop the future devices described by the ITRS roadmap. The last section discusses some practical aspects commonly used in process simulation, and gives a description of the steps required to create a structure suitable for device simulation.

27.2 PROCESS SIMULATION METHODS

Since all commercial process simulators use a combination of the finite-element (FE) and finite-volume (FV) methods, we begin with a brief introduction to the topic, including common techniques and strategies. A complete description of FE/FV methods is out of the scope of this chapter but there are many fine books[2] which describe the topic thoroughly. However, it is important to discuss requirements for process simulation for achieving accurate results. These requirements are based on the same requirements which are generic to FE/FV techniques, with an additional difficulty coming from the changes in the geometry during the simulated fabrication of the device. Process simulation uses an FE/FV mesh to compute and store the dopant and stress profiles. Each geometrical change in the simulation domain requires a new mesh which fits to the new boundaries. As will be described below, the large number of geometry-modifying steps involved and the nature of process simulation where each step depends on the cumulative results of all previous steps, make process simulation an especially challenging application of the FE/FV technique.

One of the most important results of process simulation is the dopant profile after processing. The accuracy of the profile strongly depends on maintaining a proper density of mesh points at any time during the simulation. The density of points should be just enough to resolve all dopant and defect profiles but not more, because the computational expense of solving the diffusion equations increases with the number of mesh points. A typical full-flow CMOS process simulation can have more than 50 mesh changes, and the number of mesh changes can increase dramatically if adaptive meshing is performed. For each mesh change, interpolation is used to obtain data values on the new mesh. It is important to manage the mesh changes in such a way as to avoid accuracy degradation due to interpolation error. The easiest way to do this is to always keep points once they are introduced into the mesh, but this has the drawback of producing too many mesh points, which can be computationally expensive. Maintaining a balance between interpolation error, computational expense, and minimization of required user input is important for obtaining accurate results with a minimum of computational expense. This is especially true when simulating devices in 3-D. Without careful placement of mesh, either the accuracy will suffer unacceptably, or the computational expense will be too great to be useful. Process simulation tools so far have had limited success in completely automating mesh adaptation such that no user intervention is required. This places a requirement on the user to understand meshing and how it affects simulation accuracy and run time, and burdens the user with the task of tracking mesh changes during the simulation to ensure that proper mesh is maintained.

One of the most important uses of TCAD tools is to explore new device technology where many exploratory simulations are performed in order to give the device designer a better understanding of possible benefits as well as drawbacks of a given technology. This use case demands sequential simulations with some analysis in between. In order to be useful, many simulation cycles must be run within the time allotted for exploration, putting a high priority on minimization of simulation run-time. Currently, full-flow standard CMOS simulations are most often accomplished with a combination of 1-D and 2-D simulation and take less than a few hours on a 2.6 GHz Pentium 4. To perform these simulations (from gate formation onward) in 3-D would take a minimum of 24 h for minimum-accuracy simulation. Most of the information desired from TCAD simulations can be extracted from the simplification that the device can be treated uniformly in depth (i.e., as a 2-D simulation). To include the effects device shape along the depth or to investigate implant shadowing, 3-D simulations must be performed.

27.3 ION IMPLANTATION

Ion implantation introduces dopant atoms into a wafer by ionizing the atoms or molecules containing the atoms, and these ionized species are then subjected to an electric field driving the ionized species into the wafer. The resulting depth profile for each type of implanted species depends mainly on the energy of the species and the dose. The energy is set by the process engineer by adjusting the potential difference between the ion source and the wafer, and the dose is determined by ion current and exposure time. Other factors which affect the profile are wafer orientation, and to a lesser extent, dose rate (ion current) and implantation temperature.

Physically, ionized dopants incident on the wafer will scatter if they come close to nuclei in the solid. In crystalline materials such as silicon, there are many "channeling" directions in which the nuclei are lined up in columns one behind the other, providing "channels" between the columns for the ions to travel long distances without scattering. Channeling is greatly enhanced if the incident ion beam is very nearly lined up with a channeling direction, but any orientation will give some channeling depending on the energy, dose, and species. In any case, dopants end up in exposed areas of the device, and some fraction are scattered uniformly in all directions, giving "lateral straggle" under a mask (such as resist). Figure 27.1 illustrates channeling by probing ion penetration depth vs. tilt angle. The inherent randomness of amorphous materials such as SiO_2 scatters incoming ions more often, as there are essentially no channels for ions to travel. Process engineers often use a sacrificial oxide layer over a silicon region where a shallow implanted profile is desired. When an incoming ion scatters from a nucleus, energy can be transferred to the nucleus, creating a recoil.

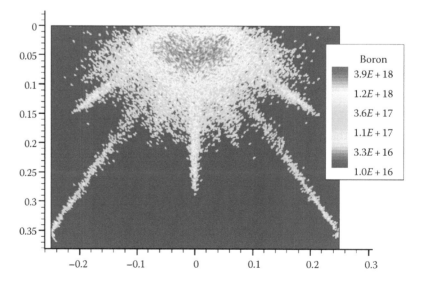

FIGURE 27.1 Figure created using Monte Carlo implant by varying the tilt from 0° to 75° to probe the channeling behavior in silicon, dramatically illustrating the channeling tails in silicon.

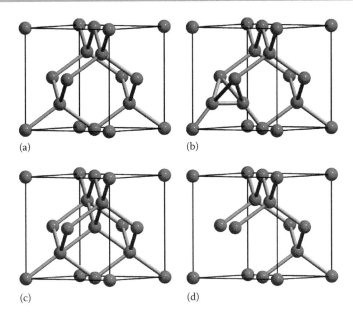

FIGURE 27.2 Atomic configuration of (a) perfect silicon lattice; (b) a split ⟨110⟩ interstitial; (c) a tetrahedral interstitial; and (d) a vacancy in the silicon crystal lattice. These configurations are not the lowest energy configuration so it is expected that there would be some adjustment of the atomic locations in the real material.

The recoil is typically a silicon atom, which has been removed from its lattice site in the crystal (creating a vacancy in the lattice). The recoil atom is scattered elsewhere—typically to an interstitial location. Interstitials and vacancies are known as point defects and when a pair is formed as part of a recoil collision, they are called a Frenkel pair. Figure 27.2 shows the atomic configuration of two interstitial types, and a vacancy compared with the defect-free silicon crystal lattice.

After an implant, the concentration of Frenkel pairs is very large compared to the concentration of ions/dopants, because the recoiling nuclei can themselves cause secondary damage in addition to the primary damage caused by ion/dopant scattering. As the implant dose is increased, fewer and fewer silicon nuclei are in their crystalline positions. Eventually the damage accumulation can cause amorphization of the material in those areas where the Frenkel pair concentration exceeds about 20% of the lattice density. After amorphization, channeling is essentially completely suppressed. This effect is sometimes exploited by process engineers by implanting an impurity such as Ge to create an amorphous layer to suppress channeling in a subsequent dopant implant. Figure 27.3 shows the effect of increasing damage on channeling. As the dose of an implant is increased, the proportion of newly arriving ions encountering a pristine lattice is reduced, hence reducing the proportion of ions which end up in channeling trajectories.

There are two common techniques for computing ion-implantation profiles available in commercial process simulators—analytic and Monte Carlo. Analytic implantation uses analytic functions to compute the dopant and defect profile resulting from an implant. The analytic functions are normally expressed in two directions: along the ion trajectory, and perpendicular to the trajectory which models scattering of ions. Because of this, it is necessary to perform a convolution integral to compute the contribution from all surface points to the concentration at every point in the bulk. Figure 27.4 shows how this is done. Because of the rapid roll-off in both directions, the integration can be limited to surface points nearby. Alternatively, Monte Carlo implantation computes the individual ion trajectories starting from a randomly chosen position in space. This technique will be described in more detail below. Because of the detailed nature of the calculation, Monte Carlo implantation can be between ten and hundred times slower, depending on the desired concentration resolution. There are a number of techniques which are typically employed to speed up the Monte Carlo implantation, but in general the analytic technique is much less computationally expensive than Monte Carlo. The trade-offs are that analytic implantation can be quite inaccurate in complicated geometries where the ion beam passes

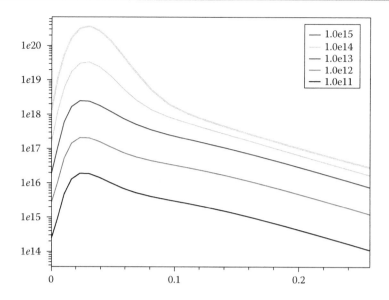

FIGURE 27.3 A series of arsenic implants for different doses, all 40 keV energy. Larger dose implants have a larger fraction of de-channeling due to increased damage accumulation. This is apparent from the increase in the peak height relative to the channeling (i.e., shoulder height).

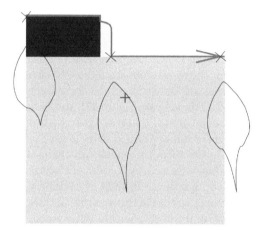

FIGURE 27.4 Schematic figure showing implantation integration scheme. The concentration distribution which results from ions impinging at one point on the surface is called a point response (constant concentration profiles in black). Analytic implantation uses functions to define the point responses. In a structure, the concentration of a point (cross) is computed by integrating the contribution from each point on the surface. However, there is no practical solution to handle multiple layers correctly with the analytic method. It would require a huge database of combinations of materials of varying thicknesses.

through multiple regions, and analytic implantation does not handle damage properly for multiple sequential implants with no annealing step in between.

27.3.1 ANALYTIC IMPLANTATION

Analytic implantation relies on a set of look-up tables which supply parameters for the profile of the implanted species, and a separate table which gives the profile for the implant damage, that is, Frenkel pair distribution. The implant parameters are a function of energy, species, dose, rotation, and tilt angles, and in addition may be a function of an overlayer thickness. Typically, Pearson IV functions are used for the primary profile and complementary error functions are used for the lateral roll-off due to scattering.

Many specific techniques have been developed to enhance the accuracy of analytic implantation for technology simulation. One common technique to reduce channeling as mentioned before is to use a sacrificial oxide. As the oxide thickness has a strong influence on scattering in the silicon layer below, special tables have been created which give the implant parameters as a function of oxide overlayer thickness. Similarly, the effect of damage accumulation on subsequent implants requires special handling. It is possible to employ some heuristics to estimate the effect of all implants since the last anneal cycle. The most important effect which must be mimicked is tracking the effect of amorphizing implant on subsequent implants. An amorphous layer has a similar impact on implant profiles as oxide, so implant parameter tables which have oxide thickness dependence can be used to estimate the resulting profiles. Amorphization is typically estimated by a threshold for Frenkel pair formation—above the threshold, the material is considered amorphous. However, experimental implant tables are normally only obtained for implants into crystalline silicon. Another difficulty comes from the fact that multiple implants may involve multiple species, and so far, because of the number of combinations possible and the expense, no experimental multiple implant tables with or without amorphization are commercially available.

In addition to the difficulty of modeling multiple implants, the geometry of devices also present challenges for analytic implantation. High-angle implants such as source drain engineering implants, are especially problematic. These implants are targeted at a corner near the gate edge and because of the high angle, the thickness of polysilicon and oxide overlayers that the ion beam passes through before hitting silicon varies dramatically there. Since the implant parameters depend on the thickness of the material covering silicon, these implants cause great difficulty in the analytic technique. In addition, modern CMOS devices have very thin layers, so a large fraction of ions can pass through layers with perhaps only a few scattering events. Modeling ions, which scatter from more than one type of material, present an enormous challenge for any type of analytic technique.

One model which has been developed to overcome these difficulties can be found in Reference 3. In this model, amorphization and damage accumulation are approximated by adjusting implant parameters to account for reduced channeling. Another technique based on the same idea was employed to adjust parameters to account for the dechanneling effect of SiO_2 over layers when there are no implant parameters available for such conditions.

27.3.2 MONTE CARLO IMPLANTATION

Monte Carlo implantation, as its name implies, explicitly computes individual ion, and optionally, recoil trajectories starting from a random location in space. Each scattering event will reduce the ion's kinetic energy, change the direction of the ion, and possibly produce secondary recoils, which may also be followed in a manner similar to the ion. The implantation damage is obtained naturally from the number of recoils produced per unit volume. Also, ions and even recoils can be tracked from one material into another so that multiple material layers can be handled accurately. A central strength of the technique is the wealth of experimental data and theoretical understanding of nuclear scattering. The ions start with a sufficiently large energy that nuclear scattering dominates the process. Only after many collisions with target atoms is the incident ion slow enough for the electronic stopping effects to become important. These effects are also fairly well understood, but there are functions used which require fitting parameters to simplify the computation. The least well-understood aspect of the technique, and an area of some ongoing research, is in differences in the detailed nature of the damage produced by different ions. In addition to the more accurate ion and damage profiles produced by the Monte Carlo method, dose loss due to backscattering events near the surface is obtained naturally.

Regarding performance, computing the scattering events is the most expensive part of the Monte Carlo implant calculation. Look-up tables can be used to compute these events, but since the number of ions incident upon a single device for a single implant can be very large (of the order of 10^7 mm^{22}) and there may be 10–30 implants for a process flow, Monte Carlo implant computations can be the most expensive step in a process simulation if they are used to simulate every implant. For typical conditions, analytic implant is of the order of ten times faster than Monte Carlo in 2-D, but for 3-D the times are somewhat closer. There are a number of techniques which can be employed to speed up the computation. One such technique is trajectory

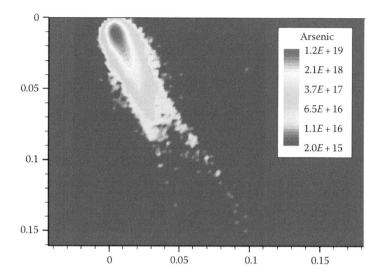

FIGURE 27.5   Point response Monte Carlo implantation simulation for a 10^{14} cm^{22}, 15 keV arsenic implant tilt 25°.

replication, where in a homogeneous part of a structure, a trajectory may be reused instead of being re-computed. Another technique commonly used is rare-event enhancement. This technique keeps track of the statistical significance of all the elements of the structure. When ions enter a region of the simulation where no data exists, they are split into multiple "pseudo-particles" with reduced weight, so that more of the simulation time is spent on the rare events instead of those events which already have good statistics.

There are a couple of things which can be done to combine the accuracy of Monte Carlo with the speed of analytic implantation. The most straightforward technique which is commonly employed by TCAD vendors and users is to calibrate the Monte Carlo implantation to a limited technology-specific set of implantation conditions. Then, it is possible to create specialized implant tables for each implant by using Monte Carlo implantation in 1-D and extracting Pearson IV parameters for the analytic tables. For a small subset of implantation conditions, this is not an enormous task and can have great benefits in terms of accuracy. This technique effectively creates process-specific ion implant tables. Another technique used to enhance analytic implantation is to create 2-D or 3-D point response functions using Monte Carlo. Instead of two perpendicular analytic functions, a 2-D or 3-D matrix is computed using the Monte Carlo technique by introducing numerous ions at a single location in a block of silicon (see Figure 27.5). The resulting intensity distribution is used as a look-up table to perform the integration over nearby surface points. This technique is used to account for the anisotropic nature of implants into crystalline materials, specifically the off-axis channeling.

27.4 DIFFUSION

One of the most important goals of process simulation is the accurate modeling of the active dopant profile evolution. In the early years of semiconductor devices, the device dimensions were huge compared to today's standards, and dopants were typically introduced through in-diffusion. This process occurs in near-equilibrium conditions and results in smooth and deep profiles by today's standards. The push toward smaller devices has fostered the development of process conditions which are further away from equilibrium. Specifically, device designers require high peak active concentrations and steep dopant profile gradients. With near-equilibrium conditions, the kinetic mechanisms of diffusion are unimportant because they are either constant or nearly constant through the whole step. As conditions are pushed further from equilibrium, the detailed kinetics of diffusion becomes more important. This can be seen in the development of physical models where the first process simulators used one equation per dopant and assumed equilibrium concentration of defects; later two defect equations were added to simulate defect kinetics but the dopant–defect pairs were still considered to be in local equilibrium; and now sometimes three equations per dopant in addition to two defect equations are solved to simulate the full dopant, defect, and dopant–defect kinetics. This effect is similarly seen in the evolution of dopant and defect cluster equations, which were not even used at first, but now are relied upon for day-to-day use. Research groups are exploring simulations which include tens to thousands of clustering species, most of which are transient as they are not typically present in significant concentrations by the end of the anneal step. The most detailed calculations are now accomplished with kinetic Monte Carlo diffusion simulations where the number of species types is essentially unlimited.

In addition to accurate models for dopants, accurate diffusion modeling strongly depends on models of point and extended defects which come mostly from ion implantation, but are also introduced during oxidation as well as a small, thermally activated concentration. After ion implantation, there is a tremendous amount of disorder or "damage" in the implanted regions. Annealing of the sample is necessary to reduce damage and activate the dopants. Reducing damage reduces electron scattering, improves electron mobility and reduces leakage current. Dopant activation will be discussed in more detail below, but involves the dopants occupying substitutional locations in the silicon lattice. The side effect of annealing is that the ion/dopant profiles redistribute or diffuse and this diffusion can strongly depend on the nature of the implanted damage.

Understanding the kinetics of diffusion is important to understanding the trends in thermal processing. But before discussing kinetics, it is first necessary to discuss the states and structures that dopants and defects can be in. Dopants can occupy substitutional silicon lattice sites, in which case they are normally considered active. Alternatively, dopants can be paired with an interstitial, vacancy, combination of interstitials and vacancies, or simply be in an interstitial location in which case they are inactive because in this state they do not affect the electron or hole concentrations. The substitutional sites are stable sites for the dopants so they are considered immobile species. On the other hand, the interstitial sites are not energetically favorable, so the dopants tend to migrate if they are in interstitial locations. When a point defect such as an interstitial or vacancy comes close to a dopant, it may change the bonding or local structure in a way to "kick out" the dopant from its substitutional site and hence allow the dopant to migrate. The reverse process whereby a dopant, paired with a point defect or simply in an interstitial location, may exchange with a silicon atom in the lattice creating an interstitial and leaving the dopant in a substitutional location. Consequently, point defects play an important role in dopant diffusion. Additionally, dopants can form immobile clusters consisting of both defects and dopants.

Interstitials play an especially important role in diffusion of ion-implanted wafers. Not only is there a large amount of damage in the form of Frenkel pairs as discussed above, but there is also an additional interstitial for every dopant atom implanted. Because of the conservation of lattice sites, an implanted ion adds an extra atom into the lattice—eventually most of the dopant atoms end up in substitutional sites, displacing one silicon atom per dopant.

The evolution of interstitial and vacancy profiles is governed by a number of physical processes. Interstitials can recombine with vacancies to reduce damage, but they might also combine with other interstitials to form interstitial clusters. Interstitial clusters form in part because of a reduction in the total strain energy as the two particles come close together. For a given size of interstitial cluster, it is possible that many different structures with different corresponding formation energies exist. Normally, models only consider the most stable form of the cluster. Additionally, the clusters can grow one interstitial at a time as other interstitials come close by and attach. The clusters can grow large enough to form extended defects visible in TEM and are generally called 311s because they are linear in shape and grow along the ⟨311⟩ direction in Si. The 311s can also grow larger and will at some point unfault into dislocation loops with hundreds to tens of thousands of interstitials each. Conversely, interstitial clusters can shrink and dissolve away as well. It is generally accepted that clusters are constantly releasing interstitials at a temperature-dependent rate and are acquiring interstitials at a rate dependent on the free interstitial concentration. Typically, right after implantation, the free interstitial concentration is very large, which favors the growth of interstitial clusters. After a while, free interstitials have recombined with vacancies, recombined with the surface, or clustered with a dopant, and the population is reduced. Once the free interstitial population is sufficiently low, the rate of release of interstitials is greater than the acquisition rate and the clusters begin shrinking. So the clusters are both sinks and sources for interstitials. The evolution of the free interstitials is responsible for an effect observed long ago called transient enhanced diffusion (TED).[4] What was observed was an initial burst of dopant migration after ion implantation, followed by much slower migration—close to what would be expected from measured bulk boron diffusivity. This effect is seen to a limited extent for all ion-implanted dopants but is most important for boron. TED was attributed to kinetics of interstitial clustering and dissolution[5] and continues to be a very challenging effect to simulate accurately.

Similar to defects, dopants can also cluster. Dopant clustering is assumed to freeze migration by trapping the dopants in immobile clusters, as is the case with defects, but activation is also reduced. Dopant clusters have been assumed to be inactive in the past, but the charge state of dopant clusters is now a topic of ongoing research.

Both boron and arsenic are known to cluster with point defects; boron with interstitials, and arsenic with vacancies. These species further complicate the TED picture because these clusters compete with 311s for interstitials and dissolve at different rates than do 311s. As for the other dopants, there is less certainty. There are some indications of phosphorus clustering with interstitials[6] and antimony with vacancies.[9] There appears to be no significant concentration of dopant–defect clusters of indium for typical process conditions.

When the concentration of dopants becomes very large, as is the case in the highly doped source and drain regions, precipitates usually in the form of silicides may form. The temperature-dependent concentration for dopants dissolved in Si (i.e., in substitutional sites) to form precipitates is called the solubility limit. All dopants are expected to have this behavior, which limits the maximum active concentration of dopants. Dopant clustering competes with precipitation and may be more important depending on the conditions. Some extremely fast anneal steps including laser annealing have shown that activation above solid solubility can be achieved, but this is a metastable state which can relax (forming clusters and precipitates) upon further annealing.

There are two main methods used to solve diffusion: methods which solve partial differential equations such as FE/FV which will be called "continuum methods" and kinetic Monte Carlo techniques. The continuum method for diffusion can handle most of today's important effects to a high degree of accuracy, and therefore has been and continues to be the method of choice. However, trends in MOS scaling have prompted the use of extremely rapid anneal steps with anneal times of the order of a few seconds and ramp rates in excess of 1000°C/s. It was found that faster thermal ramp rates at the same peak temperature can reduce dopant redistribution while still activating the dopants and reducing damage. As was mentioned before, the trend away from equilibrium makes it important to solve the detailed diffusion kinetics, which makes the continuum method for diffusion increasingly more expensive. Meanwhile, as the device dimensions shrink, kinetic Monte Carlo has become increasingly less expensive because the simulation time strongly depends on the number of diffusing species, which in turn depends on the size of the simulation domain. In addition, kinetic Monte Carlo can solve for an essentially unlimited number of species without paying any performance penalty. So it is likely that the use of kinetic Monte Carlo will some day overtake continuum diffusion.

27.4.1 CONTINUUM METHODS FOR DIFFUSION

Continuum methods for solving a diffusion/reaction system use partial differential equations which can be solved using the finite difference or finite element (FD/FE) technique. The number of equations which are solved depends on the desired accuracy. As few as one equation per dopant can be used for long-time high-temperature furnace anneals, and to simulate more detailed kinetics, as many as three equations per dopant and dozens of defect and dopant cluster equations could be used for an rapid thermal annealing (RTA) step where ramp rates exceed 250°C/s and total annealing time may only be a few seconds. The extra equations cost both memory and execution time and the effect is especially detrimental in 3-D where the simulation time is dominated by linear solvers for which simulation times scale faster than linearly with the number of equations. To reduce the number of partial differential equation (PDE's), the complete physical picture is simplified by assuming some species can be approximated using an expression instead of solving a separate PDE. Normally, this is done by making steady-state approximations. The closer the system is to equilibrium, the better these steady-state assumptions are. As computing power increases, it is expected that increasingly complete physical models will be practical and better results can be expected.

The most commonly used model in the industry today relies on one equation per dopant and two defect equations. Dopant–defect pairing is assumed to be in a steady state such that the concentration of dopant–defects is proportional to the product of the dopant and defect concentrations. Without going into all the details, the model can be formulated as follows[7]:

$$\frac{\partial C_A}{\partial t} = -\nabla \cdot J_A - R_A^{clus} \tag{27.1}$$

$$\frac{\partial C_X^{total}}{\partial t} = -\nabla \cdot J_X - \nabla \cdot J_A - R_{IV} - R_A^{clus} \tag{27.2}$$

$$C_X^{total} = C_X - C_{AX} \tag{27.3}$$

$$J_A \propto \sum_{c,X} D_{AX}^c \left(\frac{n}{n_i}\right)^{-c-z} \nabla \left(C_A^+ \left(\frac{n}{n_i}\right)^{-z}\right) \tag{27.4}$$

$$J_X \propto \sum_c D_X^c \left(\frac{n}{n_i}\right)^{-c} \nabla \left(\frac{C_X}{C_X^*}\right) \tag{27.5}$$

where

C_A is the concentration of dopant A

C_X^{total} is the total concentration of defect X (X is either interstitial or vacancy) and is the sum of the free defect X concentration and the pair concentration, C_{AX}

J_A and J_X are the particle current of dopant A and defect X, respectively

D_{AX}^C is the dopant–defect X diffusivity which depends on the charge state c

D_X^C is the diffusivity of defect X at charge state c

n is the electron concentration

n_i is the intrinsic electron concentration

R_{IV} is the bulk interstitial vacancy recombination

Finally, the various terms labeled R^{clus} are recombination/clustering terms, which depend upon what clustering models are selected. One thing to notice about the model is the strong coupling between Equations 27.1 and 27.2 which reflects the dopant diffusion dependence on defects. The extension of this model to include nonequilibrium defect pairing can be found in Reference 8. A nice comprehensive review of diffusion modeling can be found in Reference 9.

The performance of continuum diffusion solvers is affected by many factors, but the most important ones are number of mesh nodes, number of PDEs to be solved, and the "nonlinearity" of the PDE equations. Roughly speaking, the linearity of an equation refers to the time rate of change of the solution variable, given a perturbation of one of the variables being solved. If a small perturbation causes a large nonlinear change in the solution variable, then the solver must take shorter steps in order to accurately capture the time evolution of the system. Because diffusion equations are nonlinear time-dependent PDEs, the normal procedure for solving the equations is using an implicit time-stepping scheme where each time-step is solved using the Newton method. Assembly of the Jacobian (a critical part of solving nonlinear PDEs) has become a significant portion of the total time because of the complexity of the equations which are solved. Once the Jacobian is assembled, a system of linear equations must be solved. The total number of unknowns in the linear system can be estimated by the number of nodes multiplied by the number of solution variables. Because the computational expense of linear solvers in general scale faster than linearly, for very big problems with a number of nodes and PDEs, linear solve times dominate the total time, and for small problems, the assembly dominates. Another consideration for large problems is memory consumption, though with the adoption of 64-bit CPUs, this is less of a problem than it used to be.

27.4.2 KINETIC MONTE CARLO DIFFUSION

Kinetic Monte Carlo diffusion, unlike the continuum method, does not place restrictions on the complexity of the physical model. Kinetic Monte Carlo tracks the positions defects, dopants and clusters (collectively referred to as "particles") of all particles within the simulation domain. Diffusion is simulated by particles hopping from lattice site to lattice site in random directions. If diffusing particles come close to dopants or clusters, they can react to form a new species or increase the size of a cluster. Clusters can break up by losing an interstitial, and interstitials can recombine at the surface or with vacancies. All of the possible events are put into a large event table weighted with their relative event rates, and the next event is chosen with a random number. Keeping track of all available species, their locations, and their possible reactions to form new species is simply a book-keeping exercise and costs essentially nothing in terms of performance but can require a significant amount of memory. The most expensive step in kinetic Monte Carlo is determining the neighbors after a diffusing species hops. In order to speed up the computation, a regular grid is usually created which stores the list of species in each grid cell. This reduces the list of neighbor candidates considerably, but can be very costly in terms of memory usage.

The main cause of the poor performance of kinetic Monte Carlo with respect to continuum diffusion is the large number of fast-diffusing species. Right after an implant, there is an enormous amount of damage in the form of point defects. The average time step per event is extremely small, perhaps of the order of picoseconds. The total reaction rate of the system is given by $R = \sum_i r_i n_i$ where r_i is the rate of event type i and n_i is current multiplicity of event i. The Monte Carlo time step is proportional to $1/R$, which and R is very large just after an implant because of the very large number of possible diffusion events. By the end of a simulation, there are far fewer diffusing species because of various re-combination events, so the time step increases dramatically. However, once the number of diffusing species becomes very small, one can also run into finite size effects in the simulation. One important finite size effect comes from the handling of the interstitials, which in a very large simulation box would form a very smooth, deep profile, and in the active part of the device would result in a slowly decreasing concentration. This effect is not easy to reproduce accurately with a small simulation domain.

As the size of devices shrink, the cost of kinetic Monte Carlo is reduced because there are fewer species and the number of cells required for neighbor look-up is reduced. Considering the shrinking device dimension trend, the complexity of kinetic reactions which can be included using kinetic Monte Carlo, and the fact that kinetic Monte Carlo does not depend on a high-quality mesh, it is likely that at some point and for some problems, kinetic Monte Carlo will become the diffusion module of choice.

One way to take advantage of both methods even today is to use the full kinetic Monte Carlo for a very short initial time period. Essentially, the kinetic Monte Carlo would serve to compute the starting conditions for the continuum method for diffusion. Without going into too many details, there is not a good understanding of the precise starting conditions after an implant. So, for example, the proportion of dopants which end up in interstitial as opposed to substitutional sites after an implant is not known. Similarly, dopant, dopant–defect, and interstitial clusters form very quickly after diffusion starts and without solving for hundreds of species, it may be difficult to know which are the dominant species at the beginning of the anneal. So one way to get the best of both techniques may be to start with kinetic Monte Carlo and then finish the simulation using FD/FE method with a simpler physical model.

27.5 OXIDATION

Thermal oxidation of silicon forms a very high-quality dielectric, SiO_2, and is perhaps the most compelling reason that Si is the dominant material used for electronics. Very thin defect-free layers of SiO_2 are used to insulate the gate from the channel in CMOS devices. Recently, small amounts of nitrogen have been added to SiO_2-forming oxynitrides to improve reliability and reduce unwanted diffusion of boron from the gate to the channel. In addition to its use as a gate dielectric, oxidation is also used to line isolation trenches between devices, which are subsequently filled with a deposited oxide such as tetraethyl orthosilicate (TEOS).

The simulation of oxidation has two functions—one is the formation and growth of SiO_2 when Si is exposed to an oxidizing ambient, and the other is the re-distribution of dopants due to diffusion and due to the growth and flow of the oxide layer. After an initial native layer of oxide is formed on the silicon surface, the subsequent growth of the oxide layer occurs at the Si–SiO_2 interface and is caused by oxidant(s) diffusing through SiO_2 and reacting at the Si surface. Oxidants can be O_2, H_2O, O, or compounds involving N, to create oxynitrides. Figure 27.6 shows the simulation of a poly re-oxidation step. The growth rate of the SiO_2 is known to be affected by the type(s) and concentration(s) of oxidant(s), stress, and dopant concentration. In addition, the oxidation rate depends on the orientation of the oxidizing surface as is shown in Figure 27.7. One important aspect of oxidation is that SiO_2 is not naturally as tightly packed as Si, which causes compression in SiO_2 laterally and tension in Si near the interface. The lateral compression of the SiO_2 in turn causes expansion in the vertical direction through Poisson's ratio. This expansion causes motion of the entire SiO_2 layer so that both the Si–SiO_2 and the SiO_2–gas surfaces are in motion. If there are any exposed polysilicon layers, those will behave similarly to silicon. Because oxidation is always solved using FD/FE methods and moving boundaries, this boundary movement creates meshing challenges which are particularly difficult to overcome in a full 3-D simulation.

In addition to re-distributing dopants during inert thermal ramps, diffusion modules must also take into account several important effects during oxidation. The most important of those effects is oxidation-enhanced diffusion (OED). The conversion of Si to SiO_2 introduces interstitials at the Si–SiO_2 interface which then contribute to enhanced diffusion of interstitial diffusing dopants. The conversion of silicon to oxide is not by itself considered to affect the dopants, so for each step part of the dopant profile in silicon must be transferred to oxide, but at the same time silicon/oxide segregation must also be solved. In general, dopants favor either silicon or oxide—most likely as a result of chemical effects. In addition, it is critical to maintain the dose of the dopants locally. Therefore global remeshing, which is normally used to produce a clean mesh but introduces interpolation errors, must be minimized as much as possible.

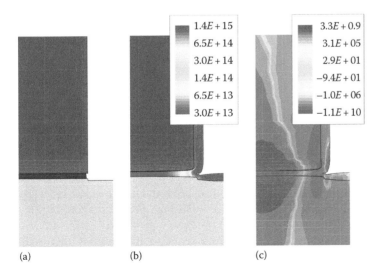

1.4E + 15		3.3E + 0.9
6.5E + 14		3.1E + 05
3.0E + 14		2.9E + 01
1.4E + 14		−9.4E + 01
6.5E + 13		−1.0E + 06
3.0E + 13		−1.1E + 10

(a)   (b)   (c)

FIGURE 27.6   An oxidation simulation showing oxidant diffusion, Si consumption, and SiO_2 expansion. (a) Light pink is silicon, brown is oxide, and magenta is polysilicon; (b) same simulation as in (a), but showing oxidant concentration in the oxide. Because the oxidant concentration does not reach far underneath the gate, oxidation occurs mostly at the edge of the poly gate; (c) same simulation as in (a), (b) but the shading shows the component of stress in the vertical direction. Stress, which is concentrated in the region where oxidation happens, is a by-product of the consumption of silicon by the oxidant-forming less-dense SiO_2.

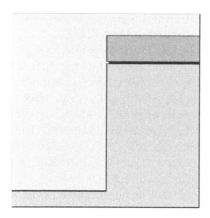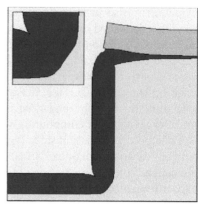

FIGURE 27.7   Simulation of trench oxidation showing orientation-dependent oxidation of a trench. Oxidation rate is assumed to be slowest along the ⟨111⟩ direction in silicon, resulting in a facet at the bottom of the trench. The wafer orientation is ⟨100⟩ and devices are normally oriented along ⟨110⟩.

27.6 ETCH AND DEPOSITION

Traditionally, TCAD has focused on front-end fabrication simulation ending with source/drain implant and anneal steps. So it has not been necessary to include a full physical etch and deposition module. The main goal of TCAD simulations is ultimately the device electrical characteristics and there are only a few device dimensions which have a significant impact on the electrical characteristics including, for example, gate oxide thickness, gate width and length, shape of the poly gate at the bottom, and spacer width. For these critical parts of the device, simple geometrical etch and deposition models have been sufficient. Additionally, process engineers developed very good etch and deposition techniques, creating fairly simple shapes. As device dimensions shrink and more complex techniques are required to realize structures, it is becoming increasingly important especially in the gate and spacer formation to take into account physical effects in etching, deposition, and even resist topography. While this has traditionally been outside the scope of process simulation, it is useful in the context of this chapter to point out the mechanisms which lead to nonidealistic device structures and techniques which have been developed to simulate those effects.

27.6.1 ETCH

For the purposes of this chapter it is sufficient to focus on two main etch types which are dominant in the industry, dry etching or plasma related etching (including reactive ion etching), and wet chemical etching. The two types of etch produce different shapes. Dry etching tends to be highly directional and is used to produce nearly square bottom holes, whereas wet chemical etching tends to be more isotropic producing rounded bottom holes and under etching (see Figure 27.8).

Reactive ion etching involves creation of a plasma of reactants and biasing of a sample in a way that ions from the plasma are incident on the sample surface. A reaction occurs at the wafer surface and the waste products are pumped away. Since the incident ions are driven by a mostly vertical electric field, the etch rate is mostly vertical and a vertical directional etching results. Material selectivity is enhanced by the use of reactive ions, which preferentially react with one material over another. The make-up of the plasma can contain reactive species and inert species. Reactive species give higher etch rates and better selectivity but tend to produce isotropic profiles, whereas inert species have lower etch-rate selectivity and produce more directional profiles. It is also possible to combine inert with reactive species to obtain a compromise.

Accurate physical simulation of these processes can be very challenging. A full simulation would include the following: ionization reactions in the plasma, the interaction of the plasma with the wafer, trajectories and energies of incoming species, surface reactions for the reactive species, surface diffusion of adsorbed reactive species, re-emission of inert, reacted, or unreacted

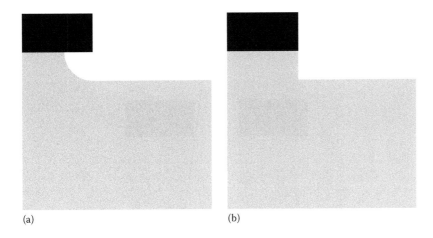

(a)   (b)

FIGURE 27.8   Basic etch types: (a) "Wet" etch normally is simulated with isotropic etching which causes material to be removed under masks. The dark material is assumed to be an etch mask; (b) "Dry" etching (reactive ion etching or plasma etching) is normally approximated with what is labeled anisotropic etch or sometimes called directional etch.

species back into the gas (which can land at other surface sites if there is direct visibility), and finally desorption of reacted species back into the gas. So in addition to many chemical reactions, a set of coupled integral equations must be solved which compute the re-emitted flux coming from other surface elements, which are in the line of sight of the current surface element (see Figure 27.9). After these computations are performed, the etch rate as a function of position on the surface is computed, and finally the boundary is moved.

27.6.2 DEPOSITION

Depending on the material to be deposited, there are several deposition techniques currently used which are mostly variants of chemical vapor deposition (CVD) or sputtering. These two classes of deposition processes, isotropic and sputter deposition, generate different layer topography, and most deposit steps are handled with one of these two types of deposition modules. Figure 27.10 shows a schematic of the two deposition types. A description of some of the physical effects responsible for the deposited geometries is given below.

For deposition of oxide, nitride, polysilicon, or silicon, it is common to use low-pressure CVD (LPCVD) or plasma-enhanced CVD (PECVD). In general, the following set of steps describes CVD: adsorption of a gas molecule, possible surface diffusion of the adsorbed species to find a favorable site (such as kinks in atomic layer steps on the surface), a reaction taking place which is either temperature- or plasma-assisted, the reaction depositing the material and releasing reacted species, and finally desorption of reacted species. Depending on the conditions in the reactor, the deposition can occur isotropically or anisotropically. In addition to the possible directionality of the incoming beam, many other processes can affect the resulting topography including re-deposition, surface diffusion, and changes in deposition conditions (partial pressures of mixed gases) inside deep trenches. TCAD simulators have traditionally stopped short of modeling gas or plasma reactions but instead have offered a set of geometrically based models which can effectively model directional, re-deposition, and surface diffusion.

For metal over-layers, sputter deposition is the most common method. With this technique, electrons incident on a target of the desired deposition metal kick out atoms, which are then deposited on the wafer. This technique results in an atomic beam which has an angular distribution. The directionality of the atomic beam can cause an instability in the deposited shape. For example, in Figure 27.10a, deposition takes place near a step in the surface. At the top corner of the step, deposition can occur on the side wall which is exposed to the incoming flux, whereas the bottom of the step is first partially shadowed and then fully shadowed as deposition progresses. This results in a shape like the one shown. For comparison, the right side of Figure 27.10 shows

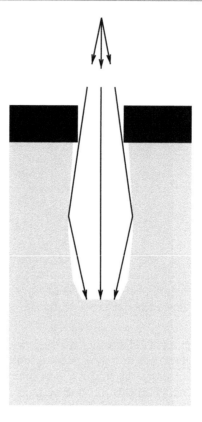

FIGURE 27.9 Schematic figure illustrating geometrical effects leading to more realistic shape. In a real dry-etch process, the incoming beam has a slight spread (exaggerated here for clarity). The beam-spread can create bowed side walls and rounded trench bottoms. The sides of the hole are only exposed to a part of the incoming beam, thus slowing the etch rate on the sides compared with the middle of the hole. In addition, the etchant flux is incident at a glancing angle on the side walls, increasing the chance of reflection re-directing part of the flux to the bottom of the hole.

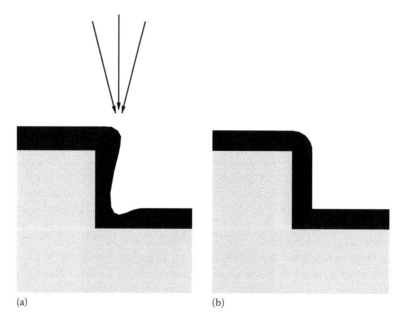

(a)   (b)

FIGURE 27.10 Schematic figure of (a) sputter deposition vs. (b) isotropic deposition. In sputter deposition, the directionality of the incoming flux leads to irregular shapes near steep surface features. The bottom of such features can be shadowed by an overhang created at the top.

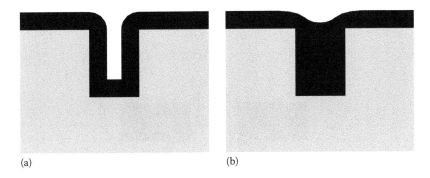

(a)   (b)

FIGURE 27.11   Schematic figure of (a) re-flow and (b) surface diffusion. Both re-flow and surface diffusion are driven by the surface free energy (flow is from high-energy to low-energy regions). Convex corners have a high energy and concave corners have a low energy.

isotropic deposition, which can result from CVD-type deposition where all areas of the surface have a uniform growth rate because the deposition rate depends on the gas concentration which is uniform everywhere along the surface.

27.6.3 REFLOW AND SURFACE DIFFUSION

Doped oxides such as TEOS or silane-oxygen doped with phosphorus (PSG), boron (BSG), or a combination of both (BPSG), at moderate to high temperatures become soft and flow like a viscous liquid. This process is most often associated with back-end processing and can be used to fill in trenches and smooth deposited features (see Figure 27.11 for a schematic illustration). PSG re-flow occurs at a fairly high temperature (>950°C) and can cause significant dopant re-distribution, whereas for BPSG the re-flow temperatures are in the range of 400°C–450°C, which would not cause dopant re-distribution. Normally, the shape of the deposited doped oxide layer is not important because it is followed by a chemical mechanical-polishing (CMP) step which flattens the top surface. So most often, an inert anneal step following a trench fill would be sufficient to model doped oxide deposition with re-flow.

High-temperature deposition can be used to smooth out instabilities caused by a directional beam because of surface diffusion effects. Topography changes due to surface diffusion and re-flow, to a certain extent, are driven by a minimization of surface energy. Surface energy is proportional to the surface curvature: convex corners have a high surface energy and concave corners have a low surface energy. So, when doped oxides are heated, re-flow will cause material to flow from convex to concave regions, which tend to smooth films and fill vias and holes. Surface diffusion normally only occurs on semiconductor and metal surfaces at higher temperatures than re-flow of TEOS and only in high vacuum, but the simulation technique is pretty much the same. Simulating re-flow or surface diffusion can be accomplished by computing and applying a normal speed function where the growth/removal rate (assumed to be normal to the surface) is a function of the local surface curvature.

27.6.4 RE-DEPOSITION

Re-deposition occurs when a deposition molecule scatters from a part of a surface and is adsorbed or deposited somewhere else (see Figure 27.12). Re-deposition can be modeled in a geometric way by using the concept of a sticking coefficient. Each location on the growing surface is exposed to an incoming flux, which depends on the visibility if the deposition is directional. The rate at which each individual surface element moves depends on the direct incoming flux plus the indirect flux coming from other parts of the surface, multiplied by the sticking coefficient. In order to account for multiple reflections, an integral equation must be solved which couples every surface element to every other surface element (some coupling may be discarded if the pair of surface elements do not see each other).

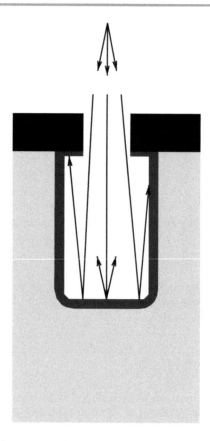

FIGURE 27.12 Schematic showing re-deposition. Some deposition flux may be reflected to be re-deposited elsewhere in the structure.

27.6.5 EPITAXY

The growth of silicon epitaxial films is similar to CVD except that the temperature is higher, and the growth rate lower. The low deposition rates allow incoming species to find favorable growth sites such as kink sites, leading to atomic layer-by-layer growth. Epitaxy is often done to prepare the starting wafer with a very flat contaminant-free surface, as well as to control the wafer doping. There are various other applications as well. Recently, SiGe epitaxy has been used in the source and drain regions to introduce stress in the channel, thereby increasing electron mobility. Epitaxy most often occurs at relatively high temperature compared with CVD of dielectric films. Often, dopants are introduced during epitaxy or it may be that dopants are already present in the structure when epitaxy is performed. For those cases, it is necessary to compute the simultaneous diffusion of dopants during the deposition process. There are two possible techniques to accomplish this. One technique involves breaking up the deposition into smaller alternating diffusion and deposition steps; the other solves a moving boundary problem like oxidation where a dopant flux is introduced into the growing boundary nodes and provisions are made to account for the change in neighboring element size and the introduction of a dopant flux.

27.6.6 SELECTIVE DEPOSITION

Because deposition occurs as a reaction with a gas species, it is possible to tailor the gas in such a way that it reacts with only one of the materials present on the surface of the wafer. For example, silicon tetrachloride ($SiCl_4$) will deposit polysilicon on silicon at temperatures above 800°C and will grow a layer of Si on Si without depositing on SiO_2 or Si_3N_4 at temperatures above 1100°C (HCl may be added to improve selectivity). Other gas mixtures have also been used to reduce the required deposition temperature. This type of deposition can be used to explore novel devices or other specific uses. Simulation of this type of deposition is fairly easy. The only concern is how to handle the edges where the active material meets an inert material. Typically, it is handled as an isotropic material-specific deposition. And similarly to standard epitaxy, if dopants are involved, diffusion must be solved simultaneously with deposition.

27.6.7 TECHNIQUES FOR MODELING ETCH AND DEPOSITION

We begin by discussing the two main methods which have been employed to compute the layer topography and topology.

27.6.7.1 LEVELSET METHOD

Levelset methods[10] solve equations that describe the evolution of surface motion given a normal speed function. Instead of solving potentially complex discrete-topology evolution problems, the levelset equation resolves the motion of a front in a continuous manner. Discretized levelset equations can be solved using FE/FV methods. Because of this, levelset methods handle changes in topology naturally and reliably. The normal speed function of the evolving surface can be quite general. It can be a function of the local surface orientation, curvature or visibility, or can be computed with an integral along the front. Another possibility is to import a speed function from an external simulator. This allows a straightforward way to couple physical etch and deposition tools to process simulators. The current simulation boundaries can be passed to an etch and deposit module which would compute the normal speed on the boundary points. This information can be passed back to the process simulation tool which would use the levelset method to move the boundaries, handle topology changes, update the mesh, and interpolate the previous values onto the current mesh.

The main drawback of the levelset methods is the difficulty in handling thin layers and sharp corners. Also, for a nontrivial normal speed function, accuracy of the evolving front may be a concern making it necessary to maintain a proper mesh density to solve the levelset function. In this case there is a trade-off between accuracy and etch, and deposit computation cost both in terms of memory use and execution speed. Discontinuities in the speed function and sharp corners have to be handled with care. Discontinuities in the speed function come from the visibility—meaning that a surface segment is either exposed to the beam or is shadowed by some other part of the surface. Sharp corners in general cannot be handled exactly—normally a specified geometrical tolerance is used to maintain the accuracy of the boundary evolution. The fundamental reason for these limitations is that the levelset function which defines the boundary location is assumed to be a continuous function (see Figure 27.13). The other main difficulty with levelset methods is in handling thin layers for reasons similar to the corner problem. In the case

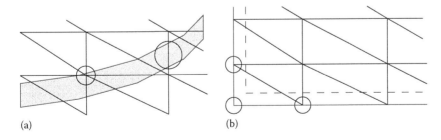

(a)   (b)

FIGURE 27.13 Difficulties with the levelset method: (a) A thin shaded region is formed during a deposition step. The levelset method defines the location of a surface as the contour of constant values of the levelset function which is a field defined on the mesh at each nodal point. This implies that there can only be one level crossing for each edge or the topology of the region will be incorrect. Inside the large circle, the double crossing will be missed whereas if there is mesh point inside the region, as with the small circle, then the topology will be correct; In (b), deposition is simulated starting with the solid thin line. However, all the edges between the circled mesh points will either have no level crossing or will have an undetectable double crossing, so the corner will be lost. Adaptive meshing can be used to solve both problems.

of thin regions, mesh edges which cross the region entirely without a node inside the region will miss the two levelset crossing inside, causing a hole in the thin layer (see Figure 27.13). Even if the layer is recognized, an accurate solution needs at least 4–5 points across the layer, which can be very computationally expensive.

27.6.7.2 STRING ALGORITHMS

There are many variations of string algorithms and other analytic techniques to move boundaries. They all involve breaking up a surface into a set of edges; the positions of the points are moved analytically and collisions must be detected and resolved. Therein lies one of the difficulties with these methods—the resolution of colliding boundaries can be very complicated. This method has not been very successful for 3-D simulation. However, for simple geometries and for 2-D simulation, string algorithms can perform better than levelset-based methods, because string algorithms are extremely accurate, can easily handle discontinuities in both the speed function and the surface itself, and are computationally inexpensive.

27.7 LITHOGRAPHY AND PHOTORESIST MODELING

Lithography/photoresist modeling is out of the scope of this chapter, but it is mentioned here for completeness. The shape of the resist will affect the final etch shape, and not just the width; but it may be important to the final structure if the resist shape has sloped and/or rough walls. A full litho/etch simulation would include photo-resist exposure, postexposure bake, and development—essentially all steps responsible for the final resist topography. In the exposure step, it is important to include diffraction effects, and allow for phase-shifting masks (PSM) and masks with optical proximity corrections (OPC). There has been some work done recently to transfer resist shapes from the lithography and resist simulators to the process simulators to get more accurate gate shapes.

27.8 SILICIDATION

The formation of silicides is simulated in a manner similar to oxidation. A metal is deposited on Si, then a native layer of the silicide is created between the metal and the Si, and finally the moving boundary problem is solved to move the Si–silicide and metal–silicide boundaries. In the silicide, two species can be present—silicon diffusing from the silicon side reacting at the metal, and metal atoms diffusing through the silicide and reacting at the Si–silicide boundary. Each reaction can either introduce compressive or tensile stresses in silicon, depending on the relative density of the silicide compared to silicon.

27.9 MECHANICS MODELING

Stress and strain modeling have become an evermore important goal of process modeling. Not only is the stress state itself important as an estimate of the likelihood of dislocations which have a deleterious effect on yield, but also the stress and strain can affect dopant diffusion and be used to improve device performance[11] through changes to the silicon bandgap, and electron mobility. In fact, stress is now used as a tool to improve electrical characteristics in the latest devices produced by many major manufacturers.[12–14] Known sources of strain come from many sources: thermal mismatch, deposited stress, oxidation, lattice mismatch, densification, and high dose implantation. A more detailed discussion of various stress sources except oxidation and silicidation, which have already been discussed separately in Sections 27.5 and 27.8, respectively, will be covered below.

Thermal mismatch is modeled as a natural by-product of thermal expansion, which varies from material to material. This is normally accomplished by referencing all expansion coefficients to the silicon substrate. The over layers expand and contract relative to the substrate, introducing stress.

The deposition of Si_3N_4 is known to introduce stress. Thermal mismatch is at least partially responsible, and chemical reactions involving Si–H and N–H bonds are thought to account for the rest. Stress increases with film thickness up to a certain level, leading to the conclusion that the dominant mechanism may be similar to lattice mismatch stress. In addition, subsequent annealing of the film can cause further stress. Studies indicate a maximum stress which can be attained during deposition depending on the temperature which has been attributed to a visco-elastic behavior of the films.[15] Because of the lack of a good model for the stress state of the deposited film, normally an experimentally determined (user-defined) constant stress is applied to the deposited film and a subsequent relaxation of the whole system is performed to give an estimate for the deposited stress state. It is quite likely that because of the renewed interest in stress and strain modeling, commercial simulators will implement more sophisticated stress models in the future.

The source of strain in strained silicon and SiGe is lattice mismatch. Typically, both of these technologies start with a Si wafer on top of which is grown a relaxed SiGe buffer layer. The buffer layer growth creates dislocations due to the build-up of stress from the difference in lattice constants between SiGe and pure Si. Typically, the Ge mole fraction is varied in the buffer layer to control the density of dislocations and ensure that they are predominantly located near the bottom of the layer. At the top of the buffer layer, the stress is minimal, so the top of the buffer layer can be thought of as a starting substrate with a variable lattice constant. A pure Si or SiGe layer with a mole fraction different from the relaxed buffer layer is grown on top of the buffer layer, so a tensile layer of Si or a tensile or compressive layer of SiGe can be produced in this manner. Because there are no good analytic models which can accurately reproduce the stress state during the buffer layer growth, and because the top part of the buffer layer is mostly or completely relaxed, it is normally sufficient to treat the lattice mismatch by assigning the lattice constant to be fixed at some point near the top of the buffer layer, and apply an external strain proportional to the concentration of Ge (see Figure 27.14).

Densification is the process where the density of a material can increase during annealing. The deposited dielectric TEOS undergoes densification due to the evaporation of residual organics from the deposition and due to chemical changes in the material induced by annealing. These reactions cause the natural density of the material to increase, normally creating a tensile stress in the material because the region boundaries are bonded to neighboring regions.

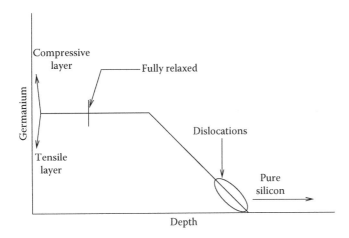

FIGURE 27.14 Schematic figure of graded buffer layer. SiGe buffer layers allow for a buffer layer with a customized lattice constant between that of Si and that of SiGe (approximately 4% larger than Si lattice constant). The formation of the buffer layer uses a grading of Ge concentration, which helps to trap dislocations deep in the graded region. Because there are no known analytic models for handling dislocations and stress relaxation, it is generally assumed in process simulation that a relaxed buffer layer of given Ge concentration can be produced. The user chooses a relaxed depth, where the lattice constant is fixed. Above the relaxed depth, no dislocations are assumed to be formed, so the strain is a simple linear function of Ge concentration.

Implantation has been shown to introduce compressive stress in Si of the order of 10^8 (N/m²).[16] Good models for the evolution of stress coming from implantation are not available currently. It seems likely even without a good model that the stress is in large part due to the Frenkel pair concentration generated during implantation, and in amorphous regions volume increase of approximately 6% have been observed leading to very large stresses. The recombination of the Frenkel pairs happens very quickly compared to dopant diffusion, so it could be that stress due to implantation is greatly reduced before significant dopant diffusion takes place. Despite this, it may still be important for extremely fast ramp rates and short anneal times which are being increasingly utilized. Stress coming from high concentrations of dopants could be significant, especially if coupled with stress-dependent diffusion models.

A final note about material properties: silicon is normally modeled as an elastic material, but this is not necessarily a good approximation if dislocations are to be considered. Dislocations can relax stresses in silicon but they allow leakage current to flow and can potentially short the transistor. However, if they are carefully controlled, as is the case with SiGe-relaxed substrates, their existence can be tolerated. SiO_2 is modeled as a visco-elastic material which can flow at high temperatures but remains elastic at low temperatures. Si_3N_4 is normally modeled as a stiff elastic material, and polysilicon is elastic like silicon. These properties influence the evolution of stress and strain during the simulation.

27.10 PUTTING IT ALL TOGETHER

There are a number of simulation strategies which are commonly used to improve simulation accuracy and performance. A very important capability for a simulator is to be able to easily change the number of spatial dimensions of the simulation (see Figure 27.15). Many initial process

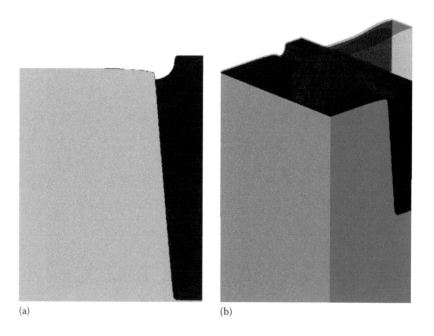

(a)   (b)

FIGURE 27.15   2-D (a) and 3-D (b) techniques. It is common to simulate in the lowest dimensionality until it is necessary to extrude to a higher dimension to save on computational resources. To avoid the necessity of creating many 3-D meshes, it is highly beneficial to create a structure containing all intermediate boundaries mesh the structure once. After this the structure that is meshed during the simulation regions only need to be turned "on" and "off" to reproduce all intermediate structures. This technique minimizes 3-D meshing difficulties and interpolation, but cannot simulate reaction/diffusion situations such as oxidation where the boundaries move at the same time the dopants are diffusing.

steps such as epitaxy, well implants, and gate implants and subsequent anneal steps can be performed in 1-D. Some shallow trench isolation (STI) trench formation and liner oxide steps can be performed in 2-D. Each increase in dimension will cost 10 times or more in CPU and memory, so staying in the lowest possible spatial dimension for as long as possible will reduce simulation time without hurting accuracy. Many structures such as CMOS devices have one- or two-fold symmetry. Simulating all process steps with either half structures in 2-D or one-fourth structures in 3-D and only building the full structure as a last step not only can cut CPU and memory by their respective fractions, but will also help improve device simulation accuracy because the mesh will reflect the symmetry of the device.

Another technique which has been employed to assist in difficulties of producing 3-D meshes is to produce one composite structure at the beginning of the simulation which contains all intermediate boundaries.[7] Once this structure is meshed, etch and deposition steps can be mimicked by changing the material of a given region (deposition would be changing a gas region to a material, and etching the reverse). This technique eliminates interpolation error from changing meshes and reduces the simulation time by reducing many mesh operations, but it cannot be used to solve any moving boundary problems such as oxidation or silicidation.

Transferring information from process simulation to device simulation requires a structure and associated data fields for active dopants and stress. Often the mesh must be recreated so it can be tailored to device simulation, and hence the process data are then interpolated onto the new mesh. For device simulation, it is necessary to create contacts in the mesh where potentials will be applied during device simulation. The contacts are simply a collection of surface elements and a name is designated for each contact.

More sophisticated device modeling has led to more information transfer. At first, all that was needed was net doping (donors–acceptors) and total doping (sum of all dopants) for mobility degradation due to scattering. If partial ionization of dopants is to be considered, then the concentration of each dopant must be sent separately. In addition, for 90 nm node and below, the deliberate introduction of stress is used to increase drive currents. Advances in mobility models and band structure models depending on the full strain/stress tensor and is now transferred from process simulators directly to device simulators. Figure 27.16 shows an example of a structure created by combination of Synopsys process simulation and mesh generation tools which is ready for device simulation.

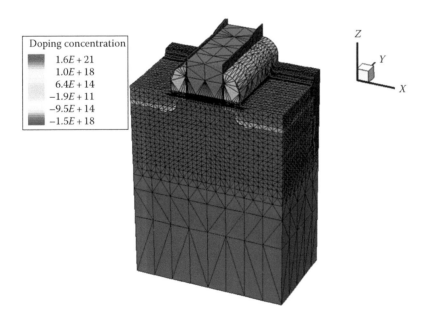

FIGURE 27.16   Final 3-D structure ready for device simulation. The half structure is reflected and contacts are added (outlined in purple on the top). The mesh lines are shown, and the doping concentration is shown in a rainbow, shaded red for n type and blue for p type.

27.11 CONCLUSIONS

Process simulation is a very powerful tool to investigate next-generation device manufacturing strategies and problems. In combination with device simulation, investigations of the benefits and trade-off with novel device structures can be explored at relatively low cost. In addition, parametric yield issues can also be probed at a fraction of the time and cost of a purely experimental approach. These trends will likely give rise to increased use and reliance on TCAD in future process development and yield improvement.

ACKNOWLEDGMENTS

I would like to thank Synopsys for support in preparing this document, and Victor Moroz and Vinay Rao for helping me review the contents. I would also like to thank Beat Sahli for contributing Figure 27.2.

REFERENCES

1. D.A. Antoniadis, S.E. Hansen, R.W. Dutton, and A.G. Gonzalez, SUPREM I-A program for IC process modeling and simulation, Technical Report No. 5019-2, Stanford Electronics Laboratories, Stanford, CA, 1978.
2. S. Selberher, *Analysis and Simulation of Semiconductor Devices*, Springer, New York, 1984; O.C. Zienkiewicz and R.L. Taylor, *The Finite Element Method*, McGraw-Hill, London, U.K., 1991.
3. Synopsys, Taurus-process reference manual release 2004.09, Mountain View, CA, 2004.
4. W.K. Hofker, H.W. Werner, D.P. Oosthoek, and N.J. Koeman, Taurus-process reference manual release 2004.09, *Appl. Phys.*, 4, 125, 1974.
5. P.A. Stolk et al., Physical mechanisms of transient enhanced dopant diffusion in ion-implanted silicon, *J. Appl. Phys.*, 81, 6031, 1997.
6. M. Uematsu, Simulation of high concentration phosphorus diffusion in silicon taking into account Ostwald Repening of defects, *Jpn. J. Appl. Phys.*, 38, 6188, 1999; P.H. Keys, R. Brindos, V. Krishnamoorthy, M. Puga-Lambers, K.S. Jones, and M.E. Law, Phosphorus/silicon interstitial annealing after ion implantation, *Mater. Res. Soc. Symp. Proc.*, 610, B6.6, 2000.
7. Synopsys, FLOOPS-ISE release10.0 manual, Mountain View, CA, 2004.
8. S.T. Dunham, A quantitative model for the coupled diffusion of phosphorus and point defects in silicon, *J. Electrochem. Soc.*, 139, 2628, 1992.
9. P. Pichler, *Intrinsic Point Defects, Impurities, and Their Diffusion in Silicon*, Springer, Wien, Austria, 2004 and references therein.
10. A. Sethian, *Level Set Methods and Fast Marching Methods*, Cambridge University Press, New York, 2002.
11. Rim, K., J.L. Hoyt, and J.F. Gibbons, Fabrication and analysis of deep submicron strained-Si N-MOSFET's, *IEEE Trans. Electron. Dev.*, 47, 1406–1415, 2000.
12. T. Ghani, M. Armstrong, C. Auth et al., A 90 nm high volume manufacturing logic technology featuring novel 45 nm gate length strained silicon CMOS transistors, *IEDM Technical Digest*, Washington, DC, December 7–10, 2003.
13. H. Shang, J. Chu, S. Bedell, E.P. Gusev, P. Jamison, Y. Zhang, J. Ott, M. Copel, D. Sadana, K. Guarini, and M. Ieong, Selectively formed high mobility strained Ge PMOSFETs for high performance CMOS, *IEDM Technical Digest*, San Francisco, CA, December 13–15, 2004.
14. S. Takagi, T. Mizuno, T. Tezuka et al., Channel structure design, fabrication and carrier transport properties of strained-Si/SiGe-on-insulator (strained-SOI) MOSFETs, *IEDM Technical Digest*, Washington, DC, December 7–10, 2003.
15. A.G. Noskov, E.B. Gorokhov, G.A. Sokolova, E.M. Trukhanov, and S.I. Stenin, Correlation between stress and structure in chemically vapour deposited silicon nitride films, *Thin Solid Films*, 162, 129, 1988.
16. C.A. Volkert, Stress and plastic flow in silicon during amorphization by ion bombardment, *J. Appl. Phys.*, 70, 3521, 1991.

Device Modeling: From Physics to Electrical Parameter Extraction

Robert W. Dutton, Chang-Hoon Choi, and Edwin C. Kan

CONTENTS

28.1 INTRODUCTION

Technology files and design rules are essential building blocks of the IC design process. Their accuracy and robustness over process technology and its variability and the operating conditions of the IC—environmental, parasitic interactions, and testing, including adverse conditions such as electrostatic discharge (ESD)—are critical in determining performance, yield, and reliability. Development of these technology and design rule files involves an iterative process that spans across the boundaries of technology and device development, product design, and quality assurance. Modeling and simulation play a critical role in supporting many aspects of this evolution process.

The goal of this chapter is to start from the physical description of integrated circuit devices, considering both the physical configuration and related device properties, and then to build the links between the broad range of physics and electrical behavior models that support circuit design. Physics-based modeling of devices, in distributed and lumped forms, is an essential part of the IC process development. It seeks to quantify the underlying understanding of the technology, and abstract that knowledge to the device design level, including extraction of the key parameters that support circuit design and statistical metrology. Although the emphasis of this chapter will be on metal-oxide-semiconductor (MOS) transistors—the workhorse of the IC industry—it is useful to overview briefly the development history of the modeling tools and methodology that has set the stage for the present state of the art.

The evolution of technology computer-aided design (TCAD)—the synergistic combination of process, device and circuit simulation, and modeling tools—has its roots in bipolar technology starting in the late 1960s, and the challenges of junction isolated, double- and triple-diffused transistors. These devices and technology were the basis of the first integrated circuits; nonetheless, many of the scaling issues and underlying physical effects are still integral to IC design, even after four decades of IC development. With these early generations of IC, process variability and parametric yield were an issue—a theme that will re-emerge as a controlling factor in future IC technology as well.

Process control issues—both for the intrinsic devices and all the associated parasitics—presented formidable challenges and mandated the development of a range of advanced physical models for process and device simulation. Starting in the late 1960s and into the 1970s, the modeling approaches exploited were dominantly one- and two-dimensional (2-D) simulators. While TCAD in these early generations showed exciting promise in addressing the physics-oriented challenges of bipolar technology, the superior scalability and power consumption of MOS technology revolutionized the IC industry. By the mid-1980s, CMOS became the dominant driver for integrated electronics. Nonetheless, these early TCAD developments [1,2] set the stage for their growth and broad deployment as an essential toolset that has leveraged technology development through the very-large and ultra-large-scale-integration (VLSI and ULSI) eras, which are now the mainstream.

IC development for more than a quarter century has been dominated by the MOS technology. In the 1970s and 1980s n-channel MOS (NMOS) was favored owing to speed and area advantages, coupled with technology limitations and concerns related to isolation, parasitic effects, and process complexity. During that era of NMOS-dominated LSI and the emergence of VLSI, the fundamental scaling laws of MOS technology were codified and broadly applied [3]. It was also during this period that TCAD reached maturity in terms of realizing robust process modeling (primarily one-dimensional) which then became an integral technology design tool, used universally across the industry [4]. At the same time device simulation, dominantly 2-D owing to the nature of MOS devices, became the workhorse of technologists in the design and scaling of devices [5,6]. The transition from NMOS to CMOS technology resulted in the necessity of tightly coupled and fully 2-D simulators for process and device simulations. This third generation of TCAD tools became critical to address the full complexity of twin-well CMOS technology (see Figure 28.3a), including issues of design rules and parasitic effects such as latch-up [7,8]. An abbreviated but prospective view of this period, through the mid-1980s, is given in [9], from the point of view of how TCAD tools were used in the design process [10].

Today the requirements for and use of TCAD cut across a very broad landscape of design automation issues, including many fundamental physical limits. At the core are still a host of process and device modeling challenges that support intrinsic device scaling and parasitic extraction. These applications include technology and design rule development, extraction of compact models and more generally design for manufacturability (DFM) [11]. The dominance of interconnects for giga-scale integration (transistor counts in O[billion]) and clocking frequencies in O(10 GHz) has mandated the development of tools and methodologies that embrace patterning by electromagnetic simulations—both for optical patterns, and electronic and optical interconnect performance modeling—as well as circuit-level modeling. This broad range of issues at the device and interconnect levels, including links to underlying patterning and processing technologies, is summarized in Figure 28.1, and provides a conceptual framework for the discussion that now follows.

Figure 28.1a depicts a hierarchy of process, device, and circuit levels of simulation tools. On each side of the boxes indicating modeling level are icons that schematically depict representative applications for TCAD. The left side gives emphasis to DFM issues such as shallow-trench isolation (STI), extra features required for phase-shift masking (PSM), and challenges for multilevel interconnects that include processing issues of chemical–mechanical polishing (CMP) and the need to consider electromagnetic effects using field solvers. The icons on the right side show the more traditional hierarchy of expected TCAD results and applications: complete process

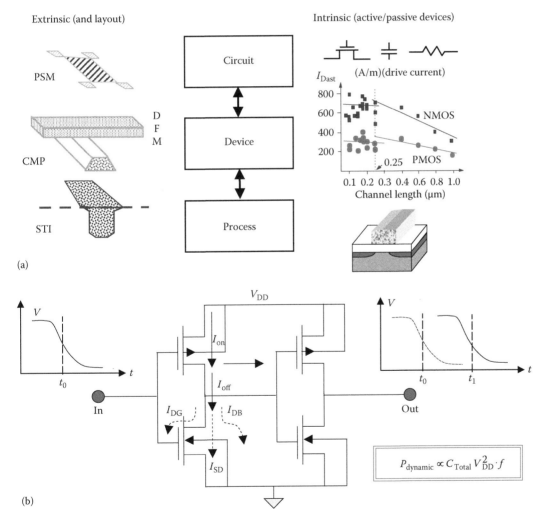

FIGURE 28.1   (a) Hierarchy of technology CAD tools building from the process level to circuits. Icons on the left show typical manufacturing issues; icons on the right reflect MOS scaling results based on TCAD. (b) Schematic view of circuit-oriented aspects of modeling that illustrate voltage-time behavior of inverters, components of "on" and "off" currents and dynamic power constraints.

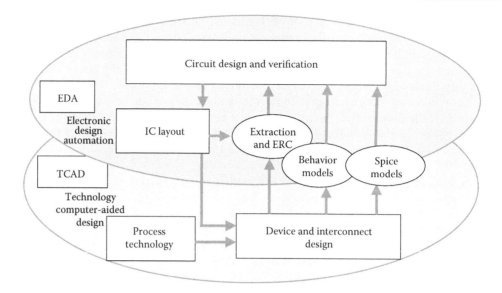

FIGURE 28.2   Information links between EDA and TCAD toolsets. Layout and process technology files support the TCAD flow which in turn outputs device (and interconnect) electrical data used for EDA models.

simulations of the intrinsic devices, predictions of drive current scaling, and extraction of technology files for the complete set of devices and parasitics.

Figure 28.2 again looks at TCAD capabilities but this time more in the context of design flow information and how this relates to the physical layers and modeling of the electronic design automation (EDA) world. Here the simulation levels of process and device modeling are considered as integral capabilities (within TCAD) that together provide the "mapping" from mask-level information to the functional capabilities needed at the EDA level, such as compact models ("technology files") and even higher-level behavioral models. Also shown is the extraction and electrical rule checking (ERC) this indicates that many of the details that to date have been embedded in analytical formulations, may in fact also be linked to the deeper TCAD level in order to support the growing complexity of technology scaling.

28.2 MOS TECHNOLOGY AND INTRINSIC DEVICE MODELING

The previous section has motivated the scope of device- and technology-level modeling with emphasis on TCAD as the toolset that provides details ranging from deep physical models of fabrication processes and device physics to the higher-level EDA interfaces, to electrical extraction—both for technology files and design rules in support of behavior-level models. Figure 28.1a points out the modeling hierarchy going from process technology to circuit modeling; the hierarchy has a dual "ladder" in terms of intrinsic and extrinsic devices. Chapter 24 considers the process modeling level of the hierarchy; Chapter 19 considers the statistical effects associated with the design process that have a profound effect on overall performance. The extrinsic devices and modeling of ICs will be considered in later sections of this chapter. This section will focus on considerations of the intrinsic MOS device, including supporting details of the variety of p–n junction effects that are inherent to the bulk CMOS processing technology.

It is useful to briefly consider the objectives of device design, both from the system and technology perspectives, prior to going into greater detail at the physical modeling level. Figure 28.1b provides a simplified conceptual view of "the big picture." This figure shows two inverter stages and the resulting input–output voltage–time plot of the circuit. Clearly, from the digital systems point of view, the key parameters of interest are timing delays, switching power, leakage current, and cross-coupling ("crosstalk") with other blocks. The voltage levels and transition speed are also of concern. The figure also shows schematically the importance of I_{on} vs. I_{off}, which in turn is related to drive current (and mobility) for the "on" device and several leakage paths for the "off" devices. Not shown explicitly in Figure 28.1b are the capacitances—both intrinsic and parasitic—that affect dynamic performance. Moreover, the power scaling which is now a major driving force in the industry is reflected in the simplified equation shown in Figure 28.1b—critical parameters are capacitance, power supply, and clocking frequency. Key parameters that relate device behavior to system performance include the threshold voltage, driving current, and subthreshold characteristics. It is the confluence of system performance issues with the underlying technology and device design variables that results in the ongoing scaling laws that we now codify as "Moore's law" [12]. With this backdrop of system performance metrics that relate to the technology perspective, we now begin the discussion of the physical (and physics-based) considerations of the MOS transistor.

The physics and modeling of integrated circuits can roughly be divided between the surface effects associated with MOS devices and a range of bulk junction effects that involve bipolar and associated leakage and parasitic effects. As mentioned in the introduction, CMOS transistors for ULSI now claim the lion's share of practical IC devices. In CMOS technology the bulk junction device effects are virtually all parasitic. In Section 28.4 some of the bipolar options, for example the use of BiCMOS for some power and RF applications will be considered. There are of course also issues of reliability engineering—for example, ESD protection circuits and devices—where substrate and parasitic devices are of pivotal importance. These effects and modeling are not considered here; the reader is referred to several excellent monographs in the area of ESD and I/O modeling [13–15]. The following discussion will use CMOS devices as the driver for discussing the physical modeling issues. Figure 28.3a shows a low-resolution cross-sectional version of a basic CMOS technology. Figure 28.3b and the subsequent discussions will focus on the n-channel MOS technology using a twin-well process and STI on a lightly doped substrate.

28.2.1 INTRINSIC MOS TRANSISTOR

The intrinsic MOS device relies on gate control to capacitively induce inversion charge in the channel between source and drain regions. Historically, polycrystalline silicon gates have been used, where they are highly doped, in concert with the source-drain regions—using the so-called self-aligned process—to control both the threshold work function of the gate and reduce gate resistance. With ever-shrinking channel lengths, a range of substrate parasitics associated with the channel region have become first-order effects in fabrication and device modeling. The shallowness and lateral abruptness of the source-drain extension (SDE) regions (see Figure 28.3b) are critical to controlling parasitic resistances and overlap capacitance. They also have a direct impact on the on-to-off current ratios—an important figure of merit. Source-drain leakage is another factor limited by various implanted doping profiles such as "halo" doping (i.e., HALO), which is introduced to prevent parasitic substrate leakage and electrostatic punch-through between source and drain. Yet as doping in these regions below the channel as well as in the source-drain regions (along with the SDE) increase, consistent with scaling laws [3], the band-to-band tunneling (BTBT) leakage and breakdown phenomena become more severe. BTBT leakage and junction breakdown have to be kept in check by accurate doping placement to form a sandwich low-doping area, where nanometer precision is often necessary. Moreover, ultra-thin gate dielectrics now also have significant leakage current. The following subsections summarize the physics and modeling needed to capture properly the essential features of these diverse physical effects.

Figure 28.3b shows the technology cross-section of a typical scaled (i.e., sub-90 nm) MOS, including highlighted information that indicates 2-D doping effects related to threshold, subthreshold and source-drain "engineering" (all the details of SDE, extrinsic resistance, capacitance, and substrate coupling). In Figure 28.3c, a simplified equivalent circuit of various intrinsic, extrinsic, and distributed substrate components is indicated. The key points to understand, both from the physical cross-section of the device and the simplified lumped equivalent circuit, are now briefly discussed. This is followed by the formulation of the modeling equations that are used for simulation and in creating technology files for circuit design.

The MOS transistor is inherently a 2-D field-effect device with gate electrode creating a vertical field that induces carriers from the source region that are extracted toward the drain. The technological complexity shown in Figure 28.3b reflects the evolution over several generations of scaling

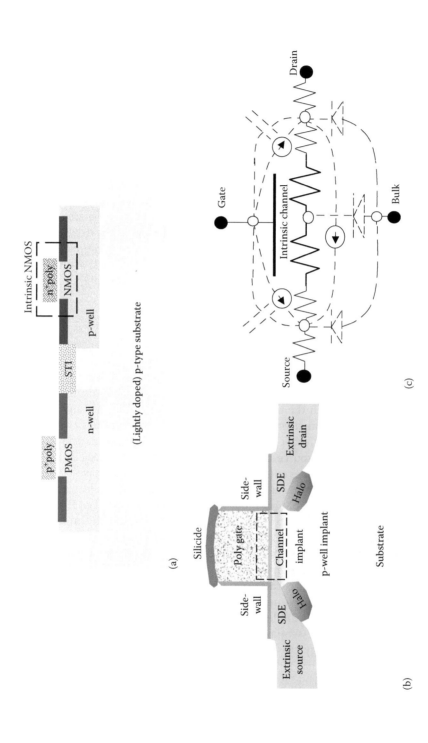

FIGURE 28.3   (a) Simplified cross-section of a twin-well CMOS process using a lightly doped p-type substrate p and n-channel MOS transistors are shown along with STI isolation. (b) Cross-section of the n-channel MOS transistor, including details of gate, sidewalls and substrate doping profiles (SDE, channel/well implants, halo doping, etc.). (c) Schematic view of MOS transistor model that reflects intrinsic channel field effects, bulk node interactions with the channel ("back gate"), S/D diodes and various R, C, and leakage paths.

that has been necessary to achieve the desired inversion charge density and at the same time reduce parasitics that include resistance, capacitance, and leakage—both source-to-drain and bulk leakage from the substrate diodes. The intrinsic part of the MOS transistor (denoted by the dashed region in Figure 28.3b) continues to be scaled using electrostatics that tries to maintain greater vertical electric field (from the gate) compared to the lateral field down the channel. However, with each generation of scaling this intrinsic portion of the device becomes smaller and contributions of the extrinsic parasitics, including source-drain contact regions, are not scaling as rapidly. The implications can be seen by considering the equivalent circuit shown in Figure 28.3c.

The simplified schematic of the MOS transistor shown in Figure 28.3c is not specifically targeted to represent the equivalent circuit used in compact modeling. It is to point out how many of the physical device effects (coming from the technology cross-section shown in Figure 28.3b) relate to intrinsic and parasitic effects. The key points to be emphasized here are as follows: (1) the channel region couples electrostatically to both gate and bulk and is a distributed effect (i.e., the channel charge can be shared with both source and drain); (2) there are several leakage paths that couple gate-to-channel (i.e., tunneling of carriers through the gate) and source-to-drain (i.e., thermionic emission and diffusion as well as quantum mechanical [QM] tunneling); (3) there are both resistive and capacitive parasitics associated with source and drain regions. As noted above, these passive parasitics are increasingly coming to dominate the device and only serve to degrade the intrinsic transistor action.

The following discussion will first address the intrinsic behavior of the MOS transistor and associated simulation and modeling issues, followed by a discussion of considerations of the extrinsic and parasitic effects. Figure 28.4a gives a first-order transistor equation that describes how the drain current depends on electrostatic effects of gate voltage and source-to-drain electric field. This simplified formulation indicates the importance of both the quantity of channel charge (capacitively induced by the gate) and the value of carrier velocity (as reflected by the mobility). The supporting equations show the voltage dependence of C_g and how the various scattering mechanisms affect the overall channel mobility. From a scaling perspective the technology approaches now being considered (and implemented in many research demonstrations) involve increasing the dielectric constant of the gate and the implementation of materials and processing that increase the mobility. Figure 28.4b shows how drive current is scaled with technology generation. The reduced slope for bulk silicon devices over the past several generations of scaling has motivated serious research efforts and industrial demonstrations of materials with increased mobility. Which options might become the mainstream are still under investigation. Models that embrace the necessary physical effects—both for conventional silicon and for advanced material options—are considered as part of the following discussion.

28.2.1.1 INVERSION-LAYER MOBILITY MODELING

The modeling of mobility and its dependence on physical parameters, ambient and operating conditions, is arguably one of the most important dependencies both for TCAD physical models and for circuit-level compact models. For mobility modeling at the TCAD level the electrical variables are the local electric field at each point in the device (i.e., components parallel and perpendicular to the planar device surface) and the various scattering mechanisms, including their technology and ambient dependencies. By contrast, the circuit-level models represent a macroscopic approximation of the physical phenomena and parameterize the effects only in terms of terminal voltages vis-à-vis internal (and microscopic) electric fields. Through the electrical extraction process the two representations can be compared and calibrated. The following discussion will consider two typical and well-known TCAD-based mobility models, show comparisons with representative measured data, and then finally contrast these physical models with a typical compact model representation for mobility.

The Lombardi surface mobility (LSM) model was first proposed in the late 1980s as an empirical model that captures the vertical field dependence and accounts for the doping dependence of bulk mobility (μ_b) as well as acoustic phonon (μ_{ac}) and surface roughness (μ_{sr}) scattering effects [16]. The overall formulation exploits the classical Matthiessen's summation rule, shown as the first equation in Figure 28.5, i.e., the inverse of the total surface mobility (μ_s) is the sum of the inverse of all constituent mobility terms. The set of relationships shown in Figure 28.5 give details of one popular implementation of the LSM model in the MEDICI simulator. It is beyond the

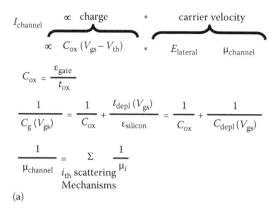

$$I_{channel} \overbrace{\qquad}^{charge} \propto charge \quad * \quad \overbrace{\qquad}^{carrier\ velocity} carrier\ velocity$$

$$\propto C_{ox}\,(V_{gs} - V_{th}) \quad * \quad E_{lateral} \quad \mu_{channel}$$

$$C_{ox} = \frac{\varepsilon_{gate}}{t_{ox}}$$

$$\frac{1}{C_g\,(V_{gs})} = \frac{1}{C_{ox}} + \frac{t_{depl}\,(V_{gs})}{\varepsilon_{silicon}} = \frac{1}{C_{ox}} + \frac{1}{C_{depl}\,(V_{gs})}$$

$$\frac{1}{\mu_{channel}} = \sum_{\substack{i_{th}\ scattering \\ Mechanisms}} \frac{1}{\mu_i}$$

(a)

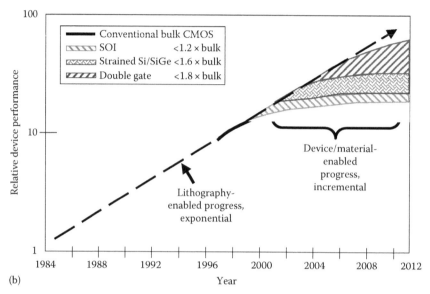

(b)

FIGURE 28.4   (a) First-order model of how MOS inversion layer channel current ($I_{channel}$) depends on gate-included charge and carrier velocity. The physical parameters are C_g and $\mu_{channel}$. (b) Relative MOS device performance occurring for conventional bulk CMOS and projects incremental improvements with new materials.

scope of this work to discuss in detail the parameters of the model, other than to point out that the key physical quantities are total doping (N_{total}), maximum mobility for the respective material ($\mu_{max,n}$ here for electrons) and perpendicular local electric field ($E \perp n$). All model parameters are indicated with capital letters and a ".LSM" suffix.

The LSM model was developed with emphasis on the vertical electric field effects, hence additional modeling terms are needed for the lateral field effects. The work of Shin et al. [17,18] is an example of a model that considers both vertical and lateral field effects, especially since it is able to match the so-called "universal mobility" curves [20] and was suitable for implementation as a local field model—a key requirement for TCAD simulator implementation. Another alternative model developed by Darwish et al. [19] considers lateral field effects and allows the model to be localized in space ($\vec{r}$). Figure 28.6 summarizes the key equations used by Darwish, again using Matthiessen's summation rule, where the subscripts refer to: b, bulk; ac, acoustic phonon; sr, surface roughness; L, lattice; and I, ionized impurity scattering events, respectively. The term involving acoustic phonons has been refined and, while still similar in form to that in the Lombardi model, is based on improved physical modeling of the velocity dependence. The perpendicular field terms have the same meaning as considered above and the form of the surface roughness term is also unchanged. Figure 28.7 shows a plot of mobility vs. normal electric field with doping as a parameter; the curves compare the Darwish model with measured data of Takagi et al. [20]. There is generally good agreement between both models and the data; the Darwish model shows

$$\mu_S = \left[\frac{1}{\mu_{ac}} + \frac{1}{\mu_b} + \frac{1}{\mu_{sr}} \right]^{-1}$$

$$\mu_{ac} = \frac{BN.LSM}{E_{\perp,n}} + \frac{CN.LSM.N_{total}^{EXN4.LSM}}{T\sqrt[3]{E_{\perp,n}}}$$

$$\mu_b = MUN0.LSM + \frac{\mu_{max,n} - MUN0.LSM}{1 + \left(\dfrac{N_{total}}{CRN.LSM} \right)^{EXN1.LSM}} - \frac{MUN1.LSM}{1 + \left(\dfrac{CSN.LSM}{N_{total}} \right)^{EXN2.LSM}}$$

$$\mu_{max,n} = MUN2.LSM \left(\frac{T}{300} \right)^{-EXN3.LSM}$$

$$\mu_{sr} = \left(\frac{DN.LSM}{E_{\perp}^{EXN8.LSM}} \right)$$

FIGURE 28.5 Lombardi surface mobility (LSM) model formulation [16], including parameterization as implemented in MEDICI. Matthiessin's used in summation of various scatting terms.

$$\frac{1}{\mu_o(\vec{r})} = \frac{1}{\mu_b(\vec{r})} + \frac{1}{\mu_{ac}(\vec{r})} + \frac{1}{\mu_{sr}(\vec{r})}$$

$$\frac{1}{\mu_b(\vec{r})} = \frac{1}{\mu_L} + \frac{1}{\mu_I(\vec{r})} \qquad \mu_{sr}(\vec{r}) = \frac{\delta}{E_{\perp}^{Y}(\vec{r})}$$

$$\mu_{ac}(\vec{r}) = \left(\frac{BT'}{E_{\perp}(\vec{r})} + \frac{CN_I^T}{E_{\perp}^{1/3}(\vec{r})} - \frac{1}{T'} \right)$$

FIGURE 28.6 Darwish mobility model expression [19] that considers local position dependence $(\vec{r})$. Comparisons of this model and data from Takagi et al. [20] are given in Figure 28.7.

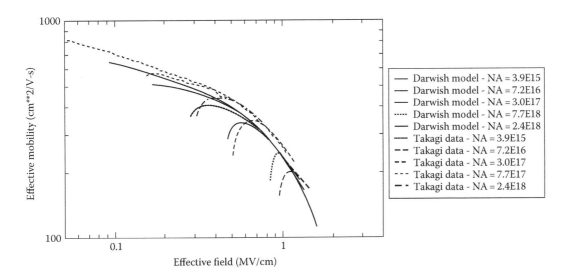

FIGURE 28.7 Effective mobility vs. effective electric field showing experimental results of Takagi et al. [20] and Darwish TCAD model formulation [19] over range of doping levels with good agreement.

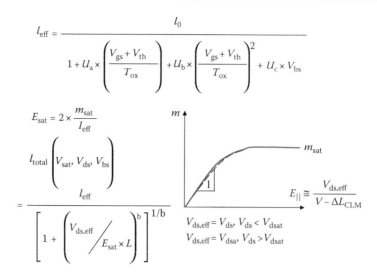

$$l_{\text{eff}} = \frac{l_0}{1 + U_a \times \left(\dfrac{V_{gs} + V_{th}}{T_{ox}}\right) + U_b \times \left(\dfrac{V_{gs} + V_{th}}{T_{ox}}\right)^2 + U_c \times V_{bs}}$$

$$E_{\text{sat}} = 2 \times \frac{m_{\text{sat}}}{l_{\text{eff}}}$$

$$\frac{l_{\text{total}}\left(V_{\text{sat}},\, V_{ds},\, V_{bs}\right)}{l_{\text{eff}}} = \frac{1}{\left[1 + \left(\dfrac{V_{ds,\text{eff}}}{E_{\text{sat}} \times L}\right)^b\right]^{1/b}}$$

$$E_{||} \cong \frac{V_{ds,\text{eff}}}{V - \Delta L_{\text{CLM}}}$$

$$V_{ds,\text{eff}} = V_{ds},\ V_{ds} < V_{dsat}$$
$$V_{ds,\text{eff}} = V_{dsa},\ V_{ds} > V_{dsat}$$

FIGURE 28.8 Compact (circuit-level) model for effective carrier mobility showing gate (V_{gs}) and bulk (V_{bs}) voltage dependencies, along with definition of critical field term, where velocity saturation occurs and simple "scaling" term used in linear region.

good agreement for larger effective electric field (E_{eff}). The notion of a "universal curve" for mobility simply refers to the fact that the channel mobility can be modeled as a function of E_{eff} over several orders of magnitude, where the substrate doping effect is represented in terms of only μ_b and E_{eff}, and the $\mu(E_{\text{eff}})$ is independent of the other device parameters such as oxide thickness and interface states.

The above discussion is representative of physical models for mobility used in TCAD simulators. These models utilize detailed information about the behavior of the inversion-layer channel in MOS transistors and parameterize it in a form suitable for implementation in numerical device simulators. There a variety of models available in TCAD tools and the interested reader should refer to the documentation provided by the respective tool developers. It is useful to look briefly at the mobility modeling used in circuit simulation and to contrast it with that used at the TCAD level. Figure 28.8 shows a typical mobility formulation used in a circuit-level compact model (i.e., BSIM3 used in HSPICE). In contrast to the TCAD model that has physical parameters such as doping level and local electric fields occurring inside the device, the compact model for mobility only considers terminal voltages and fitting parameters, including expressions used to take into account effects of velocity saturation. The inserted figure shows the carrier velocity vs. the parallel electric field plot, indicating the region where carriers become velocity-saturated. The dashed curve in Figure 28.8 shows the approximation used to smooth the transition between the low and high parallel field mobility regions as a function of the drain voltage; the last equation shows one form used to achieve that smoothing. The main point to emphasize about compact models is the fact that they require curve fitting to terminal I–V data for transistors. This can be done either with experimental data or TCAD-generated I–V curves, which in turn exploit the physical mobility models discussed above.

28.2.1.2 CHANNEL CHARGE MODELING

The channel charge and all the effects of gate and dielectric materials are as important as the mobility as indicated by the formula shown in Figure 28.4a. The basic formula indicates a simple relationship between gate capacitance and gate-to-source voltage based on the series combination of oxide capacitance (C_{ox}) and the bulk depletion layer capacitance (determined by t_{depl}). In fact, the dependence of C_g is much more complex as reflected in Figure 28.9a. This figure shows typical measured and simulated C–V characteristics for a 2 nm oxide, along with the TCAD-simulated characteristics. The idealized formula given in Figure 28.4a is shown with a broken line. For gate bias both above and below threshold (around 0 V) there are possibly poly-depletion QM effects in the channel region (bulk) and in the poly-silicon gate region [21]. Moreover, as the gate thickness

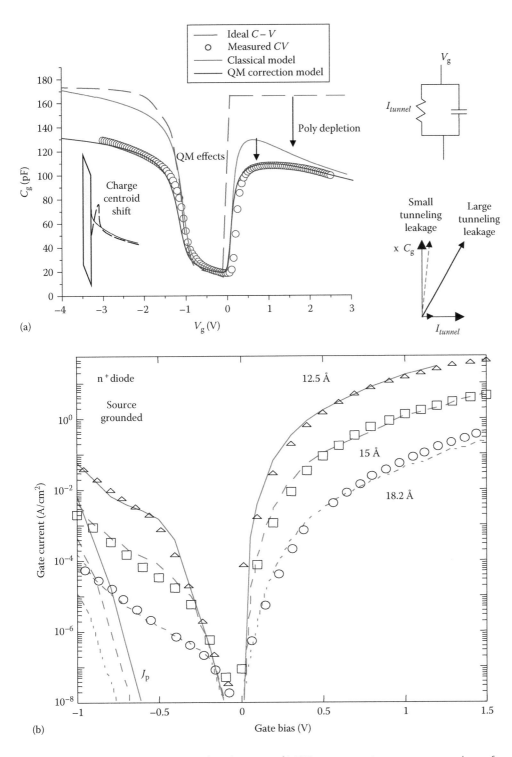

FIGURE 28.9 (a) Measured and simulated curves of MOS gate capacitance vs. gate voltage for an n-channel, poly-silicon gate with 2 nm-oxide thickness. Simulated curves compare impact of QM effects. (b) Simulated and measured data for gate current in an MOS capacitor as a function of gate oxide thickness. Simulations were performed using NEMO; data provided by Hewlett-Packard.

decreases further, tunneling of carriers through the gate region becomes distinguishable. Quantum mechanical effects in $C-V$ include the quantum repulsion (i.e., the peak of carrier concentration has to be away from the interface in inversion and accumulation) and tunneling which corrects the impedance phase during capacitance extraction. Figure 28.9b shows a semilog plot of simulated and measured gate currents for much thinner oxides as a function of bias. The simulations were performed using the NEMO simulator [22]. Tunneling currents, along with electrostatic effects of depletion in the poly-gate regions and QM confinement effects, play a key role in determining the observed $C-V$ characteristics as well as $I-V$ characteristics in scaled MOS transistors.

From both the technology scaling and circuit design perspectives, it is important to separate the effect of gate capacitance (which controls channel charge) from the gate leakage current which is a parasitic effect. In order to extract properly and model the channel charge component separately from the effects of tunneling current on the measured $C-V$ curves, an equivalent circuit extraction approach has been used as reflected in Figure 28.10a [21]. This circuit represents a large area MOS device (100 μm × 100 μm) that has been modeled as a distributed RC network, including voltage-dependent tunneling current generators extracted from the data shown in Figure 28.9b. Circuit simulation and optimization are used to fit the capacitive terms such that the simulated and measured data for the imaginary component of ac input admittance agree, as shown in Figure 28.10b. For an oxide thickness of 1.8 nm, the C_g plot shown in Figure 28.10b is not unlike that shown in Figure 28.9a. However, as the oxide thickness decreases further (i.e., 1.5 and 1.3 nm), the curves are dominated by the tunneling current component. That is, as direct (tunneling) current becomes a larger fraction of the total current (AC1DC), there is a fall-off in the apparent capacitance component. This spurious effect needs to be corrected for in order to provide the correct information to the IC designer. These results illustrate an important link between TCAD-based and compact modeling. The equivalent circuit used for extraction of the actual gate capacitance (from measured data) was parameterized using TCAD, where each component can be more independently extracted that is not separable in actual measurement. In turn, the resulting model (including tunneling current generators) is essential for circuit-level modeling to fit a wider range of technologies. Implications of tunneling current on circuit performance, using this overall modeling approach, are discussed elsewhere [23].

As stated above, there are important QM effects that impact the confinement of inversion-layer charge in the channel and gate of MOS devices. Figure 28.11 shows the comparison of electron distributions in both the bulk silicon (at the surface inversion layer) and in the poly-silicon gate regions. Figure 28.11a contrasts the classical solution for the inversion layer with that of the QM solution where there are discrete energy levels in the channel region, resulting in a solution that has zero carriers at $x = 0$. The energy levels also impose constraints in terms of the number of available states (density of states [DOS]) that can be occupied by electrons. This in turn places limits on the amount of charge that can be induced in each level and ensemble.

In Figure 28.11b, the electrons in the poly-gate have a spread-out distribution for the QM solution vs. a sharply peaked distribution at the interface for the classical solution. These QM effects have a very significant impact on both $C-V$ and $I-V$ characteristics of scaled devices. The impact on the $C-V$ curves, as demonstrated in Figure 28.9a, is a reduction in capacitance. This can be understood readily by looking at the simplified expression for C_g given in Figure 28.4a; an increase in t_{depl} in the bulk region causes the total gate capacitance to decrease. Similarly, by adding a dipole layer in the poly-silicon gate there is an additional series capacitance that again reduces C_g. As alternative silicon-on-insulator (SOI) and double-gated (DG) devices are considered in future scaling nodes, there are coupled QM effects that can potentially impact performance. For example, in ultra-thin layers, resonant (QM) tunneling between the gates becomes possible [24]. Even without considering such unusual effects, the 2-D electrostatic (QM) effects can have a major impact on scaled SOI devices.

It is beyond the scope of this chapter to delve seriously into future trends of MOS scaling. The current projections of the International Technology Roadmap for Semiconductors (ITRS) [12] outlines the ongoing needs to consider alternative device geometries and materials for channel lengths below the 45 nm technology node. As discussed above for bulk silicon devices, many of the issues related to carrier mobility and channel charge need to be revisited based on these new structures. The beauty and genius of the poly-silicon, self-aligned gate technology has been its scalability—the dimensional and electrical control of key parameters. The above discussions of

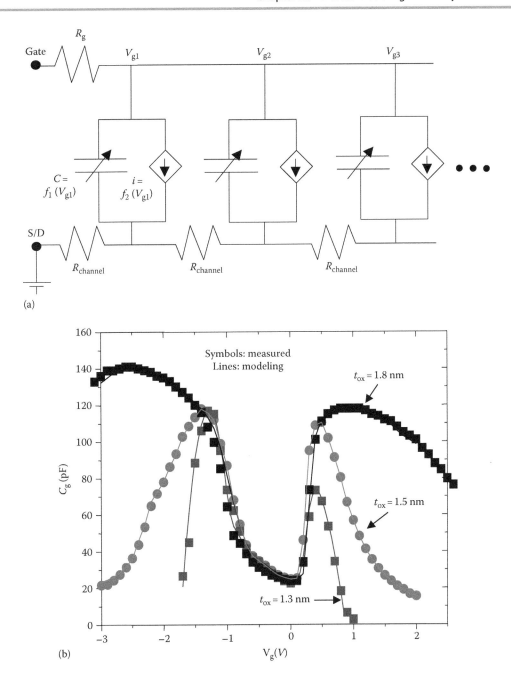

FIGURE 28.10   (a) Equivalent circuit (lumped) model for large area (100 μm × 100 μm) MOs capacitor, S/D grounded showing distributed *R–C* effects well as dependent generators representing tunneling current. (b) Results comparing measure and equivalent circuit simulations of imaginary component of input admittance for large area MOS capacitors structure with oxide thickness as a parameter.

gate leakage and QM effects in the gate now brings to the foreground the motivation for changing both dielectric materials of the gate and the gate material itself. Some of these limits are now illustrated for ultra-scaled, DG SOI devices, one of the contenders for future device scaling.

Figure 28.12 shows the impact of QM effects on both threshold and subthreshold characteristics of an idealized 20 nm device structure. Figure 28.12a shows the simulated electrical characteristics—*C–V* and *I–V* curves. For $V_{ds} = 0$ V, the main effect of the poly-QM is the shift in threshold voltage from subband formation and charge centroid repulsion at the interface. Notice that the sub-threshold slope hardly degrades by including the poly-QM effect. Especially as the

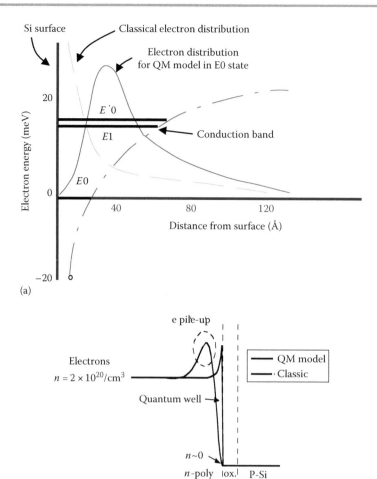

(a)

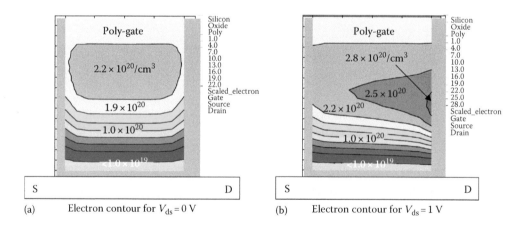

FIGURE 28.11   (a) Surface region of bulk MOS device that shows well potential created by the conduction band energy (-); discrete energy levels imposed by QM (*E*0,*E*1); electron distribution based on classical theory (- - -). (b) Electron distribution in the n-type poly-silicon gate of an NMOS transistor, comparing charge buildup based on classical theory vs. that for a quantum-based (QM) solution for charge. The peak region shows a "QM depletion" resulting in *V*th and *C*–*V* shifts.

FIGURE 28.12   (a) *C*–*V* and *I*–*V* characteristic for a 20 nm channel length DG SOI transistor, comparing effects of QM effects in the poly-gate region with idealized (metal-like) gate. Results show shifts in threshold as well as serious drain-induced current effects. (b) Simulations of QM poly depletion effects for two drain bias conditions (same device and *I*–*V* as Figure 28.12a). Results show significant gate depletion (left) and lateral effects that influence drain-induced barrier lowering (right).

drain voltage increases to $V_{ds} = 1$ V, the I_{ds} curve shows drain-induced coupling effects that occur through the poly-gate region, which increases the drive current quite significantly. Figure 28.12b shows the 2-D contours of charge in the poly-gate region for $V_{ds} = 0$ and 1 V when $V_{gs} = 0.8$ V; the poly-depletion effects are first-order in importance for such ultra-scaled devices [25]. The take-home message from these simulations is that even with improvements of channel scaling based on SOI, the scaling of the poly-silicon gates is rapidly reaching electrostatic limits. The issues of threshold voltage, its control—both electrical and with process variations—and technology options are now briefly explored.

28.2.1.3 THRESHOLD VOLTAGE MODELING

Threshold voltage is the third key parameter that affects performance of MOS devices, namely it controls the channel charge as given by the equation in Figure 28.4a. This equation for channel charge involves both the gate capacitance and "overdrive" voltage (the amount of voltage greater than the threshold voltage, V_{th}). There are many factors that affect V_{th}, including choice of gate electrode materials, device topology (bulk vs. SOI) and bias conditions of all four terminals (see Figure 28.3c). A brief discussion of technology options complimentary MOS, silicon-on-insulator, Bipolar-CMOS (i.e., bulk CMOS, SOI, BiCMOS, etc.) will be covered in Section 28.4. The next section will consider many of the threshold voltage and other parasitic effects associated with bulk CMOS. The emphasis of this subsection is to consider first-order, vertical field only, threshold behavior of bulk CMOS, and the basic silicon-on-insulator NMOS, where the "back gate" is the electrically floating bulk wafer. Alternative SOI substrate configurations, including the most advanced three-dimensional (3-D) structures, are summarized elsewhere [26].

Figure 28.13 summarizes two typical expressions for threshold voltage, along with details of the supporting equations and definitions. Threshold (V_{th}) is typically defined in terms of an electrostatic condition that results in a specified amount of inversion charge or drain current in the device. The two equations shown in Figure 28.13 represent typical formulations for bulk (Figure 28.13a) and SOI (Figure 28.13b) MOS devices, respectively; the threshold condition is defined in terms of a specific surface potential. The threshold expressions consist of several terms; these are contributions to the overall gate potential coming from effects of the gate(s) and bulk/SOI material properties, as will be discussed now. From a physical modeling point of view, process and device simulations are used to characterize the multidimensional electrostatic effects in each device structure. The threshold expressions as shown in Figure 28.13 seek to capture the key dependencies and parameterize them based on analytical approximations of the numerical simulations. The following discussion represents the long-channel results; the effects of source, drain, and substrate are discussed further in the next section. In device simulation, since the 2-D Poisson equation is explicitly solved, all electrostatic effects are included, provided that the correct DOS information is available. Lumping the effects to different parts of the device is useful for device design and compact model construction.

The material-related effects reflect electrical differences between the bulk/SOI regions and the gate(s). The V_{FB} and $\phi_{F,poly}$ terms in Figure 28.13a and b are the specific terms that capture the gate-related dependencies. The expression given in Figure 28.13a (for bulk CMOS) considers both gate-related and fixed interface charge effects, which are lumped into V_{fb}. The gate voltage needs to have the semiconductor to be in the "flat-band" condition (i.e., no vertical electric field at the interface). The expression given in Figure 28.13b considers the gate material to be heavily (n^1) doped poly-silicon; the term $\phi_{F,poly}$ represents the flat-band offset due to only that contribution (i.e., interface or other fixed charge has been neglected). As threshold voltage continues to be scaled down with channel length, these gate-related contributions pose scaling limits; changes in gate materials with a wider selection of work functions are becoming necessary to coordinate the scaling of these terms with the other threshold contributions discussed next.

The remaining terms in both Figure 28.13a and b represent capacitive coupling terms, which generally have the form Q/C_{ox}. The important trends to observe related to these terms include interdependence of doping (N_{Abulk} or $N_{A,SOI}$) with gates(s) (C_{ox} or dual gates if SOI) and substrate bias (V_{sb} for bulk CMOS or capacitively through the substrate potential as a "second gate" via Ψ_{sub}). In Figure 28.13a there are two such terms, one representing the bulk depletion layer (i.e., the term involving N_{Abulk}) and another involving an implanted dose (D_I) of charge. For bulk CMOS,

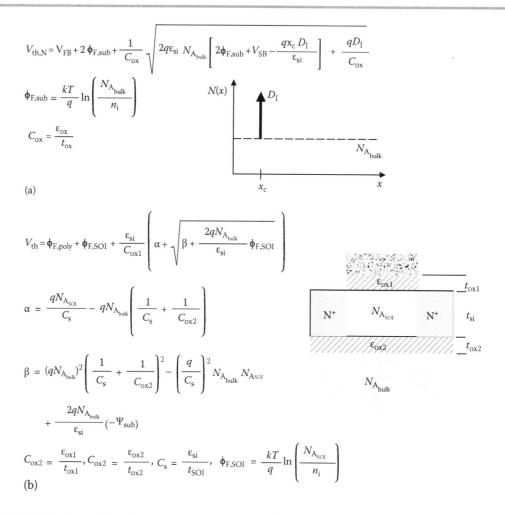

$$V_{th,N} = V_{FB} + 2\,\phi_{F,sub} + \frac{1}{C_{ox}} \sqrt{2q\varepsilon_{si}\, N_{A_{bulk}} \left[2\phi_{F,sub} + V_{SB} - \frac{qx_c D_I}{\varepsilon_{si}} \right]} + \frac{qD_I}{C_{ox}}$$

$$\phi_{F,sub} = \frac{kT}{q} \ln\left(\frac{N_{A_{bulk}}}{n_i} \right)$$

$$C_{ox} = \frac{\varepsilon_{ox}}{t_{ox}}$$

(a)

$$V_{th} = \phi_{F,poly} + \phi_{F,SOI} + \frac{\varepsilon_{si}}{C_{ox1}} \left(\alpha + \sqrt{ \beta + \frac{2qN_{A_{bulk}}}{\varepsilon_{si}} \phi_{F,SOI} } \right)$$

$$\alpha = \frac{qN_{A_{SOI}}}{C_s} - qN_{A_{bulk}} \left(\frac{1}{C_s} + \frac{1}{C_{ox2}} \right)$$

$$\beta = (qN_{A_{bulk}})^2 \left(\frac{1}{C_s} + \frac{1}{C_{ox2}} \right)^2 - \left(\frac{q}{C_s} \right)^2 N_{A_{bulk}} N_{A_{SOI}}$$

$$+ \frac{2qN_{A_{bulk}}}{\varepsilon_{si}} (-\Psi_{sub})$$

$$C_{ox2} = \frac{\varepsilon_{ox1}}{t_{ox1}}, \, C_{ox2} = \frac{\varepsilon_{ox2}}{t_{ox2}}, \, C_s = \frac{\varepsilon_{si}}{t_{SOI}}, \, \phi_{F,SOI} = \frac{kT}{q} \ln\left(\frac{N_{A_{SOI}}}{n_i} \right)$$

(b)

FIGURE 28.13 (a) Threshold voltage expression of a traditional bulk CMOS device (NMOS) with an equivalent threshold adjustment implant dose (DI) centered at X_c. The notation follows that given in [38]. (b) Threshold voltage expression for an SOI (fully depleted) NMOS, where the channel is induced at the top surface of the SOI layer (t_{si}). The definition of terms follows the notation and derivation in [27].

as shown in Figure 28.3b, the substrate doping profiles can be complex, and 2-D effects such as SDE and HALO profiles cannot be ignored. The simple threshold model shown in Figure 28.13a lumps all vertical nonuniform doping effects into an equivalent dose of dopant charge. In the case of Figure 28.13b for an SOI device, the bracketed term involving a and b terms as well as $N_{A,SOI}$ also demonstrates the general form for threshold voltage, controlled by the top gate via C_{ox1}. The α and β terms reflect the effects of the substrate ($N_{A_{bulk}}$) and bottom gate via C_{ox2}. Key points to note here are the multielectrode effects (i.e., dependence on C_{ox1}, C_{ox2}, and Ψ_{sub}) as well as the silicon layer thickness. The expression shown in Figure 28.13b is for a fully-depleted (FD) SOI in all V_{gs} operating conditions; here the thickness as well as silicon layer doping are inter-related in determining the depletion charge contribution to V_{th}. For ultra-thin SOI, the layer doping becomes less important. Details of these two threshold formulations can be found elsewhere [27,28].

The complexity of nonuniformly doped, single-gate structures and DG SOI structures leads to increasingly complex threshold expressions; the two shown in Figure 28.13 are only representative. For bulk CMOS the trade-off between flat-band voltage, surface potential and associated bulk charge pose ongoing challenges in the design of the gate stack and choices in substrate doping profiles. Silicon-on-insulator provides additional parameters to consider, particularly the semiconductor layer thickness (t_{si}) and the additional control of a DG structure. With FD, DG SOI the control of layer thickness becomes a first-order concern and layer doping becomes much less of an important controlling factor. Looking for "the optimal" device structure can be very challenging; the optimization process is largely one of economics (fabrication costs) and performance objectives. However, it is worth pointing out that threshold voltage (and its relationship to power supply level) will be a key factor in the trade-off between power and speed for large digital circuits.

28.2.2 SUBSTRATE EFFECTS ON MOS TRANSISTORS

The drain current expression given in Figure 28.4a is useful in understanding the basic issues of the MOS as an electrostatically induced conducting channel "resistor," controlled by the gate voltage with respect to the threshold voltage. For bulk CMOS the substrate terminal (the *p*- or *n*-well for the *p*- or *n*-channel device, respectively, as in Figure 28.3a) can directly influence the threshold voltage. That is, for the most basic formulation of threshold voltage used in circuit simulation, it is modulated by the substrate bias (V_{sb}) as shown in the following equation:

$$V_{th}(V_{sb}) = V_{th0} + \frac{\sqrt{2q\varepsilon_{si}N_{A\text{bulk}}}}{C_{ox}} \left[(2\phi_{F,sub} + V_{sb})^{1/2} - (2\phi_{F,sub})^{1/2} \right]$$

where the first term represents the threshold voltage with zero V_{sb} applied and the second term represents the correction of the "VTO" term (using SPICE terminology) for the additional bulk charge (Q/C_{ox}) that needs to be included. Note that the threshold expression given in Figure 28.13a does include the first term from the square-bracketed expression given above. The second term represents the component of bulk charge term that must be removed from V_{th0}, that is, it must be replaced by the first term (only) to be correct (and not "double count" the Q/C_{ox} contributions). The term in front of the square-bracketed terms is the so-called body coefficient, GAMMA = $\gamma \equiv \sqrt{2q\varepsilon_{si}N_{A\text{bulk}}}/C_{ox}$ (again using SPICE notation). This threshold expression and body coefficient are a simplified form of that given in Figure 28.13a and only serve to link the first-order SPICE terminology to the discussion given in the previous section. Additionally, this form clearly points out the fact that threshold voltage shifts as a result of applying "body bias" (V_{sb}). The assumption that no current flows below threshold is discussed next, along with other parasitic substrate effects.

In reality, drain current flows for gate voltages below "threshold"; this is the so-called subthreshold region that is important both for low-power design and in terms of understanding parasitic leakage effects. Prior to strong inversion at the MOS interface that creates the channel, which in turn pins the surface potential, there is an exponential buildup of charge at the surface due primarily to the gate-induced (V_{gs}) effects. This is the expected (and desirable) subthreshold effect. There are also electrostatic effects of V_{ds} and V_{bs} which are generally of a parasitic nature in that they shift the *I–V* curves and can induce unwanted leakage which is not controlled by the primary (front) gate. Figure 28.14a gives representative expressions for the drain current dependencies on these potentials. Figure 28.14b shows a semilog plot of I_{ds} vs. V_{gs} with V_{ds} as a parameter for three channel length devices where the channel doping profiles have purposefully not been scaled to adjust for the shorter channel effects. That is, electrostatically the long-channel device is properly scaled in terms of dopant profiles; the shorter channel-length devices allow too much coupling of drain potential with the source region.

28.2.2.1 SUBTHRESHOLD AND DRAIN-INDUCED-BARRIER-LOWERING (DIBL) CURRENTS

The first set of expressions in Figure 28.14a are for the subthreshold current regime [29] and demonstrate several key features of device performance in this region of operation. The exponential dependence of drain current on V_{gs} reflects the buildup of channel charge before the surface potential becomes pinned; the current flow is a result of diffusion rather than due to the lateral drift field (i.e., Figure 28.4a) [27]. The "*m*" factor in the exponential is determined by the ratio of the substrate depletion capacitance to the oxide capacitance, which reflects the division of potential between gate and bulk regions; if the ratio is 0 (*m* = 1) this means that the gate potential is fully coupled to the surface. The impact of this *m* factor on slope of the drain current is shown in

$$I_{ds} = \mu_{eff} C_{ox} \frac{W}{L} (m-1) \left(\frac{kT}{q}\right)^2 e^{q(V_g - V_{th})/mkT} \left(1 - e^{-qV_{ds}/kT}\right)$$

$$V_g = V_{th} + m(\psi_s - 2\phi_{F,sub})$$

$$m = 1 + \frac{\sqrt{\varepsilon_{si} q N_{A_{bulk}}/4\phi_{F,sub}}}{C_{ox}} = 1 + \frac{C_{depl}}{C_{ox}}$$

$$C_{depl} \equiv \frac{\varepsilon_{si}}{t_{depl} (\psi_s = 2\phi_{F,sub})}$$

$$I_{ds} = q D_n \frac{W \cdot Z^*}{L^*} \frac{n_i^2}{N_{A_{sub}}} e^{\left(\frac{q(\psi^* - V_s)}{kT}\right)} \left[1 - e^{\left(\frac{-qV_{ds}}{kT}\right)}\right] \psi^*(x^*, y^*) = \psi^*(V_{gs}, V_{ds}, V_{sb})$$

(a)

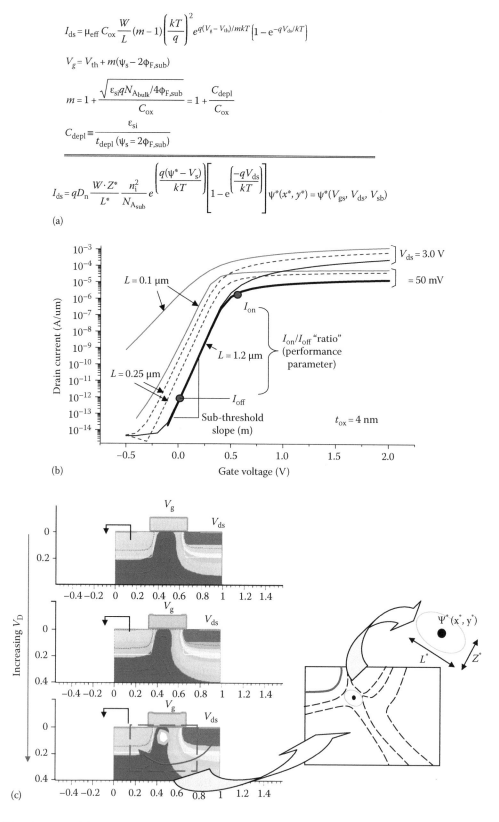

(b)

(c)

FIGURE 28.14   (a) Drain current vs. gate, source, drain-source, and bulk (barrier) potential for two regimes of sub threshold operation: (top) gate-controlled (bottom) bulk-controlled. (b) Semilog plot of drain voltage (V_{ds}) as a parameter for three channel length NMOS devices. (c) Two-dimensional plot of electrostatic potential in MOS device for three drain bias (V_{ds}) conditions. The results show typical operation as well as drain-induced barrier lowering (DIBL) for sufficiently high drain bias.

Figure 28.14b. The theoretical limit is $m = 1$ but for practical doping levels $m>1$ and increases with the square root of the doping. Careful readers may notice that the prefactor of I_D contains $(m-1)$ which makes the ideal case without sub-threshold leakage, which is nonphysical but a common problem in deriving equations by artificial separation of two operating regions. In SPICE model implementation, there will often be a fitting factor to join the equations for above threshold and sub-threshold regions with a current at the threshold voltage defined, so the $(m-1)$ prefactor does not matter even for ideal sub-threshold leakage case. This m factor determines the slope of the semilog I_{ds} vs. V_{gs} in the subthreshold region. From a circuit perspective the ratio of "on current" to "off current" is critical; this ratio is directly impacted by the m factor. This determines how large a voltage difference is required to achieve the desired on–off current ratio.

The first set of equations in Figure 28.14a does not include the drain bias effect on the threshold voltage or the sub-threshold slope, because when $V_{ds}>3kT$, it has almost no influence on the drain current. The second equation shown in Figure 28.14a represents the drain current that flows as the drain voltage overpowers the gate control—the so-called drain-induced barrier-lowering (DIBL). Notice that the most leaky path (MLP) for the sub-threshold drain current can only be determined from 2-D potential con-tours, and can be away from the surface channels depending on the doping and SOI design parameters. The form of the expression is rather similar to the previous subthreshold equation, but this apparent similarity is superficial. The DIBL expression is a bulk conduction that is controlled by the potential (Ψ^*) which occurs at a "saddle point" as indicated in Figure 28.14c. The current is controlled by the width/length ratio, Z^*/L^*, which are the TCAD-extracted parameters of the saddle point shown in the figure. The product of WZ^* (a cross-sectional area factor) in the numerator indicates that this can be a bulk conduction term in the MLP, and the difference between Ψ^* and V_s indicates the "barrier control" at the saddle point. Details of the method for extraction of these saddle-point parameters are given in [30].

The expressions given in Figure 28.14a and b represent electrostatically controlled current conditions—the first where the gate is the primary controlling factor and the second expression results from the MLP induced by the drain voltage. As stated above, the subthreshold conduction regime with strong gate control is expected and is useful in low-power design. The condition that results in DIBL-degraded sub-threshold leakage is one of several parasitic substrate conduction effects. In order to reduce this effect, the substrate doping, the HALO implant dose, or the body thickness of SOI can be scaled [31]. However, this in turn can result in parasitic conduction at the drain end of the channel. For heavily doped junctions there can be band-to-band (Zener) tunneling as well as trap-assisted tunneling due to damage introduced during the HALO implant. A simulation-based model for extraction of these tunneling currents has been proposed [32] and further refined based on empirical data from experiments using different HALO dose conditions [33]. This again provides evidence that the doping distribution details (and their statistics) are of critical importance in device design and modeling.

28.2.2.2 BAND-TO-BAND TUNNELING (ZENER) CURRENT

Figure 28.15a shows an expression for the BTBT given in [32] based on pioneering work in the 1950s by Kane and Keldysh on Zener tunneling in junctions. The rate equation (R_{btbt}) is expressed in terms of tunneling current density per unit of energy, parameterized in terms of electric field $(\vec{E})$ and energy band parameters (which are denoted with superscript "~"). The supporting equations give further details for R_{btbt} including an expression for D that is suitable for TCAD-based calculations and a modified form for E'_0 [33] that accounts for the doping dependence (that can also be extracted based on TCAD process simulations). Figure 28.15b shows schematically the MOS device, the HALO (ion implanted) region and the energy bands between the p^1 and n^1 regions where the BTBT occurs. The plot to the right in Figure 28.15b shows the experimental data and simulation results based on the formulation shown in Figure 28.15a for different HALO implantation doses [33]. There is a strong doping (dose) dependence; changing the angle of the HALO implantation also has an effect on how defects influence the leakage currents.

Band-to-band tunneling is also an important concern for drain and substrate currents in the deep sub-threshold and high V_D, where the channel region close to the drain side is driven to accumulation by the gate bias as shown in Figure 28.15c. In high V_D, this induced Zener junction can break down and carries a significant current, which has a negative transconductance usually

$$R_{btbt} - \frac{dJ_{btbt}}{d\tilde{E}}\,\vec{E}$$

$$R_{btbt} = -B\left|\vec{E}\right|^{5/2} D(\vec{E},\tilde{E},\tilde{E}_{fn},\tilde{E}_{fp})e^{\left(\frac{E'_0}{|\vec{E}|}\right)}$$

$$D(\psi,\tilde{E}_{fn},\tilde{E}_{fp})\ \frac{1}{\exp\left[(-\tilde{E}_{fp}-q\psi)/kT\right]+1} - \frac{1}{\exp\left[(-\tilde{E}_{fp}-q\psi)/kT\right]+1}$$

$$E'_0 = E_0 - a(N_A - N_0)$$

(a)

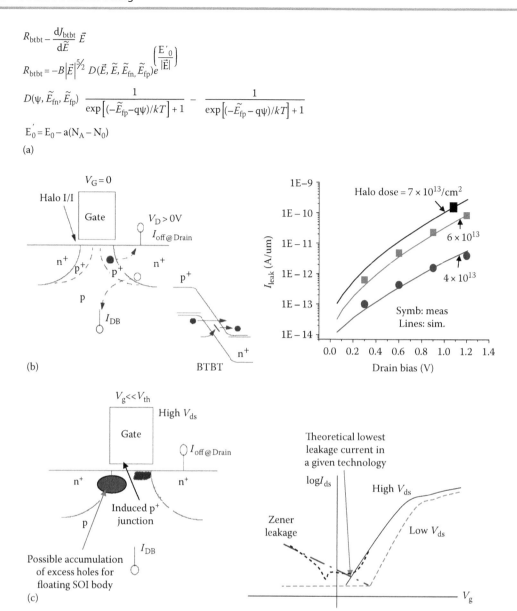

FIGURE 28.15 (a) Rate equation for band-to-band tunneling (R_{btbt}) expressed in terms of the tunneling current density per unit of energy (dJ/dE) with supporting equations for the rate equation for the rate equation [32], including a modified term for the doping dependence [33]. (b) Schematic view of the band-to-band tunneling, including trap-assisted tunneling, and plots of the modeling results (Figure 28.15a) along with experimental data for different HALO implant doses [33]. (c) Schematic view of the band-to-band tunneling in the gate-induced drain leakage (GIDL) region. The leakage current has a negative transconductance and determines the lowest possible leakage current in high V_{ds}.

called gate-induced drain leakage (GIDL). Notice that GIDL can happen even for low-doped substrates and can inject significant amounts of carriers to the floating substrate in SOI structures, similar to those created by impact ionization, until the body potential is raised high enough by the excess hole concentration. This can potentially turn on the parasitic bipolar junction transistors (BJT) in the substrate and cannot be shut off by the gate bias. Detailed models including the self-consistent body potential for modeling BTBT can again be obtained from TCAD simulation, and appropriate substrate and drain-doping engineering needs to be employed to control this parasitic effect.

Drain-induced barrier lowering, together with channel length modulation (CLM) by velocity saturation or velocity overshoot also sets the output resistance in the saturation region when

$V_{ds} > V_{dsat} \sim V_g - V_{th}$. The product of small-signal output resistance and transconductance is referred to as the self-limited gain in that operating point, which is a critical parameter in analog and RF circuit designs. In digital circuits, the self-limited gain is also important for sense amplifiers and phase-lock-loop (PLL) designs. Since DIBL and CLM are quite complicated in their multidimensional potential profile nature, detailed TCAD simulation is usually necessary to guide the design process [34].

The above discussion has considered subthreshold conduction as well as parasitic drain-to-source (DIBL) and drain junction (BTBT) leakage currents. The discussion also addressed the first-order body-bias effect in terms of threshold shift with V_{sb}. There are also 2-D effects on threshold voltage that result from the influence of source and drain potential on the depletion charge in the channel region. The next following discussion considers generally the inhomogeneous substrate depletion effects, including influence on threshold voltage and parasitic junction capacitance.

28.3 PARASITIC JUNCTION AND INHOMOGENEOUS SUBSTRATE EFFECTS

The previous discussions of threshold voltage have focused on the vertical field effects and have generally ignored the two-dimension (2-D) effects, other than the DIBL and BTBT leakage effects. From the perspective of achieving an electrostatically "well-tempered" MOS device where the vertical field dominates over lateral field effects, this is the desirable condition. However, as scaling continues for bulk CMOS there are an increasing number of doping profiles—HALO, SDE, etc.—that influence the 2-D electrostatics. Figure 28.3b shows a fairly realistic cross-section of such a scaled device (i.e., 90 nm technology node). In this section, we will briefly look at some of the other substrate effects that impact circuit performance.

Figure 28.16 considers that basic NMOS device, now showing electrostatic potential contours under that gate and junction regions (broken lines). Also shown in Figure 28.16 are several trapezoidal and triangular regions that schematically indicate regions of influence in terms of electrostatic potential and dopant distributions. The trapezoid directly under the gate represents the bulk charge dominantly controlled by the gate—the term Q_{bulk}/C_{ox} in the threshold expression. The two neighboring triangles (Q_{bulk-S} and Q_{bulk-D}) represent the portions of bulk charge that are strongly influenced by source and drain potentials. That is, electrostatically, as these potentials change they have a greater influence on the bulk charge than does the gate potential, resulting in 2-D effects on V_{th}. Also shown in Figure 28.16a are the sidewall (SW) contributions of charge of the source and drain junctions. The portion of charge within the L_{eff} region but not controlled by the gate bias has been used as the initial estimation of the short-channel effect (SCE) [27]. It can be readily seen that the source/drain charge effect has more influence in short-channel devices than in long-channel devices. While these charges may only weakly influence the threshold voltage in the "well-tempered" device design, they do have a major impact on SW capacitance as discussed below. We will not specifically consider how these 2-D bulk charge effects impact threshold voltage or the circuit-level models. Suffice it to say that TCAD simulations and electrostatic results such as those shown in Figure 28.16a are the basis for developing appropriate threshold expressions that account for these 2-D effects.

As noted above, the source-bulk and drain-bulk junctions exhibit 2-D effects as a result of various doping and other isolation-related effects along the edges. Figure 28.16a shows one example of these edge-effects resulting from the HALO implant doping profiles. Recently, technologists have begun using "strain engineering" to improve mobility in transistors [35]. There are several ways to induce this strain, for example using epitaxial layers with silicon-germanium (SiGe) or alternatively localized (i.e., ion-implanted) material changes. These material changes have localized effects on the 2-D junction parameters both in terms of dielectric constants and energy band parameters. Figure 28.16b shows the results of one set of comparisons for junction capacitance in an epitaxial SiGe process. Results of 2-D (MEDICI) device-simulated drain junction capacitance are compared with parameterized compact (circuit) models for the area and SW components of the capacitance, including different components for the SiGe and strained-silicon (S-Si) regions. Such parameterization of the various components is essential for accurate circuit

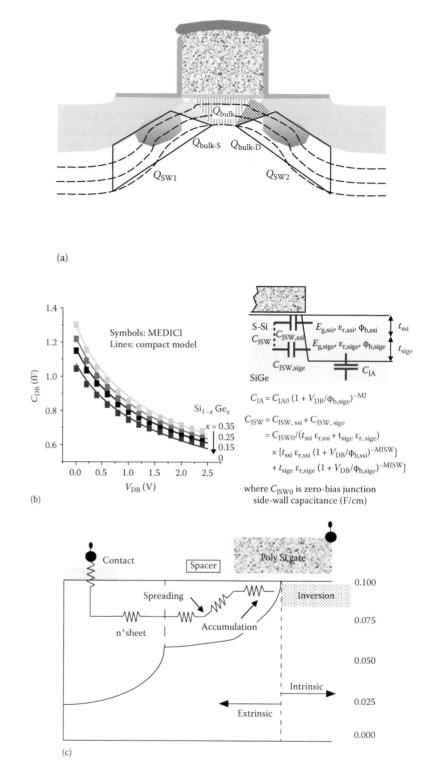

FIGURE 28.16   (a) Cross-section on NMOS (same as Figure 28.3b) showing electrostatic potential contours (broken lines) and trapezoids and triangles of charge, representing influence of gate, source and drain potential as well as side-wall (SW) effects due to doping. (b) Junction capacitance (CDB) as a function of bias (VDB) for junctions with varying degress of strain, induced by different germanium fraction (x). The components of capacitance area (CJA) and side-wall (CJSW) are indicated, along with compact model forms. (c) Cross-section of contact region (source end) of a scaled NMOS, with emphasis on the several components of resistance: SDE region, spreading resistance, bulk (n1) sheet resistance and contact resistance.

modeling; TCAD simulations provide an excellent means to perform the necessary extraction. From a technology point of view the capacitance increases with increased germanium fraction (x), leading to a design trade-off for this particular example between increased drive current (due to higher mobility) and increased capacitance as indicated in Figure 28.16b. Part of the drain junction capacitance will be reflected also as the capacitance between drain and gate, especially when $V_{ds}<V_{dsat}$. The increased C_{dg} is a serious concern in RF and analog CMOS circuit design, and has to be accounted for carefully in determining technology alternatives.

There are other substrate-related parasitic effects that influence bulk CMOS performance including edge-effects in the isolation (i.e., see STI region in Figure 28.3a), source and drain resistance (i.e., the SDE and extrinsic regions in Figure 28.3b), and bulk resistance of the substrate. Some of these issues are considered in Chapter 24 from the technology perspective. The device implications range from leakage current concerns for the isolation to system-on-chip (SoC) noise coupling resulting from substrate coupling as is discussed in Chapter 21. Parasitic source and drain resistance deserves some further discussion since it has major implications on limits of scaling and also issues related to statistics of the devices, namely, the potential drop across the source-side resistance directly subtracts from the gate over-drive that in turn reduces the drive current.

Figure 28.16c shows in detail the various components that contribute to the parasitic source resistance. These include the SDE "link up" (or overlap) region between the channel and extrinsic region, spreading resistance that results as the current goes from the confinement of the SDE into the deeper junction region, and finally the bulk junction and contact regions. As junctions become shallower, consistent with the scaling laws, the sheet resistance of these regions also increases and it becomes more difficult to fully activate the high-concentration doping profiles (see Chapter 24). In addition, the SDE "tip regions" are influenced by surface effects and the lateral diffusion and abruptness of this profile is critical in controlling devices performance—both statistically and in terms of the "on–off ratio." The determination of the parameters associated with source/drain parasitic resistance are very important; Taur has developed the "shift-and-ratio" method [27] which is broadly used in this extraction process. There are also many details related to the profile scaling that can be obtained using Taur's approach in combination with TCAD-based modeling [36].

At this point it is useful to specifically point out the importance of process modeling and its relationship to statistical effects (see Chapters 19 and 24). Namely, in the regime of sub-100 nm scaling, the issues of dopant activation and lateral diffusion are critical. That is, they increasingly tighten the constraints on thermal budget for diffusion (the product of diffusivity and time in the Gaussian solution of a diffusion equation—"Dt") which in turn directly affects junction depth. The shrinking device dimension also results in smaller volumes for a given dopant density, which in turn bring statistics into play in the modeling. This is also a good vantage point to return to the "closure" between system-level and technology-driven perspectives of device scaling as initially introduced in discussing Figure 28.1b. Namely, many of the system-level care-about issues such as leakage current, parasitic R/C effects, and generally I_{on}/I_{off} ratio come down to the complete array of process technology and device scaling issues representative of the cross-section shown in Figure 28.16c. These dopant profile and dopant activation issues, constrained by physical constants such as dopant solid solubility and other parameters of dopant kinetics, directly impact the electrical device properties, including their statistical distributions.

In this section and the previous one on intrinsic MOS device effects, we have considered the use of TCAD to understand parameter dependencies and the relationship between these physical effects and models that bridge to the circuit world. Figure 28.17 summarizes schematically the set of physical modeling effects considered to this point and how they relate to the world of compact modeling. We have purposefully used the NMOS device as the primary vehicle and have emphasized bulk CMOS technology. While there have been limited discussions of technology shifts to SOI and the use of other materials such as SiGe to induce S-Si, we have primarily discussed the ways to address known "roadblocks" that are becoming performance limitations with bulk CMOS. The next section presents a slightly broader view of other device technology concerns and points the reader toward supplemental reading that can go beyond this very cursory discussion.

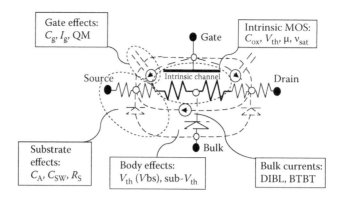

FIGURE 28.17   Summary of modeling considerations presented in Sections 28.2 and 28.3. Intrinsic MOS, gate, body/bulk effects (Section 28.2) and parasitic substrate capacitance/resistance effects (Section 28.3).

28.4 DEVICE TECHNOLOGY ALTERNATIVES

The above discussion has emphasized CMOS technology that is indeed the engine of the integrated electronics economy and will continue to be so for the indefinite future. The coordinated scaling of analog, digital and a variety of memory technologies using the same basic infrastructure is the driving force. Over the years there have been a variety of contenders seeking to change this "all MOS" landscape. As noted in the introduction, BJT preceded MOS and for some applications, especially in mixed-signals, it still has competitive advantages. For about a decade ca. the 1980s bulk BiCMOS (BJT devices integrated with CMOS) had been predicted as mainstream; however, due to limitations of power, "on voltage" and generally issues related to scalability, "all CMOS" regained its dominant position.

In certain analog areas there have been alternative devices that held out promise based on unique performance, for example the charge-coupled device (CCD) technology used for imaging. The CCD while generally compatible with bulk CMOS is finding itself squeezed for market space by the growing prevalence of CMOS-based digital cameras, due largely to the digitally dominated infrastructure for information processing. This digital signal processing (DSP) approach also has impacted wireless applications as well as data conversion and other interface circuits. What this means is that chips are dominantly digital with add-on analog or other interface capabilities only to the degree that they can be justified *economically* instead of just technically [37]. This economics-driven perspective (which is in fact the true message of Moore's law) also poses a very real challenge for SoC vs. system-in-package (SiP). The following few sub-sections will consider device options that have had some success over the past decade in terms of integration with mainstream CMOS technology into single-chip solutions, primarily from the perspective of physical modeling.

28.4.1 SILICON-ON-INSULATOR TECHNOLOGY

There is growing interest in 3-D MOS structures including SOI-like devices [26]. A DG SOI device example was discussed above (Figure 28.12), primarily to illustrate the severity of scaling limits when using poly-silicon gates. The major advantages of SOI technology come from the elimination of the underlying bulk silicon layers, associated dopant distributions needed to avoid parasitic currents, and capacitive effects. For example, having an insulator rather than silicon bulk region dramatically alters—mainly reduces—junction and body effects for the MOS devices. Additionally, there are advantages to DG SOI structures, including more complete surrounding of the channel region, that are manifest in increased gate control and greater effective channel charge. This in turn can give tighter SCE control for V_{th} parametric yield with a given gate oxide thickness. The infrastructure for SOI technology is generally compatible with the mainstream. For applications where the cost is warranted in return for performance advantages, this technology will continue to be used. An excellent discussion of the technology and scaling issues for SOI can be found in [38].

In spite of the benefits and promise of SOI-like technologies, including commercial availability, there are also some limitations. For example, the so-called FD SOI technology, which embraces a range of device topologies [26], relies on the silicon body thickness and gate work function to set threshold voltages in contrast to the bulk and gate doping levels traditionally introduced by ion implantation. This shifts the design burden of scaling, which was dominated by selective introduction of dopants into the substrate using ion implantation, to other process control issues such as deposition or growth of thin layers and control of gate work functions. Additionally, heat generation and its removal is of growing concern, not only for SOI but even for bulk CMOS; thermal management has become a fundamental limitation to scaling across the landscape of IC technologies [39]. This growing concern with thermal management, especially for SOI technologies, also leads to new metrics to be considered in technology design; one example related to use of germanium on insulator (GOI) will be considered later in this section.

28.4.2 SILICON-GERMANIUM HETEROJUNCTION BIPOLAR TRANSISTOR TECHNOLOGY

In spite of previous scaling challenges encountered by bulk BiCMOS, the introduction of SiGe bipolar transistors into the same technology with bulk CMOS has found an economic window, particularly in the telecommunications sector, both wireline and wireless applications. In contrast to earlier BiCMOS technologies that tried to exploit diffused junctions to create bipolar transistors within the fabric of a CMOS technology, the heterojunction bipolar transistor (HBT) technology uses selective epitaxial addition of SiGe layers that allow a more loosely coupled process flow. Details of the process flow and device design trade-offs can be found elsewhere [40]. This discussion will briefly contrast the physical modeling issues for bipolar vis à vis MOS devices.

In contrast to the MOS devices that are voltage-controlled and resistive in nature, bipolar transistors are current-controlled so that current gain ($\beta \equiv I_C/I_B$) is a critical parameter. With increased interest in low-power design, the use of the subthreshold regime with MOS devices results in an exponential dependence of I_d on V_{gs} with subthreshold slope (semilog plot) given by the "m factor" as discussed above. This m factor, similar in concept to the ideality factor in bipolar designs, is greater than unity due to the voltage sharing between the gate capacitance and bulk depletion capacitance as discussed above. The bipolar transistor does not suffer from this limitation so that the transconductance is simply $g_m = I_C/V_{thermal}$, where $V_{thermal} = kT/q$. Hence, the bipolar transistor offers a larger transconductance compared to the MOS device for a given current level and, for circuit configurations with low impedance and capacitance, the bipolar transistors can deliver excellent high-frequency performance. Due to the many telecommunications applications for such high-frequency devices, there have been intense scaling efforts to reduce unwanted parasitics in the HBT technology, particularly junction capacitance and resistance in the base and collector leads.

It is important to emphasize two fundamental differences between bipolar and MOS devices, besides those mentioned above. First, the bipolar transistor conducts current vertically in bulk material, under control of the base-emitter voltage. Hence, the current density tends to be uniformly spread over the emitter contact, if we for the moment ignore current crowding due to base resistance, compared to spreading resistance effects in the MOS SDE region, leading into the channel region. The second point relates to the current control vs. voltage control nature of the bipolar device. Namely, the flow of base current is necessary for the HBT operation, yet at the same time the controlling base current is an unavoidable consequence of bipolar current flow in junction diodes and it is strongly process-dependent. Interestingly enough, with the ongoing scaling of MOS devices the importance of gate current has been discussed above. Nevertheless, this gate current is truly a parasitic effect whereas for the HBT the flow of base current is unavoidable and required for bipolar transistor operation. Alternatively, the control base current can be isolated using a BiCMOS configuration, but this will inevitably reduce the layout efficiency and increase the technology difficulty.

Circuit design techniques that fully leverage the capabilities of bipolar transistors are well established. In the golden era of LSI, bipolar devices were used for both digital as well as analog design [41,42]. With ongoing scaling, the bipolar technology has been almost completely replaced by MOS in digital applications, especially due to power limitations. One important limitation of the bipolar device is that in order to have it turned on, the base-emitter voltage must be sufficiently large (i.e., $V_{BE}(on) = kT/q\ln(I_C/I_S)$, where I_S represents the saturation current parameter). By contrast, the MOS threshold voltage can be much smaller than this; the primary limitation becomes the I_{on}/I_{off} ratio as mentioned earlier. With the ongoing reduction of the supply voltage, the bipolar "on voltage" poses a nearly insurmountable and fundamental obstacle for low-voltage design with bipolar technology.

28.4.3 FUTURE TECHNOLOGY TRENDS

The next few generations of technology scaling pose a range of challenges as well as opportunities. This chapter has focused primarily on bulk CMOS as the ongoing mainstream, with previews of scaling issues that will come into play for SOI type devices. The ongoing challenges related to leakage currents—both in the substrate and through the gate—bring forward the need for rethinking the materials and device architectures used for future technologies. As pointed out in earlier discussions, going to SOI and FD, DG structures shifts the challenges of substrate doping to thickness control and further concerns about heat generation and removal. The shift to SOI will continue to be evolutionary and driven by specific application requirements—speed, reduced substrate coupling (i.e., radiation effects) or other performance related issues. By contrast to evolution toward SOI-like structures, the re-engineering of the gate-stack materials is an imminent concern.

The challenges of increased gate leakage and limitations to further scale of threshold voltages using polycrystalline silicon gates have brought the issues of alternative gate-stack materials into the foreground. The two related changes—gate dielectric and gate metal—have a profound impact on scaling and device performance. The introduction of materials with increased dielectric constant (so-called "high-K" materials) offers the possibility to increase gate capacitance (and the vertical electric field) while using thicker layers compared to the ~1.5–2 nm oxide limit discussed above. This increased dielectric capacitance improves the sub-threshold slope ("m factor") and also can, depending on processing conditions, reduce the gate tunneling current. The changes required to a metal gate, replacing the classic silicon gate technology, are mandated to achieve desired threshold voltages and to overcome gate depletion effects as discussed earlier. The complexity of these changes—moving away from the traditional silicon-dioxide and poly-silicon materials—represents one of the potentially most "disruptive" technology changes over the most recent generations of scaling. This is an area of intense ongoing technology development; one interesting benchmark study is presented from the perspective of a foundry [43].

There are a variety of more radical, speculative technologies that are on the horizon, each seeking to find a niche in providing value. The elegance and genius of the monolithic silicon scaling that has supported Moore's law has been the ability to pattern layers at ever smaller dimensions and to integrate new materials while maintaining backward compatibility with devices and circuit fabrics from earlier technology nodes. It is beyond the scope of this chapter and premature to speculate on the impact of nano-technology options, ranging from carbon nano-tubes (CNT) and nano-wires to self-assembly of layers of other materials and functional capabilities. New "fabrics" for memory and sensor technologies seem to be the most promising opportunities for nano-technology. In the context of the well-established monolithic process, the overarching challenge is how to structure these new layers to be aligned with underlying layers. This challenge equally applies to alternatives that fall into the class of new 3-D IC options.

As discussed briefly in considering SOI, various alternative structures, including geometry innovations and new semiconductor materials such as germanium, are now being seriously considered. Germanium has already been mentioned in the context of SiGe HBT technology; its benefits as an elemental semiconductor in field-effect transistor (FET) technology are also being reconsidered. Namely, germanium preceded silicon as a leading contender for making junction transistors in the 1940s and 1950s. However, the stability of silicon dioxide compared to water-soluble germanium oxide was a leading factor that secured the future for silicon. At the same time, the higher mobility for both holes and electrons in germanium, compared to silicon, has continued to be a very attractive feature. Moreover, there has been impressive progress in the fabrication of nearly single-crystal

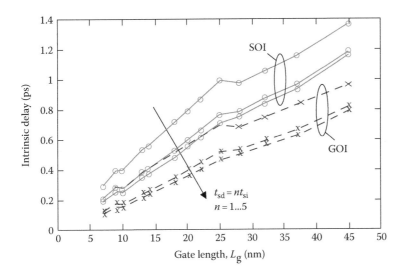

FIGURE 28.18   Self-consistent electro-thermal computational result of intrinsic delay for SOI and GOI devices. The S/D extension region thickness t_{sd} is varied, ranging from 1X to 5X the t_{si} thickness.

germanium-on-insulator (GOI) devices at low processing temperatures. There are still a host of technology challenges including the ongoing issues of gate-stack engineering as well as dopant properties to mention only two. Nonetheless, the ongoing push in scaling and increased search for new materials with improved electrical performance, have considerably broadened the playing field for innovations compared to a decade ago. An interesting point to emphasize about this quest for new materials and revisiting materials such as germanium is the fact that substantial data about the properties, including their process dependencies and statistical variations, need to be gathered in order to validate their suitability in production technologies.

To illustrate the point about how these new materials and reconsideration of device design trade-offs can lead in interesting directions, a recent simulation-based study has compared SOI and GOI devices in terms of performance [44]. This study has leveraged a detailed thermal modeling of performance, coupled with the electrical analysis. Figure 28.18 shows a plot of gate delay for the two technologies, parameterized in terms of thickness of the S/D extension regions. For the GOI device it is assumed the germanium layer thickness is $t_{Ge} = 3/4 t_{Si}$, where $t_{Si} = L_g/4$ with L_g being the device gate length. While the silicon has higher thermal conductivity, there is greater spatial falloff due to larger phonon mean free path. One key reason for the improved GOI performance compared to the SOI case, is the fact that germanium has a two-fold mobility advantage over silicon, allowing a 40% lower V_{dd} to be used and thus also reducing the power dissipation. The parameterized sets of curves show performance as a function of increasing the thickness of the raised S/D extension regions; it is shown that beyond, ~$3t_{film}$ increase provide no additional benefit in terms of increased speed. These results suggest that further scaling of SOI, GOI, and possibly other thin-film semiconductor materials will need to understand carefully the electro-thermal trade-offs. In a sense, this kind of detailed physics-based study is the prelude to the ongoing challenges of 3-D integration of active devices.

28.5 CONCLUSIONS

This chapter concludes with comments very similar to those with which it began; physics-based modeling of devices is an essential part of the development process for IC electronics. By means of examples for bulk CMOS we have attempted to quantify much of the underlying device and technology understanding, and abstract that knowledge to levels that link with circuit design and compact modeling. Although the emphasis of the chapter has been bulk MOS transistors—the workhorse of the IC industry—there have been suitable previews of evolution toward SOI-like technologies as well as limited consideration of future technology trends. Even more so than over the past three decades, future technologies will exploit new materials in concert with detailed physical understanding and modeling. As device dimensions continue to shrink and packets of charge and current reach fundamental and quantized limits, the importance of these physics-based models is critical. Hence, many of the examples used in this chapter have explored these limits and demonstrated their impact on scaling—quantized channel/gate charge, gate tunneling current and thermal limits for SOI/GOI are illustrative. Future scaling will continue to be leveraged by such modeling that will in turn provide better devices and models for design.

REFERENCES

1. H.J. DeMan and R. Mertens, SITCAP—A simulator for bipolar transistors for computer-aided circuit analysis programs, *International Solid-State Circuits Conference (ISSCC), Technical Digest,* Philadelphia, PA, February, 1973, pp. 104–105.
2. R.W. Dutton and D.A. Antoniadis, Process simulation for device design and control, *International Solid-State Circuits Conference (ISSCC), Technical Digest,* Philadelphia, PA, February, 1979, pp. 244–245.
3. R.H. Dennard, F.H. Gaensslen, H.N. Yu, V.L. Rodeout, E. Bassous, and A.R. LeBlanc, Design of ion-implanted MOSFETs with very small physical dimensions, *IEEE J. Solid State Circ.,* SC-9, 256–268, 1974.
4. R.W. Dutton and S.E. Hansen, Process modeling of integrated circuit device technology, *Proc. IEEE,* 69, 1305–1320, 1981.
5. P.E. Cottrell and E.M. Buturla, Two-dimensional static and transient simulation of mobile carrier transport in a semiconductor, *Proceedings NASECODE I (Numerical Analysis of Semiconductor Devices),* Boole Press, Dublin, OH, 1979, pp. 31–64.
6. S. Selberherr, W. Fichtner, and H.W. Potzl, Miminos—A program package to facilitate MOS device design and analysis, *Proceedings NASECODE I (Numerical Analysis of Semiconductor Devices),* Boole Press, Ireland, 1979, pp. 275–279.
7. C.S. Rafferty, M.R. Pinto, and R.W. Dutton, Iterative methods in semiconductor device simulation, *IEEE Trans. Electr. Dev.,* ED-32 (10), 2018–2027, 1985.
8. M.R. Pinto and R.W. Dutton, Accurate trigger condition analysis for CMOS latchup, *IEEE Electr. Dev. Lett.,* EDL-6(2), 100–102, 1985.
9. R.W. Dutton, Modeling and simulation for VLSI, *International Electron Devices Meeting (IEDM), Technical Digest,* Los Angeles, CA, December 1986, pp. 2–7.
10. K.M. Cham, S.-Y. Oh, D. Chin, and J.L. Moll, *Computer-Aided Design and VLSI Device Development,* Kluwer, Dordrecht, the Netherlands, 1986.
11. R.W. Dutton and A.J. Strojwas, Perspectives on technology and technology-driven CAD, *IEEE Trans. CAD-ICAS,* 19(12), 1544–1560, 2000.
12. International Technology Roadmap for Semiconductors (available from http://public.itrs.net) ITRS, 2003.
13. C. Duvvury and A. Amerasekera, ESD: A pervasive reliability concern for IC technologies, *Proc. IEEE,* 81, 690–702, 1993.
14. A. Amerasekera and C. Duvvury, *ESD in Silicon Integrated Circuits,* 2nd edn., Wiley, New York, 2002.
15. S. Dabral and T.J. Maloney, *Basic ESD and I/O Design,* Wiley, New York, 1998.
16. C. Lombardi, S. Manzini, A. Saporito, and M. Vanzi, A physically based mobility model for numerical simulation of nonplanar devices, *IEEE Trans. CAD,* 7(11), 1164–1171, 1988.
17. H. Shin, A.F. Tasch, C.M. Mazier and S.K. Banerjee, A new approach to verify and derive a transverse-field-dependent mobility model for electrons in MOS inversion layers, *IEEE Trans. Electr. Dev.,* TED-36(6), 1117–1124, 1989.
18. H. Shin, G.M. Yeric, A.F. Tasch and C.M. Mazier, Physcially-based models for effective mobility and local-field mobility of electrons in MOS inversion layers, *Solid State Electr.,* 34(6), 545–552, 1991.
19. M.N. Darwish, J.L. Lentz, M.R. Pinto, P.M. Zeitzoff, T.J. Krutsick, and J.J. Vuong, An improved electron and hole mobility model for general purpose device simulation, *IEEE Trans. Electr. Dev.,* 44(9), 1529–1538, September 1997.
20. S. Takagi, M. Iwase, and A. Toriumi, On universality of inversion-layer mobility in *n*- and *p*-channel MOSFETs, *International Electron Devices Meeting (IEDM), Technical Digest,* San Francisco, CA, December, 1988, pp. 398–401.
21. C.-H. Choi, J.-S. Goo, T.-Y. Oh, Z. Yu, R.W. Dutton, A. Bayoumi, M. Cao, P. Vande Voorde, D. Vook, and C.H. Diaz, MOS *C–V* characterization of ultrathin gate oxide thickness (1.3–1.8 nm), *IEEE Elec. Dev. Lett.,* EDL-20(6), 292–294, 1999.
22. C. Bowen, C.L. Fernando, G. Klimeck, A. Chatterjee, D. Blanks, R. Lake, J. Hu, J. Davis, M. Kulkarni, S. Hattangady, and I.-C. Chen, Physical oxide thickness extraction and verification using quantum mechanical simulation, *International Electron Devices Meeting (IEDM), Technical Digest,* Washington, DC, December, 1997, pp. 869–872.

23. C.-H. Choi, K.-Y. Nam, Z. Yu, and R.W. Dutton, Impact of gate direct tunneling current on circuit performance: a simulation study, *IEEE Trans. Electr. Dev.*, ED-48 (12), 2001.
24. C.H. Choi, Z. Yu, and R.W. Dutton, Resonant gate tunneling current in double-gate SOI: a simulation study, *IEEE Trans. Electr. Dev.*, 50(12), 2579–2581, 2003.
25. C.-H. Choi, Z. Yu, and R.W. Dutton, Two-dimensional polysilicon quantum-mechanical effects in double-gate SOI, *International Electron Devices Meeting (IEDM), Technical Digest*, San Francisco, CA, December, 2002, pp. 723–726.
26. H.-S.P. Wong, D.J. Frank, P.M. Solomon, H.-J. Wann, and J. Welser, Nanoscale CMOS, *Proc. IEEE*, 87, 537–570, 1999.
27. Y. Taur and T.H. Ning, *Fundamentals of Modern VLSI Devices*, Cambridge University Press, Cambridge, U.K., 1998.
28. J.B. Kuo and K.-W. Su, *CMOS VLSI Engineering Silicon-on-Insulator (SOI)*, Kluwer, Boston, MA, 1998.
29. R.M. Swanson and J.D. Meindl, Ion-implanted complementary MOS transistors in low-voltage circuits, *IEEE J. Solid State Circ.*, SC-7, 146, 1972.
30. J.A. Greenfield and R.W. Dutton, Nonplanar VLSI device analysis using the solution of Poisson's equation, *IEEE Trans. Electr. Dev.*, ED-27(8), 1520–1532, 1980.
31. R.-H. Yan, A. Ourmazd, and K.F. Lee, Scaling the Si MOSFET: From bulk to SOI to bulk, *IEEE Trans. Electr. Dev.*, 39(7), 1704–1710, 1992.
32. G.A.M. Hurkx, D.B.M. Klaassen, and M.P.G. Knuvers, A new recombination model for device simulation including tunneling, *IEEE Trans. Electr. Dev.*, 39(2), 331–338, 1992.
33. C.-H. Choi, S.-H. Yang, G. Pollack, S. Ekbote, P.R. Chidambaram, S. Johnson, C. Machala, and R.W. Dutton, Characterization of Zener-tunneling drain leakage current in high-dose halo implants, *2003 IEEE International Conference on Simulation of Semiconductor Processes and Devices (SISPAD), Technical Digest*, Boston, MA, September 2003, pp. 133–136.
34. R.W. Dutton, B. Troyanovsky, Z. Yu, E.C. Kan, K. Wang, and T. Chen, Advance analog circuit modeling with virtual device and instrument, *Proceedings of the ISSCC*, San Francisco, CA, February 1996, p. 78.
35. J.L. Hoyt, H.M. Nayfeh, S. Eguchi, I. Aberg, G. Xia, T. Drake, E.A. Fitzgerald, and D.A. Antoniadis, Strained silicon MOSFET technology, *International Electron Devices Meeting (IEDM), Technical Digest*, San Francisco, CA, December 2002, pp. 23–26.
36. M.Y. Kwong, R. Kasnavi, P. Griffin, J.D. Plummer, and R.W. Dutton, Impact of lateral source/drain abruptness on device performance, *IEEE Trans. Electr. Dev.*, 49(11), 1882–1890, 2002.
37. D. Buss, B.L. Evans, J. Bellay, W. Krenik, B. Haroun, D. Leipold, K. Maggio, J.-Y. Yang, and T. Moise, SOC CMOS technology for personal Internet products, *IEEE Trans. Electr. Dev.*, 50(3), 546–556, 2003.
38. J.P. Colinge, *Silicon-on-Insulator Technology*, 3rd ed., Kluwer, Dordrecht, the Netherlands, 2004.
39. E. Pop, R. Dutton, and K. Goodson, Thermal analysis of ultra-thin body device scaling, *International Electron Devices Meeting (IEDM), Technical Digest*, San Francisco, CA, December 2003, pp. 883–886.
40. G. Freeman, B. Jagannathan, S.-J. Jeng, J.-S. Rieh, A.D. Stricker, D.C. Ahlgren, and S. Subbanna, Transistor design and application considerations for .200-GHz SiGe HBTs, *IEEE Trans. Electr. Dev.*, 50(3), 645–655, 2003.
41. P.R. Gray, P.J. Hurst, S.H. Lewis, and R.G. Meyer, *Analysis and Design of Analog Integrated Circuits*, 4th edn., Wiley, New York, 2001.
42. M.I. Elmasry, *Digital Bipolar Integrated Circuits*, Wiley Interscience, New York, 1983.
43. H.C.-H. Wang, S.-J. Chen, M.-F. Wang, P.-Y. Tsai, C.-W. Tsai, T.-W. Wang, S.M. Ting et al., Low power device technology with SiGe channel, HfSiON, and poly-Si gate, *International Electron Devices Meeting (IEDM), Technical Digest*, San Francisco, CA, 2004.
44. E. Pop, C.O. Chui, S. Sinha, R. Dutton, and K. Goodson, Electro-thermal performance comparison and optimization of thin-body SOI and GOI MOSFETs, *International Electron Devices Meeting (IEDM), Technical Digest*, San Francisco, CA, December, 2004.

High-Accuracy Parasitic Extraction

Mattan Kamon and Ralph Iverson

CONTENTS

Abstract

In this chapter, we describe high-accuracy parasitic extraction by both fast integral equation methods and random-walk-based methods. For any extraction application, the appropriate simplification of Maxwell's equations must be chosen and then solved to generate an appropriate model suitable for circuit simulation. We will briefly describe these various domains of electromagnetics and then discuss integral equation approaches as they are used for capacitance, inductance, and full-wave analysis. We conclude the chapter with a discussion of random-walk methods as used for capacitance extraction.

29.1 INTRODUCTION

The extraction of parasitic circuit models is important for various aspects of physical verification such as timing, signal noise, substrate noise, and power grid analysis. As circuit speeds and densities have increased, the need to account accurately for parasitic effects for larger and more complicated interconnect structures has grown, and so has the complexity of relevant electromagnetic phenomena, from resistance and capacitance (RC) to effects that also involve inductance (RL or RLC) and now even a full electromagnetic wave propagation. This increase in complexity has also grown for the analysis of passive devices such as integrated inductors.

Electromagnetic behavior is governed by Maxwell's equations, and all methods for parasitic extraction require solving some form of Maxwell's equations. That form may be a simple analytic parallel-plate capacitance equation or may involve a full numerical solution for a complicated 3D geometry with wave propagation. Analytic formulas for simple or simplified geometry can be used where accuracy is less important than speed, but when the geometric configuration is not simple and accuracy demands do not allow simplification, a numerical solution of the appropriate form of Maxwell's equations must be employed (see Chapter 25, for details).

Maxwell's equations of the appropriate form are typically solved by one of two classes of methods. The first uses a differential form of the governing equations and requires the discretization (meshing) of the entire domain in which the electromagnetic fields reside. Two of the most common approaches in this first class are the finite difference (FD) and finite-element method (FEM). The resultant linear algebraic system (matrix) that must be solved is large but *sparse* (contains very few nonzero entries). Methods such as sparse factorization, conjugate-gradient, or multigrid methods can be used to solve these sparse linear systems, the best of which require $O(N)$ time and memory, where N is the size of the linear system. However, most problems in electronic design automation (EDA) are *open* problems, also called exterior problems, and since the fields decrease slowly toward infinity, these methods can require extremely large N.

The second class includes integral equation methods that instead require a discretization of only the *sources* of electromagnetic fields. Integral equation methods are called boundary-element methods (BEMs) when the sources exist only on 2D surfaces in 3D problems. Those sources can be physical quantities, such as the surface charge density for the capacitance problem, or mathematical abstractions resulting from the application of Green's theorem. For open problems, the sources of the field exist in a much smaller domain than the fields themselves, and thus the linear systems generated by integral equations methods are much smaller than those in FD and FEM, as illustrated for a small portion of two signal lines in Figure 29.1.

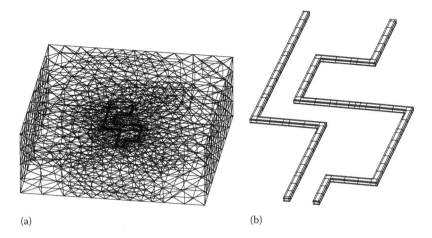

FIGURE 29.1   (a) The finite-element method mesh discretizes the space (infinite) surrounding the conductors, while (b) the boundary-element method mesh only discretizes the conductor surfaces.

Integral equation methods, however, generate *dense* (all entries are nonzero) linear systems, making such methods preferable to FD or FEM only for small problems. Dense linear systems require $O(N^2)$ memory to store and $O(N^3)$ to solve via direct Gaussian elimination or at best $O(N^2)$ to solve via an iterative method. Such growth rates are prohibitive for more than a small section of interconnect.

Much work has gone into improving both the differential and integral equation approaches, as well as random-walk methods [1]. Methods of truncating the discretization required by the FD and FEM approaches have greatly reduced the number of elements required [2,3]. Integral equation approaches have become particularly popular for interconnect extraction due to *sparsification* techniques, also sometimes called matrix compression, acceleration, or matrix-free techniques, which have brought nearly $O(N)$ growth in storage and solution time to integral equation methods [4–10].

This chapter focuses on the sparsified integral equation approaches and the random-walk methods. The sparsified integral equation techniques are typically used to solve capacitance and inductance extraction problems. The random-walk methods have become quite mature for capacitance extraction. For problems requiring the solution of the full Maxwell's equations (full wave), both differential and integral equation approaches are common.

Part I of this chapter discusses the sparsification methods in relation to the various integral forms of Maxwell's equations used in EDA. Part II discusses the random-walk approach and compares it to one of the deterministic approaches of Part I.

PART I: EXTRACTION VIA FAST INTEGRAL EQUATION METHODS

29.2 INTRODUCTION

In this part, we describe the popular detailed parasitic extraction approaches by integral equation solution of the governing electromagnetics. For any extraction application, the appropriate simplification of Maxwell's equations must be chosen and then solved to generate an appropriate model suitable for circuit simulation. We will briefly describe these various domains of electromagnetics. *Sparsified* integral equation approaches have become quite popular for their solution. We will briefly describe the sparsification techniques in the context of each of the relevant electromagnetic domains, as well as the methods of generating models appropriate for circuit simulation.

In the next section, we explore the various forms of Maxwell's equations, and then in Section 29.4, we describe sparsification methods for capacitance extraction. These fast methods can also be applied to inductance extraction as we see in Section 29.5, as well as the distributed RLC and full-wave solutions in Sections 29.6.1 and 29.6.2. We end with Section 29.7 on recent research in *variation-aware* extraction to predict the distribution of extracted parameters in the presence of random variation.

29.3 FORMS OF MAXWELL'S EQUATIONS

The term "field solver" commonly refers to any tool that numerically solves some form of Maxwell's equations. Most problems related to EDA do not require solving the full set of Maxwell's equations. In this section, we describe the simplified forms of Maxwell's equations for various applications in extraction. Table 29.1 summarizes the rest of this section.

While any parasitic behavior is described by Maxwell's equations, it would be far too inefficient to invoke the field solver many times during higher-level circuit or behavioral simulation. Instead, the simulator typically takes some responsibility for the electromagnetic behavior by assuming some network topology of parasitic elements, and then the field solver is used to *extract* parameters for the elements of the topology.

For example, parasitic extraction for timing analysis with negligible inductance would require solution of the electroquasistatic (distributed RC) form of Maxwell's equations. However, field solvers are not typically used for the electroquasistatic problem. Instead, as described in Chapter 25, static resistance and static capacitance are computed separately on small sections of interconnect and the higher-level simulator is responsible for the dynamic RC behavior. The simulator accomplishes this by connecting the R and C together in an appropriate ladder network. This permits the use of a field solver that solves the much simpler electrostatic (C) problem. The electrostatic problem is discussed in Section 29.4.

Delegation of the dynamics to the simulator for RC extraction is only possible because an accurate circuit topology for the interconnect (a ladder network) is available *a priori*. In contrast, consider analyzing the dynamic behavior of an SRAM cell that is very sensitive to the amount and distribution of coupling capacitances and resistance. Here, attempting to associate automatically a circuit topology would require heuristics that would likely incur intolerable error. Instead, the full electroquasistatic equations would be the best approach [11,12].

In contrast to parasitic RC extraction, field solvers for parasitic inductance extraction must solve the magnetoquasistatic (RL), not the magnetostatic (L) problem. The resistance cannot be separated from the inductance because it directly affects the current distribution in a frequency-dependent manner and the inductance is very dependent on this current distribution. Two examples are the *skin effect*, which causes currents to travel more toward the surface of conductors as the frequency rises, and the *return path* of signals in a ground plane or grid arrangement, which is closer to the signal as the frequency rises (in fact, the challenge for full-chip inductance extraction centers around heuristics for the return path; see Chapter 25, or [13]). Solution techniques for the inductance problem are described in Section 29.5.

TABLE 29.1   Forms of Maxwell's Equations Discussed in This Chapter and Elsewhere in This Book

Form of Maxwell's Equations	Common Circuit Description	Section	Example Application
Electrostatic	C	Sections 29.4 and 29.8–29.10	Charge dominated, $L \ll \lambda$
Conduction	R	See Chapter 26, Section 26.3.2	Line resistance, IR drop
Electroquasistatic	RC	See Chapter 26, Section 26.3.2	SRAM cell, substrate noise
Magnetoquasistatic	RL	Section 29.5	Current dominated, $L \ll \lambda l$
Electromagnetoquasistatic	RLC	Section 29.6.1	$L < \lambda, L \approx \lambda$, on-chip inductors
Full Maxwell (full-wave)	S-parameters	Section 29.6.2	Interconnect, $L > \lambda$, RF/microwave circuit passive devices

Note: L is a characteristic length of interconnect; λ is the wavelength of an electromagnetic wave at the frequency of interest.

Historically, many extraction problems could be *lumped* into inductance-dominated portions, such as the leads or wire bonds of lead-frame packages, and capacitance-dominated portions, such as sections of on-chip wiring. However, as circuit speeds have increased, more interconnect problems require a *distributed* representation of not only RC but also inductance. A good example is the on-chip inductor, which has parasitic capacitance to the substrate as well as turn-to-turn capacitance. Accurately determining the self-resonance cannot be accomplished with a single inductance extraction separate from capacitance. Solving this distributed RLC problem has sometimes been called the electromagnetoquasistatic form of Maxwell's equations [14], which will be described in Section 29.6.1.

Modeling a system as distributed is also important for capturing the finite propagation speed of electromagnetic phenomena. When such effects are important, the full Maxwell's equation (full-wave) problem must be solved as briefly described in Section 29.6.2.

A common method for determining whether a system is distributed or lumped is to compare the characteristic length of interconnect to the wavelength of electromagnetic waves at the maximum frequency of signals in the system. A rule of thumb is that interconnect longer than one-tenth the wavelength is distributed. To determine the maximum frequency for digital systems, one possible metric described in [15] is

$$(29.1) \qquad\qquad F = \frac{0.5}{T_r}$$

where
  F is in hertz
  T_r is the 10%–90% pulse rise time in seconds

For instance, a 2 mm wire bond in an encapsulant with a relative electrical permittivity of 4 should use a distributed model for rise times faster than 67 ps.

For distributed problems consisting of a handful of wires, the components of an RLC network could be computed directly without the need for a matrix solution [16]. At today's circuit densities, such methods are impractical due to the density of mutual couplings, and a field solver that directly generates a reduced-order model in the form of a few linear ordinary-differential equations is typically required, as discussed in Section 29.6.

The next three sections briefly derive the various classes of equations described earlier and discuss recent sparsification approaches to solve for the quantities of interest. For detailed derivations and descriptions of each of the regimes in the aforementioned table, see [17]. A good summary of the relationship between Maxwell's equations and circuit equations can be found in [18].

29.4 FAST FIELD SOLVERS: CAPACITANCE SOLUTION

Sparsification for solving integral equations is best illustrated through the capacitance problem. Methods of fast inductance and full-wave solution build on the fast techniques for capacitance.

The capacitance problem seeks the relation between net charge and voltage for a set of k-conducting bodies: $Q = CV$, where C is the $k \times k$ capacitance matrix, V the vector of k conductor voltages, and Q the vector of net charge on each of the k conductors. A column i of this matrix can be determined by computing the net charge on all k conductors when conductor i is set to 1 V and the rest to zero.

To compute the net charge from the voltage, we assume magnetic effects are negligible and solve the Laplace equation for the potential ϕ

$$(29.2) \qquad\qquad \nabla^2 \phi(r) = 0, \quad r \in \Omega$$

where $\phi(r)$ is the potential in the space Ω surrounding all the conductors, $\phi(r) = V$ on the surface of each conductor i, and V_i the voltage of conductor i.

Given a solution to Equation 29.2 in Ω, it can be shown that the surface charge density, ρ_s, is

(29.3)
$$\rho_s = -\varepsilon \frac{d\phi}{dn}$$

where
  n is the surface normal pointing out of a conductor
  ε is the permittivity of the space around the conductor

The net charge on any conductor i is then

(29.4)
$$Q_i = \int_{S_i} \rho_s \, da$$

where S_i is the surface of conductor i.

The Laplace equation 29.2 is the differential form for electrostatics as would be solved by an FD [19,20] or FEM [21,22] method. Many integral forms exist [23–25]. The most familiar is the first-kind formulation, also known as charge superposition,

(29.5)
$$\phi(r) = \int_S \rho_s(r') \frac{1}{4\pi\varepsilon |r - r'|} da'$$

where
  S is the union of all conductor surfaces
  $|r-r'|$ is the distance between charge point r' and evaluation point r.

Equation 29.5 may be more familiar from introductory electromagnetics as the integral form of the potential due to a set of point charges, q_i : $\phi(r) = \sum_i q_i / 4\pi\varepsilon |r - r_i|$.

Equation 29.5 directly relates the potential at a point on a conductor surface, which is known, to the unknown charge distribution on the surface. After solving this integral equation for ρ_s, one can use Equation 29.4 to compute the net charge.

This presentation assumes that the space around the conductors is a single, infinite, homogenous dielectric. Methods that solve the differential form can naturally handle arbitrary dielectrics, and methods that solve the integral forms can be augmented for piecewise-constant regions of isotropic dielectrics [24,26,27] as are common in integrated circuit and package geometries.

A common approach to solve Equation 29.5 numerically is to discretize the surface of the conductors into panels as shown on the right in Figure 29.1 and assume that the charge density is uniformly distributed on each panel. That is, ρ_s is approximated as a sum of basis functions, w_i,

(29.6)
$$\rho_s(r) \approx \sum_{i=1}^{n} q_i w_i(r)$$

where $w_i(r) = 1$, for r on panel i and 0 elsewhere. The goal is then to solve for the unknown magnitude of the charge density, q_i, on each panel.

To turn this discretization into a system of linear equations, one can require that the integral equation be satisfied at the center of each panel. Such an approach is formally called a weighted residuals [28] approach based on collocation [29]. For a problem with n panels, inserting Equation 29.6 into Equation 29.5 enforced at the n panel centers results in a linear system

(29.7)
$$Pq = v$$

where
  $q \in \mathbb{R}^n$ are the unknown charges on each panel
  $v \in \mathbb{R}^n$ the known panel voltages
  $P \in \mathbb{R}^{n \times n}$

The entries of the matrix, P_{ij}, can be computed from the integral relation

(29.8)
$$P_{ij} = \frac{1}{a_j} \int_{p_j} \frac{1}{4\pi\varepsilon |\, r_i - r'\,|} da'$$

where
 r_i is the center of panel i
 p_j is the surface of panel j
 a_j is the area of panel j

The integral can be computed analytically for quadrilateral panels [30]. Note that because the potential at any point has a contribution from the charge on *every* panel, $P_{ij} \neq 0$ for all i, j, and the matrix P is dense.

Computing each column of the capacitance matrix, C, is now a matter of solving Equation 29.7 for the charge vector, q, and summing up the charge for each conductor to arrive at the net charge vector Q. Solving Equation 29.7 by a direct matrix solution method such as Gaussian elimination requires an $O(n^3)$-time computation, which is intractable for complicated interconnect geometries requiring hundreds of thousands of panels. To avoid the $O(n^3)$ time, an iterative technique, such as GMRES [31], can be employed. The dominant cost of applying an iterative scheme is the $O(n^2)$ time of the matrix–vector product computation at each iteration (Px, where x is a vector computed at each iteration).

Sparsification approaches reduce the $O(n^2)$ time of the matrix–vector product to $O(n)$ or $O(n \log n)$ time by exploiting the fact that $y = Px$ can be interpreted as the evaluation of the potentials, y, due to a set of known charges, x. Most of these approaches hinge on the fact that the potential contribution from far away charges is much smaller than those nearby. These far-field interactions are treated approximately without significantly impacting overall accuracy.

One of the first sparsification techniques applied to capacitance extraction was the fast multipole method (FMM) [4–6] with $O(n)$ complexity in time and memory. The idea, illustrated in Figure 29.2a, is to approximate a group of distant charges within some radius R as a single charge and then use that approximation for evaluation of potentials at points a distance r away, where $r \gg R$. Using a single charge to represent this group is a monopole expansion and using a series of higher-order representations, such as a monopole, dipole, and quadrupole, is a multipole expansion. Similarly, the potential due to many multipole expansions can be expressed with a local expansion centered at a point around a cluster of evaluation points as illustrated in Figure 29.2b. By using multipole and local expansions and by keeping the ratio r/R constant for all expansions, Px can be computed in $O(n)$ time and memory for a given error tolerance. The details of applying an iterative solution algorithm and FMM sparsification to the capacitance problem can be found in [32], with a proof of $O(n)$ given in [33].

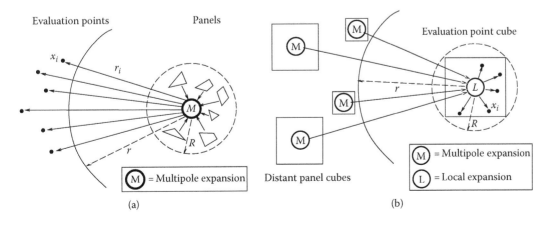

FIGURE 29.2 (a) Multipole expansions represent the potential due to a cluster of charges. (b) Local expansions represent the potential at a cluster of evaluation points due to many multipole expansions.

FMM methods reduce the computation time and memory by orders of magnitude for even modest capacitance problems. It was shown in [6] that even a small problem of roughly 6000 panels could be solved 10 times faster than a dense matrix–vector product iterative scheme and 100 times faster than direct factorization.

The drive to solve even larger problems motivates continued work on multipole-based sparsification for interconnect extraction [34,35]. In addition, other methods of sparsification such as the precorrected-FFT approach [8], which will be discussed in Section 29.6.2, and SVD-based approaches [9] further improve performance.

29.5 FAST INDUCTANCE SOLUTION

Given an interconnect network with some user-defined ports, inductance extraction is the process of computing the complex frequency-dependent port admittance matrix, Y_t, under the magnetoquasistatic approximation. In this section, we give a brief overview of fast inductance extraction.

Because of the strong dependence of the port inductance on the current distribution, the inductance problem is best understood when the governing integral equation is cast into an equivalent circuit. The resistive and inductive part of the partial element equivalent circuit (PEEC) method [36,37] is the best-known method for generating an equivalent circuit. Originally, this equivalent circuit was then used directly within a circuit simulator to model the interconnect. Unfortunately, present-day interconnect problems generate equivalent circuits that are too large for direct processing by a circuit simulator. Instead, either the equivalent circuit must be solved to extract the port admittances at a particular frequency of interest or the equivalent circuit model must be reduced to something tractable for insertion into a circuit simulator. For either approach, sparsification methods greatly increase the complexity of interconnect these methods can analyze.

In this section, we describe the governing integral equation, derive the equivalent circuit, and then describe the circuit solution methods necessary to apply sparsification methods. The model reduction approach will be described briefly at the end of Section 29.6.1.

29.5.1 INTEGRAL EQUATION AND EQUIVALENT CIRCUIT

Several integral equation–based approaches have been used to derive the Y_t associated with a given package or interconnect structure [37–40]. These integral formulations start with Maxwell's equations in the frequency domain and then apply the magnetoquasistatic assumption that the displacement current is negligible everywhere.

This leaves the current density, J, and scalar potential, Φ, as unknown quantities for the integral equation

$$(29.9) \qquad \frac{J(r)}{\sigma} + \frac{s\mu}{4\pi} \int_V \frac{J(r')}{|r-r'|} dv' = -\nabla\Phi(r), \quad r \in V$$

where
   s is the Laplace frequency
   σ is the conductivity
   μ is the permeability
   V is the union of the volumes of all conductors

The unknown quantities can be computed by combining Equation 29.9 with conservation of current

$$(29.10) \qquad \nabla \cdot J = 0$$

To create a linear system from Equation 29.9, note that current travels in the *volume* of the conductors, so the interior of all conductors must be discretized. Current within a long thin conductor can be assumed to flow parallel to its surface since the magnetoquasistatic assumption implies there is no charge accumulation on the surface. Thus, for long thin structures, such as pins of a package or signal lines on an integrated circuit, the conductor can be divided along its length into segments. In order to properly capture skin and proximity effects in these long, thin conductors, the cross section of the segments can be divided into a bundle of parallel *filaments*. It is also possible to use the filament approach for planar structures, such as ground planes, where the current distribution is 2D. In such cases, a grid of filaments must be used.

Since the current density inside each filament is assumed to be constant, the unknown current distribution can be approximated as a sum of basis functions,

$$(29.11) \qquad J(r) \approx \sum_{i=1}^{b} I_i w_i(r) \mathbf{l_i}$$

where
   b is the number of filaments
   I_i is the current inside filament i
   $\mathbf{l_i}$ a unit vector along the length of the filament
   $w_i(r)$ has a value of zero outside filament i and $1/a_i$ inside, where a_i is the cross-sectional area.

Note that unlike capacitance extraction, these basis functions are vector quantities.

By following an approach similar to that for capacitance extraction to generate a linear system, we can arrive at

$$(29.12) \qquad (R + sL)I_b = \tilde{\Phi}_A - \tilde{\Phi}_B$$

where $I_b \in \mathbb{C}^b$ is the vector of b filament currents;

$$(29.13) \qquad R_{ii} = \frac{l_i}{\sigma a_i}$$

is the $b \times b$ diagonal matrix of filament DC resistances;

$$(29.14) \qquad L_{ij} = \frac{\mu}{4\pi a_i a_j} \int\limits_{V_i} \int\limits_{V_j} \frac{\mathbf{l_i} \cdot \mathbf{l_j}}{|r - r'|} dV' dV$$

is the $b \times b$ dense, symmetric positive-definite matrix of partial inductances; V_i and V_j the volumes of filaments i and j, respectively; and $\tilde{\Phi}_A$ and $\tilde{\Phi}_B$ the averages of the potentials over the cross sections of the filament end faces. Analytic methods for computing the terms of this matrix are given in [37,41,42]. Equation 29.12 can also be written as

$$(29.15) \qquad ZI_b = V_b$$

where

   $Z = R + sL \in \mathbb{C}^{b \times b}$ is called the branch impedance matrix
   $V_b = \tilde{\Phi}_A - \tilde{\Phi}_B$ is the vector of branch voltages

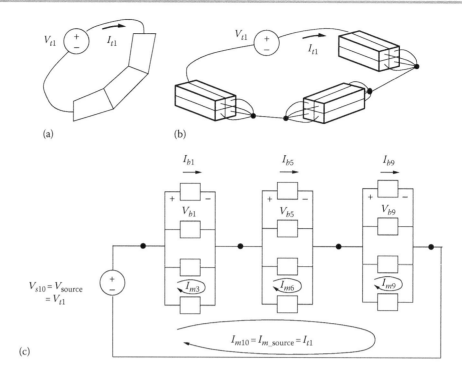

FIGURE 29.3 One conductor, (a) as piecewise-straight segments, (b) discretized into filaments, (c) modeled as a circuit with each box element representing a filament as a resistor in series with an inductor. Note that each filament has mutual inductance to every other filament. Mesh analysis quantities indicated with m suffix.

Note that intuitively one can view filament i as a resistor with resistance R_{ii} in series with an inductor with self-inductance L_{ii} and $b - 1$ mutual inductances L_{ij}, $i \neq j$, each magnetically coupling filament i to one other filament j. To enforce Equation 29.10, the interconnection of the current filaments can be represented with a planar graph, where the n nodes in the graph are associated with connection points between filaments and the b branches in the graph represent the filaments into which each conductor segment is discretized. These ideas are illustrated in Figure 29.3. The circuit obtained from the graph is the resistive and inductive part of the well-known PEEC method [36,37].

Describing filaments as inductors is not precise since inductance is a closed-loop quantity but each filament is a straight section. A filament's inductance represents only part of a loop and is thus termed the "partial" inductance. It can be shown that the correct loop inductance will be extracted if (1) all the filaments that make up the loop of current are included in the PEEC circuit and (2) all the mutual inductances between the filaments are included, that is, the full L matrix described earlier is used for extraction. A full description of the connection between the partial inductance concept and the definition of loop inductance is given in [43]. More detail is also given in Chapter 25.

The practical application of the partial inductance concept is that creating a PEEC circuit of just an on-chip signal line or just a single bonding wire is not adequate to model the inductive behavior of signals carried on that conductor. Instead, all the lines that represent the returning current and all the mutual inductances must also be included in the model to capture the true loop inductance.

29.5.2 EXTRACTION OF PORT ADMITTANCES VIA SPARSIFICATION

The creation of the PEEC circuit and the R and L values could be the end of the extraction step. The PEEC circuit network could be inserted into a circuit simulator to represent the interconnect. However, most interconnect problems generate thousands to hundreds of thousands of filaments, and thus insertion in a circuit simulator would be computationally intractable because of the dense mutual couplings between filaments. Instead, it is best to solve the PEEC circuit for the port admittances, Y_t, and use those for circuit simulation. Since the L matrix is dense, the computational time of solving will be at least $O(b^3)$ for direct methods. Instead, we wish to apply an iterative method with some sparsification approach.

A standard approach to form a linear system for the circuit would be to apply nodal analysis and generate a sparse tableau form [44]. Such a system includes both the constitutive relations Equation 29.15 and the circuit topology relations implied by Equation 29.10. Unfortunately, applying an iterative technique to the sparse tableau form requires a very large number of iterations to achieve a solution [14]. A common method of reducing the iteration count is to apply a *preconditioner* [45]; however, developing a preconditioner for this problem is difficult because the sparse tableau approach includes unknowns of different types.

Instead, we can reformulate the equations using mesh analysis and then apply an iterative method. Here, "mesh" refers to the circuit concept rather than the discretization mesh described previously. In mesh analysis [46], a mesh is any loop of branches in the circuit graph that does not enclose any other branches. Kirchhoff's voltage law, which implies that the sum of branch voltages around each mesh in the network must be zero, is represented by

$$(29.16) \qquad\qquad M V_b = V_s$$

where
  V_b is the vector of voltages across each branch except for the source branches
  V_s is the mostly zero vector of source branch voltages
  $M \in \mathbb{R}^{m \times b}$ the mesh matrix, where $m = b' - n + 1$ is the number of meshes and b' the number of
    filament branches plus the number of source branches

By defining $I_m \in \mathbb{C}^m$ as the vector of currents that circulate around each mesh as shown in Figure 29.3, it can be shown that

$$(29.17) \qquad\qquad M Z M^T I_m = V_s$$

This system has a single type of unknown quantity, I_m, and can be preconditioned to achieve $O(m^2)$ solution time with an iterative method [40].

The system is dense due to the dense mutual couplings of L from Equation 29.14. Note that Equation 29.14 shares the same *kernel*, $1/|r-r'|$, as the capacitance problem, and thus similar sparsification techniques can also be applied to inductance.

Applying sparsification techniques such as FMM for the inductance problem is not as effective as doing so for the capacitance problem for two reasons. First, the mutual inductance depends on the dot product between the vector directions of the two filaments of the coupling as shown in Equation 29.14. Since sparsification techniques represent many different filaments as a group, this representation must be done component-wise, requiring three sparsified evaluations to perform a matrix–vector product for L, compared to the single one for capacitance. Second, capturing skin effect can require the cross sections of wires to be divided into bundles of 25–100 filaments. Since the current density typically does not change along the length of a straight wire, the best mesh is often a very long bundle of densely packed filaments. Unfortunately, the long length of the wires limits the multipole radius, R, to be large, and since $r \gg R$, much less of the matrix–vector product can be sparsified.

Nonetheless, applying a mesh analysis–based iterative scheme sparsified by the FMM can result in significant speedup as shown on the left in Figure 29.4. The data are for two long traces with return paths through a finite-conductivity ground plane detailed in [40]. Using a mesh analysis–based iterative scheme was nearly two orders of magnitude faster than direct factorization for $m = 12{,}000$. Usage of FMM sparsification cut the time further by nearly a factor of 4. The memory savings usage of FMM was also over an order of magnitude for this modest problem size as shown on the right in Figure 29.4.

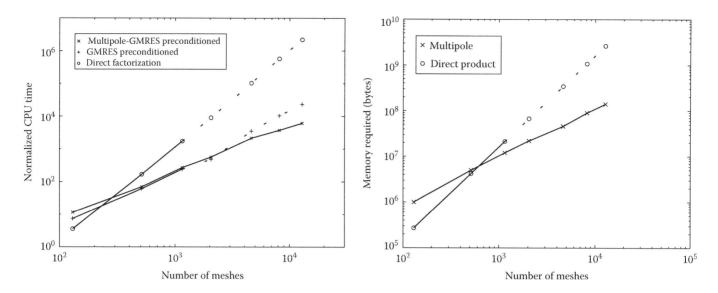

FIGURE 29.4   Comparison of the CPU time (left) and memory (right) to compute the reduced inductance matrix for two traces over a solid plane.

Some of the shortcomings of applying sparsification to inductance extraction under skin effects can be overcome by using basis functions whose shapes match more closely the actual exponentially varying current distribution, instead of using the piecewise-constant functions of Equation 29.11. Such techniques include those that use an analytic form for the basis functions [47,48], as well as numerically determined ones [49,50]. Alternately, if an integral equation could be derived that requires only a discretization of the surface yet correctly captures the interior current distribution for both low and high frequencies, then the many interior filaments could be avoided. Such a pure surface formulation for Maxwell's equations is pursued in [51,52].

29.6 DISTRIBUTED RLC AND FULL-WAVE SOLUTION

When an application requires a distributed interconnect model, the full Maxwell's equations must be considered. While both finite-element and integral equation–based approaches are common, we restrict the discussion to integral equation techniques. The integral equation commonly used extends that from Equation 29.9:

$$(29.18) \qquad \frac{J(r)}{\sigma} + \frac{s\mu}{4\pi} \int_V \frac{J(r')e^{(s/c)|r-r'|}}{|r-r'|} \mathrm{d}v' = -\nabla\Phi(r), \quad r \in V$$

$$(29.19) \qquad \Phi(r) = \frac{1}{4\pi\varepsilon} \int_S \frac{\rho_s(r')e^{(s/c)|r-r'|}}{|r-r'|} \mathrm{d}v', \quad r \in \mathbb{R}^3$$

where
  s is the Laplace frequency
  c is the speed of light
  V is the interior of all conductors
  J is the current density in V
  S is the surface of all conductors
  ρ_s is the charge density on S

Additionally, the currents and charge obey the conservation equation:

$$(29.20) \qquad \nabla \cdot \boldsymbol{J}(\boldsymbol{r}) = 0, \quad r \in V$$

$$(29.21) \qquad \boldsymbol{n} \cdot \boldsymbol{J}(\boldsymbol{r}) = -s\rho_s(\boldsymbol{r}), \quad r \in S$$

where n is the inward normal on S. These equations form the basis of many integral equation approaches. By adding capacitance-like circuit elements to the circuit in Figure 29.3 to represent Equation 29.19, we have the rPEEC method [53]. By assuming the propagation delay is negligible ($e^{(s/c)|r-r'|} \approx 1$), we arrive at the original PEEC method [16].

Other similar integral equations form the basis of codes in other fields. If all current is assumed to have a direction in the xy plane and conductors are thin in the z direction compared to their width, we arrive at the 2 1/2 D approaches common in the RF and microwave community. In 3D, if the frequency is high enough to assume current is confined to an infinitesimally thin layer on the surface of conductors, then we arrive at the high-frequency surface approaches, such as [54] often used for electromagnetic scattering and radiation calculations.

For continuity from previous sections and to highlight recent work, we will describe the PEEC-like approaches in the rest of this section.

The integral operators of Equations 29.18 and 29.19 are similar to Equation 29.5 and 29.9 described for capacitance and inductance extraction, but with the addition of a propagation delay term $e^{(s/c)|r-r'|}$. By modifying the L and P matrices appropriately and by considering the charge panels as PEEC circuit elements as well, we can follow a mesh-based circuit approach to generate a linear system for extracting the port admittances, Y_t,

$$(29.22) \qquad M Z_{EM} M^T I_m = V_s$$

where

$$(29.23) \qquad Z_{EM} = \begin{bmatrix} R + sL(s) & 0 \\ 0 & P(s)/s \end{bmatrix}$$

Note that the matrices L and P are now frequency dependent. In this context, the mesh analysis approach has similarities to the *divergence-free basis functions* used in scattering analysis [55] and more recently in RF simulation in [56].

If the frequency is low enough that $e^{(s/c)|r-r'|} \approx 1$, then L and P are no longer frequency dependent, and we arrive at the distributed RLC problem that will be discussed in Section 29.6.1. The solution to Equation 29.22 for higher frequency will be discussed in Section 29.6.2.

29.6.1 DISTRIBUTED RLC

Similar to the inductance problem from before, one would not want to insert the PEEC circuit of tens to hundreds of thousands of elements into a circuit simulator, but instead reduce Equation 29.22 to something manageable for circuit simulation.

One approach is to solve Equation 29.22 for either the admittance, Y_t, or the scattering parameters at a discrete set of frequency points. Making this solution computationally tractable again requires the application of a sparsified iterative solution technique. Because L and P are identical to the matrices from the previous sections, the FMM can be applied, provided a good preconditioner is available to keep the iteration count small.

However, the frequency dependence of distributed RLC problems is much stronger than the RL problem due to resonant behavior, and thus solving Equation 29.22 at a small set of discrete frequency points can incur a significant error. For instance, this approach can completely miss a resonant peak that falls between two adjacent frequency points. In addition, using frequency-domain discrete data for time-domain simulation requires an additional data fitting approximation step for inclusion in circuit simulation (for a discussion of this issue, see [57]).

Instead, one can apply model order reduction methods to Equation 29.22 to directly derive a model for circuit simulation. These methods require that the original system be written in a state space form such as the nth-order system

$$(29.24) \qquad\qquad s\mathcal{L}x = -\mathcal{R}x + BV_t$$

$$I_t = C^T x$$

where
  $x \in \mathbb{R}^n$ is the vector of states with dimension comparable to the number of discretization elements (panels and filaments)
  s is the Laplace frequency
  $V_t, I_t \in \mathbb{R}^t$ are the input and output vectors, respectively, and t is the number of ports
  $\mathcal{L}, \mathcal{R} \in \mathbb{R}^{n \times n}$
  $B, C \in \mathbb{R}^{n \times t}$

The idea of model reduction is to derive a much smaller qth-order system of form similar to Equation 29.24 with $q \ll n$, which still accurately models the system behavior. For circuit simulation, a qth-order system of this form can be written directly in an analog hardware description language such as Verilog-A or VHDL-AMS. Sparsification techniques for model generation are just as important as for iterative matrix solution since the run time and storage costs are similarly dominated by dense matrix–vector products.

The challenge of these model reduction methods is threefold: (1) to be numerically robust and efficient for the large systems of EDA, (2) to preserve the passivity of the original system, and (3) to generate an optimally compact model that is accurate over the desired frequency range. A brief history of these methods can be found in Chapter 18, Section 18.4.3. The PRIMA model reduction algorithm described in [58] satisfies (1) and passivity (2) is preserved if the original system meets certain structural criteria. In [59], it was shown that Equation 29.22 could be rewritten in a form suitable for the application of a multipoint version of PRIMA [60]. However, generating an optimally compact model required manual exploration. In [61–63], methods are developed to address objectives (1) and (3), and in [64], a numerically robust method is presented to address (2) and (3). Finally, the two-step procedure in [65] suggests that all three requirements could be satisfied by combining, for instance, a first step of [60] followed by a second step of [64]. Such a procedure is not fully automatic since a model order must be chosen for the first step, which accurately models the desired system behavior but is not too large for the more computationally expensive second step.

29.6.2 FULL-WAVE SOLUTION

When the propagation delay of mutual couplings is important, the full-wave problem must be solved, and $L = L(s)$ and $P = P(s)$ now have the $e^{(s/c)|r-r'|}$ oscillatory term. This frequency dependence changes the strategies necessary for both sparsification and model reduction.

Sparsification techniques, such as the FMM, are well tuned for the quasistatic cases described earlier, but to date are not as efficient for surface integral equation solutions to the full Maxwell's equations [10]. Sparsification techniques such as the precorrected-FFT approach [8], which are nearly kernel independent, can be applied instead. Instead of treating the far-field interaction

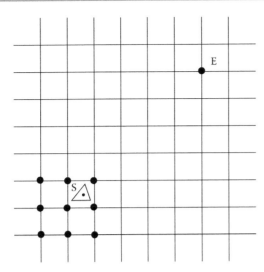

FIGURE 29.5 The precorrected-FFT approach treats the far-field interaction by projecting panel charges onto a grid (panel S projected onto nine grid points) and then uses an FFT to compute the potential everywhere on the grid. An interpolation step uses these grid potentials to compute panel potentials (not shown).

as groups of multipole expansions that must be translated in a kernel-dependent manner, the precorrected-FFT approach approximates the distribution of sources as point sources lying on a uniform grid as shown in Figure 29.5. Since the charges lie on a uniform grid, the potential they produce on the grid can be computed with a fast Fourier transform (FFT) for any translation-invariant kernel. Such an approach has been shown to have $O(N \log N)$ complexity when the geometry is reasonably homogenously distributed over a domain.

The model reduction problem for full wave becomes more difficult because the $e^{(s/c)|r-r'|}$ terms in $L(s)$ and $P(s)$ prevent Equation 29.17 from being written as the first-order linear system of Equation 29.24. Approaches for model reduction can be found in [66,67]; however, automatically generating a passive, optimally compact model is still an open problem. Under suitable approximation, the matrices generated by a FEM are not frequency dependent, and some of the recent methods described in the previous section could be applied. An approach that uses a Lanczos method is described in [57].

Similar to the inductance problem, capturing skin effect in a wide-band sense for a single formulation can lead to many current filaments. Here again, the exponentially shaped basis functions and surface formulations described at the end of Section 29.5 are advantageous for the efficient computation of the full-wave frequency response. However, such techniques typically trade size for added complexity in the frequency-dependent behavior of matrices $L(s)$ and $P(s)$. Hence, their full potential could be realized only when used in combination with a robust and efficient model order reduction procedure for frequency-dependent matrices.

29.7 VARIATION-AWARE EXTRACTION

As technology scaling has reduced the size of on-chip interconnect, the impact of process variations has become increasingly important for both device and interconnect behavior [68].

All of the techniques in the previous sections of this chapter assume that the conductor geometries are ideally smooth and of known dimension. None are inherently capable of handling the random geometric variations in the manufacturing process. In this section, we review techniques to convert the extraction solvers to "stochastic solvers." Note that this section is not about the floating random-walk (FRW) method, which is a stochastic method of capacitance extraction for a known geometry, without variation. FRW is described starting with Section 29.8.

Methods to predict the variation in the extracted parameters fall into two categories: *non-intrusive* algorithms, which can use extraction techniques of the previous sections *without* modification by solving on multiple samples but must change the discretization for each sample, and *intrusive* algorithms, which use the nominal discretization but change the representation of the unknowns and thus require specialized solution techniques.

The best-known nonintrusive technique is Monte Carlo simulation [69]. An ensemble of N_{MC} random samples are generated, which define an ensemble of geometries. The fast techniques of the previous sections are then used to solve for the impedance of each geometry in order to predict the distribution of the impedance. The difficulty is that the convergence of Monte Carlo is typically very slow, $O\left(1/\sqrt{N_{MC}}\right)$, thus requiring many thousands of samples, which is computationally infeasible for complicated interconnect structures.

Numerous authors have proposed the polynomial chaos expansion (PCE) to represent the dependence on the random variables. For instance, the nonintrusive stochastic collocation method (SCM) [70] assumes that the output impedances, $y(\eta)$, can be represented by a PCE [71] in terms of multivariate orthogonal polynomials in a vector of uncorrelated random variables, η,

$$y(\eta) \approx \sum_{i=1}^{K} y_i \Psi_i(\eta)$$

In [72,73], SCM with sparse-grid quadrature is applied to compute the capacitance distribution due to line-edge roughness and thickness variations for wiring down to 19 nm. In particular, in [72], 14 uncorrelated random variables required solving at only 421 quadrature points to compute y_i compared to thousands of points for Monte Carlo simulation.

The difficulty with SCM is that even with sparse-grid quadrature, the number of quadrature points is $O(d^k)$, where d is the number of random variables and k is the polynomial order. For the large capacitance extraction example from [74] with $d = 295$ and $k = 2$, more than 170,000 solutions with sparse-grid quadrature would be required, which is likely worse than Monte Carlo simulation.

Ideas to practically extend PCE methods for hundreds of independent random variables all exploit the fact that the solution vectors, $x(\eta)$, for various η are all structurally similar, and thus a low-order basis can be found to represent the solution. In the nonintrusive stochastic model reduction method of [75], $x \in \mathbb{R}^n$ is represented as a sum of basis vectors,

$$x_r = Uz$$

where the columns of $U \in \mathbb{R}^{n \times r}$ represent a collection of $r \ll n$ basis vectors and $z \in \mathbb{R}^r$ are the unknowns in the low-order space. After relatively few quadrature points, a low-order model created from this reduced basis represents much of the space of solutions for $x(\eta)$, and thus most of the remaining quadrature points can be solved on the low-order model instead of requiring a full solve. Since $r \ll n$, solving on the low-order model is significantly faster.

While this low-order model compresses the information about solutions in the deterministic space, it contains no information about the stochastic space. Methods of further compression as well as a comparison of the aforementioned techniques are described in [74].

PART II: STATISTICAL CAPACITANCE EXTRACTION

29.8 INTRODUCTION

A nondeterministic class of capacitance extractors involves the FRW technique. Unlike deterministic methods, a random-walk capacitance extractor can sample the exact capacitance equation, generating results that have a negligible deterministic error. Instead, the error is principally statistical in nature, which has some advantages over deterministic methods including dial-in accuracy and statistical cancellation of error.

The FRW technique, a Monte Carlo method, was used in the 1950s to calculate voltage [76]. A 3D Monte Carlo capacitance extraction using the FRW technique was developed in the early

1990s [1,77]. It has been also used for a similar problem, which is thermal analysis [78]. QuickCap, a commercial capacitance extractor incorporating this approach, has been available since the mid-1990s. Other capacitance extractors have also been developed using similar approaches [79,80]. Monte Carlo capacitance extraction has been extensively used to extract capacitance for

- All nets in test structures and critical cells
- Critical nets in full layouts
- Layout patterns (used to drive library-lookup methods)

It is especially valuable for accurate modeling of 3D physical features such as conformal and planar dielectrics, bias remaining after optical proximity correction (OPC), thickness variation due to chemical–mechanical planarization (CMP), and trapezoidal cross sections in small test structures and in large IC layouts.

Deterministic field solvers, such as those in the first part of this chapter, approximate voltage, charge, or other state variables and introduce error that is difficult to precisely quantize. In contrast, the FRW technique essentially samples an exact integral for capacitance. The statistical error can be calculated and reported. Depending on the implementation of the FRW method, the statistical error could be the only significant source of error.

Due to its statistical nature, Monte Carlo extraction has several desirable characteristics that are not available to deterministic methods:

- User-specified accuracy can be relaxed where possible, decreasing run time.
- Memory usage, orders of magnitude below that of deterministic field solvers, is not impacted when the accuracy goal is increased.
- Run time is only weakly dependent on the size of the problem.
- Capacitance values that matter collectively but not individually can be calculated to low accuracy, saving run time. A circuit simulation can be significantly more accurate than the individual capacitance values because of statistical error cancellation: The error of the sum of deterministic results is the sum of the individual errors, larger than the analogous sum of independent statistical results, where the error is the square root of the sum of the squares of the statistical errors. Because this last point is not widely recognized in the EDA industry, the accuracy goal for statistical capacitance extraction might be specified more strictly than it needs to be, resulting in much slower run times than are actually needed.

The following sections describe the theory underlying 3D Monte Carlo capacitance extraction, a description of accuracy benchmarks, important characteristics of common methods, and examples of statistical error cancellation.

29.9 THEORY

The FRW method is equivalent to Monte Carlo integration of an integral formulation for capacitance. Because the integral is exact, the error is limited only by the number of samples evaluated.

29.9.1 INTEGRAL FORMULATION FOR CAPACITANCE

The integral formulation for the capacitance associated with net i is based on the integral form of Gauss' law to find the charge on net i:

$$(29.25) \qquad Q_i = \oiint d\mathbf{A}_1\{\varepsilon\}[(\hat{\mathbf{n}} \cdot \vec{\mathbf{E}}(\mathbf{r}_1))]|$$

To develop this into an equation for capacitance, we define an operator $\vec{\nabla}_V$ that changes its scalar operand into a vector by taking the derivative with respect to the voltage on each net in turn. Applying this operator to both sides, Equation 29.25 becomes a vector equation determining all capacitance values associated with net i. In addition, $\vec{E}(\mathbf{r}_1)$ can be expressed as a surface integral of voltage weighted by a Green's function, in which the voltage can be similarly expressed. This yields an exact infinite-dimensional integral representation for column $\vec{C}_i$ of the capacitance matrix:

$$(29.26) \qquad \vec{C}_i = \oiint d\mathbf{A}_1\{\varepsilon\}[\vec{\nabla}_V(\hat{\mathbf{n}} \cdot \vec{E}(\mathbf{r}_1))]$$

$$(29.27) \qquad \vec{\nabla}_V(\hat{\mathbf{n}} \cdot \vec{E}(\mathbf{r}_1)) = \oiint d\mathbf{A}_2\{\hat{\mathbf{n}} \cdot \vec{G}_E(\mathbf{r}_2 \,|\, \mathbf{r}_1)\}[\vec{\nabla}_V\phi(\mathbf{r}_2)]$$

$$(29.28) \qquad \vec{\nabla}_V\phi(\mathbf{r}_k) = \oiint d\mathbf{A}_{k+1}\{G_\phi(\mathbf{r}_{k+1} \,|\, \mathbf{r}_k)\}[\vec{\nabla}_V\phi(\mathbf{r}_{k+1})], \;\; k = 2,\dots,\infty$$

$\vec{\nabla}_V(\hat{\mathbf{n}} \cdot \vec{E}(\mathbf{r}_1))$ in Equation 29.26 is expanded in Equation 29.27. $\vec{\nabla}_V\phi(\mathbf{r}_2)$ in Equation 29.27 is expanded in Equation 29.28. $\vec{\nabla}_V\phi(\mathbf{r}_{k+1})$ in Equation 29.28 is expanded recursively. When $\mathbf{r}$ is on net j, the term $\vec{\nabla}_V\phi(\mathbf{r})$ is an incidence vector, which is zero everywhere, except for the jth element, which is unity. Green's functions have been developed in 2D and 3D [76,81].

The integration surface in Equation 29.27 encloses $\mathbf{r}_1$, contains no nets, and is a shape for which the electric field $\vec{E}(\mathbf{r})$ can be written as a surface Green's function of voltage. $\vec{G}_E(\mathbf{r}_2 \,|\, \mathbf{r}_1)$ is the electric field (a vector) at $\mathbf{r}_1$ when the voltage on the surface is an impulse function at $\mathbf{r}_2$. For a sphere radius R centered at $\vec{G}_E(\mathbf{r}_2 \,|\, \mathbf{r}_1)$ is $3(\mathbf{r}_1-\mathbf{r}_2)/4\pi R^4$, as can be derived from the Poisson integral for a sphere [82].

Similarly, the integration surface in Equation 29.28 encloses $\mathbf{r}_k$, contains no nets, and is a shape for which the voltage $\phi(\mathbf{r}_{k+1})$ can be written as a surface Green's function. $G_\phi(\mathbf{r}_{k+1}|\mathbf{r}_k)$ is the voltage at $\mathbf{r}_k$ when the voltage on the surface is an impulse function at $\mathbf{r}_{k+1}$. For a spherical integration, surface radius R centered at $\mathbf{r}_k$, $G_\phi(\mathbf{r}_{k+1}|\mathbf{r}_k)$ is a constant, $1/4\pi R^2$. In other words, the voltage at the center of a sphere is the average of the voltage on the surface of the sphere.

26.9.2 MONTE CARLO INTEGRATION

In Monte Carlo integration, the integral of a function $f()$ is evaluated by repeatedly sampling $f()$ randomly to find the average, which is then scaled by the domain size.

A single sample of Equations 29.26 through 29.28 is equivalent to a FRW, as is illustrated in Figure 29.6. Here, a 2D sample of the integral using circles as the integration surfaces for the Green's functions starts on a surface around net i and ends on net j. Each sample is a scaled incidence vector. For the walk pictured here, only the jth value of the vector is nonzero:

- To sample Equation 29.26 for net i, $\mathbf{r}_1$ is selected at random on the closed integration surface around net i. The final incidence vector (generated when the walk hits a net) will need to be scaled by e and the perimeter of the integration surface.
- To sample Equation 29.27 at $\mathbf{r}_1$, $\mathbf{r}_2$ is selected at random on the largest circle around $\mathbf{r}_1$ that contains no nets. The incidence vector will need to be additionally scaled by $\hat{\mathbf{n}} \cdot \vec{G}_E(\mathbf{r}_2 \,|\, \mathbf{r}_1)$ (a known value) and the circumference of the circle.
- To sample Equation 29.28 at $\mathbf{r}_2$ and successive points, $\mathbf{r}_3$, $\mathbf{r}_4$, and $\mathbf{r}_5$ are similarly selected at random, each on the largest circle around the previous point. The incidence vector will need to be scaled by $G_\phi(\mathbf{r}_3|\mathbf{r}_2)$, $G_\phi(\mathbf{r}_4|\mathbf{r}_3)$, and $G_\phi(\mathbf{r}_5|\mathbf{r}_4)$ (known values) and the circumference of each circle.
- The function $\vec{\nabla}_V\phi(\mathbf{r}_5)$, here, is taken to be an incidence vector that is zero except for the jth element, which is unity. Since a random point selected on the surface of a circle will never lie exactly on a prespecified flat tangential surface, using circular Green's functions requires some level of approximation in order to complete the Monte Carlo sample. The commercial capacitance extractor QuickCap, referenced in the next section, uses Green's functions based on cubes and does not require such an approximation.

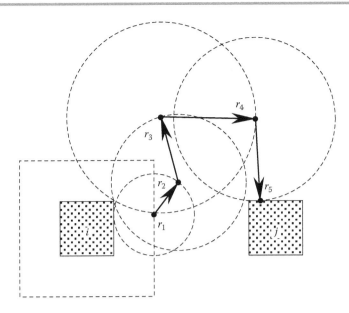

FIGURE 29.6 A 2D floating random walk for capacitance, using circles.

The series of points used to sample Equations 29.26 through 29.28 is equivalent to a FRW, where each point is selected on the perimeter of the largest homogeneous circle centered at the previous point. This example describes a simple implementation. Advanced implementations might use importance sampling or other techniques [83].

The scaled incidence vectors from many samples (or walks) are averaged together. An averaged capacitance value $\langle C_{ij} \rangle$ has a 1σ standard error δ_{ij} given by

(29.29)
$$\delta_{ij} = \sqrt{(\langle C_{ij}^2 \rangle - \langle C_{ij} \rangle^2)/n}$$

where
  $\langle C_{ij} \rangle$ is the average of the capacitance
  $\langle C_{ij}^2 \rangle$ is the average of the square of the capacitance
  n is the number of samples

As n increases, the probability distribution associated with the statistical error conforms to a normal distribution (bell curve), allowing a confidence level to be evaluated.

29.10 CHARACTERISTICS

The characteristics for a Monte Carlo extractor that follow from theory are listed here and illustrated in Sections 29.10.1 through 29.10.4. An example in Section 29.10.5 highlights a cautionary note concerning the applicability of a "gold extractor." Results are shown using QuickCap, a commercial capacitance extractor developed by then Random Logic Corporation, now acquired by Synopsys:

- *Reportable error.* The statistical error can be calculated (Equation 29.29) and reported. Section 29.10.1 shows how this can be used to calculate confidence levels.
- *Little or no bias.* No significant approximations are introduced when evaluating the capacitance integral. Benchmarks are shown in Section 29.10.2.
- *Memory efficient.* The bulk of required memory is to represent the geometric structure. The method uses no mesh. Examples are shown in Section 29.10.3.
- *Known convergence rate.* The statistical error should decrease with run time t as $t^{-1/2}$. Twice the accuracy requires four times the run time.

- *Convergence rate is a weak function of problem size.* Expanding the problem size and adding many structures should have little effect on the speed of a single walk. Examples are shown in Section 29.10.3.
- *Statistically independent errors.* Statistical cancellation occurs whenever capacitance values are effectively added, whether before or during simulation. This applies to tiled capacitance extraction, RC analysis, delay times involving multiple nets in a critical path, and cross talk involving multiple aggressor nets. This is further described in Section 29.10.4.

Because these characteristics are markedly different from those of deterministic methods, care is required when comparing the two types of extractors.

29.10.1 STANDARD ERROR

The standard error d of a result is a measure of uncertainty of that result due to statistical considerations. The standard error of a single statistical result should agree with the standard deviation of a population of results.

QuickCap reports the standard error (1σ error) associated with each capacitance value it extracts. Figure 29.7 shows a histogram of 10,000 independent Monte Carlo calculations of a single capacitance value. These 10,000 runs reported standard errors between 4.0 and 4.3 aF, agreeing with the standard deviation of the entire population, 4.01 aF.

When a result is based on many Monte Carlo samples, the probability distribution associated with the statistical error due to statistics can generally be expected to be a *normal distribution* (a bell curve), which is fully described by an average (the statistical result) and a width, δ. In such a case, the standard error can be used to calculate a confidence level. Table 29.2 lists some confidence levels.

When a statistical result is based on few samples, the standard error can be a reasonable estimate of the standard deviation of a population of results, even though the distribution of the population is not normal. The examples of statistical cancellation in Section 29.10.4 apply even when the probability distribution associated with the statistical error is not a normal distribution.

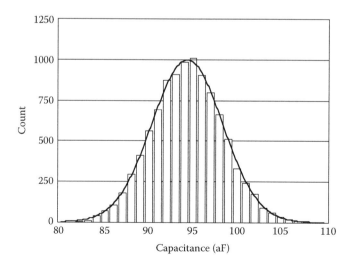

FIGURE 29.7 Histogram of 10,000 independent Monte Carlo samples. The solid line is a Gaussian characterized by the average (95.39 aF) and standard deviation (4.01 aF) of the results. (Data provided by Magma Design Automation, 1994.)

TABLE 29.2   Confidence Levels When Statistical Error Has a Normal Distribution

n	Confidence That Result Is within $\pm n\delta$	Confidence That Result Is Not More Than $n\delta$ Overestimated (Underestimated)
1	68% (2/3)	84% (5/6)
2	95% (19/20)	98% (49/50)
3	99.7% (349/350)	99.87% (749/750)
4	99.994% (1–1/15,000)	99.997% (1–1/30,000)
5	99.99994% (1–1/1,500,000)	99.99997% (1–1/3,000,000)

TABLE 29.3   QuickCap Bias for Known Problems

Problem Description	Analytic Solution	Normalized QuickCap Bias
1D, infinite parallel plates, spaced d apart	ε/d	$-0.004\% \pm 0.001\%$
2D, two infinite wires radius R, spaced D (center to center)	$\pi\varepsilon / \ln\left[\dfrac{D}{2R} + \sqrt{\left(\dfrac{D}{2R}\right)^2 - 1}\right]$	$-0.001\% \pm 0.002\%$
3D, sphere radius R in free space	$4\pi\varepsilon R$	$-0.002\% \pm 0.003\%$

Source: Data provided by Magma Design Automation, 1994.

29.10.2 BIAS

Bias is the amount of error that is not due to statistics. Bias can arise from discretization of equations (does not occur in the FRW method), from roundoff error, from boundary error, from limitations of the random-number generator, etc. Bias can be measured on a case-by-case basis when the correct answer is known.

Table 29.3 lists QuickCap bias for three problems with analytic solutions [84]: the capacitance between infinite-area parallel plates (1D electric field), the capacitance between infinite-length parallel circular wires (2D electric field), and the capacitance of a sphere in free space (3D electric field). The bias is quite small compared to the accuracy required for the analysis of IC layouts. Because the random-walk method is local (it has no mesh that depends on the problem size, and the size of each hop is limited by the *nearest* object), these bias values are expected to apply to self (total) capacitance in an environment with uniform dielectric. QuickCap's coupling capacitance bias has also been checked and has been found to be small whether in a uniform dielectric or in an environment with multiple dielectrics. Naturally, this behavior is implementation dependent.

Because the FRW method for capacitance extraction consists of Monte Carlo samples of an exact integral representation of capacitance, it can have negligible bias. The error in a method that attempts to solve for charge, voltage, or electric field by discretization is more difficult to quantify.

29.10.3 ACCURACY, MEMORY, PROBLEM SIZE, AND RUN TIME

Any capacitance extractor has trade-offs between accuracy, memory, problem size, and run time. To add perspective, Monte Carlo capacitance extraction and deterministic approaches are compared to each other, here. *Deterministic* approaches include FEMs, BEMs, and transform methods. For fairness, methods should be compared at the same accuracy levels. Also, a range of problems should be tested since each method has its own strengths and weaknesses.

The parameters that control the level of error are quite different for Monte Carlo and deterministic capacitance extraction. For the FRW approach, the error is basically controlled by the number of samples, proportional to the run time for a given problem. Since this method can report its own error, error control is straightforward.

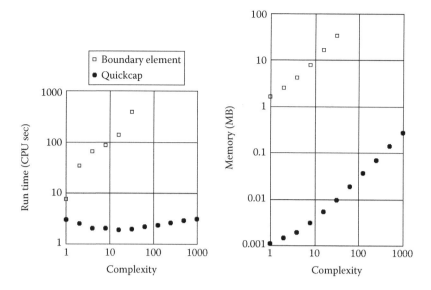

FIGURE 29.8   Characteristics of QuickCap and a boundary-element approach. The results, normalized to 2% error, are based on published data. (From Le Coz, Y.L. et al., *Solid State Electron.,* 42, 581–588, 1998.) Complexity, in arbitrary units, refers to the number of elements and to the calculation volume.

For mesh-based approaches, the error is controlled by discretization, which in turn affects both memory requirements and run time. The relationship between discretization and error is not straightforward. By varying the amount of discretization, however, the effect of discretization on the capacitance value can be observed, allowing an estimate of the error to be extrapolated.

Figure 29.8 shows how run time and memory requirements vary with problem *complexity* or the size or volume of a problem. The boundary-element results were generated with an academic version of a multipole-accelerated capacitance tool. QuickCap run time for a given level of accuracy has a weak dependence complexity, while the BEM used has an approximately linear dependency. The memory required to achieve the same level of accuracy increases linearly in both cases, though QuickCap memory requirements are many orders of magnitude below those of the BEM. Note that the error of the deterministic method, here, is that due to spatial discretization and does not include boundary error. Also note that the important feature in the run-time plot of Figure 29.8 is the slope. Not only might a commercial implementation of boundary-element extractor be significantly faster than the academic version used here, the relative speed can change with the type of problem analyzed and the convergence goal.

One characteristic not shown in Figure 29.8 is the dependence of accuracy on memory. For QuickCap, the accuracy does not depend on memory. For the BEM used here, the error is inversely proportional to the memory used. For the other method tested in [85], the error was difficult to control. It was inversely proportional to the fourth root of memory.

Sample characteristics for statistical and deterministic capacitance extractors are summarized in Table 29.4. For QuickCap, where the run time is proportional to the number of samples,

TABLE 29.4   Sample Characteristics of Statistical and Deterministic Capacitance Extractors

	Floating Random Walk	Boundary Element	Finite Element
Accuracy vs. run time	$\varepsilon \propto 1/t^{1/2}$	$\varepsilon \propto 1/t$	$\varepsilon \propto 1/t^{1/6}$
Accuracy vs. memory	Independent	$\varepsilon \propto 1/m$	$\varepsilon \propto 1/m^{1/4}$
Run time vs. complexity	Weak dependence	$t \propto C$	$t \propto C$
Memory vs. complexity	$m \propto C$	$m \propto C$	$m \propto C$

Note: From Le Coz, Y.L. et. al., *Solid State Electron.,* 42, 581–588, 1998. This summary is based on published data. Error of the deterministic capacitance does not include error introduced by boundary approximations.

the error drops slower than for the BEM of [86]. However, of primary importance is the run time in the 1%–10% range required for accurate circuit analysis. For the BEMs and FEMs, discretization error is related to memory. This is quite different from QuickCap's characteristics. Because QuickCap's run time is only weakly dependent on problem size, QuickCap will be faster than deterministic methods on large problems (for the equivalent levels of accuracy). Finally, QuickCap's memory requirements are low.

Boundary error, not addressed earlier, is the error due to neglecting objects outside of some *calculation region*. Whether one applies periodic boundary conditions, reflective boundary conditions, some *mixed* condition, or simply ignored objects outside the calculation region, the error can be calculated by finding the range of capacitance when varying the environment beyond the calculation region. Because the FRW method is memory efficient, expanding the calculation region to include additional objects is straightforward.

29.10.4 STATISTICAL CANCELLATION

When the implementation of the FRW approach has negligible bias, many advantages can be realized, including efficient tiled capacitance extraction and error reduction during circuit simulation with respect to accumulative delay time, cross talk, and the behavior of RC networks.

Statistical error cancellation occurs when adding and averaging statistical results. The sum of two statistical values, $A \pm \delta_A$ and $B \pm \delta_B$, is $(A + B) \pm \sqrt{\delta_A^2 + \delta_B^2}$. While the total error increases, the relative error decreases.

Tiled capacitance extraction can be performed with nearly 100% efficiency. In other words, the total run time and memory used can be about the same as without tiling. A tiled capacitance run can be implemented in parallel, allowing huge layouts to be analyzed in reasonable time. Consider, for example, the capacitance of the net shown in Figure 29.9. The layout is divided into two tiles, each including some of the physical data from the adjacent tile. The total capacitance is the integral (from Equation 29.6) over the dashed line. Whether this is sampled in two pieces (left) or as a single untiled run (right) makes no difference to the capacitance value or the statistical error. The only source of error is stitching error due to walks that terminate outside the valid (shaded) region. Furthermore, performing 64,000 walks on the upper part and 36,000 walks on

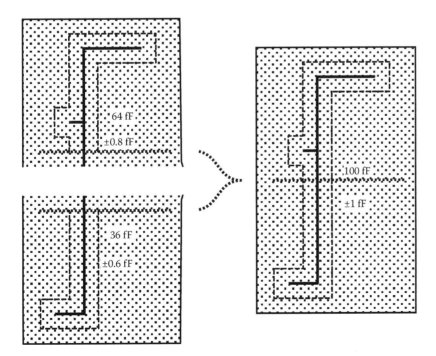

FIGURE 29.9 Tiled capacitance extraction can be almost 100% efficient while introducing negligible stitching error.

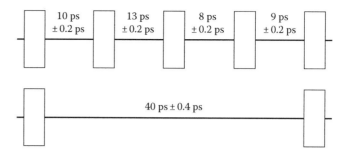

FIGURE 29.10 Statistical error is cancelled in multinet critical paths even when a timing analyzer is not aware of the error.

the lower part is approximately equivalent to performing 100,000 walks on the nontiled net in terms of the total numeric result and statistical error. Note that because of statistical cancellation, even though the relative errors associated with 64 and 36 fF, here, are ±1.25% and ±1.67%, the relative error of the sum is ±1%.

Whenever a critical path involves multiple nets, the total delay time will exhibit statistical error cancellation as long as the errors are independent, even when the analysis does not consider statistical error. In Figure 29.10, even though the individual delay times have approximately ±2% statistical error, the statistical error of the total delay time is ±1%. In order for a deterministic capacitance extractor to match this, it would need to generate ±1% results on each of the four nets in the critical path.

While small coupling capacitances might be individually negligible, they can be collectively important when many aggressor nets switch simultaneously. In Figure 29.11, each coupling capacitance is known to ±10%. Simultaneous switching results in 25% cross talk (±2% relative error or ±0.5% cross talk error). For deterministic results to match this level of accuracy, each result would need to be ±2%. Using statistical methods, the extraction time for the capacitance associated with a given victim net need not be increased when it has lots of coupling capacitance. While the individual coupling capacitance may have significant relative error, the effect of any single capacitance on circuit behavior is small, and the collective effect incorporates statistical error cancellation.

Similar statistical cancellation occurs for waveforms of RC networks. In Figure 29.12, each of 25 capacitance values is known to be ±10%. The statistical error associated with the Elmore delay time (the first-order response time) is ±2.3%, comparable to the statistical error associated with the sum (±2%). For deterministic results to match this level of accuracy, each result would need to be ±2.3%. This effect is more pronounced for more complex RC networks.

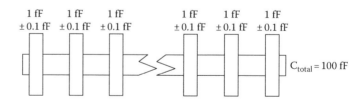

FIGURE 29.11 Statistical error cancellation occurs when aggressor nets switch simultaneously. The total coupling capacitance for 25 cross wires, here, is 25 ± 0.5 fF.

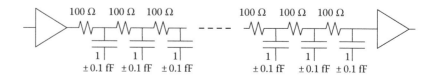

FIGURE 29.12 Statistical error cancellation occurs in resistance and capacitance (RC) networks. The total Elmore delay time due to 25 RC stages here is 32.5 ± 7.4 ps (± 2.3%).

TABLE 29.5 Statistical Cancellation Example

		Average	Std. Dev.	Min.	Max.
10,000 node caps	Absolute	458.4 aF	45.3 aF	275 aF	628 aF
	Relative		9.9%	−40%	37%
100 signal caps	Absolute	45.84 fF	0.43 fF	44.77 fF	6.89 fF
	Relative		0.94%	−2.34%	2.28%
100 delay times	Absolute	23.06 ps	0.26 ps	22.36 ps	23.77 ps
	Relative		1.14%	−3.1%	3.1%

This summary is based on QuickCap results [87]. Each signal consists of 100 nodes, and each delay time is from simulation of a 100-stage RC circuit. Relative errors are calculated by normalizing the standard deviation to the average value.

Table 29.5 summarizes results of 100 simulations of a uniform 100-stage RC line. Capacitance is extracted independently in each case. The relative statistical error of the total capacitance (100 nodes) is approximately 10× lower than for the individual node capacitances. The statistical error of the simulated delay time is similarly improved, approximately 8× to 9× better than for the individual node capacitances. This result is consistent with theory. For a long uniform line divided into a large number of stages, when the total capacitance has a relative statistical error of δ, then the Elmore delay time has a relative statistical error of $\delta * 2/\sqrt{3}$, or approximately 1.15δ.

26.10.5 TECHNOLOGY MODELING

The accuracy of a capacitance extractor is compromised when it is given bad input. To address this issue, details of a given IC fabrication technology should be examined to decide what is important. For example, the physical line widths may not match layout line widths due to bias from OPC, CMP effects can cause thickness variations, and sidewalls may not be vertical. Care must be taken when using statistical capacitance extraction or a 3D field solution as a gold standard. It can only be "gold" if given the correct input. Claims that a "fast" capacitance calculator matches a gold-standard capacitance extractor are meaningless when the gold-standard extractor is given an approximated geometry.

Figure 29.13 shows a comparison between QuickCap and silicon measurements for test structures in a 90 nm process. For the detailed OPC-bias model, most of the discrepancy can be attributed to imprecise representation of the silicon (wire width, wire spacing, dielectric values, etc.) rather than any QuickCap bias. In fact, discrepancies of almost 15% result if the technology

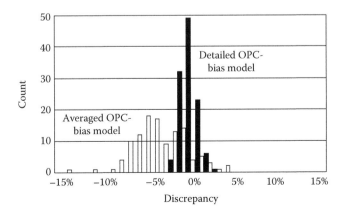

FIGURE 29.13 Comparison between silicon measurements and capacitance values extracted using two technology models. (From Data provided by Magma Design Automation, 2004.)

model uses an average OPC correction rather than one that is a function of line width and spacing. Fast methods to calculate capacitance should be based on comparisons to full 3D results that use the detailed OPC-bias model, here, not an averaged OPC-bias model.

29.11　SUMMARY

In this chapter, various regimes of electromagnetics and their governing equations as they relate to interconnect extraction and simulation were discussed. The use of integral equation techniques to solve these systems results in dense linear systems whose $O(N^3)$ computational and $O(N^2)$ memory complexity make it impractical to solve for complicated interconnect. To remedy this, we discussed sparsification techniques to bring this complexity down to $O(N)$ or $O(N \log N)$ improving the computation time and memory by orders of magnitude for capacitance, inductance, as well as distributed RLC and full-wave solution.

The FRW method for capacitance extraction was also described and has many characteristics not found in deterministic methods. In addition to reporting the statistical error (its only significant source of error), it is memory efficient and capable of handling large layouts. Statistical error cancellation results in circuit simulations that are more accurate than one might expect looking at the accuracy of individual capacitance values.

As feature sizes have shrunk, the impact of process variations has become increasingly important. Variation-aware extraction techniques were reviewed that efficiently account for this variation by directly computing the statistical distribution of the output from statistical process variations. This avoids the large number of solver invocations needed by Monte Carlo simulation.

To go beyond parasitic extraction, these computed statistical distributions can be used to predict system behavior when used within a variation-aware circuit simulator. Such a simulator must account for the variation in both the linear parasitic parameters and the nonlinear device parameters. This new research field of *uncertainty quantification* for integrated circuit and also MEMS simulation presents an even greater challenge as the system is nonlinear and time dependent and the output may not depend smoothly on the random parameters as for parasitic extraction. Recent progress is summarized in [86].

ACKNOWLEDGMENTS

For Sections 29.2–29.7, the author thanks Tom Korsmeyer, Luca Daniel, Tarek El-Moselhy, and Zheng Zhang for many suggestions and also Mike Chou, Keith Nabors, Joel Phillips, and Zhenhai Zhu for diagrams and other contributions.

REFERENCES

1. Y.L. Le Coz and R.B. Iverson, A stochastic algorithm for high speed capacitance extraction in integrated circuits, *Solid State Electron.*, 35, 1005–1012, 1992.
2. O.M. Ramahi and B. Archambeault, Adaptive absorbing boundary conditions in finite-difference time domain applications for emc simulations, *IEEE Trans. Electromagn. Comp.*, 37, 580–583, 1995.
3. J.C. Veihl and R. Mittra, An efficient implementation of berenger's perfectly matched layer (pml) for finite-difference time-domain mesh truncation, *IEEE Microw. Guided Wave Lett.*, 6, 94, 1996.
4. L. Greengard, *The Rapid Evaluation of Potential Fields in Particle Systems*, MIT Press, Cambridge, MA, 1988.
5. V. Rokhlin, Rapid solution of integral equations of classical potential theory. *J. Comput. Phys.*, 60, 187–207, 1985.
6. K. Nabors and J. White, Fastcap: A multipole accelerated 3-D capacitance extraction program, *IEEE Trans. Comput. Aided Des. Integr. Circuits Syst.*, 10, 1447–1459, 1991.
7. A. Brandt, Multilevel computations of integral transforms and particle interactions with oscillatory kernls, *Comput. Phys. Commun.*, 65, 24–38, 1991.
8. J.R. Phillips and J.K. White, A precorrected-FFT method for electrostatic analysis of complicated 3-D structures. *IEEE Trans. Comput. Aided Des. Integr. Circuits Syst.*, 16, 1059–1072, 1997.

9. S. Kapur and D.E. Long, Ies3: Efficient electrostatic and electromagnetic simulation, *IEEE Comput. Sci. Eng.*, 5, 60–67, 1998.

10. J.M. Song, C.C. Lu, W.C. Chew, and S.W. Lee, Fast Illinois solver code (FISC). *IEEE Antennas Propag. Mag.*, 40, 27–34, 1998.

11. S. Kumashiro, R. Rohrer, and A. Strojwas, A new efficient method for the transient simulation of 3-D interconnect structures, *Proceedings of International, Electron Devices Meeting*, San Francisco, CA, 1990, pp. 193–196.

12. M. Chou and J.K. White, Efficient formulation and model-order reduction for the transient simulation of three-dimensional VLSI interconnect, *IEEE Trans. Comput. Aided Des. Integr. Circuits Syst.*, 16, 1454–1476, 1997.

13. K.L. Shepard and T. Zhong, Return-limited inductances: A practical approach to on-chip inductance extraction, *IEEE Trans. Comput. Aided Des. Integr. Circuits Syst.*, 19, 425–436, 2000.

14. M. Kamon, Fast parasitic extraction and simulation of three-dimensional interconnect via quasistatic analysis, PhD dissertation, Massachusetts Institute of Technology, Electrical Engineering and Computer Science, Cambridge, MA, 1998.

15. H. Johnson and M. Graham, *High-Speed Digital Design: A Handbook of Black Magic*, Prentice-Hall, Upper Saddle River, NJ, 1993.

16. A.E. Ruehli, Equivalent circuit models for three-dimensional multiconductor systems, *IEEE Trans. Microw. Theory Tech.*, MTT-22, 216–221, 1974.

17. H.A. Haus and J.R. Melcher, *Electromagnetic Fields and Energy*, Prentice-Hall, Englewood Cliffs, NJ, 1989.

18. S. Ramo, J.R. Whinnery, and T.V Duzer, *Fields and Waves in Communication Electronics*, Wiley, New York, 1994.

19. A.H. Zemanian, R.P. Tewarson, C.P. Ju, and J.F. Jen, Three-dimensional capacitance computations for VLSI/ULSI interconnections, *IEEE Trans. Comput. Aided Des.*, 8, 1319–1326, 1989.

20. A. Seidl, H. Klose, M. Svoboda, J. Oberndorfer, and W. Rösner, CAPCAL—A 3-D capacitance solver for support of CAD systems, *IEEE Trans. Comput. Aided Des. Integr. Circuits Syst.*, 7, 549–556, 1988.

21. P.E. Cottrell and E.M. Buturla, VLSI wiring capacitance, *IBM J. Res. Dev.*, 29, 277–287, 1985.

22. T. Chou and Z.J. Cendes, Capacitance calculation of IC packages using the finite element method and planes of symmetry, *IEEE Trans. Comput. Aided Des.*, 13, 1159–1166, 1994.

23. I. Stakgold, *Boundary Value Problems of Mathematical Physics*, Vol. 2, Macmillan, New York, 1968.

24. M.A. Jaswon and G.T. Symm, *Integral Equation Methods in Potential Theory and Elastostatics*, Academic Press, London, U.K., 1977.

25. J. Tausch and J.K. White, Capacitance extraction of 3-D conductor systems in dielectric media with high-permittivity ratios, *IEEE Trans. Microw. Theory Tech.*, 47, 18–26, 1999.

26. S.M. Rao, T.K. Sarkar, and R.F. Harrington, The electrostatic field of conducting bodies in multiple di-electric media, *IEEE Trans. Microw. Theory Tech.*, MTT-32, 1441–1448, 1984.

27. K. Nabors and J. White, Multipole-accelerated capacitance extraction algorithms for 3-D structures with multiple dielectrics, *IEEE Trans. Circuits Syst. I, Fundam. Theory Appl.*, 39, 946–954, 1992.

28. S.H. Crandall, *Engineering Analysis*, McGraw-Hill, New York, 1956.

29. R.F. Harrington, *Field Computation by Moment Methods*, MacMillan, New York, 1968.

30. J.N. Newman, Distributions of sources and normal dipoles over a quadrilateral panel, *J. Eng. Math.*, 20, 113–126, 1986.

31. Y. Saad and M.H. Schultz, GMRES: A generalized minimal residual algorithm for solving nonsymmetric linear systems, *SIAM J. Sci. Stat. Comput.*, 7,856–869, 1986.

32. K. Nabors and J. White, Fast capacitance extraction of general three-dimensional structures, *IEEE Trans. Microw. Theory Tech.*, 40(7), 1496–1506, 1992.

33. K. Nabors, F.T. Korsmeyer, F.T. Leighton, and J. White, Multipole accelerated preconditioned iterative methods for three-dimensional potential integral equations of the first kind, *SIAM J. Sci. Stat. Comput.*, 15, 713–735, 1994.

34. M. Bachtold, M. Spasojevic, C. Lage, and P.B. Ljung, A system for full-chip and critical net parasitic extraction for ULSI interconnects using a fast 3-D field solver, *IEEE Trans. Comput. Aided Des. Integr. Circuits Syst.*, 19, 325–338, 2000.

35. M.W. Beattie and L.T. Pileggi, Parasitics extraction with multipole refinement, *IEEE Trans. Comput. Aided Des. Integr. Circuits Syst.*, 23, 288–292, 2004.

36. W.T. Weeks, L.L. Wu, M.F. McAllister, and A. Singh, Resistive and inductive skin effect in rectangular conductors, *IBM J. Res. Dev.*, 23, 652–660, 1979.

37. A.E. Ruehli, Inductance calculations in a complex integrated circuit environment, *IBM J. Res. Dev.*, 16, 470–481, 1972.

38. A.C. Cangellaris, J.L. Prince, and L.P. Vakanas, Frequency-dependent inductance and resistance -calculation for three-dimensional structures in high-speed interconnect systems, *IEEE Trans. Compon. Hybrids Manuf. Tech.*, 13, 154–159, 1990.

39. M.J. Tsuk and J.A. Kong, A hybrid method for the calculation of the resistance and inductance of transmission lines with arbitrary cross sections, *IEEE Trans. Microw. Theory Tech.*, 39, 1338–1347, 1991.

40. M. Kamon, M.J. Tsuk, and J. White, Fasthenry: A multipole-accelerated 3-D inductance extraction program, *IEEE Trans. Microw. Theory Tech.*, 42, 1750–1758, 1994.

41. F.W. Grover, *Inductance Calculations, Working Formulas and Tables*, Dover, New York, 1962.

42. C. Hoer and C. Love, Exact inductance equations for rectangular conductors with applications to more complicated geometries, *J. Res. Nat. Bur. Stand.*, 69C, 127–137, 1965.

43. D. Ling and A.E. Ruehli, Interconnect modeling, Chapter 11, in *Circuit Analysis, Simulation and Design, Part 2*, A.E. Ruehli, Ed., Elsevier Science, North-Holland, the Netherlands, 1987, pp. 211–332.

44. G.D. Hachtel, R.K. Brayton, and F.G. Gustavson, The sparse tableau approach to network analysis and design, *IEEE Trans. Circuit Theory*, 18, 101–113, 1971.

45. G.H. Golub and C.F. Van Loan, *Matrix Computations*, 2nd edn., The Johns Hopkins University Press, Baltimore, MD, 1989.

46. C. Desoer and E. Kuh, *Basic Circuit Theory*, McGraw-Hill, New York, 1969.

47. E. Tuncer, B.-T. Lee, and D.P. Neikirk, Interconnect series impedance determination using a surface ribbon method, *Proceedings of the IEEE Topical Meeting on Electrical Performance of Electronic Packaging*, 1994, pp. 249–252.

48. L. Daniel, A. Sangiovanni-Vincentelli, and J. White, Using conduction modes basis functions for efficient electromagnetic analysis of on-chip and off-chip interconnect, *Proceedings of the Design Automation Conference*, 2001, pp. 563–566.

49. L. Daniel, A. Sangiovanni-Vincentelli, and J. White, Proximity templates for modeling of skin and proximity effects on packages and high frequency interconnect, *Proceedings of the IEEE/ACM International Conference on Computer Aided-Design*, 2002, pp. 326–333.

50. K.M. Coperich, A.E. Ruehli, and A. Cangellaris, Enhanced skin effect for partial-element equivalent-circuit (peec) models, *IEEE Trans. Microwave Theory Tech.*, 48, 1435–1442, 2000.

51. J. Wang and J.K. White, A wide frequency range surface integral formulation for 3D RLC extraction, *Proceedings of the International Conference on Computer Aided-Design*, November 1999, pp. 453–457.

52. Z. Zhu, B. Song, and J. White, Algorithms in fastimp: A fast and wideband impedance extraction, program for complicated 3-D geometries, *Proceedings of the Design Automation Conference*, 2003, pp. 712–717.

53. H. Heeb and Albert E. Ruehli, Three-dimensional interconnect analysis using partial element equivalent circuits, *IEEE Trans. Circuits Syst. I, Fundam. Theory Appl.*, 39, 974–982, 1992.

54. S.M. Rao, D.R. Wilton, and A.W. Glisson, Electromagnetic scattering by surfaces of arbitrary shape, *IEEE Trans. Antennas Propag.*, 30, 409–418, 1982.

55. E. Arvas and R.F. Harrington, Computation of the magnetic polarizability of conducting disks and the electric polarizability of apertures, *IEEE Trans. Antennas Propag.*, 31, 719–725, 1983.

56. S. Kapur and D.E. Long, Large-Scale Full-wave simulation. *Proceedings of the 41th Design Automation Conference*, 2004, San Diego, CA, June 7–11, 2004.

57. B. Anderson, J.E. Bracken, J.B. Manges, G. Peng, and Z. Cendes, Full-wave analysis in spice via model-order reduction. *IEEE Trans. Microw. Theory Tech.*, 52, 2314–2320, 2004.

58. A. Odabasioglu, M. Celik, and L. Pileggi, Prima: Passive reduced-order interconnect macro-modeling algorithm, *CMU Report*, 1997.

59. M. Kamon, N. Marques, L.M. Silveira, and J. White, Automatic generation of accurate circuit models of 3-D interconnect. *IEEE Trans. Compon. Packag. Manuf. Technol. Part B*, 21, 225–240, 1998.

60. I.M. Elfadel and D.D. Ling, A block rational Arnoldi algorithm for multipoint passive model-order reduction of multiport RLC networks, *Proceedings of IEEE/ACM International Conference on Computer Aided-Design*, San Jose, CA, 1997.

61. J.-R. Li, F. Wang, and J. White, Efficient model reduction of interconnect via approximate system grammians, *Proceedings of International conference Computer Aided-Design*, pp. 380–383, 1999.

62. P. Rabiei and M. Pedram, Model order reduction of large circuits using balanced truncation, *Proceedings of Asia and South Pacific Design Automation Conference*, Hong Kong, China, pp. 237–240, 1999.

63. I.M. Jaimoukha and E.M. Kasenally, Krylov subspace methods for solving large Lyapunov equations, *SIAM J. Numer. Anal.*, 31, 227–251, 1994.

64. J.R. Phillips, L. Daniel, and L.M. Silveira, Guaranteed passive balancing transformations for model order reduction, *IEEE Trans. Comput. Aided Des. Integr. Circuits Syst.* 22, 1027–1041, 2003.

65. M. Kamon, F. Wang, and J.K. White, Generating nearly optimally compact models from Krylov-subspace based reduced-order models, *IEEE Trans. Circuits Syst. II, Analog Digit. Signal Process*, 47(4), 239–248, 2000.

66. J.R. Phillips, E. Chiprout, and D.D. Ling, Efficient full-wave electromagnetic analysis via model-order reduction of fast integral transforms, *Proceedings of the 33rd Design Automation Conference*, Las Vegas, NV, 1996.

67. E. Chiprout, H. Heeb, M.S. Nakhla, and A.E. Ruehli, Simulating 3-D retarded interconnect models using complex frequency hopping (CFH), *International Conference on Computer Aided-Design*, Santa Clara, CA, November 1993, pp. 6–72.

68. H. Masuda, et al., Challenge: Variability characterization and modeling for 65-to 90-nm processes, *Proceedings of CICC*, 2005, pp. 593–599.

69. N. Metropolis, A.W. Rosenbluth, M.N. Rosenbluth, A.H. Teller, and E. Teller, Equation of state calculations by fast computing machines, *J. Chem. Phys.*, 21, 1087–1092, 1953.

70. L. Mathelin and M.Y. Hussaini, A stochastic collocation algorithm, *NASA*, 153, 2003-212-153, 2003.

71. N. Wiener, The homogeneous chaos, *Am. J. Math.*, 60, 897–936, 1938.

72. H. Zhu, X. Zeng, W. Cai, J. Xue, and D. Zhou, A sparse grid based spectral stochastic collocation method for variations-aware capacitance extraction of interconnect under nanometer process technology, *Proceedings of Design Automation and Test in Europe*, Nice, France, 2007.

73. W. Yu, Q. Zhang, Z. Ye, and Z. Luo, Efficient statistical capacitance extraction of nanometer interconnects considering the on-chip line edge roughness, *Microelectron. Reliab.*, 52, 704–710, 2012.

74. T. El-Moselhy and L. Daniel, Variation-aware stochastic extraction with large parameter dimensionality: Review and comparison of state of the art intrusive and non-intrusive techniques, *Proceedings of the 12th International Symposium on Quality Electronic Design*, Santa Clara, CA, 2011.

75. T. El-Moselhy and L. Daniel, Variation-aware interconnect extraction using statistical moment preserving model order reduction, *Proceedings of Design, Automation and Test in Europe*, Dresden, Germany, 2010.

76. G.M. Brown, *Modern Mathematics for Engineers*, E.F. Beckenbach, Ed., McGraw-Hill, New York, 1956.

77. J. Jere and Y.L. Le Coz, An improved method for the multi-dielectric dirichlet problem, *IEEE Trans. Microw. Theory Tech.*, 41, 325–329, 1993.

78. Y.L. Le Coz, R.B. Iverson, T.-L. Sham, H.F. Tiersten, and M.S. Shephard, Theory of a floating random-walk algorithm for solving the steady-state heat equation in complex, materially inhomogeneous rectilinear domains, *Numer. Heat Transfer, Part B, Fundam.*, 26, 353–366, 1994.

79. M. Mascagni and N.A. Simonov, The random walk on the boundary method for calculating capacitance, *J. Comput. Phys.*, 195, 465–473, 2004.

80. M.P. Desai, CAPEM: The Capacitance Extraction Tool, http://www.ee.iitb.ac.in/~microel/download.

81. R.B. Iverson and Y.L. Le Coz, A floating random-walk algorithm for extracting electrical capacitance, *J. Math. Comput. Simul.*, 55, 59–66, 2001.

82. J.D. Jackson, *Classical Electrodynamics*, 3rd edn., Wiley, New York, 1998, p. 65, Eq. (2.19) ff.

83. W.H. Press et al., *Numerical Recipes in C*, 2nd edn., Cambridge University Press, Cambridge, U.K., pp. 316–328.

84. Data provided by Magma Design Automation, 2004.

85. Y.L. Le Coz, H.J. Greub, and R.B. Iverson, Performance of random-walk capacitance extractors for IC interconnects: A numerical study, *Solid State Electron.*, 42, 581–588, 1998.

86. Z. Zhang, X. Yang, G. Marucci, P. Maffezzoni, I. M. Elfadel, G. Karniadakis and L. Daniel, Stochastic testing simulator for integrated circuits and MEMS: Hierarchical and sparse techniques, *IEEE Custom Integrated Circuits Conference*, San Jose, CA, September 2014.

87. Data provided by Synopsys, Inc., 2014.

Index

T - #0925 - 101024 - C808 - 279/216/35 - PB - 9781138586017 - Gloss Lamination